The
World
Factbook

The
World
Factbook
1994-95

**Central
Intelligence
Agency**

BRASSEY'S
Washington·London

First Brassey's edition 1994

Brassey's, Inc., has commercially published *The World Factbook* to extend the limited audience reached through publication by the US Central Intelligence Agency.

Brassey's, Inc.

Editorial Offices	*Order Department*
Brassey's, Inc.	Brassey's Book Orders
8000 Westpark Drive	c/o Macmillan Publishing Co.
First Floor	100 Front Street, Box 500
McLean, Virginia 22102	Riverside, New Jersey 08075

Brassey's books are available at special discounts for bulk purchases for sales promotions, premiums, fund-raising, or educational use.

ISSN 0277-1527

ISBN 0-02-881052-X

The World Factbook is produced annually by the Central Intelligence Agency for the use of US government officials, and the style, format, coverage, and content are designed to meet their specific requirements. Information was provided by the Bureau of Census, CIA, Defense Intelligence Agency, Defense Nuclear Agency, Department of State, Foreign Broadcast Information Service, Maritime Administration, National Science Foundation (Polar Information Program), Naval Maritime Intelligence Center, Office of Territorial and International affairs, US Board on Geographic Names, US Coast Guard, and others.

10 9 8 7 6 5 4 3 2 1

Printed in the United States of America

Contents

Notes, Definitions,
and Abbreviations

There have been some significant changes in this edition. Czechoslovakia has been superseded by the Czech Republic and Slovakia. Eritrea gained independence from Ethiopia. The name of the Ivory Coast has been changed to Cote d'Ivoire and the Vatican City became the Holy See. New entries include Location, Map references, Abbreviation (often substituted for the country name), and Digraph (two-letter country code). Names is a new entry which includes long and short forms of both conventional and local names of countries as well as any former names. Most diacritical marks have been omitted. The electronic files used to produce the Factbook have been restructured into a database. As a result, the formats of some entries in this edition have been changed. Additional changes will occur in the 1994 Factbook. Irrigated land is a new entry with the data separate from the Land use entry. The Disputes entry is now International disputes. The GNP/GDP entry was renamed National Product and the per capita and real growth rate data placed in separate entries. Similar changes were made in the Population and Diplomatic Representation entries.

Abbreviations: (see Appendix B for international organizations and groups)

avdp.	avoirdupois
c.i.f.	cost, insurance, and freight
CY	calendar year
DWT	deadweight ton
est.	estimate
Ex-Im	Export-Import Bank of the United States
f.o.b.	free on board
FRG	Federal Republic of Germany (West Germany); used for information dated before 3 October 1990 or CY91
FY	fiscal year
GDP	gross domestic product
GDR	German Democratic Republic (East Germany); used for information dated before 3 October 1990 or CY91
GNP	gross national product
GRT	gross register ton
GWP	gross world product
km	kilometer
km^2	square kilometer
kW	kilowatt
kWh	kilowatt hour
m	meter
NA	not available
NEGL	negligible
nm	nautical mile
NZ	New Zealand
ODA	official development assistance
OOF	other official flows
PDRY	People's Democratic Republic of Yemen [Yemen (Aden) or South Yemen]; used for information dated before 22 May 1990 or CY91
UAE	United Arab Emirates
UK	United Kingdom
US	United States
USSR	Union of Soviet Socialist Republics (Soviet Union); used for information dated before 25 December 1991
YAR	Yemen Arab Republic [Yemen (Sanaa) or North Yemen]; used for information dated before 22 May 1990 or CY91

Administrative divisions: The numbers, designatory terms, and first-order administrative divisions are generally those approved by the US Board on Geographic Names (BGN). Changes that have been reported but not yet acted on by BGN are noted.

Area: Total area is the sum of all land and water areas delimited by international boundaries and/or coastlines. Land area is the aggregate of all surfaces delimited by international boundaries and/or coastlines, excluding inland water bodies (lakes, reservoirs, rivers). Comparative areas are based on total area equivalents. Most entities are compared with the entire US or one of the 50 states. The smaller entities are compared with Washington, DC (178 km^2, 69 miles^2) or The Mall in Washington, DC (0.59 km^2, 0.23 miles^2, 146 acres).

Birth rate: The average annual number of births during a year per 1,000 population at midyear; also known as crude birth rate.

Dates of information: In general, information available as of 1 January 1993 was used in the preparation of this edition. Population figures are estimates for 1 July 1993, with population growth rates estimated for calendar year 1993. Major political events have been updated through June 1993.

Death rate: The average annual number of deaths during a year per 1,000 population at midyear; also known as crude death rate.

Digraphs: The digraph is a two-letter "country code" that precisely identifies every entity without overlap, duplication, or omission. AF, for example, is the digraph for Afghanistan. It is a standardized geopolitical data element promulgated in the *Federal Information Processing Standards Publication* (FIPS) 10-3 by the National Bureau of Standards (US Department of Commerce) and maintained by the Office of the Geographer (US Department of State). The digraph is used to eliminate confusion and incompatibility in the collection, processing, and dissemination of area-specific data and is particularly useful for interchanging data between databases.

Diplomatic representation: The US Government has diplomatic relations with 180 nations. The US has diplomatic relations with 174 of the 182 UN members (excluding the Socialist Federal Republic of Yugoslavia whose status in the UN is unclear)—the exceptions are Angola, Bhutan, Cuba, Iran, Iraq, Macedonia, North Korea, and Vietnam. In addition, the US has diplomatic relations with 7 nations that are not in the UN—Andorra, Holy See, Kiribati, Nauru, Switzerland, Tonga, and Tuvalu.

Economic aid: This entry refers to bilateral commitments of official development assistance (ODA), which is defined as government grants that are administered with the promotion of economic development and welfare of LDCs as their main objective and are concessional in character and contain a grant element of at least 25%, and other official flows (OOF) or transactions by the official sector whose main objective is other than development motivated or whose grant element is below the 25% threshold for ODA. OOF transactions include official export credits (such as Ex-Im Bank credits), official equity and portfolio investment, and debt reorganization by the official sector that does not meet concessional terms. Aid is considered to have been committed when agreements are initialed by the parties involved and constitute a formal declaration of intent.

Notes, Definitions,
and Abbreviations *(continued)*

Entities: Some of the nations, dependent areas, areas of special sovereignty, and governments included in this publication are not independent, and others are not officially recognized by the US Government. "Nation" refers to a people politically organized into a sovereign state with a definite territory. "Dependent area" refers to a broad category of political entities that are associated in some way with a nation. Names used for page headings are usually the short-form names as approved by the US Board on Geographic Names. There are 266 entities in *The World Factbook* that may be categorized as follows:

NATIONS

182 UN members (excluding the Socialist Federal Republic of Yugoslavia whose status in the UN is unclear)

  8 nations that are not members of the UN—Andorra, Holy See, Kiribati, Nauru, Serbia and Montenegro, Switzerland, Tonga, Tuvalu

OTHER

  1 Taiwan

DEPENDENT AREAS

  6 Australia—Ashmore and Cartier Islands, Christmas Island, Cocos (Keeling) Islands, Coral Sea Islands, Heard Island and McDonald Islands, Norfolk Island

  2 Denmark—Faroe Islands, Greenland

16 France—Bassas da India, Clipperton Island, Europa Island, French Guiana, French Polynesia, French Southern and Antarctic Lands, Glorioso Islands, Guadeloupe, Juan de Nova Island, Martinique, Mayotte, New Caledonia, Reunion, Saint Pierre and Miquelon, Tromelin Island, Wallis and Futuna

  2 Netherlands—Aruba, Netherlands Antilles

  3 New Zealand—Cook Islands, Niue, Tokelau

  3 Norway—Bouvet Island, Jan Mayen, Svalbard

  1 Portugal—Macau

16 United Kingdom—Anguilla, Bermuda, British Indian Ocean Territory, British Virgin Islands, Cayman Islands, Falkland Islands, Gibraltar, Guernsey, Hong Kong, Jersey, Isle of Man, Montserrat, Pitcairn Islands, Saint Helena, South Georgia and the South Sandwich Islands, Turks and Caicos Islands

15 United States—American Samoa, Baker Island, Guam, Howland Island, Jarvis Island, Johnston Atoll, Kingman Reef, Midway Islands, Navassa Island, Northern Mariana Islands, Trust Territory of the Pacific Islands (Palau), Palmyra Atoll, Puerto Rico, Virgin Islands, Wake Island

MISCELLANEOUS

  6 Antarctica, Gaza Strip, Paracel Islands, Spratly Islands, West Bank, Western Sahara

OTHER ENTITIES

  4 oceans—Arctic Ocean, Atlantic Ocean, Indian Ocean, Pacific Ocean

  1 World

266 total

note: The US Government does not recognize the four so-called independent homelands of Bophuthatswana, Ciskei, Transkei, and Venda in South Africa.

Exchange rate: The value of a nation's monetary unit at a given date or over a given period of time, as expressed in units of local currency per US dollar and as determined by international market forces or official fiat.

Gross domestic product (GDP): The value of all goods and services produced domestically in a given year.

Gross national product (GNP): The value of all goods and services produced domestically in a given year, plus income earned abroad, minus income earned by foreigners from domestic production.

Gross world product (GWP): The aggregate value of all goods and services produced worldwide in a given year.

GNP/GDP methodology: In the "Economy" section, GNP/GDP dollar estimates for the OECD countries, the former Soviet republics, and the East European countries are derived from *purchasing power parity* (PPP) calculations rather than from conversions at official currency exchange rates. The PPP method normally involves the use of international dollar price weights, which are applied to the quantities of goods and services produced in a given economy. In addition to the lack of reliable data from the majority of countries, the statistician faces a major difficulty in specifying, identifying, and allowing for the quality of goods and services. The division of a PPP GNP/GDP estimate in dollars by the corresponding estimate in the local currency gives *the PPP conversion rate*. One thousand dollars will buy the same market basket of goods in the US as one thousand dollars—converted to the local currency at the PPP conversion rate—will buy in the other country. GNP/GDP estimates for the LDCs, on the other hand, are based on the conversion of GNP/GDP estimates in local currencies to dollars at the official currency exchange rates. Because currency exchange rates depend on a variety of international and domestic financial forces that often have little relation to domestic output, use of these rates is less satisfactory for calculating GNP/GDP than the PPP method. Furthermore, exchange rates may suddenly go up or down by 10% or more because of market forces or official fiat whereas real output has remained unchanged. One additional caution: the proportion of, say, defense expenditures as a percent of GNP/GDP in local currency accounts may differ substantially from the proportion when GNP/GDP accounts are expressed in PPP terms, as, for example, when an observer estimates the dollar level of Russian or Japanese military expenditures; similar problems exist when components are expressed in dollars under currency exchange rate procedures. Finally, as academic research moves forward on the PPP method, we hope to convert all GNP/GDP estimates to this method in future editions of *The World Factbook*.

Growth rate (population): The annual percent change in the population, resulting from a surplus (or deficit) of births over deaths and the balance of migrants entering and leaving a country. The rate may be positive or negative.

Illicit drugs: There are five categories of illicit drugs—narcotics, stimulants, depressants (sedatives), hallucinogens, and cannabis. These categories include many drugs legally produced and prescribed by doctors as well as those illegally produced and sold outside medical channels.
   Cannabis (Cannabis sativa) is the common hemp plant, which provides hallucinogens with some sedative properties, and includes marijuana (pot, Acapulco gold, grass, reefer), tetrahydrocannabinol (THC, Marinol), hashish (hash), and hashish oil (hash oil).

Coca (Erythroxylon coca) is a bush, and the leaves contain the stimulant cocaine. Coca is not to be confused with cocoa, which comes from cacao seeds and is used in making chocolate, cocoa, and cocoa butter.

Cocaine is a stimulant derived from the leaves of the coca bush.

Depressants (sedatives) are drugs that reduce tension and anxiety and include chloral hydrate, barbiturates (Amytal, Nembutal, Seconal, phenobarbital), benzodiazepines (Librium, Valium), methaqualone (Quaalude), glutethimide (Doriden), and others (Equanil, Placidyl, Valmid).

Drugs are any chemical substances that effect a physical, mental, emotional, or behavioral change in an individual.

Drug abuse is the use of any licit or illicit chemical substance that results in physical, mental, emotional, or behavioral impairment in an individual.

Hallucinogens are drugs that affect sensation, thinking, self-awareness, and emotion. Hallucinogens include LSD (acid, microdot), mescaline and peyote (mexc, buttons, cactus), amphetamine variants (PMA, STP, DOB), phencyclidine (PCP, angel dust, hog), phencyclidine analogues (PCE, PCPy, TCP), and others (psilocybin, psilocyn).

Hashish is the resinous exudate of the cannabis or hemp plant (Cannabis sativa).

Heroin is a semisynthetic derivative of morphine.

Marijuana is the dried leaves of the cannabis or hemp plant (Cannabis sativa).

Narcotics are drugs that relieve pain, often induce sleep, and refer to opium, opium derivatives, and synthetic substitutes. Natural narcotics include opium (paregoric, parepectolin), morphine (MS-Contin, Roxanol), codeine (Tylenol w/codeine, Empirin w/codeine, Robitussan A-C), and thebaine. Semisynthetic narcotics include heroin (horse, smack), and hydromorphone (Dilaudid). Synthetic narcotics include meperidine or Pethidine (Demerol, Mepergan), methadone (Dolophine, Methadose), and others (Darvon, Lomotil).

Opium is the milky exudate of the incised, unripe seedpod of the opium poppy.

Opium poppy (Papaver somniferum) is the source for many natural and semisynthetic narcotics.

Poppy straw concentrate is the alkaloid derived from the mature dried opium poppy.

Qat (kat, khat) is a stimulant from the buds or leaves of Catha edulis that is chewed or drunk as tea.

Stimulants are drugs that relieve mild depression, increase energy and activity, and include cocaine (coke, snow, crack), amphetamines (Desoxyn, Dexedrine), phenmetrazine (Preludin), methylphenidate (Ritalin), and others (Cylert, Sanorex, Tenuate).

Infant mortality rate: The number of deaths to infants under one year old in a given year per 1,000 live births occurring in the same year.

International disputes: This category includes a wide variety of situations that range from traditional bilateral boundary disputes to unilateral claims of one sort or another. Information regarding disputes over international boundaries and maritime boundaries has been reviewed by the Department of State. References to other situations may also be included that are border or frontier relevant, such as resource disputes, geopolitical questions, or irredentist issues. However, inclusion does not necessarily constitute official acceptance or recognition by the US Government.

Irrigated land: The figure refers to the number of km^2 that is artificially supplied with water.

Land use: Human use of the land surface is categorized as *arable land*—land cultivated for crops that are replanted after each harvest (wheat, maize, rice); *permanent crops*—land cultivated for crops that are not replanted after each harvest (citrus, coffee, rubber); *meadows and pastures*—land permanently used for herbaceous forage crops; *forest and woodland*—land under dense or open stands of trees; and *other*—any land type not specifically mentioned above (urban areas, roads, desert).

Leaders: The chief of state is the titular leader of the country who represents the state at official and ceremonial funcions but is not involved with the day-to-day activities of the government. The head of government is the administrative leader who manages the day-to-day activities of the government. In the UK, the monarch is the chief of state, and the Prime Minister is the head of government. In the US, the President is both the chief of state and the head of government.

Life expectancy at birth: The average number of years to be lived by a group of people all born in the same year, if mortality at each age remains constant in the future.

Literacy: There are no universal definitions and standards of literacy. Unless otherwise noted, all rates are based on the most common definition—the ability to read and write at a specified age. Detailing the standards that individual countries use to assess the ability to read and write is beyond the scope of this publication.

Maps: All maps will be available only in the printed version of *The World Factbook* for the foreseeable future.

Maritime claims: The proximity of neighboring states may prevent some national claims from being extended the full distance.

Merchant marine: All ships engaged in the carriage of goods. All commercial vessels (as opposed to all nonmilitary ships), which excludes tugs, fishing vessels, offshore oil rigs, etc.; also, a grouping of merchant ships by nationality or register.
   Captive register—A register of ships maintained by a territory, possession, or colony primarily or exclusively for the use of ships owned in the parent country; also referred to as an offshore register, the offshore equivalent of an internal register. Ships on a captive register will fly the same flag as the parent country, or a local variant of it, but will be subject to the maritime laws and taxation rules of the offshore territory. Although the nature of a captive register makes it especially desirable for ships owned in the parent country, just as in the internal register, the ships may also be owned abroad. The captive register then acts as a flag of convenience register, except that it is not the register of an independent state.
   Flag of convenience register—A national register offering registration to a merchant ship not owned in the flag state. The major flags of convenience (FOC) attract ships to their register by virtue of low fees, low or nonexistent taxation of profits, and liberal manning requirements. True FOC registers are characterized by having relatively few of the ships registered actually owned in the flag state. Thus, while virtually any flag can be used for ships under a given set of circumstances, an FOC register is one where the majority of the merchant fleet is owned abroad. It is also referred to as an open register.

Flag state—The nation in which a ship is registered and which holds legal jurisdiction over operation of the ship, whether at home or abroad. Differences in flag state maritime legislation determine how a ship is manned and taxed and whether a foreign-owned ship may be placed on the register.

Internal register—A register of ships maintained as a subset of a national register. Ships on the internal register fly the national flag and have that nationality but are subject to a separate set of maritime rules from those on the main national register. These differences usually include lower taxation of profits, manning by foreign nationals, and, usually, ownership outside the flag state (when it functions as an FOC register). The Norwegian International Ship Register and Danish International Ship Register are the most notable examples of an internal register. Both have been instrumental in stemming flight from the national flag to flags of convenience and in attracting foreign-owned ships to the Norwegian and Danish flags.

Merchant ship—A vessel that carries goods against payment of freight; commonly used to denote any nonmilitary ship but accurately restricted to commercial vessels only.

Register—The record of a ship's ownership and nationality as listed with the maritime authorities of a country; also, the compendium of such individual ships' registrations. Registration of a ship provides it with a nationality and makes it subject to the laws of the country in which registered (the flag state) regardless of the nationality of the ship's ultimate owner.

Money figures: All are expressed in contemporaneous US dollars unless otherwise indicated.

National product: The total output of goods and services in a country in a given year. See Gross domestic product (GDP), Gross national product (GNP), and GNP/GDP methodology.

Net migration rate: The balance between the number of persons entering and leaving a country during the year per 1,000 persons (based on midyear population). An excess of persons entering the country is referred to as net immigration (3.56 migrants/1,000 population); an excess of persons leaving the country as net emigration (-9.26 migrants/1,000 population).

Population: Figures are estimates from the Bureau of the Census based on statistics from population censuses, vital registration systems, or sample surveys pertaining to the recent past, and on assumptions about future trends.

Total fertility rate: The average number of children that would be born per woman if all women lived to the end of their childbearing years and bore children according to a given fertility rate at each age.

Years: All year references are for the calendar year (CY) unless indicated as fiscal year (FY).

Note: Information for the US and US dependencies was compiled from material in the public domain and does not represent Intelligence Community estimates. *The Handbook of International Economic Statistics,* published annually in September by the Central Intelligence Agency, contains detailed economic information for the Organization for Economic Cooperation and Development (OECD) countries, Eastern Europe, the newly independent republics of the former nations of Yugoslavia and the Soviet Union, and selected other countries. The Handbook can be obtained wherever *The World Factbook* is available.

The
World
Factbook

Afghanistan

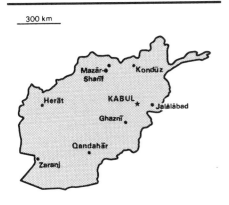

300 km

Geography

Location: South Asia, between Iran and Pakistan
Map references: Asia, Middle East, Standard Time Zones of the World
Area:
total area: 647,500 km²
land area: 647,500 km²
comparative area: slightly smaller than Texas
Land boundaries: total 5,529 km, China 76 km, Iran 936 km, Pakistan 2,430 km, Tajikistan 1,206 km, Turkmenistan 744 km, Uzbekistan 137 km
Coastline: 0 km (landlocked)
Maritime claims: none; landlocked
International disputes: periodic disputes with Iran over Helmand water rights; Iran supports clients in country, private Pakistani and Saudi sources may also be active; power struggles among various groups for control of Kabul, regional rivalries among emerging warlords, traditional tribal disputes continue; support to Islamic fighters in Tajikistan's civil war; border dispute with Pakistan (Durand Line)
Climate: arid to semiarid; cold winters and hot summers
Terrain: mostly rugged mountains; plains in north and southwest
Natural resources: natural gas, petroleum, coal, copper, talc, barites, sulphur, lead, zinc, iron ore, salt, precious and semiprecious stones
Land use:
arable land: 12%
permanent crops: 0%
meadows and pastures: 46%
forest and woodland: 3%
other: 39%
Irrigated land: 26,600 km² (1989 est.)
Environment: damaging earthquakes occur in Hindu Kush mountains; soil degradation, desertification, overgrazing, deforestation, pollution, flooding
Note: landlocked

People

Population: 16,494,145 (July 1993 est.)
Population growth rate: 2.45% (1993 est.)
Birth rate: 43.83 births/1,000 population (1993 est.)
Death rate: 19.33 deaths/1,000 population (1993 est.)
Net migration rate: 0 migrant(s)/1,000 population (1993 est.)
Infant mortality rate: 158.9 deaths/1,000 live births (1993 est.)
Life expectancy at birth:
total population: 44.41 years
male: 45.09 years
female: 43.71 years (1993 est.)
Total fertility rate: 6.34 children born/woman (1993 est.)
Nationality:
noun: Afghan(s)
adjective: Afghan
Ethnic divisions: Pashtun 38%, Tajik 25%, Uzbek 6%, Hazara 19%, minor ethnic groups (Chahar Aimaks, Turkmen, Baloch, and others)
Religions: Sunni Muslim 84%, Shi'a Muslim 15%, other 1%
Languages: Pashtu 35%, Afghan Persian (Dari) 50%, Turkic languages (primarily Uzbek and Turkmen) 11%, 30 minor languages (primarily Balochi and Pashai) 4%, much bilingualism
Literacy: age 15 and over can read and write (1990)
total population: 29%
male: 44%
female: 14%
Labor force: 4.98 million
by occupation: agriculture and animal husbandry 67.8%, industry 10.2%, construction 6.3%, commerce 5.0%, services and other 10.7% (1980 est.)

Government

Names:
conventional long form: Islamic State of Afghanistan
conventional short form: Afghanistan
former: Republic of Afghanistan
Digraph: AF
Type: transitional government
Capital: Kabul
Administrative divisions: 30 provinces (velayat, singular—velayat); Badakhshan, Badghis, Baghlan, Balkh, Bamian, Farah, Faryab, Ghazni, Ghowr, Helmand, Herat, Jowzjan, Kabol, Kandahar, Kapisa, Konar, Kondoz, Laghman, Lowgar, Nangarhar, Nimruz, Oruzgan, Paktia, Paktika, Parvan, Samangan, Sar-e Pol, Takhar, Vardak, Zabol
note: there may be a new province of Nurestan (Nuristan)
Independence: 19 August 1919 (from UK)
Constitution: the old Communist-era constitution has been suspended; a new Islamic constitution has yet to be ratified
Legal system: a new legal system has not been adopted but the transitional government has declared it will follow Islamic law (Shari'a)
National holiday: Victory of the Muslim Nation, 28 April; Remembrance Day for Martyrs and Disabled, 4 May; Independence Day, 19 August
Political parties and leaders: current political organizations include Jamiat-i-Islami (Islamic Society), Burhanuddin RABBANI, Ahmad Shah MASOOD; Hizbi Islami-Gulbuddin (Islamic Party), Gulbuddin HIKMATYAR faction; Hizbi Islami-Khalis (Islamic Party) Yunis KHALIS faction; Ittihad-i-Islami Barai Azadi Afghanistan (Islamic Union for the Liberation of Afghanistan), Abdul Rasul SAYYAF; Harakat-Inqilab-i-Islami (Islamic Revolutionary Movement), Mohammad Nabi MOHAMMADI; Jabha-i-Najat-i-Milli Afghanistan (Afghanistan National Liberation Front), Sibghatullah MOJADDEDI; Mahaz-i-Milli-Islami (National Islamic Front), Sayed Ahamad GAILANI; Hizbi Wahdat (Islamic Unity Party), Abdul Ali MAZARI; Harakat-i-Islami (Islamic Movement), Mohammed Asif MOHSENI; a new northern organization consisting of resistance and former regional figures is Jonbesh-i-Milli Islami (National Islamic Movement), Rashid DOSTAM
note: the former ruling Watan Party has been disbanded
Other political or pressure groups: the former resistance commanders are the major power brokers in the countryside; shuras (councils) of commanders are now administering most cities outside Kabul; ulema (religious scholars); tribal elders
Suffrage: undetermined; previously universal, male ages 15-50
Elections:
President: last held NA December 1992 (next to be held NA December 1994); results—Burhanuddin RABBANI was elected to a two-year term by a national shura
Executive branch: president, prime minister; Afghan leaders are still in the process of choosing a cabinet (May 1993)
Legislative branch: a unicameral parliament consisting of 205 members was chosen by the shura in January 1993; non-functioning as of June 1993
Judicial branch: an interim Chief Justice of the Supreme Court has been appointed, but a new court system has not yet been organized
Leaders:
Chief of State: President Burhanuddin RABBANI (since 2 January 1993); First Vice President Mohammad NABI Mohammadi (since NA); First Vice President Mohammad SHAH Fazli (since NA)

Afghanistan *(continued)*

Head of Government: Prime Minister-designate Gulbaddin HIKMATYAR (since NA) ; Deputy Prime Minister Sulayman GAILANI (since NA); Deputy Prime Minister Din MOHAMMAD (since NA); Deputy Prime Minister Ahmad SHAH Ahmadzai (since NA)

Member of: AsDB (has previously been a member of), CP, ECO, ESCAP, FAO, G-77, IAEA, IBRD, ICAO, IDA, IDB, IFAD, IFC, ILO, IMF, INTELSAT, IOC, ITU, LORCS, NAM, OIC, UN, UNCTAD, UNESCO, UNIDO, UPU, WFTU, WHO, WMO, WTO

Diplomatic representation in US:
chief of mission: (vacant); Charge d'Affaires Abdul RAHIM
chancery: 2341 Wyoming Avenue NW, Washington, DC 20008
telephone: (202) 234-3770 or 3771

US diplomatic representation:
chief of mission: (vacant)
embassy: Ansari Wat, Wazir Akbar Khan Mina, Kabul
mailing address: use embassy street address
telephone: 62230 through 62235 or 62436
note: US Embassy in Kabul was closed in January 1989

Flag: a new flag of unknown description reportedly has been adopted; previous flag consisted of three equal horizontal bands of black (top), red, and green, with the national coat of arms superimposed on the hoist side of the black and red bands; similar to the flag of Malawi, which is shorter and bears a radiant, rising red sun centered in the black band

---

Economy

Overview: Fundamentally, Afghanistan is an extremely poor, landlocked country, highly dependent on farming (wheat especially) and livestock raising (sheep and goats). Economic considerations have played second fiddle to political and military upheavals during more than 13 years of war, including the nearly 10-year Soviet military occupation (which ended 15 February 1989). Over the past decade, one-third of the population fled the country, with Pakistan sheltering more than 3 million refugees and Iran about 1.3 million. Another 1 million probably moved into and around urban areas within Afghanistan. Although reliable data are unavailable, gross domestic product is lower than 12 years ago because of the loss of labor and capital and the disruption of trade and transport.

National product: GDP—exchange rate conversion—$3 billion (1989 est.)

National product real growth rate: NA%

National product per capita: $200 (1989 est.)

Inflation rate (consumer prices): over 90% (1991 est.)

Unemployment rate: NA%

Budget: revenues $NA; expenditures $NA, including capital expenditures of $NA

Exports: $236 million (f.o.b., FY91 est.)
commodities: natural gas 55%, fruits and nuts 24%, handwoven carpets, wool, cotton, hides, and pelts
partners: former USSR, Pakistan

Imports: $874 million (c.i.f., FY91 est.)
commodities: food and petroleum products
partners: former USSR, Pakistan

External debt: $2.3 billion (March 1991 est.)

Industrial production: growth rate 2.3% (FY91 est.); accounts for about 25% of GDP

Electricity: 480,000 kW capacity; 1,000 million kWh produced, 60 kWh per capita (1992)

Industries: small-scale production of textiles, soap, furniture, shoes, fertilizer, and cement; handwoven carpets; natural gas, oil, coal, copper

Agriculture: largely subsistence farming and nomadic animal husbandry; cash products—wheat, fruits, nuts, karakul pelts, wool, mutton

Illicit drugs: an illicit producer of opium poppy and cannabis for the international drug trade; world's second-largest opium producer (after Burma) and a major source of hashish

Economic aid: US commitments, including Ex-Im (FY70-89), $380 million; Western (non-US) countries, ODA and OOF bilateral commitments (1970-89), $510 million; OPEC bilateral aid (1979-89), $57 million; Communist countries (1970-89), $4.1 billion; net official Western disbursements (1985-89), $270 million

Currency: 1 afghani (AF) = 100 puls

Exchange rates: afghanis (Af) per US$1—1,019 (March 1993), 900 (November 1991), 850 (1991), 700 (1989-90), 220 (1988-89); note—these rates reflect the free market exchange rates rather than the official exchange rates

Fiscal year: 21 March-20 March

Communications

Railroads: 9.6 km (single track) 1.524-meter gauge from Kushka (Turkmenistan) to Towraghondi and 15.0 km from Termez (Uzbekistan) to Kheyrabad transshipment point on south bank of Amu Darya

Highways: 21,000 km total (1984); 2,800 km hard surface, 1,650 km bituminous-treated gravel and improved earth, 16,550 km unimproved earth and tracks

Inland waterways: total navigability 1,200 km; chiefly Amu Darya, which handles vessels up to about 500 metric tons

Pipelines: petroleum products—Uzbekistan to Bagram and Turkmenistan to Shindand; natural gas 180 km

Ports: Shir Khan and Kheyrabad (river ports)

Airports:
total: 41
usable: 36
with permanent-surface runways: 9
with runways over 3,659 m: 0
with runways 2,440-3,659 m: 11
with runways 1,220-2,439 m: 16

Telecommunications: limited telephone, telegraph, and radiobroadcast services; television introduced in 1980; 31,200 telephones; broadcast stations—5 AM, no FM, 1 TV; 1 satellite earth station

---

Defense Forces

Branches: the military still does not yet exist on a national scale; some elements of the former Army, Air and Air Defense Forces, National Guard, Border Guard Forces, National Police Force (Sarandoi), and tribal militias remain intact

Manpower availability: males age 15-49 4,094,481; fit for military service 2,196,136; reach military age (22) annually 153,333 (1993 est.)

Defense expenditures: the new government has not yet adopted a defense budget

Albania

Geography

Location: Southeastern Europe, on the Balkan Peninsula between Serbia and Montenegro and Greece
Map references: Africa, Ethnic Groups in Eastern Europe, Europe, Standard Time Zones of the World
Area:
total area: 28,750 km²
land area: 27,400 km²
comparative area: slightly larger than Maryland
Land boundaries: total 720 km, Greece 282 km, Macedonia 151 km, Serbia and Montenegro 287 km (114 km with Serbia, 173 km with Montenegro)
Coastline: 362 km
Maritime claims:
continental shelf: not specified
territorial sea: 12 nm
International disputes: Kosovo question with Serbia and Montenegro; Northern Epirus question with Greece
Climate: mild temperate; cool, cloudy, wet winters; hot, clear, dry summers; interior is cooler and wetter
Terrain: mostly mountains and hills; small plains along coast
Natural resources: petroleum, natural gas, coal, chromium, copper, timber, nickel
Land use:
arable land: 21%
permanent crops: 4%
meadows and pastures: 15%
forest and woodland: 38%
other: 22%
Irrigated land: 4,230 km² (1989)
Environment: subject to destructive earthquakes; tsunami occur along southwestern coast
Note: strategic location along Strait of Otranto (links Adriatic Sea to Ionian Sea and Mediterranean Sea)

People

Population: 3,333,839 (July 1993 est.)
Population growth rate: 1.21% (1993 est.)

Birth rate: 23.24 births/1,000 population (1993 est.)
Death rate: 5.45 deaths/1,000 population (1993 est.)
Net migration rate: -5.67 migrant(s)/1,000 population (1993 est.)
Infant mortality rate: 31.8 deaths/1,000 live births (1993 est.)
Life expectancy at birth:
total population: 73 years
male: 70.01 years
female: 76.21 years (1993 est.)
Total fertility rate: 2.85 children born/woman (1993 est.)
Nationality:
noun: Albanian(s)
adjective: Albanian
Ethnic divisions: Albanian 90%, Greeks 8%, other 2% (Vlachs, Gypsies, Serbs, and Bulgarians) (1989 est.)
Religions: Muslim 70%, Greek Orthodox 20%, Roman Catholic 10%
note: all mosques and churches were closed in 1967 and religious observances prohibited; in November 1990, Albania began allowing private religious practice
Languages: Albanian (Tosk is the official dialect), Greek
Literacy: age 9 and over can read and write (1955)
total population: 72%
male: 80%
female: 63%
Labor force: 1.5 million (1987)
by occupation: agriculture 60%, industry and commerce 40% (1986)

Government

Names:
conventional long form: Republic of Albania
conventional short form: Albania
local long form: Republika e Shqiperise
local short form: Shqiperia
former: People's Socialist Republic of Albania
Digraph: AL
Type: nascent democracy
Capital: Tirane
Administrative divisions: 26 districts (rrethe, singular—rreth); Berat, Dibre, Durres, Elbasan, Fier, Gjirokaster, Gramsh, Kolonje, Korce, Kruje, Kukes, Lezhe, Librazhd, Lushnje, Mat, Mirdite, Permet, Pogradec, Puke, Sarande, Shkoder, Skrapar, Tepelene, Tirane, Tropoje, Vlore
Independence: 28 November 1912 (from Ottoman Empire)
Constitution: an interim basic law was approved by the People's Assembly on 29 April 1991; a new constitution was to be drafted for adoption in 1992, but is still in process
Legal system: has not accepted compulsory ICJ jurisdiction

National holiday: Liberation Day, 29 November (1944)
Political parties and leaders: there are at least 18 political parties; most prominent are the Albanian Socialist Party (ASP; formerly the Albania Workers Party), Fatos NANO, first secretary; Democratic Party (DP), Eduard SELAMI, chairman; Albanian Republican Party (RP), Sabri GODO; Omonia (Greek minority party), leader NA (ran in 1992 election as Unity for Human Rights Party (UHP)); Social Democratic Party (SDP), Skender GJINUSHI; Democratic Alliance Party (DAP), Spartak NGJELA, chairman
Suffrage: 18 years of age, universal and compulsory
Elections:
People's Assembly: last held 22 March 1992; results—DP 62.29%, ASP 25.57%, SDP 4.33%, RP 3.15%, UHP 2.92%, other 1.74%; seats—(140 total) DP 92, ASP 38, SDP 7, RP 1, UHP 2
Executive branch: president, prime minister of the Council of Ministers, two deputy prime ministers of the Council of Ministers
Legislative branch: unicameral People's Assembly (Kuvendi Popullor)
Judicial branch: Supreme Court
Leaders:
Chief of State: President of the Republic Sali BERISHA (since 9 April 1992)
Head of Government: Prime Minister of the Council of Ministers Aleksander Gabriel MEKSI (since 10 April 1992)
Member of: BSEC, CSCE, EBRD, ECE, FAO, IAEA, IBRD, ICAO, IDA, IFAD, IFC, IMF, INTERPOL, IOC, IOM (observer), ISO, ITU, LORCS, NACC, OIC, UN, UNCTAD, UNESCO, UNIDO, UPU, WHO, WIPO, WMO
Diplomatic representation in US:
chief of mission: Ambassador Roland BIMO
chancery: 1511 K Street, NW, Washington, DC
telephone: (202) 223-4942
FAX: (202) 223-4950
US diplomatic representation:
chief of mission: Ambassador William E. RYERSON
embassy: Rruga Labinoti 103, room 2921, Tirane
mailing address: PSC 59, Box 100 (A), APO AE 09624
telephone: 355-42-32875, 33520
FAX: 355-42-32222
Flag: red with a black two-headed eagle in the center

Economy

Overview: The Albanian economy, already providing the lowest standard of living in Europe, contracted sharply in 1991, with most industries producing at only a fraction

Albania *(continued)*

of past levels and an unemployment rate estimated at 40%. For over 40 years, the Stalinist-type economy operated on the principle of central planning and state ownership of the means of production. Fitful economic reforms begun during 1991, including the liberalization of prices and trade, the privatization of shops and transport, and land reform, were crippled by widespread civil disorder. Following its overwhelming victory in the 22 March 1992 elections, the new Democratic government announced a program of shock therapy to stabilize the economy and establish a market economy. In an effort to expand international ties, Tirane has reestablished diplomatic relations with the major republics of the former Soviet Union and the US and has joined the IMF and the World Bank. The Albanians have also passed legislation allowing foreign investment, but not foreign ownership of real estate. Albania possesses considerable mineral resources and, until 1990, was largely self-sufficient in food; however, the breakup of cooperative farms in 1991 and general economic decline forced Albania to rely on foreign aid to maintain adequate supplies. In 1992 the government tightened budgetary contols leading to another drop in domestic output. The agricultural sector is steadily gaining from the privatization process. Low domestic output is supplemented by remittances from the 200,000 Albanians working abroad.

National product: GDP—purchasing power equivalent—$2.5 billion (1992 est.)

National product real growth rate: -10% (1992 est.)

National product per capita: $760 (1992 est.)

Inflation rate (consumer prices): 210% (1992 est.)

Unemployment rate: 40% (1992 est.)

Budget: revenues $1.1 billion; expenditures $1.4 billion, including capital expenditures of $70 million (1991 est.)

Exports: $45 million (f.o.b., 1992 est.)
commodities: asphalt, metals and metallic ores, electricity, crude oil, vegetables, fruits, tobacco
partners: Italy, Macedonia, Germany, Greece, Czechoslovakia, Poland, Romania, Bulgaria, Hungary

Imports: $120 million (f.o.b., 1992 est.)
commodities: machinery, consumer goods, grains
partners: Italy, Macedonia, Germany, Czechoslovakia, Romania, Poland, Hungary, Bulgaria, Greece

External debt: $500 million (1992 est.)

Industrial production: growth rate −55% (1991 est.)

Electricity: 1,690,000 kW capacity; 5,000 million kWh produced, 1,520 kWh per capita (1992)

Industries: food processing, textiles and clothing, lumber, oil, cement, chemicals, mining, basic metals, hydropower

Agriculture: arable land per capita among lowest in Europe; over 60% of arable land now in private hands; one-half of work force engaged in farming; wide range of temperate-zone crops and livestock

Illicit drugs: transshipment point for Southwest Asian heroin transiting the Balkan route

Economic aid: recipient—$190 million humanitarian aid, $94 million in loans/guarantees/credits

Currency: 1 lek (L) = 100 qintars

Exchange rates: leke (L) per US$1—97 (January 1993), 50 (January 1992), 25 (September 1991)

Fiscal year: calendar year

Communications

Railroads: 543 km total; 509 km 1.435-meter standard gauge, single track and 34 km narrow gauge, single track (1990); line connecting Titograd (Serbia and Montenegro) and Shkoder (Albania) completed August 1986

Highways: 16,700 km total; 6,700 km highways, 10,000 km forest and agricultural cart roads (1990)

Inland waterways: 43 km plus Albanian sections of Lake Scutari, Lake Ohrid, and Lake Prespa (1990)

Pipelines: crude oil 145 km; petroleum products 55 km; natural gas 64 km (1991)

Ports: Durres, Sarande, Vlore

Merchant marine: 11 cargo ships (1,000 GRT or over) totaling 52,967 GRT/76,887 DWT

Airports:
total: 12
usable: 10
with permanent-surface runways: 3
with runways over 3,659 m: 0
with runways 2,440-3,659 m: 6
with runways 1,220-2,439 m: 4

Telecommunications: inadequate service; 15,000 telephones; broadcast stations—13 AM, 1 TV; 514,000 radios, 255,000 TVs (1987 est.)

Defense Forces

Branches: Army, Navy, Air and Air Defense Forces, Interior Ministry Troops

Manpower availability: males age 15-49 896,613; fit for military service 739,359; reach military age (19) annually 32,740 (1993 est.)

Defense expenditures: 215 million leke, NA% of GNP (1993 est.); note—conversion of defense expenditures into US dollars using the current exchange rate could produce misleading results

Algeria

Geography

Location: Northern Africa, along the Mediterranean Sea, between Morocco and Tunisia

Map references: Africa, Europe

Area:
total area: 2,381,740 km^2
land area: 2,381,740 km^2
comparative area: slightly less than 3.5 times the size of Texas

Land boundaries: total 6,343 km, Libya 982 km, Mali 1,376 km, Mauritania 463 km, Morocco 1,559 km, Niger 956 km, Tunisia 965 km, Western Sahara 42 km

Coastline: 998 km

Maritime claims:
territorial sea: 12 nm

International disputes: Libya claims part of southeastern Algeria; land boundary disputes with Tunisia under discussion

Climate: arid to semiarid; mild, wet winters with hot, dry summers along coast; drier with cold winters and hot summers on high plateau; sirocco is a hot, dust/sand-laden wind especially common in summer

Terrain: mostly high plateau and desert; some mountains; narrow, discontinuous coastal plain

Natural resources: petroleum, natural gas, iron ore, phosphates, uranium, lead, zinc

Land use:
arable land: 3%
permanent crops: 0%
meadows and pastures: 13%
forest and woodland: 2%
other: 82%

Irrigated land: 3,360 km^2 (1989 est.)

Environment: mountainous areas subject to severe earthquakes; desertification

Note: second-largest country in Africa (after Sudan)

People

Population: 27,256,252 (July 1993 est.)

Population growth rate: 2.34% (1993 est.)

Birth rate: 30.38 births/1,000 population (1993 est.)

Death rate: 6.41 deaths/1,000 population (1993 est.)

Net migration rate: -0.53 migrant(s)/1,000 population (1993 est.)

Infant mortality rate: 54 deaths/1,000 live births (1993 est.)

Life expectancy at birth:

total population: 67.35 years

male: 66.32 years

female: 68.41 years (1993 est.)

Total fertility rate: 3.96 children born/woman (1993 est.)

Nationality:

noun: Algerian(s)

adjective: Algerian

Ethnic divisions: Arab-Berber 99%, European less than 1%

Religions: Sunni Muslim (state religion) 99%, Christian and Jewish 1%

Languages: Arabic (official), French, Berber dialects

Literacy: age 15 and over can read and write (1990)

total population: 57%

male: 70%

female: 46%

Labor force: 6.2 million (1992 est.)

by occupation: government 29.5%, agriculture 22%, construction and public works 16.2%, industry 13.6%, commerce and services 13.5%, transportation and communication 5.2% (1989)

Government

Names:

conventional long form: Democratic and Popular Republic of Algeria

conventional short form: Algeria

local long form: Al Jumhuriyah al Jaza'iriyah ad Dimuqratiyah ash Shabiyah

local short form: Al Jaza'ir

Digraph: AG

Type: republic

Capital: Algiers

Administrative divisions: 48 provinces (wilayast, singular—wilaya); Adrar, Ain Defla, Ain Temouchent, Alger, Annaba, Batna, Bechar, Bejaia, Biskra, Blida, Bordj Bou Arreridj, Bouira, Boumerdes, Chlef, Constantine, Djelfa, El Bayadh, El Oued, El Tarf, Ghardaia, Guelma, Illizi, Jijel, Khenchela, Laghouat, Mascara, Medea, Mila, Mostaganem, M'Sila, Naama, Oran, Ouargla, Oum el Bouaghi, Relizane, Saida, Setif, Sidi Bel Abbes, Skikda, Souk Ahras, Tamanghasset, Tebessa, Tiaret, Tindouf, Tipaza, Tissemsilt, Tizi Ouzou, Tlemcen

Independence: 5 July 1962 (from France)

Constitution: 19 November 1976, effective 22 November 1976; revised February 1989

Legal system: socialist, based on French and Islamic law; judicial review of legislative acts in ad hoc Constitutional Council composed of various public officials, including several Supreme Court justices; has not accepted compulsory ICJ jurisdiction

National holiday: Anniversary of the Revolution, 1 November (1954)

Political parties and leaders: Islamic Salvation Front (FIS), Ali BELHADJ, Dr. Abassi MADANI, Abdelkader HACHANI (all under arrest), Rabeh KEBIR; National Liberation Front (FLN), Abdelhamid MEHRI, Secretary General; Socialist Forces Front (FFS), Hocine Ait AHMED, Secretary General

note: the government established a multiparty system in September 1989 and, as of 31 December 1990, over 30 legal parties existed

Suffrage: 18 years of age; universal

Elections:

National People's Assembly: first round held on 26 December 1991 (second round canceled by the military after President BENDJEDID resigned 11 January 1992); results—percent of vote by party NA; seats—(281 total); the fundamentalist FIS won 188 of the 231 seats contested in the first round; note—elections (municipal and wilaya) were held in June 1990, the first in Algerian history; results—FIS 55%, FLN 27.5%, other 17.5%, with 65% of the voters participating

President of the High State Committee: next election to be held December 1993

Executive branch: President of the High State Committee, prime minister, Council of Ministers (cabinet)

Legislative branch: unicameral National People's Assembly (Al-Majlis Ech-Chaabi Al-Watani)

Judicial branch: Supreme Court (Cour Supreme)

Leaders:

Chief of State: High State Committee President Ali KAFI (since 2 July 1992)

Head of Government: Prime Minister Belaid ABDESSELAM (since 8 July 1992)

Member of: ABEDA, AfDB, AFESD, AL, AMF, AMU, CCC, ECA, FAO, G-15, G-19, G-24, G-77, IAEA, IBRD, ICAO, IDA, IDB, IFAD, IFC, ILO, IMF, IMO, INMARSAT, INTELSAT, INTERPOL, IOC, ISO, ITU, LORCS, NAM, OAPEC, OAS (observer), OAU, OIC, OPEC, UN, UNAVEM II, UNCTAD, UNESCO, UNHCR, UNIDO, UNTAC, UPU, WCL, WHO, WIPO, WMO, WTO

Diplomatic representation in US:

chief of mission: Ambassador Mohamed ZARHOUNI

chancery: 2118 Kalorama Road NW, Washington, DC 20008

telephone: (202) 265-2800

US diplomatic representation:

chief of mission: Ambassador Mary Ann CASEY

embassy: 4 Chemin Cheikh Bachir El-Ibrahimi, Algiers

mailing address: B. P. Box 549, Alger-Gare, 16000 Algiers

telephone: [213] (2) 601-425 or 255, 186

FAX: [213] (2) 603979

consulate: Oran

Flag: two equal vertical bands of green (hoist side) and white with a red five-pointed star within a red crescent; the crescent, star, and color green are traditional symbols of Islam (the state religion)

Economy

Overview: The oil and natural gas sector forms the backbone of the economy, hydrocarbons accounting for nearly all export receipts, about 30% of government revenues, and nearly 25% of GDP. In 1973-74 the sharp increase in oil prices led to a booming economy and helped to finance an ambitious program of industrialization. Plunging oil and gas prices, combined with the mismanagement of Algeria's highly centralized economy, has brought the nation to its most serious social and economic crisis since full independence in 1988. The current government has put reform, including privatization of some public sector companies and an overhaul of the banking and financial system, on hold, but has continued efforts to admit private enterprise to the hydrocarbon industry.

National product: GDP—exchange rate conversion—$42 billion (1992 est.)

National product real growth rate: 2.8% (1992 est.)

National product per capita: $1,570 (1992 est.)

Inflation rate (consumer prices): 55% (1992 est.)

Unemployment rate: 35% (1992 est.)

Budget: revenues $14.4 billion; expenditures $14.6 billion, including capital expenditures of $3.5 billion (1992 est.)

Exports: $11.6 billion (f.o.b., 1992 est.)

commodities: petroleum and natural gas 97%

partners: Italy, France, US, Germany, Spain

Imports: $8.2 billion (f.o.b., 1992 est.)

commodities: capital goods 39.7%, food and beverages 21.7%, consumer goods 11.8% (1990)

partners: France, Italy, Germany, US, Spain

External debt: $26 billion (1992 est.)

Industrial production: growth rate NA%

Electricity: 6,380,000 kW capacity; 16,834 million kWh produced, 630 kWh per capita (1992)

Industries: petroleum, light industries, natural gas, mining, electrical, petrochemical, food processing

Agriculture: accounts for 10.8% of GDP

Algeria *(continued)*

(1991) and employs 22% of labor force; products—wheat, barley, oats, grapes, olives, citrus, fruits, sheep, cattle; net importer of food—grain, vegetable oil, sugar
Economic aid: US commitments, including Ex-Im (FY70-85), $1.4 billion; Western (non-US) countries, ODA and OOF bilateral commitments (1970-89), $925 million; OPEC bilateral aid (1979-89), $1.8 billion; Communist countries (1970-89), $2.7 billion; net official disbursements (1985-89), −$375 million
Currency: 1 Algerian dinar (DA) = 100 centimes
Exchange rates: Algerian dinars (DA) per US$1—22.787 (January 1993), 21.836 (1992), 18.473 (1991), 8.958 (1990), 7.6086 (1989), 5.9148 (1988)
Fiscal year: calendar year

Communications

Railroads: 4,060 km total; 2,616 km standard gauge (1.435 m), 1,188 km 1.055-meter gauge, 256 km 1.000-meter gauge; 300 km electrified; 215 km double track
Highways: 90,031 km total; 58,868 km concrete or bituminous, 31,163 km gravel, crushed stone, unimproved earth (1990)
Pipelines: crude oil 6,612 km; petroleum products 298 km; natural gas 2,948 km
Ports: Algiers, Annaba, Arzew, Bejaia, Djendjene, Ghazaouet, Jijel, Mers el Kebir, Mostaganem, Oran, Skikda
Merchant marine: 75 ships (1,000 GRT or over) totaling 903,179 GRT/1,064,211 DWT; includes 5 short-sea passenger, 27 cargo, 12 roll-on/roll-off cargo, 5 oil tanker, 9 liquefied gas, 7 chemical tanker, 9 bulk, 1 specialized tanker
Airports:
total: 141
usable: 124
with permanent-surface runways: 53
with runways over 3,659 m: 2
with runways 2,440-3,659 m: 32
with runways 1,220-2,439 m: 65
Telecommunications: excellent domestic and international service in the north, sparse in the south; 822,000 telephones; broadcast stations—26 AM, no FM, 18 TV; 1,600,000 TV sets; 5,200,000 radios; 5 submarine cables; microwave radio relay to Italy, France, Spain, Morocco, and Tunisia; coaxial cable to Morocco and Tunisia; satellite earth stations—1 Atlantic Ocean INTELSAT, 1 Indian Ocean INTELSAT, 1 Intersputnik, 1 ARABSAT, and 12 domestic; 20 additional satellite earth stations are planned

---

Defense Forces

Branches: National Popular Army, Navy, Air Force, Territorial Air Defense

Manpower availability: males age 15-49 6,610,342; fit for military service 4,063,261; reach military age (19) annually 291,685 (1993 est.)
Defense expenditures: exchange rate conversion—$1.36 billion, 2.5% of GDP (1993 est.)

American Samoa
(territory of the US)

Geography

Location: in the South Pacific Ocean, 3,700 km south-southwest of Honolulu, about halfway between Hawaii and New Zealand
Map references: Oceania
Area:
total area: 199 km²
land area: 199 km²
comparative area: slightly larger than Washington, DC
note: includes Rose Island and Swains Island
Land boundaries: 0 km
Coastline: 116 km
Maritime claims:
contiguous zone: 24 nm
continental shelf: 200 m or depth of exploitation
exclusive economic zone: 200 nm
territorial sea: 12 nm
International disputes: none
Climate: tropical marine, moderated by southeast trade winds; annual rainfall averages 124 inches; rainy season from November to April, dry season from May to October; little seasonal temperature variation
Terrain: five volcanic islands with rugged peaks and limited coastal plains, two coral atolls (Rose Island, Swains Island)
Natural resources: pumice, pumicite
Land use:
arable land: 10%
permanent crops: 5%
meadows and pastures: 0%
forest and woodland: 75%
other: 10%
Irrigated land: NA km²
Environment: typhoons common from December to March
Note: Pago Pago has one of the best natural deepwater harbors in the South Pacific Ocean, sheltered by shape from rough seas and protected by peripheral mountains from high winds; strategic location in the South Pacific Ocean

People

Population: 53,139 (July 1993 est.)
Population growth rate: 3.9% (1993 est.)
Birth rate: 37 births/1,000 population (1993 est.)
Death rate: 4 deaths/1,000 population (1993 est.)
Net migration rate: 6 migrant(s)/1,000 population (1993 est.)
Infant mortality rate: 19 deaths/1,000 live births (1993 est.)
Life expectancy at birth:
total population: 73 years
male: 71 years
female: 75 years (1993 est.)
Total fertility rate: 4.41 children born/woman (1993 est.)
Nationality:
noun: American Samoan(s)
adjective: American Samoan
Ethnic divisions: Samoan (Polynesian) 89%, Caucasian 2%, Tongan 4%, other 5%
Religions: Christian Congregationalist 50%, Roman Catholic 20%, Protestant denominations and other 30%
Languages: Samoan (closely related to Hawaiian and other Polynesian languages), English; most people are bilingual
Literacy: age 15 and over can read and write (1980)
total population: 97%
male: 97%
female: 97%
Labor force: 14,400 (1990)
by occupation: government 33%, tuna canneries 34%, other 33% (1990)

Government

Names:
conventional long form: Territory of American Samoa
conventional short form: American Samoa
Abbreviation: AS
Digraph: AQ
Type: unincorporated and unorganized territory of the US; administered by the US Department of Interior, Office of Territorial and International Affairs
Capital: Pago Pago
Administrative divisions: none (territory of the US)
Independence: none (territory of the US)
Constitution: ratified 1966, in effect 1967
Legal system: NA
National holiday: Territorial Flag Day, 17 April (1900)
Political parties and leaders: NA
Suffrage: 18 years of age; universal
Elections:
Governor: last held 3 November 1992 (next to be held NA November 1996); results—A. P. LUTALI was elected (percent of vote NA)
House of Representatives: last held 3 November 1992 (next to be held NA November 1994); results—representatives popularly elected from 17 house districts; seats—(21 total, 20 elected, and 1 nonvoting delegate from Swains Island)
Senate: last held 3 November 1992 (next to be held NA November 1996); results—senators elected by village chiefs from 12 senate districts; seats—(18 total) number of seats by party NA
US House of Representatives: last held 3 November 1992 (next to be held NA November 1994); results—Eni R. F. H. FALEOMAVAEGA reelected as delegate
Executive branch: popularly elected governor and lieutenant governor
Legislative branch: bicameral Legislative Assembly (Fono) consists of an upper house or Senate (appointed by county village chiefs) and a lower house or House of Representatives (elected)
Judicial branch: High Court
Leaders:
Chief of State: President William Jefferson CLINTON (since 20 January 1993); Vice President Albert GORE, Jr. (since 20 January 1993)
Head of Government: Governor A. P. LUTALI (since 3 January 1993); Lieutenant Governor Tauese P. SUNIA (since 3 January 1993)
Member of: ESCAP (associate), INTERPOL (subbureau), IOC, SPC
Diplomatic representation in US: none (territory of the US)
Flag: blue with a white triangle edged in red that is based on the fly side and extends to the hoist side; a brown and white American bald eagle flying toward the hoist side is carrying two traditional Samoan symbols of authority, a staff and a war club

Economy

Overview: Economic activity is strongly linked to the US, with which American Samoa does 80-90% of its foreign trade. Tuna fishing and tuna processing plants are the backbone of the private sector, with canned tuna the primary export. The tuna canneries and the government are by far the two largest employers. Other economic activities include a slowly developing tourist industry. Transfers from the US government add substantially to American Samoa's economic well-being.
National product: GDP—purchasing power equivalent—$128 million (1991)
National product real growth rate: NA%
National product per capita: $2,600 (1991)
Inflation rate (consumer prices): 7% (1990)
Unemployment rate: 12% (1991)
Budget: revenues $97,000,000 (includes $43,000,000 in local revenue and $54,000,000 in grant revenue); including capital expenditures of $NA (FY91)
Exports: $306 million (f.o.b., 1989)
commodities: canned tuna 93%
partners: US 99.6%
Imports: $360.3 million (c.i.f., 1989)
commodities: materials for canneries 56%, food 8%, petroleum products 7%, machinery and parts 6%
partners: US 62%, Japan 9%, NZ 7%, Australia 11%, Fiji 4%, other 7%
External debt: $NA
Industrial production: growth rate NA%
Electricity: 42,000 kW capacity; 100 million kWh produced, 2,020 kWh per capita (1990)
Industries: tuna canneries (largely dependent on foreign fishing vessels), meat canning, handicrafts
Agriculture: bananas, coconuts, vegetables, taro, breadfruit, yams, copra, pineapples, papayas, dairy farming
Economic aid: $21,042,650 in operational funds and $1,227,000 in construction funds for capital improvement projects from the US Department of Interior (1991)
Currency: US currency is used
Fiscal year: 1 October-30 September

Communications

Railroads: none
Highways: 350 km total; 150 km paved, 200 km unpaved
Ports: Pago Pago, Ta'u, Ofu, Auasi, Aanu'u (new construction), Faleosao
Airports:
total: 3
usable: 3
with permanent-surface runways: 3
with runways over 3,659 m: 0
with runways 2,440 to 3,659 m : 1 (international airport at Tafuna)
with runways 1,200 to 2,439 m: 0
note: small airstrips on Fituita and Ofu
Telecommunications: 8,399 telephones; broadcast stations—1 AM, 1 FM, 1 TV; good telex, telegraph, and facsimile services; 1 Pacific Ocean INTELSAT earth station, 1 COMSAT earth station

Defense Forces

Note: defense is the responsibility of the US

Andorra

Geography

Location: Western Europe, between France and Spain
Map references: Europe, Standard Time Zones of the World
Area:
total area: 450 km²
land area: 450 km²
comparative area: slightly more than 2.5 times the size of Washington, DC
Land boundaries: total 125 km, France 60 km, Spain 65 km
Coastline: 0 km (landlocked)
Maritime claims: none; landlocked
International disputes: none
Climate: temperate; snowy, cold winters and cool, dry summers
Terrain: rugged mountains dissected by narrow valleys
Natural resources: hydropower, mineral water, timber, iron ore, lead
Land use:
arable land: 2%
permanent crops: 0%
meadows and pastures: 56%
forest and woodland: 22%
other: 20%
Irrigated land: NA km²
Environment: deforestation, overgrazing
Note: landlocked

People

Population: 61,962 (July 1993 est.)
Population growth rate: 3.27% (1993 est.)
Birth rate: 13.78 births/1,000 population (1993 est.)
Death rate: 6.99 deaths/1,000 population (1993 est.)
Net migration rate: 25.92 migrant(s)/1,000 population (1993 est.)
Infant mortality rate: 8.1 deaths/1,000 live births (1993 est.)
Life expectancy at birth:
total population: 78.22 years
male: 75.35 years
female: 81.34 years (1993 est.)

Total fertility rate: 1.73 children born/woman (1993 est.)
Nationality:
noun: Andorran(s)
adjective: Andorran
Ethnic divisions: Spanish 61%, Andorran 30%, French 6%, other 3%
Religions: Roman Catholic (predominant)
Languages: Catalan (official), French, Castilian
Literacy:
total population: NA%
male: NA%
female: NA%
Labor force: NA

Government

Names:
conventional long form: Principality of Andorra
conventional short form: Andorra
local long form: Principat d'Andorra
local short form: Andorra
Digraph: AN
Type: parliamentary coprincipality under formal sovereignty of president of France and Spanish bishop of Seo de Urgel, who are represented locally by officials called veguers; to be changed to a parliamentary form of government
Capital: Andorra la Vella
Administrative divisions: 7 parishes (parroquies, singular—parroquia); Andorra, Canillo, Encamp, La Massana, Les Escaldes, Ordino, Sant Julia de Loria
Independence: 1278
Constitution: Andorra's first written constitution was drafted in 1991; adopted 14 March 1993; to take effect within 15 days
Legal system: based on French and Spanish civil codes; no judicial review of legislative acts; has not accepted compulsory ICJ jurisdiction
National holiday: Mare de Deu de Meritxell, 8 September
Political parties and leaders: political parties not yet legally recognized; traditionally no political parties but partisans for particular independent candidates for the General Council on the basis of competence, personality, and orientation toward Spain or France; various small pressure groups developed in 1972; first formal political party, Andorran Democratic Association, was formed in 1976 and reorganized in 1979 as Andorran Democratic Party
Suffrage: 18 years of age, universal
Elections:
General Council of the Valleys: last held 12 April 1992 (next to be held April 1996); results—percent of vote by party NA; seats—(28 total) number of seats by party NA
Executive branch: two co-princes (president of France, bishop of Seo de Urgel in Spain),

two designated representatives (French veguer, Episcopal veguer), two permanent delegates (French prefect for the department of Pyrenees-Orientales, Spanish vicar general for the Seo de Urgel diocese), president of government, Executive Council
Legislative branch: unicameral General Council of the Valleys (Consell General de las Valls)
Judicial branch: Supreme Court of Andorra at Perpignan (France) for civil cases, the Ecclesiastical Court of the bishop of Seo de Urgel (Spain) for civil cases, Tribunal of the Courts (Tribunal des Cortes) for criminal cases
Leaders:
Chiefs of State: French Co-Prince Francois MITTERRAND (since 21 May 1981), represented by Veguer de Franca Jean Pierre COURTOIS (since NA); Spanish Episcopal Co-Prince Mgr. Juan MARTI Alanis (since 31 January 1971), represented by Veguer Episcopal Francesc BADIA Bata
Head of Government: Executive Council President Oscar RIBAS Reig (since 10 Decmber 1989)
Member of: INTERPOL, IOC
Diplomatic representation in US: Andorra has no mission in the US
US diplomatic representation: Andorra is included within the Barcelona (Spain) Consular District, and the US Consul General visits Andorra periodically
Flag: three equal vertical bands of blue (hoist side), yellow, and red with the national coat of arms centered in the yellow band; the coat of arms features a quartered shield; similar to the flags of Chad and Romania that do not have a national coat of arms in the center

Economy

Overview: The mainstay of Andorra's economy is tourism. An estimated 13 million tourists visit annually, attracted by Andorra's duty-free status and by its summer and winter resorts. The banking sector, with its "tax haven" status, also contributes significantly to the economy. Agricultural production is limited by a scarcity of arable land, and most food has to be imported. The principal livestock activity is sheep raising. Manufacturing consists mainly of cigarettes, cigars, and furniture. Although it is a member of the EC customs union, it is unclear what effect the European Single Market will have on the advantages Andorra obtains from its duty-free status.
National product: GDP—purchasing power equivalent—$760 million (1992 est.)
National product real growth rate: NA% (1992 est.)
National product per capita: $14,000 (1992 est.)
Inflation rate (consumer prices): NA%

Angola

Unemployment rate: 0%
Budget: revenues $119.4 million; expenditures $190 million, including capital expenditures of $NA (1990)
Exports: $23 million (f.o.b., 1989)
commodities: electricity, tobacco products, furniture
partners: France, Spain
Imports: $888.7 million (f.o.b., 1989)
commodities: consumer goods, food
partners: France, Spain
External debt: $NA
Industrial production: growth rate NA%
Electricity: 35,000 kW capacity; 140 million kWh produced, 2,570 kWh per capita (1992)
Industries: tourism (particularly skiing), sheep, timber, tobacco, banking
Agriculture: sheep raising; small quantities of tobacco, rye, wheat, barley, oats, and some vegetables
Economic aid: none
Currency: the French and Spanish currencies are used
Exchange rates: French francs (F) per US$1—5.4812 (January 1993), 5.2938 (1992), 5.6421 (1991), 5.4453 (1990), 6.3801 (1989), 5.9569 (1988); Spanish pesetas (Ptas) per US$1—114.59 (January 1993), 102.38 (1992), 103.91 (1991), 101.93 (1990), 118.38 (1989), 116.49 (1988)
Fiscal year: calendar year

Communications

Highways: 96 km
Telecommunications: international digital microwave network; international landline circuits to France and Spain; broadcast stations—1 AM, no FM, no TV; 17,700 telephones

---

Defense Forces

Note: defense is the responsibility of France and Spain

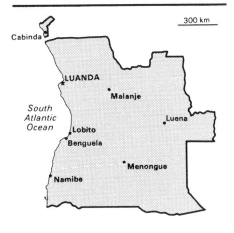

Geography

Location: Southern Africa, bordering the South Atlantic Ocean between Namibia and Zaire
Map references: Africa, Standard Time Zones of the World
Area:
total area: 1,246,700 km^2
land area: 1,246,700 km^2
comparative area: slightly less than twice the size of Texas
Land boundaries: total 5,198 km, Congo 201 km, Namibia 1,376 km, Zaire 2,511 km, Zambia 1,110 km
Coastline: 1,600 km
Maritime claims:
exclusive fishing zone: 200 nm
territorial sea: 20 nm
International disputes: civil war since independence on 11 November 1975; a ceasefire held from 31 May 1991 until October 1992, when the insurgent National Union for the Total Independence of Angola refused to accept its defeat in internationally monitored elections; fighting has since resumed across the countryside
Climate: semiarid in south and along coast to Luanda; north has cool, dry season (May to October) and hot, rainy season (November to April)
Terrain: narrow coastal plain rises abruptly to vast interior plateau
Natural resources: petroleum, diamonds, iron ore, phosphates, copper, feldspar, gold, bauxite, uranium
Land use:
arable land: 2%
permanent crops: 0%
meadows and pastures: 23%
forest and woodland: 43%
other: 32%
Irrigated land: NA km^2
Environment: locally heavy rainfall causes periodic flooding on plateau; desertification
Note: Cabinda is separated from rest of country by Zaire

People

Population: 9,545,235 (July 1993 est.)
Population growth rate: 2.67% (1993 est.)
Birth rate: 45.8 births/1,000 population (1993 est.)
Death rate: 18.96 deaths/1,000 population (1993 est.)
Net migration rate: -0.15 migrant(s)/1,000 population (1993 est.)
Infant mortality rate: 148.6 deaths/1,000 live births (1993 est.)
Life expectancy at birth:
total population: 45.26 years
male: 43.26 years
female: 47.35 years (1993 est.)
Total fertility rate: 6.54 children born/woman (1993 est.)
Nationality:
noun: Angolan(s)
adjective: Angolan
Ethnic divisions: Ovimbundu 37%, Kimbundu 25%, Bakongo 13%, Mestico 2%, European 1%, other 22%
Religions: indigenous beliefs 47%, Roman Catholic 38%, Protestant 15% (est.)
Languages: Portuguese (official), Bantu dialects
Literacy: age 15 and over can read and write (1990)
total population: 42%
male: 56%
female: 28%
Labor force: 2.783 million economically active
by occupation: agriculture 85%, industry 15% (1985 est.)

Government

Names:
conventional long form: Republic of Angola
conventional short form: Angola
local long form: Republic de Angola
local short form: Angola
former: People's Republic of Angola
Digraph: AO
Type: transitional government nominally a multiparty democracy with a strong presidential system
Capital: Luanda
Administrative divisions: 18 provinces (provincias, singular—provincia); Bengo, Benguela, Bie, Cabinda, Cuando Cubango, Cuanza Norte, Cuanza Sul, Cunene, Huambo, Huila, Luanda, Lunda Norte, Lunda Sul, Malanje, Moxico, Namibe, Uige, Zaire
Independence: 11 November 1975 (from Portugal)
Constitution: 11 November 1975; revised 7 January 1978, 11 August 1980, and 6 March 1991
Legal system: based on Portuguese civil law system and customary law; recently modified to accommodate political pluralism

and increased use of free markets
National holiday: Independence Day, 11
November (1975)
Political parties and leaders: Popular
Movement for the Liberation of Angola
(MPLA), led by Jose EDUARDO DOS
SANTOS, is the ruling party and has been in
power since 1975; National Union for the
Total Independence of Angola (UNITA), led
by Jonas SAVIMBI, remains a legal party
despite its returned to armed resistance to
the government; five minor parties have
small numbers of seats in the National
Assembly
Other political or pressure groups:
Cabindan State Liberation Front (FLEC),
NZZIA Tiago, leader
note: FLEC is waging a small-scale, highly
factionalized, armed struggle for the
independence of Cabinda Province
Suffrage: 18 years of age; universal
Elections: first nationwide, multiparty
elections were held in late September 1992
with disputed results; further elections are
being discussed
Executive branch: president, prime
minister, Council of Ministers (cabinet)
Legislative branch: unicameral National
Assembly (Assembleia Nacional)
Judicial branch: Supreme Court (Tribunal
da Relacrao)
Leaders:
Chief of State: President Jose Eduardo dos
SANTOS (since 21 September 1979)
Head of Government: Prime Minister
Marcolino Jose Carlos MOCO (since 2
December 1992)
Member of: ACP, AfDB, CCC, CEEAC
(observer), ECA, FAO, FLS, G-77, IBRD,
ICAO, IDA, IFAD, IFC, ILO, IMF, IMO,
INTELSAT, INTERPOL, IOC, IOM, ITU,
LORCS, NAM, OAU, SADC, UN,
UNCTAD, UNESCO, UNIDO, UPU, WCL,
WFTU, WHO, WIPO, WMO, WTO
Diplomatic representation in US: none
representation: Jose PATRICIO, Permanent
Observer to the Organization of American
States
address: Permanent Observer to the
Organization of American States, 1899 L
Street, NW, 5th floor, Washington, DC
20038
telephone: (202) 785-1156
FAX: (202) 785-1258
US diplomatic representation:
director: Edmund DE JARNETTE
liaison office: Rua Major Kanhangolo, Nes
132/138, Luanda
mailing address: CP6484, Luanda, Angola
(mail international); USLO Luanda,
Department of State, Washington, D.C.
20521-2550 (pouch)
telephone: [244] (2) 34-54-81
FAX: [244] (2) 39-05-15
note: the US maintains a liaison office in
Luanda accredited to the Joint Political

Military Commission that oversees
implementation of the Angola Peace
Accords; this office does not perform any
commercial or consular services; the US
does not maintain diplomatic relations with
the Government of the Republic of Angola
Flag: two equal horizontal bands of red
(top) and black with a centered yellow
emblem consisting of a five-pointed star
within half a cogwheel crossed by a machete
(in the style of a hammer and sickle)

Economy

Overview: Subsistence agriculture provides
the main livelihood for 80-90% of the
population, but accounts for less than 15%
of GDP. Oil production is vital to the
economy, contributing about 60% to GDP.
Bitter internal fighting continues to severely
affect the nonoil economy, and food needs to
be imported. For the long run, Angola has
the advantage of rich natural resources in
addition to oil, notably gold, diamonds, and
arable land. To realize its economic potential
Angola not only must secure domestic peace
but also must reform government policies
that have led to distortions and imbalances
throughout the economy.
National product: GDP—exchange rate
conversion—$5.1 billion (1991 est.)
National product real growth rate: 1.7%
(1991 est.)
National product per capita: $950 (1991
est.)
Inflation rate (consumer prices): 1,000%
(1992 est.)
Unemployment rate: NA%
Budget: revenues $2.1 billion; expenditures
$3.6 billion, including capital expenditures
of $963 million (1991 est.)
Exports: $3.7 billion (f.o.b., 1991 est.)
commodities: oil, liquefied petroleum gas,
diamonds, coffee, sisal, fish and fish
products, timber, cotton
partners: US, France, Germany, Netherlands,
Brazil
Imports: $1.5 billion (f.o.b., 1991 est.)
commodities: capital equipment (machinery
and electrical equipment), food, vehicles and
spare parts, textiles and clothing, medicines;
substantial military deliveries
partners: Portugal, Brazil, US, France, Spain
External debt: $8.0 billion (1991)
Industrial production: growth rate NA%;
accounts for about 60% of GDP, including
petroleum output
Electricity: 510,000 kW capacity; 800
million kWh produced, 84 kWh per capita
(1991)
Industries: petroleum; mining diamonds,
iron ore, phosphates, feldspar, bauxite,
uranium, and gold;, fish processing; food
processing; brewing; tobacco; sugar; textiles;
cement; basic metal products

Agriculture: cash crops—coffee, sisal, corn,
cotton, sugar cane, manioc, tobacco; food
crops—cassava, corn, vegetables, plantains,
bananas; livestock production accounts for
20%, fishing 4%, forestry 2% of total
agricultural output; disruptions caused by
civil war and marketing deficiencies require
food imports
Economic aid: US commitments, including
Ex-Im (FY70-89), $265 million; Western
(non-US) countries, ODA and OOF bilateral
commitments (1970-89), $1,105 million;
Communist countries (1970-89), $1.3 billion;
net official disbursements (1985-89), $750
million
Currency: 1 kwanza (Kz) = 100 kwei
Exchange rates: kwanza (Kz) per US$1—
4,000 (black market rate was 17,000 on 30
April 1993)
Fiscal year: calendar year

Communications

Railroads: 3,189 km total; 2,879 km
1.067-meter gauge, 310 km 0.600-meter
gauge; limited trackage in use because of
landmines still in place from the civil war;
majority of the Benguela Railroad also
closed because of civil war
Highways: 73,828 km total; 8,577 km
bituminous-surface treatment, 29,350 km
crushed stone, gravel, or improved earth,
remainder unimproved earth
Inland waterways: 1,295 km navigable
Pipelines: crude oil 179 km
Ports: Luanda, Lobito, Namibe, Cabinda
Merchant marine: 12 ships (1,000 GRT or
over) totaling 66,348 GRT/102,825 DWT;
includes 11 cargo, 1 oil tanker
Airports:
total: 302
usable: 173
with permanent-surface runways: 32
with runways over 3,659 m: 2
with runways 2,440-3,659 m: 17
with runways 1,220-2,439 m: 57
Telecommunications: limited system of
wire, microwave radio relay, and troposcatter
routes; high frequency radio used
extensively for military links; 40,300
telephones; broadcast stations—17 AM, 13
FM, 6 TV; 2 Atlantic Ocean INTELSAT
earth stations

Defense Forces

Branches: Army, Navy, Air Force/Air
Defense, People's Defense Organization and
Territorial Troops, Frontier Guard
Manpower availability: males age 15-49
2,204,155; fit for military service 1,109,292;
reach military age (18) annually 94,919
(1993 est.)
Defense expenditures: exchange rate
conversion—$NA, NA% of GDP

Anguilla
(dependent territory of the UK)

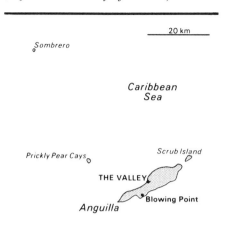

Geography

Location: in the eastern Caribbean Sea, about 270 km east of Puerto Rico
Map references: Central America and the Caribbean
Area:
total area: 91 km²
land area: 91 km²
comparative area: about half the size of Washington, DC
Land boundaries: 0 km
Coastline: 61 km
Maritime claims:
exclusive fishing zone: 200 nm
territorial sea: 3 nm
International disputes: none
Climate: tropical; moderated by northeast trade winds
Terrain: flat and low-lying island of coral and limestone
Natural resources: negligible; salt, fish, lobster
Land use:
arable land: NA%
permanent crops: NA%
meadows and pastures: NA%
forest and woodland: NA%
other: NA% (mostly rock with sparse scrub oak, few trees, some commercial salt ponds)
Irrigated land: NA km²
Environment: frequent hurricanes, other tropical storms (July to October)

People

Population: 7,006 (July 1993 est.)
Population growth rate: 0.64% (1993 est.)
Birth rate: 24.26 births/1,000 population (1993 est.)
Death rate: 8.28 deaths/1,000 population (1993 est.)
Net migration rate: -9.56 migrant(s)/1,000 population (1993 est.)
Infant mortality rate: 17.7 deaths/1,000 live births (1993 est.)
Life expectancy at birth:
total population: 73.89 years
male: 71.1 years
female: 76.7 years (1993 est.)
Total fertility rate: 3.09 children born/woman (1993 est.)
Nationality:
noun: Anguillan(s)
adjective: Anguillan
Ethnic divisions: black African
Religions: Anglican 40%, Methodist 33%, Seventh-Day Adventist 7%, Baptist 5%, Roman Catholic 3%, other 12%
Languages: English (official)
Literacy: age 12 and over can read and write (1984)
total population: 95%
male: 95%
female: 95%
Labor force: 2,780 (1984)
by occupation: NA

Government

Names:
conventional long form: none
conventional short form: Anguilla
Digraph: AV
Type: dependent territory of the UK
Capital: The Valley
Administrative divisions: none (dependent territory of the UK)
Independence: none (dependent territory of the UK)
Constitution: 1 April 1982
Legal system: based on English common law
National holiday: Anguilla Day, 30 May
Political parties and leaders: Anguilla National Alliance (ANA), Emile GUMBS; Anguilla United Party (AUP), Hubert HUGHES; Anguilla Democratic Party (ADP), Victor BANKS
Suffrage: 18 years of age; universal
Elections:
House of Assembly: last held 27 February 1989 (next to be held February 1994); results—percent of vote by party NA; seats—(11 total, 7 elected) ANA 3, AUP 2, ADP 1, independent 1
Executive branch: British monarch, governor, chief minister, Executive Council (cabinet)
Legislative branch: unicameral House of Assembly
Judicial branch: High Court
Leaders:
Chief of State: Queen ELIZABETH II (since 6 February 1952), represented by Governor Alan W. SHARE (since August 1992)
Head of Government: Chief Minister Emile GUMBS (since NA March 1984, served previously from February 1977 to May 1980)
Member of: CARICOM (observer), CDB
Diplomatic representation in US: none (dependent territory of the UK)

Flag: two horizontal bands of white (top, almost triple width) and light blue with three orange dolphins in an interlocking circular design centered in the white band; a new flag may have been in use since 30 May 1990

Economy

Overview: Anguilla has few natural resources, and the economy depends heavily on lobster fishing, offshore banking, tourism, and remittances from emigrants. In recent years the economy has benefited from a boom in tourism. Development plans center around the improvement of the infrastructure, particularly transport and tourist facilities, and also light industry.
National product: GDP—exchange rate conversion—$47.4 million (1991 est.)
National product real growth rate: 6.5% (1991 est.)
National product per capita: $6,800 (1991 est.)
Inflation rate (consumer prices): 4.6% (1991 est.)
Unemployment rate: 5% (1988 est.)
Budget: revenues $13.8 million; expenditures $15.2 million, including capital expenditures of $2.4 million (1992 est.)
Exports: $1.4 million (f.o.b., 1987)
commodities: lobster and salt
partners: NA
Imports: $10.3 million (f.o.b., 1987)
commodities: NA
partners: NA
External debt: $NA
Industrial production: growth rate NA%
Electricity: 2,000 kW capacity; 6 million kWh produced, 862 kWh per capita (1992)
Industries: tourism, boat building, salt
Agriculture: pigeon peas, corn, sweet potatoes, sheep, goats, pigs, cattle, poultry, fishing (including lobster)
Economic aid: Western (non-US) countries, ODA and OOF bilateral commitments (1970-89), $38 million
Currency: 1 EC dollar (EC$) = 100 cents
Exchange rates: East Caribbean dollars (EC$) per US$1—2.70 (fixed rate since 1976)
Fiscal year: NA

Communications

Highways: 60 km surfaced
Ports: Road Bay, Blowing Point
Airports:
total: 3
usable: 2
with permanent-surface runways: 1 (1,000 m at Wallblake Airport)
with runways over 3,659 m: 0
with runways 2,440-3,659 m: 0
with runways 1,220-2,439 m: 0

Anguilla *(continued)*

Telecommunications: modern internal telephone system; 890 telephones; broadcast stations—3 AM, 1 FM, no TV; radio relay microwave link to island of Saint Martin

---

Defense Forces

Note: defense is the responsibility of the UK

Antarctica

1000 km

South Atlantic Ocean

South Orkney Islands

ice shelf

Graham Land

Indian Ocean

• South Pole

ice shelf

Victoria Land

Wilkes Land

South Pacific Ocean

Indian Ocean

Geography

Location: continent mostly south of the Antarctic Circle
Map references: Antarctic Region
Area:
total area: 14 million km² (est.)
land area: 14 million km² (est.)
comparative area: slightly less than 1.5 times the size of the US
note: second-smallest continent (after Australia)
Land boundaries: none, but see entry on International disputes
Coastline: 17,968 km
Maritime claims: none, but see entry on International Disputes
International disputes: Antarctic Treaty defers claims (see Antarctic Treaty Summary below); sections (some overlapping) claimed by Argentina, Australia, Chile, France (Adelie Land), New Zealand (Ross Dependency), Norway (Queen Maud Land), and UK; the US and most other nations do not recognize the territorial claims of other nations and have made no claims themselves (the US and Russia reserve the right to do so); no formal claims have been made in the sector between 90 degrees west and 150 degrees west, where, because of floating ice, Antarctica is unapproachable from the sea
Climate: severe low temperatures vary with latitude, elevation, and distance from the ocean; East Antarctica is colder than West Antarctica because of its higher elevation; Antarctic Peninsula has the most moderate climate; higher temperatures occur in January along the coast and average slightly below freezing
Terrain: about 98% thick continental ice sheet and 2% barren rock, with average elevations between 2,000 and 4,000 meters; mountain ranges up to 4,897 meters high; ice-free coastal areas include parts of southern Victoria Land, Wilkes Land, the Antarctic Peninsula area, and parts of Ross Island on McMurdo Sound; glaciers form ice shelves along about half of the coastline, and

floating ice shelves constitute 11% of the area of the continent
Natural resources: none presently exploited; iron ore, chromium, copper, gold, nickel, platinum and other minerals, and coal and hydrocarbons have been found in small, uncommercial quantities
Land use:
arable land: 0%
permanent crops: 0%
meadows and pastures: 0%
forest and woodland: 0%
other: 100% (ice 98%, barren rock 2%)
Irrigated land: 0 km²
Environment: mostly uninhabitable; katabatic (gravity-driven) winds blow coastward from the high interior; frequent blizzards form near the foot of the plateau; a circumpolar ocean current flows clockwise along the coast as do cyclonic storms that form over the ocean; during summer more solar radiation reaches the surface at the South Pole than is received at the Equator in an equivalent period; in October 1991 it was reported that the ozone shield, which protects the Earth's surface from harmful ultraviolet radiation, had dwindled to the lowest level ever recorded over Antarctica; active volcanism on Deception Island and isolated areas of West Antarctica; other seismic activity rare and weak
Note: the coldest, windiest, highest, and driest continent

People

Population: no indigenous inhabitants; note—there are seasonally staffed research stations
Summer (January) population: over 4,115 total; Argentina 207, Australia 268, Belgium 13, Brazil 80, Chile 256, China NA, Ecuador NA, Finland 11, France 78, Germany 32, Greenpeace 12, India 60, Italy 210, Japan 59, South Korea 14, Netherlands 10, NZ 264, Norway 23, Peru 39, Poland NA, South Africa 79, Spain 43, Sweden 10, UK 116, Uruguay NA, US 1,666, former USSR 565 (1989-90)
Winter (July) population: over 1,046 total; Argentina 150, Australia 71, Brazil 12, Chile 73, China NA, France 33, Germany 19, Greenpeace 5, India 1, Japan 38, South Korea 14, NZ 11, Poland NA, South Africa 12, UK 69, Uruguay NA, US 225, former USSR 313 (1989-90)
Year-round stations: 42 total; Argentina 6, Australia 3, Brazil 1, Chile 3, China 2, Finland 1, France 1, Germany 1, India 1, Japan 2, South Korea 1, NZ 1, Poland 1, South Africa 3, UK 5, Uruguay 1, US 3, former USSR 6 (1990-91)
Summer only stations: over 38 total; Argentina 7, Australia 3, Chile 5, Germany 3, India 1, Italy 1, Japan 4, NZ 2, Norway 1, Peru 1, South Africa 1, Spain 1, Sweden 2,

UK 1, US numerous, former USSR 5 (1989-90); note—the disintegration of the former USSR has placed the status and future of its Antarctic facilities in doubt; stations may be subject to closings at any time because of ongoing economic difficulties

---

Government

Names:
conventional long form: none
conventional short form: Antarctica
Digraph: AY
Type:
Antarctic Treaty Summary: The Antarctic Treaty, signed on 1 December 1959 and entered into force on 23 June 1961, establishes the legal framework for the management of Antarctica. Administration is carried out through consultative member meetings—the 17th Antarctic Treaty Consultative Meeting was in Venice in November 1992. Currently, there are 41 treaty member nations: 26 consultative and 15 acceding. Consultative (voting) members include the seven nations that claim portions of Antarctica as national territory (some claims overlap) and 19 nonclaimant nations. The US and some other nations that have made no claims have reserved the right to do so. The US does not recognize the claims of others. The year in parentheses indicates when an acceding nation was voted to full consultative (voting) status, while no date indicates the country was an original 1959 treaty signatory. Claimant nations are—Argentina, Australia, Chile, France, New Zealand, Norway, and the UK. Nonclaimant consultative nations are—Belgium, Brazil (1983), China (1985), Ecuador (1990), Finland (1989), Germany (1981), India (1983), Italy (1987), Japan, South Korea (1989), Netherlands (1990), Peru (1989), Poland (1977), South Africa, Spain (1988), Sweden (1988), Uruguay (1985), the US, and Russia. Acceding (nonvoting) members, with year of accession in parentheses, are—Austria (1987), Bulgaria (1978), Canada (1988), Colombia (1988), Cuba (1984), Czechoslovakia (1962), Denmark (1965), Greece (1987), Guatemala (1991), Hungary (1984), North Korea (1987), Papua New Guinea (1981), Romania (1971), Switzerland (1990), and Ukraine (1992).
Article 1: area to be used for peaceful purposes only; military activity, such as weapons testing, is prohibited, but military personnel and equipment may be used for scientific research or any other peaceful purpose
Article 2: freedom of scientific investigation and cooperation shall continue
Article 3: free exchange of information and personnel in cooperation with the UN and other international agencies
Article 4: does not recognize, dispute, or establish territorial claims and no new claims shall be asserted while the treaty is in force
Article 5: prohibits nuclear explosions or disposal of radioactive wastes
Article 6: includes under the treaty all land and ice shelves south of 60 degrees 00 minutes south
Article 7: treaty-state observers have free access, including aerial observation, to any area and may inspect all stations, installations, and equipment; advance notice of all activities and of the introduction of military personnel must be given
Article 8: allows for jurisdiction over observers and scientists by their own states
Article 9: frequent consultative meetings take place among member nations
Article 10: treaty states will discourage activities by any country in Antarctica that are contrary to the treaty
Article 11: disputes to be settled peacefully by the parties concerned or, ultimately, by the ICJ
Article 12, 13, 14: deal with upholding, interpreting, and amending the treaty among involved nations
Other agreements: more than 170 recommendations adopted at treaty consultative meetings and ratified by governments include—Agreed Measures for the Conservation of Antarctic Fauna and Flora (1964); Convention for the Conservation of Antarctic Seals (1972); Convention on the Conservation of Antarctic Marine Living Resources (1980); a mineral resources agreement was signed in 1988 but was subsequently rejected; in 1991 the Protocol on Environmental Protection to the Antarctic Treaty was signed and awaits ratification; this agreement provides for the protection of the Antarctic environment through five specific annexes on marine pollution, fauna, and flora, environmental impact assessments, waste management, and protected areas; it also prohibits all activities relating to mineral resources except scientific research; four parties have ratified Protocol as of June 1993
Legal system: US law, including certain criminal offenses by or against US nationals, such as murder, may apply to areas not under jurisdiction of other countries. Some US laws directly apply to Antarctica. For example, the Antarctic Conservation Act, 16 U.S.C. section 2401 et seq., provides civil and criminal penalties for the following activities, unless authorized by regulation of statute: The taking of native mammals or birds; the introduction of nonindigenous plants and animals; entry into specially protected or scientific areas; the discharge or disposal of pollutants; and the importation into the US of certain items from Antarctica. Violation of the Antarctic Conservation Act carries penalties of up to $10,000 in fines and 1 year in prison. The Departments of Treasury, Commerce, Transportation, and Interior share enforcement responsibilities. Public Law 95-541, the US Antarctic Conservation Act of 1978, requires expeditions from the US to Antarctica to notify, in advance, the Office of Oceans and Polar Affairs, Room 5801, Department of State, Washington, DC 20520, which reports such plans to other nations as required by the Antarctic Treaty. For more information contact Permit Office, Office of Polar Programs, National Science Foundation, Washington, DC 20550.

Economy

Overview: No economic activity at present except for fishing off the coast and small-scale tourism, both based abroad.

Communications

Ports: none; offshore anchorage only at most coastal stations
Airports: 42 landing facilities at different locations operated by 15 national governments party to the Treaty; one additional air facility operated by commercial (nongovernmental) tourist organization; helicopter pads at 28 of these locations; runways at 10 locations are gravel, sea ice, glacier ice, or compacted snow surface suitable for wheeled fixed-wing aircraft; no paved runways; 16 locations have snow-surface skiways limited to use by ski-equipped planes—11 runways/skiways 1,000 to 3,000 m, 3 runways/skiways less than 1,000 m, 5 runways/skiways greater than 3,000 m, and 7 of unspecified or variable length; airports generally subject to severe restrictions and limitations resulting from extreme seasonal and geographic conditions; airports do not meet ICAO standards; advance approval from governments required for landing

---

Defense Forces

Note: the Antarctic Treaty prohibits any measures of a military nature, such as the establishment of military bases and fortifications, the carrying out of military maneuvers, or the testing of any type of weapon; it permits the use of military personnel or equipment for scientific research or for any other peaceful purposes

Antigua and Barbuda

Geography

Location: in the eastern Caribbean Sea, about 420 km east-southeast of Puerto Rico
Map references: Central America and the Caribbean, Standard Time Zones of the World
Area:
total area: 440 km²
land area: 440 km²
comparative area: slightly less than 2.5 times the size of Washington, DC
note: includes Redonda
Land boundaries: 0 km
Coastline: 153 km
Maritime claims:
contiguous zone: 24 nm
exclusive economic zone: 200 nm
territorial sea: 12 nm
International disputes: none
Climate: tropical marine; little seasonal temperature variation
Terrain: mostly low-lying limestone and coral islands with some higher volcanic areas
Natural resources: negligible; pleasant climate fosters tourism
Land use:
arable land: 18%
permanent crops: 0%
meadows and pastures: 7%
forest and woodland: 16%
other: 59%
Irrigated land: NA km²
Environment: subject to hurricanes and tropical storms (July to October); insufficient freshwater resources; deeply indented coastline provides many natural harbors

People

Population: 64,406 (July 1993 est.)
Population growth rate: 0.51% (1993 est.)
Birth rate: 17.51 births/1,000 population (1993 est.)
Death rate: 5.5 deaths/1,000 population (1993 est.)

Net migration rate: -6.96 migrant(s)/1,000 population (1993 est.)
Infant mortality rate: 19.2 deaths/1,000 live births (1993 est.)
Life expectancy at birth:
total population: 72.83 years
male: 70.81 years
female: 74.95 years (1993 est.)
Total fertility rate: 1.67 children born/woman (1993 est.)
Nationality:
noun: Antiguan(s), Barbudan(s)
adjective: Antiguan, Barbudan
Ethnic divisions: black African, British, Portuguese, Lebanese, Syrian
Religions: Anglican (predominant), other Protestant sects, some Roman Catholic
Languages: English (official), local dialects
Literacy: age 15 and over having completed 5 or more years of schooling (1960)
total population: 89%
male: 90%
female: 88%
Labor force: 30,000
by occupation: commerce and services 82%, agriculture 11%, industry 7% (1983)

Government

Names:
conventional long form: none
conventional short form: Antigua and Barbuda
Digraph: AC
Type: parliamentary democracy
Capital: Saint John's
Administrative divisions: 6 parishes and 2 dependencies*; Barbuda*, Redonda*, Saint George, Saint John, Saint Mary, Saint Paul, Saint Peter, Saint Philip
Independence: 1 November 1981 (from UK)
Constitution: 1 November 1981
Legal system: based on English common law
National holiday: Independence Day, 1 November (1981)
Political parties and leaders: Antigua Labor Party (ALP), Vere Cornwall BIRD, Sr., Lester BIRD; United Progressive Party (UPP), Baldwin SPENCER
Other political or pressure groups: United Progressive Party (UPP), headed by Baldwin SPENCER, a coalition of three opposition political parties—the United National Democratic Party (UNDP); the Antigua Caribbean Liberation Movement (ACLM); and the Progressive Labor Movement (PLM); Antigua Trades and Labor Union (ATLU), headed by Noel THOMAS
Suffrage: 18 years of age; universal
Elections:
House of Representatives: last held 9 March 1989 (next to be held NA 1994); results—percent of vote by party NA;

seats—(17 total) ALP 15, UPP 1, independent 1
Executive branch: British monarch, governor general, prime minister, Cabinet
Legislative branch: bicameral Parliament consists of an upper house or Senate and a lower house or House of Representatives
Judicial branch: Eastern Caribbean Supreme Court
Leaders:
Chief of State: Queen ELIZABETH II (since 6 February 1952), represented by Governor General Sir Wilfred Ebenezer JACOBS (since 1 November 1981, previously Governor since 1976)
Head of Government: Prime Minister Vere Cornwall BIRD, Sr. (since NA 1976); Deputy Prime Minister Lester BIRD (since NA)
Member of: ACP, C, CARICOM, CDB, ECLAC, FAO, G-77, GATT, IBRD, ICAO, ICFTU, IFAD, IFC, ILO, IMF, IMO, INTERPOL, IOC, ITU, NAM (observer), OAS, OECS, OPANAL, UN, UNCTAD, UNESCO, WCL, WHO, WMO
Diplomatic representation in US:
chief of mission: Ambassador Patrick Albert LEWIS
chancery: Suite 2H, 3400 International Drive NW, Washington, DC 20008
telephone: (202) 362-5211 or 5166, 5122, 5225
consulate: Miami
US diplomatic representation:
chief of mission: the US Ambassador to Barbados is accredited to Antigua and Barbuda, and, in his absence, the Embassy is headed by Charge d'Affaires Bryant J. SALTER
embassy: Queen Elizabeth Highway, Saint John's
mailing address: FPO AA 34054-0001
telephone: (809) 462-3505 or 3506
FAX: (809) 462-3516
Flag: red with an inverted isosceles triangle based on the top edge of the flag; the triangle contains three horizontal bands of black (top), light blue, and white with a yellow rising sun in the black band

Economy

Overview: The economy is primarily service oriented, with tourism the most important determinant of economic performance. During the period 1987-90, real GDP expanded at an annual average rate of about 6%. Tourism makes a direct contribution to GDP of about 13% and also affects growth in other sectors—particularly in construction, communications, and public utilities. Although Antigua and Barbuda is one of the few areas in the Caribbean experiencing a labor shortage in some sectors of the economy, it has been hurt in 1991-92 by a downturn in tourism caused by the Persian

Gulf war and the US recession.
National product: GDP—exchange rate conversion—$424 million (1991 est.)
National product real growth rate: 1.4% (1991 est.)
National product per capita: $6,600 (1991 est.)
Inflation rate (consumer prices): 6.5% (1991 est.)
Unemployment rate: 5% (1988 est.)
Budget: revenues $105 million; expenditures $161 million, including capital expenditures of $56 million (1992)
Exports: $32 million (f.o.b., 1991)
commodities: petroleum products 48%, manufactures 23%, food and live animals 4%, machinery and transport equipment 17%
partners: OECS 26%, Barbados 15%, Guyana 4%, Trinidad and Tobago 2%, US 0.3%
Imports: $317.5 million (c.i.f., 1991)
commodities: food and live animals, machinery and transport equipment, manufactures, chemicals, oil
partners: US 27%, UK 16%, Canada 4%, OECS 3%, other 50%
External debt: $250 million (1990 est.)
Industrial production: growth rate 3% (1989 est.); accounts for 5% of GDP
Electricity: 52,100 kW capacity; 95 million kWh produced, 1,482 kWh per capita (1992)
Industries: tourism, construction, light manufacturing (clothing, alcohol, household appliances)
Agriculture: accounts for 4% of GDP; expanding output of cotton, fruits, vegetables, and livestock; other crops—bananas, coconuts, cucumbers, mangoes, sugarcane; not self-sufficient in food
Economic aid: US commitments, $10 million (1985-88); Western (non-US) countries, ODA and OOF bilateral commitments (1970-89), $50 million
Currency: 1 EC dollar (EC$) = 100 cents
Exchange rates: East Caribbean dollars (EC$) per US$1—2.70 (fixed rate since 1976)
Fiscal year: 1 April-31 March

Communications

Railroads: 64 km 0.760-meter narrow gauge and 13 km 0.610-meter gauge used almost exclusively for handling sugarcane
Highways: 240 km
Ports: Saint John's
Merchant marine: 149 ships (1,000 GRT or over) totaling 529,202 GRT/778,506 DWT; includes 96 cargo, 3 refrigerated cargo, 21 container, 5 roll-on/roll-off cargo, 1 multifunction large-load carrier, 2 oil tanker, 19 chemical tanker, 2 bulk; note—a flag of convenience registry

Airports:
total: 3
usable: 3
with permanent-surface runways: 2
with runways 3,659 m: 0
with runways 2,440-3,659 m: 1
with runways 1,220-2,439 m: 0
Telecommunications: good automatic telephone system; 6,700 telephones; tropospheric scatter links with Saba and Guadeloupe; broadcast stations—4 AM, 2 FM, 2 TV, 2 shortwave; 1 coaxial submarine cable; 1 Atlantic Ocean INTELSAT earth station

---

Defense Forces

Branches: Royal Antigua and Barbuda Defense Force, Royal Antigua and Barbuda Police Force (including the Coast Guard)
Manpower availability: NA
Defense expenditures: exchange rate conversion—$1.4 million, 1% of GDP (FY90/91)

Arctic Ocean

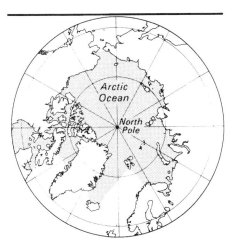

Geography

Location: body of water mostly north of the Arctic Circle
Map references: Arctic Region, Asia, North America, Standard Time Zones of the World
Area:
total area: 14.056 million km^2
comparative area: slightly more than 1.5 times the size of the US; smallest of the world's four oceans (after Pacific Ocean, Atlantic Ocean, and Indian Ocean)
note: includes Baffin Bay, Barents Sea, Beaufort Sea, Chukchi Sea, East Siberian Sea, Greenland Sea, Hudson Bay, Hudson Strait, Kara Sea, Laptev Sea, and other tributary water bodies
Coastline: 45,389 km
International disputes: some maritime disputes (see littoral states); Svalbard is the focus of a maritime boundary dispute between Norway and Russia
Climate: polar climate characterized by persistent cold and relatively narrow annual temperature ranges; winters characterized by continuous darkness, cold and stable weather conditions, and clear skies; summers characterized by continuous daylight, damp and foggy weather, and weak cyclones with rain or snow
Terrain: central surface covered by a perennial drifting polar icepack that averages about 3 meters in thickness, although pressure ridges may be three times that size; clockwise drift pattern in the Beaufort Gyral Stream, but nearly straight line movement from the New Siberian Islands (Russia) to Denmark Strait (between Greenland and Iceland); the ice pack is surrounded by open seas during the summer, but more than doubles in size during the winter and extends to the encircling land masses; the ocean floor is about 50% continental shelf (highest percentage of any ocean) with the remainder a central basin interrupted by three submarine ridges (Alpha Cordillera, Nansen Cordillera, and Lomonsov Ridge);

Arctic Ocean *(continued)*

maximum depth is 4,665 meters in the Fram Basin

Natural resources: sand and gravel aggregates, placer deposits, polymetallic nodules, oil and gas fields, fish, marine mammals (seals and whales)

Environment: endangered marine species include walruses and whales; ice islands occasionally break away from northern Ellesmere Island; icebergs calved from glaciers in western Greenland and extreme northeastern Canada; maximum snow cover in March or April about 20 to 50 centimeters over the frozen ocean and lasts about 10 months; permafrost in islands; virtually icelocked from October to June; fragile ecosystem slow to change and slow to recover from disruptions or damage

Note: major chokepoint is the southern Chukchi Sea (northern access to the Pacific Ocean via the Bering Strait); ships subject to superstructure icing from October to May; strategic location between North America and Russia; shortest marine link between the extremes of eastern and western Russia, floating research stations operated by the US and Russia

Government

Digraph: XQ

Economy

Overview: Economic activity is limited to the exploitation of natural resources, including petroleum, natural gas, fish, and seals.

Communications

Ports: Churchill (Canada), Murmansk (Russia), Prudhoe Bay (US)

Telecommunications: no submarine cables

Note: sparse network of air, ocean, river, and land routes; the Northwest Passage (North America) and Northern Sea Route (Eurasia) are important seasonal waterways

Argentina

Geography

Location: Eastern South America, bordering the South Atlantic Ocean between Chile and Uruguay

Map references: South America, Standard Time Zones of the World

Area:

total area: 2,766,890 km²

land area: 2,736,690 km²

comparative area: slightly less than three-tenths the size of the US

Land boundaries: total 9,665 km, Bolivia 832 km, Brazil 1,224 km, Chile 5,150 km, Paraguay 1,880 km, Uruguay 579 km

Coastline: 4,989 km

Maritime claims:

contiguous zone: 24 nm

continental shelf: 200 m (depth) or to depth of exploitation

exclusive economic zone: not specified

territorial sea: 200 nm; overflight and navigation permitted beyond 12 nm

International disputes: short section of the boundary with Uruguay is in dispute; short section of the boundary with Chile is indefinite; claims British-administered Falkland Islands (Islas Malvinas); claims British-administered South Georgia and the South Sandwich Islands; territorial claim in Antarctica

Climate: mostly temperate; arid in southeast; subantarctic in southwest

Terrain: rich plains of the Pampas in northern half, flat to rolling plateau of Patagonia in south, rugged Andes along western border

Natural resources: fertile plains of the pampas, lead, zinc, tin, copper, iron ore, manganese, petroleum, uranium

Land use:

arable land: 9%

permanent crops: 4%

meadows and pastures: 52%

forest and woodland: 22%

other: 13%

Irrigated land: 17,600 km² (1989 est.)

Environment: Tucuman and Mendoza areas in Andes subject to earthquakes; pamperos are violent windstorms that can strike Pampas and northeast; irrigated soil degradation; desertification; air and water pollution in Buenos Aires

Note: second-largest country in South America (after Brazil); strategic location relative to sea lanes between South Atlantic and South Pacific Oceans (Strait of Magellan, Beagle Channel, Drake Passage)

People

Population: 33,533,256 (July 1993 est.)

Population growth rate: 1.13% (1993 est.)

Birth rate: 19.75 births/1,000 population (1993 est.)

Death rate: 8.64 deaths/1,000 population (1993 est.)

Net migration rate: 0.22 migrant(s)/1,000 population (1993 est.)

Infant mortality rate: 30 deaths/1,000 live births (1993 est.)

Life expectancy at birth:

total population: 71.19 years

male: 67.91 years

female: 74.65 years (1993 est.)

Total fertility rate: 2.72 children born/woman (1993 est.)

Nationality:

noun: Argentine(s)

adjective: Argentine

Ethnic divisions: white 85%, mestizo, Indian, or other nonwhite groups 15%

Religions: nominally Roman Catholic 90% (less than 20% practicing), Protestant 2%, Jewish 2%, other 6%

Languages: Spanish (official), English, Italian, German, French

Literacy: age 15 and over can read and write (1990)

total population: 95%

male: 96%

female: 95%

Labor force: 10.9 million

by occupation: agriculture 12%, industry 31%, services 57% (1985 est.)

Government

Names:

conventional long form: Argentine Republic

conventional short form: Argentina

local long form: Republica Argentina

local short form: Argentina

Digraph: AR

Type: republic

Capital: Buenos Aires

Administrative divisions: 23 provinces (provincias, singular—provincia), and 1 federal district* (distrito federal); Buenos Aires, Catamarca, Chaco, Chubut, Cordoba, Corrientes, Distrito Federal*, Entre Rios, Formosa, Jujuy, La Pampa, La Rioja, Mendoza, Misiones, Neuquen, Rio Negro,

Salta, San Juan, San Luis, Santa Cruz, Santa Fe, Santiago del Estero, Tierra del Fuego (Territorio Nacional de la Tierra del Fuego, Antartida e Islas del Atlantico Sur), Tucuman

note: the national territory is in the process of becoming a province; the US does not recognize claims to Antarctica

Independence: 9 July 1816 (from Spain)

Constitution: 1 May 1853

Legal system: mixture of US and West European legal systems; has not accepted compulsory ICJ jurisdiction

National holiday: Revolution Day, 25 May (1810)

Political parties and leaders: Justicialist Party (JP), Carlos Saul MENEM, Peronist umbrella political organization; Radical Civic Union (UCR), Mario LOSADA, moderately left-of-center party; Union of the Democratic Center (UCD), Jorge AGUADO, conservative party; Intransigent Party (PI), Dr. Oscar ALENDE, leftist party; Dignity and Independence Political Party (MODIN), Aldo RICO, right-wing party; several provincial parties

Other political or pressure groups: Peronist-dominated labor movement; General Confederation of Labor (CGT; Peronist-leaning umbrella labor organization); Argentine Industrial Union (manufacturers' association); Argentine Rural Society (large landowners' association); business organizations; students; the Roman Catholic Church; the Armed Forces

Suffrage: 18 years of age; universal

Elections:

Chamber of Deputies: last held in three phases during late 1991 for half of 254 seats; seats (254 total)—JP 122, UCR 85, UCD 10, other 37 (1993)

President: last held 14 May 1989 (next to be held NA May 1995); results—Carlos Saul MENEM was elected

Senate: last held May 1989, but provincial elections in late 1991 set the stage for indirect elections by provincial senators for one-third of 46 seats in the national senate in May 1992; seats (46 total)—JP 27, UCR 14, others 5

Executive branch: president, vice president, Cabinet

Legislative branch: bicameral National Congress (Congreso Nacional) consists of an upper chamber or Senate (Senado) and a lower chamber or Chamber of Deputies (Camara de Diputados)

Judicial branch: Supreme Court (Corte Suprema)

Leaders:

Chief of State and Head of Government: President Carlos Saul MENEM (since 8 July 1989); Vice President (position vacant)

Member of: AG (observer), Australian Group, CCC, ECLAC, FAO, G-6, G-11, G-15, G-19, G-24, AfDB, G-77, GATT, IADB, IAEA, IBRD, ICAO, ICC, ICFTU, IDA, IFAD, IFC, ILO, IMF, IMO, INMARSAT, INTELSAT, INTERPOL, IOC, IOM, ISO, ITU, LAES, LAIA, LORCS, MERCOSUR, MINURSO, OAS, PCA, RG, UN, UNAVEM II, UNCTAD, UNESCO, UNHCR, UNIDO, UNIKOM, UNOMOZ, UNPROFOR, UNTAC, UNTSO, UPU, WCL, WFTU, WHO, WIPO, WMO, WTO

Diplomatic representation in US:

chief of mission: Ambassador Carlos ORTIZ DE ROZAS

chancery: 1600 New Hampshire Avenue NW, Washington, DC 20009

telephone: (202) 939-6400 through 6403

consulates general: Houston, Miami, New Orleans, New York, San Francisco, and San Juan (Puerto Rico)

consulates: Baltimore, Chicago, and Los Angeles

US diplomatic representation:

chief of mission: Ambassador James CHEEK (since 28 May 1993)

embassy: 4300 Colombia, 1425 Buenos Aires

mailing address: APO AA 34034

telephone: [54] (1) 774-7611 or 8811, 9911

FAX: [54] (1) 775-4205

Flag: three equal horizontal bands of light blue (top), white, and light blue; centered in the white band is a radiant yellow sun with a human face known as the Sun of May

---

Economy

Overview: Argentina is rich in natural resources and has a highly literate population, an export-oriented agricultural sector, and a diversified industrial base. Nevertheless, following decades of mismanagement and statist policies, the economy in the late 1980s was plagued with huge external debts and recurring bouts of hyperinflation. Elected in 1989, in the depths of recession, President MENEM has implemented a comprehensive economic restructuring program that shows signs of putting Argentina on a path of stable, sustainable growth. Argentina's currency has traded at par with the US dollar since April 1991, and inflation has fallen to its lowest level in 20 years. Argentines have responded to the relative price stability by repatriating flight capital and investing in domestic industry. Much remains to be done in the 1990s in dismantling the old statist barriers to growth and in solidifying the recent economic gains.

National product: GDP—exchange rate conversion—$112 billion (1992 est.)

National product real growth rate: 7% (1992 est.)

National product per capita: $3,400 (1992 est.)

Inflation rate (consumer prices): 17.7% (1992)

Unemployment rate: 6.9% (1992)

Budget: revenues $33.1 billion; expenditures $35.8 billion, including capital expenditures of $3.5 billion (1992)

Exports: $12.3 billion (f.o.b., 1992 est.)

commodities: meat, wheat, corn, oilseed, hides, wool

partners: US 12%, Brazil, Italy, Japan, Netherlands

Imports: $14.0 billion (c.i.f., 1992 est.)

commodities: machinery and equipment, chemicals, metals, fuels and lubricants, agricultural products

partners: US 22%, Brazil, Germany, Bolivia, Japan, Italy, Netherlands

External debt: $54 billion (June 1992)

Industrial production: growth rate 10% (1992 est.); accounts for 26% of GDP

Electricity: 17,911,000 kW capacity; 51,305 million kWh produced, 1,559 kWh per capita (1992)

Industries: food processing, motor vehicles, consumer durables, textiles, chemicals and petrochemicals, printing, metallurgy, steel

Agriculture: accounts for 8% of GDP (including fishing); produces abundant food for both domestic consumption and exports; among world's top five exporters of grain and beef; principal crops—wheat, corn, sorghum, soybeans, sugar beets

Illicit drugs: increasing use as a transshipment country for cocaine headed for the US and Europe

Economic aid: US commitments, including Ex-Im (FY70-89), $1.0 billion; Western (non-US) countries, ODA and OOF bilateral commitments (1970-89), $4.4 billion; Communist countries (1970-89), $718 million

Currency: 1 peso = 100 centavos

Exchange rates: pesos per US$1—0.99000 (January1993), 0.99064 (1992), 0.95355 (1991), 0.48759 (1990), 0.04233 (1989), 0.00088 (1988)

Fiscal year: calendar year

Communications

Railroads: 34,172 km total (includes 209 km electrified); includes a mixture of 1.435-meter standard gauge, 1.676-meter broad gauge, 1.000-meter narrow gauge, and 0.750-meter narrow gauge

Highways: 208,350 km total; 47,550 km paved, 39,500 km gravel, 101,000 km improved earth, 20,300 km unimproved earth

Inland waterways: 11,000 km navigable

Pipelines: crude oil 4,090 km; petroleum products 2,900 km; natural gas 9,918 km

Ports: Bahia Blanca, Buenos Aires, Comodoro Rivadavia, La Plata, Rosario, Santa Fe

Merchant marine: 60 ships (1,000 GRT or over) totaling 1,695,420 GRT/1,073,904 DWT; includes 30 cargo, 5 refrigerated

Argentina *(continued)*

cargo, 4 container, 1 railcar carrier, 14 oil tanker, 1 chemical tanker, 4 bulk, 1 roll-on/roll-off

Airports:
total: 1,700
usable: 1,451
with permanet-surface runways: 137
with runways over 3,659 m: 1
with runways 2,440-3,659 m: 31
with runways 1,220-2,439 m: 326

Telecommunications: extensive modern system; 2,650,000 telephones (12,000 public telephones); microwave widely used; broadcast stations—171 AM, no FM, 231 TV, 13 shortwave; 2 Atlantic Ocean INTELSAT earth stations; domestic satellite network has 40 earth stations

Defense Forces

Branches: Argentine Army, Navy of the Argentine Republic, Argentine Air Force, National Gendarmerie, Argentine Naval Prefecture (Coast Guard only), National Aeronautical Police Force

Manpower availability: males age 15-49 8,267,316; fit for military service 6,702,303; reach military age (20) annually 284,641 (1993 est.)

Defense expenditures: exchange rate conversion—$NA, NA% of GDP

Armenia

Geography

Location: Southeastern Europe, between Turkey and Azerbaijan

Map references: Africa, Asia, Commonwealth of Independent States—European States, Middle East, Standard Time Zones of the World

Area:
total area: 29,800 km²
land area: 28,400 km²
comparative area: slightly larger than Maryland

Land boundaries: total 1,254 km, Azerbaijan (east) 566 km, Azerbaijan (south) 221 km, Georgia 164 km, Iran 35 km, Turkey 268 km

Coastline: 0 km (landlocked)

Maritime claims: none; landlocked

International disputes: violent and longstanding dispute with Azerbaijan over ethnically Armenian exclave of Nagorno-Karabakh; some irredentism by Armenians living in southern Georgia; traditional demands on former Armenian lands in Turkey have greatly subsided

Climate: continental, hot, and subject to drought

Terrain: high Armenian Plateau with mountains; little forest land; fast flowing rivers; good soil in Aras River valley

Natural resources: small deposits of gold, copper, molybdenum, zinc, alumina

Land use:
arable land: 29%
permanent crops: 0%
meadows and pastures: 15%
forest and woodland: 0%
other: 56%

Irrigated land: 3,050 km² (1990)

Environment: pollution of Razdan and Aras Rivers; air pollution in Yerevan; energy blockade has led to deforestation as citizens scavenge for firewood, use of Lake Sevan water for hydropower has lowered lake level, threatened fish population

Note: landlocked

People

Population: 3,481,207 (July 1993 est.)

Population growth rate: 1.23% (1993 est.)

Birth rate: 25.79 births/1,000 population (1993 est.)

Death rate: 6.77 deaths/1,000 population (1993 est.)

Net migration rate: -6.76 migrant(s)/1,000 population (1993 est.)

Infant mortality rate: 28.2 deaths/1,000 live births (1993 est.)

Life expectancy at birth:
total population: 71.77 years
male: 68.36 years
female: 75.36 years (1993 est.)

Total fertility rate: 3.31 children born/woman (1993 est.)

Nationality:
noun: Armenian(s)
adjective: Armenian

Ethnic divisions: Armenian 93%, Azeri 3%, Russian 2%, other 2%

Religions: Armenian Orthodox 94%

Languages: Armenian 96%, Russian 2%, other 2%

Literacy: age 9-49 can read and write (1970)
total population: 100%
male: 100%
female: 100%

Labor force: 1.63 million
by occupation: industry and construction 42%, agriculture and forestry 18%, other 40% (1990)

Government

Names:
conventional long form: Republic of Armenia
conventional short form: Armenia
local long form: Hayastani Hanrapetut'yun
local short form: Hayastan
former: Armenian Soviet Socialist Republic; Armenian Republic

Digraph: AM

Type: republic

Capital: Yerevan

Administrative divisions: none (all rayons are under direct republic jurisdiction)

Independence: 23 September 1991 (from Soviet Union)

Constitution: adopted NA April 1978; post-Soviet constitution not yet adopted

Legal system: based on civil law system

National holiday: NA

Political parties and leaders: Armenian National Movement, Husik LAZARYAN, chairman; National Democratic Union; National Self-Determination Association; Armenian Democratic Liberal Organization, Ramkavar AZATAKAN, chairman; Dashnatktsutyan Party (Armenian Revolutionary Federation, ARF), Rouben

MIRZAKHANIN; Chairman of Parliamentary opposition—Mekhak GABRIYELYAN; Christian Democratic Union; Constitutional Rights Union; Republican Party

Suffrage: 18 years of age; universal

Elections:

President: last held 16 October 1991 (next to be held NA); results—Levon Akopovich TER-PETROSYAN 86%; radical nationalists about 7%; note—Levon TER-PETROSYAN was elected Chairman of the Armenian Supreme Soviet 4 August 1990

Supreme Soviet: last held 20 May 1990 (next to be held NA); results—percent of vote by party NA; seats—(240 total) non-aligned 149, Armenian National Movement 52, Armenian Democratic Liberal Organization 14, Dashnatktsutyan 12, National Democratic Union 9, Christian Democratic Union 1, Constitutional Rights Union 1, National Self-Determination Association 1, Republican Party 1

Executive branch: president, council of ministers, prime minister

Legislative branch: unicameral Supreme Soviet

Judicial branch: Supreme Court

Leaders:

Chief of State: President Levon Akopovich TER-PETROSYAN (since 16 October 1991), Vice President Gagik ARUTYUNYAN (since 16 October 1991)

Head of Government: Prime Minister Hrant BAGRATYAN (since NA February 1993); Supreme Soviet Chairman Babken ARARKTSYAN (since NA 1990)

Member of: BSEC, CIS, CSCE, EBRD, IBRD, ICAO, IMF, NACC, UN, UNCTAD, UNESCO, UNIDO, UPU, WHO

Diplomatic representation in US:

chief of mission: Ambassador Rouben SHUGARIAN

chancery: 122 C Street NW, Suite 360, Washington, DC 20001

telephone: (202) 628-5766

US diplomatic representation:

chief of mission: Ambassador Designate Harry GILMORE

embassy: 18 Gen Bagramian, Yerevan

mailing address: use embassy street address

telephone: (7) (885) 215-1122, 215-1144

FAX: (7) (885) 215-1122

Flag: three equal horizontal bands of red (top), blue, and gold

---

Economy

Overview: Armenia under the old centrally planned Soviet system had built up textile, machine-building, and other industries and had become a key supplier to sister republics. In turn, Armenia had depended on supplies of raw materials and energy from the other republics. Most of these supplies enter the republic by rail through Azerbaijan (85%) and Georgia (15%). The economy has been severely hurt by ethnic strife with Azerbaijan over control of the Nagorno-Karabakh Autonomous Oblast, a mostly Armenian-populated enclave within the national boundaries of Azerbaijan. In addition to outright warfare, the strife has included interdiction of Armenian imports on the Azerbaijani railroads and expensive airlifts of supplies to beleaguered Armenians in Nagorno-Karabakh. An earthquake in December 1988 destroyed about one-tenth of industrial capacity and housing, the repair of which has not been possible because the supply of funds and real resources has been disrupted by the reorganization and subsequent dismantling of the central USSR administrative apparatus. Among facilities made unserviceable by the earthquake are the Yerevan nuclear power plant, which had supplied 40% of Armenia's needs for electric power and a plant that produced one-quarter of the output of elevators in the former USSR. Armenia has some deposits of nonferrous metal ores (bauxite, copper, zinc, and molybdenum) that are largely unexploited. For the mid-term, Armenia's economic prospects seem particularly bleak because of ethnic strife and the unusually high dependence on outside areas, themselves in a chaotic state of transformation. The dramatic drop in output in 1992 is attributable largely to the cumulative impact of the blockade; of particular importance was the shutting off in the summer of 1992 of rail and road links to Russia through Georgia due to civil strife in the latter republic.

National product: GDP $NA

National product real growth rate: -34% (1992)

National product per capita: $NA

Inflation rate (consumer prices): 20% per month (first quarter 1993)

Unemployment rate: 2% of officially registered unemployed but large numbers of underemployed

Budget: revenues $NA; expenditures $NA, including capital expenditures of $NA

Exports: $30 million to outside the successor states of the former USSR (f.o.b., 1992)

commodities: machinery and transport equipment, light industrial products, processed food items (1991)

partners: NA

Imports: $300 million from outside the successor statees of the former USSR (c.i.f., 1992)

commodities: machinery, energy, consumer goods (1991)

partners: NA

External debt: $650 million (December 1991 est.)

Industrial production: growth rate −50% (1992 est.)

Electricity: 2,875,000 kW capacity; 9,000 million kWh produced, 2,585 kWh per capita (1992)

Industries: diverse, including (in percent of output of former USSR) metalcutting machine tools (5.5%), forging-pressing machines (1.9%), electric motors (9%), tires (1.5%), knitted wear (4.4%), hosiery (3.0%), shoes (2.2%), silk fabric (0.8%), washing machines (2.0%), chemicals, trucks, watches, instruments, and microelectronics (1990)

Agriculture: accounts for about 20% of GDP; only 29% of land area is arable; employs 18% of labor force; citrus, cotton, and dairy farming; vineyards near Yerevan are famous for brandy and other liqueurs

Illicit drugs: illicit producer of cannabis mostly for domestic consumption; used as a transshipment point for illicit drugs to Western Europe

Economic aid: wheat from US, Turkey

Currency: retaining Russian ruble as currency (January 1993)

Exchange rates: rubles per US$1—415 (24 December 1992) but subject to wide fluctuations

Fiscal year: calendar year

Communications

Railroads: 840 km; does not include industrial lines (1990)

Highways: 11,300 km total; 10,500 km hard surfaced, 800 km earth (1990)

Inland waterways: NA km

Pipelines: natural gas 900 km (1991)

Ports: none; landlocked

Airports:

total: 12

useable: 10

with permanent-surface runways: 6

with runways over 3,659 m: 1

with runways 2,440-3,659 m: 4

with runways 1,220-2,439 m: 3

Telecommunications: progress on installation of fiber optic cable and construction of facilities for mobile cellular phone service remains in the negotiation phase for joint venture agreement; Armenia has about 260,000 telephones, of which about 110,000 are in Yerevan; average telephone density is 8 per 100 persons; international connections to other former republics of the USSR are by landline or microwave and to other countries by satellite and by leased connection through the Moscow international gateway switch; broadcast stations—100% of population receives Armenian and Russian TV programs; satellite earth station—INTELSAT

Defense Forces

Branches: Army, Air Force, National Guard, Security Forces (internal and border troops)

Manpower availability: males age 15-49

Armenia (continued)

848,223; fit for military service 681,058; reach military age (18) annually 28,101 (1993 est.)
Defense expenditures: 250 million rubles, NA% of GDP (1992 est.); note—conversion of the military budget into US dollars using the current exchange rate could produce misleading results

Aruba
(part of the Dutch realm)

Geography

Location: in the southern Caribbean Sea, 28 km north of Venezuela and 125 km east of Colombia
Map references: Central America and the Caribbean
Area:
total area: 193 km²
land area: 193 km²
comparative area: slightly larger than Washington, DC
Land boundaries: 0 km
Coastline: 68.5 km
Maritime claims:
exclusive fishing zone: 12 nm
territorial sea: 12 nm
International disputes: none
Climate: tropical marine; little seasonal temperature variation
Terrain: flat with a few hills; scant vegetation
Natural resources: negligible; white sandy beaches
Land use:
arable land: 0%
permanent crops: 0%
meadows and pastures: 0%
forest and woodland: 0%
other: 100%
Irrigated land: NA km²
Environment: lies outside the Caribbean hurricane belt

People

Population: 65,117 (July 1993 est.)
Population growth rate: 0.66% (1993 est.)
Birth rate: 15.33 births/1,000 population (1993 est.)
Death rate: 6.05 deaths/1,000 population (1993 est.)
Net migration rate: -2.72 migrant(s)/1,000 population (1993 est.)
Infant mortality rate: 8.6 deaths/1,000 live births (1993 est.)
Life expectancy at birth:
total population: 76.3 years

male: 72.65 years
female: 80.13 years (1993 est.)
Total fertility rate: 1.83 children born/woman (1993 est.)
Nationality:
noun: Aruban(s)
adjective: Aruban
Ethnic divisions: mixed European/Caribbean Indian 80%
Religions: Roman Catholic 82%, Protestant 8%, Hindu, Muslim, Confucian, Jewish
Languages: Dutch (official), Papiamento (a Spanish, Portuguese, Dutch, English dialect), English (widely spoken), Spanish
Literacy:
total population: NA%
male: NA%
female: NA%
Labor force: NA
by occupation: most employment is in the tourist industry (1986)

Government

Names:
conventional long form: none
conventional short form: Aruba
Digraph: AA
Type: part of the Dutch realm; full autonomy in internal affairs obtained in 1986 upon separation from the Netherlands Antilles
Capital: Oranjestad
Administrative divisions: none (self-governing part of the Netherlands)
Independence: none (part of the Dutch realm; in 1990, Aruba requested and received from the Netherlands cancellation of the agreement to automatically give independence to the island in 1996)
Constitution: 1 January 1986
Legal system: based on Dutch civil law system, with some English common law influence
National holiday: Flag Day, 18 March
Political parties and leaders: Electoral Movement Party (MEP), Nelson ODUBER; Aruban People's Party (AVP), Henny EMAN; National Democratic Action (ADN), Pedro Charro KELLY; New Patriotic Party (PPN), Eddy WERLEMEN; Aruban Patriotic Party (PPA), Benny NISBET; Aruban Democratic Party (PDA), Leo BERLINSKI; Democratic Action '86 (AD '86), Arturo ODUBER; Organization for Aruban Liberty (OLA), Glenbert CROES
note: governing coalition includes the MEP, PPA, and ADN
Suffrage: 18 years of age; universal
Elections:
Legislature: last held 8 January 1993 (next to be held by NA January 1997); results—percent of vote by party NA; seats—(21 total) MEP 9, AVP 8, ADN 1, PPA 1, OLA 1, other 1

Executive branch: Dutch monarch, governor, prime minister, Council of Ministers (cabinet)
Legislative branch: unicameral legislature (Staten)
Judicial branch: Joint High Court of Justice
Leaders:
Chief of State: Queen BEATRIX Wilhelmina Armgard (since 30 April 1980), represented by Governor General Olindo KOOLMAN (since NA)
Head of Government: Prime Minister Nelson ODUBER (since NA February 1989)
Member of: ECLAC (associate), INTERPOL, IOC, UNESCO (associate), WCL, WTO (associate)
Diplomatic representation in US: none (self-governing part of the Netherlands)
Flag: blue with two narrow horizontal yellow stripes across the lower portion and a red, four-pointed star outlined in white in the upper hoist-side corner

Economy

Overview: Tourism is the mainstay of the economy, although offshore banking and oil refining and storage are also important. Hotel capacity expanded rapidly between 1985 and 1989 and nearly doubled in 1990 alone. Unemployment has steadily declined from about 20% in 1986 to about 3% in 1991. The reopening of the local oil refinery, once a major source of employment and foreign exchange earnings, promises to give the economy an additional boost.
National product: GDP—exchange rate conversion—$900 million (1991 est.)
National product real growth rate: 6% (1991 est.)
National product per capita: $14,000 (1991 est.)
Inflation rate (consumer prices): 5.6% (1991)
Unemployment rate: 3% (1991 est.)
Budget: revenues $145 million; expenditures $185 million, including capital expenditures of $42 million (1988)
Exports: $902.4 million, including oil re-exports (f.o.b., 1991)
commodities: mostly petroleum products
partners: US 64%, EC
Imports: $1,311.3 million, including oil for processing and re-export (f.o.b., 1991)
commodities: food, consumer goods, manufactures, petroleum products
partners: US 8%, EC
External debt: $81 million (1987)
Industrial production: growth rate NA%
Electricity: 310,000 kW capacity; 945 million kWh produced, 14,610 kWh per capita (1992)
Industries: tourism, transshipment facilities, oil refining
Agriculture: poor quality soils and low rainfall limit agricultural activity to the

cultivation of aloes, some livestock, and fishing
Illicit drugs: drug money laundering center
Economic aid: Western (non-US) countries ODA and OOF bilateral commitments (1980-89), $220 million
Currency: 1 Aruban florin (Af.) = 100 cents
Exchange rates: Aruban florins (Af.) per US$1—1.7900 (fixed rate since 1986)
Fiscal year: calendar year

Communications

Highways: NA km all-weather highways
Ports: Oranjestad, Sint Nicolaas
Airports:
total: 2
usable: 2
with permanent-surface runways: 2
with runways over 3,659 m: 0
with runways 2,440-3,659 m: 1
with runways 1,220-2,439 m: 0
note: government-owned airport east of Oranjestad accepts transatlantic flights
Telecommunications: generally adequate; extensive interisland microwave radio relay links; 72,168 telephones; broadcast stations—4 AM, 4 FM, 1 TV; 1 submarine cable to Sint Maarten

Defense Forces

Note: defense is the responsibility of the Netherlands

Ashmore and Cartier Islands
(territory of Australia)

Geography

Location: in the Indian Ocean, 320 km off the northwest coast of Australia, between Australia and Indonesia
Map references: Oceania, Southeast Asia
Area:
total area: 5 km²
land area: 5 km²
comparative area: about 8.5 times the size of The Mall in Washington, DC
note: includes Ashmore Reef (West, Middle, and East Islets) and Cartier Island
Land boundaries: 0 km
Coastline: 74.1 km
Maritime claims:
contiguous zone: 12 nm
continental shelf: 200 m (depth) or to depth of exploration
exclusive fishing zone: 200 nm
territorial sea: 3 nm
International disputes: none
Climate: tropical
Terrain: low with sand and coral
Natural resources: fish
Land use:
arable land: 0%
permanent crops: 0%
meadows and pastures: 0%
forest and woodland: 0%
other: 100% (all grass and sand)
Irrigated land: 0 km²
Environment: surrounded by shoals and reefs; Ashmore Reef National Nature Reserve established in August 1983

People

Population: no indigenous inhabitants; note—there are only seasonal caretakers

Government

Names:
conventional long form: Territory of Ashmore and Cartier Islands

Ashmore and Cartier Islands

(continued)

conventional short form: Ashmore and Cartier Islands
Digraph: AT
Type: territory of Australia administered by the Australian Ministry for Arts, Sports, the Environment, Tourism, and Territories
Capital: none; administered from Canberra, Australia
Administrative divisions: none (territory of Australia)
Independence: none (territory of Australia)
Legal system: relevant laws of the Northern Territory of Australia
Diplomatic representation in US: none (territory of Australia)
US diplomatic representation: none (territory of Australia)

Economy

Overview: no economic activity

Communications

Ports: none; offshore anchorage only

Defense Forces

Note: defense is the responsibility of Australia; periodic visits by the Royal Australian Navy and Royal Australian Air Force

Atlantic Ocean

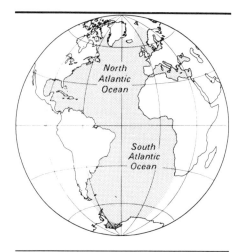

Geography

Location: body of water between the Western Hemisphere and Europe/Africa
Map references: Africa, Antarctic Region, Arctic Region, Central America and the Caribbean, Europe, North America, South America, Standard Time Zones of the World
Area:
total area: 82.217 million km²
comparative area: slightly less than nine times the size of the US; second-largest of the world's four oceans (after the Pacific Ocean, but larger than Indian Ocean or Arctic Ocean)
note: includes Baltic Sea, Black Sea, Caribbean Sea, Davis Strait, Denmark Strait, Drake Passage, Gulf of Mexico, Mediterranean Sea, North Sea, Norwegian Sea, Weddell Sea, and other tributary water bodies
Coastline: 111,866 km
International disputes: some maritime disputes (see littoral states)
Climate: tropical cyclones (hurricanes) develop off the coast of Africa near Cape Verde and move westward into the Caribbean Sea; hurricanes can occur from May to December, but are most frequent from August to November
Terrain: surface usually covered with sea ice in Labrador Sea, Denmark Strait, and Baltic Sea from October to June; clockwise warm water gyre (broad, circular system of currents) in the north Atlantic, counterclockwise warm water gyre in the south Atlantic; the ocean floor is dominated by the Mid-Atlantic Ridge, a rugged north-south centerline for the entire Atlantic basin; maximum depth is 8,605 meters in the Puerto Rico Trench
Natural resources: oil and gas fields, fish, marine mammals (seals and whales), sand and gravel aggregates, placer deposits, polymetallic nodules, precious stones

Environment: endangered marine species include the manatee, seals, sea lions, turtles, and whales; municipal sludge pollution off eastern US, southern Brazil, and eastern Argentina; oil pollution in Caribbean Sea, Gulf of Mexico, Lake Maracaibo, Mediterranean Sea, and North Sea; industrial waste and municipal sewage pollution in Baltic Sea, North Sea, and Mediterranean Sea; icebergs common in Davis Strait, Denmark Strait, and the northwestern Atlantic from February to August and have been spotted as far south as Bermuda and the Madeira Islands; icebergs from Antarctica occur in the extreme southern Atlantic
Note: ships subject to superstructure icing in extreme north Atlantic from October to May and extreme south Atlantic from May to October; persistent fog can be a hazard to shipping from May to September; major choke points include the Dardanelles, Strait of Gibraltar, access to the Panama and Suez Canals; strategic straits include the Dover Strait, Straits of Florida, Mona Passage, The Sound (Oresund), and Windward Passage; north Atlantic shipping lanes subject to icebergs from February to August; the Equator divides the Atlantic Ocean into the North Atlantic Ocean and South Atlantic Ocean

Government

Digraph: ZH

Economy

Overview: Economic activity is limited to exploitation of natural resources, especially fish, dredging aragonite sands (The Bahamas), and crude oil and natural gas production (Caribbean Sea, Gulf of Mexico, and North Sea).

Communications

Ports: Alexandria (Egypt), Algiers (Algeria), Antwerp (Belgium), Barcelona (Spain), Buenos Aires (Argentina), Casablanca (Morocco), Colon (Panama), Copenhagen (Denmark), Dakar (Senegal), Gdansk (Poland), Hamburg (Germany), Helsinki (Finland), Las Palmas (Canary Islands, Spain), Le Havre (France), Lisbon (Portugal), London (UK), Marseille (France), Montevideo (Uruguay), Montreal (Canada), Naples (Italy), New Orleans (US), New York (US), Oran (Algeria), Oslo (Norway), Piraeus (Greece), Rio de Janeiro (Brazil), Rotterdam (Netherlands), Saint Petersburg (formerly Leningrad; Russia), Stockholm (Sweden)

Australia

Geography

Location: Oceania, between Indonesia and New Zealand
Map references: Southeast Asia, Oceania, Antarctic Region, Standard Time Zones of the World
Area:
total area: 7,686,850 km²
land area: 7,617,930 km²
comparative area: slightly smaller than the US
note: includes Macquarie Island
Land boundaries: 0 km
Coastline: 25,760 km
Maritime claims:
contiguous zone: 12 nm
continental shelf: 200 m (depth) or to depth of exploitation
exclusive fishing zone: 200 nm
territorial sea: 12 nm
International disputes: territorial claim in Antarctica (Australian Antarctic Territory)
Climate: generally arid to semiarid; temperate in south and east; tropical in north
Terrain: mostly low plateau with deserts; fertile plain in southeast
Natural resources: bauxite, coal, iron ore, copper, tin, silver, uranium, nickel, tungsten, mineral sands, lead, zinc, diamonds, natural gas, petroleum
Land use:
arable land: 6%
permanent crops: 0%
meadows and pastures: 58%
forest and woodland: 14%
other: 22%
Irrigated land: 18,800 km² (1989 est.)
Environment: subject to severe droughts and floods; cyclones along coast; limited freshwater availability; irrigated soil degradation; regular, tropical, invigorating, sea breeze known as "the Doctor" occurs along west coast in summer; desertification
Note: world's smallest continent but sixth-largest country

People

Population: 17,827,204 (July 1993 est.)
Population growth rate: 1.41% (1993 est.)
Birth rate: 14.43 births/1,000 population (1993 est.)
Death rate: 7.38 deaths/1,000 population (1993 est.)
Net migration rate: 7.01 migrant(s)/1,000 population (1993 est.)
Infant mortality rate: 7.4 deaths/1,000 live births (1993 est.)
Life expectancy at birth:
total population: 77.36 years
male: 74.24 years
female: 80.63 years (1993 est.)
Total fertility rate: 1.83 children born/woman (1993 est.)
Nationality:
noun: Australian(s)
adjective: Australian
Ethnic divisions: Caucasian 95%, Asian 4%, Aboriginal and other 1%
Religions: Anglican 26.1%, Roman Catholic 26%, other Christian 24.3%
Languages: English, native languages
Literacy: age 15 and over can read and write (1980)
total population: 100%
male: 100%
female: 100%
Labor force: 8.63 million (September 1991)
by occupation: finance and services 33.8%, public and community services 22.3%, wholesale and retail trade 20.1%, manufacturing and industry 16.2%, agriculture 6.1% (1987)

Government

Names:
conventional long form: Commonwealth of Australia
conventional short form: Australia
Digraph: AS
Type: federal parliamentary state
Capital: Canberra
Administrative divisions: 6 states and 2 territories*; Australian Capital Territory*, New South Wales, Northern Territory*, Queensland, South Australia, Tasmania, Victoria, Western Australia
Dependent areas: Ashmore and Cartier Islands, Christmas Island, Cocos (Keeling) Islands, Coral Sea Islands, Heard Island and McDonald Islands, Norfolk Island
Independence: 1 January 1901 (federation of UK colonies)
Constitution: 9 July 1900, effective 1 January 1901
Legal system: based on English common law; accepts compulsory ICJ jurisdiction, with reservations
National holiday: Australia Day, 26 January

Australia (continued)

Political parties and leaders:
government: Australian Labor Party, Paul John KEATING
opposition: Liberal Party, John HEWSON; National Party, Timothy FISCHER; Australian Democratic Party, John COULTER
Other political or pressure groups: Australian Democratic Labor Party (anti-Communist Labor Party splinter group); Peace and Nuclear Disarmament Action (Nuclear Disarmament Party splinter group)
Suffrage: 18 years of age; universal and compulsory
Elections:
House of Representatives: last held 13 March 1993 (next to be held by NA May 1996); results—percent of vote by party NA; seats—(147 total) Labor 80, Liberal-National 65, independent 2
Senate: last held 13 March 1993 (next to be held by NA May 1999); results—percent of vote by party NA; seats—(76 total) Liberal-National 36, Labor 30, Australian Democrats 7, Greens 2, independents 1
Executive branch: British monarch, governor general, prime minister, deputy prime minister, Cabinet
Legislative branch: bicameral Federal Parliament consists of an upper house or Senate and a lower house or House of Representatives
Judicial branch: High Court
Leaders:
Chief of State: Queen ELIZABETH II (since 6 February 1952), represented by Governor General William George HAYDEN (since 16 February 1989)
Head of Government: Prime Minister Paul John KEATING (since 20 December 1991); Deputy Prime Minister Brian HOWE (since 4 June 1991)
Member of: AfDB, AG (observer), ANZUS, APEC, AsDB, Australia Group, BIS, C, CCC, COCOM, CP, EBRD, ESCAP, FAO, GATT, G-8, IAEA, IBRD, ICAO, ICC, ICFTU, IDA, IEA, IFAD, IFC, ILO, IMF, IMO, INMARSAT, INTELSAT, INTERPOL, IOC, IOM, ISO, ITU, LORCS, MINURSO, MTCR, NAM (guest), NEA, NSG, OECD, PCA, SPARTECA, SPC, SPF, UN, UNCTAD, UNESCO, UNHCR, UNIDO, UNOSOM, UNTAC, UNTSO, UPU, WFTU, WHO, WIPO, WMO, ZC
Diplomatic representation in US:
chief of mission: Ambassador Michael J. COOK
chancery: 1601 Massachusetts Avenue NW, Washington, DC 20036
telephone: (202) 797-3000
consulates general: Chicago, Honolulu, Houston, Los Angeles, New York, Pago Pago (American Samoa), and San Francisco
US diplomatic representation:
chief of mission: (vacant)
embassy: Moonah Place, Yarralumla, Canberra, Australian Capital Territory 2600
mailing address: APO AP 96549
telephone: [61] (6) 270-5000
FAX: [61] (6) 270-5970
consulates general: Melbourne, Perth, and Sydney
consulate: Brisbane
Flag: blue with the flag of the UK in the upper hoist-side quadrant and a large seven-pointed star in the lower hoist-side quadrant; the remaining half is a representation of the Southern Cross constellation in white with one small five-pointed star and four, larger, seven-pointed stars

---

Economy

Overview: Australia has a prosperous Western-style capitalist economy, with a per capita GDP comparable to levels in industrialized West European countries. Rich in natural resources, Australia is a major exporter of agricultural products, minerals, metals, and fossil fuels. Of the top 25 exports, 21 are primary products, so that, as happened during 1983-84, a downturn in world commodity prices can have a big impact on the economy. The government is pushing for increased exports of manufactured goods, but competition in international markets continues to be severe.
National product: GDP—purchasing power equivalent—$293.5 billion (1992)
National product real growth rate: 2.5% (1992)
National product per capita: $16,700 (1992)
Inflation rate (consumer prices): 0.8% (September 1992)
Unemployment rate: 11.3% (December 1992)
Budget: revenues $68.5 billion; expenditures $78.0 billion, including capital expenditures of $NA (FY93)
Exports: $41.7 billion (f.o.b., FY91)
commodities: coal, gold, meat, wool, alumina, wheat, machinery and transport equipment
partners: Japan 26%, US 11%, NZ 6%, South Korea 4%, Singapore 4%, UK, Taiwan, Hong Kong
Imports: $37.8 billion (f.o.b., FY91)
commodities: machinery and transport equipment, computers and office machines, crude oil and petroleum products
partners: US 24%, Japan 19%, UK 6%, FRG 7%, NZ 4% (1990)
External debt: $130.4 billion (June 1991)
Industrial production: growth rate NA%; accounts for 32% of GDP
Electricity: 40,000,000 kW capacity; 150,000 million kWh produced, 8,475 kWh per capita (1992)

Industries: mining, industrial and transportation equipment, food processing, chemicals, steel
Agriculture: accounts for 5% of GDP and 37% of export revenues; world's largest exporter of beef and wool, second-largest for mutton, and among top wheat exporters; major crops—wheat, barley, sugarcane, fruit; livestock—cattle, sheep, poultry
Illicit drugs: Tasmania is one of the world's major suppliers of licit opiate products; government maintains strict controls over areas of opium poppy cultivation and output of poppy straw concentrate
Economic aid: donor—ODA and OOF commitments (1970-89), $10.4 billion
Currency: 1 Australian dollar ($A) = 100 cents
Exchange rates: Australian dollars ($A) per US$1—1.4837 (January 1993), 1.3600 (1992), 1.2836 (1991), 1.2799 (1990), 1.2618 (1989), 1.2752 (1988)
Fiscal year: 1 July-30 June

Communications

Railroads: 40,478 km total; 7,970 km 1.600-meter gauge, 16,201 km 1.435-meter standard gauge, 16,307 km 1.067-meter gauge; 183 km dual gauge; 1,130 km electrified; government owned (except for a few hundred kilometers of privately owned track) (1985)
Highways: 837,872 km total; 243,750 km paved, 228,396 km gravel, crushed stone, or stabilized soil surface, 365,726 km unimproved earth
Inland waterways: 8,368 km; mainly by small, shallow-draft craft
Pipelines: crude oil 2,500 km; petroleum products 500 km; natural gas 5,600 km
Ports: Adelaide, Brisbane, Cairns, Darwin, Devonport, Fremantle, Geelong, Hobart, Launceston, Mackay, Melbourne, Sydney, Townsville
Merchant marine: 82 ships (1,000 GRT or over) totaling 2,347,271 GRT/3,534,926 DWT; includes 2 short-sea passenger, 8 cargo, 7 container, 8 roll-on/roll-off, 1 vehicle carrier, 17 oil tanker, 3 chemical tanker, 4 liquefied gas, 30 bulk, 2 combination bulk
Airports:
total: 481
usable: 439
with permanent-surface runways: 243
with runways over 3,659 m: 1
with runways 2,440-3,659 m: 20
with runways 1,220-2,439 m: 268
Telecommunications: good international and domestic service; 8.7 million telephones; broadcast stations—258 AM, 67 FM, 134 TV; submarine cables to New Zealand, Papua New Guinea, and Indonesia; domestic satellite service; satellite stations—4 Indian

Austria

Ocean INTELSAT, 6 Pacific Ocean
INTELSAT earth stations

Defense Forces

Branches: Australian Army, Royal
Australian Navy, Royal Australian Air Force
Manpower availability: males age 15-49
4,830,068; fit for military service 4,198,622;
reach military age (17) annually 135,591
(1993 est.)
Defense expenditures: exchange rate
conversion—$7.1 billion, 2.4% of GDP
(FY92/93)

150 km

Geography

Location: Central Europe, between Germany
and Hungary
Map references: Africa, Arctic Region,
Europe, Standard Time Zones of the World
Area:
total area: 83,850 km²
land area: 82,730 km²
comparative area: slightly smaller than
Maine
Land boundaries: total 2,496 km, Czech
Republic 362 km, Germany 784 km,
Hungary 366 km, Italy 430 km,
Liechtenstein 37 km, Slovakia 91 km,
Slovenia 262 km, Switzerland 164 km
Coastline: 0 km (landlocked)
Maritime claims: none; landlocked
International disputes: none
Climate: temperate; continental, cloudy;
cold winters with frequent rain in lowlands
and snow in mountains; cool summers with
occasional showers
Terrain: in the west and south mostly
mountains (Alps); along the eastern and
northern margins mostly flat or gently
sloping
Natural resources: iron ore, petroleum,
timber, magnesite, aluminum, lead, coal,
lignite, copper, hydropower
Land use:
arable land: 17%
permanent crops: 1%
meadows and pastures: 24%
forest and woodland: 39%
other: 19%
Irrigated land: 40 km² (1989)
Environment: population is concentrated on
eastern lowlands because of steep slopes,
poor soils, and low temperatures elsewhere
Note: landlocked; strategic location at the
crossroads of central Europe with many
easily traversable Alpine passes and valleys;
major river is the Danube

People

Population: 7,915,145 (July 1993 est.)
Population growth rate: 0.55% (1993 est.)
Birth rate: 11.54 births/1,000 population
(1993 est.)
Death rate: 10.42 deaths/1,000 population
(1993 est.)
Net migration rate: 4.42 migrant(s)/1,000
population (1993 est.)
Infant mortality rate: 7.3 deaths/1,000 live
births (1993 est.)
Life expectancy at birth:
total population: 76.4 years
male: 73.18 years
female: 79.8 years (1993 est.)
Total fertility rate: 1.47 children
born/woman (1993 est.)
Nationality:
noun: Austrian(s)
adjective: Austrian
Ethnic divisions: German 99.4%, Croatian
0.3%, Slovene 0.2%, other 0.1%
Religions: Roman Catholic 85%, Protestant
6%, other 9%
Languages: German
Literacy: age 15 and over can read and
write (1974)
total population: 99%
male: NA%
female: NA%
Labor force: 3.47 million (1989)
by occupation: services 56.4%, industry and
crafts 35.4%, agriculture and forestry 8.1%
note: an estimated 200,000 Austrians are
employed in other European countries;
foreign laborers in Austria number 177,840,
about 6% of labor force (1988)

Government

Names:
conventional long form: Republic of Austria
conventional short form: Austria
local long form: Republik Oesterreich
local short form: Oesterreich
Digraph: AU
Type: federal republic
Capital: Vienna
Administrative divisions: 9 states
(bundeslander, singular—bundesland);
Burgenland, Karnten, Niederosterreich,
Oberosterreich, Salzburg, Steiermark, Tirol,
Vorarlberg, Wien
Independence: 12 November 1918 (from
Austro-Hungarian Empire)
Constitution: 1920; revised 1929 (reinstated
1945)
Legal system: civil law system with Roman
law origin; judicial review of legislative acts
by a Constitutional Court; separate
administrative and civil/penal supreme
courts; has not accepted compulsory ICJ
jurisdiction

National holiday: National Day, 26 October (1955)

Political parties and leaders: Social Democratic Party of Austria (SPO), Franz VRANITZKY, chairman; Austrian People's Party (OVP), Erhard BUSEK, chairman; Freedom Party of Austria (FPO), Jorg HAIDER, chairman; Communist Party (KPO), Walter SILBERMAYER, chairman; Green Alternative List (GAL), Johannes VOGGENHUBER, chairman

Other political or pressure groups: Federal Chamber of Commerce and Industry; Austrian Trade Union Federation (primarily Socialist); three composite leagues of the Austrian People's Party (OVP) representing business, labor, and farmers; OVP-oriented League of Austrian Industrialists; Roman Catholic Church, including its chief lay organization, Catholic Action

Suffrage: 19 years of age, universal; compulsory for presidential elections

Elections:
President: last held 24 May 1992 (next to be held 1996); results of second ballot—Thomas KLESTIL 57%, Rudolf STREICHER 43%
National Council: last held 7 October 1990 (next to be held October 1994); results—SPO 43%, OVP 32.1%, FPO 16.6%, GAL 4.5%, KPO 0.7%, other 0.32%; seats—(183 total) SPO 80, OVP 60, FPO 33, GAL 10

Executive branch: president, chancellor, vice chancellor, Council of Ministers (cabinet)

Legislative branch: bicameral Federal Assembly (Bundesversammlung) consists of an upper council or Federal Council (Bundesrat) and a lower council or National Council (Nationalrat)

Judicial branch: Supreme Judicial Court (Oberster Gerichtshof) for civil and criminal cases, Administrative Court (Verwaltungsgerichtshof) for bureaucratic cases, Constitutional Court (Verfassungsgerichtshof) for constitutional cases

Leaders:
Chief of State: President Thomas KLESTIL (since 8 July 1992)
Head of Government: Chancellor Franz VRANITZKY (since 16 June 1986); Vice Chancellor Erhard BUSEK (since 2 July 1991)

Member of: AfDB, AG (observer), AsDB, Australia Group, BIS, CCC, CE, CEI, CERN, COCOM (cooperating country), CSCE, EBRD, ECE, EFTA, ESA, FAO, G-9, GATT, IADB, IAEA, IBRD, ICAO, ICC, ICFTU, IDA, IEA, IFAD, IFC, ILO, IMF, IMO, INTELSAT, INTERPOL, IOC, IOM, ISO, ITU, LORCS, MINURSO, MTCR, NAM (guest), NEA, NSG, OAS (observer), OECD, PCA, UN, UNCTAD, UNESCO, UNDOF, UNFICYP, UNHCR, UNIDO, UNIKOM, UNOSOM, UNTAC, UNTSO, UPU, WCL, WFTU, WHO, WIPO, WMO, WTO, ZC

Diplomatic representation in US:
chief of mission: Ambassador Friedrich HOESS
chancery: 3524 International Court NW, Washington, DC 20008-3035
telephone: (202) 895-6700
FAX: (202) 895-6750
consulates general: Chicago, Los Angeles, and New York

US diplomatic representation:
chief of mission: Ambassador Roy Michael HUFFINGTON
chancery: Boltzmanngasse 16, A-1091, Unit 27937, Vienna
mailing address: APO AE 09222
telephone: [43] (1) 31-339
FAX: [43] (1) 310-0682
consulate general: Salzburg

Flag: three equal horizontal bands of red (top), white, and red

---

Economy

Overview: Austria boasts a prosperous and stable socialist market economy with a sizable proportion of nationalized industry and extensive welfare benefits. Thanks to an excellent raw material endowment, a technically skilled labor force, and strong links to German industrial firms, Austria occupies specialized niches in European industry and services (tourism, banking) and produces almost enough food to feed itself with only 8% of the labor force in agriculture. Increased export sales resulting from German unification, continued to boost Austria's economy through 1991. However, Germany's economic difficulties in 1992 slowed Austria's GDP growth to 2% from the 3% of 1991. Austria's economy, moreover, is not expected to grow by more than 1% in 1993, and inflation is forecast to remain about 4%. Unemployment will likely remain at current levels at least until 1994. Living standards in Austria are comparable with the large industrial countries of Western Europe. Problems for the 1990s include an aging population, the high level of subsidies, and the struggle to keep welfare benefits within budgetary capabilities. The continued opening of Eastern European markets, however, will increase demand for Austrian exports. Austria, a member of the European Free Trade Association (EFTA), in 1992 ratified the European Economic Area Treaty, which will extend European Community rules on the free movement of people, goods, capital and services to the EFTA countries, and Austrians plan to hold a national referendum within the next two years to vote on EC membership.

National product: GDP—purchasing power equivalent—$141.3 billion (1992)

National product real growth rate: 1.8% (1992)

National product per capita: $18,000 (1992)

Inflation rate (consumer prices): 4% (1992 est.)

Unemployment rate: 6.4% (1992 est.)

Budget: revenues $47.8 billion; expenditures $53.0 billion, including capital expenditures of $NA (1992 est.)

Exports: $43.5 billion (1992 est.)
commodities: machinery and equipment, iron and steel, lumber, textiles, paper products, chemicals
partners: EC 65.8% (Germany 39%), EFTA 9.1%, Eastern Europe/former USSR 9.0%, Japan 1.7%, US 2.8% (1991)

Imports: $50.7 billion (1992 est.)
commodities: petroleum, foodstuffs, machinery and equipment, vehicles, chemicals, textiles and clothing, pharmaceuticals
partners: EC 67.8% (Germany 43.0%), EFTA 6.9%, Eastern Europe/former USSR 6.0%, Japan 4.8%, US 3.9% (1991)

External debt: $11.8 billion (1990 est.)

Industrial production: growth rate 2.0% (1991)

Electricity: 17,600,000 kW capacity; 49,500 million kWh produced, 6,300 kWh per capita (1992)

Industries: foods, iron and steel, machines, textiles, chemicals, electrical, paper and pulp, tourism, mining, motor vehicles

Agriculture: accounts for 3.2% of GDP (including forestry); principal crops and animals—grains, fruit, potatoes, sugar beets, sawn wood, cattle, pigs, poultry; 80-90% self-sufficient in food

Illicit drugs: transshipment point for Southwest Asian heroin transiting the Balkan route

Economic aid: donor—ODA and OOF commitments (1970-89), $2.4 billion

Currency: 1 Austrian schilling (S) = 100 groschen

Exchange rates: Austrian schillings (S) per US$1—11.363 (January 1993), 10.989 (1992), 11.676 (1991), 11.370 (1990), 13.231 (1989), 12.348 (1988)

Fiscal year: calendar year

Communications

Railroads: 5,749 km total; 5,652 km government owned and 97 km privately owned (0.760-, 1.435- and 1.000-meter gauge); 5,394 km 1.435-meter standard gauge of which 3,154 km is electrified and 1,520 km is double tracked; 339 km 0.760-meter narrow gauge of which 84 km is electrified

Highways: 95,412 km total; 34,612 km are the primary network (including 1,012 km of

Azerbaijan

autobahn, 10,400 km of federal, and 23,200 km of provincial roads); of this number, 21,812 km are paved and 12,800 km are unpaved; in addition, there are 60,800 km of communal roads (mostly gravel, crushed stone, earth)

Inland waterways: 446 km

Pipelines: crude oil 554 km; natural gas 2,611 km; petroleum products 171 km

Ports: Vienna, Linz (Danube river ports)

Merchant marine: 29 ships (1,000 GRT or over) totaling 154,159 GRT/256,765 DWT; includes 23 cargo, 1 refrigerated cargo, 1 oil tanker, 1 chemical tanker, 3 bulk

Airports:
total: 55
usable: 55
with permanent-surface runways: 20
with runways over 3,659 m: 0
with runways 2,440-3,659 m: 6
with runways 1,220-2,439 m: 4

Telecommunications: highly developed and efficient; 4,014,000 telephones; broadcast stations—6 AM, 21 (545 repeaters) FM, 47 (870 repeaters) TV; satellite ground stations for Atlantic Ocean INTELSAT, Indian Ocean INTELSAT, and EUTELSAT systems

---

Defense Forces

Branches: Army (including Flying Division)

Manpower availability: males age 15-49 2,016,464; fit for military service 1,694,140; reach military age (19) annually 50,259 (1993 est.)

Defense expenditures: exchange rate conversion—$1.7 billion, 0.9% of GDP (1993 est.)

Geography

Location: Southeastern Europe, between Armenia and Turkmenistan, bordering the Caspian Sea

Map references: Africa, Asia, Commonwealth of Independent States—Central Asian States, Commonwealth of Independent States—European States, Middle East, Standard Time Zones of the World

Area:
total area: 86,600 km²
land area: 86,100 km²
comparative area: slightly larger than Maine
note: includes the Nakhichevan' Autonomous Republic and the Nagorno-Karabakh Autonomous Oblast; region's autonomy was abolished by Azerbaijan Supreme Soviet on 26 November 1991

Land boundaries: total 2,013 km, Armenia (west) 566 km, Armenia (southwest) 221 km, Georgia 322 km, Iran (south) 432 km, Iran (southwest) 179 km, Russia 284 km, Turkey 9 km

Coastline: 0 km (landlocked)
note: Azerbaijan does border the Caspian Sea (800 km, est.)

Maritime claims: NA
note: Azerbaijani claims in Caspian Sea unknown; 10 nm fishing zone provided for in 1940 treaty regarding trade and navigation between Soviet Union and Iran

International disputes: violent and longstanding dispute with Armenia over status of Nagorno-Karabakh, lesser dispute concerns Nakhichevan; some Azerbaijanis desire absorption of and/or unification with the ethnically Azeri portion of Iran; minor irredentist disputes along Georgia border

Climate: dry, semiarid steppe; subject to drought

Terrain: large, flat Kura-Aras Lowland (much of it below sea level) with Great Caucasus Mountains to the north, Karabakh Upland in west; Baku lies on Aspheson Peninsula that juts into Caspian Sea

Natural resources: petroleum, natural gas, iron ore, nonferrous metals, alumina

Land use:
arable land: 18%
permanent crops: 0%
meadows and pastures: 25%
forest and woodland: 0%
other: 57%

Irrigated land: 14,010 km² (1990)

Environment: local scientists consider Apsheron Peninsula, including Baku and Sumgait, and the Caspian Sea to be "most ecologically devastated area in the world" because of severe air and water pollution

Note: landlocked

People

Population: 7,573,435 (July 1993 est.)

Population growth rate: 1.5% (1993 est.)

Birth rate: 24.09 births/1,000 population (1993 est.)

Death rate: 6.61 deaths/1,000 population (1993 est.)

Net migration rate: -2.45 migrant(s)/1,000 population (1993 est.)

Infant mortality rate: 35.7 deaths/1,000 live births (1993 est.)

Life expectancy at birth:
total population: 70.6 years
male: 66.77 years
female: 74.63 years (1993 est.)

Total fertility rate: 2.76 children born/woman (1993 est.)

Nationality:
noun: Azerbaijani(s)
adjective: Azerbaijani

Ethnic divisions: Azeri 82.7%, Russian 5.6%, Armenian 5.6%, Daghestanis 3.2%, other 2.9%, note—Armenian share may be less than 5.6% because many Armenians have fled the ethnic violence since 1989 census

Religions: Moslem 87%, Russian Orthodox 5.6%, Armenian Orthodox 5.6%, other 1.8%

Languages: Azeri 82%, Russian 7%, Armenian 5%, other 6%

Literacy: age 9-49 can read and write (1970)
total population: 100%
male: 100%
female: 100%

Labor force: 2.789 million
by occupation: agriculture and forestry 32%, industry and construction 26%, other 42% (1990)

Government

Names:
conventional long form: Azerbaijani Republic
conventional short form: Azerbaijan
local long form: Azarbaijchan Respublikasy
local short form: none

Azerbaijan (continued)

former: Azerbaijan Soviet Socialist Republic
Digraph: AJ
Type: republic
Capital: Baku (Baky)
Administrative divisions: 1 autonomous republic (avtomnaya respublika); Nakhichevan (administrative center at Nakhichevan)
note: all rayons except for the exclave of Nakhichevan are under direct republic jurisdiction; 1 autonomous oblast, Nagorno-Karabakh (officially abolished by Azerbaijani Supreme Soviet on 26 November 1991) has declared itself Nagorno-Karabakh Republic
Independence: 30 August 1991 (from Soviet Union)
Constitution: adopted NA April 1978; writing a new constitution mid-1993
Legal system: based on civil law system
National holiday: NA
Political parties and leaders: New Azerbaijan Party, ALIYEV; Musavat Party (Azerbaijan Popular Front—APF), Isa GAMBAROV; National Independence Party (main opposition party), Etibar MAMEDOV; Social Democratic Party (SDP), Zardusht Ali ZADE; Party of Revolutionary Revival (successor to the Communist Party), Sayad Afes OGLV, general secretary; Party of Independent Azerbaijan, SOVLEYMANOV
Other political or pressure groups: self-proclaimed Armenian Nagorno-Karabakh Republic
Suffrage: 18 years of age; universal
Elections:
President: last held 8 June 1992 (next to be held NA); results—Abdulfaz Ali ELCHIBEY, won 60% of vote
National Council: last held 30 September and 14 October 1990 for the Supreme Soviet (next expected to be held late 1993 for the National Council); seats for Supreme Soviet—(360 total) Communists 280, Democratic Bloc 45 (grouping of opposition parties), other 15, vacant 20; note—on 19 May 1992 the Supreme Soviet was disbanded in favor of a Popular Front-dominated National Council; seats—(50 total) 25 Popular Front, 25 opposition elements
Executive branch: president, council of ministers
Legislative branch: National Parliament (National Assembly or Milli Mejlis)
Judicial branch: Supreme Court
Leaders:
Chief of State: President Ebulfez ELCHIBEY (since 7 June 1992)
Head of Government: Prime Minister Penah HUSEYNOV (since 29 April 1993; resigned 7 June 1993; likely replacement—E'tibar MAMEDOV); National Parliament Chairman Isa GAMBAROV (since 19 May 1992; resigned 13 June 1993; likely replacement Geydar ALIYEV)

Member of: BSEC, CIS, CSCE, EBRD, ECO, ESCAP, IBRD, IDB, ILO, IMF, INTELSAT, ITU, NACC, OIC, UN, UNCTAD, UNESCO
Diplomatic representation in US:
chief of mission: Ambassador Hafiz PASHAYEV
chancery: 1615 L Street NW, Washington, DC 20036
telephone: NA
US diplomatic representation:
chief of mission: Ambassador Richard MILES
embassy: Hotel Intourist, Baku
mailing address: APO AE 09862
telephone: 7-8922-91-79-56
Flag: three equal horizontal bands of blue (top), red, and green; a crescent and eight-pointed star in white are centered in red band

---

Economy

Overview: Azerbaijan is less developed industrially than either Armenia or Georgia, the other Transcaucasian states. It resembles the Central Asian states in its majority Muslim population, high structural unemployment, and low standard of living. The economy's most prominent products are cotton, oil, and gas. Production from the Caspian oil and gas field has been in decline for several years. With foreign assistance, the oil industry might generate the funds needed to spur industrial development. However, civil unrest, marked by armed conflict in the Nagorno-Karabakh region between Muslim Azeris and Christian Armenians, makes foreign investors wary. Azerbaijan accounted for 1.5% to 2% of the capital stock and output of the former Soviet Union. Azerbaijan shares all the formidable problems of the ex-Soviet republics in making the transition from a command to a market economy, but its considerable energy resources brighten its propects somewhat. Old economic ties and structures have yet to be replaced. A particularly galling constraint on economic revival is the Nagorno-Karabakh conflict, said to consume 25% of Azerbaijan's economic resources.
National product: GDP $NA
National product real growth rate: -25% (1992)
National product per capita: $NA
Inflation rate (consumer prices): 20% per month (1992 est.)
Unemployment rate: 0.2% includes officially registered unemployed; also large numbers of underemployed workers
Budget: revenues $NA; expenditures $NA, including capital expenditures of $NA (1992)
Exports: $821 million to outside the successor states of the former USSR (f.o.b., 1992 est.)

commodities: oil and gas, chemicals, oilfield equipment, textiles, cotton (1991)
partners: mostly CIS and European countries
Imports: $300 million from outside the successor states of the former USSR (c.i.f., 1992 est.)
commodities: machinery and parts, consumer durables, foodstuffs, textiles (1991)
partners: European countries
External debt: $1.3 billion (1991 est.)
Industrial production: growth rate -27% (1992)
Electricity: 6,025,000 kW capacity; 22,300 million kWh produced, 2,990 kWh per capita (1992)
Industries: petroleum and natural gas, petroleum products, oilfield equipment; steel, iron ore, cement; chemicals and petrochemicals; textiles
Agriculture: cotton, grain, rice, grapes, fruit, vegetables, tea, tobacco; cattle, pigs, sheep and goats
Illicit drugs: illicit producer of cannabis and opium; mostly for CIS consumption; limited government eradication program; used as transshipment points for illicit drugs to Western Europe
Economic aid: wheat from Turkey
Currency: 1 manat (abbreviation NA) = 10 Russian rubles; ruble still used
Exchange rates: NA
Fiscal year: calendar year

Communications

Railroads: 2,090 km; does not include industrial lines (1990)
Highways: 36,700 km total (1990); 31,800 km hard surfaced; 4,900 km earth
Pipelines: crude oil 1,130 km, petroleum products 630 km, natural gas 1,240 km
Ports: inland—Baku (Baky)
Airports:
total: 65
useable: 33
with permanent-surface runways: 26
with runways over 3,659 m: 0
with runways 2,440-3,659 m: 8
with runways 1,220-2,439 m: 23
Telecommunications: domestic telephone service is of poor quality and inadequate; 644,000 domestic telephone lines (density—9 lines per 100 persons (1991)), 202,000 persons waiting for telephone installations (January 1991); connections to other former USSR republics by cable and microwave and to other countries via the Moscow international gateway switch; INTELSAT earth station installed in late 1992 in Baku with Turkish financial assistance with access to 200 countries through Turkey; domestic and Russian TV programs are received locally and Turkish and Iranian TV is received from an

The Bahamas

INTELSAT satellite through a receive-only earth station

Defense Forces

Branches: Army, Air Force, Navy, National Guard, Security Forces (internal and border troops)
Manpower availability: males age 15-49 1,842,917; fit for military service 1,497,640; reach military age (18) annually 66,928 (1993 est.).
Defense expenditures: 2,848 million rubles, NA% of GDP (1992 est.); note—conversion of the military budget into US dollars using the current exchange rate could produce misleading results

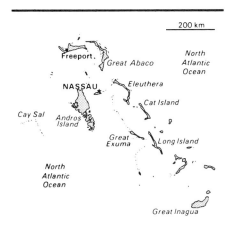

Geography

Location: in the western North Atlantic Ocean, southeast of Florida and northwest of Cuba
Map references: Central America and the Caribbean, North America, Standard Time Zones of the World
Area:
total area: 13,940 km²
land area: 10,070 km²
comparative area: slightly larger than Connecticut
Land boundaries: 0 km
Coastline: 3,542 km
Maritime claims:
continental shelf: 200 m (depth) or to depth of exploitation
exclusive fishing zone: 200 nm
territorial sea: 3 nm
International disputes: none
Climate: tropical marine; moderated by warm waters of Gulf Stream
Terrain: long, flat coral formations with some low rounded hills
Natural resources: salt, aragonite, timber
Land use:
arable land: 1%
permanent crops: 0%
meadows and pastures: 0%
forest and woodland: 32%
other: 67%
Irrigated land: NA km²
Environment: subject to hurricanes and other tropical storms that cause extensive flood damage
Note: strategic location adjacent to US and Cuba; extensive island chain

People

Population: 268,726 (July 1993 est.)
Population growth rate: 1.62% (1993 est.)
Birth rate: 18.97 births/1,000 population (1993 est.)
Death rate: 5.15 deaths/1,000 population (1993 est.)

Net migration rate: 2.42 migrant(s)/1,000 population (1993 est.)
Infant mortality rate: 31.6 deaths/1,000 live births (1993 est.)
Life expectancy at birth:
total population: 72.02 years
male: 68.19 years
female: 75.96 years (1993 est.)
Total fertility rate: 1.9 children born/woman (1993 est.)
Nationality:
noun: Bahamian(s)
adjective: Bahamian
Ethnic divisions: black 85%, white 15%
Religions: Baptist 32%, Anglican 20%, Roman Catholic 19%, Methodist 6%, Church of God 6%, other Protestant 12%, none or unknown 3%, other 2%
Languages: English, Creole, among Haitian immigrants
Literacy: age 15 and over but definition of literacy not available (1963)
total population: 90%
male: 90%
female: 89%
Labor force: 127,400
by occupation: government 30%, hotels and restaurants 25%, business services 10%, agriculture 5% (1989)

Government

Names:
conventional long form: The Commonwealth of The Bahamas
conventional short form: The Bahamas
Digraph: BF
Type: commonwealth
Capital: Nassau
Administrative divisions: 21 districts; Acklins and Crooked Islands, Bimini, Cat Island, Exuma, Freeport, Fresh Creek, Governor's Harbour, Green Turtle Cay, Harbour Island, High Rock, Inagua, Kemps Bay, Long Island, Marsh Harbour, Mayaguana, New Providence, Nichollstown and Berry Islands, Ragged Island, Rock Sound, Sandy Point, San Salvador and Rum Cay
Independence: 10 July 1973 (from UK)
Constitution: 10 July 1973
Legal system: based on English common law
National holiday: National Day, 10 July (1973)
Political parties and leaders: Progressive Liberal Party (PLP), Sir Lynden O. PINDLING; Free National Movement (FNM), Hubert Alexander INGRAHAM; Vanguard Nationalist and Socialist Party (VNPS), Lionel CAREY, chairman; People's Democratic Force (PDF), Fred MITCHELL
Other political or pressure groups: Vanguard Nationalist and Socialist Party (VNSP), a small leftist party headed by

The Bahamas *(continued)*

Lionel CAREY; Trade Union Congress (TUC), headed by Arlington MILLER
Suffrage: 18 years of age; universal
Elections:
House of Assembly: last held 19 August 1992 (next to be held by August 1997); results—percent of vote by party NA; seats—(49 total) FNM 32, PLP 17
Executive branch: British monarch, governor general, prime minister, deputy prime minister, Cabinet
Legislative branch: bicameral Parliament consists of an appointed upper house or Senate and a directly elected lower house or House of Assembly
Judicial branch: Supreme Court
Leaders:
Chief of State: Queen ELIZABETH II (since 6 February 1952), represented by Governor General Sir Clifford DARLING (since 2 January 1992)
Head of Government: Prime Minister Hubert INGRAHAM (since 19 August 1992)
Member of: ACP, C, CCC, CARICOM, CDB, ECLAC, FAO, G-77, IADB, IBRD, ICAO, ICFTU, IFC, ILO, IMF, IMO, INTELSAT, INTERPOL, IOC, ITU, LORCS, NAM, OAS, OPANAL, UN, UNCTAD, UNESCO, UNIDO, UPU, WHO, WIPO, WMO
Diplomatic representation in US:
chief of mission: Ambassador Timothy Baswell DONALDSON (since October 1992)
chancery: 2220 Massachusetts Avenue NW, Washington, DC 20008
telephone: (202) 319-2660
consulates general: Miami and New York
US diplomatic representation:
chief of mission: Ambassador Chic HECHT
embassy: Mosmar Building, Queen Street, Nassau
mailing address: P. O. Box N-8197, Nassau
telephone: (809) 322-1181 or 328-2206
FAX: (809) 328-7838
Flag: three equal horizontal bands of aquamarine (top), gold, and aquamarine with a black equilateral triangle based on the hoist side

Economy

Overview: The Bahamas is a stable, middle-income, developing nation whose economy is based primarily on tourism and offshore banking. Tourism alone provides about 50% of GDP and directly or indirectly employs about 50,000 people or 40% of the local work force. The economy has slackened in recent years, as the annual increase in the number of tourists slowed. Nonetheless, per capita GDP is one of the highest in the region.
National product: GDP—exchange rate conversion—$2.6 billion (1991 est.)

National product real growth rate: 3% (1991)
National product per capita: $10,200 (1991 est.)
Inflation rate (consumer prices): 7.2% (1991)
Unemployment rate: 16% (1991 est.)
Budget: revenues $627.5 million; expenditures $727.5 million, including capital expenditures of $100 million (1992 est.)
Exports: $306 million (f.o.b., 1991 est.)
commodities: pharmaceuticals, cement, rum, crawfish
partners: US 41%, Norway 30%, Denmark 4%
Imports: $1.14 billion (c.i.f., 1991 est.)
commodities: foodstuffs, manufactured goods, mineral fuels, crude oil
partners: US 35%, Nigeria 21%, Japan 13%, Angola 11%
External debt: $1.2 billion (December 1990)
Industrial production: growth rate 3% (1990); accounts for 15% of GDP
Electricity: 424,000 kW capacity; 929 million kWh produced, 3,599 kWh per capita (1992)
Industries: tourism, banking, cement, oil refining and transshipment, salt production, rum, aragonite, pharmaceuticals, spiral welded steel pipe
Agriculture: accounts for 5% of GDP; dominated by small-scale producers; principal products-citrus fruit, vegetables, poultry; large net importer of food
Illicit drugs: transshipment point for cocaine
Economic aid: US commitments, including Ex-Im (FY85-89), $1.0 million; Western (non-US) countries, ODA and OOF bilateral commitments (1970-89), $345 million
Currency: 1 Bahamian dollar (B$) = 100 cents
Exchange rates: Bahamian dollar (B$) per US$1—1.00 (fixed rate)
Fiscal year: calendar year

Communications

Highways: 2,400 km total; 1,350 km paved, 1,050 km gravel
Ports: Freeport, Nassau
Merchant marine: 853 ships (1,000 GRT or over) totaling 20,136,078 GRT/33,119,750 DWT; includes 53 passenger, 18 short-sea passenger, 159 cargo, 40 roll-on/roll-off cargo, 48 container, 6 vehicle carrier, 181 oil tanker, 14 liquefied gas, 22 combination ore/oil, 43 chemical tanker, 1 specialized tanker, 159 bulk, 7 combination bulk, 102 refrigerated cargo; note—a flag of convenience registry
Airports:
total: 60

usable: 55
with permanent-surface runways: 31
with runways over 3,659 m: 0
with runways 2,440-3, 659 m: 3
with runways 1,220-2,439 m: 26
Telecommunications: highly developed; 99,000 telephones in totally automatic system; tropospheric scatter and submarine cable links to Florida; broadcast stations—3 AM, 2 FM, 1 TV; 3 coaxial submarine cables; 1 Atlantic Ocean INTELSAT earth station

Defense Forces

Branches: Royal Bahamas Defense Force (Coast Guard only), Royal Bahamas Police Force
Manpower availability: males age 15-49 68,020; fit for military service NA (1993 est.)
Defense expenditures: exchange rate conversion—$65 million, 2.7% of GDP (1990)

Bahrain

Persian Gulf

Al Muharraq
MANAMA
Minā' Salmān
Sitrah
Āwāli

Gulf of
Bahrain

Hāwar Islands are
in dispute between
Bahrain and Qatar.

10 km

Geography

Location: Middle East, in the central Persian Gulf, between Saudi Arabia and Qatar
Map references: Africa, Middle East, Standard Time Zones of the World
Area:
total area: 620 km²
land area: 620 km²
comparative area: slightly less than 3.5 times the size of Washington, DC
Land boundaries: 0 km
Coastline: 161 km
Maritime claims:
contiguous zone: 24 nm
continental shelf: not specified
territorial sea: 12 nm
International disputes: territorial dispute with Qatar over the Hawar Islands; maritime boundary with Qatar
Climate: arid; mild, pleasant winters; very hot, humid summers
Terrain: mostly low desert plain rising gently to low central escarpment
Natural resources: oil, associated and nonassociated natural gas, fish
Land use:
arable land: 2%
permanent crops: 2%
meadows and pastures: 6%
forest and woodland: 0%
other: 90%
Irrigated land: 10 km² (1989 est.)
Environment: subsurface water sources being rapidly depleted (requires development of desalination facilities); dust storms; desertification
Note: close to primary Middle Eastern petroleum sources; strategic location in Persian Gulf through which much of Western world's petroleum must transit to reach open ocean

People

Population: 568,471 (July 1993 est.)
Population growth rate: 3.01% (1993 est.)

Birth rate: 26.89 births/1,000 population (1993 est.)
Death rate: 3.87 deaths/1,000 population (1993 est.)
Net migration rate: 7.04 migrant(s)/1,000 population (1993 est.)
Infant mortality rate: 20.1 deaths/1,000 live births (1993 est.)
Life expectancy at birth:
total population: 73.12 years
male: 70.72 years
female: 75.63 years (1993 est.)
Total fertility rate: 3.99 children born/woman (1993 est.)
Nationality:
noun: Bahraini(s)
adjective: Bahraini
Ethnic divisions: Bahraini 63%, Asian 13%, other Arab 10%, Iranian 8%, other 6%
Religions: Shi'a Muslim 70%, Sunni Muslim 30%
Languages: Arabic, English, Farsi, Urdu
Literacy: age 15 and over can read and write (1990)
total population: 77%
male: 82%
female: 69%
Labor force: 140,000
by occupation: industry and commerce 85%, agriculture 5%, services 5%, government 3% (1982)
note: 42% of labor force is Bahraini

Government

Names:
conventional long form: State of Bahrain
conventional short form: Bahrain
local long form: Dawlat al Bahrayn
local short form: Al Bahrayn
Digraph: BA
Type: traditional monarchy
Capital: Manama
Administrative divisions: 12 districts (manatiq, singular—mintaqah); Al Hadd, Al Manamah, Al Mintaqah al Gharbiyah, Al Mintaqah al Wusta, Al Mintaqah ash Shamaliyah, Al Muharraq, Ar Rifa'wa al Mintaqah al Janubiyah, Jidd Hafs, Madinat Hamad, Madinat 'Isa, Mintaqat Juzur Hawar, Sitrah
Independence: 15 August 1971 (from UK)
Constitution: 26 May 1973, effective 6 December 1973
Legal system: based on Islamic law and English common law
National holiday: Independence Day, 16 December
Political parties and leaders: political parties prohibited; several small, clandestine leftist and Islamic fundamentalist groups are active
Suffrage: none
Elections: none
Executive branch: amir, crown prince and heir apparent, prime minister, Cabinet

Legislative branch: unicameral National Assembly was dissolved 26 August 1975 and legislative powers were assumed by the Cabinet; appointed Advisory Council established 16 December 1992
Judicial branch: High Civil Appeals Court
Leaders:
Chief of State: Amir 'ISA bin Salman Al Khalifa (since 2 November 1961); Heir Apparent HAMAD bin 'Isa Al Khalifa (son of Amir; born 28 January 1950)
Head of Government: Prime Minister KHALIFA bin Salman Al Khalifa (since 19 January 1970)
Member of: ABEDA, AFESD, AL, AMF, ESCWA, FAO, G-77, GCC, IBRD, ICAO, IDB, ILO, IMF, IMO, INMARSAT, INTERPOL, IOC, ISO (correspondent), ITU, LORCS, NAM, OAPEC, OIC, UN, UNCTAD, UNESCO, UNIDO, UPU, WFTU, WHO, WMO
Diplomatic representation in US:
chief of mission: Ambassador 'Abd al-Rahman Faris Al KHALIFA
chancery: 3502 International Drive NW, Washington, DC 20008
telephone: (202) 342-0741 or 342-0742
consulate general: New York
US diplomatic representation:
chief of mission: Ambassador Dr. Charles W. HOSTLER
embassy: Road No. 3119 (next to Alahli Sports Club), Zinj District, Manama
mailing address: P. O. 26431, Manama, or FPO AE 09834-6210
telephone: [973] 273-300
FAX: (973) 272-594
Flag: red with a white serrated band (eight white points) on the hoist side

Economy

Overview: Petroleum production and processing account for about 80% of export receipts, 60% of government revenues, and 31% of GDP. Economic conditions have fluctuated with the changing fortunes of oil since 1985, for example, during the Gulf crisis of 1990-91. Bahrain with its highly developed communication and transport facilities is home to numerous multinational firms with business in the Gulf. A large share of exports consists of petroleum products made from imported crude.
National product: GDP—exchange rate conversion—$4.3 billion (1992 est.)
National product real growth rate: 3% (1992 est.)
National product per capita: $7,800 (1992 est.)
Inflation rate (consumer prices): 2% (1992 est.)
Unemployment rate: 8%-10% (1989)
Budget: revenues $1.2 billion; expenditures $1.32 billion, including capital expenditures of $NA (1989)

Bahrain (continued)

Exports: $3.5 billion (f.o.b., 1991)
commodities: petroleum and petroleum products 80%, aluminum 7%
partners: Japan 13%, UAE 12%, India 10%, Pakistan 8%
Imports: $3.7 billion (f.o.b., 1991)
commodities: nonoil 59%, crude oil 41%
partners: Saudi Arabia 41%, US 14%, UK 7%, Japan 5%
External debt: $1.8 billion (1991 est.)
Industrial production: growth rate 3.8% (1988); accounts for 44% of GDP
Electricity: 1,600,000 kW capacity; 4,700 million kWh produced, 8,500 kWh per capita (1992 est.)
Industries: petroleum processing and refining, aluminum smelting, offshore banking, ship repairing
Agriculture: including fishing, accounts for less than 2% of GDP; not self-sufficient in food production; heavily subsidized sector produces fruit, vegetables, poultry, dairy products, shrimp, fish; fish catch 9,000 metric tons in 1987
Economic aid: US commitments, including Ex-Im (FY70-79), $24 million; Western (non-US) countries, ODA and OOF bilateral commitments (1970-89), $45 million; OPEC bilateral aid (1979-89), $9.8 billion
Currency: 1 Bahraini dinar (BD) = 1,000 fils
Exchange rates: Bahraini dinars (BD) per US$1—0.3760 (fixed rate)
Fiscal year: calendar year

Communications

Highways: 200 km bituminous surfaced, including 25 km bridge-causeway to Saudi Arabia opened in November 1986; NA km natural surface tracks
Pipelines: crude oil 56 km; petroleum products 16 km; natural gas 32 km
Ports: Mina' Salman, Manama, Sitrah
Merchant marine: 9 ships (1,000 GRT or over) totaling 186,331 GRT/249,490 DWT; includes 5 cargo, 2 container, 1 liquefied gas, 1 bulk
Airports:
total: 3
usable: 3
with permanent-surface runways: 2
with runways over 3,659 m: 2
with runways 2,440-3,659 m: 0
with runways 1,220-2,439 m: 1
Telecommunications: modern system; good domestic services; 98,000 telephones (1 for every 6 persons); excellent international connections; tropospheric scatter to Qatar, UAE; microwave radio relay to Saudi Arabia; submarine cable to Qatar, UAE, and Saudi Arabia; satellite earth stations—1 Atlantic Ocean INTELSAT, 1 Indian Ocean INTELSAT, 1 ARABSAT; broadcast stations—2 AM, 3 FM, 2 TV

Defense Forces

Branches: Army, Navy, Air Force, Air Defense, Police Force
Manpower availability: males age 15-49 194,770; fit for military service 107,696; reach military age (15) annually 5,043 (1993 est.)
Defense expenditures: exchange rate conversion—$245 million, 6% of GDP (1990)

Baker Island
(territory of the US)

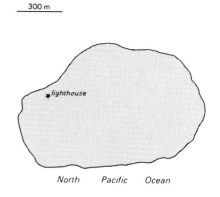

North   Pacific   Ocean

Geography

Location: in the North Pacific Ocean, just north of the Equator, 2,575 km southwest of Honolulu, about halfway between Hawaii and Australia
Map references: Oceania
Area:
total area: 1.4 km²
land area: 1.4 km²
comparative area: about 2.3 times the size of The Mall in Washington, DC
Land boundaries: 0 km
Coastline: 4.8 km
Maritime claims:
contiguous zone: 12 nm
continental shelf: 200 m or depth of exploitation
exclusive economic zone: 200 nm
territorial sea: 12 nm
International disputes: none
Climate: equatorial; scant rainfall, constant wind, burning sun
Terrain: low, nearly level coral island surrounded by a narrow fringing reef
Natural resources: guano (deposits worked until 1891)
Land use:
arable land: 0%
permanent crops: 0%
meadows and pastures: 0%
forest and woodland: 0%
other: 100%
Irrigated land: 0 km²
Environment: treeless, sparse, and scattered vegetation consisting of grasses, prostrate vines, and low growing shrubs; lacks fresh water; primarily a nesting, roosting, and foraging habitat for seabirds, shorebirds, and marine wildlife

People

Population: uninhabited; note—American civilians evacuated in 1942 after Japanese air and naval attacks during World War II; occupied by US military during

Bangladesh

World War II, but abandoned after the war; public entry is by special-use permit only and generally restricted to scientists and educators; a cemetery and cemetery ruins are located near the middle of the west coast

Government

Names:
conventional long form: none
conventional short form: Baker Island
Digraph: FQ
Type: unincorporated territory of the US administered by the Fish and Wildlife Service of the US Department of the Interior as part of the National Wildlife Refuge system
Capital: none; administered from Washington, DC

Economy

Overview: no economic activity

Communications

Ports: none; offshore anchorage only, one boat landing area along the middle of the west coast
Airports: 1 abandoned World War II runway of 1,665 m
Note: there is a day beacon near the middle of the west coast

Defense Forces

Note: defense is the responsibility of the US; visited annually by the US Coast Guard

150 km
Boundary representation is not necessarily authoritative

Geography

Location: South Asia, at the head of the Bay of Bengal, almost completely surrounded by India
Map references: Asia, Standard Time Zones of the World
Area:
total area: 144,000 km^2
land area: 133,910 km^2
comparative area: slightly smaller than Wisconsin
Land boundaries: total 4,246 km, Burma 193 km, India 4,053 km
Coastline: 580 km
Maritime claims:
contiguous zone: 18 nm
continental shelf: up to outer limits of continental margin
exclusive economic zone: 200 nm
territorial sea: 12 nm
International disputes: a portion of the boundary with India is in dispute; water-sharing problems with upstream riparian India over the Ganges
Climate: tropical; cool, dry winter (October to March); hot, humid summer (March to June); cool, rainy monsoon (June to October)
Terrain: mostly flat alluvial plain; hilly in southeast
Natural resources: natural gas, arable land, timber
Land use:
arable land: 67%
permanent crops: 2%
meadows and pastures: 4%
forest and woodland: 16%
other: 11%
Irrigated land: 27,380 km^2 (1989)
Environment: vulnerable to droughts; much of country routinely flooded during summer monsoon season; overpopulation; deforestation

People

Population: 122,254,849 (July 1993 est.)
Population growth rate: 2.35% (1993 est.)
Birth rate: 35.41 births/1,000 population (1993 est.)
Death rate: 11.94 deaths/1,000 population (1993 est.)
Net migration rate: 0 migrant(s)/1,000 population (1993 est.)
Infant mortality rate: 109.2 deaths/1,000 live births (1993 est.)
Life expectancy at birth:
total population: 54.7 years
male: 55 years
female: 54.38 years (1993 est.)
Total fertility rate: 4.55 children born/woman (1993 est.)
Nationality:
noun: Bangladeshi(s)
adjective: Bangladesh
Ethnic divisions: Bengali 98%, Biharis 250,000, tribals less than 1 million
Religions: Muslim 83%, Hindu 16%, Buddhist, Christian, other
Languages: Bangla (official), English
Literacy: age 15 and over can read and write (1990)
total population: 35%
male: 47%
female: 22%
Labor force: 35.1 million
by occupation: agriculture 74%, services 15%, industry and commerce 11% (FY86)
note: extensive export of labor to Saudi Arabia, UAE, and Oman (1991)

Government

Names:
conventional long form: People's Republic of Bangladesh
conventional short form: Bangladesh
former: East Pakistan
Digraph: BG
Type: republic
Capital: Dhaka
Administrative divisions: 64 districts (zillagulo, singular—zilla); Bagerhat, Bandarban, Barguna, Barisal, Bhola, Bogra, Brahmanbaria, Chandpur, Chapai Nawabganj, Chattagram, Chuadanga, Comilla, Cox's Bazar, Dhaka, Dinajpur, Faridpur, Feni, Gaibandha, Gazipur, Gopalganj, Habiganj, Jaipurhat, Jamalpur, Jessore, Jhalakati, Jhenaidah, Khagrachari, Khulna, Kishorganj, Kurigram, Kushtia, Laksmipur, Lalmonirhat, Madaripur, Magura, Manikganj, Meherpur, Moulavibazar, Munshiganj, Mymensingh, Naogaon, Narail, Narayanganj, Narsingdi, Nator, Netrakona, Nilphamari, Noakhali, Pabna, Panchagar, Parbattya Chattagram, Patuakhali, Pirojpur, Rajbari, Rajshahi, Rangpur, Satkhira, Shariyatpur, Sherpur, Sirajganj, Sunamganj,

Sylhet, Tangail, Thakurgaon
Independence: 16 December 1971 (from Pakistan)
Constitution: 4 November 1972, effective 16 December 1972, suspended following coup of 24 March 1982, restored 10 November 1986, amended NA March 1991
Legal system: based on English common law
National holiday: Independence Day, 26 March (1971)
Political parties and leaders: Bangladesh Nationalist Party (BNP), Khaleda ZIAur Rahman; Awami League (AL), Sheikh Hasina WAJED; Jatiyo Party (JP), Hussain Mohammad ERSHAD (in jail); Jamaat-E-Islami (JI), Ali KHAN; Bangladesh Communist Party (BCP), Saifuddin Ahmed MANIK; National Awami Party (Muzaffar); Workers Party, leader NA; Jatiyo Samajtantik Dal (JSD), Serajul ALAM KHAN; Ganotantri Party, leader NA; Islami Oikya Jote, leader NA; National Democratic Party (NDP), leader NA; Muslim League, Khan A. SABUR; Democratic League, Khondakar MUSHTAQUE Ahmed; Democratic League, Khondakar MUSHTAQUE Ahmed; United People's Party, Kazi ZAFAR Ahmed
Suffrage: 18 years of age; universal
Elections:
National Parliament: last held 27 February 1991 (next to be held NA February 1996); results—percent of vote by party NA; seats—(330 total, 300 elected and 30 seats reserved for women) BNP 168, AL 93, JP 35, JI 20, BCP 5, National Awami Party (Muzaffar) 1, Workers Party 1, JSD 1, Ganotantri Party 1, Islami Oikya Jote 1, NDP 1, independents 3
President: last held 8 October 1991 (next to be held by NA October 1996); results—Abdur Rahman BISWAS received 52.1% of parliamentary vote
Executive branch: president, prime minister, Cabinet
Legislative branch: unicameral National Parliament (Jatiya Sangsad)
Judicial branch: Supreme Court
Leaders:
Chief of State: President Abdur Rahman BISWAS (since 8 October 1991)
Head of Government: Prime Minister Khaleda ZIAur Rahman (since 20 March 1991)
Member of: AsDB, C, CCC, CP, ESCAP, FAO, G-77, GATT, IAEA, IBRD, ICAO, ICFTU, IDA, IDB, IFAD, IFC, ILO, IMF, IMO, INTELSAT, INTERPOL, IOC, IOM, ISO, ITU, LORCS, MINURSO, NAM, OIC, SAARC, UN, UNCTAD, UNESCO, UNIDO, UNIKOM, UNOMOZ, UNOSOM, UNTAC, UNPROFOR, UPU, WHO, WFTU, WIPO, WCL, WMO, WTO
Diplomatic representation in US:
chief of mission: Ambassador Abul AHSAN
chancery: 2201 Wisconsin Avenue NW,

Washington, DC 20007
telephone: (202) 342-8372 through 8376
consulate general: New York
US diplomatic representation:
chief of mission: Ambassador William B. MILAM
embassy: Diplomatic Enclave, Madani Avenue, Baridhara, Dhaka
mailing address: G. P. O. Box 323, Dhaka 1212
telephone: [880] (2) 884700-22
FAX: [880] (2) 883648
Flag: green with a large red disk slightly to the hoist side of center; green is the traditional color of Islam

---

Economy

Overview: Bangladesh is one of the world's poorest, most densely populated, and least developed nations. Its economy is overwhelmingly agricultural. Major impediments to growth include frequent cyclones and floods, government interference with the economy, a rapidly growing labor force that cannot be absorbed by agriculture, a low level of industrialization, failure to fully exploit energy resources (natural gas), and inefficient and inadequate power supplies. An excellent rice crop and expansion of the export garment industry helped growth in FY91/92. Policy reforms intended to reduce government regulation of private industry and promote public-sector efficiency have been announced but are being implemented only slowly.
National product: GDP—exchange rate conversion—$23.8 billion (FY92)
National product real growth rate: 3.8% (FY92)
National product per capita: $200 (FY92)
Inflation rate (consumer prices): 5.09% (FY92)
Unemployment rate: NA%
Budget: revenues $2.5 billion; expenditures $3.7 billion, including capital expenditures of $NA (FY92)
Exports: $2.0 billion (FY92)
commodities: garments, jute and jute goods, leather, shrimp
partners: US 28%, Western Europe 39% (FY91)
Imports: $3.4 billion (FY91/92)
commodities: capital goods, petroleum, food, textiles
partners: Japan 10.0%, Western Europe 17%, US 5.0% (FY91)
External debt: $11.8 billion (FY92 est.)
Industrial production: growth rate 4.0% (FY92 est.); accounts for less than 10% of GDP
Electricity: 2,400,000 kW capacity; 9,000 million kWh produced, 75 kWh per capita (1992)
Industries: jute manufacturing, cotton textiles, food processing, steel, fertilizer

Agriculture: accounts for about 40% of GDP, 60% of employment, and one-fifth of exports; imports 10% of food grain requirements; world's largest exporter of jute; commercial products—jute, rice, wheat, tea, sugarcane, potatoes, beef, milk, poultry; shortages include wheat, vegetable oils, cotton; fish catch 778,000 metric tons in 1986
Illicit drugs: transit country for illegal drugs produced in neighboring countries
Economic aid: US commitments, including Ex-Im (FY70-89), $3.4 billion; Western (non-US) countries, ODA and OOF bilateral commitments (1980-89), $11.65 million; OPEC bilateral aid (1979-89), $6.52 million; Communist countries (1970-89), $1.5 billion
Currency: 1 taka (Tk) = 100 paise
Exchange rates: taka (Tk) per US$1—39.000 (January 1993), 38.951 (1992), 36.596 (1991), 34.569 (1990), 32.270 (1989), 31.733 (1988)
Fiscal year: 1 July-30 June

Communications

Railroads: 2,892 km total (1986); 1,914 km 1.000 meter gauge, 978 km 1.676 meter broad gauge
Highways: 7,240 km total (1985); 3,840 km paved, 3,400 km unpaved
Inland waterways: 5,150-8,046 km navigable waterways (includes 2,575-3,058 km main cargo routes)
Pipelines: natural gas 1,220 km
Ports: Chittagong, Chalna
Merchant marine: 42 ships (1,000 GRT or over) totaling 314,228 GRT/461,607 DWT; includes 34 cargo, 2 oil tanker, 3 refrigerated cargo, 3 bulk
Airports:
total: 16
usable: 12
with permanent-surface runways: 12
with runways over 3,659 m: 0
with runways 2,440-3,659 m: 4
with runways 1,220-2,439 m: 6
Telecommunications: adequate international radio communications and landline service; fair domestic wire and microwave service; fair broadcast service; 241,250 telephones; broadcast stations—9 AM, 6 FM, 11 TV; 2 Indian Ocean INTELSAT satellite earth stations

---

Defense Forces

Branches: Army, Navy, Air Force
paramilitary forces: Bangladesh Rifles, Bangladesh Ansars, Armed Police Reserve, Defense Parties, National Cadet Corps
Manpower availability: males age 15-49 30,909,597; fit for military service 18,348,702 (1993 est.)
Defense expenditures: exchange rate conversion—$355 million, 1.5% of GDP (FY92/93)

Barbados

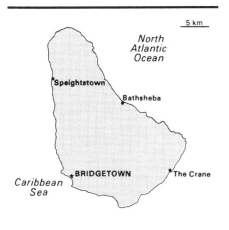

Geography

Location: in the extreme eastern Caribbean Sea, about 375 km northeast of Venezuela
Map references: Central America and the Caribbean, South America, Standard Time Zones of the World
Area:
total area: 430 km²
land area: 430 km²
comparative area: slightly less than 2.5 times the size of Washington, DC
Land boundaries: 0 km
Coastline: 97 km
Maritime claims:
exclusive economic zone: 200 nm
territorial sea: 12 nm
International disputes: none
Climate: tropical; rainy season (June to October)
Terrain: relatively flat; rises gently to central highland region
Natural resources: petroleum, fishing, natural gas
Land use:
arable land: 77%
permanent crops: 0%
meadows and pastures: 9%
forest and woodland: 0%
other: 14%
Irrigated land: NA km²
Environment: subject to hurricanes (especially June to October)
Note: easternmost Caribbean island

People

Population: 255,338 (July 1993 est.)
Population growth rate: 0.18% (1993 est.)
Birth rate: 15.78 births/1,000 population (1993 est.)
Death rate: 8.53 deaths/1,000 population (1993 est.)
Net migration rate: -5.49 migrant(s)/1,000 population (1993 est.)
Infant mortality rate: 21.3 deaths/1,000 live births (1993 est.)

Life expectancy at birth:
total population: 73.49 years
male: 70.75 years
female: 76.46 years (1993 est.)
Total fertility rate: 1.77 children born/woman (1993 est.)
Nationality:
noun: Barbadian(s)
adjective: Barbadian
Ethnic divisions: African 80%, mixed 16%, European 4%
Religions: Protestant 67% (Anglican 40%, Pentecostal 8%, Methodist 7%, other 12%), Roman Catholic 4%, none 17%, unknown 3%, other 9% (1980)
Languages: English
Literacy: age 15 and over having ever attended school (1970)
total population: 99%
male: 99%
female: 99%
Labor force: 120,900 (1991)
by occupation: services and government 37%, commerce 22%, manufacturing and construction 22%, transportation, storage, communications, and financial institutions 9%, agriculture 8%, utilities 2% (1985 est.)

Government

Names:
conventional long form: none
conventional short form: Barbados
Digraph: BB
Type: parliamentary democracy
Capital: Bridgetown
Administrative divisions: 11 parishes; Christ Church, Saint Andrew, Saint George, Saint James, Saint John, Saint Joseph, Saint Lucy, Saint Michael, Saint Peter, Saint Philip, Saint Thomas
note: the new city of Bridgetown may be given parish status
Independence: 30 November 1966 (from UK)
Constitution: 30 November 1966
Legal system: English common law; no judicial review of legislative acts
National holiday: Independence Day, 30 November (1966)
Political parties and leaders: Democratic Labor Party (DLP), Erskine SANDIFORD; Barbados Labor Party (BLP), Henry FORDE; National Democratic Party (NDP), Richie HAYNES
Other political or pressure groups: Barbados Workers Union, Leroy TROTMAN; People's Progressive Movement, Eric SEALY; Workers' Party of Barbados, Dr. George BELLE; Clement Payne Labor Union, David COMMISSIONG
Suffrage: 18 years of age; universal
Elections:
House of Assembly: last held 22 January 1991 (next to be held by January 1996);

results—DLP 49.8%; seats—(28 total) DLP 18, BLP 10
Executive branch: British monarch, governor general, prime minister, deputy prime minister, Cabinet
Legislative branch: bicameral Parliament consists of an upper house or Senate and a lower house or House of Assembly
Judicial branch: Supreme Court of Judicature
Leaders:
Chief of State: Queen ELIZABETH II (since 6 February 1952), represented by Governor General Dame Nita BARROW (since 6 June 1990)
Head of Government: Prime Minister Lloyd Erskine SANDIFORD (since 2 June 1987)
Member of: ACP, C, CARICOM, CDB, ECLAC, FAO, G-77, GATT, IADB, IBRD, ICAO, ICFTU, IFAD, IFC, ILO, IMF, IMO, INTELSAT, INTERPOL, IOC, ISO (correspondent), ITU, LAES, LORCS, NAM, OAS, OPANAL, UN, UNCTAD, UNESCO, UNIDO, UPU, WHO, WIPO, WMO
Diplomatic representation in US:
chief of mission: Ambassador Dr. Rudi WEBSTER
chancery: 2144 Wyoming Avenue NW, Washington, DC 20008
telephone: (202) 939-9200 through 9202
consulate general: New York
consulate: Los Angeles
US diplomatic representation:
chief of mission: Ambassador G. Philip HUGHES
embassy: Canadian Imperial Bank of Commerce Building, Broad Street, Bridgetown
mailing address: P. O. Box 302, Box B, FPO AA 34054
telephone: (809) 436-4950 through 4957
FAX: (809) 429-5246
Flag: three equal vertical bands of blue (hoist side), yellow, and blue with the head of a black trident centered on the gold band; the trident head represents independence and a break with the past (the colonial coat of arms contained a complete trident)

Economy

Overview: A per capita income of $7,000 gives Barbados one of the highest standards of living of all the small island states of the eastern Caribbean. Historically, the economy was based on the cultivation of sugarcane and related activities. In recent years, however, the economy has diversified into manufacturing and tourism. The tourist industry is now a major employer of the labor force and a primary source of foreign exchange. The economy slowed in 1990-91, however, and Bridgetown's declining hard currency reserves and inability to finance its

Barbados *(continued)*

deficits have caused it to adopt an austere economic reform program.
National product: GDP—exchange rate conversion—$1.8 billion (1991)
National product real growth rate: -4% (1991)
National product per capita: $7,000 (1991)
Inflation rate (consumer prices): 8.1% (1991)
Unemployment rate: 23% (1992)
Budget: revenues $547 million; expenditures $620 million (FY92-93), including capital expenditures of $60 million
Exports: $205.8 million (f.o.b., 1991)
commodities: sugar and molasses, chemicals, electrical components, clothing, rum, machinery and transport equipment
partners: CARICOM 31%, US 16%, UK 13%
Imports: $697 million (c.i.f., 1991)
commodities: foodstuffs, consumer durables, raw materials, machinery, crude oil, construction materials, chemicals
partners: US 34%, CARICOM 16%, UK 11%, Canada 6%
External debt: $750 million (1991 est.)
Industrial production: growth rate −1.3% (1991); accounts for 10% of GDP
Electricity: 152,100 kW capacity; 540 million kWh produced, 2,118 kWh per capita (1992)
Industries: tourism, sugar, light manufacturing, component assembly for export, petroleum
Agriculture: accounts for 8% of GDP; major cash crop is sugarcane; other crops—vegetables, cotton; not self-sufficient in food
Economic aid: US commitments, including Ex-Im (FY70-89), $15 million; Western (non-US) countries, ODA and OOF bilateral commitments (1970-89), $171 million
Currency: 1 Barbadian dollar (Bds$) = 100 cents
Exchange rates: Barbadian dollars (Bds$) per US$1—2.0113 (fixed rate)
Fiscal year: 1 April-31 March

Communications

Highways: 1,570 km total; 1,475 km paved, 95 km gravel and earth
Ports: Bridgetown
Merchant marine: 3 ships (1,000 GRT or over) totaling 48,710 GRT79,263 DWT; includes 1 cargo, 2 oil tanker
Airports:
total: 1
usable: 1
with permanent-surface runways: 1
with runways over 3,659 m: 0
with runways 2,440-3,659 m: 1
with runways 1,220-2,439 m: 0
Telecommunications: islandwide automatic telephone system with 89,000 telephones; tropospheric scatter link to Trinidad and Saint Lucia; broadcast stations—3 AM, 2 FM, 2 (1 is pay) TV; 1 Atlantic Ocean INTELSAT earth station

Defense Forces

Branches: Royal Barbados Defense Force, including the Ground Forces and Coast Guard, Royal Barbados Police Force
Manpower availability: males age 15-49 70,254; fit for military service 49,096 (1993 est.); no conscription
Defense expenditures: exchange rate conversion—$10 million, 0.7% of GDP (1989)

Bassas da India
(possession of France)

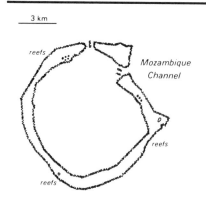

Geography

Location: Southern Africa, in the southern Mozambique Channel about halfway between Madagascar and Mozambique
Map references: Africa
Area:
total area: NA km²
land area: NA km²
comparative area: NA
Land boundaries: 0 km
Coastline: 35.2 km
Maritime claims:
contiguous zone: 24 nm
continental shelf: 200 m (depth) or to depth of exploitation
exclusive economic zone: 200 nm
territorial sea: 12 nm
International disputes: claimed by Madagascar
Climate: tropical
Terrain: a volcanic rock 2.4 meters high
Natural resources: none
Land use:
arable land: 0%
permanent crops: 0%
meadows and pastures: 0%
forest and woodland: 0%
other: 100% (all rock)
Irrigated land: 0 km²
Environment: surrounded by reefs; subject to periodic cyclones
Note: navigational hazard since it is usually under water during high tide

People

Population: uninhabited

Government

Names:
conventional long form: none
conventional short form: Bassas da India
Digraph: BS

Belarus

Type: French possession administered by Commissioner of the Republic Jacques DEWATRE (since July 1991), resident in Reunion
Capital: none; administered by France from Reunion
Independence: none (possession of France)

Economy

Overview: no economic activity

Communications

Ports: none; offshore anchorage only

Defense Forces

Note: defense is the responsibility of France

Geography

Location: Eastern Europe, between Poland and Russia
Map references: Asia, Commonwealth of Independent States—European States, Europe, Standard Time Zones of the World
Area:
total area: 207,600 km^2
land area: 207,600 km^2
comparative area: slightly smaller than Kansas
Land boundaries: total 3,098 km, Latvia 141 km, Lithuania 502 km, Poland 605 km, Russia 959 km, Ukraine 891 km
Coastline: 0 km (landlocked)
Maritime claims: none; landlocked
International disputes: none
Climate: mild and moist; transitional between continental and maritime
Terrain: generally flat and contains much marshland
Natural resources: forest land, peat deposits
Land use:
arable land: 29%
permanent crops: 0%
meadows and pastures: 15%
forest and woodland: 0%
other: 56%
Irrigated land: 1,490 km^2 (1990)
Environment: southern part of Belarus highly contaminated with fallout from 1986 nuclear reactor accident at Chornobyl'
Note: landlocked

People

Population: 10,370,269 (July 1993 est.)
Population growth rate: 0.34% (1993 est.)
Birth rate: 13.28 births/1,000 population (1993 est.)
Death rate: 11.1 deaths/1,000 population (1993 est.)
Net migration rate: 1.26 migrant(s)/1,000 population (1993 est.)
Infant mortality rate: 19.2 deaths/1,000 live births (1993 est.)

Life expectancy at birth:
total population: 70.73 years
male: 66.04 years
female: 75.66 years (1993 est.)
Total fertility rate: 1.89 children born/woman (1993 est.)
Nationality:
noun: Belarusian(s)
adjective: Belarusian
Ethnic divisions: Belarusian 77.9%, Russian 13.2%, Polish 4.1%, Ukrainian 2.9%, other 1.9%
Religions: Eastern Orthodox NA%, other NA%
Languages: Byelorussian, Russian, other
Literacy: age 9-49 can read and write (1970)
total population: 100%
male: 100%
female: 100%
Labor force: 5.418 million
by occupation: industry and construction 42%, agriculture and forestry 20%, other 38% (1990)

Government

Names:
conventional long form: Republic of Belarus
conventional short form: Belarus
local long form: Respublika Belarus
local short form: none
former: Belorussian (Byelorussian) Soviet Socialist Republic
Digraph: BO
Type: republic
Capital: Minsk
Administrative divisions: 6 oblasts (voblastsi, singular—voblasts') and one municipality* (harady, singular—horad); Brestskaya, Homyel'skaya, Minsk*, Hrodzyenskaya, Mahilyowskaya, Minskaya, Vitsyebskaya
note: each voblasts' has the same name as its administrative center
Independence: 25 August 1991 (from Soviet Union)
Constitution: adopted NA April 1978
Legal system: based on civil law system
National holiday: 24 August (1991)
Political parties and leaders: Belarusian Popular Front (BPF), Zenon PAZNYAK, chairman; United Democratic Party of Belarus (UDPB), Aleksandr DOBROVOLSKIY, chairman; Social Democratic Party of Belarus (SDBP), Mikhail TKACHEV, chairman; Belarus Workers Union, Mikhail SOBOL, Chairman; Belarus Peasants Party; Party of People's Unity, Gennadiy KARPENKO; Communist Party of Belarus
Suffrage: 18 years of age; universal
Elections:
Supreme Soviet: last held 4 April 1990 (next to be held NA); results—Communists 87%;

Belarus (continued)

seats—(360 total) number of seats by party NA; note—50 seats are for public bodies; the Communist Party obtained an overwhelming majority

Executive branch: chairman of the Supreme Soviet, chairman of the Council of Ministers; note—Belarus has approved a directly elected presidency but so far no elections have been scheduled

Legislative branch: unicameral Supreme Soviet

Judicial branch: Supreme Court

Leaders:

Chief of State: Chairman of the Supreme Soviet Stanislav S. SHUSHKEVICH (since 18 September 1991)

Head of Government: Prime Minister Vyacheslav F. KEBICH (since NA April 1990), First Deputy Prime Minister Mikhail MYASNIKOVICH (since NA 1991)

Member of: CBSS (observer), CIS, CSCE, ECE, IAEA, IBRD, ILO, IMF, INMARSAT, IOC, ITU, NACC, PCA, UN, UNCTAD, UNESCO, UNIDO, UPU, WHO, WIPO, WMO

Diplomatic representation in US:

chief of mission: Ambassador Designate Sergey Nikolayevich MARTYNOV

chancery: 1511 K Street NW, Suite 619, Washington, DC 20036

telephone: (202) 638-2954

US diplomatic representation:

chief of mission: Ambassador David H. SWARTZ

embassy: Starovilenskaya #46, Minsk

mailing address: APO AE 09862

telephone: 7-0172-34-65-37

Flag: three horizontal bands of white (top), red, and white

Economy

Overview: In many ways Belarus resembles the three Baltic states, for example, in its industrial competence, its higher-than-average standard of living, and its critical dependence on the other former Soviet states for fuels and raw materials. Belarus ranks fourth in gross output among the former Soviet republics, having produced 4% of the total GDP and employing 4% of the labor force in the old USSR. Once a mainly agricultural area, it now supplies important producer and consumer goods—sometimes as the sole producer—to the other states. Belarus had a significant share of the machine-building capacity of the former USSR. It is especially noted for production of tractors, large trucks, machine tools, and automation equipment. The soil in Belarus is not as fertile as the black earth of Ukraine, but by emphasizing favorable crops and livestock (especially pigs and chickens), Belarus has become a net exporter to the

other former republics of meat, milk, eggs, flour, and potatoes. Belarus produces only small amounts of oil and gas and receives most of its fuel from Russia through the Druzhba oil pipeline and the Northern Lights gas pipeline. These pipelines transit Belarus en route to Eastern Europe. Belarus produces petrochemicals, plastics, synthetic fibers (nearly 30% of former Soviet output), and fertilizer (20% of former Soviet output). Raw material resources are limited to potash and peat deposits. The peat (more than one-third of the total for the former Soviet Union) is used in domestic heating, as boiler fuel for electric power stations, and in the production of chemicals. The potash supports fertilizer production. In 1992 GDP fell an estimated 13%, largely because the country is highly dependent on the ailing Russian economy for raw materials and parts.

National product: GDP $NA

National product real growth rate: -13% (1992 est.)

National product per capita: $NA

Inflation rate (consumer prices): 30% per month (first quarter 1993)

Unemployment rate: 0.5% of officially registered unemployed; large numbers of underemployed workers

Budget: revenues $NA; expenditures $NA, including capital expenditures of $NA

Exports: $1.1 billion to outside of the successor states of the former USSR (f.o.b., 1992)

commodities: machinery and transport equipment, chemicals, foodstuffs

partners: NA

Imports: $751 million from outside the successor states of the former USSR (c.i.f., 1992)

commodities: machinery, chemicals, textiles

partners: NA

External debt: $2.6 billion (end of 1991)

Industrial production: growth rate −9.6%; accounts for about 50% of GDP (1992)

Electricity: 8,025,000 kW capacity; 37,600 million kWh produced, 3,626 kWh per capita (1992)

Industries: employ about 27% of labor force and produce a wide variety of products essential to the other states; products include (in percent share of total output of former Soviet Union): tractors (12%); metal-cutting machine tools (11%); off-highway dump trucks up to 110-metric-ton load capacity (100%); wheel-type earthmovers for construction and mining (100%); eight-wheel-drive, high-flotation trucks with cargo capacity of 25 metric tons for use in tundra and roadless areas (100%); equipment for animal husbandry and livestock feeding (25%); motorcycles (21.3%); television sets (11%); chemical fibers (28%); fertilizer

(18%); linen fabric (11%); wool fabric (7%); radios; refrigerators; and other consumer goods

Agriculture: accounts for almost 25% of GDP and 5.7% of total agricultural output of former Soviet Union; employs 20% of the labor force; in 1988 produced the following (in percent of total Soviet production): grain (3.6%), potatoes (12.2%), vegetables (3.0%), meat (6.0%), milk (7.0%); net exporter of meat, milk, eggs, flour, potatoes

Illicit drugs: illicit producer of opium and cannabis; mostly for the domestic market; transshipment point for illicit drugs to Western Europe

Economic aid: NA

Currency: 1 rubel (abbreviation NA) = 10 Russian rubles

note: the rubel circulates with the Russian ruble; certain purchase are made only with rubels; government has established a different, and varying, exchange rate for trade between Belarus and Russia

Exchange rates: NA

Fiscal year: calendar year

Communications

Railroads: 5,570 km; does not include industrial lines (1990)

Highways: 98,200 km total; 66,100 km hard surfaced, 32,100 km earth (1990)

Inland waterways: NA km

Pipelines: crude oil 1,470 km, refined products 1,100 km, natural gas 1,980 km (1992)

Ports: none; landlocked

Merchant marine: claims 5% of former Soviet fleet

Airports:

total: 124

useable: 55

with permanent-surface runways: 31

with runways over 3,659 m: 1

with runways 2,440-3,659 m: 28

with runways 1,220-2,439 m: 20

Telecommunications: construction of NMT-450 analog cellular network proceeding in Minsk, in addition to installation of some 300 km of fiber optic cable in the city network; telephone network has 1.7 million lines, 15% of which are switched automatically; Minsk has 450,000 lines; telephone density is approximately 17 per 100 persons; as of 1 December 1991, 721,000 applications from households for telephones were still unsatisfied; international connections to other former Soviet republics are by landline or microwave and to other countries by leased connection through the Moscow international gateway switch; Belarus has not constructed ground stations for international telecommunications via satellite to date

Belgium

Defense Forces

Branches: Army, Air Forces, Air Defense Forces, Security Forces (internal and border troops)
Manpower availability: males age 15-49 2,491,039; fit for military service 1,964,577; reach military age (18) annually 71,875 (1993 est.)
Defense expenditures: 56.5 billion rubles, NA% of GDP (1993 est.); note—conversion of the military budget into US dollars using the current exchange rate could produce misleading results

Geography

Location: Western Europe, bordering on the North Sea, between France and the Netherlands
Map references: Arctic Region, Europe, Standard Time Zones of the World
Area:
total area: 30,510 km²
land area: 30,230 km²
comparative area: slightly larger than Maryland
Land boundaries: total 1,385 km, France 620 km, Germany 167 km, Luxembourg 148 km, Netherlands 450 km
Coastline: 64 km
Maritime claims:
continental shelf: equidistant line with neighbors
exclusive fishing zone: equidistant line with neighbors (extends about 68 km from coast)
territorial sea: 12 nm
International disputes: none
Climate: temperate; mild winters, cool summers; rainy, humid, cloudy
Terrain: flat coastal plains in northwest, central rolling hills, rugged mountains of Ardennes Forest in southeast
Natural resources: coal, natural gas
Land use:
arable land: 24%
permanent crops: 1%
meadows and pastures: 20%
forest and woodland: 21%
other: 34%
Irrigated land: 10 km² (1989 est.)
Environment: air and water pollution
Note: crossroads of Western Europe; majority of West European capitals within 1,000 km of Brussels which is the seat of the EC

People

Population: 10,040,939 (July 1993 est.)
Population growth rate: 0.23% (1993 est.)
Birth rate: 11.94 births/1,000 population (1993 est.)

Death rate: 10.32 deaths/1,000 population (1993 est.)
Net migration rate: 0.7 migrant(s)/1,000 population (1993 est.)
Infant mortality rate: 7.4 deaths/1,000 live births (1993 est.)
Life expectancy at birth:
total population: 76.72 years
male: 73.41 years
female: 80.21 years (1993 est.)
Total fertility rate: 1.62 children born/woman (1993 est.)
Nationality:
noun: Belgian(s)
adjective: Belgian
Ethnic divisions: Fleming 55%, Walloon 33%, mixed or other 12%
Religions: Roman Catholic 75%, Protestant or other 25%
Languages: Flemish (Dutch) 56%, French 32%, German 1%, legally bilingual 11% divided along ethnic lines
Literacy: age 15 and over can read and write (1980)
total population: 99%
male: NA%
female: NA%
Labor force: 4.126 million
by occupation: services 63.6%, industry 28%, construction 6.1%, agriculture 2.3% (1988)

Government

Names:
conventional long form: Kingdom of Belgium
conventional short form: Belgium
local long form: Royaume de Belgique
local short form: Belgique
Digraph: BE
Type: constitutional monarchy
Capital: Brussels
Administrative divisions: 9 provinces (French: provinces, singular—province; Flemish: provincien, singular—provincie); Antwerpen, Brabant, Hainaut, Liege, Limburg, Luxembourg, Namur, Oost-Vlaanderen, West-Vlaanderen
Independence: 4 October 1830 (from the Netherlands)
Constitution: 7 February 1831, last revised 8-9 August 1980; the government is in the process of revising the Constitution with the aim of federalizing the Belgian state
Legal system: civil law system influenced by English constitutional theory; judicial review of legislative acts; accepts compulsory ICJ jurisdiction, with reservations
National holiday: National Day, 21 July (ascension of King Leopold to the throne in 1831)
Political parties and leaders: Flemish Social Christian (CVP), Herman VAN ROMPUY, president; Walloon Social

Belgium (continued)

Christian (PSC), Melchior WATHELET, president; Flemish Socialist (SP), Frank VANDENBROUCKE, president; Walloon Socialist (PS), Guy SPITAELS; Flemish Liberals and Democrats (VLD), Guy VERHOFSTADT, president; Walloon Liberal (PRL), Antoine DUQUESNE, president; Francophone Democratic Front (FDF), Georges CLERFAYT, president; Volksunie (VU), Jaak GABRIELS, president; Communist Party (PCB), Louis VAN GEYT, president; Vlaams Blok (VB), Karel VAN DILLEN, chairman; ROSSEM, Jean Pierre VAN ROSSEM; National Front (FN), Werner van STEEN; Live Differently (AGALEV; Flemish Green party), Leo COX; Ecologist (ECOLO; Francophone Green party), NA; other minor parties

Other political or pressure groups: Christian and Socialist Trade Unions; Federation of Belgian Industries; numerous other associations representing bankers, manufacturers, middle-class artisans, and the legal and medical professions; various organizations represent the cultural interests of Flanders and Wallonia; various peace groups such as the Flemish Action Committee Against Nuclear Weapons and Pax Christi

Suffrage: 18 years of age, universal and compulsory

Elections:
Senate: last held 24 November 1991 (next to be held by November 1996); results—percent of vote by party NA; seats—(184 total; of which 106 are directly elected) CVP 20, SP 14, PVV (now VLD) 13, VU 5, AGALEV 5, VB 5, ROSSEN 1, PS 18, PRL 9, PSC 9, ECOLO 6, FDF 1
Chamber of Representatives: last held 24 November 1991 (next to be held by November 1996); results—CVP 16.7%, PS 13.6%, SP 12.0%, PVV (now VLD) 11.9%, PRL 8.2%, PSC 7.8%, VB 6.6%, VU 5.9%, ECOLO 5.1%, AGALEV 4.9%, FDF 2.6%, ROSSEM 3.2%, FN 1.5%; seats—(212 total) CVP 39, PS 35, SP 28, PVV (now VLD) 26, PRL 20, PSC 18, FB 12, VU 10, ECOLO 10, AGALEV 7, FDF 3, ROSSEM 3, FN 1

Executive branch: monarch, prime minister, three deputy prime ministers, Cabinet

Legislative branch: bicameral Parliament consists of an upper chamber or Senate (Flemish—Senaat, French—Senat) and a lower chamber or Chamber of Representatives (Flemish—Kamer van Volksvertegenwoordigers, French—Chambre des Representants)

Judicial branch: Supreme Court of Justice (Flemish—Hof van Cassatie, French—Cour de Cassation)

Leaders:
Chief of State: King BAUDOUIN I (since 17 July 1951); Heir Apparent Prince

ALBERT of Liege (brother of the King; born 6 June 1934)
Head of Government: Prime Minister Jean-Luc DEHAENE (since 6 March 1992)
Member of: AG (observer), ACCT, AfDB, AsDB, Australian Group, Benelux, BIS, CCC, CE, CERN, COCOM, CSCE, EBRD, EC, ECE, EIB, ESA, FAO, G-9, G-10, GATT, IADB, IAEA, IBRD, ICAO, ICC, ICFTU, IDA, IEA, IFAD, IFC, ILO, IMF, IMO, INMARSAT, INTELSAT, INTERPOL, IOC, IOM, ISO, ITU, LORCS, MTCR, NACC, NATO, NEA, NSG, OAS (observer), OECD, PCA, UN, UNCTAD, UNESCO, UNHCR, UNIDO, UNMOGIP, UNOSOM, UNPROFOR, UNRWA, UNTAC, UNTSO, UPU, WCL, WEU, WHO, WIPO, WMO, WTO, ZC

Diplomatic representation in US:
chief of mission: Ambassador Juan CASSIERS
chancery: 3330 Garfield Street NW, Washington, DC 20008
telephone: (202) 333-6900
FAX: (202) 333-3079
consulates general: Atlanta, Chicago, Los Angeles, and New York

US diplomatic representation:
chief of mission: Ambassador Bruce S. GELB
embassy: 27 Boulevard du Regent, Brussels
mailing address: B-1000 Brussels, PSC 82, Box 002, APO AE 09724
telephone: [32] (2) 513-3830
FAX: [32] (2) 511-2725

Flag: three equal vertical bands of black (hoist side), yellow, and red; the design was based on the flag of France

---

Economy

Overview: This small private enterprise economy has capitalized on its central geographic location, highly developed transport network, and diversified industrial and commercial base. Industry is concentrated mainly in the populous Flemish area in the north, although the government is encouraging reinvestment in the southern region of Walloon. With few natural resources Belgium must import essential raw materials, making its economy closely dependent on the state of world markets. Over 70% of trade is with other EC countries. The economy grew at a strong 4% pace during the period 1988-90, but economic growth slowed to a 1% pace in 1991-92. The economy is expected to turn in another sluggish 1% performance in 1993. Belgium's public debt remains high at 120% of GDP and the government is trying to control its expenditures to bring the figure more into line with other industrialized countries.

National product: GDP—purchasing power

equivalent—$177.9 billion (1992)

National product real growth rate: 0.8% (1992)

National product per capita: $17,800 (1992)

Inflation rate (consumer prices): 2.6% (1992 est.)

Unemployment rate: 9.8% (end 1992)

Budget: revenues $97.8 billion; expenditures $109.3 billion, including capital expenditures of $NA (1989)

Exports: $118 billion (f.o.b., 1991) Belgium-Luxembourg Economic Union
commodities: iron and steel, transportation equipment, tractors, diamonds, petroleum products
partners: EC 75.5%, US 3.7%, former Communist countries 1.4% (1991)

Imports: $121 billion (c.i.f., 1991) Belgium-Luxembourg Economic Union
commodities: fuels, grains, chemicals, foodstuffs
partners: EC 73%, US 4.8%, oil-exporting less developed countries 4%, former Communist countries 1.8% (1991)

External debt: $31.3 billion (1992 est.)

Industrial production: growth rate 1.6% (1992 est.)

Electricity: 17,500,000 kW capacity; 68,000 million kWh produced, 6,790 kWh per capita (1992)

Industries: engineering and metal products, motor vehicle assembly, processed food and beverages, chemicals, basic metals, textiles, glass, petroleum, coal

Agriculture: accounts for 2.3% of GDP; emphasis on livestock production—beef, veal, pork, milk; major crops are sugar beets, fresh vegetables, fruits, grain, tobacco; net importer of farm products

Illicit drugs: source of precursor chemicals for South American cocaine processors; increasingly important gateway country for cocaine entering the European market

Economic aid: donor—ODA and OOF commitments (1970-89), $5.8 billion

Currency: 1 Belgian franc (BF) = 100 centimes

Exchange rates: Belgian francs (BF) per US$1—33.256 (January 1993), 32.150 (1992), 34.148 (1991), 33.418 (1990), 39.404 (1989), 36.768 (1988)

Fiscal year: calendar year

Communications

Railroads: Belgian National Railways (SNCB) operates 3,568 km 1.435-meter standard gauge, government owned; 2,563 km double track; 2,207 km electrified

Highways: 103,396 km total; 1,317 km limited access, divided autoroute; 11,717 km national highway; 1,362 km provincial road; about 38,000 km paved and 51,000 km unpaved rural roads

Belize

Inland waterways: 2,043 km (1,528 km in regular commercial use)
Pipelines: petroleum products 1,167 km; crude oil 161 km; natural gas 3,300 km
Ports: Antwerp, Brugge, Gent, Oostende, Zeebrugge
Merchant marine: 23 ships (1,000 GRT or over) totaling 96,949 GRT/133,658 DWT; includes 10 cargo, 5 oil tanker, 2 liquefied gas, 5 chemical tanker, 1 bulk
Airports:
total: 42
usable: 42
with permanent-surface runways: 24
with runways over 3,659 m: 0
with runways 2,440-3,659 m: 14
with runways 1,220-2,439 m: 3
Telecommunications: highly developed, technologically advanced, and completely automated domestic and international telephone and telegraph facilities; extensive cable network; limited microwave radio relay network; 4,720,000 telephones; broadcast stations—3 AM, 39 FM, 32 TV; 5 submarine cables; 2 satellite earth stations—Atlantic Ocean INTELSAT and EUTELSAT systems; nationwide mobile phone system

Defense Forces

Branches: Army, Navy, Air Force, National Gendarmerie
Manpower availability: males age 15-49 2,556,189; fit for military service 2,133,051; reach military age (19) annually 63,532 (1993 est.)
Defense expenditures: exchange rate conversion—$4 billion, 2% of GDP (1992)

Geography

Location: Central America, bordering the Caribbean Sea between Guatemala and Mexico
Map references: Central America and the Caribbean, North America, Standard Time Zones of the World
Area:
total area: 22,960 km²
land area: 22,800 km²
comparative area: slightly larger than Massachusetts
Land boundaries: total 516 km, Guatemala 266 km, Mexico 250 km
Coastline: 386 km
Maritime claims:
territorial sea: 12 nm in the north, 3 nm in the south
note: from the mouth of the Sarstoon River to Ranguana Caye, Belize's territorial sea is 3 miles; according to Belize's Maritime Areas Act, 1992, the purpose of this limitation is to provide a framework for the negotiation of a definitive agreement on territorial differences with Guatemala
International disputes: border with Guatemala in dispute; negotiations to resolve the dispute have begun
Climate: tropical; very hot and humid; rainy season (May to February)
Terrain: flat, swampy coastal plain; low mountains in south
Natural resources: arable land potential, timber, fish
Land use:
arable land: 2%
permanent crops: 0%
meadows and pastures: 2%
forest and woodland: 44%
other: 52%
Irrigated land: 20 km² (1989 est.)
Environment: frequent devastating hurricanes (September to December) and coastal flooding (especially in south); deforestation

Note: national capital moved 80 km inland from Belize City to Belmopan because of hurricanes; only country in Central America without a coastline on the North Pacific Ocean

People

Population: 203,957 (July 1993 est.)
Population growth rate: 2.42% (1993 est.)
Birth rate: 35.75 births/1,000 population (1993 est.)
Death rate: 6.15 deaths/1,000 population (1993 est.)
Net migration rate: -5.44 migrant(s)/1,000 population (1993 est.)
Infant mortality rate: 36.5 deaths/1,000 live births (1993 est.)
Life expectancy at birth:
total population: 67.85 years
male: 65.91 years
female: 69.88 years (1993 est.)
Total fertility rate: 4.53 children born/woman (1993 est.)
Nationality:
noun: Belizean(s)
adjective: Belizean
Ethnic divisions: Mestizo 44%, Creole 30%, Maya 11%, Garifuna 7%, other 8%
Religions: Roman Catholic 62%, Protestant 30% (Anglican 12%, Methodist 6%, Mennonite 4%, Seventh-Day Adventist 3%, Pentecostal 2%, Jehovah's Witnesses 1%, other 2%), none 2%, other 6% (1980)
Languages: English (official), Spanish, Maya, Garifuna (Carib)
Literacy: age 15 and over having ever attended school (1970)
total population: 91%
male: 91%
female: 91%
Labor force: 51,500
by occupation: agriculture 30%, services 16%, government 15.4%, commerce 11.2%, manufacturing 10.3%
note: shortage of skilled labor and all types of technical personnel (1985)

Government

Names:
conventional long form: none
conventional short form: Belize
former: British Honduras
Digraph: BH
Type: parliamentary democracy
Capital: Belmopan
Administrative divisions: 6 districts; Belize, Cayo, Corozal, Orange Walk, Stann Creek, Toledo
Independence: 21 September 1981 (from UK)
Constitution: 21 September 1981
Legal system: English law

Belize (continued)

National holiday: Independence Day, 21 September

Political parties and leaders: People's United Party (PUP), George PRICE, Florencio MARIN, Said MUSA; United Democratic Party (UDP), Manuel ESQUIVEL, Dean LINDO, Dean BARROW; National Alliance for Belizean Rights, leader NA

Other political or pressure groups: Society for the Promotion of Education and Research (SPEAR), Assad SHOMAN; United Workers Front, leader NA

Suffrage: 18 years of age; universal

Elections:
National Assembly: last held 4 September 1989 (next to be held September 1994); results—percent of vote by party NA; seats—(28 total) PUP 15, UDP 13; note—in January 1990 one member expelled from UDP joined PUP, making the seat count PUP 16, UDP 12

Executive branch: British monarch, governor general, prime minister, deputy prime minister, Cabinet

Legislative branch: bicameral National Assembly consists of an upper house or Senate and a lower house or House of Representatives

Judicial branch: Supreme Court

Leaders:
Chief of State: Queen ELIZABETH II (since 6 February 1952), represented by Governor General Dame Minita Elmira GORDON (since 21 September 1981)

Head of Government: Prime Minister George Cadle PRICE (since 4 September 1989)

Member of: ACP, C, CARICOM, CDB, ECLAC, FAO, G-77, GATT, IBRD, ICAO, IDA, IFAD, IFC, ILO, IMF, IMO, INTERPOL, IOC, IOM (observer), ITU, LORCS, NAM, OAS, UN, UNCTAD, UNESCO, UNIDO, UPU, WCL, WHO, WMO

Diplomatic representation in US:
chief of mission: Ambassador James V. HYDE
chancery: 2535 Massachusetts Avenue NW, Washington, DC 20008
telephone: (202) 332-9636

US diplomatic representation:
chief of mission: Ambassador Eugene L. SCASSA
embassy: Gabourel Lane and Hutson Street, Belize City
mailing address: P. O. Box 286, Belize City
telephone: [501] (2) 77161 through 77163
FAX: [501] (2) 30802

Flag: blue with a narrow red stripe along the top and the bottom edges; centered is a large white disk bearing the coat of arms; the coat of arms features a shield flanked by two workers in front of a mahogany tree with the related motto SUB UMBRA FLOREO (I Flourish in the Shade) on a scroll at the bottom, all encircled by a green garland

---

Economy

Overview: The economy is based primarily on agriculture, agro-based industry, and merchandising, with tourism and construction assuming increasing importance. Agriculture accounts for about 30% of GDP and provides 75% of export earnings, while sugar, the chief crop, accounts for almost 40% of hard currency earnings. The US, Belize's main trading partner, is assisting in efforts to reduce dependency on sugar with an agricultural diversification program.

National product: GDP—exchange rate conversion—$373 million (1990 est.)

National product real growth rate: 10% (1990)

National product per capita: $1,635 (1990 est.)

Inflation rate (consumer prices): 5.5% (1991)

Unemployment rate: 12% (1991 est.)

Budget: revenues $126.8 million; expenditures $123.1 million, including capital expenditures of $44.8 million (FY91 est.)

Exports: $95.6 million (f.o.b., 1991)
commodities: sugar, citrus, clothing, bananas, fish products, molasses
partners: US 49%, UK, EC, Mexico (1991)

Imports: $194 million (c.i.f., 1991 est.)
commodities: machinery and transportation equipment, food, manufactured goods, fuels, chemicals, pharmaceuticals
partners: US 60%, UK, EC, Mexico (1991)

External debt: $143.7 million (1991)

Industrial production: growth rate 3.7% (1990); accounts for 12% of GDP

Electricity: 34,532 kW capacity; 90 million kWh produced, 393 kWh per capita (1992)

Industries: garment production, citrus concentrates, sugar refining, rum, beverages, tourism

Agriculture: accounts for 22% of GDP (including fish and forestry); commercial crops include sugarcane, bananas, coca, citrus fruits; expanding output of lumber and cultured shrimp; net importer of basic foods

Illicit drugs: an illicit producer of cannabis for the international drug trade; eradication program cut marijuana production from 200 metric tons in 1987 to about 50 metric tons in 1991; transshipment point for cocaine

Economic aid: US commitments, including Ex-Im (FY70-89), $104 million; Western (non-US) countries, ODA and OOF bilateral commitments (1970-89), $215 million

Currency: 1 Belizean dollar (Bz$) = 100 cents

Exchange rates: Belizean dollars (Bz$) per US$1—2.00 (fixed rate)

Fiscal year: 1 April-31 March

Communications

Highways: 2,710 km total; 500 km paved, 1,600 km gravel, 300 km improved earth, and 310 km unimproved earth

Inland waterways: 825 km river network used by shallow-draft craft; seasonally navigable

Ports: Belize City; additional ports for shallow draught craft include Corozol, Punta Gorda, Big Creek

Merchant marine: 4 ships (1,000 GRT or over) totaling 9,768 GRT/12,721 DWT; includes 3 cargo, 1 roll-on/roll-off

Airports:
total: 42
usable: 32
with permanent-surface runways: 3
with runways over 3,659 m: 0
with runways 2,440-3,659 m: 1
with runways 1,229-2,439 mr: 2

Telecommunications: 8,650 telephones; above-average system based on microwave radio relay; broadcast stations—6 AM, 5 FM, 1 TV, 1 shortwave; 1 Atlantic Ocean INTELSAT earth station

---

Defense Forces

Branches: British Forces Belize, Belize Defense Force (including Army, Navy, Air Force, and Volunteer Guard), Belize National Police

Manpower availability: males age 15-49 47,135; fit for military service 28,070; reach military age (18) annually 2,066 (1993 est.)

Defense expenditures: exchange rate conversion—$5.4 million, 2% of GDP (1992)

Benin

150 km

Bight of Benin

Geography

Location: Western Africa, bordering the North Atlantic Ocean between Nigeria and Togo
Map references: Africa, Standard Time Zones of the World
Area:
total area: 112,620 km²
land area: 110,620 km²
comparative area: slightly smaller than Pennsylvania
Land boundaries: total 1,989 km, Burkina 306 km, Niger 266 km, Nigeria 773 km, Togo 644 km
Coastline: 121 km
Maritime claims:
territorial sea: 200 nm
International disputes: none
Climate: tropical; hot, humid in south; semiarid in north
Terrain: mostly flat to undulating plain; some hills and low mountains
Natural resources: small offshore oil deposits, limestone, marble, timber
Land use:
arable land: 12%
permanent crops: 4%
meadows and pastures: 4%
forest and woodland: 35%
other: 45%
Irrigated land: 60 km² (1989 est.)
Environment: hot, dry, dusty harmattan wind may affect north in winter; deforestation; desertification
Note: recent droughts have severely affected marginal agriculture in north; no natural harbors

People

Population: 5,166,735 (July 1993 est.)
Population growth rate: 3.33% (1993 est.)
Birth rate: 48.09 births/1,000 population (1993 est.)
Death rate: 14.8 deaths/1,000 population (1993 est.)

Net migration rate: 0 migrant(s)/1,000 population (1993 est.)
Infant mortality rate: 112.7 deaths/1,000 live births (1993 est.)
Life expectancy at birth:
total population: 51.31 years
male: 49.51 years
female: 53.16 years (1993 est.)
Total fertility rate: 6.86 children born/woman (1993 est.)
Nationality:
noun: Beninese (singular and plural)
adjective: Beninese
Ethnic divisions: African 99% (42 ethnic groups, most important being Fon, Adja, Yoruba, Bariba), Europeans 5,500
Religions: indigenous beliefs 70%, Muslim 15%, Christian 15%
Languages: French (official), Fon and Yoruba (most common vernaculars in south), tribal languages (at least six major ones in north)
Literacy: age 15 and over can read and write (1990)
total population: 23%
male: 32%
female: 16%
Labor force: 1.9 million (1987)
by occupation: agriculture 60%, transport, commerce, and public services 38%, industry less than 2%
note: 49% of population of working age (1985)

Government

Names:
conventional long form: Republic of Benin
conventional short form: Benin
local long form: Republique Populaire du Benin
local short form: Benin
former: Dahomey
Digraph: BN
Type: republic under multiparty democratic rule dropped Marxism-Leninism December 1989; democratic reforms adopted February 1990; transition to multiparty system completed 4 April 1991
Capital: Porto-Novo
Administrative divisions: 6 provinces; Atakora, Atlantique, Borgou, Mono, Oueme, Zou
Independence: 1 August 1960 (from France)
Constitution: 2 December 1990
Legal system: based on French civil law and customary law; has not accepted compulsory ICJ jurisdiction
National holiday: National Day, 1 August (1990)
Political parties and leaders: Alliance of the Democratic Union for the Forces of Progress (UDFP), Timothee ADANLIN; Movement for Democracy and Social

Progress (MDPS), Jean-Roger AHOYO; Union for Liberty and Development (ULD), Marcellin DEGBE; Alliance of the National Party for Democracy and Development (PNDD) and the Democratic Renewal Party (PRD), Pascal Chabi KAO; Alliance of the Social Democratic Party (PSD) and the National Union for Solidarity and Progress (UNSP), Bruno AMOUSSOU; Our Common Cause (NCC), Albert TEVOEDJRE; National Rally for Democracy (RND), Joseph KEKE; Alliance of the National Movement for Democracy and Development (MNDD), leader NA; Movement for Solidarity, Union, and Progress (MSUP), Adebo ADENIYI; Union for Democracy and National Reconstruction (UDRN), Azaria FAKOREDE; Union for Democracy and National Solidarity (UDS), Mama Amadou N'DIAYE; Assembly of Liberal Democrats for National Reconstruction (RDL), Severin ADJOVI; Alliance of the Alliance for Social Democracy (ASD), Robert DOSSOU; Bloc for Social Democracy (BSD), Michel MAGNIDE; Alliance of the Alliance for Democracy and Progress (ADP), Akindes ADEKPEDJOU; Democratic Union for Social Renewal (UDRS), Bio Gado Seko N'GOYE; National Union for Democracy and Progress (UNDP), Robert TAGNON; Party for Progress and Democracy, Theophile NATA; numerous other small parties
Suffrage: 18 years of age; universal
Elections:
National Assembly: last held 10 and 24 March 1991; results—percent of vote by party NA; seats—(64 total) UDFP-MDPS-ULD 12, PNDD/PRD 9, PSD/UNSP 8, NCC 7, RND 7, MNDD/MSUP/UDRN 6, UDS 5, RDL 4, ASD/BSD 3, ADP/UDRS 2, UNDP 1
President: last held 10 and 24 March 1991; results—Nicephore SOGLO 68%, Mathieu KEREKOU 32%
Executive branch: president, cabinet
Legislative branch: unicameral National Assembly (Assemblee Nationale)
Judicial branch: Supreme Court (Cour Supreme)
Leaders:
Chief of State and Head of Government: President Nicephore SOGLO (since 4 April 1991)
Member of: ACCT, ACP, AfDB, CEAO, ECA, ECOWAS, Entente, FAO, FZ, G-77, GATT, IBRD, ICAO, IDA, IDB, IFAD, IFC, ILO, IMF, IMO, INTELSAT, INTERPOL, IOC, ITU, LORCS, NAM, OAU, OIC, UN, UNCTAD, UNESCO, UNIDO, UPU, WADB, WCL, WHO, WIPO, WMO, WTO
Diplomatic representation in US:
chief of mission: Ambassador Candide AHOUANSOU
chancery: 2737 Cathedral Avenue NW, Washington, DC 20008
telephone: (202) 232-6656

Benin (continued)

US diplomatic representation:
chief of mission: Ambassador Ruth A. DAVIS
embassy: Rue Caporal Anani Bernard, Cotonou
mailing address: B. P. 2012, Cotonou
telephone: [229] 30-06-50, 30-05-13, 30-17-92
FAX: [229] 30-14-39 and 30-19-74
Flag: two equal horizontal bands of yellow (top) and red with a vertical green band on the hoist side

---

Economy

Overview: Benin is one of the least developed countries in the world because of limited natural resources and a poorly developed infrastructure. Agriculture accounts for about 35% of GDP, employs about 60% of the labor force, and generates a major share of foreign exchange earnings. The industrial sector contributes only about 15% to GDP and employs 2% of the work force. Low prices in recent years have kept down hard currency earnings from Benin's major exports of agricultural products and crude oil.
National product: GDP—exchange rate conversion—$2 billion (1991)
National product real growth rate: 3% (1991)
National product per capita: $410 (1991)
Inflation rate (consumer prices): 3.4% (1990)
Unemployment rate: NA%
Budget: revenues $194 million; expenditures $390 million, including capital expenditures of $104 million (1990 est.)
Exports: $263.3 million (f.o.b., 1990 est.)
commodities: crude oil, cotton, palm products, cocoa
partners: FRG 36%, France 16%, Spain 14%, Italy 8%, UK 4%
Imports: $428 million (f.o.b., 1990 est.)
commodities: foodstuffs, beverages, tobacco, petroleum products, intermediate goods, capital goods, light consumer goods
partners: France 34%, Netherlands 10%, Japan 7%, Italy 6%, US 4%
External debt: $1.0 billion (December 1990 est.)
Industrial production: growth rate −0.7% (1988); accounts for 15% of GDP
Electricity: 30,000 kW capacity; 25 million kWh produced, 5 kWh per capita (1991)
Industries: textiles, cigarettes, construction materials, beverages, food production, petroleum
Agriculture: accounts for 35% of GDP; small farms produce 90% of agricultural output; production is dominated by food crops—corn, sorghum, cassava, beans, rice; cash crops include cotton, palm oil, peanuts; poultry and livestock output has not kept up with consumption
Economic aid: US commitments, including Ex-Im (FY70-89), $46 million; Western (non-US) countries, ODA and OOF bilateral commitments (1970-89), $1,300 million; OPEC bilateral aid (1979-89), $19 million; Communist countries (1970-89), $101 million
Currency: 1 CFA franc (CFAF) = 100 centimes
Exchange rates: Communaute Financiere Africaine francs (CFAF) per US$1—274.06 (January 1993), 264.69 (1992), 282.11 (1991), 272.26 (1990), 319.01 (1989), 297.85 (1988)
Fiscal year: calendar year

Communications

Railroads: 578 km, all 1.000-meter gauge, single track
Highways: 5,050 km total; 920 km paved, 2,600 laterite, 1,530 km improved earth
Inland waterways: navigable along small sections, important only locally
Ports: Cotonou
Airports:
total: 7
usable: 5
with permanent-surface runways: 1
with runways over 3,659 m: 0
with runways 2,439-3,659 m: 1
with runways 1,220-2,439 m: 2
Telecommunications: fair system of open wire, submarine cable, and radio relay microwave; broadcast stations—2 AM, 2 FM, 2 TV; 1 Atlantic Ocean INTELSAT earth station

---

Defense Forces

Branches: Armed Forces (including Army, Navy, Air Force), National Gendarmerie
Manpower availability: males age 15-49 1,075,053; females age 15-49 1,170,693; males fit for military service 550,645; females fit for military service 591,506; males reach military age (18) annually 56,872; females reach military age (18) annually 55,141 (1993 est.); both sexes are liable for military service
Defense expenditures: exchange rate conversion—$29 million, 1.7% of GDP (1988 est.)

Bermuda
(dependent territory of the UK)

Geography

Location: in the western North Atlantic Ocean, 1,050 km east of North Carolina
Map references: North America
Area:
total area: 50 km²
land area: 50 km²
comparative area: about 0.3 times the size of Washington, DC
Land boundaries: 0 km
Coastline: 103 km
Maritime claims:
exclusive fishing zone: 200 nm
territorial sea: 12 nm
International disputes: none
Climate: subtropical; mild, humid; gales, strong winds common in winter
Terrain: low hills separated by fertile depressions
Natural resources: limestone, pleasant climate fostering tourism
Land use:
arable land: 0%
permanent crops: 0%
meadows and pastures: 0%
forest and woodland: 20%
other: 80%
Irrigated land: NA km²
Environment: ample rainfall, but no rivers or freshwater lakes; consists of about 360 small coral islands
Note: some reclaimed land leased by US Government

People

Population: 60,686 (July 1993 est.)
Population growth rate: 0.78% (1993 est.)
Birth rate: 15.21 births/1,000 population (1993 est.)
Death rate: 7.3 deaths/1,000 population (1993 est.)
Net migration rate: -0.13 migrant(s)/1,000 population (1993 est.)
Infant mortality rate: 13.16 deaths/1,000 live births (1993 est.)

Life expectancy at birth:
total population: 75.03 years
male: 73.36 years
female: 76.97 years (1993 est.)
Total fertility rate: 1.82 children
born/woman (1993 est.)
Nationality:
noun: Bermudian(s)
adjective: Bermudian
Ethnic divisions: black 61%, white and
other 39%
Religions: Anglican 37%, Roman Catholic
14%, African Methodist Episcopal (Zion)
10%, Methodist 6%, Seventh-Day Adventist
5%, other 28%
Languages: English
Literacy: age 15 and over can read and
write (1970)
total population: 98%
male: 98%
female: 99%
Labor force: 32,000
by occupation: clerical 25%, services 22%,
laborers 21%, professional and technical
13%, administrative and managerial 10%,
sales 7%, agriculture and fishing 2% (1984)

Government

Names:
conventional long form: none
conventional short form: Bermuda
Digraph: BD
Type: dependent territory of the UK
Capital: Hamilton
Administrative divisions: 9 parishes and 2
municipalities*; Devonshire, Hamilton,
Hamilton*, Paget, Pembroke, Saint George*,
Saint Georges, Sandys, Smiths,
Southampton, Warwick
Independence: none (dependent territory of
the UK)
Constitution: 8 June 1968
Legal system: English law
National holiday: Bermuda Day, 22 May
Political parties and leaders: United
Bermuda Party (UBP), John W. D. SWAN;
Progressive Labor Party (PLP), Frederick
WADE; National Liberal Party (NLP),
Gilbert DARRELL
Other political or pressure groups:
Bermuda Industrial Union (BIU), Ottiwell
SIMMONS
Suffrage: 21 years of age; universal
Elections:
House of Assembly: last held 9 February
1989 (next to be held by February 1994);
results—percent of vote by party NA;
seats—(40 total) UBP 23, PLP 15, NLP 1,
other 1
Executive branch: British monarch,
governor, deputy governor, premier, deputy
premier, Executive Council (cabinet)
Legislative branch: bicameral Parliament
consists of an upper house or Senate and a

lower house or House of Assembly
Judicial branch: Supreme Court
Leaders:
Chief of State: Queen ELIZABETH II (since
6 February 1952), represented by Governor
Lord David WADDINGTON (since NA)
Head of Government: Premier John William
David SWAN (since NA January 1982)
Member of: CARICOM (observer), CCC,
ICFTU, INTERPOL (subbureau), IOC
Diplomatic representation in US: as a
dependent territory of the UK, Bermuda's
interests in the US are represented by the
UK
US diplomatic representation:
chief of mission: Consul General L. Ebersole
GAINES
consulate general: Crown Hill, 16 Middle
Road, Devonshire, Hamilton
mailing address: P. O. Box HM325,
Hamilton HMBX; PSC 1002, FPO AE
09727-1002
telephone: (809) 295-1342
FAX: (809) 295-1592
Flag: red with the flag of the UK in the
upper hoist-side quadrant and the Bermudian
coat of arms (white and blue shield with a
red lion holding a scrolled shield showing
the sinking of the ship Sea Venture off
Bermuda in 1609) centered on the outer half
of the flag

Economy

Overview: Bermuda enjoys one of the
highest per capita incomes in the world,
having successfully exploited its location by
providing luxury tourist facilities and
financial services. The tourist industry
attracts more than 90% of its business from
North America. The industrial sector is
small, and agriculture is severely limited by
a lack of suitable land. About 80% of food
needs are imported.
National product: GDP—purchasing power
equivalent—$1.3 billion (1991)
National product real growth rate: -1.5%
(1991)
National product per capita: $22,000
(1991)
Inflation rate (consumer prices): 4.4%
(1991)
Unemployment rate: 6% (1991)
Budget: revenues $327.5 million;
expenditures $308.9 million, including
capital expenditures of $35.4 million (FY91
est.)
Exports: $50 million (f.o.b., FY89)
commodities: semitropical produce, light
manufactures, re-exports of pharmaceuticals
partners: US 55%, UK 32%, Canada 11%,
other 2%
Imports: 527.2 million (f.o.b., FY89)
commodities: fuel, foodstuffs, machinery
partners: US 60%, UK 8%, Venezuela 7%,

Canada 5%, Japan 5%, other 15%
External debt: $NA
Industrial production: growth rate NA%
Electricity: 154,000 kW capacity; 504
million kWh produced, 8,370 kWh per
capita (1992)
Industries: tourism, finance, structural
concrete products, paints, pharmaceuticals,
ship repairing
Agriculture: accounts for less than 1% of
GDP; most basic foods must be imported;
produces bananas, vegetables, citrus fruits,
flowers, dairy products
Economic aid: US commitments, including
Ex-Im (FY70-81), $34 million; Western
(non-US) countries, ODA and OOF bilateral
commitments (1970-89), $277 million
Currency: 1 Bermudian dollar (Bd$) = 100
cents
Exchange rates: Bermudian dollar (Bd$)
per US$1—1.0000 (fixed rate)
Fiscal year: 1 April-31 March

Communications

Highways: 210 km public roads, all paved
(about 400 km of private roads)
Ports: Freeport, Hamilton, Saint George
Merchant marine: 72 ships (1,000 GRT or
over) totaling 3,451.099 GRT/5,937,636
DWT; includes 5 cargo, 5 refrigerated cargo,
5 container, 7 roll-on/roll-off, 21 oil tanker,
13 liquefied gas, 16 bulk; note—a flag of
convenience registry
Airports:
total: 1
usable: 1
with permanent-surface runways: 1
with runways over 3,659 m: 0
with runways 2,440-3,659 m: 1
with runways 1,220-2,439 m: 0
Telecommunications: modern with fully
automatic telephone system; 52,670
telephones; broadcast stations—5 AM, 3
FM, 2 TV; 3 submarine cables; 2 Atlantic
Ocean INTELSAT earth stations

Defense Forces

Branches: Bermuda Regiment, Bermuda
Police Force, Bermuda Reserve Constabulary
Note: defense is the responsibility of the UK

Bhutan

75 km

Geography

Location: South Asia, in the Himalayas, between China and India
Map references: Asia, Standard Time Zones of the World
Area:
total area: 47,000 km²
land area: 47,000 km²
comparative area: slightly more than half the size of Indiana
Land boundaries: total 1,075 km, China 470 km, India 605 km
Coastline: 0 km (landlocked)
Maritime claims: none; landlocked
International disputes: none
Climate: varies; tropical in southern plains; cool winters and hot summers in central valleys; severe winters and cool summers in Himalayas
Terrain: mostly mountainous with some fertile valleys and savanna
Natural resources: timber, hydropower, gypsum, calcium carbide, tourism potential
Land use:
arable land: 2%
permanent crops: 0%
meadows and pastures: 5%
forest and woodland: 70%
other: 23%
Irrigated land: 340 km² (1989 est.)
Environment: violent storms coming down from the Himalayas were the source of the country name which translates as Land of the Thunder Dragon
Note: landlocked; strategic location between China and India; controls several key Himalayan mountain passes

People

Population: 700,000 (July 1993 est.)
Population growth rate: 2.33% (1993 est.)
Birth rate: 39.59 births/1,000 population (1993 est.)
Death rate: 16.26 deaths/1,000 population (1993 est.)

Net migration rate: 0 migrant(s)/1,000 population (1993 est.)
Infant mortality rate: 123.3 deaths/1,000 live births (1993 est.)
Life expectancy at birth:
total population: 50.17 years
male: 50.74 years
female: 49.58 years (1993 est.)
Total fertility rate: 5.45 children born/woman (1993 est.)
Nationality:
noun: Bhutanese (singular and plural)
adjective: Bhutanese
Ethnic divisions: Bhote 50%, ethnic Nepalese 35%, indigenous or migrant tribes 15%
Religions: Lamaistic Buddhism 75%, Indian- and Nepalese-influenced Hinduism 25%
Languages: Dzongkha (official), Bhotes speak various Tibetan dialects; Nepalese speak various Nepalese dialects
Literacy:
total population: NA%
male: NA%
female: NA%
Labor force: NA
by occupation: agriculture 93%, services 5%, industry and commerce 2%
note: massive lack of skilled labor

Government

Names:
conventional long form: Kingdom of Bhutan
conventional short form: Bhutan
Digraph: BT
Type: monarchy; special treaty relationship with India
Capital: Thimphu
Administrative divisions: 18 districts (dzongkhag, singular and plural); Bumthang, Chhukha, Chirang, Daga, Geylegphug, Ha, Lhuntshi, Mongar, Paro, Pemagatsel, Punakha, Samchi, Samdrup Jongkhar, Shemgang, Tashigang, Thimphu, Tongsa, Wangdi Phodrang
Independence: 8 August 1949 (from India)
Constitution: no written constitution or bill of rights
Legal system: based on Indian law and English common law; has not accepted compulsory ICJ jurisdiction
National holiday: National Day, 17 December (1907) (Ugyen Wangchuck became first hereditary king)
Political parties and leaders: no legal parties
Other political or pressure groups: Buddhist clergy; Indian merchant community; ethnic Nepalese organizations leading militant antigovernment campaign
Suffrage: each family has one vote in village-level elections
Elections: no national elections

Executive branch: monarch, chairman of the Royal Advisory Council, Royal Advisory Council (Lodoi Tsokde), chairman of the Council of Ministers, Council of Ministers (Lhengye Shungtsog)
Legislative branch: unicameral National Assembly (Tshogdu)
Judicial branch: High Court
Leaders:
Chief of State and Head of Government: King Jigme Singye WANGCHUCK (since 24 July 1972)
Member of: AsDB, CP, ESCAP, FAO, G-77, IBRD, ICAO, IDA, IFAD, IMF, INTELSAT, IOC, ITU, NAM, SAARC, UN, UNCTAD, UNESCO, UNIDO, UPU, WHO
Diplomatic representation in US: no formal diplomatic relations; the Bhutanese mission to the UN in New York has consular jurisdiction in the US
US diplomatic representation: no formal diplomatic relations, although informal contact is maintained between the Bhutanese and US Embassies in New Delhi (India)
Flag: divided diagonally from the lower hoist side corner; the upper triangle is orange and the lower triangle is red; centered along the dividing line is a large black and white dragon facing away from the hoist side

Economy

Overview: The economy, one of the world's least developed, is based on agriculture and forestry, which provide the main livelihood for 90% of the population and account for about 50% of GDP. Rugged mountains dominate the terrain and make the building of roads and other infrastructure difficult and expensive. The economy is closely aligned with that of India through strong trade and monetary links. The industrial sector is small and technologically backward, with most production of the cottage industry type. Most development projects, such as road construction, rely on Indian migrant labor. Bhutan's hydropower potential and its attraction for tourists are its most important natural resources; however, the government limits the number of tourists to 3,000/year to minimize foreign influence.
National product: GDP—exchange rate conversion—$320 million (1991 est.)
National product real growth rate: 3.1% (1991 est.)
National product per capita: $200 (1991 est.)
Inflation rate (consumer prices): 10% (FY91 est.)
Unemployment rate: NA%
Budget: revenues $112 million; expenditures $121 million, including capital expenditures of $58 million (FY91 est.)
Exports: $74 million (f.o.b., FY91 est.)
commodities: cardamon, gypsum, timber,

Bolivia

handicrafts, cement, fruit, electricity (to India)
partners: India 90%
Imports: $106.4 million (c.i.f., FY91 est.)
commodities: fuel and lubricants, grain, machinery and parts, vehicles, fabrics
partners: India 83%
External debt: $120 million (June 91)
Industrial production: growth rate NA%; accounts for 18% of GDP; primarily cottage industry and home based handicrafts
Electricity: 336,000 kW capacity; 1,542.2 million kWh produced, 2,203 kWh per capita (25.8% is exported to India, leaving only 1,633 kWh per capita) (1990-91)
Industries: cement, wood products, processed fruits, alcoholic beverages, calcium carbide
Agriculture: accounts for 45% of GDP; based on subsistence farming and animal husbandry; self-sufficient in food except for foodgrains; other production—rice, corn, root crops, citrus fruit, dairy products, eggs
Economic aid: Western (non-US) countries, ODA and OOF bilateral commitments (1970-89), $115 million; OPEC bilateral aid (1979-89), $11 million
Currency: 1 ngultrum (Nu) = 100 chetrum; note—Indian currency is also legal tender
Exchange rates: ngultrum (Nu) per US$1—26.156 (January 1993), 25.918 (1992), 22.742 (1991), 17.504 (1990), 16.226 (1989), 13.917 (1988); note—the Bhutanese ngultrum is at par with the Indian rupee
Fiscal year: 1 July-30 June

Communications

Highways: 2,165 km total; 1,703 km surfaced
Airports:
total: 2
usable: 2
with permanent-surface runways: 1
with runways over 3,659 m: 0
with runways 2,440-3,659 m: 0
with runways 1,220-2,439 m: 2
Telecommunications: domestic telephone service is very poor with very few telephones in use; international telephone and telegraph service is by land line through India; a satellite earth station was planned (1990); broadcast stations—1 AM, 1 FM, no TV (1990)

Defense Forces

Branches: Royal Bhutan Army, Palace Guard, Militia
Manpower availability: males age 15-49 415,315; fit for military service 222,027; reach military age (18) annually 17,344 (1993 est.)
Defense expenditures: exchange rate conversion—$NA, NA% of GDP

Geography

Location: Central South America, between Brazil and Chile
Map references: South America, Standard Time Zones of the World
Area:
total area: 1,098,580 km²
land area: 1,084,390 km²
comparative area: slightly less than three times the size of Montana
Land boundaries: total 6,743 km, Argentina 832 km, Brazil 3,400 km, Chile 861 km, Paraguay 750 km, Peru 900 km
Coastline: 0 km (landlocked)
Maritime claims: none; landlocked
International disputes: has wanted a sovereign corridor to the South Pacific Ocean since the Atacama area was lost to Chile in 1884; dispute with Chile over Rio Lauca water rights
Climate: varies with altitude; humid and tropical to cold and semiarid
Terrain: rugged Andes Mountains with a highland plateau (Altiplano), hills, lowland plains of the Amazon basin
Natural resources: tin, natural gas, petroleum, zinc, tungsten, antimony, silver, iron ore, lead, gold, timber
Land use:
arable land: 3%
permanent crops: 0%
meadows and pastures: 25%
forest and woodland: 52%
other: 20%
Irrigated land: 1,650 km² (1989 est.)
Environment: cold, thin air of high plateau is obstacle to efficient fuel combustion; overgrazing; soil erosion; desertification
Note: landlocked; shares control of Lago Titicaca, world's highest navigable lake, with Peru

People

Population: 7,544,099 (July 1993 est.)
Population growth rate: 2.31% (1993 est.)

Birth rate: 32.83 births/1,000 population (1993 est.)
Death rate: 8.63 deaths/1,000 population (1993 est.)
Net migration rate: -1.06 migrant(s)/1,000 population (1993 est.)
Infant mortality rate: 76.7 deaths/1,000 live births (1993 est.)
Life expectancy at birth:
total population: 62.77 years
male: 60.34 years
female: 65.33 years (1993 est.)
Total fertility rate: 4.31 children born/woman (1993 est.)
Nationality:
noun: Bolivian(s)
adjective: Bolivian
Ethnic divisions: Quechua 30%, Aymara 25%, mixed 25-30%, European 5-15%
Religions: Roman Catholic 95%, Protestant (Evangelical Methodist)
Languages: Spanish (official), Quechua (official), Aymara (official)
Literacy: age 15 and over can read and write (1990)
total population: 78%
male: 85%
female: 71%
Labor force: 1.7 million
by occupation: agriculture 50%, services and utilities 26%, manufacturing 10%, mining 4%, other 10% (1983)

Government

Names:
conventional long form: Republic of Bolivia
conventional short form: Bolivia
local long form: Republica de Bolivia
local short form: Bolivia
Digraph: BL
Type: republic
Capital: La Paz (seat of government); Sucre (legal capital and seat of judiciary)
Administrative divisions: 9 departments (departamentos, singular—departamento); Chuquisaca, Cochabamba, Beni, La Paz, Oruro, Pando, Potosi, Santa Cruz, Tarija
Independence: 6 August 1825 (from Spain)
Constitution: 2 February 1967
Legal system: based on Spanish law and Code Napoleon; has not accepted compulsory ICJ jurisdiction
National holiday: Independence Day, 6 August (1825)
Political parties and leaders: Movement of the Revolutionary Left (MIR), Jaime PAZ Zamora; Nationalist Democratic Action (ADN), Hugo BANZER Suarez; Nationalist Revolutionary Movement (MNR), Gonzalo SANCHEZ de Lozada; Civic Solidarity Union (UCS), Max FERNANDEZ Rojas; Conscience of the Fatherland (CONDEPA), Carlos PALENQUE Aviles; Christian Democratic Party (PDC), Jorge AGREDO;

Bolivia (continued)

Free Bolivia Movement (MBL), Antonio ARANIBAR; United Left (IU), a coalition of leftist parties that includes Patriotic National Convergency Axis (EJE-P), Walter DELGADILLO and Bolivian Communist Party (PCB), Humberto RAMIREZ; Revolutionary Vanguard—9th of April (VR-9), Carlos SERRATE Reich
Suffrage: 18 years of age; universal and compulsory (married) 21 years of age; universal and compulsory (single)
Elections:
Chamber of Deputies: last held 7 May 1989 (next to be held 6 June 1993); results—percent of vote by party NA; note—legislative and presidential candidates run on a unified slate, so vote percentages are the same as in section on presidential election results; seats—(130 total) MNR 40, ADN 35, MIR 33, IU 10, CONDEPA 9, PDC 3
Chamber of Senators: last held 7 May 1989 (next to be held 6 June 1993); results—percent of vote by party NA; note—legislative and presidential candidates run on a unified slate, so vote percentages are the same as in section on presidential election results; seats—(27 total) MNR 9, ADN 7, MIR 8, CONDEPA 2, PDC 1
President: last held 7 May 1989 (next to be held 6 June 1993); results—Gonzalo SANCHEZ de Lozada (MNR) 23%, Hugo BANZER Suarez (ADN) 22%, Jaime PAZ Zamora (MIR) 19%; no candidate received a majority of the popular vote; Jaime PAZ Zamora (MIR) formed a coalition with Hugo BANZER (ADN); with ADN support, PAZ Zamora won the congressional runoff election on 4 August and was inaugurated on 6 August 1989
Executive branch: president, vice president, Cabinet
Legislative branch: bicameral National Congress (Congreso Nacional) consists of an upper chamber or Chamber of Senators (Camara de Senadores) and a lower chamber or Chamber of Deputies (Camara de Diputados)
Judicial branch: Supreme Court (Corte Suprema)
Leaders:
Chief of State and Head of Government: President Jaime PAZ Zamora (since 6 August 1989); Vice President Luis OSSIO Sanjines (since 6 August 1989)
Member of: AG, ECLAC, FAO, GATT, G-11, G-77, IADB, IAEA, IBRD, ICAO, IDA, IFAD, IFC, ILO, IMF, IMO, INTELSAT, INTERPOL, IOC, IOM, ITU, LAES, LAIA, LORCS, NAM, OAS, OPANAL, PCA, RG, UN, UNCTAD, UNESCO, UNIDO, UPU, WCL, WFTU, WHO, WMO, WTO
Diplomatic representation in US:
chief of mission: Ambassador Jorge CRESPO

chancery: 3014 Massachusetts Avenue NW, Washington, DC 20008
telephone: (202) 483-4410 through 4412
consulates general: Los Angeles, Miami, New York, and San Francisco
US diplomatic representation:
chief of mission: Ambassador Charles R. BOWERS
embassy: Banco Popular del Peru Building, corner of Calles Mercado y Colon, La Paz
mailing address: P. O. Box 425, La Paz, or APO AA 34032
telephone: [591] (2) 350251 or 350120
FAX: [591] (2) 359875
Flag: three equal horizontal bands of red (top), yellow, and green with the coat of arms centered on the yellow band; similar to the flag of Ghana, which has a large black five-pointed star centered in the yellow band

Economy

Overview: With its long history of semifeudalistic social controls, dependence on volatile prices for its mineral exports, and bouts of hyperinflation, Bolivia has remained one of the poorest and least developed Latin American countries. Since August 1989, President PAZ Zamora, despite his Marxist origins, has maintained a moderate policy of repressing domestic terrorism, containing inflation, and achieving annual GDP growth of 3 to 4%. For many farmers, who constitute half of the country's work force, the main cash crop is coca, which is sold for cocaine processing.
National product: GDP—exchange rate conversion—$4.9 billion (1992)
National product real growth rate: 3.8% (1992)
National product per capita: $670 (1992)
Inflation rate (consumer prices): 10.5% (December 1992)
Unemployment rate: 5% (1992)
Budget: revenues $1.5 billion; expenditures $1.57 billion, including capital expenditures of $627 million (1993 est.)
Exports: $609 million (f.o.b., 1992)
commodities: metals 46%, hydrocarbons 21%, other 33% (coffee, soybeans, sugar, cotton, timber)
partners: US 15%, Argentina
Imports: 1.185 billion (c.i.f., 1992)
commodities: food, petroleum, consumer goods, capital goods
partners: US 22%
External debt: 3.7 billion (December 1992)
Industrial production: growth rate 7% (1992); accounts for almost 32% of GDP
Electricity: 865,000 kW capacity; 1,834 million kWh produced, 250 kWh per capita (1992)
Industries: mining, smelting, petroleum, food and beverage, tobacco, handicrafts,

clothing; illicit drug industry reportedly produces 15% of its revenues
Agriculture: accounts for about 21% of GDP (including forestry and fisheries); principal commodities—coffee, coca, cotton, corn, sugarcane, rice, potatoes, timber; self-sufficient in food
Illicit drugs: world's second-largest producer of coca (after Peru) with an estimated 47,900 hectares under cultivation; voluntary and forced eradication program unable to prevent production from rising to 82,000 metric tons in 1992 from 74,700 tons in 1989; government considers all but 12,000 hectares illicit; intermediate coca products and cocaine exported to or through Colombia and Brazil to the US and other international drug markets
Economic aid: US commitments, including Ex-Im (FY70-89), $990 million; Western (non-US) countries, ODA and OOF bilateral commitments (1970-89), $2,025 million; Communist countries (1970-89), $340 million
Currency: 1 boliviano ($B) = 100 centavos
Exchange rates: bolivianos ($B) per US$1—3.9437 (August 1992), 3.85 (1992), 3.5806 (1991), 3.1727 (1990), 2.6917 (1989), 2.3502 (1988), 2.0549 (1987)
Fiscal year: calendar year

Communications

Railroads: 3,684 km total, all narrow gauge; 3,652 km 1.000-meter gauge and 32 km 0.760-meter gauge, all government owned, single track
Highways: 38,836 km total; 1,300 km paved, 6,700 km gravel, 30,836 km improved and unimproved earth
Inland waterways: 10,000 km of commercially navigable waterways
Pipelines: crude oil 1,800 km; petroleum products 580 km; natural gas 1,495 km
Ports: none; maritime outlets are Arica and Antofagasta in Chile, Matarani and Ilo in Peru
Merchant marine: 2 cargo ships (1,000 GRT or over) totaling 14,051 GRT/22,155 DWT
Airports:
total: 1,225
usable: 1,043
with permanent-surface runways: 9
with runways over 3,659 m: 2
with runways 2,440-3,659 m: 7
with runways 1,220-2,439 m: 161
Telecommunications: microwave radio relay system being expanded; improved international services; 144,300 telephones; broadcast stations—129 AM, no FM, 43 TV, 68 shortwave; 1 Atlantic Ocean INTELSAT earth station

Bosnia and Herzegovina

Defense Forces

Branches: Army (Ejercito Boliviano), Navy includes Marines (Fuerza Navala), Air Force (Fuerza Aereo de Bolivia), National Police Force (Boliviano Policia Nacional)
Manpower availability: males age 15-49 1,786,137; fit for military service 1,162,160; reach military age (19) annually 78,125 (1993 est.)
Defense expenditures: exchange rate conversion—$80 million, 1.6% of GDP (1990 est.)

Note: Bosnia and Herzegovina is suffering from interethnic civil strife which began in March 1992 after the Bosnian Government held a referendum on independence. Bosnia's Serbs—supported by neighboring Serbia—responded with armed resistance aimed at partitioning the republic along ethnic lines and joining Serb held areas to a "greater Serbia". Since the onset of the conflict, which has driven approximately half of the pre-war population of 4.4 million from their homes, both the Bosnian Serbs and the Bosnian Croats have asserted control of more than three-quarters of the territory formerly under the control of the Bosnian Government. The UN and the EC are continuing to try to mediate a plan for peace.

Geography

Location: Southeastern Europe, on the Balkan Peninsula, between Croatia and Serbia and Montenegro
Map references: Africa, Arctic Region, Ethnic Groups in Eastern Europe, Europe, Standard Time Zones of the World
Area:
total area: 51,233 km^2
land area: 51,233 km^2
comparative area: slightly larger than Tennessee
Land boundaries: total 1,369 km, Croatia (northwest) 751 km, Croatia (south) 91 km, Serbia and Montenegro 527 km (312 km with Serbia; 215 km with Montenegro)
Coastline: 20 km
Maritime claims:
continental shelf: 200 m depth
exclusive economic zone: 12 nm
exclusive fishing zone: 12 nm
territorial sea: 12 nm
International disputes: Serbia and Montenegro and Croatia seek to cantonize Bosnia and Herzegovina; Muslim majority being forced from many areas
Climate: hot summers and cold winters; areas of high elevation have short, cool summers and long, severe winters; mild, rainy winters along coast
Terrain: mountains and valleys
Natural resources: coal, iron, bauxite, manganese, timber, wood products, copper, chromium, lead, zinc
Land use:
arable land: 20%
permanent crops: 2%
meadows and pastures: 25%
forest and woodland: 36%
other: 17%
Irrigated land: NA km^2
Environment: air pollution from metallurgical plants; water scarce; sites for disposing of urban waste are limited; subject to frequent and destructive earthquakes

People

Population: 4,618,804 (July 1993 est.)
note: all data dealing with population is subject to considerable error because of the dislocations caused by military action and ethnic cleansing
Population growth rate: 0.72% (1993 est.)
Birth rate: 13.54 births/1,000 population (1993 est.)
Death rate: 6.38 deaths/1,000 population (1993 est.)
Net migration rate: 0 migrant(s)/1,000 population (1993 est.)
Infant mortality rate: 13.2 deaths/1,000 live births (1993 est.)
Life expectancy at birth:
total population: 74.8 years
male: 72.11 years
female: 77.67 years (1993 est.)
Total fertility rate: 1.62 children born/woman (1993 est.)
Nationality:
noun: Bosnian(s), Herzegovinian(s)
adjective: Bosnian, Herzegovinian
Ethnic divisions: Muslim 44%, Serb 31%, Croat 17%, other 8%
Religions: Muslim 40%, Orthodox 31%, Catholic 15%, Protestant 4%, other 10%
Languages: Serbo-Croatian 99%
Literacy:
total population: NA%
male: NA%
female: NA%
Labor force: 1,026,254
by occupation: agriculture 2%, industry, mining 45% (1991 est.)

Government

Names:
conventional long form: Republic of Bosnia and Herzegovina
conventional short form: Bosnia and Herzegovina
local long form: Republika Bosna i Hercegovina

Bosnia and Herzegovina

(continued)

local short form: Bosna i Hercegovina
Digraph: BK
Type: emerging democracy
Capital: Sarajevo
Administrative divisions: 109 districts
(opcine, singular—opcina) Banovici, Banja
Luka, Bihac, Bijeljina, Bileca, Bosanska
Dubica, Bosanska Graaiskia, Bosanska
Krupa, Bosanski Brod, Bosanski Novi,
Bosanski Petrovac, Bosanski Samac,
Bosansko Grahovo, Bratunac, Brcko, Breza,
Bugojno, Busovaca, Cazin, Cajilice,
Capljina, Celinac, Citluk, Derventa, Duboj,
Donji Vakuf, Foca, Fojnica, Gacko, Glamoc,
Gorazde Gornji Vakuf, Gracanica, Gradacac,
Grude, Han Pijesak Jablanica, Jajce, Kakanj,
Kalesija, Kalinovik, Kiseljak, Kladanj,
Kljuc, Konjic, Kotor Varos, Kresevo,
Kupres, Laktasi, Listica, Livno, Lopare,
Lukavac, Ljubinje, Ljubuski, Maglaj,
Modrica, Mostar, Mrkonjic Grad, Neum,
Nevesinje, Odzak, Olovo, Orasje, Posusje,
Prijedor, Prnjavor, Prozor, (Pucarevo) Novi
Travnik, Rogatica, Rudo, Sanski Most,
Sarajevo-Centar, Sarajevo-Hadzici,
Sarajevo-Ilidza, Sarajevo-Ilijas,
Sarajevo-Novi Grad, Sarajevo-Novo,
Sarajevo-Pale, Sarajevo-Stari Grad,
Sarajevo-Trnovo, Sarajevo-Vogosca, Skender
Vakuf, Sokolac, Srbac, Srebrenica,
Srebrenik, Stoloc, Sekovici, Sipovo, Teslic,
Tesanj, (Titov Drvar) Drvar, Duvno, Travnik,
Trebinje, Tuzla, Ugljevik, Vare, Velika
Kladusa, Visoko, Visegrad, Vitez Vlasenica,
Zavidovici, Zenica, Zvornik, Zepce, Zivinice
note: currently under negotiation with the
assistance of international mediators
Independence: NA April 1992 (from
Yugoslavia)
Constitution: NA
Legal system: based on civil law system
National holiday: NA
Political parties and leaders: Party of
Democratic Action (SDA), Mirsad CEMAN;
Croatian Democratic Union of Bosnia and
Herzegovina (HDZ BiH), Mate BOBAN;
Serbian Democratic Party of Bosnia and
Herzegovina (SDS BiH), Radovan
KARADZIC, president; Muslim-Bosnian
Organization (MBO), Adil
ZULFIKARPASIC, president; Democratic
Party of Socialists (DSS), Nijaz
DURAKOVIC, president; Party of
Democratic Changes, leader NA; Serbian
Movement for Renewal (SPO), Milan
TRIVUNCIC; Alliance of Reform Forces of
Yugoslavia for Bosnia and Herzegovina
(SRSJ BiH), Dr. Nenad KECMANOVIC,
president; Democratic League of Greens
(DSZ), Drazen PETROVIC; Liberal Party
(LS), Rasim KADIC, president
Other political or pressure groups: NA
Suffrage: 16 years of age, if employed; 18
years of age, universal

Elections:
Chamber of Municipalities: last held
November-December 1990 (next to be held
NA); seats—(110 total) SDA 43, SDS BiH
38, HDZ BiH 23, Party of Democratic
Changes 4, DSS 1, SPO 1
Chamber of Citizens: last held NA 1990
(next to be held NA); seats—(130 total)
SDA 43, SDS BiH 34, HDZ BiH 21, Party
of Democratic Changes 15, SRSJ BiH 12,
MBO 2, DSS 1, DSZ 1, LS 1
Executive branch: collective presidency,
prime minister, deputy prime ministers,
cabinet
Legislative branch: bicameral National
Assembly consists of an upper house or
Chamber of Municipalities (Vijece Opeina)
and a lower house or Chamber of Citizens
(Vijece Gradanstvo)
Judicial branch: Supreme Court,
Constitutional Court
Leaders:
Chief of State: President Alija
IZETBEGOVIC (since NA December 1990),
other members of the collective presidency:
Ejup GANIC (since NA), Miro LASIC
(since NA December 1992), Mirko
PEJANOVIC (since NA), Tatjana
LJUJIC-MIJATOVIC (since NA December
1992), Fikret ABDIC
Head of Government: Prime Minister Mile
AKMADZIC (since NA October 1992);
Deputy Prime Minister Zlatko
LAGUMDZIJA (since NA); Deputy Prime
Minister Miodrag SIMOVIC (since NA);
Deputy Prime Minister Hadzo EFENDIC
(since NA)
Member of: CEI, CSCE, ECE, UN,
UNCTAD, WHO
Diplomatic representation in US:
chief of mission: NA
chancery: NA
telephone: NA
US diplomatic representation: the US
maintains full diplomatic relations with
Bosnia and Herzegovina but has not yet
established an embassy in Sarajevo
Flag: white with a large blue shield; the
shield contains white Roman crosses with a
white diagonal band running from the upper
hoist corner to the lower fly side

Economy

Overview: Bosnia and Herzegovina ranked
next to Macedonia as the poorest republic in
the old Yugoslav federation. Although
agriculture has been almost all in private
hands, farms have been small and
inefficient, and the republic traditionally has
been a net importer of food. Industry has
been greatly overstaffed, one reflection of
the rigidities of Communist central planning
and management. Tito had pushed the
development of military industries in the

republic with the result that Bosnia hosted a
large share of Yugoslavia's defense plants.
As of March 1993, Bosnia and Herzegovina
was being torn apart by the continued bitter
interethnic warfare that has caused
production to plummet, unemployment and
inflation to soar, and human misery to
multiply. No reliable economic statistics for
1992 are available, although output clearly
fell below the already depressed 1991 level.
National product: GDP—purchasing power
equivalent—$14 billion (1991 est.)
National product real growth rate: -37%
(1991 est.)
National product per capita: $3,200 (1991
est.)
Inflation rate (consumer prices): 80% per
month (1991)
Unemployment rate: 28% (February 1992
est.)
Budget: revenues $NA; expenditures $NA,
including capital expenditures of $NA
Exports: $2,054 million (1990)
commodities: manufactured goods 31%,
machinery and transport equipment 20.8%,
raw materials 18%, miscellaneous
manufactured articles 17.3%, chemicals
9.4%, fuel and lubricants 1.4%, food and
live animals 1.2%
partners: principally the other former
Yugoslav republics
Imports: $1,891 million (1990)
commodities: fuels and lubricants 32%,
machinery and transport equipment 23.3%,
other manufactures 21.3%, chemicals 10%,
raw materials 6.7%, food and live animals
5.5%, beverages and tobacco 1.9%
partners: principally the other former
Yugoslav republics
External debt: $NA
Industrial production: growth rate NA%,
but production is sharply down because of
interethnic and interrepublic warfare
(1991-92)
Electricity: 3,800,000 kW capacity; 7,500
million kWh produced, 1,700 kWh per
capita (1992)
Industries: steel production, mining (coal,
iron ore, lead, zinc, manganese, and
bauxite), manufacturing (vehicle assembly,
textiles, tobacco products, wooden furniture,
40% of former Yugoslavia's armaments
including tank and aircraft assembly,
domestic appliances), oil refining
Agriculture: accounted for 9.0% of GDP in
1989; regularly produces less than 50% of
food needs; the foothills of northern Bosnia
support orchards, vineyards, livestock, and
some wheat and corn; long winters and
heavy precipitation leach soil fertility
reducing agricultural output in the
mountains; farms are mostly privately held,
small, and not very productive
Illicit drugs: NA
Economic aid: $NA

Botswana

Currency: Croatian dinar used in ethnic Croat areas, "Yugoslav" dinar used in all other areas
Exchange rates: NA
Fiscal year: calendar year

Communications

Railroads: NA km
Highways: 21,168 km total (1991); 11,436 km paved, 8,146 km gravel, 1,586 km earth; note—highways now disrupted
Inland waterways: NA km
Pipelines: crude oil 174 km, natural gas 90 km (1992); note—pipelines now disrupted
Ports: coastal—none; inland—Bosanski Brod on the Sava River
Airports:
total: 27
useable: 22
with permanent-surface runways: 8
with runways over 3659: 0
with runways 2440-3659 m: 4
with runways 1220-2439 m: 5
Telecommunications: telephone and telegraph network is in need of modernization and expansion, many urban areas being below average compared with services in other former Yugoslav republics; 727,000 telephones; broadcast stations—9 AM, 2 FM, 6 TV; 840,000 radios; 1,012,094 TVs; NA submarine coaxial cables; satellite ground stations—none

Defense Forces

Branches: Army
Manpower availability: males age 15-49 1,283,576; fit for military service 1,045,512; reach military age (19) annually 37,827 (1993 est.)
Defense expenditures: $NA, NA% of GDP

Geography

Location: Southern Africa, north of South Africa
Map references: Africa, Standard Time Zones of the World
Area:
total area: 600,370 km²
land area: 585,370 km²
comparative area: slightly smaller than Texas
Land boundaries: total 4,013 km, Namibia 1,360 km, South Africa 1,840 km, Zimbabwe 813 km
Coastline: 0 km (landlocked)
Maritime claims: none; landlocked
International disputes: short section of boundary with Namibia is indefinite; disputed island with Namibia in the Chobe River; quadripoint with Namibia, Zambia, and Zimbabwe is in disagreement; recent dispute with Namibia over uninhabited Sidudu Island in Linyanti River
Climate: semiarid; warm winters and hot summers
Terrain: predominately flat to gently rolling tableland; Kalahari Desert in southwest
Natural resources: diamonds, copper, nickel, salt, soda ash, potash, coal, iron ore, silver
Land use:
arable land: 2%
permanent crops: 0%
meadows and pastures: 75%
forest and woodland: 2%
other: 21%
Irrigated land: 20 km² (1989 est.)
Environment: overgrazing, desertification
Note: landlocked

People

Population: 1,325,920 (July 1993 est.)
Population growth rate: 2.53% (1993 est.)
Birth rate: 33.39 births/1,000 population (1993 est.)
Death rate: 8.05 deaths/1,000 population (1993 est.)

Net migration rate: 0 migrant(s)/1,000 population (1993 est.)
Infant mortality rate: 40.6 deaths/1,000 live births (1993 est.)
Life expectancy at birth:
total population: 62.54 years
male: 59.52 years
female: 65.65 years (1993 est.)
Total fertility rate: 4.25 children born/woman (1993 est.)
Nationality:
noun: Motswana (singular), Batswana (plural)
adjective: Motswana (singular), Batswana (plural)
Ethnic divisions: Batswana 95%, Kalanga, Basarwa, and Kgalagadi 4%, white 1%
Religions: indigenous beliefs 50%, Christian 50%
Languages: English (official), Setswana
Literacy: age 15 and over able to read and write simple sentences (1990)
total population: 72%
male: 67%
female: 74%
Labor force: 400,000
by occupation: 198,500 formal sector employees, most others are engaged in cattle raising and subsistence agriculture (1990 est.); 14,600 are employed in various mines in South Africa (1990)

Government

Names:
conventional long form: Republic of Botswana
conventional short form: Botswana
former: Bechuanaland
Digraph: BC
Type: parliamentary republic
Capital: Gaborone
Administrative divisions: 10 districts; Central, Chobe, Ghanzi, Kgalagadi, Kgatleng, Kweneng, Ngamiland, North-East, South-East, Southern; in addition, there are 4 town councils—Francistown, Gaborone, Lobaste, Selebi-Phikwe
Independence: 30 September 1966 (from UK)
Constitution: March 1965, effective 30 September 1966
Legal system: based on Roman-Dutch law and local customary law; judicial review limited to matters of interpretation; has not accepted compulsory ICJ jurisdiction
National holiday: Independence Day, 30 September (1966)
Political parties and leaders: Botswana Democratic Party (BDP), Sir Ketumile MASIRE; Botswana National Front (BNF), Kenneth KOMA; Boswana People's Party (BPP), Knight MARIPE; Botswana Independence Party (BIP), Motsamai MPHO
Suffrage: 21 years of age; universal

Botswana (continued)

Elections:

National Assembly: last held 7 October 1989 (next to be held October 1994); results—percent of vote by party NA; seats—(38 total, 34 elected) BDP 35, BNF 3
President: last held 7 October 1989 (next to be held October 1994); results—President Sir Ketumile MASIRE was reelected by the National Assembly
Executive branch: president, vice president, Cabinet
Legislative branch: bicameral National Assembly consists of an upper house or House of Chiefs and a lower house or National Assembly
Judicial branch: High Court, Court of Appeal
Leaders:
Chief of State and Head of Government: President Sir Ketunile MASIRE (since 13 July 1980); Vice President Festus MOGAE (since 9 March 1992)
Member of: ACP, AfDB, C, CCC, ECA, FAO, FLS, G-77, GATT, IBRD, ICAO, ICFTU, IDA, IFAD, IFC, ILO, IMF, INTERPOL, IOC, ITU, LORCS, NAM, OAU, SACU, SADC, UN, UNCTAD, UNESCO, UNIDO, UNOMOZ, UPU, WCL, WHO, WMO
Diplomatic representation in US:
chief of mission: Ambassador Botsweletse Kingsley SEBELE
chancery: Suite 7M, 3400 International Drive NW, Washington, DC 20008
telephone: (202) 244-4990 or 4991
US diplomatic representation:
chief of mission: Ambassador David PASSAGE
embassy: address NA, Gaborone
mailing address: P. O. Box 90, Gaborone
telephone: [267] 353-982
FAX: [267] 356-947
Flag: light blue with a horizontal white-edged black stripe in the center

---

Economy

Overview: The economy has historically been based on cattle raising and crops. Agriculture today provides a livelihood for more than 80% of the population, but produces only about 50% of food needs. The driving force behind the rapid economic growth of the 1970s and 1980s has been the mining industry. This sector, mostly on the strength of diamonds, has gone from generating 25% of GDP in 1980 to 50% in 1991. No other sector has experienced such growth, especially not agriculture, which is plagued by erratic rainfall and poor soils. The unemployment rate remains a problem at 25%. Although diamond production was down slightly in 1992, substantial gains in coal output and manufacturing helped boost the economy

National product: GDP—purchasing power equivalent—$3.6 billion (FY92 est.)
National product real growth rate: 5.8% (FY92 est.)
National product per capita: $2,450 (FY92 est.)
Inflation rate (consumer prices): 16.5% (December 1992)
Unemployment rate: 25% (1989)
Budget: revenues $1.7 billion; expenditures $1.99 billion, including capital expenditures of $652 million (FY94)
Exports: $1.6 billion (f.o.b. 1991)
commodities: diamonds 78%, copper and nickel 8%, meat 4%
partners: Switzerland, UK, SACU (Southern African Customs Union)
Imports: $1.7 billion (c.i.f., 1991)
commodities: foodstuffs, vehicles and transport equipment, textiles, petroleum products
partners: Switzerland, SACU (Southern African Customs Union), UK, US
External debt: $344 million (December 1991)
Industrial production: growth rate 6.9% (1991); accounts for about 53% of GDP, including mining
Electricity: 220,000 kW capacity; 1,123 million kWh produced, 846 kWh per capita (1991)
Industries: mining of diamonds, copper, nickel, coal, salt, soda ash, potash; livestock processing
Agriculture: accounts for only 5% of GDP; subsistence farming predominates; cattle raising supports 50% of the population; must import up to of 80% of food needs
Economic aid: US aid, $13 million (1992); US commitments, including Ex-Im (FY70-89), $257 million; Western (non-US) countries, ODA and OOF bilateral commitments (1970-89), $1,875 million; OPEC bilateral aid (1979-89), $43 million; Communist countries (1970-89), $29 million; in 1992: Norway (largest donor) $16 million, Sweden $15.5 million, Germany $3.6 million, EC/Lome-IV $3-6 million in grants, $28.7 million in long-term projects
Currency: 1 pula (P) = 100 thebe
Exchange rates: pula (P) per US$1—2.31 (February 1993), 2.1327 (1992), 2.0173 (1991), 1.8601 (1990), 2.0125 (1989), 1.8159 (1988)
Fiscal year: 1 April-31 March

Communications

Railroads: 712 km 1.067-meter gauge
Highways: 11,514 km total; 1,600 km paved; 1,700 km crushed stone or gravel, 5,177 km improved earth, 3,037 km unimproved earth
Airports:
total: 100

usable: 87
with permanent-surface runways: 8
with runways over 3,659 m: 0
with runways 2,440-3,659 m: 1
with runways 1,220-2,439 m: 29
Telecommunications: the small system is a combination of open-wire lines, microwave radio relay links, and a few radio-communications stations; 26,000 telephones; broadcast stations—7 AM, 13 FM, no TV; 1 Indian Ocean INTELSAT earth station

---

Defense Forces

Branches: Botswana Defense Force (including Army and Air Wing), Botswana National Police
Manpower availability: males age 15-49 282,885; fit for military service 148,895; reach military age (18) annually 14,868 (1993 est.)
Defense expenditures: exchange rate conversion—$196 million, 4.9% of GDP (FY93/94)

Bouvet Island
(territory of Norway)

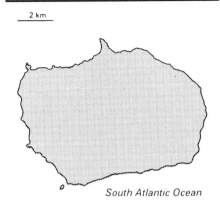

2 km

South Atlantic Ocean

Geography

Location: in the south Atlantic Ocean, 2,575 km south-southwest of the Cape of Good Hope (South Africa)
Map references: Antarctic Region
Area:
total area: 58 km²
land area: 58 km²
comparative area: about 0.3 times the size of Washington, DC
Land boundaries: 0 km
Coastline: 29.6 km
Maritime claims:
territorial sea: 4 nm
International disputes: none
Climate: antarctic
Terrain: volcanic; maximum elevation about 800 meters; coast is mostly inaccessible
Natural resources: none
Land use:
arable land: 0%
permanent crops: 0%
meadows and pastures: 0%
forest and woodland: 0%
other: 100% (all ice)
Irrigated land: 0 km²
Environment: covered by glacial ice
Note: located in the South Atlantic Ocean

People

Population: uninhabited

Government

Names:
conventional long form: none
conventional short form: Bouvet Island
Digraph: BV
Type: territory of Norway
Capital: none; administered from Oslo, Norway
Independence: none (territory of Norway)

Economy

Overview: no economic activity

Communications

Ports: none; offshore anchorage only
Telecommunications: automatic meteorological station

Defense Forces

Note: defense is the responsibility of Norway

Brazil

North Atlantic Ocean
Belém
Manaus
São Luís
Rio Branco
Recife
Salvador
Cuiabá
BRASÍLIA
Belo Horizonte
Corumbá
Rio de Janeiro
São Paulo
South Atlantic Ocean
Boundary representation is not necessarily authoritative
1000 km
Pôrto Alegre

Geography

Location: Eastern South America, bordering the Atlantic Ocean
Map references: South America, Standard Time Zones of the World
Area:
total area: 8,511,965 km²
land area: 8,456,510 km²
comparative area: slightly smaller than the US
note: includes Arquipelago de Fernando de Noronha, Atol das Rocas, Ilha da Trindade, Ilhas Martin Vaz, and Penedos de Sao Pedro e Sao Paulo
Land boundaries: total 14,691 km, Argentina 1,224 km, Bolivia 3,400 km, Colombia 1,643 km, French Guiana 673 km, Guyana 1,119 km, Paraguay 1,290 km, Peru 1,560 km, Suriname 597 km, Uruguay 985 km, Venezuela 2,200 km
Coastline: 7,491 km
Maritime claims:
contiguous zone: 24 nm
continental shelf: 200 m depth or to depth of exploitation
exclusive economic zone: 200 nm
territorial sea: 12 nm
International disputes: short section of the boundary with Paraguay (just west of Guaira Falls on the Rio Parana) is in dispute; two short sections of boundary with Uruguay are in dispute—Arrio Invernada (Arroyo de la Invernada) area of the Rio Quarai (Rio Cuareim) and the islands at the confluence of the Rio Quarai (Rio Cuareim) and the Uruguay
Climate: mostly tropical, but temperate in south
Terrain: mostly flat to rolling lowlands in north; some plains, hills, mountains, and narrow coastal belt
Natural resources: iron ore, manganese, bauxite, nickel, uranium, phosphates, tin, hydropower, gold, platinum, petroleum, timber

Brazil (continued)

Land use:
arable land: 7%
permanent crops: 1%
meadows and pastures: 19%
forest and woodland: 67%
other: 6%
Irrigated land: 27,000 km² (1989 est.)
Environment: recurrent droughts in northeast; floods and frost in south; deforestation in Amazon basin; air and water pollution in Rio de Janeiro, Sao Paulo, and several other large cities
Note: largest country in South America; shares common boundaries with every South American country except Chile and Ecuador

People

Population: 156,664,223 (July 1993 est.)
Population growth rate: 1.35% (1993 est.)
Birth rate: 21.77 births/1,000 population (1993 est.)
Death rate: 8.3 deaths/1,000 population (1993 est.)
Net migration rate: 0 migrant(s)/1,000 population (1993 est.)
Infant mortality rate: 61.7 deaths/1,000 live births (1993 est.)
Life expectancy at birth:
total population: 62.7 years
male: 58.28 years
female: 67.33 years (1993 est.)
Total fertility rate: 2.49 children born/woman (1993 est.)
Nationality:
noun: Brazilian(s)
adjective: Brazilian
Ethnic divisions: Portuguese, Italian, German, Japanese, Amerindian, black 6%, white 55%, mixed 38%, other 1%
Religions: Roman Catholic (nominal) 90%
Languages: Portuguese (official), Spanish, English, French
Literacy: age 15 and over can read and write (1990)
total population: 81%
male: 82%
female: 80%
Labor force: 57 million (1989 est.)
by occupation: services 42%, agriculture 31%, industry 27%

Government

Names:
conventional long form: Federative Republic of Brazil
conventional short form: Brazil
local long form: Republica Federativa do Brasil
local short form: Brasil
Digraph: BR
Type: federal republic
Capital: Brasilia
Administrative divisions: 26 states (estados, singular—estado) and 1 federal district* (distrito federal); Acre, Alagoas, Amapa, Amazonas, Bahia, Ceara, Distrito Federal*, Espirito Santo, Goias, Maranhao, Mato Grosso, Mato Grosso do Sul, Minas Gerais, Para, Paraiba, Parana, Pernambuco, Piaui, Rio de Janeiro, Rio Grande do Norte, Rio Grande do Sul, Rondonia, Roraima, Santa Catarina, Sao Paulo, Sergipe, Tocantins
Independence: 7 September 1822 (from Portugal)
Constitution: 5 October 1988
Legal system: based on Roman codes; has not accepted compulsory ICJ jurisdiction
National holiday: Independence Day, 7 September (1822)
Political parties and leaders: National Reconstruction Party (PRN), Daniel TOURINHO, president; Brazilian Democratic Movement Party (PMDB), Roberto ROLLEMBERG, president; Liberal Front Party (PFL), Jose Mucio MONTEIRO, president; Workers' Party (PT), Luis Ignacio (Lula) da SILVA, president; Brazilian Labor Party (PTB), Luiz GONZAGA de Paiva Muniz, president; Democratic Labor Party (PDT), Leonel BRIZOLA, president; Democratic Social Party (PPS), Paulo MALUF, president; Brazilian Social Democracy Party (PSDB), Tasso JEREISSATI, president; Popular Socialist Party (PPS), Roberto FREIRE, president; Communist Party of Brazil (PCdoB), Joao AMAZONAS, secretary general; Christian Democratic Party (PDC), Siqueira CAMPOS, president
Other political or pressure groups: left wing of the Catholic Church and labor unions allied to leftist Worker's Party are critical of government's social and economic policies
Suffrage: voluntary between 16 and 18 years of age and over 70; compulsory over 18 and under 70 years of age
Elections:
Chamber of Deputies: last held 3 October 1990 (next to be held November 1994); results—PMDB 21%, PFL 17%, PDT 9%, PDS 8%, PRN 7.9%, PTB 7%, PT 7%, other 23.1%; seats—(503 total as of 3 February 1991) PMDB 108, PFL 87, PDT 46, PDS 43, PRN 40, PTB 35, PT 35, other 109
Federal Senate: last held 3 October 1990 (next to be held November 1994); results—percent of vote by party NA; seats—(81 total as of 3 February 1991) PMDB 27, PFL 15, PSDB 10, PTB 8, PDT 5, other 16
President: last held 15 November 1989, with runoff on 17 December 1989 (next to be held November 1994); results—Fernando COLLOR de Mello 53%, Luis Inacio da SILVA 47%; note—first free, direct presidential election since 1960
Executive branch: president, vice president, Cabinet
Legislative branch: bicameral National Congress (Congresso Nacional) consists of an upper chamber or Federal Senate (Senado Federal) and a lower chamber or Chamber of Deputies (Camara dos Deputados)
Judicial branch: Supreme Federal Tribunal
Leaders:
Chief of State and Head of Government: President Itamar FRANCO (since 29 December 1992)
Member of: AfDB, AG (observer), CCC, ECLAC, FAO, G-11, G-15, G-19, G-24, G-77, GATT, IADB, IAEA, IBRD, ICAO, ICC, ICFTU, IDA, IFAD, IFC, ILO, IMF, IMO, INMARSAT, INTELSAT, INTERPOL, IOC, IOM (observer), ISO, ITU, LAES, LAIA, LORCS, MERCOSUR, NAM (observer), OAS, ONUSAL, OPANAL, PCA, RG, UN, UNAVEM II, UNCTAD, UNESCO, UNHCR, UNIDO, UNOMOZ, UNPROFOR, UPU, WCL, WHO, WFTU, WIPO, WMO, WTO
Diplomatic representation in US:
chief of mission: Ambassador Rubens RICUPERO
chancery: 3006 Massachusetts Avenue NW, Washington, DC 20008
telephone: (202) 745-2700
consulates general: Chicago, Los Angeles, Miami, New Orleans, and New York
consulates: Dallas, Houston, and San Francisco
US diplomatic representation:
chief of mission: Ambassador Richard MELTON
embassy: Avenida das Nacoes, Lote 3, Brasilia, Distrito Federal
mailing address: APO AA 34030
telephone: [55] (61) 321-7272
FAX: [55] (61) 225-9136
consulates general: Rio de Janeiro, Sao Paulo
consulates: Porto Alegre, Recife
Flag: green with a large yellow diamond in the center bearing a blue celestial globe with 23 white five-pointed stars (one for each state) arranged in the same pattern as the night sky over Brazil; the globe has a white equatorial band with the motto ORDEM E PROGRESSO (Order and Progress)

Economy

Overview: The economy, with large agrarian, mining, and manufacturing sectors, entered the 1990s with declining real growth, runaway inflation, an unserviceable foreign debt of $122 billion, and a lack of policy direction. In addition, the economy remained highly regulated, inward-looking, and protected by substantial trade and investment barriers. Ownership of major industrial and mining facilities is divided among private interests—including several multinationals—and the government. Most large agricultural holdings are private, with the government channeling financing to this sector. Conflicts between large landholders and landless peasants have produced intermittent violence. The COLLOR government, which assumed office in March 1990, launched an ambitious reform program that sought to modernize and reinvigorate the economy by stabilizing prices, deregulating the economy, and opening it to increased foreign competition. The government also obtained an IMF standby loan in January 1992 and reached agreements with commercial bankers on the repayment of interest arrears and on the reduction of debt and debt service payments. Galloping inflation—the rate doubled in 1992—continues to undermine economic stability. Itamar FRANCO, who assumed the presidency following President COLLOR'S resignation in December 1992, has promised to support the basic premises of COLLOR'S reform program but has yet to define clearly his economic policies. Brazil's natural resources remain a major, long-term economic strength.

National product: GDP—exchange rate conversion—$369 billion (1992)

National product real growth rate: -0.2% (1992)

National product per capita: $2,350 (1992)

Inflation rate (consumer prices): 1,174% (1992)

Unemployment rate: 5.9% (1992)

Budget: revenues $164.3 billion; expenditures $170.6 billion, including capital expenditures of $32.9 billion (1990)

Exports: $35.0 billion (1992)
commodities: iron ore, soybean bran, orange juice, footwear, coffee, motor vehicle parts
partners: EC 32.3%, US 20.3%, Latin America 11.6%, Japan 9% (1991)

Imports: $20.0 billion (1992)
commodities: crude oil, capital goods, chemical products, foodstuffs, coal
partners: Middle East 12.4%, US 23.5%, EC 21.8%, Latin America 18.8%, Japan 6% (1991)

External debt: $123.3 billion (December 1992)

Industrial production: growth rate −3.8% (1992); accounts for 39% of GDP

Electricity: 63,765,000 kW capacity; 242,184 million kWh produced, 1,531 kWh per capita (1992)

Industries: textiles and other consumer goods, shoes, chemicals, cement, lumber, iron ore, steel, motor vehicles and auto parts, metalworking, capital goods, tin

Agriculture: accounts for 11% of GDP; world's largest producer and exporter of coffee and orange juice concentrate and second-largest exporter of soybeans; other products—rice, corn, sugarcane, cocoa, beef; self-sufficient in food, except for wheat

Illicit drugs: illicit producer of cannabis and coca, mostly for domestic consumption; government has a modest eradication program to control cannabis and coca cultivation; important transshipment country for Bolivian and Colombian cocaine headed for the US and Europe

Economic aid: US commitments, including Ex-Im (FY70-89), $2.5 billion; Western (non-US) countries, ODA and OOF bilateral commitments (1970-89), $10.2 million; OPEC bilateral aid (1979-89), $284 million; former Communist countries (1970-89), $1.3 billion

Currency: 1 cruzeiro (Cr$) = 100 centavos

Exchange rates: cruzeiros (Cr$) per US$1—13,827.06 (January 1993), 4,506.45 (1992), 406.61 (1991), 68.300 (1990), 2.834 (1989), 0.26238 (1988)

Fiscal year: calendar year

Communications

Railroads: 28,828 km total; 24,864 km 1.000-meter gauge, 3,877 km 1.600-meter gauge, 74 km mixed 1.600-1.000-meter gauge, 13 km 0.760-meter gauge; 2,360 km electrified

Highways: 1,448,000 km total; 48,000 km paved, 1,400,000 km gravel or earth

Inland waterways: 50,000 km navigable

Pipelines: crude oil 2,000 km; petroleum products 3,804 km; natural gas 1,095 km

Ports: Belem, Fortaleza, Ilheus, Manaus, Paranagua, Porto Alegre, Recife, Rio de Janeiro, Rio Grande, Salvador, Santos

Merchant marine: 232 ships (1,000 GRT or over) totaling 5,335,234 GRT/8,986,734 DWT; includes 5 passenger-cargo, 42 cargo, 1 refrigerated cargo, 10 container, 11 roll-on/roll-off, 58 oil tanker, 15 chemical tanker, 12 combination ore/oil, 65 bulk, 2 combination bulk, 11 vehicle carrier; in addition, 1 naval tanker is sometimes used commercially

Airports:
total: 3,613
usable: 3,031
with permanent-surface runways: 431
with runways over 3,659 m: 2
with runways 2,440-3,659 m: 22
with runways 1,220-2,439 m: 584

Telecommunications: good system; extensive microwave radio relay facilities; 9.86 million telephones; broadcast stations—1,223 AM, no FM, 112 TV, 151 shortwave; 3 coaxial submarine cables, 3 Atlantic Ocean INTELSAT earth stations and 64 domestic satellite earth stations

---

Defense Forces

Branches: Brazilian Army, Navy of Brazil (including Marines), Brazilian Air Force, Military Police (paramilitary)

Manpower availability: males age 15-49 42,623,934; fit for military service 28,721,849; reach military age (18) annually 1,655,918 (1993 est.)

Defense expenditures: exchange rate conversion—$1.1 billion, 3% of GDP (1990)

British Indian Ocean Territory
(dependent territory of the UK)

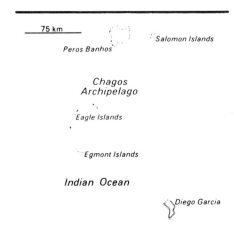

Geography

Location: in the Indian Ocean, south of India about halfway between Africa and Indonesia
Map references: Standard Time Zones of the World
Area:
total area: 60 km²
land area: 60 km²
comparative area: about 0.3 times the size of Washington, DC
note: includes the island of Diego Garcia
Land boundaries: 0 km
Coastline: 698 km
Maritime claims:
exclusive fishing zone: 200 nm
territorial sea: 3 nm
International disputes: the entire Chagos Archipelago is claimed by Mauritius
Climate: tropical marine; hot, humid, moderated by trade winds
Terrain: flat and low (up to 4 meters in elevation)
Natural resources: coconuts, fish
Land use:
arable land: 0%
permanent crops: 0%
meadows and pastures: 0%
forest and woodland: 0%
other: 100%
Irrigated land: 0 km²
Environment: archipelago of 2,300 islands
Note: Diego Garcia, largest and southernmost island, occupies strategic location in central Indian Ocean; island is site of joint US-UK military facility

People

Population: no indigenous inhabitants
note: there are UK-US military personnel; civilian inhabitants, known as the Ilois, evacuated to Mauritius before construction of UK-US military facilities

Government

Names:
conventional long form: British Indian Ocean Territory
conventional short form: none
Abbreviation: BIOT
Digraph: IO
Type: dependent territory of the UK
Capital: none
Independence: none (dependent territory of the UK)
Leaders:
Chief of State: Queen ELIZABETH II (since 6 February 1952)
Head of Government: Commissioner Mr. T. G. HARRIS (since NA); Administrator Mr. R. G. WELLS (since NA 1991); note—both reside in the UK
Diplomatic representation in US: none (dependent territory of UK)
Flag: white with the flag of the UK in the upper hoist-side quadrant and six blue wavy horizontal stripes bearing a palm tree and yellow crown centered on the outer half of the flag

Economy

Overview: All economic activity is concentrated on the largest island of Diego Garcia, where joint UK-US defense facilities are located. Construction projects and various services needed to support the military installations are done by military and contract employees from the UK, Mauritius, the Philippines, and the US. There are no industrial or agricultural activities on the islands.
Electricity: provided by the US military

Communications

Highways: short stretch of paved road between port and airfield on Diego Garcia
Ports: Diego Garcia
Airports:
total: 1
usable: 1
with permanent-surface runways: 1
with runways over 3,659 m: 1 on Diego Garcia
with runways 2,439-3,659 m: 0
with runways 1,229-2,439 m: 0
Telecommunications: minimal facilities; broadcast stations (operated by US Navy)—1 AM, 1 FM, 1 TV; 1 Atlantic Ocean INTELSAT earth station

Defense Forces

Note: defense is the responsibility of the UK

British Virgin Islands
(dependent territory of the UK)

Geography

Location: in the eastern Caribbean Sea, about 110 km east of Puerto Rico
Map references: Central America and the Caribbean
Area:
total area: 150 km²
land area: 150 km²
comparative area: about 0.8 times the size of Washington, DC
note: includes the island of Anegada
Land boundaries: 0 km
Coastline: 80 km
Maritime claims:
exclusive fishing zone: 200 nm
territorial sea: 3 nm
International disputes: none
Climate: subtropical; humid; temperatures moderated by trade winds
Terrain: coral islands relatively flat; volcanic islands steep, hilly
Natural resources: negligible
Land use:
arable land: 20%
permanent crops: 7%
meadows and pastures: 33%
forest and woodland: 7%
other: 33%
Irrigated land: NA km²
Environment: subject to hurricanes and tropical storms from July to October
Note: strong ties to nearby US Virgin Islands and Puerto Rico

People

Population: 12,707 (July 1993 est.)
Population growth rate: 1.22% (1993 est.)
Birth rate: 20.37 births/1,000 population (1993 est.)
Death rate: 6.11 deaths/1,000 population (1993 est.)
Net migration rate: -2.1 migrant(s)/1,000 population (1993 est.)
Infant mortality rate: 19.68 deaths/1,000 live births (1993 est.)

Life expectancy at birth:
total population: 72.62 years
male: 70.77 years
female: 74.6 years (1993 est.)
Total fertility rate: 2.28 children born/woman (1993 est.)
Nationality:
noun: British Virgin Islander(s)
adjective: British Virgin Islander
Ethnic divisions: black 90%, white, Asian
Religions: Protestant 86% (Methodist 45%, Anglican 21%, Church of God 7%, Seventh-Day Adventist 5%, Baptist 4%, Jehovah's Witnesses 2%, other 2%), Roman Catholic 6%, none 2%, other 6% (1981)
Languages: English (official)
Literacy: age 15 and over can read and write (1970)
total population: 98%
male: 98%
female: 98%
Labor force: 4,911 (1980)
by occupation: NA

Government

Names:
conventional long form: none
conventional short form: British Virgin Islands
Abbreviation: BVI
Digraph: VI
Type: dependent territory of the UK
Capital: Road Town
Administrative divisions: none (dependent territory of the UK)
Independence: none (dependent territory of the UK)
Constitution: 1 June 1977
Legal system: English law
National holiday: Territory Day, 1 July
Political parties and leaders: United Party (UP), Conrad MADURO; Virgin Islands Party (VIP), H. Lavity STOUTT; Independent Progressive Movement (IPM), Cyril B. ROMNEY
Suffrage: 18 years of age; universal
Elections:
Legislative Council: last held 12 November 1990 (next to be held by November 1995); results—percent of vote by party NA; seats—(9 total) VIP 6, IPM 1, independents 2
Executive branch: British monarch, governor, chief minister, Executive Council (cabinet)
Legislative branch: unicameral Legislative Council
Judicial branch: Eastern Caribbean Supreme Court
Leaders:
Chief of State: Queen ELIZABETH II (since 6 February 1952), represented by Governor Peter Alfred PENFOLD (since NA 1991)
Head of Government: Chief Minister H. Lavity STOUTT (since NA 1986)

Member of: CARICOM (associate), CDB, ECLAC (associate), IOC, OECS (associate), UNESCO (associate)
Diplomatic representation in US: none (dependent territory of UK)
Flag: blue with the flag of the UK in the upper hoist-side quadrant and the Virgin Islander coat of arms centered in the outer half of the flag; the coat of arms depicts a woman flanked on either side by a vertical column of six oil lamps above a scroll bearing the Latin word VIGILATE (Be Watchful)

Economy

Overview: The economy, one of the most prosperous in the Caribbean area, is highly dependent on the tourist industry, which generates about 21% of the national income. In 1985 the government offered offshore registration to companies wishing to incorporate in the islands, and, in consequence, incorporation fees generated about $2 million in 1987. The economy slowed in 1991 because of the poor performances of the tourist sector and tight commercial bank credit. Livestock raising is the most significant agricultural activity. The islands' crops, limited by poor soils, are unable to meet food requirements.
National product: GDP—purchasing power equivalent—$133 million (1991)
National product real growth rate: 2% (1991)
National product per capita: $10,600 (1991)
Inflation rate (consumer prices): 2.5% (1990 est.)
Unemployment rate: NEGL% (1992)
Budget: revenues $51 million; expenditures $88 million, including capital expenditures of $38 million (1991)
Exports: $2.7 million (f.o.b., 1988)
commodities: rum, fresh fish, gravel, sand, fruits, animals
partners: Virgin Islands (US), Puerto Rico, US
Imports: $11.5 million (c.i.f., 1988)
commodities: building materials, automobiles, foodstuffs, machinery
partners: Virgin Islands (US), Puerto Rico, US
External debt: $4.5 million (1985)
Industrial production: growth rate 4.0% (1985)
Electricity: 10,500 kW capacity; 43 million kWh produced, 3,510 kWh per capita (1990)
Industries: tourism, light industry, construction, rum, concrete block, offshore financial center
Agriculture: livestock (including poultry), fish, fruit, vegetables
Economic aid: NA
Currency: US currency is used

Exchange rates: US currency is used
Fiscal year: 1 April-31 March

Communications

Highways: 106 km motorable roads (1983)
Ports: Road Town
Airports:
total: 3
usable: 3
with permanent-surface runways: 2
with runways over 3,659 m: 0
with runways 2,440-3,659 m: 0
with runways 1,220-2,439 m: 0
Telecommunications: 3,000 telephones; worldwide external telephone service; submarine cable communication links to Bermuda; broadcast stations—1 AM, no FM, 1 TV

Defense Forces

Note: defense is the responsibility of the UK

Brunei

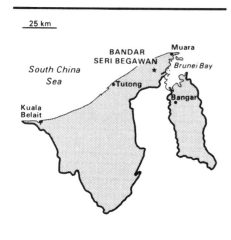

Geography

Location: Southeast Asia, on the northern coast of Borneo almost completely surrounded by Malaysia
Map references: Asia, Oceania, Southeast Asia, Standard Time Zones of the World
Area:
total area: 5,770 km²
land area: 5,270 km²
comparative area: slightly larger than Delaware
Land boundaries: total 381 km, Malysia 381 km
Coastline: 161 km
Maritime claims:
exclusive fishing zone: 200 nm
territorial sea: 12 nm
International disputes: may wish to purchase the Malaysian salient that divides the country; all of the Spratly Islands are claimed by China, Taiwan, and Vietnam; parts of them are claimed by Malaysia and the Philippines; in 1984, Brunei established an exclusive fishing zone that encompasses Louisa Reef, but has not publicly claimed the island
Climate: tropical; hot, humid, rainy
Terrain: flat coastal plain rises to mountains in east; hilly lowland in west
Natural resources: petroleum, natural gas, timber
Land use:
arable land: 1%
permanent crops: 1%
meadows and pastures: 1%
forest and woodland: 79%
other: 18%
Irrigated land: 10 km² (1989 est.)
Environment: typhoons, earthquakes, and severe flooding are rare
Note: close to vital sea lanes through South China Sea linking Indian and Pacific Oceans; two parts physically separated by Malaysia; almost an enclave of Malaysia

People

Population: 276,984 (July 1993 est.)
Population growth rate: 2.77% (1993 est.)
Birth rate: 26.55 births/1,000 population (1993 est.)
Death rate: 5.02 deaths/1,000 population (1993 est.)
Net migration rate: 6.15 migrant(s)/1,000 population (1993 est.)
Infant mortality rate: 25.7 deaths/1,000 live births (1993 est.)
Life expectancy at birth:
total population: 70.94 years
male: 69.27 years
female: 72.65 years (1993 est.)
Total fertility rate: 3.45 children born/woman (1993 est.)
Nationality:
noun: Bruneian(s)
adjective: Bruneian
Ethnic divisions: Malay 64%, Chinese 20%, other 16%
Religions: Muslim (official) 63%, Buddhism 14%, Christian 8%, indigenous beliefs and other 15% (1981)
Languages: Malay (official), English, Chinese
Literacy: age 15 and over can read and write (1981)
total population: 77%
male: 85%
female: 69%
Labor force: 89,000 (includes members of the Army)
by occupation: government 47.5%, production of oil, natural gas, services, and construction 41.9%, agriculture, forestry, and fishing 3.8% (1986)
note: 33% of labor force is foreign (1988)

Government

Names:
conventional long form: Negara Brunei Darussalam
conventional short form: Brunei
Digraph: BX
Type: constitutional sultanate
Capital: Bandar Seri Begawan
Administrative divisions: 4 districts (daerah-daerah, singular—daerah); Belait, Brunei and Muara, Temburong, Tutong
Independence: 1 January 1984 (from UK)
Constitution: 29 September 1959 (some provisions suspended under a State of Emergency since December 1962, others since independence on 1 January 1984)
Legal system: based on Islamic law
National holiday: 23 February (1984)
Political parties and leaders: Brunei United National Party (inactive), Anak HASANUDDIN, chairman; Brunei National Democratic Party (the first legal political party and now banned), leader NA

Suffrage: none
Elections:
Legislative Council: last held in March 1962; in 1970 the Council was changed to an appointive body by decree of the sultan and no elections are planned
Executive branch: sultan, prime minister, Council of Cabinet Ministers
Legislative branch: unicameral Legislative Council (Majlis Masyuarat Megeri)
Judicial branch: Supreme Court
Leaders:
Chief of State and Head of Government: Sultan and Prime Minister His Majesty Paduka Seri Baginda Sultan Haji HASSANAL Bolkiah Mu'izzaddin Waddaulah (since 5 October 1967)
Member of: APEC, ASEAN, C, ESCAP, FAO, G-77, ICAO, IDB, IMO, INTERPOL, IOC, ISO (correspondent), ITU, OIC, UN, UNCTAD, UPU, WHO, WMO
Diplomatic representation in US:
chief of mission: Ambassador Mohamed KASSIM bin Haji Mohamed Daud
chancery: 2600 Virginia Avenue NW, Suite 3000, Washington, DC 20037
telephone: (202) 342-0159
US diplomatic representation:
chief of mission: Ambassador Donald Burnham ENSENAT
embassy: Third Floor, Teck Guan Plaza, Jalan Sultan, Bandar Seri Begawan
mailing address: American Embassy Box B, APO AP 96440
telephone: [673] (2) 229-670
FAX: [673] (2) 225-293
Flag: yellow with two diagonal bands of white (top, almost double width) and black starting from the upper hoist side; the national emblem in red is superimposed at the center; the emblem includes a swallow-tailed flag on top of a winged column within an upturned crescent above a scroll and flanked by two upraised hands

Economy

Overview: The economy is a mixture of foreign and domestic entrepreneurship, government regulation and welfare measures, and village tradition. It is almost totally supported by exports of crude oil and natural gas, with revenues from the petroleum sector accounting for more than 50% of GDP. Per capita GDP of $8,800 is among the highest in the Third World, and substantial income from overseas investment supplements domestic production. The government provides for all medical services and subsidizes food and housing.
National product: GDP—exchange rate conversion—$3.5 billion (1990 est.)
National product real growth rate: 1% (1990 est.)

National product per capita: $8,800 (1990 est.)
Inflation rate (consumer prices): 1.3% (1989)
Unemployment rate: 3.7% (1989)
Budget: revenues $1.3 billion; expenditures $1.5 billion, including capital expenditures of $255 million (1989 est.)
Exports: $2.2 billion (f.o.b., 1990 est.)
commodities: crude oil, liquefied natural gas, petroleum products
partners: Japan 53%, UK 12%, South Korea 9%, Thailand 7%, Singapore 5% (1990)
Imports: $1.7 billion (c.i.f., 1990 est.)
commodities: machinery and transport equipment, manufactured goods, food, chemicals
partners: Singapore 35%, UK 26%, Switzerland 9%, US 9%, Japan 5% (1990)
External debt: none
Industrial production: growth rate 12.9% (1987); accounts for 52.4% of GDP
Electricity: 310,000 kW capacity; 890 million kWh produced, 3,300 kWh per capita (1990)
Industries: petroleum, petroleum refining, liquefied natural gas, construction
Agriculture: imports about 80% of its food needs; principal crops and livestock include rice, cassava, bananas, buffaloes, and pigs
Economic aid: US commitments, including Ex-Im (FY70-87), $20.6 million; Western (non-US) countries, ODA and OOF bilateral commitments (1970-89), $153 million
Currency: 1 Bruneian dollar (B$) = 100 cents
Exchange rates: Bruneian dollars (B$) per US$1—1.6531 (January 1993), 1.6290 (1992), 1.7276 (1991), 1.8125 (1990), 1.9503 (1989), 2.0124 (1988); note—the Bruneian dollar is at par with the Singapore dollar
Fiscal year: calendar year

Communications

Railroads: 13 km 0.610-meter narrow-gauge private line
Highways: 1,090 km total; 370 km paved (bituminous treated) and another 52 km under construction, 720 km gravel or unimproved
Inland waterways: 209 km; navigable by craft drawing less than 1.2 meters
Pipelines: crude oil 135 km; petroleum products 418 km; natural gas 920 km
Ports: Kuala Belait, Muara
Merchant marine: 7 liquefied gas carriers (1,000 GRT or over) totaling 348,476 GRT/340,635 DWT
Airports:
total: 2
usable: 2
with permanent-surface runways: 1
with runway over 3,659 m: 1

with runway 2,440-3,659 m: 0
with runway 1,220-2,439 m: 1
Telecommunications: service throughout country is adequate for present needs; international service good to adjacent Malaysia; radiobroadcast coverage good; 33,000 telephones (1987); broadcast stations—4 AM/FM, 1 TV; 74,000 radio receivers (1987); satellite earth stations—1 Indian Ocean INTELSAT and 1 Pacific Ocean INTELSAT

Defense Forces

Branches: Ground Force, Navy, Air Force, Royal Brunei Police
Manpower availability: males age 15-49 77,407; fit for military service 45,112; reach military age (18) annually 2,676 (1993 est.)
Defense expenditures: exchange rate conversion—$300 million, 9% of GDP (1990)

Bulgaria

Geography

Location: Southeastern Europe, bordering the Black Sea, between Romania and Turkey
Map references: Africa, Arctic Region, Ethnic Groups in Eastern Europe, Europe, Middle East, Standard Time Zones of the World
Area:
total area: 110,910 km²
land area: 110,550 km²
comparative area: slightly larger than Tennessee
Land boundaries: total 1,808 km, Greece 494 km, Macedonia 148 km, Romania 608 km, Serbia and Montenegro 318 km (all with Serbia), Turkey 240 km
Coastline: 354 km
Maritime claims:
contiguous zone: 24 nm
exclusive economic zone: 200 nm
territorial sea: 12 nm
International disputes: Macedonia question with Greece and Macedonia
Climate: temperate; cold, damp winters; hot, dry summers
Terrain: mostly mountains with lowlands in north and south
Natural resources: bauxite, copper, lead, zinc, coal, timber, arable land
Land use:
arable land: 34%
permanent crops: 3%
meadows and pastures: 18%
forest and woodland: 35%
other: 10%
Irrigated land: 10 km² (1989 est.)
Environment: subject to earthquakes, landslides; deforestation; air pollution
Note: strategic location near Turkish Straits; controls key land routes from Europe to Middle East and Asia

People

Population: 8,831,168 (July 1993 est.)
Population growth rate: -0.39% (1993 est.)

Bulgaria *(continued)*

Birth rate: 11.69 births/1,000 population (1993 est.)

Death rate: 11.54 deaths/1,000 population (1993 est.)

Net migration rate: -4.05 migrant(s)/1,000 population (1993 est.)

Infant mortality rate: 12.6 deaths/1,000 live births (1993 est.)

Life expectancy at birth:
total population: 72.82 years
male: 69.55 years
female: 76.26 years (1993 est.)

Total fertility rate: 1.71 children born/woman (1993 est.)

Nationality:
noun: Bulgarian(s)
adjective: Bulgarian

Ethnic divisions: Bulgarian 85.3%, Turk 8.5%, Gypsy 2.6%, Macedonian 2.5%, Armenian 0.3%, Russian 0.2%, other 0.6%

Religions: Bulgarian Orthodox 85%, Muslim 13%, Jewish 0.8%, Roman Catholic 0.5%, Uniate Catholic 0.2%, Protestant, Gregorian-Armenian, and other 0.5%

Languages: Bulgarian; secondary languages closely correspond to ethnic breakdown

Literacy: age 15 and over can read and write (1970)
total population: 93%
male: NA%
female: NA%

Labor force: 4.3 million
by occupation: industry 33%, agriculture 20%, other 47% (1987)

Government

Names:
conventional long form: Republic of Bulgaria
conventional short form: Bulgaria

Digraph: BU

Type: emerging democracy

Capital: Sofia

Administrative divisions: 9 provinces (oblasti, singular—oblast); Burgas, Grad Sofiya, Khaskovo, Lovech, Mikhaylovgrad, Plovdiv, Razgrad, Sofiya, Varna

Independence: 22 September 1908 (from Ottoman Empire)

Constitution: adopted 12 July 1991

Legal system: based on civil law system, with Soviet law influence; has accepted compulsory ICJ jurisdiction

National holiday: 3 March (1878)

Political parties and leaders: Union of Democratic Forces (UDF), Filip DIMITROV, chairman, an alliance of approximately 20 pro-Democratic parties including United Democratic Center, Democratic Party, Radical Democratic Party, Christian Democratic Union, Alternative Social Liberal Party, Republican Party, Civic Initiative Movement, Union of the Repressed, and about a dozen other groups;

Movement for Rights and Freedoms (ethnic Turkish party) (MRF), Ahmed DOGAN, chairman; Bulgarian Socialist Party (BSP), Zhan VIDENOV, chairman

Other political or pressure groups: Ecoglasnost; Podkrepa (Support) Labor Confederation; Fatherland Union; Bulgarian Democratic Youth (formerly Communist Youth Union); Confederation of Independent Trade Unions of Bulgaria (KNSB); Nationwide Committee for Defense of National Interests; Peasant Youth League; Bulgarian Agrarian National Union—United (BZNS); Bulgarian Democratic Center; "Nikola Petkov" Bulgarian Agrarian National Union; Internal Macedonian Revolutionary Organization—Union of Macedonian Societies (IMRO-UMS); numerous regional, ethnic, and national interest groups with various agendas

Suffrage: 18 years of age; universal and compulsory

Elections:
President: last held January 1992; results—Zhelyu ZHELEV was elected by popular vote
National Assembly: last held 13 October 1991; results—UDF 34%, BSP 33%, MRF 7.5%; seats—(240 total) UDF 110, BSP 106, Movement for Rights and Freedoms 24

Executive branch: president, chairman of the Council of Ministers (prime minister), three deputy chairmen of the Council of Ministers, Council of Ministers

Legislative branch: unicameral National Assembly (Narodno Sobranie)

Judicial branch: Supreme Court, Constitutional Court

Leaders:
Chief of State: President Zhelyu Mitev ZHELEV (since 1 August 1990); Vice President Blaga Nikolova DIMITROVA (since NA)
Head of Government: Chairman of the Council of Ministers (Prime Minister) Lyuben Borisov BEROV (since 30 December 1992); Deputy Chairmen of the Council of Ministers (Deputy Prime Ministers) Valentin KARABASHEV, Neycho NEEV, and Evgeniy MATINCHEV (since 30 December 1992)

Member of: BIS, BSEC, CCC, CE, CSCE, EBRD, ECE, FAO, G-9, IAEA, IBRD, ICAO, ICFTU, IFC, ILO, IMF, IMO, INMARSAT, INTERPOL, IOC, IOM (observer), ISO, ITU, LORCS, NACC, NAM (guest), NSG, PCA, UN, UNCTAD, UNESCO, UNIDO, UNTAC, UPU, WHO, WIPO, WMO, WTO, ZC

Diplomatic representation in US:
chief of mission: Ambassador Ognyan Raytchev PISHEV
chancery: 1621 22nd Street NW, Washington, DC 20008
telephone: (202) 387-7969

FAX: (202) 234-7973

US diplomatic representation:
chief of mission: Ambassador Hugh Kenneth HILL
embassy: 1 Alexander Stamboliski Boulevard, Sofia, Unit 25402
mailing address: APO AE 09213-5740
telephone: [359] (2) 88-48-01 through 05
FAX: [359] (2) 80-19-77

Flag: three equal horizontal bands of white (top), green, and red; the national emblem formerly on the hoist side of the white stripe has been removed—it contained a rampant lion within a wreath of wheat ears below a red five-pointed star and above a ribbon bearing the dates 681 (first Bulgarian state established) and 1944 (liberation from Nazi control)

Economy

Overview: Growth in the lackluster Bulgarian economy fell to the 2% annual level in the 1980s. By 1990, Sofia's foreign debt had skyrocketed to over $10 billion—giving a debt-service ratio of more than 40% of hard currency earnings and leading the regime to declare a moratorium on its hard currency payments. The post-Communist government faces major problems of renovating an aging industrial plant; keeping abreast of rapidly unfolding technological developments; investing in additional energy capacity (the portion of electric power from nuclear energy reached over one-third in 1990); and motivating workers, in part by giving them a share in the earnings of their enterprises. Political bickering in Sofia and the collapse of the DIMITROV government in October 1992 have slowed the economic reform process. New Prime Minister BEROV, however, has pledged to continue the reforms initiated by the previous government. He has promised to continue cooperation with the World Bank and IMF, advance negotiations on rescheduling commercial debt, and push ahead with privatization. BEROV's government—whose main parliamentary supporters are the former Communist Bulgarian Socialist Party (BSP)—nonetheless appears likely to pursue more interventionist tactics in overcoming the country's economic problems.

National product: GDP—purchasing power equivalent—$34.1 billion (1992)

National product real growth rate: -7.7% (1992)

National product per capita: $3,800 (1992)

Inflation rate (consumer prices): 80% (1992)

Unemployment rate: 15% (1992)

Budget: revenues $8 billion; expenditures $5 billion, including capital expenditures of $NA (1991 est.)

Exports: $3.5 billion (f.o.b., 1991)
commodities: machinery and equipment 30.6%; agricultural products 24%; manufactured consumer goods 22.2%; fuels, minerals, raw materials, and metals 10.5%; other 12.7% (1991)
partners: former CEMA countries 57.7% (USSR 48.6%, Poland 2.1%, Czechoslovakia 0.9%); developed countries 26.3% (Germany 4.8%, Greece 2.2%); less developed countries 15.9% (Libya 2.1%, Iran 0.7%) (1991)
Imports: $2.8 billion (f.o.b., 1991)
commodities: fuels, minerals, and raw materials 58.7%; machinery and equipment 15.8%; manufactured consumer goods 4.4%; agricultural products 15.2%; other 5.9%
partners: former CEMA countries 51.0% (former USSR 43.2%, Poland 3.7%); developed countries 32.8% (Germany 7.0%, Austria 4.7%); less developed countries 16.2% (Iran 2.8%, Libya 2.5%)
External debt: $12 billion (1991)
Industrial production: growth rate −21% (1992 est.); accounts for about 37% of GDP (1990)
Electricity: 11,500,000 kW capacity; 45,000 million kWh produced, 5,070 kWh per capita (1992)
Industries: machine building and metal working, food processing, chemicals, textiles, building materials, ferrous and nonferrous metals
Agriculture: accounts for 22% of GDP (1990); climate and soil conditions support livestock raising and the growing of various grain crops, oilseeds, vegetables, fruits, and tobacco; more than one-third of the arable land devoted to grain; world's fourth-largest tobacco exporter; surplus food producer
Illicit drugs: transshipment point for southwest Asian heroin transiting the Balkan route
Economic aid: donor—$1.6 billion in bilateral aid to non-Communist less developed countries (1956-89)
Currency: 1 lev (Lv) = 100 stotinki
Exchange rates: leva (Lv) per US$1—24.56 (January 1993),17.18 (January 1992), 16.13 (March 1991), 0.7446 (November 1990), 0.84 (1989), 0.82 (1988), 0.90 (1987); note—floating exchange rate since February 1991
Fiscal year: calendar year

Communications

Railroads: 4,300 km total, all government owned (1987); 4,055 km 1.435-meter standard gauge, 245 km narrow gauge; 917 km double track; 2,640 km electrified
Highways: 36,908 km total; 33,535 km hard surface (including 242 km superhighways); 3,373 km earth roads (1987)
Inland waterways: 470 km (1987)

Pipelines: crude oil 193 km; petroleum products 525 km; natural gas 1,400 km (1992)
Ports: coastal—Burgas, Varna, Varna West; inland—Ruse, Vidin, and Lom on the Danube
Merchant marine: 112 ships (1,000 GRT and over) totaling 1,262,320 GRT/1,887,729 DWT; includes 2 short-sea passenger, 30 cargo, 2 container, 1 passenger-cargo training, 6 roll-on/roll-off, 15 oil tanker, 4 chemical carrier, 2 railcar carrier, 50 bulk; Bulgaria owns 1 ship (1,000 GRT or over) totaling 8,717 DWT operating under Liberian registry
Airports:
total: 380
usable: 380
with permanent-surface runways: 120
with runways over 3659 m: 0
with runways 2,440-3,659 m: 20
with runways 1,220-2,439 m: 20
Telecommunications: extensive but antiquated transmission system of coaxial cable and mirowave radio relay; 2.6 million telephones; direct dialing to 36 countries; phone density is 29 phones per 100 persons (1992); almost two-thirds of the lines are residential; 67% of Sofia households have phones (November 1988); telephone service is available in most villages; broadcast stations—20 AM, 15 FM, and 29 TV, with 1 Soviet TV repeater in Sofia; 2.1 million TV sets (1990); 92% of country receives No. 1 television program (May 1990); 1 satellite ground station using Intersputnik; INTELSAT is used through a Greek earth station

---

Defense Forces

Branches: Army, Navy, Air and Air Defense Forces, Frontier Troops, Internal Troops
Manpower availability: males age 15-49 2,178,136; fit for military service 1,819,901; reach military age (19) annually 69,495 (1993 est.)
Defense expenditures: 5.77 billion leva, NA% of GDP (1993 est.); note—conversion of defense expenditures into US dollars using the current exchange rate could produce misleading results

Burkina

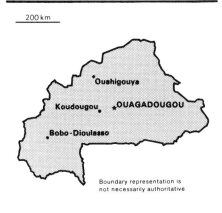

200 km

Boundary representation is not necessarily authoritative

Geography

Location: Western Africa, between Ghana and Mali
Map references: Africa, Standard Time Zones of the World
Area:
total area: 274,200 km²
land area: 273,800 km²
comparative area: slightly larger than Colorado
Land boundaries: total 3,192 km, Benin 306 km, Ghana 548 km, Cote d'Ivoire 584 km, Mali 1,000 km, Niger 628 km, Togo 126 km
Coastline: 0 km (landlocked)
Maritime claims: none; landlocked
International disputes: the disputed international boundary between Burkina and Mali was submitted to the International Court of Justice (ICJ) in October 1983 and the ICJ issued its final ruling in December 1986, which both sides agreed to accept; Burkina and Mali are proceeding with boundary demarcation, including the tripoint with Niger
Climate: tropical; warm, dry winters; hot, wet summers
Terrain: mostly flat to dissected, undulating plains; hills in west and southeast
Natural resources: manganese, limestone, marble; small deposits of gold, antimony, copper, nickel, bauxite, lead, phosphates, zinc, silver
Land use:
arable land: 10%
permanent crops: 0%
meadows and pastures: 37%
forest and woodland: 26%
other: 27%
Irrigated land: 160 km² (1989 est.)
Environment: recent droughts and desertification severely affecting marginal agricultural activities, population distribution, economy; overgrazing; deforestation
Note: landlocked

Burkina *(continued)*

People

Population: 9,852,529 (July 1993 est.)
Population growth rate: 2.83% (1993 est.)
Birth rate: 48.8 births/1,000 population (1993 est.)
Death rate: 18.19 deaths/1,000 population (1993 est.)
Net migration rate: -2.28 migrant(s)/1,000 population (1993 est.)
Infant mortality rate: 119.8 deaths/1,000 live births (1993 est.)
Life expectancy at birth:
total population: 47.47 years
male: 46.66 years
female: 48.3 years (1993 est.)
Total fertility rate: 7 children born/woman (1993 est.)
Nationality:
noun: Burkinabe (singular and plural)
adjective: Burkinabe
Ethnic divisions: Mossi (about 2.5 million), Gurunsi, Senufo, Lobi, Bobo, Mande, Fulani
Religions: indigenous beliefs 65%, Muslim 25%, Christian (mainly Roman Catholic) 10%
Languages: French (official), tribal languages belong to Sudanic family, spoken by 90% of the population
Literacy: age 15 and over can read and write (1990)
total population: 18%
male: 28%
female: 9%
Labor force: 3.3 million residents; 30,000 are wage earners
by occupation: agriculture 82%, industry 13%, commerce, services, and government 5%
note: 20% of male labor force migrates annually to neighboring countries for seasonal employment (1984); 44% of population of working age (1985)

Government

Names:
conventional long form: Burkina Faso
conventional short form: Burkina
former: Upper Volta
Digraph: UV
Type: parliamentary
Capital: Ouagadougou
Administrative divisions: 30 provinces; Bam, Bazega, Bougouriba, Boulgou, Boulkiemde, Ganzourgou, Gnagna, Gourma, Houet, Kadiogo, Kenedougou, Komoe, Kossi, Kouritenga, Mouhoun, Namentenga, Naouri, Oubritenga, Oudalan, Passore, Poni, Sanguie, Sanmatenga, Seno, Sissili, Soum, Sourou, Tapoa, Yatenga, Zoundweogo
Independence: 5 August 1960 (from France)
Constitution: June 1991

Legal system: based on French civil law system and customary law
National holiday: Anniversary of the Revolution, 4 August (1983)
Political parties and leaders: Organization for People's Democracy-Labor Movement (ODP-MT), ruling party, Marc Christian Roch KABORE; National Convention of Progressive Patriots-Social Democratic Party (CNPP-PSD), Pierre TAPSOBA; African Democratic Assembly (RDA), Gerard Kango OUEDRAOGO; Alliance for Democracy and Federation (ADF), Herman YAMEOGO
Other political or pressure groups: committees for the defense of the revolution; watchdog/political action groups throughout the country in both organizations and communities
Suffrage: none
Elections:
President: last held December 1991
Assembly of People's Deputies: last held 24 May 1992 (next to be held NA); results—percent of vote by party NA; seats—(107 total), ODP-MT 78, CNPP-PSD 12, RDA 6, ADF 4, other 7
Executive branch: president, Council of Ministers
Legislative branch: Assembly of People's Deputies
note: the current law also provides for a second consultative chamber, which had not been formally constituted as of 1 July 1992
Judicial branch: Appeals Court
Leaders:
Chief of State and Head of Government: President Captain Blaise COMPAORE (since 15 October 1987)
Member of: ACCT, ACP, AfDB, CCC, CEAO, ECA, ECOWAS, Entente, FAO, FZ, G-77, GATT, IBRD, ICAO, ICC, ICFTU, IDA, IDB, IFAD, IFC, ILO, IMF, INTELSAT, INTERPOL, IOC, ITU, LORCS, NAM, OAU, OIC, PCA, UN, UNCTAD, UNESCO, UNIDO, UPU, WADB, WCL, WFTU, WHO, WIPO, WMO, WTO
Diplomatic representation in US:
chief of mission: (vacant)
chancery: 2340 Massachusetts Avenue NW, Washington, DC 20008
telephone: (202) 332-5577 or 6895
US diplomatic representation:
chief of mission: Ambassador Edward P. BYRNN
embassy: Avenue Raoul Follerau, Ouagadougou
mailing address: 01 B. P. 35, Ouagadougou
telephone: [226] 30-67- 23 through 25
FAX: [226] 31-23-68
Flag: two equal horizontal bands of red (top) and green with a yellow five-pointed star in the center; uses the popular pan-African colors of Ethiopia

Economy

Overview: One of the poorest countries in the world, Burkina has a high population density, few natural resources, and relatively infertile soil. Economic development is hindered by a poor communications network within a landlocked country. Agriculture provides about 40% of GDP and is entirely of a subsistence nature. Industry, dominated by unprofitable government-controlled corporations, accounts for about 15% of GDP.
National product: GDP—exchange rate conversion—$3.3 billion (1991)
National product real growth rate: 1.3% (1990 est.)
National product per capita: $350 (1991)
Inflation rate (consumer prices): -1% (1990)
Unemployment rate: NA%
Budget: revenues $495 million; expenditures $786 million, including capital expenditures of $NA (1991)
Exports: $304.8 million (f.o.b., 1990)
commodities: cotton, gold, animal products
partners: EC 45%, Taiwan 15%, Cote d'Ivoire 15% (1987)
Imports: $593 million (f.o.b., 1990)
commodities: machinery, food products, petroleum
partners: EC 51%, Africa 25%, US 6% (1987)
External debt: $865 million (December 1991 est.)
Industrial production: growth rate 5.7% (1990 est.), accounts for about 23% of GDP (1989)
Electricity: 120,000 kW capacity; 320 million kWh produced, 40 kWh per capita (1991)
Industries: cotton lint, beverages, agricultural processing, soap, cigarettes, textiles, gold mining and extraction
Agriculture: accounts for about 30% of GDP; cash crops—peanuts, shea nuts, sesame, cotton; food crops—sorghum, millet, corn, rice; livestock; not self-sufficient in food grains
Economic aid: US commitments, including Ex-Im (FY70-89), $294 million; Western (non-US) countries, ODA and OOF bilateral commitments (1970-89), $2.9 billion; Communist countries (1970-89), $113 million
Currency: 1 CFA franc (CFAF) = 100 centimes
Exchange rates: CFA francs (CFAF) per US$1—274.06 (January 1993), 264.69 (1992), 282.11 (1991), 272.26 (1990), 319.01 (1989), 297.85 (1988)
Fiscal year: calendar year

Burma

Communications

Railroads: 620 km total; 520 km Ouagadougou to Cote d'Ivoire border and 100 km Ouagadougou to Kaya; all 1.00-meter gauge and single track
Highways: 16,500 km total; 1,300 km paved, 7,400 km improved, 7,800 km unimproved (1985)
Airports:
total: 48
usable: 38
with permanent-surface runways: 2
with runways over 3,659 m: 0
with runways 2,440-3,659 m: 2
with runways 1,220-2,439 m: 8
Telecommunications: all services only fair; microwave radio relay, wire, and radio communication stations in use; broadcast stations—2 AM, 1 FM, 2 TV; 1 Atlantic Ocean INTELSAT earth station

Defense Forces

Branches: Army, Air Force, National Gendarmerie, National Police, People's Militia
Manpower availability: males age 15-49 1,947,935; fit for military service 995,532 (1993 est.); no conscription
Defense expenditures: exchange rate conversion—$NA, NA% of GDP

Geography

Location: Southeast Asia, bordering the Bay of Bengal, between Bangladesh and Thailand
Map references: Asia, Southeast Asia, Standard Time Zones of the World
Area:
total area: 678,500 km²
land area: 657,740 km²
comparative area: slightly smaller than Texas
Land boundaries: total 5,876 km, Bangladesh 193 km, China 2,185 km, India 1,463 km, Laos 235 km, Thailand 1,800 km
Coastline: 1,930 km
Maritime claims:
contiguous zone: 24 nm
continental shelf: 200 nm or to the edge of continental margin
exclusive economic zone: 200 nm
territorial sea: 12 nm
International disputes: none
Climate: tropical monsoon; cloudy, rainy, hot, humid summers (southwest monsoon, June to September); less cloudy, scant rainfall, mild temperatures, lower humidity during winter (northeast monsoon, December to April)
Terrain: central lowlands ringed by steep, rugged highlands
Natural resources: petroleum, timber, tin, antimony, zinc, copper, tungsten, lead, coal, some marble, limestone, precious stones, natural gas
Land use:
arable land: 15%
permanent crops: 1%
meadows and pastures: 1%
forest and woodland: 49%
other: 34%
Irrigated land: 10,180 km² (1989)
Environment: subject to destructive earthquakes and cyclones; flooding and landslides common during rainy season (June to September); deforestation
Note: strategic location near major Indian Ocean shipping lanes

People

Population: 43,455,953 (July 1993 est.)
Population growth rate: 1.88% (1993 est.)
Birth rate: 28.88 births/1,000 population (1993 est.)
Death rate: 10.05 deaths/1,000 population (1993 est.)
Net migration rate: 0 migrant(s)/1,000 population (1993 est.)
Infant mortality rate: 65.7 deaths/1,000 live births (1993 est.)
Life expectancy at birth:
total population: 59.5 years
male: 57.5 years
female: 61.63 years (1993 est.)
Total fertility rate: 3.7 children born/woman (1993 est.)
Nationality:
noun: Burmese (singular and plural)
adjective: Burmese
Ethnic divisions: Burman 68%, Shan 9%, Karen 7%, Rakhine 4%, Chinese 3%, Mon 2%, Indian 2%, other 5%
Religions: Buddhist 89%, Christian 4% (Baptist 3%, Roman Catholic 1%), Muslim 4%, animist beliefs 1%, other 2%
Languages: Burmese; minority ethnic groups have their own languages
Literacy: age 15 and over can read and write (1990)
total population: 81%
male: 89%
female: 72%
Labor force: 16.007 million (1992)
by occupation: agriculture 65.2%, industry 14.3%, trade 10.1%, government 6.3%, other 4.1% (FY89 est.)

Government

Names:
conventional long form: Union of Burma
conventional short form: Burma
local long form: Pyidaungzu Myanma Naingngandaw (translated by the US Government as Union of Myanma and by the Burmese as Union of Myanmar)
local short form: Myanma Naingngandaw
former: Socialist Republic of the Union of Burma
Digraph: BM
Type: military regime
Capital: Rangoon (sometimes translated as Yangon)
Administrative divisions: 7 divisions* (yin-mya, singular—yin) and 7 states (pyine-mya, singular—pyine); Chin State, Irrawaddy*, Kachin State, Karan State, Kayah State, Magwe*, Mandalay*, Mon State, Pegu*, Rakhine State, Rangoon*, Sagaing*, Shan State, Tenasserim*
Independence: 4 January 1948 (from UK)
Constitution: 3 January 1974 (suspended since 18 September 1988); National

Burma (continued)

Convention started on 9 January 1993 to draft chapter headings for a new constitution
Legal system: has not accepted compulsory ICJ jurisdiction
National holiday: Independence Day, 4 January (1948)
Political parties and leaders: National Unity Party (NUP; proregime), THA KYAW; National League for Democracy (NLD), U AUNG SHWE; National Coalition of Union of Burma (NCGUB), SEIN WIN (which consists of individuals legitimately elected to parliament, but not recognized by military regime) fled to border area and joined with insurgents in December 1990 to form a parallel government
Other political or pressure groups: Kachin Independence Army (KIA); United Wa State Army (UWSA); Karen National Union (KNU—the only non-drug group); several Shan factions, including the Mong Tai Army (MTA)
Suffrage: 18 years of age; universal
Elections:
People's Assembly: last held 27 May 1990, but Assembly never convened; results—NLD 80%; seats—(485 total) NLD 396, the regime-favored NUP 10, other 79
Executive branch: chairman of the State Law and Order Restoration Council, State Law and Order Restoration Council
Legislative branch: unicameral People's Assembly (Pyithu Hluttaw) was dissolved after the coup of 18 September 1988
Judicial branch: none; Council of People's Justices was abolished after the coup of 18 September 1988
Leaders:
Chief of State and Head of Government: Chairman of the State Law and Order Restoration Council Gen. THAN SHWE (since 23 April 1992)
Member of: AsDB, CCC, CP, ESCAP, FAO, G-77, GATT, IAEA, IBRD, ICAO, IDA, IFAD, IFC, ILO, IMF, IMO, INTERPOL, IOC, ITU, LORCS, UN, UNCTAD, UNESCO, UNIDO, UPU, WHO, WMO
Diplomatic representation in US:
chief of mission: Ambassador U THAUNG
chancery: 2300 S Street NW, Washington, DC 20008
telephone: (202) 332-9044 through 9046
consulate general: New York
US diplomatic representation:
chief of mission: (vacant); Deputy Chief of Mission, Charge d'Affaires Franklin P. HUDDLE, Jr.
embassy: 581 Merchant Street, Rangoon
mailing address: GPO Box 521, AMEMB Box B, APO AP 96546
telephone: [95] (1) 82055, 82181
FAX: [95] (1) 80409
Flag: red with a blue rectangle in the upper hoist-side corner bearing, all in white, 14 five-pointed stars encircling a cogwheel containing a stalk of rice; the 14 stars represent the 14 administrative divisions

Economy

Overview: Burma is a poor Asian country, with a per capita GDP of about $660. The nation has been unable to achieve any substantial improvement in export earnings because of falling prices for many of its major commodity exports. For rice, traditionally the most important export, the drop in world prices has been accompanied by shrinking markets and a smaller volume of sales. In 1985 teak replaced rice as the largest export and continues to hold this position. The economy is heavily dependent on the agricultural sector, which generates about 40% of GDP and provides employment for 65% of the work force. Burma has been largely isolated from international economic forces and has been trying to encourage foreign investment, so far with little success.
National product: GDP—exchange rate conversion—$28 billion (1992)
National product real growth rate: 1.3% (1992)
National product per capita: $660 (1992)
Inflation rate (consumer prices): 50% (1992)
Unemployment rate: 9.6% (FY89 est.) in urban areas
Budget: revenues $8.1 billion; expenditures $11.6 billion, including capital expenditures of $NA (1992)
Exports: $535.1 million (FY92)
commodities: teak, rice, oilseed, metals, rubber, gems
partners: China, India, Thailand, Singapore
Imports: $907.0 million (FY92)
commodities: machinery, transport equipment, chemicals, food products
partners: Japan, China, Singapore
External debt: $4.0 billion (1992)
Industrial production: growth rate 2.6% (FY90 est.); accounts for 10% of GDP
Electricity: 1,100,000 kW capacity; 2,800 million kWh produced, 65 kWh per capita (1992)
Industries: agricultural processing; textiles and footwear; wood and wood products; petroleum refining; mining of copper, tin, tungsten, iron; construction materials; pharmaceuticals; fertilizer
Agriculture: accounts for 40% of GDP (including fish and forestry); self-sufficient in food; principal crops—paddy rice, corn, oilseed, sugarcane, pulses; world's largest stand of hardwood trees; rice and teak account for 55% of export revenues
Illicit drugs: world's largest illicit producer of opium poppy and minor producer of cannabis for the international drug trade; opium production has nearly doubled since the collapse of Rangoon's antinarcotic programs
Economic aid: US commitments, including Ex-Im (FY70-89), $158 million; Western (non-US) countries, ODA and OOF bilateral commitments (1970-89), $3.9 billion; Communist countries (1970-89), $424 million
Currency: 1 kyat (K) = 100 pyas
Exchange rates: kyats (K) per US$1—6.0963 (January 1992), 6.2837 (1991), 6.3386 (1990), 6.7049 (1989), 6.46 (1988), 6.6535 (1987); unofficial—105
Fiscal year: 1 April-31 March

Communications

Railroads: 3,991 km total, all government owned; 3,878 km 1.000-meter gauge, 113 km narrow-gauge industrial lines; 362 km double track
Highways: 27,000 km total; 3,200 km bituminous, 17,700 km improved earth or gravel, 6,100 km unimproved earth
Inland waterways: 12,800 km; 3,200 km navigable by large commercial vessels
Pipelines: crude oil 1,343 km; natural gas 330 km
Ports: Rangoon, Moulmein, Bassein
Merchant marine: 62 ships (1,000 GRT or over) totaling 940,264 GRT/1,315,156 DWT; includes 3 passenger-cargo, 18 cargo, 5 refrigerated cargo, 4 vehicle carrier, 2 container, 2 oil tanker, 3 chemical, 1 combination ore/oil, 23 bulk, 1 combination bulk
Airports:
total: 83
usable: 78
with permanent-surface runways: 26
with runways over 3,659 m: 0
with runways 2,440-3,659 m: 3
with runways 1,220-2,439 m: 38
Telecommunications: meets minimum requirements for local and intercity service for business and government; international service is good; 53,000 telephones (1986); radiobroadcast coverage is limited to the most populous areas; broadcast stations—2 AM, 1 FM, 1 TV (1985); 1 Indian Ocean INTELSAT earth station

Defense Forces

Branches: Army, Navy, Air Force
Manpower availability: males age 15-49 11,004,419; females age 15-49 10,945,899; males fit for military service 5,894,514; females fit for military service 5,847,958; males reach military age (18) annually 435,030; females reach military age (18) annually 420,487 (1993 est.); both sexes are liable for military service
Defense expenditures: exchange rate conversion—$NA, NA% of GDP (1992)

Burundi

Geography

Location: Central Africa, between Tanzania and Zaire
Map references: Africa, Standard Time Zones of the World
Area:
total area: 27,830 km²
land area: 25,650 km²
comparative area: slightly larger than Maryland
Land boundaries: total 974 km, Rwanda 290 km, Tanzania 451 km, Zaire 233 km
Coastline: 0 km (landlocked)
Maritime claims: none; landlocked
International disputes: none
Climate: temperate; warm; occasional frost in uplands
Terrain: mostly rolling to hilly highland; some plains
Natural resources: nickel, uranium, rare earth oxide, peat, cobalt, copper, platinum (not yet exploited), vanadium
Land use:
arable land: 43%
permanent crops: 8%
meadows and pastures: 35%
forest and woodland: 2%
other: 12%
Irrigated land: 720 km² (1989 est.)
Environment: soil exhaustion; soil erosion; deforestation
Note: landlocked; straddles crest of the Nile-Congo watershed

People

Population: 5,985,308 (July 1993 est.)
Population growth rate: 2.34% (1993 est.)
Birth rate: 44.69 births/1,000 population (1993 est.)
Death rate: 21.25 deaths/1,000 population (1993 est.)
Net migration rate: 0 migrant(s)/1,000 population (1993 est.)
Infant mortality rate: 115.6 deaths/1,000 live births (1993 est.)

Life expectancy at birth:
total population: 40.75 years
male: 38.79 years
female: 42.76 years (1993 est.)
Total fertility rate: 6.76 children born/woman (1993 est.)
Nationality:
noun: Burundian(s)
adjective: Burundi
Ethnic divisions:
Africans: Hutu (Bantu) 85%, Tutsi (Hamitic) 14%, Twa (Pygmy) 1% (other Africans include about 70,000 refugees, mostly Rwandans and Zairians)
non-Africans: Europeans 3,000, South Asians 2,000
Religions: Christian 67% (Roman Catholic 62%, Protestant 5%), indigenous beliefs 32%, Muslim 1%
Languages: Kirundi (official), French (official), Swahili (along Lake Tanganyika and in the Bujumbura area)
Literacy: age 15 and over can read and write (1990)
total population: 50%
male: 61%
female: 40%
Labor force: 1.9 million (1983 est.)
by occupation: agriculture 93.0%, government 4.0%, industry and commerce 1.5%, services 1.5%
note: 52% of population of working age (1985)

Government

Names:
conventional long form: Republic of Burundi
conventional short form: Burundi
local long form: Republika y'u Burundi
local short form: Burundi
Digraph: BY
Type: republic
Capital: Bujumbura
Administrative divisions: 15 provinces; Bubanza, Bujumbura, Bururi, Cankuzo, Cibitoke, Gitega, Karuzi, Kayanza, Kirundo, Makamba, Muramvya, Muyinga, Ngozi, Rutana, Ruyigi
Independence: 1 July 1962 (from UN trusteeship under Belgian administration)
Constitution: 13 March 1992 draft provides for establishment of plural political system
Legal system: based on German and Belgian civil codes and customary law; has not accepted compulsory ICJ jurisdiction
National holiday: Independence Day, 1 July (1962)
Political parties and leaders: only party—National Party of Unity and Progress (UPRONA), Nicolas MAYUGI, secretary general;
note: although Burundi is still officially a one-party state, at least four political parties were formed in 1991 and set the precedent

for constitutional reform in 1992—Burundi Democratic Front (FRODEBU), Organization of the People of Burundi (RPB), Socialist Party of Burundi (PSB), Royalist Parliamentary Party (PRP)—the most significant opposition party is FRODEBU, led by Melchior NDADAYE; the Party for the Liberation of the Hutu People (PALIPEHUTU), formed in exile in the early 1980s, is an ethnically based political party dedicated to majority rule; the government has long accused PALIPEHUTU of practicing divesive ethnic politics and fomenting violence against the state; PALIPEHUTU's exclusivist charter makes it an unlikely candidate for legalization under the new constitution that will require party membership open to all ethnic groups
Suffrage: universal adult at age NA
Elections:
National Assembly: note—The National Unity Charter outlining the principles for constitutional government was adopted by a national referendum on 5 February 1991; new elections to the National Assembly are to take place 29 June 1993; presidential elections are to take place 1 June 1993
Executive branch: president; chairman of the Central Committee of the National Party of Unity and Progress (UPRONA), prime minister
Legislative branch: unicameral National Assembly (Assemblee Nationale) was dissolved following the coup of 3 September 1987; at an extraordinary party congress held from 27 to 29 December 1990, the Central Committee of the National Party of Unity and Progress (UPRONA) replaced the Military Committee for National Salvation, and became the supreme governing body during the transition to constitutional government
Judicial branch: Supreme Court (Cour Supreme)
Leaders:
Chief of State: President Major Pierre BUYOYA (since 9 September 1987)
Head of Government: Prime Minister Adrien SIBOMANA (since 26 October 1988)
Member of: ACCT, ACP, AfDB, CCC, CEEAC, CEPGL, ECA, FAO, G-77, GATT, IBRD, ICAO, IDA, IFAD, IFC, ILO, IMF, INTERPOL, ITU, LORCS, NAM, OAU, UN, UNCTAD, UNESCO, UNIDO, UPU, WHO, WIPO, WMO, WTO
Diplomatic representation in US:
chief of mission: Ambassador Julien KAVAKURE
chancery: Suite 212, 2233 Wisconsin Avenue NW, Washington, DC 20007
telephone: (202) 342-2574
US diplomatic representation:
chief of mission: Ambassador Cynthia Shepherd PERRY
embassy: Avenue des Etats-Unis, Bujumbura
mailing address: B. P. 1720, Bujumbura

Burundi *(continued)*

telephone: [257] (223) 454
FAX: [257] (222) 926
Flag: divided by a white diagonal cross into red panels (top and bottom) and green panels (hoist side and outer side) with a white disk superimposed at the center bearing three red six-pointed stars outlined in green arranged in a triangular design (one star above, two stars below)

Economy

Overview: A landlocked, resource-poor country in an early stage of economic development, Burundi is predominately agricultural with only a few basic industries. Its economic health depends on the coffee crop, which accounts for an average 90% of foreign exchange earnings each year. The ability to pay for imports therefore continues to rest largely on the vagaries of the climate and the international coffee market. As part of its economic reform agenda, launched in February 1991 with IMF and World Bank support, Burundi is trying to diversify its agricultural exports and attract foreign investment in industry. Several state-owned coffee companies were privatized via public auction in September 1991.
National product: GDP—exchange rate conversion—$1.23 billion (1991 est.)
National product real growth rate: 5% (1991 est.)
National product per capita: $205 (1991 est.)
Inflation rate (consumer prices): 9% (1991 est.)
Unemployment rate: NA%
Budget: revenues $318 million; expenditures $326 million, including capital expenditures of $150 million (1991 est.)
Exports: $91.7 million (f.o.b., 1991)
commodities: coffee 81%, tea, hides, and skins
partners: EC 83%, US 5%, Asia 2%
Imports: $246 million (c.i.f., 1991)
commodities: capital goods 31%, petroleum products 15%, foodstuffs, consumer goods
partners: EC 57%, Asia 23%, US 3%
External debt: $1.0 billion (1990 est.)
Industrial production: real growth rate 11.0% (1991 est.); accounts for about 5% of GDP
Electricity: 55,000 kW capacity; 105 million kWh produced, 20 kWh per capita (1991)
Industries: light consumer goods such as blankets, shoes, soap; assembly of imports; public works construction; food processing
Agriculture: accounts for 60% of GDP; 90% of population dependent on subsistence farming; marginally self-sufficient in food production; cash crops—coffee, cotton, tea; food crops—corn, sorghum, sweet potatoes, bananas, manioc; livestock—meat, milk, hides and skins

Economic aid: US commitments, including Ex-Im (FY70-89), $71 million; Western (non-US) countries, ODA and OOF bilateral commitments (1970-89), $10.2 billion; OPEC bilateral aid (1979-89), $32 million; Communist countries (1970-89), $175 million
Currency: 1 Burundi franc (FBu) = 100 centimes
Exchange rates: Burundi francs (FBu) per US$1—235.75 (January 1993), 208.30 (1992), 181.51 (1991), 171.26 (1990), 158.67 (1989), 140.40 (1988)
Fiscal year: calendar year

Communications

Highways: 5,900 km total; 400 km paved, 2,500 km gravel or laterite, 3,000 km improved or unimproved earth
Inland waterways: Lake Tanganyika
Ports: Bujumbura (lake port) connects to transportation systems of Tanzania and Zaire
Airports:
total: 5
usable: 4
with permanent-surface runways: 1
with runways over 3,659 m: 0
with runways 2,440-3,659 m: 1
with runways 1,220-2,439 m: 4
Telecommunications: sparse system of wire, radiocommunications, and low-capacity microwave radio relay links; 8,000 telephones; broadcast stations—2 AM, 2 FM, 1 TV; 1 Indian Ocean INTELSAT earth station

Defense Forces

Branches: Army (includes naval and air units), paramilitary Gendarmerie
Manpower availability: males age 15-49 1,283,308; fit for military service 670,381; reach military age (16) annually 62,700 (1993 est.)
Defense expenditures: exchange rate conversion—$28 million, 3.7% of GDP (1989)

Cambodia

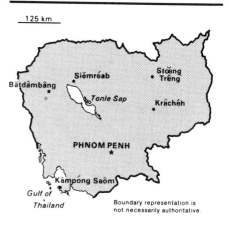

Boundary representation is not necessarily authoritative.

Geography

Location: Southeast Asia, bordering the Gulf of Thailand, between Thailand and Vietnam
Map references: Asia, Southeast Asia, Standard Time Zones of the World
Area:
total area: 181,040 km²
land area: 176,520 km²
comparative area: slightly smaller than Oklahoma
Land boundaries: total 2,572 km, Laos 541 km, Thailand 803 km, Vietnam 1,228 km
Coastline: 443 km
Maritime claims:
contiguous zone: 24 nm
continental shelf: 200 nm
exclusive economic zone: 200 nm
territorial sea: 12 nm
International disputes: offshore islands and three sections of the boundary with Vietnam are in dispute; maritime boundary with Vietnam not defined
Climate: tropical; rainy, monsoon season (May to October); dry season (December to March); little seasonal temperature variation
Terrain: mostly low, flat plains; mountains in southwest and north
Natural resources: timber, gemstones, some iron ore, manganese, phosphates, hydropower potential
Land use:
arable land: 16%
permanent crops: 1%
meadows and pastures: 3%
forest and woodland: 76%
other: 4%
Irrigated land: 920 km² (1989 est.)
Environment: a land of paddies and forests dominated by Mekong River and Tonle Sap
Note: buffer between Thailand and Vietnam

People

Population: 9,898,900 (July 1993 est.)
Population growth rate: 4.41% (1993 est.)

Birth rate: 45.52 births/1,000 population (1993 est.)
Death rate: 16.57 deaths/1,000 population (1993 est.)
Net migration rate: 15.15 migrant(s)/1,000 population (1993 est.)
Infant mortality rate: 111.5 deaths/1,000 live births (1993 est.)
Life expectancy at birth:
total population: 49.06 years
male: 47.6 years
female: 50.6 years (1993 est.)
Total fertility rate: 5.81 children born/woman (1993 est.)
Nationality:
noun: Cambodian(s)
adjective: Cambodian
Ethnic divisions: Khmer 90%, Vietnamese 5%, Chinese 1%, other 4%
Religions: Theravada Buddhism 95%, other 5%
Languages: Khmer (official), French
Literacy: age 15 and over can read and write (1990)
total population: 35%
male: 48%
female: 22%
Labor force: 2,500,000 to 3,000,000
by occupation: agriculture 80% (1988 est.)

Government

Names:
conventional long form: none
conventional short form: Cambodia
Digraph: CB
Type: transitional government currently administered by the Supreme National Council (SNC), a body set up under United Nations' auspices, in preparation for an internationally supervised election in 1993 and including representatives from each of the country's four political factions
Capital: Phnom Penh
Administrative divisions: 20 provinces (khet, singular and plural); Banteay Meanchey, Batdambang, Kampong Cham, Kampong Chhnang, Kampong Spoe, Kampong Thum, Kampot, Kandal, Kaoh Kong, Kracheh, Mondol Kiri, Phnum Penh, Pouthisat, Preah Vihear, Prey Veng, Rotanokiri, Siemreab-Otdar Meanchey, Stoeng Treng, Svay Rieng, Takev
Independence: 9 November 1949 (from France)
Constitution: a new constitution will be drafted after the national election in 1993
Legal system: NA
National holiday:
NGC: Independence Day, 17 April (1975)
SOC: Liberation Day, 7 January (1979)
Political parties and leaders: Democratic Kampuchea (DK, also known as the Khmer Rouge) under KHIEU SAMPHAN; Cambodian Pracheachon Party or

Cambodian People's Party (CPP) under CHEA SIM; Khmer People's National Liberation Front (KPNLF) under SON SANN; National United Front for an Independent, Neutral, Peaceful, and Cooperative Cambodia (FUNCINPEC) under Prince NORODOM RANARIDDH; Liberal Democratic Party (LDP) under SAK SUTSAKHAN
Suffrage: 18 years of age; universal
Elections: UN-supervised election for a 120-member constituent assembly based on proportional representation within each province is scheduled for 23-27 May 1993; the assembly will draft and approve a constitution and then transform itself into a legislature that will create a new Cambodian Government
Executive branch: a 12-member Supreme National Council (SNC), chaired by Prince NORODOM SIHANOUK, composed of representatives from each of the four political factions; faction names and delegation leaders are: State of Cambodia (SOC)—HUN SEN; Democratic Kampuchea (DK or Khmer Rouge)—KHIEU SAMPHAN; Khmer People's National Liberation Front (KPNLF)—SON SANN; National United Front for an Independent, Peaceful, Neutral, and Cooperative Cambodia (FUNCINPEC)—Prince NORODOM RANARIDDH
Legislative branch: pending a national election in 1993, the incumbent SOC faction's unicameral National Assembly is the only functioning national legislative body
Judicial branch: Supreme People's Court pending a national election in 1993, the incumbent SOC faction's Supreme People's Court is the only functioning national judicial body
Leaders:
Chief of State: SNC—Chairman Prince NORODOM SIHANOUK, under UN supervision
Head of Government: NGC—vacant, but will be determined following the national election in 1993; SOC—Chairman of the Council of Ministers HUN SEN (since 14 January 1985)
Member of: AsDB, CP, ESCAP, FAO, G-77, IAEA, IBRD, ICAO, IDA, IFAD, ILO, IMF, IMO, INTERPOL, ITU, LORCS, NAM, PCA, UN, UNCTAD, UNESCO, UPU, WFTU, WHO, WMO, WTO
Diplomatic representation in US: the Supreme National Council (SNC) represents Cambodia in international organizations
US diplomatic representation:
US representative: Charles TWINNING
mission: 27 EO Street 240, Phnom Penh
mailing address: Box P, APO AP 96546
telephone: (855) 23-26436 or (855) 23-26438

FAX: (855) 23-26437
Flag: SNC—blue background with white map of Cambodia in middle; SOC—two equal horizontal bands of red (top) and blue with a gold stylized five-towered temple representing Angkor Wat in the center

Economy

Overview: Cambodia remains a desperately poor country whose economic recovery is held hostage to continued political unrest and factional hostilities. The country's immediate economic challenge is an acute financial crisis that is undermining monetary stability and preventing disbursement of foreign development assistance. Cambodia is still recovering from an abrupt shift in 1990 to free-market economic mechanisms and a cutoff in aid from former Soviet bloc countries; these changes have severely impacted on public sector revenues and performance. The country's infrastructure of roads, bridges, and power plants has been severely degraded, now having only 40-50% of prewar capacity. The economy remains essentially rural, with 90% of the population living in the countryside and dependent mainly on subsistence agriculture. Statistical data on the economy continue to be sparse and unreliable.
National product: GDP—exchange rate conversion—$2 billion (1991 est.)
National product real growth rate: NA%
National product per capita: $280 (1991 est.)
Inflation rate (consumer prices): 250-300% (1992 est.)
Unemployment rate: NA%
Budget: revenues $120 million; expenditures $NA, including capital expenditures of $NA (1992 est.)
Exports: $59 million (f.o.b., 1990 est.)
commodities: natural rubber, rice, pepper, wood
partners: Vietnam, USSR, Eastern Europe, Japan, India
Imports: $170 million (c.i.f., 1990 est.)
commodities: international food aid; fuels, consumer goods, machinery
partners: Vietnam, USSR, Eastern Europe, Japan, India
External debt: $717 million (1990)
Industrial production: growth rate NA%
Electricity: 35,000 kW capacity; 70 million kWh produced, 9 kWh per capita (1990)
Industries: rice milling, fishing, wood and wood products, rubber, cement, gem mining
Agriculture: mainly subsistence farming except for rubber plantations; main crops—rice, rubber, corn; food shortages—rice, meat, vegetables, dairy products, sugar, flour
Economic aid: US commitments, including Ex-Im (FY70-89), $725 million; Western

Cambodia *(continued)*

(non-US countries) (1970-89), $300 million;
Communist countries (1970-89), $1.8 billion
Currency: 1 riel (CR) = 100 sen
Exchange rates: riels (CR) per
US$1—2,800 (September 1992), 500
(December 1991), 560 (1990), 159.00
(1988), 100.00 (1987)
Fiscal year: calendar year

Communications

Railroads: 612 km 1.000-meter gauge,
government owned
Highways: 13,351 km total; 2,622 km
bituminous; 7,105 km crushed stone, gravel,
or improved earth; 3,624 km unimproved
earth; some roads in disrepair
Inland waterways: 3,700 km navigable all
year to craft drawing 0.6 meters; 282 km
navigable to craft drawing 1.8 meters
Ports: Kampong Saom, Phnom Penh
Airports:
total: 15
usable: 9
with permanent-surface runways: 5
with runways over 3,659 m: 0
with runways 2,440-3,659 m: 2
with runways 1,220-2,439 m: 4
Telecommunications: service barely
adequate for government requirements and
virtually nonexistent for general public;
international service limited to Vietnam and
other adjacent countries; broadcast
stations—1 AM, no FM, 1 TV

Defense Forces

Branches:
SOC: Cambodian People's Armed Forces
(CPAF)
Communist resistance forces: National Army
of Democratic Kampuchea (Khmer Rouge)
non-Communist resistance forces: Armee
National Kampuchea Independent (ANKI)
which is sometimes anglicized as National
Army of Independent Cambodia (NAIC),
Khmer People's National Liberation Armed
Forces (KPNLAF)
Manpower availability: males age 15-49
1,883,679; fit for military service 1,033,168;
reach military age (18) annually 74,585
(1993 est.)
Defense expenditures: exchange rate
conversion—$NA, NA% of GDP

Cameroon

300 km

Lake Chad
Maroua
Garoua
Ngaoundéré
Bafoussam
Nkongsamba
Douala
YAOUNDÉ
Kribi
Gulf of Guinea

Geography

Location: Western Africa, bordering the
North Atlantic Ocean between Equatorial
Guinea and Nigeria
Map references: Africa, Standard Time
Zones of the World
Area:
total area: 475,440 km^2
land area: 469,440 km^2
comparative area: slightly larger than
California
Land boundaries: total 4,591 km, Central
African Republic 797 km, Chad 1,094 km,
Congo 523 km, Equatorial Guinea 189 km,
Gabon 298 km, Nigeria 1,690 km
Coastline: 402 km
Maritime claims:
territorial sea: 50 nm
International disputes: demarcation of
international boundaries in Lake Chad, the
lack of which has led to border incidents in
the past, is completed and awaiting
ratification by Cameroon, Chad, Niger, and
Nigeria; boundary commission, created with
Nigeria to discuss unresolved land and
maritime boundaries, has not yet convened
Climate: varies with terrain from tropical
along coast to semiarid and hot in north
Terrain: diverse, with coastal plain in
southwest, dissected plateau in center,
mountains in west, plains in north
Natural resources: petroleum, bauxite, iron
ore, timber, hydropower potential
Land use:
arable land: 13%
permanent crops: 2%
meadows and pastures: 18%
forest and woodland: 54%
other: 13%
Irrigated land: 280 km^2 (1989 est.)
Environment: recent volcanic activity with
release of poisonous gases; deforestation;
overgrazing; desertification
Note: sometimes referred to as the hinge of
Africa

People

Population: 12,755,873 (July 1993 est.)
Population growth rate: 2.9% (1993 est.)
Birth rate: 40.66 births/1,000 population
(1993 est.)
Death rate: 11.63 deaths/1,000 population
(1993 est.)
Net migration rate: 0 migrant(s)/1,000
population (1993 est.)
Infant mortality rate: 78.8 deaths/1,000
live births (1993 est.)
Life expectancy at birth:
total population: 56.66 years
male: 54.65 years
female: 58.74 years (1993 est.)
Total fertility rate: 5.88 children
born/woman (1993 est.)
Nationality:
noun: Cameroonian(s)
adjective: Cameroonian
Ethnic divisions: Cameroon Highlanders
31%, Equatorial Bantu 19%, Kirdi 11%,
Fulani 10%, Northwestern Bantu 8%,
Eastern Nigritic 7%, other African 13%,
non-African less than 1%
Religions: indigenous beliefs 51%, Christian
33%, Muslim 16%
Languages: 24 major African language
groups, English (official), French (official)
Literacy: age 15 and over can read and
write (1990)
total population: 54%
male: 66%
female: 43%
Labor force: NA
by occupation: agriculture 74.4%, industry
and transport 11.4%, other services 14.2%
(1983)
note: 50% of population of working age
(15-64 years) (1985)

Government

Names:
conventional long form: Republic of
Cameroon
conventional short form: Cameroon
former: French Cameroon
Digraph: CM
Type: unitary republic; multiparty
presidential regime (opposition parties
legalized 1990)
Capital: Yaounde
Administrative divisions: 10 provinces;
Adamaoua, Centre, Est, Extreme-Nord,
Littoral, Nord, Nord-Ouest, Ouest, Sud,
Sud-Ouest
Independence: 1 January 1960 (from UN
trusteeship under French administration)
Constitution: 20 May 1972
Legal system: based on French civil law
system, with common law influence; has not
accepted compulsory ICJ jurisdiction
National holiday: National Day, 20 May
(1972)

Political parties and leaders: Cameroon People's Democratic Movement (CPDM), Paul BIYA, president, is government-controlled and was formerly the only party, but opposition parties were legalized in 1990
major opposition parties: National Union for Democracy and Progress (UNDP)
major oppositon parties: Social Democratic Front (SDF)
major opposition parties: Cameroonian Democratic Union (UDC); Union of Cameroonian Populations (UPC)
Other political or pressure groups: NA
Suffrage: 20 years of age; universal
Elections:
National Assembly: last held 1 March 1992 (next scheduled for March 1997); results—(180 seats) CPDM 88, UNDP 68, UPC 18, MDR 6
President: last held 11 October 1992; results—President Paul BIYA reelected with about 40% of the vote amid widespread allegations of fraud; SDF candidate John FRU NDI got 36% of the vote; UNDP candidate Bello Bouba MAIGARI got 19% of the vote
Executive branch: president, Cabinet
Legislative branch: unicameral National Assembly (Assemblee Nationale)
Judicial branch: Supreme Court
Leaders:
Chief of State: President Paul BIYA (since 6 November 1982)
Head of Government: Prime Minister Simon ACHIDI ACHU (since 9 April 1992)
Member of: ACCT (associate), ACP, AfDB, BDEAC, CCC, CEEAC, ECA, FAO, FZ, G-19, G-77, GATT, IAEA, IBRD, ICAO, ICC, IDA, IDB, IFAD, IFC, ILO, IMF, IMO, INMARSAT, INTELSAT, INTERPOL, IOC, ITU, LORCS, NAM, OAU, OIC, PCA, UDEAC, UN, UNCTAD, UNESCO, UNIDO, UNTAC, UPU, WCL, WHO, WIPO, WMO, WTO
Diplomatic representation in US:
chief of mission: Ambassador Paul PONDI
chancery: 2349 Massachusetts Avenue NW, Washington, DC 20008
telephone: (202) 265-8790 through 8794
US diplomatic representation:
chief of mission: Ambassador Harriet ISOM
embassy: Rue Nachtigal, Yaounde
mailing address: B. P. 817, Yaounde
telephone: [237] 234-014
FAX: [237] 230-753
consulate: Douala
Flag: three equal vertical bands of green (hoist side), red, and yellow with a yellow five-pointed star centered in the red band; uses the popular pan-African colors of Ethiopia

Economy

Overview: Because of its offshore oil resources, Cameroon has one of the highest incomes per capita in tropical Africa. Still, it faces many of the serious problems facing other underdeveloped countries, such as political instability, a top-heavy civil service, and a generally unfavorable climate for business enterprise. The development of the oil sector led rapid economic growth between 1970 and 1985. Growth came to an abrupt halt in 1986 precipitated by steep declines in the prices of major exports: coffee, cocoa, and petroleum. Export earnings were cut by almost one-third, and inefficiencies in fiscal management were exposed. In 1990-92, with support from the IMF and World Bank, the government has begun to introduce reforms designed to spur business investment, increase efficiency in agriculture, and recapitalize the nation's banks. Nationwide strikes organized by opposition parties in 1991, however, undermined these efforts.
National product: GDP—exchange rate conversion—$11.5 billion (1990 est.)
National product real growth rate: 3% (1990 est.)
National product per capita: $1,040 (1990 est.)
Inflation rate (consumer prices): 3% (1990 est.)
Unemployment rate: 25% (1990 est.)
Budget: revenues $1.7 billion; expenditures $2.4 billion, including capital expenditures of $422 million (FY90 est.)
Exports: $1.8 billion (f.o.b., 1991)
commodities: petroleum products 51%, coffee, beans, cocoa, aluminum products, timber
partners: EC (particularly France) about 50%, US, African countries
Imports: $1.2 billion (c.i.f., 1991)
commodities: machines and electrical equipment, food, consumer goods, transport equipment
partners: EC about 60%, France 41%, Germany 9%, African countries, Japan, US 4%
External debt: $6 billion (1991)
Industrial production: growth rate 6.4% (FY87); accounts for 30% of GDP
Electricity: 755,000 kW capacity; 2,190 million kWh produced, 190 kWh per capita (1991)
Industries: petroleum production and refining, food processing, light consumer goods, textiles, sawmills
Agriculture: the agriculture and forestry sectors provide employment for the majority of the population, contributing nearly 25% to GDP and providing a high degree of self-sufficiency in staple foods; commercial and food crops include coffee, cocoa, timber, cotton, rubber, bananas, oilseed, grains, livestock, root starches
Economic aid: US commitments, including Ex-Im (FY70-90), $479 million; Western (non-US) countries, ODA and OOF bilateral commitments (1970-90), $4.75 billion; OPEC bilateral aid (1979-89), $29 million; Communist countries (1970-89), $125 million
Currency: 1 CFA franc (CFAF) = 100 centimes
Exchange rates: Communaute Financiere Africaine francs (CFAF) per US$1—274.06 (January 1993), 264.69 (1992), 282.11 (1991), 272.26 (1990), 319.01 (1989), 297.85 (1988)
Fiscal year: 1 July-30 June

Communications

Railroads: 1,003 km total; 858 km 1.000-meter gauge, 145 km 0.600-meter gauge
Highways: about 65,000 km total; includes 2,682 km paved, 32,318 km gravel and improved earth, and 30,000 km of unimproved earth
Inland waterways: 2,090 km; of decreasing importance
Ports: Douala
Merchant marine: 2 cargo ships (1,000 GRT or over) totaling 24,122 GRT/33,509 DWT
Airports:
total: 59
usable: 51
with permanent-surface runways: 11
with runways over 3,659 m: 0
with runways 2,440-3,659 m: 6
with runways 1,220-2,439 m: 51
Telecommunications: good system of open wire, cable, troposcatter, and microwave radio relay; 26,000 telephones, 2 telephones per 1,000 persons, available only to business and government; broadcast stations—11 AM, 11 FM, 1 TV; 2 Atlantic Ocean INTELSAT earth stations

Defense Forces

Branches: Army, Navy (including Naval Infantry), Air Force, National Gendarmerie, Presidential Guard
Manpower availability: males age 15-49 2,844,280; fit for military service 1,432,563; reach military age (18) annually 125,453 (1993 est.)
Defense expenditures: exchange rate conversion—$219 million, less than 2% of GDP (1990 est.)

Canada

Geography

Location: Northern North America, bordering the North Atlantic Ocean and North Pacific Ocean north of the US
Map references: Arctic Region, North America, Standard Time Zones of the World
Area:
total area: 9,976,140 km²
land area: 9,220,970 km²
comparative area: slightly larger than US
Land boundaries: total 8,893 km, US 8,893 km (includes 2,477 km with Alaska)
Coastline: 243,791 km
Maritime claims:
continental shelf: 200 m depth or to depth of exploitation
exclusive fishing zone: 200 nm
territorial sea: 12 nm
International disputes: maritime boundary disputes with the US; Saint Pierre and Miquelon is focus of maritime boundary dispute between Canada and France
Climate: varies from temperate in south to subarctic and arctic in north
Terrain: mostly plains with mountains in west and lowlands in southeast
Natural resources: nickel, zinc, copper, gold, lead, molybdenum, potash, silver, fish, timber, wildlife, coal, petroleum, natural gas
Land use:
arable land: 5%
permanent crops: 0%
meadows and pastures: 3%
forest and woodland: 35%
other: 57%
Irrigated land: 8,400 km² (1989 est.)
Environment: 80% of population concentrated within 160 km of US border; continuous permafrost in north a serious obstacle to development
Note: second-largest country in world (after Russia); strategic location between Russia and US via north polar route

People

Population: 27,769,993 (July 1993 est.)
Population growth rate: 1.28% (1993 est.)
Birth rate: 14.48 births/1,000 population (1993 est.)
Death rate: 7.35 deaths/1,000 population (1993 est.)
Net migration rate: 5.68 migrant(s)/1,000 population (1993 est.)
Infant mortality rate: 7 deaths/1,000 live births (1993 est.)
Life expectancy at birth:
total population: 77.98 years
male: 74.54 years
female: 81.6 years (1993 est.)
Total fertility rate: 1.84 children born/woman (1993 est.)
Nationality:
noun: Canadian(s)
adjective: Canadian
Ethnic divisions: British Isles origin 40%, French origin 27%, other European 20%, indigenous Indian and Eskimo 1.5%
Religions: Roman Catholic 46%, United Church 16%, Anglican 10%, other 28%
Languages: English (official), French (official)
Literacy: age 15 and over can read and write (1981)
total population: 99%
male: NA%
female: NA%
Labor force: 13.38 million
by occupation: services 75%, manufacturing 14%, agriculture 4%, construction 3%, other 4% (1988)

Government

Names:
conventional long form: none
conventional short form: Canada
Digraph: CA
Type: confederation with parliamentary democracy
Capital: Ottawa
Administrative divisions: 10 provinces and 2 territories*; Alberta, British Columbia, Manitoba, New Brunswick, Newfoundland, Northwest Territories*, Nova Scotia, Ontario, Prince Edward Island, Quebec, Saskatchewan, Yukon Territory*
Independence: 1 July 1867 (from UK)
Constitution: amended British North America Act 1867 patriated to Canada 17 April 1982; charter of rights and unwritten customs
Legal system: based on English common law, except in Quebec, where civil law system based on French law prevails; accepts compulsory ICJ jurisdiction, with reservations
National holiday: Canada Day, 1 July (1867)

Political parties and leaders: Progressive Conservative Party, Brian MULRONEY; Liberal Party, Jean CHRETIEN; New Democratic Party, Audrey McLAUGHLIN; Reform Party, Preston MANNING; Bloc Quebecois, Lucien BOUCHARD
Suffrage: 18 years of age; universal
Elections:
House of Commons: last held 21 November 1988 (next to be held by November 1993); results—Progressive Conservative Party 43%, Liberal Party 32%, New Democratic Party 20%, other 5%; seats—(295 total) Progressive Conservative Party 159, Liberal Party 80, New Democratic Party 44, Bloc Quebecois 9, independents 3
Executive branch: British monarch, governor general, prime minister, deputy prime minister, Cabinet
Legislative branch: bicameral Parliament (Parlement) consists of an upper house or Senate (Senat) and a lower house or House of Commons (Chambre des Communes)
Judicial branch: Supreme Court
Leaders:
Chief of State: Queen ELIZABETH II (since 6 February 1952), represented by Governor General Raymond John HNATYSHYN (since 29 January 1990)
Head of Government: Prime Minister Kim CAMBELL was chosen to replace Brian MULRONEY on 13 June 1993
Member of: ACCT, AfDB, AG (observer), APEC, AsDB, Australia Group, BIS, C, CCC, CDB (non-regional), COCOM, CP, CSCE, EBRD, ECE, ECLAC, ESA (cooperating state), FAO, G-7, G-8, G-10, GATT, IADB, IAEA, IBRD, ICAO, ICC, ICFTU, IDA, IEA, IFAD, IFC, ILO, IMF, IMO, INMARSAT, INTELSAT, INTERPOL, IOC, IOM, ISO, ITU, LORCS, MINURSO, MTCR, NACC, NAM (guest), NATO, NEA, NSG, OAS, OECD, ONUSAL, PCA, UN, UNAVEM II, UNCTAD, UNDOF, UNESCO, UNFICYP, UNHCR, UNIDO, UNIKOM, UNOMOZ, UNOSOM, UNPROFOR, UNTAC, UNTSO, UPU, WCL, WHO, WMO, WIPO, WTO, ZC
Diplomatic representation in US:
chief of mission: Ambassador John DE CHASTELAIN
chancery: 501 Pennsylvania Avenue NW, Washington, DC 20001
telephone: (202) 682-1740
FAX: (202) 682-7726
consulates general: Atlanta, Boston, Buffalo, Chicago, Cleveland, Dallas, Detroit, Los Angeles, Minneapolis, New York, Philadelphia, San Francisco, and Seattle
US diplomatic representation:
chief of mission: Ambassador-designate Governor James J. BLANCHARD
embassy: 100 Wellington Street, K1P 5T1, Ottawa
mailing address: P. O. Box 5000, Ogdensburg, NY 13669-0430

telephone: (613) 238-5335 or (613) 238-4470
FAX: (613) 238-5720
consulates general: Calgary, Halifax, Montreal, Quebec, Toronto, and Vancouver
Flag: three vertical bands of red (hoist side), white (double width, square), and red with a red maple leaf centered in the white band

Economy

Overview: As an affluent, high-tech industrial society, Canada today closely resembles the US in per capita output, market-oriented economic system, and pattern of production. Since World War II the impressive growth of the manufacturing, mining, and service sectors has transformed the nation from a largely rural economy into one primarily industrial and urban. In the 1980s, Canada registered one of the highest rates of real growth among the OECD nations, averaging about 3.2%. With its great natural resources, skilled labor force, and modern capital plant, Canada has excellent economic prospects. However, the continuing constitutional impasse between English- and French-speaking areas has observers discussing a possible split in the confederation; foregn investors have become edgy.
National product: GDP—purchasing power equivalent—$537.1 billion (1992)
National product real growth rate: 0.9% (1992)
National product per capita: $19,600 (1992)
Inflation rate (consumer prices): 1.5% (1992)
Unemployment rate: 11.5% (December 1992)
Budget: revenues $111.8 billion; expenditures $138.3 billion, including capital expenditures of $NA (FY90 est.)
Exports: $124.0 billion (f.o.b., 1991)
commodities: newsprint, wood pulp, timber, crude petroleum, machinery, natural gas, aluminum, motor vehicles and parts; telecommunications equipment
partners: US, Japan, UK, Germany, South Korea, Netherlands, China
Imports: $118 billion (c.i.f., 1991)
commodities: crude oil, chemicals, motor vehicles and parts, durable consumer goods, electronic computers; telecommunications equipment and parts
partners: US, Japan, UK, Germany, France, Mexico, Taiwan, South Korea
External debt: $247 billion (1987)
Industrial production: growth rate 1% (1992); accounts for 34% of GDP
Electricity: 109,340,000 kW capacity; 493,000 million kWh produced, 17,900 kWh per capita (1992)

Industries: processed and unprocessed minerals, food products, wood and paper products, transportation equipment, chemicals, fish products, petroleum and natural gas
Agriculture: accounts for about 3% of GDP; one of the world's major producers and exporters of grain (wheat and barley); key source of US agricultural imports; large forest resources cover 35% of total land area; commercial fisheries provide annual catch of 1.5 million metric tons, of which 75% is exported
Illicit drugs: illicit producer of cannabis for the domestic drug market; use of hydroponics technology permits growers to plant large quantities of high-quality marijuana indoors; growing role as a transit point for heroin and cocaine entering the US market
Economic aid: donor—ODA and OOF commitments (1970-89), $7.2 billion
Currency: 1 Canadian dollar (Can$) = 100 cents
Exchange rates: Canadian dollars (Can$) per US$1—1.2776 (January 1993), 1.2087 (1992), 1.1457 (1991), 1.1668 (1990), 1.1840 (1989), 1.2307 (1988)
Fiscal year: 1 April-31 March

Communications

Railroads: 146,444 km total; two major transcontinental freight railway systems—Canadian National (government owned) and Canadian Pacific Railway; passenger service—VIA (government operated); 158 km is electrified
Highways: 884,272 km total; 712,936 km surfaced (250,023 km paved), 171,336 km earth
Inland waterways: 3,000 km, including Saint Lawrence Seaway
Pipelines: crude and refined oil 23,564 km; natural gas 74,980 km
Ports: Halifax, Montreal, Quebec, Saint John (New Brunswick), Saint John's (Newfoundland), Toronto, Vancouver
Merchant marine: 63 ships (1,000 GRT or over) totaling 454,582 GRT/646,329 DWT; includes 1 passenger, 3 short-sea passenger, 2 passenger-cargo, 8 cargo, 2 railcar carrier, 1 refrigerated cargo, 7 roll-on/roll-off, 1 container, 24 oil tanker, 4 chemical tanker, 1 specialized tanker, 9 bulk; note—does not include ships used exclusively in the Great Lakes
Airports:
total: 1,420
useable: 1,142
with permanent-surface runways: 457
with runways over 3,659 m: 4
with runways 2,440-3,659 m: 30
with runways 1,220-2,439 m: 330

Telecommunications: excellent service provided by modern media; 18.0 million telephones; broadcast stations—900 AM, 29 FM, 53 (1,400 repeaters) TV; 5 coaxial submarine cables; over 300 earth stations operating in INTELSAT (including 4 Atlantic Ocean and 1 Pacific Ocean) and domestic systems

Defense Forces

Branches: Canadian Armed Forces (including Land Forces Command, Maritime Command, Air Command, Communications Command, Training Command), Royal Canadian Mounted Police (RCMP)
Manpower availability: males age 15-49 7,444,767; fit for military service 6,440,927; reach military age (17) annually 191,884 (1993 est.)
Defense expenditures: exchange rate conversion—$11.3 billion, 2% of GDP (FY92/93)

Cape Verde

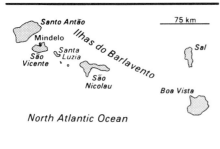

Geography

Location: in the southeastern North Atlantic Ocean, 500 km west of Senegal in Western Africa
Map references: Africa, Standard Time Zones of the World
Area:
total area: 4,030 km²
land area: 4,030 km²
comparative area: slightly larger than Rhode Island
Land boundaries: 0 km
Coastline: 965 km
Maritime claims: measured from claimed archipelagic baselines
exclusive economic zone: 200 nm
territorial sea: 12 nm
International disputes: none
Climate: temperate; warm, dry, summer; precipitation very erratic
Terrain: steep, rugged, rocky, volcanic
Natural resources: salt, basalt rock, pozzolana, limestone, kaolin, fish
Land use:
arable land: 9%
permanent crops: 0%
meadows and pastures: 6%
forest and woodland: 0%
other: 85%
Irrigated land: 20 km² (1989 est.)
Environment: subject to prolonged droughts; harmattan wind can obscure visibility; volcanically and seismically active; deforestation; overgrazing
Note: strategic location 500 km from west coast of Africa near major north-south sea routes; important communications station; important sea and air refueling site

People

Population: 410,535 (July 1993 est.)
Population growth rate: 3.03% (1993 est.)
Birth rate: 47.02 births/1,000 population (1993 est.)
Death rate: 9.43 deaths/1,000 population (1993 est.)

Net migration rate: -7.31 migrant(s)/1,000 population (1993 est.)
Infant mortality rate: 59.6 deaths/1,000 live births (1993 est.)
Life expectancy at birth:
total population: 62.18 years
male: 60.3 years
female: 64.15 years (1993 est.)
Total fertility rate: 6.41 children born/woman (1993 est.)
Nationality:
noun: Cape Verdean(s)
adjective: Cape Verdean
Ethnic divisions: Creole (mulatto) 71%, African 28%, European 1%
Religions: Roman Catholicism fused with indigenous beliefs
Languages: Portuguese, Crioulo, a blend of Portuguese and West African words
Literacy: age 15 and over can read and write (1989)
total population: 66%
male: NA
female: NA
Labor force: 102,000 (1985 est.)
by occupation: agriculture (mostly subsistence) 57%, services 29%, industry 14% (1981)
note: 51% of population of working age (1985)

Government

Names:
conventional long form: Republic of Cape Verde
conventional short form: Cape Verde
local long form: Republica de Cabo Verde
local short form: Cabo Verde
Digraph: CV
Type: republic
Capital: Praia
Administrative divisions: 14 districts (concelhos, singular—concelho); Boa Vista, Brava, Fogo, Maio, Paul, Praia, Porto Novo, Ribeira Grande, Sal, Santa Catarina, Santa Cruz, Sao Nicolau, Sao Vicente, Tarrafal
Independence: 5 July 1975 (from Portugal)
Constitution: 7 September 1980; amended 12 February 1981, December 1988, and 28 September 1990 (legalized opposition parties)
Legal system: NA
National holiday: Independence Day, 5 July (1975)
Political parties and leaders: Movement for Democracy (MPD), Prime Minister Carlos VEIGA, founder and chairman; African Party for Independence of Cape Verde (PAICV), Pedro Verona Rodrigues PIRES, chairman
Suffrage: 18 years of age; universal
Elections:
People's National Assembly: last held 13 January 1991 (next to be held January

1996); results—percent of vote by party NA; seats—(79 total) MPD 56, PAICV 23; note—this multiparty Assembly election ended 15 years of single-party rule
President: last held 17 February 1991 (next to be held February 1996); results—Antonio Monteiro MASCARENHAS (MPD) received 72.6% of vote
Executive branch: president, prime minister, deputy minister, secretaries of state, Council of Ministers (cabinet)
Legislative branch: unicameral People's National Assembly (Assembleia Nacional Popular)
Judicial branch: Supreme Tribunal of Justice (Supremo Tribunal de Justia)
Leaders:
Chief of State: President Antonio Monteiro MASCARENHAS (since 22 March 1991)
Head of Government: Prime Minister Carlos Alberto Wahnon de Carvalho VEIGA (since 13 January 1991)
Member of: ACP, AfDB, ECA, ECOWAS, FAO, G-77, IBRD, ICAO, IDA, IFAD, IFC, ILO, IMF, IMO, INTELSAT, INTERPOL, IOM (observer), ITU, LORCS, NAM, OAU, UN (Cape Verde assumed a nonpermanent seat on the Security Council on 1 January 1992), UNCTAD, UNESCO, UNIDO, UNOMOZ, UPU, WCL, WHO, WMO
Diplomatic representation in US:
chief of mission: Ambassador Carlos Alberto Santos SILVA
chancery: 3415 Massachusetts Avenue NW, Washington, DC 20007
telephone: (202) 965-6820
consulate general: Boston
US diplomatic representation:
chief of mission: Ambassador Joseph SEGARS
embassy: Rua Hoji Ya Henda 81, Praia
mailing address: C. P. 201, Praia
telephone: [238] 61-56-16 or 61-56-17
FAX: [238] 61-13-55
Flag: a new flag of unknown description reportedly has been adopted; previous flag consisted of two equal horizontal bands of yellow (top) and green with a vertical red band on the hoist side; in the upper portion of the red band is a black five-pointed star framed by two corn stalks and a yellow clam shell; uses the popular pan-African colors of Ethiopia; similar to the flag of Guinea-Bissau, which is longer and has an unadorned black star centered in the red band

Economy

Overview: Cape Verde's low per capita GDP reflects a poor natural resource base, a serious, long-term drought, and a high birthrate. The economy is service oriented, with commerce, transport, and public services accounting for 60% of GDP.

Although nearly 70% of the population lives in rural areas, agriculture's share of GDP is only 16%; the fishing sector accounts for 4%. About 90% of food must be imported. The fishing potential, mostly lobster and tuna, is not fully exploited. In 1988 fishing represented only 3.5% of GDP. Cape Verde annually runs a high trade deficit, financed by remittances from emigrants and foreign aid. Economic reforms launched by the new democratic government in February 1991 are aimed at developing the private sector and attracting foreign investment to diversify the economy.

National product: GDP—exchange rate conversion—$310 million (1990 est.)
National product real growth rate: 4% (1990 est.)
National product per capita: $800 (1990 est.)
Inflation rate (consumer prices): 8.7% (1991 est.)
Unemployment rate: 25% (1988)
Budget: revenues $104 million; expenditures $133 million, including capital expenditures of $72 million (1991 est.)
Exports: $5.7 million (f.o.b., 1990 est.)
commodities: fish, bananas, hides and skins
partners: Portugal 40%, Algeria 31%, Angola, Netherlands (1990 est.)
Imports: $120 million (c.i.f., 1990 est.)
commodities: foodstuffs, consumer goods, industrial products, transport equipment
partners: Sweden 33%, Spain 11%, Germany 5%, Portugal 3%, France 3%, Netherlands, US (1990 est.)
External debt: $156 million (1991)
Industrial production: growth rate 18% (1988 est.); accounts for 4% of GDP
Electricity: 15,000 kW capacity; 15 million kWh produced, 40 kWh per capita (1991)
Industries: fish processing, salt mining, clothing factories, ship repair, construction materials, food and beverage production
Agriculture: accounts for 20% of GDP (including fishing); largely subsistence farming; bananas are the only export crop; other crops—corn, beans, sweet potatoes, coffee; growth potential of agricultural sector limited by poor soils and scanty rainfall; annual food imports required; fish catch provides for both domestic consumption and small exports
Economic aid: US commitments, including Ex-Im (FY75-90), $93 million; Western (non-US) countries, ODA and OOF bilateral commitments (1970-90), $586 million; OPEC bilateral aid (1979-89), $12 million; Communist countries (1970-89), $36 million
Currency: 1 Cape Verdean escudo (CVEsc) = 100 centavos
Exchange rates: Cape Verdean escudos (CVEsc) per US$1—75.47 (January 1993), 73.10 (1992), 71.41 (1991), 64.10 (November 1990), 74.86 (December 1989), 72.01 (1988)
Fiscal year: calendar year

Communications

Ports: Mindelo, Praia
Merchant marine: 7 cargo ships (1,000 GRT or over) totaling 11,717 GRT/19,000 DWT
Airports:
total: 6
usable: 6
with permanent-surface runways: 6
with runways over 3,659 m: 0
with runways 2,440-3,659 m: 1
with runways 1,220-2,439 m: 2
Telecommunications: interisland microwave radio relay system, high-frequency radio to Senegal and Guinea-Bissau; over 1,700 telephones; broadcast stations—1 AM, 6 FM, 1 TV; 2 coaxial submarine cables; 1 Atlantic Ocean INTELSAT earth station

Defense Forces

Branches: People's Revolutionary Armed Forces (FARP) (including Army and Navy), Security Service
Manpower availability: males age 15-49 75,431; fit for military service 44,358 (1993 est.)
Defense expenditures: exchange rate conversion—$NA, NA% of GDP

Cayman Islands
(dependent territory of the UK)

Geography

Location: in the northwestern Caribbean Sea, nearly halfway between Cuba and Honduras
Map references: Central America and the Caribbean
Area:
total area: 260 km²
land area: 260 km²
comparative area: slightly less than 1.5 times the size of Washington, DC
Land boundaries: 0 km
Coastline: 160 km
Maritime claims:
exclusive fishing zone: 200 nm
territorial sea: 3 nm
International disputes: none
Climate: tropical marine; warm, rainy summers (May to October) and cool, relatively dry winters (November to April)
Terrain: low-lying limestone base surrounded by coral reefs
Natural resources: fish, climate and beaches that foster tourism
Land use:
arable land: 0%
permanent crops: 0%
meadows and pastures: 8%
forest and woodland: 23%
other: 69%
Irrigated land: NA km²
Environment: within the Caribbean hurricane belt
Note: important location between Cuba and Central America

People

Population: 30,440 (July 1993 est.)
Population growth rate: 4.35% (1993 est.)
Birth rate: 15.32 births/1,000 population (1993 est.)
Death rate: 4.98 deaths/1,000 population (1993 est.)
Net migration rate: 33.2 migrant(s)/1,000 population (1993 est.)

Cayman Islands (continued)

Infant mortality rate: 8.4 deaths/1,000 live births (1993 est.)

Life expectancy at birth:
total population: 77.1 years
male: 75.37 years
female: 78.81 years (1993 est.)

Total fertility rate: 1.48 children born/woman (1993 est.)

Nationality:
noun: Caymanian(s)
adjective: Caymanian

Ethnic divisions: mixed 40%, white 20%, black 20%, expatriates of various ethnic groups 20%

Religions: United Church (Presbyterian and Congregational), Anglican, Baptist, Roman Catholic, Church of God, other Protestant denominations

Languages: English

Literacy: age 15 and over having ever attended school (1970)
total population: 98%
male: 98%
female: 98%

Labor force: 8,061
by occupation: service workers 18.7%, clerical 18.6%, construction 12.5%, finance and investment 6.7%, directors and business managers 5.9% (1979)

Government

Names:
conventional long form: none
conventional short form: Cayman Islands

Digraph: CJ

Type: dependent territory of the UK

Capital: George Town

Administrative divisions: 8 districts; Creek, Eastern, Midland, South Town, Spot Bay, Stake Bay, West End, Western

Independence: none (dependent territory of the UK)

Constitution: 1959, revised 1972

Legal system: British common law and local statutes

National holiday: Constitution Day (first Monday in July)

Political parties and leaders: no formal political parties

Suffrage: 18 years of age; universal

Elections:
Legislative Assembly: last held November 1992 (next to be held November 1996); results—percent of vote by party NA; seats—(15 total, 12 elected)

Executive branch: British monarch, governor, Executive Council (cabinet)

Legislative branch: unicameral Legislative Assembly

Judicial branch: Grand Court, Cayman Islands Court of Appeal

Leaders:
Chief of State: Queen ELIZABETH II (since 6 February 1952)

Head of Government: Governor and President of the Executive Council Michael GORE (since NA May 1992)

Member of: CARICOM (observer), CDB, INTERPOL (subbureau), IOC

Diplomatic representation in US: as a dependent territory of the UK, Caymanian interests in the US are represented by the UK

Flag: blue, with the flag of the UK in the upper hoist-side quadrant and the Caymanian coat of arms on a white disk centered on the outer half of the flag; the coat of arms includes a pineapple and turtle above a shield with three stars (representing the three islands) and a scroll at the bottom bearing the motto HE HATH FOUNDED IT UPON THE SEAS

Economy

Overview: The economy depends heavily on tourism (70% of GDP and 75% of export earnings) and offshore financial services, with the tourist industry aimed at the luxury market and catering mainly to visitors from North America. About 90% of the islands' food and consumer goods needs must be imported. The Caymanians enjoy one of the highest standards of living in the region.

National product: GDP—exchange rate conversion—$670 million (1991 est.)

National product real growth rate: 4.4% (1991)

National product per capita: $23,000 (1991 est.)

Inflation rate (consumer prices): 8% (1990 est.)

Unemployment rate: 7% (1992)

Budget: revenues $141.5 million; expenditures $160.7 million, including capital expenditures of $NA (1991)

Exports: $1.5 million (f.o.b., 1987 est.)
commodities: turtle products, manufactured consumer goods
partners: mostly US

Imports: $136 million (c.i.f., 1987 est.)
commodities: foodstuffs, manufactured goods
partners: US, Trinidad and Tobago, UK, Netherlands Antilles, Japan

External debt: $15 million (1986)

Industrial production: growth rate NA%

Electricity: 74,000 kW capacity; 256 million kWh produced, 8,780 kWh per capita (1992)

Industries: tourism, banking, insurance and finance, construction, building materials, furniture making

Agriculture: minor production of vegetables, fruit, livestock; turtle farming

Economic aid: US commitments, including Ex-Im (FY70-89), $26.7 million; Western (non-US) countries, ODA and OOF bilateral commitments (1970-89), $35 million

Currency: 1 Caymanian dollar (CI$) = 100 cents

Exchange rates: Caymanian dollars (CI$) per US$1—1.20 (fixed rate)

Fiscal year: 1 April-31 March

Communications

Highways: 160 km of main roads

Ports: George Town, Cayman Brac

Merchant marine: 29 ships (1,000 GRT or over) totaling 307,738 GRT/468,659 DWT; includes 1 passenger-cargo, 8 cargo, 8 roll-on/roll-off cargo, 3 oil tanker, 2 chemical tanker, 1 liquefied gas carrier, 4 bulk, 2 combination bulk; note—a flag of convenience registry

Airports:
total: 3
usable: 3
with permanent-surface runways: 2
with runways over 3,659 m: 0
with runways 2,440-3,659 m: 0
with runways 1,220-2,439 m: 2

Telecommunications: 35,000 telephones; telephone system uses 1 submarine coaxial cable and 1 Atlantic Ocean INTELSAT earth station to link islands and access international services; broadcast stations—2 AM, 1 FM, no TV

Defense Forces

Branches: Royal Cayman Islands Police Force (RCIPF)

Note: defense is the responsibility of the UK

Central African Republic

Geography

Location: Central Africa, between Chad and Zaire
Map references: Africa, Standard Time Zones of the World
Area:
total area: 622,980 km²
land area: 622,980 km²
comparative area: slightly smaller than Texas
Land boundaries: total 5,203 km, Cameroon 797 km, Chad 1,197 km, Congo 467 km, Sudan 1,165 km, Zaire 1,577 km
Coastline: 0 km (landlocked)
Maritime claims: none; landlocked
International disputes: none
Climate: tropical; hot, dry winters; mild to hot, wet summers
Terrain: vast, flat to rolling, monotonous plateau; scattered hills in northeast and southwest
Natural resources: diamonds, uranium, timber, gold, oil
Land use:
arable land: 3%
permanent crops: 0%
meadows and pastures: 5%
forest and woodland: 64%
other: 28%
Irrigated land: NA km²
Environment: hot, dry, dusty harmattan winds affect northern areas; poaching has diminished reputation as one of last great wildlife refuges; desertification
Note: landlocked; almost the precise center of Africa

People

Population: 3,073,979 (July 1993 est.)
Population growth rate: 2.23% (1993 est.)
Birth rate: 42.77 births/1,000 population (1993 est.)
Death rate: 20.49 deaths/1,000 population (1993 est.)
Net migration rate: 0 migrant(s)/1,000 population (1993 est.)
Infant mortality rate: 138.7 deaths/1,000 live births (1993 est.)
Life expectancy at birth:
total population: 42.94 years
male: 41.46 years
female: 44.45 years (1993 est.)
Total fertility rate: 5.47 children born/woman (1993 est.)
Nationality:
noun: Central African(s)
adjective: Central African
Ethnic divisions: Baya 34%, Banda 27%, Sara 10%, Mandjia 21%, Mboum 4%, M'Baka 4%, Europeans 6,500 (including 3,600 French)
Religions: indigenous beliefs 24%, Protestant 25%, Roman Catholic 25%, Muslim 15%, other 11%
note: animistic beliefs and practices strongly influence the Christian majority
Languages: French (official), Sangho (lingua franca and national language), Arabic, Hunsa, Swahili
Literacy: age 15 and over can read and write (1990)
total population: 27%
male: 33%
female: 15%
Labor force: 775,413 (1986 est.)
by occupation: agriculture 85%, commerce and services 9%, industry 3%, government 3%
note: about 64,000 salaried workers; 55% of population of working age (1985)

Government

Names:
conventional long form: Central African Republic
conventional short form: none
local long form: Republique Centrafricaine
local short form: none
former: Central African Empire
Abbreviation: CAR
Digraph: CT
Type: republic; one-party presidential regime since 1986
Capital: Bangui
Administrative divisions: 14 prefectures (prefectures, singular—prefecture), 2 economic prefectures* (prefectures economiques, singular—prefecture economique), and 1 commune**; Bamingui-Bangoran, Bangui** Basse-Kotto, Gribingui*, Haute-Kotto, Haute-Sangha, Haut-Mbomou, Kemo-Gribingui, Lobaye, Mbomou, Nana-Mambere, Ombella-Mpoko, Ouaka, Ouham, Ouham-Pende, Sangha*, Vakaga
Independence: 13 August 1960 (from France)
Constitution: 21 November 1986
Legal system: based on French law
National holiday: National Day, 1 December (1958) (proclamation of the republic)
Political parties and leaders: Central African Democratic Party (RDC), the government party, Laurent GOMINA-PAMPALI; Council of Moderates Coalition includes; Union of the People for Economic and Social Development (UPDS), Katossy SIMANI; Liberal Republican Party (PARELI), Augustin M'BOE; Central African Socialist Movement (MSCA), Michel BENGUE; Concerted Democratic Forces (CFD), a coalition of 13 parties, including; Alliance for Democracy and Progress (ADP), Francois PEHOUA; Central African Republican party (PRC), Ruth ROLLAND; Social Democratic Party (PSD), Enoch DERANT-LAKOUE; Civic Forum (FC), Gen. Timothee MALENDOMA; Liberal Democratic Party (PLD), Nestor KOMBOT-NAGUEMON
Suffrage: 21 years of age; universal
Elections:
President: last held 25 October 1992; widespread irregularities at some polls led to dismissal of results by Supreme Court; elections are rescheduled for 17 October 1993
National Assembly: last held 25 October 1992; widespread irregularities at some polls led to dismissal of results by Supreme Court; elections are rescheduled for 17 October 1993
Executive branch: president, prime minister, Council of Ministers (cabinet)
Legislative branch: unicameral National Assembly (Assemblee Nationale) advised by the Economic and Regional Council (Conseil Economique et Regional); when they sit together this is known as the Congress (Congres)
Judicial branch: Supreme Court (Cour Supreme)
Leaders:
Chief of State: President Andre-Dieudonne KOLINGBA (since 1 September 1981)
Head of Government: Prime Minister Enoch DERANT-LAKOUE (since 2 March 1993)
Member of: ACCT, ACP, AfDB, BDEAC, CCC, CEEAC, ECA, FAO, FZ, G-77, GATT, IBRD, ICAO, ICFTU, IDA, IFAD, IFC, ILO, IMF, INTELSAT, INTERPOL, IOC, ITU, LORCS, NAM, OAU, UDEAC, UN, UNCTAD, UNESCO, UNIDO, UPU, WCL, WHO, WIPO, WMO
Diplomatic representation in US:
chief of mission: Ambassador Jean-Pierre SOHAHONG-KOMBET
chancery: 1618 22nd Street NW, Washington, DC 20008
telephone: (202) 483-7800 or 7801
US diplomatic representation:
chief of mission: Ambassador Robert E. GRIBBIN

Central African Republic
(continued)

embassy: Avenue David Dacko, Bangui
mailing address: B. P. 924, Bangui
telephone: [236] 61-02-00, 61-25-78, 61-43-33, 61-02-10
FAX: [236] 61-44-94
Flag: four equal horizontal bands of blue (top), white, green, and yellow with a vertical red band in center; there is a yellow five-pointed star on the hoist side of the blue band

---

Economy

Overview: Subsistence agriculture, including forestry, is the backbone of the CAR economy, with more than 70% of the population living in the countryside. In 1988 the agricultural sector generated about 40% of GDP. Agricultural products accounted for about 60% of export earnings and the diamond industry for 30%. Important constraints to economic development include the CAR's landlocked position, a poor transportation system, and a weak human resource base. Multilateral and bilateral development assistance, particularly from France, plays a major role in providing capital for new investment.
National product: GDP—exchange rate conversion—$1.3 billion (1990 est.)
National product real growth rate: -3% (1990 est.)
National product per capita: $440 (1990 est.)
Inflation rate (consumer prices): -3% (1990 est.)
Unemployment rate: 30% (1988 est.) in Bangui
Budget: revenues $175 million; expenditures $312 million, including capital expenditures of $122 million (1991 est.)
Exports: $138 million (1991 est.)
commodities: diamonds, cotton, coffee, timber, tobacco
partners: France, Belgium, Italy, Japan, US
Imports: $205 million (1991 est.)
commodities: food, textiles, petroleum products, machinery, electrical equipment, motor vehicles, chemicals, pharmaceuticals, consumer goods, industrial products
partners: France, other EC countries, Japan, Algeria
External debt: $859 million (1991)
Industrial production: growth rate 4% (1990 est.); accounts for 14% of GDP
Electricity: 40,000 kW capacity; 95 million kWh produced, 30 kWh per capita (1991)
Industries: diamond mining, sawmills, breweries, textiles, footwear, assembly of bicycles and motorcycles
Agriculture: accounts for 40% of GDP; self-sufficient in food production except for grain; commercial crops—cotton, coffee, tobacco, timber; food crops—manioc, yams, millet, corn, bananas

Economic aid: US commitments, including Ex-Im (FY70-90), $52 million; Western (non-US) countries, ODA and OOF bilateral commitments (1970-90), $1.6 billion; OPEC bilateral aid (1979-89), $6 million; Communist countries (1970-89), $38 million
Currency: 1 CFA franc (CFAF) = 100 centimes
Exchange rates: Communaute Financiere Africaine francs (CFAF) per US$1—274.06 (January 1993), 264.69 (1992), 282.11 (1991), 272.26 (1990), 319.01 (1989), 297.85 (1988)
Fiscal year: calendar year

Communications

Highways: 22,000 km total; 458 km bituminous, 10,542 km improved earth, 11,000 unimproved earth
Inland waterways: 800 km; traditional trade carried on by means of shallow-draft dugouts; Oubangui is the most important river
Airports:
total: 66
usable: 51
with permanent-surface runways: 3
with runways over 3,659 m: 0
with runways 2,440-3,659 m: 2
with runways 1,220-2,439 m: 20
Telecommunications: fair system; network relies primarily on radio relay links, with low-capacity, low-powered radiocommunication also used; broadcast stations—1 AM, 1 FM, 1 TV; 1 Atlantic Ocean INTELSAT earth station

---

Defense Forces

Branches: Central African Army (including Republican Guard), Air Force, National Gendarmerie, Police Force
Manpower availability: males age 15-49 685,575; fit for military service 358,836 (1993 est.)
Defense expenditures: exchange rate conversion—$23 million, 1.8% of GDP (1989 est.)

Chad

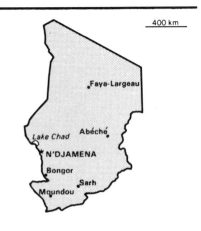

400 km

Geography

Location: Central Africa, between the Central African Republic and Libya
Map references: Africa, Standard Time Zones of the World
Area:
total area: 1.284 million km²
land area: 1,259,200 km²
comparative area: slightly more than three times the size of California
Land boundaries: total 5,968 km, Cameroon 1,094 km, Central African Republic 1,197 km, Libya 1,055 km, Niger 1,175 km, Nigeria 87 km, Sudan 1,360 km
Coastline: 0 km (landlocked)
Maritime claims: none; landlocked
International disputes: Libya claims and occupies the 100,000 km² Aozou Strip in the far north; demarcation of international boundaries in Lake Chad, the lack of which has led to border incidents in the past, is completed and awaiting ratification by Cameroon, Chad, Niger, and Nigeria
Climate: tropical in south, desert in north
Terrain: broad, arid plains in center, desert in north, mountains in northwest, lowlands in south
Natural resources: petroleum (unexploited but exploration under way), uranium, natron, kaolin, fish (Lake Chad)
Land use:
arable land: 2%
permanent crops: 0%
meadows and pastures: 36%
forest and woodland: 11%
other: 51%
Irrigated land: 100 km² (1989 est.)
Environment: hot, dry, dusty harmattan winds occur in north; drought and desertification adversely affecting south; subject to plagues of locusts
Note: landlocked; Lake Chad is the most significant water body in the Sahel

People

Population: 5,350,971 (July 1993 est.)
Population growth rate: 2.13% (1993 est.)
Birth rate: 42.21 births/1,000 population (1993 est.)
Death rate: 20.93 deaths/1,000 population (1993 est.)
Net migration rate: 0 migrant(s)/1,000 population (1993 est.)
Infant mortality rate: 134 deaths/1,000 live births (1993 est.)
Life expectancy at birth:
total population: 40.41 years
male: 39.36 years
female: 41.5 years (1993 est.)
Total fertility rate: 5.33 children born/woman (1993 est.)
Nationality:
noun: Chadian(s)
adjective: Chadian
Ethnic divisions:
north and center: Muslims (Arabs, Toubou, Hadjerai, Fulbe, Kotoko, Kanembou, Baguirmi, Boulala, Zaghawa, and Maba)
south: non-Muslims (Sara, Ngambaye, Mbaye, Goulaye, Moundang, Moussei, Massa) nonindigenous 150,000, of whom 1,000 are French
Religions: Muslim 44%, Christian 33%, indigenous beliefs, animism 23%
Languages: French (official), Arabic (official), Sara (in south), Sango (in south), more than 100 different languages and dialects are spoken
Literacy: age 15 and over can read and write French or Arabic (1990)
total population: 30%
male: 42%
female: 18%
Labor force: NA
by occupation: agriculture 85% (engaged in unpaid subsistence farming, herding, and fishing)

Government

Names:
conventional long form: Republic of Chad
conventional short form: Chad
local long form: Republique du Tchad
local short form: Tchad
Digraph: CD
Type: republic
Capital: N'Djamena
Administrative divisions: 14 prefectures (prefectures, singular—prefecture); Batha, Biltine, Borkou-Ennedi-Tibesti, Chari-Baguirmi, Guera, Kanem, Lac, Logone Occidental, Logone Oriental, Mayo-Kebbi, Moyen-Chari, Ouaddai, Salamat, Tandjile
Independence: 11 August 1960 (from France)
Constitution: 22 December 1989, suspended 3 December 1990; Provisional National

Charter 1 March 1991; national conference drafting new constitution to submit to referendum January 1993
Legal system: based on French civil law system and Chadian customary law; has not accepted compulsory ICJ jurisdiction
National holiday: 11 August
Political parties and leaders: Patriotic Salvation Movement (MPS; former dissident group), Idriss DEBY, chairman
note: President DEBY has promised political pluralism, a new constitution, and free elections by September 1993; numerous dissident groups; 26 opposition political parties
Other political or pressure groups: NA
Suffrage: universal at age NA
Elections:
National Consultative Council: last held 8 July 1990; disbanded 3 December 1990
President: last held 10 December 1989 (next to be held NA); results—President Hissein HABRE was elected without opposition; note—the government of then President HABRE fell on 1 December 1990, and Idriss DEBY seized power on 3 December 1990; national conference opened 15 January 1993; election to follow by end of year
Executive branch: president, Council of State (cabinet)
Legislative branch: unicameral National Consultative Council (Conseil National Consultatif) was disbanded 3 December 1990 and replaced by the Provisional Council of the Republic, with 30 members appointed by President DEBY on 8 March 1991
Judicial branch: Court of Appeal
Leaders:
Chief of State: Col. Idriss DEBY (since 4 December 1990)
Head of Government: Prime Minister Joseph YODOYMAN (since NA August 1992)
Member of: ACCT, ACP, AfDB, BDEAC, CEEAC, ECA, FAO, FZ, G-77, GATT, IBRD, ICAO, ICFTU, IDA, IDB, IFAD, ILO, IMF, INTELSAT, INTERPOL, IOC, ITU, LORCS, NAM, OAU, OIC, UDEAC, UN, UNCTAD, UNESCO, UNIDO, UPU, WCL, WHO, WIPO, WMO, WTO
Diplomatic representation in US:
chief of mission: Ambassador Kombaria Loumaye MEKONYO
chancery: 2002 R Street NW, Washington, DC 20009
telephone: (202) 462-4009
US diplomatic representation:
chief of mission: Ambassador Richard W. BOGOSIAN
embassy: Avenue Felix Eboue, N'Djamena
mailing address: B. P. 413, N'Djamena
telephone: [235] (51) 62-18, 40-09, or 51-62-11
FAX: [235] 51-33-72

Flag: three equal vertical bands of blue (hoist side), yellow, and red; similar to the flag of Romania; also similar to the flag of Andorra, which has a national coat of arms featuring a quartered shield centered in the yellow band; design was based on the flag of France

Economy

Overview: The climate, geographic location, and lack of infrastructure and natural resources make Chad one of the most underdeveloped countries in the world. Its economy is burdened by the ravages of civil war, conflict with Libya, drought, and food shortages. In 1986 real GDP returned to its 1977 level, with cotton, the major cash crop, accounting for 48% of exports. Over 80% of the work force is employed in subsistence farming and fishing. Industry is based almost entirely on the processing of agricultural products, including cotton, sugarcane, and cattle. Chad is highly dependent on foreign aid, with its economy in trouble and many regions suffering from shortages. Oil companies are exploring areas north of Lake Chad and in the Doba basin in the south. Good crop weather led to 8.4% growth in 1991.
National product: GDP—exchange rate conversion—$1.1 billion (1991 est.)
National product real growth rate: 8.4% (1991 est.)
National product per capita: $215 (1991 est.)
Inflation rate (consumer prices): 2%-3% (1991 est.)
Unemployment rate: NA%
Budget: revenues $115 million; expenditures $412 million, including capital expenditures of $218 million (1991 est.)
Exports: $193.9 million (f.o.b., 1991)
commodities: cotton 48%, cattle 35%, textiles 5%, fish
partners: France, Nigeria, Cameroon
Imports: $294.1 million (f.o.b., 1991)
commodities: machinery and transportation equipment 39%, industrial goods 20%, petroleum products 13%, foodstuffs 9%; note—excludes military equipment
partners: US, France, Nigeria, Cameroon
External debt: $492 million (December 1990 est.)
Industrial production: growth rate 12.9% (1989 est.); accounts for nearly 15% of GDP
Electricity: 40,000 kW capacity; 70 million kWh produced, 15 kWh per capita (1991)
Industries: cotton textile mills, slaughterhouses, brewery, natron (sodium carbonate), soap, cigarettes
Agriculture: accounts for about 45% of GDP; largely subsistence farming; cotton most important cash crop; food crops include sorghum, millet, peanuts, rice,

Chad (continued)

potatoes, manioc; livestock—cattle, sheep, goats, camels; self-sufficient in food in years of adequate rainfall
Economic aid: US commitments, including Ex-Im (FY70-89), $198 million; Western (non-US) countries, ODA and OOF bilateral commitments (1970-89), $1.5 billion; OPEC bilateral aid (1979-89), $28 million; Communist countries (1970-89), $80 million
Currency: 1 CFA franc (CFAF) = 100 centimes
Exchange rates: Communaute Financiere Africaine Francs (CFAF) per US$1—274.06 (January 1993), 264.69 (1992), 282.11 (1991), 272.26 (1990), 319.01 (1989), 297.85 (1988)
Fiscal year: calendar year

Communications

Highways: 31,322 km total; 32 km bituminous; 7,300 km gravel and laterite; remainder unimproved earth
Inland waterways: 2,000 km navigable
Airports:
total: 69
usable: 55
with permanent-surface runways: 5
with runways over 3,659 m: 0
with runways 2,440-3,659 m: 4
with runways 1,220-2,439 m: 24
Telecommunications: fair system of radiocommunication stations for intercity links; broadcast stations—6 AM, 1 FM, limited TV service; many facilities are inoperative; 1 Atlantic Ocean INTELSAT earth station

Defense Forces

Branches: Army (includes Ground Forces, Air Force, and Gendarmerie), Republican Guard
Manpower availability: males age 15-49 1,246,617; fit for military service 647,908; reach military age (20) annually 52,870 (1993 est.)
Defense expenditures: exchange rate conversion—$58 million, 5.6% of GDP (1989)

Chile

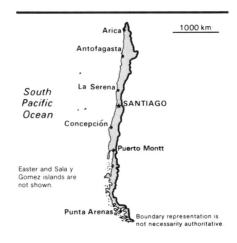

Geography

Location: Western South America, bordering the South Pacific Ocean between Argentina and Peru
Map references: South America, Standard Time Zones of the World
Area:
total area: 756,950 km²
land area: 748,800 km²
comparative area: slightly smaller than twice the size of Montana
note: includes Isla de Pascua (Easter Island) and Isla Sala y Gomez
Land boundaries: total 6,171 km, Argentina 5,150 km, Bolivia 861 km, Peru 160 km
Coastline: 6,435 km
Maritime claims:
contiguous zone: 24 nm
continental shelf: 200 nm
exclusive economic zone: 200 nm
territorial sea: 12 nm
International disputes: short section of the southern boundary with Argentina is indefinite; Bolivia has wanted a sovereign corridor to the South Pacific Ocean since the Atacama area was lost to Chile in 1884; dispute with Bolivia over Rio Lauca water rights; territorial claim in Antarctica (Chilean Antarctic Territory) partially overlaps Argentine and British claims
Climate: temperate; desert in north; cool and damp in south
Terrain: low coastal mountains; fertile central valley; rugged Andes in east
Natural resources: copper, timber, iron ore, nitrates, precious metals, molybdenum
Land use:
arable land: 7%
permanent crops: 0%
meadows and pastures: 16%
forest and woodland: 21%
other: 56%
Irrigated land: 12,650 km² (1989 est.)
Environment: subject to severe earthquakes, active volcanism, tsunami; Atacama Desert

one of world's driest regions; desertification
Note: strategic location relative to sea lanes between Atlantic and Pacific Oceans (Strait of Magellan, Beagle Channel, Drake Passage)

People

Population: 13,739,759 (July 1993 est.)
Population growth rate: 1.54% (1993 est.)
Birth rate: 20.9 births/1,000 population (1993 est.)
Death rate: 5.55 deaths/1,000 population (1993 est.)
Net migration rate: 0 migrant(s)/1,000 population (1993 est.)
Infant mortality rate: 15.9 deaths/1,000 live births (1993 est.)
Life expectancy at birth:
total population: 74.15 years
male: 71.16 years
female: 77.29 years (1993 est.)
Total fertility rate: 2.51 children born/woman (1993 est.)
Nationality:
noun: Chilean(s)
adjective: Chilean
Ethnic divisions: European and European-Indian 95%, Indian 3%, other 2%
Religions: Roman Catholic 89%, Protestant 11%, Jewish
Languages: Spanish
Literacy: age 15 and over can read and write (1990)
total population: 93%
male: 94%
female: 93%
Labor force: 4.728 million
by occupation: services 38.3% (includes government 12%), industry and commerce 33.8%, agriculture, forestry, and fishing 19.2%, mining 2.3%, construction 6.4% (1990)

Government

Names:
conventional long form: Republic of Chile
conventional short form: Chile
local long form: Republica de Chile
local short form: Chile
Digraph: CI
Type: republic
Capital: Santiago
Administrative divisions: 13 regions (regiones, singular—region); Aisen del General Carlos Ibanez del Campo, Antofagasta, Araucania, Atacama, Bio-Bio, Coquimbo, Libertador General Bernardo O'Higgins, Los Lagos, Magallanes y de la Antartica Chilena, Maule, Region Metropolitana, Tarapaca, Valparaiso
note: the US does not recognize claims to Antarctica

Independence: 18 September 1810 (from Spain)

Constitution: 11 September 1980, effective 11 March 1981; amended 30 July 1989

Legal system: based on Code of 1857 derived from Spanish law and subsequent codes influenced by French and Austrian law; judicial review of legislative acts in the Supreme Court; has not accepted compulsory ICJ jurisdiction

National holiday: Independence Day, 18 September (1810)

Political parties and leaders: Concertation of Parties for Democracy consists mainly of four parties: PDC, PPD, PR, PS; Christian Democratic Party (PDC), Eduardo FREI Ruiz-Tagle; Party for Democracy (PPD), Sergio BITAR; Radical Party (PR), Carlos GONZALEZ Marquez; Sociaistl Party (PS), German CORREA; Independent Democratic Union (UDI), Jovino NOVOA; National Renovation (RN), Andree ALLAMAND; Center-Center Union (UCC), Francisco Juner ERRAZURIZ; Communist Party of Chile (PCCh), Volodia TEITELBOIM; Allende Leftist Democratic Movement (MIDA), Mario PALESTRO

Other political or pressure groups: revitalized university student federations at all major universities dominated by opposition political groups; labor—United Labor Central (CUT) includes trade unionists from the country's five largest labor confederations; Roman Catholic Church

Suffrage: 18 years of age; universal and compulsory

Elections:
Chamber of Deputies: last held 14 December 1989 (next to be held December 1993); results—percent of vote by party NA; seats—(120 total) Concertation of Parties for Democracy 71 (PDC 38, PPD 17, PR 5, other 11), RN 29, UDI 11, right-wing independents 9

President: last held 14 December 1989 (next to be held December 1993); results—Patricio AYLWIN (PDC) 55.2%, Hernan BUCHI 29.4%, other 15.4%

Senate: last held 14 December 1989 (next to be held December 1993); results—percent of vote by party NA; seats—(46 total, 38 elected) Concertation of Parties for Democracy 22 (PDC 13, PPD 5, PR 2, PSD 1, PRSD 1), RN 6, UDI 2, right-wing independents 8

Executive branch: president, Cabinet

Legislative branch: bicameral National Congress (Congreso Nacional) consisting of an upper house or Senate (Senado) and a lower house or Chamber of Deputies (Camara de Diputados)

Judicial branch: Supreme Court (Corte Suprema)

Leaders:
Chief of State and Head of Government: President Patricio AYLWIN Azocar (since 11 March 1990)

Member of: CCC, ECLAC, FAO, G-11, G-77, GATT, IADB, IAEA, IBRD, ICAO, ICFTU, IDA, IFAD, IFC, ILO, IMF, IMO, INMARSAT, INTELSAT, INTERPOL, IOC, IOM, ISO, ITU, LAES, LAIA, LORCS, OAS, OPANAL, PCA, RG, UN, UNCTAD, UNESCO, UNIDO, UNIKOM, UNMOGIP, UNTAC, UNTSO, UPU, WCL, WFTU, WHO, WIPO, WMO, WTO

Diplomatic representation in US:
chief of mission: Ambassador Patricio SILVA Echenique
chancery: 1732 Massachusetts Avenue NW, Washington, DC 20036
telephone: (202) 785-1746
consulates general: Houston, Los Angeles, Miami, New York, Philadelphia, and San Francisco

US diplomatic representation:
chief of mission: Ambassador Curtis W. KAMMAN
embassy: Codina Building, 1343 Agustinas, Santiago
mailing address: APO AA 34033
telephone: [56] (2) 671-0133
FAX: [56] (2) 699-1141

Flag: two equal horizontal bands of white (top) and red; there is a blue square the same height as the white band at the hoist-side end of the white band; the square bears a white five-pointed star in the center; design was based on the US flag

Economy

Overview: The government of President AYLWIN, which took power in 1990, retained the economic policies of PINOCHET, although the share of spending for social welfare has risen steadily. In 1991 growth in GDP recovered to 6% (led by consumer spending) after only 2% growth in 1990. The pace accelerated in 1992 as the result of strong investment and export growth, and GDP rose 10.4%. Nonetheless, inflation fell further, to 12.7%, compared with 27.3% in 1990 and 18.7% in 1991. The buoyant economy spurred a 25% growth in imports, and the trade surplus fell in 1992, although international reserves increased. Inflationary pressures are not expected to ease much in 1993, and economic growth is likely to approach 7%.

National product: GDP—exchange rate conversion—$34.7 billion (1992 est.)

National product real growth rate: 10.4% (1992)

National product per capita: $2,550 (1992)

Inflation rate (consumer prices): 12.7% (1992)

Unemployment rate: 4.9% (1992)

Budget: revenues $10.9 billion; expenditures $10.9 billion, including capital expenditures of $1.2 billion (1993)

Exports: $10 billion (f.o.b., 1992)
commodities: copper 41%, other metals and minerals 8.7%, wood products 7.1%, fish and fishmeal 9.8%, fruits 8.4% (1991)
partners: EC 32%, US 18%, Japan 18%, Brazil 5% (1991)

Imports: $9.2 billion (f.o.b., 1992)
commodities: capital goods 25.2%, spare parts 24.8%, raw materials 15.4%, petroleum 10%, foodstuffs 5.7%
partners: US 21%, EC 18%, Brazil 9%, Japan 8% (1991)

External debt: $16.9 billion (year end 1991)

Industrial production: growth rate 14.56% (1992); accounts for 34% of GDP

Electricity: 5,769,000 kW capacity; 22,010 million kWh produced, 1,630 kWh per capita (1992)

Industries: copper, other minerals, foodstuffs, fish processing, iron and steel, wood and wood products, transport equipment, cement, textiles

Agriculture: accounts for about 9% of GDP (including fishing and forestry); major exporter of fruit, fish, and timber products; major crops—wheat, corn, grapes, beans, sugar beets, potatoes, deciduous fruit; livestock products—beef, poultry, wool; self-sufficient in most foods; 1991 fish catch of 6.6 million metric tons; net agricultural importer

Economic aid: US commitments, including Ex-Im (FY70-89), $521 million; Western (non-US) countries, ODA and OOF bilateral commitments (1970-89), $1.6 billion; Communist countries (1970-89), $386 million

Currency: 1 Chilean peso (Ch$) = 100 centavos

Exchange rates: Chilean pesos (Ch$) per US$1—384.04 (January 1993), 362.59 (1992), 349.37 (1991), 305.06 (1990), 267.16 (1989), 245.05 (1988)

Fiscal year: calendar year

Communications

Railroads: 7,766 km total; 3,974 km 1.676-meter gauge, 150 km 1.435-meter standard gauge, 3,642 km 1.000-meter gauge; 1,865 km 1.676-meter gauge and 80 km 1.000-meter gauge electrified

Highways: 79,025 km total; 9,913 km paved, 33,140 km gravel, 35,972 km improved and unimproved earth (1984)

Inland waterways: 725 km

Pipelines: crude oil 755 km; petroleum products 785 km; natural gas 320 km

Ports: Antofagasta, Iquique, Puerto Montt, Punta Arenas, Valparaiso, San Antonio, Talcahuano, Arica

Chile *(continued)*

Merchant marine: 31 ships (1,000 GRT or over) totaling 445,330 GRT/756,018 DWT; includes 8 cargo, 1 refrigerated cargo, 3 roll-on/roll-off cargo, 2 oil tanker, 3 chemical tanker, 3 liquefied gas tanker, 3 combination ore/oil, 8 bulk; note—in addition, 1 naval tanker and 1 military transport are sometimes used commercially

Airports:
total: 396
usable: 351
with permanent-surface runways: 48
with runways over 3,659 m: 0
with runways 2,440-3,659 m: 13
with runways 1,220-2,439 m: 57

Telecommunications: modern telephone system based on extensive microwave radio relay facilities; 768,000 telephones; broadcast stations—159 AM, no FM, 131 TV, 11 shortwave; satellite ground stations—2 Atlantic Ocean INTELSAT and 3 domestic

---

Defense Forces

Branches: Army of the Nation, National Navy (including Naval Air, Coast Guard, and Marines), Air Force of the Nation, Carabineros of Chile (National Police), Investigative Police

Manpower availability: males age 15-49 3.653 million; fit for military service 2,722,479; reach military age (19) annually 119,434 (1993 est.)

Defense expenditures: exchange rate conversion—$1 billion, 3.4% of GDP (1991 est.)

China
(also see separate Taiwan entry)

1200 km

Boundary representation is not necessarily authoritative.

Geography

Location: East Asia, between India and Mongolia

Map references: Asia, Southeast Asia, Standard Time Zones of the World

Area:
total area: 9,596,960 km²
land area: 9,326,410 km²
comparative area: slightly larger than the US

Land boundaries: total 22,143.34 km, Afghanistan 76 km, Bhutan 470 km, Burma 2,185 km, Hong Kong 30 km, India 3,380 km, Kazakhstan 1,533 km, North Korea 1,416 km, Kyrgyzstan 858 km, Laos 423 km, Macau 0.34 km, Mongolia 4,673 km, Nepal 1,236 km, Pakistan 523 km, Russia (northeast) 3,605 km, Russia (northwest) 40 km, Tajikistan 414 km, Vietnam 1,281 km

Coastline: 14,500 km

Maritime claims:
continental shelf: claim to shallow areas of East China Sea and Yellow Sea
territorial sea: 12 nm

International disputes: boundary with India; bilateral negotiations are under way to resolve disputed sections of the boundary with Russia; boundary with Tajikistan under dispute; a short section of the boundary with North Korea is indefinite; involved in a complex dispute over the Spratly Islands with Malaysia, Philippines, Taiwan, Vietnam, and possibly Brunei; maritime boundary dispute with Vietnam in the Gulf of Tonkin; Paracel Islands occupied by China, but claimed by Vietnam and Taiwan; claims Japanese-administered Senkaku-shoto, as does Taiwan, (Senkaku Islands/Diaoyu Tai)

Climate: extremely diverse; tropical in south to subarctic in north

Terrain: mostly mountains, high plateaus, deserts in west; plains, deltas, and hills in east

Natural resources: coal, iron ore, petroleum, mercury, tin, tungsten, antimony, manganese, molybdenum, vanadium, magnetite, aluminum, lead, zinc, uranium, world's largest hydropower potential

Land use:
arable land: 10%
permanent crops: 0%
meadows and pastures: 31%
forest and woodland: 14%
other: 45%

Irrigated land: 478,220 km² (1991—Chinese statistic)

Environment: frequent typhoons (about five times per year along southern and eastern coasts), damaging floods, tsunamis, earthquakes; deforestation; soil erosion; industrial pollution; water pollution; air pollution; desertification

Note: world's third-largest country (after Russia and Canada)

People

Population: 1,177,584,537 (July 1993 est.)

Population growth rate: 1.1% (1993 est.)

Birth rate: 18.29 births/1,000 population (1993 est.)

Death rate: 7.34 deaths/1,000 population (1993 est.)

Net migration rate: 0 migrant(s)/1,000 population (1993 est.)

Infant mortality rate: 52.1 deaths/1,000 live births (1993 est.)

Life expectancy at birth:
total population: 67.74 years
male: 66.78 years
female: 68.8 years (1993 est.)

Total fertility rate: 1.85 children born/woman (1993 est.)

Nationality:
noun: Chinese (singular and plural)
adjective: Chinese

Ethnic divisions: Han Chinese 91.9%, Zhuang, Uygur, Hui, Yi, Tibetan, Miao, Manchu, Mongol, Buyi, Korean, and other nationalities 8.1%

Religions: Daoism (Taoism), Buddhism, Muslim 2-3%, Christian 1% (est.)
note: officially atheist, but traditionally pragmatic and eclectic

Languages: Standard Chinese (Putonghua) or Mandarin (based on the Beijing dialect), Yue (Cantonese), Wu (Shanghainese), Minbei (Fuzhou), Minnan (Hokkien-Taiwanese), Xiang, Gan, Hakka dialects, minority languages (see Ethnic divisions entry)

Literacy: age 15 and over can read and write (1990)
total population: 73%
male: 84%
female: 62%

Labor force: 567.4 million
by occupation: agriculture and forestry 60%, industry and commerce 25%, construction and mining 5%, social services 5%, other 5% (1990 est.)

Government

Names:
conventional long form: People's Republic of China
conventional short form: China
local long form: Zhonghua Renmin Gongheguo
local short form: Zhong Guo
Abbreviation: PRC
Digraph: CH
Type: Communist state
Capital: Beijing
Administrative divisions: 23 provinces (sheng, singular and plural), 5 autonomous regions* (zizhiqu, singular and plural), and 3 municipalities** (shi, singular and plural); Anhui, Beijing Shi**, Fujian, Gansu, Guangdong, Guangxi*, Guizhou, Hainan, Hebei, Heilongjiang, Henan, Hubei, Hunan, Jiangsu, Jiangxi, Jilin, Liaoning, Nei Mongol*, Ningxia*, Qinghai, Shaanxi, Shandong, Shanghai Shi**, Shanxi, Sichuan, Tianjin Shi**, Xinjiang*, Xizang* (Tibet), Yunnan, Zhejiang
note: China considers Taiwan its 23rd province
Independence: 221 BC (unification under the Qin or Ch'in Dynasty 221 BC; Qing or Ch'ing Dynasty replaced by the Republic on 12 February 1912; People's Republic established 1 October 1949)
Constitution: most recent promulgated 4 December 1982
Legal system: a complex amalgam of custom and statute, largely criminal law; rudimentary civil code in effect since 1 January 1987; new legal codes in effect since 1 January 1980; continuing efforts are being made to improve civil, administrative, criminal, and commercial law
National holiday: National Day, 1 October (1949)
Political parties and leaders: Chinese Communist Party (CCP), JIANG Zemin, general secretary of the Central Committee (since 24 June 1989); eight registered small parties controlled by CCP
Other political or pressure groups: such meaningful opposition as exists consists of loose coalitions, usually within the party and government organization, that vary by issue
Suffrage: 18 years of age; universal
Elections:
National People's Congress: last held March 1993 (next to be held March 1998); results—CCP is the only party but there are also independents; seats—(2,977 total) (elected at county or xian level)
President: last held 27 March 1993 (next to be held NA 1998); results—JIANG Zemin was nominally elected by the Eighth National People's Congress
Executive branch: president, vice president, premier, four vice premiers, State Council

Legislative branch: unicameral National People's Congress (Quanguo Renmin Daibiao Dahui)
Judicial branch: Supreme People's Court
Leaders:
Chief of State: President JIANG Zemin (since 27 March 1993); Vice President RONG Yiren (since 27 March 1993)
Chief of State and Head of Government (de facto): DENG Xiaoping (since NA 1977)
Head of Government: Premier LI Peng (Acting Premier since 24 November 1987, Premier since 9 April 1988) Vice Premier ZHU Rongji (since 8 April 1991); Vice Premier ZOU Jiahua (since 8 April 1991); Vice Premier QIAN Qichen (since 29 March 1993); Vice Premier LI Lanqing (29 March 1993)
Member of: AfDB, APEC, AsDB, CCC, ESCAP, FAO, IAEA, IBRD, ICAO, IDA, IFAD, IFC, ILO, IMF, IMO, INMARSAT, INTELSAT, INTERPOL, IOC, ISO, ITU, LORCS, MINURSO, NAM (observer), PCA, UN, UNCTAD, UNESCO, UNHCR, UNIDO, UNIKOM, UN Security Council, UNTAC, UNTSO, UN Trusteeship Council, UPU, WHO, WIPO, WMO, WTO
Diplomatic representation in US:
chief of mission: Ambassador LI Daoyu
chancery: 2300 Connecticut Avenue NW, Washington, DC 20008
telephone: (202) 328-2500 through 2502
consulates general: Chicago, Houston, Los Angeles, New York, and San Francisco
US diplomatic representation:
chief of mission: Ambassador J. Stapleton ROY
embassy: Xiu Shui Bei Jie 3, Beijing
mailing address: 100600, PSC 461, Box 50, Beijing or FPO AP 96521-0002
telephone: [86] (1) 532-3831
FAX: [86] (1) 532-3178
consulates general: Chengdu, Guangzhou, Shanghai, Shenyang
Flag: red with a large yellow five-pointed star and four smaller yellow five-pointed stars (arranged in a vertical arc toward the middle of the flag) in the upper hoist-side corner

Economy

Overview: Beginning in late 1978 the Chinese leadership has been trying to move the economy from the sluggish Soviet-style centrally planned economy to a more productive and flexible economy with market elements, but still within the framework of monolithic Communist control. To this end the authorities have switched to a system of household responsibility in agriculture in place of the old collectivization, increased the authority of local officials and plant managers in industry, permitted a wide variety of

small-scale enterprise in services and light manufacturing, and opened the foreign economic sector to increased trade and joint ventures. The most gratifying result has been a strong spurt in production, particularly in agriculture in the early 1980s. Industry also has posted major gains, especially in coastal areas near Hong Kong and opposite Taiwan, where foreign investment and modern production methods have helped spur production of both domestic and export goods. Aggregate output has more than doubled since 1978. On the darker side, the leadership has often experienced in its hybrid system the worst results of socialism (bureaucracy, lassitude, corruption) and of capitalism (windfall gains and stepped-up inflation). Beijing thus has periodically backtracked, retightening central controls at intervals and thereby lessening the credibility of the reform process. In 1991, and again in 1992, output rose substantially, particularly in the favored coastal areas. Popular resistance, changes in central policy, and loss of authority by rural cadres have weakened China's population control program, which is essential to the nation's long-term economic viability.
National product: GNP $NA
National product real growth rate: 12.8% (1992 est.)
National product per capita: $NA
Inflation rate (consumer prices): 5.4% (1992)
Unemployment rate: 2.3% in urban areas (1992)
Budget: deficit $16.3 billion (1992)
Exports: $85.0 billion (f.o.b., 1992)
commodities: textiles, garments, telecommunications and recording equipment, petroleum, minerals
partners: Hong Kong and Macau, Japan, US, Germany, South Korea, Russia (1992)
Imports: $80.6 billion (c.i.f., 1992)
commodities: specialized industrial machinery, chemicals, manufactured goods, steel, textile yarn, fertilizer
partners: Hong Kong and Macau, Japan, US, Taiwan, Germany, Russia (1992)
External debt: $69.3 billion (1992)
Industrial production: growth rate 20.8% (1992)
Electricity: 158,690,000 kW capacity; 740,000 million kWh produced, 630 kWh per capita (1992)
Industries: iron and steel, coal, machine building, armaments, textiles, petroleum, cement, chemical fertilizers, consumer durables, food processing
Agriculture: accounts for 26% of GNP; among the world's largest producers of rice, potatoes, sorghum, peanuts, tea, millet, barley, and pork; commercial crops include cotton, other fibers, and oilseeds; produces variety of livestock products; basically

China (continued)

self-sufficient in food; fish catch of 13.35 million metric tons (including fresh water and pond raised) (1991)
Illicit drugs: illicit producer of opium in at least 18 provinces and administrative regions; bulk of production is in Yunnan Province; transshipment point for heroin produced in the Golden Triangle
Economic aid: donor—to less developed countries (1970-89) $7.0 billion; US commitments, including Ex-Im (FY70-87), $220.7 million; Western (non-US) countries, ODA and OOF bilateral commitments (1970-87), $13.5 billion
Currency: 1 yuan (¥) = 10 jiao
Exchange rates: yuan (¥) per US$1—5.7640 (January 1993), 5.5146 (1992), 5.3234 (1991), 4.7832 (1990), 3.7651 (1989), 3.7221 (1988)
Fiscal year: calendar year

Communications

Railroads: total about 64,000 km; 54,000 km of common carrier lines, of which 53,400 km are 1.435-meter gauge (standard) and 600 km are 1.000-meter gauge (narrow); 11,200 km of standard gauge common carrier route are double tracked and 6,900 km are electrified (1990); an additional 10,000 km of varying gauges (0.762 to 1.067-meter) are dedicated industrial lines
Highways: about 1,029,000 km (1990) total; 170,000 km (est.) paved roads, 648,000 km (est.) gravel/improved earth roads, 211,000 km (est.) unimproved earth roads and tracks
Inland waterways: 138,600 km; about 109,800 km navigable
Pipelines: crude oil 9,700 km (1990); petroleum products 1,100 km; natural gas 6,200 km
Ports: Dalian, Guangzhou, Huangpu, Qingdao, Qinhuangdao, Shanghai, Xingang, Zhanjiang, Ningbo, Xiamen, Tanggu, Shantou
Merchant marine: 1,478 ships (1,000 GRT or over) totaling 14,029,320 GRT/21,120,522 DWT; includes 25 passenger, 42 short-sea passenger, 18 passenger-cargo, 6 cargo/training, 811 cargo, 11 refrigerated cargo, 81 container, 18 roll-on/roll-off cargo, 1 multifunction/barge carrier, 177 oil tanker, 11 chemical tanker, 263 bulk, 3 liquefied gas, 1 vehicle carrier, 9 combination bulk, 1 barge carrier; note—China beneficially owns an additional 227 ships (1,000 GRT or over) totaling approximately 6,187,117 DWT that operate under Panamanian, British, Hong Kong, Maltese, Liberian, Vanuatu, Cypriot, Saint Vincent, Bahamian, and Romanian registry
Airports:
total: 330
usable: 330
with permanent-surface runways: 260

with runways over 3,500 m: fewer than 10
with runways 2,440-3,659 m: 90
with runways 1,220-2,439 m: 200
Telecommunications: domestic and international services are increasingly available for private use; unevenly distributed internal system serves principal cities, industrial centers, and most townships; 11,000,000 telephones (December 1989); broadcast stations—274 AM, unknown FM, 202 (2,050 repeaters) TV; more than 215 million radio receivers; 75 million TVs; satellite earth stations—4 Pacific Ocean INTELSAT, 1 Indian Ocean INTELSAT, 1 INMARSAT, and 55 domestic

Defense Forces

Branches: People's Liberation Army (PLA), PLA Navy (including Marines), PLA Air Force
Manpower availability: males age 15-49 343,361,925; fit for military service 190,665,512; reach military age (18) annually 10,844,047 (1993 est.)
Defense expenditures: exchange rate conversion—$NA, NA% of GNP

Christmas Island
(territory of Australia)

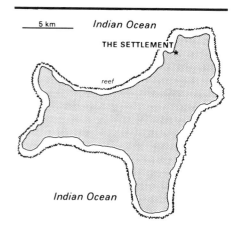

Geography

Location: in the Indian Ocean, between Australia and Indonesia
Map references: Southeast Asia
Area:
total area: 135 km^2
land area: 135 km^2
comparative area: about 0.8 times the size of Washington, DC
Land boundaries: 0 km
Coastline: 138.9 km
Maritime claims:
contiguous zone: 12 nm
exclusive fishing zone: 200 nm
territorial sea: 3 nm
International disputes: none
Climate: tropical; heat and humidity moderated by trade winds
Terrain: steep cliffs along coast rise abruptly to central plateau
Natural resources: phosphate
Land use:
arable land: 0%
permanent crops: 0%
meadows and pastures: 0%
forest and woodland: 0%
other: 100%
Irrigated land: NA km^2
Environment: almost completely surrounded by a reef
Note: located along major sea lanes of Indian Ocean

People

Population: 1,685 (July 1993 est.)
Population growth rate: -2.44% (1993 est.)
Birth rate: NA births/1,000 population
Death rate: NA deaths/1,000 population
Net migration rate: NA migrant(s)/1,000 population
Infant mortality rate: NA deaths/1,000 live births
Life expectancy at birth:
total population: NA years
male: NA years

female: NA years
Total fertility rate: NA children born/woman
Nationality:
noun: Christmas Islander(s)
adjective: Christmas Island
Ethnic divisions: Chinese 61%, Malay 25%, European 11%, other 3%, no indigenous population
Religions: Buddhist 36.1%, Muslim 25.4%, Christian 17.7% (Roman Catholic 8.2%, Church of England 3.2%, Presbyterian 0.9%, Uniting Church 0.4%, Methodist 0.2%, Baptist 0.1%, and other 4.7%), none 12.7%, unknown 4.6%, other 3.5% (1981)
Languages: English
Literacy:
total population: NA%
male: NA%
female: NA%
Labor force: NA
by occupation: all workers are employees of the Phosphate Mining Company of Christmas Island, Ltd.

Government

Names:
conventional long form: Territory of Christmas Island
conventional short form: Christmas Island
Digraph: KT
Type: territory of Australia
Capital: The Settlement
Administrative divisions: none (territory of Australia)
Independence: none (territory of Australia)
Constitution: Christmas Island Act of 1958
Legal system: under the authority of the governor general of Australia
National holiday: NA
Political parties and leaders: none
Executive branch: British monarch, governor general of Australia, administrator, Advisory Council (cabinet)
Legislative branch: none
Judicial branch: none
Leaders:
Chief of State: Queen ELIZABETH II (since 6 February 1952)
Head of Government: Administrator M. J. GRIMES (since NA)
Member of: none
Diplomatic representation in US: none (territory of Australia)
US diplomatic representation: none (territory of Australia)
Flag: the flag of Australia is used

Economy

Overview: Phosphate mining had been the only significant economic activity, but in December 1987 the Australian Government closed the mine as no longer economically

viable. Plans have been under way to reopen the mine and also to build a casino and hotel to develop tourism.
National product: $NA
National product real growth rate: NA%
National product per capita: $NA
Inflation rate (consumer prices): NA%
Unemployment rate: NA%
Budget: revenues $NA; expenditures $NA, including capital expenditures of $NA
Exports: $NA
commodities: phosphate
partners: Australia, NZ
Imports: $NA
commodities: consumer goods
partners: principally Australia
External debt: $NA
Industrial production: growth rate NA%
Electricity: 11,000 kW capacity; 30 million kWh produced, 17,800 kWh per capita (1990)
Industries: phosphate extraction (near depletion)
Agriculture: NA
Economic aid: none
Currency: 1 Australian dollar ($A) = 100 cents
Exchange rates: Australian dollars ($A) per US$1—1.4837 (January 1993), 1.3600 (1992), 1.2836 (1991), 1.2799 (1990), 1.2618 (1989), 1.2752 (1988)
Fiscal year: 1 July-30 June

Communications

Highways: adequate road system
Ports: Flying Fish Cove
Airports:
total: 1
useable: 1
with permanent-surface runways: 1
with runways over 3,659 m: 0
with runways 2,440-3,659 m: 0
with runways 1,220-2,439: 1
Telecommunications: 4,000 radios (1982); broadcasting stations—1 AM, 1 TV

Defense Forces

Note: defense is the responsibility of Australia

Clipperton Island
(possession of France)

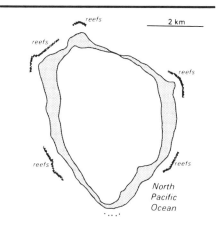

Geography

Location: in the North Pacific Ocean, 1,120 km southwest of Mexico
Map references: World
Area:
total area: 7 km²
land area: 7 km²
comparative area: about 12 times the size of the Mall in Washington, DC
Land boundaries: 0 km
Coastline: 11.1 km
Maritime claims:
exclusive economic zone: 200 nm
territorial sea: 12 nm
International disputes: claimed by Mexico
Climate: tropical
Terrain: coral atoll
Natural resources: none
Land use:
arable land: 0%
permanent crops: 0%
meadows and pastures: 0%
forest and woodland: 0%
other: 100% (all coral)
Irrigated land: 0 km²
Environment: reef about 8 km in circumference

People

Population: uninhabited

Government

Names:
conventional long form: none
conventional short form: Clipperton Island
local long form: none
local short form: Ile Clipperton
former: sometimes called Ile de la Passion
Digraph: IP
Type: French possession administered by France from French Polynesia by High Commissioner of the Republic
Capital: none; administered by France from French Polynesia

Clipperton Island *(continued)*

Independence: none (possession of France)

Economy

Overview: The only economic activity is a tuna fishing station.

Communications

Ports: none; offshore anchorage only

Defense Forces

Note: defense is the responsibility of France

Cocos (Keeling) Islands
(territory of Australia)

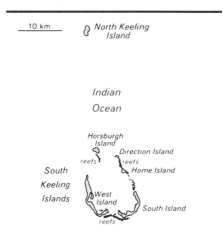

Geography

Location: in the Indian Ocean, 1,070 km southwest of Indonesia, about halfway between Australia and Sri Lanka
Map references: Southeast Asia
Area:
total area: 14 km²
land area: 14 km²
comparative area: about 24 times the size of the Mall in Washington, DC
note: includes the two main islands of West Island and Home Island
Land boundaries: 0 km
Coastline: 2.6 km
Maritime claims:
exclusive fishing zone: 200 nm
territorial sea: 3 nm
International disputes: none
Climate: pleasant, modified by the southeast trade wind for about nine months of the year; moderate rain fall
Terrain: flat, low-lying coral atolls
Natural resources: fish
Land use:
arable land: 0%
permanent crops: 0%
meadows and pastures: 0%
forest and woodland: 0%
other: 100%
Irrigated land: NA km²
Environment: two coral atolls thickly covered with coconut palms and other vegetation

---

People

Population: 593 (July 1993 est.)
Population growth rate: -0.53% (1993 est.)
Birth rate: NA births/1,000 population
Death rate: NA deaths/1,000 population
Net migration rate: NA migrant(s)/1,000 population
Infant mortality rate: NA deaths/1,000 live births
Life expectancy at birth:
total population: NA years

male: NA years
female: NA years
Total fertility rate: NA children born/women
Nationality:
noun: Cocos Islander(s)
adjective: Cocos Islander
Ethnic divisions:
West Island: Europeans
Home Island: Cocos Malays
Religions: Sunni Muslims
Languages: English
Literacy:
total population: NA%
male: NA%
female: NA%
Labor force: NA

Government

Names:
conventional long form: Territory of Cocos (Keeling) Islands
conventional short form: Cocos (Keeling) Islands
Digraph: CK
Type: territory of Australia
Capital: West Island
Administrative divisions: none (territory of Australia)
Independence: none (territory of Australia)
Constitution: Cocos (Keeling) Islands Act of 1955
Legal system: based upon the laws of Australia and local laws
National holiday: NA
Political parties and leaders: NA
Suffrage: NA
Elections: NA
Executive branch: British monarch, governor general of Australia, administrator, chairman of the Islands Council
Legislative branch: unicameral Islands Council
Judicial branch: Supreme Court
Leaders:
Chief of State: Queen ELIZABETH II (since 6 February 1952)
Head of Government: Administrator B. CUNNINGHAM (since NA); Chairman of the Islands Council Haji WAHIN bin Bynie (since NA)
Member of: none
Diplomatic representation in US: none (territory of Australia)
US diplomatic representation: none (territory of Australia)
Flag: the flag of Australia is used

Economy

Overview: Grown throughout the islands, coconuts are the sole cash crop. Copra and fresh coconuts are the major export earners. Small local gardens and fishing contribute to

Colombia

the food supply, but additional food and most other necessities must be imported from Australia.
National product: GDP $NA
National product real growth rate: NA%
National product per capita: $NA
Inflation rate (consumer prices): NA%
Budget: revenues $NA; expenditures $NA, including capital expenditures of $NA
Exports: $NA
commodities: copra
partners: Australia
Imports: $NA
commodities: foodstuffs
partners: Australia
External debt: $NA
Industrial production: growth rate NA%
Electricity: 1,000 kW capacity; 2 million kWh produced, 2,980 kWh per capita (1990)
Industries: copra products
Agriculture: gardens provide vegetables, bananas, pawpaws, coconuts
Economic aid: none
Currency: 1 Australian dollar ($A) = 100 cents
Exchange rates: Australian dollars ($A) per US$1—1.4837 (January 1993), 1.3600 (1992), 1.2836 (1991), 1.2799 (1990), 1.2618 (1989), 1.2752 (1988)
Fiscal year: 1 July-30 June

Communications

Ports: none; lagoon anchorage only
Airports:
total: 1
useable: 1
with permanent-surface runways: 1
with runways over 3,659 m: 0
with runways 2,440-3,659 m: 0
with runways 1,220-2,439 m: 1
Telecommunications: 250 radios (1985); linked by telephone, telex, and facsimile communications via satellite with Australia; broadcast stations—1 AM, no FM, no TV

Defense Forces

Note: defense is the responsibility of Australia

Providencia, Malpelo, and San Andrés islands are not shown.

Geography

Location: Northern South America, between Panama and Venezuela
Map references: Central America and the Caribbean, South America, Standard Time Zones of the World
Area:
total area: 1,138,910 km²
land area: 1,038,700 km²
comparative area: slightly less than three times the size of Montana
note: includes Isla de Malpelo, Roncador Cay, Serrana Bank, and Serranilla Bank
Land boundaries: total 7,408 km, Brazil 1,643 km, Ecuador 590 km, Panama 225 km, Peru 2,900 km, Venezuela 2,050 km
Coastline: 3,208 km (Caribbean Sea 1,760 km, North Pacific Ocean 1,448 km)
Maritime claims:
continental shelf: not specified
exclusive economic zone: 200 nm
territorial sea: 12 nm
International disputes: maritime boundary dispute with Venezuela in the Gulf of Venezuela; territorial dispute with Nicaragua over Archipelago de San Andres y Providencia and Quita Sueno Bank
Climate: tropical along coast and eastern plains; cooler in highlands
Terrain: flat coastal lowlands, central highlands, high Andes mountains, eastern lowland plains
Natural resources: petroleum, natural gas, coal, iron ore, nickel, gold, copper, emeralds
Land use:
arable land: 4%
permanent crops: 2%
meadows and pastures: 29%
forest and woodland: 49%
other: 16%
Irrigated land: 5,150 km² (1989 est.)
Environment: highlands subject to volcanic eruptions; deforestation; soil damage from overuse of pesticides; periodic droughts
Note: only South American country with coastlines on both North Pacific Ocean and Caribbean Sea

People

Population: 34,942,767 (July 1993 est.)
Population growth rate: 1.83% (1993 est.)
Birth rate: 23.4 births/1,000 population (1993 est.)
Death rate: 4.82 deaths/1,000 population (1993 est.)
Net migration rate: -0.25 migrant(s)/1,000 population (1993 est.)
Infant mortality rate: 29.7 deaths/1,000 live births (1993 est.)
Life expectancy at birth:
total population: 71.72 years
male: 68.99 years
female: 74.53 years (1993 est.)
Total fertility rate: 2.54 children born/woman (1993 est.)
Nationality:
noun: Colombian(s)
adjective: Colombian
Ethnic divisions: mestizo 58%, white 20%, mulatto 14%, black 4%, mixed black-Indian 3%, Indian 1%
Religions: Roman Catholic 95%
Languages: Spanish
Literacy: age 15 and over can read and write (1990)
total population: 87%
male: 88%
female: 86%
Labor force: 12 million (1990)
by occupation: services 46%, agriculture 30%, industry 24% (1990)

Government

Names:
conventional long form: Republic of Colombia
conventional short form: Colombia
local long form: Republica de Colombia
local short form: Colombia
Digraph: CO
Type: republic; executive branch dominates government structure
Capital: Bogota
Administrative divisions: 23 departments (departamentos, singular—departamento), 5 commissariats* (comisarias, singular—comisaria), 4 intendancies** (intendencias, singular—intendencia), and 1 special district*** (distrito especial); Amazonas*, Antioquia, Arauca**, Atlantico, Bogota***, Bolivar, Boyaca, Caldas, Caqueta, Casanare**, Cauca, Cesar, Choco, Cordoba, Cundinamarca, Guainia*, Guaviare*, Huila, La Guajira, Magdalena, Meta, Narino, Norte de Santander, Putumayo**, Quindio, Risaralda, San Andres y Providencia**, Santander, Sucre, Tolima, Valle del Cauca, Vaupes*, Vichada*
note: the Constitution of 5 July 1991 states that the commissariats and intendancies are to become full departments and a capital

Colombia (continued)

district (distrito capital) of Santa Fe de Bogota is to be established by 1997
Independence: 20 July 1810 (from Spain)
Constitution: 5 July 1991
Legal system: based on Spanish law; judicial review of executive and legislative acts; accepts compulsory ICJ jurisdiction, with reservations
National holiday: Independence Day, 20 July (1810)
Political parties and leaders: Liberal Party (PL), Cesar GAVIRIA Trujillo, president; Social Conservative Party (PCS), Misael PASTRANA Borrero; National Salvation Movement (MSN), Alvaro GOMEZ Hurtado; Democratic Alliance M-19 (AD/M-19) is headed by 19th of April Movement (M-19) leader Antonio NAVARRO Wolf, coalition of small leftist parties and dissident liberals and conservatives; Patriotic Union (UP) is a legal political party formed by Revolutionary Armed Forces of Colombia (FARC) and Colombian Communist Party (PCC), Carlos ROMERO
Other political or pressure groups: three insurgent groups are active in Colombia—Revolutionary Armed Forces of Colombia (FARC), Manuel MARULANDA and Alfonso CANO; National Liberation Army (ELN), Manuel PEREZ; and dissidents of the recently demobilized People's Liberation Army (EPL), Francisco CARABALLO
Suffrage: 18 years of age; universal and compulsory
Elections:
President: last held 27 May 1990 (next to be held May 1994); results—Cesar GAVIRIA Trujillo (Liberal) 47%, Alvaro GOMEZ Hurtado (National Salvation Movement) 24%, Antonio NAVARRO Wolff (M-19) 13%, Rodrigo LLOREDA (Conservative) 12%
Senate: last held 27 October 1991 (next to be held March 1994); results—percent of vote by party NA; seats—(102 total) Liberal 58, Conservative 22, AD/M-19 9, MSN 5, UP 1, other 7
House of Representatives: last held 27 October 1991 (next to be held March 1994); results—percent of vote by party NA; seats—(161 total) Liberal 87, Conservative 31, AD/M-19 13, MSN 10, UP 3, other 17
Executive branch: president, presidential designate, Cabinet
Legislative branch: bicameral Congress (Congreso) consists of a nationally elected upper chamber or Senate (Senado) and a nationally elected lower chamber or House of Representatives (Camara de Representantes)
Judicial branch: Supreme Court of Justice (Corte Suprema de Justical), Constitutional Court, Council of State

Leaders:
Chief of State and Head of Government: President Cesar GAVIRIA Trujillo (since 7 August 1990)
Member of: AG, CDB, CG, ECLAC, FAO, G-3, G-11, G-24, G-77, GATT, IADB, IAEA, IBRD, ICAO, ICC, ICFTU, IDA, IFAD, IFC, ILO, IMF, IMO, INMARSAT, INTELSAT, INTERPOL, IOC, IOM, ISO, ITU, LAES, LAIA, LORCS, NAM, OAS, ONUSAL, OPANAL, PCA, RG, UN, UNCTAD, UNESCO, UNHCR, UNIDO, UNPROFOR, UPU, WCL, WFTU, WHO, WIPO, WMO, WTO
Diplomatic representation in US:
chief of mission: Ambassador Jaime GARCIA Parra
chancery: 2118 Leroy Place NW, Washington, DC 20008
telephone: (202) 387-8338
consulates general: Chicago, Houston, Miami, New Orleans, New York, San Francisco, and San Juan (Puerto Rico)
consulates: Atlanta, Boston, Detroit, Los Angeles, and Tampa
US diplomatic representation:
chief of mission: Ambassador Morris D. BUSBY
embassy: Calle 38, No. 8-61, Bogota
mailing address: P. O. Box A. A. 3831, Bogota or APO AA 34038
telephone: [57] (1) 285-1300 or 1688
FAX: [57] (1) 288-5687
consulate: Barranquilla
Flag: three horizontal bands of yellow (top, double-width), blue, and red; similar to the flag of Ecuador, which is longer and bears the Ecuadorian coat of arms superimposed in the center

---

Economy

Overview: Economic development has slowed gradually since 1986, but growth rates remain high by Latin American standards. Conservative economic policies have kept inflation and unemployment near 30% and 10%, respectively. The rapid development of oil, coal, and other nontraditional industries in recent years has helped to offset the decline in coffee prices—Colombia's major export. The collapse of the International Coffee Agreement in the summer of 1989, a troublesome rural insurgency, energy rationing, and drug-related violence have dampened growth. The level of violence, in Bogota in particular, surged to higher levels in the first quarter of 1993, further delaying the economic resurgence expected from government reforms. These reforms center on fiscal restraint, trade and investment liberalization, financial and labor reform, and privatization of state utilities and commercial banks.

National product: GDP—exchange rate conversion—$51 billion (1992 est.)
National product real growth rate: 3.3% (1992 est.)
National product per capita: $1,500 (1992 est.)
Inflation rate (consumer prices): 25% (1992)
Unemployment rate: 10% (1992)
Budget: revenues $5.0 billion; current expenditures $5.1 billion, capital expenditures $964 million (1991 est.)
Exports: $7.4 billion (f.o.b., 1992 est.)
commodities: petroleum, coffee, coal, bananas, fresh cut flowers
partners: US 44%, EC 21%, Japan 5%, Netherlands 4%, Sweden 3% (1991)
Imports: $5.5 billion (c.i.f., 1992 est.)
commodities: industrial equipment, transportation equipment, consumer goods, chemicals, paper products
partners: US 36%, EC 16%, Brazil 4%, Venezuela 3%, Japan 3% (1991)
External debt: $17.0 billion (1992)
Industrial production: growth rate −0.5% (1991); accounts for 20% of GDP
Electricity: 10,193,000 kW capacity; 36,000 million kWh produced, 1,050 kWh per capita (1992)
Industries: textiles, food processing, oil, clothing and footwear, beverages, chemicals, metal products, cement; mining—gold, coal, emeralds, iron, nickel, silver, salt
Agriculture: growth rate 3% (1991 est.) accounts for 22% of GDP; crops make up two-thirds and livestock one-third of agricultural output; climate and soils permit a wide variety of crops, such as coffee, rice, tobacco, corn, sugarcane, cocoa beans, oilseeds, vegetables; forest products and shrimp farming are becoming more important
Illicit drugs: illicit producer of cannabis, coca, and opium; about 37,500 hectares of coca under cultivation; the world's largest processor of coca derivatives into cocaine; supplier of cocaine to the US and other international drug markets
Economic aid: US commitments, including Ex-Im (FY70-89), $1.6 billion; Western (non-US) countries, ODA and OOF bilateral commitments (1970-89), $3.3 billion, Communist countries (1970-89), $399 million
Currency: 1 Colombian peso (Col$) = 100 centavos
Exchange rates: Colombian pesos (Col$) per US$1—820.08 (January 1993), 759.28 (1992), 633.05 (1991), 502.26 (1990), 382.57 (1989), 299.17 (1988)
Fiscal year: calendar year

Communications

Railroads: 3,386 km; 3,236 km 0.914-meter gauge, single track (2,611 km in use), 150

Comoros

km 1.435-meter gauge

Highways: 75,450 km total; 9,350 km paved, 66,100 km earth and gravel surfaces

Inland waterways: 14,300 km, navigable by river boats

Pipelines: crude oil 3,585 km; petroleum products 1,350 km; natural gas 830 km; natural gas liquids 125 km

Ports: Barranquilla, Buenaventura, Cartagena, Covenas, San Andres, Santa Marta, Tumaco

Merchant marine: 27 ships (1,000 GRT or over) totaling 227,719 GRT/356,665 DWT; includes 9 cargo, 3 oil tanker, 8 bulk, 7 container

Airports:
total: 1,233
usable: 1,059
with permanent-surface: 69
with runways over 3,659 m: 1
with runways 2,440-2,459 m: 9
with runways 1,220-2,439 m: 200

Telecommunications: nationwide radio relay system; 1,890,000 telephones; broadcast stations—413 AM, no FM, 33 TV, 28 shortwave; 2 Atlantic Ocean INTELSAT earth stations and 11 domestic satellite earth stations

Defense Forces

Branches: Army (Ejercito Nacional), Navy (Armada Nacional, including Marines), Air Force (Fuerza Aerea Colombiana), National Police (Policia Nacional)

Manpower availability: males age 15-49 9,428,358; fit for military service 6,375,944; reach military age (18) annually 356,993 (1993 est.)

Defense expenditures: exchange rate conversion—$630 million, 1.3% of GDP (1993 est.)

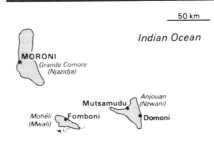

Geography

Location: in the extreme northern Mozambique Channel, about two-thirds of the way between northern Madagascar and northern Mozambique

Map references: Africa, Standard Time Zones of the World

Area:
total area: 2,170 km²
land area: 2,170 km²
comparative area: slightly more than 12 times the size of Washington, DC

Land boundaries: 0 km

Coastline: 340 km

Maritime claims:
exclusive economic zone: 200 nm
territorial sea: 12 nm

International disputes: claims French-administered Mayotte

Climate: tropical marine; rainy season (November to May)

Terrain: volcanic islands, interiors vary from steep mountains to low hills

Natural resources: negligible

Land use:
arable land: 35%
permanent crops: 8%
meadows and pastures: 7%
forest and woodland: 16%
other: 34%

Irrigated land: NA km²

Environment: soil degradation and erosion; deforestation; cyclones possible during rainy season

Note: important location at northern end of Mozambique Channel

People

Population: 511,651 (July 1993 est.)

Population growth rate: 3.54% (1993 est.)

Birth rate: 46.75 births/1,000 population (1993 est.)

Death rate: 11.31 deaths/1,000 population (1993 est.)

Net migration rate: 0 migrant(s)/1,000 population (1993 est.)

Infant mortality rate: 81.8 deaths/1,000 live births (1993 est.)

Life expectancy at birth:
total population: 57.35 years
male: 55.23 years
female: 59.55 years (1993 est.)

Total fertility rate: 6.86 children born/woman (1993 est.)

Nationality:
noun: Comoran(s)
adjective: Comoran

Ethnic divisions: Antalote, Cafre, Makoa, Oimatsaha, Sakalava

Religions: Sunni Muslim 86%, Roman Catholic 14%

Languages: Arabic (official), French (official), Comoran (a blend of Swahili and Arabic)

Literacy: age 15 and over can read and write (1980)
total population: 48%
male: 56%
female: 40%

Labor force: 140,000 (1982)
by occupation: agriculture 80%, government 3%
note: 51% of population of working age (1985)

Government

Names:
conventional long form: Federal Islamic Republic of the Comoros
conventional short form: Comoros
local long form: Republique Federale Islamique des Comores
local short form: Comores

Digraph: CN

Type: independent republic

Capital: Moroni

Administrative divisions: three islands; Njazidja (Grand Comore), Nzwani (Anjouan), and Mwali (Moheli)
note: there are also four municipalities named Domoni, Fomboni, Moroni, and Mutsamudu

Independence: 6 July 1975 (from France)

Constitution: 7 June 1992

Legal system: French and Muslim law in a new consolidated code

National holiday: Independence Day, 6 July (1975)

Political parties and leaders: over 20 political parties are currently active, the most important of which are; Comoran Union for Progress (UDZIMA), Omar TAMOU; Islands' Fraternity and Unity Party (CHUMA), Said Ali KEMAL; Comoran Party for Democracy and Progress (PCDP), Ali MROUDJAE; Realizing Freedom's Capability (UWEZO), Mouazair ABDALLAH; Democratic Front of the Comoros (FDR), Moustapha CHELKH; Dialogue Proposition Action

Comoros (continued)

(DPA/MWANGAZA), Said MCHAWGAMA; Rally for Change and Democracy (RACHADE), Hassan HACHIM; Union for Democracy and Decentralization (UNDC), Mohamed Taki Halidi IBRAHAM; Maecha Bora, leader NA; MDP/NGDC (expansion NA), leader NA; Comoran Popular Front (FPC), Mohamed HASSANALI, Mohamed El Arif OUKACHA, Abdou MOUSTAKIM (Secretary General)

Suffrage: 18 years of age; universal

Elections:

Federal Assembly: last held November-December 1992 (next to be held NA March 1997); results—percent of vote by party NA; seats—(42 total) UNDC 7, CHUMA 3, ADP 2, MDP/NGDC 5, FDC 2, MAECHA BORA 2, FPC 2, RACHADE 1, UWEZO 1, MWANGAZA 1, 16 other seats to smaller parties

President: last held 11 March 1990 (next to be held March 1996); results—Said Mohamed DJOHAR (UDZIMA) 55%, Mohamed TAKI Abdulkarim (UNDC) 45%

Executive branch: president, Council of Ministers (cabinet), prime minister

Legislative branch: unicameral Federal Assembly (Assemblee Federale)

Judicial branch: Supreme Court (Cour Supreme)

Leaders:

Chief of State and Head of Government: President Said Mohamed DJOHAR (since 11 March 1990); Prime Minister Ibrahim HALIDI (since 1 January 1992)

Member of: ACCT, ACP, AfDB, ECA, FAO, FZ, G-77, IBRD, ICAO, IDA, IDB, IFAD, IFC, ILO, IMF, ITU, NAM, OAU, OIC, UN, UNCTAD, UNESCO, UNIDO, UPU, WHO, WMO

Diplomatic representation in US:

chief of mission: Ambassador Amini Ali MOUMIN

chancery: (temporary) at the Comoran Permanent Mission to the UN, 336 East 45th Street, 2nd Floor, New York, NY 10017

telephone: (212) 972-8010

US diplomatic representation:

chief of mission: Ambassador Kenneth N. PELTIER

embassy: address NA, Moroni

mailing address: B. P. 1318, Moroni

telephone: [269] 73-22-03, 73-29-22

FAX: no service available at this time

Flag: green with a white crescent placed diagonally (closed side of the crescent points to the upper hoist-side corner of the flag); there are four white five-pointed stars placed in a line between the points of the crescent; the crescent, stars, and color green are traditional symbols of Islam; the four stars represent the four main islands of the archipelago—Mwali, Njazidja, Nzwani, and Mayotte (which is a territorial collectivity of France, but claimed by the Comoros)

Economy

Overview: One of the world's poorest countries, Comoros is made up of several islands that have poor transportation links, a young and rapidly increasing population, and few natural resources. The low educational level of the labor force contributes to a low level of economic activity, high unemployment, and a heavy dependence on foreign grants and technical assistance. Agriculture, including fishing, hunting, and forestry, is the leading sector of the economy. It contributes 40% to GDP, employs 80% of the labor force, and provides most of the exports. The country is not self-sufficient in food production, and rice, the main staple, accounts for 90% of imports. During the period 1982-86 the industrial sector grew at an annual average rate of 5.3%, but its contribution to GDP was only 5% in 1988. Despite major investment in the tourist industry, which accounts for about 25% of GDP, growth has stagnated since 1983. A sluggish growth rate of 1.5% during 1985-90 has led to large budget deficits, declining incomes, and balance-of-payments difficulties. Preliminary estimates for FY92 show a moderate increase in the growth rate based on increased exports, tourism, and government investment outlays.

National product: GDP—exchange rate conversion—$260 million (1991 est.)

National product real growth rate: 2.7% (1991 est.)

National product per capita: $540 (1991 est.)

Inflation rate (consumer prices): 4% (1991 est.)

Unemployment rate: over 16% (1988 est.)

Budget: revenues $96 million; expenditures $88 million, including capital expenditures of $33 million (1991 est.)

Exports: $16 million (f.o.b., 1990 est.)

commodities: vanilla, cloves, perfume oil, copra, ylang-ylang

partners: US 53%, France 41%, Africa 4%, FRG 2% (1988)

Imports: $41 million (f.o.b., 1990 est.)

commodities: rice and other foodstuffs, cement, petroleum products, consumer goods

partners: Europe 62% (France 22%), Africa 5%, Pakistan, China (1988)

External debt: $196 million (1991 est.)

Industrial production: growth rate –6.5% (1989 est.); accounts for 10% of GDP

Electricity: 16,000 kW capacity; 25 million kWh produced, 50 kWh per capita (1991)

Industries: perfume distillation, textiles, furniture, jewelry, construction materials, soft drinks

Agriculture: accounts for 40% of GDP; most of population works in subsistence agriculture and fishing; plantations produce cash crops for export—vanilla, cloves, perfume essences, copra; principal food crops—coconuts, bananas, cassava; world's leading producer of essence of ylang-ylang (for perfumes) and second-largest producer of vanilla; large net food importer

Economic aid: US commitments, including Ex-Im (FY80-89), $10 million; Western (non-US) countries, ODA and OOF bilateral commitments (1970-89), $435 million; OPEC bilateral aid (1979-89), $22 million; Communist countries (1970-89), $18 million

Currency: 1 Comoran franc (CF) = 100 centimes

Exchange rates: Comoran francs (CF) per US$1—274.06 (January 1993), 264.69 (1992), 282.11 (1991), 272.26 (1990), 319.01 (1989), 297.85 (1988),; note—linked to the French franc at 50 to 1 French franc

Fiscal year: calendar year

Communications

Highways: 750 km total; about 210 km bituminous, remainder crushed stone or gravel

Ports: Mutsamudu, Moroni

Airports:

total: 4

usable: 4

with permanent-surface runways: 4

with runways over 3,659 m: 0

with runways 2,440-3,659 m: 1

with runways 1,220-2,439 m: 3

Telecommunications: sparse system of radio relay and high-frequency radio communication stations for interisland and external communications to Madagascar and Reunion; over 1,800 telephones; broadcast stations—2 AM, 1 FM, no TV

Defense Forces

Branches: Comoran Defense Force (FDC)

Manpower availability: males age 15-49 108,867; fit for military service 65,106 (1993 est.)

Defense expenditures: $NA, NA% of GDP

Congo

200 km

Geography

Location: Western Africa, bordering the South Atlantic Ocean between Gabon and Zaire
Map references: Africa, Standard Time Zones of the World
Area:
total area: 342,000 km²
land area: 341,500 km²
comparative area: slightly smaller than Montana
Land boundaries: total 5,504 km, Angola 201 km, Cameroon 523 km, Central African Republic 467 km, Gabon 1,903 km, Zaire 2,410 km
Coastline: 169 km
Maritime claims:
territorial sea: 200 nm
International disputes: long section with Zaire along the Congo River is indefinite (no division of the river or its islands has been made)
Climate: tropical; rainy season (March to June); dry season (June to October); constantly high temperatures and humidity; particularly enervating climate astride the Equator
Terrain: coastal plain, southern basin, central plateau, northern basin
Natural resources: petroleum, timber, potash, lead, zinc, uranium, copper, phosphates, natural gas
Land use:
arable land: 2%
permanent crops: 0%
meadows and pastures: 29%
forest and woodland: 62%
other: 7%
Irrigated land: 40 km² (1989)
Environment: deforestation; about 70% of the population lives in Brazzaville, Pointe Noire, or along the railroad between them

People

Population: 2,388,667 (July 1993 est.)
Population growth rate: 2.44% (1993 est.)

Birth rate: 40.68 births/1,000 population (1993 est.)
Death rate: 16.28 deaths/1,000 population (1993 est.)
Net migration rate: 0 migrant(s)/1,000 population (1993 est.)
Infant mortality rate: 112.7 deaths/1,000 live births (1993 est.)
Life expectancy at birth:
total population: 48.04 years
male: 46.3 years
female: 49.84 years (1993 est.)
Total fertility rate: 5.38 children born/woman (1993 est.)
Nationality:
noun: Congolese (singular and plural)
adjective: Congolese or Congo
Ethnic divisions:
south: Kongo 48%
north: Sangha 20%, M'Bochi 12%
center: Teke 17%, Europeans 8,500 (mostly French)
Religions: Christian 50%, animist 48%, Muslim 2%
Languages: French (official), African languages (Lingala and Kikongo are the most widely used)
Literacy: age 15 and over can read and write (1990)
total population: 57%
male: 70%
female: 44%
Labor force: 79,100 wage earners
by occupation: agriculture 75%, commerce, industry, and government 25%
note: 51% of population of working age; 40% of population economically active (1985)

Government

Names:
conventional long form: Republic of the Congo
conventional short form: Congo
local long form: Republique Populaire du Congo
local short form: Congo
former: Congo/Brazzaville
Digraph: CF
Type: republic
Capital: Brazzaville
Administrative divisions: 9 regions (regions, singular—region) and 1 commune*; Bouenza, Brazzaville*, Cuvette, Kouilou, Lekoumou, Likouala, Niari, Plateaux, Pool, Sangha
Independence: 15 August 1960 (from France)
Constitution: 8 July 1979, currently being modified
Legal system: based on French civil law system and customary law
National holiday: Congolese National Day, 15 August (1960)

Political parties and leaders: Congolese Labor Party (PCT), headed by former president Denis SASSOU-NGUESSO; Union for Democratic Renewal (URD)—a coalition of opposition parties; Panafrican Union for Social Development (UPADS)
Other political or pressure groups: Union of Congolese Socialist Youth (UJSC); Congolese Trade Union Congress (CSC); Revolutionary Union of Congolese Women (URFC); General Union of Congolese Pupils and Students (UGEEC)
Suffrage: 18 years of age; universal
Elections:
President: last held 2-16 August 1992 (next to be held August 1997); results—President Pascal LISSOUBA won with 61% of the vote
National Assembly: last held 24 June-19 July 1992; results—(125 total) UPADS 39, MCDDI (part of URD coalition) 29, PCT 19; more than a dozen smaller parties split the remaining 38 seats
note: National Assembly dissolved in November 1992; next election to be held May 1993
Executive branch: president, prime minister, Council of Ministers (cabinet)
Legislative branch: unicameral National Assembly (Assemblee Nationale) was dissolved on NA November 1992
Judicial branch: Supreme Court (Cour Supreme)
Leaders:
Chief of State: President Pascal LISSOUBA (since August 1992)
Head of Government: Prime Minister Claude Antoine DA COSTA (since December 1992)
Member of: ACCT, ACP, AfDB, BDEAC, CCC, CEEAC, ECA, FAO, FZ, G-77, GATT, IBRD, ICAO, IDA, IFAD, IFC, ILO, IMF, IMO, INTELSAT, INTERPOL, IOC, ITU, LORCS, NAM, OAU, UDEAC, UN, UNAVEM II, UNCTAD, UNESCO, UNIDO, UNTAC, UPU, WFTU, WHO, WIPO, WMO, WTO
Diplomatic representation in US:
chief of mission: Ambassador Roger ISSOMBO
chancery: 4891 Colorado Avenue NW, Washington, DC 20011
telephone: (202) 726-5500
US diplomatic representation:
chief of mission: Ambassador James Daniel PHILLIPS
embassy: Avenue Amilcar Cabral, Brazzaville
mailing address: B. P. 1015, Brazzaville, or Box C, APO AE 09828
telephone: (242) 83-20-70
FAX: [242] 83-63-38
Flag: red, divided diagonally from the lower hoist side by a yellow band; the upper triangle (hoist side) is green and the lower triangle is red; uses the popular pan-African colors of Ethiopia

Congo (continued)

Economy

Overview: Congo's economy is a mixture of village agriculture and handicrafts, a beginning industrial sector based largely on oil, supporting services, and a government characterized by budget problems and overstaffing. A reform program, supported by the IMF and World Bank, ran into difficulties in 1990-91 because of problems in changing to a democratic political regime and a heavy debt-servicing burden. Oil has supplanted forestry as the mainstay of the economy, providing about two-thirds of government revenues and exports. In the early 1980s rapidly rising oil revenues enabled Congo to finance large-scale development projects with growth averaging 5% annually, one of the highest rates in Africa. During the period 1987-91, however, growth has slowed to an average of roughly 1.5% annually, only half the population growth rate. The new government, responding to pressure from businessmen and the electorate, has promised to reduce the bureaucracy and government regulation but little has been accomplished as of early 1993.

National product: GDP—exchange rate conversion—$2.5 billion (1991 est.)

National product real growth rate: 0.6% (1991 est.)

National product per capita: $1,070 (1991 est.)

Inflation rate (consumer prices): -0.6% (1991 est.)

Unemployment rate: NA%

Budget: revenues $765 million; expenditures $952 million, including capital expenditures of $65 million (1990)

Exports: $1.1 billion (f.o.b., 1990)
commodities: crude oil 72%, lumber, plywood, coffee, cocoa, sugar, diamonds
partners: US, France, other EC countries

Imports: $704 million (c.i.f., 1990)
commodities: foodstuffs, consumer goods, intermediate manufactures, capital equipment
partners: France, Italy, other EC countries, US, Germany, Spain, Japan, Brazil

External debt: $4.1 billion (1991)

Industrial production: growth rate 1.2% (1989); accounts for 33% of GDP; includes petroleum

Electricity: 140,000 kW capacity; 315 million kWh produced, 135 kWh per capita (1991)

Industries: petroleum, cement, lumbering, brewing, sugar milling, palm oil, soap, cigarette

Agriculture: accounts for 13% of GDP (including fishing and forestry); cassava accounts for 90% of food output; other crops—rice, corn, peanuts, vegetables; cash crops include coffee and cocoa; forest products important export earner; imports over 90% of food needs

Economic aid: US commitments, including Ex-Im (FY70-90), $63 million; Western (non-US) countries, ODA and OOF bilateral commitments (1970-90), $2.5 billion; OPEC bilateral aid (1979-89), $15 million; Communist countries (1970-89), $338 million

Currency: 1 CFA franc (CFAF) = 100 centimes

Exchange rates: Communaute Financiere Africaine francs (CFAF) per US$1—274.06 (January 1993), 264.69 (1992), 282.11 (1991), 272.26 (1990), 319.01 (1989), 297.85 (1988)

Fiscal year: calendar year

Communications

Railroads: 797 km, 1.067-meter gauge, single track (includes 285 km that are privately owned)

Highways: 11,960 km total; 560 km paved; 850 km gravel and laterite; 5,350 km improved earth; 5,200 km unimproved earth

Inland waterways: the Congo and Ubangi (Oubangui) Rivers provide 1,120 km of commercially navigable water transport; the rest are used for local traffic only

Pipelines: crude oil 25 km

Ports: Pointe-Noire (ocean port), Brazzaville (river port)

Airports:
total: 44
usable: 41
with permanent-surface runways: 5
with runways over 3,659 m: 0
with runways 2,440-3,659 m: 1
with runways 1,220-2,439 m: 16

Telecommunications: services adequate for government use; primary network is composed of radio relay routes and coaxial cables; key centers are Brazzaville, Pointe-Noire, and Loubomo; 18,100 telephones; broadcast stations—4 AM, 1 FM, 4 TV; 1 Atlantic Ocean satellite earth station

Defense Forces

Branches: Army, Navy (including Marines), Air Force, National Police

Manpower availability: males age 15-49 534,802; fit for military service 272,051; reach military age (20) annually 24,190 (1993 est.)

Defense expenditures: exchange rate conversion—$NA, NA% of GDP

Cook Islands
(free association with New Zealand)

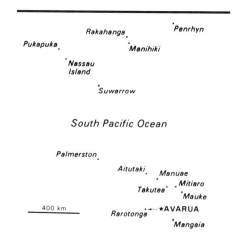

Geography

Location: Oceania, 4,500 km south of Hawaii in the South Pacific Ocean, about halfway between Hawaii and New Zealand

Map references: Oceania

Area:
total area: 240 km²
land area: 240 km²
comparative area: slightly less than 1.3 times the size of Washington, DC

Land boundaries: 0 km

Coastline: 120 km

Maritime claims:
continental shelf: 200 nm or the edge of continental margin
exclusive economic zone: 200 nm
territorial sea: 12 nm

International disputes: none

Climate: tropical; moderated by trade winds

Terrain: low coral atolls in north; volcanic, hilly islands in south

Natural resources: negligible

Land use:
arable land: 4%
permanent crops: 22%
meadows and pastures: 0%
forest and woodland: 0%
other: 74%

Irrigated land: NA km²

Environment: subject to typhoons from November to March

People

Population: 18,903 (July 1993 est.)

Population growth rate: 1.18% (1993 est.)

Birth rate: 23.4 births/1,000 population (1993 est.)

Death rate: 5.2 deaths/1,000 population (1993 est.)

Net migration rate: -6.45 migrant(s)/1,000 population (1993 est.)

Infant mortality rate: 24.7 deaths/1,000 live births (1993 est.)

Life expectancy at birth:
total population: 71.14 years

male: 69.2 years
female: 73.1 years (1993 est.)
Total fertility rate: 3.32 children born/woman (1993 est.)
Nationality:
noun: Cook Islander(s)
adjective: Cook Islander
Ethnic divisions: Polynesian (full blood) 81.3%, Polynesian and European 7.7%, Polynesian and other 7.7%, European 2.4%, other 0.9%
Religions: Christian (majority of populace members of Cook Islands Christian Church)
Languages: English (official), Maori
Literacy:
total population: NA%
male: NA%
female: NA%
Labor force: 5,810
by occupation: agriculture 29%, government 27%, services 25%, industry 15%, other 4% (1981)

Government

Names:
conventional long form: none
conventional short form: Cook Islands
Digraph: CW
Type: self-governing parliamentary government in free association with New Zealand; Cook Islands is fully responsible for internal affairs; New Zealand retains responsibility for external affairs, in consultation with the Cook Islands
Capital: Avarua
Administrative divisions: none
Independence: none (became self-governing in free association with New Zealand on 4 August 1965 and has the right at any time to move to full independence by unilateral action)
Constitution: 4 August 1965
Legal system: NA
National holiday: Constitution Day, 4 August
Political parties and leaders: Cook Islands Party, Geoffrey HENRY; Democratic Tumu Party, Vincent INGRAM; Democratic Party, Terepai MAOATE; Cook Islands Labor Party, Rena JONASSEN; Cook Islands People's Party, Sadaraka SADARAKA
Suffrage: universal adult at age NA
Elections:
Parliament: last held 19 January 1989 (next to be held by January 1994); results—percent of vote by party NA; seats—(24 total) Cook Islands Party 12, Democratic Tumu Party 2, opposition coalition (including Democratic Party) 9, independent 1
Executive branch: British monarch, representative of the UK, representative of New Zealand, prime minister, deputy prime minister, Cabinet

Legislative branch: unicameral Parliament; note—the House of Arikis (chiefs) advises on traditional matters, but has no legislative powers
Judicial branch: High Court
Leaders:
Chief of State: Queen ELIZABETH II (since 6 February 1952); Representative of the UK Sir Tangaroa TANGAROA (since NA); Representative of New Zealand Adrian SINCOCK (since NA)
Head of Government: Prime Minister Geoffrey HENRY (since 1 February 1989); Deputy Prime Minister Inatio AKARURU (since NA February 1989)
Member of: AsDB, ESCAP (associate), ICAO, IOC, SPARTECA, SPC, SPF, UNESCO, WHO
Diplomatic representation in US: none (self-governing in free association with New Zealand)
US diplomatic representation: none (self-governing in free association with New Zealand)
Flag: blue, with the flag of the UK in the upper hoist-side quadrant and a large circle of 15 white five-pointed stars (one for every island) centered in the outer half of the flag

Economy

Overview: Agriculture provides the economic base. The major export earners are fruit, copra, and clothing. Manufacturing activities are limited to a fruit-processing plant and several clothing factories. Economic development is hindered by the isolation of the islands from foreign markets and a lack of natural resources and good transportation links. A large trade deficit is annually made up for by remittances from emigrants and from foreign aid. Current economic development plans call for exploiting the tourism potential and expanding the fishing industry.
National product: GDP—exchange rate conversion—$40 million (1988 est.)
National product real growth rate: 5.3% (1986-88 est.)
National product per capita: $2,200 (1988 est.)
Inflation rate (consumer prices): 8% (1988)
Unemployment rate: NA%
Budget: revenues $33.8 million; expenditures $34.4 million, including capital expenditures of $NA (1990 est.)
Exports: $4.0 million (f.o.b., 1988)
commodities: copra, fresh and canned fruit, clothing
partners: NZ 80%, Japan
Imports: $38.7 million (c.i.f., 1988)
commodities: foodstuffs, textiles, fuels, timber
partners: NZ 49%, Japan, Australia, US

External debt: $NA
Industrial production: growth rate NA%
Electricity: 14,000 kW capacity; 21 million kWh produced, 1,170 kWh per capita (1990)
Industries: fruit processing, tourism
Agriculture: export crops—copra, citrus fruits, pineapples, tomatoes, bananas; subsistence crops—yams, taro
Economic aid: Western (non-US) countries, ODA and OOF bilateral commitments (1970-89), $128 million
Currency: 1 New Zealand dollar (NZ$) = 100 cents
Exchange rates: New Zealand dollars (NZ$) per US$1—1.9490 (January 1993), 1.8584 (1992), 1.7266 (1991), 1.6750 (1990), 1.6711 (1989), 1.5244 (1988)
Fiscal year: 1 April-31 March

Communications

Highways: 187 km total (1980); 35 km paved, 35 km gravel, 84 km improved earth, 33 km unimproved earth
Ports: Avatiu
Merchant marine: 1 cargo ship (1,000 or over) totaling 1,464 GRT/2,181 DWT
Airports:
total: 7
usable: 7
with permanent-surface runways: 1
with runways over 3,659 m: 0
with runways 2,440-3,659 m: 0
with runways 1,220-2,439 m: 5
Telecommunications: broadcast stations—1 AM, 1 FM, 1 TV; 11,000 radio receivers; 17,000 TV receivers (1989); 2,052 telephones; 1 Pacific Ocean INTELSAT earth station

Defense Forces

Note: defense is the responsibility of New Zealand

Coral Sea Islands
(territory of Australia)

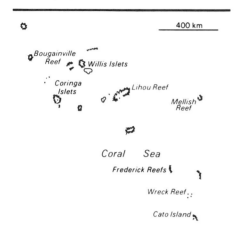

Geography

Location: Oceania, just off the northeast coast of Australia in the Coral Sea
Map references: Oceania
Area:
total area: less than 3 km²
land area: less than 3 km²
comparative area: NA
note: includes numerous small islands and reefs scattered over a sea area of about 1 million km², with Willis Islets the most important
Land boundaries: 0 km
Coastline: 3,095 km
Maritime claims:
exclusive fishing zone: 200 nm
territorial sea: 3 nm
International disputes: none
Climate: tropical
Terrain: sand and coral reefs and islands (or cays)
Natural resources: negligible
Land use:
arable land: 0%
permanent crops: 0%
meadows and pastures: 0%
forest and woodland: 0%
other: 100% (mostly grass or scrub cover)
Irrigated land: 0 km²
Environment: subject to occasional tropical cyclones; no permanent fresh water; important nesting area for birds and turtles

People

Population: no indigenous inhabitants; note—there are 3 meteorologists

Government

Names:
conventional long form: Coral Sea Islands Territory
conventional short form: Coral Sea Islands
Digraph: CR

Type: territory of Australia administered by the Ministry for Arts, Sport, the Environment, Tourism, and Territories
Capital: none; administered from Canberra, Australia
Independence: none (territory of Australia)
Flag: the flag of Australia is used

Economy

Overview: no economic activity

Communications

Ports: none; offshore anchorages only

Defense Forces

Note: defense is the responsibility of Australia; visited regularly by the Royal Australian Navy; Australia has control over the activities of visitors

Costa Rica

Isla del Coco
is not shown.

Geography

Location: Central America, between Nicaragua and Panama
Map references: Central America and the Caribbean, South America
Area:
total area: 51,100 km²
land area: 50,660 km²
comparative area: slightly smaller than West Virginia
note: includes Isla del Coco
Land boundaries: total 639 km, Nicaragua 309 km, Panama 330 km
Coastline: 1,290 km
Maritime claims:
continental shelf: 200 nm
exclusive economic zone: 200 nm
territorial sea: 12 nm
International disputes: none
Climate: tropical; dry season (December to April); rainy season (May to November)
Terrain: coastal plains separated by rugged mountains
Natural resources: hydropower potential
Land use:
arable land: 6%
permanent crops: 7%
meadows and pastures: 45%
forest and woodland: 34%
other: 8%
Irrigated land: 1,180 km² (1989 est.)
Environment: subject to occasional earthquakes, hurricanes along Atlantic coast; frequent flooding of lowlands at onset of rainy season; active volcanoes; deforestation; soil erosion

People

Population: 3,264,776 (July 1993 est.)
Population growth rate: 2.38% (1993 est.)
Birth rate: 26.07 births/1,000 population (1993 est.)
Death rate: 3.57 deaths/1,000 population (1993 est.)
Net migration rate: 1.26 migrant(s)/1,000 population (1993 est.)

Infant mortality rate: 11.6 deaths/1,000 live births (1993 est.)
Life expectancy at birth:
total population: 77.49 years
male: 75.56 years
female: 79.52 years (1993 est.)
Total fertility rate: 3.11 children born/woman (1993 est.)
Nationality:
noun: Costa Rican(s)
adjective: Costa Rican
Ethnic divisions: white (including mestizo) 96%, black 2%, Indian 1%, Chinese 1%
Religions: Roman Catholic 95%
Languages: Spanish (official), English; spoken around Puerto Limon
Literacy: age 15 and over can read and write (1990)
total population: 93%
male: 93%
female: 93%
Labor force: 868,300
by occupation: industry and commerce 35.1%, government and services 33%, agriculture 27%, other 4.9% (1985 est.)

Government

Names:
conventional long form: Republic of Costa Rica
conventional short form: Costa Rica
local long form: Republica de Costa Rica
local short form: Costa Rica
Digraph: CS
Type: democratic republic
Capital: San Jose
Administrative divisions: 7 provinces (provincias, singular—provincia); Alajuela, Cartago, Guanacaste, Heredia, Limon, Puntarenas, San Jose
Independence: 15 September 1821 (from Spain)
Constitution: 9 November 1949
Legal system: based on Spanish civil law system; judicial review of legislative acts in the Supreme Court; has not accepted compulsory ICJ jurisdiction
National holiday: Independence Day, 15 September (1821)
Political parties and leaders: National Liberation Party (PLN), Carlos Manuel CASTILLO Morales; Social Christian Unity Party (PUSC), Rafael Angel CALDERON Fournier; Marxist Popular Vanguard Party (PVP), Humberto VARGAS Carbonell; New Republic Movement (MNR), Sergio Erick ARDON Ramirez; Progressive Party (PP), Isaac Felipe AZOFEIFA Bolanos; People's Party of Costa Rica (PPC), Lenin CHACON Vargas; Radical Democratic Party (PRD), Juan Jose ECHEVERRIA Brealey
Other political or pressure groups: Costa Rican Confederation of Democratic Workers (CCTD; Liberation Party affiliate);

Confederated Union of Workers (CUT; Communist Party affiliate); Authentic Confederation of Democratic Workers (CATD; Communist Party affiliate); Chamber of Coffee Growers; National Association for Economic Development (ANFE); Free Costa Rica Movement (MCRL; rightwing militants); National Association of Educators (ANDE)
Suffrage: 18 years of age; universal and compulsory
Elections:
Legislative Assembly: last held 4 February 1990 (next to be held February 1994); results—percent of vote by party NA; seats—(57 total) PUSC 29, PLN 25, PVP/PPC 1, regional parties 2
President: last held 4 February 1990 (next to be held February 1994); results—Rafael Angel CALDERON Fournier 51%, Carlos Manuel CASTILLO 47%
Executive branch: president, two vice presidents, Cabinet
Legislative branch: unicameral Legislative Assembly (Asamblea Legislativa)
Judicial branch: Supreme Court (Corte Suprema)
Leaders:
Chief of State and Head of Government: President Rafael Angel CALDERON Fournier (since 8 May 1990); First Vice President German SERRANO Pinto (since 8 May 1990); Second Vice President Arnoldo LOPEZ Echandi (since 8 May 1990)
Member of: AG (observer), BCIE, CACM, ECLAC, FAO, G-77, GATT, IADB, IAEA, IBRD, ICAO, ICFTU, IDA, IFAD, IFC, ILO, IMF, IMO, INTELSAT, INTERPOL, IOC, IOM, ITU, LAES, LAIA (observer), LORCS, NAM (observer), OAS, OPANAL, UN, UNCTAD, UNESCO, UNIDO, UPU, WCL, WFTU, WHO, WIPO, WMO
Diplomatic representation in US:
chief of mission: Ambassador Gonzalo FACIO Segreda
chancery: Suite 211, 1825 Connecticut Avenue NW, Washington, DC 20009
telephone: (202) 234-2945 through 2947
consulates general: Albuquerque, Houston, Los Angeles, Miami, New Orleans, New York, San Diego, San Francisco, and San Juan (Puerto Rico)
consulate: Buffalo
US diplomatic representation:
chief of mission: Ambassador Luis GUINOT, Jr.
embassy: Pavas Road, San Jose
mailing address: APO AA 34020
telephone: [506] 20-39-39
FAX: (506) 20-2305
Flag: five horizontal bands of blue (top), white, red (double width), white, and blue, with the coat of arms in a white disk on the hoist side of the red band

Economy

Overview: In 1992 the economy grew at an estimated 5.4%, up from the 2.5% gain of 1991 and the gain of 1990. Increases in agricultural production (on the strength of good coffee and banana crops) and in nontraditional exports are responsible for much of the growth. In 1992 consumer prices rose by 17%, below the 27% of 1991. The trade deficit of $100 million was substantially below the 1991 deficit of $270 million. Unemployment is officially reported at 4.0%, but much underemployment remains. External debt, on a per capita basis, is among the world's highest.
National product: GDP—exchange rate conversion—$6.4 billion (1992 est.)
National product real growth rate: 5.4% (1992 est.)
National product per capita: $2,000 (1992 est.)
Inflation rate (consumer prices): 17% (1992 est.)
Unemployment rate: 4% (1992)
Budget: revenues $1.1 billion; expenditures $1.34 billion, including capital expenditures of $110 million (1991 est.)
Exports: $1.7 billion (f.o.b., 1992 est.)
commodities: coffee, bananas, textiles, sugar
partners: US 75%, Germany, Guatemala, Netherlands, UK, Japan
Imports: $1.8 billion (c.i.f., 1992 est.)
commodities: raw materials, consumer goods, capital equipment, petroleum
partners: US 45%, Japan, Guatemala, Germany
External debt: $3.2 billion (1991)
Industrial production: growth rate 1.0% (1991); accounts for 19% of GDP
Electricity: 927,000 kW capacity; 3,612 million kWh produced, 1,130 kWh per capita (1992)
Industries: food processing, textiles and clothing, construction materials, fertilizer, plastic products
Agriculture: accounts for 17% of GDP and 70% of exports; cash commodities—coffee, beef, bananas, sugar; other food crops include corn, rice, beans, potatoes; normally self-sufficient in food except for grain; depletion of forest resources resulting in lower timber output
Illicit drugs: illicit production of cannabis on small scattered plots; transshipment country for cocaine from South America
Economic aid: US commitments, including Ex-Im (FY70-89), $1.4 billion; Western (non-US) countries, ODA and OOF bilateral commitments (1970-89), $935 million; Communist countries (1971-89), $27 million
Currency: 1 Costa Rican colon (C) = 100 centimos
Exchange rates: Costa Rican colones (C) per US$1—137.72 (January 1993), 134.51

Costa Rica *(continued)*

(1992), 122.43 (1991), 91.58 (1990), 81.504
(1989), 75.805 (1988)
Fiscal year: calendar year

Communications

Railroads: 950 km total, all 1.067-meter
gauge; 260 km electrified
Highways: 15,400 km total; 7,030 km
paved, 7,010 km gravel, 1,360 km
unimproved earth
Inland waterways: about 730 km,
seasonally navigable
Pipelines: petroleum products 176 km
Ports: Puerto Limon, Caldera, Golfito,
Moin, Puntarenas
Merchant marine: 1 cargo ship (1,000 GRT
or over) totaling 2,878 GRT/4,506 DWT
Airports:
total: 162
usable: 144
with permanent-surface runways: 28
with runways over 3,659 m: 0
with runways 2,440-3,659 m: 2
with runways 1,220-2,439 m: 8
Telecommunications: very good domestic
telephone service; 292,000 telephones;
connection into Central American
Microwave System; broadcast stations—71
AM, no FM, 18 TV, 13 shortwave; 1
Atlantic Ocean INTELSAT earth station

Defense Forces

Branches: Civil Guard, Rural Assistance
Guard
note: constitution prohibits armed forces
Manpower availability: males age 15-49
851,713; fit for military service 573,854;
reach military age (18) annually 31,987
(1993 est.)
Defense expenditures: exchange rate
conversion—$22 million, 0.5% of GDP
(1989)

Cote d'Ivoire
(also known as Ivory Coast)

Geography

Location: Western Africa, bordering the
North Atlantic Ocean between Ghana and
Liberia
Map references: Africa, Standard Time
Zones of the World
Area:
total area: 322,460 km²
land area: 318,000 km²
comparative area: slightly larger than New
Mexico
Land boundaries: total 3,110 km, Burkina
584 km, Ghana 668 km, Guinea 610 km,
Liberia 716 km, Mali 532 km
Coastline: 515 km
Maritime claims:
continental shelf: 200 m depth
exclusive economic zone: 200 nm
territorial sea: 12 nm
International disputes: none
Climate: tropical along coast, semiarid in
far north; three seasons—warm and dry
(November to March), hot and dry (March
to May), hot and wet (June to October)
Terrain: mostly flat to undulating plains;
mountains in northwest
Natural resources: petroleum, diamonds,
manganese, iron ore, cobalt, bauxite, copper
Land use:
arable land: 9%
permanent crops: 4%
meadows and pastures: 9%
forest and woodland: 26%
other: 52%
Irrigated land: 620 km² (1989 est.)
Environment: coast has heavy surf and no
natural harbors; severe deforestation

People

Population: 13,808,447 (July 1993 est.)
Population growth rate: 3.5% (1993 est.)
Birth rate: 46.88 births/1,000 population
(1993 est.)
Death rate: 15.07 deaths/1,000 population
(1993 est.)

Net migration rate: 3.15 migrant(s)/1,000
population (1993 est.)
Infant mortality rate: 97 deaths/1,000 live
births (1993 est.)
Life expectancy at birth:
total population: 48.97 years
male: 46.98 years
female: 51.03 years (1993 est.)
Total fertility rate: 6.73 children
born/woman (1993 est.)
Nationality:
noun: Ivorian(s)
adjective: Ivorian
Ethnic divisions: Baoule 23%, Bete 18%,
Senoufou 15%, Malinke 11%, Agni, foreign
Africans (mostly Burkinabe about 2 million),
non-Africans 130,000 to 330,000 (French
30,000 and Lebanese 100,000 to 300,000)
Religions: indigenous 63%, Muslim 25%,
Christian 12%
Languages: French (official), 60 native
dialects Dioula is the most widely spoken
Literacy: age 15 and over can read and
write (1990)
total population: 54%
male: 67%
female: 40%
Labor force: 5.718 million
by occupation: over 85% of population
engaged in agriculture, forestry, livestock
raising; about 11% of labor force are wage
earners, nearly half in agriculture and the
remainder in government, industry,
commerce, and professions
note: 54% of population of working age
(1985)

Government

Names:
conventional long form: Republic of Cote
d'Ivoire
conventional short form: Cote d'Ivoire
local long form: Republique de Cote
d'Ivoire
local short form: Cote d'Ivoire
former: Ivory Coast
Digraph: IV
Type: republic multiparty presidential
regime established 1960
Capital: Yamoussoukro
note: although Yamoussoukro has been the
capital since 1983, Abidjan remains the
administrative center; foreign governments,
including the United States, maintain
presence in Abidjan
Administrative divisions: 49 departments
(departementes, singular—(departement);
Abengourou, Abidjan, Aboisso, Adzope,
Agboville, Bangolo, Beoumi, Biankouma,
Bondoukou, Bongouanou, Bouafle, Bouake,
Bouna, Boundiali, Dabakala, Daloa, Danane,
Daoukro, Dimbokro, Divo, Duekoue,
Ferkessedougou, Gagnoa, Grand-Lahou,
Guiglo, Issia, Katiola, Korhogo, Lakota,

Man, Mankono, Mbahiakro, Odienne, Oume, Sakassou, San-Pedro, Sassandra, Seguela, Sinfra, Soubre, Tabou, Tanda, Tingrela, Tiassale, Touba, Toumodi, Vavoua, Yamoussoukro, Zuenoula

Independence: 7 August 1960 (from France)

Constitution: 3 November 1960

Legal system: based on French civil law system and customary law; judicial review in the Constitutional Chamber of the Supreme Court; has not accepted compulsory ICJ jurisdiction

National holiday: National Day, 7 December

Political parties and leaders: Democratic Party of the Cote d'Ivoire (PDCI), Dr. Felix HOUPHOUET-BOIGNY; Ivorian Popular Front (FPI), Laurent GBAGBO; Ivorian Worker's Party (PIT), Francis WODIE; Ivorian Socialist Party (PSI), Morifere BAMBA; over 20 smaller parties

Suffrage: 21 years of age; universal

Elections:

President: last held 28 October 1990 (next to be held October 1995); results—President Felix HOUPHOUET-BOIGNY received 81% of the vote in his first contested election; he is currently serving his seventh consecutive five-year term

National Assembly: last held 25 November 1990 (next to be held November 1995); results—percent of vote by party NA; seats—(175 total) PDCI 163, FPI 9, PIT 1, independents 2

Executive branch: president, Council of Ministers (cabinet)

Legislative branch: unicameral National Assembly (Assemblee Nationale)

Judicial branch: Supreme Court (Cour Supreme)

Leaders:

Chief of State: President Dr. Felix HOUPHOUET-BOIGNY (since 27 November 1960)

Head of Government: Prime Minister Alassane OUATTARA (since 7 November 1990)

Member of: ACCT, ACP, AfDB, CCC, CEAO, ECA, ECOWAS, Entente, FAO, FZ, G-24, G-77, GATT, IAEA, IBRD, ICAO, ICC, IDA, IFAD, IFC, ILO, IMF, IMO, INTELSAT, INTERPOL, IOC, ISO, ITU, LORCS, NAM, OAU, UN, UNCTAD, UNESCO, UNIDO, UPU, WADB, WCL, WHO, WIPO, WMO, WTO

Diplomatic representation in US:

chief of mission: Ambassador Charles GOMIS

chancery: 2424 Massachusetts Avenue NW, Washington, DC 20008

telephone: (202) 797-0300

US diplomatic representation:

chief of mission: Ambassador Hume A. HORAN

embassy: 5 Rue Jesse Owens, Abidjan

mailing address: 01 B. P. 1712, Abidjan

telephone: [225] 21-09-79 or 21-46-72

FAX: [225] 22-32-59

Flag: three equal vertical bands of orange (hoist side), white, and green; similar to the flag of Ireland, which is longer and has the colors reversed—green (hoist side), white, and orange; also similar to the flag of Italy, which is green (hoist side), white, and red; design was based on the flag of France

Economy

Overview: Cote d'Ivoire is among the world's largest producers and exporters of coffee, cocoa beans, and palm-kernel oil. Consequently, the economy is highly sensitive to fluctuations in international prices for coffee and cocoa and to weather conditions. Despite attempts by the government to diversify, the economy is still largely dependent on agriculture and related industries. The agricultural sector accounts for over one-third of GDP and about 80% of export earnings and employs about 85% of the labor force. A collapse of world cocoa and coffee prices in 1986 threw the economy into a recession, from which the country had not recovered by 1990. Continuing low prices for commodity exports, an overvalued exchange rate, a bloated public-sector wage bill, and a large foreign debt hindered economic recovery in 1991. The government, which has sponsored various economic reform programs, especially in agriculture, projected an increase of 1.6% in GNP in 1992.

National product: GDP—exchange rate conversion—$10 billion (1991)

National product real growth rate: -0.6% (1991)

National product per capita: $800 (1991)

Inflation rate (consumer prices): 1% (1991 est.)

Unemployment rate: 14% (1985)

Budget: revenues $2.3 billion; expenditures $3.6 billion, including capital expenditures of $274 million (1990 est.)

Exports: $2.8 billion (f.o.b., 1990)

commodities: cocoa 30%, coffee 20%, tropical woods 11%, petroleum, cotton, bananas, pineapples, palm oil, cotton

partners: France, FRG, Netherlands, US, Belgium, Spain (1985)

Imports: $1.6 billion (f.o.b., 1990)

commodities: food, capital goods, consumer goods, fuel

partners: France 29%, other EC 29%, Nigeria 16%, US 4%, Japan 3% (1989)

External debt: $15.0 billion (1990 est.)

Industrial production: growth rate 6% (1990); accounts for 11% of GDP

Electricity: 1,210,000 kW capacity; 1,970 million kWh produced, 150 kWh per capita (1991)

Industries: foodstuffs, wood processing, oil refinery, automobile assembly, textiles, fertilizer, beverage

Agriculture: most important sector, contributing one-third to GDP and 80% to exports; cash crops include coffee, cocoa beans, timber, bananas, palm kernels, rubber; food crops—corn, rice, manioc, sweet potatoes; not self-sufficient in bread grain and dairy products

Illicit drugs: illicit producer of cannabis; mostly for local consumption; some international drug trade; transshipment point for Southwest Asian heroin to Europe

Economic aid: US commitments, including Ex-Im (FY70-89), $356 million; Western (non-US) countries, ODA and OOF bilateral commitments (1970-88), $5.2 billion

Currency: 1 CFA franc (CFAF) = 100 centimes

Exchange rates: Communaute Financiere Africaine francs (CFAF) per US$1—274.06 (January 1993), 264.69 (1992), 282.11 (1991), 272.26 (1990), 319.01 (1989), 297.85 (1988)

Fiscal year: calendar year

Communications

Railroads: 660 km (Burkina border to Abidjan, 1.00-meter gauge, single track, except 25 km Abidjan-Anyama section is double track)

Highways: 46,600 km total; 3,600 km paved; 32,000 km gravel, crushed stone, laterite, and improved earth; 11,000 km unimproved

Inland waterways: 980 km navigable rivers, canals, and numerous coastal lagoons

Ports: Abidjan, San-Pedro

Merchant marine: 7 ships (1,000 GRT or over) totaling 71,945 GRT/ 90,684 DWT; includes 1 oil tanker, 1 chemical tanker, 3 container, 2 roll-on/roll-off

Airports:

total: 42

usable: 37

with permanent-surface runways: 7

with runways over 3,659 m: 0

with runways 2,440-3,659 m: 3

with runways 1,220-2,439 m: 15

Telecommunications: well-developed by African standards but operating well below capacity; consists of open-wire lines and radio relay microwave links; 87,700 telephones; broadcast stations—3 AM, 17 FM, 13 TV, 1 Atlantic Ocean and 1 Indian Ocean INTELSAT earth station; 2 coaxial submarine cables

Defense Forces

Branches: Army, Navy, Air Force, paramilitary Gendarmerie, Republican Guard, Military Fire Group

Cote d'Ivoire *(continued)*

Manpower availability: males age 15-49 3,131,016; fit for military service 1,624,401; reach military age (18) annually 145,827 (1993 est.)
Defense expenditures: exchange rate conversion—$200 million, 2.3% of GDP (1988)

Croatia

Geography

Location: Southeastern Europe, on the Balkan Peninsula, bordering the Adriatic Sea, between Slovenia and Bosnia and Herzegovina
Map references: Africa, Ethnic Groups in Eastern Europe, Europe, Standard Time Zones of the World
Area:
total area: 56,538 km²
land area: 56,410 km²
comparative area: slightly smaller than West Virginia
Land boundaries: total 1,843 km, Bosnia and Herzegovina (east) 751 km, Bosnia and Herzegovina (southeast) 91 km, Hungary 292 km, Serbia and Montenegro 254 km (239 km with Serbia; 15 km with Montenego), Slovenia 455 km
Coastline: 5,790 km (mainland 1,778 km, islands 4,012 km)
Maritime claims:
continental shelf: 200 m depth or to depth of exploitation
exclusive economic zone: 12 nm
exclusive fishing zone: 12 nm
territorial sea: 12 nm
International disputes: Serbian enclaves in eastern Croatia and along the western Bosnia and Herzegovinian border; dispute with Slovenia over fishing rights in Adriatic
Climate: Mediterranean and continental; continental climate predominant with hot summers and cold winters; mild winters, dry summers along coast
Terrain: geographically diverse; flat plains along Hungarian border, low mountains and highlands near Adriatic coast, coastline, and islands
Natural resources: oil, some coal, bauxite, low-grade iron ore, calcium, natural asphalt, silica, mica, clays, salt
Land use:
arable land: 32%
permanent crops: 20%
meadows and pastures: 18%
forest and woodland: 15%
other: 15%
Irrigated land: NA km²
Environment: air pollution from metallurgical plants; damaged forest; coastal pollution from industrial and domestic waste; subject to frequent and destructive earthquakes
Note: controls most land routes from Western Europe to Aegean Sea and Turkish Straits

People

Population: 4,694,398 (July 1993 est.)
Population growth rate: 0.07% (1993 est.)
Birth rate: 11.38 births/1,000 population (1993 est.)
Death rate: 10.73 deaths/1,000 population (1993 est.)
Net migration rate: 0 migrant(s)/1,000 population (1993 est.)
Infant mortality rate: 9 deaths/1,000 live births (1993 est.)
Life expectancy at birth:
total population: 73.19 years
male: 69.7 years
female: 76.89 years (1993 est.)
Total fertility rate: 1.66 children born/woman (1993 est.)
Nationality:
noun: Croat(s)
adjective: Croatian
Ethnic divisions: Croat 78%, Serb 12%, Muslim 0.9%, Hungarian 0.5%, Slovenian 0.5%, others 8.1%
Religions: Catholic 76.5%, Orthodox 11.1%, Slavic Muslim 1.2%, Protestant 1.4%, others and unknown 9.8%
Languages: Serbo-Croatian 96%, other 4%
Literacy:
total population: NA%
male: NA%
female: NA%
Labor force: 1,509,489
by occupation: industry and mining 37%, agriculture 16% (1981 est.), government NA%, other

Government

Names:
conventional long form: Republic of Croatia
conventional short form: Croatia
local long form: Republika Hrvatska
local short form: Hrvatska
Digraph: HR
Type: parliamentary democracy
Capital: Zagreb
Administrative divisions: 100 districts (opcine, singular—opcina) Beli Manastir, Biograd (Biograd Na Moru), Bielovar, Bjelovar, Brac, Buje, Buzet, Cabar, Cakovec, Cazma, Cres Losinj, Crikvenica, Daruvar, Delnice, Djakovo (Dakovo), Donja Stubica,

Donji Lapac, Dordevac, Drnis, Dubrovnik, Duga Resa, Dugo Selo, Dvor, Garesnica, Glina, Gospic, Gracac, Grubisno Polje, Hvar, Imotski, Ivanec, Ivanic-Grad, Jastrebarsko, Karlovac, Klanjec, Knin, Koprivnica, Korcula, Kostajnica, Krapina, Krizevci, Krk, Kutina, Labin, Lastovo, Ludbreg, Makarska, Metkovic, Nova Gradiska, Novi Marof, Novska, Obrovac, Ogulin, Omis, Opatija, Orahovica, Osijek, Otocac, Ozalj, Pag, Pazin, Petrinja, Ploce (Kardeljevo), Podravska Slatina, Porec, Pregrada, Pukrac, Pula, Rab, Rijeka, Rovinj, Samobor (part of Zagreb), Senj, Sesvete, Sibenik, Sinj, Sisak, Slavonska Pozega, Slavonski Brod, Slunj, Split (Solin, Kastela), Titova Korenica, Trogir, Valpovo, Varazdin, Vinkovci, Virovitica, Vukovar, Vis, Vojnic, Vrborsko, Vrbovec, Vrgin-Most, Vrgorac, Zabok, Zadar, Zagreb (Grad Zagreb), Zelina (Sveti Ivan Zelina), Zlatar Bistrica, Zupanja

Independence: NA June 1991 (from Yugoslavia)

Constitution: adopted on 2 December 1991

Legal system: based on civil law system

National holiday: Statehood Day, 30 May (1990)

Political parties and leaders: Croatian Democratic Union (HDZ), Stjepan MESIC, chairman of the executive council; Croatian People's Party (HNS), Savka DABCEVIC-KUCAR, president; Croatian Christian Democratic Party (HKDS), Ivan CESAR, president; Croatian Party of Rights, Dobroslav PARAGA; Croatian Social Liberal Party (HSLS), Drazen BUDISA, president; Croatian Peasant Party (HSS), leader NA; Istrian Democratic Assembly (IDS), leader NA; Social-Democratic Party (SDP), leader NA; Croatian National Party (PNS), leader NA

Other political or pressure groups: NA

Suffrage: 16 years of age, if employed; 18 years of age, universal

Elections:
President: last held 4 August 1992 (next to be held NA); Franjo TUDJMAN reelected with about 56% of the vote; Dobroslav PARAGA 5%
House of Parishes: last held 7 February 1993 (next to be held NA February 1997); seats—(68 total; 63 elected, 5 presidentially appointed) HDZ 37, HSLS 16, HSS 5, IDS 3, SDP 1, PNS 1
Chamber of Deputies: last held NA August 1992 (next to be held NA August 1996); seats—(138 total) 87 HDZ

Executive branch: president, prime minister, deputy prime ministers, cabinet

Legislative branch: bicameral Parliament consists of an upper house or House of Parishes (Zupanije Dom) and a lower house or Chamber of Deputies (Predstavnicke Dom)

Judicial branch: Supreme Court, Constitutional Court

Leaders:
Chief of State: President Franjo TUDJMAN (since 30 May 1990)
Head of Government: Prime Minister Nikica VALENTIC (since NA April 1993); Deputy Prime Ministers Mate GRANIC, Vladimir SEKS, Borislav SKEGRO (since NA)

Member of: CEI, CSCE, ECE, ICAO, IMO, IOM (observer), UN, UNCTAD, UNESCO, UNIDO, UPU, WHO

Diplomatic representation in US:
chief of mission: Ambassador Peter A. SARCEVIC
chancery: 2356 Massachusetts Avenue, NW, Washington, DC 20036
telephone: (202) 543-5586

US diplomatic representation:
chief of mission: (vacant)
embassy: Andrije Hebranga 2, Zagreb
mailing address: AMEMB Unit 25402, APO AE 09213-5080
telephone: [38] (41) 444-800
FAX: [38] (41) 440-235

Flag: red, white, and blue horizontal bands with Croatian coat of arms (red and white checkered)

Economy

Overview: Before the dissolution of Yugoslavia, the republic of Croatia, after Slovenia, was the most prosperous and industrialized area, with a per capita output roughly comparable to that of Portugal and perhaps one-third above the Yugoslav average. Croatian Serb Nationalists control approximately one third of the Croatian territory, and one of the overriding determinants of Croatia's long-term political and economic prospects will be the resolution of this territorial dispute. Croatia faces monumental problems stemming from: the legacy of longtime Communist mismanagement of the economy; large foreign debt; damage during the fighting to bridges, factories, powerlines, buildings, and houses; the large refugee population, both Croatian and Bosnian; and the disruption of economic ties to Serbia and the other former Yugoslav republics, as well as within its own territory. At the minimum, extensive Western aid and investment, especially in the tourist and oil industries, would seem necessary to salvage a desperate economic situation. However, peace and political stability must come first. As of June 1993, fighting continues among Croats, Serbs, and Muslims, and national boundaries and final political arrangements are still in doubt.

National product: GDP—purchasing power equivalent—$26.3 billion (1991 est.)

National product real growth rate: -25% (1991 est.)

National product per capita: $5,600 (1991 est.)

Inflation rate (consumer prices): 50% (monthly rate, December 1992)

Unemployment rate: 20% (December 1991 est.)

Budget: revenues $NA; expenditures $NA, including capital expenditures of $NA

Exports: $2.9 billion (1990)
commodities: machinery and transport equipment 30%, other manufacturers 37%, chemicals 11%, food and live animals 9%, raw materials 6.5%, fuels and lubricants 5%
partners: principally the other former Yugoslav republics

Imports: $4.4 billion (1990)
commodities: machinery and transport equipment 21%, fuels and lubricants 19%, food and live animals 16%, chemicals 14%, manufactured goods 13%, miscellaneous manufactured articles 9%, raw materials 6.5%, beverages and tobacco 1%
partners: principally other former Yugoslav republics

External debt: $2.6 billion (will assume some part of foreign debt of former Yugoslavia)

Industrial production: growth rate −29% (1991 est.)

Electricity: 3,570,000 kW capacity; 11,500 million kWh produced, 2,400 kWh per capita (1992)

Industries: chemicals and plastics, machine tools, fabricated metal, electronics, pig iron and rolled steel products, aluminum reduction, paper, wood products (including furniture), building materials (including cement), textiles, shipbuilding, petroleum and petroleum refining, food processing and beverages

Agriculture: Croatia normally produces a food surplus; most agricultural land in private hands and concentrated in Croat-majority districts in Slavonia and Istria; much of Slavonia's land has been put out of production by fighting; wheat, corn, sugar beets, sunflowers, alfalfa, and clover are main crops in Slavonia; central Croatian highlands are less fertile but support cereal production, orchards, vineyards, livestock breeding, and dairy farming; coastal areas and offshore islands grow olives, citrus fruits, and vegetables

Economic aid: $NA

Currency: 1 Croatian dinar (CD) = 100 paras

Exchange rates: Croatian dinar per US $1—60.00 (April 1992)

Fiscal year: calendar year

Communications

Railroads: 2,592 km of standard guage (1.435 m) of which 864 km are electrified (1992); note—disrupted by territorial dispute

Highways: 32,071 km total; 23,305 km paved, 8,439 km gravel, 327 km earth

Croatia *(continued)*

(1990); note—key highways note disrupted because of territorial dispute
Inland waterways: 785 km perennially navigable
Pipelines: crude oil 670 km, petroleum products 20 km, natural gas 310 km (1992); note—now disrupted because of territorial dispute
Ports: coastal—Rijeka, Split, Kardeljevo (Ploce); inland—Vukovar, Osijek, Sisak, Vinkovci
Merchant marine: 18 ships (1,000 GRT or over) totaling 77,074 GRT/93,052 DWT; includes 4 cargo, 1 roll-on/roll-off, 10 passenger ferries, 2 bulk, 1 oil tanker; note—also controlled by Croatian shipowners are 198 ships (1,000 GRT or over) under flags of convenience—primarily Malta and St. Vincent—totaling 2,602,678 GRT/4,070,852 DWT; includes 89 cargo, 9 roll-on/ roll-off, 6 refrigerated cargo, 14 container, 3 multifunction large load carriers, 51 bulk, 5 passenger, 11 oil tanker, 4 chemical tanker, 6 service vessel
Airports:
total: 75
usable: 72
with permanent-surface runways: 15
with runways over 3,659 m: 0
with runways 2,440-3,659 m: 10
with runways 1,220-2,439 m: 5
Telecommunications: 350,000 telephones; broadcast stations—14 AM, 8 FM, 12 (2 repeaters) TV; 1,100,000 radios; 1,027,000 TVs; NA submarine coaxial cables; satellite ground stations—none

Defense Forces

Branches: Ground Forces, Naval Forces, Air and Air Defense Forces
Manpower availability: males age 15-49 1,177,029; fit for military service 943,259; reach military age (19) annually 32,873 (1993 est.)
Defense expenditures: 337-393 billion Croatian dinars, NA% of GDP (1993 est.); note—conversion of defense expenditures into US dollars using the current exchange rate could produce misleading results

Cuba

Geography

Location: in the northern Caribbean Sea, 145 km south of Key West (Florida)
Map references: Central America and the Caribbean, North America, Standard Time Zones of the World
Area:
total area: 110,860 km²
land area: 110,860 km²
comparative area: slightly smaller than Pennsylvania
Land boundaries: total 29 km, US Naval Base at Guantanamo 29 km
note: Guantanamo is leased and as such remains part of Cuba
Coastline: 3,735 km
Maritime claims:
exclusive economic zone: 200 nm
territorial sea: 12 nm
International disputes: US Naval Base at Guantanamo is leased to US and only mutual agreement or US abandonment of the area can terminate the lease
Climate: tropical; moderated by trade winds; dry season (November to April); rainy season (May to October)
Terrain: mostly flat to rolling plains with rugged hills and mountains in the southeast
Natural resources: cobalt, nickel, iron ore, copper, manganese, salt, timber, silica, petroleum
Land use:
arable land: 23%
permanent crops: 6%
meadows and pastures: 23%
forest and woodland: 17%
other: 31%
Irrigated land: 8,960 km² (1989)
Environment: averages one hurricane every other year
Note: largest country in Caribbean

People

Population: 10,957,088 (July 1993 est.)
Population growth rate: 1% (1993 est.)

Birth rate: 17.08 births/1,000 population (1993 est.)
Death rate: 6.5 deaths/1,000 population (1993 est.)
Net migration rate: -0.63 migrant(s)/1,000 population (1993 est.)
Infant mortality rate: 10.5 deaths/1,000 live births (1993 est.)
Life expectancy at birth:
total population: 76.72 years
male: 74.59 years
female: 78.99 years (1993 est.)
Total fertility rate: 1.83 children born/woman (1993 est.)
Nationality:
noun: Cuban(s)
adjective: Cuban
Ethnic divisions: mulatto 51%, white 37%, black 11%, Chinese 1%
Religions: nominally Roman Catholic 85% prior to Castro assuming power
Languages: Spanish
Literacy: age 15 and over can read and write (1990)
total population: 94%
male: 95%
female: 93%
Labor force: 4,620,800 economically active population (1988); 3,578,800 in state sector
by occupation: services and government 30%, industry 22%, agriculture 20%, commerce 11%, construction 10%, transportation and communications 7% (June 1990)

Government

Names:
conventional long form: Republic of Cuba
conventional short form: Cuba
local long form: Republica de Cuba
local short form: Cuba
Digraph: CU
Type: Communist state
Capital: Havana
Administrative divisions: 14 provinces (provincias, singular—provincia) and 1 special municipality* (municipio especial); Camaguey, Ciego de Avila, Cienfuegos, Ciudad de La Habana, Granma, Guantanamo, Holguin, Isla de la Juventud*, La Habana, Las Tunas, Matanzas, Pinar del Rio, Sancti Spiritus, Santiago de Cuba, Villa Clara
Independence: 20 May 1902 (from Spain 10 December 1898; administered by the US from 1898 to 1902)
Constitution: 24 February 1976
Legal system: based on Spanish and American law, with large elements of Communist legal theory; does not accept compulsory ICJ jurisdiction
National holiday: Rebellion Day, 26 July (1953)

Political parties and leaders: only party—Cuban Communist Party (PCC), Fidel CASTRO Ruz, first secretary
Suffrage: 16 years of age; universal
Elections:
National Assembly of People's Power: last held December 1986 (next to be held February 1993); results—PCC is the only party; seats—(510 total; after the February election, the National Assembly will have 590 seats) indirectly elected from slates approved by special candidacy commissions
Executive branch: president of the Council of State, first vice president of the Council of State, Council of State, president of the Council of Ministers, first vice president of the Council of Ministers, Executive Committee of the Council of Ministers, Council of Ministers
Legislative branch: unicameral National Assembly of the People's Power (Asamblea Nacional del Poder Popular)
Judicial branch: People's Supreme Court (Tribunal Supremo Popular)
Leaders:
Chief of State and Head of Government: President of the Council of State and President of the Council of Ministers Fidel CASTRO Ruz (Prime Minister from February 1959 until 24 February 1976 when office was abolished; President since 2 December 1976); First Vice President of the Council of State and First Vice President of the Council of Ministers Gen. Raul CASTRO Ruz (since 2 December 1976)
Member of: CCC, ECLAC, FAO, G-77, GATT, IAEA, ICAO, IFAD, ILO, IMO, INMARSAT, INTERPOL, IOC, ISO, ITU, LAES, LAIA (observer), LORCS, NAM, OAS (excluded from formal participation since 1962), PCA, UN, UNCTAD, UNESCO, UNIDO, UPU, WCL, WFTU, WHO, WIPO, WMO, WTO
Diplomatic representation in US:
chief of mission: Principal Officer Alfonso FRAGA Perez (since August 1992)
chancery: 2630 and 2639 16th Street NW, US Interests Section, Swiss Embassy, Washington, DC 20009
telephone: (202) 797-8518 or 8519, 8520, 8609, 8610
US diplomatic representation:
chief of mission: Principal Officer Alan H. FLANIGAN
US Interests Section: USINT, Swiss Embassy, Calzada entre L Y M, Vedado Seccion, Havana
mailing address: USINT, Swiss Embassy, Calzada Entre L Y M, Vedado, Havava
telephone: 32-0051, 32-0543
FAX: no service available at this time
note: protecting power in Cuba is Switzerland—US Interests Section, Swiss Embassy

Flag: five equal horizontal bands of blue (top and bottom) alternating with white; a red equilateral triangle based on the hoist side bears a white five-pointed star in the center

---

Economy

Overview: Since Castro's takeover of Cuba in 1959, the economy has been run in the Soviet style of government ownership of substantially all the means of production and government planning of all but the smallest details of economic activity. Thus, Cuba, like the former Warsaw Pact nations, has remained in the backwater of economic modernization. The economy contracted by about one-third between 1989 and 1992 as it absorbed the loss of $4 billion of annual economic aid from the former Soviet Union and much smaller amounts from Eastern Europe. The government implemented numerous energy conservation measures and import substitution schemes to cope with a large decline in imports. To reduce fuel consumption, Havana has cut back bus service and imported approximately 1 million bicycles from China, domesticated nearly 200,000 oxen to replace tractors, and halted a large amount of industrial production. The government has prioritized domestic food production and promoted herbal medicines since 1990 to compensate for lower imports. Havana also has been shifting its trade away from the former Soviet republics and Eastern Europe toward the industrialized countries of Latin America and the OECD.
National product: GNP—exchange rate conversion—$14.9 billion (1992 est.)
National product real growth rate: -15% (1992 est.)
National product per capita: $1,370 (1992 est.)
Inflation rate (consumer prices): NA%
Unemployment rate: NA%
Budget: revenues $12.46 billion; expenditures $14.45 billion, including capital expenditures of $NA (1990 est.)
Exports: $2.1 billion (f.o.b., 1992 est.)
commodities: sugar, nickel, shellfish, tobacco, medical products, citrus, coffee
partners: Russia 30%, Canada 10%, China 9%, Japan 6%, Spain 4% (1992 est.)
Imports: $2.2 billion (c.i.f., 1992 est.)
commodities: petroleum, food, machinery, chemicals
partners: Russia 10%, China 9%, Spain 9%, Mexico 5%, Italy 5%, Canada 4%, France 4% (1992 est.)
External debt: $6.8 billion (convertible currency, July 1989)
Industrial production: NA
Electricity: 3,889,000 kW capacity; 16,248 million kWh produced, 1,500 kWh per capita (1992)

Industries: sugar milling and refining, petroleum refining, food and tobacco processing, textiles, chemicals, paper and wood products, metals (particularly nickel), cement, fertilizers, consumer goods, agricultural machinery
Agriculture: accounts for 11% of GNP (including fishing and forestry); key commercial crops—sugarcane, tobacco, and citrus fruits; other products—coffee, rice, potatoes, meat, beans; world's largest sugar exporter; not self-sufficient in food (excluding sugar); sector hurt by growing shortages of fuels and parts
Economic aid: Western (non-US) countries, ODA and OOF bilateral commitments (1970-89), $710 million; Communist countries (1970-89), $18.5 billion
Currency: 1 Cuban peso (Cu$) = 100 centavos
Exchange rates: Cuban pesos (Cu$) per US$1—1.0000 (linked to the US dollar)
Fiscal year: calendar year

Communications

Railroads: 12,947 km total; Cuban National Railways operates 5,053 km of 1.435-meter gauge track; 151.7 km electrified; 7,742 km of sugar plantation lines of 0.914-m and 1.435-m gauge
Highways: 26,477 km total; 14,477 km paved, 12,000 km gravel and earth surfaced (1989 est.)
Inland waterways: 240 km
Ports: Cienfuegos, Havana, Mariel, Matanzas, Santiago de Cuba; 7 secondary, 35 minor
Merchant marine: 73 ships (1,000 GRT or over) totaling 511,522 GRT/720,270 DWT; includes 42 cargo, 10 refrigerated cargo, 1 cargo/training, 11 oil tanker, 1 chemical tanker, 4 liquefied gas, 4 bulk; note—Cuba beneficially owns an additional 38 ships (1,000 GRT and over) totaling 529,090 DWT under the registry of Panama, Cyprus, and Malta
Airports:
total: 186
usable: 166
with permanent-surface runways: 73
with runways over 3,659 m: 3
with runways 2,440-3,659 m: 12
with runways 1,220-2,439 m: 19
Telecommunications: broadcast stations—150 AM, 5 FM, 58 TV; 1,530,000 TVs; 2,140,000 radios; 229,000 telephones; 1 Atlantic Ocean INTELSAT earth station

---

Defense Forces

Branches: Revolutionary Armed Forces (FAR)—including Ground Forces, Revolutionary Navy (MGR), Air and Air Defense Force (DAAFAR), Ministry of the

Cuba *(continued)*

Armed Forces Special Troops, Border Guard Troops, Territorial Militia Troops (MTT), Youth Labor Army (EJT)
Manpower availability: males age 15-49 3,087,255; females age 15-49 3,064,663; males fit for military service 1,929,698; females fit for military service 1,910,733; males reach military age (17) annually 90,409; females reach military age (17) annually 87,274 (1993 est.)
Defense expenditures: exchange rate conversion—$1.2-1.4 billion; 10% of GNP in 1990 plan was for defense and internal security
Note: the breakup of the Soviet Union, the key military supporter and supplier of Cuba, has resulted in substantially less outside help for Cuba's defense forces

Cyprus

Geography

Location: in the eastern Mediterreanean Sea, 97 km west of Syria and 64 km west of Turkey
Map references: Africa, Middle East, Standard Time Zones of the World
Area:
total area: 9,250 km²
land area: 9,240 km²
comparative area: about 0.7 times the size of Connecticut
Land boundaries: 0 km
Coastline: 648 km
Maritime claims:
continental shelf: 200 m depth or to depth of exploitation
territorial sea: 12 nm
International disputes: 1974 hostilities divided the island into two de facto autonomous areas, a Greek area controlled by the Cypriot Government (60% of the island's land area) and a Turkish-Cypriot area (35% of the island) that are separated by a narrow UN buffer zone; in addition, there are two UK sovereign base areas (about 5% of the island's land area)
Climate: temperate, Mediterranean with hot, dry summers and cool, wet winters
Terrain: central plain with mountains to north and south
Natural resources: copper, pyrites, asbestos, gypsum, timber, salt, marble, clay earth pigment
Land use:
arable land: 40%
permanent crops: 7%
meadows and pastures: 10%
forest and woodland: 18%
other: 25%
Irrigated land: 350 km² (1989)
Environment: moderate earthquake activity; water resource problems (no natural reservoir catchments, seasonal disparity in rainfall, and most potable resources concentrated in the Turkish-Cypriot area)

People

Population: 723,371 (July 1993 est.)
Population growth rate: 0.94% (1993 est.)
Birth rate: 17.14 births/1,000 population (1993 est.)
Death rate: 7.74 deaths/1,000 population (1993 est.)
Net migration rate: 0 migrant(s)/1,000 population (1993 est.)
Infant mortality rate: 9.3 deaths/1,000 live births (1993 est.)
Life expectancy at birth:
total population: 75.98 years
male: 73.75 years
female: 78.31 years (1993 est.)
Total fertility rate: 2.34 children born/woman (1993 est.)
Nationality:
noun: Cypriot(s)
adjective: Cypriot
Ethnic divisions: Greek 78%, Turkish 18%, other 4%
Religions: Greek Orthodox 78%, Muslim 18%, Maronite, Armenian, Apostolic, and other 4%
Languages: Greek, Turkish, English
Literacy: age 15 and over can read and write (1987)
total population: 94%
male: 98%
female: 91%
Labor force:
Greek area: 282,000
by occupation: services 57%, industry 29%, agriculture 14% (1991)
Turkish area: 72,000
by occupation: services 57%, industry 22%, agriculture 21% (1991)

Government

Names:
conventional long form: Republic of Cyprus
conventional short form: Cyprus
Digraph: CY
Type: republic
note: a disaggregation of the two ethnic communities inhabiting the island began after the outbreak of communal strife in 1963; this separation was further solidified following the Turkish invasion of the island in July 1974, which gave the Turkish Cypriots de facto control in the north; Greek Cypriots control the only internationally recognized government; on 15 November 1983 Turkish Cypriot President Rauf DENKTASH declared independence and the formation of a "Turkish Republic of Northern Cyprus" (TRNC), which has been recognized only by Turkey; both sides publicly call for the resolution of intercommunal differences and creation of a new federal system of government
Capital: Nicosia

Administrative divisions: 6 districts; Famagusta, Kyrenia, Larnaca, Limassol, Nicosia, Paphos

Independence: 16 August 1960 (from UK)

Constitution: 16 August 1960; negotiations to create the basis for a new or revised constitution to govern the island and to better relations between Greek and Turkish Cypriots have been held intermittently; in 1975 Turkish Cypriots created their own Constitution and governing bodies within the "Turkish Federated State of Cyprus," which was renamed the "Turkish Republic of Northern Cyprus" in 1983; a new Constitution for the Turkish area passed by referendum in May 1985

Legal system: based on common law, with civil law modifications

National holiday: Independence Day, 1 October (15 November is celebrated as Independence Day in the Turkish area)

Political parties and leaders:
Greek Cypriot: Progressive Party of the Working People (AKEL; Communist Party), Dimitrios CHRISTOFIAS; Democratic Rally (DISY), Glafkos CLERIDES; Democratic Party (DIKO), Spyros KYPRIANOU; United Democratic Union of the Center (EDEK), Vassos LYSSARIDIS; Socialist Democratic Renewal Movement (ADISOK), Mikhalis PAPAPETROU; Liberal Party, Nikos ROLANDIS; Free Democrats, George VASSILIOU

Turkish area: National Unity Party (UBP), Dervis EROGLU; Communal Liberation Party (TKP), Mustafa AKINCI; Republican Turkish Party (CTP), Ozker OZGUR; New Cyprus Party (YKP), Alpay DURDURAN; Social Democratic Party (SDP), Ergun VEHBI; New Birth Party (YDP), Ali Ozkan ALTINISHIK; Free Democratic Party (HDP), Ismet KOTAK; Nationalist Justice Party (MAP), Zorlu TORE; United Sovereignty Party, Arif Salih KIRDAG; Democratic Party (DP), Hakki ATUN; Fatherland Party (VP), Orhan UCOK; CTP, TKP, and YDP joined in the coalition Democratic Struggle Party (DMP) for the 22 April 1990 legislative election; the CTP and TKP boycotted the byelection of 13 October 1991, in which 12 seats were at stake; the DMP was dissolved after the 1990 election

Other political or pressure groups: United Democratic Youth Organization (EDON; Communist controlled); Union of Cyprus Farmers (EKA; Communist controlled); Cyprus Farmers Union (PEK; pro-West); Pan-Cyprian Labor Federation (PEO; Communist controlled); Confederation of Cypriot Workers (SEK; pro-West); Federation of Turkish Cypriot Labor Unions (Turk-Sen); Confederation of Revolutionary Labor Unions (Dev-Is)

Suffrage: 18 years of age; universal

Elections:
President: last held 14 February 1993 (next to be held February 1998); results—Glafkos CLERIDES 50.3%, George VASSILIOU 49.7%

House of Representatives: last held 19 May 1991; results—DISY 35.8%, AKEL (Communist) 30.6%, DIKO 19.5%, EDEK 10. 9%; others 3.2%; seats—(56 total) DISY 20, AKEL (Communist) 18, DIKO 11, EDEK 7

Turkish Area: President: last held 22 April 1990 (next to be held April 1995); results—Rauf R. DENKTASH 66%, Ismail BOZKURT 32.05%

Turkish Area: Assembly of the Republic: last held 6 May 1990 (next to be held May 1995); results—UBP (conservative) 54.4%, DMP 44.4% YKP 0.9%; seats—(50 total) UBP (conservative) 45, SDP 1, HDP 2, YDP 2; note—by-election of 13 October 1991 was for 12 seats; DP delegates broke away from the UBP and formed their own party after the last election; seats as of July 1992 UBP 34, SPD 1, HDP 1, YDP 2, DP 10, independents 2

Executive branch: president, Council of Ministers (cabinet); note—there is a president, prime minister, and Council of Ministers (cabinet) in the Turkish area

Legislative branch: unicameral House of Representatives (Vouli Antiprosopon); note—there is a unicameral Assembly of the Republic (Cumhuriyet Meclisi) in the Turkish area

Judicial branch: Supreme Court; note—there is also a Supreme Court in the Turkish area

Leaders:
Chief of State and Head of Government: President Glafkos CLERIDES (since 28 February 1993)

note: Rauf R. DENKTASH has been president of the Turkish area since 13 February 1975; Dervish EROGLU has been prime minister of the Turkish area since 20 July 1985

Member of: C, CCC, CE, CSCE, EBRD, ECE, FAO, G-77, GATT, IAEA, IBRD, ICAO, ICC, ICFTU, IDA, IFAD, IFC, ILO, IMF, IMO, INMARSAT, INTELSAT, INTERPOL, IOC, IOM, ISO, ITU, NAM, OAS (observer), UN, UNCTAD, UNESCO, UNIDO, UPU, WCL, WFTU, WHO, WIPO, WMO, WTO

Diplomatic representation in US:
chief of mission: Ambassador Michael E. SHERIFIS

chancery: 2211 R Street NW, Washington, DC 20008

telephone: (202) 462-5772

consulate general: New York

note: Representative of the Turkish area in the US is Namik KORMAN, office at 1667 K Street, NW, Washington DC, telephone (202) 887-6198

US diplomatic representation:
chief of mission: Ambassador Robert E. LAMB

embassy: corner of Therissos Street and Dositheos Street, Nicosia

mailing address: APO AE 09836

telephone: [357] (2) 465151

FAX: [357] (2) 459-571

Flag: white with a copper-colored silhouette of the island (the name Cyprus is derived from the Greek word for copper) above two green crossed olive branches in the center of the flag; the branches symbolize the hope for peace and reconciliation between the Greek and Turkish communities

note: the Turkish cypriot flag has a horizontal red stripe at the top and bottom with a red crescent and red star on a white field

---

Economy

Overview: The Greek Cypriot economy is small, diversified, and prosperous. Industry contributes 16.5% to GDP and employs 29% of the labor force, while the service sector contributes 62% to GDP and employs 57% of the labor force. Rapid growth in exports of agricultural and manufactured products and in tourism have played important roles in the average 6.8% rise in GDP between 1986 and 1990. This progress was temporarily checked in 1991, because of the adverse effects of the Gulf War on tourism. Nevertheless in mid-1991, the World Bank "graduated" Cyprus off its list of developing countries. In contrast to the bright picture in the south, the Turkish Cypriot economy has less than half the per capita GDP and suffered a series of reverses in 1991. Crippled by the effects of the Gulf war, the collapse of the fruit-to-electronics conglomerate, Polly Peck, Ltd., and a drought, the Turkish area in late 1991 asked for a multibillion-dollar grant from Turkey to help ease the burden of the economic crisis. In addition, the Turkish government extended a $100 million loan in November 1992 to be used for economic development projects in 1993. Turkey normally underwrites a substantial portion of the Turkish Cypriot economy.

National product:
Greek area: GDP—purchasing power equivalent—$6.3 billion (1992)
Turkish area: GDP—purchasing power equivalent—$600 million (1990)

National product real growth rate:
Greek area: 6.5% (1992)
Turkish area: 5.9% (1990)

National product per capita:
Greek area: $11,000 (1992)
Turkish area: $4,000 (1990)

Inflation rate (consumer prices):
Greek area: 5.1% (1991)

Cyprus *(continued)*

Turkish area: 69.4% (1990)
Unemployment rate:
Greek area: 2.4% (1991)
Turkish area: 1.5% (1991)
Budget: revenues $1.7 billion; expenditures $2.2 billion, including capital expenditures of $350 million (1993)
Exports: $875 million (f.o.b., 1991)
commodities: citrus, potatoes, grapes, wine, cement, clothing and shoes
partners: UK 23%, Greece 10%, Lebanon 10%, Germany 5%
Imports: $2.4 billion (f.o.b., 1991)
commodities: consumer goods, petroleum and lubricants, food and feed grains, machinery
partners: UK 13%, Japan 12%, Italy 10%, Germany 9.1%
External debt: $1.9 billion (1991)
Industrial production: growth rate 0.4% (1991); accounts for 16.5% of GDP
Electricity: 620,000 kW capacity; 1,770 million kWh produced, 2,530 kWh per capita (1991)
Industries: food, beverages, textiles, chemicals, metal products, tourism, wood products
Agriculture: contributes 6% to GDP and employs 14% of labor force in the south; major crops—potatoes, vegetables, barley, grapes, olives, citrus fruits; vegetables and fruit provide 25% of export revenues
Illicit drugs: transit point for heroin via air routes and container traffic to Europe, especially from Lebanon and Turkey
Economic aid: US commitments, including Ex-Im (FY70-89), $292 million; Western (non-US) countries, ODA and OOF bilateral commitments (1970-89), $250 million; OPEC bilateral aid (1979-89), $62 million; Communist countries (1970-89), $24 million
Currency: 1 Cypriot pound (£C) = 100 cents; 1 Turkish lira (TL) = 100 kurus
Exchange rates: NA
Fiscal year: calendar year

Communications

Highways: 10,780 km total; 5,170 km paved; 5,610 km gravel, crushed stone, and earth
Ports: Famagusta, Kyrenia, Larnaca, Limassol, Paphos
Merchant marine: 1,299 ships (1,000 GRT or over) totaling 21,045,037 GRT/37,119,933 DWT; includes 10 short-sea passenger, 1 passenger-cargo, 463 cargo, 77 refrigerated cargo, 24 roll-on/roll-off, 70 container, 4 multifunction large load carrier, 110 oil tanker, 3 specialized tanker, 3 liquefied gas, 26 chemical tanker, 32 combination ore/oil, 422 bulk, 3 vehicle carrier, 48 combination bulk, 1 railcar carrier, 2 passenger; note—a flag of convenience registry; Cuba owns 27 of these ships, Russia owns 36, Latvia also has 7 ships, Croatia owns 2, and Romania 5

Airports:
total: 13
usable: 13
with permanent-surface runways: 10
with runways over 3,659 m: 0
with runways 2,440-3,659 m: 7
with runways 1,220-2,439 m: 1
Telecommunications: excellent in both the area controlled by the Cypriot Government (Greek area), and in the Turkish-Cypriot administered area; 210,000 telephones; largely open-wire and microwave radio relay; broadcast stations—11 AM, 8 FM, 1 (34 repeaters) TV in Greek sector and 2 AM, 6 FM and 1 TV in Turkish sector; international service by tropospheric scatter, 3 submarine cables, and satellite earth stations—1 Atlantic Ocean INTELSAT, 1 Indian Ocean INTELSAT and EUTELSAT earth stations

Defense Forces

Branches:
Greek area: Greek Cypriot National Guard (GCNG; including air and naval elements), Greek Cypriot Police
Turkish area: Turkish Cypriot Security Force
Manpower availability: males age 15-49 185,371; fit for military service 127,536; reach military age (18) annually 5,085 (1993 est.)
Defense expenditures: exchange rate conversion—$209 million, 5% of GDP (1990 est.)

Czech Republic

150 km

Geography

Location: Eastern Europe, between Germany and Slovakia
Map references: Ethnic Groups in Eastern Europe, Europe, Standard Time Zones of the World
Area:
total area: 78,703 km²
land area: 78,645 km²
comparative area: slightly smaller than South Carolina
Land boundaries: total 1,880 km, Austria 362 km, Germany 646 km, Poland 658 km, Slovakia 214 km
Coastline: 0 km (landlocked)
Maritime claims: none; landlocked
International disputes: Liechtenstein claims 620 square miles of Czech territory confiscated from its royal family in 1918; the Czech Republic insists that restitution does not go back before February 1948, when the Communists seized power; unresolved property dispute issues with Slovakia over redistribution of Czech and Slovak Federal Republic's property; establishment of international border between Czech Republic and Slovakia
Climate: temperate; cool summers; cold, cloudy, humid winters
Terrain: two main regions: Bohemia in the west, consisting of rolling plains, hills, and plateaus surrounded by low mountains; and Moravia in the east, consisting of very hilly country
Natural resources: hard coal, kaolin, clay, graphite
Land use:
arable land: NA%
permanent crops: NA%
meadows and pastures: NA%
forest and woodland: NA%
other: NA%
Irrigated land: NA km²
Environment: NA
Note: landlocked; strategically located astride some of oldest and most significant land routes in Europe; Moravian Gate is a

traditional military corridor between the North European Plain and the Danube in central Europe

People

Population: 10,389,256 (July 1993 est.)
Population growth rate: 0.16% (1993 est.)
Birth rate: 13 births/1,000 population (1993 est.)
Death rate: 11.44 deaths/1,000 population (1993 est.)
Net migration rate: 0 migrant(s)/1,000 population (1993 est.)
Infant mortality rate: 9.7 deaths/1,000 live births (1993 est.)
Life expectancy at birth:
total population: 72.64 years
male: 68.9 years
female: 76.58 years (1993 est.)
Total fertility rate: 1.85 children born/woman (1993 est.)
Nationality:
noun: Czech(s)
adjective: Czech
Ethnic divisions: Czech 94.4%, Slovak 3%, Polish 0.6%, German 0.5%, Gypsy 0.3%, Hungarian 0.2%, other 1%
Religions: atheist 39.8%, Roman Catholic 39.2%, Protestant 4.6%, Orthodox 3%, other 13.4%
Languages: Czech, Slovak
Literacy:
total population: NA%
male: NA%
female: NA%
Labor force: 5.389 million
by occupation: industry 37.9%, agriculture 8.1%, construction 8.8%, communications and other 45.2% (1990)

Government

Names:
conventional long form: Czech Republic
conventional short form: none
local long form: Ceska Republika
local short form: Cechy
Digraph: EZ
Type: parliamentary democracy
Capital: Prague
Administrative divisions: 7 regions (kraje, kraj—singular); Severocesky, Zapadocesky, Jihocesky, Vychodocesky, Praha, Severomoravsky, Jihomoravsky
Independence: 1 January 1993 (from Czechoslovakia)
Constitution: ratified 16 December 1992; effective 1 January 1993
Legal system: civil law system based on Austro-Hungarian codes; has not accepted compulsory ICJ jurisdiction; legal code modified to bring it in line with Conference on Security and Cooperation in Europe (CSCE) obligations and to expunge Marxist-Leninist legal theory

National holiday: NA
Political parties and leaders: Civic Democratic Party, Vaclav KLAUS, chairman; Christian Democratic Union, leader NA; Civic Democratic Alliance, Jan KALVODA, chairman; Christian Democratic Party, Vaclav BENDA, chairman; Czech People's Party, Josef LUX; Czechoslovak Social Democracy, Milos ZEMAN, chairman; Left Bloc, leader NA; Republican Party, Miroslav SLADEK, chairman; Movement for Self-Governing Democracy for Moravia and Silesia, Jan STRYCER, chairman; Liberal Social Union, leader NA; Assembly for the Republic, leader NA
Other political or pressure groups: Czech Democratic Left Movement; Civic Movement
Suffrage: 18 years of age; universal
Elections:
President: last held 26 January 1993 (next to be held NA January 1998); results—Vaclav HAVEL elected by the National Council
Senate: elections not yet held; seats (81 total)
Chamber of Deputies: last held 5-6 June 1992 (next to be held NA 1996); results—percent of vote by party NA; seats—(200 total) Civic Democratic Party/Christian Democratic Party 76, Left Bloc 35, Czechoslovak Social Democracy 16, Liberal Social Union 16, Christian Democratic Union/Czech People's Party 15, Assembly for the Republic/Republican Party 14, Civic Democratic Alliance 14, Movement for Self-Governing Democracy for Moravia and Silesia 14
Executive branch: president, prime minister, Cabinet
Legislative branch: bicameral National Council (Narodni rada) will consist of an upper house or Senate (which has not yet been established) and a lower house or Chamber of Deputies
Judicial branch: Supreme Court, Constitutional Court
Leaders:
Chief of State: President Vaclav HAVEL (since 26 January 1993)
Head of Government: Prime Minister Vaclav KLAUS (since NA June 1992); Deputy Prime Ministers Ivan KOCARNIK, Josef LUX, Jan KALVODA (since NA June 1992)
Member of: BIS, CCC, CE, CEI, CERN, CSCE, EBRD, ECE, FAO, GATT, IAEA, IBRD, ICAO, IDA, IFC, IFCTU, ILO, IMF, IMO, INMARSAT, INTELSAT, INTERPOL, IOC, IOM (observer), ISO, ITU, LORCS, NACC, NAM (guest), NSG, PCA, UN (as of 8 January 1993), UNAVEM II, UNCTAD, UNESCO, UNIDO, UNOSOM, UNPROFOR, UPU, WHO, WIPO, WMO, WTO, ZC
Diplomatic representation in US:
chief of mission: Ambassador Michael ZANTOVSKY

chancery: 3900 Spring of Freedom Street NW, Washington, DC 20008
telephone: (202) 363-6315 or 6316
US diplomatic representation:
chief of mission: Ambassador Adrian A. BASORA
embassy: Trziste 15, 125 48, Prague 1
mailing address: Unit 25402; APO AE 09213-5630
telephone: [42] (2) 536-641/6
FAX: [42] (2) 532-457
Flag: two equal horizontal bands of white (top) and red with a blue isosceles triangle based on the hoist side

Economy

Overview: The dissolution of Czechoslovakia into two independent nation states—the Czech Republic and Slovakia—on 1 January 1993 has complicated the task of moving toward a more open and decentralized economy. The old Czechoslovakia, even though highly industrialized by East European standards, suffered from an aging capital plant, lagging technology, and a deficiency in energy and many raw materials. In January 1991, approximately one year after the end of communist control of Eastern Europe, theCzech and Slovak Federal Republic launched a sweeping program to convert its almost entirely state-owned and controlled economy to a market system. In 1991-92 these measures resulted in privatization of some medium- and small-scale economic activity and the setting of more than 90% of prices by the market—but at a cost in inflation, unemployment, and lower output. For Czechoslovakia as a whole inflation in 1991 was roughly 50% and output fell 15%. In 1992, in the Czech lands, inflation dropped to an estimated 12.5% and GDP was down a more moderate 5%. For 1993 the government of the Czech Republic anticipates inflation of 15-20% and a rise in unemployment to perhaps 12% as some large-scale enterprises go into bankruptcy; GDP may drop as much as 3%, mainly because of the disruption of trade links with Slovakia. Although the governments of the Czech Republic and Slovakia had envisaged retaining the koruna as a common currency, at least in the short term, the two countries ended the currency union in February 1993.
National product: GDP—purchasing power equivalent—$75.3 billion (1992 est.)
National product real growth rate: -5% (1992 est.)
National product per capita: $7,300 (1992 est.)
Inflation rate (consumer prices): 12.5% (1992 est.)
Unemployment rate: 3.1% (1992 est.)
Budget: revenues $NA; expenditures $NA, including capital expenditures of $NA

Czech Republic *(continued)*

Exports: $8.2 billion (f.o.b., 1992)
commodities: manufactured goods, machinery and transport equipment, chemicals, fuels, minerals, and metals
partners: Slovakia, Germany, Poland, Austria, Hungary, Italy, France, US, UK, CIS republics
Imports: $8.9 billion (f.o.b., 1992)
commodities: machinery and transport equipment, fuels and lubricants, manfactured goods, raw materials, chemicals, agricultural products
partners: Slovakia, CIS republics, Germany Austria, Poland, Switzerland, Hungary, UK, Italy
External debt: $3.8 billion, hard currency indebtedness (December 1992)
Industrial production: growth rate –4% (November 1992 over November 1991); accounts for over 60% of GDP
Electricity: 16,500,000 kW capacity; 62,200 million kWh produced, 6,030 kWh per capita (1992)
Industries: fuels, ferrous metallurgy, machinery and equipment, coal, motor vehicles, glass, armaments
Agriculture: largely self-sufficient in food production; diversified crop and livestock production, including grains, potatoes, sugar beets, hops, fruit, hogs, cattle, and poultry; exporter of forest products
Illicit drugs: the former Czechoslovakia was a transshipment point for Southwest Asian heroin and was emerging as a transshipment point for Latin American cocaine (1992)
Economic aid: the former Czechoslovakia was a donor—$4.2 billion in bilateral aid to non-Communist less developed countries (1954-89)
Currency: 1 koruna (Kc) = 100 haleru
Exchange rates: koruny (Kcs) per US$1—28.59 (December 1992), 28.26 (1992), 29.53 (1991), 17.95 (1990), 15.05 (1989), 14.36 (1988), 13.69 (1987)
Fiscal year: calendar year

Communications

Railroads: 9,434 km total (1988)
Highways: 55,890 km total (1988)
Inland waterways: NA km; the Elbe (Labe) is the principal river
Pipelines: natural gas 5,400 km
Ports: coastal outlets are in Poland (Gdynia, Gdansk, Szczecin), Croatia (Rijeka), Slovenia (Koper), Germany (Hamburg, Rostock); principal river ports are Prague on the Vltava, Decin on the Elbe (Labe)
Merchant marine: the former Czechoslovakia had 22 ships (1,000 GRT or over) totaling 290,185 GRT/437,291 DWT; includes 13 cargo, 9 bulk; may be shared with Slovakia
Airports:
total: 75

usable: 75
with permanent-surface runways: 8
with runways over 3,659 m: 0
with runways 2,440-3,659 m: 2
with runways 1,220-2,439 m: 4
Telecommunications: NA

Defense Forces

Branches: Army, Air and Air Defense Forces, Civil Defense, Railroad Units
Manpower availability: males age 15-49 2,736,657; fit for military service 2,083,555; reach military age (18) annually 95,335 (1993 est.)
Defense expenditures: 23 billion koruny, NA% of GNP (1993 est.); note—conversion of defense expenditures into US dollars using the current exchange rate could produce misleading results

Denmark

Geography

Location: Northwestern Europe, bordering the North Sea on a peninsula north of Germany
Map references: Arctic Region, Europe, Standard Time Zones of the World
Area:
total area: 43,070 km²
land area: 42,370 km²
comparative area: slightly more than twice the size of Massachusetts
note: includes the island of Bornholm in the Baltic Sea and the rest of metropolitan Denmark, but excludes the Faroe Islands and Greenland
Land boundaries: total 68 km, Germany 68 km
Coastline: 3,379 km
Maritime claims:
contiguous zone: 4 nm
continental shelf: 200 m depth or to depth of exploitation
exclusive fishing zone: 200 nm
territorial sea: 3 nm
International disputes: Rockall continental shelf dispute involving Iceland, Ireland, and the UK (Ireland and the UK have signed a boundary agreement in the Rockall area); dispute between Denmark and Norway over maritime boundary in Arctic Ocean between Greenland and Jan Mayen is before the International Court of Justice
Climate: temperate; humid and overcast; mild, windy winters and cool summers
Terrain: low and flat to gently rolling plains
Natural resources: petroleum, natural gas, fish, salt, limestone
Land use:
arable land: 61%
permanent crops: 0%
meadows and pastures: 6%
forest and woodland: 12%
other: 21%
Irrigated land: 4,300 km² (1989 est.)
Environment: air and water pollution

Note: controls Danish Straits linking Baltic and North Seas

People

Population: 5,175,922 (July 1993 est.)
Population growth rate: 0.23% (1993 est.)
Birth rate: 12.5 births/1,000 population (1993 est.)
Death rate: 11.42 deaths/1,000 population (1993 est.)
Net migration rate: 1.24 migrant(s)/1,000 population (1993 est.)
Infant mortality rate: 7.1 deaths/1,000 live births (1993 est.)
Life expectancy at birth:
total population: 75.51 years
male: 72.63 years
female: 78.56 years (1993 est.)
Total fertility rate: 1.68 children born/woman (1993 est.)
Nationality:
noun: Dane(s)
adjective: Danish
Ethnic divisions: Scandinavian, Eskimo, Faroese, German
Religions: Evangelical Lutheran 91%, other Protestant and Roman Catholic 2%, other 7% (1988)
Languages: Danish, Faroese, Greenlandic (an Eskimo dialect), German (small minority)
Literacy: age 15 and over can read and write (1980)
total population: 99%
male: NA%
female: NA%
Labor force: 2,553,900
by occupation: private services 37.1%, government services 30.4%, manufacturing and mining 20%, construction 6.3%, agriculture, forestry, and fishing 5.6%, electricity/gas/water 0.6% (1991)

Government

Names:
conventional long form: Kingdom of Denmark
conventional short form: Denmark
local long form: Kongeriget Danmark
local short form: Danmark
Digraph: DA
Type: constitutional monarchy
Capital: Copenhagen
Administrative divisions: metropolitan Denmark—14 counties (amter, singular—amt) and 1 city* (stad); Arhus, Bornholm, Frederiksborg, Fyn, Kbenhavn, Nordjylland, Ribe, Ringkbing, Roskilde, Snderjylland, Staden Kbenhavn*, Storstrm, Vejle, Vestsjaelland, Viborg
note: see separate entries for the Faroe Islands and Greenland, which are part of the Danish realm and self-governing administrative divisions

Independence: 1849 (became a constitutional monarchy)
Constitution: 5 June 1953
Legal system: civil law system; judicial review of legislative acts; accepts compulsory ICJ jurisdiction, with reservations
National holiday: Birthday of the Queen, 16 April (1940)
Political parties and leaders: Social Democratic Party, Poul Nyrup RASMUSSEN; Conservative Party, Poul SCHLUETER; Liberal Party, Uffe ELLEMANN-JENSEN; Socialist People's Party, Holger K. NIELSEN; Progress Party, Pia KJAERSGAARD; Center Democratic Party, Mimi Stilling JAKOBSEN; Radical Liberal Party, Marianne JELVED; Christian People's Party, Jann SJURSEN; Common Course, Preben Moller HANSEN; Danish Workers' Party
Suffrage: 21 years of age; universal
Elections:
Parliament: last held 12 December 1990 (next to be held by December 1994); results—Social Democratic Party 37.4%, Conservative Party 16.0%, Liberal 15.8%, Socialist People's Party 8.3%, Progress Party 6.4%, Center Democratic Party 5.1%, Radical Liberal Party 3.5%, Christian People's Party 2.3%, other 5.2%; seats—(179 total; includes 2 from Greenland and 2 from the Faroe Islands) Social Democratic 69, Conservative 30, Liberal 29, Socialist People's 15, Progress Party 12, Center Democratic 9, Radical Liberal 7, Christian People's 4
Executive branch: monarch, heir apparent, prime minister, Cabinet
Legislative branch: unicameral parliament (Folketing)
Judicial branch: Supreme Court
Leaders:
Chief of State: Queen MARGRETHE II (since NA January 1972); Heir Apparent Crown Prince FREDERIK, elder son of the Queen (born 26 May 1968)
Head of Government: Prime Minister Poul Nyrup RASMUSSEN (since NA January 1993)
Member of: AfDB, AG (observer), AsDB, Australia Group, BIS, CBSS, CCC, CE, CERN, COCOM, CSCE, EBRD, EC, ECE, EIB, ESA, FAO, G-9, GATT, IADB, IAEA, IBRD, ICAO, ICC, ICFTU, IDA, IEA, IFAD, IFC, ILO, IMF, IMO, INMARSAT, INTELSAT, INTERPOL, IOC, IOM, ISO, ITU, LORCS, MTCR, NACC, NATO, NC, NEA, NIB, NSG, OECD, PCA, UN, UNCTAD, UNESCO, UNFICYP, UNHCR, UNIDO, UNIKOM, UNMOGIP, UNPROFOR, UNTSO, UPU, WHO, WIPO, WMO, ZC
Diplomatic representation in US:
chief of mission: Ambassador Peter Pedersen DYVIG

chancery: 3200 Whitehaven Street NW, Washington, DC 20008
telephone: (202) 234-4300
FAX: (202) 328-1470
consulates general: Chicago, Los Angeles, and New York
US diplomatic representation:
chief of mission: Ambassador Richard B. STONE
embassy: Dag Hammarskjolds Alle 24, 2100 Copenhagen O
mailing address: APO AE 09716
telephone: [45] (31) 42-31-44
FAX: [45] (35) 43-0223
Flag: red with a white cross that extends to the edges of the flag; the vertical part of the cross is shifted to the hoist side, and that design element of the DANNEBROG (Danish flag) was subsequently adopted by the other Nordic countries of Finland, Iceland, Norway, and Sweden

Economy

Overview: This modern economy features high-tech agriculture, up-to-date small-scale and corporate industry, extensive government welfare measures, comfortable living standards, and high dependence on foreign trade. Denmark's new center-left coalition government will concentrate on reducing the persistent high unemployment rate and the budget deficit as well as following the previous government's policies of maintaining low inflation and a current account surplus. In the face of recent international market pressure on the Danish krone, the coalition has also vowed to maintain a stable currency. The coalition hopes to lower marginal income taxes while maintaining overall tax revenues; boost industrial competitiveness through labor market and tax reforms and increased research and development funds; and improve welfare services for the neediest while cutting paperwork and delays. Prime Minister RASMUSSEN's reforms will focus on adapting Denmark to EC's economic and monetary union (EMU) criteria by 1999, although Copenhagen won from the EC the right to opt out of the EMU if a national referendum rejects it. Denmark is, in fact, one of the few EC countries likely to fit into the EMU on time. Denmark is weathering the current worldwide slump better than many West European countries. As the EC's single market (formally established on 1 January 1993) gets underway, Danish economic growth is expected to pickup to around 2% in 1993. Expected Danish approval of the Maastricht treaty on EC political and economic union in May 1993 would almost certainly reverse the drop in investment, further boosting growth. The current account surplus remains strong as

Denmark (continued)

limitations on wage increases and low inflation—expected to be around 1% in 1993—improve export competitiveness. Although unemployment is high, it remains stable compared to most European countries.
National product: GDP—purchasing power equivalent—$94.2 billion (1992)
National product real growth rate: 1% (1992)
National product per capita: $18,200 (1992)
Inflation rate (consumer prices): 1.5% (1992)
Unemployment rate: 11.4% (1992)
Budget: revenues $48.8 billion; expenditures $55.3 billion, including capital expenditures of $NA (1992)
Exports: $37.3 billion (f.o.b., 1992)
commodities: meat and meat products, dairy products, transport equipment (shipbuilding), fish, chemicals, industrial machinery
partners: EC 54.3% (Germany 23.6%, UK 10.1%, France 5.7%), Sweden 10.5%, Norway 5.8%, US 4.9%, Japan 3.6% (1992)
Imports: $30.3 billion (c.i.f., 1992)
commodities: petroleum, machinery and equipment, chemicals, grain and foodstuffs, textiles, paper
partners: EC 53.4% (Germany 23.1%, UK 8.2%, France 5.6%), Sweden 10.8%, Norway 5.4%, US 5.7%, Japan 4.1% (1992)
External debt: $40 billion (1992 est.)
Industrial production: growth rate 1.9% (1992)
Electricity: 11,215,000 kW capacity; 34,170 million kWh produced, 6,610 kWh per capita (1992)
Industries: food processing, machinery and equipment, textiles and clothing, chemical products, electronics, construction, furniture, and other wood products, shipbuilding
Agriculture: accounts for 4% of GDP and employs 5.6% of labor force (includes fishing and forestry); farm products account for nearly 15% of export revenues; principal products—meat, dairy, grain, potatoes, rape, sugar beets, fish; self-sufficient in food production
Economic aid: donor—ODA and OOF commitments (1970-89) $5.9 billion
Currency: 1 Danish krone (DKr) = 100 re
Exchange rates: Danish kroner (DKr) per US$1—6.236 (January 1993), 6.036 (1992), 6.396 (1991), 6.189 (1990), 7.310 (1989), 6.732 (1988)
Fiscal year: calendar year

Communications

Railroads: 2,770 km; Danish State Railways (DSB) operate 2,120 km (1,999 km rail line and 121 km rail ferry services); 188 km electrified, 730 km double tracked; 650 km of standard-gauge lines are privately owned and operated
Highways: 66,482 km total; 64,551 km concrete, bitumen, or stone block; 1,931 km gravel, crushed stone, improved earth
Inland waterways: 417 km
Pipelines: crude oil 110 km; petroleum products 578 km; natural gas 700 km
Ports: Alborg, Arhus, Copenhagen, Esbjerg, Fredericia; numerous secondary and minor ports
Merchant marine: 328 ships (1,000 GRT or over) totaling 5,043,277 GRT/7,230,634 DWT; includes 13 short-sea passenger, 102 cargo, 19 refrigerated cargo, 47 container, 37 roll-on/roll-off, 1 railcar carrier, 33 oil tanker, 18 chemical tanker, 36 liquefied gas, 4 livestock carrier, 17 bulk, 1 combination bulk; note—Denmark has created its own internal register, called the Danish International Ship register (DIS); DIS ships do not have to meet Danish manning regulations, and they amount to a flag of convenience within the Danish register; by the end of 1990, 258 of the Danish-flag ships belonged to the DIS
Airports:
total: 118
usable: 109
with permanent-surface runways: 28
with runways over 3,659 m: 0
with runways 2,440-3,659 m: 9
with runways 1,220-2,439 m: 7
Telecommunications: excellent telephone, telegraph, and broadcast services; 4,509,000 telephones; buried and submarine cables and microwave radio relay support trunk network; broadcast stations—3 AM, 2 FM, 50 TV; 19 submarine coaxial cables; 7 earth stations operating in INTELSAT, EUTELSAT, and INMARSAT

Defense Forces

Branches: Royal Danish Army, Royal Danish Navy, Royal Danish Air Force, Home Guard
Manpower availability: males age 15-49 1,368,211; fit for military service 1,176,559; reach military age (20) annually 37,248 (1993 est.)
Defense expenditures: exchange rate conversion—$2.8 billion, 2% of GDP (1992)

Djibouti

Geography

Location: Eastern Africa, at the entrance to the Red Sea between Ethiopia and Somalia
Map references: Africa, Middle East, Standard Time Zones of the World
Area:
total area: 22,000 km²
land area: 21,980 km²
comparative area: slightly larger than Massachusetts
Land boundaries: total 508 km, Erithea 113 km, Ethiopia 337 km, Somalia 58 km
Coastline: 314 km
Maritime claims:
contiguous zone: 24 nm
exclusive economic zone: 200 nm
territorial sea: 12 nm
International disputes: possible claim by Somalia based on unification of ethnic Somalis
Climate: desert; torrid, dry
Terrain: coastal plain and plateau separated by central mountains
Natural resources: geothermal areas
Land use:
arable land: 0%
permanent crops: 0%
meadows and pastures: 9%
forest and woodland: 0%
other: 91%
Irrigated land: NA km²
Environment: vast wasteland
Note: strategic location near world's busiest shipping lanes and close to Arabian oilfields; terminus of rail traffic into Ethiopia

People

Population: 401,579 (July 1993 est.)
Population growth rate: 2.7% (1993 est.)
Birth rate: 43.05 births/1,000 population (1993 est.)
Death rate: 16.06 deaths/1,000 population (1993 est.)
Net migration rate: 0 migrant(s)/1,000 population (1993 est.)

Infant mortality rate: 113.2 deaths/1,000 live births (1993 est.)

Life expectancy at birth:

total population: 48.78 years

male: 47.01 years

female: 50.59 years (1993 est.)

Total fertility rate: 6.27 children born/woman (1993 est.)

Nationality:

noun: Djiboutian(s)

adjective: Djiboutian

Ethnic divisions: Somali 60%, Afar 35%, French, Arab, Ethiopian, and Italian 5%

Religions: Muslim 94%, Christian 6%

Languages: French (official), Arabic (official), Somali, Afar

Literacy: age 15 and over can read and write (1990)

total population: 48%

male: 63%

female: 34%

Labor force: NA

by occupation: a small number of semiskilled laborers at the port and 3,000 railway workers

note: 52% of population of working age (1983)

Government

Names:

conventional long form: Republic of Djibouti

conventional short form: Djibouti

former: French Territory of the Afars and Issas French Somaliland

Digraph: DJ

Type: republic

Capital: Djibouti

Administrative divisions: 5 districts (cercles, singular—cercle); 'Ali Sabih, Dikhil, Djibouti, Obock, Tadjoura

Independence: 27 June 1977 (from France)

Constitution: multiparty constitution approved in referendum September 1992

Legal system: based on French civil law system, traditional practices, and Islamic law

National holiday: Independence Day, 27 June (1977)

Political parties and leaders:

ruling party: People's Progress Assembly (RPP), Hassan GOULED Aptidon

other parties: Democratic Renewal Party (PRD), Mohamed Jama ELABE; Democratic National Party (PND), ADEN Robleh Awaleh

Other political or pressure groups: Front for the Restoration of Unity and Democracy (FRUD) and affiliates; Movement for Unity and Democracy (MUD)

Suffrage: universal adult at age NA

Elections:

National Assembly: last held 18 December 1992; results—RPP is the only party; seats—(65 total) RPP 65

President: last held 24 April 1987 (next to be held April 1993); results—President Hassan GOULED Aptidon was reelected without opposition

Executive branch: president, prime minister, Council of Ministers

Legislative branch: unicameral Chamber of Deputies (Chambre des Deputes)

Judicial branch: Supreme Court (Cour Supreme)

Leaders:

Chief of State: President HASSAN GOULED Aptidon (since 24 June 1977)

Head of Government: Prime Minister BARKAT Gourad Hamadou (since 30 September 1978)

Member of: ACCT, ACP, AfDB, AFESD, AL, ECA, FAO, G-77, IBRD, ICAO, IDA, IDB, IFAD, IFC, IGADD, ILO, IMF, IMO, INTERPOL, IOC, ITU, LORCS, NAM, OAU, OIC, UN, UNESCO, UNCTAD, UNIDO, UPU, WHO, WMO

Diplomatic representation in US:

chief of mission: Ambassador Roble OLHAYE

chancery: Suite 515, 1156 15th Street NW, Washington, DC 20005

telephone: (202) 331-0270

US diplomatic representation:

chief of mission: Ambassador Charles R. BAQUET III

embassy: Plateau du Serpent, Boulevard Marechal Joffre, Djibouti

mailing address: B. P. 185, Djibouti

telephone: [253] 35-39-95

FAX: [253] 35-39-40

Flag: two equal horizontal bands of light blue (top) and light green with a white isosceles triangle based on the hoist side bearing a red five-pointed star in the center

Economy

Overview: The economy is based on service activities connected with the country's strategic location and status as a free trade zone in northeast Africa. Djibouti provides services as both a transit port for the region and an international transshipment and refueling center. It has few natural resources and little industry. The nation is, therefore, heavily dependent on foreign assistance to help support its balance of payments and to finance development projects. An unemployment rate of over 30% continues to be a major problem. Per capita consumption dropped an estimated 35% over the last five years because of recession and a high population growth rate (including immigrants and refugees).

National product: GDP—exchange rate conversion—$358 million (1990 est.)

National product real growth rate: 1.2% (1990 est.)

National product per capita: $1,030 (1990 est.)

Inflation rate (consumer prices): 7.7% (1991 est.)

Unemployment rate: over 30% (1989)

Budget: revenues $170 million; expenditures $203 million, including capital expenditures of $70 million (1991 est.)

Exports: $186 million (f.o.b., 1991 est.)

commodities: hides and skins, coffee (in transit)

partners: Africa 50%, Middle East 40%, Western Europe 9%

Imports: $360 million (f.o.b., 1991 est.)

commodities: foods, beverages, transport equipment, chemicals, petroleum products

partners: Western Europe 54%, Middle East 20%, Asia 19%

External debt: $355 million (December 1990)

Industrial production: growth rate 10.0% (1990); manufacturing accounts for 11% of GDP

Electricity: 115,000 kW capacity; 200 million kWh produced, 580 kWh per capita (1991)

Industries: limited to a few small-scale enterprises, such as dairy products and mineral-water bottling

Agriculture: accounts for only 3% of GDP; scanty rainfall limits crop production to mostly fruit and vegetables; half of population pastoral nomads herding goats, sheep, and camels; imports bulk of food needs

Economic aid: US commitments, including Ex-Im (FY78-89), $39 million; Western (non-US) countries, including ODA and OOF bilateral commitments (1970-89), $1.1 billion; OPEC bilateral aid (1979-89), $149 million; Communist countries (1970-89), $35 million

Currency: 1 Djiboutian franc (DF) = 100 centimes

Exchange rates: Djiboutian francs (DF) per US$1—177.721 (fixed rate since 1973)

Fiscal year: calendar year

Communications

Railroads: the Ethiopian-Djibouti railroad extends for 97 km through Djibouti

Highways: 2,900 km total; 280 km paved; 2,620 km improved or unimproved earth (1982)

Ports: Djibouti

Merchant marine: 1 cargo ship (1,000 GRT or over) totaling 1,369 GRT/3,030 DWT

Airports:

total: 13

usable: 11

with permanent-surface runways: 2

with runways over 3,659 m: 0

with runways 2,440-3,659 m: 2

with runways 1,220-2,439 m: 5

Telecommunications: telephone facilities in the city of Djibouti are adequate as are the

Djibouti (continued)

microwave radio relay connections to outlying areas of the country; international connections via submarine cable to Saudi Arabia and by satellite to other countries; one ground station each for Indian Ocean INTELSAT and ARABSAT; broadcast stations—2 AM, 2 FM, 1 TV

Defense Forces

Branches: Djibouti National Army (including Navy and Air Force), National Security Force (Force Nationale de Securite), National Police Force
Manpower availability: males age 15-49 97,943; fit for military service 57,187 (1993 est.)
Defense expenditures: exchange rate conversion—$26 million, NA% of GDP (1989)

Dominica

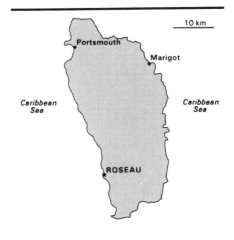

Geography

Location: in the eastern Caribbean, about halfway between Puerto Rico and Trinidad and Tobago
Map references: Central America and the Caribbean, South America, Standard Time Zones of the World
Area:
total area: 750 km^2
land area: 750 km^2
comparative area: slightly more than four times the size of Washington, DC
Land boundaries: 0 km
Coastline: 148 km
Maritime claims:
contiguous zone: 24 nm
exclusive economic zone: 200 nm
territorial sea: 12 nm
International disputes: none
Climate: tropical; moderated by northeast trade winds; heavy rainfall
Terrain: rugged mountains of volcanic origin
Natural resources: timber
Land use:
arable land: 9%
permanent crops: 13%
meadows and pastures: 3%
forest and woodland: 41%
other: 34%
Irrigated land: NA km^2
Environment: flash floods a constant hazard; occasional hurricanes

People

Population: 86,547 (July 1993 est.)
Population growth rate: 1.31% (1993 est.)
Birth rate: 20.82 births/1,000 population (1993 est.)
Death rate: 5.06 deaths/1,000 population (1993 est.)
Net migration rate: -2.63 migrant(s)/1,000 population (1993 est.)
Infant mortality rate: 10.7 deaths/1,000 live births (1993 est.)

Life expectancy at birth:
total population: 76.72 years
male: 73.89 years
female: 79.71 years (1993 est.)
Total fertility rate: 2.03 children born/woman (1993 est.)
Nationality:
noun: Dominican(s)
adjective: Dominican
Ethnic divisions: black, Carib Indians
Religions: Roman Catholic 77%, Protestant 15% (Methodist 5%, Pentecostal 3%, Seventh-Day Adventist 3%, Baptist 2%, other 2%), none 2%, unknown 1%, other 5%
Languages: English (official), French patois
Literacy: age 15 and over having ever attended school (1970)
total population: 94%
male: 94%
female: 94%
Labor force: 25,000
by occupation: agriculture 40%, industry and commerce 32%, services 28% (1984)

Government

Names:
conventional long form: Commonwealth of Dominica
conventional short form: Dominica
Digraph: DO
Type: parliamentary democracy
Capital: Roseau
Administrative divisions: 10 parishes; Saint Andrew, Saint David, Saint George, Saint John, Saint Joseph, Saint Luke, Saint Mark, Saint Patrick, Saint Paul, Saint Peter
Independence: 3 November 1978 (from UK)
Constitution: 3 November 1978
Legal system: based on English common law
National holiday: Independence Day, 3 November (1978)
Political parties and leaders: Dominica Freedom Party (DFP), (Mary) Eugenia CHARLES; Dominica Labor Party (DLP), Rosie DOUGLAS; United Workers Party (UWP), Edison JAMES
Other political or pressure groups: Dominica Liberation Movement (DLM), a small leftist group
Suffrage: 18 years of age; universal
Elections:
House of Assembly: last held 28 May 1990 (next to be held May 1995); results—percent of vote by party NA; seats—(30 total; 9 appointed senators and 21 elected representatives) DFP 11, UWP 6, DLP 4
President: last held 20 December 1988 (next to be held December 1993); results—President Sir Clarence Augustus SEIGNORET was reelected by the House of Assembly
Executive branch: president, prime minister, Cabinet

Legislative branch: unicameral House of Assembly
Judicial branch: Eastern Caribbean Supreme Court
Leaders:
Chief of State: President Sir Clarence Augustus SEIGNORET (since 19 December 1983)
Head of Government: Prime Minister (Mary) Eugenia CHARLES (since 21 July 1980, elected for a third term 28 May 1990)
Member of: ACCT, ACP, C, CARICOM, CDB, ECLAC, FAO, G-77, IBRD, ICFTU, IDA, IFAD, IFC, ILO, IMF, IMO, INTERPOL, LORCS, NAM (observer), OAS, OECS, OPANAL, UN, UNCTAD, UNESCO, UNIDO, UPU, WCL, WHO, WMO
Diplomatic representation in US: there is no chancery in the US
US diplomatic representation: no official presence since the Ambassador resides in Bridgetown (Barbados), but travels frequently to Dominica
Flag: green with a centered cross of three equal bands—the vertical part is yellow (hoist side), black, and white—the horizontal part is yellow (top), black, and white; superimposed in the center of the cross is a red disk bearing a sisserou parrot encircled by 10 green five-pointed stars edged in yellow; the 10 stars represent the 10 administrative divisions (parishes)

Economy

Overview: The economy is dependent on agriculture and thus is highly vulnerable to climatic conditions. Agriculture accounts for about 30% of GDP and employs 40% of the labor force. Principal products include bananas, citrus, mangoes, root crops, and coconuts. In 1991, GDP grew by 2.1%. The tourist industry remains undeveloped because of a rugged coastline and the lack of an international airport.
National product: GDP—purchasing power equivalent—$174 million (1991 est.)
National product real growth rate: 2.1% (1991 est.)
National product per capita: $2,100 (1991 est.)
Inflation rate (consumer prices): 4.5% (1991)
Unemployment rate: 15% (1991)
Budget: revenues $70 million; expenditures $84 million, including capital expenditures of $26 million (FY91 est.)
Exports: $66.0 million (c.i.f., 1991)
commodities: bananas, soap, bay oil, vegetables, grapefruit, oranges
partners: UK 50%, CARICOM countries, US, Italy
Imports: $110.0 million (c.i.f., 1991)
commodities: manufactured goods, machinery and equipment, food, chemicals

partners: US 27%, CARICOM, UK, Canada
External debt: $87 million (1991)
Industrial production: growth rate 4.5% in manufacturing (1988 est.); accounts for 18% of GDP
Electricity: 7,000 kW capacity; 16 million kWh produced, 185 kWh per capita (1992)
Industries: soap, coconut oil, tourism, copra, furniture, cement blocks, shoes
Agriculture: accounts for 26% of GDP; principal crops—bananas, citrus, mangoes, root crops, coconuts; bananas provide the bulk of export earnings; forestry and fisheries potential not exploited
Economic aid: Western (non-US) countries, ODA and OOF bilateral commitments (1970-89), $120 million
Currency: 1 EC dollar (EC$) = 100 cents
Exchange rates: East Caribbean dollars (EC$) per US$1—2.70 (fixed rate since 1976)
Fiscal year: 1 July-30 June

Communications

Highways: 750 km total; 370 km paved, 380 km gravel and earth
Ports: Roseau, Portsmouth
Airports:
total: 2
usable: 2
with permanent-surface runways: 2
with runways over 3,659 m: 0
with runways 2,440-3,659 m: 0
with runways 1,220-2,439 m: 1
Telecommunications: 4,600 telephones in fully automatic network; VHF and UHF link to Saint Lucia; new SHF links to Martinique and Guadeloupe; broadcast stations—3 AM, 2 FM, 1 cable TV

Defense Forces

Branches: Commonwealth of Dominica Police Force
Manpower availability: NA
Defense expenditures: exchange rate conversion—$NA, NA% of GDP

Dominican Republic

North Atlantic Ocean

100 km

Hispaniola

Monte Cristo · Puerto Plata · Santiago · La Vega · Bahía de Samaná · Elías Piña · Lago Enriquillo · SANTO DOMINGO · Higüey · Haina · San Pedro de Macorís · Barahona · Pedernales

Caribbean Sea

Geography

Location: in the northern Caribbean Sea, about halfway between Cuba and Puerto Rico
Map references: Central America and the Caribbean, Standard Time Zones of the World
Area:
total area: 48,730 km²
land area: 48,380 km²
comparative area: slightly more than twice the size of New Hampshire
Land boundaries: total 275 km, Haiti 275 km
Coastline: 1,288 km
Maritime claims:
contiguous zone: 24 nm
continental shelf: 200 nm or the outer edge of continental margin
exclusive economic zone: 200 nm
territorial sea: 6 nm
International disputes: none
Climate: tropical maritime; little seasonal temperature variation
Terrain: rugged highlands and mountains with fertile valleys interspersed
Natural resources: nickel, bauxite, gold, silver
Land use:
arable land: 23%
permanent crops: 7%
meadows and pastures: 43%
forest and woodland: 13%
other: 14%
Irrigated land: 2,250 km² (1989)
Environment: subject to occasional hurricanes (July to October); deforestation
Note: shares island of Hispaniola with Haiti (western one-third is Haiti, eastern two-thirds is the Dominican Republic)

People

Population: 7,683,940 (July 1993 est.)
Population growth rate: 1.86% (1993 est.)
Birth rate: 25.68 births/1,000 population (1993 est.)

Dominican Republic (continued)

Death rate: 6.38 deaths/1,000 population (1993 est.)

Net migration rate: -0.68 migrant(s)/1,000 population (1993 est.)

Infant mortality rate: 53.6 deaths/1,000 live births (1993 est.)

Life expectancy at birth:
total population: 67.98 years
male: 65.87 years
female: 70.21 years (1993 est.)

Total fertility rate: 2.89 children born/woman (1993 est.)

Nationality:
noun: Dominican(s)
adjective: Dominican

Ethnic divisions: mixed 73%, white 16%, black 11%

Religions: Roman Catholic 95%

Languages: Spanish

Literacy: age 15 and over can read and write (1990)
total population: 83%
male: 85%
female: 82%

Labor force: 2,300,000 to 2,600,000
by occupation: agriculture 49%, services 33%, industry 18% (1986)

Government

Names:
conventional long form: Dominican Republic
conventional short form: none
local long form: Republica Dominicana
local short form: none

Digraph: DR

Type: republic

Capital: Santo Domingo

Administrative divisions: 29 provinces (provincias, singular—provincia) and 1 district* (distrito); Azua, Baoruco, Barahona, Dajabon, Distrito Nacional*, Duarte, Elias Pina, El Seibo, Espaillat, Hato Mayor, Independencia, La Altagracia, La Romana, La Vega, Maria Trinidad Sanchez, Monsenor Nouel, Monte Cristi, Monte Plata, Pedernales, Peravia, Puerto Plata, Salcedo, Samana, Sanchez Ramirez, San Cristobal, San Juan, San Pedro De Macoris, Santiago, Santiago Rodriguez, Valverde

Independence: 27 February 1844 (from Haiti)

Constitution: 28 November 1966

Legal system: based on French civil codes

National holiday: Independence Day, 27 February (1844)

Political parties and leaders:
Major parties: Social Christian Reformist Party (PRSC), Joaquin BALAGUER Ricardo; Dominican Liberation Party (PLD), Juan BOSCH Gavino; Dominican Revolutionary Party (PRD), Jose Franciso PENA Gomez; Independent Revolutionary Party (PRI), Jacobo MAJLUTA
Minor parties: National Veterans and

Civilian Party (PNVC), Juan Rene BEAUCHAMPS Javier; Liberal Party of the Dominican Republic (PLRD), Andres Van Der HORST; Democratic Quisqueyan Party (PQD), Elias WESSIN Chavez; National Progressive Force (FNP), Marino VINICIO Castillo; Popular Christian Party (PPC), Rogelio DELGADO Bogaert; Dominican Communist Party (PCD), Narciso ISA Conde; Dominican Workers' Party (PTD), Ivan RODRIGUEZ; Anti-Imperialist Patriotic Union (UPA), Ignacio RODRIGUEZ Chiappini; Alliance for Democracy Party (APD), Maximilano Rabelais PUIG Miller, Nelsida MARMOLEJOS, Vicente BENGOA
note: in 1983 several leftist parties, including the PCD, joined to form the Dominican Leftist Front (FID); however, they still retain individual party structures

Other political or pressure groups: Collective of Popular Organzations (COP), leader NA

Suffrage: 18 years of age; universal and compulsory or married persons regardless of age
note: members of the armed forces and police cannot vote

Elections:
Chamber of Deputies: last held 16 May 1990 (next to be held May 1994); results—percent of vote by party NA; seats—(120 total) PLD 44, PRSC 41, PRD 33, PRI 2
President: last held 16 May 1990 (next to be held May 1994); results—Joaquin BALAGUER (PRSC) 35.7%, Juan BOSCH Gavino (PLD) 34.4%
Senate: last held 16 May 1990 (next to be held May 1994); results—percent of vote by party NA; seats—(30 total) PRSC 16, PLD 12, PRD 2

Executive branch: president, vice president, Cabinet

Legislative branch: bicameral National Congress (Congreso Nacional) consists of an upper chamber or Senate (Senado) and lower chamber or Chamber of Deputies (Camara de Diputados)

Judicial branch: Supreme Court (Corte Suprema)

Leaders:
Chief of State and Head of Government: President Joaquin BALAGUER Ricardo (since 16 August 1986, fifth elected term began 16 August 1990); Vice President Carlos A. MORALES Troncoso (since 16 August 1986)

Member of: ACP, CARICOM (observer), ECLAC, FAO, G-11, G-77, GATT, IADB, IAEA, IBRD, ICAO, ICFTU, IDA, IFAD, IFC, ILO, IMF, IMO, INTELSAT, INTERPOL, IOC, IOM, ITU, LAES, LAIA (observer), LORCS, NAM (guest), OAS, OPANAL, PCA, UN, UNCTAD, UNESCO,

UNIDO, UPU, WCL, WFTU, WHO, WIPO, WMO, WTO

Diplomatic representation in US:
chief of mission: Ambassador Jose del Carmen ARIZA Gomez
chancery: 1715 22nd Street NW, Washington, DC 20008
telephone: (202) 332-6280
consulates general: Boston, Chicago, Los Angeles, Mayaguez (Puerto Rico), Miami, New Orleans, New York, Philadelphia, San Juan (Puerto Rico)
consulates: Charlotte Amalie (Virgin Islands), Detroit, Houston, Jacksonville, Minneapolis, Mobile, Ponce (Puerto Rico), and San Francisco

US diplomatic representation:
chief of mission: Ambassador Robert S. PASTORINO
embassy: corner of Calle Cesar Nicolas Penson and Calle Leopoldo Navarro, Santo Domingo
mailing address: APO AA 34041-0008
telephone: (809) 541-2171 and 541-8100
FAX: (809) 686-7437

Flag: a centered white cross that extends to the edges, divides the flag into four rectangles—the top ones are blue (hoist side) and red, the bottom ones are red (hoist side) and blue; a small coat of arms is at the center of the cross

Economy

Overview: The economy is largely dependent on trade; imported components average 60% of the value of goods consumed in the domestic market. Rapid growth of free trade zones has established a significant expansion of manufacturing for export, especially wearing apparel. Over the past decade, tourism has also increased in importance and is a major earner of foreign exchange and a source of new jobs. Agriculture remains a key sector of the economy. The principal commercial crop is sugarcane, followed by coffee, cotton, cocoa, and tobacco. Domestic industry is based on the processing of agricultural products, oil refining, minerals, and chemicals. Unemployment is officially reported at about 30%, but there is considerable underemployment.

National product: GDP—exchange rate conversion—$8.4 billion (1992 est.)

National product real growth rate: 5% (1992 est.)

National product per capita: $1,120 (1992 est.)

Inflation rate (consumer prices): 6% (1992 est.)

Unemployment rate: 30% (1992 est.)

Budget: revenues $1.4 billion; expenditures $1.8 billion, including capital expenditures of $NA (1993 est.)

Exports: $600 million (f.o.b., 1992)
commodities: ferronickel, sugar, gold, coffee, cocoa
partners: US 60%, EC 19%, Puerto Rico 8% (1990)
Imports: $2 billion (c.i.f., 1992 est.)
commodities: foodstuffs, petroleum, cotton and fabrics, chemicals and pharmaceuticals
partners: US 50%
External debt: $4.7 billion (1992 est.)
Industrial production: growth rate −1.5% (1991); accounts for 20% of GDP
Electricity: 2,283,000 kW capacity; 5,000 million kWh produced, 660 kWh per capita (1992)
Industries: tourism, sugar processing, ferronickel and gold mining, textiles, cement, tobacco
Agriculture: accounts for 15% of GDP and employs 49% of labor force; sugarcane is the most important commercial crop, followed by coffee, cotton, cocoa, and tobacco; food crops—rice, beans, potatoes, corn, bananas; animal output—cattle, hogs, dairy products, meat, eggs; not self-sufficient in food
Illicit drugs: transshipment point for South American drugs destined for the US
Economic aid: US commitments, including Ex-Im (FY85-89), $575 million; Western (non-US) countries, ODA and OOF bilateral commitments (1970-89), $655 million
Currency: 1 Dominican peso (RD$) = 100 centavos
Exchange rates: Dominican pesos (RD$) per US$1—12.7 (1992), 12.692 (1991), 8.525 (1990), 6.340 (1989), 6.113 (1988)
Fiscal year: calendar year

Communications

Railroads: 1,655 km total in numerous segments; 4 different gauges from 0.558 m to 1.435 m
Highways: 12,000 km total; 5,800 km paved, 5,600 km gravel and improved earth, 600 km unimproved
Pipelines: crude oil 96 km; petroleum products 8 km
Ports: Santo Domingo, Haina, San Pedro de Macoris, Puerto Plata
Merchant marine: 1 cargo ship (1,000 GRT or over) totaling 1,587 GRT/1,165 DWT
Airports:
total: 36
usable: 30
with permanent-surface runways: 12
with runways over 3,659 m: 0
with runways 2,440-3,659 m: 4
with runways 1,220-2,439 m: 8
Telecommunications: relatively efficient domestic system based on islandwide microwave relay network; 190,000 telephones; broadcast stations—120 AM, no FM, 18 TV, 6 shortwave; 1 coaxial submarine cable; 1 Atlantic Ocean INTELSAT earth station

Defense Forces

Branches: Army, Navy, Air Force, National Police
Manpower availability: males age 15-49 2,064,244; fit for military service 1,302,644; reach military age (18) annually 80,991 (1993 est.)
Defense expenditures: exchange rate conversion—$110 million, 0.7% of GDP (1993 est.)

Ecuador

Galapagos Islands

Geography

Location: Western South America, bordering the Pacific Ocean at the Equator between Colombia and Peru
Map references: South America, Standard Time Zones of the World
Area:
total area: 283,560 km²
land area: 276,840 km²
comparative area: slightly smaller than Nevada
note: includes Galapagos Islands
Land boundaries: total 2,010 km, Colombia 590 km, Peru 1,420 km
Coastline: 2,237 km
Maritime claims:
continental shelf: claims continental shelf between mainland and Galapagos Islands
territorial sea: 200 nm
International disputes: three sections of the boundary with Peru are in dispute
Climate: tropical along coast becoming cooler inland
Terrain: coastal plain (Costa), inter-Andean central highlands (Sierra), and flat to rolling eastern jungle (Oriente)
Natural resources: petroleum, fish, timber
Land use:
arable land: 6%
permanent crops: 3%
meadows and pastures: 17%
forest and woodland: 51%
other: 23%
Irrigated land: 5,500 km² (1989 est.)
Environment: subject to frequent earthquakes, landslides, volcanic activity; deforestation; desertification; soil erosion; periodic droughts
Note: Cotopaxi in Andes is highest active volcano in world

People

Population: 10,461,072 (July 1993 est.)
Population growth rate: 2.07% (1993 est.)
Birth rate: 26.54 births/1,000 population (1993 est.)

Ecuador *(continued)*

Death rate: 5.8 deaths/1,000 population (1993 est.)
Net migration rate: 0 migrant(s)/1,000 population (1993 est.)
Infant mortality rate: 40.8 deaths/1,000 live births (1993 est.)
Life expectancy at birth:
total population: 69.61 years
male: 67.09 years
female: 72.25 years (1993 est.)
Total fertility rate: 3.19 children born/woman (1993 est.)
Nationality:
noun: Ecuadorian(s)
adjective: Ecuadorian
Ethnic divisions: mestizo (mixed Indian and Spanish) 55%, Indian 25%, Spanish 10%, black 10%
Religions: Roman Catholic 95%
Languages: Spanish (official), Indian languages (especially Quechua)
Literacy: age 15 and over can read and write (1990)
total population: 86%
male: 88%
female: 84%
Labor force: 2.8 million
by occupation: agriculture 35%, manufacturing 21%, commerce 16%, services and other activities 28% (1982)

Government

Names:
conventional long form: Republic of Ecuador
conventional short form: Ecuador
local long form: Republica del Ecuador
local short form: Ecuador
Digraph: EC
Type: republic
Capital: Quito
Administrative divisions: 21 provinces (provincias, singular—provincia); Azuay, Bolivar, Canar, Carchi, Chimborazo, Cotopaxi, El Oro, Esmeraldas, Galapagos, Guayas, Imbabura, Loja, Los Rios, Manabi, Morona-Santiago, Napo, Pastaza, Pichincha, Sucumbios, Tungurahua, Zamora-Chinchipe
Independence: 24 May 1822 (from Spain)
Constitution: 10 August 1979
Legal system: based on civil law system; has not accepted compulsory ICJ jurisdiction
National holiday: Independence Day, 10 August (1809) (independence of Quito)
Political parties and leaders:
Center-Right parties: Social Christian Party (PSC), Jaime NEBOT Saadi, president; Republican Unity Party (PUR), President Sixto DURAN-BALLEN, leader; Conservative Party (CE), Vice President Alberto DAHIK, president
Center-Left parties: Democratic Left (ID), Andres VALLEJO Arcos, Rodrigo BORJA Cevallos, leaders; Popular Democracy (DP),

Jamil MANUAD Witt, president; Ecuadorian Radical Liberal Party (PLRE), Carlos Luis PLAZA Aray, director; Radical Alfarista Front (FRA), Jaime ASPIAZU Seminario, director
Populist parties: Roldista Party (PRE), Abdala BUCARAM Ortiz, director; Concentration of Popular Forces (CFP), Rafael SANTELICES, director; Popular Revolutionary Action (APRE), Frank VARGAS Passos, leader; Assad Bucaram Party (PAB), Avicena BUCARAM, leader; People, Change, and Democracy (PCD), Raul AULESTIA, director
Far-Left parties: Popular Democratic Movement (MPD), Jorge Fausto MORENO, director; Ecuadorian Socialist Party (PSE), Leon ROLDOS, leader; Broad Leftist Front (FADI), Jose Xavier GARAYCOA, president; Ecuadorian National Liberation (LN), Alfredo CASTILLO, director
Communists: Communist Party of Ecuador (PCE, pro-North Korea), Rene Leon Mague MOSWUERRA, secretary general (5,00 members); Communist Party of Ecuador/Marxist-Leninist (PCMLE, Maoist), leader NA (3,000 members)
Suffrage: 18 years of age; universal, compulsory for literate persons ages 18-65, optional for other eligible voters
Elections:
President: runoff election held 5 July 1992 (next to be held NA 1996); results—Sixto DURAN-BALLEN elected as president and Alberto DAHIK elected as vice president
National Congress: last held 17 May 1992 (next to be held NA May 1994); results—percent of vote by party NA; seats—(77 total) PSC 20, PRE 15, PUR 12, ID 7, PC 6, DP 5, PSE 3, MPD 3, PLRE 2, CFP 2, FRA 1, APRE 1
Executive branch: president, vice president, Cabinet
Legislative branch: unicameral National Congress (Congreso Nacional)
Judicial branch: Supreme Court (Corte Suprema)
Leaders:
Chief of State and Head of Government: President Sixto DURAN-BALLEN (since 10 August 1992); Vice President Alberto DAHIK (since 10 August 1992)
Member of: AG, ECLAC, FAO, G-11, G-77, IADB, IAEA, IBRD, ICAO, ICC, ICFTU, IDA, IFAD, IFC, ILO, IMF, IMO, INTELSAT, INTERPOL, IOC, IOM, ITU, LAES, LAIA, LORCS, NAM, OAS, ONUSAL, OPANAL, PCA, RG, UN, UNCTAD, UNESCO, UNIDO, UPU, WCL, WFTU, WHO, WIPO, WMO, WTO
Diplomatic representation in US:
chief of mission: Ambassador Edgar TERAN
chancery: 2535 15th Street NW, Washington, DC 20009
telephone: (202) 234-7200

consulates general: Chicago, Houston, Los Angeles, Miami, New Orleans, New York, and San Francisco
consulate: San Diego
US diplomatic representation:
chief of mission: (vacant); Charge d'Affaires James F. MACK
embassy: Avenida 12 de Octubre y Avenida Patria, Quito
mailing address: P. O. Box 538, Quito, or APO AA 34039-3420
telephone: [593] (2) 562-890
FAX: [593] (2) 502-052
consulate general: Guayaquil
Flag: three horizontal bands of yellow (top, double width), blue, and red with the coat of arms superimposed at the center of the flag; similar to the flag of Colombia that is shorter and does not bear a coat of arms

---

Economy

Overview: Ecuador has substantial oil resources and rich agricultural areas. Growth has been uneven because of natural disasters, fluctuations in global oil prices, and government policies designed to curb inflation. Banana exports, second only to oil, have suffered as a result of EC import quotas and banana blight. The new President Sixto DURAN-BALLEN, has a much more favorable attitude toward foreign investment than did his predecessor. Ecuador has implemented trade agreements with Colombia, Peru, Bolivia, and Venezuela and has applied for GATT membership. At the end of 1991, Ecuador received a standby IMF loan of $105 million, which will permit the country to proceed with the rescheduling of Paris Club debt. In September 1992, the government launched a new, macroeconomic program that gives more play to market forces; as of March 1993, the program seemed to be paying off.
National product: GDP—exchange rate conversion—$11.8 billion (1992)
National product real growth rate: 3% (1992)
National product per capita: $1,100 (1992)
Inflation rate (consumer prices): 70% (1992)
Unemployment rate: 8% (1992)
Budget: revenues $1.9 billion; expenditures $1.9 billion, including capital expenditures of $NA (1992)
Exports: $3.0 billion (f.o.b., 1992)
commodities: petroleum 42%, bananas, shrimp, cocoa, coffee
partners: US 53.4%, Latin America, Caribbean, EC countries
Imports: $2.4 billion (f.o.b., 1992)
commodities: transport equipment, vehicles, machinery, chemicals
partners: US 32.7%, Latin America, Caribbean, EC countries, Japan

External debt: $12.7 billion (1992)
Industrial production: growth rate 3.9% (1991); accounts for almost 40% of GDP, including petroleum
Electricity: 2,921,000 kW capacity; 7,676 million kWh produced, 700 kWh per capita (1992)
Industries: petroleum, food processing, textiles, metal works, paper products, wood products, chemicals, plastics, fishing, timber
Agriculture: accounts for 18% of GDP and 35% of labor force (including fishing and forestry); leading producer and exporter of bananas and balsawood; other exports—coffee, cocoa, fish, shrimp; crop production—rice, potatoes, manioc, plantains, sugarcane; livestock sector—cattle, sheep, hogs, beef, pork, dairy products; net importer of foodgrains, dairy products, and sugar
Illicit drugs: minor illicit producer of coca following the successful eradication campaign of 1985-87; significant transit country, however, for derivatives of coca originating in Colombia, Bolivia, and Peru; importer of precursor chemicals used in production of illicit narcotics; important money-laundering hub
Economic aid: US commitments, including Ex-Im (FY70-89), $498 million; Western (non-US) countries, ODA and OOF bilateral commitments (1970-89), $2.15 billion; Communist countries (1970-89), $64 million
Currency: 1 sucre (S/) = 100 centavos
Exchange rates: sucres (S/) per US$1—1,453.8 (August 1992), 1,046.25 (1991), 869.54 (December 1990), 767.75 (1990), 526.35 (1989), 301.61 (1988)
Fiscal year: calendar year

Communications

Railroads: 965 km total; all 1.067-meter-gauge single track
Highways: 28,000 km total; 3,600 km paved, 17,400 km gravel and improved earth, 7,000 km unimproved earth
Inland waterways: 1,500 km
Pipelines: crude oil 800 km; petroleum products 1,358 km
Ports: Guayaquil, Manta, Puerto Bolivar, Esmeraldas
Merchant marine: 45 ships (1,000 GRT or over) totaling 333,380 GRT/483,862 DWT; includes 2 passenger, 4 cargo, 17 refrigerated cargo, 4 container, 1 roll-on/roll-off, 15 oil tanker, 1 liquefied gas, 1 bulk
Airports:
total: 174
usable: 173
with permanent-surface runways: 52
with runway over 3,659 m: 1
with runways 2,440-3,659 m: 6
with runways 1,220-2,439 m: 21

Telecommunications: domestic facilities generally adequate; 318,000 telephones; broadcast stations—272 AM, no FM, 33 TV, 39 shortwave; 1 Atlantic Ocean INTELSAT earth station

---

Defense Forces

Branches: Army (Ejercito Ecuatoriano), Navy (Armada Ecuatoriana), Air Force (Fuerza Aerea Ecuatoriana), National Police
Manpower availability: males age 15-49 2,655,520; fit for military service 1,798,122; reach military age (20) annually 109,413 (1993 est.)
Defense expenditures: exchange rate conversion—$NA, NA% of GDP

Egypt

Boundary representation is not necessarily authoritative.

Geography

Location: Northern Africa, bordering the Mediterranean Sea and the Red Sea, between Sudan and Libya
Map references: Africa, Middle East, Standard Time Zones of the World
Area:
total area: 1,001,450 km^2
land area: 995,450 km^2
comparative area: slightly more than three times the size of New Mexico
Land boundaries: total 2,689 km, Gaza Strip 11 km, Israel 255 km, Libya 1,150 km, Sudan 1,273 km
Coastline: 2,450 km
Maritime claims:
contiguous zone: 24 nm
continental shelf: 200 m depth or to depth of exploitation
exclusive economic zone: not specified
territorial sea: 12 nm
International disputes: administrative boundary with Sudan does not coincide with international boundary, creating the "Hala'ib Triangle," a barren area of 20,580 km^2, the dispute over this area escalated in 1993
Climate: desert; hot, dry summers with moderate winters
Terrain: vast desert plateau interrupted by Nile valley and delta
Natural resources: petroleum, natural gas, iron ore, phosphates, manganese, limestone, gypsum, talc, asbestos, lead, zinc
Land use:
arable land: 3%
permanent crops: 2%
meadows and pastures: 0%
forest and woodland: 0%
other: 95%
Irrigated land: 25,850 km^2 (1989 est.)
Environment: Nile is only perennial water source; increasing soil salinization below Aswan High Dam; hot, driving windstorm called khamsin occurs in spring; water pollution; desertification

Egypt (continued)

Note: controls Sinai Peninsula, only land bridge between Africa and remainder of Eastern Hemisphere; controls Suez Canal, shortest sea link between Indian Ocean and Mediterranean; size and juxtaposition to Israel establish its major role in Middle Eastern geopolitics

People

Population: 59,585,529 (July 1993 est.)
Population growth rate: 2.3% (1993 est.); note—the US Bureau of the Census has lowered its 1993 estimate of growth to 2.0% on the basis of a 1992 Egyptian government survey, whereas estimates of other observers go as high as 2.9%
Birth rate: 33 births/1,000 population (1993 est.)
Death rate: 9 deaths/1,000 population (1993 est.)
Net migration rate: NEGL migrant(s)/1,000 population (1993 est.)
Infant mortality rate: 78.3 deaths/1,000 live births (1993 est.)
Life expectancy at birth:
total population: 60.46 years
male: 58.61 years
female: 62.41 years (1993 est.)
Total fertility rate: 4.35 children born/woman (1993 est.)
Nationality:
noun: Egyptian(s)
adjective: Egyptian
Ethnic divisions: Eastern Hamitic stock 90%, Greek, Italian, Syro-Lebanese 10%
Religions: Muslim (mostly Sunni) 94% (official estimate), Coptic Christian and other 6% (official estimate)
Languages: Arabic (official), English and French widely understood by educated classes
Literacy: age 15 and over can read and write (1990)
total population: 48%
male: 63%
female: 34%
Labor force: 15 million (1989 est.)
by occupation: government, public sector enterprises, and armed forces 36%, agriculture 34%, privately owned service and manufacturing enterprises 20% (1984)
note: shortage of skilled labor; 2,500,000 Egyptians work abroad, mostly in Saudi Arabia and the Gulf Arab states (1988 est.)

Government

Names:
conventional long form: Arab Republic of Egypt
conventional short form: Egypt
local long form: Jumhuriyat Misr al-Arabiyah
local short form: none
former: United Arab Republic (with Syria)

Digraph: EG
Type: republic
Capital: Cairo
Administrative divisions: 26 governorates (muhafazat, singular—muhafazah); Ad Daqahliyah, Al Bahr al Ahmar, Al Buhayrah, Al Fayyum, Al Gharbiyah, Al Iskandariyah, Al Isma'iliyah, Al Jizah, Al Minufiyah, Al Minya, Al Qahirah, Al Qalyubiyah, Al Wadi al Jadid, Ash Sharqiyah, As Suways, Aswan, Asyu't, Bani Suwayf, Bur Sa'id, Dumyat, Janub Sina, Kafr ash Shaykh, Matruh, Qina, Shamal Sina, Suhaj
Independence: 28 February 1922 (from UK)
Constitution: 11 September 1971
Legal system: based on English common law, Islamic law, and Napoleonic codes; judicial review by Supreme Court and Council of State (oversees validity of administrative decisions); accepts compulsory ICJ jurisdiction, with reservations
National holiday: Anniversary of the Revolution, 23 July (1952)
Political parties and leaders: National Democratic Party (NDP), President Mohammed Hosni MUBARAK, leader, is the dominant party; legal opposition parties are Socialist Liberal Party (SLP), Kamal MURAD; Socialist Labor Party, Ibrahim SHUKRI; National Progressive Unionist Grouping (NPUG), Khalid MUHYI-AL-DIN; Umma Party, Ahmad al-SABAHI; New Wafd Party (NWP), Fu'ad SIRAJ AL-DIN; Misr al-Fatah Party (Young Egypt Party), Ali al-Din SALIH; The Greens Party, Hasan RAJABD; Nasserist Arab Democratic Party, Muhammad Rif'at al-MUHAMI; Democratic Unionist Party, Mohammed 'Abd-al-Mun'im TURK; Democratic Peoples' Party, Anwar AFISI
note: formation of political parties must be approved by government
Other political or pressure groups: Islamic groups are illegal, but the largest one, the Muslim Brotherhood, is tolerated by the government; trade unions and professional associations are officially sanctioned
Suffrage: 18 years of age; universal and compulsory
Elections:
Advisory Council: last held 8 June 1989 (next to be held June 1995); results—NDP 100%; seats—(258 total, 172 elected) NDP 172
People's Assembly: last held 29 November 1990 (next to be held November 1995); results—NDP 78.4%, NPUG 1.4%, independents 18.7%; seats—(437 total, 444 elected) NDP 348, NPUG 6, independents 83; note—most opposition parties boycotted
President: last held 5 October 1987 (next to be held October 1993); results—President Hosni MUBARAK was reelected

Executive branch: president, prime minister, Cabinet
Legislative branch: unicameral People's Assembly (Majlis al-Cha'b); note—there is an Advisory Council (Majlis al-Shura) that functions in a consultative role
Judicial branch: Supreme Constitutional Court
Leaders:
Chief of State: President Mohammed Hosni MUBARAK (was made acting President on 6 October 1981 upon the assassination of President SADAT and sworn in as president on 14 October 1981)
Head of Government: Prime Minister Atef Mohammed Najib SEDKY (since 12 November 1986)
Member of: ABEDA, ACC, ACCT (associate), AfDB, AFESD, AG (observer), AL, AMF, CAEU, CCC, EBRD, ECA, ESCWA, FAO, G-15, G-19, G-24, G-77, GATT, IAEA, IBRD, ICAO, ICC, IDA, IDB, IFAD, IFC, ILO, IMF, IMO, INMARSAT, INTELSAT, INTERPOL, IOC, IOM, ISO, ITU, LORCS, MINURSO, NAM, OAPEC, OAS (observer), OAU, OIC, PCA, UN, UNAVEM II, UNCTAD, UNESCO, UNIDO, UNOMOZ, UNOSOM, UNPROFOR, UPU, UNRWA, WHO, WIPO, WMO, WTO
Diplomatic representation in US:
chief of mission: Ambassador Ahmed MAHER El Sayed
chancery: 2310 Decatur Place NW, Washington, DC 20008
telephone: (202) 232-5400
consulates general: Chicago, Houston, New York, and San Francisco
US diplomatic representation:
chief of mission: Ambassador Robert PELLETREAU
embassy: Lazougi Street, Garden City, Cairo
mailing address: APO AE 09839
telephone: [20] (2) 355-7371
FAX: [20] (2) 355-7375
consulate general: Alexandria
Flag: three equal horizontal bands of red (top), white, and black with the national emblem (a shield superimposed on a golden eagle facing the hoist side above a scroll bearing the name of the country in Arabic) centered in the white band; similar to the flag of Yemen, which has a plain white band; also similar to the flag of Syria that has two green stars and to the flag of Iraq, which has three green stars (plus an Arabic inscription) in a horizontal line centered in the white band

Economy

Overview: Egypt has one of the largest public sectors of all the Third World economies, most industrial plants being owned by the government. Overregulation holds back technical modernization and foreign investment. Even so, the economy grew rapidly during the late 1970s and early 1980s, but in 1986 the collapse of world oil prices and an increasingly heavy burden of debt servicing led Egypt to begin negotiations with the IMF for balance-of-payments support. Egypt's first IMF standby arrangement concluded in mid-1987 was suspended in early 1988 because of the government's failure to adopt promised reforms. Egypt signed a follow-on program with the IMF and also negotiated a structural adjustment loan with the World Bank in 1991. In 1991-92 the government made solid progress on administrative reforms such as liberalizing exchange and interest rates but resisted implementing major structural reforms like streamlining the public sector. As a result, the economy has not gained momentum and unemployment has become a growing problem. In 1992-93 tourism has plunged 20% or so because of sporadic attacks by Islamic extremists on tourist groups. President MUBARAK has cited population growth as the main cause of the country's economic troubles. The addition of about 1.4 million people a year to the already huge population of 60 million exerts enormous pressure on the 5% of the land area available for agriculture.

National product: GDP—exchange rate conversion—$41.2 billion (1992 est.)

National product real growth rate: 2.1% (1992 est.)

National product per capita: $730 (1992 est.)

Inflation rate (consumer prices): 21% (1992 est.)

Unemployment rate: 20% (1992 est.)

Budget: revenues $12.6 billion; expenditures $15.2 billion, including capital expenditures of $4 billion (FY92 est.)

Exports: $3.6 billion (f.o.b., FY92 est.)
commodities: crude oil and petroleum products, cotton yarn, raw cotton, textiles, metal products, chemicals
partners: EC, Eastern Europe, US, Japan

Imports: $10.0 billion (c.i.f., FY92 est.)
commodities: machinery and equipment, foods, fertilizers, wood products, durable consumer goods, capital goods
partners: EC, US, Japan, Eastern Europe

External debt: $38 billion (December 1991 est.)

Industrial production: growth rate 7.3% (FY89 est.); accounts for 18% of GDP

Electricity: 14,175,000 kW capacity; 47,000 million kWh produced, 830 kWh per capita (1992)

Industries: textiles, food processing, tourism, chemicals, petroleum, construction, cement, metals

Agriculture: accounts for 20% of GDP and employs more than one-third of labor force; dependent on irrigation water from the Nile; world's sixth-largest cotton exporter; other crops produced include rice, corn, wheat, beans, fruit, vegetables; not self-sufficient in food for a rapidly expanding population; livestock—cattle, water buffalo, sheep, goats; annual fish catch about 140,000 metric tons

Illicit drugs: a transit point for Southwest Asian and Southeast Asian heroin and opium moving to Europe and the US; popular transit stop for Nigerian couriers; large domestic consumption of hashish and heroin from Lebanon and Syria

Economic aid: US commitments, including Ex-Im (FY70-89), $15.7 billion; Western (non-US) countries, ODA and OOF bilateral commitments (1970-88), $10.1 billion; OPEC bilateral aid (1979-89), $2.9 billion; Communist countries (1970-89), $2.4 billion

Currency: 1 Egyptian pound (£E) = 100 piasters

Exchange rates: Egyptian pounds (£E) per US$1—3.345 (November 1992), 2.7072 (1990), 2.5171 (1989), 2.2233 (1988), 1.5183 (1987)

Fiscal year: 1 July-30 June

Communications

Railroads: 5,110 km total; 4,763 km 1,435-meter standard gauge, 347 km 0.750-meter gauge; 951 km double track; 25 km electrified

Highways: 51,925 km total; 17,900 km paved, 2,500 km gravel, 13,500 km improved earth, 18,025 km unimproved earth

Inland waterways: 3,500 km (including the Nile, Lake Nasser, Alexandria-Cairo Waterway, and numerous smaller canals in the delta); Suez Canal, 193.5 km long (including approaches), used by oceangoing vessels drawing up to 16.1 meters of water

Pipelines: crude oil 1,171 km; petroleum products 596 km; natural gas 460 km

Ports: Alexandria, Port Said, Suez, Bur Safajah, Damietta

Merchant marine: 168 ships (1,000 GRT or over) totaling 1,097,707 GRT/1,592,885 DWT; includes 25 passenger, 6 short-sea passenger, 2 passenger-cargo, 88 cargo, 3 refrigerated cargo, 14 roll-on/roll-off, 13 oil tanker, 16 bulk, 1 container

Airports:
total: 92
usable: 82
with permanent-surface runways: 66
with runways over 3,659 m: 2
with runways 2,440-3,659 m: 44
with runways 1,220-2,439 m: 24

Telecommunications: large system by Third World standards but inadequate for present requirements and undergoing extensive upgrading; about 600,000 telephones (est.)—11 telephones per 1,000 persons; principal centers at Alexandria, Cairo, Al Mansurah, Ismailia Suez, and Tanta are connected by coaxial cable and microwave radio relay; international traffic is carried by satellite—one earth station for each of Atlantic Ocean INTELSAT, Indian Ocean INTELSAT, ARABSAT and INMARSAT; by 5 coaxial submarine cables, microwave troposcatter (to Sudan), and microwave radio relay (to Libya, Israel, and Jordan); broadcast stations—39 AM, 6 FM, and 41 TV

Defense Forces

Branches: Army, Navy, Air Force, Air Defense Command

Manpower availability: males age 15-49 14,513,752; fit for military service 9,434,020; reach military age (20) annually 581,858 (1993 est.)

Defense expenditures: exchange rate conversion—$2.05 billion, 5% of GDP (FY92/93)

El Salvador

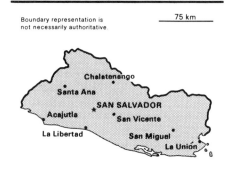

Boundary representation is
not necessarily authoritative.

75 km

Chalatenango
Santa Ana
SAN SALVADOR
Acajutla
San Vicente
La Libertad
San Miguel
La Unión

North Pacific Ocean

Geography

Location: Central America, bordering the
North Pacific Ocean between Guatemala and
Honduras
Map references: Central America and the
Caribbean, North America, Standard Time
Zones of the World
Area:
total area: 21,040 km²
land area: 20,720 km²
comparative area: slightly smaller than
Massachusetts
Land boundaries: total 545 km, Guatemala
203 km, Honduras 342 km
Coastline: 307 km
Maritime claims:
territorial sea: 200 nm; overflight and
navigation permitted beyond 12 nm
International disputes: land boundary
dispute with Honduras mostly resolved by
11 September 1992 International Court of
Justice (ICJ) decision; ICJ referred the
maritime boundary in the Golfo de Fonseca
to an earlier agreement in this century and
advised that some tripartite resolution among
El Salvador, Honduras and Nicaragua likely
would be required
Climate: tropical; rainy season (May to
October); dry season (November to April)
Terrain: mostly mountains with narrow
coastal belt and central plateau
Natural resources: hydropower, geothermal
power, petroleum
Land use:
arable land: 27%
permanent crops: 8%
meadows and pastures: 29%
forest and woodland: 6%
other: 30%
Irrigated land: 1,200 km² (1989)
Environment: the Land of Volcanoes;
subject to frequent and sometimes very
destructive earthquakes; deforestation; soil
erosion; water pollution
Note: smallest Central American country
and only one without a coastline on
Caribbean Sea

People

Population: 5,636,524 (July 1993 est.)
Population growth rate: 2.04% (1993 est.)
Birth rate: 33.12 births/1,000 population
(1993 est.)
Death rate: 6.53 deaths/1,000 population
(1993 est.)
Net migration rate: -6.21 migrant(s)/1,000
population (1993 est.)
Infant mortality rate: 42.8 deaths/1,000
live births (1993 est.)
Life expectancy at birth:
total population: 66.5 years
male: 63.93 years
female: 69.2 years (1993 est.)
Total fertility rate: 3.87 children
born/woman (1993 est.)
Nationality:
noun: Salvadoran(s)
adjective: Salvadoran
Ethnic divisions: mestizo 94%, Indian 5%,
white 1%
Religions: Roman Catholic 75%
note: Roman Catholic about 75%; there is
extensive activity by Protestant groups
throughout the country; by the end of 1992,
there were an estimated 1 million Protestant
evangelicals in El Salvador
Languages: Spanish, Nahua (among some
Indians)
Literacy: age 15 and over can read and
write (1990)
total population: 73%
male: 76%
female: 70%
Labor force: 1.7 million (1982 est.)
by occupation: agriculture 40%, commerce
16%, manufacturing 15%, government 13%,
financial services 9%, transportation 6%,
other 1%
note: shortage of skilled labor and a large
pool of unskilled labor, but manpower
training programs improving situation (1984
est.)

Government

Names:
conventional long form: Republic of El
Salvador
conventional short form: El Salvador
local long form: Republica de El Salvador
local short form: El Salvador
Digraph: ES
Type: republic
Capital: San Salvador
Administrative divisions: 14 departments
(departamentos, singular—departamento);
Ahuachapan, Cabanas, Chalatenango,
Cuscatlan, La Libertad, La Paz, La Union,
Morazan, San Miguel, San Salvador, Santa
Ana, San Vicente, Sonsonate, Usulutan
Independence: 15 September 1821 (from
Spain)

Constitution: 20 December 1983
Legal system: based on civil and Roman
law, with traces of common law; judicial
review of legislative acts in the Supreme
Court; accepts compulsory ICJ jurisdiction,
with reservations
National holiday: Independence Day, 15
September (1821)
Political parties and leaders: National
Republican Alliance (Arena), Armando
CALDERON Sol, president; Christian
Democratic Party (PDC), Fidel CHAVEZ
Mena, secretary general; National
Conciliation Party (PCN), Ciro CRUZ
Zepeda, president; Democratic Convergence
(CD) is a coalition of three parties—the
Social Democratic Party (PSD), Carlos Diaz
BARRERA, secretary general; Democratic
Nationalist Union (UDN), Mario
AGUINADA Carranza, secretary general;
and the Popular Social Christian Movement
(MPSC), Dr. Ruben Ignacio ZAMORA
Rivas; Authentic Christian Movement
(MAC), Guillermo Antonia GUEVARA
Lacayo, president; Farabundo Marti National
Liberation Front (FMLM), Jorge Shafik
HANDAL, general coordinator, has five
factions—Popular Liberation Forces (FPL),
Salvador SANCHEZ Ceren; Armed Forces
of National Resistance (FARN), Ferman
CIENFUEGOS; People's Revolutionary
Army (ERP), Joaquin VILLA LOBOS
Huezo; Salvadoran Communist Party/Armed
Forces of Liberation (PCES/FAL), Jorge
Shafik HANDAL; and Central American
Workers' Revolutionary Party
(PRTC)/Popular Liberation Revolutionary
Aermed Forces (FARLP), Francisco JOVEL
Other political or pressure groups:
FMLN labor front organizations: National
Union of Salvadoran Workers (UNTS),
leftist umbrella front group, leads FMLN
front network; National Federation of
Salvadoran Workers (FENASTRAS), best
organized of front groups and controlled by
FMLN's National Resistance (RN); Social
Security Institute Workers Union (STISSS),
one of the most militant fronts, is controlled
by FMLN's Armed Forces of National
Resistance (FARN) and RN; Association of
Telecommunications Workers (ASTTEL);
Unitary Federation of Salvadoran Unions
(FUSS), leftist; Treasury Ministry
Employees (AGEMHA)
FMLN nonlabor front organizations:
Committee of Mothers and Families of
Political Prisoners, Disappeared Persons, and
Assassinated of El Salvador (COMADRES);
Nongovernmental Human Rights
Commission (CDHES); Committee of
Dismissed and Unemployed of El Salvador
(CODYDES); General Association of
Salvadoran University Students (AGEUS);
National Association of Salvadoran
Educators (ANDES-21 DE JUNIO);

Salvadoran Revolutionary Student Front (FERS), associated with the Popular Forces of Liberation (FPL); Association of National University Educators (ADUES); Salvadoran University Students Front (FEUS); Christian Committee for the Displaced of El Salvador (CRIPDES), an FPL front; The Association for Communal Development in El Salvador (PADECOES), controlled by the People's Revolutionary Army (ERP); Confederation of Cooperative Associations of El Salvador (COACES)

labor organizations: Federation of Construction and Transport Workers Unions (FESINCONSTRANS), independent; Salvadoran Communal Union (UCS), peasant association; Democratic Workers Central (CTD), moderate; General Confederation of Workers (CGT), moderate; National Union of Workers and Peasants (UNOC), moderate labor coalition of democratic labor organizations; United Workers Front (FUT)

business organizations: National Association of Private Enterprise (ANEP), conservative; Productive Alliance (AP), conservative; National Federation of Salvadoran Small Businessmen (FENAPES), conservative

Suffrage: 18 years of age; universal

Elections:

Legislative Assembly: last held 10 March 1991 (next to be held March 1994); results—ARENA 44.3%, PDC 27.96%, CD 12.16%, PCN 8.99%, MAC 3.23%, UDN 2.68%; seats—(84 total) ARENA 39, PDC 26, PCN 9, CD 8, UDN 1, MAC 1

President: last held 19 March 1989 (next to be held March 1994); results—Alfredo CRISTIANI (ARENA) 53.8%, Fidel CHAVEZ Mena (PDC) 36.6%, other 9.6%

Executive branch: president, vice president, Council of Ministers (cabinet)

Legislative branch: unicameral Legislative Assembly (Asamblea Legislativa)

Judicial branch: Supreme Court (Corte Suprema)

Leaders:

Chief of State and Head of Government: President (Felix) Alfredo CRISTIANI Buchard (since 1 June 1989); Vice President (Jose) Francisco MERINO Lopez (since 1 June 1989)

Member of: BCIE, CACM, ECLAC, FAO, G-77, GATT, IADB, IAEA, IBRD, ICAO, ICFTU, IDA, IFAD, IFC, ILO, IMF, IMO, INTELSAT, IOC, IOM, ITU, LAES, LAIA (observer), LORCS, NAM (observer), OAS, OPANAL, PCA, UN, UNCTAD, UNESCO, UNIDO, UPU, WCL, WFTU, WHO, WIPO, WMO

Diplomatic representation in US:

chief of mission: Ambassador Miguel Angel SALAVERRIA

chancery: 2308 California Street NW, Washington, DC 20008

telephone: (202) 265-9671 through 3482

consulates general: Houston, Los Angeles, Miami, New Orleans, New York, and San Francisco

US diplomatic representation:

chief of mission: Charge d'Affaires Peter F. ROMERO

embassy: Final Boulevard, Station Antigua Cuscatlan, San Salvador

mailing address: APO AA 34023

telephone: [503] 78-4444

FAX: [503] 78-6011

Flag: three equal horizontal bands of blue (top), white, and blue with the national coat of arms centered in the white band; the coat of arms features a round emblem encircled by the words REPUBLICA DE EL SALVADOR EN LA AMERICA CENTRAL; similar to the flag of Nicaragua, which has a different coat of arms centered in the white band—it features a triangle encircled by the words REPUBLICA DE NICARAGUA on top and AMERICA CENTRAL on the bottom; also similar to the flag of Honduras, which has five blue stars arranged in an X pattern centered in the white band

---

Economy

Overview: The agricultural sector accounts for 24% of GDP, employs about 40% of the labor force, and contributes about 66% to total exports. Coffee is the major commercial crop, accounting for 45% of export earnings. The manufacturing sector, based largely on food and beverage processing, accounts for 18% of GDP and 15% of employment. Economic losses because of guerrilla sabotage total more than $2 billion since 1979. The costs of maintaining a large military seriously constrain the government's efforts to provide essential social services. Nevertheless, growth in national output during the period 1990-92 exceeded growth in population for the first time since 1987.

National product: GDP—exchange rate conversion—$5.9 billion (1992 est.)

National product real growth rate: 4.6% (1992 est.)

National product per capita: $1,060 (1992 est.)

Inflation rate (consumer prices): 17% (1992 est.)

Unemployment rate: 7.5% (1991)

Budget: revenues $846 million; expenditures $890 million, including capital expenditures of $NA (1992 est.)

Exports: $693 million (f.o.b., 1992 est.)

commodities: coffee 45%, sugar, shrimp, cotton

partners: US 33%, Guatemala, Germany, Costa Rica

Imports: $1.47 billion (c.i.f., 1992 est.)

commodities: raw materials, consumer goods, capital goods

partners: US 43%, Guatemala, Mexico, Venezuela, Germany

External debt: $2.6 billion (December 1992)

Industrial production: growth rate 4.7% (1991); accounts for 22% of GDP

Electricity: 713,800 kW capacity; 2,190 million kWh produced, 390 kWh per capita (1992)

Industries: food processing, beverages, petroleum, nonmetallic products, tobacco, chemicals, textiles, furniture

Agriculture: accounts for 24% of GDP and 40% of labor force (including fishing and forestry); coffee most important commercial crop; other products—sugarcane, corn, rice, beans, oilseeds, beef, dairy products, shrimp; not self-sufficient in food

Illicit drugs: transshipment point for cocaine

Economic aid: US commitments, including Ex-Im (FY70-90), $2.95 billion, plus $250 million for 1992-96; Western (non-US) countries, ODA and OOF bilateral commitments (1970-89), $525 million

Currency: 1 Salvadoran colon (C) = 100 centavos

Exchange rates: Salvadoran colones (C) per US$1—8.7600 (January 1993), 9.1700 (1992), 8.0300 (1991), fixed rate of 5.000 (1986-1989)

Fiscal year: calendar year

Communications

Railroads: 602 km 0.914-meter gauge, single track; 542 km in use

Highways: 10,000 km total; 1,500 km paved, 4,100 km gravel, 4,400 km improved and unimproved earth

Inland waterways: Rio Lempa partially navigable

Ports: Acajutla, Cutuco

Airports:

total: 105

usable: 74

with permanent-surface runways: 5

with runways over 3,659 m: 0

with runways 2,440-3,659 m: 1

with runways 1,220-2,439 m: 5

Telecommunications: nationwide trunk microwave radio relay system; connection into Central American Microwave System; 116,000 telephones (21 telephones per 1,000 persons); broadcast stations—77 AM, no FM, 5 TV, 2 shortwave; 1 Atlantic Ocean INTELSAT earth station

El Salvador *(continued)*

Defense Forces

Branches: Army, Navy, Air Force
Manpower availability: males age 15-49 1,305,853; fit for military service 836,192; reach military age (18) annually 71,101 (1993 est.)
Defense expenditures: exchange rate conversion—$104 million, 3%-4% of GDP (1993 est.)

Equatorial Guinea

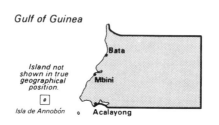

Geography

Location: Western Africa, bordering the North Atlantic Ocean between Cameroon and Gabon
Map references: Africa, Standard Time Zones of the World
Area:
total area: 28,050 km^2
land area: 28,050 km^2
comparative area: slightly larger than Maryland
Land boundaries: total 539 km, Cameroon 189 km, Gabon 350 km
Coastline: 296 km
Maritime claims:
exclusive economic zone: 200 nm
territorial sea: 12 nm
International disputes: maritime boundary dispute with Gabon because of disputed sovereignty over islands in Corisco Bay
Climate: tropical; always hot, humid
Terrain: coastal plains rise to interior hills; islands are volcanic
Natural resources: timber, petroleum, small unexploited deposits of gold, manganese, uranium
Land use:
arable land: 8%
permanent crops: 4%
meadows and pastures: 4%
forest and woodland: 51%
other: 33%
Irrigated land: NA km^2
Environment: subject to violent windstorms
Note: insular and continental regions rather widely separated

People

Population: 399,055 (July 1993 est.)
Population growth rate: 2.6% (1993 est.)
Birth rate: 41.1 births/1,000 population (1993 est.)
Death rate: 15.11 deaths/1,000 population (1993 est.)
Net migration rate: 0 migrant(s)/1,000 population (1993 est.)

Infant mortality rate: 104.9 deaths/1,000 live births (1993 est.)
Life expectancy at birth:
total population: 51.63 years
male: 49.56 years
female: 53.76 years (1993 est.)
Total fertility rate: 5.33 children born/woman (1993 est.)
Nationality:
noun: Equatorial Guinean(s) or Equatoguinean(s)
adjective: Equatorial Guinean or Equatoguinean
Ethnic divisions: Bioko (primarily Bubi, some Fernandinos), Rio Muni (primarily Fang), Europeans less than 1,000, mostly Spanish
Religions: nominally Christian and predominantly Roman Catholic, pagan practices
Languages: Spanish (official), pidgin English, Fang, Bubi, Ibo
Literacy: age 15 and over can read and write (1990)
total population: 50%
male: 64%
female: 37%
Labor force: 172,000 (1986 est.)
by occupation: agriculture 66%, services 23%, industry 11% (1980)
note: labor shortages on plantations; 58% of population of working age (1985)

Government

Names:
conventional long form: Republic of Equatorial Guinea
conventional short form: Equatorial Guinea
local long form: Republica de Guinea Ecuatorial
local short form: Guinea Ecuatorial
former: Spanish Guinea
Digraph: EK
Type: republic in transition to multiparty democracy
Capital: Malabo
Administrative divisions: 7 provinces (provincias, singular—provincia); Annobon, Bioko Norte, Bioko Sur, Centro Sur, Kie-Ntem, Litoral, Wele-Nzas
Independence: 12 October 1968 (from Spain)
Constitution: new constitution 17 November 1991
Legal system: partly based on Spanish civil law and tribal custom
National holiday: Independence Day, 12 October (1968)
Political parties and leaders: ruling—Democratic Party for Equatorial Guinea (PDGE), Brig. Gen. (Ret.) Teodoro OBIANG NGUEMA MBASOGO, party leader
Suffrage: universal adult at age NA

Elections:
President: last held 25 June 1989 (next to be held 25 June 1996); results—President Brig. Gen. (Ret.) Teodoro OBIANG NGUEMA MBASOGO was reelected without opposition
Chamber of People's Representatives: last held 10 July 1988 (next to be held 10 July 1993); results—PDGE is the only party; seats—(41 total) PDGE 41
Executive branch: president, prime minister, deputy prime minister, Council of Ministers (cabinet)
Legislative branch: unicameral House of Representatives of the People (Camara de Representantes del Pueblo)
Judicial branch: Supreme Tribunal
Leaders:
Chief of State: President Brig. Gen. (Ret.) Teodoro OBIANG NGUEMA MBASOGO (since 3 August 1979)
Head of Government: Prime Minister Silvestre SIALE BILEKA (since 17 January 1992); Deputy Prime Minister Miguel OYONO NDONG MIFUMU (since 22 January 1992)
Member of: ACCT, ACP, AfDB, BDEAC, CEEAC, ECA, ECOWAS, FAO, FZ, G-77, IBRD, ICAO, IDA, IFAD, IFC, ILO, IMF, IMO, INTERPOL, IOC, ITU, LORCS (associate), NAM, OAS (observer), OAU, UDEAC, UN, UNCTAD, UNESCO, UNIDO, UPU, WHO
Diplomatic representation in US:
chief of mission: Ambassador Damaso OBIANG NDONG
chancery: (temporary) 57 Magnolia Avenue, Mount Vernon, NY 10553
telephone: (914) 667-9664
US diplomatic representation:
chief of mission: Ambassador John E. BENNETT
embassy: Calle de Los Ministros, Malabo
mailing address: P.O. Box 597, Malabo
telephone: [240] (9) 2185
FAX: [240] (9) 2164
Flag: three equal horizontal bands of green (top), white, and red with a blue isosceles triangle based on the hoist side and the coat of arms centered in the white band; the coat of arms has six yellow six-pointed stars (representing the mainland and five offshore islands) above a gray shield bearing a silk-cotton tree and below which is a scroll with the motto UNIDAD, PAZ, JUSTICIA (Unity, Peace, Justice)

Economy

Overview: The economy, devastated during the regime of former President Macias NGUEMA, is based on agriculture, forestry, and fishing, which account for about half of GDP and nearly all exports. Subsistence agriculture predominates, with cocoa, coffee, and wood products providing income, foreign exchange, and government revenues. There is little industry. Commerce accounts for about 8% of GDP and the construction, public works, and service sectors for about 38%. Undeveloped natural resources include titanium, iron ore, manganese, uranium, and alluvial gold. Oil exploration, taking place under concessions offered to US, French, and Spanish firms, has been moderately successful. Increased production from recently discovered natural gas deposits will provide a greater share of exports by 1995.
National product: GDP—exchange rate conversion—$144 million (1991 est.)
National product real growth rate: -1% (1991 est.)
National product per capita: $380 (1991 est.)
Inflation rate (consumer prices): 1.4% (1990)
Unemployment rate: NA%
Budget: revenues $26 million; expenditures $30 million, including capital expenditures of $3 million (1991 est.)
Exports: $37 million (f.o.b., 1990 est.)
commodities: coffee, timber, cocoa beans
partners: Spain 38.2%, Italy 12.2%, Netherlands 11.4%, FRG 6.9%, Nigeria 12.4% (1988)
Imports: $63.0 million (c.i.f., 1990)
commodities: petroleum, food, beverages, clothing, machinery
partners: France 25.9%, Spain 21.0%, Italy 16%, US 12.8%, Netherlands 8%, FRG 3.1%, Gabon 2.9%, Nigeria 1.8% (1988)
External debt: $213 million (1990)
Industrial production: growth rate 6.8% (1990 est.)
Electricity: 23,000 kW capacity; 60 million kWh produced, 160 kWh per capita (1991)
Industries: fishing, sawmilling
Agriculture: cash crops—timber and coffee from Rio Muni, cocoa from Bioko; food crops—rice, yams, cassava, bananas, oil palm nuts, manioc, livestock
Economic aid: US commitments, including Ex-Im (FY81-89), $14 million; Western (non-US) countries, ODA and OOF bilateral commitments (1970-89) $130 million; Communist countries (1970-89), $55 million
Currency: 1 CFA franc (CFAF) = 100 centimes
Exchange rates: Communaute Financiere Africaine francs (CFAF) per US$1—274.06 (January 1993), 264.69 (1992), 282.11 (1991), 272.26 (1990), 319.01 (1989), 297.85 (1988)
Fiscal year: 1 April-31 March

Communications

Highways: Rio Muni—2,460 km; Bioko—300 km
Ports: Malabo, Bata

Merchant marine: 2 ships (1,000 GRT or over) totaling 6,413 GRT/6,699 DWT; includes 1 cargo and 1 passenger-cargo
Airports:
total: 3
usable: 3
with permanent-surface runways: 2
with runways over 3,659 m: 0
with runways 2,440-3,659 m: 1
with runways 1,220-2,439 m: 1
Telecommunications: poor system with adequate government services; international communications from Bata and Malabo to African and European countries; 2,000 telephones; broadcast stations—2 AM, no FM, 1 TV; 1 Indian Ocean INTELSAT earth station

Defense Forces

Branches: Army, Navy, Air Force, National Guard, National Police
Manpower availability: males age 15-49 84,323; fit for military service 42,812 (1993 est.)
Defense expenditures: exchange rate conversion—$NA, NA% of GDP

Eritrea

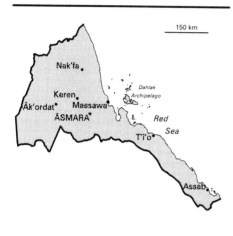

Geography

Location: Eastern Africa, bordering the Red Sea between Djibouti and Sudan
Map references: Africa, Standard Time Zones of the World
Area:
total area: 121,320 km²
land area: 121,320 km²
comparative area: slightly larger than Pennsylvania
Land boundaries: total 1,630 km, Djibouti 113 km, Ethiopia 912 km, Sudan 605 km
Coastline: 1,151 km (land and island coastline is 2,234 km)
Maritime claims:
territorial sea: 12 nm
International disputes: none
Climate: hot, dry desert strip along Red Sea coast; cooler and wetter in the central highlands (up to 61 cm of rainfall annually); semiarid in western hills and lowlands; rainfall heaviest during June-September except on coast desert
Terrain: dominated by extension of Ethiopian north-south trending highlands, descending on the east to a coastal desert plan, on the northwest to hilly terrain and on the southwest to flat-to-rolling plains
Natural resources: gold, potash, zinc, copper, salt, probably oil, fish
Land use:
arable land: 3%
permanent crops: 2% (coffee)
meadows and pastures: 40%
forest and woodland: 5%
other: 50%
Irrigated land: NA km²
Environment: frequent droughts, famine; deforestation; soil eroision; overgrazing; loss of infrastructure from civil warfare
Note: strategic geopolitical position along world's busiest shipping lanes and close to Arabian oilfields, Eritrea retained the entire coastline of Ethiopia along the Red Sea upon de jure independence from Ethiopia on 27 April 1993

People

Population: 3,467,087 (July 1993 est.)
Population growth rate: 3.46% (1993 est.)
Birth rate: NA births/1,000 population
Death rate: NA deaths/1,000 population
Net migration rate: NA migrant(s)/1,000 population
Infant mortality rate: NA deaths/1,000 live births
Life expectancy at birth:
total population: NA years
male: NA years
female: NA years
Total fertility rate: NA children born/woman
Nationality:
noun: Eritrean(s)
adjective: Eritrean
Ethnic divisions: ethnic Tigrays 50%, Tigre and Kunama 40%, Afar 4%, Saho (Red Sea coast dwellers) 3%
Religions: Muslim, Coptic Christian, Roman Catholic, Protestant
Languages: Tigre and Kunama, Cushitic dialects, Tigre, Nora Bana, Arabic
Literacy: NA%
Labor force: NA

Government

Names:
conventional long form: State of Eritrea
conventional short form: Eritrea
local long form: none
local short form: none
former: Eritrea Autonomous Region in Ethiopia
Digraph: ER
Type: transitional government
note: on 29 May 1991 ISSAIAS Afeworke, secretary general of the Eritrean People's Liberation Front (EPLF), announced the formation of the Provisional Government in Eritrea (PGE), in preparation for the 23-25 April 1993 referendum on independence for the autonomous region of Eritrea; the result was a landslide vote for independence that was announced on 27 April 1993
Capital: Asmara (formerly Asmera)
Administrative divisions: NA
Independence: 27 April 1993 (from Ethiopia; formerly the Eritrea Autonomous Region)
Constitution: transitional "constitution" decreed 19 May 1993
Legal system: NA
National holiday: National Day (independence from Ethiopia), 24 May (1993)
Political parties and leaders: Eritrean People's Liberation Front (EPLF) (Christian Muslim), ISSAIAS Afeworke, PETROS Soloman; Eritrean Liberation Front (ELF) (Muslim), ABDULLAH Muhammed;

Eritrean Liberation Front—United Organization (ELF-UO), leader NA
Other political or pressure groups: Oromo Liberation Front (OLF); Ethiopian People's Revolutionary Party (EPRP); numerous small, ethnic-based groups have formed since Mengistu's resignation, including several Islamic militant groups
Suffrage: NA
Elections: multinational election before 20 May 1997
Executive branch: president, Eritrean National Council
Legislative branch: National Assembly
Judicial branch: Judiciary
Leaders:
Chief of State and Head of Government: President ISSAIAS Aferworke
Member of: OAU, UN, UNCTAD
Diplomatic representation in US:
chief of mission: NA
chancery: NA
telephone: NA
US diplomatic representation:
chief of mission: Joseph P. O'NEILL
embassy: NA
mailing address: NA
telephone: 251-4-113-720
FAX: NA
Flag: red isosceles triangle (based on the hoist side) dividing the flag into two right triangles; the upper triangle is green, the lower one is blue; a gold wreath encircling a gold olive branch is centered on the hoist side of the red triangle

Economy

Overview: With independence from Ethiopia on 27 April 1993, Eritrea faces the bitter economic problems of a small, desperately poor African country. Most of the population will continue to depend on subsistence farming. Domestic output is substantially augmented by worker remittances from abroad. Government revenues come from custom duties and income and sales taxes. Eritrea has inherited the entire coastline of Ethiopia and has long-term prospects for revenues from the devlopment of offshore oil, offshore fishing and tourist development. For the time being, Ethiopia will be largely dependent on Eritrean ports for its foreign trade.
National product: GDP—exchange rate conversion—$400 million (1992 est.)
National product real growth rate: NA%
National product per capita: $115 (1992 est.)
Inflation rate (consumer prices): NA%
Unemployment rate: NA%
Budget: revenues $NA; expenditures $NA, including capital expenditures of $NA
Exports: $NA
commodities: NA

Estonia

partners: NA
Imports: $NA
commodities: NA
partners: NA
External debt: $NA
Industrial production: growth rate NA%
Electricity: NA kW capacity; NA kWh produced, NA kWh per capita
Industries: food processing, beverages, clothing and textiles
Agriculture: NA
Economic aid: NA
Currency: NA
Exchange rates: NA
Fiscal year: NA

Communications

Railroads: 307 km total; 307 km 1.000-meter gauge; 307 km 0.950-meter gauge (nonoperational) linking Ak'ordat and Asmara (formerly Asmera) with the port of Massawa (formerly Mits'iwa) (1993 est.)
Highways: 3,845 km total; 807 km paved, 840 km gravel, 402 km improved earth, 1,796 km unimproved earth
Ports: Assab (formerly Aseb), Massawa (formerly Mits'iwa)
Merchant marine: 14 ships (1,000 GRT or over) totaling 71,837 GRT/90,492 DWT; includes 9 cargo, 1 roll-on/roll off, 1 livestock carrier, 2 oil tanker, 1 refrigerated cargo
Airports:
total: 5
usable: 5
with permanent-surface runways: 2
with runways over 3,659 m: 0
with runways 2,440-3,659 m: 2
with runways 1,220-2,439 m: 2
Telecommunications: NA

Defense Forces

Branches: Eritrean People's Liberation Front (EPLF)
Manpower availability: males age 15-49 NA; fit for military service NA; reach military age (18) annually NA
Defense expenditures: exchange rate conversion—$NA, NA% of GDP

150 km

Final boundaries of Estonia, Latvia, and Lithuania with the former Soviet Union are expected to be confirmed by agreement.

Geography

Location: Northeastern Europe, bordering the Baltic Sea, between Sweden and Russia
Map references: Arctic Region, Asia, Europe, Standard Time Zones of the World
Area:
total area: 45,100 km²
land area: 43,200 km²
comparative area: slightly larger than New Hampshire and Vermont combined
note: includes 1,520 islands in the Baltic Sea
Land boundaries: total 557 km, Latvia 267 km, Russia 290 km
Coastline: 1,393 km
Maritime claims:
territorial sea: 12 nm
International disputes: international small border strips along the northern (Narva) and southern (Petseri) sections of eastern border with Russia ceded to Russia in 1945 by the Estonian SSR
Climate: maritime, wet, moderate winters
Terrain: marshy, lowlands
Natural resources: shale oil, peat, phosphorite, amber
Land use:
arable land: 22%
permanent crops: 0%
meadows and pastures: 11%
forest and woodland: 31%
other: 36%
Irrigated land: 110 km² (1990)
Environment: air heavily polluted with sulphur dioxide from oil-shale burning power plants in northeast; radioactive wastes dumped in open reservoir in Sillamae, a few dozen meters from Baltic Sea; contamination of soil and ground water with petroleum products, chemicals at military bases

---

People

Population: 1,608,469 (July 1993 est.)
Population growth rate: 0.52% (1993 est.)
Birth rate: 14.05 births/1,000 population (1993 est.)

Death rate: 12.13 deaths/1,000 population (1993 est.)
Net migration rate: 3.28 migrant(s)/1,000 population (1993 est.)
Infant mortality rate: 19.5 deaths/1,000 live births (1993 est.)
Life expectancy at birth:
total population: 69.75 years
male: 64.75 years
female: 74.99 years (1993 est.)
Total fertility rate: 2.01 children born/woman (1993 est.)
Nationality:
noun: Estonian(s)
adjective: Estonian
Ethnic divisions: Estonian 61.5%, Russian 30.3%, Ukrainian 3.17%, Belarusian 1.8%, Finn 1.1%, other 2.13% (1989)
Religions: Lutheran
Languages: Estonian (official), Latvian, Lithuanian, Russian, other
Literacy: age 9-49 can read and write (1970)
total population: 100%
male: 100%
female: 100%
Labor force: 796,000
by occupation: industry and construction 42%, agriculture and forestry 20%, other 38% (1990)

Government

Names:
conventional long form: Republic of Estonia
conventional short form: Estonia
local long form: Eesti Vabariik
local short form: Eesti
former: Estonian Soviet Socialist Republic
Digraph: EN
Type: republic
Capital: Tallinn
Administrative divisions: none (all districts are under direct republic jurisdiction)
Independence: 6 September 1991 (from Soviet Union)
Constitution: adopted 28 June 1992
Legal system: based on civil law system; no judicial review of legislative acts
National holiday: Independence Day, 24 February (1918)
Political parties and leaders: Popular Front of Estonia (Rahvarinne), NA chairman; Estonian Christian Democratic Party, Aivar KALA, chairman; Estonian Christian Democratic Union, Illar HALLASTE, chairman; Estonian Heritage Society (EMS), Trivimi VELLISTE, chairman; Estonian National Independence Party (ENIP), Lagle PAREK, chairman; Estonian Social Democratic Party, Marju LAURISTIN, chairman; Estonian Green Party, Tonu OJA; Independent Estonian Communist Party, Vaino VALJAS; People's Centrist Party, Edgar SAVISAAR, chairman; Estonian

Estonia (continued)

Royalist Party (ERP), Kalle KULBOK, chairman; Entrpreneurs' Party (EP), Tiit MADE; Estonian Fatherland Party, Mart LAAR, chairman; Safe Home; Moderates; Estonian Citizen
Suffrage: 18 years of age; universal
Elections:
President: last held 20 September 1992; (next to be held NA); results—no candidate received majority; newly elected Parliament elected Lennart MERI (NA October 1992)
Parliament: last held 20 September 1992; (next to be held NA); results—Fatherland 21%, Safe Home 14%, Popular Front 13%, Moderates 10%, Estonian National Independence Party 8%, Royalists 7%, Estonian Citizen 7%, Estonian Entrepreneurs 2%, other 18%; seats—(101 total) Fatherland 29, Safe Home 18, Popular Front 15, Moderates 12, ENIP 10, Royalists 8, Estonian Citizen 8, Estonian Entrepreneurs 1
Congress of Estonia: last held March 1990 (next to be held NA); note—Congress of Estonia was a quasi-governmental structure which disbanded itself October 1992 after the new Parliament and government were installed
Executive branch: president, prime minister, cabinet
Legislative branch: unicameral Parliament (Riigikogu)
Judicial branch: Supreme Court
Leaders:
Chief of State: President Lennart MERI (since NA October 1992)
Head of Government: Prime Minister Mart LAAR (since NA October 1992)
Member of: CBSS, CSCE, EBRD, ECE, FAO, IAEA, IBRD, ICAO, ICFTU, ILO, IMF, IMO, NACC, UN, UNCTAD, UNESCO, UPU
Diplomatic representation in US:
chief of mission: Ambassador Toomas Hendrik IIVES
chancery: (temporary) 630 Fifth Avenue, Suite 2415, New York, NY 10111
telephone: (212) 247-2131
consulate general: New York
US diplomatic representation:
chief of mission: Ambassador Robert C. FRASURE
embassy: Kentmanni 20, Tallin EE 0001
mailing address: use embassy street address
telephone: 011-[358] (49) 303-182 (cellular)
FAX: [358] (49) 306-817 (cellular)
note: dialing to Baltics still requires use of an international operator unless you use the cellular phone lines
Flag: pre-1940 flag restored by Supreme Soviet in May 1990—three equal horizontal bands of blue (top), black, and white

Economy

Overview: As of June 1993 Estonia ranks first among the 15 former Soviet republics in moving from its obsolete command economy to a modern market economy. Yet serious problems remain. In contrast to the estimated 30% drop in output in 1992, GDP should grow by a small percent in 1993. Of key importance has been the introduction of the kroon in August 1993 and the subsequent reductions in inflation to 1%-2% per month. Starting in July 1991, under a new law on private ownership, small enterprises, such as retail shops and restaurants, were sold to private owners. The auctioning of large-scale enterprises is progressing with the proceeds being held in escrow until the prior ownership (that is, Estonian or the Commonwealth of Independent States) can be established. Estonia ranks first in per capita consumption among the former Soviet republics. Agriculture is well developed, especially meat production, and provides a surplus for export. Only about one-fifth of the work force is in agriculture. The major share of the work force engages in manufacturing both capital and consumer goods based on raw materials and intermediate products from the other former Soviet republics. These manufactures are of high quality by ex-Soviet standards and are exported to the other republics. Estonia's mineral resources are limited to major deposits of shale oil (60% of the old Soviet total) and phosphorites (400 million tons). Estonia has a large, relatively modern port and produces more than half of its own energy needs at highly polluting shale oil power plants. It has advantages in the transition, not having suffered so long under the Soviet yoke and having better chances of developing profitable ties to the Nordic and West European countries. Like Latvia, but unlike Lithuania, the large portion of ethnic Russians (30%) in the population poses still another difficulty in the transition to an independent market economy.
National product: GDP $NA
National product real growth rate: -30% (1992 est.)
National product per capita: $NA
Inflation rate (consumer prices): 1%-2% per month (first quarter 1993)
Unemployment rate: 3% (March 1993); but large number of underemployed workers
Budget: revenues $223 million; expenditures $142 million, including capital expenditures of $NA (1992)
Exports: $NA
commodities: textile 11%, wood products and timber 9%, dairy products 9%
partners: Russia and the other former Soviet republics 50%, West 50% (1992)
Imports: $NA
commodities: machinery 45%, oil 13%, chemicals 12%
partners: Finland 15%, Russia 18%

External debt: $650 million (end of 1991)
Industrial production: growth rate –40% (1992)
Electricity: 3,700,000 kW capacity; 22,900 million kWh produced, 14,245 kWh per capita (1992)
Industries: accounts for 30% of labor force; oil shale, shipbuilding, phosphates, electric motors, excavators, cement, furniture, clothing, textiles, paper, shoes, apparel
Agriculture: employs 20% of work force; very efficient; net exports of meat, fish, dairy products, and potatoes; imports of feedgrains for livestock; fruits and vegetables
Illicit drugs: transshipment point for illicit drugs from Central and Southwest Asia to Western Europe; limited illicit opium producer; mostly for domestic production
Economic aid: US commitments, including Ex-Im (1992), $10 million
Currency: 1 Estonian kroon (EEK) = 100 NA; (introduced in August 1992)
Exchange rates: kroons (EEK) per US$1—12 (January 1993)
Fiscal year: calendar year

Communications

Railroads: 1,030 km (includes NA km electrified); does not include industrial lines (1990)
Highways: 30,300 km total (1990); 29,200 km hard surfaced; 1,100 km earth
Inland waterways: 500 km perennially navigable
Pipelines: natural gas 420 km (1992)
Ports: coastal—Tallinn, Novotallin, Parnu; inland—Narva
Merchant marine: 68 ships (1,000 GRT or over) totaling 394,501 GRT/526,502 DWT; includes 52 cargo, 6 roll-on/roll-off, 2 short-sea passenger, 6 bulk, 2 container
Airports:
total: 29
useable: 18
with permanent-surface runways: 11
with runways over 3,659 m: 0
with runways 2,440-3,659 m: 10
with runways 1,220-2,439 m: 8
Telecommunications: 300,000 telephone subscribers in 1990 with international direct dial service available to Finland, Germany, Austria, UK and France; 21 telephone lines per 100 persons as of 1991; broadcast stations—3 TV (provide Estonian programs as well as Moscow Ostenkino's first and second programs); international traffic is carried to the other former USSR republics by landline or microwave and to other countries by leased connection to the Moscow international gateway switch via 19 incoming/20 outgoing international channels, by the Finnish cellular net, and by an old copper submarine cable to Finland soon to

Ethiopia

be replaced by an undersea fiber optic cable system; there is also a new international telephone exchange in Tallinn handling 60 channels via Helsinki; 2 analog mobile cellular networks with international roaming capability to Scandinavia are operating in major cities

Defense Forces

Branches: Ground Forces, Maritime Border Guard, National Guard (Kaitseliit), Security Forces (internal and border troops)
Manpower availability: males age 15-49 387,733; fit for military service 306,056; reach military age (18) annually 11,570 (1993 est.)
Defense expenditures: 124.4 million kroons, NA% of GDP (forecast for 1993); note—conversion of the military budget into US dollars using the current exchange rate could produce misleading results

Geography

Location: Eastern Africa, between Somalia and Sudan
Map references: Africa, Standard Time Zones of the World
Area:
total area: 1,127,127 km²
land area: 1,119,683 km²
comparative area: slightly less than twice the size of Texas
Land boundaries: total 5,311 km, Djibouti 337 km, Erithea 912 km, Kenya 830 km, Somalia 1,626 km, Sudan 1,606 km
Coastline: none—landlocked
Maritime claims: none—landlocked
International disputes: southern half of the boundary with Somalia is a Provisional Administrative Line; possible claim by Somalia based on unification of ethnic Somalis; territorial dispute with Somalia over the Ogaden
Climate: tropical monsoon with wide topographic-induced variation; some areas prone to extended droughts
Terrain: high plateau with central mountain range divided by Great Rift Valley
Natural resources: small reserves of gold, platinum, copper, potash
Land use:
arable land: 12%
permanent crops: 1%
meadows and pastures: 41%
forest and woodland: 24%
other: 22%
Irrigated land: 1,620 km² (1989 est.)
Environment: geologically active Great Rift Valley susceptible to earthquakes, volcanic eruptions; deforestation; overgrazing; soil erosion; desertification; frequent droughts; famine
Note: landlocked—entire coastline along the Red Sea was lost with the de jure independence of Eritrea on 27 April 1993

People

Population: 53,278,446 (July 1993 est.)
note: Ethiopian demographic data, except population and population growth rate, include Eritrea
Population growth rate: 3.41% (1993 est.)
Birth rate: 45.37 births/1,000 population (1993 est.)
Death rate: 14.23 deaths/1,000 population (1993 est.)
Net migration rate: 2.94 migrant(s)/1,000 population (1993 est.)
Infant mortality rate: 108.8 deaths/1,000 live births (1993 est.)
Life expectancy at birth:
total population: 52.21 years
male: 50.6 years
female: 53.88 years (1993 est.)
Total fertility rate: 6.88 children born/woman (1993 est.)
Nationality:
noun: Ethiopian(s)
adjective: Ethiopian
Ethnic divisions: Oromo 40%, Amhara and Tigrean 32%, Sidamo 9%, Shankella 6%, Somali 6%, Afar 4%, Gurage 2%, other 1%
Religions: Muslim 45-50%, Ethiopian Orthodox 35-40%, animist 12%, other 5%
Languages: Amharic (official), Tigrinya, Orominga, Guaraginga, Somali, Arabic, English (major foreign language taught in schools)
Literacy: age 10 and over can read and write (1983)
total population: 62%
male: NA%
female: NA%
Labor force: 18 million
by occupation: agriculture and animal husbandry 80%, government and services 12%, industry and construction 8% (1985)

Government

Names:
conventional long form: none
conventional short form: Ethiopia
local long form: none
local short form: Ityop'iya
Digraph: ET
Type: transitional government
note: on 28 May 1991 the Ethiopian People's Revolutionary Democratic Front (EPRDF) toppled the authoritarian government of MENGISTU Haile-Mariam and took control in Addis Ababa; the Transitional Government of Ethiopia (TGE), announced a two-year transitional period
Capital: Addis Ababa
Administrative divisions: 14 administrative regions (astedader akababiwach, singular—astedader akababi) Addis Ababa, Afar, Amhara, Benishangul, Gambela, Gurage-Hadiya-Kambata, Harer, Kefa, Omo,

Ethiopia *(continued)*

Oromo, Sidamo, Somali, Tigray, Wolayta
Independence: oldest independent country in Africa and one of the oldest in the world—at least 2,000 years
Constitution: to be redrafted by 1993
Legal system: NA
National holiday: National Day, 28 May (1991) (defeat of Mengistu regime)
Political parties and leaders: NA
Other political or pressure groups: Oromo Liberation Front (OLF); Ethiopian People's Revolutionary Party (EPRP); numerous small, ethnic-based groups have formed since Mengistu's resignation, including several Islamic militant groups
Suffrage: 18 years of age; universal
Elections:
President: last held 10 September 1987; next election planned after new constitution drafted; results—MENGISTU Haile-Mariam elected by the now defunct National Assembly, but resigned and left Ethiopia on 21 May 1991
Constituent Assembly: now planned for January 1994 (to ratify constitution to be drafted by end of 1993)
Executive branch: president, prime minister, Council of Ministers
Legislative branch: unicameral Constituent Assembly
Judicial branch: Supreme Court
Leaders:
Chief of State: President MELES Zenawi (since 1 June 1991)
Head of Government: Prime Minister TAMIRAT Layne (since 6 June 1991)
Member of: ACP, AfDB, CCC, ECA, FAO, G-24, G-77, IAEA, IBRD, ICAO, IDA, IFAD, IFC, IGADD, ILO, IMF, IMO, INTELSAT, INTERPOL, IOC, ISO, ITU, LORCS, NAM, OAU, UN, UNCTAD, UNESCO, UNHCR, UNIDO, UPU, WFTU, WHO, WMO, WTO
Diplomatic representation in US:
chief of mission: Ambassador BERHANE Gebre-Christos
chancery: 2134 Kalorama Road NW, Washington, DC 20008
telephone: (202) 234-2281 or 2282
US diplomatic representation:
chief of mission: Ambassador Marc A. BAAS
embassy: Entoto Street, Addis Ababa
mailing address: P. O. Box 1014, Addis Ababa
telephone: [251] (1) 550-666
FAX: [251] (1) 551-166
Flag: three equal horizontal bands of green (top), yellow, and red; Ethiopia is the oldest independent country in Africa, and the colors of her flag were so often adopted by other African countries upon independence that they became known as the pan-African colors

Economy

Overview: With the independence of Eritrea on 27 April 1993, Ethiopia continues to face difficult economic problems as one of the poorest and least developed countries in Africa. (The accompanying analysis and figures predate the independence of Eritrea.) Its economy is based on subsistence agriculture, which accounts for about 45% of GDP, 90% of exports, and 80% of total employment; coffee generates 60% of export earnings. The manufacturing sector is heavily dependent on inputs from the agricultural sector. Over 90% of large-scale industry, but less than 10% of agriculture, is state run; the government is considering selling off a portion of state-owned plants. Favorable agricultural weather largely explains the 4.5% growth in output in FY89, whereas drought and deteriorating internal security conditions prevented growth in FY90. In 1991 the lack of law and order, particularly in the south, interfered with economic development and growth. In 1992, because of some easing of civil strife and aid from the outside world, the economy substantially improved.
National product: GDP—exchange rate conversion—$6.6 billion (FY92 est.)
National product real growth rate: 6% (FY92 est.)
National product per capita: $130 (FY92 est.)
Inflation rate (consumer prices): 7.8% (1989)
Unemployment rate: NA%
Budget: revenues $1.4 billion; expenditures $2.3 billion, including capital expenditures of $565 million (FY91)
Exports: $276 million (f.o.b., FY90)
commodities: coffee, leather products, gold, petroleum products
partners: EC, Djibouti, Japan, Saudi Arabia, US
Imports: $1.0 billion (c.i.f., FY90)
commodities: capital goods, consumer goods, fuel
partners: EC, Eastern Europe, Japan, US
External debt: $3.48 billion (1991)
Industrial production: growth rate 2.3% (FY89 est.); accounts for 12% of GDP
Electricity: 330,000 kW capacity; 650 million kWh produced, 10 kWh per capita (1991)
Industries: food processing, beverages, textiles, chemicals, metals processing, cement
Agriculture: accounts for 47% of GDP and is the most important sector of the economy even though frequent droughts and poor cultivation practices keep farm output low; famines not uncommon; export crops of coffee and oilseeds grown partly on state farms; estimated 50% of agricultural

production at subsistence level; principal crops and livestock—cereals, pulses, coffee, oilseeds, sugarcane, potatoes and other vegetables, hides and skins, cattle, sheep, goats
Illicit drugs: transit hub for heroin originating in Southwest and Southeast Asia and destined for Europe and North America; cultivates qat (chat) for local use and regional export
Economic aid: US commitments, including Ex-Im (FY70-89), $504 million; Western (non-US) countries, ODA and OOF bilateral commitments (1970-89), $3.4 billion; OPEC bilateral aid (1979-89), $8 million; Communist countries (1970-89), $2.0 billion
Currency: 1 birr (Br) = 100 cents
Exchange rates: birr (Br) per US$1—5.0000 (fixed rate)
Fiscal year: 8 July-7 July

Communications

Railroads: 781 km total; 781 km 1.000-meter gauge; 307 km 0.950-meter gauge linking Addis Ababa (Ethiopia) to Djibouti; control of railroad is shared between Djibouti and Ethiopia
Highways: 39,150 km total; 2,776 km paved, 7,504 km gravel, 2,054 km improved earth, 26,816 km unimproved earth (1993 est.)
Ports: none; landlocked
Merchant marine: none; landlocked
Airports:
total: 121
usable: 82
with permanent-surface runways: 9
with runways over 3,659 m: 1
with runways 2,440-3,659 m: 13
with runways 1,220-2,439 m: 83 (1993 est.)
Telecommunications: open-wire and radio relay system adequate for government use; open-wire to Sudan and Djibouti; microwave radio relay to Kenya and Djibouti; broadcast stations—4 AM, no FM, 1 TV; 100,000 TV sets; 9,000,000 radios; satellite earth stations—1 Atlantic Ocean INTELSAT and 2 Pacific Ocean INTELSAT

Defense Forces

Branches: Ethiopian People's Revolutionary Democratic Front (EPRDF)
Manpower availability: males age 15-49 12,793,340; fit for military service 6,640,616; reach military age (18) annually 576,329 (1993 est.)
Defense expenditures: exchange rate conversion—$NA, NA% of GDP

Europa Island

(possession of France)

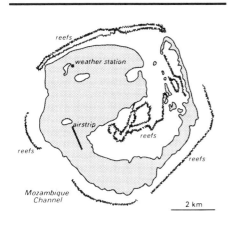

Geography

Location: Southern Africa, in the southern Mozambique Channel about halfway between Madagascar and Mozambique
Map references: Africa
Area:
total area: 28 km²
land area: 28 km²
comparative area: about 0.2 times the size of Washington, DC
Land boundaries: 0 km
Coastline: 22.2 km
Maritime claims:
exclusive economic zone: 200 nm
territorial sea: 12 nm
International disputes: claimed by Madagascar
Climate: tropical
Terrain: NA
Natural resources: negligible
Land use:
arable land: NA%
permanent crops: NA%
meadows and pastures: NA%
forest and woodland: NA%
other: NA% (heavily wooded)
Irrigated land: 0 km²
Environment: wildlife sanctuary

People

Population: uninhabited

Government

Names:
conventional long form: none
conventional short form: Europa Island
local long form: none
local short form: Ile Europa
Digraph: EU
Type: French possession administered by Commissioner of the Republic; resident in Reunion

Capital: none; administered by France from Reunion
Independence: none (possession of France)

Economy

Overview: no economic activity

Communications

Ports: none; offshore anchorage only
Airports:
total: 1
usable: 1
with permanent-surface runways: 0
with runways over 3,659 m: 0
with runways 2,439-3,659 m: 0
with runways 1,220-2,439 m: 1
Telecommunications: 1 meteorological station

Defense Forces

Note: defense is the responsibility of France

Falkland Islands (Islas Malvinas)

(dependent territory of the UK)

Geography

Location: in the South Atlantic Ocean, off the southern coast of Argentina
Map references: Antarctic Region, South America
Area:
total area: 12,170 km²
land area: 12,170 km²
comparative area: slightly smaller than Connecticut
note: includes the two main islands of East and West Falkland and about 200 small islands
Land boundaries: 0 km
Coastline: 1,288 km
Maritime claims:
continental shelf: 100 m depth
exclusive fishing zone: 200 nm
territorial sea: 12 nm
International disputes: administered by the UK, claimed by Argentina
Climate: cold marine; strong westerly winds, cloudy, humid; rain occurs on more than half of days in year; occasional snow all year, except in January and February, but does not accumulate
Terrain: rocky, hilly, mountainous with some boggy, undulating plains
Natural resources: fish, wildlife
Land use:
arable land: 0%
permanent crops: 0%
meadows and pastures: 99%
forest and woodland: 0%
other: 1%
Irrigated land: NA km²
Environment: poor soil fertility and a short growing season
Note: deeply indented coast provides good natural harbors

People

Population: 2,206 (July 1993 est.)
Population growth rate: 2.43% (1993 est.)
Birth rate: NA births/1,000 population
Death rate: NA deaths/1,000 population

Falkland Islands (Islas Malvinas)

(continued)

Net migration rate: NA migrant(s)/1,000 population
Infant mortality rate: NA deaths/1,000 population
Life expectancy at birth:
total population: NA years
male: NA years
female: NA years
Total fertility rate: NA children born/woman
Nationality:
noun: Falkland Islander(s)
adjective: Falkland Island
Ethnic divisions: British
Religions: primarily Anglican, Roman Catholic, United Free Church, Evangelist Church, Jehovah's Witnesses, Lutheran, Seventh-Day Adventist
Languages: English
Literacy:
total population: NA%
male: NA%
female: NA%
Labor force: 1,100 (est.)
by occupation: agriculture 95% (mostly sheepherding)

Government

Names:
conventional long form: Colony of the Falkland Islands
conventional short form: Falkland Islands (Islas Malvinas)
Digraph: FA
Type: dependent territory of the UK
Capital: Stanley
Administrative divisions: none (dependent territory of the UK)
Independence: none (dependent territory of the UK)
Constitution: 3 October 1985
Legal system: English common law
National holiday: Liberation Day, 14 June (1982)
Suffrage: 18 years of age; universal
Elections:
Legislative Council: last held 11 October 1989 (next to be held October 1994); results—percent of vote by party NA; seats—(10 total, 8 elected) number of seats by party NA
Executive branch: British monarch, governor, Executive Council
Legislative branch: unicameral Legislative Council
Judicial branch: Supreme Court
Leaders:
Chief of State: Queen ELIZABETH II (since 6 February 1952)
Head of Government: Governor David Everard TATHAM (since August 1992)
Member of: ICFTU
Diplomatic representation in US: none (dependent territory of the UK)

US diplomatic representation: none (dependent territory of the UK)
Flag: blue with the flag of the UK in the upper hoist-side quadrant and the Falkland Island coat of arms in a white disk centered on the outer half of the flag; the coat of arms contains a white ram (sheep raising is the major economic activity) above the sailing ship Desire (whose crew discovered the islands) with a scroll at the bottom bearing the motto DESIRE THE RIGHT

Economy

Overview: The economy is based on sheep farming, which directly or indirectly employs most of the work force. A few dairy herds are kept to meet domestic consumption of milk and milk products, and crops grown are primarily those for providing winter fodder. Exports feature shipments of high-grade wool to the UK and the sale of postage stamps and coins. Rich stocks of fish in the surrounding waters are not presently exploited by the islanders. So far, efforts to establish a domestic fishing industry have been unsuccessful. In 1987 the government began selling fishing licenses to foreign trawlers operating within the Falklands exclusive fishing zone. These license fees amount to more than $40 million per year and are a primary source of income for the government. To encourage tourism, the Falkland Islands Development Corporation has built three lodges for visitors attracted by the abundant wildlife and trout fishing.
National product: GDP $NA
National product real growth rate: NA%
National product per capita: $NA
Inflation rate (consumer prices): 7.4% (1980-87 average)
Unemployment rate: NA%; labor shortage
Budget: revenues $62.7 million; expenditures $41.8 million, including capital expenditures of $NA (FY90)
Exports: at least $14.7 million
commodities: wool, hides and skins, and meat
partners: UK, Netherlands, Japan (1987 est.)
Imports: at least $13.9 million
commodities: food, clothing, fuels, and machinery
partners: UK, Netherlands Antilles (Curacao), Japan (1987 est.)
External debt: $NA
Industrial production: growth rate NA%
Electricity: 9,200 kW capacity; 17 million kWh produced, 8,940 kWh per capita (1992)
Industries: wool and fish processing
Agriculture: predominantly sheep farming; small dairy herds; some fodder and vegetable crops
Economic aid: Western (non-US) countries, ODA and OOF bilateral commitments (1970-89), $277 million

Currency: 1 Falkland pound (£F) = 100 pence
Exchange rates: Falkland pound (£F) per US$1—0.6527 (January 1993), 0.5664 (1992), 0.5652 (1991), 0.5604 (1990), 0.6099 (1989), 0.5614 (1988); note—the Falkland pound is at par with the British pound
Fiscal year: 1 April-31 March

Communications

Highways: 510 km total; 30 km paved, 80 km gravel, and 400 km unimproved earth
Ports: Stanley
Airports:
total: 5
usable: 5
with permanent-surface runways: 2
with runways over 3,659 m: 0
with runways 2,440-3,659 m: 1
with runways 1,220-2,439 m: 0
Telecommunications: government-operated radiotelephone and private VHF/CB radio networks provide effective service to almost all points on both islands; 590 telephones; broadcast stations—2 AM, 3 FM, no TV; 1 Atlantic Ocean INTELSAT earth station with links through London to other countries

Defense Forces

Branches: British Forces Falkland Islands (including Army, Royal Air Force, Royal Navy, and Royal Marines), Police Force
Note: defense is the responsibility of the UK

Faroe Islands
(part of the Danish realm)

Geography

Location: in the north Atlantic Ocean, located half way between Norway and Iceland

Map references: Arctic Region

Area:

total area: 1,400 km²

land area: 1,400 km²

comparative area: slightly less than eight times the size of Washington, DC

Land boundaries: 0 km

Coastline: 764 km

Maritime claims:

exclusive fishing zone: 200 nm

territorial sea: 3 nm

International disputes: none

Climate: mild winters, cool summers; usually overcast; foggy, windy

Terrain: rugged, rocky, some low peaks; cliffs along most of coast

Natural resources: fish

Land use:

arable land: 2%

permanent crops: 0%

meadows and pastures: 0%

forest and woodland: 0%

other: 98%

Irrigated land: NA km²

Environment: precipitous terrain limits habitation to small coastal lowlands; archipelago of 18 inhabited islands and a few uninhabited islets

Note: strategically located along important sea lanes in northeastern Atlantic

People

Population: 48,065 (July 1993 est.)

Population growth rate: 0.67% (1993 est.)

Birth rate: 18.45 births/1,000 population (1993 est.)

Death rate: 7.57 deaths/1,000 population (1993 est.)

Net migration rate: -4.2 migrant(s)/1,000 population (1993 est.)

Infant mortality rate: 8.3 deaths/1,000 live births (1993 est.)

Life expectancy at birth:

total population: 77.92 years

male: 74.51 years

female: 81.45 years (1993 est.)

Total fertility rate: 2.52 children born/woman (1993 est.)

Nationality:

noun: Faroese (singular and plural)

adjective: Faroese

Ethnic divisions: Scandinavian

Religions: Evangelical Lutheran

Languages: Faroese (derived from Old Norse), Danish

Literacy:

total population: NA%

male: NA%

female: NA%

Labor force: 17,585

by occupation: largely engaged in fishing, manufacturing, transportation, and commerce

Government

Names:

conventional long form: none

conventional short form: Faroe Islands

local long form: none

local short form: Foroyar

Digraph: FO

Type: part of the Danish realm; self-governing overseas administrative division of Denmark

Capital: Torshavn

Administrative divisions: none (self-governing overseas administrative division of Denmark)

Independence: none (part of the Danish realm; self-governing overseas administrative division of Denmark)

Constitution: Danish

Legal system: Danish

National holiday: Birthday of the Queen, 16 April (1940)

Political parties and leaders:

three-party ruling coalition: Social Democratic Party, Marita PETERSEN; Republican Party, Signer HANSEN; Home Rule Party, Hilmar KASS

opposition: Cooperation Coalition Party, Pauli ELLEFSEN; Progressive and Fishing Industry Party-Christian People's Party (PFIP-CPP), leader NA; Progress Party, leader NA; People's Party, Jogvan SUND-STEIN

Suffrage: 20 years of age; universal

Elections:

Danish Parliament: last held on 12 December 1990 (next to be held by December 1994); results—percent of vote by party NA; seats—(2 total) Social Democratic 1, People's Party 1; note—the Faroe Islands elects two representatives to the Danish Parliament

Faroese Parliament: last held 17 November 1990 (next to be held November 1994); results—Social Democratic 27.4%, People's Party 21.9%, Cooperation Coalition Party 18.9%, Republican Party 14.7%, Home Rule 8.8%, PFIP-CPP 5.9%, other 2.4%; seats—(32 total) two-party coalition 17 (Social Democratic 10, People's Party 7), Cooperation Coalition Party 6, Republican Party 4, Home Rule 3, PFIP-CPP 2

Executive branch: Danish monarch, high commissioner, prime minister, deputy prime minister, Cabinet (Landsstyri)

Legislative branch: unicameral Parliament (Lgting)

Judicial branch: none

Leaders:

Chief of State: Queen MARGRETHE II (since 14 January 1972), represented by High Commissioner Bent KLINTE (since NA)

Head of Government: Prime Minister Marita PETERSEN (since 18 January 1993)

Member of: none

Diplomatic representation in US: none (self-governing overseas administrative division of Denmark)

US diplomatic representation: none (self-governing overseas administrative division of Denmark)

Flag: white with a red cross outlined in blue that extends to the edges of the flag; the vertical part of the cross is shifted to the hoist side in the style of the DANNEBROG (Danish flag)

Economy

Overview: The Faroese, who have long enjoyed the affluent living standards of the Danes and other Scandinavians, now must cope with the decline of the all-important fishing industry and one of the world's heaviest per capita external debts of nearly $30,000. When the nations of the world extended their fishing zones to 200 nautical miles in the early 1970s, the Faroese no longer could continue their traditional long-distance fishing and subsequently depleted their own nearby fishing areas. The government's tight controls on fish stocks and its austerity measures have caused a recession, and subsidy cuts will force nationalization in the fishing industry, which has already been plagued with bankruptcies. Copenhagen has threatened to withhold its annual subsidy of $130 million—roughly one-third of the islands' budget revenues—unless the Faroese make significant efforts to balance their budget. To this extent the Faroe government is expected to continue its tough policies, including introducing a 20% VAT in 1993, and has agreed to an IMF economic-political stabilization plan. In addition to its annual subsidy, the Danish government has bailed out the second largest Faroe bank to the tune

Faroe Islands *(continued)*

National product: GDP—purchasing power equivalent—$662 million (1989 est.)
National product real growth rate: 3% (1989 est.)
National product per capita: $14,000 (1989 est.)
Inflation rate (consumer prices): 2% (1988)
Unemployment rate: 5%-6% (1991 est.)
Budget: revenues $425 million; expenditures $480 million, including capital expenditures of $NA (1991 est.)
Exports: $386 million (f.o.b., 1990 est.)
commodities: fish and fish products 88%, animal feedstuffs, transport equipment (ships) (1989)
partners: Denmark 20%, Germany 18.3%, UK 14.2%, France 11.2%, Spain 7.9%, US 4.5%
Imports: $322 million (c.i.f., 1990 est.)
commodities: machinery and transport equipment 24.4%, manufactures 24%, food and livestock 19%, fuels 12%, chemicals 6.5%
partners: Denmark 43.8%, Norway 19.8%, Sweden 4.9%, Germany 4.2%, US 1.3%
External debt: $1.3 billion (1991)
Industrial production: growth rate NA%
Electricity: 80,000 kW capacity; 280 million kWh produced, 5,760 kWh per capita (1992)
Industries: fishing, shipbuilding, handicrafts
Agriculture: accounts for 27% of GDP and employs 27% of labor force; principal crops—potatoes and vegetables; livestock—sheep; annual fish catch about 360,000 metric tons
Economic aid: receives an annual subsidy from Denmark of about $130 million
Currency: 1 Danish krone (DKr) = 100 ore
Exchange rates: Danish kroner (DKr) per US$1—6.236 (January 1993), 6.036 (1992), 6.396 (1991), 6.189 (1990), 7.310 (1989), 6.732 (1988)
Fiscal year: 1 April-31 March

Communications

Highways: 200 km
Ports: Torshavn, Tvoroyri
Merchant marine: 10 ships (1,000 GRT or over) totaling 22,015 GRT/24,007 DWT; includes 1 short-sea passenger, 5 cargo, 2 roll-on/roll-off, 2 refrigerated cargo; note—a subset of the Danish register
Airports:
total: 1
useable: 1
with permanent-surface runways: 1
with runways over 3659 m: 0
with runways 2440-3659 m: 0
with runways 1220-2439 m: 1
Telecommunications: good international communications; fair domestic facilities; 27,900 telephones; broadcast stations—1 AM, 3 (10 repeaters) FM, 3 (29 repeaters) TV; 3 coaxial submarine cables

---

Defense Forces

Branches: small Police Force, no organized native military forces
Note: defense is the responsibility of Denmark

Fiji

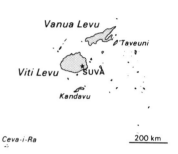

Geography

Location: Oceania, 2,500 km north of New Zealand in the South Pacific Ocean
Map references: Oceania, Standard Time Zones of the World
Area:
total area: 18,270 km²
land area: 18,270 km²
comparative area: slightly smaller than New Jersey
Land boundaries: 0 km
Coastline: 1,129 km
Maritime claims: (measured from claimed archipelagic baselines)
continental shelf: 200 m depth or to depth of exploitation; rectilinear shelf claim added
exclusive economic zone: 200 nm
territorial sea: 12 nm
International disputes: none
Climate: tropical marine; only slight seasonal temperature variation
Terrain: mostly mountains of volcanic origin
Natural resources: timber, fish, gold, copper, offshore oil potential
Land use:
arable land: 8%
permanent crops: 5%
meadows and pastures: 3%
forest and woodland: 65%
other: 19%
Irrigated land: 10 km² (1989 est.)
Environment: subject to hurricanes from November to January; includes 332 islands of which approximately 110 are inhabited

---

People

Population: 756,762 (July 1993 est.)
Population growth rate: 0.95% (1993 est.)
Birth rate: 24.74 births/1,000 population (1993 est.)
Death rate: 6.59 deaths/1,000 population (1993 est.)
Net migration rate: -8.65 migrant(s)/1,000 population (1993 est.)

Infant mortality rate: 18.4 deaths/1,000 live births (1993 est.)
Life expectancy at birth:
total population: 64.86 years
male: 62.62 years
female: 67.21 years (1993 est.)
Total fertility rate: 2.98 children born/woman (1993 est.)
Nationality:
noun: Fijian(s)
adjective: Fijian
Ethnic divisions: Fijian 49%, Indian 46%, European, other Pacific Islanders, overseas Chinese, and other 5%
Religions: Christian 52% (Methodist 37%, Roman Catholic 9%), Hindu 38%, Muslim 8%, other 2%
note: Fijians are mainly Christian, Indians are Hindu, and there is a Muslim minority (1986)
Languages: English (official), Fijian, Hindustani
Literacy: age 15 and over can read and write (1985)
total population: 86%
male: 90%
female: 81%
Labor force: 235,000
by occupation: subsistence agriculture 67%, wage earners 18%, salary earners 15% (1987)

Government

Names:
conventional long form: Republic of Fiji
conventional short form: Fiji
Digraph: FJ
Type: republic
note: military coup leader Maj. Gen. Sitiveni RABUKA formally declared Fiji a republic on 6 October 1987
Capital: Suva
Administrative divisions: 4 divisions and 1 dependency*; Central, Eastern, Northern, Rotuma*, Western
Independence: 10 October 1970 (from UK)
Constitution: 10 October 1970 (suspended 1 October 1987); a new Constitution was proposed on 23 September 1988 and promulgated on 25 July 1990; the 1990 Constitution is currently still under review (February 1993)
Legal system: based on British system
National holiday: Independence Day, 10 October (1970)
Political parties and leaders: Fijian Political Party (SVT—primarily Fijian), leader Maj. Gen. Sitivini RABUKA; National Federation Party (NFP; primarily Indian), Siddiq KOYA; Christian Fijian Nationalist Party (CFNP), Sakeasi BUTADROKA; Fiji Labor Party (FLP), Jokapeci KOROI; All National Congress (ANC), Apisai TORA; General Voters Party

(GVP), Max OLSSON; Fiji Conservative Party (FCP), Isireli VUIBAU; Conservative Party of Fiji (CPF), Jolale ULUDOLE and Viliame SAVU; Fiji Indian Liberal Party, Swami MAHARAJ; Fiji Indian Congress Party, Ishwari BAJPAI; Fiji Independent Labor (Muslim), leader NA; Four Corners Party, David TULVANUAVOU
Suffrage: none
Elections:
House of Representatives: last held 23-29 May 1992 (next to be held NA 1997); results—percent of vote by party NA; seats—(70 total, with ethnic Fijians allocated 37 seats, ethnic Indians 27 seats, and independents and other 6 seats) number of seats by party NA
Executive branch: president, prime minister, Cabinet, Great Councils of Chiefs (highest ranking members of the traditional chiefly system)
Legislative branch: the bicameral Parliament, consisting of an upper house or Senate and a lower house or House of Representatives, was dissolved following the coup of 14 May 1987; the Constitution of 23 September 1988 provides for a bicameral Parliament
Judicial branch: Supreme Court
Leaders:
Chief of State: President Ratu Sir Penaia Kanatabatu GANILAU (since 5 December 1987); Vice President Ratu Sir Kamisese MARA (since 14 April 1992); Vice President Ratu Sir Josaia TAIVAIQIA (since 14 April 1992)
Head of Government: Prime Minister Sitiveni RABUKA (since 2 June 1992); Deputy Prime Minister Filipe BOLE (since 11 June 1992); Deputy Prime Minister Ratu Timoci VESIKULA (since 11 June 1993)
Member of: ACP, AsDB, CP, ESCAP, FAO, G-77, IBRD, ICAO, ICFTU, IDA, IFAD, IFC, ILO, IMF, IMO, INTELSAT, INTERPOL, IOC, ITU, LORCS, PCA, SPARTECA, SPC, SPF, UN, UNCTAD, UNESCO, UNIDO, UNIFIL, UNIKOM, UNOSOM, UPU, WHO, WIPO, WMO
Diplomatic representation in US:
chief of mission: Ambassador Pita Kewa NACUVA
chancery: Suite 240, 2233 Wisconsin Avenue NW, Washington, DC 20007
telephone: (202) 337-8320
consulate: New York
US diplomatic representation:
chief of mission: Ambassador Evelyn I. H. TEEGEN
embassy: 31 Loftus Street, Suva
mailing address: P. O. Box 218, Suva
telephone: [679] 314-466
FAX: [679] 300-081
Flag: light blue with the flag of the UK in the upper hoist-side quadrant and the Fijian shield centered on the outer half of the flag;

the shield depicts a yellow lion above a white field quartered by the cross of Saint George featuring stalks of sugarcane, a palm tree, bananas, and a white dove

Economy

Overview: Fiji's economy is primarily agricultural, with a large subsistence sector. Sugar exports are a major source of foreign exchange, and sugar processing accounts for one-third of industrial output. Industry, including sugar milling, contributes 13% to GDP. Fiji traditionally had earned considerable sums of hard currency from the 250,000 tourists who visited each year. In 1987, however, after two military coups, the economy went into decline. GDP dropped by 7.8% in 1987 and by another 2.5% in 1988; political uncertainty created a drop in tourism, and the worst drought of the century caused sugar production to fall sharply. In contrast, sugar and tourism turned in strong performances in 1989, and the economy rebounded vigorously. In 1990 the economy received a setback from cyclone Sina, which cut sugar output by an estimated 21%. Sugar exports recovered in 1991-92.
National product: GDP—exchange rate conversion—$1.4 billion (1992 est.)
National product real growth rate: 3% (1992 est.)
National product per capita: $1,900 (1992 est.)
Inflation rate (consumer prices): 5% (1992 est.)
Unemployment rate: 5.9% (1991 est.)
Budget: revenues $455 million; expenditures $546 million, including capital expenditures of $NA (1993 est.)
Exports: $435 million (f.o.b., 1991)
commodities: sugar 40%, gold, clothing, copra, processed fish, lumber
partners: EC 31%, Australia 21%, Japan 8%, US 6%
Imports: $553 million (c.i.f., 1991)
commodities: machinery and transport equipment 32%, food 15%, petroleum products, consumer goods, chemicals
partners: Australia 30%, NZ 17%, Japan 13%, EC 6%, US 6%
External debt: $428 million (December 1990 est.)
Industrial production: growth rate 8.4% (1991 est.); accounts for 13% of GDP
Electricity: 215,000 kW capacity; 420 million kWh produced, 560 kWh per capita (1992)
Industries: sugar, tourism, copra, gold, silver, clothing, lumber, small cottage industries
Agriculture: accounts for 23% of GDP; principal cash crop is sugarcane; coconuts, cassava, rice, sweet potatoes, bananas; small

Fiji *(continued)*

livestock sector includes cattle, pigs, horses, and goats; fish catch nearly 33,000 tons (1989)
Economic aid: Western (non-US) countries, ODA and OOF bilateral commitments (1980-89), $815 million
Currency: 1 Fijian dollar (F$) = 100 cents
Exchange rates: Fijian dollars (F$) per US$1—1.5809 (January 1993), 1.5029 (1992), 1.4756 (1991), 1.4809 (1990), 1.4833 (1989), 1.4303 (1988)
Fiscal year: calendar year

Communications

Railroads: 644 km 0.610-meter narrow gauge, belonging to the government-owned Fiji Sugar Corporation
Highways: 3,300 km total; 1,590 km paved; 1,290 km gravel, crushed stone, or stabilized soil surface; 420 unimproved earth (1984)
Inland waterways: 203 km; 122 km navigable by motorized craft and 200-metric-ton barges
Ports: Lambasa, Lautoka, Savusavu, Suva
Merchant marine: 7 ships (1,000 GRT or over) totaling 40,072 GRT/47,187 DWT; includes 2 roll-on/roll-off, 2 container, 1 oil tanker, 1 chemical tanker, 1 cargo
Airports:
total: 25
usable: 22
with permanent-surface runways: 2
with runways over 3,659 m: 0
with runways 2,440-3,659 m: 1
with runways 1,220-2,439 m: 2
Telecommunications: modern local, interisland, and international (wire/radio integrated) public and special-purpose telephone, telegraph, and teleprinter facilities; regional radio center; important COMPAC cable link between US-Canada and New Zealand-Australia; 53,228 telephones (71 telephones per 1,000 persons); broadcast stations—7 AM, 1 FM, no TV; 1 Pacific Ocean INTELSAT earth station

Defense Forces

Branches: Fiji Military Force (FMF; including a naval division, police)
Manpower availability: males age 15-49 194,634; fit for military service 107,304; reach military age (18) annually 7,834 (1993 est.)
Defense expenditures: exchange rate conversion—$22.4 million, about 2% of GDP (FY91/92)

Finland

Geography

Location: Northern Europe, bordering the Baltic Sea between Sweden and Russia
Map references: Arctic Region, Europe, Standard Time Zones of the World
Area:
total area: 337,030 km²
land area: 305,470 km²
comparative area: slightly smaller than Montana
Land boundaries: total 2,628 km, Norway 729 km, Sweden 586 km, Russia 1,313 km
Coastline: 1,126 km (excludes islands and coastal indentations)
Maritime claims:
contiguous zone: 6 nm
continental shelf: 200 m depth or to depth of exploitation
exclusive fishing zone: 12 nm
territorial sea: 4 nm
International disputes: none
Climate: cold temperate; potentially subarctic, but comparatively mild because of moderating influence of the North Atlantic Current, Baltic Sea, and more than 60,000 lakes
Terrain: mostly low, flat to rolling plains interspersed with lakes and low hills
Natural resources: timber, copper, zinc, iron ore, silver
Land use:
arable land: 8%
permanent crops: 0%
meadows and pastures: 0%
forest and woodland: 76%
other: 16%
Irrigated land: 620 km² (1989 est.)
Environment: permanently wet ground covers about 30% of land; population concentrated on small southwestern coastal plain
Note: long boundary with Russia; Helsinki is northernmost national capital on European continent

People

Population: 5,050,942 (July 1993 est.)
Population growth rate: 0.37% (1993 est.)
Birth rate: 12.61 births/1,000 population (1993 est.)
Death rate: 9.91 deaths/1,000 population (1993 est.)
Net migration rate: 1.04 migrant(s)/1,000 population (1993 est.)
Infant mortality rate: 5.4 deaths/1,000 live births (1993 est.)
Life expectancy at birth:
total population: 75.65 years
male: 71.85 years
female: 79.62 years (1993 est.)
Total fertility rate: 1.79 children born/woman (1993 est.)
Nationality:
noun: Finn(s)
adjective: Finnish
Ethnic divisions: Finn, Swede, Lapp, Gypsy, Tatar
Religions: Evangelical Lutheran 89%, Greek Orthodox 1%, none 9%, other 1%
Languages: Finnish 93.5% (official), Swedish 6.3% (official), small Lapp- and Russian-speaking minorities
Literacy: age 15 and over can read and write (1980)
total population: 100%
male: NA%
female: NA%
Labor force: 2.533 million
by occupation: public services 30.4%, industry 20.9%, commerce 15.0%, finance, insurance, and business services 10.2%, agriculture and forestry 8.6%, transport and communications 7.7%, construction 7.2%

Government

Names:
conventional long form: Republic of Finland
conventional short form: Finland
local long form: Suomen Tasavalta
local short form: Suomi
Digraph: FI
Type: republic
Capital: Helsinki
Administrative divisions: 12 provinces (laanit, singular—laani); Ahvenanmaa, Hame, Keski-Suomi, Kuopio, Kymi, Lappi, Mikkeli, Oulu, Pohjois-Karjala, Turku ja Pori, Uusimaa, Vaasa
Independence: 6 December 1917 (from Soviet Union)
Constitution: 17 July 1919
Legal system: civil law system based on Swedish law; Supreme Court may request legislation interpreting or modifying laws; accepts compulsory ICJ jurisdiction, with reservations
National holiday: Independence Day, 6 December (1917)

Political parties and leaders:
government coalition: Center Party, Esko AHO; National Coalition (conservative) Party, Perti SALOLAINEN; Swedish People's Party, (Johan) Ole NORRBACK; Finnish Christian League, Toimi KANKAANNIEMI
other parties: Social Democratic Party, Antero KEKKONEN, Acting Chairman; Leftist Alliance (Communist) People's Democratic League and Democratic Alternative, Claes ANDERSON; Green League, Pekka SAURI; Rural Party, Tina MAKELA; Liberal People's Party, Kalle MAATTA
Other political or pressure groups: Finnish Communist Party-Unity, Yrjo HAKANEN; Constitutional Rightist Party; Finnish Pensioners Party; Communist Workers Party, Timo LAHDENMAKI
Suffrage: 18 years of age; universal
Elections:
President: last held 31 January—1 February and 15 February 1988 (next to be held January 1994); results—Mauno KOIVISTO 48%, Paavo VAYRYNEN 20%, Harri HOLKERI 18%
Parliament: last held 17 March 1991 (next to be held March 1995); results—Center Party 24.8%, Social Democratic Party 22.1%, National Coalition (Conservative) Party 19.3%, Leftist Alliance (Communist) 10.1%, Green League 6.8%, Swedish People's Party 5.5%, Rural 4.8%, Finnish Christian League 3.1%, Liberal People's Party 0.8%; seats—(200 total) Center Party 55, Social Democratic Party 48, National Coalition (Conservative) Party 40, Leftist Alliance (Communist) 19, Swedish People's Party 12, Green League 10, Finnish Christian League 8, Rural 7, Liberal People's Party 1
Executive branch: president, prime minister, deputy prime minister, Council of State (Valtioneuvosto)
Legislative branch: unicameral Parliament (Eduskunta)
Judicial branch: Supreme Court (Korkein Oikeus)
Leaders:
Chief of State: President Mauno KOIVISTO (since 27 January 1982)
Head of Government: Prime Minister Esko AHO (since 26 April 1991); Deputy Prime Minister Ilkka KANERVA (since 26 April 1991)
Member of: AfDB, AG (observer), AsDB, Australia Group, BIS, CBSS, CCC, CE, CERN, COCOM (cooperating country), CSCE, EBRD, ECE, EFTA, ESA (associate), FAO, G-9, GATT, IADB, IAEA, IBRD, ICAO, ICC, ICFTU, IDA, IFAD, IFC, ILO, IMF, IMO, INMARSAT, INTELSAT, INTERPOL, IOC, IOM, ISO, ITU, LORCS, MTCR, NAM (guest), NC, NEA, NIB,

NSG, OAS (observer), OECD, PCA, UN, UNCTAD, UNDOF, UNESCO, UNFICYP, UNHCR, UNIDO, UNIFIL, UNIKOM, UNMOGIP, UNOSOM, UNPROFOR, UNTSO, UPU, WHO, WIPO, WMO, WTO, ZC
Diplomatic representation in US:
chief of mission: Ambassador Jukka VALTASAARI
chancery: 3216 New Mexico Avenue NW, Washington, DC 20016
telephone: (202) 363-2430
FAX: (202) 363-8233
consulates general: Los Angeles and New York
consulates: Chicago and Houston
US diplomatic representation:
chief of mission: Ambassador John H. KELLY
embassy: Itainen Puistotie 14A, SF-00140, Helsinki
mailing address: APO AE 09723
telephone: [358] (0) 171931
FAX: [358] (0) 174681
Flag: white with a blue cross that extends to the edges of the flag; the vertical part of the cross is shifted to the hoist side in the style of the DANNEBROG (Danish flag)

Economy

Overview: Finland has a highly industrialized, largely free market economy, with per capita output two-thirds of the US figure. Its key economic sector is manufacturing—principally the wood, metals, and engineering industries. Trade is important, with the export of goods representing about 30% of GDP. Except for timber and several minerals, Finland depends on imports of raw materials, energy, and some components for manufactured goods. Because of the climate, agricultural development is limited to maintaining self-sufficiency in basic products. The economy, which experienced an average of 4.9% annual growth between 1987 and 1989, sank into deep recession in 1991 as growth contracted by 6.5%. The recession—which continued in 1992 with growth contracting by 3.5%—has been caused by economic overheating, depressed foreign markets, and the dismantling of the barter system between Finland and the former Soviet Union under which Soviet oil and gas had been exchanged for Finnish manufactured goods. The Finnish Government has proposed efforts to increase industrial competitiveness and efficiency by an increase in exports to Western markets, cuts in public expenditures, partial privatization of state enterprises, and changes in monetary policy. In June 1991 Helsinki had tied the markka to the EC's European Currency Unit (ECU) to promote

stability. Ongoing speculation resulting from a lack of confidence in the government's policies forced Helsinki to devalue the markka by about 12% in November 1991 and to indefinitely break the link in September 1992. By boosting the competitiveness of Finnish exports, these measures presumably have kept the economic downturn from being even more severe. Unemployment probably will remain a serious problem during the next few years—monthly figures in early 1993 are approaching 20%—with the majority of Finnish firms facing a weak domestic market and the troubled German and Swedish export markets. Declining revenues, increased transfer payments, and extensive funding to bail out the banking system are expected to push the central government's budget deficit to nearly 13% in 1993. Helsinki continues to harmonize its economic policies with those of the EC during Finland's current EC membership bid.
National product: GDP—purchasing power equivalent—$79.4 billion (1992)
National product real growth rate: -3.5% (1992)
National product per capita: $15,900 (1992)
Inflation rate (consumer prices): 2.1% (1992)
Unemployment rate: 13.1% (1992)
Budget: revenues $26.8 billion; expenditures $40.6 billion, including capital expenditures of $NA (1992)
Exports: $24.0 billion (f.o.b., 1992)
commodities: timber, paper and pulp, ships, machinery, clothing and footwear
partners: EC 53.2% (Germany 15.6%, UK 10.7%), EFTA 19.5% (Sweden 12.8%), US 5.9%, Japan 1.3%, Russia 2.8% (1992)
Imports: $21.2 billion (c.i.f., 1992)
commodities: foodstuffs, petroleum and petroleum products, chemicals, transport equipment, iron and steel, machinery, textile yarn and fabrics, fodder grains
partners: EC 47.2% (Germany 16.9%, UK 8.7%), EFTA 19.0% (Sweden 11.7%), US 6.1%, Japan 5.5%, Russia 7.1% (1992)
External debt: $25 billion (1992)
Industrial production: growth rate 7.6% (1992 est.)
Electricity: 13,500,000 kW capacity; 55,300 million kWh produced, 11,050 kWh per capita (1992)
Industries: metal products, shipbuilding, forestry and wood processing (pulp, paper), copper refining, foodstuffs, chemicals, textiles, clothing
Agriculture: accounts for 5% of GDP (including forestry); livestock production, especially dairy cattle, predominates; forestry is an important export earner and a secondary occupation for the rural population; main crops—cereals, sugar beets,

Finland *(continued)*

potatoes; 85% self-sufficient, but short of foodgrains and fodder grains; annual fish catch about 160,000 metric tons
Economic aid: donor—ODA and OOF commitments (1970-89), $2.7 billion
Currency: 1 markkaa (FMk) or Finmark = 100 pennia
Exchange rates: markkaa (FMk) per US$1—5.4193 (January 1993), 4.4794 (1992), 4.0440 (1991), 3.8235 (1990), 4.2912 (1989), 4.1828 (1988)
Fiscal year: calendar year

Communications

Railroads: 5,924 km total; Finnish State Railways (VR) operate a total of 5,863 km 1.524-meter gauge, of which 480 km are multiple track and 1,445 km are electrified
Highways: about 103,000 km total, including 35,000 km paved (bituminous, concrete, bituminous-treated surface) and 38,000 km unpaved (stabilized gravel, gravel, earth); additional 30,000 km of private (state-subsidized) roads
Inland waterways: 6,675 km total (including Saimaa Canal); 3,700 km suitable for steamers
Pipelines: natural gas 580 km
Ports: Helsinki, Oulu, Pori, Rauma, Turku
Merchant marine: 87 ships (1,000 GRT or over) totaling 935,260 GRT/973,995 DWT; includes 3 passenger, 11 short-sea passenger, 17 cargo, 1 refrigerated cargo, 26 roll-on/roll-off, 14 oil tanker, 6 chemical tanker, 2 liquefied gas, 7 bulk
Airports:
total: 160
usable: 157
with permanent-surface runways: 66
with runways over 3,659 m: 0
with runways 2,440-3,659 m: 25
with runways 1,220-2,439 m: 22
Telecommunications: good service from cable and microwave radio relay network; 3,140,000 telephones; broadcast stations—6 AM, 105 FM, 235 TV; 1 submarine cable; INTELSAT satellite transmission service via Swedish earth station and a receive-only INTELSAT earth station near Helsinki

---

Defense Forces

Branches: Army, Navy, Air Force, Frontier Guard (including Coast Guard)
Manpower availability: males age 15-49 1,323,381; fit for military service 1,091,613; reach military age (17) annually 33,828 (1993 est.)
Defense expenditures: exchange rate conversion—$1.93 billion, about 2% of GDP (1992)

France

Geography

Location: Western Europe, bordering the North Atlantic Ocean between Spain and Germany
Map references: Europe, Standard Time Zones of the World
Area:
total area: 547,030 km²
land area: 545,630 km²
comparative area: slightly more than twice the size of Colorado
note: includes Corsica and the rest of metropolitan France, but excludes the overseas administrative divisions
Land boundaries: total 2,892.4 km, Andorra 60 km, Belgium 620 km, Germany 451 km, Italy 488 km, Luxembourg 73 km, Monaco 4.4 km, Spain 623 km, Switzerland 573 km
Coastline: 3,427 km (mainland 2,783 km, Corsica 644 km)
Maritime claims:
contiguous zone: 12-24 nm
exclusive economic zone: 200 nm
territorial sea: 12 nm
International disputes: Madagascar claims Bassas da India, Europa Island, Glorioso Islands, Juan de Nova Island, and Tromelin Island; Comoros claims Mayotte; Mauritius claims Tromelin Island; Seychelles claims Tromelin Island; Suriname claims part of French Guiana; Mexico claims Clipperton Island; territorial claim in Antarctica (Adelie Land); Saint Pierre and Miquelon is focus of maritime boundary dispute between Canada and France
Climate: generally cool winters and mild summers, but mild winters and hot summers along the Mediterranean
Terrain: mostly flat plains or gently rolling hills in north and west; remainder is mountainous, especially Pyrenees in south, Alps in east
Natural resources: coal, iron ore, bauxite, fish, timber, zinc, potash

Land use:
arable land: 32%
permanent crops: 2%
meadows and pastures: 23%
forest and woodland: 27%
other: 16%
Irrigated land: 11,600 km² (1989 est.)
Environment: most of large urban areas and industrial centers in Rhone, Garonne, Seine, or Loire River basins; occasional warm tropical wind known as mistral
Note: largest West European nation

People

Population: 57,566,091 (July 1993 est.)
Population growth rate: 0.48% (1993 est.)
Birth rate: 13.24 births/1,000 population (1993 est.)
Death rate: 9.3 deaths/1,000 population (1993 est.)
Net migration rate: 0.87 migrant(s)/1,000 population (1993 est.)
Infant mortality rate: 6.8 deaths/1,000 live births (1993 est.)
Life expectancy at birth:
total population: 78 years
male: 74.04 years
female: 82.16 years (1993 est.)
Total fertility rate: 1.8 children born/woman (1993 est.)
Nationality:
noun: Frenchman(men), Frenchwoman(women)
adjective: French
Ethnic divisions: Celtic and Latin with Teutonic, Slavic, North African, Indochinese, Basque minorities
Religions: Roman Catholic 90%, Protestant 2%, Jewish 1%, Muslim (North African workers) 1%, unaffiliated 6%
Languages: French 100%, rapidly declining regional dialects and languages (Provencal, Breton, Alsatian, Corsican, Catalan, Basque, Flemish)
Literacy: age 15 and over can read and write (1980)
total population: 99%
male: NA%
female: NA%
Labor force: 24.17 million
by occupation: services 61.5%, industry 31.3%, agriculture 7.2% (1987)

Government

Names:
conventional long form: French Republic
conventional short form: France
local long form: Republique Francaise
local short form: France
Digraph: FR
Type: republic
Capital: Paris

Administrative divisions: 22 regions (regions, singular—region); Alsace, Aquitaine, Auvergne, Basse-Normandie, Bourgogne, Bretagne, Centre, Champagne-Ardenne, Corse, Franche-Comte, Haute-Normandie, Ile-de-France, Languedoc-Roussillon, Limousin, Lorraine, Midi-Pyrenees, Nord-Pas-de-Calais, Pays de la Loire, Picardie, Poitou-Charentes, Provence-Alpes-Cote d'Azur, Rhone-Alpes
note: the 22 regions are subdivided into 96 departments; see separate entries for the overseas departments (French Guiana, Guadeloupe, Martinique, Reunion) and the territorial collectivities (Mayotte, Saint Pierre and Miquelon)
Dependent areas: Bassas da India, Clipperton Island, Europa Island, French Polynesia, French Southern and Antarctic Lands, Glorioso Islands, Juan de Nova Island, New Caledonia, Tromelin Island, Wallis and Futuna
note: the US does not recognize claims to Antarctica
Independence: 486 (unified by Clovis)
Constitution: 28 September 1958, amended concerning election of president in 1962, ammended to comply with provisions of EC Maastricht Treaty in 1992
Legal system: civil law system with indigenous concepts; review of administrative but not legislative acts
National holiday: National Day, Taking of theBastille, 14 July (1789)
Political parties and leaders: Rally for the Republic (RPR), Jacques CHIRAC; Union for French Democracy (UDF, federation of UREI, UC, RDE), Valery Giscard d'ESTAING; Republican Party (PR), Gerard LONGUET; Center for Social Democrats (CDS), Pierre MEHAIGNERIE; Radical (RAD), Yves GALLAND; Socialist Party (PS), Michel ROCARD; Left Radical Movement (MRG), Emile ZUCCARELLI; Communist Party (PCF), Georges MARCHAIS; National Front (FN), Jean-Marie LE PEN; Union of Republican and Independents (UREI); Centrist Union (UC); (RDE)
Other political or pressure groups: Communist-controlled labor union (Confederation Generale du Travail) nearly 2.4 million members (claimed); Socialist-leaning labor union (Confederation Francaise Democratique du Travail or CFDT) about 800,000 members est.; independent labor union (Force Ouvriere) 1 million members (est.); independent white-collar union (Confederation Generale des Cadres) 340,000 members (claimed); National Council of French Employers (Conseil National du Patronat Francais—CNPF or Patronat)
Suffrage: 18 years of age; universal

Elections:
President: last held 8 May 1988 (next to be held by May 1995); results—Second Ballot Francois MITTERRAND 54%, Jacques CHIRAC 46%
Senate: last held NA September 1992 (next to be held September 1995—nine-year term, elected by thirds every three years); results—percent of vote by party NA; seats—(321 total; 296 metropolitan France, 13 for overseas departments and territories, and 12 for French nationals abroad) RPR 91, UDF 142 (UREI 51, UC 68, RDE 23), PS 66, PCF 16, independents 2, other 4
National Assembly: last held 21 and 28 March 1993 (next to be held NA 1998); results—percent of vote by party NA; seats—(577 total) RPR 247, UDF 213, PS 67, PCF 24, independents 26
Executive branch: president, prime minister, Council of Ministers (cabinet)
Legislative branch: bicameral Parliament (Parlement) consists of an upper house or Senate (Senat) and a lower house or National Assembly (Assemblee Nationale)
Judicial branch: Constitutional Court (Cour Constitutionnelle)
Leaders:
Chief of State: President Francois MITTERRAND (since 21 May 1981)
Head of Government: Prime Minister Edouard BALLADUR (since 29 March 1993)
Member of: ACCT, AfDB, AG (observer), AsDB, Australia Group, BDEAC, BIS, CCC, CDB (non-regional), CE, CERN, COCOM, CSCE, EBRD, EC, ECA (associate), ECE, ECLAC, EIB, ESA, ESCAP, FAO, FZ, GATT, G-5, G-7, G-10, IADB, IAEA, IBRD, ICAO, ICC, ICFTU, IDA, IFAD, IFC, ILO, IMF, IMO, INMARSAT, INTELSAT, INTERPOL, IOC, IOM, ISO, ITU, LORCS, MINURSO, MTCR, NACC, NATO, NEA, NSG, OAS (observer), OECD, PCA, SPC, UN, UNCTAD, UNESCO, UNHCR, UNIDO, UNIFIL, UNIKOM, UNPROFOR, UNRWA, UN Security Council, UNTAC, UN Trusteeship Council, UNTSO, UPU, WCL, WEU, WFTU, WHO, WIPO, WMO, WTO, ZC
Diplomatic representation in US:
chief of mission: Ambassador Jacques ANDREANI
chancery: 4101 Reservoir Road NW, Washington, DC 20007
telephone: (202) 944-6000
consulates general: Atlanta, Boston, Chicago, Honolulu, Houston, Los Angeles, Miami, New Orleans, New York, San Francisco, and San Juan (Puerto Rico)
US diplomatic representation:
chief of mission: Ambassador Pamela HARRIMAN
embassy: 2 Avenue Gabriel, 75382 Paris Cedex 08, Unit 21551

mailing address: APO AE 09777
telephone: [33] (1) 4296-12-02 or 4261-80-75
FAX: [33] (1) 4266-9783
consulates general: Bordeaux, Marseille, Strasbourg
Flag: three equal vertical bands of blue (hoist side), white, and red; known as the French Tricouleur (Tricolor); the design and colors have been the basis for a number of other flags, including those of Belgium, Chad, Ireland, Cote d'Ivoire, and Luxembourg; the official flag for all French dependent areas

---

Economy

Overview: One of the world's most developed economies, France has substantial agricultural resources and a highly diversified modern industrial sector. Large tracts of fertile land, the application of modern technology, and subsidies have combined to make it the leading agricultural producer in Western Europe. France is largely self-sufficient in agricultural products and is a major exporter of wheat and dairy products. The industrial sector generates about one-quarter of GDP, and the growing services sector has become crucial to the economy. The French economy is entering its fourth consecutive year of sluggish growth after a strong expansion in the late 1980s. Growth averaged only 1.3% in 1990-92 and is expected to drop to between zero and –0.5% in 1993. The government budget deficit rose to 3.2% of GDP in 1992 and is expected to be far larger than planned in the 1993 budget. Paris remains committed to maintaining the franc-deutsch mark parity, which has kept French interest rates high despite France's low inflation. Although the pace of economic integration within the European Community has slowed down, integration presumably will remain a major force shaping the fortunes of the various economic sectors.
National product: GDP—purchasing power equivalent—$1.08 trillion (1992)
National product real growth rate: 1.1% (1992)
National product per capita: $18,900 (1992)
Inflation rate (consumer prices): 2.1% (1992 est.)
Unemployment rate: 10.5% (end 1992)
Budget: revenues $220.5 billion; expenditures $249.1 billion, including capital expenditures of $47 billion (1993 budget)
Exports: $212.7 billion (f.o.b., 1991)
commodities: machinery and transportation equipment, chemicals, foodstuffs, agricultural products, iron and steel products, textiles and clothing
partners: Germany 18.6%, Italy 11.0%, Spain 11.0%, Belgium-Luxembourg 9.1%,

France *(continued)*

UK 8.8%, Netherlands 7.9%, US 6.4%, Japan 2.0%, former USSR 0.7% (1991 est.)
Imports: $230.3 billion (c.i.f., 1991)
commodities: crude oil, machinery and equipment, agricultural products, chemicals, iron and steel products
partners: Germany 17.8%, Italy 10.9%, US 9.5%, Netherlands 8.9%, Spain 8.8%, Belgium-Luxembourg 8.5%, UK 7.5%, Japan 4.1%, former USSR 1.3% (1991 est.)
External debt: $270 billion (December 1992)
Industrial production: growth rate 0.2% (1992 est.)
Electricity: 110,000,000 kW capacity; 426,000 million kWh produced, 7,430 kWh per capita (1992)
Industries: steel, machinery, chemicals, automobiles, metallurgy, aircraft, electronics, mining, textiles, food processing, tourism
Agriculture: accounts for 4% of GDP (including fishing and forestry); one of the world's top five wheat producers; other principal products—beef, dairy products, cereals, sugar beets, potatoes, wine grapes; self-sufficient for most temperate-zone foods; shortages include fats and oils and tropical produce, but overall net exporter of farm products; fish catch of 850,000 metric tons ranks among world's top 20 countries and is all used domestically
Economic aid: donor—ODA and OOF commitments (1970-89), $75.1 billion
Currency: 1 French franc (F) = 100 centimes
Exchange rates: French francs (F) per US$1—5.4812 (January 1993), 5.2938 (1992), 5.6421 (1991), 5.4453 (1990), 6.3801 (1989), 5.9569 (1988)
Fiscal year: calendar year

Communications

Railroads: French National Railways (SNCF) operates 34,322 km 1.435-meter standard gauge; 12,434 km electrified, 15,132 km double or multiple track; 99 km of various gauges (1.000-meter), privately owned and operated
Highways: 1,551,400 km total; 33,400 km national highway; 347,000 km departmental highway; 421,000 km community roads; 750,000 km rural roads; 5,401 km of controlled-access divided autoroutes; about 803,000 km paved
Inland waterways: 14,932 km; 6,969 km heavily traveled
Pipelines: crude oil 3,059 km; petroleum products 4,487 km; natural gas 24,746 km
Ports: coastal—Bordeaux, Boulogne, Brest, Cherbourg, Dunkerque, Fos-Sur-Mer, Le Havre, Marseille, Nantes, Sete, Toulon; inland—Rouen
Merchant marine: 130 ships (1,000 GRT or over) totaling 3,224,945 GRT/5,067,252

DWT; includes 7 short-sea passenger, 10 cargo, 20 container, 1 multifunction large-load carrier, 27 roll-on/roll-off, 36 oil tanker, 11 chemical tanker, 6 liquefied gas, 2 specialized tanker, 10 bulk; note—France also maintains a captive register for French-owned ships in the Kerguelen Islands (French Southern and Antarctic Lands) and French Polynesia
Airports:
total: 471
usable: 461
with permanent-surface runways: 256
with runways over 3,659 m: 3
with runways 2,440-3,659 m: 37
with runways 1,220-2,439 m: 136
Telecommunications: highly developed; extensive cable and microwave radio relay networks; large-scale introduction of optical-fiber systems; satellite systems for domestic traffic; 39,200,000 telephones; broadcast stations—41 AM, 800 (mostly repeaters) FM, 846 (mostly repeaters) TV; 24 submarine coaxial cables; 2 INTELSAT earth stations (with total of 5 antennas—2 for the Indian Ocean INTELSAT and 3 for the Atlantic Ocean INTELSAT); HF radio communications with more than 20 countries; INMARSAT service; EUTELSAT TV service

Defense Forces

Branches: Army, Navy (including Naval Air), Air Force, National Gendarmerie
Manpower availability: males age 15-49 14,662,761; fit for military service 12,247,950; reach military age (18) annually 386,504 (1993 est.)
Defense expenditures: exchange rate conversion—$36.6 billion, 3.1% of GDP (1993 est.)

French Guiana
(overseas department of France)

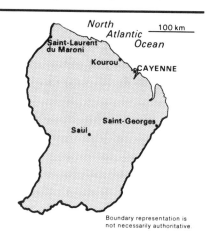

Boundary representation is not necessarily authoritative.

Geography

Location: northern South America, bordering on the North Atlantic Ocean between Suriname and Brazil
Map references: South America, Standard Time Zones of the World
Area:
total area: 91,000 km²
land area: 89,150 km²
comparative area: slightly smaller than Indiana
Land boundaries: total 1,183 km, Brazil 673 km, Suriname 510 km
Coastline: 378 km
Maritime claims:
exclusive economic zone: 200 nm
territorial sea: 12 nm
International disputes: Suriname claims area between Riviere Litani and Riviere Marouini (both headwaters of the Lawa)
Climate: tropical; hot, humid; little seasonal temperature variation
Terrain: low-lying coastal plains rising to hills and small mountains
Natural resources: bauxite, timber, gold (widely scattered), cinnabar, kaolin, fish
Land use:
arable land: 0%
permanent crops: 0%
meadows and pastures: 0%
forest and woodland: 82%
other: 18%
Irrigated land: NA km²
Environment: mostly an unsettled wilderness

People

Population: 133,376 (July 1993 est.)
Population growth rate: 4.42% (1993 est.)
Birth rate: 26.46 births/1,000 population (1993 est.)
Death rate: 4.72 deaths/1,000 population (1993 est.)
Net migration rate: 22.49 migrant(s)/1,000 population (1993 est.)

Infant mortality rate: 16.6 deaths/1,000 live births (1993 est.)

Life expectancy at birth:

total population: 74.87 years

male: 71.59 years

female: 78.32 years (1993 est.)

Total fertility rate: 3.54 children born/woman (1993 est.)

Nationality:

noun: French Guianese (singular and plural)

adjective: French Guianese

Ethnic divisions: black or mulatto 66%, Caucasian 12%, East Indian, Chinese, Amerindian 12%, other 10%

Religions: Roman Catholic

Languages: French

Literacy: age 15 and over can read and write (1982)

total population: 82%

male: 81%

female: 83%

Labor force: 23,265

by occupation: services, government, and commerce 60.6%, industry 21.2%, agriculture 18.2% (1980)

Government

Names:

conventional long form: Department of Guiana

conventional short form: French Guiana

local long form: none

local short form: Guyane

Digraph: FG

Type: overseas department of France

Capital: Cayenne

Administrative divisions: none (overseas department of France)

Independence: none (overseas department of France)

Constitution: 28 September 1958 (French Constitution)

Legal system: French legal system

National holiday: National Day, Taking of the Bastille, 14 July (1789)

Political parties and leaders: Guianese Socialist Party (PSG), Gerard HOLDER; Rally for the Republic (RPR), Paulin BRUNE; Union of the Center Rally (URC); Union for French Democracy (UDF), Claude Ho A CHUCK; Guyana Democratic Front (FDG), Georges OTHILY

Suffrage: 18 years of age; universal

Elections:

French National Assembly: last held 24 September 1989 (next to be held March 1993); results—percent of vote by party NA; seats—(2 total) PSG 1, RPR 1

French Senate: last held 24 September 1989 (next to be held September 1998); results—percent of vote by party NA; seats—(1 total) PSG 1

Regional Council: last held 22 March 1992 (next to be held NA); results—percent of

vote by party NA; seats—(31 total) PSG 16

Executive branch: French president, commissioner of the republic

Legislative branch: unicameral General Council and a unicameral Regional Council

Judicial branch: Court of Appeals (highest local court based in Martinique with jurisdiction over Martinique, Guadeloupe, and French Guiana)

Leaders:

Chief of State: President Francois MITTERRAND (since 21 May 1981)

Head of Government: Prefect Jean-Francois CORDET (since NA 1992)

Member of: FZ, WCL

Diplomatic representation in US: as an overseas department of France, the interests of French Guiana are represented in the US by France

US diplomatic representation: none (overseas department of France)

Flag: the flag of France is used

Economy

Overview: The economy is tied closely to that of France through subsidies and imports. Besides the French space center at Kourou, fishing and forestry are the most important economic activities, with exports of fish and fish products (mostly shrimp) accounting for more than 60% of total revenue in 1987. The large reserves of tropical hardwoods, not fully exploited, support an expanding sawmill industry that provides sawn logs for export. Cultivation of crops—rice, cassava, bananas, and sugarcane—is limited to the coastal area, where the population is largely concentrated. French Guiana is heavily dependent on imports of food and energy. Unemployment is a serious problem, particularly among younger workers.

National product: GDP—exchange rate conversion—$421 million (1986)

National product real growth rate: NA%

National product per capita: $4,390 (1986)

Inflation rate (consumer prices): 4.1% (1987)

Unemployment rate: 13% (1990)

Budget: revenues $735 million; expenditures $735 million, including capital expenditures of $NA (1987)

Exports: $64.8 million (f.o.b., 1990)

commodities: shrimp, timber, rum, rosewood essence

partners: France 36%, US 14%, Japan 6% (1990)

Imports: $435 million (c.i.f., 1990)

commodities: food (grains, processed meat), other consumer goods, producer goods, petroleum

partners: France 62%, Trinidad and Tobago 9%, US 4%, FRG 3% (1987)

External debt: $1.2 billion (1988)

Industrial production: growth rate NA%

Electricity: 92,000 kW capacity; 185 million kWh produced, 1,450 kWh per capita (1992)

Industries: construction, shrimp processing, forestry products, rum, gold mining

Agriculture: some vegetables for local consumption; rice, corn, manioc, cocoa, bananas, sugar; livestock—cattle, pigs, poultry

Economic aid: Western (non-US) countries, ODA and OOF bilateral commitments (1970-89), $1.51 billion

Currency: 1 French franc (F) = 100 centimes

Exchange rates: French francs (F) per US$1—5.4812 (January 1993), 5.2938 (1992), 5.6421 (1991), 5.4453 (1990), 6.3801 (1989), 5.9569 (1988)

Fiscal year: calendar year

Communications

Highways: 680 km total; 510 km paved, 170 km improved and unimproved earth

Inland waterways: 460 km, navigable by small oceangoing vessels and river and coastal steamers; 3,300 km navigable by native craft

Ports: Cayenne

Airports:

total: 10

usable: 10

with permanent-surface runways: 4

with runways over 3,659 m: 0

with runways 2,440-3,659 m: 1

with runways 1,220-2,439 m: 1

Telecommunications: fair open-wire and microwave radio relay system; 18,100 telephones; broadcast stations—5 AM, 7 FM, 9 TV; 1 Atlantic Ocean INTELSAT earth station

Defense Forces

Branches: French Forces, Gendarmerie

Manpower availability: males 15-49 39,005; fit for military service 25,477 (1993 est.)

Defense expenditures: $NA, NA% of GDP

Note: defense is the responsibility of France

French Polynesia
(overseas territory of France)

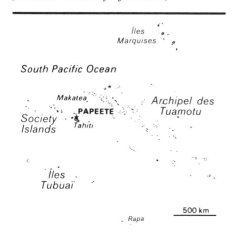

Geography

Location: Oceania, halfway between Australia and South America
Map references: Oceania
Area:
total area: 3,941 km²
land area: 3,660 km²
comparative area: slightly less than one-third the size of Connecticut
Land boundaries: 0 km
Coastline: 2,525 km
Maritime claims:
exclusive economic zone: 200 nm
territorial sea: 12 nm
International disputes: none
Climate: tropical, but moderate
Terrain: mixture of rugged high islands and low islands with reefs
Natural resources: timber, fish, cobalt
Land use:
arable land: 1%
permanent crops: 19%
meadows and pastures: 5%
forest and woodland: 31%
other: 44%
Irrigated land: NA km²
Environment: occasional cyclonic storm in January; includes five archipelagoes
Note: Makatea in French Polynesia is one of the three great phosphate rock islands in the Pacific Ocean—the others are Banaba (Ocean Island) in Kiribati and Nauru

People

Population: 210,333 (July 1993 est.)
Population growth rate: 2.26% (1993 est.)
Birth rate: 27.89 births/1,000 population (1993 est.)
Death rate: 5.27 deaths/1,000 population (1993 est.)
Net migration rate: 0 migrant(s)/1,000 population (1993 est.)
Infant mortality rate: 15 deaths/1,000 live births (1993 est.)
Life expectancy at birth:
total population: 70.33 years

male: 67.95 years
female: 72.84 years (1993 est.)
Total fertility rate: 3.32 children born/woman (1993 est.)
Nationality:
noun: French Polynesian(s)
adjective: French Polynesian
Ethnic divisions: Polynesian 78%, Chinese 12%, local French 6%, metropolitan French 4%
Religions: Protestant 54%, Roman Catholic 30%, other 16%
Languages: French (official), Tahitian (official)
Literacy: age 14 and over but definition of literacy not available (1977)
total population: 98%
male: 98%
female: 98%
Labor force: 76,630 employed (1988)

Government

Names:
conventional long form: Territory of French Polynesia
conventional short form: French Polynesia
local long form: Territoire de la Polynesie Francaise
local short form: Polynesie Francaise
Digraph: FP
Type: overseas territory of France since 1946
Capital: Papeete
Administrative divisions: none (overseas territory of France); there are no first-order administrative divisions as defined by the US Government, but there are 5 archipelagic divisions named Archipel des Marquises, Archipel des Tuamotu, Archipel des Tubuai, Iles du Vent, and Iles Sous-le-Vent
note: Clipperton Island is administered by France from French Polynesia
Independence: none (overseas territory of France)
Constitution: 28 September 1958 (French Constitution)
Legal system: based on French system
National holiday: National Day, Taking of the Bastille, 14 July (1789)
Political parties and leaders: People's Rally (Tahoeraa Huiraatira; Gaullist), Gaston FLOSSE; Polynesian Union Party (Te Tiarama; centrist), Alexandre LEONTIEFF; New Fatherland Party (Ai'a Api), Emile VERNAUDON; Polynesian Liberation Front (Tavini Huiraatira), Oscar TEMARU; other small parties
Suffrage: 18 years of age; universal
Elections:
French National Assembly: last held 5 and 12 June 1988 (next to be held 21 and 28 March 1993); results—percent of vote by party NA; seats—(2 total) People's Rally (Gaullist) 1, New Fatherland Party 1

French Senate: last held 24 September 1989 (next to be held September 1998); results—percent of vote by party NA; seats—(1 total) party NA
Territorial Assembly: last held 17 March 1991 (next to be held March 1996); results—percent of vote by party NA; seats—(41 total) People's Rally (Gaullist) 18, Polynesian Union Party 14, New Fatherland Party 5, other 4
Executive branch: French president, high commissioner of the republic, president of the Council of Ministers, vice president of the Council of Ministers, Council of Ministers
Legislative branch: unicameral Territorial Assembly
Judicial branch: Court of Appeal, Court of the First Instance, Court of Administrative Law
Leaders:
Chief of State: President Francois MITTERRAND (since 21 May 1981); High Commissioner of the Republic Michel JAU (since NA 1992)
Head of Government: President of the Council of Ministers Gaston FLOSSE (since 10 May 1991); Vice President of the Council of Ministers Joel BUILLARD (since 12 September 1991)
Member of: ESCAP (associate), FZ, ICFTU, SPC, WMO
Diplomatic representation in US: as an overseas territory of France, French Polynesian interests are represented in the US by France
US diplomatic representation: none (overseas territory of France)
Flag: the flag of France is used

Economy

Overview: Since 1962, when France stationed military personnel in the region, French Polynesia has changed from a subsistence economy to one in which a high proportion of the work force is either employed by the military or supports the tourist industry. Tourism accounts for about 20% of GDP and is a primary source of hard currency earnings.
National product: GDP—exchange rate conversion—$1.2 billion (1991 est.)
National product real growth rate: NA%
National product per capita: $6,000 (1991 est.)
Inflation rate (consumer prices): 2.9% (1989)
Unemployment rate: 14.9% (1988 est.)
Budget: revenues $614 million; expenditures $957 million, including capital expenditures of $NA (1988)
Exports: $88.9 million (f.o.b., 1989)
commodities: coconut products 79%, mother-of-pearl 14%, vanilla, shark meat

partners: France 54%, US 17%, Japan 17%
Imports: $765 million (c.i.f., 1989)
commodities: fuels, foodstuffs, equipment
partners: France 53%, US 11%, Australia 6%, NZ 5%
External debt: $NA
Industrial production: growth rate NA%
Electricity: 75,000 kW capacity; 275 million kWh produced, 1,330 kWh per capita (1992)
Industries: tourism, pearls, agricultural processing, handicrafts
Agriculture: coconut and vanilla plantations; vegetables and fruit; poultry, beef, dairy products
Economic aid: Western (non-US) countries, ODA and OOF bilateral commitments (1970-88), $3.95 billion
Currency: 1 CFP franc (CFPF) = 100 centimes
Exchange rates: Comptoirs Francais du Pacifique francs (CFPF) per US$1—99.65 (January 1993), 96.24 (1992), 102.57 (1991), 99.00 (1990), 115.99 (1989), 108.30 (1988); note—linked at the rate of 18.18 to the French franc
Fiscal year: calendar year

Communications

Highways: 600 km (1982)
Ports: Papeete, Bora-bora
Merchant marine: 3 ships (1,000 GRT or over) totaling 4,127 GRT/6,710 DWT; includes 2 passenger-cargo, 1 refrigerated cargo; note—a captive subset of the French register
Airports:
total: 43
usable: 41
with permanent-surface runways: 23
with runways over 3,659 m: 0
with runways 2,440-3,659 m: 2
with runways 1,220-2,439 m: 12
Telecommunications: 33,200 telephones; 84,000 radio receivers; 26,400 TV sets; broadcast stations—5 AM, 2 FM, 6 TV; 1 Pacific Ocean INTELSAT earth station

Defense Forces

Branches: French forces (including Army, Navy, Air Force), Gendarmerie
Note: defense is responsibility of France

Geography

Location: in the southern Indian Ocean, about equidistant between Africa, Antarctica, and Australia
Map references: Antarctic Region, Standard Time Zones of the World
Area:
total area: 7,781 km^2
land area: 7,781 km^2
comparative area: slightly less than 1.5 times the size of Delaware
note: includes Ile Amsterdam, Ile Saint-Paul, Iles Kerguelen, and Iles Crozet; excludes Terre Adelie claim of about 500,000 km^2 in Antarctica that is not recognized by the US
Land boundaries: 0 km
Coastline: 1,232 km
Maritime claims:
exclusive economic zone: 200 nm from Iles Kerguelen only
territorial sea: 12 nm
International disputes: Terre Adelie claim in Antarctica is not recognized by the US
Climate: antarctic
Terrain: volcanic
Natural resources: fish, crayfish
Land use:
arable land: 0%
permanent crops: 0%
meadows and pastures: 0%
forest and woodland: 0%
other: 100%
Irrigated land: 0 km^2
Environment: Ile Amsterdam and Ile Saint-Paul are extinct volcanoes
Note: remote location in the southern Indian Ocean

People

Population: no indigenous inhabitants; note—there are researchers whose numbers vary from 150 in winter (July) to 200 in summer (January)

Government

Names:
conventional long form: Territory of the French Southern and Antarctic Lands
conventional short form: French Southern and Antarctic Lands
local long form: Territoire des Terres Australes et Antarctiques Francaises
local short form: Terres Australes et Antarctiques Francaises
Digraph: FS
Type: overseas territory of France since 1955; governed by High Administrator Bernard de GOUTTES (since May 1990), who is assisted by a 7-member Consultative Council and a 12-member Scientific Council
Capital: none; administered from Paris, France
Administrative divisions: none (overseas territory of France); there are no first-order administrative divisions as defined by the US Government, but there are 3 districts named Ile Crozet, Iles Kerguelen, and Iles Saint-Paul et Amsterdam; excludes Terre Adelie claim in Antarctica that is not recognized by the US
Independence: none (overseas territory of France)
Flag: the flag of France is used

Economy

Overview: Economic activity is limited to servicing meteorological and geophysical research stations and French and other fishing fleets. The fishing catches landed on Iles Kerguelen by foreign ships are exported to France and Reunion.
Budget: revenues $17.5 million; expenditures $NA, including capital expenditures of $NA (1992)

Communications

Ports: none; offshore anchorage only
Merchant marine: 16 ships (1,000 GRT or over) totaling 292,490 GRT/514,389 DWT; includes 2 cargo, 4 refrigerated cargo, 4 roll-on/roll-off cargo, 2 oil tanker, 3 bulk, 1 multifunction large load carrier; note—a captive subset of the French register
Telecommunications: NA

Defense Forces

Note: defense is the responsibility of France

137

Gabon

Geography

Location: Western Africa, bordering the Atlantic Ocean at the Equator between the Congo and Equatorial Guinea
Map references: Africa, Standard Time Zones of the World
Area:
total area: 267,670 km²
land area: 257,670 km²
comparative area: slightly smaller than Colorado
Land boundaries: total 2,551 km, Cameroon 298 km, Congo 1,903 km, Equatorial Guinea 350 km
Coastline: 885 km
Maritime claims:
contiguous zone: 24 nm
exclusive economic zone: 200 nm
territorial sea: 12 nm
International disputes: maritime boundary dispute with Equatorial Guinea because of disputed sovereignty over islands in Corisco Bay
Climate: tropical; always hot, humid
Terrain: narrow coastal plain; hilly interior; savanna in east and south
Natural resources: petroleum, manganese, uranium, gold, timber, iron ore
Land use:
arable land: 1%
permanent crops: 1%
meadows and pastures: 18%
forest and woodland: 78%
other: 2%
Irrigated land: NA km²
Environment: deforestation

People

Population: 1,122,550 (July 1993 est.)
Population growth rate: 1.45% (1993 est.)
Birth rate: 28.63 births/1,000 population (1993 est.)
Death rate: 14.08 deaths/1,000 population (1993 est.)
Net migration rate: 0 migrant(s)/1,000 population (1993 est.)

Infant mortality rate: 97.3 deaths/1,000 live births (1993 est.)
Life expectancy at birth:
total population: 54.19 years
male: 51.46 years
female: 57.01 years (1993 est.)
Total fertility rate: 4.02 children born/woman (1993 est.)
Nationality:
noun: Gabonese (singular and plural)
adjective: Gabonese
Ethnic divisions: Bantu tribes including four major tribal groupings (Fang, Eshira, Bapounou, Bateke), Africans and Europeans 100,000, including 27,000 French
Religions: Christian 55-75%, Muslim less than 1%, animist
Languages: French (official), Fang, Myene, Bateke, Bapounou/Eschira, Bandjabi
Literacy: age 15 and over can read and write (1990)
total population: 61%
male: 74%
female: 48%
Labor force: 120,000 salaried
by occupation: agriculture 65.0%, industry and commerce 30.0%, services 2.5%, government 2.5%
note: 58% of population of working age (1983)

Government

Names:
conventional long form: Gabonese Republic
conventional short form: Gabon
local long form: Republique Gabonaise
local short form: Gabon
Digraph: GB
Type: republic; multiparty presidential regime (opposition parties legalized 1990)
Capital: Libreville
Administrative divisions: 9 provinces; Estuaire, Haut-Ogooue, Moyen-Ogooue, Ngounie, Nyanga, Ogooue-Ivindo, Ogooue-Lolo, Ogooue-Maritime, Woleu-Ntem
Independence: 17 August 1960 (from France)
Constitution: 21 February 1961, revised 15 April 1975
Legal system: based on French civil law system and customary law; judicial review of legislative acts in Constitutional Chamber of the Supreme Court; compulsory ICJ jurisdiction not accepted
National holiday: Renovation Day, 12 March (1968) (Gabonese Democratic Party established)
Political parties and leaders: Gabonese Democratic Party (PDG, former sole party), El Hadj Omar BONGO, president; National Recovery Movement-Lumberjacks (Morena-Bucherons); Gabonese Party for Progress (PGP); National Recovery

Movement (Morena-Original); Association for Socialism in Gabon (APSG); Gabonese Socialist Union (USG); Circle for Renewal and Progress (CRP); Union for Democracy and Development (UDD)
Suffrage: 21 years of age; universal
Elections:
National Assembly: last held on 28 October 1990 (next to be held by NA); results—percent of vote NA; seats—(120 total, 111 elected) PDG 62, National Recovery Movement—Lumberjacks (Morena-Bucherons) 19, PGP 18, National Recovery Movement (Morena-Original) 7, APSG 6, USG 4, CRP 1, independents 3
President: last held on 9 November 1986 (next to be held December 1993); results—President Omar BONGO was reelected without opposition
Executive branch: president, prime minister, Cabinet
Legislative branch: unicameral National Assembly (Assemblee Nationale)
Judicial branch: Supreme Court (Cour Supreme)
Leaders:
Chief of State: President El Hadj Omar BONGO (since 2 December 1967)
Head of Government: Prime Minister Casimir OYE-MBA (since 3 May 1990)
Member of: ACCT, ACP, AfDB, BDEAC, CCC, CEEAC, ECA, FAO, FZ, G-24, G-77, GATT, IAEA, IBRD, ICAO, ICC, IDA, IDB, IFAD, IFC, ILO, IMF, IMO, INMARSAT, INTELSAT, INTERPOL, IOC, ITU, LORCS (associate), NAM, OAU, OIC, OPEC, UDEAC, UN, UNCTAD, UNESCO, UNIDO, UPU, WCL, WHO, WIPO, WMO, WTO
Diplomatic representation in US:
chief of mission: (vacant)
chancery: 2034 20th Street NW, Washington, DC 20009
telephone: (202) 797-1000
US diplomatic representation:
chief of mission: Ambassador John C. WILSON IV
embassy: Boulevard de la Mer, Libreville
mailing address: B. P. 4000, Libreville
telephone: (241) 762003/4, or 743492
FAX: [241] 745-507
Flag: three equal horizontal bands of green (top), yellow, and blue

Economy

Overview: The economy, dependent on timber and manganese until the early 1970s, is now dominated by the oil sector. In 1981-85, oil accounted for about 45% of GDP, 80% of export earnings, and 65% of government revenues on average. The high oil prices of the early 1980s contributed to a substantial increase in per capita national income, stimulated domestic demand, reinforced migration from rural to urban areas, and raised the level of real wages to among the highest in Sub-Saharan Africa. The subsequent slide of Gabon's economy, which began with falling oil prices in 1985, was reversed in 1989-90, but debt servicing obligations continue to limit prospects for further domestic development. Real growth in 1991-92 was weak because of a combination of an overstaffed bureaucracy, a large budget deficit, and the continued underdevelopment of the whole economy outside the petroleum sector.

National product: GDP—exchange rate conversion—$4.6 billion (1991)
National product real growth rate: 13% (1990 est.)
National product per capita: $4,200 (1991 est.)
Inflation rate (consumer prices): 0.7% (1991 est.)
Unemployment rate: NA%
Budget: revenues $1.4 billion; expenditures $1.4 billion, including capital expenditures of $247 million (1990 est.)
Exports: $2.2 billion (f.o.b., 1991)
commodities: crude oil 80%, manganese 7%, wood 7%, uranium 2%
partners: France 48%, US 15%, Germany 2%, Japan 2%
Imports: $702 million (c.i.f., 1991 est.)
commodities: foodstuffs, chemical products, petroleum products, construction materials, manufactures, machinery
partners: France 64%, African countries 7%, US 5%, Japan 3%
External debt: $4.4 billion (1991)
Industrial production: growth rate—10% (1988 est.); accounts for 45% of GDP, including petroleum
Electricity: 315,000 kW capacity; 995 million kWh produced, 920 kWh per capita (1991)
Industries: petroleum, food and beverages, lumbering and plywood, textiles, mining—manganese, uranium, gold, cement
Agriculture: accounts for 10% of GDP (including fishing and forestry); cash crops—cocoa, coffee, palm oil; livestock not developed; importer of food; small fishing operations provide a catch of about 20,000 metric tons; okoume (a tropical softwood) is the most important timber product
Economic aid: US commitments, including Ex-Im (FY70-90), $68 million; Western (non-US) countries, ODA and OOF bilateral commitments (1970-90), $2,342 million; Communist countries (1970-89), $27 million
Currency: 1 CFA franc (CFAF) = 100 centimes
Exchange rates: Communaute Financiere Africaine francs (CFAF) per US$1—274.06 (January 1993), 264.69 (1992), 282.11 (1991), 272.26 (1990), 319.01 (1989), 297.85 (1988)
Fiscal year: calendar year

Communications

Railroads: 649 km 1.437-meter standard-gauge single track (Transgabonese Railroad)
Highways: 7,500 km total; 560 km paved, 960 km laterite, 5,980 km earth
Inland waterways: 1,600 km perennially navigable
Pipelines: crude oil 270 km; petroleum products 14 km
Ports: Owendo, Port-Gentil, Libreville
Merchant marine: 2 cargo ships (1,000 GRT or over) totaling 18,563 GRT/25,330 DWT
Airports:
total: 68
usable: 56
with permanent-surface runways: 10
with runways over 3,659 m: 0
with runways 2,440-3,659 m: 2
with runways 1,220-2,439 m: 22
Telecommunications: adequate system of cable, radio relay, tropospheric scatter links and radiocommunication stations; 15,000 telephones; broadcast stations—6 AM, 6 FM, 3 (5 repeaters) TV; satellite earth stations—3 Atlantic Ocean INTELSAT and 12 domestic satellite

Defense Forces

Branches: Army, Navy, Air Force, Presidential Guard, National Gendarmerie, National Police
Manpower availability: males age 15-49 269,066; fit for military service 135,836; reach military age (20) annually 9,680 (1993 est.)
Defense expenditures: exchange rate conversion—$102 million, 3.2% of GDP (1990 est.)

The Gambia

Boundary representation is not necessarily authoritative.

Geography

Location: Western Africa, bordering the North Atlantic Ocean almost completely surrounded by Senegal
Map references: Africa, Standard Time Zones of the World
Area:
total area: 11,300 km²
land area: 10,000 km²
comparative area: slightly more than twice the size of Delaware
Land boundaries: total 740 km, Senegal 740 km
Coastline: 80 km
Maritime claims:
contiguous zone: 18 nm
continental shelf: not specified
exclusive fishing zone: 200 nm
territorial sea: 12 nm
International disputes: short section of boundary with Senegal is indefinite
Climate: tropical; hot, rainy season (June to November); cooler, dry season (November to May)
Terrain: flood plain of the Gambia River flanked by some low hills
Natural resources: fish
Land use:
arable land: 16%
permanent crops: 0%
meadows and pastures: 9%
forest and woodland: 20%
other: 55%
Irrigated land: 120 km² (1989 est.)
Environment: deforestation
Note: almost an enclave of Senegal; smallest country on the continent of Africa

People

Population: 930,249 (July 1993 est.)
Population growth rate: 3.07% (1993 est.)
Birth rate: 46.85 births/1,000 population (1993 est.)
Death rate: 16.1 deaths/1,000 population (1993 est.)

The Gambia (continued)

Net migration rate: 0 migrant(s)/1,000 population (1993 est.)
Infant mortality rate: 126.3 deaths/1,000 live births (1993 est.)
Life expectancy at birth:
total population: 49.61 years
male: 47.41 years
female: 51.87 years (1993 est.)
Total fertility rate: 6.35 children born/woman (1993 est.)
Nationality:
noun: Gambian(s)
adjective: Gambian
Ethnic divisions: African 99% (Mandinka 42%, Fula 18%, Wolof 16%, Jola 10%, Serahuli 9%, other 4%), non-Gambian 1%
Religions: Muslim 90%, Christian 9%, indigenous beliefs 1%
Languages: English (official), Mandinka, Wolof, Fula, other indigenous vernaculars
Literacy: age 15 and over can read and write (1990)
total population: 27%
male: 39%
female: 16%
Labor force: 400,000 (1986 est.)
by occupation: agriculture 75.0%, industry, commerce, and services 18.9%, government 6.1%
note: 55% population of working age (1983)

Government

Names:
conventional long form: Republic of The Gambia
conventional short form: The Gambia
Digraph: GA
Type: republic under multiparty democratic rule
Capital: Banjul
Administrative divisions: 5 divisions and 1 city*; Banjul*, Lower River, MacCarthy Island, North Bank, Upper River, Western
Independence: 18 February 1965 (from UK; The Gambia and Senegal signed an agreement on 12 December 1981 that called for the creation of a loose confederation to be known as Senegambia, but the agreement was dissolved on 30 September 1989)
Constitution: 24 April 1970
Legal system: based on a composite of English common law, Koranic law, and customary law; accepts compulsory ICJ jurisdiction, with reservations
National holiday: Independence Day, 18 February (1965)
Political parties and leaders: People's Progressive Party (PPP), Dawda K. JAWARA, secretary general; National Convention Party (NCP), Sheriff DIBBA; Gambian People's Party (GPP), Hassan Musa CAMARA; United Party (UP), leader NA; People's Democratic Organization of Independence and Socialism (PDOIS), leader

NA; People's Democratic Party (PDP), Jabel SALLAH
Suffrage: 21 years of age; universal
Elections:
House of Representatives: last held on 11 March 1987 (next to be held by March 1992); results—PPP 56.6%, NCP 27.6%, GPP 14.7%, PDOIS 1%; seats—(43 total, 36 elected) PPP 31, NCP 5
President: last held on 11 March 1987 (next to be held March 1992); results—Sir Dawda JAWARA (PPP) 61.1%, Sherif Mustapha DIBBA (NCP) 25.2%, Assan Musa CAMARA (GPP) 13.7%
Executive branch: president, vice president, Cabinet
Legislative branch: unicameral House of Representatives
Judicial branch: Supreme Court
Leaders:
Chief of State and Head of Government: President Alhaji Sir Dawda Kairaba JAWARA (since 24 April 1970); Vice President Saihou SABALLY (since NA)
Member of: ACP, AfDB, C, CCC, ECA, ECOWAS, FAO, G-77, GATT, IBRD, ICAO, ICFTU, IDA, IDB, IFAD, IFC, IMF, IMO, INTERPOL, IOC, ITU, LORCS, NAM, OAU, OIC, UN, UNCTAD, UNESCO, UNIDO, UPU, WCL, WFTU, WHO, WIPO, WMO, WTO
Diplomatic representation in US:
chief of mission: Ambassador Ousman A. SALLAH
chancery: Suite 720, 1030 15th Street NW, Washington, DC 20005
telephone: (202) 842-1356 or 842-1359
US diplomatic representation:
chief of mission: Ambassador Arlene RENDER
embassy: Pipeline Road (Kairaba Avenue), Fajara, Banjul
mailing address: P. M. B. No. 19, Banjul
telephone: [220] 92856 or 92858, 91970, 91971
FAX: (220) 92475
Flag: three equal horizontal bands of red (top), blue with white edges, and green

Economy

Overview: The Gambia has no important mineral or other natural resources and has a limited agricultural base. It is one of the world's poorest countries with a per capita income of about $325. About 75% of the population is engaged in crop production and livestock raising, which contribute 30% to GDP. Small-scale manufacturing activity—processing peanuts, fish, and hides—accounts for less than 10% of GDP. Tourism is a growing industry. The Gambia imports one-third of its food, all fuel, and most manufactured goods. Exports are concentrated on peanut products (about 75% of total value).

National product: GDP—exchange rate conversion—$292 million (1991 est.)
National product real growth rate: 3% (1991)
National product per capita: $325 (1991 est.)
Inflation rate (consumer prices): 12% (1992 est.)
Unemployment rate: NA%
Budget: revenues $94 million; expenditures $80 million, including capital expenditures of $25 million (FY91 est.)
Exports: $133 million (f.o.b., FY91 est.)
commodities: peanuts and peanut products, fish, cotton lint, palm kernels
partners: Japan 60%, Europe 29%, Africa 5%, US 1%, other 5% (1989)
Imports: $174 million (f.o.b., FY91 est.)
commodities: foodstuffs, manufactures, raw materials, fuel, machinery and transport equipment
partners: Europe 57%, Asia 25%, USSR and Eastern Europe 9%, US 6%, other 3% (1989)
External debt: $336 million (December 1990 est.)
Industrial production: growth rate 6.7%; accounts for 5.8% of GDP (FY90)
Electricity: 30,000 kW capacity; 65 million kWh produced, 75 kWh per capita (1991)
Industries: peanut processing, tourism, beverages, agricultural machinery assembly, woodworking, metalworking, clothing
Agriculture: accounts for 30% of GDP and employs about 75% of the population; imports one-third of food requirements; major export crop is peanuts; other principal crops—millet, sorghum, rice, corn, cassava, palm kernels; livestock—cattle, sheep, goats; forestry and fishing resources not fully exploited
Economic aid: US commitments, including Ex-Im (FY70-89), $93 million; Western (non-US) countries, ODA and OOF bilateral commitments (1970-89), $535 million; Communist countries (1970-89), $39 million
Currency: 1 dalasi (D) = 100 bututs
Exchange rates: dalasi (D) per US$1—8.673 (October 1992), 8.803 (1991), 7.883 (1990), 7.5846 (1989), 6.7086 (1988), 7.0744 (1987)
Fiscal year: 1 July-30 June

Communications

Highways: 3,083 km total; 431 km paved, 501 km gravel/laterite, and 2,151 km unimproved earth
Inland waterways: 400 km
Ports: Banjul
Airports:
total: 1
usable: 1
with permanent-surface runways: 1
with runways over 3,659 m: 0

Gaza Strip

with runways 2,440-3,659 m: 1
with runways 1,220-2,439 m: 0
Telecommunications: adequate network of radio relay and wire; 3,500 telephones; broadcast stations—3 AM, 2 FM; 1 Atlantic Ocean INTELSAT earth station

Defense Forces

Branches: Army, Navy, National Gendarmerie, National Police
Manpower availability: males age 15-49 201,026; fit for military service 101,642 (1993 est.)
Defense expenditures: exchange rate conversion—$NA, NA% of GDP

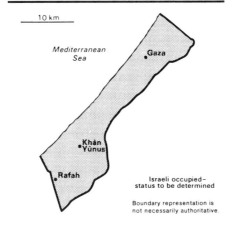

Note: The war between Israel and the Arab states in June 1967 ended with Israel in control of the West Bank and the Gaza Strip, the Sinai, and the Golan Heights. As stated in the 1978 Camp David accords and reaffirmed by President Bush's post-Gulf crisis peace initiative, the final status of the West Bank and the Gaza Strip, their relationship with their neighbors, and a peace treaty between Israel and Jordan are to be negotiated among the concerned parties. Camp David further specifies that these negotiations will resolve the respective boundaries. Pending the completion of this process, it is US policy that the final status of the West Bank and the Gaza Strip has yet to be determined. In the US view, the term West Bank describes all of the area west of the Jordan River under Jordanian administration before the 1967 Arab-Israeli war. With respect to negotiations envisaged in the framework agreement, however, it is US policy that a distinction must be made between Jerusalem and the rest of the West Bank because of the city's special status and circumstances. Therefore, a negotiated solution for the final status of Jerusalem could be different in character from that of the rest of the West Bank.

Geography

Location: Middle East, bordering the eastern Mediterranean Sea, between Egypt and Israel
Map references: Middle East
Area:
total area: 380 km²
land area: 380 km²
comparative area: slightly more than twice the size of Washington, DC
Land boundaries: total 62 km, Egypt 11 km, Israel 51 km
Coastline: 40 km
Maritime claims: Israeli occupied with status to be determined
International disputes: Israeli occupied with status to be determined

Climate: temperate, mild winters, dry and warm to hot summers
Terrain: flat to rolling, sand- and dune-covered coastal plain
Natural resources: negligible
Land use:
arable land: 13%
permanent crops: 32%
meadows and pastures: 0%
forest and woodland: 0%
other: 55%
Irrigated land: 200 km²
Environment: desertification

People

Population: 705,834 (July 1993 est.)
note: in addition, there are 4,000 Jewish settlers in the Gaza Strip (1993 est.)
Population growth rate: 3.56% (1993 est.)
Birth rate: 45.66 births/1,000 population (1993 est.)
Death rate: 5.71 deaths/1,000 population (1993 est.)
Net migration rate: -4.35 migrant(s)/1,000 population (1993 est.)
Infant mortality rate: 38.9 deaths/1,000 live births (1993 est.)
Life expectancy at birth:
total population: 67.26 years
male: 66.01 years
female: 68.57 years (1993 est.)
Total fertility rate: 7.51 children born/woman (1993 est.)
Nationality:
noun: NA
adjective: NA
Ethnic divisions: Palestinian Arab and other 99.8%, Jewish 0.2%
Religions: Muslim (predominantly Sunni) 99%, Christian 0.7%, Jewish 0.3%
Languages: Arabic, Hebrew (spoken by Israeli settlers), English (widely understood)
Literacy:
total population: NA%
male: NA%
female: NA%
Labor force: NA
by occupation: small industry, commerce and business 32.0%, construction 24.4%, service and other 25.5%, agriculture 18.1% (1984)
note: excluding Israeli Jewish settlers

Government

Note: The Gaza Strip is currently governed by Israeli military authorities and Israeli civil administration. It is US policy that the final status of the Gaza Strip will be determined by negotiations among the concerned parties. These negotiations will determine how this area is to be governed.
Names:
conventional long form: none

Gaza Strip *(continued)*

conventional short form: Gaza Strip
local long form: none
local short form: Qita Ghazzah
Digraph: GZ

Economy

Overview: In 1990 roughly 40% of Gaza Strip workers were employed across the border by Israeli industrial, construction, and agricultural enterprises, with worker remittances accounting for about one-third of GNP. The construction, agricultural, and industrial sectors account for about 15%, 12%, and 8% of GNP, respectively. Gaza depends upon Israel for some 90% of its external trade. Unrest in the territory in 1988-93 (intifadah) has raised unemployment and substantially lowered the standard of living of Gazans. The Persian Gulf crisis and its aftershocks also have dealt severe blows to Gaza since August 1990. Worker remittances from the Gulf states have plunged, unemployment has increased, and exports have fallen dramatically. The area's economic outlook remains bleak.
National product: GNP—exchange rate conversion—$380 million (1991 est.)
National product real growth rate: -30% (1991 est.)
National product per capita: $590 (1991 est.)
Inflation rate (consumer prices): 9% (1991 est.)
Unemployment rate: 20% (1990 est.)
Budget: revenues $33.8 million; expenditures $33.3 million, including capital expenditures of $NA (FY88)
Exports: $30 million (f.o.b., 1989)
commodities: citrus
partners: Israel, Egypt
Imports: $255 million (c.i.f., 1989)
commodities: food, consumer goods, construction materials
partners: Israel, Egypt
External debt: $NA
Industrial production: growth rate 10% (1989); accounts for about 8% of GNP
Electricity: power supplied by Israel
Industries: generally small family businesses that produce textiles, soap, olive-wood carvings, and mother-of-pearl souvenirs; the Israelis have established some small-scale modern industries in an industrial center
Agriculture: accounts for about 12% of GNP; olives, citrus and other fruits, vegetables, beef, dairy products
Economic aid: NA
Currency: 1 new Israeli shekel (NIS) = 100 new agorot
Exchange rates: new Israeli shekels (NIS) per US$1—2.6480 (November 1992), 2.4591 (1992), 2.2791 (1991), 2.0162 (1990), 1.9164 (1989), 1.5989 (1988), 1.5946 (1987)
Fiscal year: calendar year (since 1 January 1992)

Communications

Railroads: one line, abandoned and in disrepair, some trackage remains
Highways: small, poorly developed indigenous road network
Ports: facilities for small boats to service the city of Gaza
Airports:
total: 1
useable: 1
with permanent-surface runways: 0
with runways over 3,659 m: 0
with runways 2,440-3,659 m: 0
with runways 1,220-2,439 m: 0
Telecommunications: broadcast stations—no AM, no FM, no TV

Defense Forces

Branches: NA
Manpower availability: males age 15-49 136,311; fit for military service NA (1993 est.)
Defense expenditures: exchange rate conversion—$NA, NA% of GDP

Georgia

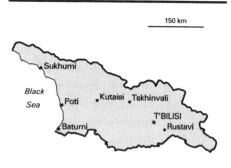

Note: Georgia is currently besieged by conflicts driven by separatists in its Abkazian and South Ossetian enclaves, and supporters of ousted President GAMAKHURDIA control much of western Georgia

Geography

Location: Southeastern Europe, bordering the Black Sea, between Turkey and Russia
Map references: Africa, Asia, Middle East, Standard Time Zones of the World
Area:
total area: 69,700 km²
land area: 69,700 km²
comparative area: slightly larger than South Carolina
Land boundaries: total 1,461 km, Armenia 164 km, Azerbaijan 322 km, Russia 723 km, Turkey 252 km
Coastline: 310 km
Maritime claims:
note: 12 nm in 1973 USSR-Turkish Protocol concerning the sea boundary between the two states in the Black Sea; Georgia claims the coastline along the Black Sea as its international waters, although it cannot control this area and the Russian navy and commercial ships transit freely
International disputes: none
Climate: warm and pleasant; Mediterranean-like on Black Sea coast
Terrain: largely mountainous with Great Caucasus Mountains in the north and Lesser Caucasus Mountains in the south; Kolkhida Lowland opens to the Black Sea in the west; Kura River Basin in the east; good soils in river valley flood plains, foothills of Kolkhida lowland
Natural resources: forest lands, hydropower, manganese deposits, iron ores, copper, minor coal and oil deposits; coastal climate and soils allow for important tea and citrus growth
Land use:
arable land: NA%

permanent crops: NA%
meadows and pastures: NA%
forest and woodland: NA%
other: NA%
Irrigated land: 4,660 km² (1990)
Environment: air pollution, particularly in Rustavi; heavy pollution of Kura River, Black Sea

People

Population: 5,634,296 (July 1993 est.)
Population growth rate: 0.85% (1993 est.)
Birth rate: 16.48 births/1,000 population (1993 est.)
Death rate: 8.68 deaths/1,000 population (1993 est.)
Net migration rate: 0.64 migrant(s)/1,000 population (1993 est.) note—this data may be low because of movement of Ossetian, Russian, and Abkhaz refugees due to ongoing conflicts
Infant mortality rate: 24.2 deaths/1,000 live births (1993 est.)
Life expectancy at birth:
total population: 72.58 years
male: 68.89 years
female: 76.46 years (1993 est.)
Total fertility rate: 2.21 children born/woman (1993 est.)
Nationality:
noun: Georgian(s)
adjective: Georgian
Ethnic divisions: Georgian 70.1%, Armenian 8.1%, Russian 6.3%, Azeri 5.7%, Ossetian 3%, Abkhaz 1.8%, other 5%
Religions: Georgian Orthodox 65%, Russian Orthodox 10%, Muslim 11%, Armenian Orthodox 8%, unknown 6%
Languages: Armenian 7%, Azerbaijani 6%, Georgian 71% (official), Russian 9%, other 7%
Literacy: age 9-49 can read and write (1970)
total population: 100%
male: 100%
female: 100%
Labor force: 2.763 million
by occupation: industry and construction 31%, agriculture and forestry 25%, other 44% (1990)

Government

Names:
conventional long form: Republic of Georgia
conventional short form: Georgia
local long form: Sakartvelo Respublika
local short form: Sakartvelo
former: Georgian Soviet Socialist Republic
Digraph: GG
Type: republic
Capital: T'bilisi (Tbilisi)
Administrative divisions: 2 autonomous republics (avtomnoy respubliki, singular—avtom respublika); Abkhazia

(Sukhumi), Ajaria (Batumi)
note: the administrative centers of the autonomous republics are included in parentheses; there are no oblasts—the rayons around T'bilisi are under direct republic jurisdiction; also included is the South Ossetia Autonomous Oblast
Independence: 9 April 1991 (from Soviet Union)
Constitution: adopted NA 1921; currently amending constitution for Parliamentary and popular review by late 1995
Legal system: based on civil law system
National holiday: Independence Day, 9 April 1991
Political parties and leaders: All-Georgian Merab Kostava Society, Vazha ADAMIA, chairman; All-Georgian Traditionalists' Union, Akakiy ASATIANI, chairman; Georgian National Front—Radical Union, Ruslan GONGADZE, chairman; Georgian Social Democratic Party, Guram MUCHAIDZE, chairman; Green Party, Zurab ZHVANIA, chairman; Monarchist-Conservative Party (MCP), Temur ZHORZHOLIANI, chairman; Georgian Popular Front (GPF), Nodar NATADZE, chairman; National Democratic Party (NDP), Georgi CHANTURIA, chairman; National Independence Party (NIP), Irakli TSERETELI and Irakli BATIASHVILI, chairmen; Charter 1991 Party, Tedo PAATASHVILI, chairman; Democratic Georgia Party, Georgiy SHENGELAYA, Chairman; Peace Bloc; Unity; October 11
Other political or pressure groups: supporters of ousted President GAMSAKHURDIA boycotted the October elections and remain an important source of opposition and instability
Suffrage: 18 years of age; universal
Elections:
Chairman of Parliament: last held NA October 1992 (next to be held NA); results—Eduard SHEVARDNADZE 95%
Georgian Parliament (Supreme Soviet): last held 11 October 1992 (next to be held NA); results—percent of vote by party NA; seats—(225 total) number of seats by party NA; note—representatives of 26 parties elected; Peace Bloc, October 11, Unity, National Democratic Party, and the Greens Party won the largest representation
Executive branch: chairman of Parliament, Council of Ministers, prime minister
Legislative branch: unicameral Parliament
Judicial branch: Supreme Court
Leaders:
Chief of State: Chairman of Parliament Eduard Amvrosiyevich SHEVARDNADZE (since 10 March 1992)
Head of Government: Prime Minister Tengiz SIGUA (since NA January 1992); First Deputy Prime Minister Roman

GOTSIRIDZE (since NA); Deputy Prime Ministers Aleksandr KAVADZE, Avtandil MARGIANI, Zurab KERVALISHVILI (since NA)
Member of: BSEC, CSCE, EBRD, IBRD, IMF, NACC, UN, UNCTAD, UNESCO, WHO
Diplomatic representation in US:
chief of mission: NA
chancery: NA
telephone: NA
US diplomatic representation:
chief of mission: Ambassador Kent N. BROWN
embassy: #25 Antoneli Street, T'bilisi
mailing address: APO AE 09862
telephone: (7) 8832-74-46-23
Flag: maroon field with small rectangle in upper hoist side corner; rectangle divided horizontally with black on top, white below

Economy

Overview: Among the former Soviet republics, Georgia has been noted for its Black Sea tourist industry, its large output of citrus fruits and tea, and an industrial sector that accounted, however, for less than 2% of the USSR's output. Another salient characteristic of the economy has been a flourishing private sector (compared with the other republics). About 25% of the labor force is employed in agriculture. Mineral resources consist of manganese and copper, and, to a lesser extent, molybdenum, arsenic, tungsten, and mercury. Except for very small quantities of domestic oil, gas, and coal, fuel must be imported from neighboring republics. Oil and its products have been delivered by pipeline from Azerbaijan to the port of Batumi for export and local refining. Gas has been supplied in pipelines from Krasnodar and Stavropol'. The dismantling of central economic controls has been delayed by political factionalism, marked by bitter armed struggles. In early 1993 the Georgian economy was operating at well less than half capacity due to disruptions in fuel supplies and vital transportation links as a result of conflicts in Abkhazia and South Ossetia, antigovernment activity in Western Georgia, and Azerbaijani pressure against Georgian assistance for Armenia. To restore economic viability, Georgia must establish domestic peace and must maintain economic ties to the other former Soviet republics while developing new links to the West.
National product: GDP $NA
National product real growth rate: -35% (1992 est.)
National product per capita: $NA
Inflation rate (consumer prices): 50% per month (January 1993 est.)
Unemployment rate: 3% but large numbers of underemployed workers

Georgia *(continued)*

Budget: revenues $NA; expenditures $NA, including capital expenditures of $NA
Exports: $NA
commodities: citrus fruits, tea, other agricultural products; diverse types of machinery; ferrous and nonferrous metals; textiles
partners: Russia, Turkey, Armenia, Azerbaijan (1992)
Imports: $NA
commodities: machinery and parts, fuel, transport equipment, textiles
partners: Russia, Ukraine (1992)
External debt: $650 million (1991 est.)
Industrial production: growth rate −50% (1992)
Electricity: 4,875,000 kW capacity; 15,800 million kWh produced, about 2,835 kWh per capita (1992)
Industries: heavy industrial products include raw steel, rolled steel, cement, lumber; machine tools, foundry equipment, electric mining locomotives, tower cranes, electric welding equipment, machinery for food preparation, meat packing, dairy, and fishing industries; air-conditioning electric motors up to 100 kW in size, electric motors for cranes, magnetic starters for motors; devices for control of industrial processes; trucks, tractors, and other farm machinery; light industrial products, including cloth, hosiery, and shoes
Agriculture: accounted for 97% of former USSR citrus fruits and 93% of former USSR tea; berries and grapes; sugar; vegetables, grains, potatoes; cattle, pigs, sheep, goats, poultry; tobacco
Illicit drugs: illicit producers of cannabis and opium; mostly for domestic consumption; used as transshipment point for illicit drugs to Western Europe
Economic aid: NA
Currency: coupons introduced in April 1993 to be followed by introduction of the lari at undetermined future date; Russian ruble remains official currency until introduction of the lari
Exchange rates: rubles per US$1—415 (24 December 1992) but subject to wide fluctuations
Fiscal year: calendar year

Communications

Railroads: 1,570 km, does not include industrial lines (1990)
Highways: 33,900 km total; 29,500 km hard surfaced, 4,400 km earth (1990)
Pipelines: crude oil 370 km, refined products 300 km, natural gas 440 km (1992)
Ports: coastal—Batumi, Poti, Sukhumi
Merchant marine: 47 ships (1,000 GRT or over) totaling 658,192 GRT/1,014,056 DWT; includes 16 bulk cargo, 30 oil tanker, and 1 specialized liquid carrier

Airports:
total: 37
useable: 26
with permanent-surface runways: 19
with runways over 3,659 m: 0
with runways 2,440-3,659 m: 10
with runways 1,220-2,439 m: 9
Telecommunications: poor telephone service; as of 1991, 672,000 republic telephone lines providing 12 lines per 100 persons; 339,000 unsatisfied applications for telephones (31 January 1992); international links via landline to CIS members and Turkey; low capacity satellite earth station and leased international connections via the Moscow international gateway switch; international electronic mail and telex service established
Note: transportation network is disrupted by ethnic conflict, criminal activities, and fuel shortages

Defense Forces

Branches: Army, National Guard, Interior Ministry Troops
Manpower availability: males age 15-49 1,338,606; fit for military service 1,066,309; reach military age (18) annually 43,415 (1993 est.)
Defense expenditures: $NA, NA% of GNP
Note: Georgian forces are poorly organized and not fully under the government's control

Germany

Geography

Location: Western Europe, bordering the North Sea between France and Poland
Map references: Arctic Region, Europe, Standard Time Zones of the World
Area:
total area: 356,910 km²
land area: 349,520 km²
comparative area: slightly smaller than Montana
note: includes the formerly separate Federal Republic of Germany, the German Democratic Republic, and Berlin following formal unification on 3 October 1990
Land boundaries: total 3,621 km, Austria 784 km, Belgium 167 km, Czech Republic 646 km, Denmark 68 km, France 451 km, Luxembourg 138 km, Netherlands 577 km, Poland 456 km, Switzerland 334 km
Coastline: 2,389 km
Maritime claims:
continental shelf: 200 m depth or to depth of exploitation
exclusive fishing zone: 200 nm
territorial sea: 3 nm in North Sea and Schleswig-Holstein coast of Baltic Sea (extends, at one point, to 16 nm in the Helgolander Bucht); 12 nm in remainder of Baltic Sea
International disputes: none
Climate: temperate and marine; cool, cloudy, wet winters and summers; occasional warm, tropical foehn wind; high relative humidity
Terrain: lowlands in north, uplands in center, Bavarian Alps in south
Natural resources: iron ore, coal, potash, timber, lignite, uranium, copper, natural gas, salt, nickel
Land use:
arable land: 34%
permanent crops: 1%
meadows and pastures: 16%
forest and woodland: 30%
other: 19%
Irrigated land: 4,800 km² (1989 est.)

Environment: air and water pollution; groundwater, lakes, and air quality in eastern Germany are especially bad; significant deforestation in the eastern mountains caused by air pollution and acid rain
Note: strategic location on North European Plain and along the entrance to the Baltic Sea

People

Population: 80,767,591 (July 1993 est.)
Population growth rate: 0.4% (1993 est.)
Birth rate: 11 births/1,000 population (1993 est.)
Death rate: 11 deaths/1,000 population (1993 est.)
Net migration rate: 4 migrant(s)/1,000 population (1993 est.)
Infant mortality rate: 7 deaths/1,000 live births (1993 est.)
Life expectancy at birth:
total population: 76 years
male: 73 years
female: 79 years (1993 est.)
Total fertility rate: 1.4 children born/woman (1993 est.)
Nationality:
noun: German(s)
adjective: German
Ethnic divisions: German 95.1%, Turkish 2.3%, Italians 0.7%, Greeks 0.4%, Poles 0.4%, other 1.1% (made up largely of people fleeing the war in the former Yugoslavia)
Religions: Protestant 45%, Roman Catholic 37%, unaffiliated or other 18%
Languages: German
Literacy: age 15 and over can read and write (1977 est.)
total population: 99%
male: NA%
female: NA%
Labor force: 36.75 million
by occupation: industry 41%, agriculture 6%, other 53% (1987)

Government

Names:
conventional long form: Federal Republic of Germany
conventional short form: Germany
local long form: Bundesrepublik Deutschland
local short form: Deutschland
Digraph: GM
Type: federal republic
Capital: Berlin
note: the shift from Bonn to Berlin will take place over a period of years with Bonn retaining many administrative functions and several ministries
Administrative divisions: 16 states (laender, singular—land); Baden-Wuerttemberg, Bayern, Berlin, Brandenburg, Bremen, Hamburg, Hessen, Mecklenburg-Vorpommern, Niedersachsen, Nordrhein-Westfalen, Rheinland-Pfalz, Saarland, Sachsen, Sachsen-Anhalt, Schleswig-Holstein, Thuringen
Independence: 18 January 1871 (German Empire unification); divided into four zones of occupation (UK, US, USSR, and later, France) in 1945 following World War II; Federal Republic of Germany (FRG or West Germany) proclaimed 23 May 1949 and included the former UK, US, and French zones; German Democratic Republic (GDR or East Germany) proclaimed 7 October 1949 and included the former USSR zone; unification of West Germany and East Germany took place 3 October 1990; all four power rights formally relinquished 15 March 1991
Constitution: 23 May 1949, provisional constitution known as Basic Law
Legal system: civil law system with indigenous concepts; judicial review of legislative acts in the Federal Constitutional Court; has not accepted compulsory ICJ jurisdiction
National holiday: German Unity Day, 3 October (1990)
Political parties and leaders: Christian Democratic Union (CDU), Helmut KOHL, chairman; Christian Social Union (CSU), Theo WAIGEL, chairman; Free Democratic Party (FDP), Klaus KINKEL, chairman; Social Democratic Party (SPD); Green Party, Ludger VOLMER, Christine WEISKE, co-chairmen (after the 2 December 1990 election the East and West German Green Parties united); Alliance 90 united to form one party in September 1991, Petra MORAWE, chairwoman; Party of Democratic Socialism (PDS), Gregor GYSI, chairman; Republikaner, Franz SCHOENHUBER; National Democratic Party (NPD), Walter BACHMANN; Communist Party (DKP), Rolf PRIEMER
Other political or pressure groups: expellee, refugee, and veterans groups
Suffrage: 18 years of age; universal
Elections:
Federal Diet: last held 2 December 1990 (next to be held October 1994); results—CDU 36.7%, SPD 33.5%, FDP 11.0%, CSU 7.1%, Green Party (West Germany) 3.9%, PDS 2.4%, Republikaner 2.1%, Alliance 90/Green Party (East Germany) 1.2%, other 2.1%; seats—(662 total, 656 statutory with special rules to allow for slight expansion) CDU 268, SPD 239, FDP 79, CSU 51, PDS 17, Alliance 90/Green Party (East Germany) 8; note—special rules for this election allowed former East German parties to win seats if they received at least 5% of vote in eastern Germany

Executive branch: president, chancellor, Cabinet
Legislative branch: bicameral parliament (no official name for the two chambers as a whole) consists of an upper chamber or Federal Council (Bundesrat) and a lower chamber or Federal Diet (Bundestag)
Judicial branch: Federal Constitutional Court (Bundesverfassungsgericht)
Leaders:
Chief of State: President Dr. Richard von WEIZSACKER (since 1 July 1984)
Head of Government: Chancellor Dr. Helmut KOHL (since 4 October 1982)
Member of: AfDB, AG (observer), AsDB, Australian Group, BDEAC, BIS, CBSS, CCC, CDB (non-regional), CE, CERN, COCOM, CSCE, EBRD, EC, ECE, EIB, ESA, FAO, G-5, G-7, G-10, GATT, IADB, IAEA, IBRD, ICAO, ICC, ICFTU, IDA, IEA, IFAD, IFC, ILO, IMF, IMO, INMARSAT, INTELSAT, INTERPOL, IOC, IOM, ISO, ITU, LORCS, MTCR, NACC, NAM (guest), NATO, NEA, NSG, OAS (observer), OECD, PCA, UN, UNCTAD, UNESCO, UNIDO, UNHCR, UNTAC, UPU, WEU, WHO, WIPO, WMO, WTO, ZC
Diplomatic representation in US:
chief of mission: Ambassador Juergen RUHFUS
chancery: 4645 Reservoir Road NW, Washington, DC 20007
telephone: (202) 298-4000
consulates general: Atlanta, Boston, Chicago, Detroit, Houston, Los Angeles, Miami, New York, San Francisco, Seattle
consulates: Manila (Trust Territories of the Pacific Islands) and Wellington (America Samoa)
US diplomatic representation:
chief of mission: Ambassador Robert M. KIMMITT
embassy: Deichmanns Avenue, 5300 Bonn 2, Unit 21701
mailing address: APO AE 09080
telephone: [49] (228) 3391
FAX: [49] (228) 339-2663
branch office: Berlin
consulates general: Frankfurt, Hamburg, Leipzig, Munich, and Stuttgart
Flag: three equal horizontal bands of black (top), red, and yellow

Economy

Overview: With the collapse of communism in Eastern Europe in 1989, prospects seemed bright for a fairly rapid incorporation of East Germany into the highly successful West German economy. The Federal Republic, however, continues to experience difficulties in integrating and modernizing eastern Germany, and the tremendous costs of unification have sunk western Germany

Germany (continued)

deeper into recession. The western German economy grew by less than 1% in 1992 as the Bundesbank set high interest rates to offset the inflationary effects of large government deficits and high wage settlements. Eastern Germany grew by 6.8% in 1992 but this was from a shrunken base. Despite government transfers to the east amounting to nearly $110 billion annually, a self-sustaining economy in the region is still some years away. The bright spots are eastern Germany's construction, transportation, telecommunications, and service sectors, which have experienced strong growth. Western Germany has an advanced market economy and is a world leader in exports. It has a highly urbanized and skilled population that enjoys excellent living standards, abundant leisure time, and comprehensive social welfare benefits. Western Germany is relatively poor in natural resources, coal being the most important mineral. Western Germany's world-class companies manufacture technologically advanced goods. The region's economy is mature: services and manufacturing account for the dominant share of economic activity, and raw materials and semimanufactured goods constitute a large portion of imports. In recent years, manufacturing has accounted for about 31% of GDP, with other sectors contributing lesser amounts. Gross fixed investment in 1992 accounted for about 21.5% of GDP. GDP in the western region is now $20,000 per capita, or 85% of US per capita GDP. Eastern Germany's economy appears to be changing from one anchored on manufacturing into a more service-oriented economy. The German government, however, is intent on maintaining a manufacturing base in the east and is considering a policy for subsidizing industrial cores in the region. Eastern Germany's share of all-German GDP is only 7% and eastern productivity is just 30% that of the west even though eastern wages are at roughly 70% of western levels. The privatization agency for eastern Germany, Treuhand, has privatized more than four -fifths of the almost 12,000 firms under its control and will likely wind down operations in 1994. Private investment in the region continues to be lackluster, resulting primarily from the deepening recession in western Germany and excessively high eastern wages. Eastern Germany has one of the world's largest reserves of low-grade lignite coal but little else in the way of mineral resources. The quality of statistics from eastern Germany is improving, yet many gaps remain; the federal government began producing all-German data for select economic statistics at the start of 1992. The most challenging economic problem is

promoting eastern Germany's economic reconstruction—specifically, finding the right mix of fiscal, monetary, regulatory, and tax policies that will spur investment in eastern Germany—without destabilizing western Germany's economy or damaging relations with West European partners. The government hopes a "solidarity pact" among labor unions, business, state governments, and the SPD opposition will provide the right mix of wage restraints, investment incentives, and spending cuts to stimulate eastern recovery. Finally, the homogeneity of the German economic culture has been changed by the admission of large numbers of immigrants.

National product:
Germany: GDP—purchasing power equivalent—$1.398 trillion (1992)
western: GDP—purchasing power equivalent—$1.294 trillion (1992)
eastern: GDP—purchasing power equivalent—$104 billion (1992)

National product real growth rate:
Germany: 1.5% (1992)
western: 0.9% (1992)
eastern: 8% (1992)

National product per capita:
Germany: $17,400 (1992)
western: $20,000 (1992)
eastern: $6,500 (1992)

Inflation rate (consumer prices):
western: 4% (1992)
eastern: NA%

Unemployment rate:
western: 7.1% (1992)
eastern: 13.5% (December 1992)

Budget:
western (federal, state, local): revenues $684 billion; expenditures $704 billion, including capital expenditures $NA (1990)
eastern: revenues $NA; expenditures $NA, including capital expenditures of $NA

Exports: $378.0 billion (f.o.b., 1991)
commodities: manufactures 86.6% (including machines and machine tools, chemicals, motor vehicles, iron and steel products), agricultural products 4.9%, raw materials 2.3%, fuels 1.3%
partners: EC countries 54.3% (France 12.9%, Italy 9.3%, Netherlands 8.3%, UK 7.7%, Belgium-Luxembourg 7.4%), other Western Europe 17.0%, US 6.4%, Eastern Europe 5.6%, OPEC 3.4% (1992)

Imports: $354.5 billion (f.o.b., 1991)
commodities: manufactures 68.5%, agricultural products 12.0%, fuels 9.7%, raw materials 7.1%
partners: EC countries 52.0% (France 12.0%, Netherlands 9.6%, Italy 9.2%, Belgium-Luxembourg 7.0%, UK 6.8%), other Western Europe 15.2%, US 6.6%, Eastern Europe 5.5%, OPEC 2.4% (1992)

External debt: $NA

Industrial production:
western: growth rates −5% (1992 est.)
eastern: $NA

Electricity: 134,000,000 kW capacity; 580,000 million kWh produced, 7,160 kWh per capita (1992)

Industries:
western: among world's largest producers of iron, steel, coal, cement, chemicals, machinery, vehicles, machine tools, electronics; food and beverages
eastern: metal fabrication, chemicals, brown coal, shipbuilding, machine building, food and beverages, textiles, petroleum refining

Agriculture:
western: accounts for about 2% of GDP (including fishing and forestry); diversified crop and livestock farming; principal crops and livestock include potatoes, wheat, barley, sugar beets, fruit, cabbage, cattle, pigs, poultry; net importer of food; fish catch of 202,000 metric tons in 1987
eastern: accounts for about 10% of GDP (including fishing and forestry); principal crops—wheat, rye, barley, potatoes, sugar beets, fruit; livestock products include pork, beef, chicken, milk, hides and skins; net importer of food; fish catch of 193,600 metric tons in 1987

Illicit drugs: source of precursor chemicals for South American cocaine processors

Economic aid:
western: donor—ODA and OOF commitments (1970-89), $75.5 billion
eastern: donor—$4.0 billion extended bilaterally to non-Communist less developed countries (1956-89)

Currency: 1 deutsche mark (DM) = 100 pfennige

Exchange rates: deutsche marks (DM) per US$1—1.6158 (January 1993), 1.5617 (1992), 1.6595 (1991), 1.6157 (1990), 1.8800 (1989), 1.7562 (1988)

Fiscal year: calendar year

Communications

Railroads:
western: 31,443 km total; 27,421 km government owned, 1.435-meter standard gauge (12,491 km double track, 11,501 km electrified); 4,022 km nongovernment owned, including 3,598 km 1.435-meter standard gauge (214 km electrified) and 424 km 1.000-meter gauge (186 km electrified)
eastern: 14,025 km total; 13,750 km 1.435-meter standard gauge, 275 km 1.000-meter or other narrow gauge; 3,830 (est.) km 1.435-meter standard gauge double-track; 3,475 km overhead electrified (1988)

Highways:
western: 466,305 km total; 169,568 km primary, includes 6,435 km autobahn, 32,460

km national highways (Bundesstrassen), 65,425 km state highways (Landesstrassen), 65,248 km county roads (Kreisstrassen); 296,737 km of secondary communal roads (Gemeindestrassen)

eastern: 124,604 km total; 47,203 km concrete, asphalt, stone block, of which 1,855 km are autobahn and limited access roads, 11,326 km are trunk roads, and 34,022 km are regional roads; 77,401 km municipal roads (1988)

Inland waterways:

western: 5,222 km, of which almost 70% are usable by craft of 1,000-metric-ton capacity or larger; major rivers include the Rhine and Elbe; Kiel Canal is an important connection between the Baltic Sea and North Sea

eastern: 2,319 km (1988)

Pipelines: crude oil 3,644 km; petroleum products 3,946 km; natural gas 97,564 km (1988)

Ports: coastal—Bremerhaven, Brunsbuttel, Cuxhaven, Emden, Bremen, Hamburg, Kiel, Lubeck, Wilhelmshaven, Rostock, Wismar, Stralsund, Sassnitz; inland—31 major on Rhine and Elbe rivers

Merchant marine: 565 ships (1,000 GRT or over) totaling 4,928,759 GRT/6,292,193 DWT; includes 5 short-sea passenger, 3 passenger, 303 cargo, 10 refrigerated cargo, 134 container, 28 roll-on/roll-off cargo, 5 railcar carrier, 7 barge carrier, 9 oil tanker, 21 chemical tanker, 17 liquefied gas tanker, 5 combination ore/oil, 6 combination bulk, 12 bulk; note—the German register includes ships of the former East and West Germany; during 1991 the fleet underwent major restructuring as surplus ships were sold off

Airports:

total: 499

usable: 492

with permanent-surface runways: 271

with runways over 3,659 m: 5

with runways 2,440-3,659 m: 59

with runways 1,220-2,439 m: 67

Telecommunications:

western: highly developed, modern telecommunication service to all parts of the country; fully adequate in all respects; 40,300,000 telephones; intensively developed, highly redundant cable and microwave radio relay networks, all completely automatic; broadcast stations—80 AM, 470 FM, 225 (6,000 repeaters) TV; 6 submarine coaxial cables; satellite earth stations—12 Atlantic Ocean INTELSAT antennas, 2 Indian Ocean INTELSAT antennas, EUTELSAT, and domestic systems; 2 HF radiocommunication centers; tropospheric links

eastern: badly needs modernization; 3,970,000 telephones; broadcast stations—23 AM, 17 FM, 21 TV (15 Soviet TV repeaters); 6,181,860 TVs; 6,700,000 radios; 1 satellite earth station operating in INTELSAT and Intersputnik systems

Defense Forces

Branches: Army, Navy, Air Force

Manpower availability: males age 15-49 20,295,655; fit for military service 17,577,570; reach military age (18) annually 411,854 (1993 est.)

Defense expenditures: exchange rate conversion—$42.4 billion, 2.2% of GDP (1992)

Ghana

Geography

Location: Western Africa, bordering the North Atlantic Ocean between Cote d'Ivoire and Togo

Map references: Africa, Standard Time Zones of the World

Area:

total area: 238,540 km²

land area: 230,020 km²

comparative area: slightly smaller than Oregon

Land boundaries: total 2,093 km, Burkina 548 km, Cote d'Ivoire 668 km, Togo 877 km

Coastline: 539 km

Maritime claims:

contiguous zone: 24 nm

continental shelf: 200 nm

exclusive economic zone: 200 nm

territorial sea: 12 nm

International disputes: none

Climate: tropical; warm and comparatively dry along southeast coast; hot and humid in southwest; hot and dry in north

Terrain: mostly low plains with dissected plateau in south-central area

Natural resources: gold, timber, industrial diamonds, bauxite, manganese, fish, rubber

Land use:

arable land: 5%

permanent crops: 7%

meadows and pastures: 15%

forest and woodland: 37%

other: 36%

Irrigated land: 80 km² (1989)

Environment: recent drought in north severely affecting marginal agricultural activities; deforestation; overgrazing; soil erosion; dry, northeasterly harmattan wind (January to March)

Note: Lake Volta is the world's largest artificial lake

People

Population: 16,699,105 (July 1993 est.)

Population growth rate: 3.12% (1993 est.)

Birth rate: 44.66 births/1,000 population (1993 est.)

Death rate: 12.52 deaths/1,000 population (1993 est.)

Net migration rate: -1 migrant(s)/1,000 population (1993 est.)

Infant mortality rate: 84.5 deaths/1,000 live births (1993 est.)

Life expectancy at birth:
total population: 55.19 years
male: 53.27 years
female: 57.17 years (1993 est.)

Total fertility rate: 6.21 children born/woman (1993 est.)

Nationality:
noun: Ghanaian(s)
adjective: Ghanaian

Ethnic divisions: black African 99.8% (major tribes—Akan 44%, Moshi-Dagomba 16%, Ewe 13%, Ga 8%), European and other 0.2%

Religions: indigenous beliefs 38%, Muslim 30%, Christian 24%, other 8%

Languages: English (official), African languages (including Akan, Moshi-Dagomba, Ewe, and Ga)

Literacy: age 15 and over can read and write (1990)
total population: 60%
male: 70%
female: 51%

Labor force: 3.7 million
by occupation: agriculture and fishing 54.7%, industry 18.7%, sales and clerical 15.2%, services, transportation, and communications 7.7%, professional 3.7%
note: 48% of population of working age (1983)

Government

Names:
conventional long form: Republic of Ghana
conventional short form: Ghana
former: Gold Coast

Digraph: GH

Type: constitutional democracy

Capital: Accra

Administrative divisions: 10 regions; Ashanti, Brong-Ahafo, Central, Eastern, Greater Accra, Northern, Upper East, Upper West, Volta, Western

Independence: 6 March 1957 (from UK)

Constitution: new constitution approved 28 April 1992

Legal system: based on English common law and customary law; has not accepted compulsory ICJ jurisdiction

National holiday: Independence Day, 6 March (1957)

Political parties and leaders: National Democratic Congress, Jerry John Rawlings; New Patriotic Party, Albert Adu BOAHEN; People's Heritage Party, Alex Erskine; various other smaller parties

Suffrage: universal at 18

Elections:
President: last held 3 November 1992 (next to be held NA)
National Assembly: last held 29 December 1992 (next to be held NA)

Executive branch: president, cabinet

Legislative branch: unicameral National Assembly

Judicial branch: Supreme Court

Leaders:
Chief of State and Head of Government: President Jerry John RAWLINGS (since 3 November 1992)

Member of: ACP, AfDB, C, CCC, ECA, ECOWAS, FAO, G-24, G-77, GATT, IAEA, IBRD, ICAO, IDA, IFAD, IFC, ILO, IMF, IMO, INTELSAT, INTERPOL, IOC, IOM (observer), ISO, ITU, LORCS, MINURSO, NAM, OAU, UN, UNCTAD, UNESCO, UNIDO, UNIFIL, UNIKOM, UNPROFOR, UNTAC, UPU, WCL, WHO, WIPO, WMO, WTO

Diplomatic representation in US:
chief of mission: Ambassador Dr. Joseph ABBEY
chancery: 3512 International Drive NW, Washington, DC 20008
telephone: (202) 686-4520
consulate general: New York

US diplomatic representation:
chief of mission: Ambassador Kenneth L. BROWN
embassy: Ring Road East, East of Danquah Circle, Accra
mailing address: P. O. Box 194, Accra
telephone: [233] (21) 775348, 775349, 775295 or 775298
FAX: [233] (21) 776008

Flag: three equal horizontal bands of red (top), yellow, and green with a large black five-pointed star centered in the gold band; uses the popular pan-African colors of Ethiopia; similar to the flag of Bolivia, which has a coat of arms centered in the yellow band

---

Economy

Overview: Supported by substantial international assistance, Ghana has been implementing a steady economic rebuilding program since 1983, including moves toward privatization and relaxation of government controls. Heavily dependent on cocoa, gold, and timber exports, economic growth so far has not spread substantially to other areas of the economy. The costs of sending peacekeeping forces to Liberia and preparing for the transition to a democratic government have boosted government expenditures and undercut structural adjustment reforms. Ghana opened a stock exchange in 1990. Meanwhile, declining world commodity prices for Ghana's exports has placed the government under severe financial pressure.

National product: GDP—exchange rate conversion—$6.6 billion (1992 est.)

National product real growth rate: 3.9% (1992 est.)

National product per capita: $410 (1992 est.)

Inflation rate (consumer prices): 10% (1992 est.)

Unemployment rate: 10% (1991)

Budget: revenues $1.0 billion; expenditures $905 million, including capital expenditures of $200 million (1991 est.)

Exports: $1.1 billion (f.o.b., 1992)
commodities: cocoa 45%, gold, timber, tuna, bauxite, and aluminum
partners: Germany 29%, UK 12%, US 12%, Japan 5%

Imports: $1.4 billion (c.i.f., 1992 est.)
commodities: petroleum 16%, consumer goods, foods, intermediate goods, capital equipment
partners: UK 23%, US 11%, Germany 10%, Japan 6%

External debt: $4.6 billion (1992 est.)

Industrial production: growth rate 4.6% in manufacturing (1991); accounts for almost 15% of GDP

Electricity: 1,180,000 kW capacity; 4,490 million kWh produced, 290 kWh per capita (1991)

Industries: mining, lumbering, light manufacturing, aluminum, food processing

Agriculture: accounts for about 50% of GDP (including fishing and forestry); the major cash crop is cocoa; other principal crops—rice, coffee, cassava, peanuts, corn, shea nuts, timber; normally self-sufficient in food

Illicit drugs: illicit producer of cannabis for the international drug trade

Economic aid: US commitments, including Ex-Im (FY70-89), $455 million; Western (non-US) countries, ODA and OOF bilateral commitments (1970-89), $2.6 billion; OPEC bilateral aid (1979-89), $78 million; Communist countries (1970-89), $106 million

Currency: 1 cedi (C) = 100 pesewas

Exchange rates: ceolis per US$1—437 (July 1992)

Fiscal year: calendar year

Communications

Railroads: 953 km, all 1.067-meter gauge; 32 km double track; railroads undergoing major renovation

Highways: 32,250 km total; 6,084 km concrete or bituminous surface, 26,166 km gravel, laterite, and improved earth surfaces

Inland waterways: Volta, Ankobra, and Tano Rivers provide 168 km of perennial navigation for launches and lighters; Lake

Volta provides 1,125 km of arterial and feeder waterways
Pipelines: none
Ports: Tema, Takoradi
Merchant marine: 6 ships (1,000 GRT or over) totaling 59,293 GRT/78,246 DWT; includes 5 cargo, 1 refrigerated cargo
Airports:
total: 10
usable: 9
with permanent-surface runways: 5
with runways over 3,659 m: 0
with runways 2,440-3,659 m: 2
with runways 1,220-2,439 m: 6
Telecommunications: poor to fair system handled primarily by microwave radio relay links; 42,300 telephones; broadcast stations—4 AM, 1 FM, 4 (8 translators) TV; 1 Atlantic Ocean INTELSAT earth station

---

Defense Forces

Branches: Army, Navy, Air Force, Police Force, Civil Defense
Manpower availability: males age 15-49 3,766,073; fit for military service 2,105,865; reach military age (18) annually 171,145 (1993 est.)
Defense expenditures: exchange rate conversion—$30 million, less than 1% of GDP (1989 est.)

Gibraltar
(dependent territory of the UK)

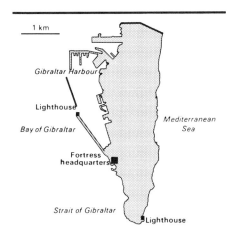

Geography

Location: Southwestern Europe, bordering the Strait of Gibraltar, which links the North Atlantic Ocean and the Mediterranean Sea, on the southern coast of Spain
Map references: Africa, Europe
Area:
total area: 6.5 km²
land area: 6.5 km²
comparative area: about 11 times the size of the Mall in Washington, DC
Land boundaries: total 1.2 km, Spain 1.2 km
Coastline: 12 km
Maritime claims:
exclusive fishing zone: 3 nm
territorial sea: 3 nm
International disputes: source of occasional friction between Spain and the UK
Climate: Mediterranean with mild winters and warm summers
Terrain: a narrow coastal lowland borders The Rock
Natural resources: negligible
Land use:
arable land: 0%
permanent crops: 0%
meadows and pastures: 0%
forest and woodland: 0%
other: 100%
Irrigated land: NA km²
Environment: natural freshwater sources are meager, so large water catchments (concrete or natural rock) collect rain water
Note: strategic location on Strait of Gibraltar that links the North Atlantic Ocean and Mediterranean Sea

---

People

Population: 31,508 (July 1993 est.)
Population growth rate: 0.53% (1993 est.)
Birth rate: 15.68 births/1,000 population (1993 est.)
Death rate: 8.89 deaths/1,000 population (1993 est.)

Net migration rate: -1.46 migrant(s)/1,000 population (1993 est.)
Infant mortality rate: 8.4 deaths/1,000 live births (1993 est.)
Life expectancy at birth:
total population: 76.06 years
male: 73.18 years
female: 78.91 years (1993 est.)
Total fertility rate: 2.37 children born/woman (1993 est.)
Nationality:
noun: Gibraltarian(s)
adjective: Gibraltar
Ethnic divisions: Italian, English, Maltese, Portuguese, Spanish
Religions: Roman Catholic 74%, Protestant 11% (Church of England 8%, other 3%), Moslem 8%, Jewish 2%, none or other 5% (1981)
Languages: English (used in schools and for official purposes), Spanish, Italian, Portuguese, Russian
Literacy:
total population: NA%
male: NA%
female: NA%
Labor force: 14,800 (including non-Gibraltar laborers)
note: UK military establishments and civil government employ nearly 50% of the labor force

---

Government

Names:
conventional long form: none
conventional short form: Gibraltar
Digraph: GI
Type: dependent territory of the UK
Capital: Gilbraltar
Administrative divisions: none (dependent territory of the UK)
Independence: none (dependent territory of the UK)
Constitution: 30 May 1969
Legal system: English law
National holiday: Commonwealth Day (second Monday of March)
Political parties and leaders: Socialist Labor Party (SL), Joe BOSSANO; Gibraltar Labor Party/Association for the Advancement of Civil Rights (GCL/AACR), leader NA; Gibraltar Social Democrats, Peter CARUANA; Gibraltar National Party, Joe GARCIA
Other political or pressure groups: Housewives Association; Chamber of Commerce; Gibraltar Representatives Organization
Suffrage: 18 years of age; universal, plus other UK subjects resident six months or more
Elections:
House of Assembly: last held on 16 January 1992 (next to be held January 1996);

Gibraltar (continued)

results—SL 73.3%; seats—(18 total, 15 elected) number of seats by party NA
Executive branch: British monarch, governor, chief minister, Gibraltar Council, Council of Ministers (cabinet)
Legislative branch: unicameral House of Assembly
Judicial branch: Supreme Court, Court of Appeal
Leaders:
Chief of State: Queen ELIZABETH II (since 6 February 1952), represented by Governor and Commander in Chief Adm. Sir Derek REFFELL (since NA 1989)
Head of Government: Chief Minister Joe BOSSANO (since 25 March 1988)
Member of: INTERPOL (subbureau)
Diplomatic representation in US: none (dependent territory of the UK)
US diplomatic representation: none (dependent territory of the UK)
Flag: two horizontal bands of white (top, double width) and red with a three-towered red castle in the center of the white band; hanging from the castle gate is a gold key centered in the red band

Economy

Overview: The economy depends heavily on British defense expenditures, revenue from tourists, fees for services to shipping, and revenues from banking and finance activities. Because more than 70% of the economy is in the public sector, changes in government spending have a major impact on the level of employment. Construction workers are particularly affected when government expenditures are cut.
National product: GNP—exchange rate conversion—$182 million (FY87)
National product real growth rate: 5% (FY87)
National product per capita: $4,600 (FY87)
Inflation rate (consumer prices): 3.6% (1988)
Unemployment rate: NA%
Budget: revenues $136 million; expenditures $139 million, including capital expenditures of $NA (FY88)
Exports: $82 million (f.o.b., 1988)
commodities: (principally reexports) petroleum 51%, manufactured goods 41%, other 8%
partners: UK, Morocco, Portugal, Netherlands, Spain, US, FRG
Imports: $258 million (c.i.f., 1988)
commodities: fuels, manufactured goods, and foodstuffs
partners: UK, Spain, Japan, Netherlands
External debt: $318 million (1987)
Industrial production: growth rate NA%
Electricity: 47,000 kW capacity; 200 million kWh produced, 6,740 kWh per capita (1992)

Industries: tourism, banking and finance, construction, commerce; support to large UK naval and air bases; transit trade and supply depot in the port; light manufacturing of tobacco, roasted coffee, ice, mineral waters, candy, beer, and canned fish
Agriculture: none
Economic aid: US commitments, including Ex-Im (FY70-88), $0.8 million; Western (non-US) countries, ODA and OOF bilateral commitments (1970-89), $188 million
Currency: 1 Gibraltar pound (£G) = 100 pence
Exchange rates: Gibraltar pounds (£G) per US$1—0.6527 (January 1993), 0.5664 (1992), 0.5652 (1991), 0.5603 (1990), 0.6099 (1989), 0.5614 (1988); note—the Gibraltar pound is at par with the British pound
Fiscal year: 1 July-30 June

Communications

Railroads: 1.000-meter-gauge system in dockyard area only
Highways: 50 km, mostly good bitumen and concrete
Pipelines: none
Ports: Gibraltar
Merchant marine: 32 ships (1,000 GRT or over) totaling 642,446 GRT/1,141,592 DWT; includes 4 cargo, 2 refrigerated cargo, 1 container, 18 oil tanker, 2 chemical tanker, 5 bulk; note—a flag of convenience registry
Airports:
total: 1
useable: 1
with permanent surface runways: 1
with runways over 3,659 m: 0
with runways 2,440-3,659 m: 0
with runways 1,220-2,439 m: 1
Telecommunications: adequate, automatic domestic system and adequate international radiocommunication and microwave facilities; 9,400 telephones; broadcast stations—1 AM, 6 FM, 4 TV; 1 Atlantic Ocean INTELSAT earth station

Defense Forces

Branches: British Army, Royal Navy, Royal Air Force
Note: defense is the responsibility of the UK

Glorioso Islands
(possession of France)

Geography

Location: Southern Africa, in the Indian Ocean just north of Madagascar
Map references: Africa
Area:
total area: 5 km²
land area: 5 km²
comparative area: about 8.5 times the size of the Mall in Washington, DC
note: includes Ile Glorieuse, Ile du Lys, Verte Rocks, Wreck Rock, and South Rock
Land boundaries: 0 km
Coastline: 35.2 km
Maritime claims:
contiguous zone: 12 nm
exclusive economic zone: 200 nm
territorial sea: 12 nm
International disputes: claimed by Madagascar
Climate: tropical
Terrain: NA
Natural resources: guano, coconuts
Land use:
arable land: 0%
permanent crops: 0%
meadows and pastures: 0%
forest and woodland: 0%
other: 100% (all lush vegetation and coconut palms)
Irrigated land: 0 km²
Environment: subject to periodic cyclones

People

Population: unihabited

Government

Names:
conventional long form: none
conventional short form: Glorioso Islands
local long form: none
local short form: Iles Glorieuses
Digraph: GO

Greece

Type: French possession administered by Commissioner of the Republic, resident in Reunion
Capital: none; administered by France from Reunion
Independence: none (possession of France)

Economy

Overview: no economic activity

Communications

Ports: none; offshore anchorage only
Airports:
total: 1
usable: 1
with permanent-surface runways: 0
with runsways over 3,6359 m: 0
with runways 2,440-3,659 m: 0
with runways 1,220-2,439 m: 1

Defense Forces

Note: defense is the responsibility of France

Geography

Location: Southern Europe, bordering the Mediterranean Sea between Turkey and Bulgaria
Map references: Africa, Europe, Standard Time Zones of the World
Area:
total area: 131,940 km²
land area: 130,800 km²
comparative area: slightly smaller than Alabama
Land boundaries: total 1,210 km, Albania 282 km, Bulgaria 494 km, Turkey 206 km, Macedonia 228 km
Coastline: 13,676 km
Maritime claims:
continental shelf: 200 m depth or to depth of exploitation
territorial sea: 6 nm, but Greece has threatened to claim 12 nm
International disputes: air, continental shelf, and territorial water disputes with Turkey in Aegean Sea; Cyprus question; northern Epirus question with Albania; Macedonia question with Bulgaria and Macedonia
Climate: temperate; mild, wet winters; hot, dry summers
Terrain: mostly mountains with ranges extending into sea as peninsulas or chains of islands
Natural resources: bauxite, lignite, magnesite, petroleum, marble
Land use:
arable land: 23%
permanent crops: 8%
meadows and pastures: 40%
forest and woodland: 20%
other: 9%
Irrigated land: 11,900 km² (1989 est.)
Environment: subject to severe earthquakes; air pollution
Note: strategic location dominating the Aegean Sea and southern approach to Turkish Straits; a peninsular country, possessing an archipelago of about 2,000 islands

People

Population: 10,470,460 (July 1993 est.)
Population growth rate: 0.95% (1993 est.)
Birth rate: 10.42 births/1,000 population (1993 est.)
Death rate: 9.36 deaths/1,000 population (1993 est.)
Net migration rate: 8.46 migrant(s)/1,000 population (1993 est.)
Infant mortality rate: 8.9 deaths/1,000 live births (1993 est.)
Life expectancy at birth:
total population: 77.5 years
male: 75.02 years
female: 80.12 years (1993 est.)
Total fertility rate: 1.44 children born/woman (1993 est.)
Nationality:
noun: Greek(s)
adjective: Greek
Ethnic divisions: Greek 98%, other 2%
note: the Greek Government states there are no ethnic divisions in Greece
Religions: Greek Orthodox 98%, Muslim 1.3%, other 0.7%
Languages: Greek (official), English, French
Literacy: age 15 and over can read and write (1990)
total population: 93%
male: 98%
female: 89%
Labor force: 3,966,900
by occupation: services 45%, agriculture 27%, industry 28% (1990)

Government

Names:
conventional long form: Hellenic Republic
conventional short form: Greece
local long form: Elliniki Dhimokratia
local short form: Ellas
former: Kingdom of Greece
Digraph: GR
Type: presidential parliamentary government; monarchy rejected by referendum 8 December 1974
Capital: Athens
Administrative divisions: 52 prefectures (nomoi, singular—nomos); Aitolia kai Akarnania, Akhaia, Argolis, Arkadhia, Arta, Attiki, Dhodhekanisos, Dhrama, Evritania, Evros, Evvoia, Florina, Fokis, Fthiotis, Grevena, Ilia, Imathia, Ioannina, Iraklion, Kardhitsa, Kastoria, Kavala, Kefallinia, Kerkira, Khalkidhiki, Khania, Khios, Kikladhes, Kilkis, Korinthia, Kozani, Lakonia, Larisa, Lasithi, Lesvos, Levkas, Magnisia, Messinia, Pella, Pieria, Piraievs, Preveza, Rethimni, Rodhopi, Samos, Serrai, Thesprotia, Thessaloniki, Trikala, Voiotia, Xanthi, Zakinthos, autonomous region: Agion Oros (Mt. Athos)

Greece (continued)

Independence: 1829 (from the Ottoman Empire)

Constitution: 11 June 1975

Legal system: based on codified Roman law; judiciary divided into civil, criminal, and administrative courts

National holiday: Independence Day, 25 March (1821) (proclamation of the war of independence)

Political parties and leaders: New Democracy (ND; conservative), Konstantinos MITSOTAKIS; Panhellenic Socialist Movement (PASOK), Andreas PAPANDREOU; Left Alliance, Maria DAMANAKI; Democratic Renewal (DIANA), Konstantinos STEFANOPOULOS; Communist Party (KKE), Aleka PAPARIGA; Ecologist-Alternative List, leader rotates

Suffrage: 18 years of age; universal and compulsory

Elections:

President: last held 4 May 1990 (next to be held May 1995); results—Konstantinos KARAMANLIS was elected by Parliament

Chamber of Deputies: last held 8 April 1990 (next must be held by May 1994); results—ND 46.89%, PASOK 38.62%, Left Alliance 10.27%, PASOK/Left Alliance 1.02%, Ecologist-Alternative List 0.77%, DIANA 0.67%, Muslim independents 0.5%; seats—(300 total) ND 150, PASOK 123, Left Alliance 19, PASOK-Left Alliance 4, Muslim independents 2, DEANA 1, Ecologist-Alternative List 1

note: deputies shifting from one party to another and the dissolution of party coalitions have resulted in the following seating arrangement: ND 152, PASOK 124, Left Alliance 14, KKE 7, Muslim deputies 2, Ecologist-Alternative List 1

Executive branch: president, prime minister, Cabinet

Legislative branch: unicameral Greek Chamber of Deputies (Vouli ton Ellinon)

Judicial branch: Supreme Judicial Court, Special Supreme Tribunal

Leaders:

Chief of State: President Konstantinos KARAMANLIS (since 5 May 1990)

Head of Government: Prime Minister Konstantinos MITSOTAKIS (since 11 April 1990)

Member of: Australian Group, BIS, BSEC, CCC, CE, CERN, COCOM, CSCE, EBRD, EC, ECE, EIB, FAO, G-6, GATT, IAEA, IBRD, ICAO, ICC, ICFTU, IDA, IEA, IFAD, IFC, ILO, IMF, IMO, INMARSAT, INTELSAT, INTERPOL, IOC, IOM, ISO, ITU, LORCS, MINURSO, MTCR, NACC, NAM (guest), NATO, NEA, NSG, OAS (observer), OECD, PCA, UN, UNCTAD, UNESCO, UNHCR, UNIDO, UNIKOM, UPU, WEU (observer), WHO, WIPO, WMO, WTO, ZC

Diplomatic representation in US:

chief of mission: Ambassador Christos ZACHARAKIS

chancery: 2221 Massachusetts Avenue NW, Washington, DC 20008

telephone: (202) 939-5800

FAX: (202) 939-5824

consulates general: Atlanta, Boston, Chicago, Los Angeles, New York, and San Francisco

consulate: New Orleans

US diplomatic representation:

chief of mission: (vacant); Charge d'Affaires James A. WILLIAMS

embassy: 91 Vasilissis Sophias Boulevard, 10160 Athens

mailing address: PSC 108, Box 56, APO AE 09842

telephone: [30] (1) 721-2951 or 721-8401

FAX: [30] (1) 645-6282

consulate general: Thessaloniki

Flag: nine equal horizontal stripes of blue alternating with white; there is a blue square in the upper hoist-side corner bearing a white cross; the cross symbolizes Greek Orthodoxy, the established religion of the country

---

Economy

Overview: Greece has a mixed capitalist economy with the basic entrepreneurial system overlaid in 1981-89 by a socialist system that enlarged the public sector from 55% of GDP in 1981 to about 70% when Prime Minister MITSOTAKIS took office. Tourism continues as a major source of foreign exchange, and agriculture is self-sufficient except for meat, dairy products, and animal feedstuffs. Since 1986, real GDP growth has averaged only 1.6% a year, compared with the Europen Community average of 3%. The MITSOTAKIS government has made little progress during its two and one-half years in power in coming to grips with Greece's main economic problems: an inflation rate still four times the EC average, a large public sector deficit, and a fragile current account position. In early 1991, the government secured a three-year, $2.5 billion assistance package from the EC under the strictest terms yet imposed on a member country, as the EC finally ran out of patience with Greece's failure to put its financial affairs in order. On the advice of the EC Commission, Greece delayed applying for the second installment until 1993 because of the failure of the government to meet the 1992 targets. Although MITSOTAKIS faced down the unions in mid-1992 in a dispute over privatization plans, social security reform, and tax and price increases, and his new economics czar, Stephanos MANOS, is a respected economist committed to renovating the ailing economy. However, a national elections due by May 1994 will probably prompt MITSOTAKIS to backtrack on economic reform. In 1993, the GDP growth rate likely will remain low; the inflation rate probably will continue to fall, while remaining the highest in the EC.

National product: GDP—purchasing power equivalent—$82.9 billion (1992)

National product real growth rate: 1.2% (1992)

National product per capita: $8,200 (1992)

Inflation rate (consumer prices): 15.6% (1992)

Unemployment rate: 9.1% (1992)

Budget: revenues $37.6 billion; expenditures $45.1 billion, including capital expenditures of $5.4 billion (1993)

Exports: $6.8 billion (f.o.b., 1991)

commodities: manufactured goods 53%, foodstuffs 31%, fuels 9%

partners: Germany 24%, France 18%, Italy 17%, UK 7%, US 6%

Imports: $21.5 billion (c.i.f., 1991)

commodities: manufactured goods 71%, foodstuffs 14%, fuels 10%

partners: Germany 20%, Italy 14%, France 8%, UK 5%, US 4%

External debt: $23.7 billion (1991)

Industrial production: growth rate −1.0% (1991); accounts for 20% of GDP

Electricity: 10,500,000 kW capacity; 36,400 million kWh produced, 3,610 kWh per capita (1992)

Industries: food and tobacco processing, textiles, chemicals, metal products, tourism, mining, petroleum

Agriculture: including fishing and forestry, accounts for 15% of GDP and 27% of the labor force; principal products—wheat, corn, barley, sugar beets, olives, tomatoes, wine, tobacco, potatoes; self-sufficient in food except meat, dairy products, and animal feedstuffs; fish catch of 116,600 metric tons in 1988

Illicit drugs: illicit producer of cannabis and limited opium; mostly for domestic production; serves as a gateway to Europe for traffickers smuggling cannabis and heroin from the Middle East and Southwest Asia to the West and precursor chemicals to the East; transshipment point for Southwest Asian heroin transiting the Balkan route

Economic aid: US commitments, including Ex-Im (FY70-81), $525 million; Western (non-US) countries, ODA and OOF bilateral commitments (1970-89), $1,390 million

Currency: 1 drachma (Dr) = 100 lepta

Exchange rates: drachma (Dr) per US$1—215.82 (January 1993), 190.62 (1992), 182.27 (1991), 158.51 (1990), 162.42 (1989), 141.86 (1988)

Fiscal year: calendar year

Greenland
(part of the Danish realm)

Communications

Railroads: 2,479 km total; 1,565 km 1.435-meter standard gauge, of which 36 km electrified and 100 km double track; 892 km 1.000-meter gauge; 22 km 0.750-meter narrow gauge; all government owned
Highways: 38,938 km total; 16,090 km paved, 13,676 km crushed stone and gravel, 5,632 km improved earth, 3,540 km unimproved earth
Inland waterways: 80 km; system consists of three coastal canals; including the Corinth Canal (6 km) which crosses the Isthmus of Corinth connecting the Gulf of Corinth with the Saronic Gulf and shortens the sea voyage from the Adriatic to Piraievs (Piraeus) by 325 km; and three unconnected rivers
Pipelines: crude oil 26 km; petroleum products 547 km
Ports: Piraievs (Piraeus), Thessaloniki
Merchant marine: 998 ships (1,000 GRT or over) totaling 25,483,768 GRT/47,047,285 DWT; includes 14 passenger, 66 short-sea passenger, 2 passenger-cargo, 128 cargo, 26 container, 15 roll-on/roll-off cargo, 14 refrigerated cargo, 1 vehicle carrier, 214 oil tanker, 19 chemical tanker, 7 liquefied gas, 42 combination ore/oil, 3 specialized tanker, 424 bulk, 22 combination bulk, 1 livestock carrier; note—ethnic Greeks also own large numbers of ships under the registry of Liberia, Panama, Cyprus, Malta, and The Bahamas
Airports:
total: 78
usable: 77
with permanent-surface runways: 63
with runways over 3,659 m: 0
with runways 2,440-3,659 m: 20
with runways 1,220-2,439 m: 24
Telecommunications: adequate, modern networks reach all areas; 4,080,000 telephones; microwave radio relay carries most traffic; extensive open-wire network; submarine cables to off-shore islands; broadcast stations—29 AM, 17 (20 repeaters) FM, 361 TV; tropospheric links, 8 submarine cables; 1 satellite earth station operating in INTELSAT (1 Atlantic Ocean and 1 Indian Ocean antenna), and EUTELSAT systems

Defense Forces

Branches: Hellenic Army, Hellenic Navy, Hellenic Air Force, National Guard, Police
Manpower availability: males age 15-49 2,606,267; fit for military service 1,996,835; reach military age (21) annually 73,541 (1993 est.)
Defense expenditures: exchange rate conversion—$4.2 billion, 5.1% of GDP (1992)

Geography

Location: in the North Atlantic Ocean, between Canada and Norway
Map references: Arctic Region, North America, Standard Time Zones of the World
Area:
total area: 2,175,600 km²
land area: 341,700 km² (ice free)
comparative area: slightly more than three times the size of Texas
Land boundaries: 0 km
Coastline: 44,087 km
Maritime claims:
exclusive fishing zone: 200 nm
territorial sea: 3 nm
International disputes: Denmark has challenged Norway's maritime claims between Greenland and Jan Mayen
Climate: arctic to subarctic; cool summers, cold winters
Terrain: flat to gradually sloping icecap covers all but a narrow, mountainous, barren, rocky coast
Natural resources: zinc, lead, iron ore, coal, molybdenum, cryolite, uranium, fish
Land use:
arable land: 0%
permanent crops: 0%
meadows and pastures: 1%
forest and woodland: 0%
other: 99%
Irrigated land: NA km²
Environment: sparse population confined to small settlements along coast; continuous permafrost over northern two-thirds of the island
Note: dominates North Atlantic Ocean between North America and Europe

People

Population: 56,533 (July 1993 est.)
Population growth rate: 0.84% (1993 est.)
Birth rate: 19.62 births/1,000 population (1993 est.)
Death rate: 7.66 deaths/1,000 population (1993 est.)

Net migration rate: -3.54 migrant(s)/1,000 population (1993 est.)
Infant mortality rate: 28.4 deaths/1,000 live births (1993 est.)
Life expectancy at birth:
total population: 66.19 years
male: 61.79 years
female: 70.6 years (1993 est.)
Total fertility rate: 2.33 children born/woman (1993 est.)
Nationality:
noun: Greenlander(s)
adjective: Greenlandic
Ethnic divisions: Greenlander 86% (Eskimos and Greenland-born Caucasians), Danish 14%
Religions: Evangelical Lutheran
Languages: Eskimo dialects, Danish
Literacy:
total population: NA%
male: NA%
female: NA%
Labor force: 22,800
by occupation: largely engaged in fishing, hunting, sheep breeding

Government

Names:
conventional long form: none
conventional short form: Greenland
local long form: none
local short form: Kalaallit Nunaat
Digraph: GL
Type: part of the Danish realm; self-governing overseas administrative division
Capital: Nuuk (Godthab)
Administrative divisions: 3 municipalities (kommuner, singular—kommun); Nordgronland, Ostgronland, Vestgronland
Independence: none (part of the Danish realm; self-governing overseas administrative division)
Constitution: Danish
Legal system: Danish
National holiday: Birthday of the Queen, 16 April (1940)
Political parties and leaders: two-party ruling coalition; Siumut (a moderate socialist party that advocates more distinct Greenlandic identity and greater autonomy from Denmark), Lars Emil JOHANSEN, chairman; Inuit Ataqatigiit (IA; a Marxist-Leninist party that favors complete independence from Denmark rather than home rule), Arqaluk LYNGE; Atassut Party (a more conservative party that favors continuing close relations with Denmark), leader NA; Polar Party (conservative-Greenland nationalist), Lars CHEMNITZ; Center Party (a new nonsocialist protest party), leader NA
Suffrage: 18 years of age; universal
Elections:
Danish Folketing: last held on 12 December

Greenland (continued)

1990 (next to be held by December 1994); Greenland elects two representatives to the Folketing; results—percent of vote by party NA; seats—(2 total) Siumut 1, Atassut 1
Landsting: last held on 5 March 1991 (next to be held 5 March 1995); results—percent of vote by party NA; seats—(27 total) Siumut 11, Atassut Party 8, Inuit Ataqatigiit 5, Center Party 2, Polar Party 1
Executive branch: Danish monarch, high commissioner, home rule chairman, prime minister, Cabinet (Landsstyre)
Legislative branch: unicameral Parliament (Landsting)
Judicial branch: High Court (Landsret)
Leaders:
Chief of State: Queen MARGRETHE II (since 14 January 1972), represented by High Commissioner Torben Hede PEDERSEN (since NA)
Head of Government: Home Rule Chairman Lars Emil JOHANSEN (since 15 March 1991)
Diplomatic representation in US: none (self-governing overseas administrative division of Denmark)
US diplomatic representation: none (self-governing overseas administrative division of Denmark)
Flag: two equal horizontal bands of white (top) and red with a large disk slightly to the hoist side of center—the top half of the disk is red, the bottom half is white

Economy

Overview: Greenland's economic situation at present is difficult and unemployment increases. Prospects for economic growth in the immediate future are not bright. The Home Rule Government's economic restraint measures introduced in the late 1980s have assisted in shifting red figures into a balance in the public budget. Foreign trade produced a surplus in 1989 and 1990, but has now returned to a deficit. Following the closing of the Black Angel lead and zinc mine in 1989, Greenland today is fully dependent on fishing and fish processing, this sector accounting for 95% of exports. Prospects for fisheries are not bright, as the important shrimp catches will at best stabilize and cod catches have dropped. Resumption of mining and hydrocarbon activities is not around the corner, thus leaving only tourism with some potential for the near future. The public sector in Greenland, i.e. the HRG and its commercial entities and the municipalities, plays a dominant role in Greenland accounting for about two thirds of total employment. About half the government's revenues come from grants from the Danish Government.
National product: GNP—purchasing power equivalent—$500 million (1988)
National product real growth rate: -10% (1990)

National product per capita: $9,000 (1988)
Inflation rate (consumer prices): 1.6% (1991)
Unemployment rate: 9% (1990 est.)
Budget: revenues $381 million; expenditures $381 million, including capital expenditures of $36 million (1989)
Exports: $340.6 million (f.o.b., 1991)
commodities: fish and fish products 95%
partners: Denmark 79%, Benelux 9%, Germany 5%
Imports: $403 million (c.i.f., 1991)
commodities: manufactured goods 28%, machinery and transport equipment 24%, food and live animals 12.4%, petroleum products 12%
partners: Denmark 65%, Norway 8.8%, US 4.6%, Germany 3.8%, Japan 3.8%, Sweden 2.4%
External debt: $480 million (1990 est.)
Industrial production: growth rate NA%
Electricity: 84,000 kW capacity; 176 million kWh produced, 3,060 kWh per capita (1992)
Industries: fish processing (mainly shrimp), lead and zinc mining, handicrafts, some small shipyards, potential for platinum and gold mining
Agriculture: sector dominated by fishing and sheep raising; crops limited to forage and small garden vegetables; 1988 fish catch of 133,500 metric tons
Economic aid: none
Currency: 1 Danish krone (DKr) = 100 re
Exchange rates: Danish kroner (DKr) per US$1—6.236 (January 1993), 6.036 (1992), 6.396 (1991), 6.189 (1990), 7.310 (1989), 6.732 (1988)
Fiscal year: calendar year

Communications

Highways: 80 km
Ports: Kangerluarsoruseq (Faeringehavn), Paamiut (Frederikshaab), Nuuk (Godthaab), Sisimiut (Holsteinsborg), Julianehaab, Maarmorilik, North Star Bay
Airports:
total: 11
usable: 8
with permanent-surface runways: 5
with runways over 3,659 m: 0
with runways 2,440-3,659 m: 2
with runways 1,220-2,439 m: 2
Telecommunications: adequate domestic and international service provided by cables and microwave radio relay; 17,900 telephones; broadcast stations—5 AM, 7 (35 repeaters) FM, 4 (9 repeaters) TV; 2 coaxial submarine cables; 1 Atlantic Ocean INTELSAT earth station

Defense Forces

Note: defense is responsibility of Denmark

Grenada

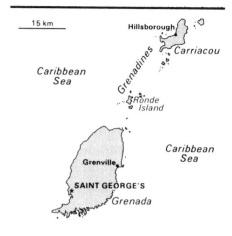

Geography

Location: in the eastern Caribbean Sea, about 150 im north of Trinidad and Tobago
Map references: Central America and the Caribbean, South America, Standard Time Zones of the World
Area:
total area: 340 km²
land area: 340 km²
comparative area: slightly less than twice the size of Washington, DC
Land boundaries: 0 km
Coastline: 121 km
Maritime claims:
exclusive economic zone: 200 nm
territorial sea: 12 nm
International disputes: none
Climate: tropical; tempered by northeast trade winds
Terrain: volcanic in origin with central mountains
Natural resources: timber, tropical fruit, deepwater harbors
Land use:
arable land: 15%
permanent crops: 26%
meadows and pastures: 3%
forest and woodland: 9%
other: 47%
Irrigated land: NA km²
Environment: lies on edge of hurricane belt; hurricane season lasts from June to November
Note: islands of the Grenadines group are divided politically with Saint Vincent and the Grenadines

People

Population: 93,830 (July 1993 est.)
Population growth rate: 0.24% (1993 est.)
Birth rate: 30.85 births/1,000 population (1993 est.)
Death rate: 6.46 deaths/1,000 population (1993 est.)
Net migration rate: -21.95 migrant(s)/1,000 population (1993 est.)

Infant mortality rate: 12.7 deaths/1,000 live births (1993 est.)

Life expectancy at birth:
total population: 70.15 years
male: 67.79 years
female: 72.54 years (1993 est.)

Total fertility rate: 4 children born/woman (1993 est.)

Nationality:
noun: Grenadian(s)
adjective: Grenadian

Ethnic divisions: black African

Religions: Roman Catholic, Anglican, other Protestant sects

Languages: English (official), French patois

Literacy: age 15 and over having ever attended school (1970)
total population: 98%
male: 98%
female: 98%

Labor force: 36,000
by occupation: services 31%, agriculture 24%, construction 8%, manufacturing 5%, other 32% (1985)

Government

Names:
conventional long form: none
conventional short form: Grenada

Digraph: GJ

Type: parliamentary democracy

Capital: Saint George's

Administrative divisions: 6 parishes and 1 dependency*; Carriacou and Petit Martinique*, Saint Andrew, Saint David, Saint George, Saint John, Saint Mark, Saint Patrick

Independence: 7 February 1974 (from UK)

Constitution: 19 December 1973

Legal system: based on English common law

National holiday: Independence Day, 7 February (1974)

Political parties and leaders: National Democratic Congress (NDC), Nicholas BRATHWAITE; Grenada United Labor Party (GULP), Sir Eric GAIRY; The National Party (TNP), Ben JONES; New National Party (NNP), Keith MITCHELL; Maurice Bishop Patriotic Movement (MBPM), Terrence MARRYSHOW; New Jewel Movement (NJM), Bernard COARD

Suffrage: 18 years of age; universal

Elections:
House of Representatives: last held on 13 March 1990 (next to be held by NA March 1996); results—percent of vote by party NA; seats—(15 total) NDC 8, GULP 3, TNP 2, NNP 2

Executive branch: British monarch, governor general, prime minister, Ministers of Government (cabinet)

Legislative branch: bicameral Parliament consists of an upper house or Senate and a lower house or House of Representatives

Judicial branch: Supreme Court

Leaders:
Chief of State: Queen ELIZABETH II (since 6 February 1952), represented by Governor General Reginald Oswald PALMER (since 6 August 1992)
Head of Government: Prime Minister Nicholas BRATHWAITE (since 13 March 1990)

Member of: ACP, C, CARICOM, CDB, ECLAC, FAO, G-77, IBRD, ICAO, ICFTU, IDA, IFAD, IFC, ILO, IMF, INTERPOL, IOC, ITU, LAES, LORCS, NAM, OAS, OECS, OPANAL, UN, UNCTAD, UNESCO, UNIDO, UPU, WCL, WHO, WTO

Diplomatic representation in US:
chief of mission: Ambassador Denneth MODESTE
chancery: 1701 New Hampshire Avenue NW, Washington, DC 20009
telephone: (202) 265-2561
consulate general: New York

US diplomatic representation:
chief of mission: Charge d'Affaires Annette T. VELER
embassy: Ross Point Inn, Saint George's
mailing address: P. O. Box 54, Saint George's
telephone: (809) 444-1173 through 1178
FAX: (809) 444-4820

Flag: a rectangle divided diagonally into yellow triangles (top and bottom) and green triangles (hoist side and outer side) with a red border around the flag; there are seven yellow five-pointed stars with three centered in the top red border, three centered in the bottom red border, and one on a red disk superimposed at the center of the flag; there is also a symbolic nutmeg pod on the hoist-side triangle (Grenada is the world's second-largest producer of nutmeg, after Indonesia); the seven stars represent the seven administrative divisions

---

Economy

Overview: The economy is essentially agricultural and centers on the traditional production of spices and tropical plants. Agriculture accounts for about 16% of GDP and 80% of exports and employs 24% of the labor force. Tourism is the leading foreign exchange earner, followed by agricultural exports. Manufacturing remains relatively undeveloped, but is expected to grow, given a more favorable private investment climate since 1983. The economy achieved an impressive average annual growth rate of 5.5% in 1986-91 but stalled in 1992. Unemployment remains high at about 25%.

National product: GDP—purchasing power equivalent—$250 million (1992 est.)

National product real growth rate: -0.4% (1992 est.)

National product per capita: $3,000 (1992 est.)

Inflation rate (consumer prices): 2.6% (1991 est.)

Unemployment rate: 25% (1992 est.)

Budget: revenues $78 million; expenditures $51 million, including capital expenditures of $22 million (1991 est.)

Exports: $30 million (f.o.b., 1991 est.)
commodities: nutmeg 36%, cocoa beans 9%, bananas 14%, mace 8%, textiles 5%
partners: US 12%, UK, FRG, Netherlands, Trinidad and Tobago (1989)

Imports: $110 million (f.o.b., 1991 est.)
commodities: food 25%, manufactured goods 22%, machinery 20%, chemicals 10%, fuel 6% (1989)
partners: US 29%, UK, Trinidad and Tobago, Japan, Canada (1989)

External debt: $104 million (1990 est.)

Industrial production: growth rate 5.8% (1989 est.); accounts for 9% of GDP

Electricity: 12,500 kW capacity; 26 million kWh produced, 310 kWh per capita (1992)

Industries: food and beverage, textile, light assembly operations, tourism, construction

Agriculture: accounts for 16% of GDP and 80% of exports; bananas, cocoa, nutmeg, and mace account for two-thirds of total crop production; world's second-largest producer and fourth-largest exporter of nutmeg and mace; small-size farms predominate, growing a variety of citrus fruits, avocados, root crops, sugarcane, corn, and vegetables

Economic aid: US commitments, including Ex-Im (FY84-89), $60 million; Western (non-US) countries, ODA and OOF bilateral commitments (1970-89), $70 million; Communist countries (1970-89), $32 million

Currency: 1 EC dollar (EC$) = 100 cents

Exchange rates: East Caribbean dollars (EC$) per US$1—2.70 (fixed rate since 1976)

Fiscal year: calendar year

Communications

Highways: 1,000 km total; 600 km paved, 300 km otherwise improved; 100 km unimproved

Ports: Saint George's

Airports:
total: 3
usable: 3
with permanent-surface runways: 2
with runways over 3,659 m: 0
with runways 2,440-3,659 m: 1
with runways 1,220-2,439 m: 1

Telecommunications: automatic, islandwide telephone system with 5,650 telephones; new SHF radio links to Trinidad and Tobago and Saint Vincent; VHF and UHF radio links to Trinidad and Carriacou; broadcast stations—1 AM, no FM, 1 TV

Grenada *(continued)*

Defense Forces

Branches: Royal Grenada Police Force, Coast Guard
Manpower availability: NA
Defense expenditures: $NA, NA% of GDP

Guadeloupe
(overseas department of France)

Geography

Location: in the Caribbean Sea, 500 km southeast of Puerto Rico
Map references: Central America and the Caribbean
Area:
total area: 1,780 km²
land area: 1,760 km²
comparative area: 10 times the size of Washington, DC
Land boundaries: 0 km
Coastline: 306 km
Maritime claims:
exclusive economic zone: 200 nm
territorial sea: 12 nm
International disputes: none
Climate: subtropical tempered by trade winds; relatively high humidity
Terrain: Basse-Terre is volcanic in origin with interior mountains; Grand-Terre is low limestone formation
Natural resources: cultivable land, beaches and climate that foster tourism
Land use:
arable land: 18%
permanent crops: 5%
meadows and pastures: 13%
forest and woodland: 40%
other: 24%
Irrigated land: 30 km² (1989 est.)
Environment: subject to hurricanes (June to October); La Soufriere is an active volcano

People

Population: 422,114 (July 1993 est.)
Population growth rate: 1.67% (1993 est.)
Birth rate: 18.18 births/1,000 population (1993 est.)
Death rate: 5.94 deaths/1,000 population (1993 est.)
Net migration rate: 4.42 migrant(s)/1,000 population (1993 est.)
Infant mortality rate: 9.2 deaths/1,000 live births (1993 est.)
Life expectancy at birth:
total population: 76.72 years

male: 73.67 years
female: 79.9 years (1993 est.)
Total fertility rate: 2.08 children born/woman (1993 est.)
Nationality:
noun: Guadeloupian(s)
adjective: Guadeloupe
Ethnic divisions: black or mulatto 90%, white 5%, East Indian, Lebanese, Chinese less than 5%
Religions: Roman Catholic 95%, Hindu and pagan African 5%
Languages: French, creole patois
Literacy: age 15 and over can read and write (1982)
total population: 90%
male: 90%
female: 91%
Labor force: 120,000
by occupation: services, government, and commerce 53.0%, industry 25.8%, agriculture 21.2%

Government

Names:
conventional long form: Department of Guadeloupe
conventional short form: Guadeloupe
local long form: Departement de la Guadeloupe
local short form: Guadeloupe
Digraph: GP
Type: overseas department of France
Capital: Basse-Terre
Administrative divisions: none (overseas department of France)
Independence: none (overseas department of France)
Constitution: 28 September 1958 (French Constitution)
Legal system: French legal system
National holiday: National Day, Taking of the Bastille, 14 July (1789)
Political parties and leaders: Rally for the Republic (RPR), Marlene CAPTANT; Communist Party of Guadeloupe (PCG), Christian Medard CELESTE; Socialist Party (PS), Dominique LARIFLA; Popular Union for the Liberation of Guadeloupe (UPLG); Independent Republicans; Union for French Democracy (UDF); Union for the Center Rally (URC coalition of the PS, RPR, and UDF); Guadeloupe Objective (OG), Lucette MICHAUX-CHEVRY
Other political or pressure groups: Popular Union for the Liberation of Guadeloupe (UPLG); Popular Movement for Independent Guadeloupe (MPGI); General Union of Guadeloupe Workers (UGTG); General Federation of Guadeloupe Workers (CGT-G); Christian Movement for the Liberation of Guadeloupe (KLPG)
Suffrage: 18 years of age; universal

Elections:

French National Assembly: last held on 5 and 12 June 1988 (next to be held March 1993); Guadeloupe elects four representatives; results—percent of vote by party NA; seats—(4 total) PS 2 seats, RPR 1 seat, PCG 1 seat

French Senate: last held in September 1986 (next to be held September 1995); Guadeloupe elects two representatives; results—percent of vote by party NA; seats—(2 total) PCG 1, PS 1

General Council: last held 25 September and 8 October 1988 (next to be held by NA 1992); results—percent of vote by party NA; seats—(42 total) PS 26, URC 16

Regional Council: last held on 22 March 1992 (next to be held by 16 March 1998); results—OG 33.1%, PSG 28.7%, PCG 23.8%, UDF 10.7%, other 3.7%; seats—(41 total) OG 15, PSG 12, PCG 10, UDF 4

Executive branch: government commissioner

Legislative branch: unicameral General Council and unicameral Regional Council

Judicial branch: Court of Appeal (Cour d'Appel) with jurisdiction over Guadeloupe, French Guiana, and Martinique

Leaders:

Chief of State: President Francois MITTERRAND (since 21 May 1981)

Head of Government: Prefect Franck PERRIEZ (since NA 1992)

Member of: FZ, WCL

Diplomatic representation in US: as an overseas department of France, the interests of Guadeloupe are represented in the US by France

US diplomatic representation: none (overseas department of France)

Flag: the flag of France is used

Economy

Overview: The economy depends on agriculture, tourism, light industry, and services. It is also dependent upon France for large subsidies and imports. Tourism is a key industry, with most tourists from the US. In addition, an increasingly large number of cruise ships visit the islands. The traditionally important sugarcane crop is slowly being replaced by other crops, such as bananas (which now supply about 50% of export earnings), eggplant, and flowers. Other vegetables and root crops are cultivated for local consumption, although Guadeloupe is still dependent on imported food, which comes mainly from France. Light industry consists mostly of sugar and rum production. Most manufactured goods and fuel are imported. Unemployment is especially high among the young.

National product: GDP—exchange rate

conversion—$1.5 billion (1989)

National product real growth rate: NA%

National product per capita: $4,700 (1989)

Inflation rate (consumer prices): 3.7% (1990)

Unemployment rate: 31.3% (1990)

Budget: revenues $333 million; expenditures $671 million, including capital expenditures of $NA (1989)

Exports: $168 million (f.o.b., 1988)

commodities: bananas, sugar, rum

partners: France 68%, Martinique 22% (1987)

Imports: $1.2 billion (c.i.f., 1988)

commodities: vehicles, foodstuffs, clothing and other consumer goods, construction materials, petroleum products

partners: France 64%, Italy, FRG, US (1987)

External debt: $NA

Industrial production: growth rate NA%

Electricity: 171,500 kW capacity; 441 million kWh produced, 1,080 kWh per capita (1992)

Industries: construction, cement, rum, sugar, tourism

Agriculture: cash crops—bananas, sugarcane; other products include tropical fruits and vegetables; livestock—cattle, pigs, goats; not self-sufficient in food

Economic aid: US commitments, including Ex-Im (FY70-88), $4 million; Western (non-US) countries, ODA and OOF bilateral commitments (1970-89), $8.235 billion

Currency: 1 French franc (F) = 100 centimes

Exchange rates: French francs (F) per US$1—5.4812 (January 1993), 5.2938 (1992), 5.6421 (1991), 5.4453 (1990), 6.3801 (1989), 5.9569 (1988)

Fiscal year: calendar year

Communications

Railroads: privately owned, narrow-gauge plantation lines

Highways: 1,940 km total; 1,600 km paved, 340 km gravel and earth

Ports: Pointe-a-Pitre, Basse-Terre

Airports:

total: 9

usable: 9

with permanent-surface runways: 8

with runways over 3,659 m: 0

with runways 2,440-3,659 m: 1

with runways 1,220-2,439 m: 1

Telecommunications: domestic facilities inadequate; 57,300 telephones; interisland microwave radio relay to Antigua and Barbuda, Dominica, and Martinique; broadcast stations—2 AM, 8 FM (30 private stations licensed to broadcast FM), 9 TV; 1 Atlantic Ocean INTELSAT ground station

---

Defense Forces

Branches: French Forces, Gendarmerie

Manpower availability: males age 15-49 98,069; fit for military service NA (1993 est.)

Note: defense is responsibility of France

Guam

(territory of the US)

Geography

Location: in the North Pacific Ocean, 5,955 km west-southwest of Honolulu, about three-quarters of the way between Hawaii and the Philippines

Map references: Oceania

Area:

total area: 541.3 km²

land area: 541.3 km²

comparative area: slightly more than three times the size of Washington, DC

Land boundaries: 0 km

Coastline: 125.5 km

Maritime claims:

contiguous zone: 24 nm

continental shelf: 200 m or depth of exploitation

exclusive economic zone: 200 nm

territorial sea: 12 nm

International disputes: none

Climate: tropical marine; generally warm and humid, moderated by northeast trade winds; dry season from January to June, rainy season from July to December; little seasonal temperature variation

Terrain: volcanic origin, surrounded by coral reefs; relatively flat coraline limestone plateau (source of most fresh water) with steep coastal cliffs and narrow coastal plains in north, low-rising hills in center, mountains in south

Natural resources: fishing (largely undeveloped), tourism (especially from Japan)

Land use:

arable land: 11%

permanent crops: 11%

meadows and pastures: 15%

forest and woodland: 18%

other: 45%

Irrigated land: NA km²

Environment: frequent squalls during rainy season; subject to relatively rare, but potentially very destructive typhoons (especially in August)

Note: largest and southernmost island in the Mariana Islands archipelago; strategic location in western North Pacific Ocean

People

Population: 145,935 (July 1993 est.)

Population growth rate: 2.53% (1993 est.)

Birth rate: 26.16 births/1,000 population (1993 est.)

Death rate: 3.86 deaths/1,000 population (1993 est.)

Net migration rate: 3 migrant(s)/1,000 population (1993 est.)

Infant mortality rate: 15.17 deaths/1,000 live births (1993 est.)

Life expectancy at birth:

total population: 74.29 years

male: 72.42 years

female: 76.13 years (1993 est.)

Total fertility rate: 2.44 children born/woman (1993 est.)

Nationality:

noun: Guamanian(s)

adjective: Guamanian

Ethnic divisions: Chamorro 47%, Filipino 25%, Caucasian 10%, Chinese, Japanese, Korean, and other 18%

Religions: Roman Catholic 98%, other 2%

Languages: English, Chamorro, Japanese

Literacy: age 15 and over can read and write (1980)

total population: 96%

male: 96%

female: 96%

Labor force: 46,930 (1990)

by occupation: federal and territorial government 40%, private 60% (trade 18%, services 15.6%, construction 13.8%, other 12.6%) (1990)

Government

Names:

conventional long form: Territory of Guam

conventional short form: Guam

Digraph: GQ

Type: organized, unincorporated territory of the US with policy relations between Guam and the US under the jurisdiction of the Office of Territorial and International Affairs, US Department of the Interior

Capital: Agana

Administrative divisions: none (territory of the US)

Independence: none (territory of the US)

Constitution: Organic Act of 1 August 1950

Legal system: modeled on US; federal laws apply

National holiday: Guam Discovery Day (first Monday in March); Liberation Day, 21 July

Political parties and leaders: Democratic Party (controls the legislature); Republican Party (party of the Governor)

Suffrage: 18 years of age; universal; US citizens, but do not vote in US presidential elections

Elections:

Governor: last held on 6 November 1990 (next to be held NA November 1994); results—Joseph F. ADA reelected

Legislature: last held on 9 November 1992 (next to be held NA November 1994); results—percent of vote by party NA; seats—(21 total) Democratic 14, Republican 7

US House of Representatives: last held 9 November 1992 (next to be held NA November 1994); Guam elects one delegate; results—Robert UNDERWOOD was elected as delegate; seats—(1 total) Democrat 1

Executive branch: US president, governor, lieutenant governor, Cabinet

Legislative branch: unicameral Legislature

Judicial branch: Federal District Court, Territorial Superior Court

Leaders:

Chief of State: President William Jefferson CLINTON (since 20 January 1993); Vice President Albert GORE, Jr. (since 20 January 1993)

Head of Government: Governor Joseph A. ADA (since November 1986); Lieutenant Governor Frank F. BLAS (since NA)

Member of: ESCAP (associate), IOC, SPC

Diplomatic representation in US: none (territory of the US)

Flag: territorial flag is dark blue with a narrow red border on all four sides; centered is a red-bordered, pointed, vertical ellipse containing a beach scene, outrigger canoe with sail, and a palm tree with the word GUAM superimposed in bold red letters; US flag is the national flag

Economy

Overview: The economy depends mainly on US military spending and on revenues from tourism. Over the past 20 years the tourist industry has grown rapidly, creating a construction boom for new hotels and the expansion of older ones. Visitors numbered about 900,000 in 1992. About 60% of the labor force works for the private sector and the rest for government. Most food and industrial goods are imported, with about 75% from the US.

National product: GNP—purchasing power equivalent—$2 billion (1991 est.)

National product real growth rate: NA%

National product per capita: $14,000 (1991 est.)

Inflation rate (consumer prices): 4% (1992 est.)

Unemployment rate: 2% (1992 est.)

Budget: revenues $525 million; expenditures $395 million, including capital expenditures of $NA

Guatemala

Exports: $34 million (f.o.b., 1984)
commodities: mostly transshipments of refined petroleum products, construction materials, fish, food and beverage products
partners: US 25%, Trust Territory of the Pacific Islands 63%, other 12%
Imports: $493 million (c.i.f., 1984)
commodities: petroleum and petroleum products, food, manufactured goods
partners: US 23%, Japan 19%, other 58%
External debt: $NA
Industrial production: growth rate NA%
Electricity: 500,000 kW capacity; 2,300 million kWh produced, 16,300 kWh per capita (1990)
Industries: US military, tourism, construction, transshipment services, concrete products, printing and publishing, food processing, textiles
Agriculture: relatively undeveloped with most food imported; fruits, vegetables, eggs, pork, poultry, beef, copra
Economic aid: although Guam receives no foreign aid, it does receive large transfer payments from the general revenues of the US Federal Treasury into which Guamanians pay no income or excise taxes; under the provisions of a special law of Congress, the Guamanian Treasury, rather than the US Treasury, receives federal income taxes paid by military and civilian Federal employees stationed in Guam
Currency: US currency is used
Fiscal year: 1 October-30 September

Communications

Highways: 674 km all-weather roads
Ports: Apra Harbor
Airports:
total: 5
usable: 4
with permanent-surface runways: 3
with runways over 3,659 m: 0
with runways 2,440-3,659 m: 3
with runways 1,200-2,439 m: 0
Telecommunications: 26,317 telephones (1989); broadcast stations—3 AM, 3 FM, 3 TV; 2 Pacific Ocean INTELSAT ground stations

Defense Forces

Note: defense is the responsibility of the US

Geography

Location: Central America, between Honduras and Mexico
Map references: Central America and the Caribbean, North America, Standard Time Zones of the World
Area:
total area: 108,890 km²
land area: 108,430 km²
comparative area: slightly smaller than Tennessee
Land boundaries: total 1,687 km, Belize 266 km, El Salvador 203 km, Honduras 256 km, Mexico 962 km
Coastline: 400 km
Maritime claims:
continental shelf: the outer edge of the continental shelf
exclusive economic zone: 200 nm
territorial sea: 12 nm
International disputes: border with Belize in dispute; negotiations to resolve the dispute have begun
Climate: tropical; hot, humid in lowlands; cooler in highlands
Terrain: mostly mountains with narrow coastal plains and rolling limestone plateau (Peten)
Natural resources: petroleum, nickel, rare woods, fish, chicle
Land use:
arable land: 12%
permanent crops: 4%
meadows and pastures: 12%
forest and woodland: 40%
other: 32%
Irrigated land: 780 km² (1989 est.)
Environment: numerous volcanoes in mountains, with frequent violent earthquakes; Caribbean coast subject to hurricanes and other tropical storms; deforestation; soil erosion; water pollution
Note: no natural harbors on west coast

People

Population: 10,446,015 (July 1993 est.)
Population growth rate: 2.63% (1993 est.)
Birth rate: 36.19 births/1,000 population (1993 est.)
Death rate: 7.74 deaths/1,000 population (1993 est.)
Net migration rate: -2.18 migrant(s)/1,000 population (1993 est.)
Infant mortality rate: 55.6 deaths/1,000 live births (1993 est.)
Life expectancy at birth:
total population: 63.99 years
male: 61.46 years
female: 66.65 years (1993 est.)
Total fertility rate: 4.9 children born/woman (1993 est.)
Nationality:
noun: Guatemalan(s)
adjective: Guatemalan
Ethnic divisions: Ladino 56% (mestizo—mixed Indian and European ancestry), Indian 44%
Religions: Roman Catholic, Protestant, traditional Mayan
Languages: Spanish 60%, Indian language 40% (18 Indian dialects, including Quiche, Cakchiquel, Kekchi)
Literacy: age 15 and over can read and write (1990)
total population: 55%
male: 63%
female: 47%
Labor force: 2.5 million
by occupation: agriculture 60%, services 13%, manufacturing 12%, commerce 7%, construction 4%, transport 3%, utilities 0.8%, mining 0.4% (1985)

Government

Names:
conventional long form: Republic of Guatemala
conventional short form: Guatemala
local long form: Republica de Guatemala
local short form: Guatemala
Digraph: GT
Type: republic
Capital: Guatemala
Administrative divisions: 22 departments (departamentos, singular—departamento); Alta Verapaz, Baja Verapaz, Chimaltenango, Chiquimula, El Progreso, Escuintla, Guatemala, Huehuetenango, Izabal, Jalapa, Jutiapa, Peten, Quetzaltenango, Quiche, Retalhuleu, Sacatepequez, San Marcos, Santa Rosa, Solola, Suchitepequez, Totonicapan, Zacapa
Independence: 15 September 1821 (from Spain)
Constitution: 31 May 1985, effective 14 January 1986

Guatemala (continued)

note: suspended on 25 May 1993 by President SERRANO; reinstated on 5 June 1993 following ouster of president

Legal system: civil law system; judicial review of legislative acts; has not accepted compulsory ICJ jurisdiction

National holiday: Independence Day, 15 September (1821)

Political parties and leaders: National Centrist Union (UCN), Jorge CARPIO Nicolle; Solidarity Action Movement (MAS), Jorge SERRANO Elias; Christian Democratic Party (DCG), Alfonso CABRERA Hidalgo; National Advancement Party (PAN), Alvaro ARZU Irigoyen; National Liberation Movement (MLN), Mario SANDOVAL Alarcon; Social Democratic Party (PSD), Mario SOLARZANO Martinez; Popular Alliance 5 (AP-5), Max ORLANDO Molina; Revolutionary Party (PR), Carlos CHAVARRIA; National Authentic Center (CAN), Hector MAYORA Dawe; Democratic Institutional Party (PID), Oscar RIVAS; Nationalist United Front (FUN), Gabriel GIRON; Guatemalan Republican Front (FRG), Efrain RIOS Montt

Other political or pressure groups: Federated Chambers of Commerce and Industry (CACIF); Mutual Support Group (GAM); Agrarian Owners Group (UNAGRO); Committee for Campesino Unity (CUC); leftist guerrilla movement known as Guatemalan National Revolutionary Union (URNG) has four main factions—Guerrilla army of the Poor (EGP); Revolutionary Organization of the People in Arms (ORPA); Rebel Armed Forces (FAR); Guatemalan Labor Party (PGT/O)

Suffrage: 18 years of age; universal

Elections:
Congress: last held on 11 November 1990 (next to be held 11 November 1995); results—UCN 25.6%, MAS 24.3%, DCG 17.5%, PAN 17.3%, MLN 4.8%, PSD/AP-5 3.6%, PR 2.1%; seats—(116 total) UCN 38, DCG 27, MAS 18, PAN 12, Pro-Rios Montt 10, MLN 4, PR 1, PSD/AP-5 1, independent 5
President: runoff held on 11 January 1991 (next to be held 11 November 1995); results—Jorge SERRANO Elias (MAS) 68.1%, Jorge CARPIO Nicolle (UCN) 31.9%

note: President SERRANO resigned on 1 June 1993 shortly after dissolving Congress and the judiciary; on 6 June 1993, Ramiro DE LEON Carpio was chosen as the new president by a vote of Congress; he will finish off the remainder of SERRANO's five-year term which expires in 1995

Executive branch: president, vice president, Council of Ministers (cabinet)

Legislative branch: unicameral Congress of the Republic (Congreso de la Republica)

Judicial branch: Supreme Court of Justice (Corte Suprema de Justicia)

Leaders:
Chief of State and Head of Government: President Ramiro DE LEON Carpio (since 6 June 1993); Vice President Arturo HERBRUGER (since 18 June 1993)

Member of: BCIE, CACM, CCC, ECLAC, FAO, G-24, G-77, GATT, IADB, IAEA, IBRD, ICAO, ICFTU, IDA, IFAD, IFC, ILO, IMF, IMO, INTELSAT, INTERPOL, IOC, IOM, ITU, LAES, LAIA (observer), LORCS, NAM, OAS, OPANAL, PCA, UN, UNCTAD, UNESCO, UNIDO, UPU, WCL, WFTU, WHO, WIPO, WMO

Diplomatic representation in US:
chief of mission: Ambassador Juan Jose CASO-FANJUL
chancery: 2220 R Street NW, Washington, DC 20008
telephone: (202) 745-4952 through 4954
consulates general: Chicago, Houston, Los Angeles, Miami, New Orleans, New York, and San Francisco

US diplomatic representation:
chief of mission: Ambassador Marilyn MCAFEE (since 28 May 1993)
embassy: 7-01 Avenida de la Reforma, Zone 10, Guatemala City
mailing address: APO AA 34024
telephone: [502] (2) 31-15-41
FAX: [502] (2) 318855

Flag: three equal vertical bands of light blue (hoist side), white, and light blue with the coat of arms centered in the white band; the coat of arms includes a green and red quetzal (the national bird) and a scroll bearing the inscription LIBERTAD 15 DE SEPTIEMBRE DE 1821 (the original date of independence from Spain) all superimposed on a pair of crossed rifles and a pair of crossed swords and framed by a wreath

---

Economy

Overview: The economy is based on family and corporate agriculture, which accounts for 26% of GDP, employs about 60% of the labor force, and supplies two-thirds of exports. Manufacturing, predominantly in private hands, accounts for about 18% of GDP and 12% of the labor force. In both 1990 and 1991, the economy grew by 3%, the fourth and fifth consecutive years of mild growth. In 1992 growth picked up to 4% as government policies favoring competition and foreign trade and investment took stronger hold.

National product: GDP—exchange rate conversion—$12.6 billion (1992 est.)

National product real growth rate: 4.2% (1992)

National product per capita: $1,300 (1992 est.)

Inflation rate (consumer prices): 14% (1992 est.)

Unemployment rate: 6.5% (1991 est.), with 30-40% underemployment

Budget: revenues $604 million; expenditures $808 million, including capital expenditures of $134 million (1990 est.)

Exports: $1.3 billion (f.o.b., 1992)
commodities: coffee 26%, sugar 13%, bananas 7%, beef 3%
partners: US 36%, El Salvador, Costa Rica, Germany, Honduras

Imports: $1.8 billion (c.i.f., 1992)
commodities: fuel and petroleum products, machinery, grain, fertilizers, motor vehicles
partners: US 40%, Mexico, Venezuela, Japan, Germany

External debt: $2.5 billion (December 1992 est.)

Industrial production: growth rate 1.9% (1991 est.); accounts for 18% of GDP

Electricity: 847,600 kW capacity; 2,500 million kWh produced, 260 kWh per capita (1992)

Industries: sugar, textiles and clothing, furniture, chemicals, petroleum, metals, rubber, tourism

Agriculture: accounts for 26% of GDP; most important sector of economy; contributes two-thirds of export earnings; principal crops—sugarcane, corn, bananas, coffee, beans, cardamom; livestock—cattle, sheep, pigs, chickens; food importer

Illicit drugs: illicit producer of opium poppy and cannabis for the international drug trade; the government has an active eradication program for cannabis and opium poppy; transit country for cocaine shipments

Economic aid: US commitments, including Ex-Im (FY70-90), $1.1 billion; Western (non-US) countries, ODA and OOF bilateral commitments (1970-89), $7.92 billion

Currency: 1 quetzal (Q) = 100 centavos

Exchange rates: free market quetzales (Q) per US$1—5.2850 (December 1993), 5.1706 (1992), 5.0289 (1991), 2.8161 (1989), 2.6196 (1988); note—black-market rate 2.800 (May 1989)

Fiscal year: calendar year

Communications

Railroads: 1,019 km 0.914-meter gauge, single track; 917 km government owned, 102 km privately owned

Highways: 26,429 km total; 2,868 km paved, 11,421 km gravel, and 12,140 unimproved

Inland waterways: 260 km navigable year round; additional 730 km navigable during high-water season

Pipelines: crude oil 275 km

Ports: Puerto Barrios, Puerto Quetzal, Santo Tomas de Castilla

Merchant marine: 1 cargo ship (1,000 GRT or over) totaling 4,129 GRT/6,450 DWT

Airports:
total: 474
usable: 418
with permanent-surface runways: 11
with runways over 3,659 m: 0
with runways 2,440-3,659 m: 3
with runways 1,220-2,439 m: 21
Telecommunications: fairly modern network centered in Guatemala [city]; 97,670 telephones; broadcast stations—91 AM, no FM, 25 TV, 15 shortwave; connection into Central American Microwave System; 1 Atlantic Ocean INTELSAT earth station

---

Defense Forces

Branches: Army, Navy, Air Force
Manpower availability: males age 15-49 2,410,760; fit for military service 1,576,569; reach military age (18) annually 115,178 (1993 est.)
Defense expenditures: exchange rate conversion—$121 million, 1% of GDP (1993)

Guernsey
(British crown dependency)

Geography

Location: in the English Channel, 52 km west of France between UK and France
Map references: Europe
Area:
total area: 194 km²
land area: 194 km²
comparative area: slightly larger than Washington, DC
note: includes Alderney, Guernsey, Herm, Sark, and some other smaller islands
Land boundaries: 0 km
Coastline: 50 km
Maritime claims:
exclusive fishing zone: 200 nm
territorial sea: 3 nm
International disputes: none
Climate: temperate with mild winters and cool summers; about 50% of days are overcast
Terrain: mostly level with low hills in southwest
Natural resources: cropland
Land use:
arable land: NA%
permanent crops: NA%
meadows and pastures: NA%
forest and woodland: NA%
other: NA%
Irrigated land: NA km²
Environment: large, deepwater harbor at Saint Peter Port

People

Population: 63,075 (July 1993 est.)
Population growth rate: 1.02% (1993 est.)
Birth rate: 13.1 births/1,000 population (1993 est.)
Death rate: 10.08 deaths/1,000 population (1993 est.)
Net migration rate: 7.23 migrant(s)/1,000 population (1993 est.)
Infant mortality rate: 6.7 deaths/1,000 live births (1993 est.)
Life expectancy at birth:
total population: 77.96 years

male: 75.27 years
female: 80.68 years (1993 est.)
Total fertility rate: 1.66 children born/woman (1993 est.)
Nationality:
noun: Channel Islander(s)
adjective: Channel Islander
Ethnic divisions: UK and Norman-French descent
Religions: Anglican, Roman Catholic, Presbyterian, Baptist, Congregational, Methodist
Languages: English, French; Norman-French dialect spoken in country districts
Literacy:
total population: NA%
male: NA%
female: NA%
Labor force: NA

Government

Names:
conventional long form: Bailiwick of Guernsey
conventional short form: Guernsey
Digraph: GK
Type: British crown dependency
Capital: Saint Peter Port
Administrative divisions: none (British crown dependency)
Independence: none (British crown dependency)
Constitution: unwritten; partly statutes, partly common law and practice
Legal system: English law and local statute; justice is administered by the Royal Court
National holiday: Liberation Day, 9 May (1945)
Political parties and leaders: none; all independents
Suffrage: 18 years of age; universal
Elections:
Assembly of the States: last held NA (next to be held NA); results—no percent of vote by party since all are independents; seats—(60 total, 33 elected), all independents
Executive branch: British monarch, lieutenant governor, bailiff, deputy bailiff
Legislative branch: unicameral Assembly of the States
Judicial branch: Royal Court
Leaders:
Chief of State: Queen ELIZABETH II (since 6 February 1952)
Head of Government: Lieutenant Governor and Commander in Chief Lt. Gen. Sir Michael WILKINS (since NA 1990); Bailiff Mr. Graham Martyn DOREY (since February 1992)
Member of: none
Diplomatic representation in US: none (British crown dependency)
US diplomatic representation: none (British crown dependency)

Guernsey *(continued)*

Flag: white with the red cross of Saint George (patron saint of England) extending to the edges of the flag

Economy

Overview: Tourism is a major source of revenue. Other economic activity includes financial services, breeding the world-famous Guernsey cattle, and growing tomatoes and flowers for export.
National product: GDP—$NA
National product real growth rate: 9% (1987)
National product per capita: $NA
Inflation rate (consumer prices): 7% (1988)
Unemployment rate: NA%
Budget: revenues $208.9 million; expenditures $173.9 million, including capital expenditures of $NA (1988)
Exports: $NA
commodities: tomatoes, flowers and ferns, sweet peppers, eggplant, other vegetables
partners: UK (regarded as internal trade)
Imports: $NA
commodities: coal, gasoline, and oil
partners: UK (regarded as internal trade)
External debt: $NA
Industrial production: growth rate NA%
Electricity: 173,000 kW capacity; 525 million kWh produced, 9,060 kWh per capita (1992)
Industries: tourism, banking
Agriculture: tomatoes, flowers (mostly grown in greenhouses), sweet peppers, eggplant, other vegetables, fruit; Guernsey cattle
Economic aid: none
Currency: 1 Guernsey (£G) pound = 100 pence
Exchange rates: Guernsey pounds (£G) per US$1—0.6527 (January 1993), 0.5664 (1992), 0.5652 (1991), 0.5603 (1990), 0.6099 (1989), 0.5614 (1988); note—the Guernsey pound is at par with the British pound
Fiscal year: calendar year

Communications

Ports: Saint Peter Port, Saint Sampson
Airports:
total: 2
useable: 2
with permanent-surface runways: 2
with runways over 3,659 m: 0
with runways 2,440-3,659 m: 0
with runways 1,220-2,439 m: 1
Telecommunications: broadcast stations—1 AM, no FM, 1 TV; 41,900 telephones; 1 submarine cable

Defense Forces

Note: defense is the responsibility of the UK

Guinea

Geography

Location: Western Africa, bordering the North Atlantic Ocean between Guinea-Bissau and Sierra Leone
Map references: Africa, Standard Time Zones of the World
Area:
total area: 245,860 km²
land area: 245,860 km²
comparative area: slightly smaller than Oregon
Land boundaries: total 3,399 km, Guinea-Bissau 386 km, Cote d'Ivoire 610 km, Liberia 563 km, Mali 858 km, Senegal 330 km, Sierra Leone 652 km
Coastline: 320 km
Maritime claims:
exclusive economic zone: 200 nm
territorial sea: 12 nm
International disputes: none
Climate: generally hot and humid; monsoonal-type rainy season (June to November) with southwesterly winds; dry season (December to May) with northeasterly harmattan winds
Terrain: generally flat coastal plain, hilly to mountainous interior
Natural resources: bauxite, iron ore, diamonds, gold, uranium, hydropower, fish
Land use:
arable land: 6%
permanent crops: 0%
meadows and pastures: 12%
forest and woodland: 42%
other: 40%
Irrigated land: 240 km² (1989 est.)
Environment: hot, dry, dusty harmattan haze may reduce visibility during dry season; deforestation

People

Population: 6,236,506 (July 1993 est.)
Population growth rate: 2.46% (1993 est.)
Birth rate: 44.76 births/1,000 population (1993 est.)

Death rate: 20.13 deaths/1,000 population (1993 est.)
Net migration rate: 0 migrant(s)/1,000 population (1993 est.)
Infant mortality rate: 141.7 deaths/1,000 live births (1993 est.)
Life expectancy at birth:
total population: 43.68 years
male: 41.49 years
female: 45.93 years (1993 est.)
Total fertility rate: 5.9 children born/woman (1993 est.)
Nationality:
noun: Guinean(s)
adjective: Guinean
Ethnic divisions: Fulani 35%, Malinke 30%, Soussou 20%, indigenous tribes 15%
Religions: Muslim 85%, Christian 8%, indigenous beliefs 7%
Languages: French (official); each tribe has its own language
Literacy: age 15 and over can read and write (1990)
total population: 24%
male: 35%
female: 13%
Labor force: 2.4 million (1983)
by occupation: agriculture 82.0%, industry and commerce 11.0%, services 5.4%
note: 88,112 civil servants (1987); 52% of population of working age (1985)

Government

Names:
conventional long form: Republic of Guinea
conventional short form: Guinea
local long form: Republique de Guinee
local short form: Guinee
former: French Guinea
Digraph: GV
Type: republic
Capital: Conakry
Administrative divisions: 33 administrative regions (regions administratives, singular—region administrative); Beyla, Boffa, Boke, Conakry, Coyah, Dabola, Dalaba, Dinguiraye, Faranah, Forecariah, Fria, Gaoual, Gueckedou, Kankan, Kerouane, Kindia, Kissidougou, Koubia, Koundara, Kouroussa, Labe, Lelouma, Lola, Macenta, Mali, Mamou, Mandiana, Nzerekore, Pita, Siguiri, Telimele, Tougue, Yomou
Independence: 2 October 1958 (from France)
Constitution: 23 December 1990 (Loi Fundamentale)
Legal system: based on French civil law system, customary law, and decree; legal codes currently being revised; has not accepted compulsory ICJ jurisdiction
National holiday: Anniversary of the Second Republic, 3 April (1984)
Political parties and leaders: political parties were legalized on 1 April 1992;

pro-government: Party for Unity and Progress (PUP), leader NA
other: Rally for the Guinean People (RPG), Alpha CONDE; Union for a New Republic (UNR), Mamadon BAH; Party for Renewal and Progress (PRP), Siradion DIALLO
Suffrage: none
Elections: none
Executive branch: president, Transitional Committee for National Recovery (Comite Transitionale de Redressement National or CTRN) replaced the Military Committee for National Recovery (Comite Militaire de Redressement National or CMRN); Council of Ministers (cabinet)
Legislative branch: unicameral People's National Assembly (Assemblee Nationale Populaire) was dissolved after the 3 April 1984 coup; framework established in December 1991 for a new National Assembly with 114 seats
Judicial branch: Court of Appeal (Cour d'Appel)
Leaders:
Chief of State and Head of Government: Gen. Lansana CONTE (since 5 April 1984)
Member of: ACCT, ACP, AfDB, CCC, CEAO (observer), ECA, ECOWAS, FAO, G-77, IBRD, ICAO, IDA, IDB, IFAD, IFC, ILO, IMF, IMO, INTELSAT, INTERPOL, IOC, ISO (correspondent), ITU, LORCS, MINURSO, NAM, OAU, OIC, UN, UNCTAD, UNESCO, UNIDO, UPU, WCL, WHO, WIPO, WMO, WTO
Diplomatic representation in US:
chief of mission: (vacant); Charge d'Affaires ad interim Ansoumane CAMARA
chancery: 2112 Leroy Place NW, Washington, DC 20008
telephone: (202) 483-9420
US diplomatic representation:
chief of mission: Ambassador Dane F. SMITH, Jr.
embassy: 2nd Boulevard and 9th Avenue, Conakry
mailing address: B. P. 603, Conakry
telephone: (224) 44-15-20 through 24
FAX: (224) 44-15-22
Flag: three equal vertical bands of red (hoist side), yellow, and green; uses the popular pan-African colors of Ethiopia; similar to the flag of Rwanda, which has a large black letter R centered in the yellow band

Economy

Overview: Although possessing many natural resources and considerable potential for agricultural development, Guinea is one of the poorest countries in the world. The agricultural sector contributes about 40% to GDP and employs more than 80% of the work force, while industry accounts for 27% of GDP. Guinea possesses over 25% of the world's bauxite reserves; exports of bauxite and alumina accounted for about 70% of total exports in 1989.
National product: GDP—exchange rate conversion—$3 billion (1990 est.)
National product real growth rate: 4.3% (1990 est.)
National product per capita: $410 (1990 est.)
Inflation rate (consumer prices): 19.6% (1990 est.)
Unemployment rate: NA%
Budget: revenues $449 million; expenditures $708 million, including capital expenditures of $361 million (1990 est.)
Exports: $788 million (f.o.b., 1990 est.)
commodities: alumina, bauxite, diamonds, coffee, pineapples, bananas, palm kernels
partners: US 33%, EC 33%, USSR and Eastern Europe 20%, Canada
Imports: $692 million (c.i.f., 1990 est.)
commodities: petroleum products, metals, machinery, transport equipment, foodstuffs, textiles, and other grain
partners: US 16%, France, Brazil
External debt: $2.6 billion (1990 est.)
Industrial production: growth rate NA%; accounts for 27% of GDP
Electricity: 113,000 kW capacity; 300 million kWh produced, 40 kWh per capita (1989)
Industries: bauxite mining, alumina, gold, diamond mining, light manufacturing and agricultural processing industries
Agriculture: accounts for 40% of GDP (includes fishing and forestry); mostly subsistence farming; principal products—rice, coffee, pineapples, palm kernels, cassava, bananas, sweet potatoes, timber; livestock—cattle, sheep and goats; not self-sufficient in food grains
Economic aid: US commitments, including Ex-Im (FY70-89), $227 million; Western (non-US) countries, ODA and OOF bilateral commitments (1970-89), $1,465 million; OPEC bilateral aid (1979-89), $120 million; Communist countries (1970-89), $446 million
Currency: 1 Guinean franc (FG) = 100 centimes
Exchange rates: Guinean francs (FG) per US$1—675 (1990), 618 (1989), 515 (1988), 440 (1987), 383 (1986)
Fiscal year: calendar year

Communications

Railroads: 1,045 km; 806 km 1.000-meter gauge, 239 km 1.435-meter standard gauge
Highways: 30,100 km total; 1,145 km paved, 12,955 km gravel or laterite (of which barely 4,500 km are currently all-weather roads), 16,000 km unimproved earth (1987)
Inland waterways: 1,295 km navigable by shallow-draft native craft

Ports: Conakry, Kamsar
Airports:
total: 15
usable: 15
with permanent-surface runways: 4
with runways over 3,659 m: 0
with runways 2,440-3,659 m: 3
with runways 1,220-2,439 m: 10
Telecommunications: poor to fair system of open-wire lines, small radiocommunication stations, and new radio relay system; 15,000 telephones; broadcast stations—3 AM 1 FM, 1 TV; 65,000 TV sets; 200,000 radio receivers; 1 Atlantic Ocean INTELSAT earth station

Defense Forces

Branches: Army, Navy (acts primarily as a coast guard), Air Force, Presidential Guard, Republican Guard, paramilitary National Gendarmerie, National Police Force
Manpower availability: males age 15-49 1,403,776; fit for military service 708,078 (1993 est.)
Defense expenditures: exchange rate conversion—$29 million, 1.2% of GDP (1988)

Guinea-Bissau

100 km

North Atlantic Ocean

Geography

Location: Western Africa, bordering the North Atlantic Ocean between Guinea and Senegal
Map references: Africa, Standard Time Zones of the World
Area:
total area: 36,120 km^2
land area: 28,000 km^2
comparative area: slightly less than three times the size of Connecticut
Land boundaries: total 724 km, Guinea 386 km, Senegal 338 km
Coastline: 350 km
Maritime claims:
exclusive economic zone: 200 nm
territorial sea: 12 nm
International disputes: the International Court of Justice (ICJ) on 12 November 1991 rendered its decision on the Guinea-Bissau/Senegal maritime boundary in favor of Senegal
Climate: tropical; generally hot and humid; monsoonal-type rainy season (June to November) with southwesterly winds; dry season (December to May) with northeasterly harmattan winds
Terrain: mostly low coastal plain rising to savanna in east
Natural resources: unexploited deposits of petroleum, bauxite, phosphates, fish, timber
Land use:
arable land: 11%
permanent crops: 1%
meadows and pastures: 43%
forest and woodland: 38%
other: 7%
Irrigated land: NA km^2
Environment: hot, dry, dusty harmattan haze may reduce visibility during dry season

People

Population: 1,072,439 (July 1993 est.)
Population growth rate: 2.38% (1993 est.)
Birth rate: 41.26 births/1,000 population (1993 est.)

Death rate: 17.45 deaths/1,000 population (1993 est.)
Net migration rate: 0 migrant(s)/1,000 population (1993 est.)
Infant mortality rate: 122.1 deaths/1,000 live births (1993 est.)
Life expectancy at birth:
total population: 47.03 years
male: 45.38 years
female: 48.73 years (1993 est.)
Total fertility rate: 5.6 children born/woman (1993 est.)
Nationality:
noun: Guinea-Bissauan(s)
adjective: Guinea-Bissauan
Ethnic divisions: African 99% (Balanta 30%, Fula 20%, Manjaca 14%, Mandinga 13%, Papel 7%), European and mulatto less than 1%
Religions: indigenous beliefs 65%, Muslim 30%, Christian 5%
Languages: Portuguese (official), Criolo, African languages
Literacy: age 15 and over can read and write (1990)
total population: 36%
male: 50%
female: 24%
Labor force: 403,000 (est.)
by occupation: agriculture 90%, industry, services, and commerce 5%, government 5%
note: population of working age 53% (1983)

Government

Names:
conventional long form: Republic of Guinea-Bissau
conventional short form: Guinea-Bissau
local long form: Republica de Guine-Bissau
local short form: Guine-Bissau
former: Portuguese Guinea
Digraph: PU
Type: republic highly centralized multiparty since mid-1991; the African Party for the Independence of Guinea-Bissau and Cape Verde (PAIGC) held an extraordinary party congress in December 1990 and established a two-year transition program during which the constitution will be revised, allowing for multiple political parties and a presidential election in 1993
Capital: Bissau
Administrative divisions: 9 regions (regioes, singular—regiao); Bafata, Biombo, Bissau, Bolama, Cacheu, Gabu, Oio, Quinara, Tombali
Independence: 10 September 1974 (from Portugal)
Constitution: 16 May 1984
Legal system: NA
National holiday: Independence Day, 10 September (1974)
Political parties and leaders: African Party for the Independence of Guinea-Bissau and

Cape Verde (PAIGC), President Joao Bernardo VIEIRA, leader; Democratic Social Front (FDS), Rafael BARBOSA, leader; Bafata Movement, Domingos Fernandes GARNER, leader; Democratic Front, Aristides MENEZES, leader
note: PAIGC is still the major party (of 10 parties) and controls all aspects of the government
Suffrage: 15 years of age; universal
Elections:
National People's Assembly: last held 15 June 1989 (next to be held 15 June 1994); results—PAIGC is the only party; seats—(150 total) PAIGC 150, appointed by Regional Councils
President of Council of State: last held 19 June 1989 (next to be held NA 1993); results—Gen. Joao Bernardo VIEIRA was reelected without opposition by the National People's Assembly
Executive branch: president of the Council of State, vice presidents of the Council of State, Council of State, Council of Ministers (cabinet)
Legislative branch: unicameral National People's Assembly (Assembleia Nacional Popular)
Judicial branch: none; there is a Ministry of Justice in the Council of Ministers
Leaders:
Chief of State and Head of Government: President of the Council of State Gen. Joao Bernardo VIEIRA (assumed power 14 November 1980 and elected President of Council of State on 16 May 1984)
Member of: ACCT (associate), ACP, AfDB, ECA, ECOWAS, FAO, G-77, IBRD, ICAO, IDA, IDB, IFAD, IFC, ILO, IMF, IMO, IOM (observer), ITU, LORCS, NAM, OAU, OIC, UN, UNAVEM II, UNCTAD, UNESCO, UNIDO, UNOMOZ, UPU, WFTU, WHO, WIPO, WMO, WTO
Diplomatic representation in US:
chief of mission: Ambassador Alfredo Lopes CABRAL
chancery: 918 16th Street NW, Mezzanine Suite, Washington, DC 20006
telephone: (202) 872-4222
US diplomatic representation:
chief of mission: Ambassador Roger A. MAGUIRE
embassy: 17 Avenida Domingos Ramos, Bissau
mailing address: 1067 Bissau Codex, Bissau
telephone: [245] 20-1139, 20-1145, 20-1113
FAX: [245] 20-1159
Flag: two equal horizontal bands of yellow (top) and green with a vertical red band on the hoist side; there is a black five-pointed star centered in the red band; uses the popular pan-African colors of Ethiopia; similar to the flag of Cape Verde, which has the black star raised above the center of the red band and is framed by two corn stalks and a yellow clam shell

Economy

Overview: Guinea-Bissau ranks among the poorest countries in the world, with a per capita GDP of roughly $200. Agriculture and fishing are the main economic activities. Cashew nuts, peanuts, and palm kernels are the primary exports. Exploitation of known mineral deposits is unlikely at present because of a weak infrastructure and the high cost of development. The government's four-year plan (1988-91) targeted agricultural development as the top priority.
National product: GDP—exchange rate conversion—$210 million (1991 est.)
National product real growth rate: 2.3% (1991 est.)
National product per capita: $210 (1991 est.)
Inflation rate (consumer prices): 55% (1991 est.)
Unemployment rate: NA%
Budget: revenues $33.6 million; expenditures $44.8 million, including capital expenditures of $.57 million (1991 est.)
Exports: $20.4 million (f.o.b., 1991 est.)
commodities: cashews, fish, peanuts, palm kernels
partners: Portugal, Senegal, France, The Gambia, Netherlands, Spain
Imports: $63.5 million (f.o.b., 1991 est.)
commodities: capital equipment, consumer goods, semiprocessed goods, foods, petroleum
partners: Portugal, Netherlands, Senegal, USSR, Germany
External debt: $462 million (December 1990 est.)
Industrial production: growth rate 1.0% (1989 est.); accounts for 10% of GDP (1989 est.)
Electricity: 22,000 kW capacity; 30 million kWh produced, 30 kWh per capita (1991)
Industries: agricultural processing, beer, soft drinks
Agriculture: accounts for over 50% of GDP, nearly 100% of exports, and 90% of employment; rice is the staple food; other crops include corn, beans, cassava, cashew nuts, peanuts, palm kernels, and cotton; not self-sufficient in food; fishing and forestry potential not fully exploited
Economic aid: US commitments, including Ex-Im (FY70-89), $49 million; Western (non-US) countries, ODA and OOF bilateral commitments (1970-89), $615 million; OPEC bilateral aid (1979-89), $41 million; Communist countries (1970-89), $68 million
Currency: 1 Guinea-Bissauan peso (PG) = 100 centavos
Exchange rates: Guinea-Bissauan pesos (PG) per US$1—1987.2 (1989), 1363.6 (1988), 851.65 (1987), 238.98 (1986)
Fiscal year: calendar year

Communications

Highways: 3,218 km; 2,698 km bituminous, remainder earth
Inland waterways: scattered stretches are important to coastal commerce
Ports: Bissau
Airports:
total: 33
usable: 15
with permanent-surface runways: 4
with runways over 3,659 m: 0
with runways 2,440-3,659 m: 1
with runways 1,220-2,439 m: 5
Telecommunications: poor system of radio relay, open-wire lines, and radiocommunications; 3,000 telephones; broadcast stations—2 AM, 3 FM, 1 TV

Defense Forces

Branches: People's Revolutionary Armed Force (FARP; including Army, Navy, Air Force), paramilitary force
Manpower availability: males age 15-49 235,931; fit for military service 134,675 (1993 est.)
Defense expenditures: exchange rate conversion—$9.3 million, 5%-6% of GDP (1987)

Guyana

Geography

Location: Northern South America, bordering the North Atlantic Ocean between Suriname and Venezuela
Map references: South America, Standard Time Zones of the World
Area:
total area: 214,970 km²
land area: 196,850 km²
comparative area: slightly smaller than Idaho
Land boundaries: total 2,462 km, Brazil 1,119 km, Suriname 600 km, Venezuela 743 km
Coastline: 459 km
Maritime claims:
continental shelf: 200 nm or the outer edge of continental margin
exclusive fishing zone: 200 nm
territorial sea: 12 nm
International disputes: all of the area west of the Essequibo River claimed by Venezuela; Suriname claims area between New (Upper Courantyne) and Courantyne/Koetari Rivers (all headwaters of the Courantyne)
Climate: tropical; hot, humid, moderated by northeast trade winds; two rainy seasons (May to mid-August, mid-November to mid-January)
Terrain: mostly rolling highlands; low coastal plain; savanna in south
Natural resources: bauxite, gold, diamonds, hardwood timber, shrimp, fish
Land use:
arable land: 3%
permanent crops: 0%
meadows and pastures: 6%
forest and woodland: 83%
other: 8%
Irrigated land: 1,300 km² (1989 est.)
Environment: flash floods a constant threat during rainy seasons; water pollution

Guyana *(continued)*

People

Population: 734,640 (July 1993 est.)
Population growth rate: -0.68% (1993 est.)
Birth rate: 20.47 births/1,000 population (1993 est.)
Death rate: 7.39 deaths/1,000 population (1993 est.)
Net migration rate: -19.89 migrant(s)/1,000 population (1993 est.)
Infant mortality rate: 49.3 deaths/1,000 live births (1993 est.)
Life expectancy at birth:
total population: 64.7 years
male: 61.46 years
female: 68.1 years (1993 est.)
Total fertility rate: 2.35 children born/woman (1993 est.)
Nationality:
noun: Guyanese (singular and plural)
adjective: Guyanese
Ethnic divisions: East Indian 51%, black and mixed 43%, Amerindian 4%, European and Chinese 2%
Religions: Christian 57%, Hindu 33%, Muslim 9%, other 1%
Languages: English, Amerindian dialects
Literacy: age 15 and over having ever attended scool (1990)
total population: 95%
male: 98%
female: 96%
Labor force: 268,000
by occupation: industry and commerce 44.5%, agriculture 33.8%, services 21.7%
note: public-sector employment amounts to 60-80% of the total labor force (1985)

Government

Names:
conventional long form: Co-operative Republic of Guyana
conventional short form: Guyana
former: British Guiana
Digraph: GY
Type: republic
Capital: Georgetown
Administrative divisions: 10 regions; Barima-Waini, Cuyuni-Mazaruni, Demerara-Mahaica, East Berbice-Corentyne, Essequibo Islands-West Demerara, Mahaica-Berbice, Pomeroon-Supenaam, Potaro-Siparuni, Upper Demerara-Berbice, Upper Takutu-Upper Essequibo
Independence: 26 May 1966 (from UK)
Constitution: 6 October 1980
Legal system: based on English common law with certain admixtures of Roman-Dutch law; has not accepted compulsory ICJ jurisdiction
National holiday: Republic Day, 23 February (1970)
Political parties and leaders: People's National Congress (PNC), Hugh Desmond HOYTE; People's Progressive Party (PPP), Cheddi JAGAN; Working People's Alliance (WPA), Eusi KWAYANA, Rupert ROOPNARINE; Democratic Labor Movement (DLM), Paul TENNASSEE; People's Democratic Movement (PDM), Llewellyn JOHN; National Democratic Front (NDF), Joseph BACCHUS; The United Force (TUF), Manzoor NADIR; United Republican Party (URP), Leslie RAMSAMMY; National Republican Party (NRP), Robert GANGADEEN; Guyana Labor Party (GLP), Nanda GOPAUL
Other political or pressure groups: Trades Union Congress (TUC); Guyana Council of Indian Organizations (GCIO); Civil Liberties Action Committee (CLAC)
note: the latter two organizations are small and active but not well organized
Suffrage: 18 years of age; universal
Elections:
Executive President: last held on 5 October 1992; results—Cheddi JAGAN was elected president since he was leader of the party with the most votes in the National Assembly elections
National Assembly: last held on 5 October 1992 (next to be held in 1997); results—PPP 53.4%, PNC 42.3%, WPA 2%, TUF 1.2%; seats—(65 total, 53 elected) PPP 36, PNC 26, WPA 2, TUF 1
Executive branch: executive president, first vice president, prime minister, first deputy prime minister, Cabinet
Legislative branch: unicameral National Assembly
Judicial branch: Supreme Court of Judicature
Leaders:
Chief of State: Executive President Cheddi JAGAN (since 5 October 1992); First Vice President Sam HINDS (since 5 October 1992)
Head of Government: Prime Minister Sam HINDS (since 5 October 1992)
Member of: ACP, C, CARICOM, CCC, CDB, ECLAC, FAO, G-77, GATT, IADB, IBRD, ICAO, ICFTU, IDA, IFAD, IFC, ILO, IMF, IMO, INTERPOL, IOC, ITU, LAES, LORCS, NAM, OAS, UN, UNCTAD, UNESCO, UNIDO, UPU, WCL, WFTU, WHO, WMO
Diplomatic representation in US:
chief of mission: Ambassador Dr. Odeen ISHMAEL
chancery: 2490 Tracy Place NW, Washington, DC 20008
telephone: (202) 265-6900
consulate general: New York
US diplomatic representation:
chief of mission: Ambassador George Jones
embassy: 99-100 Young and Duke Streets, Georgetown
mailing address: P. O. Box 10507, Georgetown

telephone: [592] (2) 54900 through 54909 and 57960 through 57969
FAX: [592] (2) 58497
Flag: green with a red isosceles triangle (based on the hoist side) superimposed on a long yellow arrowhead; there is a narrow black border between the red and yellow, and a narrow white border between the yellow and the green

Economy

Overview: Guyana is one of the world's poorest countries with a per capita income less than one-fifth the South American average. After growing on average at less than 1% a year in 1986-87, GDP dropped by 5% a year in 1988-90. The decline resulted from bad weather, labor trouble in the cane fields, and flooding and equipment problems in the bauxite industry. Consumer prices rose about 100% in 1989 and 75% in 1990, and the current account deficit widened substantially as sugar and bauxite exports fell. Moreover, electric power has been in short supply and constitutes a major barrier to future gains in national output. The government, in association with international financial agencies, seeks to reduce its payment arrears and to raise new funds. The government's stabilization program—aimed at establishing realistic exchange rates, reasonable price stability, and a resumption of growth—requires considerable public administrative abilities and continued patience by consumers during a long incubation period. Buoyed by a recovery in mining and agriculture, the economy posted 6% growth in 1991 and 7% growth in 1992, according to official figures. A large volume of illegal and quasi-legal economic activity is not captured in estimates of the country's total output.
National product: GDP—exchange rate conversion—$267.5 million (1992 est.)
National product real growth rate: 7% (1992 est.)
National product per capita: $370 (1992 est.)
Inflation rate (consumer prices): 15% (1992)
Unemployment rate: 12%-15% (1991 est.)
Budget: revenues $121 million; expenditures $225 million, including capital expenditures of $50 million (1990 est.)
Exports: $268 million (f.o.b., 1992 est.)
commodities: sugar, bauxite/alumina, rice, gold, shrimp, molasses, timber, rum
partners: UK 28%, US 25%, FRG 8%, Canada 7%, Japan 6% (1989)
Imports: $242.4 million (f.o.b., 1990 est.)
commodities: manufactures, machinery, food, petroleum
partners: US 40%, Trinidad & Tobago 13%, UK 11%, Japan 5%, Netherland Antilles 3% (1989)

External debt: $2.0 billion, including arrears (1990)
Industrial production: growth rate 12% (1990 est.); accounts for about 24% of GDP
Electricity: 253,500 kW capacity; 276 million kWh produced, 370 kWh per capita (1992)
Industries: bauxite mining, sugar, rice milling, timber, fishing (shrimp), textiles, gold mining
Agriculture: most important sector, accounting for 25% of GDP and about half of exports; sugar and rice are key crops; development potential exists for fishing and forestry; not self-sufficient in food, especially wheat, vegetable oils, and animal products
Economic aid: US commitments, including Ex-Im (FY70-89), $116 million; Western (non-US) countries, ODA and OOF bilateral commitments (1970-89), $325 million; Communist countries 1970-89, $242 million
Currency: 1 Guyanese dollar (G$) = 100 cents
Exchange rates: Guyanese dollars (G$) per US$1—125.8 (January 1993) 125.0 (1992), 111.8 (1991), 39.533 (1990), 27.159 (1989), 10.000 (1988)
Fiscal year: calendar year

Communications

Railroads: 187 km total, all single track 0.914-meter gauge
Highways: 7,665 km total; 550 km paved, 5,000 km gravel, 1,525 km earth, 590 km unimproved
Inland waterways: 6,000 km total of navigable waterways; Berbice, Demerara, and Essequibo Rivers are navigable by oceangoing vessels for 150 km, 100 km, and 80 km, respectively
Ports: Georgetown, New Amsterdam
Merchant marine: 1 cargo ship (1,000 GRT or over) totaling 1,317 GRT/2,558 DWT
Airports:
total: 53
usable: 48
with permanent-surface runways: 5
with runways over 3,659 m: 0
with runways 2,440-3,659 m: 0
with runways 1,220-2,439 m: 13
Telecommunications: fair system with radio relay network; over 27,000 telephones; tropospheric scatter link to Trinidad; broadcast stations—4 AM, 3 FM, no TV, 1 shortwave; 1 Atlantic Ocean INTELSAT earth station

Defense Forces

Branches: Guyana Defense Force (GDF; including the Ground Forces, Coast Guard and Air Corps), Guyana People's Militia (GPM), Guyana National Service (GNS)
Manpower availability: males age 15-49 196,960; fit for military service 149,583 (1993 est.)
Defense expenditures: $NA, NA% of GDP

Haiti

Geography

Location: in the northern Caribbean Sea, about 90 km southeast of Cuba
Map references: Central America and the Caribbean, Standard Time Zones of the World
Area:
total area: 27,750 km²
land area: 27,560 km²
comparative area: slightly larger than Maryland
Land boundaries: total 275 km, Dominican Republic 275 km
Coastline: 1,771 km
Maritime claims:
contiguous zone: 24 nm
continental shelf: to depth of exploitation
exclusive economic zone: 200 nm
territorial sea: 12 nm
International disputes: claims US-administered Navassa Island
Climate: tropical; semiarid where mountains in east cut off trade winds
Terrain: mostly rough and mountainous
Natural resources: bauxite
Land use:
arable land: 20%
permanent crops: 13%
meadows and pastures: 18%
forest and woodland: 4%
other: 45%
Irrigated land: 750 km² (1989 est.)
Environment: lies in the middle of the hurricane belt and subject to severe storms from June to October; occasional flooding and earthquakes; deforestation; soil erosion
Note: shares island of Hispaniola with Dominican Republic (western one-third is Haiti, eastern two-thirds is the Dominican Republic)

People

Population: 6,384,877 (July 1993 est.)
Population growth rate: 1.68% (1993 est.)
Birth rate: 40.77 births/1,000 population

Haiti (continued)

(1993 est.)

Death rate: 18.88 deaths/1,000 population (1993 est.)

Net migration rate: -5.04 migrant(s)/1,000 population (1993 est.)

Infant mortality rate: 109.5 deaths/1,000 live births (1993 est.)

Life expectancy at birth:

total population: 45.45 years

male: 43.88 years

female: 47.11 years (1993 est.)

Total fertility rate: 6.05 children born/woman (1993 est.)

Nationality:

noun: Haitian(s)

adjective: Haitian

Ethnic divisions: black 95%, mulatto and European 5%

Religions: Roman Catholic 80% (of which an overwhelming majority also practice Voodoo), Protestant 16% (Baptist 10%, Pentecostal 4%, Adventist 1%, other 1%), none 1%, other 3% (1982)

Languages: French (official) 10%, Creole

Literacy: age 15 and over can read and write (1990)

total population: 53%

male: 59%

female: 47%

Labor force: 2.3 million

by occupation: agriculture 66%, services 25%, industry 9%

note: shortage of skilled labor, unskilled labor abundant (1982)

Government

Names:

conventional long form: Republic of Haiti

conventional short form: Haiti

local long form: Republique d'Haiti

local short form: Haiti

Digraph: HA

Type: republic

Capital: Port-au-Prince

Administrative divisions: 9 departments, (departements, singular—departement); Artibonite, Centre, Grand'Anse, Nord, Nord-Est, Nord-Ouest, Ouest, Sud, Sud-Est

Independence: 1 January 1804 (from France)

Constitution: 27 August 1983, suspended February 1986; draft constitution approved March 1987, suspended June 1988, most articles reinstated March 1989; October 1991, government claims to be observing the Constitution

Legal system: based on Roman civil law system; accepts compulsory ICJ jurisdiction

National holiday: Independence Day, 1 January (1804)

Political parties and leaders: National Front for Change and Democracy (FNCD), including National Congress of Democratic Movements (CONACOM), Victor BENOIT, and National Cooperative Action Movement (MKN), Volvick Remy JOSEPH; Movement for the Installation of Democracy in Haiti (MIDH), Marc BAZIN; National Progressive Revolutionary Party (PANPRA), Serge GILLES; National Patriotic Movement of November 28 (MNP-28), Dejean BELIZAIRE; National Agricultural and Industrial Party (PAIN), Louis DEJOIE; Movement for National Reconstruction (MRN), Rene THEODORE; Haitian Christian Democratic Party (PDCH), Joseph DOUZE; Assembly of Progressive National Democrats (RDNP), Leslie MANIGAT; National Party of Labor (PNT), Thomas DESULME; Mobilization for National Development (MDN), Hubert DE RONCERAY; Democratic Movement for the Liberation of Haiti (MODELH), Francois LATORTUE; Haitian Social Christian Party (PSCH), Gregoire EUGENE; Movement for the Organization of the Country (MOP), Gesner COMEAU and Jean MOLIERE

Other political or pressure groups: Democratic Unity Confederation (KID); Roman Catholic Church; Confederation of Haitian Workers (CTH); Federation of Workers Trade Unions (FOS); Autonomous Haitian Workers (CATH); National Popular Assembly (APN)

Suffrage: 18 years of age; universal

Elections:

Chamber of Deputies: last held 16 December 1990, with runoff held 20 January 1991 (next to be held by December 1994); results—percent of vote NA; seats—(83 total) FNCD 27, ANDP 17, PDCH 7, PAIN 6, RDNP 6, MDN 5, PNT 3, MKN 2, MODELH 2, MRN 1, independents 5, other 2

President: last held 16 December 1990 (next election to be held by December 1995); results—Rev. Jean-Bertrand ARISTIDE 67.5%, Marc BAZIN 14.2%, Louis DEJOIE 4.9%

Senate: last held 18 January 1993, widely condemned as illegitimate (next to be held December 1994); results—percent of vote NA; seats—(27 total) FNCD 12, ANDP 8, PAIN 2, MRN 1, RDNP 1, PNT 1, independent 2

Executive branch: president, Council of Ministers (cabinet)

Legislative branch: bicameral National Assembly (Assemblee Nationale) consisting of an upper house or Senate and a lower house or Chamber of Deputies

Judicial branch: Court of Appeal (Cour de Cassation)

Leaders:

Chief of State: President Jean-Bertrand ARISTIDE (since 7 February 1991), ousted in a coup in September 1991, but still recognized by international community as Chief of State

Head of Government: de facto Prime Minister Marc BAZIN (since NA June 1992)

Member of: ACCT, ACP, CARICOM (observer), CCC, ECLAC, FAO, G-77, GATT, IADB, IAEA, IBRD, ICAO, IDA, IFAD, IFC, ILO, IMF, IMO, INTELSAT, INTERPOL, IOC, ITU, LAES, LORCS, OAS, OPANAL, PCA, UN, UNCTAD, UNESCO, UNIDO, UPU, WCL, WFTU, WHO, WIPO, WMO, WTO

Diplomatic representation in US:

chief of mission: Ambassador Jean CASIMIR

chancery: 2311 Massachusetts Avenue NW, Washington, DC 20008

telephone: (202) 332-4090 through 4092

consulates general: Boston, Chicago, Miami, New York, and San Juan (Puerto Rico)

US diplomatic representation:

chief of mission: Special Charge d'Affaires Charles REDMAN

embassy: Harry Truman Boulevard, Port au-Prince

mailing address: P. O. Box 1761, Port-au-Prince

telephone: [509] 22-0354, 22-0368, 22-0200, or 22-0612

FAX: [509] 23-9007

Flag: two equal horizontal bands of blue (top) and red with a centered white rectangle bearing the coat of arms, which contains a palm tree flanked by flags and two cannons above a scroll bearing the motto L'UNION FAIT LA FORCE (Union Makes Strength)

Economy

Overview: About 75% of the population live in abject poverty. Agriculture is mainly small-scale subsistence farming and employs nearly three-fourths of the work force. The majority of the population does not have ready access to safe drinking water, adequate medical care, or sufficient food. Few social assistance programs exist, and the lack of employment opportunities remains one of the most critical problems facing the economy, along with soil erosion and political instability. Trade sanctions applied by the Organization of American States in response to the September 1991 coup against President ARISTIDE have further damaged the economy.

National product: GDP—exchange rate conversion—$2.2 billion (1991 est.)

National product real growth rate: -4% (FY91 est.)

National product per capita: $340 (1991 est.)

Inflation rate (consumer prices): 20% (FY91 est.)

Unemployment rate: 25-50% (1991)

Budget: revenues $300 million; expenditures $416 million, including capital expenditures of $145 million (1990 est.)

Exports: $146 million (f.o.b., 1991 est.)
commodities: light manufactures 65%, coffee 19%, other agriculture 8%, other 8%
partners: US 84%, Italy 4%, France 3%, other industrial countries 6%, less developed countries 3% (1987)
Imports: $252 million (f.o.b., 1991 est.)
commodities: machines and manufactures 34%, food and beverages 22%, petroleum products 14%, chemicals 10%, fats and oils 9%
partners: US 64%, Netherlands Antilles 5%, Japan 5%, France 4%, Canada 3%, Germany 3% (1987)
External debt: $838 million (December 1990)
Industrial production: growth rate −2.0% (1991 est.); accounts for 15% of GDP
Electricity: 217,000 kW capacity; 480 million kWh produced, 75 kWh per capita (1992)
Industries: sugar refining, textiles, flour milling, cement manufacturing, tourism, light assembly industries based on imported parts
Agriculture: accounts for 28% of GDP and employs around 70% of work force; mostly small-scale subsistence farms; commercial crops—coffee, mangoes, sugarcane, wood; staple crops—rice, corn, sorghum; shortage of wheat flour
Illicit drugs: transshipment point for cocaine
Economic aid: US commitments, including Ex-Im (1970-89), $700 million; Western (non-US) countries, ODA and OOF bilateral commitments (1970-89), $770 million
Currency: 1 gourde (G) = 100 centimes
Exchange rates: gourdes (G) per US$1—8.4 (December 1991), fixed rate of 5.000 through second quarter of 1991)
Fiscal year: 1 October-30 September

Communications

Railroads: 40 km 0.760-meter narrow gauge, single-track, privately owned industrial line
Highways: 4,000 km total; 950 km paved, 900 km otherwise improved, 2,150 km unimproved
Inland waterways: negligible; less than 100 km navigable
Ports: Port-au-Prince, Cap-Haitien
Airports:
total: 13
usable: 10
with permanent-surface runways: 3
with runways over 3,659 m: 0
with runways 2,440-3,659 m: 1
with runways 1,220-2,439 m: 3
Telecommunications: domestic facilities barely adequate, international facilities slightly better; 36,000 telephones; broadcast stations—33 AM, no FM, 4 TV, 2 shortwave; 1 Atlantic Ocean INTELSAT earth station

Defense Forces

Branches: Army (including Police), Navy, Air Force
Manpower availability: males age 15-49 1,289,310; fit for military service 695,997; reach military age (18) annually 60,588 (1993 est.)
Defense expenditures: exchange rate conversion—$34 million, 1.5% of GDP (1988 est.)

Heard Island and McDonald Islands
(territory of Australia)

Geography

Location: in the Indian Ocean, 4,100 km southwest of Australia
Map references: Antarctic Region
Area:
total area: 412 km²
land area: 412 km²
comparative area: slightly less than 2.5 times the size of Washington, DC
Land boundaries: 0 km
Coastline: 101.9 km
Maritime claims:
exclusive fishing zone: 200 nm
territorial sea: 3 nm
International disputes: none
Climate: antarctic
Terrain: Heard Island—bleak and mountainous, with an extinct volcano; McDonald Islands—small and rocky
Natural resources: none
Land use:
arable land: 0%
permanent crops: 0%
meadows and pastures: 0%
forest and woodland: 0%
other: 100%
Irrigated land: 0 km²
Environment: primarily used for research stations

People

Population: uninhabited

Government

Names:
conventional long form: Territory of Heard Island and McDonald Islands
conventional short form: Heard Island and McDonald Islands
Digraph: HM
Type: territory of Australia administered by the Ministry for Arts, Sport, the Environment, Tourism and Territories

Heard Island and McDonald Islands (continued)

Capital: none; administered from Canberra, Australia
Independence: none (territory of Australia)

Economy

Overview: no economic activity

Communications

Ports: none; offshore anchorage only

Defense Forces

Note: defense is the responsibility of Australia

Holy See (Vatican City)

250 meters

Geography

Location: Southern Europe, an enclave of Rome—central Italy
Map references: Europe
Area:
total area: 0.44 km²
land area: 0.44 km²
comparative area: about 0.7 times the size of The Mall in Washington, DC
Land boundaries: total 3.2 km, Italy 3.2 km
Coastline: 0 km (landlocked)
Maritime claims: none; landlocked
International disputes: none
Climate: temperate; mild, rainy winters (September to mid-May) with hot, dry summers (May to September)
Terrain: low hill
Natural resources: none
Land use:
arable land: 0%
permanent crops: 0%
meadows and pastures: 0%
forest and woodland: 0%
other: 100%
Irrigated land: 0 km²
Environment: urban
Note: landlocked; enclave of Rome, Italy; world's smallest state; outside the Vatican City, 13 buildings in Rome and Castel Gandolfo (the pope's summer residence) enjoy extraterritorial rights

People

Population: 811 (July 1993 est.)
Population growth rate: 1.15% (1993 est.)
Birth rate: NA births/1,000 population
Death rate: NA deaths/1,000 population
Net migration rate: NA migrant(s)/1,000 population
Infant mortality rate: NA deaths/1,000 live births
Life expectancy at birth:
total population: NA years
male: NA years
female: NA years
Total fertility rate: NA children born/woman
Nationality:
noun: none
adjective: none
Ethnic divisions: Italians, Swiss
Religions: Roman Catholic
Languages: Italian, Latin, various other languages
Literacy:
total population: NA%
male: NA%
female: NA%
Labor force: NA
by occupation: dignitaries, priests, nuns, guards, and 3,000 lay workers who live outside the Vatican

Government

Names:
conventional long form: The Holy See (State of the Vatican City)
conventional short form: Holy See (Vatican City)
local long form: Santa Sede (Stato della Citta del Vaticano)
local short form: Santa Sede (Citta del Vaticano)
Digraph: VT
Type: monarchical-sacerdotal state
Capital: Vatican City
Independence: 11 February 1929 (from Italy)
Constitution: Apostolic Constitution of 1967 (effective 1 March 1968)
Legal system: NA
National holiday: Installation Day of the Pope, 22 October (1978) (John Paul II)
note: Pope John Paul II was elected on 16 October 1978
Political parties and leaders: none
Other political or pressure groups: none (exclusive of influence exercised by church officers)
Suffrage: limited to cardinals less than 80 years old
Elections:
Pope: last held 16 October 1978 (next to be held after the death of the current pope); results—Karol WOJTYLA was elected for life by the College of Cardinals
Executive branch: pope
Legislative branch: unicameral Pontifical Commission
Judicial branch: none; normally handled by Italy
Leaders:
Chief of State: Pope JOHN PAUL II (Karol WOJTYLA; since 16 October 1978)
Head of Government: Secretary of State Archbishop Angelo Cardinal SODANO (since NA)

Member of: CSCE, IAEA, ICFTU, IMF (observer), INTELSAT, IOM (observer), ITU, OAS (observer), UN (observer), UNCTAD, UNHCR, UPU, WIPO, WTO (observer)

Diplomatic representation in US:
chief of mission: Apostolic Pro-Nuncio Archbishop Agostino CACCIAVILLAN
chancery: 3339 Massachusetts Avenue NW, Washington, DC 20008
telephone: (202) 333-7121

US diplomatic representation:
chief of mission: Ambassador Raymond L. FLYNN
embassy: Villino Pacelli, Via Aurelia 294, 00165 Rome
mailing address: PSC 59, APO AE 09624
telephone: [396] 46741
FAX: [396] 638-0159

Flag: two vertical bands of yellow (hoist side) and white with the crossed keys of Saint Peter and the papal tiara centered in the white band

Economy

Overview: This unique, noncommercial economy is supported financially by contributions (known as Peter's Pence) from Roman Catholics throughout the world, the sale of postage stamps and tourist mementos, fees for admission to museums, and the sale of publications. The incomes and living standards of lay workers are comparable to, or somewhat better than, those of counterparts who work in the city of Rome.

Budget: revenues $86 million; expenditures $178 million, including capital expenditures of $NA (1993 est.)

Electricity: 5,000 kW standby capacity (1992); power supplied by Italy

Industries: printing and production of a small amount of mosaics and staff uniforms; worldwide banking and financial activities

Currency: 1 Vatican lira (VLit) = 100 centesimi

Exchange rates: Vatican lire (VLit) per US$1—1,482.5 (January 1993), 1,232.4 (1992), 1,240.6 (1991), 1,198.1 (1990), 1,372.1 (1989), 1,301.6 (1988); note—the Vatican lira is at par with the Italian lira which circulates freely

Fiscal year: calendar year

Communications

Railroads: 850 m, 750 mm gauge (links with Italian network near the Rome station of Saint Peter's)

Highways: none; all city streets

Telecommunications: broadcast stations—3 AM, 4 FM, no TV; 2,000-line automatic telephone exchange; no communications satellite systems

Defense Forces

Note: defense is the responsibility of Italy; Swiss Papal Guards are posted at entrances to the Vatican City

Honduras

Boundary representation is not necessarily authoritative

Geography

Location: Central America, between Guatemala and Nicaragua

Map references: Central America and the Caribbean, North America, Standard Time Zones of the World

Area:
total area: 112,090 km²
land area: 111,890 km²
comparative area: slightly larger than Tennessee

Land boundaries: total 1,520 km, Guatemala 256 km, El Salvador 342 km, Nicaragua 922 km

Coastline: 820 km

Maritime claims:
contiguous zone: 24 nm
continental shelf: 200 m depth or to depth of exploitation
exclusive economic zone: 200 nm
territorial sea: 12 nm

International disputes: land boundary dispute with El Salvador mostly resolved by 11 September 1992 International Court of Justice (ICJ) decision; ICJ referred the maritime boundary in the Golfo de Fonseca to an earlier agreement in this century and advised that some tripartite resolution among El Salvador, Honduras and Nicaragua likely would be required

Climate: subtropical in lowlands, temperate in mountains

Terrain: mostly mountains in interior, narrow coastal plains

Natural resources: timber, gold, silver, copper, lead, zinc, iron ore, antimony, coal, fish

Land use:
arable land: 14%
permanent crops: 2%
meadows and pastures: 30%
forest and woodland: 34%
other: 20%

Irrigated land: 900 km² (1989 est.)

Honduras (continued)

Environment: subject to frequent, but generally mild, earthquakes; damaging hurricanes and floods along Caribbean coast; deforestation; soil erosion

People

Population: 5,170,108 (July 1993 est.)
Population growth rate: 2.8% (1993 est.)
Birth rate: 35.82 births/1,000 population (1993 est.)
Death rate: 6.44 deaths/1,000 population (1993 est.)
Net migration rate: -1.43 migrant(s)/1,000 population (1993 est.)
Infant mortality rate: 47.2 deaths/1,000 live births (1993 est.)
Life expectancy at birth:
total population: 67.17 years
male: 64.82 years
female: 69.62 years (1993 est.)
Total fertility rate: 4.87 children born/woman (1993 est.)
Nationality:
noun: Honduran(s)
adjective: Honduran
Ethnic divisions: mestizo (mixed Indian and European) 90%, Indian 7%, black 2%, white 1%
Religions: Roman Catholic 97%, Protestant minority
Languages: Spanish, Indian dialects
Literacy: age 15 and over can read and write (1990)
total population: 73%
male: 76%
female: 71%
Labor force: 1.3 million
by occupation: agriculture 62%, services 20%, manufacturing 9%, construction 3%, other 6% (1985)

Government

Names:
conventional long form: Republic of Honduras
conventional short form: Honduras
local long form: Republica de Honduras
local short form: Honduras
Digraph: HO
Type: republic
Capital: Tegucigalpa
Administrative divisions: 18 departments (departamentos, singular—departamento); Atlantida, Choluteca, Colon, Comayagua, Copan, Cortes, El Paraiso, Francisco Morazan, Gracias a Dios, Intibuca, Islas de la Bahia, La Paz, Lempira, Ocotepeque, Olancho, Santa Barbara, Valle, Yoro
Independence: 15 September 1821 (from Spain)
Constitution: 11 January 1982, effective 20 January 1982

Legal system: rooted in Roman and Spanish civil law; some influence of English common law; accepts ICJ jurisdiction, with reservations
National holiday: Independence Day, 15 September (1821)
Political parties and leaders: Liberal Party (PLH), Carlos Roberto REINA, presidential candidate, Rafael PINEDA Ponce, president; National Party (PN) has two factions: Movimiento Nacional de Reivindication Callejista (Monarca), Rafael Leonardo CALLEJAS, and Oswaldista, Oswaldo RAMOS SOTO, presidential candidate; National Innovation and Unity Party (PINU), German LEITZELAR, president; Christian Democratic Party (PDCH), Efrain DIAZ Arrivillaga, president
Other political or pressure groups: National Association of Honduran Campesinos (ANACH); Honduran Council of Private Enterprise (COHEP); Confederation of Honduran Workers (CTH); National Union of Campesinos (UNC); General Workers Confederation (CGT); United Federation of Honduran Workers (FUTH); Committee for the Defense of Human Rights in Honduras (CODEH); Coordinating Committee of Popular Organizations (CCOP)
Suffrage: 18 years of age; universal and compulsory
Elections:
President: last held on 26 November 1989 (next to be held November 1993); results—Rafael Leonardo CALLEJAS (PNH) 51%, Carlos FLORES Facusse (PLH) 43.3%, other 5.7%
National Congress: last held on 26 November 1989 (next to be held November 1993); results—PNH 51%, PLH 43%, PDCH 1.9%, PINU-SD 1.5%, other 2.6%; seats—(128 total) PNH 71, PLH 55, PINU-SD 2
Executive branch: president, Council of Ministers (cabinet)
Legislative branch: unicameral National Congress (Congreso Nacional)
Judicial branch: Supreme Court of Justice (Corte Suprema de Justica)
Leaders:
Chief of State and Head of Government: President Rafael Leonardo CALLEJAS Romero (since 26 January 1990)
Member of: BCIE, CACM, ECLAC, FAO, G-77, IADB, IBRD, ICAO, ICFTU, IDA, IFAD, IFC, ILO, IMF, IMO, INTELSAT, INTERPOL, IOC, IOM, ITU, LAES, LAIA (observer), LORCS, OAS, OPANAL, PCA, UN, UNCTAD, UNESCO, UNIDO, UPU, WCL, WFTU, WHO, WIPO, WMO
Diplomatic representation in US:
chief of mission: Ambassador Rene Arturo BENDANA-VALENZUELA
chancery: 3007 Tilden Street NW, Washington, DC 20008

telephone: (202) 966-7702
consulates general: Chicago, Los Angeles, Miami, New Orleans, New York, and San Francisco
consulates: Baton Rouge, Boston, Detroit, Houston, and Jacksonville
US diplomatic representation:
chief of mission: Ambassador William Bryce (since 28 May 1993)
embassy: Avenida La Paz, Tegucigalpa
mailing address: APO AA 34022, Tegucigalpa
telephone: [504] 32-3120
FAX: [504] 32-0027
Flag: three equal horizontal bands of blue (top), white, and blue with five blue five-pointed stars arranged in an X pattern centered in the white band; the stars represent the members of the former Federal Republic of Central America—Costa Rica, El Salvador, Guatemala, Honduras, and Nicaragua; similar to the flag of El Salvador, which features a round emblem encircled by the words REPUBLICA DE EL SALVADOR EN LA AMERICA CENTRAL centered in the white band; also similar to the flag of Nicaragua, which features a triangle encircled by the word REPUBLICA DE NICARAGUA on top and AMERICA CENTRAL on the bottom, centered in the white band

Economy

Overview: Honduras is one of the poorest countries in the Western Hemisphere. Agriculture, the most important sector of the economy, accounts for more than 25% of GDP, employs 62% of the labor force, and produces two-thirds of exports. Productivity remains low. Industry, still in its early stages, employs nearly 9% of the labor force, accounts for 15% of GDP, and generates 20% of exports. The service sectors, including public administration, account for 50% of GDP and employ nearly 20% of the labor force. Basic problems facing the economy include rapid population growth, high unemployment, a lack of basic services, a large and inefficient public sector, and the dependence of the export sector mostly on coffee and bananas, which are subject to sharp price fluctuations. A far-reaching reform program initiated by President CALLEJAS in 1990 is beginning to take hold.
National product: GDP—exchange rate conversion—$5.5 billion (1992 est.)
National product real growth rate: 3.6% (1992 est.)
National product per capita: $1,090 (1992 est.)
Inflation rate (consumer prices): 8% (1992 est.)
Unemployment rate: 15% (30-40% underemployed) (1989)

Budget: revenues $1.4 billion; expenditures $1.9 billion, including capital expenditures of $511 million (1990 est.)
Exports: $1.0 billion (f.o.b., 1991)
commodities: bananas, coffee, shrimp, lobster, minerals, meat, lumber
partners: US 65%, Germany 9%, Japan 8%, Belgium 7%
Imports: $1.3 billion (c.i.f. 1991)
commodities: machinery and transport equipment, chemical products, manufactured goods, fuel and oil, foodstuffs
partners: US 45%, Japan 9%, Netherlands 7%, Mexico 7%, Venezuela 6%
External debt: $2.8 billion (1990)
Industrial production: growth rate 0.8% (1990 est.); accounts for 15% of GDP
Electricity: 575,000 kW capacity; 2,000 million kWh produced, 390 kWh per capita (1992)
Industries: agricultural processing (sugar and coffee), textiles, clothing, wood products
Agriculture: most important sector, accounting for more than 25% of GDP, more than 60% of the labor force, and two-thirds of exports; principal products include bananas, coffee, timber, beef, citrus fruit, shrimp; importer of wheat
Illicit drugs: illicit producer of cannabis, cultivated on small plots and used principally for local consumption; transshipment point for cocaine
Economic aid: US commitments, including Ex-Im (FY70-89), $1.4 billion; Western (non-US) countries, ODA and OOF bilateral commitments (1970-89), $1.1 billion
Currency: 1 lempira (L) = 100 centavos
Exchange rates: lempiras (L) per US$1—5.4 (fixed rate); 5.70 parallel black-market rate (November 1990); the lempira was allowed to float in 1992; current rate about US$1—5.65
Fiscal year: calendar year

Communications

Railroads: 785 km total; 508 km 1.067-meter gauge, 277 km 0.914-meter gauge
Highways: 8,950 km total; 1,700 km paved, 5,000 km otherwise improved, 2,250 km unimproved earth
Inland waterways: 465 km navigable by small craft
Ports: Puerto Castilla, Puerto Cortes, San Lorenzo
Merchant marine: 252 ships (1,000 GRT or over) totaling 819,100 GRT/1,195,276 DWT; includes 2 passenger-cargo, 162 cargo, 20 refrigerated cargo, 10 container, 6 roll-on/roll-off cargo, 22 oil tanker, 1 chemical tanker, 2 specialized tanker, 22 bulk, 3 passenger, 2 short-sea passenger; note—a flag of convenience registry; Russia owns 10 ships under the Honduran flag

Airports:
total: 165
usable: 137
with permanent-surface runways: 11
with runways over 3,659 m: 0
with runways 2,440-3,659 m: 4
with runways 1,220-2,439 m: 14
Telecommunications: inadequate system with only 7 telephones per 1,000 persons; international services provided by 2 Atlantic Ocean INTELSAT earch stations and the Central American microwave radio relay system; broadcast stations—176 AM, no FM, 7 SW, 28 TV

Defense Forces

Branches: Army, Navy (including Marines), Air Force, Public Security Forces (FUSEP)
Manpower availability: males age 15-49 1,185,072; fit for military service 706,291; reach military age (18) annually 58,583 (1993 est.)
Defense expenditures: exchange rate conversion—$45 million, about 1% of GDP (1993 est.)

Hong Kong
(dependent territory of the UK)

15 km

Lema Channel

Geography

Location: East Asia, on the southeast coast of China bordering the South China Sea
Map references: Asia, Southeast Asia, Standard Time Zones of the World
Area:
total area: 1,040 km²
land area: 990 km²
comparative area: slightly less than six times the size of Washington, DC
Land boundaries: total 30 km, China 30 km
Coastline: 733 km
Maritime claims:
exclusive fishing zone: 3 nm
territorial sea: 3 nm
International disputes: none
Climate: tropical monsoon; cool and humid in winter, hot and rainy from spring through summer, warm and sunny in fall
Terrain: hilly to mountainous with steep slopes; lowlands in north
Natural resources: outstanding deepwater harbor, feldspar
Land use:
arable land: 7%
permanent crops: 1%
meadows and pastures: 1%
forest and woodland: 12%
other: 79%
Irrigated land: 20 km² (1989)
Environment: more than 200 islands; occasional typhoons

People

Population: 5,552,965 (July 1993 est.)
Population growth rate: -0.06% (1993 est.)
Birth rate: 12.27 births/1,000 population (1993 est.)
Death rate: 5.68 deaths/1,000 population (1993 est.)
Net migration rate: -7.2 migrant(s)/1,000 population (1993 est.)
Infant mortality rate: 5.9 deaths/1,000 live births (1993 est.)

Hong Kong (continued)

Life expectancy at birth:
total population: 79.99 years
male: 76.55 years
female: 83.64 years (1993 est.)
Total fertility rate: 1.34 children
born/woman (1993 est.)
Nationality:
noun: Chinese
adjective: Chinese
Ethnic divisions: Chinese 98%, other 2%
Religions: eclectic mixture of local religions
90%, Christian 10%
Languages: Chinese (Cantonese), English
Literacy: age 15 and over can read and
write (1971)
total population: 77%
male: 90%
female: 64%
Labor force: 2.8 million (1990)
by occupation: manufacturing 28.5%,
wholesale and retail trade, restaurants, and
hotels 27.9%, services 17.7%, financing,
insurance, and real estate 9.2%, transport
and communications 4.5%, construction
2.5%, other 9.7% (1989)

Government

Names:
conventional long form: none
conventional short form: Hong Kong
Abbreviation: HK
Digraph: HK
Type: dependent territory of the UK
scheduled to revert to China in 1997
Capital: Victoria
Administrative divisions: none (dependent
territory of the UK)
Independence: none (dependent territory of
the UK; the UK signed an agreement with
China on 19 December 1984 to return Hong
Kong to China on 1 July 1997; in the joint
declaration, China promises to respect Hong
Kong's existing social and economic systems
and lifestyle)
Constitution: unwritten; partly statutes,
partly common law and practice; new Basic
Law approved in March 1990 in preparation
for 1997
Legal system: based on English common
law
National holiday: Liberation Day, 29
August (1945)
Political parties and leaders: United
Democrats of Hong Kong, Martin LEE,
chairman; Democratic Alliance for the
Betterment of Hong Kong; Hong Kong
Democratic Foundation
Other political or pressure groups:
Cooperative Resources Center, Allen LEE,
chairman; Meeting Point, Anthony
CHEUNG, chairman; Association of
Democracy and People's Livelihood,
Frederick FUNG Kin Kee, chairman; Liberal
Democratic Federation, HEUNG Yee Kuk;

Federation of Trade Unions (pro-China);
Hong Kong and Kowloon Trade Union
Council (pro-Taiwan); Confederation of
Trade Unions (prodemocracy); Hong Kong
General Chamber of Commerce; Chinese
General Chamber of Commerce (pro-China);
Federation of Hong Kong Industries;
Chinese Manufacturers' Association of Hong
Kong; Hong Kong Professional Teachers'
Union; Hong Kong Alliance in Support of
the Patriotic Democratic Movement in China
Suffrage: direct election 21 years of age;
universal as a permanent resident living in
the territory of Hong Kong for the past
seven years indirect election limited to about
100,000 professionals of electoral college
and functional constituencies
Elections:
Legislative Council: indirect elections last
held 12 September 1991 and direct elections
were held for the first time 15 September
1991 (next to be held in September 1995
when the number of directly-elected seats
increases to 20); results—percent of vote by
party NA; seats—(60 total; 21 indirectly
elected by functional constituencies, 18
directly elected, 18 appointed by governor, 3
ex officio members); indirect
elections—number of seats by functional
constituency NA; direct elections—UDHK
12, Meeting Point 3, ADPL 1, other 2
Executive branch: British monarch,
governor, chief secretary of the Executive
Council
Legislative branch: unicameral Legislative
Council
Judicial branch: Supreme Court
Leaders:
Chief of State: Queen ELIZABETH II (since
6 February 1952)
Head of Government: Governor Chris
PATTEN (since NA July 1992); Chief
Secretary Sir David Robert FORD (since
NA February 1987)
Member of: APEC, AsDB, CCC, ESCAP
(associate), GATT, ICFTU, IMO (associate),
INTERPOL (subbureau), IOC, ISO
(correspondent), WCL, WMO
Diplomatic representation in US: as a
dependent territory of the UK, the interests
of Hong Kong in the US are represented by
the UK
US diplomatic representation:
chief of mission: Consul General Richard L.
WILLIAMS
embassy: Consulate General at 26 Garden
Road, Hong Kong
mailing address: Box 30, Hong Kong, or
FPO AP 96522-0002
telephone: [852] 239-011
Flag: blue with the flag of the UK in the
upper hoist-side quadrant with the Hong
Kong coat of arms on a white disk centered
on the outer half of the flag; the coat of
arms contains a shield (bearing two junks

below a crown) held by a lion (representing
the UK) and a dragon (representing China)
with another lion above the shield and a
banner bearing the words HONG KONG
below the shield

Economy

Overview: Hong Kong has a bustling free
market economy with few tariffs or nontariff
barriers. Natural resources are limited, and
food and raw materials must be imported.
Manufacturing accounts for about 18% of
GDP, employs 28% of the labor force, and
exports about 90% of its output. Real GDP
growth averaged a remarkable 8% in
1987-88, slowed to 3.0% in 1989-90, and
picked up to 4.2% in 1991 and 5.9% in
1992. Unemployment, which has been
declining since the mid-1980s, is now about
2%. A shortage of labor continues to put
upward pressure on prices and the cost of
living. Short-term prospects remain bright so
long as major trading partners continue to be
reasonably prosperous.
National product: GDP—exchange rate
conversion—$86 billion (1992 est.)
National product real growth rate: 5.9%
(1992)
National product per capita: $14,600
(1992 est.)
Inflation rate (consumer prices): 9.4%
(1992)
Unemployment rate: 2% (1992 est.)
Budget: revenues $17.4 billion; expenditures
$14.7 billion, including capital expenditures
of $NA (FY92)
Exports: $118 billion, including reexports of
$85.1 billion (f.o.b., 1992 est.)
commodities: clothing, textiles, yarn and
fabric, footwear, electrical appliances,
watches and clocks, toys
partners: US 29%, China 21%, Germany
8%, UK 6%, Japan 5% (1990)
Imports: $120 billion (c.i.f., 1992 est.)
commodities: foodstuffs, transport
equipment, raw materials, semimanufactures,
petroleum
partners: China 37%, Japan 16%, Taiwan
9%, US 8% (1990)
External debt: $9.5 billion (December 1990
est.)
Industrial production: growth rate NA%
Electricity: 9,566,000 kW capacity; 29,400
million kWh produced, 4,980 kWh per
capita (1992)
Industries: textiles, clothing, tourism,
electronics, plastics, toys, watches, clocks
Agriculture: minor role in the economy;
rice, vegetables, dairy products; less than
20% self-sufficient; shortages of rice, wheat,
water
Illicit drugs: a hub for Southeast Asian
heroin trade; transshipment and major
financial and money-laundering center

Economic aid: US commitments, including Ex-Im (FY70-87), $152 million; Western (non-US) countries, ODA and OOF bilateral commitments (1970-89), $923 million
Currency: 1 Hong Kong dollar (HK$) = 100 cents
Exchange rates: Hong Kong dollars (HK$) per US$—7.800 (1992), 7.771 (1991), 7.790 (1990), 7.800 (1989), 7.810 (1988), 7.760 (1987); note—linked to the US dollar at the rate of about 7.8 HK$ per 1 US$ since 1985
Fiscal year: 1 April—31 March

Communications

Railroads: 35 km 1.435-meter standard gauge, government owned
Highways: 1,100 km total; 794 km paved, 306 km gravel, crushed stone, or earth
Ports: Hong Kong
Merchant marine: 176 ships (1,000 GRT or over), totaling 5,870,007 GRT/10,006,390 DWT; includes 1 passenger, 1 short-sea passenger, 20 cargo, 6 refrigerated cargo, 29 container, 15 oil tanker, 3 chemical tanker, 6 combination ore/oil, 5 liquefied gas, 88 bulk, 2 combination bulk; note—a flag of convenience registry; ships registered in Hong Kong fly the UK flag, and an estimated 500 Hong Kong-owned ships are registered elsewhere
Airports:
total: 2
useable: 2
with permanent-surface runways: 2
with runways over 3,659 m: 0
with runways 2,440-3,659 m: 1
with runways 1,220-2,439 m: 0
Telecommunications: modern facilities provide excellent domestic and international services; 3,000,000 telephones; microwave transmission links and extensive optical fiber transmission network; broadcast stations—6 AM, 6 FM, 4 TV; 1 British Broadcasting Corporation (BBC) repeater station and 1 British Forces Broadcasting Service repeater station; 2,500,000 radio receivers; 1,312,000 TV sets (1,224,000 color TV sets); satellite earth stations—1 Pacific Ocean INTELSAT and 2 Indian Ocean INTELSAT; coaxial cable to Guangzhou, China; links to 5 international submarine cables providing access to ASEAN member nations, Japan, Taiwan, Australia, Middle East, and Western Europe

Defense Forces

Branches: Headquarters of British Forces, Royal Navy, Royal Air Force, Royal Hong Kong Auxiliary Air Force, Royal Hong Kong Police Force
Manpower availability: males age 15-49 1,635,516; fit for military service 1,256,057; reach military age (18) annually 43,128 (1993 est.)

Defense expenditures: exchange rate conversion—$300 million, 0.5% of GDP (1989 est.); this represents one-fourth of the total cost of defending itself, the remainder being paid by the UK
Note: defense is the responsibility of the UK

Howland Island
(territory of the US)

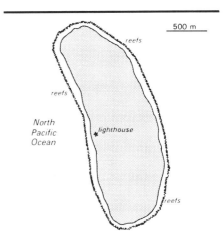

Geography

Location: in the North Pacific Ocean, 2,575 km southwest of Honolulu, just north of the Equator, about halfway between Hawaii and Australia
Map references: Oceania
Area:
total area: 1.6 km²
land area: 1.6 km²
comparative area: about 2.7 times the size of the Mall in Washington, DC
Land boundaries: 0 km
Coastline: 6.4 km
Maritime claims:
contiguous zone: 24 nm
continental shelf: 200 m or the depth of exploitation
exclusive economic zone: 200 nm
territorial sea: 12 nm
International disputes: none
Climate: equatorial; scant rainfall, constant wind, burning sun
Terrain: low-lying, nearly level, sandy, coral island surrounded by a narrow fringing reef; depressed central area
Natural resources: guano (deposits worked until late 1800s)
Land use:
arable land: 0%
permanent crops: 0%
meadows and pastures: 0%
forest and woodland: 5%
other: 95%
Irrigated land: 0 km²
Environment: almost totally covered with grasses, prostrate vines, and low-growing shrubs; small area of trees in the center; lacks fresh water; primarily a nesting, roosting, and foraging habitat for seabirds, shorebirds, and marine wildlife; feral cats

People

Population: uninhabited; note—American civilians evacuated in 1942 after Japanese air and naval attacks during World War II;

175

Howland Island (continued)

occupied by US military during World War II, but abandoned after the war; public entry is by special-use permit only and generally restricted to scientists and educators

Government

Names:
conventional long form: none
conventional short form: Howland Island
Digraph: HQ
Type: unincorporated territory of the US administered by the Fish and Wildlife Service of the US Department of the Interior as part of the National Wildlife Refuge System
Capital: none; administered from Washington, DC

Economy

Overview: no economic activity

Communications

Ports: none; offshore anchorage only, one boat landing area along the middle of the west coast
Airports: airstrip constructed in 1937 for scheduled refueling stop on the round-the-world flight of Amelia Earhart and Fred Noonan—they left Lae, New Guinea, for Howland Island, but were never seen again; the airstrip is no longer serviceable
Note: Earhart Light is a day beacon near the middle of the west coast that was partially destroyed during World War II, but has since been rebuilt in memory of famed aviatrix Amelia Earhart

Defense Forces

Note: defense is the responsibility of the US; visited annually by the US Coast Guard

Hungary

125 km

Geography

Location: Eastern Europe, between Slovakia and Romania
Map references: Ethnic Groups in Eastern Europe, Europe
Area:
total area: 93,030 km²
land area: 92,340 km²
comparative area: slightly smaller than Indiana
Land boundaries: total 1,952 km, Austria 366 km, Croatia 292 km, Romania 443 km, Serbia and Montenegro 151 km (all with Serbia), Slovakia 515 km, Slovenia 82 km, Ukraine 103 km
Coastline: 0 km (landlocked)
Maritime claims: none; landlocked
International disputes: Gabcikovo Dam dispute with Slovakia; Vojvodina taken from Hungary and awarded to the former Yugoslavia by treaty of Trianon in 1920
Climate: temperate; cold, cloudy, humid winters; warm summers
Terrain: mostly flat to rolling plains
Natural resources: bauxite, coal, natural gas, fertile soils
Land use:
arable land: 50.7%
permanent crops: 6.1%
meadows and pastures: 12.6%
forest and woodland: 18.3%
other: 12.3%
Irrigated land: 1,750 km² (1989)
Environment: levees are common along many streams, but flooding occurs almost every year
Note: landlocked; strategic location astride main land routes between Western Europe and Balkan Peninsula as well as between Ukraine and Mediterranean basin

People

Population: 10,324,018 (July 1993 est.)
Population growth rate: -0.07% (1993 est.)
Birth rate: 12.33 births/1,000 population (1993 est.)

Death rate: 13.02 deaths/1,000 population (1993 est.)
Net migration rate: 0 migrant(s)/1,000 population (1993 est.)
Infant mortality rate: 13.1 deaths/1,000 live births (1993 est.)
Life expectancy at birth:
total population: 70.86 years
male: 66.81 years
female: 75.12 years (1993 est.)
Total fertility rate: 1.83 children born/woman (1993 est.)
Nationality:
noun: Hungarian(s)
adjective: Hungarian
Ethnic divisions: Hungarian 89.9%, Gypsy 4%, German 2.6%, Serb 2%, Slovak 0.8%, Romanian 0.7%
Religions: Roman Catholic 67.5%, Calvinist 20%, Lutheran 5%, atheist and other 7.5%
Languages: Hungarian 98.2%, other 1.8%
Literacy: age 15 and over can read and write (1980)
total population: 99%
male: 99%
female: 98%
Labor force: 5.4 million
by occupation: services, trade, government, and other 44.8%, industry 29.7%, agriculture 16.1%, construction 7.0% (1991)

Government

Names:
conventional long form: Republic of Hungary
conventional short form: Hungary
local long form: Magyar Koztarsasag
local short form: Magyarorszag
Digraph: HU
Type: republic
Capital: Budapest
Administrative divisions: 38 counties (megyek, singular—megye) and 1 capital city* (fovaros); Bacs-Kiskun, Baranya, Bekes, Bekescsaba, Borsod-Abauj-Zemplen, Budapest*, Csongrad, Debrecen, Dunaujvaros, Eger, Fejer, Gyor, Gyor-Moson-Sopron, Hajdu-Bihar, Heves, Hodmezovasarhely, Jasz-Nagykun-Szolnok, Kaposvar, Kecskemet, Komarom-Esztergom, Miskolc, Nagykanizsa, Nograd, Nyiregyhaza, Pecs, Pest, Somogy, Sopron, Szabolcs-Szatmar-Bereg, Szeged, Szekesfehervar, Szolnok, Szombathely, Tatabanya, Tolna, Vas, Veszprem, Zala, Zalaegerszeg
Independence: 1001 (unification by King Stephen I)
Constitution: 18 August 1949, effective 20 August 1949, revised 19 April 1972; 18 October 1989 revision ensured legal rights for individuals and constitutional checks on the authority of the prime minister and also established the principle of parliamentary oversight

Legal system: in process of revision, moving toward rule of law based on Western model

National holiday: October 23 (1956) (commemorates the Hungarian uprising)

Political parties and leaders: Democratic Forum, Jozsef ANTALL, chairman, Dr. Lajos FUR, executive chairman; Independent Smallholders (FKGP), Jozsef TORGYAN, president; Hungarian Socialist Party (MSZP), Gyula HORN, chairman; Christian Democratic People's Party (KDNP), Dr. Lazlo SURJAN, president; Federation of Young Democrats (FIDESZ), Viktor ORBAN, chairman; Alliance of Free Democrats (SZDSZ), Ivan PETO, chairman *note:* the Hungarian Socialist (Communist) Workers' Party (MSZMP) renounced Communism and became the Hungarian Socialist Party (MSP) in October 1989; there is still a small (fringe) MSZMP

Suffrage: 18 years of age; universal

Elections:

President: last held 3 August 1990 (next to be held NA 1995); results—President GONCZ elected by parliamentary vote; note—President GONCZ was elected by the National Assembly with a total of 295 votes out of 304 as interim President from 2 May 1990 until elected President

National Assembly: last held on 25 March 1990 (first round, with the second round held 8 April 1990); results—percent of vote by party NA; seats—(386 total) Democratic Forum 162, Free Democrats 90, Independent Smallholders 45, Hungarian Socialist Party (MSP) 33, Young Democrats 22, Christian Democrats 21, independents or jointly sponsored candidates 13

Executive branch: president, prime minister

Legislative branch: unicameral National Assembly (Orszaggyules)

Judicial branch: Constitutional Court

Leaders:

Chief of State: President Arpad GONCZ (since 3 August 1990; previously interim president from 2 May 1990)

Head of Government: Prime Minister Jozsef ANTALL (since 21 May 1990)

Member of: Australian Group, BIS, CCC, CE, CEI, CERN, CSCE, EBRD, ECE, FAO, G-9, GATT, IAEA, IBRD, ICAO, IDA, IFC, ILO, IMF, IMO, INTERPOL, IOC, IOM, ISO, ITU, LORCS, MTCR, NACC, NAM (guest), NSG, PCA, UN, UNAVEM II, UNCTAD, UNESCO, UNHCR, UNIDO, UNIKOM, UNOMOZ, UPU, WHO, WIPO, WMO, WTO, ZC

Diplomatic representation in US:

chief of mission: Ambassador Pal TAR

chancery: 3910 Shoemaker Street NW, Washington DC 20008

telephone: (202) 362-6730

FAX: (202) 966-8135

consulate general: New York

US diplomatic representation:

chief of mission: Ambassador Charles H. THOMAS

embassy: V. Szabadsag Ter 12, Budapest

mailing address: Am Embassy, Unit 25402, APO AE 09213-5270

telephone: [36] (1) 112-6450

FAX: [36] (1) 132-8934

Flag: three equal horizontal bands of red (top), white, and green

---

Economy

Overview: Hungary is in the midst of a difficult transition from a command to a market economy. Agriculture is an important sector, providing sizable export earnings and meeting domestic food needs. Industry accounts for about 40% of GDP and 30% of employment. Hungary claims that less than 25% of foreign trade is now with former CEMA countries, while about 70% is with OECD members. Hungary's economic reform programs during the Communist era gave it a head start in creating a market economy and attracting foreign investment. In 1991, Hungary received 60% of all foreign investment in Eastern Europe, and in 1992 received the largest single share. The growing private sector accounts for about one-third of national output according to unofficial estimates. Privatization of state enterprises is progressing, although excessive red tape, bureaucratic oversight, and uncertainties about pricing have slowed the process. Escalating unemployment and high rates of inflation may impede efforts to speed up privatization and budget reform, while Hungary's heavy foreign debt will make the government reluctant to introduce full convertibility of the forint before 1994 and to rein in inflation. The government is projecting an end to the 5-year recession in 1993, and GDP is forecast to grow 0%-3%.

National product: GDP—purchasing power equivalent—$55.4 billion (1992 est.)

National product real growth rate: -5% (1992 est.)

National product per capita: $5,380 (1992 est.)

Inflation rate (consumer prices): 23% (1992 est.)

Unemployment rate: 12.3% (1992)

Budget: revenues $13.2 billion; expenditures $15.4 billion, including capital expenditures $NA (1993 est.)

Exports: $10.9 billion (f.o.b., 1992 est.)

commodities: raw materials, semi-finished goods, chemicals 35.5%, machinery 13.5%, light industry 23.3%, food and agricultural 24.8%, fuels and energy 2.8%

partners: OECD 70.7%, (EC 50.1%, EFTA 15.0%), LDCs 5.1%, former CEMA members 23.2%, others 1.0% (1991)

Imports: $11.7 billion (f.o.b., 1992 est.)

commodities: fuels and energy 14.9%, raw materials, semi-finished goods, chemicals 37.6%, machinery 19.7%, light industry 21.5%, food and agricultural 6.3%

partners: OECD 71.0%, (EC 45.4%, EFTA 20.0%), LDCs 3.9%, former CEMA members 23.9%, others 1.2% (1991)

External debt: $23.5 billion (September 1992)

Industrial production: growth rate -10% (1992)

Electricity: 7,200,000 kW capacity; 30,000 million kWh produced, 3,000 kWh per capita (1992)

Industries: mining, metallurgy, construction materials, processed foods, textiles, chemicals (especially pharmaceuticals), buses, automobiles

Agriculture: including forestry, accounts for 15% of GDP and 16% of employment; highly diversified crop and livestock farming; principal crops—wheat, corn, sunflowers, potatoes, sugar beets; livestock—hogs, cattle, poultry, dairy products; self-sufficient in food output

Illicit drugs: transshipment point for Southeast Asia heroin transiting the Balkan route

Economic aid: recipient—$9.1 billion in assistance from OECD countries (from 1st quarter 1990 to end of 2nd quarter 1991)

Currency: 1 forint (Ft) = 100 filler

Exchange rates: forints per US$1—83.97 (December 1992), 78.99 (1992), 74.74 (1991), 63.21 (1990), 59.07 (1989), 50.41 (1988)

Fiscal year: calendar year

Communications

Railroads: 7,765 km total; 7,508 km 1.435-meter standard gauge, 222 km narrow gauge (mostly 0.760-meter), 35 km 1.520-meter broad gauge; 1,236 km double track, 2,249 km electrified; all government owned (1990)

Highways: 130,218 km total; 29,919 km national highway system (27,212 km asphalt, 126 km concrete, 50 km stone and road brick, 2,131 km macadam, 400 km unpaved); 58,495 km country roads (66% unpaved), and 41,804 km other roads (70% unpaved) (1988)

Inland waterways: 1,622 km (1988)

Pipelines: crude oil 1,204 km; natural gas 4,387 km (1991)

Ports: Budapest and Dunaujvaros are river ports on the Danube; coastal outlets are Rostock (Germany), Gdansk (Poland), Gdynia (Poland), Szczecin (Poland), Galati (Romania), and Braila (Romania)

Merchant marine: 12 cargo ships (1,000 GRT or over) and 1 bulk totaling 83,091 GRT/115,950 DWT

Hungary *(continued)*

Airports:
total: 92
usable: 92
with permanent-surface runways: 25
with runways over 3,659 m: 1
with runways 2,440-3,659 m: 20
with runways 1,220-2,439 m: 28
Telecommunications: automatic telephone network based on microwave radio relay system; 1,128,800 phones (1991); telephone density is at 19.4 per 100 inhabitants; 49% of all phones are in Budapest; 608,000 telephones on order (1991); 12-15 year wait for a phone; 14,213 telex lines (1991); broadcast stations—32 AM, 15 FM, 41 TV (8 Soviet TV repeaters); 4.2 million TVs (1990); 1 satellite ground station using INTELSAT and Intersputnik

Defense Forces

Branches: Ground Forces, Air and Air Defense Forces, Border Guard, Territorial Defense
Manpower availability: males age 15-49 2,630,552; fit for military service 2,101,637; reach military age (18) annually 91,979 (1993 est.)
Defense expenditures: 66.5 billion forints, NA% of GNP (1993 est.); note—conversion of defense expenditures into US dollars using the current exchange rate could produce misleading results

Iceland

Geography

Location: in the North Atlantic Ocean, between Greenland and Norway
Map references: Arctic Region, Europe, North America, Standard Time Zones of the World
Area:
total area: 103,000 km²
land area: 100,250 km²
comparative area: slightly smaller than Kentucky
Land boundaries: 0 km
Coastline: 4,988 km
Maritime claims:
continental shelf: 200 nm or the edge of continental margin
exclusive economic zone: 200 nm
territorial sea: 12 nm
International disputes: Rockall continental shelf dispute involving Denmark, Ireland, and the UK (Ireland and the UK have signed a boundary agreement in the Rockall area)
Climate: temperate; moderated by North Atlantic Current; mild, windy winters; damp, cool summers
Terrain: mostly plateau interspersed with mountain peaks, icefields; coast deeply indented by bays and fiords
Natural resources: fish, hydropower, geothermal power, diatomite
Land use:
arable land: 1%
permanent crops: 0%
meadows and pastures: 20%
forest and woodland: 1%
other: 78%
Irrigated land: NA km²
Environment: subject to earthquakes and volcanic activity
Note: strategic location between Greenland and Europe; westernmost European country; more land covered by glaciers than in all of continental Europe

People

Population: 261,270 (July 1993 est.)
note: population data estimates based on average growth rate may differ slightly from official population data because of volatile migration rates
Population growth rate: 0.88% (1993 est.)
Birth rate: 16.99 births/1,000 population (1993 est.)
Death rate: 6.74 deaths/1,000 population (1993 est.)
Net migration rate: -1.47 migrant(s)/1,000 population (1993 est.)
Infant mortality rate: 4 deaths/1,000 live births (1993 est.)
Life expectancy at birth:
total population: 78.69 years
male: 76.45 years
female: 81.04 years (1993 est.)
Total fertility rate: 2.16 children born/woman (1993 est.)
Nationality:
noun: Icelander(s)
adjective: Icelandic
Ethnic divisions: homogeneous mixture of descendants of Norwegians and Celts
Religions: Evangelical Lutheran 96%, other Protestant and Roman Catholic 3%, none 1% (1988)
Languages: Icelandic
Literacy: age 15 and over can read and write (1976)
total population: 100%
male: NA%
female: NA%
Labor force: 127,900
by occupation: commerce, transportation, and services 60.0%, manufacturing 12.5%, fishing and fish processing 11.8%, construction 10.8%, agriculture 4.0% (1990)

Government

Names:
conventional long form: Republic of Iceland
conventional short form: Iceland
local long form: Lyoveldio Island
local short form: Island
Digraph: IC
Type: republic
Capital: Reykjavik
Administrative divisions: 23 counties (syslar, singular—sysla) and 14 independent towns* (kaupstadhir, singular—kaupstadhur); Akranes*, Akureyri*, Arnessysla, Austur-Bardhastrandarsysla, Austur-Hunavatnssysla, Austur-Skaftafellssysla, Borgarfjardharsysla, Dalasysla, Eyjafjardharsysla, Gullbringusysla, Hafnarfjordhur*, Husavik*,

Isafjordhur*, Keflavik*, Kjosarsysla, Kopavogur*, Myrasysla, Neskaupstadhur*, Nordhur-Isafjardharsysla, Nordhur-Mulasys-la, Nordhur-Thingeyjarsysla, Olafsfjordhur*, Rangarvallasysla, Reykjavik*, Saudharkrokur*, Seydhisfjordhur*, Siglufjordhur*, Skagafjardharsysla, Snaefellsnes-og Hnappadalssysla, Strandasysla, Sudhur-Mulasysla, Sudhur-Thingeyjarsysla, Vesttmannaeyjar*, Vestur-Bardhastrandarsysla, Vestur-Hunavatnssysla, Vestur-Isafjardharsysla, Vestur-Skaftafellssysla

Independence: 17 June 1944 (from Denmark)
Constitution: 16 June 1944, effective 17 June 1944
Legal system: civil law system based on Danish law; does not accept compulsory ICJ jurisdiction
National holiday: Anniversary of the Establishment of the Republic, 17 June (1944)
Political parties and leaders: Independence Party (conservative), David ODDSSON; Progressive Party, Steingrimur HERMANNSSON; Social Democratic Party, Jon Baldvin HANNIBALSSON; People's Alliance (left socialist), Olafur Ragnar GRIMSSON; Women's List
Suffrage: 18 years of age; universal
Elections:
President: last held on 29 June 1988 (next scheduled for June 1996); results—there was no election in 1992 as President Vigdis FINNBOGADOTTIR was unopposed
Althing: last held on 20 April 1991 (next to be held by April 1995); results—Independence Party 38.6%, Progressive Party 18.9%, Social Democratic Party 15.5%, People's Alliance 14.4%, Womens List 8.3%, Liberals 1.2%, other 3.1%; seats—(63 total) Independence 26, Progressive 13, Social Democratic 10, People's Alliance 9, Womens List 5
Executive branch: president, prime minister, Cabinet
Legislative branch: unicameral Parliament (Althing)
Judicial branch: Supreme Court (Haestirettur)
Leaders:
Chief of State: President Vigdis FINNBOGADOTTIR (since 1 August 1980)
Head of Government: Prime Minister David ODDSSON (since 30 April 1991)
Member of: Australian Group, BIS, CCC, CE, CSCE, EBRD, ECE, EFTA, FAO, GATT, IAEA, IBRD, ICAO, ICC, ICFTU, IDA, IFC, ILO, IMF, IMO, INMARSAT, INTELSAT, INTERPOL, IOC, ISO (correspondent), ITU, LORCS, MTCR, NACC, NATO, NC, NEA, NIB, OECD,

PCA, UN, UNCTAD, UNESCO, UPU, WEU (associate), WHO, WIPO, WMO
Diplomatic representation in US:
chief of mission: (vacant)
chancery: 2022 Connecticut Avenue NW, Washington DC 20008
telephone: (202) 265-6653 through 6655
FAX: (202) 265-6656
consulate general: New York
US diplomatic representation:
chief of mission: (vacant); Charge d'Affaires Jon GUNDERSEN
embassy: Laufasvegur 21, Box 40, Reykjavik
mailing address: USEMB, PSC 1003, Box 40, FPO AE 09728-0340
telephone: [354] (1) 29100
FAX: [354] (1) 29139
Flag: blue with a red cross outlined in white that extends to the edges of the flag; the vertical part of the cross is shifted to the hoist side in the style of the Dannebrog (Danish flag)

Economy

Overview: Iceland's Scandinavian-type economy is basically capitalistic, but with an extensive welfare system, relatively low unemployment, and comparatively even distribution of income. The economy is heavily dependent on the fishing industry, which provides nearly 75% of export earnings and employs 12% of the workforce. In the absence of other natural resources—except energy—Iceland's economy is vulnerable to changing world fish prices. Iceland's economy has been in recession since 1988. The recession deepened in 1992 due to severe cutbacks in fishing quotas and falling world prices for the country's main exports: fish and fish products, aluminum, and ferrosilicon. Real GDP declined 3.3% in 1992 and is forecast to contract another 1.5% in 1993. The center-right government's economic goals include reducing the budget and current account deficits, limiting foreign borrowing, containing inflation, revising agricultural and fishing policies, diversifying the economy, and privatizing state-owned industries. The recession has led to a wave of bankruptcies and mergers throughout the economy, as well as the highest unemployment of the post-World War II period. The national unemployment rate reached 5% in early 1993, with some parts of the country experiencing unemployment in the 9-10% range. Inflation, previously a serious problem, declined from double digit rates in the 1980s to only 3.7% in 1992.
National product: GDP—purchasing power equivalent—$4.5 billion (1992)
National product real growth rate: -3.3% (1992)

National product per capita: $17,400 (1992)
Inflation rate (consumer prices): 3.7% (1992 est.)
Unemployment rate: 5% (first quarter 1993)
Budget: revenues $1.8 billion; expenditures $1.9 billion, including capital expenditures of $191 million (1992)
Exports: $1.5 billion (f.o.b., 1992)
commodities: fish and fish products, animal products, aluminum, ferrosilicon, diatomite
partners: EC 68% (UK 25%, Germany 12%), US 11%, Japan 8% (1992)
Imports: $1.5 billion (c.i.f., 1992)
commodities: machinery and transportation equipment, petroleum products, foodstuffs, textiles
partners: EC 53% (Germany 14%, Denmark 10%, UK 9%), Norway 14%, US 9% (1992)
External debt: $3.9 billion (1992 est.)
Industrial production: growth rate 1.75% (1991 est.)
Electricity: 1,063,000 kW capacity; 5,165 million kWh produced, 19,940 kWh per capita (1992)
Industries: fish processing, aluminum smelting, ferro-silicon production, geothermal power
Agriculture: accounts for about 25% of GDP; fishing is most important economic activity, contributing nearly 75% to export earnings; principal crops—potatoes, turnips; livestock—cattle, sheep; self-sufficient in crops; fish catch of about 1.4 million metric tons in 1989
Economic aid: US commitments, including Ex-Im (FY70-81), $19.1 million
Currency: 1 Icelandic krona (IKr) = 100 aurar
Exchange rates: Icelandic kronur (IKr) per US$1—63.789 (January 1993), 57.546 (1992), 58.996 (1991), 58.284 (1990), 57.042 (1989), 43.014 (1988)
Fiscal year: calendar year

Communications

Highways: 11,543 km total; 2,690 km hard surfaced, 8,853 km gravel and earth
Ports: Reykjavik, Akureyri, Hafnarfjordhur, Keflavik, Seydhisfjordhur, Siglufjordhur, Vestmannaeyjar
Merchant marine: 10 ships (1,000 GRT or over) totaling 35,832 GRT/53,037 DWT; includes 3 cargo, 3 refrigerated cargo, 2 roll-on/roll-off cargo, 1 oil tanker, 1 chemical tanker
Airports:
total: 90
usable: 84
with permanent-surface runways: 8
with runways over 3,659 m: 0
with runways 2,440-3,659 m: 1
with runways 1,220-2,439 m: 12

Iceland (continued)

Telecommunications: adequate domestic service; coaxial and fiber-optical cables and microwave radio relay for trunk network; 140,000 telephones; broadcast stations—5 AM, 147 (transmitters and repeaters) FM, 202 (transmitters and repeaters) TV; 2 submarine cables; 1 Atlantic Ocean INTELSAT earth station carries all international traffic; a second INTELSAT earth station is scheduled to be operational in 1993

Defense Forces

Branches: Police, Coast Guard
note: no armed forces, Iceland's defense is provided by the US-manned Icelandic Defense Force (IDF) headquartered at Keflavik
Manpower availability: males age 15-49 69,499; fit for military service 61,798 (1993 est.); no conscription or compulsory military service
Defense expenditures: none

India

Geography

Location: South Asia, bordering the Arabian Sea and the Bay of Bengal, between Bangladesh and Pakistan
Map references: Asia, Standard Time Zones of the World
Area:
total area: 3,287,590 km²
land area: 2,973,190 km²
comparative area: slightly more than one-third the size of the US
Land boundaries: total 14,103 km, Bangladesh 4,053 km, Bhutan 605 km, Burma 1,463 km, China 3,380 km, Nepal 1,690 km, Pakistan 2,912 km
Coastline: 7,000 km
Maritime claims:
contiguous zone: 24 nm
continental shelf: 200 nm or the edge of continental margin
exclusive economic zone: 200 nm
territorial sea: 12 nm
International disputes: boundaries with Bangladesh and China; status of Kashmir with Pakistan; water-sharing problems with downstream riparians, Bangladesh over the Ganges and Pakistan over the Indus
Climate: varies from tropical monsoon in south to temperate in north
Terrain: upland plain (Deccan Plateau) in south, flat to rolling plain along the Ganges, deserts in west, Himalayas in north
Natural resources: coal (fourth-largest reserves in the world), iron ore, manganese, mica, bauxite, titanium ore, chromite, natural gas, diamonds, petroleum, limestone
Land use:
arable land: 55%
permanent crops: 1%
meadows and pastures: 4%
forest and woodland: 23%
other: 17%
Irrigated land: 430,390 km² (1989)
Environment: droughts, flash floods, severe thunderstorms common; deforestation; soil erosion; overgrazing; air and water pollution; desertification

Note: dominates South Asian subcontinent; near important Indian Ocean trade routes

People

Population: 903,158,968 (July 1993 est.)
Population growth rate: 1.86% (1993 est.)
Birth rate: 29.11 births/1,000 population (1993 est.)
Death rate: 10.52 deaths/1,000 population (1993 est.)
Net migration rate: 0 migrant(s)/1,000 population (1993 est.)
Infant mortality rate: 80.5 deaths/1,000 live births (1993 est.)
Life expectancy at birth:
total population: 58.12 years
male: 57.69 years
female: 58.59 years (1993 est.)
Total fertility rate: 3.57 children born/woman (1993 est.)
Nationality:
noun: Indian(s)
adjective: Indian
Ethnic divisions: Indo-Aryan 72%, Dravidian 25%, Mongoloid and other 3%
Religions: Hindu 82.6%, Muslim 11.4%, Christian 2.4%, Sikh 2%, Buddhist 0.7%, Jains 0.5%, other 0.4%
Languages: English enjoys associate status but is the most important language for national, political, and commercial communication, Hindi the national language and primary tongue of 30% of the people, Bengali (official), Telugu (official), Marathi (official), Tamil (official), Urdu (official), Gujarati (official), Malayalam (official), Kannada (official), Oriya (official), Punjabi (official), Assamese (official), Kashmiri (official), Sindhi (official), Sanskrit (official), Hindustani a popular variant of Hindu/Urdu, is spoken widely throughout northern India
note: 24 languages each spoken by a million or more persons; numerous other languages and dialects, for the most part mutually unintelligible
Literacy: age 15 and over can read and write (1990)
total population: 48%
male: 62%
female: 34%
Labor force: 284.4 million
by occupation: agriculture 67% (FY85)

Government

Names:
conventional long form: Republic of India
conventional short form: India
Digraph: IN
Type: federal republic
Capital: New Delhi
Administrative divisions: 25 states and 7 union territories*; Andaman and Nicobar Islands*, Andhra Pradesh, Arunachal

Pradesh, Assam, Bihar, Chandigarh*, Dadra and Nagar Haveli*, Daman and Diu*, Delhi*, Goa, Gujarat, Haryana, Himachal Pradesh, Jammu and Kashmir, Karnataka, Kerala, Lakshadweep*, Madhya Pradesh, Maharashtra, Manipur, Meghalaya, Mizoram, Nagaland, Orissa, Pondicherry*, Punjab, Rajasthan, Sikkim, Tamil Nadu, Tripura, Uttar Pradesh, West Bengal

Independence: 15 August 1947 (from UK)
Constitution: 26 January 1950
Legal system: based on English common law; limited judicial review of legislative acts; accepts compulsory ICJ jurisdiction, with reservations
National holiday: Anniversary of the Proclamation of the Republic, 26 January (1950)
Political parties and leaders: Congress (I) Party, P. V. Narasimha RAO, president; Bharatiya Janata Party, M. M. JOSHI; Janata Dal Party; Communist Party of India/Marxist (CPI/M), Harkishan Singh SURJEET; Communist Party of India (CPI), C. Rajeswara RAO; Telugu Desam (a regional party in Andhra Pradesh), N. T. Rama RAO; All-India Anna Dravida Munnetra Kazagham (AIADMK; a regional party in Tamil Nadu), JAYALALITHA Jeyaram; Samajwadi Janata Party, CHANDRA SHEKHAR; Shiv Sena, Bal THACKERAY; Revolutionary Socialist Party (RSP), Tridip CHOWDHURY; Bahujana Samaj Party (BSP), Kanshi RAM; Congress (S) Party, leader NA; Communist Party of India/Marxist-Leninist (CPI/ML), Satyanarayan SINGH; Dravida Munnetra Kazagham (a regional party in Tamil Nadu), M. KARUNANIDHI; Akali Dal factions representing Sikh religious community in the Punjab; National Conference (NC; a regional party in Jammu and Kashmir), Farooq ABDULLAH; Asom Gana Parishad (a regional party in Assam), Prafulla MAHANTA
Other political or pressure groups: various separatist groups seeking greater communal and/or regional autonomy; numerous religious or militant/chauvinistic organizations, including Adam Sena, Ananda Marg, Vishwa Hindu Parishad, and Rashtriya Swayamsevak Sangh
Suffrage: 18 years of age; universal
Elections:
People's Assembly: last held 21 May, 12 and 15 June 1991 (next to be held by November 1996); results—percent of vote by party NA; seats—(545 total, 543 elected, 2 appointed) Congress (I) Party 245, Bharatiya Janata Party 119, Janata Dal Party 39, Janata Dal (Ajit Singh) 20, CPI/M 35, CPI 14, Telugu Desam 13, AIADMK 11, Samajwadi Janata Party 5, Shiv Sena 4, RSP 4, BSP 1, Congress (S) Party 1, other 23, vacant 9
Executive branch: president, vice president, prime minister, Council of Ministers

Legislative branch: bicameral Parliament (Sansad) consists of an upper house or Council of States (Rajya Sabha) and a lower house or People's Assembly (Lok Sabha)
Judicial branch: Supreme Court
Leaders:
Chief of State: President Shankar Dayal SHARMA (since 25 July 1992); Vice President K.R. NARAYANAN (since 21 August 1992)
Head of Government: Prime Minister P. V. Narasimha RAO (since 21 June 1991)
Member of: AG (observer), AsDB, C, CCC, CP, ESCAP, FAO, G-6, G-15, G-19, AfDB, G-24, G-77, GATT, IAEA, IBRD, ICAO, ICC, ICFTU, IDA, IFAD, IFC, ILO, IMF, IMO, INMARSAT, INTELSAT, INTERPOL, IOC, IOM (observer), ISO, ITU, LORCS, NAM, ONUSAL, PCA, SAARC, UN, UNAVEM II, UNCTAD, UNESCO, UNIDO, UNIKOM, UNOMOZ, UNTAC, UPU, WFTU, WHO, WIPO, WMO, WTO
Diplomatic representation in US:
chief of mission: Ambassador Siddhartha Shankar RAY
chancery: 2107 Massachusetts Avenue NW, Washington, DC 20008
telephone: (202) 939-7000
consulates general: Chicago, New York, and San Francisco
US diplomatic representation:
chief of mission: Ambassador Thomas R. Pickering
embassy: Shanti Path, Chanakyapuri 110021, New Delhi
mailing address: use embassy street address
telephone: [91] (11) 600651
FAX: [91] (11) 687-2028, 687-2391
consulates general: Bombay, Calcutta, Madras
Flag: three equal horizontal bands of orange (top), white, and green with a blue chakra (24-spoked wheel) centered in the white band; similar to the flag of Niger, which has a small orange disk centered in the white band

---

Economy

Overview: India's economy is a mixture of traditional village farming, modern agriculture, handicrafts, a wide range of modern industries, and a multitude of support services. Faster economic growth in the 1980s permitted a significant increase in real per capita private consumption. A large share of the population, perhaps as much as 40%, remains too poor to afford an adequate diet. Financial strains in 1990 and 1991 prompted government austerity measures that slowed industrial growth but permitted India to meet its international payment obligations without rescheduling its debt. Policy reforms since 1991 have extended

earlier economic liberalization and greatly reduced government controls on production, trade, and investment.
National product: GDP—exchange rate conversion—$240 billion (FY93 est.)
National product real growth rate: 4% (FY93 est.)
National product per capita: $270 (FY93 est.)
Inflation rate (consumer prices): 11.9% (1992 est.)
Unemployment rate: NA%
Budget: revenues $39.2 billion; expenditures $41.06 billion, including capital expenditures of $10.2 billion (FY92)
Exports: $19.8 billion (f.o.b., FY93 est.)
commodities: gems and jewelry, clothing, engineering goods, leather manufactures, cotton yarn, and fabric
partners: USSR 16.1%, US 14.7%, West Germany 7.8% (FY91)
Imports: $25.5 billion (c.i.f., FY93 est.)
commodities: crude oil and petroleum products, gems, fertilizer, chemicals, machinery
partners: US 12.1%, West Germany 8.0%, Japan 7.5% (FY91)
External debt: $73.0 billion (March 1992)
Industrial production: growth rate 2.5% (FY93 est.); accounts for about 25% of GDP
Electricity: 82,000,000 kW capacity; 310,000 million kWh produced, 340 kWh per capita (1992)
Industries: textiles, chemicals, food processing, steel, transportation equipment, cement, mining, petroleum, machinery
Agriculture: accounts for about 30% of GDP and employs 67% of labor force; principal crops—rice, wheat, oilseeds, cotton, jute, tea, sugarcane, potatoes; livestock—cattle, buffaloes, sheep, goats, poultry; fish catch of about 3 million metric tons ranks India among the world's top 10 fishing nations
Illicit drugs: licit producer of opium poppy for the pharmaceutical trade, but some opium is diverted to illicit international drug markets; major transit country for illicit narcotics produced in neighboring countries; illicit producer of hashish
Economic aid: US commitments, including Ex-Im (FY70-89), $4.4 billion; Western (non-US) countries, ODA and OOF bilateral commitments (1980-89), $31.7 billion; OPEC bilateral aid (1979-89), $315 million; USSR (1970-89), $11.6 billion; Eastern Europe (1970-89), $105 million
Currency: 1 Indian rupee (Re) = 100 paise
Exchange rates: Indian rupees (Rs) per US$1—26.156 (January 1993), 25.918 (1992), 22.742 (1991), 17.504 (1990), 16.226 (1989), 13.917 (1988)
Fiscal year: 1 April-31 March

India (continued)

Communications

Railroads: 61,850 km total (1986); 33,553 km 1.676-meter broad gauge, 24,051 km 1.000-meter gauge, 4,246 km narrow gauge (0.762 meter and 0.610 meter); 12,617 km is double track; 6,500 km is electrified
Highways: 1,970,000 km total (1989); 960,000 km surfaced and 1,010,000 km gravel, crushed stone, or earth
Inland waterways: 16,180 km; 3,631 km navigable by large vessels
Pipelines: crude oil 3,497 km; petroleum products 1,703 km; natural gas 902 km (1989)
Ports: Bombay, Calcutta, Cochin, Kandla, Madras, New Mangalore, Port Blair (Andaman Islands)
Merchant marine: 306 ships (1,000 GRT or over) totaling 6,278,672 GRT/10,446,073 DWT; includes 1 short-sea passenger, 6 passenger-cargo, 87 cargo, 1 roll-on/roll-off, 8 container, 63 oil tanker, 10 chemical tanker, 8 combination ore/oil, 114 bulk, 2 combination bulk, 6 liquefied gas
Airports:
total: 336
usable: 285
with permanent-surface runways: 205
with runways over 3,659 m: 2
with runways 2,440-3,659 m: 58
with runways 1,220-2,439 m: 90
Telecommunications: domestic telephone system is poor providing only one telephone for about 200 persons on average; long distance telephoning has been improved by a domestic satellite system which also carries TV; international service is provided by 3 Indian Ocean INTELSAT earth stations and by submarine cables to Malaysia and the United Arab Emirates; broadcast stations—96 AM, 4 FM, 274 TV (government controlled)

Defense Forces

Branches: Army, Navy, Air Force, Security or Paramilitary Forces (including Border Security Force, Assam Rifles, and Coast Guard)
Manpower availability: males age 15-49 242,866,053; fit for military service 143,008,471; about 9,466,323 reach military age (17) annually (1993 est.)
Defense expenditures: exchange rate conversion—$5.8 billion, 2.4% of GDP (FY93/94)

Indian Ocean

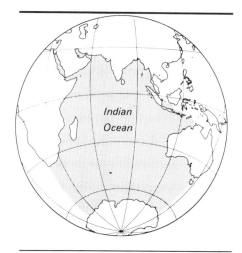

Geography

Location: body of water between Africa, Asia, Australia, and Antarctica
Map references: Southeast Asia, Standard Time Zones of the World
Area:
total area: 73.6 million km²
comparative area: slightly less than eight times the size of the US; third-largest ocean (after the Pacific Ocean and Atlantic Ocean, but larger than the Arctic Ocean)
note: includes Arabian Sea, Bass Straight, Bay of Bengal, Java Sea, Persian Gulf, Red Sea, Straight of Malacca, Timor Sea, and other tributary water bodies
Coastline: 66,526 km
International disputes: some maritime disputes (see littoral states)
Climate: northeast monsoon (December to April), southwest monsoon (June to October); tropical cyclones occur during May/June and October/November in the north Indian Ocean and January/February in the south Indian Ocean
Terrain: surface dominated by counterclockwise gyre (broad, circular system of currents) in the south Indian Ocean; unique reversal of surface currents in the north Indian Ocean, low atmospheric pressure over southwest Asia from hot, rising, summer air results in the southwest monsoon and southwest-to-northeast winds and currents, while high pressure over northern Asia from cold, falling, winter air results in the northeast monsoon and northeast-to-southwest winds and currents; ocean floor is dominated by the Mid-Indian Ocean Ridge and subdivided by the Southeast Indian Ocean Ridge, Southwest Indian Ocean Ridge, and Ninety East Ridge; maximum depth is 7,258 meters in the Java Trench
Natural resources: oil and gas fields, fish, shrimp, sand and gravel aggregates, placer deposits, polymetallic nodules

Environment: endangered marine species include the dugong, seals, turtles, and whales; oil pollution in the Arabian Sea, Persian Gulf, and Red Sea
Note: major chokepoints include Bab el Mandeb, Strait of Hormuz, Strait of Malacca, southern access to the Suez Canal, and the Lombok Strait; ships subject to superstructure icing in extreme south near Antarctica from May to October

Government

Digraph: XO

Economy

Overview: The Indian Ocean provides major sea routes connecting the Middle East, Africa, and East Asia with Europe and the Americas. It carries a particularly heavy traffic of petroleum and petroleum products from the oil fields of the Persian Gulf and Indonesia. Its fish are of great and growing importance to the bordering countries for domestic consumption and export. Fishing fleets from Russia, Japan, Korea, and Taiwan also exploit the Indian Ocean, mainly for shrimp and tuna. Large reserves of hydrocarbons are being tapped in the offshore areas of Saudi Arabia, Iran, India, and Western Australia. An estimated 40% of the world's offshore oil production comes from the Indian Ocean. Beach sands rich in heavy minerals and offshore placer deposits are actively exploited by bordering countries, particularly India, South Africa, Indonesia, Sri Lanka, and Thailand.
Industries: based on exploitation of natural resources, particularly marine life, minerals, oil and gas production, fishing, sand and gravel aggregates, placer deposits

Communications

Ports: Bombay (India), Calcutta (India), Madras (India), Colombo (Sri Lanka), Durban (South Africa), Fremantle (Australia), Jakarta (Indonesia), Melbourne (Australia), Richard's Bay (South Africa)
Telecommunications: submarine cables from India to United Arab Emirates and Malaysia, and from Sri Lanka to Djibouti and Indonesia

Indonesia

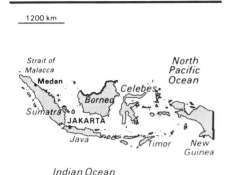

1200 km

Strait of Malacca
North Pacific Ocean
Medan
Celebes
Borneo
Sumatra
JAKARTA
Java
Timor
New Guinea
Indian Ocean

Geography

Location: Southeast Asia, between Malaysia and Australia
Map references: Oceania, Southeast Asia, Standard Time Zones of the World
Area:
total area: 1,919,440 km²
land area: 1,826,440 km²
comparative area: slightly less than three times the size of Texas
Land boundaries: total 2,602 km, Malaysia 1,782 km, Papua New Guinea 820 km
Coastline: 54,716 km
Maritime claims: measured from claimed archipelagic baselines
exclusive economic zone: 200 nm
territorial sea: 12 nm
International disputes: sovereignty over Timor Timur (East Timor Province) disputed with Portugal and not recognized by the UN; two islands in dispute with Malaysia
Climate: tropical; hot, humid; more moderate in highlands
Terrain: mostly coastal lowlands; larger islands have interior mountains
Natural resources: petroleum, tin, natural gas, nickel, timber, bauxite, copper, fertile soils, coal, gold, silver
Land use:
arable land: 8%
permanent crops: 3%
meadows and pastures: 7%
forest and woodland: 67%
other: 15%
Irrigated land: 75,500 km² (1989 est.)
Environment: archipelago of 13,500 islands (6,000 inhabited); occasional floods, severe droughts, and tsunamis; deforestation
Note: straddles Equator; strategic location astride or along major sea lanes from Indian Ocean to Pacific Ocean

People

Population: 197,232,428 (July 1993 est.)
Population growth rate: 1.61% (1993 est.)

Birth rate: 24.84 births/1,000 population (1993 est.)
Death rate: 8.73 deaths/1,000 population (1993 est.)
Net migration rate: 0 migrant(s)/1,000 population (1993 est.)
Infant mortality rate: 69.6 deaths/1,000 live births (1993 est.)
Life expectancy at birth:
total population: 60.26 years
male: 58.28 years
female: 62.34 years (1993 est.)
Total fertility rate: 2.86 children born/woman (1993 est.)
Nationality:
noun: Indonesian(s)
adjective: Indonesian
Ethnic divisions: Javanese 45%, Sundanese 14%, Madurese 7.5%, coastal Malays 7.5%, other 26%
Religions: Muslim 87%, Protestant 6%, Roman Catholic 3%, Hindu 2%, Buddhist 1%, other 1% (1985)
Languages: Bahasa Indonesia (modified form of Malay; official), English, Dutch, local dialects the most widely spoken of which is Javanese
Literacy: age 15 and over can read and write (1990)
total population: 77%
male: 84%
female: 68%
Labor force: 67 million
by occupation: agriculture 55%, manufacturing 10%, construction 4%, transport and communications 3% (1985 est.)

Government

Names:
conventional long form: Republic of Indonesia
conventional short form: Indonesia
local long form: Republik Indonesia
local short form: Indonesia
former name: Netherlands East Indies; Dutch East Indies
Digraph: ID
Type: republic
Capital: Jakarta
Administrative divisions: 24 provinces (propinsi-propinsi, singular—propinsi), 2 special regions* (daerah-daerah istimewa, singular—daerah istimewa), and 1 special capital city district** (daerah khusus ibukota); Aceh*, Bali, Bengkulu, Irian Jaya, Jakarta Raya**, Jambi, Jawa Barat, Jawa Tengah, Jawa Timur, Kalimantan Barat, Kalimantan Selatan, Kalimantan Tengah, Kalimantan Timur, Lampung, Maluku, Nusa Tenggara Barat, Nusa Tenggara Timur, Riau, Sulawesi Selatan, Sulawesi Tengah, Sulawesi Tenggara, Sulawesi Utara, Sumatera Barat, Sumatera Selatan, Sumatera Utara, Timor Timur, Yogyakarta*

Independence: 17 August 1945 (proclaimed independence; on 27 December 1949, Indonesia became legally independent from the Netherlands)
Constitution: August 1945, abrogated by Federal Constitution of 1949 and Provisional Constitution of 1950, restored 5 July 1959
Legal system: based on Roman-Dutch law, substantially modified by indigenous concepts and by new criminal procedures code; has not accepted compulsory ICJ jurisdiction
National holiday: Independence Day, 17 August (1945)
Political parties and leaders: GOLKAR (quasi-official party based on functional groups), Lt. Gen. (Ret.) WAHONO, general chairman; Indonesia Democracy Party (PDI—federation of former Nationalist and Christian Parties), SOERYADI, chairman; Development Unity Party (PPP, federation of former Islamic parties), Ismail Hasan METAREUM, chairman
Suffrage: 17 years of age; universal and married persons regardless of age
Elections:
House of Representatives: last held on 8 June 1992 (next to be held NA 1997); results—GOLKAR 68%, PPP 17%, PDI 15%; seats—(500 total, 400 elected, 100 appointed) GOLKAR 282, PPP 62, PDI 56
Executive branch: president, vice president, Cabinet
Legislative branch: unicameral House of Representatives (Dewan Perwakilan Rakyat or DPR); note - the People's Consultative Assembly (Majelis Permusyawaratan Rakyat or MPR) includes the DPR plus 500 indirectly elected members who meet every five years to elect the president and vice president and, theoretically, to determine national policy
Judicial branch: Supreme Court (Mahkamah Agung)
Leaders:
Chief of State and Head of Government: President Gen. (Ret.) SOEHARTO (since 27 March 1968); Vice President Gen. (Ret.) Try SUTRISNO (since 11 March 1993)
Member of: APEC, AsDB, ASEAN, CCC, CP, ESCAP, FAO, G-15, G-19, G-77, GATT, IAEA, IBRD, ICAO, ICC, ICFTU, IDA, IDB, IFAD, IFC, ILO, IMF, IMO, INMARSAT, INTELSAT, INTERPOL, IOC, IOM (observer), ISO, ITU, LORCS, NAM, OIC, OPEC, UN, UNCTAD, UNESCO, UNIDO, UNIKOM, UNOSOM, UNTAC, UPU, WCL, WFTU, WHO, WIPO, WMO, WTO
Diplomatic representation in US:
chief of mission: Ambassador Abdul Rachman RAMLY
chancery: 2020 Massachusetts Avenue NW, Washington, DC 20036
telephone: (202) 775-5200

Indonesia *(continued)*

consulates general: Houston, New York, and Los Angeles
consulates: Chicago and San Francisco
US diplomatic representation:
chief of mission: Ambassador Robert L. BARRY
embassy: Medan Merdeka Selatan 5, Jakarta
mailing address: APO AP 96520
telephone: [62] (21) 360-360
FAX: [62] (21) 360-644
consulates: Medan, Surabaya
Flag: two equal horizontal bands of red (top) and white; similar to the flag of Monaco, which is shorter; also similar to the flag of Poland, which is white (top) and red

Economy

Overview: Indonesia is a mixed economy with many socialist institutions and central planning but with a recent emphasis on deregulation and private enterprise. Indonesia has extensive natural wealth, yet, with a large and rapidly increasing population, it remains a poor country. Real GDP growth in 1985-92 averaged about 6%, quite impressive, but not sufficient to both slash underemployment and absorb the 2.3 million workers annually entering the labor force. Agriculture, including forestry and fishing, is an important sector, accounting for almost 20% of GDP and over 50% of the labor force. The staple crop is rice. Once the world's largest rice importer, Indonesia is now nearly self-sufficient. Plantation crops—rubber and palm oil—and textiles and plywood are being encouraged for both export and job generation. Industrial output now accounts for almost 40% of GDP and is based on a supply of diverse natural resources, including crude oil, natural gas, timber, metals, and coal. Of these, the oil sector dominates the external economy, generating more than 20% of the government's revenues and 40% of export earnings in 1989. However, the economy's growth is highly dependent on the continuing expansion of nonoil exports. Japan remains Indonesia's most important customer and supplier of aid. Rapid growth in the money supply in 1989-90 prompted Jakarta to implement a tight monetary policy in 1991, forcing the private sector to go to foreign banks for investment financing. Real interest rates remained above 10% and off-shore commercial debt grew. The growth in off-shore debt prompted Jakarta to limit foreign borrowing beginning in late 1991. Despite the continued problems in moving toward a more open financial system and the persistence of a fairly tight credit situation, GDP growth in 1992 is estimated to have stayed at 6%.
National product: GDP—exchange rate conversion—$133 billion (1992 est.)

National product real growth rate: 6% (1992 est.)
National product per capita: $680 (1992 est.)
Inflation rate (consumer prices): 8% (1992 est.)
Unemployment rate: 3% ; underemployment 45% (1991 est.)
Budget: revenues $17.2 billion; expenditures $23.4 billion, including capital expenditures of $8.9 billion (FY91)
Exports: $29.4 billion (f.o.b., 1991)
commodities: petroleum and liquefied natural gas 40%, timber 15%, textiles 7%, rubber 5%, coffee 3%
partners: Japan 37%, Europe 13%, US 12%, Singapore 8% (1991)
Imports: $24.6 billion (f.o.b., 1991)
commodities: machinery 39%, chemical products 19%, manufactured goods 16%
partners: Japan 25%, Europe 23%, US 13%, Singapore 5% (1991)
External debt: $50.5 billion (1992 est.)
Industrial production: growth rate 11.6% (1989 est.); accounts for almost 40% of GDP
Electricity: 11,600,000 kW capacity; 38,000 million kWh produced, 200 kWh per capita (1990)
Industries: petroleum and natural gas, textiles, mining, cement, chemical fertilizers, plywood, food, rubber
Agriculture: accounts for almost 20% of GDP; subsistence food production; small-holder and plantation production for export; main products are rice, cassava, peanuts, rubber, cocoa, coffee, palm oil, copra, other tropical products, poultry, beef, pork, eggs
Illicit drugs: illicit producer of cannabis for the international drug trade, but not a major player; government actively eradicating plantings and prosecuting traffickers
Economic aid: US commitments, including Ex-Im (FY70-89), $4.4 billion; Western (non-US) countries, ODA and OOF bilateral commitments (1970-89), $25.9 billion; OPEC bilateral aid (1979-89), $213 million; Communist countries (1970-89), $175 million
Currency: 1 Indonesian rupiah (Rp) = 100 sen (sen no longer used)
Exchange rates: Indonesian rupiahs (Rp) per US$1—2,064.7 (January 1993), 2,029.9 (1992), 1,950.3 (1991), 1,842.8 (1990), 1,770.1 (1989), 1,685.7 (1988)
Fiscal year: 1 April-31 March

Communications

Railroads: 6,964 km total; 6,389 km 1.067-meter gauge, 497 km 0.750-meter gauge, 78 km 0.600-meter gauge; 211 km double track; 101 km electrified; all government owned

Highways: 119,500 km total; 11,812 km state, 34,180 km provincial, and 73,508 km district roads
Inland waterways: 21,579 km total; Sumatra 5,471 km, Java and Madura 820 km, Kalimantan 10,460 km, Celebes 241 km, Irian Jaya 4,587 km
Pipelines: crude oil 2,505 km; petroleum products 456 km; natural gas 1,703 km (1989)
Ports: Cilacap, Cirebon, Jakarta, Kupang, Palembang, Ujungpandang, Semarang, Surabaya
Merchant marine: 401 ships (1,000 GRT or over) totaling 1,766,201 GRT/2,642,529 DWT; includes 6 short-sea passenger, 13 passenger-cargo, 238 cargo, 10 container, 4 roll-on/roll-off cargo, 4 vehicle carrier, 78 oil tanker, 6 chemical tanker, 6 liquefied gas, 7 specialized tanker, 1 livestock carrier, 26 bulk, 2 passenger
Airports:
total: 435
usable: 411
with permanent-surface runways: 119
with runways over 3,659 m: 1
with runways 2,440-3,659 m: 11
with runways 1,220-2,439 m: 67
Telecommunications: interisland microwave system and HF police net; domestic service fair, international service good; radiobroadcast coverage good; 763,000 telephones (1986); broadcast stations—618 AM, 38 FM, 9 TV; satellite earth stations—1 Indian Ocean INTELSAT earth station and 1 Pacific Ocean INTELSAT earth station; and 1 domestic satellite communications system

Defense Forces

Branches: Army, Navy, Air Force, National Police
Manpower availability: males age 15-49 53,160,364; fit for military service 31,395,254; reach military age (18) annually 2,148,927 (1993 est.)
Defense expenditures: exchange rate conversion—$2.1 billion, 1.5% of GNP (FY93/94 est.)

Iran

Geography

Location: Middle East, between the Persian Gulf and the Caspian Sea
Map references: Asia, Middle East, Standard Time Zones of the World
Area:
total area: 1.648 million km²
land area: 1.636 million km²
comparative area: slightly larger than Alaska
Land boundaries: total 5,440 km, Afghanistan 936 km, Armenia 35 km, Azerbaijan (north) 432 km, Azerbaijan (northwest) 179 km, Iraq 1,458 km, Pakistan 909 km, Turkey 499 km, Turkmenistan 992 km
Coastline: 2,440 km
note: Iran also borders the Caspian Sea (740 km)
Maritime claims:
continental shelf: not specified
exclusive fishing zone: 50 nm in the Gulf of Oman; continental shelf limit, continental shelf boundaries, or median lines in the Persian Gulf
territorial sea: 12 nm
International disputes: Iran and Iraq restored diplomatic relations in 1990 but are still trying to work out written agreements settling outstanding disputes from their eight-year war concerning border demarcation, prisoners-of-war, and freedom of navigation and sovereignty over the Shatt al Arab waterway; Iran occupies two islands in the Persian Gulf claimed by the UAE: Tunb as Sughra (Arabic), Jazireh-ye Tonb-e Kuchek (Persian) or Lesser Tunb, and Tunb al Kubra (Arabic), Jazireh-ye Tonb-e Bozorg (Persian) or Greater Tunb; it jointly administers with the UAE an island in the Persian Gulf claimed by the UAE, Abu Musa (Arabic) or Jazireh-ye Abu Musa (Persian); in 1992 the dispute over Abu Musa and the Tunb Islands became more acute when Iran unilaterally tried to control the entry of third country nationals into the UAE portion of Abu Musa island, Tehran subsequently backed off in the face of significant diplomatic support for the UAE in the region; periodic disputes with Afghanistan over Helmand water rights,
Climate: mostly arid or semiarid, subtropical along Caspian coast
Terrain: rugged, mountainous rim; high, central basin with deserts, mountains; small, discontinuous plains along both coasts
Natural resources: petroleum, natural gas, coal, chromium, copper, iron ore, lead, manganese, zinc, sulfur
Land use:
arable land: 8%
permanent crops: 0%
meadows and pastures: 27%
forest and woodland: 11%
other: 54%
Irrigated land: 57,500 km² (1989 est.)
Environment: deforestation; overgrazing; desertification

People

Population: 63,369,809 (July 1993 est.)
Population growth rate: 3.49% (1993 est.)
Birth rate: 43 births/1,000 population (1993 est.)
Death rate: 8.06 deaths/1,000 population (1993 est.)
Net migration rate: 0 migrant(s)/1,000 population (1993 est.)
Infant mortality rate: 62.1 deaths/1,000 live births (1993 est.)
Life expectancy at birth:
total population: 65.26 years
male: 64.37 years
female: 66.19 years (1993 est.)
Total fertility rate: 6.4 children born/woman (1993 est.)
Nationality:
noun: Iranian(s)
adjective: Iranian
Ethnic divisions: Persian 51%, Azerbaijani 24%, Gilaki and Mazandarani 8%, Kurd 7%, Arab 3%, Lur 2%, Baloch 2%, Turkmen 2%, other 1%
Religions: Shi'a Muslim 95%, Sunni Muslim 4%, Zoroastrian, Jewish, Christian, and Baha'i 1%
Languages: Persian and Persian dialects 58%, Turkic and Turkic dialects 26%, Kurdish 9%, Luri 2%, Baloch 1%, Arabic 1%, Turkish 1%, other 2%
Literacy: age 15 and over can read and write (1990)
total population: 54%
male: 64%
female: 43%
Labor force: 15.4 million
by occupation: agriculture 33%, manufacturing 21%
note: shortage of skilled labor (1988 est.)

Government

Names:
conventional long form: Islamic Republic of Iran
conventional short form: Iran
local long form: Jomhuri-ye Eslami-ye Iran
local short form: Iran
Digraph: IR
Type: theocratic republic
Capital: Tehran
Administrative divisions: 24 provinces (ostanha, singular—ostan); Azarbayjan-e Bakhtari, Azarbayjan-e Khavari, Bakhtaran, Bushehr, Chahar Mahall va Bakhtiari, Esfahan, Fars, Gilan, Hamadan, Hormozgan, Ilam, Kerman, Khorasan, Khuzestan, Kohkiluyeh va Buyer Ahmadi, Kordestan, Lorestan, Markazi, Mazandaran, Semnan, Sistan va Baluchestan, Tehran, Yazd, Zanjan
Independence: 1 April 1979 (Islamic Republic of Iran proclaimed)
Constitution: 2-3 December 1979; revised 1989 to expand powers of the presidency and eliminate the prime ministership
Legal system: the Constitution codifies Islamic principles of government
National holiday: Islamic Republic Day, 1 April (1979)
Political parties and leaders: there are at least 18 licensed parties; the three most important are—Tehran Militant Clergy Association, Mohammad Reza MAHDAVI-KANI; Militant Clerics Association, Mehdi MAHDAVI-KARUBI and Mohammad Asqar MUSAVI-KHOINIHA; Fedaiyin Islam Organization, Sadeq KHALKHALI
Other political or pressure groups: groups that generally support the Islamic Republic include Hizballah, Hojjatiyeh Society, Mojahedin of the Islamic Revolution, Muslim Students Following the Line of the Imam; armed political groups that have been almost completely repressed by the government include Mojahedin-e Khalq Organization (MEK), People's Fedayeen, Kurdish Democratic Party; the Society for the defense of Freedom
Suffrage: 15 years of age; universal
Elections:
President: last held July 1989 (next to be held 11 June 1993); results—Ali Akbar HASHEMI-RAFSANJANI was elected with only token opposition
Islamic Consultative Assembly: last held 8 April 1992 (next to be held April 1996); results—percent of vote by party NA; seats—(270 seats total) number of seats by party NA
Executive branch: supreme leader (velay-t-e faqih), president, Council of Ministers
Legislative branch: unicameral Islamic Consultative Assembly (Majles-e-Shura-ye-Eslami)

185

Iran (continued)

Judicial branch: Supreme Court
Leaders:
Supreme Leader and functional Chief of State: Leader of the Islamic Revolution Ayatollah Ali HOSEINI-KHAMENEI (since 4 June 1989)
Head of Government: President Ali Akbar HASHEMI-RAFSANJANI (since 3 August 1989)
Member of: CCC, CP, ESCAP, ECO, FAO, G-19, G-24, G-77, IAEA, IBRD, ICAO, ICC, IDA, IDB, IFAD, IFC, ILO, IMF, IMO, INMARSAT, INTELSAT, INTERPOL, IOC, ISO, ITU, LORCS, NAM, OIC, OPEC, PCA, UN, UNCTAD, UNESCO, UNHCR, UNIDO, UPU, WFTU, WHO, WIPO, WMO, WTO
Diplomatic representation in US:
chief of mission: Iran has an Interests Section in the Pakistani Embassy in Washington, DC
chancery: Iranian Interests Section, 2209 Wisconsin Ave. NW, Washington, DC 20007
telephone: (202) 965-4990
US diplomatic representation: protecting power in Iran is Switzerland
Flag: three equal horizontal bands of green (top), white, and red; the national emblem (a stylized representation of the word Allah) in red is centered in the white band; Allah Alkbar (God is Great) in white Arabic script is repeated 11 times along the bottom edge of the green band and 11 times along the top edge of the red band

Economy

Overview: Iran's economy is a mixture of central planning, state ownership of oil and other large enterprises, village agriculture, and small-scale private trading and service ventures. After a decade of economic decline, Iran's real GDP grew by 10% in FY90 and 6% in FY91, according to Iranian Government statistics. An oil windfall in 1990 combined with a substantial increase in imports contributed to Iran's recent economic growth. Iran has also begun implementing a number of economic reforms to reduce government intervention (including subsidies) and has allocated substantial resources to development projects in the hope of stimulating the economy. Lower oil revenues in 1991—oil accounts for more than 90% of export revenues—together with a surge in imports greatly weakened Iran's international financial position. By mid-1992 Iran was unable to meet its obligations to foreign creditors. Subsequently the government has tried to boost oil exports, curb imports (especially of consumer goods), and renegotiate terms of its foreign debts.
National product: GNP—exchange rate conversion—$90 billion (FY92)
National product real growth rate: 6% (FY91)

National product per capita: $1,500 (FY91)
Inflation rate (consumer prices): 23.7% (September 1991-September 1992)
Unemployment rate: 30% (1991 est.)
Budget: revenues $63 billion; expenditures $80 billion, including capital expenditures of $23 billion (FY90 est.)
Exports: $17.2 billion (f.o.b., FY91 est.)
commodities: petroleum 90%, carpets, fruits, nuts, hides
partners: Japan, Italy, France, Netherlands, Belgium/Luxembourg, Spain, and Germany
Imports: $21.0 billion (c.i.f., FY91 est.)
commodities: machinery, military supplies, metal works, foodstuffs, pharmaceuticals, technical services, refined oil products
partners: Germany, Japan, Italy, UK, France
External debt: $17 billion (FY91 est.)
Industrial production: growth rate 12% (1990 est.); accounts for almost 30% of GDP, including petroleum
Electricity: 15,649,000 kW capacity; 43,600 million kWh produced, 710 kWh per capita (1992)
Industries: petroleum, petrochemicals, textiles, cement and other building materials, food processing (particularly sugar refining and vegetable oil production), metal fabricating
Agriculture: accounts for about 20% of GDP; principal products—wheat, rice, other grains, sugar beets, fruits, nuts, cotton, dairy products, wool, caviar; not self-sufficient in food
Illicit drugs: illicit producer of opium poppy for the domestic and international drug trade; transshipment point for Southwest Asian heroin to Europe
Economic aid: US commitments, including Ex-Im (FY70-80), $1.0 billion; Western (non-US) countries, ODA and OOF bilateral commitments (1970-89), $1.675 billion; Communist countries (1970-89), $976 million; note—aid fell sharply following the 1979 revolution
Currency: 1 Iranian rial (IR) = 10 tomans
Exchange rates: Iranian rials (IR) per US$1—67.095 (January 1993), 65.552 (1992), 67.505 (1991), 68.096 (1990), 72.015 (1989), 68.683 (1988); black-market rate 1,400 (January 1991); note—in March 1993 the Iranian government announced a new single-parity exchange rate system with a new official rate of 1,538 rials per dollar
Fiscal year: 21 March-20 March

Communications

Railroads: 4,852 km total; 4,760 km 1.432-meter gauge, 92 km 1.676-meter gauge; 480 km under construction from Bafq to Bandar-e Abbas, rail construction from Bafq to Sirjan has been completed and is operational; section from Sirjan to Bandar-e

Abbas still under construction
Highways: 140,200 km total; 42,694 km paved surfaces; 46,866 km gravel and crushed stone; 49,440 km improved earth; 1,200 km (est.) rural road network
Inland waterways: 904 km; the Shatt al Arab is usually navigable by maritime traffic for about 130 km; channel has been dredged to 3 meters and is in use
Pipelines: crude oil 5,900 km; petroleum products 3,900 km; natural gas 4,550 km
Ports: Abadan (largely destroyed in fighting during 1980-88 war), Bandar Beheshti, Bandar-e Abbas, Bandar-e Bushehr, Bandar-e Khomeyni, Bandar-e Torkeman (Caspian Sea port), Khorramshahr (repaired after being largely destroyed in fighting during 1980-88 war) has been in limited operation since November 1992
Merchant marine: 135 ships (1,000 GRT or over) totaling 4,480,726 GRT/8,332,593 DWT; includes 39 cargo, 6 roll-on/roll-off cargo, 32 oil tanker, 4 chemical tanker, 3 refrigerated cargo, 48 bulk, 2 combination bulk, 1 liquefied gas
Airports:
total: 219
usable: 194
with permanent-surface runways: 83
with runways over 3,659 m: 16
with runways 2,440-3,659 m: 20
with runways 1,220-2,439 m: 70
Telecommunications: microwave radio relay extends throughout country; system centered in Tehran; 2,143,000 telephones (35 telephones per 1,000 persons); broadcast stations—77 AM, 3 FM, 28 TV; satellite earth stations—2 Atlantic Ocean INTELSAT and 1 Indian Ocean INTELSAT; HF radio and microwave radio relay to Turkey, Pakistan, Syria, Kuwait, Tajikistan, and Uzbekistan; submarine fiber optic cable to UAE

Defense Forces

Branches: Islamic Republic of Iran Ground Forces, Navy, Air and Air Defense Force, Revolutionary Guards (including Basij militia and own ground, air, and naval forces), Law Enforcement Forces
Manpower availability: males age 15-49 13,812,367; fit for military service 8,218,286; reach military age (21) annually 575,392 (1993 est.)
Defense expenditures: hard currency expenditures on defense are 7-10% of total hard currency expenditures; rial expenditures on defense are 8-13% of total rial expenditures (1992 est.)
note: conversion of rial expenditures into US dollars using the prevailing exchange rate could produce misleading results

Iraq

Geography

Location: Middle East, between Iran and Saudi Arabia

Map references: Middle East, Standard Time Zones of the World

Area:
total area: 437,072 km²
land area: 432,162 km²
comparative area: slightly more than twice the size of Idaho

Land boundaries: total 3,631 km, Iran 1,458 km, Jordan 181 km, Kuwait 242 km, Saudi Arabia 814 km, Syria 605 km, Turkey 331 km

Coastline: 58 km

Maritime claims:
continental shelf: not specified
territorial sea: 12 nm

International disputes: Iran and Iraq restored diplomatic relations in 1990 but are still trying to work out written agreements settling outstanding disputes from their eight-year war concerning border demarcation, prisoners-of-war, and freedom of navigation and sovereignty over the Shatt al Arab waterway; in April 1991 official Iraqi acceptance of UN Security Council Resolution 687, which demands that Iraq accept the inviolability of the boundary set forth in its 1963 agreement with Kuwait, ending earlier claims to Bubiyan and Warbah Islands or to all of Kuwait; the 20 May 1993 final report of the UN Iraq/Kuwait Boundary Demarcation Commission was welcomed by the Security Council in Resolution 833 of 27 May 1993, which also reaffirmed that the decisions of the commission on the boundary were final, bringing to a completion the official demarcation of the Iraq-Kuwait boundary; Iraqi officials still make public statements claiming Kuwait; periodic disputes with upstream riparian Syria over Euphrates water rights; potential dispute over water development plans by Turkey for the Tigris and Euphrates Rivers

Climate: mostly desert; mild to cool winters with dry, hot, cloudless summers; northernmost regions along Iranian and Turkish borders experience cold winters with occasionally heavy snows

Terrain: mostly broad plains; reedy marshes in southeast; mountains along borders with Iran and Turkey

Natural resources: petroleum, natural gas, phosphates, sulfur

Land use:
arable land: 12%
permanent crops: 1%
meadows and pastures: 9%
forest and woodland: 3%
other: 75%

Irrigated land: 25,500 km² (1989 est)

Environment: development of Tigris-Euphrates Rivers system contingent upon agreements with upstream riparians (Syria, Turkey); air and water pollution; soil degradation (salinization) and erosion; desertification

People

Population: 19,161,956 (July 1993 est.)

Population growth rate: 3.73% (1993 est.)

Birth rate: 44.57 births/1,000 population (1993 est.)

Death rate: 7.71 deaths/1,000 population (1993 est.)

Net migration rate: 0.42 migrant(s)/1,000 population (1993 est.)

Infant mortality rate: 71.8 deaths/1,000 live births (1993 est.)

Life expectancy at birth:
total population: 64.96 years
male: 64.2 years
female: 65.76 years (1993 est.)

Total fertility rate: 6.86 children born/woman (1993 est.)

Nationality:
noun: Iraqi(s)
adjective: Iraqi

Ethnic divisions: Arab 75-80%, Kurdish 15-20%, Turkoman, Assyrian or other 5%

Religions: Muslim 97% (Shi'a 60-65%, Sunni 32-37%), Christian or other 3%

Languages: Arabic, Kurdish (official in Kurdish regions), Assyrian, Armenian

Literacy: age 15 and over can read and write (1990)
total population: 60%
male: 70%
female: 49%

Labor force: 4.4 million (1989)
by occupation: services 48%, agriculture 30%, industry 22%
note: severe labor shortage; expatriate labor force was about 1,600,000 (July 1990); since then, it has declined substantially

Government

Names:
conventional long form: Republic of Iraq
conventional short form: Iraq
local long form: Al Jumhuriyah al Iraqiyah
local short form: Al Iraq

Digraph: IZ

Type: republic

Capital: Baghdad

Administrative divisions: 18 provinces (muhafazat, singular—muhafazah); Al Anbar, Al Basrah, Al Muthanna, Al Qadisiyah, An Najaf, Arbil, As Sulaymaniyah, At Ta'mim, Babil, Baghdad, Dahuk, Dhi Qar, Diyala, Karbala', Maysan, Ninawa, Salah ad Din, Wasit

Independence: 3 October 1932 (from League of Nations mandate under British administration)

Constitution: 22 September 1968, effective 16 July 1970 (interim Constitution); new constitution drafted in 1990 but not adopted

Legal system: based on Islamic law in special religious courts, civil law system elsewhere; has not accepted compulsory ICJ jurisdiction

National holiday: Anniversary of the Revolution, 17 July (1968)

Political parties and leaders: Ba'th Party

Other political or pressure groups: political parties and activity severely restricted; possibly some opposition to regime from disaffected members of the regime, Army officers, and Shi'a religious and Kurdish ethnic dissidents; the Green Party (government-controlled)

Suffrage: 18 years of age; universal

Elections:
National Assembly: last held on 1 April 1989 (next to be held NA); results—Sunni Arabs 53%, Shi'a Arabs 30%, Kurds 15%, Christians 2% est.; seats—(250 total) number of seats by party NA
note: in northern Iraq, a "Kurdish Assembly" was elected in May 1992 and calls for Kurdish self-determination within a federated Iraq

Executive branch: president, vice president, chairman of the Revolutionary Command Council, vice chairman of the Revolutionary Command Council, prime minister, first deputy prime minister, Council of Ministers

Legislative branch: unicameral National Assembly (Majlis al-Watani)

Judicial branch: Court of Cassation

Leaders:
Chief of State: President SADDAM Husayn (since 16 July 1979); Vice President Taha Muhyi al-Din MA'RUF (since 21 April 1974); Vice President Taha Yasin RAMADAN (since 23 March 1991)
Head of Government: Prime Minister Muhammad Hamza al-ZUBAYDI (since 13 September 1991); Deputy Prime Minister Tariq 'AZIZ (since NA 1979)

Iraq *(continued)*

Member of: ABEDA, ACC, AFESD, AL, AMF, CAEU, CCC, ESCWA, FAO, G-19, G-77, IAEA, IBRD, ICAO, IDA, IDB, IFAD, IFC, ILO, IMF, IMO, INMARSAT, INTELSAT, INTERPOL, IOC, ISO, ITU, LORCS, NAM, OAPEC, OIC, OPEC, PCA, UN, UNCTAD, UNESCO, UNIDO, UPU, WFTU, WHO, WIPO, WMO, WTO

Diplomatic representation in US:
chief of mission: Iraq has an Interest Section in the Algerian Embassy in Washington, DC
chancery: Iraqi Interests Section, 1801 P Street NW, Washington, DC 20036
telephone: (202) 483-7500
FAX: (202) 462-5066

US diplomatic representation:
chief of mission: (vacant); note—operations have been temporarily suspended; a US interests section is located in Poland's embassy in Baghdad
embassy: Masbah Quarter (opposite the Foreign Ministry Club), Baghdad
mailing address: P. O. Box 2447 Alwiyah, Baghdad
telephone: [964] (1) 719-6138 or 719-6139, 718-1840, 719-3791

Flag: three equal horizontal bands of red (top), white, and black with three green five-pointed stars in a horizontal line centered in the white band; the phrase ALLAHU AKBAR (God is Great) in green Arabic script—Allahu to the right of the middle star and Akbar to the left of the middle star—was added in January 1991 during the Persian Gulf crisis; similar to the flag of Syria that has two stars but no script and the flag of Yemen that has a plain white band; also similar to the flag of Egypt that has a symbolic eagle centered in the white band

Economy

Overview: The Ba'thist regime engages in extensive central planning and management of industrial production and foreign trade while leaving some small-scale industry and services and most agriculture to private enterprise. The economy has been dominated by the oil sector, which has traditionally provided about 95% of foreign exchange earnings. In the 1980s, financial problems caused by massive expenditures in the eight-year war with Iran and damage to oil export facilities by Iran, led the government to implement austerity measures and to borrow heavily and later reschedule foreign debt payments. After the end of hostilities in 1988, oil exports gradually increased with the construction of new pipelines and restoration of damaged facilities. Agricultural development remained

hampered by labor shortages, salinization, and dislocations caused by previous land reform and collectivization programs. The industrial sector, although accorded high priority by the government, also was under financial constraints. Iraq's seizure of Kuwait in August 1990, subsequent international economic embargoes, and military action by an international coalition beginning in January 1991 drastically changed the economic picture. Industrial and transportation facilities suffered severe damage and have been only partially restored. Oil exports remain at less than 10% of the previous level. Shortages of spare parts continue. Living standards deteriorated even further in 1992 and early 1993; consumer prices at least tripled in 1992. The UN-sponsored economic embargo has reduced exports and imports and has contributed to the sharp rise in prices. The government's policies of supporting large military and internal security forces and of allocating resources to key supporters of the regime have exacerbated shortages. In brief, per capita output in early 1993 is far below the 1989-90 level, but no reliable estimate is available.

National product: GNP—exchange rate conversion—$35 billion (1989 est.)
National product real growth rate: 10% (1989 est.)
National product per capita: $1,940 (1989 est.)
Inflation rate (consumer prices): 200% (1992 est.)
Unemployment rate: less than 5% (1989 est.)
Budget: revenues $NA; expenditures $NA, including capital expenditures of $NA
Exports: $10.4 billion (f.o.b., 1990)
commodities: crude oil and refined products, fertilizer, sulfur
partners: US, Brazil, Turkey, Japan, Netherlands, Spain (1990)
Imports: $6.6 billion (c.i.f., 1990)
commodities: manufactures, food
partners: Germany, US, Turkey, France, UK (1990)
External debt: $45 billion (1989 est.), excluding debt of about $35 billion owed to Arab Gulf states
Industrial production: NA%; manufacturing accounts for 10% of GNP (1989)
Electricity: 7,300,000 kW available out of 9,902,000 kW capacity due to Gulf war; 12,900 million kWh produced, 700 kWh per capita (1992)
Industries: petroleum production and refining, chemicals, textiles, construction materials, food processing
Agriculture: accounts for 11% of GNP and 30% of labor force; principal products—wheat, barley, rice, vegetables,

dates, other fruit, cotton, wool; livestock—cattle, sheep; not self-sufficient in food output
Economic aid: US commitments, including Ex-Im (FY70-80), $3 million; Western (non-US) countries, ODA and OOF bilateral commitments (1970-89), $647 million; Communist countries (1970-89), $3.9 billion
Currency: 1 Iraqi dinar (ID) = 1,000 fils
Exchange rates: Iraqi dinars (ID) per US$1—3.2 (fixed official rate since 1982); black-market rate (April 1993) US$1 = 53.5 Iraqi dinars
Fiscal year: calendar year

Communications

Railroads: 2,457 km 1.435-meter standard gauge
Highways: 34,700 km total; 17,500 km paved, 5,500 km improved earth, 11,700 km unimproved earth
Inland waterways: 1,015 km; Shatt al Arab is usually navigable by maritime traffic for about 130 km; channel has been dredged to 3 meters and is in use; Tigris and Euphrates Rivers have navigable sections for shallow-draft watercraft; Shatt al Basrah canal was navigable by shallow-draft craft before closing in 1991 because of the Persian Gulf war
Pipelines: crude oil 4,350 km; petroleum products 725 km; natural gas 1,360 km
Ports: Umm Qasr, Khawr az Zubayr, Al Basrah (closed since 1980)
Merchant marine: 41 ships (1,000 GRT or over) totaling 930,780 GRT/1,674,878 DWT; includes 1 passenger, 1 passenger-cargo, 15 cargo, 1 refrigerated cargo, 3 roll-on/roll-off cargo, 19 oil tanker, 1 chemical tanker; note—none of the Iraqi flag merchant fleet was trading internationally as of 1 January 1993
Airports:
total: 114
usable: 99
with permanent-surface runways: 74
with runways over 3,659 m: 9
with runways 2,440-3,659 m: 52
with runways 1,220-2,439 m: 12
Telecommunications: reconstitution of damaged telecommunication facilities began after Desert Storm, most damaged facilities have been rebuilt; the network consists of coaxial cables and microwave radio relay links; 632,000 telephones; broadcast stations—16 AM, 1 FM, 13 TV; satellite earth stations—1 Atlantic Ocean INTELSAT, 1 Indian Ocean INTELSAT, 1 Atlantic Ocean GORIZONT in the Intersputnik system and 1 ARABSAT; coaxial cable and microwave radio relay to Jordan, Kuwait, Syria, and Turkey, Kuwait line is probably non-operational

Ireland

Defense Forces

Branches: Army and Republican Guard, Navy, Air Force, Air Defense Force, Border Guard Force, Internal Security Forces
Manpower availability: males age 15-49 4,235,321; fit for military service 2,379,999; reach military age (18) annually 211,776 (1993 est.)
Defense expenditures: exchange rate conversion—$NA, NA% of GNP

Geography

Location: in the North Atlantic Ocean, across the Irish Sea from Great Britain
Map references: Europe, Standard Time Zones of the World
Area:
total area: 70,280 km²
land area: 68,890 km²
comparative area: slightly larger than West Virginia
Land boundaries: total 360 km, UK 360 km
Coastline: 1,448 km
Maritime claims:
continental shelf: not specified
exclusive fishing zone: 200 nm
territorial sea: 12 nm
International disputes: Northern Ireland question with the UK; Rockall continental shelf dispute involving Denmark, Iceland, and the UK (Ireland and the UK have signed a boundary agreement in the Rockall area)
Climate: temperate maritime; modified by North Atlantic Current; mild winters, cool summers; consistently humid; overcast about half the time
Terrain: mostly level to rolling interior plain surrounded by rugged hills and low mountains; sea cliffs on west coast
Natural resources: zinc, lead, natural gas, petroleum, barite, copper, gypsum, limestone, dolomite, peat, silver
Land use:
arable land: 14%
permanent crops: 0%
meadows and pastures: 71%
forest and woodland: 5%
other: 10%
Irrigated land: NA km²
Environment: deforestation
Note: strategic location on major air and sea routes between North American and northern Europe

People

Population: 3,529,566 (July 1993 est.)
Population growth rate: 0.26% (1993 est.)
Birth rate: 14.39 births/1,000 population (1993 est.)
Death rate: 8.71 deaths/1,000 population (1993 est.)
Net migration rate: -3.13 migrant(s)/1,000 population (1993 est.)
Infant mortality rate: 7.6 deaths/1,000 live births (1993 est.)
Life expectancy at birth:
total population: 75.38 years
male: 72.56 years
female: 78.36 years (1993 est.)
Total fertility rate: 2.02 children born/woman (1993 est.)
Nationality:
noun: Irishman(men), Irishwoman(men), Irish (collective plural)
adjective: Irish
Ethnic divisions: Celtic, English
Religions: Roman Catholic 93%, Anglican 3%, none 1%, unknown 2%, other 1% (1981)
Languages: Irish (Gaelic), spoken mainly in areas located along the western seaboard, English is the language generally used
Literacy: age 15 and over can read and write (1981)
total population: 98%
male: NA%
female: NA%
Labor force: 1.37 million
by occupation: services 57.0%, manufacturing and construction 28%, agriculture, forestry, and fishing 13.5%, energy and mining 1.5% (1992)

Government

Names:
conventional long form: none
conventional short form: Ireland
Digraph: EI
Type: republic
Capital: Dublin
Administrative divisions: 26 counties; Carlow, Cavan, Clare, Cork, Donegal, Dublin, Galway, Kerry, Kildare, Kilkenny, Laois, Leitrim, Limerick, Longford, Louth, Mayo, Meath, Monaghan, Offaly, Roscommon, Sligo, Tipperary, Waterford, Westmeath, Wexford, Wicklow
Independence: 6 December 1921 (from UK)
Constitution: 29 December 1937; adopted 1937
Legal system: based on English common law, substantially modified by indigenous concepts; judicial review of legislative acts in Supreme Court; has not accepted compulsory ICJ jurisdiction

Ireland (continued)

National holiday: Saint Patrick's Day, 17 March

Political parties and leaders: Democratic Left, Proinsias DE ROSSA; Fianna Fail, Albert REYNOLDS; Labor Party, Richard SPRING; Fine Gael, John BRUTON; Communist Party of Ireland, Michael O'RIORDAN; Sinn Fein, Gerry ADAMS; Progressive Democrats, Desmond O'MALLEY

note: Prime Minister REYNOLDS heads a coalition consisting of the Fianna Fail and the Labor Party

Suffrage: 18 years of age; universal

Elections:

President: last held 9 November 1990 (next to be held November 1997); results—Mary Bourke ROBINSON 52.8%, Brian LENIHAN 47.2%

Senate: last held on NA February 1992 (next to be held February 1997); results—percent of vote by party NA; seats—(60 total, 49 elected) Fianna Fail 26, Fine Gael 16, Labor 9, Progressive Democrats 2, Democratic Left 1, independents 6

House of Representatives: last held on 25 November 1992 (next to be held by June 1995); results—Fianna Fail 39.1%, Fine Gael 24.5%, Labor Party 19.3%, Progressive Democrats 4.7%, Democratic Left 2.8%, Sinn Fein 1.6%, Workers' Party 0.7%, independents 5.9%; seats—(166 total) Fianna Fail 68, Fine Gael 45, Labor Party 33, Progressive Democrats 10, Democratic Left 4, Greens 1, independents 5

Executive branch: president, prime minister, deputy prime minister, Cabinet

Legislative branch: bicameral Parliament (Oireachtas) consists of an upper house or Senate (Seanad Eireann) and a lower house or House of Representatives (Dail Eireann)

Judicial branch: Supreme Court

Leaders:

Chief of State: President Mary Bourke ROBINSON (since 9 November 1990)

Head of Government: Prime Minister Albert REYNOLDS (since 11 February 1992)

Member of: Australian Group, BIS, CCC, CE, COCOM (cooperating country), CSCE, EBRD, EC, ECE, EIB, ESA, FAO, GATT, IAEA, IBRD, ICAO, ICC, IDA, IEA, IFAD, IFC, ILO, IMF, IMO, INTELSAT, INTERPOL, IOC, ISO, ITU, LORCS, MINURSO, MTCR, NEA, NSG, OECD, ONUSAL, UN, UNAVEM II, UNCTAD, UNESCO, UNFICYP, UNIDO, UNIFIL, UNIKOM, UNPROFRO, UNTAC, UNTSO, UPU, WHO, WIPO, WMO, ZC

Diplomatic representation in US:

chief of mission: Ambassador Dermot A. GALLAGHER

chancery: 2234 Massachusetts Avenue NW, Washington DC 20008

telephone: (202) 462-3939

consulates general: Boston, Chicago, New York, and San Francisco

US diplomatic representation:

chief of mission: Ambassador William Henry G. FITZGERALD; Ambassador Designate Jean Kennedy SMITH (17 March 1993)

embassy: 42 Elgin Road, Ballsbridge, Dublin

mailing address: use embassy street address

telephone: [353] (1) 687122

FAX: [353] (1) 689946

Flag: three equal vertical bands of green (hoist side), white, and orange; similar to the flag of the Cote d'Ivoire, which is shorter and has the colors reversed—orange (hoist side), white, and green; also similar to the flag of Italy, which is shorter and has colors of green (hoist side), white, and red

---

Economy

Overview: The economy is small and trade dependent. Agriculture, once the most important sector, is now dwarfed by industry, which accounts for 37% of GDP, about 80% of exports, and employs 28% of the labor force. Since 1987, real GDP growth, led by exports, has averaged 4% annually. Over the same period, inflation has fallen sharply and chronic trade deficits have been transformed into annual surpluses. Unemployment, at 22.7% remains a serious problem, however, and job creation is the main focus of government policy. To ease unemployment, Dublin aggressively courts foreign investors and recently created a new industrial development agency to aid small indigenous firms. Government assistance is constrained by Dublin's continuing deficit reduction measures. After five years of fiscal restraint, total government debt still exceeds GDP. Growth probably will moderate in 1993 as the heavily indebted and trade-dependent economy is highly sensitive to changes in exchange rates and world interest rates. Exports to the UK, Ireland's major export market, probably will be hurt by the recent appreciation of the Irish currency against sterling—for the first time since 1979 the value of the Irish pound exceeds that of its British counterpart.

National product: GDP—purchasing power equivalent—$42.4 billion (1992)

National product real growth rate: 2% (1992)

National product per capita: $12,000 (1992)

Inflation rate (consumer prices): 3.5% (1992)

Unemployment rate: 22.7% (1992)

Budget: revenues $16.0 billion; expenditures $16.6 billion, including capital expenditures of $1.6 billion (1992 est.)

Exports: $28.3 billion (f.o.b., 1992)

commodities: chemicals, data processing equipment, industrial machinery, live animals, animal products

partners: EC 75% (UK 32%, Germany 13%, France 10%), US 9%

Imports: $23.3 billion (c.i.f., 1992)

commodities: food, animal feed, data processing equipment, petroleum and petroleum products, machinery, textiles, clothing

partners: EC 66% (UK 41%, Germany 8%, Netherlands 4%), US 15%

External debt: $15.0 billion (1990)

Industrial production: growth rate 8.0% (1992 est.); accounts for 37% of GDP

Electricity: 5,000,000 kW capacity; 14,500 million kWh produced, 4,120 kWh per capita (1992)

Industries: food products, brewing, textiles, clothing, chemicals, pharmaceuticals, machinery, transportation equipment, glass and crystal

Agriculture: accounts for 11% of GDP and 13% of the labor force; principal crops—turnips, barley, potatoes, sugar beets, wheat; livestock—meat and dairy products; 85% self-sufficient in food; food shortages include bread grain, fruits, vegetables

Economic aid: donor—ODA commitments (1980-89), $90 million

Currency: 1 Irish pound (£Ir) = 100 pence

Exchange rates: Irish pounds (£Ir) per US$1—0.6118 (January 1993), 0.5864 (1992), 0.6190 (1991), 0.6030 (1990), 0.7472 (1989), 0.6553 (1988)

Fiscal year: calendar year

Communications

Railroads: Irish National Railways (CIE) operates 1,947 km 1.602-meter gauge, government owned; 485 km double track; 37 km electrified

Highways: 92,294 km total; 87,422 km paved, 4,872 km gravel or crushed stone

Inland waterways: limited for commercial traffic

Pipelines: natural gas 225 km

Ports: Cork, Dublin, Waterford

Merchant marine: 57 ships (1,000 GRT or over) totaling 154,647 GRT/186,432 DWT; includes 4 short-sea passenger, 33 cargo, 2 refrigerated cargo, 4 container, 3 oil tanker, 3 specialized tanker, 3 chemical tanker, 5 bulk

Airports:

total: 40

usable: 39

with permanent-surface runways: 13

with runways over 3,659 m: 0

with runways 2,440-3,659 m: 2

with runways 1,220-2,439 m: 6

Telecommunications: modern system using cable and digital microwave circuits; 900,000 telephones; broadcast stations—9 AM, 45 FM, 86 TV; 2 coaxial submarine cables; 1 Atlantic Ocean INTELSAT earth station

Israel

(also see separate Gaza Strip and West Bank entries)

Defense Forces

Branches: Army (including Naval Service and Air Corps), National Police (Garda Siochana)

Manpower availability: males age 15-49 903,536; fit for military service 731,085; reach military age (17) annually 33,932 (1993 est.)

Defense expenditures: exchange rate conversion—$569 million, 1-2% of GDP (1993 est.)

Note: The Arab territories occupied by Israel since the 1967 war are not included in the data below. As stated in the 1978 Camp David Accords and reaffirmed by President Bush's post-Gulf crisis peace initiative, the final status of the West Bank and Gaza Strip, their relationship with their neighbors, and a peace treaty between Israel and Jordan are to be negotiated among the concerned parties. The Camp David Accords further specify that these negotiations will resolve the location of the respective boundaries. Pending the completion of this process, it is US policy that the final status of the West Bank and Gaza Strip has yet to be determined (see West Bank and Gaza Strip entries). On 25 April 1982, Israel relinquished control of the Sinai to Egypt. Statistics for the Israeli-occupied Golan Heights are included in the Syria entry.

Geography

Location: Middle East, bordering the eastern Mediterranean Sea, between Egypt and Lebanon

Map references: Africa, Middle East, Standard Time Zones of the World

Area:
total area: 20,770 km^2
land area: 20,330 km^2
comparative area: slightly larger than New Jersey

Land boundaries: total 1,006 km, Egypt 255 km, Gaza Strip 51 km, Jordan 238 km, Lebanon 79 km, Syria 76 km, West Bank 307 km

Coastline: 273 km

Maritime claims:
continental shelf: to depth of exploitation
territorial sea: 12 nm

International disputes: separated from Lebanon, Syria, and the West Bank by the 1949 Armistice Line; differences with Jordan over the location of the 1949 Armistice Line that separates the two countries; West Bank

and Gaza Strip are Israeli occupied with status to be determined; Golan Heights is Israeli occupied; Israeli troops in southern Lebanon since June 1982; water-sharing issues with Jordan

Climate: temperate; hot and dry in southern and eastern desert areas

Terrain: Negev desert in the south; low coastal plain; central mountains; Jordan Rift Valley

Natural resources: copper, phosphates, bromide, potash, clay, sand, sulfur, asphalt, manganese, small amounts of natural gas and crude oil

Land use:
arable land: 17%
permanent crops: 5%
meadows and pastures: 40%
forest and woodland: 6%
other: 32%

Irrigated land: 2,140 km^2 (1989)

Environment: sandstorms may occur during spring and summer; limited arable land and natural water resources pose serious constraints; deforestation

Note: there are 175 Jewish settlements in the West Bank, 38 in the Israeli-occupied Golan Heights, 18 in the Gaza Strip, and 14 Israeli-built Jewish neighborhoods in East Jerusalem

People

Population: 4,918,946 (July 1993 est.)
note: includes 102,000 Jewish settlers in the West Bank, 14,000 in the Israeli-occupied Golan Heights, 4,000 in the Gaza Strip, and 134,000 in East Jerusalem (1993 est.)

Population growth rate: 3.08% (1993 est.)

Birth rate: 20.72 births/1,000 population (1993 est.)

Death rate: 6.45 deaths/1,000 population (1993 est.)

Net migration rate: 16.51 migrant(s)/1,000 population (1993 est.)

Infant mortality rate: 8.9 deaths/1,000 live births (1993 est.)

Life expectancy at birth:
total population: 77.77 years
male: 75.72 years
female: 79.93 years (1993 est.)

Total fertility rate: 2.86 children born/woman (1993 est.)

Nationality:
noun: Israeli(s)
adjective: Israeli

Ethnic divisions: Jewish 83%, non-Jewish 17% (mostly Arab)

Religions: Judaism 82%, Islam 14% (mostly Sunni Muslim), Christian 2%, Druze and other 2%

Languages: Hebrew (official), Arabic used officially for Arab minority, English most commonly used foreign language

Literacy: age 15 and over can read and write (1983)

Israel (continued)

total population: 92%
male: 95%
female: 89%
Labor force: 1.4 million (1984 est.)
by occupation: public services 29.3%,
industry, mining, and manufacturing 22.8%,
commerce 12.8%, finance and business
9.5%, transport, storage, and
communications 6.8%, construction and
public works 6.5%, personal and other
services 5.8%, agriculture, forestry, and
fishing 5.5%, electricity and water 1.0%
(1983)

Government

Names:
conventional long form: State of Israel
conventional short form: Israel
local long form: Medinat Yisra'el
local short form: Yisra'el
Digraph: IS
Type: republic
Capital: Jerusalem
note: Israel proclaimed Jerusalem its capital
in 1950, but the US, like nearly all other
countries, maintains its Embassy in Tel Aviv
Administrative divisions: 6 districts
(mehozot, singular—mehoz); Central, Haifa,
Jerusalem, Northern, Southern, Tel Aviv
Independence: 14 May 1948 (from League
of Nations mandate under British
administration)
Constitution: no formal constitution; some
of the functions of a constitution are filled
by the Declaration of Establishment (1948),
the basic laws of the parliament (Knesset),
and the Israeli citizenship law
Legal system: mixture of English common
law, British Mandate regulations, and, in
personal matters, Jewish, Christian, and
Muslim legal systems; in December 1985,
Israel informed the UN Secretariat that it
would no longer accept compulsory ICJ
jurisdiction
National holiday: Independence Day, 14
May 1948 (Israel declared independence on
14 May 1948, but the Jewish calendar is
lunar and the holiday may occur in April or
May)
Political parties and leaders:
members of the government: Labor Party,
Prime Minister Yitzhak RABIN; MERETZ,
Minister of Education Shulamit ALONI;
SHAS, Minister of Interior Arieh DERI
opposition parties: Likud Party, Binyamin
NETANYAHU; Tzomet, Rafael EITAN;
National Religious Party, Zevulun
HAMMER; United Torah Jewry, Avraham
SHAPIRA; Democratic Front for Peace and
Equality (Hadash), Hashim MAHAMID;
Moledet, Rehavam ZEEVI; Arab Democratic
Party, Abd al Wahab DARAWSHAH
note: Israel currently has a coalition
government comprising 3 parties that hold

62 seats of the Knesset's 120 seats
Other political or pressure groups: Gush
Emunim, Jewish nationalists advocating
Jewish settlement on the West Bank and
Gaza Strip; Peace Now, critical of
government's West Bank/Gaza Strip and
Lebanon policies
Suffrage: 18 years of age; universal
Elections:
President: last held 24 March 1993 (next to
be held NA March 1999); results—Ezer
WEIZMAN elected by Knesset
Knesset: last held June 1992 (next to be held
by NA); results—percent of vote by party
NA; seats—(120 total) Labor Party 44,
Likud bloc 32, Meretz 12, Tzomet 8,
National Religious Party 6, Shas 6, United
Torah Jewry 4, Democratic Front for Peace
and Equality 3, Moledet 3, Arab Democratic
Party 2
Executive branch: president, prime
minister, vice prime minister, Cabinet
Legislative branch: unicameral parliament
(Knesset)
Judicial branch: Supreme Court
Leaders:
Chief of State: President Ezer WEIZMAN
(since 13 May 1993)
Head of Government: Prime Minister
Yitzhak RABIN (since July 1992)
Member of: AG (observer), CCC, CERN
(oberver), EBRD, ECE, FAO, GATT, IADB,
IAEA, IBRD, ICAO, ICC, ICFTU, IDA,
IFAD, IFC, ILO, IMF, IMO, INMARSAT,
INTELSAT, INTERPOL, IOC, IOM, ISO,
ITU, OAS (observer), PCA, UN, UNCTAD,
UNESCO, UNHCR, UNIDO, UPU, WHO,
WIPO, WMO, WTO
Diplomatic representation in US:
chief of mission: Ambassador Itamar
RABINOVICH
chancery: 3514 International Drive NW,
Washington, DC 20008
telephone: (202) 364-5500
consulates general: Atlanta, Boston,
Chicago, Houston, Los Angeles, Miami,
New York, Philadelphia, and San Francisco
US diplomatic representation:
chief of mission: Acting Ambassador
William BROWN
embassy: 71 Hayarkon Street, Tel Aviv
mailing address: APO AE 09830
telephone: [972] (3) 654338
FAX: [972] (3) 663449
consulate general: Jerusalem
Flag: white with a blue hexagram
(six-pointed linear star) known as the Magen
David (Shield of David) centered between
two equal horizontal blue bands near the top
and bottom edges of the flag

Economy

Overview: Israel has a market economy
with substantial government participation. It
depends on imports of crude oil, grains, raw

materials, and military equipment. Despite
limited natural resources, Israel has
intensively developed its agricultural and
industrial sectors over the past 20 years.
Industry employs about 20% of Israeli
workers, agriculture 5%, and services most
of the rest. Diamonds, high-technology
equipment, and agricultural products (fruits
and vegetables) are leading exports. Israel
usually posts balance-of-payments deficits,
which are covered by large transfer
payments from abroad and by foreign loans.
Roughly half of the government's $17
billion external debt is owed to the United
States, which is its major source of
economic and military aid. To earn needed
foreign exchange, Israel has been targeting
high-technology niches in international
markets, such as medical scanning
equipment. The influx of Jewish immigrants
from the former USSR, which topped
400,000 during the period 1990-92, has
increased unemployment, intensified housing
problems, and widened the government
budget deficit. At the same time, a
considerable number of the immigrants bring
to the economy valuable scientific and
professional expertise.
National product: GDP—purchasing power
equivalent—$57.4 billion (1992 est.)
National product real growth rate: 6.4%
(1992 est.)
National product per capita: $12,100
(1992 est.)
Inflation rate (consumer prices): 10%
(1992 est.)
Unemployment rate: 11% (1992 est.)
Budget: revenues $33.9 billion; expenditures
$36.8 billion, including capital expenditures
of $9.3 billion (FY93)
Exports: $11.8 billion (f.o.b., 1992 est.)
commodities: polished diamonds, citrus and
other fruits, textiles and clothing, processed
foods, fertilizer and chemical products,
military hardware, electronics
partners: US, EC, Japan, Hong Kong,
Switzerland
Imports: $19.6 billion (c.i.f., 1992 est.)
commodities: military equipment, rough
diamonds, oil, chemicals, machinery, iron
and steel, cereals, textiles, vehicles, ships,
aircraft
partners: US, EC, Switzerland, Japan, South
Africa, Canada, Hong Kong
External debt: $25 billion, of which
government debt is $17 billion (December
1992 est.)
Industrial production: growth rate 9.4%
(1992 est.); accounts for about 20% of GDP
Electricity: 5,835,000 kW capacity; 21,840
million kWh produced, 4,600 kWh per
capita (1992)
Industries: food processing, diamond
cutting and polishing, textiles, clothing,
chemicals, metal products, military
equipment, transport equipment, electrical

equipment, miscellaneous machinery, potash mining, high-technology electronics, tourism

Agriculture: accounts for about 3% of GDP; largely self-sufficient in food production, except for grains; principal products—citrus and other fruits, vegetables, cotton; livestock products—beef, dairy, poultry

Economic aid: US commitments, including Ex-Im (FY70-90), $18.2 billion; Western (non-US) countries, ODA and OOF bilateral commitments (1970-89), $2.8 billion

Currency: 1 new Israeli shekel (NIS) = 100 new agorot

Exchange rates: new Israeli shekels (NIS) per US$1—2.8000 (December 1992), 2.4591 (1992), 2.2791 (1991), 2.0162 (1990), 1.9164 (1989), 1.5989 (1988), 1.5946 (1987)

Fiscal year: calendar year (since 1 January 1992)

Communications

Railroads: 600 km 1.435-meter gauge, single track; diesel operated

Highways: 4,750 km; majority is bituminous surfaced

Pipelines: crude oil 708 km; petroleum products 290 km; natural gas 89 km

Ports: Ashdod, Haifa

Merchant marine: 35 ships (1,000 GRT or over) totaling 678,584 GRT/785,220 DWT; includes 8 cargo, 24 container, 2 refrigerated cargo, 1 roll-on/roll-off; note—Israel also maintains a significant flag of convenience fleet, which is normally at least as large as the Israeli flag fleet; the Israeli flag of convenience fleet typically includes all of its oil tankers

Airports:
total: 53
usable: 46
with permanent-surface runways: 28
with runways over 3,659 m: 0
with runways 2,440-3,659 m: 7
with runways 1,220-2,439 m: 12

Telecommunications: most highly developed in the Middle East although not the largest; good system of coaxial cable and microwave radio relay; 1,800,000 telephones; broadcast stations—14 AM, 21 FM, 20 TV; 3 submarine cables; satellite earth stations—2 Atlantic Ocean INTELSAT and 1 Indian Ocean INTELSAT

Defense Forces

Branches: Israel Defense Forces (including ground, naval, and air components)
note: historically, there have been no separate Israeli military services

Manpower availability: males age 15-49 1,240,757; females age 15-49 1,218,610; males fit for military service 1,018,212; females fit for military service 996,089;

males reach military age (18) annually 46,131; females reach military age (18) annually 44,134 (1993 est.); both sexes are liable for military service

Defense expenditures: exchange rate conversion—$12.5 billion, 18% of GDP (1993 est.)

Italy

Geography

Location: Southern Europe, a peninsula in the central Mediterranean Sea

Map references: Africa, Europe, Standard Time Zones of the World

Area:
total area: 301,230 km²
land area: 294,020 km²
comparative area: slightly larger than Arizona
note: includes Sardinia and Sicily

Land boundaries: total 1,899.2 km, Austria 430 km, France 488 km, Holy See (Vatican City) 3.2 km, San Marino 39 km, Slovenia 199 km, Switzerland 740 km

Coastline: 4,996 km

Maritime claims:
continental shelf: 200 m depth or to depth of exploitation
territorial sea: 12 nm

International disputes: small vocal minority in northern Italy seeks the return of parts of southwestern Slovenia

Climate: predominantly Mediterranean; Alpine in far north; hot, dry in south

Terrain: mostly rugged and mountainous; some plains, coastal lowlands

Natural resources: mercury, potash, marble, sulfur, dwindling natural gas and crude oil reserves, fish, coal

Land use:
arable land: 32%
permanent crops: 10%
meadows and pastures: 17%
forest and woodland: 22%
other: 19%

Irrigated land: 31,000 km² (1989 est.)

Environment: regional risks include landslides, mudflows, snowslides, earthquakes, volcanic eruptions, flooding, pollution; land sinkage in Venice

Note: strategic location dominating central Mediterranean as well as southern sea and air approaches to Western Europe

Italy (continued)

People

Population: 58,018,540 (July 1993 est.)
Population growth rate: 0.2% (1993 est.)
Birth rate: 10.65 births/1,000 population (1993 est.)
Death rate: 9.66 deaths/1,000 population (1993 est.)
Net migration rate: 1.03 migrant(s)/1,000 population (1993 est.)
Infant mortality rate: 7.8 deaths/1,000 live births (1993 est.)
Life expectancy at birth:
total population: 77.43 years
male: 74.22 years
female: 80.85 years (1993 est.)
Total fertility rate: 1.37 children born/woman (1993 est.)
Nationality:
noun: Italian(s)
adjective: Italian
Ethnic divisions: Italian (includes small clusters of German-, French-, and Slovene-Italians in the north and Albanian-Italians and Greek-Italians in the south), Sicilians, Sardinians
Religions: Roman Catholic 100%
Languages: Italian, German (parts of Trentino-Alto Adige region are predominantly German speaking), French (small French-speaking minority in Valle d'Aosta region), Slovene (Slovene-speaking minority in the Trieste-Gorizia area)
Literacy: age 15 and over can read and write (1990)
total population: 97%
male: 98%
female: 96%
Labor force: 23.988 million
by occupation: services 58%, industry 32.2%, agriculture 9.8% (1988)

Government

Names:
conventional long form: Italian Republic
conventional short form: Italy
local long form: Repubblica Italiana
local short form: Italia
former: Kingdom of Italy
Digraph: IT
Type: republic
Capital: Rome
Administrative divisions: 20 regions (regioni, singular—regione); Abruzzi, Basilicata, Calabria, Campania, Emilia-Romagna, Friuli-Venezia Giulia, Lazio, Liguria, Lombardia, Marche, Molise, Piemonte, Puglia, Sardegna, Sicilia, Toscana, Trentino-Alto Adige, Umbria, Valle d'Aosta, Veneto
Independence: 17 March 1861 (Kingdom of Italy proclaimed)
Constitution: 1 January 1948

Legal system: based on civil law system, with ecclesiastical law influence; appeals treated as trials de novo; judicial review under certain conditions in Constitutional Court; has not accepted compulsory ICJ jurisdiction
National holiday: Anniversary of the Republic, 2 June (1946)
Political parties and leaders: Christian Democratic Party (DC), Fermo Mino MARTINAZZOLI, general secretary; Rosa Russo JERVOLINO, president; Socialist Party (PSI), Giorgio BENVENUTO, party secretary; Social Democratic Party (PSDI), Enrico FERRI, party secretary; Liberal Party (PLI); Democratic Party of the Left (PDS—was Communist Party, or PCI, until January 1991), Achille OCCHETTO, secretary general; Italian Social Movement (MSI), Gianfranco FINI, national secretary; Republican Party (PRI), Giorgio BOGI, political secretary; Lega Nord (Northern League), Umberto BOSSI, president; Communist Renewal (RC), Sergio GARAVINI
Other political or pressure groups: the Roman Catholic Church; three major trade union confederations (CGIL—formerly Communist dominated, CISL—Christian Democratic, and UIL—Social Democratic, Socialist, and Republican); Italian manufacturers association (Confindustria); organized farm groups (Confcoltivatori, Confagricoltura)
Suffrage: 18 years of age, universal (except in senatorial elections, where minimum age is 25)
Elections:
Senate: last held 5-6 April 1992 (next to be held by April 1997); results—DC 27.3%, PDS 17.0%, PSI 13.6%, Northern Leagues 8.2%, other 33.9%; seats—(326 total; 315 elected, 11 appointed senators-for-life) DC 107, PDS 64, PSI 49, Leagues 25, other 70
Chamber of Deputies: last held 5-6 April 1992 (next to be held by April 1997); results—DC 29.7%, PDS 16.1%, PSI 13.6%, Northern Leagues 8.7%, RC 5.6%, MSI 5.4%, PRI 4.4%, PLI 2.8%, PSDI 2.7%, other 11%; seats—(630 total) DC 206, PDS 107, PSI 92, Northern Leagues 55, RC 35, MSI 34, PRI 27, PLI 17, PSDI 16, other 41
Executive branch: president, prime minister (president of the Council of Ministers)
Legislative branch: bicameral Parliament (Parlamento) consists of an upper chamber or Senate of the Republic (Senato della Repubblica) and a lower chamber or Chamber of Deputies (Camera dei Deputati)
Judicial branch: Constitutional Court (Corte Costituzionale)
Leaders:
Chief of State: President Oscar Luigi SCALFARO (since 28 May 1992)

Head of Government: Prime Minister Carlo Azeglio CIAMPI (29 April 1993)
Member of: AfDB, AG (observer), Australia Group, AsDB, BIS, CCC, CDB (non-regional), CE, CEI, CERN, COCOM, CSCE, EBRD, EC, ECE, ECLAC, EIB, ESA, FAO, G-7, G-10, GATT, IADB, IAEA, IBRD, ICAO, ICC, ICFTU, IDA, IFAD, IEA, IFC, ILO, IMF, IMO, INMARSAT, INTELSAT, INTERPOL, IOC, IOM, ISO, ITU, LAIA (observer), LORCS, MINURSO, MTCR, NACC, NATO, NEA, NSG, OAS (observer), OECD, PCA, UN, UNCTAD, UNESCO, UNHCR, UNIDO, UNIFIL, UNIKOM, UNMOGIP, UNOMOZ, UNTSO, UPU, WCL, WEU, WHO, WIPO, WMO, WTO, ZC
Diplomatic representation in US:
chief of mission: Ambassador Boris BIANCHERI CHIAPPORI
chancery: 1601 Fuller Street NW, Washington DC 20009
telephone: (202) 328-5500
consulates general: Boston, Chicago, Houston, Miami, New York, Los Angeles, Philadelphia, San Francisco
consulates: Detroit, New Orleans, and Newark (New Jersey)
US diplomatic representation:
chief of mission: (vacant)
embassy: Via Veneto 119/A, 00187, Rome
mailing address: PSC 59, Box 100, APO AE 09624
telephone: [39] (6) 46741
FAX: [39] (6) 488-2672
consulates general: Florence, Genoa, Milan, Naples, Palermo (Sicily)
Flag: three equal vertical bands of green (hoist side), white, and red; similar to the flag of Ireland, which is longer and is green (hoist side), white, and orange; also similar to the flag of the Cote d'Ivoire, which has the colors reversed—orange (hoist side), white, and green

Economy

Overview: Since World War II the economy has changed from one based on agriculture into a ranking industrial economy, with approximately the same total and per capita output as France and the UK. The country is still divided into a developed industrial north, dominated by private companies, and an undeveloped agricultural south, dominated by large public enterprises. Services account for 48% of GDP, industry 35%, agriculture 4%, and public administration 13%. Most raw materials needed by industry and over 75% of energy requirements must be imported. After growing at an annual average rate of 3% in 1983-90, growth slowed to about 1% in 1991 and 1992. In the second half of 1992, Rome became unsettled by the prospect of not qualifying to participate in EC plans for economic and monetary union later in the decade; thus it finally began to address its huge fiscal imbalances. Thanks to the determination of Prime Minister AMATO, the government adopted a fairly stringent budget for 1993, abandoned its highly inflationary wage indexation system, and started to scale back its extremely generous social welfare programs, including pension and health care benefits. Monetary officials, who were forced to withdraw the lira from the European monetary system in September 1992 when it came under extreme pressure in currency markets, remain committed to bringing the currency back into the grid as soon as conditions warrant. For the 1990s, Italy faces the problems of refurbishing a tottering communications system, curbing pollution in major industrial centers, and adjusting to the new competitive forces accompanying the ongoing economic integration of the European Community.

National product: GDP—purchasing power equivalent—$1.012 trillion (1992)

National product real growth rate: 0.9% (1992)

National product per capita: $17,500 (1992)

Inflation rate (consumer prices): 5.4% (1992)

Unemployment rate: 11% (1992 est.)

Budget: revenues $447 billion; expenditures $581 billion, including capital expenditures of $46 billion (1992 est.)

Exports: $168.8 million (f.o.b., 1991)
commodities: textiles, wearing apparel, metals, production machinery, motor vehicles, transportation equipment, chemicals, other
partners: EC 58.3%, US 6.8%, OPEC 5.1% (1992)

Imports: $169.7 million (f.o.b., 1991)
commodities: petroleum, industrial machinery, chemicals, metals, food, agricultural products
partners: EC 58.8%, OPEC 6.1%, US 5.5% (1992)

External debt: $42 billion (September 1992)

Industrial production: growth rate −0.5% (1992 est.); accounts for almost 35% of GDP

Electricity: 58,000,000 kW capacity; 235,000 million kWh produced, 4,060 kWh per capita (1992)

Industries: machinery, iron and steel, chemicals, food processing, textiles, motor vehicles, clothing, footwear, ceramics

Agriculture: accounts for about 4% of GDP and about 10% of the work force; self-sufficient in foods other than meat, dairy products, and cereals; principal crops—fruits, vegetables, grapes, potatoes, sugar beets, soybeans, grain, olives; fish

catch of 525,000 metric tons in 1990

Illicit drugs: increasingly important gateway country for Latin American cocaine entering the European market

Economic aid: donor—ODA and OOF commitments (1970-89), $25.9 billion

Currency: 1 Italian lira (Lit) = 100 centesimi

Exchange rates: Italian lire (Lit) per US$1—1,482.5 (January 1993), 1,232.4 (1992), 1,240.6 (1991), 1,198.1 (1990), 1,372.1 (1989), 1,301.6 (1988)

Fiscal year: calendar year

Communications

Railroads: 20,011 km total; 16,066 km 1.435-meter government-owned standard gauge (8,999 km electrified); 3,945 km privately owned—2,100 km 1.435-meter standard gauge (1,155 km electrified) and 1,845 km 0.950-meter narrow gauge (380 km electrified)

Highways: 298,000 km total; autostrada (expressway) 6,000 km, state highways 46,000 km, provincial highways 103,000 km, communal highways 143,000 km; 270,000 km paved, 23,000 km gravel and crushed stone, 5,000 km earth

Inland waterways: 2,400 km for various types of commercial traffic, although of limited overall value

Pipelines: crude oil 1,703 km; petroleum products 2,148 km; natural gas 19,400 km

Ports: Cagliari (Sardinia), Genoa, La Spezia, Livorno, Naples, Palermo (Sicily), Taranto, Trieste, Venice

Merchant marine: 536 ships (1,000 GRT or over) totaling 6,788,938 GRT/10,128,468 DWT; includes 15 passenger, 36 short-sea passenger, 87 cargo, 4 refrigerated cargo, 21 container, 69 roll-on/roll-off cargo, 8 vehicle carrier, 1 multifunction large-load carrier, 138 oil tanker, 34 chemical tanker, 45 liquefied gas, 10 specialized tanker, 9 combination ore/oil, 57 bulk, 2 combination bulk

Airports:
total: 137
usable: 133
with permanent-surface runways: 92
with runways over 3,659 m: 2
with runways 2,440-3,659 m: 36
with runways 1,220-2,439 m: 39

Telecommunications: modern, well-developed, fast; 25,600,000 telephones; fully automated telephone, telex, and data services; high-capacity cable and microwave radio relay trunks; broadcast stations—135 AM, 28 (1,840 repeaters) FM, 83 (1,000 repeaters) TV; international service by 21 submarine cables, 3 satellite earth stations operating in INTELSAT with 3 Atlantic Ocean antennas and 2 Indian Ocean antennas; also participates in INMARSAT and EUTELSAT systems

---

Defense Forces

Branches: Army, Navy, Air Force, Carabinieri

Manpower availability: males age 15-49 14,898,913; fit for military service 12,989,142; reach military age (18) annually 425,286 (1993 est.)

Defense expenditures: exchange rate conversion—$24.5 billion, 2% of GDP (1992)

Jamaica

50 km

Caribbean Sea

Negril · Montego Bay · · Ocho Rios
Black River · Mandeville · · Port Antonio
· KINGSTON
· Morant Bay

Caribbean Sea

Geography

Location: in the northern Caribbean Sea, about 160 km south of Cuba
Map references: Central America and the Caribbean, North America, Standard Time Zones of the World
Area:
total area: 10,990 km²
land area: 10,830 km²
comparative area: slightly smaller than Connecticut
Land boundaries: 0 km
Coastline: 1,022 km
Maritime claims:
exclusive economic zone: 200 nm
territorial sea: 12 nm
International disputes: none
Climate: tropical; hot, humid; temperate interior
Terrain: mostly mountains with narrow, discontinuous coastal plain
Natural resources: bauxite, gypsum, limestone
Land use:
arable land: 19%
permanent crops: 6%
meadows and pastures: 18%
forest and woodland: 28%
other: 29%
Irrigated land: 350 km² (1989 est.)
Environment: subject to hurricanes (especially July to November); deforestation; water pollution
Note: strategic location between Cayman Trench and Jamaica Channel, the main sea lanes for Panama Canal

People

Population: 2,529,981 (July 1993 est.)
Population growth rate: 0.96% (1993 est.)
Birth rate: 22.24 births/1,000 population (1993 est.)
Death rate: 5.72 deaths/1,000 population (1993 est.)
Net migration rate: -6.95 migrant(s)/1,000 population (1993 est.)

Infant mortality rate: 17.5 deaths/1,000 live births (1993 est.)
Life expectancy at birth:
total population: 74.09 years
male: 71.92 years
female: 76.36 years (1993 est.)
Total fertility rate: 2.47 children born/woman (1993 est.)
Nationality:
noun: Jamaican(s)
adjective: Jamaican
Ethnic divisions: African 76.3%, Afro-European 15.1%, East Indian and Afro-East Indian 3%, white 3.2%, Chinese and Afro-Chinese 1.2%, other 1.2%
Religions: Protestant 55.9% (Church of God 18.4%, Baptist 10%, Anglican 7.1%, Seventh-Day Adventist 6.9%, Pentecostal 5.2%, Methodist 3.1%, United Church 2.7%, other 2.5%), Roman Catholic 5%, other, including some spiritual cults 39.1% (1982)
Languages: English, Creole
Literacy: age 15 and over having ever attended school (1990)
total population: 98%
male: 98%
female: 99%
Labor force: 1,062,100
by occupation: services 41%, agriculture 22.5%, industry 19%, unemployed 17.5% (1989)

Government

Names:
conventional long form: none
conventional short form: Jamaica
Digraph: JM
Type: parliamentary democracy
Capital: Kingston
Administrative divisions: 14 parishes; Clarendon, Hanover, Kingston, Manchester, Portland, Saint Andrew, Saint Ann, Saint Catherine, Saint Elizabeth, Saint James, Saint Mary, Saint Thomas, Trelawny, Westmoreland
Independence: 6 August 1962 (from UK)
Constitution: 6 August 1962
Legal system: based on English common law; has not accepted compulsory ICJ jurisdiction
National holiday: Independence Day (first Monday in August)
Political parties and leaders: People's National Party (PNP) P. J. PATTERSON; Jamaica Labor Party (JLP), Edward SEAGA
Other political or pressure groups: Rastafarians (black religious/racial cultists, pan-Africanists)
Suffrage: 18 years of age; universal
Elections:
House of Representatives: last held 30 March 1993 (next to be held by February 1998); results—percent of vote by party NA; seats—(60 total) PNP 52, JLP 8

Executive branch: British monarch, governor general, prime minister, Cabinet
Legislative branch: bicameral Parliament consists of an upper house or Senate and a lower house or House of Representatives
Judicial branch: Supreme Court
Leaders:
Chief of State: Queen ELIZABETH II (since 6 February 1952), represented by Governor General Sir Howard COOKE (since 1 August 1991)
Head of Government: Prime Minister P. J. PATTERSON (since 30 March 1992)
Member of: ACP, C, CARICOM, CCC, CDB, ECLAC, FAO, G-19, G-77, GATT, G-15, IADB, IAEA, IBRD, ICAO, ICFTU, IFAD, IFC, ILO, IMF, IMO, INTELSAT, INTERPOL, IOC, ISO, ITU, LAES, LORCS, NAM, OAS, OPANAL, UN, UNCTAD, UNESCO, UNIDO, UPU, WCL, WFTU, WHO, WIPO, WMO, WTO
Diplomatic representation in US:
chief of mission: Ambassador Richard BERNAL
chancery: Suite 355, 1850 K Street NW, Washington, DC 20006
telephone: (202) 452-0660
consulates general: Miami and New York
US diplomatic representation:
chief of mission: Ambassador Glen A. HOLDEN
embassy: Kingston
mailing address: 3rd Floor, Jamaica Mutual Life Center, 2 Oxford Road, Kingston
telephone: (809) 929-4850 through 4859
FAX: (809) 926-6743
Flag: diagonal yellow cross divides the flag into four triangles—green (top and bottom) and black (hoist side and fly side)

Economy

Overview: The economy is based on sugar, bauxite, and tourism. In 1985 it suffered a setback with the closure of some facilities in the bauxite and alumina industry, a major source of hard currency earnings. Since 1986 an economic recovery has been under way. In 1987 conditions began to improve for the bauxite and alumina industry because of increases in world metal prices. The recovery has also been supported by growth in the manufacturing and tourism sectors. In September 1988, Hurricane Gilbert inflicted severe damage on crops and the electric power system, a sharp but temporary setback to the economy. By October 1989 the economic recovery from the hurricane was largely complete, and real growth was up about 3% for 1989. In 1991, however, growth dropped to 0.2% as a result of the US recession, lower world bauxite prices, and monetary instability. In 1992, growth was 1.5%, supported by a recovery in tourism and stabilization of the Jamaican

dollar in the second half of 1992.
National product: GDP—exchange rate conversion—$3.7 billion (1992 est.)
National product real growth rate: 1.5% (1992 est.)
National product per capita: $1,500 (1992 est.)
Inflation rate (consumer prices): 52% (1992 est.)
Unemployment rate: 15.4% (1992)
Budget: revenues $600 million; expenditures $736 million, including capital expenditures of $NA (FY91 est.)
Exports: $1.2 billion (f.o.b., 1991)
commodities: alumina, bauxite, sugar, bananas, rum
partners: US 39%, UK 14%, Canada 12%, Netherlands 8%, Norway 7%
Imports: $1.6 billion (f.o.b., 1991)
commodities: fuel, other raw materials, construction materials, food, transport equipment, other machinery and equipment
partners: US 51%, UK 6%, Venezuela 5%, Canada 5%, Japan 4.5%
External debt: $4.4 billion (1991 est.)
Industrial production: growth rate 2.0% (1990); accounts for almost 25% of GDP
Electricity: 1,127,000 kW capacity; 2,736 million kWh produced, 1,090 kWh per capita (1992)
Industries: tourism, bauxite mining, textiles, food processing, light manufactures
Agriculture: accounts for about 9% of GDP, 22% of work force, and 17% of exports; commercial crops—sugarcane, bananas, coffee, citrus, potatoes, vegetables; livestock and livestock products include poultry, goats, milk; not self-sufficient in grain, meat, and dairy products
Illicit drugs: illicit cultivation of cannabis; transshipment point for cocaine from Central and South America to North America; government has an active cannabis eradication program
Economic aid: US commitments, including Ex-Im (FY70-89), $1.2 billion; other countries, ODA and OOF bilateral commitments (1970-89), $1.6 billion
Currency: 1 Jamaican dollar (J$) = 100 cents
Exchange rates: Jamaican dollars (J$) per US$1—22.173 (September 1992), 12.116 (1991), 7.184 (1990), 5.7446 (1989), 5.4886 (1988), 5.4867 (1987)
Fiscal year: 1 April-31 March

Communications

Railroads: 294 km, all 1.435-meter standard gauge, single track
Highways: 18,200 km total; 12,600 km paved, 3,200 km gravel, 2,400 km improved earth
Pipelines: petroleum products 10 km
Ports: Kingston, Montego Bay, Port Antonio

Merchant marine: 4 ships (1,000 GRT or over) totaling 9,619 GRT/16,302 DWT; includes 1 roll-on/roll-off cargo, 1 oil tanker, 2 bulk
Airports:
total: 36
usable: 23
with permanent-surface runways: 10
with runways over 3,659 m: 0
with runways 2,440-3,659 m: 2
with runways 1,220-2,439 m: 1
Telecommunications: fully automatic domestic telephone network; 127,000 telephones; broadcast stations—10 AM, 17 FM, 8 TV; 2 Atlantic Ocean INTELSAT earth stations; 3 coaxial submarine cables

Defense Forces

Branches: Jamaica Defense Force (including Ground Forces, Coast Guard and Air Wing), Jamaica Constabulary Force
Manpower availability: males age 15-49 651,931; fit for military service 461,980 (1993 est.); no conscription; 26,445 reach minimum volunteer age (18) annually
Defense expenditures: exchange rate conversion—$19.3 million, 1% of GDP (FY91/92)

Jan Mayen
(territory of Norway)

Geography

Location: in the North Atlantic Ocean, north of the Arctic Circle about 590 km north-northeast of Iceland, between the Greenland Sea and the Norwegian Sea
Map references: Arctic Region
Area:
total area: 373 km²
land area: 373 km²
comparative area: slightly more than twice the size of Washington, DC
Land boundaries: 0 km
Coastline: 124.1 km
Maritime claims:
contiguous zone: 10 nm
continental shelf: 200 m depth or to depth of exploitation
exclusive economic zone: 200 nm
territorial sea: 4 nm
International disputes: Denmark has challenged Norway's maritime claims between Greenland and Jan Mayen
Climate: arctic maritime with frequent storms and persistent fog
Terrain: volcanic island, partly covered by glaciers; Beerenberg is the highest peak, with an elevation of 2,277 meters
Natural resources: none
Land use:
arable land: 0%
permanent crops: 0%
meadows and pastures: 0%
forest and woodland: 0%
other: 100%
Irrigated land: 0 km²
Environment: barren volcanic island with some moss and grass; volcanic activity resumed in 1970

People

Population: no permanent inhabitants; note—there are personnel who man the LORAN C base and the weather and coastal services radio station

Jan Mayen (continued)

Government

Names:
conventional long form: none
conventional short form: Jan Mayen
Digraph: JN
Type: territory of Norway
Capital: none; administered from Oslo, Norway, through a governor (sysselmann) resident in Longyearbyen (Svalbard)
Independence: none (territory of Norway)

Economy

Overview: Jan Mayen is a volcanic island with no exploitable natural resources. Economic activity is limited to providing services for employees of Norway's radio and meteorological stations located on the island.
Electricity: 15,000 kW capacity; 40 million kWh produced, NA kWh per capita (1992)

Communications

Ports: none; offshore anchorage only
Airports:
total: 1
useable: 1
with permanent-surface runways: 0
with runways over 3,659 m: 0
with runways 2,440-3,659 m: 0
with runways 1,220-2,439 m: 1
Telecommunications: radio and meteorological station

Defense Forces

Note: defense is the responsibility of Norway

Japan

Geography

Location: Northeast Asia, off the southeast coast of Russia and east of the Korean peninsula
Map references: Asia, Standard Time Zones of the World
Area:
total area: 377,835 km²
land area: 374,744 km²
comparative area: slightly smaller than California
note: includes Bonin Islands (Ogasawara-gunto), Daito-shoto, Minami-jima, Okinotori-shima, Ryukyu Islands (Nansei-shoto), and Volcano Islands (Kazan-retto)
Land boundaries: 0 km
Coastline: 29,751 km
Maritime claims:
exclusive fishing zone: 200 nm
territorial sea: 12 nm 3 nm in the international straits—La Perouse or Soya, Tsugaru, Osumi, and Eastern and Western channels of the Korea or Tsushima Strait
International disputes: Etorofu, Kunashiri, and Shikotan Islands and the Habomai island group occupied by the Soviet Union in 1945, now administered by Russia, claimed by Japan; Liancourt Rocks disputed with South Korea; Senkaku-shoto (Senkaku Islands) claimed by China and Taiwan
Climate: varies from tropical in south to cool temperate in north
Terrain: mostly rugged and mountainous
Natural resources: negligible mineral resources, fish
Land use:
arable land: 13%
permanent crops: 1%
meadows and pastures: 1%
forest and woodland: 67%
other: 18%
Irrigated land: 28,680 km² (1989)
Environment: many dormant and some active volcanoes; about 1,500 seismic occurrences (mostly tremors) every year; subject to tsunamis

Note: strategic location in northeast Asia

People

Population: 124,711,551 (July 1993 est.)
Population growth rate: 0.32% (1993 est.)
Birth rate: 10.31 births/1,000 population (1993 est.)
Death rate: 7.17 deaths/1,000 population (1993 est.)
Net migration rate: 0 migrant(s)/1,000 population (1993 est.)
Infant mortality rate: 4.3 deaths/1,000 live births (1993 est.)
Life expectancy at birth:
total population: 79.18 years
male: 76.35 years
female: 82.15 years (1993 est.)
Total fertility rate: 1.54 children born/woman (1993 est.)
Nationality:
noun: Japanese (singular and plural)
adjective: Japanese
Ethnic divisions: Japanese 99.4%, other 0.6% (mostly Korean)
Religions: Shinto 95.8%, Buddhist 76.3%, Christian 1.4%, other 12%
note: most Japanese observe both Shinto and Buddhist rites so the percentages add to` more than 100%
Languages: Japanese
Literacy: age 15 and over can read and write (1970)
total population: 99%
male: NA%
female: NA%
Labor force: 63.33 million
by occupation: trade and services 54%, manufacturing, mining, and construction 33%, agriculture, forestry, and fishing 7%, government 3% (1988)

Government

Names:
conventional long form: none
conventional short form: Japan
Digraph: JA
Type: constitutional monarchy
Capital: Tokyo
Administrative divisions: 47 prefectures; Aichi, Akita, Aomori, Chiba, Ehime, Fukui, Fukuoka, Fukushima, Gifu, Gumma, Hiroshima, Hokkaido, Hyogo, Ibaraki, Ishikawa, Iwate, Kagawa, Kagoshima, Kanagawa, Kochi, Kumamoto, Kyoto, Mie, Miyagi, Miyazaki, Nagano, Nagasaki, Nara, Niigata, Oita, Okayama, Okinawa, Osaka, Saga, Saitama, Shiga, Shimane, Shizuoka, Tochigi, Tokushima, Tokyo, Tottori, Toyama, Wakayama, Yamagata, Yamaguchi, Yamanashi
Independence: 660 BC (traditional founding by Emperor Jimmu)
Constitution: 3 May 1947

Legal system: modled after European civil law system with English-American influence; judicial review of legislative acts in the Supreme Court; accepts compulsory ICJ jurisdiction, with reservations

National holiday: Birthday of the Emperor, 23 December (1933)

Political parties and leaders: Liberal Democratic Party (LDP), Kiichi MIYAZAWA, president; Seiroku KAJIYAMA, secretary general; Social Democratic Party of Japan (SDPJ), Sadao YAMAHANA, Chairman; Democratic Socialist Party (DSP), Keizo OUCHI, chairman; Japan Communist Party (JCP), Tetsuzo FUWA, Presidium chairman; Komeito (Clean Government Party, CGP), Koshiro ISHIDA, chairman; Japan New Party (JNP), Morihiro HOSOKAWA, chairman

Suffrage: 20 years of age; universal

Elections:
House of Councillors: last held on 26 July 1992 (next to be held NA July 1995); results—percent of vote by party NA; seats—(252 total) LDP 106, SDPJ 73, CGP 24, DSP 12, JCP 11, JNP 4, other 22
House of Representatives: last held on 18 February 1990 (next to be held by NA February 1994); results—percent of vote by party NA; seats—(512 total) LDP 274, SDPJ 137, CGP 46, JCP 16, DSP 13, others 5, independents 6, vacant 15

Executive branch: Emperor, prime minister, Cabinet

Legislative branch: bicameral Diet (Kokkai) consists of an upper house or House of Councillors (Sangi-in) and a lower house or House of Representatives (Shugi-in)

Judicial branch: Supreme Court

Leaders:
Chief of State: Emperor AKIHITO (since 7 January 1989)
Head of Government: Prime Minister Kiichi MIYAZAWA (since 5 November 1991)

Member of: AfDB, AG (observer), Australia Group, APEC, AsDB, BIS, CCC, COCOM, CP, CSCE (observer), EBRD, ESCAP, FAO, G-2, G-5, G-7, G-8, G-10, GATT, IADB, IAEA, IBRD, ICAO, ICC, ICFTU, IDA, IEA, IFAD, IFC, ILO, IMF, IMO, INMARSAT, INTELSAT, INTERPOL, IOC, IOM (observer), ISO, ITU, LORCS, MTCR, NEA, NSG, OAS (observer), OECD, PCA, UN, UNCTAD, UNESCO, UNHCR, UNIDO, UNOMOZ, UNRWA, UPU, WFTU, WHO, WIPO, WMO, WTO, ZC

Diplomatic representation in US:
chief of mission: Ambassador Takakazu KURIYAMA
chancery: 2520 Massachusetts Avenue NW, Washington, DC 20008
telephone: (202) 939-6700

consulates general: Agana (Guam), Anchorage, Atlanta, Boston, Chicago, Honolulu, Houston, Kansas City (Missouri), Los Angeles, New Orleans, New York, San Francisco, Seattle, and Portland (Oregon)
consulates: Saipan (Northern Mariana Islands)

US diplomatic representation:
chief of mission: Ambassador Michael H. ARMACOST
embassy: 10-5, Akasaka 1-chome, Minato-ku (107), Tokyo
mailing address: APO AP 96337-0001
telephone: [81] (3) 3224-5000
FAX: [81] (3) 3505-1862
consulates general: Naha (Okinawa), Osaka-Kobe, Sapporo
consulate: Fukuoka

Flag: white with a large red disk (representing the sun without rays) in the center

---

Economy

Overview: Government-industry cooperation, a strong work ethic, and a comparatively small defense allocation have helped Japan advance with extraordinary rapidity, notably in high-technology fields. Industry, the most important sector of the economy, is heavily dependent on imported raw materials and fuels. Self-sufficient in rice, Japan must import about 50% of its requirements of other grain and fodder crops. Japan maintains one of the world's largest fishing fleets and accounts for nearly 15% of the global catch. Overall economic growth has been spectacular: a 10% average in the 1960s, a 5% average in the 1970s and 1980s. Economic growth slowed markedly in 1992 largely because of contractionary domestic policies intended to wring speculative excesses from the stock and real estate markets. At the same time, the stronger yen and slower global growth are containing export growth. Unemployment and inflation remain low at 2%. Japan continues to run a huge trade surplus—$107 billion in 1992, up nearly 40% from the year earlier—which supports extensive investment in foreign assets. The crowding of its habitable land area and the aging of its population are two major long-run problems.

National product: GDP—purchasing power equivalent—$2.468 trillion (1992)

National product real growth rate: 1.5% (1992)

National product per capita: $19,800 (1992)

Inflation rate (consumer prices): 2.1% (1992)

Unemployment rate: 2.2% (1992)

Budget: revenues $490 billion; expenditures $579 billion, including capital expenditures (public works only) of about $68 billion (FY93)

Exports: $339.7 billion (f.o.b., 1992)
commodities: manufactures 97% (including machinery 40%, motor vehicles 18%, consumer electronics 10%)
partners: Southeast Asia 31%, US 29%, Western Europe 23%, Communist countries 4%, Middle East 3%

Imports: $232.7 billion (c.i.f., 1992)
commodities: manufactures 44%, fossil fuels 33%, foodstuffs and raw materials 23%
partners: Southeast Asia 25%, US 22%, Western Europe 17%, Middle East 12%, former Communist countries and China 8%

External debt: $NA

Industrial production: growth rate −6.0% (1992); accounts for 30% of GDP

Electricity: 196,000,000 kW capacity; 835,000 million kWh produced, 6,700 kWh per capita (1992)

Industries: steel and non-ferrous metallurgy, heavy electrical equipment, construction and mining equipment, motor vehicles and parts, electronic and telecommunication equipment and components, machine tools and automated production systems, locomotives and railroad rolling stock, shipbuilding, chemicals, textiles, food processing

Agriculture: accounts for only 2% of GDP; highly subsidized and protected sector, with crop yields among highest in world; principal crops—rice, sugar beets, vegetables, fruit; animal products include pork, poultry, dairy and eggs; about 50% self-sufficient in food production; shortages of wheat, corn, soybeans; world's largest fish catch of 10 million metric tons in 1991

Economic aid: donor—ODA and OOF commitments (1970-89), $83.2 billion; ODA outlay of $9.1 billion in 1990 (est.)

Currency: 1 yen (¥) = 100 sen

Exchange rates: yen (¥) per US$1—125.01 (January 1993), 126.65 (1992), 134.71 (1991), 144.79 (1990), 137.96 (1989), 128.15 (1988)

Fiscal year: 1 April-31 March

Communications

Railroads: 27,327 km total; 2,012 km 1.435-meter standard gauge and 25,315 km predominantly 1.067-meter narrow gauge; 5,724 km doubletrack and multitrack sections, 9,038 km 1.067-meter narrow-gauge electrified, 2,012 km 1.435-meter standard-gauge electrified (1987)

Highways: 1,111,974 km total; 754,102 km paved, 357,872 km gravel, crushed stone, or unpaved; 4,400 km national expressways; 46,805 km national highways; 128,539 km prefectural roads; and 930,230 km city, town, and village roads, 6,400 km other

Inland waterways: about 1,770 km; seagoing craft ply all coastal inland seas

Pipelines: crude oil 84 km; petroleum products 322 km; natural gas 1,800 km

Japan *(continued)*

Ports: Chiba, Muroran, Kitakyushu, Kobe, Tomakomai, Nagoya, Osaka, Tokyo, Yokkaichi, Yokohama, Kawasaki, Niigata, Fushiki-Toyama, Shimizu, Himeji, Wakayama-Shimozu, Shimonoseki, Tokuyama-Shimomatsu

Merchant marine: 950 ships (1,000 GRT or over) totaling 21,080,149 GRT/32,334,270 DWT; includes 10 passenger, 39 short-sea passenger, 1 passenger cargo, 81 cargo, 43 container, 43 roll-on/roll-off cargo, 87 refrigerated cargo, 97 vehicle carrier, 240 oil tanker, 11 chemical tanker, 39 liquefied gas, 9 combination ore/oil, 2 specialized tanker, 247 bulk, 1 multi-function large load carrier; note—Japan also owns a large flag of convenience fleet, including up to 44% of the total number of ships under the Panamanian flag

Airports:
total: 162
usable: 159
with permanent-surface runways: 132
with runways over 3,659 m: 2
with runways 2,440-3,659 m: 32
with runways 1,220-2,439 m: 50

Telecommunications: excellent domestic and international service; 64,000,000 telephones; broadcast stations—318 AM, 58 FM, 12,350 TV (196 major—1 kw or greater); satellite earth stations—4 Pacific Ocean INTELSAT and 1 Indian Ocean INTELSAT; submarine cables to US (via Guam), Philippines, China, and Russia

Defense Forces

Branches: Japan Ground Self-Defense Force (Army), Japan Maritime Self-Defense Force (Navy), Japan Air Self-Defense Force (Air Force), Maritime Safety Agency (Coast Guard)

Manpower availability: males age 15-49 32,134,496; fit for military service 27,689,029; reach military age (18) annually 1,002,998 (1993 est.)

Defense expenditures: exchange rate conversion—$37 billion, 0.94% of GDP (FY93/94 est.)

Jarvis Island
(territory of the US)

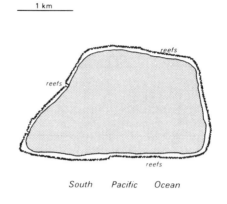

South   Pacific   Ocean

Geography

Location: in the South Pacific Ocean, 2,090 km south of Honolulu, just south of the Equator, about halfway between Hawaii and the Cook Islands

Map references: Oceania

Area:
total area: 4.5 km²
land area: 4.5 km²
comparative area: about 7.5 times the size of the Mall in Washington, DC

Land boundaries: 0 km

Coastline: 8 km

Maritime claims:
contiguous zone: 24 nm
continental shelf: 200 m or depth of exploitation
exclusive economic zone: 200 nm
territorial sea: 12 nm

International disputes: none

Climate: tropical; scant rainfall, constant wind, burning sun

Terrain: sandy, coral island surrounded by a narrow fringing reef

Natural resources: guano (deposits worked until late 1800s)

Land use:
arable land: 0%
permanent crops: 0%
meadows and pastures: 0%
forest and woodland: 0%
other: 100%

Irrigated land: 0 km²

Environment: sparse bunch grass, prostrate vines, and low-growing shrubs; lacks fresh water; primarily a nesting, roosting, and foraging habitat for seabirds, shorebirds, and marine wildlife; feral cats

People

Population: uninhabited; note—Millersville settlement on western side of island occasionally used as a weather station from 1935 until World War II, when it was abandoned; reoccupied in 1957 during the

International Geophysical Year by scientists who left in 1958; public entry is by special-use permit only and generally restricted to scientists and educators

Government

Names:
conventional long form: none
conventional short form: Jarvis Island
Digraph: DQ
Type: unincorporated territory of the US administered by the Fish and Wildlife Service of the US Department of the Interior as part of the National Wildlife Refuge System
Capital: none; administered from Washington, DC

Economy

Overview: no economic activity

Communications

Ports: none; offshore anchorage only—one boat landing area in the middle of the west coast and another near the southwest corner of the island
Note: there is a day beacon near the middle of the west coast

Defense Forces

Note: defense is the responsibility of the US; visited annually by the US Coast Guard

Jersey
(British crown dependency)

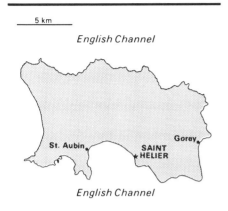

English Channel

St. Aubin • Gorey
SAINT
★HELIER

English Channel

5 km

Geography

Location: Western Europe, 27 km from France in the English Channel
Map references: Europe
Area:
total area: 117 km²
land area: 117 km²
comparative area: about 0.7 times the size of Washington, DC
Land boundaries: 0 km
Coastline: 70 km
Maritime claims:
exclusive fishing zone: 200 nm
territorial sea: 3 nm
International disputes: none
Climate: temperate; mild winters and cool summers
Terrain: gently rolling plain with low, rugged hills along north coast
Natural resources: agricultural land
Land use:
arable land: 57%
permanent crops: NA%
meadows and pastures: NA%
forest and woodland: NA%
other: NA%
Environment: about 30% of population concentrated in Saint Helier
Note: largest and southernmost of Channel Islands

People

Population: 85,450 (July 1993 est.)
Population growth rate: 0.7% (1993 est.)
Birth rate: 12.79 births/1,000 population (1993 est.)
Death rate: 10.23 deaths/1,000 population (1993 est.)
Net migration rate: 4.42 migrant(s)/1,000 population (1993 est.)
Infant mortality rate: 4.7 deaths/1,000 live births (1993 est.)
Life expectancy at birth:
total population: 76.4 years
male: 73.28 years

female: 79.86 years (1993 est.)
Total fertility rate: 1.42 children born/woman (1993 est.)
Nationality:
noun: Channel Islander(s)
adjective: Channel Islander
Ethnic divisions: UK and Norman-French descent
Religions: Anglican, Roman Catholic, Baptist, Congregational New Church, Methodist, Presbyterian
Languages: English (official), French (official), Norman-French dialect spoken in country districts
Literacy:
total population: NA%
male: NA%
female: NA%
Labor force: NA

Government

Names:
conventional long form: Bailiwick of Jersey
conventional short form: Jersey
Digraph: JE
Type: British crown dependency
Capital: Saint Helier
Administrative divisions: none (British crown dependency)
Independence: none (British crown dependency)
Constitution: unwritten; partly statutes, partly common law and practice
Legal system: English law and local statute
National holiday: Liberation Day, 9 May (1945)
Political parties and leaders: none; all independents
Suffrage: universal adult at age NA
Elections:
Assembly of the States: last held NA (next to be held NA); results—no percent of vote by party since all are independents; seats—(56 total, 52 elected) 52 independents
Executive branch: British monarch, lieutenant governor, bailiff
Legislative branch: unicameral Assembly of the States
Judicial branch: Royal Court
Leaders:
Chief of State: Queen ELIZABETH II (since 6 February 1952)
Head of Government: Lieutenant Governor and Commander in Chief Air Marshal Sir John SUTTON (since NA 1990); Bailiff Sir Peter J. CRILL (since NA)
Member of: none
Diplomatic representation in US: none (British crown dependency)
US diplomatic representation: none (British crown dependency)
Flag: white with the diagonal red cross of Saint Patrick (patron saint of Ireland) extending to the corners of the flag

Economy

Overview: The economy is based largely on financial services, agriculture, and tourism. Potatoes, cauliflower, tomatoes, and especially flowers are important export crops, shipped mostly to the UK. The Jersey breed of dairy cattle is known worldwide and represents an important export earner. Milk products go to the UK and other EC countries. In 1986 the finance sector overtook tourism as the main contributor to GDP, accounting for 40% of the island's output. In recent years the government has encouraged light industry to locate in Jersey, with the result that an electronics industry has developed alongside the traditional manufacturing of knitwear. All raw material and energy requirements are imported, as well as a large share of Jersey's food needs.
National product: GDP $NA
National product real growth rate: 8% (1987 est.)
National product per capita: $NA
Inflation rate (consumer prices): 8% (1988 est.)
Unemployment rate: NA%
Budget: revenues $308.0 million; expenditures $284.4 million, including capital expenditures of $NA (1985)
Exports: $NA
commodities: light industrial and electrical goods, foodstuffs, textiles
partners: UK
Imports: $NA
commodities: machinery and transport equipment, manufactured goods, foodstuffs, mineral fuels, chemicals
partners: UK
External debt: $NA
Industrial production: growth rate NA%
Electricity: 50,000 kW standby capacity (1992); power supplied by France
Industries: tourism, banking and finance, dairy
Agriculture: potatoes, cauliflowers, tomatoes; dairy and cattle farming
Economic aid: none
Currency: 1 Jersey pound (£J) = 100 pence
Exchange rates: Jersey pounds (£J) per US$1—0.6527 (January 1993), 0.5664 (1992), 0.5652 (1991), 0.5603 (1990), 0.6099 (1989), 0.5614 (1988); the Jersey pound is at par with the British pound
Fiscal year: 1 April-31 March

Communications

Ports: Saint Helier, Gorey, Saint Aubin
Airports:
total: 1
useable: 1
with permanent-surface runways: 1
with runways over 3,659 m: 0
with runways 2,440-3,659 m: 0
with runways 1,220-2,439 m: 1

Jersey *(continued)*

Telecommunications: 63,700 telephones; broadcast stations—1 AM, no FM, 1 TV; 3 submarine cables

Defense Forces

Note: defense is the responsibility of the UK

Johnston Atoll
(territory of the US)

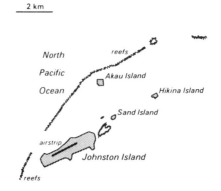

2 km

North Pacific Ocean

reefs

Akau Island

Hikina Island

Sand Island

airstrip

Johnston Island

reefs

Geography

Location: in the North Pacific Ocean, 1,430 km west-southwest of Honolulu, about one-third of the way between Hawaii and the Marshall Islands
Map references: Oceania
Area:
total area: 2.8 km²
land area: 2.8 km²
comparative area: about 4.7 times the size of the Mall in Washington, DC
Land boundaries: 0 km
Coastline: 10 km
Maritime claims:
contiguous zone: 24 nm
continental shelf: 200 m or depth of exploitation
exclusive economic zone: 200 nm
territorial sea: 12 nm
International disputes: none
Climate: tropical, but generally dry; consistent northeast trade winds with little seasonal temperature variation
Terrain: mostly flat with a maximum elevation of 4 meters
Natural resources: guano (deposits worked until about 1890)
Land use:
arable land: 0%
permanent crops: 0%
meadows and pastures: 0%
forest and woodland: 0%
other: 100%
Irrigated land: 0 km²
Environment: some low-growing vegetation
Note: strategic location in the North Pacific Ocean; Johnston Island and Sand Island are natural islands; North Island (Akau) and East Island (Hikina) are manmade islands formed from coral dredging; closed to the public; former nuclear weapons test site; site of Johnston Atoll Chemical Agent Disposal System (JACADS)

People

Population: no indigenous inhabitants; note—there are 1,400 US Government personnel and contractors

Government

Names:
conventional long form: none
conventional short form: Johnston Atoll
Digraph: JQ
Type: unincorporated territory of the US administered by the US Defense Nuclear Agency (DNA) and managed cooperatively by DNA and the Fish and Wildlife Service of the US Department of the Interior as part of the National Wildlife Refuge system
Capital: none; administered from Washington, DC
Diplomatic representation in US: none (territory of the US)
Flag: the flag of the US is used

Economy

Overview: Economic activity is limited to providing services to US military personnel and contractors located on the island. All food and manufactured goods must be imported.
Electricity: supplied by the management and operations contractor

Communications

Airports:
total: 1
usable: 1
with permanent-surface runways: 1
with runways over 3,659 m: 0
with runways 2,440 to 3,659 m: 1 with TACAN and beacon
with runways 1,220 to 2,439 m: 0
Telecommunications: excellent system including 60-channel submarine cable, Autodin/SRT terminal, digital telephone switch, Military Affiliated Radio System (MARS station), commercial satellite television system, and UHF/VHF air-ground radio

Defense Forces

Note: defense is the responsibility of the US

Jordan
(also see separate West Bank entry)

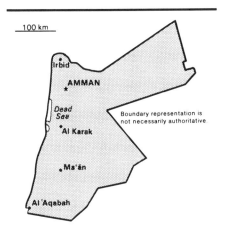

Geography

Location: Middle East, between Israel and Saudi Arabia
Map references: Africa, Middle East, Standard Time Zones of the World
Area:
total area: 89,213 km²
land area: 88,884 km²
comparative area: slightly smaller than Indiana
Land boundaries: total 1,619 km, Iraq 181 km, Israel 238 km, Saudi Arabia 728 km, Syria 375 km, West Bank 97 km
Coastline: 26 km
Maritime claims:
territorial sea: 3 nm
International disputes: differences with Israel over the location of the 1949 Armistice Line that separates the two countries; water-sharing issues with Israel
Climate: mostly arid desert; rainy season in west (November to April)
Terrain: mostly desert plateau in east, highland area in west; Great Rift Valley separates East and West Banks of the Jordan River
Natural resources: phosphates, potash, shale oil
Land use:
arable land: 4%
permanent crops: 0.5%
meadows and pastures: 1%
forest and woodland: 0.5%
other: 94%
Irrigated land: 570 km² (1989 est.)
Environment: lack of natural water resources; deforestation; overgrazing; soil erosion; desertification

People

Population: 3,823,636 (July 1993 est.)
Population growth rate: 3.57% (1993 est.)
Birth rate: 39.48 births/1,000 population (1993 est.)
Death rate: 4.32 deaths/1,000 population (1993 est.)

Net migration rate: 0.51 migrant(s)/1,000 population (1993 est.)
Infant mortality rate: 33.3 deaths/1,000 live births (1993 est.)
Life expectancy at birth:
total population: 71.61 years
male: 69.83 years
female: 73.51 years (1993 est.)
Total fertility rate: 5.79 children born/woman (1993 est.)
Nationality:
noun: Jordanian(s)
adjective: Jordanian
Ethnic divisions: Arab 98%, Circassian 1%, Armenian 1%
Religions: Sunni Muslim 92%, Christian 8%
Languages: Arabic (official), English widely understood among upper and middle classes
Literacy: age 15 and over can read and write (1990)
total population: 80%
male: 89%
female: 70%
Labor force: 572,000 (1988)
by occupation: agriculture 20%, manufacturing and mining 20% (1987 est.)

Government

Names:
conventional long form: Hashemite Kingdom of Jordan
conventional short form: Jordan
local long form: Al Mamlakah al Urduniyah al Hashimiyah
local short form: Al Urdun
former: Transjordan
Digraph: JO
Type: constitutional monarchy
Capital: Amman
Administrative divisions: 8 governorates (muhafazat, singular—muhafazah); Al Balqa', Al Karak, Al Mafraq, 'Amman, At Tafilah, Az Zarqa', Irbid, Ma'an
Independence: 25 May 1946 (from League of Nations mandate under British administration)
Constitution: 8 January 1952
Legal system: based on Islamic law and French codes; judicial review of legislative acts in a specially provided High Tribunal; has not accepted compulsory ICJ jurisdiction
National holiday: Independence Day, 25 May (1946)
Political parties and leaders: approximately 24 parties have been formed since the National Charter, but the number fluctuates; after the 1989 parliamentary elections, King Hussein promised to allow the formation of political parties; a national charter that sets forth the ground rules for democracy in Jordan—including the creation of political parties—was approved in principle by the special National Conference on 9 June 1991, but its specific provisions have yet to be passed by National Assembly

Suffrage: 20 years of age; universal
Elections:
House of Representatives: last held 8 November 1989 (next to be held November 1993); results—percent of vote by party NA; seats—(80 total) Muslim Brotherhood (fundamentalist) 22, Independent Islamic bloc (generally traditionalist) 6, Democratic bloc (mostly leftist) 9, Constitutionalist bloc (traditionalist) 17, Nationalist bloc (traditionalist) 16, independent 10
Executive branch: monarch, prime minister, deputy prime minister, Cabinet
Legislative branch: bicameral National Assembly (Majlis al-'Umma) consists of an upper house or House of Notables (Majlis al-A'ayan) and a lower house or House of Representatives (Majlis al-Nuwaab); note—the House of Representatives has been convened and dissolved by the King several times since 1974 and in November 1989 the first parliamentary elections in 22 years were held
Judicial branch: Court of Cassation
Leaders:
Chief of State: King HUSSEIN Ibn Talal Al Hashemi (since 11 August 1952)
Head of Government: Prime Minister Zayd bin SHAKIR (since 21 November 1991)
Member of: ABEDA, ACC, AFESD, AL, AMF, CAEU, CCC, ESCWA, FAO, G-77, IAEA, IBRD, ICAO, ICC, IDA, IDB, IFAD, IFC, ILO, IMF, IMO, INTELSAT, INTERPOL, IOC, IOM (observer), ISO (correspondent), ITU, LORCS, NAM, OIC, PCA, UN, UNAVEM II, UNCTAD, UNESCO, UNIDO, UNOSOM, UNRWA, UNPROFOR, UPU, WFTU, WHO, WIPO, WMO, WTO
Diplomatic representation in US:
chief of mission: Ambassador Fayez A. TARAWNEH
chancery: 3504 International Drive NW, Washington, DC 20008
telephone: (202) 966-2664
US diplomatic representation:
chief of mission: Ambassador Roger Gram HARRISON
embassy: Jebel Amman, Amman
mailing address: P. O. Box 354, Amman, or APO AE 09892
telephone: [962] (6) 644-371
Flag: three equal horizontal bands of black (top), white, and green with a red isosceles triangle based on the hoist side bearing a small white seven-pointed star; the seven points on the star represent the seven fundamental laws of the Koran

Economy

Overview: Jordan benefited from increased Arab aid during the oil boom of the late 1970s and early 1980s, when its annual GNP growth averaged more than 10%. In the

Jordan (continued)

remainder of the 1980s, however, reductions in both Arab aid and worker remittances slowed economic growth to an average of roughly 2% per year. Imports—mainly oil, capital goods, consumer durables, and food—have been outstripping exports, with the difference covered by aid, remittances, and borrowing. In mid-1989, the Jordanian Government began debt-rescheduling negotiations and agreed to implement an IMF program designed to gradually reduce the budget deficit and implement badly needed structural reforms. The Persian Gulf crisis that began in August 1990, however, aggravated Jordan's already serious economic problems, forcing the government to shelve the IMF program, stop most debt payments, and suspend rescheduling negotiations. Aid from Gulf Arab states and worker remittances have plunged, and refugees have flooded the country, straining government resources. Economic recovery is unlikely without substantial foreign aid, debt relief, and economic reform.
National product: GDP—exchange rate conversion—$3.6 billion (1991 est.)
National product real growth rate: 3% (1991 est.)
National product per capita: $1,100 (1991 est.)
Inflation rate (consumer prices): 9% (1991 est.)
Unemployment rate: 40% (1991 est.)
Budget: revenues $1.3 billion; expenditures $1.9 billion, including capital expenditures of $440 million (1992 est.)
Exports: $1.0 billion (f.o.b., 1991 est.)
commodities: phosphates, fertilizers, potash, agricultural products, manufactures
partners: India, Iraq, Saudi Arabia, Indonesia, Ethiopia, UAE, China
Imports: $2.3 billion (c.i.f., 1991 est.)
commodities: crude oil, machinery, transport equipment, food, live animals, manufactured goods
partners: EC countries, US, Iraq, Saudi Arabia, Japan, Turkey
External debt: $9 billion (December 1991 est.)
Industrial production: growth rate 1% (1991 est.); accounts for 20% of GDP
Electricity: 1,030,000 kW capacity; 3,814 million kWh produced, 1,070 kWh per capita (1992)
Industries: phosphate mining, petroleum refining, cement, potash, light manufacturing
Agriculture: accounts for about 7% of GDP; principal products are wheat, barley, citrus fruit, tomatoes, melons, olives; livestock—sheep, goats, poultry; large net importer of food
Economic aid: US commitments, including Ex-Im (FY70-89), $1.7 billion; Western (non-US) countries, ODA and OOF bilateral commitments (1970-89), $1.5 billion; OPEC bilateral aid (1979-89), $9.5 billion; Communist countries (1970-89), $44 million
Currency: 1 Jordanian dinar (JD) = 1,000 fils
Exchange rates: Jordanian dinars (JD) per US$1—0.6890 (January 1993), 0.6797 (1992), 0.6808 (1991), 0.6636 (1990), 0.5704 (1989), 0.3709 (1988)
Fiscal year: calendar year

Communications

Railroads: 789 km 1.050-meter gauge, single track
Highways: 7,500 km; 5,500 km asphalt, 2,000 km gravel and crushed stone
Pipelines: crude oil 209 km
Ports: Al 'Aqabah
Merchant marine: 2 ships (1,000 GRT or over) totaling 60,378 GRT/113,557 DWT; includes 1 cargo and 1 oil tanker
Airports:
total: 19
usable: 15
with permanent-surface runways: 14
with runways over 3,659 m: 1
with runways 2,440-3,659 m: 13
with runways 1,220-2,439 m: 0
Telecommunications: adequate telephone system of microwave, cable, and radio links; 81,500 telephones; broadcast stations—5 AM, 7 FM, 8 TV; satellite earth stations—1 Atlantic Ocean INTELSAT, 1 Indian Ocean INTELSAT, 1 ARABSAT, 1 domestic TV receive-only; coaxial cable and microwave to Iraq, Saudi Arabia, and Syria; microwave link to Lebanon is inactive; participant in MEDARABTEL, a microwave radio relay network linking Syria, Jordan, Egypt, Libya, Tunisia, Algeria, and Morocco

Defense Forces

Branches: Royal Jordanian Land Force, Royal Jordanian Air Force, Royal Naval Force, Public Security Force
Manpower availability: males age 15-49 936,213; fit for military service 664,095; reach military age (18) annually 42,093 (1993 est.)
Defense expenditures: exchange rate conversion—$434.8 million, 7.9% of GDP (1993 est.)

Juan de Nova Island
(possession of France)

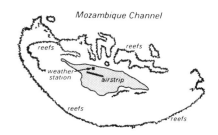

Geography

Location: Southern Africa, in the central Mozambique Channel about one-third of the way between Madagascar and Mozambique
Map references: Africa
Area:
total area: 4.4 km²
land area: 4.4 km²
comparative area: about 7.5 times the size of the Mall in Washington, DC
Land boundaries: 0 km
Coastline: 24.1 km
Maritime claims:
contiguous zone: 12 nm
continental shelf: 200 m depth or to depth of exploitation
exclusive economic zone: 200 nm
territorial sea: 12 nm
International disputes: claimed by Madagascar
Climate: tropical
Terrain: NA
Natural resources: guano deposits and other fertilizers
Land use:
arable land: 0%
permanent crops: 0%
meadows and pastures: 0%
forest and woodland: 90%
other: 10%
Irrigated land: 0 km²
Environment: subject to periodic cyclones; wildlife sanctuary

People

Population: uninhibited

Government

Names:
conventional long form: none
conventional short form: Juan de Nova Island
local long form: none
local short form: Ile Juan de Nova

Kazakhstan

Digraph: JU
Type: French possession administered by Commissioner of the Republic, resident in Reunion
Capital: none; administered by France from Reunion
Independence: none (possession of France)

Economy

Overview: no economic activity

Communications

Railroads: short line going to a jetty
Ports: none; offshore anchorage only
Airports:
total: 1
usable: 1
with permament-surface runways: 0
with runways over 3,659 m: 0
with runways 2,439-3,659 m: 0
with runways 1,220-2,439 m: 1

Defense Forces

Note: defense is the responsibility of France

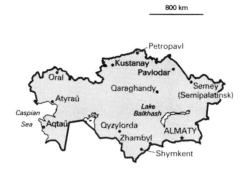

800 km

Geography

Location: South Asia, between Russia and Uzbekistan, bordering on the Caspian Sea and the Aral Sea
Map references: Asia, Commonwealth of Independent States—Central Asian States, Standard Time Zones of the World
Area:
total area: 2,717,300 km²
land area: 2,669,800 km²
comparative area: slightly less than four times the size of Texas
Land boundaries: total 12,012 km, China 1,533 km, Kyrgyzstan 1,051 km, Russia 6,846 km, Turkmenistan 379 km, Uzbekistan 2,203 km
Coastline: 0 km
note: Kazakhstan borders the Aral Sea (1,015 km) and the Caspian Sea (1,894 km)
Maritime claims: landlocked, but boundaries with Uzbekistan in the Sea of Azov and with Russia, Azerbaijan, and Turkmenistan in the Caspian Sea are yet to be determined
International disputes: none
Climate: continental, arid and semiarid
Terrain: extends from the Volga to the Altai Mountains and from the plains in western Siberia to oasis and desert in Central Asia
Natural resources: petroleum, coal, iron, manganese, chrome, nickel, cobalt, copper, molybdenum, lead, zinc, bauxite, gold, uranium, iron
Land use:
arable land: 15%
permanent crops: 0%
meadows and pastures: 57%
forest and woodland: 4%
other: 24%
Irrigated land: 23,080 km² (1990)
Environment: drying up of Aral Sea is causing increased concentrations of chemical pesticides and natural salts; industrial pollution
Note: landlocked

People

Population: 17,156,370 (July 1993 est.)
Population growth rate: 0.65% (1993 est.)
Birth rate: 19.55 births/1,000 population (1993 est.)
Death rate: 7.95 deaths/1,000 population (1993 est.)
Net migration rate: -5.06 migrant(s)/1,000 population (1993 est.)
Infant mortality rate: 41.8 deaths/1,000 live births (1993 est.)
Life expectancy at birth:
total population: 67.83 years
male: 63.17 years
female: 72.73 years (1993 est.)
Total fertility rate: 2.45 children born/woman (1993 est.)
Nationality:
noun: Kazakhstani(s)
adjective: Kazakhstani
Ethnic divisions: Kazakh (Qazaq) 41.9%, Russian 37%, Ukrainian 5.2%, German 4.7%, Uzbek 2.1%, Tatar 2%, other 7.1%
Religions: Muslim 47%, Russian Orthodox 15%, Protestant 2%, other 36%
Languages: Kazakh (Qazaq; official language), Russian (language of interethnic communication)
Literacy: age 9-49 can read and write (1970)
total population: 100%
male: 100%
female: 100%
Labor force: 7.563 million
by occupation: industry and construction 32%, agriculture and forestry 23%, other 45% (1990)

Government

Names:
conventional long form: Republic of Kazakhstan
conventional short form: Kazakhstan
local long form: Kazakhstan Respublikasy
local short form: none
former: Kazakh Soviet Socialist Republic
Digraph: KZ
Type: republic
Capital: Almaty (Alma-Ata)
Administrative divisions: 19 oblasts (oblystar, singular—oblys) and 1 city (qalalar, singular—qala)*; Almaty*, Almaty, Aqmola, Aqtobe, Atyrau, Batys Qazaqstan, Kokshetau, Mangghystau, Ongtustik Qazaqstan, Qaraghandy, Qostanay, Qyzylorda, Pavlodar, Semey, Shyghys Qazaqstan, Soltustik Qazaqstan, Taldyqorghan, Torghay, Zhambyl, Zhezqazghan,
Independence: 16 December 1991 (from the Soviet Union)
Constitution: adopted 18 January 1993
Legal system: based on civil law system

205

Kazakhstan (continued)

National holiday: Independence Day, 16 December

Political parties and leaders: Peoples Congress, Olzhas SULEYMENOV and Mukhtar SHAKHANOV, co-chairmen; Kazakh Socialist Party (former Communist Party), Nursultan NAZARBAYEV, chairman; December (Zheltoksan) Movement, Khasan KOZHAKMETOV, chairman; Freedom (AZAT) Party, Kamal ORMANTAYEV, chairman

Other political or pressure groups: Independent Trade Union Center (Birlesu; an association of independent trade union and business associations), Leonid SOLOMIN, president

Suffrage: 18 years of age; universal

Elections:
President: last held 1 December 1991 (next to be held NA 1996); percent of vote by party NA; Nursultan A. NAZARBAYEV ran unopposed
Supreme Council: last held NA April 1990 (next to be held NA December 1994); results—percent of vote by party NA; seats—(358 total) Socialist Party 338

Executive branch: president, cabinet of ministers, prime minister

Legislative branch: unicameral Supreme Soviet

Judicial branch: Supreme Court

Leaders:
Chief of State: President Nursultan A. NAZARBAYEV (since NA April 1990); Vice President Yerik ASANBAYEV (since 1 December 1991)
Head of Government: Prime Minister Sergey TERESHENKO (since 14 October 1991); First Deputy Prime Minister Davlat SEMBAYEV (since NA November 1990); Supreme Council Chairman Serikbolsyn ABDILDIN (since NA July 1991)

Member of: CIS, CSCE, EBRD, ECO, IBRD, IDA, IMF, OIC, UN, UNCTAD, UNESCO, UPU

Diplomatic representation in US:
chief of mission: Ambassador Alim S. DJAMBOURCHINE
chancery: 3421 Massachusetts Ave., NW, Washington, DC 20007
telephone: (202) 333-4504

US diplomatic representation:
chief of mission: Ambassador William H. COURTNEY
embassy: Furumanova 99/97, Almaty
mailing address: US Department of State, Washington, D.C. 20521-7030
telephone: (3272) 63-24-26

Flag: sky blue background representing the endless sky and a gold sun with 32 rays soaring above a golden steppe eagle in the center; on the hoist side is a "national ornamentation" in yellow

Economy

Overview: The second-largest in area of the 15 former Soviet republics, Kazakhstan has vast oil, coal, and agricultural resources. Kazakhstan is highly dependent on trade with Russia, exchanging its natural resources for finished consumer and industrial goods. Kazakhstan now finds itself with serious pollution problems, backward technology, and little experience in foreign markets. The government in 1992 continued to push privatization of the economy and freed many prices. Output in 1992 dropped because of problems common to the ex-Soviet Central Asian republics, especially the cumulative effects of the disruption of old supply channels and the slow process of creating new economic institutions. Kazakhstan lacks the funds, technology, and managerial skills for a quick recovery of output. US firms have been enlisted to increase oil output but face formidable obstacles; for example, oil can now reach Western markets only through pipelines that run across independent former Soviet republics. Finally, the end of monolithic Communist control has brought ethnic grievances into the open. The 6 million Russians in the republic, formerly the favored class, now face the hostility of a society dominated by Muslims. Ethnic rivalry will be just one of the formidable obstacles to the prioritization of national objectives and the creation of a productive, technologically advancing society.

National product: GDP $NA

National product real growth rate: -15% (1992 est.)

National product per capita: $NA

Inflation rate (consumer prices): 28% per month (first quarter 1993)

Unemployment rate: 0.4% includes only officially registered unemployed; also large numbers of underemployed workers

Budget: revenues $NA; expenditures $NA, including capital expenditures of $1.76 billion (1991)

Exports: $1.5 billion to outside the successor states of the former USSR (1992)
commodities: oil, ferrous and nonferrous metals, chemicals, grain, wool, meat (1991)
partners: Russia, Ukraine, Uzbekistan

Imports: $500 million from outside the successor states of the former USSR (1992)
commodities: machinery and parts, industrial materials
partners: Russia and other former Soviet republics, China

External debt: $2.6 billion (1991 est.)

Industrial production: growth rate −15% (1992 est.); accounts for 30% of net material product

Electricity: 19,135,000 kW capacity; 81,300 million kWh produced, 4,739 kWh per capita (1992)

Industries: extractive industries (oil, coal, iron ore, manganese, chromite, lead, zinc, copper, titanium, bauxite, gold, silver, phosphates, sulfur), iron and steel, nonferrous metal, tractors and other agricultural machinery, electric motors, construction materials

Agriculture: accounts for almost 40% of net material product; employs about 25% of the labor force; grain, mostly spring wheat; meat, cotton, wool

Illicit drugs: illicit producers of cannabis and opium; mostly for CIS consumption; limited government eradication program; used as transshipment point for illicit drugs to Western Europe

Economic aid: recipient of limited foreign aid (1992)

Currency: retaining Russian ruble as currency (May 1993)

Exchange rates: rubles per US$1—415 (24 December 1992) but subject to wide fluctuations

Fiscal year: calendar year

Communications

Railroads: 14,460 km (all 1.520-meter gauge); does not include industrial lines (1990)

Highways: 189,000 km total; 108,100 km hard surfaced (paved or gravel), 80,900 km earth (1990)

Inland waterways: Syr Darya

Pipelines: crude oil 2,850 km, refined products 1,500 km, natural gas 3,480 km (1992)

Ports: inland—Atyrau (Guryev; on Caspian Sea)

Airports:
total: 365
useable: 152
with permanent-surface runways: 49
with runways over 3,659 m: 8
with runways 2,440-3,659 m: 38
with runways 1,220-2,439 m: 71

Telecommunications: telephone service is poor, with only about 6 telephones for each 100 persons; of the approximately 1 million telephones, Almaty (Alma-Ata) has 184,000; international traffic with other former USSR republics and China carried by landline and microwave, and with other countries by satellite and through 8 international telecommunications circuits at the Moscow international gateway switch; satellite earth stations—INTELSAT and Orbita (TV receive only); new satellite ground station established at Almaty with Turkish financial help (December 1992) with 2500 channel band width

Kenya

Defense Forces

Branches: Army, Navy, National Guard, Security Forces (internal and border troops)
Manpower availability: males age 15-49 4,349,509; fit for military service 3,499,718; reach military age (18) annually 154,727 (1993 est.)
Defense expenditures: 69,326 million rubles, NA% of GDP (forecast for 1993); note—conversion of the military budget into US dollars using the current exchange rate could produce misleading results

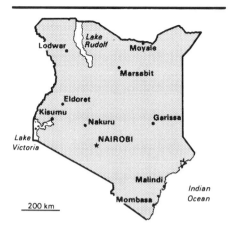

200 km

Geography

Location: Eastern Africa, bordering the northwestern India Ocean between Tanzania and Somalia
Map references: Africa, Standard Time Zones of the World
Area:
total area: 582,650 km²
land area: 569,250 km²
comparative area: slightly more than twice the size of Nevada
Land boundaries: total 3,446 km, Ethiopia 830 km, Somalia 682 km, Sudan 232 km, Tanzania 769 km, Uganda 933 km
Coastline: 536 km
Maritime claims:
exclusive economic zone: 200 nm
territorial sea: 12 nm
International disputes: administrative boundary with Sudan does not coincide with international boundary; possible claim by Somalia based on unification of ethnic Somalis
Climate: varies from tropical along coast to arid in interior
Terrain: low plains rise to central highlands bisected by Great Rift Valley; fertile plateau in west
Natural resources: gold, limestone, soda ash, salt barytes, rubies, fluorspar, garnets, wildlife
Land use:
arable land: 3%
permanent crops: 1%
meadows and pastures: 7%
forest and woodland: 4%
other: 85%
Irrigated land: 520 km² (1989)
Environment: unique physiography supports abundant and varied wildlife of scientific and economic value; deforestation; soil erosion; desertification; glaciers on Mt. Kenya
Note: the Kenyan Highlands comprise one of the most successful agricultural production regions in Africa

People

Population: 27,372,266 (July 1993 est.)
Population growth rate: 3.18% (1993 est.)
Birth rate: 43.18 births/1,000 population (1993 est.)
Death rate: 11.41 deaths/1,000 population (1993 est.)
Net migration rate: 0 migrant(s)/1,000 population (1993 est.)
Infant mortality rate: 74.7 deaths/1,000 live births (1993 est.)
Life expectancy at birth:
total population: 54.07 years
male: 52.27 years
female: 55.92 years (1993 est.)
Total fertility rate: 6.06 children born/woman (1993 est.)
Nationality:
noun: Kenyan(s)
adjective: Kenyan
Ethnic divisions: Kikuyu 21%, Luhya 14%, Luo 13%, Kalenjin 11%, Kamba 11%, Kisii 6%, Meru 6%, Asian, European, and Arab 1%
Religions: Roman Catholic 28%, Protestant (including Anglican) 26%, indigenous beliefs 18%, Muslim 6%
Languages: English (official), Swahili (official), numerous indigenous languages
Literacy: age 15 and over can read and write (1990)
total population: 69%
male: 80%
female: 58%
Labor force: 9.2 million (includes unemployed); the total employed is 1,370,000 (14.8% of the labor force)
by occupation: services 54.8%, industry 26.2%, agriculture 19.0% (1989)

Government

Names:
conventional long form: Republic of Kenya
conventional short form: Kenya
former: British East Africa
Digraph: KE
Type: republic
Capital: Nairobi
Administrative divisions: 8 provinces; Central, Coast, Eastern, Nairobi, North Eastern, Nyanza, Rift Valley, Western
Independence: 12 December 1963 (from UK)
Constitution: 12 December 1963, amended as a republic 1964; reissued with amendments 1979, 1983, 1986, 1988, 1991, and 1992
Legal system: based on English common law, tribal law, and Islamic law; judicial review in High Court; accepts compulsory ICJ jurisdiction, with reservations; constitutional amendment of 1982 making Kenya a de jure one-party state repealed in 1991

Kenya (continued)

National holiday: Independence Day, 12 December (1963)

Political parties and leaders: ruling party is Kenya African National Union (KANU), Daniel T. arap MOI, president; opposition parties include Forum for the Restoration of Democracy (FORD-Kenya), Oginga ODINGA; FORD-Asili, Kenneth MATIBA; Democratic Party of Kenya (DP), Mwai KIBAKI; Kenya National Congress (KNC), Titus MBATHI; Kenya Social Congress (KSC), George ANYONA; Kenya National Democratic Alliance (KENYA), Mukara NG'ANG'A; Party for Independent Candidates of Kenya (PKK), Otieno OTOERA

Other political or pressure groups: labor unions; exile opposition—Mwakenya and other groups

Suffrage: 18 years of age; universal

Elections:
President: last held on 29 December 1992; results—President Daniel T. arap MOI was reelected with 37% of the vote; Kenneth Matiba (FORD-ASILI) 26%; Mwai Kibaki (SP) 19%, Oginga Odinga (FORD-Kenya) 17%
National Assembly: last held on 29 December 1992; results—(188 total) KANU 100, FORD-Kenya 31, FORD-Asili 31, DP 23, smaller parties 3; president nominates 12 additional members
note: first multiparty election since repeal of one-party state law

Executive branch: president, vice president, Cabinet

Legislative branch: unicameral National Assembly (Bunge)

Judicial branch: Court of Appeal, High Court

Leaders:
Chief of State and Head of Government: President Daniel Teroitich arap MOI (since 14 October 1978); Vice President George SAITOTI (since 10 May 1989)

Member of: ACP, AfDB, C, CCC, EADB, ECA, FAO, G-77, GATT, IAEA, IBRD, ICAO, IDA, IFAD, IFC, IGADD, ILO, IMF, IMO, INTELSAT, INTERPOL, IOC, IOM, ISO, ITU, LORCS, MINURSO, NAM, OAU, UN, UNCTAD, UNESCO, UNIDO, UNIKOM, UNPROFOR, UPU, WCL, WHO, WIPO, WMO, WTO

Diplomatic representation in US:
chief of mission: Ambassador Denis Daudi AFANDE
chancery: 2249 R Street NW, Washington, DC 20008
telephone: (202) 387-6101
consulates general: Los Angeles and New York

US diplomatic representation:
chief of mission: Ambassador Smith HEMPSTONE, Jr.
embassy: corner of Moi Avenue and Haile Selassie Avenue, Nairobi

mailing address: P. O. Box 30137, Nairobi or APO AE 09831
telephone: [254] (2) 334141
FAX: [254] (2) 340838
consulate: Mombasa

Flag: three equal horizontal bands of black (top), red, and green; the red band is edged in white; a large warrior's shield covering crossed spears is superimposed at the center

Economy

Overview: Kenya's 3.6% annual population growth rate—one of the highest in the world—presents a serious problem for the country's economy. In the meantime, GDP growth in the near term has kept slightly ahead of population—annually averaging 4.9% in the 1986-90 period. Undependable weather conditions and a shortage of arable land hamper long-term growth in agriculture, the leading economic sector. In 1991, deficient rainfall, stagnant export volume, and sagging export prices held economic growth below the all-important population growth figure, and in 1992 output fell.

National product: GDP—exchange rate conversion—$8.3 billion (1992 est.)

National product real growth rate: -1% (1992 est.)

National product per capita: $320 (1992 est.)

Inflation rate (consumer prices): 30% (1992 est.)

Unemployment rate: NA%

Budget: revenues $2.4 billion; expenditures $2.8 billion, including capital expenditures of $0.74 billion (FY90)

Exports: $1.0 billion (f.o.b., 1992 est.)
commodities: tea 25%, coffee 18%, petroleum products 11% (1990)
partners: EC 44%, Africa 25%, Asia 5%, US 5%, Middle East 4% (1990)

Imports: $2.05 billion (f.o.b., 1992 est.)
commodities: machinery and transportation equipment 29%, petroleum and petroleum products 15%, iron and steel 7%, raw materials, food and consumer goods (1989)
partners: EC 45%, Asia 11%, Middle East 12%, US 5% (1988)

External debt: $7.0 billion (1992 est.)

Industrial production: growth rate 5.4% (1989 est.); accounts for 13% of GDP

Electricity: 730,000 kW capacity; 2,540 million kWh produced, 100 kWh per capita (1990)

Industries: small-scale consumer goods (plastic, furniture, batteries, textiles, soap, cigarettes, flour), agricultural processing, oil refining, cement, tourism

Agriculture: most important sector, accounting for 25% of GDP and 65% of exports; cash crops—coffee, tea, sisal, pineapple; food products—corn, wheat, sugarcane, fruit, vegetables, dairy products,

beef, pork, poultry, eggs; food output not keeping pace with population growth, and crop production has been extended into marginal land

Illicit drugs: widespread wild, small-plot cultivation of marijuana and qat; most locally consumed; transit country for Southwest Asian heroin moving to West Africa and onward to Europe and North America; Indian methaqualone also transits on way to South Africa

Economic aid: US commitments, including Ex-Im (FY70-89), $839 million; Western (non-US) countries, ODA and OOF bilateral commitments (1970-89), $7,490 million; OPEC bilateral aid (1979-89), $74 million; Communist countries (1970-89), $83 million

Currency: 1 Kenyan shilling (KSh) = 100 cents

Exchange rates: Kenyan shillings (KSh) per US$1 −36.227 (January 1993), 32.217 (1992), 27.508 (1991), 22.915 (1990), 20.572 (1989), 17.747 (1988)

Fiscal year: 1 July-30 June

Communications

Railroads: 2,040 km 1.000-meter gauge

Highways: 64,590 km total; 7,000 km paved, 4,150 km gravel, remainder improved earth

Inland waterways: part of Lake Victoria system is within boundaries of Kenya

Pipelines: petroleum products 483 km

Ports: coastal—Mombasa, Lamu; inland—Kisumu

Merchant marine: 1 oil tanker ship (1,000 GRT or over) totaling 3,727 GRT/5,558 DWT

Airports:
total: 247
usable: 208
with permanent-surface runways: 18
with runways over 3,659 m: 2
with runways 2,440-3,659 m: 3
with runways 1,220-2,439 m: 43

Telecommunications: in top group of African systems; consists primarily of radio relay links; over 260,000 telephones; broadcast stations—16 AM; 4 FM, 6 TV; satellite earth stations—1 Atlantic Ocean INTELSAT and 1 Indian Ocean INTELSAT

Defense Forces

Branches: Army, Navy, Air Force, paramilitary General Service Unit of the Police

Manpower availability: males age 15-49 5,912,744; fit for military service 3,654,738 (1993 est.); no conscription

Defense expenditures: exchange rate conversion—$294 million, 4.9% of GDP (FY88/89 est.)

Kingman Reef

(territory of the US)

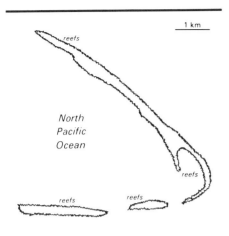

Geography

Location: in the North Pacific Ocean, 1,600 km south-southwest of Honolulu, about halfway between Hawaii and American Samoa
Map references: Oceania
Area:
total area: 1 km²
land area: 1 km²
comparative area: about 1.7 times the size of the Mall in Washington, DC
Land boundaries: 0 km
Coastline: 3 km
Maritime claims:
contiguous zone: 24 nm
continental shelf: 200 m or depth of exploitation
exclusive economic zone: 200 nm
territorial sea: 12 nm
International disputes: none
Climate: tropical, but moderated by prevailing winds
Terrain: low and nearly level with a maximum elevation of about 1 meter
Natural resources: none
Land use:
arable land: 0%
permanent crops: 0%
meadows and pastures: 0%
forest and woodland: 0%
other: 100%
Irrigated land: 0 km²
Environment: barren coral atoll with deep interior lagoon; wet or awash most of the time
Note: maximum elevation of about 1 meter makes this a navigational hazard; closed to the public

People

Population: uninhabited

Government

Names:
conventional long form: none
conventional short form: Kingman Reef
Digraph: KQ
Type: unincorporated territory of the US administered by the US Navy
Capital: none; administered from Washington, DC

Economy

Overview: no economic activity

Communications

Ports: none; offshore anchorage only
Airports: lagoon was used as a halfway station between Hawaii and American Samoa by Pan American Airways for flying boats in 1937 and 1938

Defense Forces

Note: defense is the responsibility of the US

Kiribati

Geography

Location: Oceania, straddling the equator in the Pacific Ocean, about halfway between Hawaii and Australia
Map references: Oceania
Area:
total area: 717 km²
land area: 717 km²
comparative area: slightly more than four times the size of Washington, DC
note: includes three island groups—Gilbert Islands, Line Islands, Phoenix Islands
Land boundaries: 0 km
Coastline: 1,143 km
Maritime claims:
exclusive economic zone: 200 nm
territorial sea: 12 nm
International disputes: none
Climate: tropical; marine, hot and humid, moderated by trade winds
Terrain: mostly low-lying coral atolls surrounded by extensive reefs
Natural resources: phosphate (production discontinued in 1979)
Land use:
arable land: 0%
permanent crops: 51%
meadows and pastures: 0%
forest and woodland: 3%
other: 46%
Irrigated land: NA km²
Environment: typhoons can occur any time, but usually November to March; 20 of the 33 islands are inhabited
Note: Banaba (Ocean Island) in Kiribati is one of the three great phosphate rock islands in the Pacific Ocean—the others are Makatea in French Polynesia and Nauru

People

Population: 76,320 (July 1993 est.)
Population growth rate: 2.03% (1993 est.)
Birth rate: 32.03 births/1,000 population (1993 est.)

Kiribati (continued)

Death rate: 12.31 deaths/1,000 population (1993 est.)
Net migration rate: 0.56 migrant(s)/1,000 population (1993 est.)
Infant mortality rate: 98.4 deaths/1,000 live births (1993 est.)
Life expectancy at birth:
total population: 54.16 years
male: 52.56 years
female: 55.78 years (1993 est.)
Total fertility rate: 3.82 children born/woman (1993 est.)
Nationality:
noun: I-Kiribati (singular and plural)
adjective: I-Kiribati
Ethnic divisions: Micronesian
Religions: Roman Catholic 52.6%, Protestant (Congregational) 40.9%, Seventh-Day Adventist, Baha'i, Church of God, Mormon 6% (1985)
Languages: English (official), Gilbertese
Literacy:
total population: NA%
male: NA%
female: NA%
Labor force: 7,870 economically active, not including subsistence farmers (1985 est.)

Government

Names:
conventional long form: Republic of Kiribati
conventional short form: Kiribati
former: Gilbert Islands
Digraph: KR
Type: republic
Capital: Tarawa
Administrative divisions: 3 units; Gilbert Islands, Line Islands, Phoenix Islands
note: a new administrative structure of 6 districts (Banaba, Central Gilberts, Line Islands, Northern Gilberts, Southern Gilberts, Tarawa) may have been changed to 21 island councils (one for each of the inhabited islands) named Abaiang, Abemama, Aranuka, Arorae, Banaba, Beru, Butaritari, Canton, Kiritimati, Kuria, Maiana, Makin, Marakei, Nikunau, Nonouti, Onotoa, Tabiteuea, Tabuaeran, Tamana, Tarawa, Teraina
Independence: 12 July 1979 (from UK)
Constitution: 12 July 1979
Legal system: NA
National holiday: Independence Day, 12 July (1979)
Political parties and leaders: National Progressive Party, Teatao TEANNAKI; Christian Democratic Party, Teburoro TITO; New Movement Party, leader NA; Liberal Party, Tewareka TENTOA; Maneaba Party, Roniti TEIWAKI
note: there is no tradition of formally organized political parties in Kiribati; they more closely resemble factions or interest groups because they have no party

headquarters, formal platforms, or party structures
Suffrage: 18 years of age; universal
Elections:
President: last held on 8 July 1991 (next to be held by NA 1996); results—Teatao TEANNAKI 52%, Roniti TEIWAKI 28%
House of Assembly: last held on 8 May 1991 (next to be held by NA 1996); results—percent of vote by party NA; seats—(40 total; 39 elected) percent of seats by party NA
Executive branch: president (Beretitenti), vice president (Kauoman-ni-Beretitenti), Cabinet
Legislative branch: unicameral House of Assembly (Maneaba Ni Maungatabu)
Judicial branch: Court of Appeal, High Court
Leaders:
Chief of State and Head of Government: President Teatao TEANNAKI (since 8 July 1991); Vice President Taomati IUTA (since 8 July 1991)
Member of: ACP, AsDB, C, ESCAP, IBRD, ICAO, ICFTU, IDA, IFC, IMF, INTERPOL, ITU, SPARTECA, SPC, SPF, UNESCO, UPU, WHO
Diplomatic representation in US:
chief of mission: (vacant)
US diplomatic representation: the ambassador to Fiji is accredited to Kiribati
Flag: the upper half is red with a yellow frigate bird flying over a yellow rising sun, and the lower half is blue with three horizontal wavy white stripes to represent the ocean

Economy

Overview: The country has few national resources. Commercially viable phosphate deposits were exhausted at the time of independence in 1979. Copra and fish now represent the bulk of production and exports. The economy has fluctuated widely in recent years. Real GDP declined about 8% in 1987, as the fish catch fell sharply to only one-fourth the level of 1986 and copra production was hampered by repeated rains. Output rebounded strongly in 1988, with real GDP growing by 17%. The upturn in economic growth came from an increase in copra production and a good fish catch. Following the strong surge in output in 1988, GNP increased 1% in both 1989 and 1990.
National product: GDP—exchange rate conversion—$36.8 million (1990 est.)
National product real growth rate: 1% (1990 est.)
National product per capita: $525 (1990 est.)
Inflation rate (consumer prices): 4.8% (1991 est.)

Unemployment rate: NA%
Budget: revenues $29.9 million; expenditures $16.3 million, including capital expenditures of $14.0 million (1990 est.)
Exports: $5.8 million (f.o.b., 1990 est.)
commodities: copra 18%, fish 17%, seaweed 13%
partners: EC 50%, Fiji 22%, US 18% (1990)
Imports: $26.7 million (c.i.f., 1990 est.)
commodities: foodstuffs, machinery and equipment, miscellaneous manufactured goods, fuel
partners: Australia 33%, Japan 24%, Fiji 19%, NZ 6%, US 6% (1990)
External debt: $2.0 million (December 1989 est.)
Industrial production: growth rate 0% (1988 est.); accounts for less than 4% of GDP
Electricity: 5,000 kW capacity; 13 million kWh produced, 190 kWh per capita (1990)
Industries: fishing, handicrafts
Agriculture: accounts for 15% of GDP (including fishing); copra and fish contribute about 95% to exports; subsistence farming predominates; food crops—taro, breadfruit, sweet potatoes, vegetables; not self-sufficient in food
Economic aid: Western (non-US) countries, ODA and OOF bilateral commitments (1970-89), $273 million
Currency: 1 Australian dollar ($A) = 100 cents
Exchange rates: Australian dollars ($A) per US$1—1.4837 (January 1993), 1.3600 (1992), 1.2835 (1991), 1.2799 (1990), 1.2618 (1989), 1.2752 (1988)
Fiscal year: NA

Communications

Highways: 640 km of motorable roads
Inland waterways: small network of canals, totaling 5 km, in Line Islands
Ports: Banaba and Betio (Tarawa)
Airports:
total: 21
useable: 20
with permanent-surface runways: 4
with runways over 3,659 m : 0
with runways 2,440-3,659 m: 0
with runways 1,220-2,439 m: 5
Telecommunications: 1,400 telephones; broadcast stations—1 AM, no FM, no TV; 1 Pacific Ocean INTELSAT earth station

Defense Forces

Branches: Police Force (carries out law enforcement functions and paramilitary duties; there are small police posts on all islands); no military force is maintained
Manpower availability: NA
Defense expenditures: $NA, NA% of GDP

Korea, North

150 km

Boundary representation is not necessarily authoritative.

Najin
Ch'ŏngjin
Hyesan
Kanggye
Kimch'aek
Sinŭiju
Hamhŭng
Sunch'ŏn
Sea of Japan
P'YŎNGYANG
Wŏnsan
Namp'o
Sariwŏn
Yellow Sea
Kaesŏng P'anmunjŏm

Geography

Location: Northeast Asia, between China and South Korea
Map references: Asia, Standard Time Zones of the World
Area:
total area: 120,540 km²
land area: 120,410 km²
comparative area: slightly smaller than Mississippi
Land boundaries: total 1,673 km, China 1,416 km, South Korea 238 km, Russia 19 km
Coastline: 2,495 km
Maritime claims:
territorial sea: 12 nm
exclusive economic zone: 200 nm
military boundary line: 50 nm in the Sea of Japan and the exclusive economic zone limit in the Yellow Sea where all foreign vessels and aircraft without permission are banned
International disputes: short section of boundary with China is indefinite; Demarcation Line with South Korea
Climate: temperate with rainfall concentrated in summer
Terrain: mostly hills and mountains separated by deep, narrow valleys; coastal plains wide in west, discontinuous in east
Natural resources: coal, lead, tungsten, zinc, graphite, magnesite, iron ore, copper, gold, pyrites, salt, fluorspar, hydropower
Land use:
arable land: 18%
permanent crops: 1%
meadows and pastures: 0%
forest and woodland: 74%
other: 7%
Irrigated land: 14,000 km² (1989)
Environment: mountainous interior is isolated, nearly inaccessible, and sparsely populated; late spring droughts often followed by severe flooding
Note: strategic location bordering China, South Korea, and Russia

People

Population: 22,645,811 (July 1993 est.)
Population growth rate: 1.86% (1993 est.)
Birth rate: 24.09 births/1,000 population (1993 est.)
Death rate: 5.52 deaths/1,000 population (1993 est.)
Net migration rate: 0 migrant(s)/1,000 population (1993 est.)
Infant mortality rate: 28.6 deaths/1,000 live births (1993 est.)
Life expectancy at birth:
total population: 69.51 years
male: 66.42 years
female: 72.75 years (1993 est.)
Total fertility rate: 2.4 children born/woman (1993 est.)
Nationality:
noun: Korean(s)
adjective: Korean
Ethnic divisions: racially homogeneous
Religions: Buddhism and Confucianism, some Christianity and syncretic Chondogyo
note: autonomous religious activities now almost nonexistent; government-sponsored religious groups exist to provide illusion of religious freedom
Languages: Korean
Literacy: age 15 and over can read and write (1990)
total population: 99%
male: 99%
female: 99%
Labor force: 9.615 million
by occupation: agricultural 36%, nonagricultural 64%
note: shortage of skilled and unskilled labor (mid-1987 est.)

Government

Names:
conventional long form: Democratic People's Republic of Korea
conventional short form: North Korea
local long form: Choson-minjujuui-inmin-konghwaguk
local short form: none
Abbreviation: DPRK
Digraph: KN
Type: Communist state; Stalinist dictatorship
Capital: P'yongyang
Administrative divisions: 9 provinces (do, singular and plural) and 3 special cities* (jikhalsi, singular and plural); Chagang-do (Chagang Province), Hamgyong-bukto (North Hamgyong Province), Hamgyong-namdo (South Hamgyong Province), Hwanghae-bukto (North Hwanghae Province), Hwanghae-namdo (South Hwanghae Province), Kaesong-si* (Kaesong City), Kangwon-do (Kangwon Province), Namp'o-si* (Namp'o City), P'yongan-bukto (North P'yongan Province), P'yongan-namdo (South P'yongan Province), P'yongyang-si* (P'yongyang City), Yanggang-do (Yanggang Province)
Independence: 9 September 1948
note: 15 August 1945, date of independence from the Japanese and celebrated in North Korea as National Liberation Day
Constitution: adopted 1948, completely revised 27 December 1972, revised again in April 1992
Legal system: based on German civil law system with Japanese influences and Communist legal theory; no judicial review of legislative acts; has not accepted compulsory ICJ jurisdiction
National holiday: DPRK Foundation Day, 9 September (1948)
Political parties and leaders: major party—Korean Workers' Party (KWP), KIM Il-song, general secretary, and his son, KIM Chong-il, secretary, Central Committee; Korean Social Democratic Party, KIM Yong-ho, vice-chairman; Chondoist Chongu Party, CHONG Sin-hyok, chairman
Suffrage: 17 years of age; universal
Elections:
President: last held 24 May 1990 (next to be held by NA 1994); results—President KIM Il-song was reelected without opposition
Supreme People's Assembly: last held on 7-9 April 1993 (next to be held NA); results—percent of vote by party NA; seats—(687 total) the KWP approves a single list of candidates who are elected without opposition; minor parties hold a few seats
Executive branch: president, two vice presidents, premier, ten vice premiers, State Administration Council (cabinet)
Legislative branch: unicameral Supreme People's Assembly (Ch'oego Inmin Hoeui)
Judicial branch: Central Court
Leaders:
Chief of State: President KIM Il-song (national leader since 1948, president since 28 December 1972); designated successor KIM Chong-il (son of president, born 16 February 1942)
Head of Government: Premier KANG Song-san (since December 1992)
Member of: ESCAP, FAO, G-77, IAEA, ICAO, IFAD, IMF (observer), IMO, IOC, ISO, ITU, LORCS, NAM, UN, UNCTAD, UNESCO, UNIDO, UPU, WFTU, WHO, WIPO, WMO, WTO
Diplomatic representation in US: none
US diplomatic representation: none
Flag: three horizontal bands of blue (top), red (triple width), and blue; the red band is edged in white; on the hoist side of the red band is a white disk with a red five-pointed star

Economy

Overview: More than 90% of this command economy is socialized; agricultural land is collectivized; and state-owned industry produces 95% of manufactured goods. State control of economic affairs is unusually tight even for a Communist country because of the small size and homogeneity of the society and the strict rule of KIM Il-song and his son, KIM Chong-il. Economic growth during the period 1984-88 averaged 2-3%, but output declined by 3-5% annually during 1989-92 because of systemic problems and disruptions in socialist-style economic relations with the former USSR and China. In 1992, output dropped sharply, by perhaps 10-15%, as the economy felt the cumulative effect of the reduction in outside support. The leadership insisted in maintaining its high level of military outlays from a shrinking economic pie. Moreover, a serious drawdown in inventories and critical shortages in the energy sector have led to increasing interruptions in industrial production. Abundant mineral resources and hydropower have formed the basis of industrial development since WWII. Output of the extractive industries includes coal, iron ore, magnesite, graphite, copper, zinc, lead, and precious metals. Manufacturing is centered on heavy industry, including military industry, with light industry lagging far behind. Despite the use of improved seed varieties, expansion of irrigation, and the heavy use of fertilizers, North Korea has not yet become self-sufficient in food production. Five consecutive years of poor harvests, coupled with distribution problems, have led to chronic food shortages. North Korea remains far behind South Korea in economic development and living standards.
National product: GNP—purchasing power equivalent—$22 billion (1992 est.)
National product real growth rate: -10% to −15% (1992 est.)
National product per capita: $1,000 (1992 est.)
Inflation rate (consumer prices): NA%
Unemployment rate: NA%
Budget: revenues $18.5 billion; expenditures $18.4 billion, including capital expenditures of $NA (1992)
Exports: $1.3 billion (f.o.b., 1992 est.)
commodities: minerals, metallurgical products, agricultural and fishery products, manufactures (including armaments)
partners: China, Japan, Russia, South Korea, Germany, Hong Kong, Mexico
Imports: $1.9 billion (f.o.b., 1992 est.)
commodities: petroleum, grain, coking coal, machinery and equipment, consumer goods
partners: China, Russia, Japan, Hong Kong, Germany, Singapore
External debt: $8 billion (1992 est.)

Industrial production: growth rate −15% (1992 est.)
Electricity: 7,300,000 kW capacity; 26,000 million kWh produced, 1,160 kWh per capita (1992)
Industries: machine building, military products, electric power, chemicals, mining, metallurgy, textiles, food processing
Agriculture: accounts for about 25% of GNP and 36% of work force; principal crops—rice, corn, potatoes, soybeans, pulses; livestock and livestock products—cattle, hogs, pork, eggs; not self-sufficient in grain; fish catch estimated at 1.7 million metric tons in 1987
Economic aid: Communist countries, $1.4 billion a year in the 1980s
Currency: 1 North Korean won (Wn) = 100 chon
Exchange rates: North Korean won (Wn) per US$1—2.13 (May 1992), 2.14 (September 1991), 2.1 (January 1990), 2.3 (December 1989), 2.13 (December 1988), 0.94 (March 1987)
Fiscal year: calendar year

Communications

Railroads: 4,915 km total; 4,250 km 1.435-meter standard gauge, 665 km 0.762-meter narrow gauge; 159 km double track; 3,084 km electrified; government owned (1989)
Highways: about 30,000 km (1991); 92.5% gravel, crushed stone, or earth surface; 7.5% paved
Inland waterways: 2,253 km; mostly navigable by small craft only
Pipelines: crude oil 37 km
Ports: primary—Ch'ongjin, Hungnam (Hamhung), Najin, Namp'o, Wonsan; secondary—Haeju, Kimchaek, Kosong, Sinuiju, Songnim, Sonbong (formerly Unggi), Ungsang
Merchant marine: 80 ships (1,000 GRT and over) totaling 675,666 GRT/1,057,815 DWT; includes 1 passenger, 1 short-sea passenger, 2 passenger-cargo, 67 cargo, 2 oil tanker, 5 bulk, 1 combination bulk, 1 container
Airports:
total: 55
usable : 55 (est.)
with permanent-surface runways: about 30
with runways over 3,659 m: fewer than 5
with runways 2,440-3,659 m: 20
with runways 1,220-2,439 m: 30
Telecommunications: broadcast stations—18 AM, no FM, 11 TV; 300,000 TV sets (1989); 3,500,000 radio receivers; 1 Indian Ocean INTELSAT earth station

Defense Forces

Branches: Korean People's Army (including the Army, Navy, Air Force), Civil Security Forces

Manpower availability: males age 15-49 6,567,684; fit for military service 3,996,893; reach military age (18) annually 208,132 (1993 est.)
Defense expenditures: exchange rate conversion—about $5 billion, 20-25% of GNP (1991 est.); note—the officially announced but suspect figure is $1.9 billion (1991), 8% of GNP (1991 est.)

Korea, South

150 km

SEOUL
Inch'ŏn
Wŏnju
Kangnŭng
Ullŭng-do
Yellow
Sea
Taejŏn
Taegu
Sea of
Japan
Ulsan
Kwangju
Pusan
Korea
Strait
Cheju-do

Boundary representation is
not necessarily authoritative.

Geography

Location: Northeast Asia, between North
Korea and Japan
Map references: Asia, Standard Time Zones
of the World
Area:
total area: 98,480 km²
land area: 98,190 km²
comparative area: slightly larger than
Indiana
Land boundaries: total 238 km, North
Korea 238 km
Coastline: 2,413 km
Maritime claims:
continental shelf: not specified
territorial sea: 12 nm; 3 nm in the Korea
Strait
International disputes: Demarcation Line
with North Korea; Liancourt Rocks claimed
by Japan
Climate: temperate, with rainfall heavier in
summer than winter
Terrain: mostly hills and mountains; wide
coastal plains in west and south
Natural resources: coal, tungsten, graphite,
molybdenum, lead, hydropower
Land use:
arable land: 21%
permanent crops: 1%
meadows and pastures: 1%
forest and woodland: 67%
other: 10%
Irrigated land: 13,530 km² (1989)
Environment: occasional typhoons bring
high winds and floods; earthquakes in
southwest; air pollution in large cities

People

Population: 44,613,993 (July 1993 est.)
Population growth rate: 1.05% (1993 est.)
Birth rate: 15.72 births/1,000 population
(1993 est.)
Death rate: 6.16 deaths/1,000 population
(1993 est.)
Net migration rate: 0.91 migrant(s)/1,000
population (1993 est.)

Infant mortality rate: 22.5 deaths/1,000
live births (1993 est.)
Life expectancy at birth:
total population: 70.29 years
male: 67.1 years
female: 73.68 years (1993 est.)
Total fertility rate: 1.64 children
born/woman (1993 est.)
Nationality:
noun: Korean(s)
adjective: Korean
Ethnic divisions: homogeneous (except for
about 20,000 Chinese)
Religions: Christianity 48.6%, Buddhism
47.4%, Confucianism 3%, pervasive folk
religion (Shamanism), Chondogyo (religion
of the heavenly way) 0.2%
Languages: Korean, English widely taught
in high school
Literacy: age 15 and over can read and
write (1990)
total population: 96%
male: 99%
female: 94%
Labor force: 19 million
by occupation: services and other 52%,
mining and manufacturing 27%, agriculture,
fishing, forestry 21% (1991)

Government

Names:
conventional long form: Republic of Korea
conventional short form: South Korea
local long form: Taehan-min'guk
local short form: none
Abbreviation: ROK
Digraph: KS
Type: republic
Capital: Seoul
Administrative divisions: 9 provinces (do,
singular and plural) and 6 special cities*
(jikhalsi, singular and plural); Cheju-do,
Cholla-bukto, Cholla-namdo,
Ch'ungch'ong-bukto, Ch'ungch'ong-namdo,
Inch'on-jikhalsi*, Kangwon-do,
Kwangju-jikhalsi*, Kyonggi-do,
Kyongsang-bukto, Kyongsang-namdo,
Pusan-jikhalsi*, Soul-t'ukpyolsi*,
Taegu-jikhalsi*, Taejon-jikhalsi*
Independence: 15 August 1948
Constitution: 25 February 1988
Legal system: combines elements of
continental European civil law systems,
Anglo-American law, and Chinese classical
thought
National holiday: Independence Day, 15
August (1948)
Political parties and leaders:
majority party: Democratic Liberal Party
(DLP), KIM Young Sam, president
opposition: Democratic Party (DP), LEE Ki
Taek, executive chairman; United People's
Party (UPP), CHUNG Ju Yung, chairman;
several smaller parties

note: the DLP resulted from a merger of the
Democratic Justice Party (DJP),
Reunification Democratic Party (RDP), and
New Democratic Republican Party (NDRP)
on 9 February 1990
Other political or pressure groups: Korean
National Council of Churches; National
Democratic Alliance of Korea; National
Federation of Student Associations; National
Federation of Farmers' Associations;
National Council of Labor Unions;
Federation of Korean Trade Unions; Korean
Veterans' Association; Federation of Korean
Industries; Korean Traders Association
Suffrage: 20 years of age; universal
Elections:
President: last held on 18 December 1992
(next to be held NA December 1997);
results—KIM Young Sam (DLP) 41.9%,
KIM Dae Jung (DP) 33.8%, CHUNG Ju
Yung (UPP) 16.3%, other 8%
National Assembly: last held on 24 March
1992; results—DLP 38.5%, DP 29.2%,
Unification National Party (UNP) 17.3%
(name later changed to UPP), other 15%;
seats—(299 total) DLP 149, DP 97, UNP
31, other 22; the distribution of seats as of
May 1993 was DLP 167, DP 95, UPP 14,
other 23
note: the change in the distribution of seats
reflects the fluidity of the current situation
where party members are constantly
switching from one party to another
Executive branch: president, prime
minister, two deputy prime ministers, State
Council (cabinet)
Legislative branch: unicameral National
Assembly (Kuk Hoe)
Judicial branch: Supreme Court
Leaders:
Chief of State: President KIM Young Sam
(since 25 February 1993)
Head of Government: Prime Minister
HWANG In Sung (since 25 February 1993);
Deputy Prime Minister LEE Kyung Shick
(since 25 February 1993) and Deputy Prime
Minister HAN Wan Sang (since 25 February
1993)
Member of: AfDB, APEC, AsDB, CCC,
COCOM (cooperating country), CP, EBRD,
ESCAP, FAO, G-77, GATT, IAEA, IBRD,
ICAO, ICC, ICFTU, IDA, IFAD, IFC, ILO,
IMF, IMO, INMARSAT, INTELSAT,
INTERPOL, IOC, IOM, ISO, ITU, LORCS,
OAS (observer), UN, UNCTAD, UNESCO,
UNIDO, UPU, WHO, WIPO, WMO, WTO
Diplomatic representation in US:
chief of mission: Ambassador HAN Seung
Soo
chancery: 2370 Massachusetts Avenue NW,
Washington, DC 20008
telephone: (202) 939-5600
consulates general: Agana (Guam),
Anchorage, Atlanta, Chicago, Honolulu,
Houston, Los Angeles, New York, San
Francisco, and Seattle

Kuwait

US diplomatic representation:
chief of mission: (vacant), Charge d'Affaires
Raymond BURGHARDT
embassy: 82 Sejong-Ro, Chongro-ku, Seoul,
AMEMB, Unit 15550
mailing address: APO AP 96205-0001
telephone: [82] (2) 732-2601 through 2618
FAX: [82] (2) 738-8845
consulate: Pusan
Flag: white with a red (top) and blue
yin-yang symbol in the center; there is a
different black trigram from the ancient I
Ching (Book of Changes) in each corner of
the white field

Economy

Overview: The driving force behind the
economy's dynamic growth has been the
planned development of an export-oriented
economy in a vigorously entrepreneurial
society. Real GNP increased more than 10%
annually between 1986 and 1991. This
growth ultimately led to an overheated
situation characterized by a tight labor
market, strong inflationary pressures, and a
rapidly rising current account deficit. As a
result, in 1992, focusing attention on
slowing the growth rate of inflation and
reducing the deficit is leading to a
slow-down in growth. The economy remains
the envy of the great majority of the world's
peoples.
National product: GNP—purchasing power
equivalent—$287 billion (1992 est.)
National product real growth rate: 5%
(1992 est.)
National product per capita: $6,500 (1992
est.)
Inflation rate (consumer prices): 4.5%
(1992 est.)
Unemployment rate: 2.4% (1992 est.)
Budget: revenues $48.4 billion; expenditures
$48.4 billion, including capital expenditures
of $NA (1993)
Exports: $76.8 billion (f.o.b., 1992)
commodities: textiles, clothing, electronic
and electrical equipment, footwear,
machinery, steel, automobiles, ships, fish
partners: US 24%, Japan 15% (1992)
Imports: $81.7 billion (c.i.f., 1992)
commodities: machinery, electronics and
electronic equipment, oil, steel, transport
equipment, textiles, organic chemicals,
grains
partners: Japan 24%, US 22% (1992)
External debt: $42.0 billion (1992)
Industrial production: growth rate 5.0%
(1992 est.); accounts for about 45% of GNP
Electricity: 24,000,000 kW capacity;
105,000 million kWh produced, 2,380 kWh
per capita (1992)
Industries: textiles, clothing, footwear, food
processing, chemicals, steel, electronics,
automobile production, shipbuilding
Agriculture: accounts for 8% of GNP and

employs 21% of work force (including
fishing and forestry); principal crops—rice,
root crops, barley, vegetables, fruit; livestock
and livestock products—cattle, hogs,
chickens, milk, eggs; self-sufficient in food,
except for wheat; fish catch of 2.9 million
metric tons, seventh-largest in world
Economic aid: US commitments, including
Ex-Im (FY70-89), $3.9 billion; non-US
countries (1970-89), $3.0 billion
Currency: 1 South Korean won (W) = 100
chon (theoretical)
Exchange rates: South Korean won (W) per
US$1—791.99 (January 1993), 780.65
(1992), 733.35 (1991), 707.76 (1990),
671.46 (1989), 731.47 (1988)
Fiscal year: calendar year

Communications

Railroads: 3,091 km total (1991); 3,044 km
1.435 meter standard gauge, 47 km
0.610-meter narrow gauge, 847 km double
track; 525 km electrified, government owned
Highways: 63,201 km total (1991); 1,551
expressways, 12,190 km national highway,
49,460 km provincial and local roads
Inland waterways: 1,609 km; use restricted
to small native craft
Pipelines: petroleum products 455 km
Ports: Pusan, Inchon, Kunsan, Mokpo,
Ulsan
Merchant marine: 431 ships (1,000 GRT or
over) totaling 6,689,227 GRT/11,016,014
DWT; includes 2 short-sea passenger, 138
cargo, 61 container, 11 refrigerated cargo, 9
vehicle carrier, 45 oil tanker, 12 chemical
tanker, 13 liquefied gas, 2 combination
ore/oil, 135 bulk, 2 combination bulk, 1
multifunction large-load carrier
Airports:
total: 103
usable: 93
with permanent-surface runways: 59
with runways over 3,659 m: 0
with runways 2,440-3,659 m: 22
with runways 1,220-2,439 m: 18
Telecommunications: excellent domestic
and international services; 13,276,449
telephone subscribers; broadcast stations—79
AM, 46 FM, 256 TV (57 of 1 kW or
greater); satellite earth stations—2 Pacific
Ocean INTELSAT and 1 Indian Ocean
INTELSAT

Defense Forces

Branches: Army, Navy, Marine Corps, Air
Force
Manpower availability: males age 15-49
13,286,969; fit for military service
8,542,640; reach military age (18) annually
432,434 (1993 est.)
Defense expenditures: exchange rate
conversion—$12.2 billion, 3.6% of GNP
(1993 est.)

Geography

Location: Middle East, at the head of the
Persian Gulf, between Iraq and Saudi Arabia
Map references: Africa, Middle East,
Standard Time Zones of the World
Area:
total area: 17,820 km²
land area: 17,820 km²
comparative area: slightly smaller than New
Jersey
Land boundaries: total 464 km, Iraq 242
km, Saudi Arabia 222 km
Coastline: 499 km
Maritime claims:
continental shelf: not specified
territorial sea: 12 nm
International disputes: in April 1991 Iraq
officially accepted UN Security Council
Resolution 687, which demands that Iraq
accept the inviolability of the boundary set
forth in its 1963 agreement with Kuwait,
ending earlier claims to Bubiyan and Warbah
Islands, or to all of Kuwait; the 20 May
1993 final report of the UN Iraq/Kuwait
Boundary Demarcation Commission was
welcomed by the Security Council in
Resolution 833 of 27 May 1993, which also
reaffirmed that the decisions of the
commission on the boundary were final,
bringing to a completion the official
demarcation of the Iraq-Kuwait boundary;
Iraqi officials still make public statements
claiming Kuwait; ownership of Qaruh and
Umm al Maradim Islands disputed by Saudi
Arabia
Climate: dry desert; intensely hot summers;
short, cool winters
Terrain: flat to slightly undulating desert
plain
Natural resources: petroleum, fish, shrimp,
natural gas
Land use:
arable land: 0%
permanent crops: 0%
meadows and pastures: 8%
forest and woodland: 0%
other: 92%

Irrigated land: 20 km² (1989 est.)
Environment: some of world's largest and most sophisticated desalination facilities provide most of water; air and water pollution; desertification
Note: strategic location at head of Persian Gulf

People

Population: 1,698,077 (July 1993 est.)
Population growth rate: 8.67% (1993 est.)
Birth rate: 30.29 births/1,000 population (1993 est.)
Death rate: 2.39 deaths/1,000 population (1993 est.)
Net migration rate: 58.74 migrant(s)/1,000 population (1993 est.)
Infant mortality rate: 13.1 deaths/1,000 live births (1993 est.)
Life expectancy at birth:
total population: 74.62 years
male: 72.47 years
female: 76.87 years (1993 est.)
Total fertility rate: 4.11 children born/woman (1993 est.)
Nationality:
noun: Kuwaiti(s)
adjective: Kuwaiti
Ethnic divisions: Kuwaiti 45%, other Arab 35%, South Asian 9%, Iranian 4%, other 7%
Religions: Muslim 85% (Shi'a 30%, Sunni 45%, other 10%), Christian, Hindu, Parsi, and other 15%
Languages: Arabic (official), English widely spoken
Literacy: age 15 and over can read and write (1990)
total population: 73%
male: 77%
female: 67%
Labor force: 566,000 (1986)
by occupation: services 45.0%, construction 20.0%, trade 12.0%, manufacturing 8.6%, finance and real estate 2.6%, agriculture 1.9%, power and water 1.7%, mining and quarrying 1.4%
note: 70% of labor force was non-Kuwaiti (1986)

Government

Names:
conventional long form: State of Kuwait
conventional short form: Kuwait
local long form: Dawlat al Kuwayt
local short form: Al Kuwayt
Digraph: KU
Type: nominal constitutional monarchy
Capital: Kuwait
Administrative divisions: 5 governorates (mu'hafaz'at, singular—muh'afaz'ah); Al Ah'madi, Al Jahrah, Al Kuwayt, 'Hawalli; Farwaniyah
Independence: 19 June 1961 (from UK)

Constitution: 16 November 1962 (some provisions suspended since 29 August 1962)
Legal system: civil law system with Islamic law significant in personal matters; has not accepted compulsory ICJ jurisdiction
National holiday: National Day, 25 February
Political parties and leaders: none
Other political or pressure groups: 40,000 Palestinian community; small, clandestine leftist and Shi'a fundamentalist groups are active; several groups critical of government policies are active
Suffrage: adult males who resided in Kuwait before 1920 and their male descendants at age 21
note: out of all citizens, only 10% are eligible to vote and only 5% actually vote
Elections:
National Assembly: dissolved 3 July 1986; new elections were held on 5 October 1992 with a second election in the 14th and 16th constituencies scheduled for 15 February 1993
Executive branch: amir, prime minister, deputy prime minister, Council of Ministers (cabinet)
Legislative branch: unicameral National Assembly (Majlis al 'umma) dissolved 3 July 1986; elections for new Assembly held 5 October 1992
Judicial branch: High Court of Appeal
Leaders:
Chief of State: Amir Shaykh JABIR al-Ahmad al-Jabir al-Sabah (since 31 December 1977)
Head of Government: Prime Minister and Crown Prince SA'D al-'Abdallah al-Salim al-Sabah (since 8 February 1978); Deputy Prime Minister SABAH al-Ahmad al-Jabir al-Sabah (since 17 October 1992)
Member of: ABEDA, AfDB, AFESD, AL, AMF, BDEAC, CAEU, ESCWA, FAO, G-77, GATT, GCC, IAEA, IBRD, ICAO, IDA, IDB, IFAD, IFC, ILO, IMF, IMO, INMARSAT, INTELSAT, INTERPOL, IOC, ISO (correspondent), ITU, LORCS, NAM, OAPEC, OIC, OPEC, UN, UNCTAD, UNESCO, UNIDO, UPU, WFTU, WHO, WMO, WTO
Diplomatic representation in US:
chief of mission: Ambassador Muhammad al-Sabah al-Salim al-SABAH
chancery: 2940 Tilden Street NW, Washington, DC 20008
telephone: (202) 966-0702
US diplomatic representation:
chief of mission: Ambassador Edward (Skip) GNEHM, Jr.
embassy: Bneid al-Gar (opposite the Kuwait International Hotel), Kuwait City
mailing address: P.O. Box 77 SAFAT, 13001 SAFAT, Kuwait; APO AE 09880
telephone: [965] 242-4151 through 4159
FAX: [956] 244-2855

Flag: three equal horizontal bands of green (top), white, and red with a black trapezoid based on the hoist side

Economy

Overview: Kuwait is a small and relatively open economy with proven crude oil reserves of about 94 billion barrels—10% of world reserves. Kuwait is rebuilding its war-ravaged petroleum sector and the increase in crude oil production to nearly 2.0 million barrels per day by the end of 1992 led to an enormous increase in GDP for the year. The government ran a cumulative fiscal deficit of approximately $70 billion over its last two fiscal years, reducing its foreign asset position and increasing its public debt to roughly $40 billion. Petroleum accounts for nearly half of GDP and over 90% of export and government revenue.
National product: GDP—exchange rate conversion—$15.3 billion (1992 est.)
National product real growth rate: 80% (1992 est.)
National product per capita: $11,100 (1992 est.)
Inflation rate (consumer prices): 5% (1992 est.)
Unemployment rate: NEGL% (1992 est.)
Budget: revenues $7.1 billion; expenditures $10.5 billion, including capital expenditures of $3.1 billion (FY88)
Exports: $750 million (f.o.b., 1991 est.)
commodities: oil
partners: France 16%, Italy 15%, Japan 12%, UK 11%
Imports: $4.7 billion (f.o.b., 1991 est.)
commodities: food, construction materials, vehicles and parts, clothing
partners: US 35%, Japan 12%, UK 9%, Canada 9%
External debt: $7.2 billion (December 1989 est.)
note: external debt has grown substantially in 1991 and 1992 to pay for restoration of war damage
Industrial production: growth rate NA%; accounts for NA% of GDP
Electricity: 6,873,000 kW available out of 7,398,000 kW capacity due to Persian Gulf war; 12,264 million kWh produced, 8,890 kWh per capita (1992)
Industries: petroleum, petrochemicals, desalination, food processing, building materials, salt, construction
Agriculture: practically none; dependent on imports for food; about 75% of potable water must be distilled or imported
Economic aid: donor—pledged $18.3 billion in bilateral aid to less developed countries (1979-89)
Currency: 1 Kuwaiti dinar (KD) = 1,000 fils

Kuwait (continued)

US$1—0.3044 (January 1993), 0.2934 (1992), 0.2843 (1991), 0.2915 (1990), 0.2937 (1989), 0.2790 (1988)
Fiscal year: 1 July—30 June

Communications

Railroads: none
Highways: 3,900 km total; 3,000 km bituminous; 900 km earth, sand, light gravel
Pipelines: crude oil 877 km; petroleum products 40 km; natural gas 165 km
Ports: Ash Shu'aybah, Ash Shuwaykh, Mina' al 'Ahmadi
Merchant marine: 42 ships (1,000 GRT or over), totaling 1,996,052 GRT/3,373,088 DWT; includes 7 cargo, 4 livestock carrier, 24 oil tanker, 4 liquefied gas, 3 container
Airports:
total: 7
usable: 4
with permanent-surface runways: 4
with runways over 3,659 m: 0
with runways 2,440-3,659 m: 4
with runways 1,220-2,439 m: 0
Telecommunications: civil network suffered extensive damage as a result of Desert Storm and reconstruction is still under way with some restored international and domestic capabilities; broadcast stations—3 AM, 0 FM, 3 TV; satellite earth stations—destroyed during Persian Gulf War and not rebuilt yet; temporary mobile satellite ground stations provide international telecommunications; coaxial cable and microwave radio relay to Saudi Arabia; service to Iraq is nonoperational

Defense Forces

Branches: Army, Navy, Air Force, National Police Force, National Guard
Manpower availability: males age 15-49 498,254; fit for military service 298,865; reach military age (18) annually 14,459 (1993 est.)
Defense expenditures: exchange rate conversion—$2.5 billion, 7.3% of GDP (FY92/93)

Kyrgyzstan

150 km

Geography

Location: South Asia, between China and Kazakhstan
Map references: Asia, Commonwealth of Independent States—Central Asian States, Standard Time Zones of the World
Area:
total area: 198,500 km²
land area: 191,300 km²
comparative area: slightly smaller than South Dakota
Land boundaries: total 3,878 km, China 858 km, Kazakhstan 1,051 km, Tajikistan 870 km, Uzbekistan 1,099 km
Coastline: 0 km (landlocked)
Maritime claims: none; landlocked
International disputes: territorial dispute with Tajikistan on southern boundary in Isfara Valley area
Climate: dry continental to polar in high Tien Shan; subtropical in south (Fergana Valley)
Terrain: peaks of Tien Shan rise to 7,000 meters, and associated valleys and basins encompass entire nation
Natural resources: small amounts of coal, natural gas, oil, nepheline, rare earth metals, mercury, bismuth, gold, lead, zinc, hydroelectric power
Land use:
arable land: NA%
permanent crops: NA%
meadows and pastures: NA%
forest and woodland: NA%
other: NA%
Irrigated land: 10,320 km² (1990)
Environment: NA
Note: landlocked

People

Population: 4,625,954 (July 1993 est.)
Population growth rate: 1.56% (1993 est.)
Birth rate: 26.69 births/1,000 population (1993 est.)
Death rate: 7.45 deaths/1,000 population (1993 est.)

Net migration rate: -3.62 migrant(s)/1,000 population (1993 est.)
Infant mortality rate: 47.8 deaths/1,000 live births (1993 est.)
Life expectancy at birth:
total population: 67.71 years
male: 63.47 years
female: 72.15 years (1993 est.)
Total fertility rate: 3.39 children born/woman (1993 est.)
Nationality:
noun: Kirghiz(s)
adjective: Kirghiz
Ethnic divisions: Kirghiz 52.4%, Russian 21.5%, Uzbek 12.9%, Ukrainian 2.5%, German 2.4%, other 8.3%
Religions: Muslim 70%, Russian Orthodox NA%
Languages: Kirghiz (Kyrgyz)—official language, Russian
Literacy: age 9-49 can read and write (1970)
total population: 100%
male: 100%
female: 100%
Labor force: 1.748 million
by occupation: agriculture and forestry 33%, industry and construction 28%, other 39% (1990)

Government

Names:
conventional long form: Kyrgyz Republic
conventional short form: Kyrgyzstan
local long form: Kyrgyzstan Respublikasy
local short form: none
former: Kirghiz Soviet Socialist Republic
Digraph: KG
Type: republic
Capital: Bishkek (Frunze)
Administrative divisions: 6 oblasts (oblastey, singular—oblast'); Chu, Jalal-Abad, Ysyk-Kul', Naryn, Osh, Talas
Independence: 31 August 1991 (from Soviet Union)
Constitution: adopted 5 May 1993
Legal system: based on civil law system
National holiday: National Day, 2 December
Political parties and leaders: Kyrgyz Democratic Movement, Kazat AKMAKOV, chairman; Civic Accord, Coalition representing nonnative minority groups; National Revived Asaba (Banner) Party, Asan ORMUSHEV, chairman; Communist Party was banned but has registered as political party 18 September 1992
Other political or pressure groups: National Unity Democratic Movement; Peasant Party; Council of Free Trade Unions; Union of Entrepreneurs
Suffrage: 18 years of age; universal

Elections:
President: last held 12 October 1991 (next to be held NA 1996); results—Askar AKAYEV won in uncontested election with 95% of vote with 90% of electorate voting; note—president elected by Supreme Soviet 28 October 1990, then by popular vote 12 October 1991
Zhogorku Keneshom: last held 25 February 1990 for the Supreme Soviet (next to be held no later than NA November 1994 for the Zhgorku Keneshom); results—Commnunists 90%; seats—(350 total) Communists 310
Executive branch: president, Cabinet of Ministers, prime minister
Legislative branch: unicameral Zhogorku Keneshom
Judicial branch: Supreme Court
Leaders:
Chief of State: President Askar AKAYEV (since 28 October 1990); Vice President Feliks KULOV (since 12 October 1992)
Head of Government: Prime Minister Tursenbek CHYNGYSHEV (since 2 March 1992); Deputy Prime Minister Abdygani ERKEBAYEV; Supreme Soviet Chairman Medetkan SHERIMKULOV (since NA)
Member of: CIS, CSCE, EBRD, ECO, ESCAP, IBRD, IDA, ILO, IMF, NACC, PCA, UN, UNCTAD, UNESCO, WHO
Diplomatic representation in US:
chief of mission: Ambassador Roza OTUNBAYEVA
chancery: 1511 K Street, NW, Washington, DC
telephone: (202) 347-5029
US diplomatic representation:
chief of mission: Ambassador Edward HURWITZ
embassy: (temporary) Erkindik Prospekt #66, Bishkek
mailing address: APO AE 09721
telephone: 7-3312 22-26-93, 22-35-51, 22-29-20
FAX: 7-3312 22-35-51
Flag: red field with a yellow sun in the center having 40 rays representing the 40 Krygyz tribes; on the obverse side the rays run counterclockwise, on the reverse, clockwise; in the center of the sun is a red ring crossed by two sets of three lines, a stylized representation of the roof of the traditional Kyrgyz yurt

Economy

Overview: Kyrgyzstan's small economy (less than 1% of the total for the former Soviet Union) is oriented toward agriculture, producing mainly livestock such as goats and sheep, as well as cotton, grain, and tobacco. Industry, concentrated around Bishkek, produces small quantities of electric motors, livestock feeding equipment, washing machines, furniture, cement, paper, and bricks. Mineral extraction is small, the most important minerals being coal, rare earth metals and gold. Kyrgyzstan is a net importer of many types of food and fuel but is a net exporter of electricity. In 1992, the Kirghiz leadership made progress on reform, primarily by privatizing business, granting life-long tenure to farmers, and freeing most prices. Nonetheless, in 1992 overall industrial and livestock output declined because of acute fuel shortages and a widespread lack of spare parts.
National product: GDP $NA
National product real growth rate: -25% (1992 est.)
National product per capita: $NA
Inflation rate (consumer prices): 29% per month (first quarter 1993)
Unemployment rate: 0.1% includes officially registered unemployed; also large numbers of underemployed workers
Budget: revenues $NA; expenditures $NA, including capital expenditures of $NA
Exports: $NA
commodities: wool, chemicals, cotton, ferrous and nonferrous metals, shoes, machinery, tobacco
partners: Russia 70%, Ukraine, Uzbekistan, Kazakhstan, and others
Imports: $NA
commodities: lumber, industrial products, ferrous metals, fuel, machinery, textiles, footwear
partners: other CIS republics
External debt: $650 million (1991)
Industrial production: growth rate NA% (1992)
Electricity: 4,100,000 kW capacity; 11,800 million kWh produced, 2,551 kWh per capita (1992)
Industries: small machinery, textiles, food-processing industries, cement, shoes, sawn logs, refrigerators, furniture, electric motors, gold, and rare earth metals
Agriculture: wool, tobacco, cotton, livestock (sheep, goats, cattle), vegetables, meat, grapes, fruits and berries, eggs, milk, potatoes
Illicit drugs: illicit producer of cannabis and opium; mostly for CIS consumption; limited government eradication program; used as transshipment point for illicit drugs to Western Europel
Economic aid: $300 million official and commitments by foreign donors (1992)
Currency: introduced national currency, the som (10 May 1993)
Exchange rates: rubles per US$1—415 (24 December 1992) but subject to wide fluctuations
Fiscal year: calendar year

Communications

Railroads: 370 km; does not include industrial lines (1990)
Highways: 30,300 km total; 22,600 km paved or graveled, 7,700 km earth(1990)
Pipelines: natural gas 200 km
Ports: none; landlocked
Airports:
total: 52
useable: 27
with permanent-surface runways: 12
with runways over 3,659 m: 1
with runways 2,440-3,659 m: 4
with runways 1,220-2,439 m: 13
Telecommunications: poorly developed; 56 telephones per 1000 persons (December 1990); connections with other CIS countries by landline or microwave and with other countries by leased connections with Moscow international gateway switch; satellite earth stations—Orbita and INTELSAT (TV receive only); new intelsat earth station provide TV receive-only capability for Turkish broadcasts

Defense Forces

Branches: National Guard, Security Forces (internal and border troops), Civil Defense
Manpower availability: males age 15-49 1,093,694; fit for military service 890,961 (1993 est.)
Defense expenditures: $NA, NA% of GDP

Laos

Geography

Location: Southeast Asia, between Vietnam and Thailand
Map references: Southeast Asia, Standard Time Zones of the World
Area:
total area: 236,800 km²
land area: 230,800 km²
comparative area: slightly larger than Utah
Land boundaries: total 5,083 km, Burma 235 km, Cambodia 541 km, China 423 km, Thailand 1,754 km, Vietnam 2,130 km
Coastline: 0 km (landlocked)
Maritime claims: none; landlocked
International disputes: boundary dispute with Thailand
Climate: tropical monsoon; rainy season (May to November); dry season (December to April)
Terrain: mostly rugged mountains; some plains and plateaus
Natural resources: timber, hydropower, gypsum, tin, gold, gemstones
Land use:
arable land: 4%
permanent crops: 0%
meadows and pastures: 3%
forest and woodland: 58%
other: 35%
Irrigated land: 1,200 km² (1989 est.)
Environment: deforestation; soil erosion; subject to floods
Note: landlocked

People

Population: 4,569,327 (July 1993 est.)
Population growth rate: 2.86% (1993 est.)
Birth rate: 43.82 births/1,000 population (1993 est.)
Death rate: 15.22 deaths/1,000 population (1993 est.)
Net migration rate: 0 migrant(s)/1,000 population (1993 est.)
Infant mortality rate: 104.4 deaths/1,000 live births (1993 est.)

Life expectancy at birth:
total population: 51.18 years
male: 49.67 years
female: 52.77 years (1993 est.)
Total fertility rate: 6.16 children born/woman (1993 est.)
Nationality:
noun: Lao(s) or Laotian(s)
adjective: Lao or Laotian
Ethnic divisions: Lao 50%, Phoutheung (Kha) 15%, tribal Thai 20%, Meo, Hmong, Yao, and other 15%
Religions: Buddhist 85%, animist and other 15%
Languages: Lao (official), French, English
Literacy: age 15-45 can read and write (1985)
total population: 84%
male: 92%
female: 76%
Labor force: 1-1.5 million
by occupation: agriculture 85-90% (est.)

Government

Names:
conventional long form: Lao People's Democratic Republic
conventional short form: Laos
local long form: Sathalanalat Paxathipatai Paxaxon Lao
local short form: none
Digraph: LA
Type: Communist state
Capital: Vientiane
Administrative divisions: 16 provinces (khoueng, singular and plural) and 1 municipality* (kampheng nakhon, singular and plural); Attapu, Bokeo, Bolikhamsai, Champasak, Houaphan, Khammouan, Louang Namtha, Louangphrabang, Oudomxai, Phongsali, Saravan, Savannakhet, Sekong, Vientiane, Vientiane*, Xaignabouri, Xiangkhoang
Independence: 19 July 1949 (from France)
Constitution: promulgated August 1991
Legal system: based on civil law system; has not accepted compulsory ICJ jurisdiction
National holiday: National Day, 2 December (1975) (proclamation of the Lao People's Democratic Republic)
Political parties and leaders: Lao People's Revolutionary Party (LPRP), KHAMTAI Siphandon, party president; includes Lao Front for National Construction (LFNC); other parties moribund
Other political or pressure groups: non-Communist political groups moribund; most leaders fled the country in 1975
Suffrage: 18 years of age; universal
Elections:
Third National Assembly: last held on 20 December 1992 (next to be held NA); results—percent of vote by party NA; seats—(85 total) number of seats by party NA

Executive branch: president, prime minister and two deputy prime ministers, Council of Ministers (cabinet)
Legislative branch: National Assembly
Judicial branch: Supreme People's Court
Leaders:
Chief of State: President NOUHAK Phoumsavan (since 25 November 1992)
Head of Government: Prime Minister Gen. KHAMTAI Siphandon (since 15 August 1991)
Member of: ACCT (associate), AsDB, CP, ESCAP, FAO, G-77, IBRD, ICAO, IDA, IFAD, IFC, ILO, IMF, INTERPOL, IOC, ITU, LORCS, NAM, PCA, UN, UNCTAD, UNESCO, UNIDO, UPU, WFTU, WHO, WMO, WTO
Diplomatic representation in US:
chief of mission: Ambassador HIEM Phommachanh
chancery: 2222 S Street NW, Washington, DC 20008
telephone: (202) 332-6416 or 6417
US diplomatic representation:
chief of mission: Ambassador Charles B. SALMON, Jr.
embassy: Rue Bartholonie, Vientiane
mailing address: B. P. 114, Vientiane, or AMEMB, Box V, APO AP 96546
telephone: (856) 2220, 2357, 2384
FAX: (856) 4675
Flag: three horizontal bands of red (top), blue (double width), and red with a large white disk centered in the blue band

Economy

Overview: One of the world's poorest nations, Laos has had a Communist centrally planned economy with government ownership and control of productive enterprises of any size. In recent years, however, the government has been decentralizing control and encouraging private enterprise. Laos is a landlocked country with a primitive infrastructure; that is, it has no railroads, a rudimentary road system, limited external and internal telecommunications, and electricity available in only a limited area. Subsistence agriculture is the main occupation, accounting for over 60% of GDP and providing about 85-90% of total employment. The predominant crop is rice. For the foreseeable future the economy will continue to depend for its survival on foreign aid from the IMF and other international sources; aid from the former USSR and Eastern Europe has been cut sharply.
National product: GDP—exchange rate conversion—$900 million (1991)
National product real growth rate: 4% (1991)
National product per capita: $200 (1991)

Inflation rate (consumer prices): 10%
(1991)
Unemployment rate: 21% (1989 est.)
Budget: revenues $83 million; expenditures
$188.5 million, including capital
expenditures of $94 million (1990 est.)
Exports: $72 million (f.o.b., 1990 est.)
commodities: electricity, wood products,
coffee, tin
partners: Thailand, Malaysia, Vietnam,
USSR, US, China
Imports: $238 million (c.i.f., 1990 est.)
commodities: food, fuel oil, consumer goods,
manufactures
partners: Thailand, USSR, Japan, France,
Vietnam, China
External debt: $1.1 billion (1990 est.)
Industrial production: growth rate 12%
(1991 est.); accounts for about 18% of GDP
(1991 est.)
Electricity: 226,000 kW capacity; 990
million kWh produced, 220 kWh per capita
(1992)
Industries: tin and gypsum mining, timber,
electric power, agricultural processing,
construction
Agriculture: accounts for 60% of GDP and
employs most of the work force; subsistence
farming predominates; normally
self-sufficient in nondrought years; principal
crops—rice (80% of cultivated land), sweet
potatoes, vegetables, corn, coffee, sugarcane,
cotton; livestock—buffaloes, hogs, cattle,
poultry
Illicit drugs: illicit producer of cannabis,
opium poppy for the international drug trade,
third-largest opium producer
Economic aid: US commitments, including
Ex-Im (FY70-79), $276 million; Western
(non-US) countries, ODA and OOF bilateral
commitments (1970-89), $605 million;
Communist countries (1970-89), $995
million
Currency: 1 new kip (NK) = 100 at
Exchange rates: new kips (NK) per
US$1—710 (May 1992), 710 (December
1991), 700 (September 1990), 576 (1989),
385 (1988), 200 (1987)
Fiscal year: 1 July-30 June

Communications

Railroads: none
Highways: about 27,527 km total; 1,856 km
bituminous or bituminous treated; 7,451 km
gravel, crushed stone, or improved earth;
18,220 km unimproved earth and often
impassable during rainy season mid-May to
mid-September
Inland waterways: about 4,587 km,
primarily Mekong and tributaries; 2,897
additional kilometers are sectionally
navigable by craft drawing less than 0.5 m
Pipelines: petroleum products 136 km
Ports: none

Airports:
total: 54
usable: 41
with permanent-surface runways: 8
with runways over 3,659 m: 0
with runways 2,440-3,659 m: 1
with runways 1,220-2,439 m: 15
Telecommunications: service to general
public practically non-existant; radio
communications network provides generally
erratic service to government users; 7,390
telephones (1986); broadcast stations—10
AM, no FM, 1 TV; 1 satellite earth station

Defense Forces

Branches: Lao People's Army (LPA;
including naval, aviation, and militia
elements), Air Force, National Police
Department
Manpower availability: males age 15-49
980,274; fit for military service 528,450;
reach military age (18) annually 43,849
(1993 est.)
Defense expenditures: exchange rate
conversion—$NA, NA% of GDP

Latvia

Final boundaries of Estonia, Latvia, and Lithuania
with the former Soviet Union are expected to be
confirmed by agreement

150 km

Geography

Location: Eastern Europe, bordering on the
Baltic Sea, between Sweden and Russia
Map references: Arctic Region, Asia,
Europe, Standard Time Zones of the World
Area:
total area: 64,100 km²
land area: 64,100 km²
comparative area: slightly larger than West
Virginia
Land boundaries: total 1,078 km, Belarus
141 km, Estonia 267 km, Lithuania 453 km,
Russia 217 km
Coastline: 531 km
Maritime claims:
exclusive economic zone: 200 nm
territorial sea: 12 nm
International disputes: the Abrene section
of border ceded by the Latvian Soviet
Socialist Republic to Russia in 1944
Climate: maritime; wet, moderate winters
Terrain: low plain
Natural resources: minimal; amber, peat,
limestone, dolomite
Land use:
arable land: 27%
permanent crops: 0%
meadows and pastures: 13%
forest and woodland: 39%
other: 21%
Irrigated land: 160 km² (1990)
Environment: heightened levels of air and
water pollution because of a lack of waste
conversion equipment; Gulf of Riga and
Daugava River heavily polluted;
contamination of soil and groundwater with
chemicals and petroleum products at military
bases

People

Population: 2,735,573 (July 1993 est.)
Population growth rate: 0.5% (1993 est.)
Birth rate: 13.99 births/1,000 population
(1993 est.)
Death rate: 12.73 deaths/1,000 population
(1993 est.)

Latvia (continued)

Net migration rate: 3.72 migrant(s)/1,000 population (1993 est.)
Infant mortality rate: 22 deaths/1,000 live births (1993 est.)
Life expectancy at birth:
total population: 69.23 years
male: 64.15 years
female: 74.55 years (1993 est.)
Total fertility rate: 2 children born/woman (1993 est.)
Nationality:
noun: Latvian(s)
adjective: Latvian
Ethnic divisions: Latvian 51.8%, Russian 33.8%, Belarusian 4.5%, Ukrainian 3.4%, Polish 2.3%, other 4.2%
Religions: Lutheran, Roman Catholic, Russian Orthodox
Languages: Latvian (official), Lithuanian, Russian, other
Literacy: age 9-49 can read and write (1970)
total population: 100%
male: 100%
female: 100%
Labor force: 1.407 million
by occupation: industry and construction 41%, agriculture and forestry 16%, other 43% (1990)

Government

Names:
conventional long form: Republic of Latvia
conventional short form: Latvia
local long form: Latvijas Republika
local short form: Latvija
former: Latvian Soviet Socialist Republic
Digraph: LG
Type: republic
Capital: Riga
Administrative divisions: none (all districts are under direct republic jurisdiction)
Independence: 6 September 1991 (from Soviet Union)
Constitution: adopted NA May 1922, considering rewriting constitution
Legal system: based on civil law system
National holiday: Independence Day, 18 November (1918)
Political parties and leaders: Democratic Labor Party of Latvia, Juris BOJARS, chairman; Inter-Front of the Working People of Latvia, Igor LOPATIN, chairman (Inter-Front was banned after the coup); Latvian National Movement for Independence, Eduards BERKLAVS, chairman; Latvian Democratic Party, Janis DINEVICS, chairman; Latvian Social Democratic Workers' Party, Uldis BERZINS, chairman; Latvian People's Front, Uldis AUGST-KALNS, chairman; Latvian Liberal Party, Georg LANSMANIS, chairman
Suffrage: 18 years of age; universal

Elections:
President: last held October 1988 (next to be held NA); note—Anatolijs V. GORBUNOVS elected by Supreme Soviet; elected to restyled post of Chairman of the Supreme Council on 3 May 1990; new elections have not been scheduled
Supreme Council: last held 18 March 1990 for the Supreme Soviet (next to be held 5-6 June 1993 for the Saeima); results—percent of vote by party NA; seats—(234 total) Latvian Communist Party 59, Latvian Democratic Workers Party 31, Social Democratic Party of Latvia 4, Green Party of Latvia 7, Latvian Farmers Union 7, Latvian Popular Front 126; note—the Supreme Council is an interim 201-seats legislative body; a new parliament or Saiema to be elected in June 1993
Congress of Latvia: last held April 1990 (next to be held NA); results—percent of vote by party NA; seats—(231 total) number of seats by party NA; note—the Congress of Latvia is a quasi-governmental structure
Executive branch: Chairman of Supreme Council (president), prime minister, cabinet
Legislative branch: unicameral Supreme Council
Judicial branch: Supreme Court
Leaders:
Chief of State: Chairman Supreme Council Anatolijs V. GORBUNOVS (since NA October 1988)
Head of Government: Prime Minister Ivars GODMANIS (since NA May 1990)
Member of: CBSS, CSCE, EBRD, ECE, FAO, IBRD, ICAO, IDA, ILO, IMF, IOM (observer), ITU, NACC, UN, UNCTAD, UNESCO, UNIDO, UPU, WHO
Diplomatic representation in US:
chief of mission: Ambassador Ojars KALNINS
chancery: 4325 17th Street NW, Washington, DC 20011
telephone: (202) 726-8213 and 8214
US diplomatic representation:
chief of mission: Ambassador Ints M, SILINS;
embassy: Raina Boulevard 7, Riga 226050
mailing address: APO AE 09862
telephone: 0-11 [358] (49) 311-348 (cellular)
FAX: [358] (49) 314-665 (cellular), (7) (01-32) 220-502
note: dialing to the Baltics still requires use of an international operator, unless you use the cellular phone lines
Flag: two horizontal bands of maroon (top and bottom), white (middle, narrower than other two bands)

Economy

Overview: Latvia is in the process of reforming the centrally planned economy inherited from the former USSR into a market economy. Prices have been freed, and privatization of shops and farms has begun. Latvia lacks natural resources, aside from its arable land and small forests. Its most valuable economic asset is its work force, which is better educated and disciplined than in most of the former Soviet republics. Industrial production is highly diversified, with products ranging from agricultural machinery to consumer electronics. One conspicuous vulnerability: Latvia produces only 10% of its electric power needs. Latvia in the near term must retain key commercial ties to Russia, Belarus, and Ukraine while moving in the long run toward joint ventures with technological support from, and trade ties to the West. Because of the efficiency of its mostly individual farms, Latvians enjoy a diet that is higher in meat, vegetables, and dairy products and lower in grain and potatoes than diets in the 12 non-Baltic republics of the former USSR. Good relations with Russia are threatened by animosity between ethnic Russians (34% of the population) and native Latvians. The cumulative difficulties in replacing old sources of supply and old markets, together with the phasing out of the Russian ruble as the medium of exchange, help account for the sharp 30% drop in GDP in 1992.
National product: GDP $NA
National product real growth rate: -30% (1992)
National product per capita: $NA
Inflation rate (consumer prices): 2% per month (first quarter 1993)
Unemployment rate: 3.6% (March 1993); but large numbers of underemployed workers
Budget: revenues $NA; expenditures $NA, including capital expenditures of $NA
Exports: $NA
commodities: NA
partners: NA
Imports: $NA
commodities: NA
partners: NA
External debt: $650 million (1991 est.)
Industrial production: growth rate −35% (1992 est.)
Electricity: 2,140,000 kW capacity; 5,800 million kWh produced, 2,125 kWh per capita (1992)
Industries: employs 33% of labor force; highly diversified; dependent on imports for energy, raw materials, and intermediate products; produces buses, vans, street and railroad cars, synthetic fibers, agricultural machinery, fertilizers, washing machines, radios, electronics, pharmaceuticals, processed foods, textiles
Agriculture: employs 16% of labor force; principally dairy farming and livestock feeding; products—meat, milk, eggs, grain,

Lebanon

sugar beets, potatoes, vegetables; fishing and fish packing
Illicit drugs: transshipment point for illicit drugs from Central and Southwest Asia to Western Europe; limited producer of illicit opium; mostly for domestic consumption; also produces illicit amphetamines for export
Economic aid: NA
Currency: 1 lat = 100 NA; introduced NA March 1993
Exchange rates: lats per US$1—1.32 (March 1993)
Fiscal year: calendar year

Communications

Railroads: 2,400 km; does not include industrial lines (1990)
Highways: 59,500 km total; 33,000 km hard surfaced 26,500 km earth (1990)
Inland waterways: 300 km perennially navigable
Pipelines: crude oil 750 km, refined products 780 km, natural gas 560 km (1992)
Ports: coastal—Riga, Ventspils, Liepaja; inland—Daugavpils
Merchant marine: 96 ships (1,000 GRT or over) totaling 905,006 GRT/1,178,844 DWT; includes 14 cargo, 27 refrigerated cargo, 2 container, 9 roll-on/roll-off, 44 oil tanker
Airports:
total: 50
useable: 15
with permanent-surface runways: 11
with runways over 3,659 m: 0
with runways 2,440-3,659 m: 7
with runways 1,220-2,439 m: 7
Telecommunications: NMT-450 analog cellular network is operational covering Riga, Ventspils, Daugavpils, Rezekne, and Valmiera; broadcast stations—NA; international traffic carried by leased connection to the Moscow international gateway switch and through new independent international automatic telephone exchange in Riga and the Finnish cellular net

Defense Forces

Branches: Ground Forces, Navy, Air Force, Security Forces (internal and border troops), Border Guard, Home Guard (Zemessardze)
Manpower availability: males age 15-49 648,273; fit for military service 511,297; reach military age (18) annually 18,767 (1993 est.)
Defense expenditures: 176 million rubles, 3-5% of GDP; note—conversion of the military budget into US$ using the current exchange rate could produce misleading results

Note: Lebanon has made progress toward rebuilding its political institutions and regaining its national sovereignty since the end of the devastating 16-year civil war in October 1990. Under the Ta'if accord—the blueprint for national reconciliation—the Lebanese have established a more equitable political system, particularly by giving Muslims a greater say in the political process. Since December 1990, the Lebanese have formed three cabinets and conducted the first legislative election in 20 years. Most of the militias have been weakened or disbanded. The Lebanese Armed Forces (LAF) has seized vast quantities of weapons used by the militias during the war and extended central government authority over about one-half of the country. Hizballah, the radical Sh'ia party, is the only significant group that retains most of its weapons. Foreign forces still occupy areas of Lebanon. Israel continues to support a proxy militia, The Army of South Lebanon (ASL), along a narrow stretch of territory contiguous to its border. The ASL's enclave encompasses this self-declared security zone and about 20 kilometers north to the strategic town of Jazzine. As of December 1992, Syria maintained about 30,000 troops in Lebanon. These troops are based mainly in Beirut, North Lebanon, and the Bekaa Valley. Syria's deployment was legitimized by the Arab League early in Lebanon's civil war and in the Ta'if accord. Citing the continued weakness of the LAF, Beirut's requests, and failure of the Lebanese Government to implement all of the constitutional reforms in the Ta'if accord, Damascus has so far refused to withdraw its troops from Beirut.

Geography

Location: Middle East, in the eastern Mediterranean Sea, between Israel and Syria
Map references: Africa, Middle East, Standard Time Zones of the World

Area:
total area: 10,400 km²
land area: 10,230 km²
comparative area: about 0.8 times the size of Connecticut
Land boundaries: total 454 km, Israel 79 km, Syria 375 km
Coastline: 225 km
Maritime claims:
territorial sea: 12 nm
International disputes: separated from Israel by the 1949 Armistice Line; Israeli troops in southern Lebanon since June 1982; Syrian troops in northern, central, and eastern Lebanon since October 1976
Climate: Mediterranean; mild to cool, wet winters with hot, dry summers; Lebanon mountians experience heavy winter snows
Terrain: narrow coastal plain; Al Biqa' (Bekaa Valley) separates Lebanon and Anti-Lebanon Mountains
Natural resources: limestone, iron ore, salt, water-surplus state in a water-deficit region
Land use:
arable land: 21%
permanent crops: 9%
meadows and pastures: 1%
forest and woodland: 8%
other: 61%
Irrigated land: 860 km² (1989 est.)
Environment: rugged terrain historically helped isolate, protect, and develop numerous factional groups based on religion, clan, ethnicity; deforestation; soil erosion; air and water pollution; desertification
Note: Nahr al Litani only major river in Near East not crossing an international boundary

People

Population: 3,552,369 (July 1993 est.)
Population growth rate: 1.81% (1993 est.)
Birth rate: 27.86 births/1,000 population (1993 est.)
Death rate: 6.66 deaths/1,000 population (1993 est.)
Net migration rate: -3.1 migrant(s)/1,000 population (1993 est.)
Infant mortality rate: 41 deaths/1,000 live births (1993 est.)
Life expectancy at birth:
total population: 69.01 years
male: 66.63 years
female: 71.52 years (1993 est.)
Total fertility rate: 3.47 children born/woman (1993 est.)
Nationality:
noun: Lebanese (singular and plural)
adjective: Lebanese
Ethnic divisions: Arab 95%, Armenian 4%, other 1%
Religions: Islam 70% (5 legally recognized Islamic groups—Alawite or Nusayri, Druze, Isma'ilite, Shi'a, Sunni), Christian 30% (11 legally recognized Christian groups—4 Orthodox Christian, 6 Catholic, 1 Protestant), Judaism NEGL%

Lebanon (continued)

Languages: Arabic (official), French (official), Armenian, English
Literacy: age 15 and over can read and write (1990)
total population: 80%
male: 88%
female: 73%
Labor force: 650,000
by occupation: industry, commerce, and services 79%, agriculture 11%, government 10% (1985)

Government

Names:
conventional long form: Republic of Lebanon
conventional short form: Lebanon
local long form: Al Jumhuriyah al Lubnaniyah
local short form: none
Digraph: LE
Type: republic
Capital: Beirut
Administrative divisions: 5 governorates (muhafazat, singular—muhafazah); Al Biqa, 'Al Janub, Ash Shamal, Bayrut, Jabal Lubnan
Independence: 22 November 1943 (from League of Nations mandate under French administration)
Constitution: 26 May 1926 (amended)
Legal system: mixture of Ottoman law, canon law, Napoleonic code, and civil law; no judicial review of legislative acts; has not accepted compulsory ICJ jurisdiction
National holiday: Independence Day, 22 November (1943)
Political parties and leaders: political party activity is organized along largely sectarian lines; numerous political groupings exist, consisting of individual political figures and followers motivated by religious, clan, and economic considerations
Suffrage: 21 years of age; compulsory for all males; authorized for women at age 21 with elementary education
Elections:
National Assembly: Lebanon's first legislative election in 20 years was held in the summer of 1992; the National Assembly is composed of 128 deputies, one-half Christian and one-half Muslim; its mandate expires in 1996
Executive branch: president, prime minister, Cabinet; note—by custom, the president is a Maronite Christian, the prime minister is a Sunni Muslim, and the speaker of the legislature is a Shi'a Muslim
Legislative branch: unicameral National Assembly (Arabic—Majlis Alnuwab, French—Assemblee Nationale)
Judicial branch: four Courts of Cassation (three courts for civil and commercial cases and one court for criminal cases)

Leaders:
Chief of State: President Ilyas HARAWI (since 24 November 1989)
Head of Government: Prime Minister Rafiq HARIRI (since 22 October 1992)
Member of: ABEDA, AFESD, AL, AMF, CCC, ESCWA, FAO, G-24, G-77, IAEA, IBRD, ICAO, ICC, ICFTU, IDA, IDB, IFAD, IFC, ILO, IMF, IMO, INTELSAT, INTERPOL, IOC, ITU, LORCS, NAM, OIC, PCA, UN, UNCTAD, UNESCO, UNHCR, UNIDO, UNRWA, UPU, WFTU, WHO, WIPO, WMO, WTO
Diplomatic representation in US:
chief of mission: Ambassador Simon KARAM
chancery: 2560 28th Street NW, Washington, DC 20008
telephone: (202) 939-6300
consulates general: Detroit, New York, and Los Angeles
US diplomatic representation:
chief of mission: Ambassador Ryan C. CROCKER
mailing embassy: Antelias, Beirut
address: P. O. Box 70-840, Beirut, or Box B, FPO AE 09836
telephone: [961] 417774 or 415802, 415803, 402200, 403300
Flag: three horizontal bands of red (top), white (double width), and red with a green and brown cedar tree centered in the white band

Economy

Overview: Since 1975 civil war has seriously damaged Lebanon's economic infrastructure, cut national output by half, and all but ended Lebanon's position as a Middle Eastern entrepot and banking hub. Following October 1990, however, a tentative peace has enabled the central government to begin restoring control in Beirut, collect taxes, and regain access to key port and government facilities. The battered economy has also been propped up by a financially sound banking system and resilient small- and medium-scale manufacturers. Family remittances, banking transactions, manufactured and farm exports, the narcotics trade, and international emergency aid are main sources of foreign exchange. In the relatively settled year of 1991, industrial production, agricultural output, and exports showed substantial gains. The further rebuilding of the war-ravaged country was delayed in 1992 because of an upturn in political wrangling. Hope for restoring economic momentum in 1993 rests with the new, business-oriented Prime Minister HARIRI.

National product: GDP—exchange rate conversion—$4.8 billion (1991 est.)
National product real growth rate: NA%
National product per capita: $1,400 (1991 est.)
Inflation rate (consumer prices): 100% (1992 est.)
Unemployment rate: 35% (1991 est.)
Budget: revenues $533 million; expenditures $1.3 billion, including capital expenditures of $NA (1991 est.)
Exports: $490 million (f.o.b., 1991)
commodities: agricultural products, chemicals, textiles, precious and semiprecious metals and jewelry, metals and metal products
partners: Saudi Arabia 21%, Switzerland 9.5%, Jordan 6%, Kuwait 12%, US 5%
Imports: $3.7 billion (c.i.f., 1991)
commodities: Consumer goods, machinery and transport equipment, petroleum products
partners: Italy 14%, France 12%, US 6%, Turkey 5%, Saudi Arabia 3%
External debt: $400 million (1992 est.)
Industrial production: growth rate NA%
Electricity: 1,300,000 kW capacity; 3,413 million kWh produced, 990 kWh per capita (1992)
Industries: banking, food processing, textiles, cement, oil refining, chemicals, jewelry, some metal fabricating
Agriculture: accounts for about one-third of GDP; principal products—citrus fruits, vegetables, potatoes, olives, tobacco, hemp (hashish), sheep, goats; not self-sufficient in grain
Illicit drugs: illicit producer of opium, hashish, and heroin for the international drug trade; opium poppy production in Al Biqa almost completely eradicated this year; hashish production is shipped to Western Europe, Israel, US, the Middle East, and South America
Economic aid: US commitments, including Ex-Im (FY70-88), $356 million; Western (non-US) countries, ODA and OOF bilateral commitments (1970-89), $664 million; OPEC bilateral aid (1979-89), $962 million; Communist countries (1970-89), $9 million
Currency: 1 Lebanese pound (£L) = 100 piasters
Exchange rates: Lebanese pounds (£L) per US$1—1,742.00 (April 1993), 1,712.80 (1992), 928.23 (1991), 695.09 (1990), 496.69 (1989), 409.23 (1988)
Fiscal year: calendar year

Communications

Railroads: system in disrepair, considered inoperable
Highways: 7,300 km total; 6,200 km paved, 450 km gravel and crushed stone, 650 km improved earth
Pipelines: crude oil 72 km (none in operation)

Lesotho

Ports: Beirut, Tripoli, Ra'Sil'ata, Juniyah, Sidon, Az Zahrani, Tyre, Jubayl, Shikka Jadidah
Merchant marine: 63 ships (1,000 GRT or over) totaling 270,505 GRT/403,328 DWT; includes 39 cargo, 1 refrigerated cargo, 2 vehicle carrier, 3 roll-on/roll-off, 1 container, 9 livestock carrier, 2 chemical tanker, 1 specialized tanker, 4 bulk, 1 combination bulk
Airports:
total: 9
usable: 8
with permanent-surface runways: 6
with runways over 3,659 m: 0
with runways 2,440-3,659 m: 3
with runways 1,220-2,439 m: 2
Telecommunications: telecommunications system severely damaged by civil war; rebuilding still underway; 325,000 telephones (95 telephones per 1,000 persons); domestic traffic carried primarily by microwave radio relay and a small amount of cable; international traffic by satellite—1 Indian Ocean INTELSAT earth station and 1 Atlantic Ocean INTELSAT earth station (erratic operations), coaxial cable to Syria; microwave radio relay to Syria but inoperable beyond Syria to Jordan, 3 submarine coaxial cables; broadcast stations—5 AM, 3 FM, 13 TV (numerous AM and FM stations are operated sporadically by various factions)

Defense Forces

Branches: Lebanese Armed Forces (LAF; including Army, Navy, and Air Force)
Manpower availability: males age 15-49 798,299; fit for military service 495,763 (1993 est.)
Defense expenditures: exchange rate conversion—$271 million, 8.2% of GDP (1992 budget)

Geography

Location: Southern Africa, an enclave of South Africa
Map references: Africa, Standard Time Zones of the World
Area:
total area: 30,350 km²
land area: 30,350 km²
comparative area: slightly larger than Maryland
Land boundaries: total 909 km, South Africa 909 km
Coastline: 0 km (landlocked)
Maritime claims: none; landlocked
International disputes: none
Climate: temperate; cool to cold, dry winters; hot, wet summers
Terrain: mostly highland with some plateaus, hills, and mountains
Natural resources: some diamonds and other minerals, water, agricultural and grazing land
Land use:
arable land: 10%
permanent crops: 0%
meadows and pastures: 66%
forest and woodland: 0%
other: 24%
Irrigated land: NA km²
Environment: population pressure forcing settlement in marginal areas results in overgrazing, severe soil erosion, soil exhaustion; desertification
Note: landlocked; surrounded by South Africa; Highlands Water Project will control, store, and redirect water to South Africa

People

Population: 1,896,484 (July 1993 est.)
Population growth rate: 2.52% (1993 est.)
Birth rate: 34.64 births/1,000 population (1993 est.)
Death rate: 9.44 deaths/1,000 population (1993 est.)
Net migration rate: 0 migrant(s)/1,000 population (1993 est.)

Infant mortality rate: 71.5 deaths/1,000 live births (1993 est.)
Life expectancy at birth:
total population: 61.73 years
male: 59.91 years
female: 63.6 years (1993 est.)
Total fertility rate: 4.6 children born/woman (1993 est.)
Nationality:
noun: Mosotho (singular), Basotho (plural)
adjective: Basotho
Ethnic divisions: Sotho 99.7%, Europeans 1,600, Asians 800
Religions: Christian 80%, rest indigenous beliefs
Languages: Sesotho (southern Sotho), English (official), Zulu, Xhosa
Literacy: age 15 and over can read and write (1966)
total population: 59%
male: 44%
female: 68%
Labor force: 689,000 economically active
by occupation: 86.2% of resident population engaged in subsistence agriculture; roughly 60% of active male labor force works in South Africa

Government

Names:
conventional long form: Kingdom of Lesotho
conventional short form: Lesotho
former: Basutoland
Digraph: LT
Type: constitutional monarchy
Capital: Maseru
Administrative divisions: 10 districts; Berea, Butha-Buthe, Leribe, Mafeteng, Maseru, Mohale's Hoek, Mokhotlong, Qacha's Nek, Quthing, Thaba-Tseka
Independence: 4 October 1966 (from UK)
Constitution: 4 October 1966, suspended January 1970
Legal system: based on English common law and Roman-Dutch law; judicial review of legislative acts in High Court and Court of Appeal; has not accepted compulsory ICJ jurisdiction
National holiday: Independence Day, 4 October (1966)
Political parties and leaders: Basotho National Party (BNP), Evaristus SEKHONYANA; Basutoland Congress Party (BCP), Ntsu MOKHEHLE; National Independent Party (NIP), A. C. MANYELI; Marematlou Freedom Party (MFP), Vincent MALEBO; United Democratic Party, Charles MOFELI; Communist Party of Lesotho (CPL), JCOB M. KENA
Suffrage: 21 years of age; universal
Elections:
National Assembly: dissolved following the military coup in January 1986; military has pledged elections will take place in March 1993

Lesotho *(continued)*

Executive branch: monarch, chairman of the Military Council, Military Council, Council of Ministers (cabinet)

Legislative branch: none—the bicameral Parliament was dissolved following the military coup in January 1986; note—a National Constituent Assembly convened in June 1990 to rewrite the constitution and debate issues of national importance, but it has no legislative authority

Judicial branch: High Court, Court of Appeal

Leaders:
Chief of State: King LETSIE III (since 12 November 1990 following dismissal of his father, exiled King MOSHOESHOE II, by Maj. Gen. LEKHANYA)
Head of Government: Chairman of the Military Council Gen. Elias Phisoana RAMAEMA (since 30 April 1991)

Member of: ACP, AfDB, C, CCC, ECA, FAO, G-77, GATT, IBRD, ICAO, ICFTU, IDA, IFAD, IFC, ILO, IMF, INTERPOL, IOC, ITU, LORCS, NAM, OAU, SACU, SADC, UN, UNCTAD, UNESCO, UNHCR, UNIDO, UPU, WCL, WHO, WIPO, WMO, WTO

Diplomatic representation in US:
chief of mission: Ambassador Designate Teboho KITLEI
chancery: 2511 Massachusetts Avenue NW, Washington, DC 20008
telephone: (202) 797-5534

US diplomatic representation:
chief of mission: Ambassador Leonard H.O. SPEARMAN, Sr.
embassy: address NA, Maseru
mailing address: P. O. Box 333, Maseru 100 Lesotho
telephone: [266] 312-666
FAX: (266) 310-116

Flag: divided diagonally from the lower hoist side corner; the upper half is white bearing the brown silhouette of a large shield with crossed spear and club; the lower half is a diagonal blue band with a green triangle in the corner

---

Economy

Overview: Small, landlocked, and mountainous, Lesotho has no important natural resources other than water. Its economy is based on agriculture, light manufacturing, and remittances from laborers employed in South Africa ($439 million in 1991). The great majority of households gain their livelihoods from subsistence farming and migrant labor. Manufacturing depends largely on farm products to support the milling, canning, leather, and jute industries; other industries include textile, clothing, and construction (in particular, a major water improvement project which will permit the sale of water

to South Africa). Industry's share of GDP rose from 6% in 1982 to 15% in 1989. Political and economic instability in South Africa raises uncertainty for Lesotho's economy, especially with respect to migrant worker remittances—recently the equivalent of nearly three-fourths of domestic output.

National product: GDP—exchange rate conversion—$620 million (1991 est.)
note: GNP of $1.0 billion (1991 est.)

National product real growth rate: 5.3% (1991 est.); GNP 2.2% (1991 est.)

National product per capita: $340 (1991 est.); GNP $570 (1991 est.)

Inflation rate (consumer prices): 17.9% (1991)

Unemployment rate: at least 55% among adult males (1991 est.)

Budget: revenues $388 million; expenditures $399 million, including capital expenditures of $132 million (FY93)

Exports: $57 million (f.o.b., 1991)
commodities: wool, mohair, wheat, cattle, peas, beans, corn, hides, skins, baskets
partners: South Africa 53%, EC 30%, North and South America 13% (1989)

Imports: $805 million (c.i.f., 1991)
commodities: mainly corn, building materials, clothing, vehicles, machinery, medicines, petroleum
partners: South Africa 95%, EC 2% (1989)

External debt: $358 million (for public sector) (December 1990/91 est.)

Industrial production: growth rate 5.0% (1991 est.); accounts for 11% of GDP

Electricity: power supplied by South Africa

Industries: food, beverages, textiles, handicrafts, tourism

Agriculture: accounts for 19% of GDP (1990 est.) and employs 60-70% of all households; exceedingly primitive, mostly subsistence farming and livestock; principal crops corn, wheat, pulses, sorghum, barley

Economic aid: US commitments, including Ex-Im (FY70-89), $268 million; US, $10.3 million (1992), $10.1 million (1993 est.); Western (non-US) countries, ODA and OOF bilateral commitments (1970-89), $819 million; OPEC bilateral aid (1979-89), $4 million; Communist countries (1970-89), $14 million

Currency: 1 loti (L) = 100 lisente

Exchange rates: maloti (M) per US$1—3.1576 (May 1993), 2.8497 (1992), 2.7563 (1991), 2.5863 (1990), 2.6166 (1989), 2.2611 (1988); note—the Basotho loti is at par with the South African rand

Fiscal year: 1 April-31 March

Communications

Railroads: 2.6 km; owned, operated by, and included in the statistics of South Africa
Highways: 7,215 km total; 572 km paved; 2,337 km crushed stone, gravel, or stabilized

soil; 1,806 km improved earth, 2,500 km unimproved earth

Airports:
total: 28
usable: 28
with permanent-surface runways: 3
with runways over 3,659 m: 0
with runways 2,440-3,659 m: 1
with runways 1,220-2,439 m: 2

Telecommunications: rudimentary system consisting of a few landlines, a small microwave system, and minor radio communications stations; 5,920 telephones; broadcast stations—3 AM, 2 FM, 1 TV; 1 Atlantic Ocean INTELSAT earth station

---

Defense Forces

Branches: Royal Lesotho Defense Force (RLDF; including Army, Air Wing), Royal Lesotho Mounted Police

Manpower availability: males age 15-49 422,802; fit for military service 228,102 (1993 est.)

Defense expenditures: exchange rate conversion—$55 million, 13% of GDP (1990 est.)

Liberia

Geography

Location: Western Africa, bordering the North Pacific Ocean between Cote d'Ivoire and Sierra Leone
Map references: Africa, Standard Time Zones of the World
Area:
total area: 111,370 km²
land area: 96,320 km²
comparative area: slightly larger than Tennessee
Land boundaries: total 1,585 km, Guinea 563 km, Cote d'Ivoire 716 km, Sierra Leone 306 km
Coastline: 579 km
Maritime claims:
continental shelf: 200 m depth or to depth of exploitation
territorial sea: 200 nm
International disputes: none
Climate: tropical; hot, humid; dry winters with hot days and cool to cold nights; wet, cloudy summers with frequent heavy showers
Terrain: mostly flat to rolling coastal plains rising to rolling plateau and low mountains in northeast
Natural resources: iron ore, timber, diamonds, gold
Land use:
arable land: 1%
permanent crops: 3%
meadows and pastures: 2%
forest and woodland: 39%
other: 55%
Irrigated land: 20 km² (1989 est.)
Environment: West Africa's largest tropical rain forest, subject to deforestation

People

Population: 2,874,881 (July 1993 est.)
Population growth rate: 3.37% (1993 est.)
Birth rate: 43.9 births/1,000 population (1993 est.)
Death rate: 12.38 deaths/1,000 population (1993 est.)

Net migration rate: 2.15 migrant(s)/1,000 population (1993 est.)
Infant mortality rate: 115.9 deaths/1,000 live births (1993 est.)
Life expectancy at birth:
total population: 57.28 years
male: 54.88 years
female: 59.76 years (1993 est.)
Total fertility rate: 6.42 children born/woman (1993 est.)
Nationality:
noun: Liberian(s)
adjective: Liberian
Ethnic divisions: indigenous African tribes 95% (including Kpelle, Bassa, Gio, Kru, Grebo, Mano, Krahn, Gola, Gbandi, Loma, Kissi, Vai, and Bella), Americo-Liberians 5% (descendants of repatriated slaves)
Religions: traditional 70%, Muslim 20%, Christian 10%
Languages: English 20% (official), Niger-Congo language group about 20 local languages come from this group
Literacy: age 15 and over can read and write (1990)
total population: 40%
male: 50%
female: 29%
Labor force: 510,000 including 220,000 in the monetary economy
by occupation: agriculture 70.5%, services 10.8%, industry and commerce 4.5%, other 14.2%
note: non-African foreigners hold about 95% of the top-level management and engineering jobs; 52% of population of working age

Government

Names:
conventional long form: Republic of Liberia
conventional short form: Liberia
Digraph: LI
Type: republic
Capital: Monrovia
Administrative divisions: 13 counties; Bomi, Bong, Grand Bassa, Cape Mount, Grand Gedeh, Grand Kru, Lofa, Margibi, Maryland, Montserrado, Nimba, River Cess, Sinoe
Independence: 26 July 1847
Constitution: 6 January 1986
Legal system: dual system of statutory law based on Anglo-American common law for the modern sector and customary law based on unwritten tribal practices for indigenous sector
National holiday: Independence Day, 26 July (1847)
Political parties and leaders: National Democratic Party of Liberia (NDPL), Augustus CAINE, chairman; Liberian Action Party (LAP), Emmanuel KOROMAH, chairman; Unity Party (UP), Carlos SMITH, chairman; United People's Party (UPP), Gabriel Baccus MATTHEWS, chairman

Suffrage: 18 years of age; universal
Elections:
President: last held on 15 October 1985 (next to be held NA); results—Gen. Dr. Samuel Kanyon DOE (NDPL) 50.9%, Jackson DOE (LAP) 26.4%, other 22.7%; note—President Doe was killed by rebel forces on 9 September 1990
Senate: last held on 15 October 1985 (next to be held NA); results—percent of vote by party NA; seats—(26 total) NDPL 21, LAP 3, UP 1, UPP 1
House of Representatives: last held on 15 October 1985 (next to be held NA); results—percent of vote by party NA; seats—(64 total) NDPL 51, LAP 8, UP 3, UPP 2
Executive branch: president, vice president, Cabinet
Legislative branch: bicameral National Assembly consists of an upper house or Senate and a lower house or House of Representatives
Judicial branch: People's Supreme Court
Leaders:
Chief of State and Head of Government: interim President Dr. Amos SAWYER (since 15 November 1990)
note: this is an interim government appointed by the Economic Community of West African States (ECOWAS) that will be replaced after elections are held under a West African-brokered peace plan; a rebel faction led by Charles TAYLOR is challenging the SAWYER government's legitimacy; former president, Gen. Dr. Samuel Kanyon DOE, was killed on 9 September 1990 by Prince Y. JOHNSON
Member of: ACP, AfDB, CCC, ECA, ECOWAS, FAO, G-77, IAEA, IBRD, ICAO, ICFTU, IDA, IFAD, IFC, ILO, IMF, IMO, INMARSAT, INTERPOL, IOC, ITU, LORCS, NAM, OAU, UN, UNCTAD, UNESCO, UNIDO, UPU, WHO, WIPO, WMO
Diplomatic representation in US:
chief of mission: Ambassador James TARPEH
chancery: 5201 16th Street NW, Washington, DC 20011
telephone: (202) 723-0437 through 0440
consulate general: New York
US diplomatic representation:
chief of mission: Ambassador William H. TWADDELL
embassy: 111 United Nations Drive, Monrovia
mailing address: P. O. Box 98, Monrovia, or APO AE 09813
telephone: [231] 222991 through 222994
FAX: (231) 223710
Flag: 11 equal horizontal stripes of red (top and bottom) alternating with white; there is a white five-pointed star on a blue square in the upper hoist-side corner; the design was based on the US flag

Liberia *(continued)*

Economy

Overview: Civil war since 1990 has destroyed much of Liberia's economy, especially the infrastructure in and around Monrovia. Businessmen have fled the country, taking capital and expertise with them. Many will not return. Richly endowed with water, mineral resources, forests, and a climate favorable to agriculture, Liberia had been a producer and exporter of basic products, while local manufacturing, mainly foreign owned, had been small in scope. Political instability threatens prospects for economic reconstruction and repatriation of some 750,000 Liberian refugees who have fled to neighboring countries. The political impasse between the interim government and rebel leader Charles Taylor has prevented restoration of normal economic life, including the re-establishment of a strong central government with effective economic development programs.
National product: GDP—exchange rate conversion—$988 million (1988)
National product real growth rate: 1.5% (1988)
National product per capita: $400 (1988)
Inflation rate (consumer prices): 12% (1989)
Unemployment rate: 43% urban (1988)
Budget: revenues $242.1 million; expenditures $435.4 million, including capital expenditures of $29.5 million (1989)
Exports: $505 million (f.o.b., 1989 est.)
commodities: iron ore 61%, rubber 20%, timber 11%, coffee
partners: US, EC, Netherlands
Imports: $394 million (c.i.f., 1989 est.)
commodities: rice, mineral fuels, chemicals, machinery, transportation equipment, other foodstuffs
partners: US, EC, Japan, China, Netherlands, ECOWAS
External debt: $1.6 billion (December 1990 est.)
Industrial production: growth rate 1.5% in manufacturing (1987); accounts for 22% of GDP
Electricity: 410,000 kW capacity; 750 million kWh produced, 275 kWh per capita (1991)
Industries: rubber processing, food processing, construction materials, furniture, palm oil processing, mining (iron ore, diamonds)
Agriculture: accounts for about 40% of GDP (including fishing and forestry); principal products—rubber, timber, coffee, cocoa, rice, cassava, palm oil, sugarcane, bananas, sheep, goats; not self-sufficient in food, imports 25% of rice consumption
Economic aid: US commitments, including Ex-Im (FY70-89), $665 million; Western

(non-US) countries, ODA and OOF bilateral commitments (1970-89), $870 million; OPEC bilateral aid (1979-89), $25 million; Communist countries (1970-89), $77 million
Currency: 1 Liberian dollar (L$) = 100 cents
Exchange rates: Liberian dollars (L$) per US$1—1.00 (fixed rate since 1940); unofficial parallel exchange rate of L$7 = US$1, January 1992
Fiscal year: calendar year

Communications

Railroads: 480 km total; 328 km 1.435-meter standard gauge, 152 km 1.067-meter narrow gauge; all lines single track; rail systems owned and operated by foreign steel and financial interests in conjunction with Liberian Government
Highways: 10,087 km total; 603 km bituminous treated, 2,848 km all weather, 4,313 km dry weather; there are also 2,323 km of private, laterite-surfaced roads open to public use, owned by rubber and timber companies
Ports: Monrovia, Buchanan, Greenville, Harper (or Cape Palmas)
Merchant marine: 1,618 ships (1,000 GRT or over) totaling 57,769,476 DWT/ 101,391,576 DWT; includes 20 passenger, 1 short-sea passenger, 132 cargo, 56 refrigerated cargo, 21 roll-on/roll-off, 58 vehicle carrier, 97 container, 3 barge carrier, 499 oil tanker, 108 chemical, 68 combination ore/oil, 62 liquefied gas, 6 specialized tanker, 456 bulk, 31 combination bulk; note—a flag of convenience registry; all ships are foreign owned; the top 4 owning flags are US 16%, Japan 14%, Norway 11%, and Hong Kong 9%
Airports:
total: 59
usable: 41
with permanent-surface runways: 2
with runways over 3,659 m: 0
with runways 2,440-3,659 m: 1
with runways 1,220-2,439 m: 4
Telecommunications: telephone and telegraph service via radio relay network; main center is Monrovia; broadcast stations—3 AM, 4 FM, 5 TV; 1 Atlantic Ocean INTELSAT earth station; most telecommunications services inoperable due to insurgency movement

Defense Forces

Branches: the ultimate structure of the Liberian military force will depend on who is the victor in the ongoing civil war
Manpower availability: males age 15-49 684,681; fit for military service 365,518 (1993 est.)
Defense expenditures: exchange rate conversion—$NA, NA% of GDP

Libya

Geography

Location: Northern Africa, on the southern coast of the Mediterranean Sea, between Egypt and Tunisia
Map references: Africa, Standard Time Zones of the World
Area:
total area: 1,759,540 km²
land area: 1,759,540 km²
comparative area: slightly larger than Alaska
Land boundaries: total 4,383 km, Algeria 982 km, Chad 1,055 km, Egypt 1,150 km, Niger 354 km, Sudan 383 km, Tunisia 459 km
Coastline: 1,770 km
Maritime claims:
territorial sea: 12 nm
Gulf of Sidra closing line: 32 degrees 30 minutes north
International disputes: claims and occupies the Aozou Strip in northern Chad; maritime boundary dispute with Tunisia; Libya claims part of northern Niger and part of southeastern Algeria
Climate: Mediterranean along coast; dry, extreme desert interior
Terrain: mostly barren, flat to undulating plains, plateaus, depressions
Natural resources: petroleum, natural gas, gypsum
Land use:
arable land: 2%
permanent crops: 0%
meadows and pastures: 8%
forest and woodland: 0%
other: 90%
Irrigated land: 2,420 km² (1989 est.)
Environment: hot, dry, dust-laden ghibli is a southern wind lasting one to four days in spring and fall; desertification; sparse natural surface-water resources
Note: the Great Manmade River Project, the largest water development scheme in the world, is being built to bring water from large aquifers under the Sahara to coastal cities

People

Population: 4,872,598 (July 1993 est.)
Population growth rate: 3.73% (1993 est.)
Birth rate: 45.66 births/1,000 population (1993 est.)
Death rate: 8.37 deaths/1,000 population (1993 est.)
Net migration rate: 0 migrant(s)/1,000 population (1993 est.)
Infant mortality rate: 65.5 deaths/1,000 live births (1993 est.)
Life expectancy at birth:
total population: 63.47 years
male: 61.35 years
female: 65.7 years (1993 est.)
Total fertility rate: 6.44 children born/woman (1993 est.)
Nationality:
noun: Libyan(s)
adjective: Libyan
Ethnic divisions: Berber and Arab 97%, Greeks, Maltese, Italians, Egyptians, Pakistanis, Turks, Indians, Tunisians
Religions: Sunni Muslim 97%
Languages: Arabic, Italian, English, all are widely understood in the major cities
Literacy: age 15 and over can read and write (1990)
total population: 64%
male: 75%
female: 50%
Labor force: 1 million includes about 280,000 resident foreigners
by occupation: industry 31%, services 27%, government 24%, agriculture 18%

Government

Names:
conventional long form: Socialist People's Libyan Arab Jamahiriya
conventional short form: Libya
local long form: Al Jumahiriyah al Arabiyah al Libiyah ash Shabiyah al Ishirakiyah
local short form: none
Digraph: LY
Type: Jamahiriya (a state of the masses) in theory, governed by the populace through local councils; in fact, a military dictatorship
Capital: Tripoli
Administrative divisions: 25 municipalities (baladiyah, singular—baladiyat); Ajdabiya, Al 'Aziziyah, Al Fatih, Al Jabal al Akhdar, Al Jufrah, Al Khums, Al Kufrah, An Nuqat al Khams, Ash Shati', Awbari, Az Zawiyah, Banghazi, Darnah, Ghadamis, Gharyan, Misratah, Murzuq, Sabha, Sawfajjin, Surt, Tarabulus, Tarhunah, Tubruq, Yafran, Zlitan
Independence: 24 December 1951 (from Italy)
Constitution: 11 December 1969, amended 2 March 1977
Legal system: based on Italian civil law system and Islamic law; separate religious courts; no constitutional provision for judicial review of legislative acts; has not accepted compulsory ICJ jurisdiction
National holiday: Revolution Day, 1 September (1969)
Political parties and leaders: none
Other political or pressure groups: various Arab nationalist movements and the Arab Socialist Resurrection (Ba'th) party with almost negligible memberships may be functioning clandestinely, as well as some Islamic elements
Suffrage: 18 years of age; universal and compulsory
Elections: national elections are indirect through a hierarchy of peoples' committees
Executive branch: revolutionary leader, chairman of the General People's Committee (premier), General People's Committee (cabinet)
Legislative branch: unicameral General People's Congress
Judicial branch: Supreme Court
Leaders:
Chief of State: Revolutionary Leader Col. Mu'ammar Abu Minyar al-QADHAFI (since 1 September 1969)
Head of Government: Chairman of the General People's Committee (Premier) Abu Zayd 'umar DURDA (since 7 October 1990)
Member of: ABEDA, AfDB, AFESD, AL, AMF, AMU, CAEU, CCC, ECA, FAO, G-77, IAEA, IBRD, ICAO, IDA, IDB, IFAD, IFC, ILO, IMF, IMO, INTELSAT, INTERPOL, IOC, ITU, LORCS, NAM, OAPEC, OAU, OIC, OPEC, UN, UNCTAD, UNESCO, UNIDO, UPU, WHO, WIPO, WMO, WTO
Diplomatic representation in US: none
US diplomatic representation: none
Flag: plain green; green is the traditional color of Islam (the state religion)

Economy

Overview: The socialist-oriented economy depends primarily upon revenues from the oil sector, which contributes practically all export earnings and about one-third of GDP. In 1990 per capita GDP was the highest in Africa at $5,410, but GDP growth rates have slowed and fluctuate sharply in response to changes in the world oil market. Import restrictions and inefficient resource allocations have led to shortages of basic goods and foodstuffs, although the reopening of the Libyan-Tunisian border in April 1988 and the Libyan-Egyptian border in December 1989 have eased shortages. Austerity budgets and a lack of trained technicians have undermined the government's ability to implement a number of planned infrastructure development projects. Windfall revenues from the hike in world oil prices in late 1990 improved the foreign payments position and resulted in a current account surplus for the first time in five years. The nonoil manufacturing and construction sectors, which account for about 20% of GDP, have expanded from processing mostly agricultural products to include petrochemicals, iron, steel, and aluminum. Although agriculture accounts for only 5% of GDP, it employs about 20% of the labor force. Climatic conditions and poor soils severely limit farm output, and Libya imports about 75% of its food requirements.
National product: GDP—exchange rate conversion—$26.1 billion (1992 est.)
National product real growth rate: 0.2% (1992 est.)
National product per capita: $5,800 (1992 est.)
Inflation rate (consumer prices): 7% (1991 est.)
Unemployment rate: NA%
Budget: revenues $8.1 billion; expenditures $9.8 billion, including capital expenditures of $3.1 billion (1989 est.)
Exports: $9.71 billion (f.o.b., 1992)
commodities: crude oil, refined petroleum products, natural gas
partners: Italy, former USSR, Germany, Spain, France, Belgium/Luxembourg, Turkey
Imports: $8.66 billion (f.o.b., 1992)
commodities: machinery, transport equipment, food, manufactured goods
partners: Italy, former USSR, Germany, UK, Japan, Korea
External debt: $3.5 billion, excluding military debt (1991 est.)
Industrial production: growth rate 10.5%; accounts for 7.6% of GDP (not including oil) (1990)
Electricity: 4,935,000 kW capacity; 14,385 million kWh produced, 2,952 kWh per capita (1992)
Industries: petroleum, food processing, textiles, handicrafts, cement
Agriculture: 5% of GNP; cash crops—wheat, barley, olives, dates, citrus fruits, peanuts; 75% of food is imported
Economic aid: Western (non-US) countries, ODA and OOF bilateral commitments (1970-87), $242 million; no longer a recipient
Currency: 1 Libyan dinar (LD) = 1,000 dirhams
Exchange rates: Libyan dinars (LD) per US$1—0.2998 (January 1993), 0.3013 (1992), 0.2684 (1991), 0.2699 (1990), 0.2922 (1989), 0.2853 (1988)
Fiscal year: calendar year

Communications

Railroads: Libya has had no railroad in operation since 1965, all previous systems having been dismantled; current plans are to construct a standard gauge (1.435 m) line

Libya (continued)

from the Tunisian frontier to Tripoli and Misratah, then inland to Sabha, center of a mineral rich area, but there has been no progress; other plans made jointly with Egypt would establish a rail line from As Sallum, Egypt to Tobruk with completion set for mid-1994, progress unknown
Highways: 19,300 km total; 10,800 km bituminous/bituminous treated, 8,500 km crushed stone or earth
Inland waterways: none
Pipelines: crude oil 4,383 km; natural gas 1,947 km; petroleum products 443 km (includes liquified petroleum gas 256 km)
Ports: Tobruk, Tripoli, Banghazi, Misratah, Marsa al Burayqah, Ra's Lanuf, Ra's al Unif
Merchant marine: 32 ships (1,000 GRT or over) totaling 694,883 GRT/1,215,494 DWT; includes 4 short-sea passenger, 11 cargo, 4 roll-on/roll-off, 10 oil tanker, 1 chemical tanker, 2 liquefied gas
Airports:
total: 138
usable: 124
with permanent-surface runways: 56
with runways over 3,659 m: 9
with runways 2,440-3,659 m: 27
with runways 1,220-2,439 m: 47
Telecommunications: modern telecommunications system using radio relay, coaxial cable, tropospheric scatter, and domestic satellite stations; 370,000 telephones; broadcast stations—17 AM, 3 FM, 12 TV; satellite earth stations—1 Atlantic Ocean INTELSAT, 1 Indian Ocean INTELSAT, and 14 domestic; submarine cables to France and Italy; radio relay to Tunisia and Egypt; tropospheric scatter to Greece; planned ARABSAT and Intersputnik satellite stations

Defense Forces

Branches: Armed Peoples of the Libyan Arab Jamahiriyah (including Army, Navy, Air and Air Defense Command)
Manpower availability: males age 15-49 1,058,134; fit for military service 628,285; reach military age (17) annually 50,997 (1993 est.); conscription now being implemented
Defense expenditures: exchange rate conversion—$3.3 billion, 15% of GDP (1989 est.)

Liechtenstein

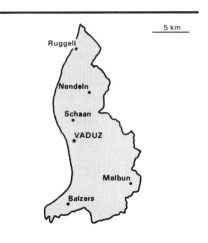

Geography

Location: Western Europe, between Austria and Switzerland
Map references: Europe, Standard Time Zones of the World
Area:
total area: 160 km^2
land area: 160 km^2
comparative area: about 0.9 times the size of Washington, DC
Land boundaries: total 78 km, Austria 37 km, Switzerland 41 km
Coastline: 0 km (landlocked)
Maritime claims: none; landlocked
International disputes: claims 620 square miles of Czech territory confiscated from its royal family in 1918; the Czech Republic insists that restitution does not go back before February 1948, when the Communists seized power
Climate: continental; cold, cloudy winters with frequent snow or rain; cool to moderately warm, cloudy, humid summers
Terrain: mostly mountainous (Alps) with Rhine Valley in western third
Natural resources: hydroelectric potential
Land use:
arable land: 25%
permanent crops: 0%
meadows and pastures: 38%
forest and woodland: 19%
other: 18%
Irrigated land: NA km^2
Environment: variety of microclimatic variations based on elevation
Note: landlocked

People

Population: 29,894 (July 1993 est.)
Population growth rate: 1.32% (1993 est.)
Birth rate: 13.15 births/1,000 population (1993 est.)
Death rate: 6.62 deaths/1,000 population (1993 est.)
Net migration rate: 6.66 migrant(s)/1,000 population (1993 est.)

Infant mortality rate: 5.3 deaths/1,000 live births (1993 est.)
Life expectancy at birth:
total population: 77.29 years
male: 73.65 years
female: 80.9 years (1993 est.)
Total fertility rate: 1.45 children born/woman (1993 est.)
Nationality:
noun: Liechtensteiner(s)
adjective: Liechtenstein
Ethnic divisions: Alemannic 95%, Italian and other 5%
Religions: Roman Catholic 87.3%, Protestant 8.3%, unknown 1.6%, other 2.8% (1988)
Languages: German (official), Alemannic dialect
Literacy: age 10 and over can read and write (1981)
total population: 100%
male: 100%
female: 100%
Labor force: 19,905 of which 11,933 are foreigners; 6,885 commute from Austria and Switzerland to work each day
by occupation: industry, trade, and building 53.2%, services 45%, agriculture, fishing, forestry, and horticulture 1.8% (1990)

Government

Names:
conventional long form: Principality of Liechtenstein
conventional short form: Liechtenstein
local long form: Furstentum Liechtenstein
local short form: Liechtenstein
Digraph: LS
Type: hereditary constitutional monarchy
Capital: Vaduz
Administrative divisions: 11 communes (gemeinden, singular—gemeinde); Balzers, Eschen, Gamprin, Mauren, Planken, Ruggell, Schaan, Schellenberg, Triesen, Triesenberg, Vaduz
Independence: 23 January 1719 (Imperial Principality of Liechtenstein established)
Constitution: 5 October 1921
Legal system: local civil and penal codes; accepts compulsory ICJ jurisdiction, with reservations
National holiday: Assumption Day, 15 August
Political parties and leaders: Fatherland Union (VU), Dr. Otto HASLER; Progressive Citizens' Party (FBP), Emanuel VOGT; Free Electoral List (FL)
Suffrage: 18 years of age; universal
Elections:
Diet: last held on 7 February 1993 (next to be held by March 1997); results—percent of vote by party NA; seats—(25 total) FBP 12, VU 11, FL 2

Executive branch: reigning prince, hereditary prince, head of government, deputy head of government
Legislative branch: unicameral Diet (Landtag)
Judicial branch: Supreme Court (Oberster Gerichtshof) for criminal cases, Superior Court (Obergericht) for civil cases
Leaders:
Chief of State: Prince Hans ADAM II (since 13 November 1989; assumed executive powers 26 August 1984); Heir Apparent Prince ALOIS von und zu Liechtenstein (born 11 June 1968)
Head of Government: Markus BUECHEL (since 7 February 1993); Deputy Head of Government Dr. Herbert WILLE (since 2 February 1986)
Member of: CE, CSCE, EBRD, ECE, EFTA, IAEA, INTELSAT, INTERPOL, IOC, ITU, LORCS, UN, UNCTAD, UPU, WCL, WIPO
Diplomatic representation in US: in routine diplomatic matters, Liechtenstein is represented in the US by the Swiss Embassy
US diplomatic representation: the US has no diplomatic or consular mission in Liechtenstein, but the US Consul General at Zurich (Switzerland) has consular accreditation at Vaduz
Flag: two equal horizontal bands of blue (top) and red with a gold crown on the hoist side of the blue band

Economy

Overview: The prosperous economy is based primarily on small-scale light industry and tourism. Industry accounts for 53% of total employment, the service sector 45% (mostly based on tourism), and agriculture and forestry 2%. The sale of postage stamps to collectors is estimated at $10 million annually. Low business taxes (the maximum tax rate is 20%) and easy incorporation rules have induced about 25,000 holding or so-called letter box companies to establish nominal offices in Liechtenstein. Such companies, incorporated solely for tax purposes, provide 30% of state revenues. The economy is tied closely to Switzerland's economy in a customs union, and incomes and living standards parallel those of the more prosperous Swiss groups.
National product: GDP—purchasing power equivalent—$630 million (1990 est.)
National product real growth rate: NA%
National product per capita: $22,300 (1990 est.)
Inflation rate (consumer prices): 5.4% (1990)
Unemployment rate: 1.5% (1990)
Budget: revenues $259 million; expenditures $292 million, including capital expenditures of $NA (1990)
Exports: $1.6 billion

commodities: small specialty machinery, dental products, stamps, hardware, pottery
partners: EFTA countries 20.9% (Switzerland 15.4%), EC countries 42.7%, other 36.4% (1990)
Imports: $NA
commodities: machinery, metal goods, textiles, foodstuffs, motor vehicles
partners: NA
External debt: $NA
Industrial production: growth rate NA%
Electricity: 23,000 kW capacity; 150 million kWh produced, 5,230 kWh per capita (1992)
Industries: electronics, metal manufacturing, textiles, ceramics, pharmaceuticals, food products, precision instruments, tourism
Agriculture: livestock, vegetables, corn, wheat, potatoes, grapes
Economic aid: none
Currency: 1 Swiss franc, franken, or franco (SwF) = 100 centimes, rappen, or centesimi
Exchange rates: Swiss francs, franken, or franchi (SwF) per US$1—1.4781 (January 1993), 1.4062 (1992), 1.4340 (1991), 1.3892 (1990), 1.6359 (1989), 1.4633 (1988)
Fiscal year: calendar year

Communications

Railroads: 18.5 km 1.435-meter standard gauge, electrified; owned, operated, and included in statistics of Austrian Federal Railways
Highways: 130.66 km main roads, 192.27 km byroads
Airports: none
Telecommunications: limited, but sufficient automatic telephone system; 25,400 telephones; linked to Swiss networks by cable and radio relay for international telephone, radio, and TV services

Defense Forces

Note: defense is responsibility of Switzerland

Lithuania

Final boundaries of Estonia, Latvia, and Lithuania with the former Soviet Union are expected to be confirmed by agreement.

150 km

Geography

Location: Eastern Europe, bordering the Baltic Sea, between Sweden and Russia
Map references: Asia, Europe, Standard Time Zones of the World
Area:
total area: 65,200 km²
land area: 65,200 km²
comparative area: slightly larger than West Virginia
Land boundaries: total 1,273 km, Belarus 502 km, Latvia 453 km, Poland 91 km, Russia (Kaliningrad) 227 km
Coastline: 108 km
Maritime claims:
territorial sea: 12 nm
International disputes: dispute with Russia (Kaliningrad Oblast) over the position of the Neman River border presently located on the Lithuanian bank and not in midriver as by international standards
Climate: maritime; wet, moderate winters
Terrain: lowland, many scattered small lakes, fertile soil
Natural resources: peat
Land use:
arable land: 49.1%
permanent crops: 0%
meadows and pastures: 22.2%
forest and woodland: 16.3%
other: 12.4%
Irrigated land: 430 km² (1990)
Environment: risk of accidents from the two Chernobyl-type reactors at the Ignalina Nuclear Power Plant; contamination of soil and groundwater with petroleum products and chemicals at military bases

People

Population: 3,819,638 (July 1993 est.)
Population growth rate: 0.76% (1993 est.)
Birth rate: 14.95 births/1,000 population (1993 est.)
Death rate: 10.94 deaths/1,000 population (1993 est.)

Lithuania *(continued)*

Net migration rate: 3.62 migrant(s)/1,000 population (1993 est.)
Infant mortality rate: 16.9 deaths/1,000 live births (1993 est.)
Life expectancy at birth:
total population: 71.12 years
male: 66.39 years
female: 76.08 years (1993 est.)
Total fertility rate: 2.03 children born/woman (1993 est.)
Nationality:
noun: Lithuanian(s)
adjective: Lithuanian
Ethnic divisions: Lithuanian 80.1%, Russian 8.6%, Polish 7.7%, Belarusian 1.5%, other 2.1%
Religions: Roman Catholic, Lutheran, other
Languages: Lithuanian (official), Polish, Russian
Literacy: age 9-49 can read and write (1970)
total population: 100%
male: 100%
female: 100%
Labor force: 1.836 million
by occupation: industry and construction 42%, agriculture and forestry 18%, other 40% (1990)

Government

Names:
conventional long form: Republic of Lithuania
conventional short form: Lithuania
local long form: Lietuvos Respublika
local short form: Lietuva
former: Lithuanian Soviet Socialist Republic
Digraph: LH
Type: republic
Capital: Vilnius
Administrative divisions: NA districts
Independence: 6 September 1991 (from Soviet Union)
Constitution: adopted 25 October 1992
Legal system: based on civil law system; no judicial review of legislative acts
National holiday: Independence Day, 16 February
Political parties and leaders: Christian Democratic Party, Egidijus KLUMBYS, chairman; Democratic Labor Party of Lithuania, Algirdas Mykolas BRAZAUSKAS, chairman; Lithuanian Democratic Party, Saulius PECELIUNAS, chairman; Lithuanian Green Party, Irena IGNATAVICIENE, chairwoman; Lithuanian Humanism Party, Vytautas KAZLAUSKAS, chairman; Lithuanian Independence Party, Virgilijus CEPAITIS, chairman; Lithuanian Liberty League, Antanas TERLECKAS; Lithuanian Liberal Union, Vytautus RADZVILAS, chairman; Lithuanian Nationalist Union, Rimantas SMETONA, chairman; Lithuanian Social Democratic

Party, Aloizas SAKALAS, chairman; Union of the Motherland, Vytavtas LANDSBERGIS, chairman
Other political or pressure groups: Sajudis; Lithuanian Future Forum; Farmers Union
Suffrage: 18 years of age; universal
Elections:
President: last held 14 February 1993 (next to be held NA); results—Algirdas BRAZAUSKAS was elected
Seimas (parliament): last held 26 October and 25 November 1992 (next to be held NA); results—Democratic Labor Party 51%; seats—(141 total) Democratic Labor Party 73
Executive branch: president, prime minister, cabinet
Legislative branch: unicameral Seimas (parliament)
Judicial branch: Supreme Court, Court of Appeals
Leaders:
Chief of State: Seimas Chairman and Acting President Algirdas Mykolas BRAZAUSKAS (since 15 November 1992); Deputy Seimas Chairmen Aloyzas SAKALAS (since NA December 1992) and Egidius BICKAUSKAS (since NA December 1992)
Head of Government: Premier Adolfas SLEZEVICIUS (since NA)
Member of: CBSS, CSCE, EBRD, ECE, FAO, IBRD, ICAO, ILO, IMF, INTERPOL, ITU, NACC, UN, UNCTAD, UNESCO, UNIDO, UPU, WHO, WIPO
Diplomatic representation in US:
chief of mission: Ambassador Stasys LOZORAITIS, Jr.
chancery: 2622 16th St. NW, Washington, DC 20009
telephone: (202) 234-5860, 2639
FAX: (202) 328-0466
consulate general: New York
US diplomatic representation:
chief of mission: Ambassador Darryl N. JOHNSON
embassy: Akmenu 6, Vilnius 232600
mailing address: APO AE 09723
telephone: 011 [7] (012-2) 222-031
FAX: 011 [7] (012-2) 222-779
Flag: three equal horizontal bands of yellow (top), green, and red

---

Economy

Overview: Lithuania is striving to become an independent privatized economy. Although it was substantially above average in living standards and technology in the old USSR, Lithuania historically lagged behind Latvia and Estonia in economic development. The country has no important natural resources aside from its arable land and strategic location. Industry depends entirely on imported materials that have

come from the republics of the former USSR. Lithuania benefits from its ice-free port at Klaipeda on the Baltic Sea and its rail and highway hub at Vilnius, which provides land communication between Eastern Europe and Russia, Latvia, Estonia, and Belarus. Industry produces a small assortment of high-quality products, ranging from complex machine tools to sophisticated consumer electronics. Because of nuclear power, Lithuania is presently self-sufficient in electricity, exporting its surplus to Latvia and Belarus; the nuclear facilities inherited from the USSR, however, have come under world scrutiny as seriously deficient in safety standards. Agriculture is efficient compared with most of the former Soviet Union. Lithuania held first place in per capita consumption of meat, second place for eggs and potatoes, and fourth place for milk and dairy products. Grain must be imported to support the meat and dairy industries. Lithuania is pressing ahead with plans to privatize at least 60% of state-owned property (industry, agriculture, and housing), having already sold almost all housing and many small enterprises using a voucher system. Other government priorities include encouraging foreign investment by protecting the property rights of foreign firms and redirecting foreign trade away from Eastern markets to the more competitive Western markets. For the moment, Lithuania will remain highly dependent on Russia for energy, raw materials, grains, and markets for its products. In 1992, output plummeted by 30% because of cumulative problems with inputs and with markets, problems that were accentuated by the phasing out of the Russian ruble as the medium of exchange.
National product: GDP $NA
National product real growth rate: -30% (1992 est.)
National product per capita: $NA
Inflation rate (consumer prices): 10%-20% per month (first quarter 1993)
Unemployment rate: 1% (February 1993); but large numbers of underemployed workers
Budget: revenues $258.5 million; expenditures $270.2 million, including capital expenditures of $NA (1992 est.)
Exports: $NA
commodities: electronics 18%, petroleum products 5%, food 10%, chemicals 6% (1989)
partners: Russia 40%, Ukraine 16%, other former Soviet republics 32%, West 12%
Imports: $NA
commodities: oil 24%, machinery 14%, chemicals 8%, grain NA% (1989)
partners: Russia 62%, Belarus 18%, former Soviet republics 10%, West 10%
External debt: $650 million (1991 est.)

Industrial production: growth rate –50% (1992 est.)

Electricity: 5,925,000 kW capacity; 25,000 million kWh produced, 6,600 kWh per capita (1992)

Industries: employs 25% of the labor force; shares in the total production of the former USSR are: metal-cutting machine tools 6.6%; electric motors 4.6%; television sets 6.2%; refrigerators and freezers 5.4%; other branches: petroleum refining, shipbuilding (small ships), furniture making, textiles, food processing, fertilizers, agricultural machinery, optical equipment, electronic components, computers, and amber

Agriculture: employs around 20% of labor force; sugar, grain, potatoes, sugarbeets, vegetables, meat, milk, dairy products, eggs, fish; most developed are the livestock and dairy branches, which depend on imported grain; net exporter of meat, milk, and eggs

Illicit drugs: transshipment point for illicit drugs from Central and Southwest Asia to Western Europe; limited producer of illicit opium; mostly for domestic consumption

Economic aid: US commitments, including Ex-Im (1992), $10 million; Western (non-US) countries, ODA and OOF bilateral commitments (1970-86), $NA million; Communist countries (1971-86), $NA million

Currency: using talonas as temporary currency (March 1993), but planning introduction of convertible litas (late 1993)

Exchange rates: NA

Fiscal year: calendar year

Communications

Railroads: 2,100 km; does not include industrial lines (1990)

Highways: 44,200 km total 35,500 km hard surfaced, 8,700 km earth (1990)

Inland waterways: 600 km perennially navigable

Pipelines: crude oil 105 km, natural gas 760 km (1992)

Ports: coastal—Klaipeda; inland—Kaunas

Merchant marine: 46 ships (1,000 GRT or over) totaling 282,633 GRT/332,447 DWT; includes 31 cargo, 3 railcar carrier, 1 roll-on/roll-off, 11 combination bulk

Airports:
total: 96
useable: 19
with permanent-surface runways: 12
with runways over 3,659 m: 0
with runways 2,440-3,659 m: 5
with runways 1,220-2,439 m: 11

Telecommunications: better developed than in most other former USSR republics; operational NMT-450 analog cellular network in Vilnius; fiber optic cable installed beween Vilnius and Kaunas; 224 telephones per 1000 persons; broadcast stations—13 AM, 26 FM, 1 SW, 1 LW, 3 TV; landlines or microwave to former USSR republics; leased connection to the Moscow international switch for traffic with other countries; satellite earth stations—(8 channels to Norway); new international digital telephone exchange in Kaunas for direct access to 13 countries via satellite link out of Copenhagen, Denmark

---

Defense Forces

Branches: Ground Forces, Navy, Air Force, Security Forces (internal and border troops), National Guard (Skat)

Manpower availability: males age 15-49 933,245; fit for military service 739,400; reach military age (18) annually 27,056 (1993 est.)

Defense expenditures: exchange rate conversion—$NA, 5.5% of GDP (1993 est.)

Luxembourg

Geography

Location: Western Europe, between Belgium and Germany

Map references: Europe, Standard Time Zones of the World

Area:
total area: 2,586 km²
land area: 2,586 km²
comparative area: slightly smaller than Rhode Island

Land boundaries: total 359 km, Belgium 148 km, France 73 km, Germany 138 km

Coastline: 0 km (landlocked)

Maritime claims: none; landlocked

International disputes: none

Climate: modified continental with mild winters, cool summers

Terrain: mostly gently rolling uplands with broad, shallow valleys; uplands to slightly mountainous in the north; steep slope down to Moselle floodplain in the southeast

Natural resources: iron ore (no longer exploited)

Land use:
arable land: 24%
permanent crops: 1%
meadows and pastures: 20%
forest and woodland: 21%
other: 34%

Irrigated land: NA km²

Environment: deforestation

Note: landlocked

People

Population: 398,220 (July 1993 est.)

Population growth rate: 1.04% (1993 est.)

Birth rate: 12.96 births/1,000 population (1993 est.)

Death rate: 9.56 deaths/1,000 population (1993 est.)

Net migration rate: 6.97 migrant(s)/1,000 population (1993 est.)

Infant mortality rate: 6.9 deaths/1,000 live births (1993 est.)

Luxembourg (continued)

Life expectancy at birth:
total population: 76.43 years
male: 72.71 years
female: 80.3 years (1993 est.)
Total fertility rate: 1.63 children born/woman (1993 est.)
Nationality:
noun: Luxembourger(s)
adjective: Luxembourg
Ethnic divisions: Celtic base (with French and German blend), Portuguese, Italian, and European (guest and worker residents)
Religions: Roman Catholic 97%, Protestant and Jewish 3%
Languages: Luxembourgisch, German, French, English
Literacy: age 15 and over can read and write (1980)
total population: 100%
male: 100%
female: 100%
Labor force: 177,300 one-third of labor force is foreign workers, mostly from Portugal, Italy, France, Belgium, and Germany
by occupation: services 65%, industry 31.6%, agriculture 3.4% (1988)

Government

Names:
conventional long form: Grand Duchy of Luxembourg
conventional short form: Luxembourg
local long form: Grand-Duche de Luxembourg
local short form: Luxembourg
Digraph: LU
Type: constitutional monarchy
Capital: Luxembourg
Administrative divisions: 3 districts; Diekirch, Grevenmacher, Luxembourg
Independence: 1839
Constitution: 17 October 1868, occasional revisions
Legal system: based on civil law system; accepts compulsory ICJ jurisdiction
National holiday: National Day, 23 June (1921) (public celebration of the Grand Duke's birthday)
Political parties and leaders: Christian Social Party (CSV), Jacques SANTER; Socialist Workers Party (LSAP), Jacques POOS; Liberal (DP), Colette FLESCH; Communist (KPL), Andre HOFFMANN; Green Alternative (GAP), Jean HUSS
Other political or pressure groups: group of steel companies representing iron and steel industry; Centrale Paysanne representing agricultural producers; Christian and Socialist labor unions; Federation of Industrialists; Artisans and Shopkeepers Federation
Suffrage: 18 years of age; universal and compulsory

Elections:
Chamber of Deputies: last held on 18 June 1989 (next to be held by June 1994); results—CSV 31.7%, LSAP 27.2%, DP 16.2%, Greens 8.4%, PAC 7.3%, KPL 5.1%, other 4.1%; seats—(60 total) CSV 22, LSAP 18, DP 11, Greens 4, PAC 4, KPL 1
Executive branch: grand duke, prime minister, vice prime minister, Council of Ministers (cabinet)
Legislative branch: unicameral Chamber of Deputies (Chambre des Deputes); note—the Council of State (Conseil d'Etat) is an advisory body whose views are considered by the Chamber of Deputies
Judicial branch: Superior Court of Justice (Cour Superieure de Justice)
Leaders:
Chief of State: Grand Duke JEAN (since 12 November 1964); Heir Apparent Prince HENRI (son of Grand Duke Jean, born 16 April 1955)
Head of Government: Prime Minister Jacques SANTER (since 21 July 1984); Vice Prime Minister Jacques F. POOS (since 21 July 1984)
Member of: ACCT, Australia Group, Benelux, CCC, CE, COCOM, CSCE, EBRD, EC, ECE, EIB, FAO, GATT, IAEA, IBRD, ICAO, ICC, ICFTU, IDA, IEA, IFAD, IFC, ILO, IMF, IMO, INTELSAT, INTERPOL, IOC, IOM, ITU, LORCS, MTCR, NACC, NATO, NEA, NSG, OECD, PCA, UN, UNCTAD, UNESCO, UNIDO, UNPROFOR, UPU, WCL, WEU, WHO, WIPO, WMO, ZC
Diplomatic representation in US:
chief of mission: Ambassador Alphonse BERNS
chancery: 2200 Massachusetts Avenue NW, Washington, DC 20008
telephone: (202) 265-4171
FAX: (202) 328-8270
consulates general: New York and San Francisco
US diplomatic representation:
chief of mission: Ambassador Edward M. ROWELL
embassy: 22 Boulevard Emmanuel-Servais, 2535 Luxembourg City
mailing address: PSC 11, APO AE 09132-5380
telephone: [352] 460123
FAX: [352] 461401
Flag: three equal horizontal bands of red (top), white, and light blue; similar to the flag of the Netherlands, which uses a darker blue and is shorter; design was based on the flag of France

Economy

Overview: The stable economy features moderate growth, low inflation, and negligible unemployment. Agriculture is based on small but highly productive family-owned farms. The industrial sector, until recently dominated by steel, has become increasingly more diversified, particularly toward high-technology firms. During the past decade, growth in the financial sector has more than compensated for the decline in steel. Services, especially banking, account for a growing proportion of the economy. Luxembourg participates in an economic union with Belgium on trade and most financial matters and is also closely connected economically to the Netherlands.
National product: GDP—purchasing power equivalent—$8.5 billion (1992)
National product real growth rate: 2.5% (1992)
National product per capita: $21,700 (1992)
Inflation rate (consumer prices): 3.6% (1992)
Unemployment rate: 1.4% (1991)
Budget: revenues $3.5 billion; expenditures $3.5 billion, including capital expenditures of $NA (1992)
Exports: $6.4 billion (f.o.b., 1991 est.)
commodities: finished steel products, chemicals, rubber products, glass, aluminum, other industrial products
partners: EC 76%, US 5%
Imports: $8.3 billion (c.i.f., 1991 est.)
commodities: minerals, metals, foodstuffs, quality consumer goods
partners: Belgium 37%, FRG 31%, France 12%, US 2%
External debt: $131.6 million (1989 est.)
Industrial production: growth rate −0.5% (1990); accounts for 25% of GDP
Electricity: 1,238,750 kW capacity; 1,375 million kWh produced, 3,450 kWh per capita (1990)
Industries: banking, iron and steel, food processing, chemicals, metal products, engineering, tires, glass, aluminum
Agriculture: accounts for less than 3% of GDP (including forestry); principal products—barley, oats, potatoes, wheat, fruits, wine grapes; cattle raising widespread
Illicit drugs: money-laundering hub
Economic aid: none
Currency: 1 Luxembourg franc (LuxF) = 100 centimes
Exchange rates: Luxembourg francs (LuxF) per US$1—33.256 (January 1993), 32.150 (1992), 34.148 (1991), 33.418 (1990), 39.404 (1989), 36.768 (1988); note—the Luxembourg franc is at par with the Belgian franc, which circulates freely in Luxembourg
Fiscal year: calendar year

Communications

Railroads: Luxembourg National Railways (CFL) operates 272 km 1.435-meter standard gauge; 178 km double track; 178 km electrified

Highways: 5,108 km total; 4,995 km paved, 57 km gravel, 56 km earth; about 80 km limited access divided highway
Inland waterways: 37 km; Moselle River
Pipelines: petroleum products 48 km
Ports: Mertert (river port)
Merchant marine: 53 ships (1,000 GRT or over) totaling 1,570,466 GRT/2,614,154 DWT; includes 2 cargo, 5 container, 5 roll-on/roll-off, 6 oil tanker, 4 chemical tanker, 3 combination ore/oil, 8 liquefied gas, 2 passenger, 8 bulk, 6 combination bulk, 4 refrigerated cargo
Airports:
total: 2
usable: 2
with permanent-surface runways: 1
with runways over 3,659 m: 1
with runways 2,440-3,659 m: 0
with runways 1,220-2,439 m: 1
Telecommunications: highly developed, completely automated and efficient system, mainly buried cables; 230,000 telephones; broadcast stations—2 AM, 3 FM, 3 TV; 3 channels leased on TAT-6 coaxial submarine cable; 1 direct-broadcast satellite earth station; nationwide mobile phone system

---

Defense Forces

Branches: Army, National Gendarmerie
Manpower availability: males age 15-49 103,607; fit for military service 86,003; reach military age (19) annually 2,227 (1993 est.)
Defense expenditures: exchange rate conversion—$100 million, 1.2% of GDP (1992)

Macau
(overseas territory of Portugal)

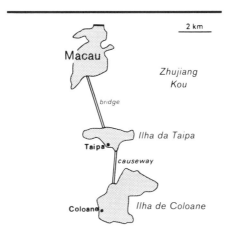

Geography

Location: East Asia, 27 km west-southwest of Hong Kong on the southeast coast of China bordering the South China Sea
Map references: Asia, Oceania, Southeast Asia, Standard Time Zones of the World
Area:
total area: 16 km²
land area: 16 km²
comparative area: about 0.1 times the size of Washington, DC
Land boundaries: total 0.34 km, China 0.34 km
Coastline: 40 km
Maritime claims: not specified
International disputes: none
Climate: subtropical; marine with cool winters, warm summers
Terrain: generally flat
Natural resources: negligible
Land use:
arable land: 0%
permanent crops: 0%
meadows and pastures: 0%
forest and woodland: 0%
other: 100%
Irrigated land: NA km²
Environment: essentially urban; one causeway and one bridge connect the two islands to the peninsula on mainland

---

People

Population: 477,850 (July 1993 est.)
Population growth rate: 1.44% (1993 est.)
Birth rate: 14.99 births/1,000 population (1993 est.)
Death rate: 4.05 deaths/1,000 population (1993 est.)
Net migration rate: 3.45 migrant(s)/1,000 population (1993 est.)
Infant mortality rate: 5.5 deaths/1,000 live births (1993 est.)
Life expectancy at birth:
total population: 79.64 years
male: 77.24 years
female: 82.17 years (1993 est.)

Total fertility rate: 1.44 children born/woman (1993 est.)
Nationality:
noun: Macanese (singular and plural)
adjective: Macau
Ethnic divisions: Chinese 95%, Portuguese 3%, other 2%
Religions: Buddhist 45%, Roman Catholic 7%, Protestant 1%, none 45.8%, other 1.2% (1981)
Languages: Portuguese (official), Cantonese is the language of commerce
Literacy: age 15 and over can read and write (1981)
total population: 90%
male: 93%
female: 86%
Labor force: 180,000 (1986)
by occupation: NA

Government

Names:
conventional long form: none
conventional short form: Macau
local long form: none
local short form: Ilha de Macau
Digraph: MC
Type: overseas territory of Portugal scheduled to revert to China in 1999
Capital: Macau
Administrative divisions: 2 districts (concelhos, singular—concelho); Ilhas, Macau
Independence: none (territory of Portugal; Portugal signed an agreement with China on 13 April 1987 to return Macau to China on 20 December 1999; in the joint declaration, China promises to respect Macau's existing social and economic systems and lifestyle for 50 year after transition)
Constitution: 17 February 1976, Organic Law of Macau; basic law drafted primarily by Beijing awaiting final approval
Legal system: Portuguese civil law system
National holiday: Day of Portugal, 10 June
Political parties and leaders: Association to Defend the Interests of Macau; Macau Democratic Center; Group to Study the Development of Macau; Macau Independent Group
Other political or pressure groups: wealthy Macanese and Chinese representing local interests, wealthy pro-Communist merchants representing China's interests; in January 1967 the Macau Government acceded to Chinese demands that gave China veto power over administration
Suffrage: 18 years of age; universal
Elections:
Legislative Assembly: last held on 10 March 1991; results—percent of vote by party NA; seats—(23 total; 8 elected by universal suffrage, 8 by indirect suffrage, and 7 appointed by the governor) number of seats by party NA

Macau *(continued)*

Executive branch: president of Portugal, governor, Consultative Council (cabinet)
Legislative branch: unicameral Legislative Assembly
Judicial branch: Supreme Court
Leaders:
Chief of State: President (of Portugal) Mario Alberto SOARES (since 9 March 1986)
Head of Government: Governor Gen. Vasco Joachim Rocha VIEIRA (since 20 March 1991)
Member of: ESCAP (associate), GATT, IMO (associate), WTO (associate)
Diplomatic representation in US: as Chinese territory under Portuguese administration, Macanese interests in the US are represented by Portugal
US diplomatic representation: the US has no offices in Macau, and US interests are monitored by the US Consulate General in Hong Kong
Flag: the flag of Portugal is used

Economy

Overview: The economy is based largely on tourism (including gambling) and textile and fireworks manufacturing. Efforts to diversify have spawned other small industries—toys, artificial flowers, and electronics. The tourist sector has accounted for roughly 25% of GDP, and the clothing industry has provided about two-thirds of export earnings; the gambling industry represented well over 40% of GDP in 1992. Macau depends on China for most of its food, fresh water, and energy imports. Japan and Hong Kong are the main suppliers of raw materials and capital goods.
National product: GDP—exchange rate conversion—$3.1 billion (1991)
National product real growth rate: 3.1% (1991)
National product per capita: $6,700 (1991)
Inflation rate (consumer prices): 8.2% (1991 est.)
Unemployment rate: 2% (1991 est.)
Budget: revenues $305 million; expenditures $298 million, including capital expenditures of $NA (1989)
Exports: $1.8 billion (1992 est.)
commodities: textiles, clothing, toys
partners: US 36%, Hong Kong 13%, Germany 12%, France 8% (1991)
Imports: $2.0 billion (1992 est.)
commodities: raw materials, foodstuffs, capital goods
partners: Hong Kong 35%, China 22%, Japan 17% (1991)
External debt: $91 million (1985)
Industrial production: NA
Electricity: 258,000 kW capacity; 855 million kWh produced, 1,806 kWh per capita (1992)

Industries: clothing, textiles, toys, plastic products, furniture, tourism
Agriculture: rice, vegetables; food shortages—rice, vegetables, meat; depends mostly on imports for food requirements
Economic aid: none
Currency: 1 pataca (P) = 100 avos
Exchange rates: patacas (P) per US$1—8.034 (1991), 8.024 (1990), 8.030 (1989), 8.044 (1988), 7.993 (1987); note—linked to the Hong Kong dollar at the rate of 1.03 patacas per Hong Kong dollar
Fiscal year: calendar year

Communications

Highways: 42 km paved
Ports: Macau
Airports: none useable, 1 under construction; 1 seaplane station
Telecommunications: fairly modern communication facilities maintained for domestic and international services; 52,000 telephones; broadcast stations—4 AM, 3 FM, no TV (TV programs received from Hong Kong); 115,000 radio receivers (est.); international high-frequency radio communication facility; access to international communications carriers provided via Hong Kong and China; 1 Indian Ocean INTELSAT earth station

Defense Forces

Manpower availability: males age 15-49 137,738; fit for military service 77,159 (1993 est.)
Note: defense is responsibility of Portugal

Macedonia

Macedonia has proclaimed independent statehood but has not been formally recognized as a state by the United States.

Note: Macedonia has proclaimed independent statehood but has not been formally recognized as a state by the United States.

Geography

Location: Southern Europe, between Serbia and Montenegro and Greece
Map references: Ethnic Groups in Eastern Europe, Europe, Standard Time Zones of the World
Area:
total area: 25,333 km²
land area: 24,856 km²
comparative area: slightly larger than Vermont
Land boundaries: total 748 km, Albania 151 km, Bulgaria 148 km, Greece 228 km, Serbia and Montenegro 221 km (all with Serbia)
Coastline: 0 km (landlocked)
Maritime claims: none; landlocked
International disputes: Greece claims republic's name implies territorial claims against Aegean Macedonia
Climate: hot, dry summers and autumns and relatively cold winters with heavy snowfall
Terrain: mountainous territory covered with deep basins and valleys; there are three large lakes, each divided by a frontier line
Natural resources: chromium, lead, zinc, manganese, tungsten, nickel, low-grade iron ore, asbestos, sulphur, timber
Land use:
arable land: 5%
permanent crops: 5%
meadows and pastures: 20%
forest and woodland: 30%
other: 40%
Irrigated land: NA km²
Environment: Macedonia suffers from high seismic hazard; air pollution from metallurgical plants
Note: landlocked; major transportation corridor from Western and Central Europe to Aegean Sea and Southern Europe to Western Europe

People

Population: 2,193,951 (July 1993 est.)
Population growth rate: 0.91% (1993 est.)
Birth rate: 15.91 births/1,000 population (1993 est.)
Death rate: 6.79 deaths/1,000 population (1993 est.)
Net migration rate: 0 migrant(s)/1,000 population (1993 est.)
Infant mortality rate: 29.7 deaths/1,000 live births (1993 est.)
Life expectancy at birth:
total population: 73.19 years
male: 71.15 years
female: 75.41 years (1993 est.)
Total fertility rate: 2 children born/woman (1993 est.)
Nationality:
noun: Macedonian(s)
adjective: Macedonian
Ethnic divisions: Macedonian 67%, Albanian 21%, Turkish 4%, Serb 2%, other 6%
Religions: Eastern Orthodox 59%, Muslim 26%, Catholic 4%, Protestant 1%, other 10%
Languages: Macedonian 70%, Albanian 21%, Turkish 3%, Serbo-Croatian 3%, other 3%
Literacy:
total population: NA%
male: NA%
female: NA%
Labor force: 507,324
by occupation: agriculture 8%, manufacturing and mining 40% (1990)

Government

Names:
conventional long form: Republic of Macedonia
conventional short form: Macedonia
local long form: Republika Makedonija
local short form: Makedonija
Digraph: MK
Type: emerging democracy
Capital: Skopje
Administrative divisions: 34 districts (opcine, singular—opcina) Berovo, Bitola, Brod, Debar, Delcevo, Demir Hisar, Gevgelija, Gostivar, Kavadarci, Kicevo, Kocani, Kratovo, Kriva Palanka, Krusevo, Kumanovo, Negotino, Ohrid, Prilep, Probistip, Radovis, Resen, Skopje-Centar, Skopje-Cair, Skopje-Karpos, Skopje-Kisela Voda, Skopje-Gazi Baba, Stip, Struga, Strumica, Sveti Nikole, Tetovo, Titov Veles, Valandovo, Vinica
Independence: 20 November 1991 (from Yugoslavia)
Constitution: adopted 17 November 1991, effective 20 November 1991
Legal system: based on civil law system; judicial review of legislative acts

National holiday: NA
Political parties and leaders:
Social-Democratic League of Macedonia (SDSM; former Communist Party), Branko CRVENKOVSKI, president; Party for Democratic Prosperity in Macedonia (PDPM), Nevzat HALILI, president; National Democratic Party (PDP), Ilijas HALINI, president; Alliance of Reform Forces of Macedonia (SRSM), Stojan ANDOV, president; Socialist Party of Macedonia (SPM), Kiro POPOVSKI, president; Internal Macedonian Revolutionary Organization—Democratic Party for Macedonian National Unity (VMRO-DPMNE), Ljupco GEORGIEVSKI, president; Party of Yugoslavs in Macedonia (SJM), Milan DURCINOV, president
Other political or pressure groups:
Movement for All Macedonian Action (MAAK); League for Democracy; Albanian Democratic Union-Liberal Party
Suffrage: 18 years of age; universal
Elections:
President: last held 27 January 1991 (next to be held NA); results—Kiro GLIGOROV was elected by the Assembly
Assembly: last held 11 and 25 November and 9 December 1990 (next to be held NA); results—percent of vote by party NA; seats—(120 total) VMRO-DPMNE 37, SDSM 31, PDPM 25, SRSM 17, SJM 1, SPM 5, others 4
Executive branch: president, Council of Ministers, prime minister
Legislative branch: unicameral Assembly (Sobranje)
Judicial branch: Constitutional Court, Judicial Court of the Republic
Leaders:
Chief of State: President Kiro GLIGOROV (since 27 January 1991)
Head of Government: Prime Minister Branko CRVENKOVSKI (since NA September 1992), Deputy Prime Ministers Jovan ANDONOV (since NA March 1991), Stevo CRVENKOVSKI (since NA September 1992), and Becir ZUTA (since NA March 1991)
Member of: EBRD, ICAO, IMF, UN, UNCTAD, WMO
Diplomatic representation in US: none; US does not recognize Macedonia
US diplomatic representation: none; US does not recognize Macedonia
Flag: 16-point gold sun (Vergino, Sun) centered on a red field

Economy

Overview: Macedonia, although the poorest among the six republics of a dissolved Yugoslav federation, can meet basic food and energy needs through its own agricultural and coal resources. It will, however, move down toward a bare subsistence level of life unless economic ties are reforged or enlarged with its neighbors Serbia and Montenegro, Albania, Greece, and Bulgaria. The economy depends on outside sources for all of its oil and gas and its modern machinery and parts. Continued political turmoil, both internally and in the region as a whole, prevents any swift readjustments of trade patterns and economic programs. Inflation in early 1992 was out of control, the result of fracturing trade links, the decline in economic activity, and general uncertainties about the future status of the country; prices rose 38% in March 1992 alone. In August 1992, Greece, angry at the use of "Macedonia" as the republic's name, imposed a partial blockade for several months. This blockade, combined with the effects of the UN sanctions on Serbia and Montenegro, cost the economy approximately $1 billion in 1992 according to official figures. Macedonia's geographical isolation, technological backwardness, and potential political instability place it far down the list of countries of interest to Western investors. Resolution of the dispute with Greece and an internal commitment to economic reform would help to encourage foreign investment over the long run. In the immediate future, the worst scenario for the economy would be the spread of fighting across its borders.
National product: GDP—purchasing power equivalent—$7.1 billion (1991 est.)
National product real growth rate: -18% (1991 est.)
National product per capita: $3,110 (1991 est.)
Inflation rate (consumer prices): 114.9% (1991 est.)
Unemployment rate: 20% (1991 est.)
Budget: revenues $NA; expenditures $NA, including capital expenditures of $NA
Exports: $578 million (1990)
commodities: manufactured goods 40%, machinery and transport equipment 14%, miscellaneous manufactured articles 23%, raw materials 7.6%, food (rice) and live animals 5.7%, beverages and tobacco 4.5%, chemicals 4.7%
partners: principally Serbia and Montenegro and the other former Yugoslav republics, Germany, Greece, Albania
Imports: $1,112 million (1990)
commodities: fuels and lubricants 19%, manufactured goods 18%, machinery and transport equipment 15%, food and live animals 14%, chemicals 11.4%, raw materials 10%, miscellaneous manufactured articles 8.0%, beverages and tobacco 3.5%
partners: other former Yugoslav republics, Greece, Albania, Germany, Bulgaria
External debt: $845.8 million
Industrial production: growth rate −18% (1991 est.)

Macedonia *(continued)*

Electricity: 1,600,000 kw capacity; 6,300 million kWh produced, 2,900 kWh per capita (1992)
Industries: low levels of technology predominate, such as, oil refining by distillation only; produces basic liquid fuels, coal, metallic chromium, lead, zinc, and ferronickel; light industry produces basic textiles, wood products, and tobacco
Agriculture: provides 12% of GDP and meets the basic need for food; principal crops are rice, tobacco, wheat, corn, and millet; also grown are cotton, sesame, mulberry leaves, citrus fruit, and vegetables; Macedonia is one of the seven legal cultivators of the opium poppy for the world pharmaceutical industry, including some exports to the US; agricultural production is highly labor intensive
Illicit drugs: NA
Economic aid: $10 million from the US for humanitarian and technical assistance; EC promised a 100 ECU million economic aid package
Currency: 1 denar (abbreviation NA) = 100 NA
Exchange rates: denar per US$1—240 (January 1991)
Fiscal year: calendar year

Communications

Railroads: NA
Highways: 10,591 km total (1991); 5,091 km paved, 1,404 km gravel, 4,096 km earth
Inland waterways: NA km
Pipelines: none
Ports: none; landlocked
Airports:
total: 17
useable: 17
with permanent-surface runways: 9
with runways over 3,659 m: 0
with runways 2,440-3,659 m: 2
with runways 1,220-2,439 m: 2
Telecommunications: 125,000 telephones; broadcast stations—6 AM, 2 FM, 5 (2 relays) TV; 370,000 radios, 325,000 TV; satellite communications ground stations—none

Defense Forces

Branches: Army, Navy, Air and Air Defense Force, Police Force
Manpower availability: males age 15-49 597,024; fit for military service 484,701; reach military age (19) annually 18,979 (1993 est.)
Defense expenditures: 7 billion denars, NA% of GNP (1993 est.); note—conversion of the military budget into US dollars using the current exchange rate could produce misleading results

Madagascar

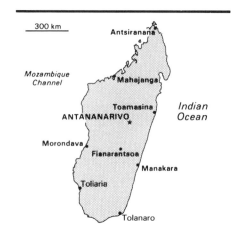

Geography

Location: in the western Indian Ocean, 430 km east of Mozambique in Southern Africa
Map references: Africa, Standard Time Zones of the World
Area:
total area: 587,040 km²
land area: 581,540 km²
comparative area: slightly less than twice the size of Arizona
Land boundaries: 0 km
Coastline: 4,828 km
Maritime claims:
exclusive economic zone: 200 nm
territorial sea: 12 nm
International disputes: claims Bassas da India, Europa Island, Glorioso Islands, Juan de Nova Island, and Tromelin Island (all administered by France)
Climate: tropical along coast, temperate inland, arid in south
Terrain: narrow coastal plain, high plateau and mountains in center
Natural resources: graphite, chromite, coal, bauxite, salt, quartz, tar sands, semiprecious stones, mica, fish
Land use:
arable land: 4%
permanent crops: 1%
meadows and pastures: 58%
forest and woodland: 26%
other: 11%
Irrigated land: 9,000 km² (1989 est.)
Environment: subject to periodic cyclones; deforestation; overgrazing; soil erosion; desertification
Note: world's fourth-largest island; strategic location along Mozambique Channel

People

Population: 13,005,989 (July 1993 est.)
Population growth rate: 3.2% (1993 est.)
Birth rate: 45.66 births/1,000 population (1993 est.)
Death rate: 13.71 deaths/1,000 population (1993 est.)

Net migration rate: 0 migrant(s)/1,000 population (1993 est.)
Infant mortality rate: 91 deaths/1,000 live births (1993 est.)
Life expectancy at birth:
total population: 53.52 years
male: 51.65 years
female: 55.45 years (1993 est.)
Total fertility rate: 6.75 children born/woman (1993 est.)
Nationality:
noun: Malagasy (singular and plural)
adjective: Malagasy
Ethnic divisions: Malayo-Indonesian (Merina and related Betsileo), Cotiers (mixed African, Malayo-Indonesian, and Arab ancestry—Betsimisaraka, Tsimihety, Antaisaka, Sakalava), French, Indian, Creole, Comoran
Religions: indigenous beliefs 52%, Christian 41%, Muslim 7%
Languages: French (official), Malagasy (official)
Literacy: age 15 and over can read and write (1990)
total population: 80%
male: 88%
female: 73%
Labor force: 4.9 million 90% nonsalaried family workers engaged in subsistence agriculture; 175,000 wage earners
by occupation: agriculture 26%, domestic service 17%, industry 15%, commerce 14%, construction 11%, services 9%, transportation 6%, other 2%
note: 51% of population of working age (1985)

Government

Names:
conventional long form: Republic of Madagascar
conventional short form: Madagascar
local long form: Republique de Madagascar
local short form: Madagascar
former: Malagasy Republic
Digraph: MA
Type: republic
Capital: Antananarivo
Administrative divisions: 6 provinces—Antananarivo, Antsiranana, Fianarantsoa, Mahajanga, Toamasina, Toliary
Independence: 26 June 1960 (from France)
Constitution: 12 September 1992
Legal system: based on French civil law system and traditional Malagasy law; has not accepted compulsory ICJ jurisdiction
National holiday: Independence Day, 26 June (1960)
Political parties and leaders: some 30 political parties now exist in Madagascar, the most important of which are Advance Guard of the Malagasy Revolution (AREMA), Didier RATSIRAKA; Congress

Party for Malagasy Independence (AKFM), RAKOTOVAO-ANDRIATIANA; Movement for National Unity (VONJY), Dr. Marojama RAZANABAHINY; Malagasy Christian Democratic Union (UDECMA), Norbert ANDRIAMORASATA; Militants for the Establishment of a Proletarian Regime (MFM), Manandafy RAKOTONIRINA; National Movement for the Independence of Madagascar (MONIMA), Monja JAONA; National Union for the Defense of Democracy (UNDD), Albert ZAFY

Other political or pressure groups: National Council of Christian Churches (FFKM), leader NA; Federalist Movement, leader NA

Suffrage: 18 years of age; universal

Elections:

President: last held on 10 February 1993 (next to be held 1998); results—Albert ZAFY (UNDD), 67%; Didier RATSIRAKA (AREMA), 33%

Popular National Assembly: last held on 28 May 1989 (next to be held May 1993); results—AREMA 88.2%, MFM 5.1%, AKFM 3.7%, VONJY 2.2%, other 0.8%; seats—(137 total) AREMA 120, MFM 7, AKFM 5, VONJY 4, MONIMA 1

Executive branch: president, prime minister, Council of Ministers

Legislative branch: unicameral Popular National Assembly (Assemblee Nationale Populaire); note—the National Assembly has suspended its operations during 1992 and early 1993 in preparation for new legislative elections. In its place, an interim High Authority of State and a Social and Economic Recovery Council have been established

Judicial branch: Supreme Court (Cour Supreme), High Constitutional Court (Haute Cour Constitutionnelle)

Leaders:

Chief of State: President Adm. Didier RATSIRAKA (since 15 June 1975)

Head of Government: Prime Minister Guy RAZANAMASY (since 8 August 1991)

Member of: ACCT, ACP, AfDB, CCC, ECA, FAO, G-77, GATT, IAEA, IBRD, ICAO, ICC, ICFTU, IDA, IFAD, IFC, ILO, IMF, IMO, INTELSAT, INTERPOL, IOC, ITU, LORCS, NAM, OAU, UN, UNCTAD, UNESCO, UNHCR, UNIDO, UPU, WCL, WFTU, WHO, WIPO, WMO, WTO

Diplomatic representation in US:

chief of mission: Ambassador Pierrot Jocelyn RAJAONARIVELO

chancery: 2374 Massachusetts Avenue NW, Washington, DC 20008

telephone: (202) 265-5525 or 5526

consulate general: New York

US diplomatic representation:

chief of mission: Ambassador Douglas BARRETT

embassy: 14 and 16 Rue Rainitovo, Antsahavola, Antananarivo

mailing address: B. P. 620, Antananarivo

telephone: [261] (2) 212-57, 209-56, 200-89, 207-18

FAX: 261-234-539

Flag: two equal horizontal bands of red (top) and green with a vertical white band of the same width on hoist side

---

Economy

Overview: Madagascar is one of the poorest countries in the world. Agriculture, including fishing and forestry, is the mainstay of the economy, accounting for over 30% of GDP and contributing to more than 70% of total export earnings. Industry is largely confined to the processing of agricultural products and textile manufacturing; in 1991 it accounted for only 13% of GDP. In 1986 the government introduced a five-year development plan that stressed self-sufficiency in food (mainly rice) by 1990, increased production for exports, and reduced energy imports. After mid-1991, however, output dropped sharply because of protracted antigovernment strikes and demonstrations for political reform.

National product: GDP—exchange rate conversion—$2.5 billion (1992 est.)

National product real growth rate: 1% (1992 est.)

National product per capita: $200 (1992 est.)

Inflation rate (consumer prices): 20% (1992 est.)

Unemployment rate: NA%

Budget: revenues $250 million; expenditures $265 million, including capital expenditures of $180 million (1991)

Exports: $312 million (f.o.b., 1991 est.)

commodities: coffee 45%, vanilla 20%, cloves 11%, sugar, petroleum products

partners: France, Japan, Italy, Germany, US

Imports: $350 million (f.o.b., 1992 est.)

commodities: intermediate manufactures 30%, capital goods 28%, petroleum 15%, consumer goods 14%, food 13%

partners: France, Germany, UK, other EC, US

External debt: $4.4 billion (1991)

Industrial production: growth rate 5.2% (1990 est.); accounts for 13% of GDP

Electricity: 125,000 kW capacity; 450 million kWh produced, 35 kWh per capita (1991)

Industries: agricultural processing (meat canneries, soap factories, breweries, tanneries, sugar refining plants), light consumer goods industries (textiles, glassware), cement, automobile assembly plant, paper, petroleum

Agriculture: accounts for 31% of GDP; cash crops—coffee, vanilla, sugarcane, cloves, cocoa; food crops—rice, cassava, beans, bananas, peanuts; cattle raising

widespread; almost self-sufficient in rice

Illicit drugs: illicit producer of cannabis (cultivated and wild varieties) used mostly for domestic consumption

Economic aid: US commitments, including Ex-Im (FY70-89), $136 million; Western (non-US) countries, ODA and OOF bilateral commitments (1970-89), $3,125 million; Communist countries (1970-89), $491 million

Currency: 1 Malagasy franc (FMG) = 100 centimes

Exchange rates: Malagasy francs (FMG) per US$1—1,910.2 (December 1992), 1,867.9 (1992), 1,835.4 (1991), 1,454.6 (December 1990), 1,603.4 (1989), 1,407.1 (1988), 1,069.2 (1987)

Fiscal year: calendar year

Communications

Railroads: 1,020 km 1.000-meter gauge

Highways: 40,000 km total; 4,694 km paved, 811 km crushed stone, gravel, or stabilized soil, 34,495 km improved and unimproved earth (est.)

Inland waterways: of local importance only; isolated streams and small portions of Canal des Pangalanes

Ports: Toamasina, Antsiranana, Mahajanga, Toliara

Merchant marine: 11 ships (1,000 GRT or over) totaling 35,359 GRT/48,772 DWT; includes 6 cargo, 2 roll-on/roll-off cargo, 1 oil tanker, 1 chemical tanker, 1 liquefied gas

Airports:

total: 146

usable: 103

with permanent-surface runways: 30

with runways over 3,659 m: 0

with runways 2,440-3,659 m: 3

with runways 1,220-2,439 m: 36

Telecommunications: above average system includes open-wire lines, coaxial cables, radio relay, and troposcatter links; submarine cable to Bahrain; satellite earth stations—1 Indian Ocean INTELSAT and broadcast stations—17 AM, 3 FM, 1 (36 repeaters) TV

Defense Forces

Branches: Popular Armed Forces (including Intervention Forces, Development Forces, Aeronaval Forces—including Navy and Air Force), Gendarmerie, Presidential Security Regiment

Manpower availability: males age 15-49 2,826,018; fit for military service 1,681,553; reach military age (20) annually 118,233 (1993 est.)

Defense expenditures: exchange rate conversion—$37 million, 2.2% of GDP (1991 est.)

Malawi

Geography

Location: Southern Africa, between Mozambique and Zambia
Map references: Africa, Standard Time Zones of the World
Area:
total area: 118,480 km²
land area: 94,080 km²
comparative area: slightly larger than Pennsylvania
Land boundaries: total 2,881 km, Mozambique 1,569 km, Tanzania 475 km, Zambia 837 km
Coastline: 0 km (landlocked)
Maritime claims: none; landlocked
International disputes: dispute with Tanzania over the boundary in Lake Nyasa (Lake Malawi)
Climate: tropical; rainy season (November to May); dry season (May to November)
Terrain: narrow elongated plateau with rolling plains, rounded hills, some mountains
Natural resources: limestone, unexploited deposits of uranium, coal, and bauxite
Land use:
arable land: 25%
permanent crops: 0%
meadows and pastures: 20%
forest and woodland: 50%
other: 5%
Irrigated land: 200 km² (1989 est.)
Environment: deforestation
Note: landlocked

People

Population: 9,831,935 (July 1993 est.)
Population growth rate: -0.95% (1993 est.)
Birth rate: 51.1 births/1,000 population (1993 est.)
Death rate: 22.87 deaths/1,000 population (1993 est.)
Net migration rate: -37.71 migrant(s)/1,000 population (1993 est.)
Infant mortality rate: 141.9 deaths/1,000 live births (1993 est.)

Life expectancy at birth:
total population: 40.48 years
male: 39.61 years
female: 41.37 years (1993 est.)
Total fertility rate: 7.5 children born/woman (1993 est.)
Nationality:
noun: Malawian(s)
adjective: Malawian
Ethnic divisions: Chewa, Nyanja, Tumbuko, Yao, Lomwe, Sena, Tonga, Ngoni, Ngonde, Asian, European
Religions: Protestant 55%, Roman Catholic 20%, Muslim 20%, traditional indigenous beliefs
Languages: English (official), Chichewa (official), other languages important regionally
Literacy: age 15 and over can read and write (1966)
total population: 22%
male: 34%
female: 12%
Labor force: 428,000 wage earners
by occupation: agriculture 43%, manufacturing 16%, personal services 15%, commerce 9%, construction 7%, miscellaneous services 4%, other permanently employed 6% (1986)

Government

Names:
conventional long form: Republic of Malawi
conventional short form: Malawi
former: Nyasaland
Digraph: MI
Type: one-party republic
note: a referendum to determine whether Malawi should remain a one-party state is scheduled to be held on 14 June 1993
Capital: Lilongwe
Administrative divisions: 24 districts; Blantyre, Chikwawa, Chiradzulu, Chitipa, Dedza, Dowa, Karonga, Kasungu, Lilongwe, Machinga (Kasupe), Mangochi, Mchinji, Mulanje, Mwanza, Mzimba, Ntcheu, Nkhata Bay, Nkhotakota, Nsanje, Ntchisi, Rumphi, Salima, Thyolo, Zomba
Independence: 6 July 1964 (from UK)
Constitution: 6 July 1964; republished as amended January 1974
Legal system: based on English common law and customary law; judicial review of legislative acts in the Supreme Court of Appeal; has not accepted compulsory ICJ jurisdiction
National holiday: Independence Day, 6 July (1964)
Political parties and leaders: only party—Malawi Congress Party (MCP), Wadson DELEZA, administrative secretary; John TEMBO, treasurer general; top party position of secretary general vacant since 1983

Other political or pressure groups: Alliance for Democracy (AFORD), Chakufwa CHIHANA; United Democratic Front (UDF) Bakili MULUZI; Malawi Democratic People (MDP), leader NA
Suffrage: 21 years of age; universal
Elections:
President: President BANDA sworn in as President for Life on 6 July 1971
National Assembly: last held 26-27 June 1987 (next to be held by June 1997); results—MCP is the only party; seats—(141 total, 136 elected) MCP 141
Executive branch: president, Cabinet
Legislative branch: unicameral National Assembly
Judicial branch: High Court, Supreme Court of Appeal
Leaders:
Chief of State and Head of Government: President Dr. Hastings Kamuzu BANDA (since 6 July 1966; sworn in as President for Life 6 July 1971)
Member of: ACP, AfDB, C, CCC, ECA, FAO, G-77, GATT, IBRD, ICAO, ICFTU, IDA, IFAD, IFC, ILO, IMF, IMO, INTELSAT, INTERPOL, IOC, ISO (correspondent), ITU, LORCS, NAM, OAU, SADC, UN, UNCTAD, UNESCO, UNIDO, UPU, WHO, WIPO, WMO, WTO
Diplomatic representation in US:
chief of mission: Ambassador Robert B. MBAYA
chancery: 2408 Massachusetts Avenue NW, Washington, DC 20008
telephone: (202) 797-1007
US diplomatic representation:
chief of mission: Ambassador Michael T. F. PISTOR
embassy: address NA, in new capital city development area in Lilongwe
mailing address: P. O. Box 30016, Lilongwe
telephone: [265] 730-166
FAX: [265] 732-282
Flag: three equal horizontal bands of black (top), red, and green with a radiant, rising, red sun centered in the black band; similar to the flag of Afghanistan, which is longer and has the national coat of arms superimposed on the hoist side of the black and red bands

Economy

Overview: Landlocked Malawi ranks among the world's least developed countries. The economy is predominately agricultural, with about 90% of the population living in rural areas. Agriculture accounts for 40% of GDP and 90% of export revenues. After two years of weak performance, economic growth improved significantly in 1988-91 as a result of good weather and a broadly based economic adjustment effort by the government. Drought cut overall output

sharply in 1992. The economy depends on substantial inflows of economic assistance from the IMF, the World Bank, and individual donor nations.
National product: GDP—exchange rate conversion—$1.9 billion (1992 est.)
National product real growth rate: -7.7% (1992 est.)
National product per capita: $200 (1992 est.)
Inflation rate (consumer prices): 21% (1992 est.)
Unemployment rate: NA%
Budget: revenues $398 million; expenditures $510 million, including capital expenditures of $154 million (FY91 est.)
Exports: $400 million (f.o.b., 1991 est.)
commodities: tobacco, tea, sugar, coffee, peanuts, wood products
partners: US, UK, Zambia, South Africa, Germany
Imports: $660 million (c.i.f., 1991 est.)
commodities: food, petroleum products, semimanufactures, consumer goods, transportation equipment
partners: South Africa, Japan, US, UK, Zimbabwe
External debt: $1.8 billion (December 1991 est.)
Industrial production: growth rate 4.0% (1990 est.); accounts for about 18% of GDP (1988)
Electricity: 190,000 kW capacity; 620 million kWh produced, 65 kWh per capita (1992)
Industries: agricultural processing (tea, tobacco, sugar), sawmilling, cement, consumer goods
Agriculture: accounts for 40% of GDP; cash crops—tobacco, sugarcane, cotton, tea, and corn; subsistence crops—potatoes, cassava, sorghum, pulses; livestock—cattle, goats
Economic aid: US commitments, including Ex-Im (FY70-89), $215 million; Western (non-US) countries, ODA and OOF bilateral commitments (1970-89), $2,150 million
Currency: 1 Malawian kwacha (MK) = 100 tambala
Exchange rates: Malawian kwacha (MK) per US$1—4.3418 (November 1992), 2.8033 (1991), 2.7289 (1990), 2.7595 (1989), 2.5613 (1988), 2.2087 (1987)
Fiscal year: 1 April-31 March

Communications

Railroads: 789 km 1.067-meter gauge
Highways: 13,135 km total; 2,364 km paved; 251 km crushed stone, gravel, or stabilized soil; 10,520 km earth and improved earth
Inland waterways: Lake Nyasa (Lake Malawi); Shire River, 144 km

Ports: Chipoka, Monkey Bay, Nkhata Bay, and Nkotakota—all on Lake Nyasa (Lake Malawi)
Airports:
total: 47
usable: 41
with permanent-surface runways: 5
with runways over 3,659 m: 0
with runways 2,440-3,659 m: 1
with runways 1,220-2,439 m: 10
Telecommunications: fair system of open-wire lines, radio relay links, and radio communications stations; 42,250 telephones; broadcast stations—10 AM, 17 FM, no TV; satellite earth stations—1 Indian Ocean INTELSAT and 1 Atlantic Ocean INTELSAT
Note: a majority of exports would normally go through Mozambique on the Beira, Nacala, and Limgogo railroads, but now most go through South Africa because of insurgent activity and damage to rail lines

---

Defense Forces

Branches: Army (including Air Wing and Naval Detachment), Police (including paramilitary Mobile Force Unit), paramilitary Malawi Young Pioneers
Manpower availability: males age 15-49 2,059,509; fit for military service 1,048,986 (1993 est.)
Defense expenditures: exchange rate conversion—$22 million, 1.6% of GDP (1989 est.)

Malawi

Geography

Location: Southeast Asia, bordering the South China Sea, between Vietnam and Indonesia
Map references: Asia, Oceania, Southeast Asia, Standard Time Zones of the World
Area:
total area: 329,750 km^2
land area: 328,550 km^2
comparative area: slightly larger than New Mexico
Land boundaries: total 2,669 km, Brunei 381 km, Indonesia 1,782 km, Thailand 506 km
Coastline: 4,675 km (Peninsular Malaysia 2,068 km, East Malaysia 2,607 km)
Maritime claims:
continental shelf: 200 m depth or to depth of exploitation; specified boundary in the South China Sea
exclusive fishing zone: 200 nm
exclusive economic zone: 200 nm
territorial sea: 12 nm
International disputes: involved in a complex dispute over the Spratly Islands with China, Philippines, Taiwan, Vietnam, and possibly Brunei; State of Sabah claimed by the Philippines; Brunei may wish to purchase the Malaysian salient that divides Brunei into two parts; two islands in dispute with Singapore; two islands in dispute with Indonesia
Climate: tropical; annual southwest (April to October) and northeast (October to February) monsoons
Terrain: coastal plains rising to hills and mountains
Natural resources: tin, petroleum, timber, copper, iron ore, natural gas, bauxite
Land use:
arable land: 3%
permanent crops: 10%
meadows and pastures: 0%
forest and woodland: 63%
other: 24%
Irrigated land: 3,420 km^2 (1989 est.)

Malaysia (continued)

Environment: subject to flooding; air and water pollution
Note: strategic location along Strait of Malacca and southern South China Sea

People

Population: 18,845,340 (July 1993 est.)
Population growth rate: 2.32% (1993 est.)
Birth rate: 28.93 births/1,000 population (1993 est.)
Death rate: 5.77 deaths/1,000 population (1993 est.)
Net migration rate: 0 migrant(s)/1,000 population (1993 est.)
Infant mortality rate: 26.5 deaths/1,000 live births (1993 est.)
Life expectancy at birth:
total population: 68.82 years
male: 65.96 years
female: 71.81 years (1993 est.)
Total fertility rate: 3.54 children born/woman (1993 est.)
Nationality:
noun: Malaysian(s)
adjective: Malaysian
Ethnic divisions: Malay and other indigenous 59%, Chinese 32%, Indian 9%
Religions:
Peninsular Malaysia: Muslim (Malays) Buddhist (Chinese), Hindu (Indians)
Sabah: Muslim 38% Christian 17%, other 45%
Sarawak: tribal religion 35% Buddhist and Confucianist 24%, Muslim 20%, Christian 16%, other 5%
Languages:
Peninsular Malaysia: Malay (official) English, Chinese dialects, Tamil
State of Sabah: English Malay, numerous tribal dialects, Chinese (Mandarin and Hakka dialects predominate)
State of Sarawak: English Malay, Mandarin, numerous tribal languages,
Literacy: age 15 and over can read and write (1990)
total population: 78%
male: 86%
female: 70%
Labor force: 7.258 million (1991 est.)

Government

Names:
conventional long form: none
conventional short form: Malaysia
former: Malayan Union
Digraph: MY
Type: constitutional monarchy
note: Federation of Malaysia formed 9 July 1963; nominally headed by the paramount ruler (king) and a bicameral Parliament; Peninsular Malaysian states—hereditary rulers in all but Melaka, where governors are appointed by Malaysian Pulau Pinang

Government; powers of state governments are limited by federal Constitution; Sabah—self-governing state, holds 20 seats in House of Representatives, with foreign affairs, defense, internal security, and other powers delegated to federal government; Sarawak—self-governing state within Malaysia, holds 27 seats in House of Representatives, with foreign affairs, defense, internal security, and other powers delegated to federal government
Capital: Kuala Lumpur
Administrative divisions: 13 states (negeri-negeri, singular—negeri) and 2 federal territories* (wilayah-wilayah persekutuan, singular—wilayah persekutuan); Johor, Kedah, Kelantan, Labuan*, Melaka, Negeri Sembilan, Pahang, Perak, Perlis, Pulau Pinang, Sabah, Sarawak, Selangor, Terengganu, Wilayah Persekutuan*
Independence: 31 August 1957 (from UK)
Constitution: 31 August 1957, amended 16 September 1963
Legal system: based on English common law; judicial review of legislative acts in the Supreme Court at request of supreme head of the federation; has not accepted compulsory ICJ jurisdiction
National holiday: National Day, 31 August (1957)
Political parties and leaders:
Peninsular Malaysia: National Front, a confederation of 13 political parties dominated by United Malays National Organization Baru (UMNO Baru), MAHATHIR bin Mohamad; Malaysian Chinese Association (MCA), LING Liong Sik; Gerakan Rakyat Malaysia, Datuk LIM Keng Yaik; Malaysian Indian Congress (MIC), Datuk S. Samy VELLU
Sabah: Berjaya Party, Datuk Haji Mohammed NOOR Mansor; Bersatu Sabah (PBS), Joseph Pairin KITINGAN; United Sabah National Organizaton (USNO), leader NA
Sarawak: coalition Sarawak National Front composed of the Party Pesaka Bumiputra Bersatu (PBB), Datuk Patinggi Amar Haji Abdul TAIB Mahmud; Sarawak United People's Party (SUPP), Datuk Amar James WONG Soon Kai; Sarawak National Party (SNAP), Datuk Amar James WONG; Parti Bansa Dayak Sarawak (PBDS), Datuk Leo MOGGIE; major opposition parties are Democratic Action Party (DAP), LIM Kit Siang; and Pan-Malaysian Islamic Party (PAS), Fadzil NOOR
Suffrage: 21 years of age; universal
Elections:
House of Representatives: last held 21 October 1990 (next to be held by August 1995); results—National Front 52%, other 48%; seats—(180 total) National Front 127, DAP 20, PAS 7, independents 4, other 22; note—within the National Front, UMNO got 71 seats and MCA 18 seats

Executive branch: paramount ruler, deputy paramount ruler, prime minister, deputy prime minister, Cabinet
Legislative branch: bicameral Parliament (Parlimen) consists of an upper house or Senate (Dewan Negara) and a lower house or House of Representatives (Dewan Rakyat)
Judicial branch: Supreme Court
Leaders:
Chief of State: Paramount Ruler AZLAN Muhibbuddin Shah ibni Sultan Yusof Izzudin (since 26 April 1989); Deputy Paramount Ruler JA'AFAR ibni Abdul Rahman (since 26 April 1989)
Head of Government: Prime Minister Dr. MAHATHIR bin Mohamad (since 16 July 1981); Deputy Prime Minister Abdul GHAFAR Bin Baba (since 7 May 1986)
Member of: APEC, AsDB, ASEAN, C, CCC, CP, ESCAP, FAO, G-15, G-77, GATT, IAEA, IBRD, ICAO, ICFTU, IDA, IDB, IFAD, IFC, ILO, IMF, IMO, INMARSAT, INTELSAT, INTERPOL, IOC, ISO, ITU, LORCS, MINURSO, NAM, OIC, UN, UNAVEM II, UNCTAD, UNESCO, UNIDO, UNIKOM, UNOMOZ, UNTAC, UPU, WCL, WHO, WIPO, WMO, WTO
Diplomatic representation in US:
chief of mission: Ambassador Abdul MAJID Mohamed
chancery: 2401 Massachusetts Avenue NW, Washington, DC 20008
telephone: (202) 328-2700
consulates general: Los Angeles and New York
US diplomatic representation:
chief of mission: Ambassador John S. WOLF
embassy: 376 Jalan Tun Razak, 50400 Kuala Lumpur
mailing address: P. O. Box No. 10035, 50700 Kuala Lumpur
telephone: [60] (3) 248-9011
FAX: [60] (3) 242-2207
Flag: fourteen equal horizontal stripes of red (top) alternating with white (bottom); there is a blue rectangle in the upper hoist-side corner bearing a yellow crescent and a yellow fourteen-pointed star; the crescent and the star are traditional symbols of Islam; the design was based on the flag of the US

Economy

Overview: The Malaysian economy, a mixture of private enterprise and a soundly managed public sector, has posted a remarkable record of 8%-9% average growth in 1987-92. This growth has resulted in a substantial reduction in poverty and a marked rise in real wages. Despite sluggish growth in the major world economies in 1992, demand for Malaysian goods remained strong and foreign investors continued to commit large sums in the economy. The government is aware of the inflationary potential of this rapid development and is closely monitoring fiscal and monetary policies.

National product: GDP—exchange rate conversion—$54.5 billion (1992 est.)

National product real growth rate: 8% (1992 est.)

National product per capita: $2,960 (1992 est.)

Inflation rate (consumer prices): 4.7% (1992 est.)

Unemployment rate: 4.1% (1992 est.)

Budget: revenues $15.6 billion; expenditures $18.0 billion, including capital expenditures of $4.5 billion (1992 est.)

Exports: $39.8 billion (f.o.b., 1992) *commodities:* electronic equipment, palm oil, petroleum and petroleum products, wood and wood products, rubber, textiles *partners:* Singapore 23%, US 18.6%, Japan 13.2%, UK 4%, Germany 4%

Imports: $39.1 billion (f.o.b., 1992) *commodities:* food, consumer goods, petroleum products, chemicals, capital equipment *partners:* Japan 26%, US 15.8%, Singapore 15.7%, Taiwan 5.6%, Germany 4.2%

External debt: $25.7 billion (1992 est.)

Industrial production: growth rate 13% (1992); accounts for NA% of GDP

Electricity: 8,000,000 kW capacity; 30,000 million kWh produced, 1,610 kWh per capita (1992)

Industries:
Peninsular Malaysia: rubber and oil palm processing and manufacturing, light manufacturing industry, electronics, tin mining and smelting, logging and processing timber
Sabah: logging, petroleum production
Sarawak: agriculture processing, petroleum production and refining, logging

Agriculture: accounts for 20% of GDP
Peninsular Malaysia: natural rubber, palm oil, rice
Sabah: mainly subsistence, but also rubber, timber, coconut, rice
Sarawak: rubber, timber, pepper; deficit of rice in all areas; fish catch of 608,000 metric tons in 1987

Illicit drugs: transit point for Golden Triangle heroin going to the US, Western Europe, and the Third World

Economic aid: US commitments, including Ex-Im (FY70-84), $170 million; Western (non-US) countries, ODA and OOF bilateral commitments (1970-89), $4.7 million; OPEC bilateral aid (1979-89), $42 million

Currency: 1 ringgit (M$) = 100 sen

Exchange rates: ringgits (M$) per US$1—2.6238 (January 1993), 2.5475 (1992), 2.7501 (1991), 1.7048 (1990), 2.7088 (1989), 2.6188 (1988)

Fiscal year: calendar year

Communications

Railroads:
Peninsular Malaysia: 1,665 km 1.04-meter gauge; 13 km double track, government owned
Sabah: 136 km 1.000-meter gauge
Sarawak: none

Highways:
Peninsular Malaysia: 23,600 km; 19,352 km hard surfaced, mostly bituminous surface treatment, and 4,248 km unpaved
Sabah: 3,782 km
Sarawak: 1,644 km

Inland waterways:
Peninsular Malaysia: 3,209 km
Sabah: 1,569 km
Sarawak: 2,518 km

Pipelines: crude oil 1,307 km; natural gas 379 km

Ports: Tanjong Kidurong, Kota Kinabalu, Kuching, Pasir Gudang, Penang, Port Kelang, Sandakan, Tawau

Merchant marine: 184 ships (1,000 GRT or over) totaling 1,869,817 GRT/2,786,765 DWT; includes 1 passenger-cargo, 2 short-sea passenger, 71 cargo, 28 container, 2 vehicle carrier, 2 roll-on/roll-off, 1 livestock carrier, 38 oil tanker, 6 chemical tanker, 6 liquefied gas, 27 bulk

Airports:
total: 111
usable: 102
with permanent-surface runways: 32
with runways over 3,659 m: 1
with runways 2,440-3,659 m: 7
with runways 1,220-2,439 m: 18

Telecommunications: good intercity service provided on Peninsular Malaysia mainly by microwave radio relay; adequate intercity microwave radio relay network between Sabah and Sarawak via Brunei; international service good; good coverage by radio and television broadcasts; 994,860 telephones (1984); broadcast stations—28 AM, 3 FM, 33 TV; submarine cables extend to India and Sarawak; SEACOM submarine cable links to Hong Kong and Singapore; satellite earth stations—1 Indian Ocean INTELSAT, 1 Pacific Ocean INTELSAT, and 2 domestic

Defense Forces

Branches: Malaysian Army, Royal Malaysian Navy, Royal Malaysian Air Force, Royal Malaysian Police Force, Marine Police, Sarawak Border Scouts

Manpower availability: males age 15-49 4,837,256; fit for military service 2,941,577; reach military age (21) annually 181,435 (1993 est.)

Defense expenditures: exchange rate conversion—$2.4 billion, about 5% of GDP (1992)

Maldives

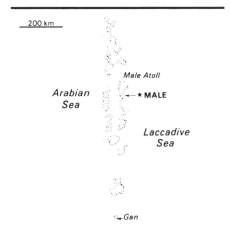

Geography

Location: South Asia, in the Indian Ocean off the southwest coast of India

Map references: Asia, Standard Time Zones of the World

Area:
total area: 300 km²
land area: 300 km²
comparative area: slightly more than 1.5 times the size of Washington, DC

Land boundaries: 0 km

Coastline: 644 km

Maritime claims:
exclusive economic zone: 35-310 nm as defined by geographic coordinates; segment of zone coincides with maritime boundary with India
territorial sea: 12 nm

International disputes: none

Climate: tropical; hot, humid; dry, northeast monsoon (November to March); rainy, southwest monsoon (June to August)

Terrain: flat with elevations only as high as 2.5 meters

Natural resources: fish

Land use:
arable land: 10%
permanent crops: 0%
meadows and pastures: 3%
forest and woodland: 3%
other: 84%

Irrigated land: NA km²

Environment: 1,200 coral islands grouped into 19 atolls

Note: archipelago of strategic location astride and along major sea lanes in Indian Ocean

People

Population: 243,094 (July 1993 est.)

Population growth rate: 3.64% (1993 est.)

Birth rate: 44.34 births/1,000 population (1993 est.)

Death rate: 7.91 deaths/1,000 population (1993 est.)

Maldives (continued)

Net migration rate: 0 migrant(s)/1,000 population (1993 est.)
Infant mortality rate: 57.6 deaths/1,000 live births (1993 est.)
Life expectancy at birth:
total population: 63.86 years
male: 62.5 years
female: 65.28 years (1993 est.)
Total fertility rate: 6.36 children born/woman (1993 est.)
Nationality:
noun: Maldivian(s)
adjective: Maldivian
Ethnic divisions: Sinhalese, Dravidian, Arab, African
Religions: Sunni Muslim
Languages: Divehi (dialect of Sinhala; script derived from Arabic), English spoken by most government officials
Literacy: age 15 and over can read and write (1985)
total population: 92%
male: 92%
female: 92%
Labor force: 66,000 (est.)
by occupation: fishing industry 25%

Government

Names:
conventional long form: Republic of Maldives
conventional short form: Maldives
Digraph: MV
Type: republic
Capital: Male
Administrative divisions: 19 districts (atolls); Aliff, Baa, Daalu, Faafu, Gaafu Aliff, Gaafu Daalu, Haa Aliff, Haa Daalu, Kaafu, Laamu, Laviyani, Meemu, Naviyani, Noonu, Raa, Seenu, Shaviyani, Thaa, Waavu
Independence: 26 July 1965 (from UK)
Constitution: 4 June 1964
Legal system: based on Islamic law with admixtures of English common law primarily in commercial matters; has not accepted compulsory ICJ jurisdiction
National holiday: Independence Day, 26 July (1965)
Political parties and leaders: no organized political parties; country governed by the Didi clan for the past eight centuries
Suffrage: 21 years of age; universal
Elections:
President: last held 23 September 1988 (next to be held September 1993); results—President Maumoon Abdul GAYOOM reelected
Citizens' Council: last held on 7 December 1989 (next to be held 7 December 1994); results—percent of vote NA; seats—(48 total, 40 elected)
Executive branch: president, Cabinet
Legislative branch: unicameral Citizens' Council (Majlis)
Judicial branch: High Court

Leaders:
Chief of State and Head of Government: President Maumoon Abdul GAYOOM (since 11 November 1978)
Member of: AsDB, C, CP, ESCAP, FAO, G-77, GATT, IBRD, ICAO, IDA, IDB, IFAD, IFC, IMF, IMO, INTERPOL, IOC, ITU, NAM, OIC, SAARC, UN, UNCTAD, UNESCO, UNIDO, UPU, WHO, WMO, WTO
Diplomatic representation in US: Maldives does not maintain an embassy in the US, but does have a UN mission in New York
US diplomatic representation:
chief of mission: the US Ambassador to Sri Lanka is accredited to Maldives and makes periodic visits there
consular agency: Midhath Hilmy, Male
telephone: 2581
Flag: red with a large green rectangle in the center bearing a vertical white crescent; the closed side of the crescent is on the hoist side of the flag

Economy

Overview: The economy is based on fishing, tourism, and shipping. Agriculture is limited to the production of a few subsistence crops that provide only 10% of food requirements. Fishing is the largest industry, employing 25% of the work force and accounting for over 60% of exports; it is also an important source of government revenue. During the 1980s tourism became one of the most important and highest growth sectors of the economy. In 1988 industry accounted for about 5% of GDP. Real GDP is officially estimated to have increased by about 10% annually during the period 1974-90.
National product: GDP—exchange rate conversion—$140 million (1991 est.)
National product real growth rate: 4.7% (1991 est.)
National product per capita: $620 (1991 est.)
Inflation rate (consumer prices): 11.5% (1991 est.)
Unemployment rate: NEGL%
Budget: revenues $52 million (excluding foreign transfers); expenditures $83 million, including capital expenditures of $39 million (1991 est.)
Exports: $53.7 million (f.o.b., 1991)
commodities: fish, clothing
partners: US, UK, Sri Lanka
Imports: $150.9 million (c.i.f., 1991)
commodities: consumer goods, intermediate and capital goods, petroleum products
partners: Singapore, Germany, Sri Lanka, India
External debt: $90 million (1991)
Industrial production: growth rate 24.0% (1990); accounts for 6% of GDP

Electricity: 5,000 kW capacity; 11 million kWh produced, 50 kWh per capita (1990)
Industries: fishing and fish processing, tourism, shipping, boat building, some coconut processing, garments, woven mats, coir (rope), handicrafts
Agriculture: accounts for almost 25% of GDP (including fishing); fishing more important than farming; limited production of coconuts, corn, sweet potatoes; most staple foods must be imported; fish catch of 67,000 tons (1990 est.)
Economic aid: US commitments, including Ex-Im (FY70-88), $28 million; Western (non-US) countries, ODA and OOF bilateral commitments (1970-89), $125 million; OPEC bilateral aid (1979-89), $14 million
Currency: 1 rufiyaa (Rf) = 100 laaris
Exchange rates: rufiyaa (Rf) per US$1—10.506 (January 1993), 10.569 (1992), 10.253 (1991), 9.509 (1990), 9.0408 (1989), 8.7846 (1988)
Fiscal year: calendar year

Communications

Highways: Male has 9.6 km of coral highways within the city
Ports: Male, Gan
Merchant marine: 14 ships (1,000 GRT or over) totaling 38,848 GRT/58,496 DWT; includes 12 cargo, 1 container, 1 oil tanker
Airports:
total: 2
useable: 2
with permanent-surface runways: 2
with runways over 3,659 m: 0
with runways 2,440-3,659 m: 2
with runways 1,220-2,439 m: 0
Telecommunications: minimal domestic and international facilities; 2,804 telephones; broadcast stations—2 AM, 1 FM, 1 TV; 1 Indian Ocean INTELSAT earth station

Defense Forces

Branches: National Security Service (paramilitary police force)
Manpower availability: males age 15-49 53,730; fit for military service 30,014 (1993 est.)
Defense expenditures: exchange rate conversion—$NA, NA% of GDP

Mali

Boundary representation is not necessarily authoritative

Geography

Location: Western Africa, between Mauritania and Niger
Map references: Africa, Standard Time Zones of the World
Area:
total area: 1.24 million km²
land area: 1.22 million km²
comparative area: slightly less than twice the size of Texas
Land boundaries: total 7,243 km, Algeria 1,376 km, Burkina 1,000 km, Guinea 858 km, Cote d'Ivoire 532 km, Mauritania 2,237 km, Niger 821 km, Senegal 419 km
Coastline: 0 km (landlocked)
Maritime claims: none; landlocked
International disputes: the disputed international boundary between Burkina and Mali was submitted to the International Court of Justice (ICJ) in October 1983 and the ICJ issued its final ruling in December 1986, which both sides agreed to accept; Burkina and Mali are proceeding with boundary demarcation, including the tripoint with Niger
Climate: subtropical to arid; hot and dry February to June; rainy, humid, and mild June to November; cool and dry November to February
Terrain: mostly flat to rolling northern plains covered by sand; savanna in south, rugged hills in northeast
Natural resources: gold, phosphates, kaolin, salt, limestone, uranium, bauxite, iron ore, manganese, tin, and copper deposits are known but not exploited
Land use:
arable land: 2%
permanent crops: 0%
meadows and pastures: 25%
forest and woodland: 7%
other: 66%
Irrigated land: 50 km² (1989 est.)
Environment: hot, dust-laden harmattan; haze common during dry seasons; desertification
Note: landlocked

People

Population: 8,868,617 (July 1993 est.)
Population growth rate: 2.66% (1993 est.)
Birth rate: 51.73 births/1,000 population (1993 est.)
Death rate: 20.81 deaths/1,000 population (1993 est.)
Net migration rate: -4.35 migrant(s)/1,000 population (1993 est.)
Infant mortality rate: 108 deaths/1,000 live births (1993 est.)
Life expectancy at birth:
total population: 45.45 years
male: 43.89 years
female: 47.06 years (1993 est.)
Total fertility rate: 7.33 children born/woman (1993 est.)
Nationality:
noun: Malian(s)
adjective: Malian
Ethnic divisions: Mande 50% (Bambara, Malinke, Sarakole), Peul 17%, Voltaic 12%, Songhai 6%, Tuareg and Moor 10%, other 5%
Religions: Muslim 90%, indigenous beliefs 9%, Christian 1%
Languages: French (official), Bambara 80%, numerous African languages
Literacy: age 15 and over can read and write (1990)
total population: 32%
male: 41%
female: 24%
Labor force: 2.666 million (1986 est.)
by occupation: agriculture 80%, services 19%, industry and commerce 1% (1981)
note: 50% of population of working age (1985)

Government

Names:
conventional long form: Republic of Mali
conventional short form: Mali
local long form: Republique de Mali
local short form: Mali
former: French Sudan
Digraph: ML
Type: republic
Capital: Bamako
Administrative divisions: 8 regions (regions, singular—region); Gao, Kayes, Kidal, Koulikoro, Mopti, Segou, Sikasso, Tombouctou
Independence: 22 September 1960 (from France)
Constitution: new constitution adopted in constitutional referendum in January 1992
Legal system: based on French civil law system and customary law; judicial review of legislative acts in Constitutional Section of Court of State; has not accepted compulsory ICJ jurisdiction
National holiday: Anniversary of the

Proclamation of the Republic, 22 September (1960)
Political parties and leaders: Alliance for Democracy (Adema), Alpha Oumar KONARE; National Committee for Democratic Initiative (CNID), Mountaga TALL; Sudanese Union/African Democratic Rally (US/RAD), Baba Hakib HAIDARA and Treoule Mamadon KONATE; Popular Movement for the Development of the Republic of West Africa; Rally for Democracy and Progress (RDP), Almamy SYLLA; Union for Democracy and Development (UDD), Moussa Balla COULIBALY; Rally for Democracy and Labor (RDT); Union of Democratic Forces for Progress (UFDP), Col. Youssouf TRAORE; Party for Democracy and Progress (PDP), Idrissa TRAORE; Malian Union for Democracy and Development (UMDD)
Suffrage: 21 years of age; universal
Elections:
President: last held in April 1992; Alpha KONARE was elected in runoff race against Montaga TALL
National Assembly: last held on 8 March 1992 (next to be held NA); results—percent of vote by party NA; seats—(total 116) Adema 76, CNID 9, US/RAD 8, Popular Movement for the Development of the Republic of West Africa 6, RDP 4, UDD 4, RDT 3, UFDP 3, PDP 2, UMDD 1
Executive branch: Transition Committee for the Salvation of the People (CTSP) composed of 25 members, predominantly civilian
Legislative branch: unicameral National Assembly
Judicial branch: Supreme Court (Cour Supreme)
Leaders:
Chief of State: President Alpha Oumar KONARE (since 8 June 1992)
Head of Government: Prime Minister Younoussi TOURE (since 8 June 1992)
Member of: ACCT, ACP, AfDB, CCC, CEAO, ECA, ECOWAS, FAO, FZ, G-77, IAEA, IBRD, ICAO, IDA, IDB, IFAD, IFC, ILO, IMF, INTELSAT, INTERPOL, IOC, ITU, LORCS, NAM, OAU, OIC, UN, UNCTAD, UNESCO, UNIDO, UPU, WADB, WCL, WHO, WIPO, WMO, WTO
Diplomatic representation in US:
chief of mission: Ambassador Siragatou Ibrahim CISSE
chancery: 2130 R Street NW, Washington, DC 20008
telephone: (202) 332-2249 or 939-8950
US diplomatic representation:
chief of mission: Ambassador Herbert Donald GELBER
embassy: Rue Rochester NY and Rue Mohamed V., Bamako
mailing address: B. P. 34, Bamako
telephone: [223] 225470
FAX: [233] 228059

Mali (continued)

Flag: three equal vertical bands of green (hoist side), yellow, and red; uses the popular pan-African colors of Ethiopia

Economy

Overview: Mali is among the poorest countries in the world, with about 70% of its land area desert or semidesert. Economic activity is largely confined to the riverine area irrigated by the Niger. About 10% of the population live as nomads and some 80% of the labor force is engaged in agriculture and fishing. Industrial activity is concentrated on processing farm commodities. In consultation with international lending agencies, the government has adopted a structural adjustment program for 1992-95, aiming at GDP annual growth of 4.6%, inflation of no more than 2.5% on average, and a substantial reduction in the external current account deficit.

National product: GDP—exchange rate conversion—$2.3 billion (1991 est.)

National product real growth rate: -0.2% (1991 est.)

National product per capita: $265 (1991 est.)

Inflation rate (consumer prices): 1.4% (1991 est.)

Unemployment rate: NA%

Budget: revenues $329 million; expenditures $519 million, including capital expenditures of $178 (1989 est.)

Exports: $320 million (f.o.b., 1991 est.)
commodities: livestock, peanuts, dried fish, cotton, skins
partners: mostly franc zone and Western Europe

Imports: $390 million (f.o.b., 1991 est.)
commodities: textiles, vehicles, petroleum products, machinery, sugar, cereals
partners: mostly franc zone and Western Europe

External debt: $2.6 billion (1991 est.)

Industrial production: growth rate 15.0% (1990 est.); accounts for 10.0% of GDP

Electricity: 260,000 kW capacity; 750 million kWh produced, 90 kWh per capita (1991)

Industries: small local consumer goods and processing, construction, phosphate, gold, fishing

Agriculture: accounts for 50% of GDP; most production based on small subsistence farms; cotton and livestock products account for over 70% of exports; other crops—millet, rice, corn, vegetables, peanuts; livestock—cattle, sheep, goats

Economic aid: US commitments, including Ex-Im (FY70-89), $349 million; Western (non-US) countries, ODA and OOF bilateral commitments (1970-89), $3,020 million; OPEC bilateral aid (1979-89), $92 million;

Communist countries (1970-89), $190 million

Currency: 1 CFA franc (CFAF) = 100 centimes

Exchange rates: Communaute Financiere Africaine francs (CFAF) per US$1—274.06 (January 1993), 264.69 (1992), 282.11 (1991), 272.26 (1990), 319.01 (1989), 297.85 (1988)

Fiscal year: calendar year

Communications

Railroads: 642 km 1.000-meter gauge; linked to Senegal's rail system through Kayes

Highways: about 15,700 km total; 1,670 km paved, 3,670 km gravel and improved earth, 10,360 km unimproved earth

Inland waterways: 1,815 km navigable

Airports:
total: 34
usable: 27
with permanent-surface runways: 8
with runways over 3,659 m: 0
with runways 2,440-3,659 m: 5
with runways 1,220-2,439 m: 10

Telecommunications: domestic system poor but improving; provides only minimal service with radio relay, wire, and radio communications stations; expansion of radio relay in progress; 11,000 telephones; broadcast stations—2 AM, 2 FM, 2 TV; satellite earth stations—1 Atlantic Ocean INTELSAT and 1 Indian Ocean INTELSAT

Defense Forces

Branches: Army, Air Force, Gendarmerie, Republican Guard, National Police (Surete Nationale)

Manpower availability: males age 15-49 1,749,662; fit for military service 995,554 (1993 est.); no conscription

Defense expenditures: exchange rate conversion—$41 million, 2% of GDP (1989)

Malta

Geography

Location: in the central Mediterranean Sea, 93 km south of Sicily (Italy), 290 km north of Libya

Map references: Europe, Standard Time Zones of the World

Area:
total area: 320 km²
land area: 320 km²
comparative area: slightly less than twice the size of Washington, DC

Land boundaries: 0 km

Coastline: 140 km

Maritime claims:
contiguous zone: 24 nm
continental shelf: 200 m depth or to depth of exploitation
exclusive fishing zone: 25 nm
territorial sea: 12 nm

International disputes: none

Climate: Mediterranean with mild, rainy winters and hot, dry summers

Terrain: mostly low, rocky, flat to dissected plains; many coastal cliffs

Natural resources: limestone, salt

Land use:
arable land: 38%
permanent crops: 3%
meadows and pastures: 0%
forest and woodland: 0%
other: 59%

Irrigated land: 10 km² (1989)

Environment: numerous bays provide good harbors; fresh water very scarce; increasing reliance on desalination

Note: the country comprises an archipelago, with only the 3 largest islands (Malta, Gozo, and Comino) being inhabited

People

Population: 363,791 (July 1993 est.)

Population growth rate: 0.84% (1993 est.)

Birth rate: 13.9 births/1,000 population (1993 est.)

Death rate: 7.52 deaths/1,000 population (1993 est.)

Net migration rate: 1.98 migrant(s)/1,000 population (1993 est.)

Infant mortality rate: 8.2 deaths/1,000 live births (1993 est.)

Life expectancy at birth:

total population: 76.52 years

male: 74.32 years

female: 78.9 years (1993 est.)

Total fertility rate: 1.97 children born/woman (1993 est.)

Nationality:

noun: Maltese (singular and plural)

adjective: Maltese

Ethnic divisions: Arab, Sicilian, Norman, Spanish, Italian, English

Religions: Roman Catholic 98%

Languages: Maltese (official), English (official)

Literacy: age 15 and over can read and write (1985)

total population: 84%

male: 86%

female: 82%

Labor force: 127,200

by occupation: government (excluding job corps) 37%, services 26%, manufacturing 22%, training programs 9%, construction 4%, agriculture 2% (1990)

Government

Names:

conventional long form: Republic of Malta

conventional short form: Malta

Digraph: MT

Type: parliamentary democracy

Capital: Valletta

Administrative divisions: none (administration directly from Valletta)

Independence: 21 September 1964 (from UK)

Constitution: 26 April 1974, effective 2 June 1974

Legal system: based on English common law and Roman civil law; has accepted compulsory ICJ jurisdiction, with reservations

National holiday: Independence Day, 21 September

Political parties and leaders: Nationalist Party (NP), Edward FENECH ADAMI; Malta Labor Party (MLP), Alfred SANT

Suffrage: 18 years of age; universal

Elections:

House of Representatives: last held on 22 February 1992 (next to be held by February 1997); results—NP 51.8%, MLP 46.5%; seats—(usually 65 total) MLP 36, NP 29; note—additional seats are given to the party with the largest popular vote to ensure a legislative majority; current total 69 (MLP 33, NP 36 after adjustment)

Executive branch: president, prime minister, deputy prime minister, Cabinet

Legislative branch: unicameral House of Representatives

Judicial branch: Constitutional Court, Court of Appeal

Leaders:

Chief of State: President Vincent (Censu) TABONE (since 4 April 1989)

Head of Government: Prime Minister Dr. Edward (Eddie) FENECH ADAMI (since 12 May 1987); Deputy Prime Minister Dr. Guido DE MARCO (since 14 May 1987)

Member of: C, CCC, CE, CSCE, EBRD, ECE, FAO, G-77, GATT, IBRD, ICAO, ICFTU, IFAD, ILO, IMF, IMO, INMARSAT, INTERPOL, IOC, IOM (observer), ITU, NAM, PCA, UN, UNCTAD, UNESCO, UNIDO, UPU, WHO, WIPO, WMO, WTO

Diplomatic representation in US:

chief of mission: Ambassador Albert BORG OLIVIER DE PUGET

chancery: 2017 Connecticut Avenue NW, Washington, DC 20008

telephone: (202) 462-3611 or 3612

FAX: (202) 387-5470

consulate: New York

US diplomatic representation:

chief of mission: (vacant)

embassy: 2nd Floor, Development House, Saint Anne Street, Floriana, Valletta

mailing address: P. O. Box 535, Valletta

telephone: [356] 240424, 240425, 243216, 243217, 243653, 223654

FAX: same as telephone numbers

Flag: two equal vertical bands of white (hoist side) and red; in the upper hoist-side corner is a representation of the George Cross, edged in red

---

Economy

Overview: Significant resources are limestone, a favorable geographic location, and a productive labor force. Malta produces only about 20% of its food needs, has limited freshwater supplies, and has no domestic energy sources. Consequently, the economy is highly dependent on foreign trade and services. Manufacturing and tourism are the largest contributors to the economy. Manufacturing accounts for about 27% of GDP, with the electronics and textile industries major contributors and the state-owned Malta drydocks employing about 4,300 people. In 1991, about 900,000 tourists visited the island. Per capita GDP at $7,600 places Malta in the middle-income range of the world's nations.

National product: GDP—exchange rate conversion—$2.7 billion (1991 est.)

National product real growth rate: 5.9% (1991)

National product per capita: $7,600 (1991 est.)

Inflation rate (consumer prices): 2.9% (1991)

Unemployment rate: 3.6% (1992)

Budget: revenues $1.1 billion; expenditures $1.1 billion, including capital expenditures of $161 million (1992 est.)

Exports: $l.2 billion (f.o.b., 1991)

commodities: clothing, textiles, footwear, ships

partners: Italy 30%, Germany 22%, UK 11%

Imports: $2.1 billion (f.o.b., 1991)

commodities: food, petroleum, machinery and semimanufactured goods

partners: Italy 30%, UK 16%, Germany 13%, US 4%

External debt: $127 million (1990 est.)

Industrial production: growth rate 19.0% (1990); accounts for 27% of GDP

Electricity: 328,000 kW capacity; 1,110 million kWh produced, 3,000 kWh per capita (1992)

Industries: tourism, electronics, ship repair yard, construction, food manufacturing, textiles, footwear, clothing, beverages, tobacco

Agriculture: accounts for 3% of GDP and 2.5% of the work force (1992); overall, 20% self-sufficient; main products—potatoes, cauliflower, grapes, wheat, barley, tomatoes, citrus, cut flowers, green peppers, hogs, poultry, eggs; generally adequate supplies of vegetables, poultry, milk, pork products; seasonal or periodic shortages in grain, animal fodder, fruits, other basic foodstuffs

Economic aid: US commitments, including Ex-Im (FY70-81), $172 million; Western (non-US) countries, ODA and OOF bilateral commitments (1970-89), $336 million; OPEC bilateral aid (1979-89), $76 million; Communist countries (1970-88), $48 million

Currency: 1 Maltese lira (LM) = 100 cents

Exchange rates: Maltese liri (LM) per US$1—0.3687 (January 1993), 0.3178 (1992), 0.3226 (1991), 0.3172 (1990), 0.3483 (1989), 0.3306 (1988)

Fiscal year: 1 April-31 March

Communications

Highways: 1,291 km total; 1,179 km paved (asphalt), 77 km crushed stone or gravel, 35 km improved and unimproved earth

Ports: Valletta, Marsaxlokk

Merchant marine: 789 ships (1,000 GRT or over) totaling 11,059,874 GRT/18,758,969 DWT; includes 6 passenger, 17 short-sea passenger, 272 cargo, 26 container, 2 passenger-cargo, 20 roll-on/roll-off, 2 vehicle carrier, 3 barge carrier, 17 refrigerated cargo, 19 chemical tanker, 15 combination ore/oil, 3 specialized tanker, 3 liquefied gas, 131 oil tanker, 223 bulk, 26 combination bulk, 3 multifunction large load carrier, 1 railcar carrier; note—a flag of convenience registry; China owns 2 ships, Russia owns 52 ships, Cuba owns 10, Vietnam owns 6, Croatia owns 37, Romania owns 3

Malta *(continued)*

Airports:
total: 1
useable: 1
with permanent-surface runways: 1
with runways over 3,659 m: 0
with runways 2,440-3,659 m: 1
with runways 1,220-2,439 m: 0
Telecommunications: automatic system satisfies normal requirements; 153,000 telephones; excellent service by broadcast stations—8 AM, 4 FM, and 2 TV; submarine cable and microwave radio relay between islands; international service by 1 submarine cable and 1 Atlantic Ocean INTELSAT earth station

Defense Forces

Branches: Armed Forces, Maltese Police Force
Manpower availability: males age 15-49 97,446; fit for military service 77,481 (1993 est.)
Defense expenditures: exchange rate conversion—$21.9 million, 1.3% of GDP (1989 est.)

Man, Isle of
(British crown dependency)

Geography

Location: in the Irish Sea, between Ireland and Great Britain
Map references: Europe
Area:
total area: 588 km²
land area: 588 km²
comparative area: nearly 3.5 times the size of Washington, DC
Land boundaries: 0 km
Coastline: 113 km
Maritime claims:
exclusive fishing zone: 200 nm
territorial sea: 3 nm
International disputes: none
Climate: cool summers and mild winters; humid; overcast about half the time
Terrain: hills in north and south bisected by central valley
Natural resources: lead, iron ore
Land use:
arable land: NA%
permanent crops: NA%
meadows and pastures: NA%
forest and woodland: NA%
other: NA% (extensive arable land and forests)
Irrigated land: NA km²
Environment: strong westerly winds prevail
Note: one small islet, the Calf of Man, lies to the southwest, and is a bird sanctuary

People

Population: 71,263 (July 1993 est.)
Population growth rate: 1.07% (1993 est.)
Birth rate: 13.57 births/1,000 population (1993 est.)
Death rate: 12.87 deaths/1,000 population (1993 est.)
Net migration rate: 9.99 migrant(s)/1,000 population (1993 est.)
Infant mortality rate: 8.5 deaths/1,000 live births (1993 est.)
Life expectancy at birth:
total population: 75.98 years

male: 73.25 years
female: 78.92 years (1993 est.)
Total fertility rate: 1.8 children born/woman (1993 est.)
Nationality:
noun: Manxman, Manxwoman
adjective: Manx
Ethnic divisions: Manx (Norse-Celtic descent), Briton
Religions: Anglican, Roman Catholic, Methodist, Baptist, Presbyterian, Society of Friends
Languages: English, Manx Gaelic
Literacy:
total population: NA%
male: NA%
female: NA%
Labor force: 25,864 (1981)
by occupation: NA

Government

Names:
conventional long form: none
conventional short form: Isle of Man
Digraph: IM
Type: British crown dependency
Capital: Douglas
Administrative divisions: none (British crown dependency)
Independence: none (British crown dependency)
Constitution: 1961, Isle of Man Constitution Act
Legal system: English law and local statute
National holiday: Tynwald Day, 5 July
Political parties and leaders: there is no party system and members sit as independents
Suffrage: 21 years of age; universal
Elections:
House of Keys: last held in 1991 (next to be held NA 1996); results—percent of vote NA; no party system; seats—(24 total) independents 24
Executive branch: British monarch, lieutenant governor, president, prime minister, Council of Ministers (cabinet)
Legislative branch: bicameral Tynwald consists of an upper house or Legislative Council and a lower house or House of Keys
Judicial branch: Court of Tynwald
Leaders:
Chief of State: Lord of Mann Queen ELIZABETH II (since 6 February 1952), represented by Lieutenant Governor Air Marshal Sir Laurence JONES (since NA 1990)
Head of Government: President of the Legislative Council Sir Charles KERRUISH (since NA 1990)
Member of: none
Diplomatic representation in US: none (British crown dependency)

US diplomatic representation: none (British crown dependency)
Flag: red with the Three Legs of Man emblem (Trinacria), in the center; the three legs are joined at the thigh and bent at the knee; in order to have the toes pointing clockwise on both sides of the flag, a two-sided emblem is used

Economy

Overview: Offshore banking, manufacturing, and tourism are key sectors of the economy. The government's policy of offering incentives to high-technology companies and financial institutions to locate on the island has paid off in expanding employment opportunities in high-income industries. As a result, agriculture and fishing, once the mainstays of the economy, have declined in their shares of GNP. Banking now contributes over 20% to GNP and manufacturing about 15%. Trade is mostly with the UK. The Isle of Man enjoys free access to European Community markets.
National product: GNP—exchange rate conversion—$490 million (1988)
National product real growth rate: NA%
National product per capita: $7,500 (1988)
Inflation rate (consumer prices): 7% (1992 est.)
Unemployment rate: 1% (1992 est.)
Budget: revenues $130.4 million; expenditures $114.4 million, including capital expenditures of $18.1 million (FY85 est.)
Exports: $NA
commodities: tweeds, herring, shellfish, meat
partners: UK
Imports: $NA
commodities: timber, fertilizers, fish
partners: UK
External debt: $NA
Industrial production: growth rate NA%
Electricity: 61,000 kW capacity; 190 million kWh produced, 2,965 kWh per capita (1992)
Industries: an important offshore financial center; financial services, light manufacturing, tourism
Agriculture: cereals and vegetables; cattle, sheep, pigs, poultry
Economic aid: NA
Currency: 1 Manx pound (£M) = 100 pence
Exchange rates: Manx pounds (£M) per US$1—0.6527 (January 1993), 0.5664 (1992), 0.5652 (1991), 0.5603 (1990), 0.6099 (1989), 0.5614 (1988); the Manx pound is at par with the British pound
Fiscal year: 1 April-31 March

Communications

Railroads: 60 km; 36 km electric track, 24 km steam track
Highways: 640 km motorable roads
Ports: Douglas, Ramsey, Peel
Merchant marine: 59 ships (1,000 GRT or over) totaling 1,363,502 GRT/2,363,502 DWT; includes 10 cargo, 6 container, 9 roll-on/roll-off, 14 oil tanker, 4 chemical tanker, 4 liquefied gas, 12 bulk; note—a captive register of the United Kingdom, although not all ships on the register are British owned
Airports:
total: 1
useable: 1
with permanent-surface runways: 1
with runways over 3,659 m: 0
with runways 2,440-3,659 m: 0
with runways 1,220-2,439 m: 1
Telecommunications: 24,435 telephones; broadcast stations—1 AM, 4 FM, 4 TV

Defense Forces

Note: defense is the responsibility of the UK

Marshall Islands

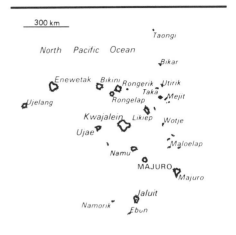

Geography

Location: Oceania, in the North Pacific Ocean, about two-thirds of the way between Hawaii and Papua New Guinea
Map references: Oceania, Standard Time Zones of the World
Area:
total area: 181.3 km²
land area: 181.3 km²
comparative area: slightly larger than Washington, DC
note: includes the atolls of Bikini, Eniwetok, and Kwajalein
Land boundaries: 0 km
Coastline: 370.4 km
Maritime claims:
contiguous zone: 24 nm
exclusive economic zone: 200 nm
territorial sea: 12 nm
International disputes: claims US territory of Wake Island
Climate: wet season May to November; hot and humid; islands border typhoon belt
Terrain: low coral limestone and sand islands
Natural resources: phosphate deposits, marine products, deep seabed minerals
Land use:
arable land: 0%
permanent crops: 60%
meadows and pastures: 0%
forest and woodland: 0%
other: 40%
Irrigated land: NA km²
Environment: occasionally subject to typhoons; two archipelagic island chains of 30 atolls and 1,152 islands
Note: Bikini and Eniwetok are former US nuclear test sites; Kwajalein, the famous World War II battleground, is now used as a US missile test range

People

Population: 51,982 (July 1993 est.)
Population growth rate: 3.87% (1993 est.)

Marshall Islands *(continued)*

Birth rate: 46.65 births/1,000 population (1993 est.)

Death rate: 7.91 deaths/1,000 population (1993 est.)

Net migration rate: 0 migrant(s)/1,000 population (1993 est.)

Infant mortality rate: 50.5 deaths/1,000 live births (1993 est.)

Life expectancy at birth:
total population: 62.79 years
male: 61.27 years
female: 64.38 years (1993 est.)

Total fertility rate: 6.99 children born/woman (1993 est.)

Nationality:
noun: Marshallese (singular and plural)
adjective: Marshallese

Ethnic divisions: Micronesian

Religions: Christian (mostly Protestant)

Languages: English (universally spoken and is the official language), two major Marshallese dialects from the Malayo-Polynesian family, Japanese

Literacy: age 15 and over can read and write (1980)
total population: 93%
male: 100%
female: 88%

Labor force: 4,800 (1986)
by occupation: NA

Government

Names:
conventional long form: Republic of the Marshall Islands
conventional short form: Marshall Islands
former: Marshall Islands District (Trust Territory of the Pacific Islands)

Digraph: RM

Type: constitutional government in free association with the US; the Compact of Free Association entered into force 21 October 1986

Capital: Majuro

Administrative divisions: none

Independence: 21 October 1986 (from the US-administered UN trusteeship)

Constitution: 1 May 1979

Legal system: based on adapted Trust Territory laws, acts of the legislature, municipal, common, and customary laws

National holiday: Proclamation of the Republic of the Marshall Islands, 1 May (1979)

Political parties and leaders: no formal parties; President KABUA is chief political (and traditional) leader

Suffrage: 18 years of age; universal

Elections:
President: last held 6 January 1992 (next to be held NA; results—President Amata KABUA was reelected
Parliament: last held 18 November 1991 (next to be held November 1995);

results—percent of vote NA; seats—(33 total)

Executive branch: president, Cabinet

Legislative branch: unicameral Nitijela (parliament)

Judicial branch: Supreme Court

Leaders:
Chief of State and Head of Government: President Amata KABUA (since 1979)

Member of: AsDB, ESCAP, IBRD, ICAO, IFC, IMF, INTERPOL, SPARTECA, SPC, SPF, UN, UNCTAD, WHO

Diplomatic representation in US:
chief of mission: Ambassador Wilfred I. KENDALL
chancery: 2433 Massachusetts Avenue, NW, Washington, DC 20008
telephone: (202) 234-5414

US diplomatic representation:
chief of mission: Ambassador David C. FIELDS
embassy: NA address, Majuro
mailing address: P. O. Box 1379, Majuro, Republic of the Marshall Islands 96960-1379
telephone: (011) 692-4011
FAX: (011) 692-4012

Flag: blue with two stripes radiating from the lower hoist-side corner—orange (top) and white; there is a white star with four large rays and 20 small rays on the hoist side above the two stripes

Economy

Overview: Agriculture and tourism are the mainstays of the economy. Agricultural production is concentrated on small farms, and the most important commercial crops are coconuts, tomatoes, melons, and breadfruit. A few cattle ranches supply the domestic meat market. Small-scale industry is limited to handicrafts, fish processing, and copra. The tourist industry is the primary source of foreign exchange and employs about 10% of the labor force. The islands have few natural resources, and imports far exceed exports. In 1987 the US Government provided grants of $40 million out of the Marshallese budget of $55 million.

National product: GDP—exchange rate conversion—$63 million (1989 est.)

National product real growth rate: NA%

National product per capita: $1,500 (1989 est.)

Inflation rate (consumer prices): NA%

Unemployment rate: NA%

Budget: revenues $55 million; expenditures $NA, including capital expenditures of $NA (1987 est.)

Exports: $2.5 million (f.o.b., 1985)
commodities: copra, copra oil, agricultural products, handicrafts
partners: NA

Imports: $29.2 million (c.i.f., 1985)
commodities: foodstuffs, beverages, building materials

partners: NA

External debt: $NA

Industrial production: growth rate NA%

Electricity: 42,000 kW capacity; 80 million kWh produced, 1,840 kWh per capita (1990)

Industries: copra, fish, tourism; craft items from shell, wood, and pearls; offshore banking (embryonic)

Agriculture: coconuts, cacao, taro, breadfruit, fruits, pigs, chickens

Economic aid: under the terms of the Compact of Free Association, the US is to provide approximately $40 million in aid annually

Currency: US currency is used

Exchange rates: US currency is used

Fiscal year: 1 October-30 September

Communications

Highways: paved roads on major islands (Majuro, Kwajalein), otherwise stone-, coral-, or laterite-surfaced roads and tracks

Ports: Majuro

Merchant marine: 29 ships (1,000 GRT or over) totaling 1,786,070 GRT/3,498,895 DWT; includes 2 cargo, 1 container, 9 oil tanker, 15 bulk carrier, 2 combination ore/oil; note—a flag of convenience registry

Airports:
total: 16
usable: 16
with permanent-surface runways: 4
with runways over 3,659m: 0
with runways 2,440-3,659 m: 0
with runways 1,220-2,439 m: 8

Telecommunications: telephone network—570 lines (Majuro) and 186 (Ebeye); telex services; islands interconnected by shortwave radio (used mostly for government purposes); broadcast stations—1 AM, 2 FM, 1 TV, 1 shortwave; 2 Pacific Ocean INTELSAT earth stations; US Government satellite communications system on Kwajalein

Defense Forces

Note: defense is the responsibility of the US

Martinique
(overseas department of France)

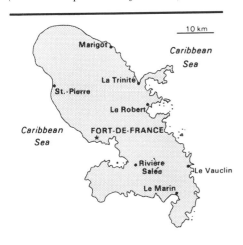

Geography

Location: in the Caribbean Sea, off the coast of Venezuela
Map references: Central America and the Caribbean, South America
Area:
total area: 1,100 km²
land area: 1,060 km²
comparative area: slightly more than six times the size of Washington, DC
Land boundaries: 0 km
Coastline: 290 km
Maritime claims:
exclusive economic zone: 200 nm
territorial sea: 12 nm
International disputes: none
Climate: tropical; moderated by trade winds; rainy season (June to October)
Terrain: mountainous with indented coastline; dormant volcano
Natural resources: coastal scenery and beaches, cultivable land
Land use:
arable land: 10%
permanent crops: 8%
meadows and pastures: 30%
forest and woodland: 26%
other: 26%
Irrigated land: 60 km² (1989 est.)
Environment: subject to hurricanes, flooding, and volcanic activity that result in an average of one major natural disaster every five years

People

Population: 387,656 (July 1993 est.)
Population growth rate: 1.21% (1993 est.)
Birth rate: 18.07 births/1,000 population (1993 est.)
Death rate: 5.94 deaths/1,000 population (1993 est.)
Net migration rate: 0 migrant(s)/1,000 population (1993 est.)
Infant mortality rate: 10.7 deaths/1,000 live births (1993 est.)

Life expectancy at birth:
total population: 77.82 years
male: 74.68 years
female: 81.01 years (1993 est.)
Total fertility rate: 1.94 children born/woman (1993 est.)
Nationality:
noun: Martiniquais (singular and plural)
adjective: Martiniquais
Ethnic divisions: African and African-Caucasian-Indian mixture 90%, Caucasian 5%, East Indian, Lebanese, Chinese less than 5%
Religions: Roman Catholic 95%, Hindu and pagan African 5%
Languages: French, Creole patois
Literacy: age 15 and over can read and write (1982)
total population: 93%
male: 92%
female: 93%
Labor force: 100,000
by occupation: service industry 31.7%, construction and public works 29.4%, agriculture 13.1%, industry 7.3%, fisheries 2.2%, other 16.3%

Government

Names:
conventional long form: Department of Martinique
conventional short form: Martinique
local long form: Departement de la Martinique
local short form: Martinique
Digraph: MB
Type: overseas department of France
Capital: Fort-de-France
Administrative divisions: none (overseas department of France)
Independence: none (overseas department of France)
Constitution: 28 September 1958 (French Constitution)
Legal system: French legal system
National holiday: National Day, Taking of the Bastille, 14 July (1789)
Political parties and leaders: Rally for the Republic (RPR); Union for a Martinique of Progress (UMP); Martinique Progressive Party (PPM); Socialist Federation of Martinique (FSM); Martinique Communist Party (PCM); Martinique Patriots (PM); Union for French Democracy (UDF)
Other political or pressure groups: Proletarian Action Group (GAP); Alhed Marie-Jeanne Socialist Revolution Group (GRS); Martinique Independence Movement (MIM); Caribbean Revolutionary Alliance (ARC); Central Union for Martinique Workers (CSTM), Marc PULVAR; Frantz Fanon Circle; League of Workers and Peasants
Suffrage: 18 years of age; universal

Elections:
French Senate: last held 24 September 1989 (next to be held NA); results—percent of vote by party NA; seats—(2 total) UDF 1, PPM 1
French National Assembly: last held on 5 and 12 June 1988 (next to be held June 1993); results—percent of vote by party NA; seats—(4 total) PPM 1, FSM 1, RPR 1, UDF 1
General Council: last held in 25 September and 8 October 1988 (next to be held by NA); results—percent of vote by party NA; seats—(44 total) number of seats by party NA; note—a leftist coalition obtained a one-seat margin
Regional Assembly: last held on NA March 1992 (next to be held by March 1998); results—percent of vote by party NA; seats—(41 total) UMP 16
Executive branch: government commissioner
Legislative branch: unicameral General Council
Judicial branch: Supreme Court
Leaders:
Chief of State: President Francois MITTERRAND (since 21 May 1981)
Head of Government: Government Commissioner Jean Claude ROURE (since 5 May 1989); President of the General Council Emile MAURICE (since NA 1988)
Member of: FZ, WCL
Diplomatic representation in US: as an overseas department of France, Martiniquais interests are represented in the US by France
US diplomatic representation:
chief of mission: Consul General Raymond G. ROBINSON
embassy: Consulate General at 14 Rue Blenac, Fort-de-France
mailing address: B. P. 561, Fort-de-France 97206
telephone: [596] 63-13-03
Flag: the flag of France is used

Economy

Overview: The economy is based on sugarcane, bananas, tourism, and light industry. Agriculture accounts for about 10% of GDP and the small industrial sector for 10%. Sugar production has declined, with most of the sugarcane now used for the production of rum. Banana exports are increasing, going mostly to France. The bulk of meat, vegetable, and grain requirements must be imported, contributing to a chronic trade deficit that requires large annual transfers of aid from France. Tourism has become more important than agricultural exports as a source of foreign exchange. The majority of the work force is employed in the service sector and in administration. Banana workers launched protests late in

Martinique (continued)

1992 because of falling banana prices and fears of greater competition in the European market from other producers.
National product: GDP—exchange rate conversion—$2 billion (1988)
National product real growth rate: NA%
National product per capita: $6,000 (1988)
Inflation rate (consumer prices): 3.9% (1990)
Unemployment rate: 32.1% (1990)
Budget: revenues $268 million; expenditures $268 million, including capital expenditures of $NA (1989 est.)
Exports: $196 million (f.o.b., 1988)
commodities: refined petroleum products, bananas, rum, pineapples
partners: France 65%, Guadeloupe 24%, Germany (1987)
Imports: $1.3 billion (c.i.f., 1988)
commodities: petroleum products, crude oil, foodstuffs, construction materials, vehicles, clothing and other consumer goods
partners: France 65%, UK, Italy, Germany, Japan, US (1987)
External debt: $NA
Industrial production: growth rate NA%
Electricity: 113,100 kW capacity; 588 million kWh produced, 1,580 kWh per capita (1992)
Industries: construction, rum, cement, oil refining, sugar, tourism
Agriculture: including fishing and forestry, accounts for about 12% of GDP; principal crops—pineapples, avocados, bananas, flowers, vegetables, sugarcane for rum; dependent on imported food, particularly meat and vegetables
Economic aid: Western (non-US) countries, ODA and OOF bilateral commitments (1970-89), $10.1 billion
Currency: 1 French franc (F) = 100 centimes
Exchange rates: French francs (F) per US$1—5.4812 (January 1993), 5.2938 (1992), 5.6421 (1991), 5.4453 (1990), 6.3801 (1989), 5.9569 (1988)
Fiscal year: calendar year

Communications

Highways: 1,680 km total; 1,300 km paved, 380 km gravel and earth
Ports: Fort-de-France
Airports:
total: 2
useable: 2
with permanent-surface runways: 1
with runways over 3,659 m: 0
with runways 2,440-3,659 m: 1
with runways 1,220-2,439 m: 0
Telecommunications: domestic facilities are adequate; 68,900 telephones; interisland microwave radio relay links to Guadeloupe, Dominica, and Saint Lucia; broadcast stations—1 AM, 6 FM, 10 TV; 2 Atlantic Ocean INTELSAT earth stations

Defense Forces

Branches: French Forces, Gendarmerie
Note: defense is the responsibility of France

Mauritania

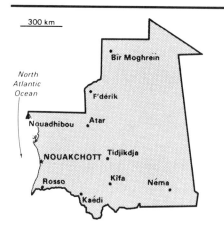

Geography

Location: Northern Africa, along the North Atlantic Ocean, between Western Sahara and Senegal
Map references: Africa, Standard Time Zones of the World
Area:
total area: 1,030,700 km²
land area: 1,030,400 km²
comparative area: slightly larger than three times the size of New Mexico
Land boundaries: total 5,074 km, Algeria 463 km, Mali 2,237 km, Senegal 813 km, Western Sahara 1,561 km
Coastline: 754 km
Maritime claims:
contiguous zone: 24 nm
continental shelf: 200 nm or the edge of continental margin
exclusive economic zone: 200 nm
territorial sea: 12 nm
International disputes: boundary with Senegal
Climate: desert; constantly hot, dry, dusty
Terrain: mostly barren, flat plains of the Sahara; some central hills
Natural resources: iron ore, gypsum, fish, copper, phosphate
Land use:
arable land: 1%
permanent crops: 0%
meadows and pastures: 38%
forest and woodland: 5%
other: 56%
Irrigated land: 120 km² (1989 est.)
Environment: hot, dry, dust/sand-laden sirocco wind blows primarily in March and April; desertification; only perennial river is the Senegal

People

Population: 2,124,792 (July 1993 est.)
Population growth rate: 3.14% (1993 est.)
Birth rate: 47.97 births/1,000 population (1993 est.)

Death rate: 16.54 deaths/1,000 population (1993 est.)

Net migration rate: 0 migrant(s)/1,000 population (1993 est.)

Infant mortality rate: 87 deaths/1,000 live births (1993 est.)

Life expectancy at birth:
total population: 47.59 years
male: 44.81 years
female: 50.48 years (1993 est.)

Total fertility rate: 7.05 children born/woman (1993 est.)

Nationality:
noun: Mauritanian(s)
adjective: Mauritanian

Ethnic divisions: mixed Maur/black 40%, Maur 30%, black 30%

Religions: Muslim 100%

Languages: Hasaniya Arabic (official), Pular, Soninke, Wolof (official)

Literacy: age 10 and over can read and write (1990)
total population: 34%
male: 47%
female: 21%

Labor force: 465,000 (1981 est.); 45,000 wage earners (1980)
by occupation: agriculture 47%, services 29%, industry and commerce 14%, government 10%
note: 53% of population of working age (1985)

Government

Names:
conventional long form: Islamic Republic of Mauritania
conventional short form: Mauritania
local long form: Al Jumhuriyah al Islamiyah al Muritaniyah
local short form: Muritaniyah

Digraph: MR

Type: republic

Capital: Nouakchott

Administrative divisions: 12 regions(regions, singular—region); Adrar, Assaba, Brakna, Dakhlet Nouadhibou, Gorgol, Guidimaka, Hodh ech Chargui, Hodh el Gharbi, Inchiri, Tagant, Tiris Zemmour, Trarza
note: there may be a new capital district of Nouakchott

Independence: 28 November 1960 (from France)

Constitution: 12 July 1991

Legal system: three-tier system: Islamic (Shari'a) courts, special courts, state security courts (in the process of being eliminated)

National holiday: Independence Day, 28 November (1960)

Political parties and leaders: legalized by constitution passed 12 July 1991, however, politics continue to be tribally based;

emerging parties include Democratic and Social Republican Party (PRDS), led by President Col. Maaouya Ould Sid'Ahmed TAYA; Union of Democratic Forces—New Era (UFD/NE), headed by Ahmed Ould DADDAH; Assembly for Democracy and Unity (RDU), Ahmed Ould SIDI BABA; Popular Social and Democratic Union (UPSD), Mohamed Mahmoud Ould MAH; Mauritanian Party for Renewal (PMR), Hameida BOUCHRAYA; National Avant-Garde Party (PAN), Khattry Ould JIDDOU; Mauritanian Party of the Democratic Center (PCDM), Bamba Ould SIDI BADI

Other political or pressure groups: Mauritanian Workers Union (UTM)

Suffrage: 18 years of age; universal

Elections:
President: last held January 1992 (next to be held January 1998); results—President Col. Maaouya Ould Sid 'Ahmed TAYA elected
Senate: last held 3 and 10 April 1992 (one-third of the seats up for re-election in 1994)
National Assembly: last held 6 and 13 March 1992 (next to be held March 1997)

Executive branch: president

Legislative branch: bicameral legislature consists of an upper house or Senate (Majlis al-Shuyukh) and a lower house or National Assembly (Majlis al-Watani)

Judicial branch: Supreme Court (Cour Supreme)

Leaders:
Chief of State and Head of Government: President Col. Maaouya Ould Sid'Ahmed TAYA (since 12 December 1984)

Member of: ABEDA, ACCT (associate), ACP, AfDB, AFESD, AL, AMF, AMU, CAEU, CCC, CEAO, ECA, ECOWAS, FAO, G-77, GATT, IBRD, ICAO, IDA, IDB, IFAD, IFC, ILO, IMF, IMO, INTELSAT, INTERPOL, IOC, ITU, LORCS, NAM, OAU, OIC, UN, UNCTAD, UNESCO, UNIDO, UPU, WHO, WIPO, WMO, WTO

Diplomatic representation in US:
chief of mission: Ambassador Mohamed Fall OULD AININA
chancery: 2129 Leroy Place NW, Washington, DC 20008
telephone: (202) 232-5700

US diplomatic representation:
chief of mission: Ambassador Gordon S. BROWN
embassy: address NA, Nouakchott
mailing address: B. P. 222, Nouakchott
telephone: [222] (2) 526-60 or 526-63
FAX: [222] (2) 525-89

Flag: green with a yellow five-pointed star above a yellow, horizontal crescent; the closed side of the crescent is down; the crescent, star, and color green are traditional symbols of Islam

Economy

Overview: A majority of the population still depends on agriculture and livestock for a livelihood, even though most of the nomads and many subsistence farmers were forced into the cities by recurrent droughts in the 1970s and 1980s. Mauritania has extensive deposits of iron ore, which account for almost 50% of total exports. The decline in world demand for this ore, however, has led to cutbacks in production. The nation's coastal waters are among the richest fishing areas in the world, but overexploitation by foreigners threatens this key source of revenue. The country's first deepwater port opened near Nouakchott in 1986. In recent years, the droughts, the endemic conflict with Senegal, rising energy costs, and economic mismanagement have resulted in a substantial buildup of foreign debt. The government has begun the second stage of an economic reform program in consultation with the World Bank, the IMF, and major donor countries. But the reform process suffered a major setback following the Gulf war of early 1991. Because of Mauritania's support of SADDAM Husayn, bilateral aid from its two top donors, Saudi Arabia and Kuwait, was suspended, and multilateral aid was reduced.

National product: GDP—exchange rate conversion—$1.1 billion (1991 est.)

National product real growth rate: 3% (1991 est.)

National product per capita: $555 (1991 est.)

Inflation rate (consumer prices): 6.2% (1991 est.)

Unemployment rate: 20% (1991 est.)

Budget: revenues $280 million; expenditures $346 million, including capital expenditures of $61 million (1989 est.)

Exports: $447 million (f.o.b., 1990)
commodities: iron ore, processed fish, small amounts of gum arabic and gypsum; unrecorded but numerically significant cattle exports to Senegal
partners: EC 43%, Japan 27%, USSR 11%, Cote d'Ivoire 3%

Imports: $385 million (c.i.f., 1990)
commodities: foodstuffs, consumer goods, petroleum products, capital goods
partners: EC 60%, Algeria 15%, China 6%, US 3%

External debt: $1.9 billion (1990)

Industrial production: growth rate 4.4% (1988 est.); accounts for almost 33% of GDP

Electricity: 190,000 kW capacity; 135 million kWh produced, 70 kWh per capita (1991)

Industries: fish processing, mining of iron ore and gypsum

Mauritania (continued)

Agriculture: accounts for 50% of GDP (including fishing); largely subsistence farming and nomadic cattle and sheep herding except in Senegal river valley; crops —dates, millet, sorghum, root crops; fish products number-one export; large food deficit in years of drought
Economic aid: US commitments, including Ex-Im (FY70-89), $168 million; Western (non-US) countries, ODA and OOF bilateral commitments (1970-89), $1.3 billion; OPEC bilateral aid (1979-89), $490 million; Communist countries (1970-89), $277 million; Arab Development Bank (1991), $20 million
Currency: 1 ouguiya (UM) = 5 khoums
Exchange rates: ouguiya (UM) per US$1—116.990 (February 1993), 87.082 (1992), 81.946 (1991), 80.609 (1990), 83.051 (1989), 75.261 (1988)
Fiscal year: calendar year

Communications

Railroads: 690 km 1.435-meter (standard) gauge, single track, owned and operated by government mining company
Highways: 7,525 km total; 1,685 km paved; 1,040 km gravel, crushed stone, or otherwise improved; 4,800 km unimproved roads, trails, tracks
Inland waterways: mostly ferry traffic on the Senegal River
Ports: Nouadhibou, Nouakchott
Merchant marine: 1 cargo ship (1,000 GRT or over) totaling 1,290 GRT/1,840 DWT
Airports:
total: 29
usable: 29
with permanent-surface runways: 9
with runways over 3,659 m: 1
with runways 2,440-3,659 m: 5
with runways 1,220-2,439 m: 16
Telecommunications: poor system of cable and open-wire lines, minor microwave radio relay links, and radio communications stations (improvements being made); broadcast stations—2 AM, no FM, 1 TV; satellite earth stations—1 Atlantic Ocean INTELSAT and 2 ARABSAT, with six planned

Defense Forces

Branches: Army, Navy, Air Force, National Gendarmerie, National Guard, National Police, Presidential Guard
Manpower availability: males age 15-49 452,008; fit for military service 220,717 (1993 est.); conscription law not implemented
Defense expenditures: exchange rate conversion—$40 million, 4.2% of GDP (1989)

Mauritius

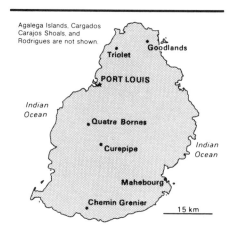

Agalega Islands, Cargados Carajos Shoals, and Rodrigues are not shown.

Geography

Location: Southern Africa, in the western Indian Ocean, 900 km east of Madagascar
Map references: Africa, Standard Time Zones of the World
Area:
total area: 1,860 km²
land area: 1,850 km²
comparative area: slightly less than 10.5 times the size of Washington, DC
note: includes Agalega Islands, Cargados Carajos Shoals (Saint Brandon), and Rodrigues
Land boundaries: 0 km
Coastline: 177 km
Maritime claims:
contiguous zone: 24 nm
continental shelf: 200 nm or the edge of continental margin
exclusive economic zone: 200 nm
territorial sea: 12 nm
International disputes: claims UK-administered Chagos Archipelago, which includes the island of Diego Garcia in UK-administered British Indian Ocean Territory; claims French-administered Tromelin Island
Climate: tropical modified by southeast trade winds; warm, dry winter (May to November); hot, wet, humid summer (November to May)
Terrain: small coastal plain rising to discontinuous mountains encircling central plateau
Natural resources: arable land, fish
Land use:
arable land: 54%
permanent crops: 4%
meadows and pastures: 4%
forest and woodland: 31%
other: 7%
Irrigated land: 170 km² (1989 est.)
Environment: subject to cyclones (November to April); almost completely surrounded by reefs

People

Population: 1,106,516 (July 1993 est.)
Population growth rate: 0.95% (1993 est.)
Birth rate: 19.67 births/1,000 population (1993 est.)
Death rate: 6.44 deaths/1,000 population (1993 est.)
Net migration rate: -3.71 migrant(s)/1,000 population (1993 est.)
Infant mortality rate: 19 deaths/1,000 live births (1993 est.)
Life expectancy at birth:
total population: 70.24 years
male: 66.34 years
female: 74.3 years (1993 est.)
Total fertility rate: 2.23 children born/woman (1993 est.)
Nationality:
noun: Mauritian(s)
adjective: Mauritian
Ethnic divisions: Indo-Mauritian 68%, Creole 27%, Sino-Mauritian 3%, Franco-Mauritian 2%
Religions: Hindu 52%, Christian 28.3% (Roman Catholic 26%, Protestant 2.3%), Muslim 16.6%, other 3.1%
Languages: English (official), Creole, French, Hindi, Urdu, Hakka, Bojpoori
Literacy: age 13 and over can read and write (1962)
total population: 61%
male: 72%
female: 50%
Labor force: 335,000
by occupation: government services 29%, agriculture and fishing 27%, manufacturing 22%, other 22%
note: 43% of population of working age (1985)

Government

Names:
conventional long form: Republic of Mauritius
conventional short form: Mauritius
Digraph: MP
Type: parliamentary democracy
Capital: Port Louis
Administrative divisions: 9 districts and 3 dependencies*; Agalega Islands*, Black River, Cargados Carajos*, Flacq, Grand Port, Moka, Pamplemousses, Plaines Wilhems, Port Louis, Riviere du Rempart, Rodrigues*, Savanne
Independence: 12 March 1968 (from UK)
Constitution: 12 March 1968
Legal system: based on French civil law system with elements of English common law in certain areas
National holiday: Independence Day, 12 March (1968)
Political parties and leaders:
government coalition: Militant Socialist Movement (MSM), A. JUGNAUTH

Mauritian Militant Movement (MMM), Paul BERENGER; Organization of the People of Rodrigues (OPR), Louis Serge CLAIR; Democratic Labor Movement (MTD), Anil BAICHOO
opposition: Mauritian Labor Party (MLP), Navin RAMGOOLMAN Socialist Workers Front, Sylvio MICHEL; Mauritian Social Democratic Party (PMSD), X. DUVAL
Other political or pressure groups: various labor unions
Suffrage: 18 years of age; universal
Elections:
Legislative Assembly: last held on 15 September 1991 (next to be held by 15 September 1996); results—MSM/MMM 53%, MLP/PMSD 38%; seats—(70 total, 62 elected) MSM/MMM alliance 59 (MSM 29, MMM 26, OPR 2, MTD 2); MLP/PMSD 3
Executive branch: president, vice president, prime minister, deputy prime minister, Council of Ministers (cabinet)
Legislative branch: unicameral Legislative Assembly
Judicial branch: Supreme Court
Leaders:
Chief of State: President Cassam UTEEM (since 1 July 1992); Vice President Robin Dranooth GHURBURRON (since 1 July 1992)
Head of Government: Prime Minister Sir Aneerood JUGNAUTH (since 12 June 1982); Deputy Prime Minister Prem NABABSING (since 26 September 1990)
Member of: ACCT, ACP, AfDB, C, CCC, ECA, FAO, G-77, GATT, IAEA, IBRD, ICAO, ICFTU, IDA, IFAD, IFC, ILO, IMF, IMO, INTELSAT, INTERPOL, IOC, ISO (correspondent), ITU, LORCS, NAM, OAU, PCA, UN, UNCTAD, UNESCO, UNIDO, UPU, WCL, WFTU, WHO, WIPO, WMO, WTO
Diplomatic representation in US:
chief of mission: Ambassador Chitmansing JESSERAMSING
chancery: Suite 134, 4301 Connecticut Avenue NW, Washington, DC 20008
telephone: (202) 244-1491 or 1492
US diplomatic representation:
chief of mission: Ambassador vacant
embassy: 4th Floor, Rogers House, John Kennedy Street, Port Louis
mailing address: 4th Floor, Rogers House, John Kennedy Street, Port Louis
telephone: [230] 208-9763 through 208-9767
FAX: [230] 208-9534
Flag: four equal horizontal bands of red (top), blue, yellow, and green

Economy

Overview: The economy is based on sugar, manufacturing (mainly textiles), and tourism. Sugarcane is grown on about 90% of the cultivated land area and accounts for 40% of

export earnings. The government's development strategy is centered on industrialization (with a view to exports), agricultural diversification, and tourism. Economic performance in FY91 was impressive, with 6% real growth and low unemployment.
National product: GDP—exchange rate conversion—$2.5 billion (FY91 est.)
National product real growth rate: 6.1% (FY91 est.)
National product per capita: $2,300 (FY91 est.)
Inflation rate (consumer prices): 7% (FY91)
Unemployment rate: 2.4% (1991 est.)
Budget: revenues $557 million; expenditures $607 million, including capital expenditures of $111 million (FY90)
Exports: $1.2 billion (f.o.b., 1990)
commodities: textiles 44%, sugar 40%, light manufactures 10%
partners: EC and US have preferential treatment, EC 77%, US 15%
Imports: $1.6 billion (f.o.b., 1990)
commodities: manufactured goods 50%, capital equipment 17%, foodstuffs 13%, petroleum products 8%, chemicals 7%
partners: EC, US, South Africa, Japan
External debt: $869 million (1991 est.)
Industrial production: growth rate 7% (1990); accounts for 25% of GDP
Electricity: 235,000 kW capacity; 630 million kWh produced, 570 kWh per capita (1992)
Industries: food processing (largely sugar milling), textiles, wearing apparel, chemicals, metal products, transport equipment, nonelectrical machinery, tourism
Agriculture: accounts for 10% of GDP; about 90% of cultivated land in sugarcane; other products—tea, corn, potatoes, bananas, pulses, cattle, goats, fish; net food importer, especially rice and fish
Illicit drugs: illicit producer of cannabis for the international drug trade
Economic aid: US commitments, including Ex-Im (FY70-89), $76 million; Western (non-US) countries (1970-89), $709 million; Communist countries (1970-89), $54 million
Currency: 1 Mauritian rupee (MauR) = 100 cents
Exchange rates: Mauritian rupees (MauRs) per US$1—16.982 (January 1993), 15.563 (1992), 15.652 (1991), 14.839 (1990), 15.250 (1989), 13.438 (1988)
Fiscal year: 1 July-30 June

Communications

Highways: 1,800 km total; 1,640 km paved, 160 km earth
Ports: Port Louis
Merchant marine: 7 ships (1,000 GRT or over) totaling 103,328 GRT/163,142 DWT; includes 3 cargo, 1 liquefied gas, 3 bulk

Airports:
total: 5
usable: 4
with permanent-surface runways: 2
with runways over 3,659 m: 0
with runways 2,440-3,659 m: 1
with runways 1,220-2,439 m: 0
Telecommunications: small system with good service utilizing primarily microwave radio relay; new microwave link to Reunion; high-frequency radio links to several countries; over 48,000 telephones; broadcast stations—2 AM, no FM, 4 TV; 1 Indian Ocean INTELSAT earth station

Defense Forces

Branches: National Police Force (including the paramilitary Special Mobile Force (SMF), Special Support Units (SSU), and National Coast Guard
Manpower availability: males age 15-49 312,056; fit for military service 159,408 (1993 est.)
Defense expenditures: exchange rate conversion—$5 million, 0.2% of GDP (FY89)

253

Mayotte
(territorial collectivity of France)

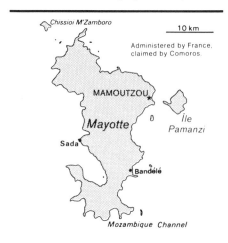

Chissioi M'Zamboro

10 km

Administered by France, claimed by Comoros.

MAMOUTZOU ★

Mayotte

Île Pamanzi

Sada

Bandélé

Mozambique Channel

Geography

Location: Southern Africa, in the northern Mozambique Channel about halfway between Madagascar and Mozambique
Map references: Africa
Area:
total area: 375 km²
land area: 375 km²
comparative area: slightly more than twice the size of Washington, DC
Land boundaries: 0 km
Coastline: 185.2 km
Maritime claims:
exclusive economic zone: 200 nm
territorial sea: 12 nm
International disputes: claimed by Comoros
Climate: tropical; marine; hot, humid, rainy season during northeastern monsoon (November to May); dry season is cooler (May to November)
Terrain: generally undulating with ancient volcanic peaks, deep ravines
Natural resources: negligible
Land use:
arable land: NA%
permanent crops: NA%
meadows and pastures: NA%
forest and woodland: NA%
other: NA%
Irrigated land: NA km²
Environment: subject to cyclones during rainy season
Note: part of Comoro Archipelago

People

Population: 89,983 (July 1993 est.)
Population growth rate: 3.8% (1993 est.)
Birth rate: 49.22 births/1,000 population (1993 est.)
Death rate: 11.22 deaths/1,000 population (1993 est.)
Net migration rate: 0 migrant(s)/1,000 population (1993 est.)
Infant mortality rate: 81.8 deaths/1,000 live births (1993 est.)

Life expectancy at birth:
total population: 57.35 years
male: 55.23 years
female: 59.55 years (1993 est.)
Total fertility rate: 6.84 children born/woman (1993 est.)
Nationality:
noun: Mahorais (singular and plural)
adjective: Mahoran
Ethnic divisions: NA
Religions: Muslim 99%, Christian (mostly Roman Catholic)
Languages: Mahorian (a Swahili dialect), French
Literacy:
total population: NA
male: NA
female: NA
Labor force: NA

Government

Names:
conventional long form: Territorial Collectivity of Mayotte
conventional short form: Mayotte
Digraph: MF
Type: territorial collectivity of France
Capital: Mamoutzou
Administrative divisions: none (territorial collectivity of France)
Independence: none (territorial collectivity of France)
Constitution: 28 September 1958 (French Constitution)
Legal system: French law
National holiday: Taking of the Bastille, 14 July (1789)
Political parties and leaders: Mahoran Popular Movement (MPM), Younoussa BAMANA; Party for the Mahoran Democratic Rally (PRDM), Daroueche MAOULIDA; Mahoran Rally for the Republic (RMPR), Mansour KAMARDINE; Union of the Center (UDC)
Suffrage: 18 years of age; universal
Elections:
General Council: last held March 1991 (next to be held March 1996); results—percent of vote by party NA; seats—(17 total) MPM 12, RPR 5
French Senate: last held on 24 September 1989 (next to be held September 1993); results—percent of vote by party NA; seats—(1 total) MPM 1
French National Assembly: last held 5 and 12 June 1988 (next to be held June 1993); results—percent of vote by party NA; seats—(1 total) UDC 1
Executive branch: government commissioner
Legislative branch: unicameral General Council (Conseil General)
Judicial branch: Supreme Court (Tribunal Superieur d'Appel)

Leaders:
Chief of State: President Francois MITTERRAND (since 21 May 1981)
Head of Government: Commissioner, Representative of the French Government Jean-Paul COSTE (since NA 1991); President of the General Council Youssouf BAMANA (since NA 1976)
Member of: FZ
Diplomatic representation in US: as a territorial collectivity of France, Mahoran interests are represented in the US by France
Flag: the flag of France is used

Economy

Overview: Economic activity is based primarily on the agricultural sector, including fishing and livestock raising. Mayotte is not self-sufficient and must import a large portion of its food requirements, mainly from France. The economy and future development of the island are heavily dependent on French financial assistance.
National product: GDP $NA
National product real growth rate: NA%
National product per capita: $NA
Inflation rate (consumer prices): NA%
Unemployment rate: NA%
Budget: revenues $NA; expenditures $37.3 million, including capital expenditures of $NA (1985)
Exports: $4.0 million (f.o.b., 1984)
commodities: ylang-ylang, vanilla
partners: France 79%, Comoros 10%, Reunion 9%
Imports: $21.8 million (f.o.b., 1984)
commodities: building materials, transportation equipment, rice, clothing, flour
partners: France 57%, Kenya 16%, South Africa 11%, Pakistan 8%
External debt: $NA
Industrial production: growth rate NA%
Electricity: NA kW capacity; NA million kWh produced, NA kWh per capita
Industries: newly created lobster and shrimp industry
Agriculture: most important sector; provides all export earnings; crops—vanilla, ylang-ylang, coffee, copra; imports major share of food needs
Economic aid: Western (non-US) countries, ODA and OOF bilateral commitments (1970-89), $402 million
Currency: 1 French franc (F) = 100 centimes
Exchange rates: French francs (F) per US$1—5.4812 (January 1993), 5.2938 (1992), 5.6421 (1991), 5.4453 (1990), 6.3801 (1989), 5.9569 (1988)
Fiscal year: calendar year

Mexico

Communications

Highways: 42 km total; 18 km bituminous
Ports: Dzaoudzi
Airports:
total: 1
usable: 1
with permanet-surface runways: 1
with runways over 3,659 m: 0
with runways 2,440-3,659 m: 0
with runways 1,220-2,439 m: 1
Telecommunications: small system administered by French Department of Posts and Telecommunications; includes radio relay and high-frequency radio communications for links to Comoros and international communications; 450 telephones; broadcast stations—1 AM, no FM, no TV

Defense Forces

Note: defense is the responsibility of France

1000 km

Geography

Location: Central America, between Guatemala and the US
Map references: North America, Standard Time Zones of the World
Area:
total area: 1,972,550 km²
land area: 1,923,040 km²
comparative area: slightly less than three times the size of Texas
Land boundaries: total 4,538 km, Belize 250 km, Guatemala 962 km, US 3,326 km
Coastline: 9,330 km
Maritime claims:
contiguous zone: 24 nm
continental shelf: 200 nm or the natural prolongation of continental margin
exclusive economic zone: 200 nm
territorial sea: 12 nm
International disputes: claims Clipperton Island (French possession)
Climate: varies from tropical to desert
Terrain: high, rugged mountains, low coastal plains, high plateaus, and desert
Natural resources: petroleum, silver, copper, gold, lead, zinc, natural gas, timber
Land use:
arable land: 12%
permanent crops: 1%
meadows and pastures: 39%
forest and woodland: 24%
other: 24%
Irrigated land: 51,500 km² (1989 est.)
Environment: subject to tsunamis along the Pacific coast and destructive earthquakes in the center and south; natural water resources scarce and polluted in north, inaccessible and poor quality in center and extreme southeast; deforestation; erosion widespread; desertification; serious air pollution in Mexico City and urban centers along US-Mexico border
Note: strategic location on southern border of US

People

Population: 90,419,606 (July 1993 est.)
Population growth rate: 1.97% (1993 est.)
Birth rate: 27.67 births/1,000 population (1993 est.)
Death rate: 4.82 deaths/1,000 population (1993 est.)
Net migration rate: -3.15 migrant(s)/1,000 population (1993 est.)
Infant mortality rate: 28.8 deaths/1,000 live births (1993 est.)
Life expectancy at birth:
total population: 72.55 years
male: 68.99 years
female: 76.3 years (1993 est.)
Total fertility rate: 3.25 children born/woman (1993 est.)
Nationality:
noun: Mexican(s)
adjective: Mexican
Ethnic divisions: mestizo (Indian-Spanish) 60%, Amerindian or predominantly Amerindian 30%, Caucasian or predominantly Caucasian 9%, other 1%
Religions: nominally Roman Catholic 89%, Protestant 6%
Languages: Spanish, various Mayan dialects
Literacy: age 15 and over can read and write (1990)
total population: 87%
male: 90%
female: 85%
Labor force: 26.2 million (1990)
by occupation: services 31.7%, agriculture, forestry, hunting, and fishing 28%, commerce 14.6%, manufacturing 11.1%, construction 8.4%, transportation 4.7%, mining and quarrying 1.5%

Government

Names:
conventional long form: United Mexican States
conventional short form: Mexico
local long form: Estados Unidos Mexicanos
local short form: Mexico
Digraph: MX
Type: federal republic operating under a centralized government
Capital: Mexico
Administrative divisions: 31 states (estados, singular—estado) and 1 federal district* (distrito federal); Aguascalientes, Baja California, Baja California Sur, Campeche, Chiapas, Chihuahua, Coahuila, Colima, Distrito Federal*, Durango, Guanajuato, Guerrero, Hidalgo, Jalisco, Mexico, Michoacan, Morelos, Nayarit, Nuevo Leon, Oaxaca, Puebla, Queretaro, Quintana Roo, San Luis Potosi, Sinaloa, Sonora, Tabasco, Tamaulipas, Tlaxcala, Veracruz, Yucatan, Zacatecas

Mexico *(continued)*

Independence: 16 September 1810 (from Spain)

Constitution: 5 February 1917

Legal system: mixture of US constitutional theory and civil law system; judicial review of legislative acts; accepts compulsory ICJ jurisdiction, with reservations

National holiday: Independence Day, 16 September (1810)

Political parties and leaders: (recognized parties) Institutional Revolutionary Party (PRI), Fernando Ortiz Arana; National Action Party (PAN), Carlos CASTILLO; Popular Socialist Party (PPS), Indalecio SAYAGO Herrera; Democratic Revolutionary Party (PRD), Roberto ROBLES Garnica; Cardenist Front for the National Reconstruction Party (PFCRN), Rafael AGUILAR Talamantes; Authentic Party of the Mexican Revolution (PARM), Carlos Enrique CANTU Rosas; Democratic Forum Party (PFD), Pablo Emilio MADERO; Mexican Ecologist Party (PEM), Jorge GONZALEZ Torres

Other political or pressure groups: Roman Catholic Church; Confederation of Mexican Workers (CTM); Confederation of Industrial Chambers (CONCAMIN); Confederation of National Chambers of Commerce (CONCANACO); National Peasant Confederation (CNC); Revolutionary Workers Party (PRT); Revolutionary Confederation of Workers and Peasants (CROC); Regional Confederation of Mexican Workers (CROM); Confederation of Employers of the Mexican Republic (COPARMEX); National Chamber of Transformation Industries (CANACINTRA); Coordinator for Foreign Trade Business Organizations (COECE); Federation of Unions Provding Goods and Services (FESEBES)

Suffrage: 18 years of age; universal and compulsory (but not enforced)

Elections:

President: last held on 6 July 1988 (next to be held August 1994); results—Carlos SALINAS de Gortari (PRI) 50.74%, Cuauhtemoc CARDENAS Solorzano (FDN) 31.06%, Manuel CLOUTHIER (PAN) 16.81%; other 1.39%; note—several of the smaller parties ran a common candidate under a coalition called the National Democratic Front (FDN)

Senate: last held on 18 August 1991 (next to be held midyear 1994); results—percent of vote by party NA; seats in full Senate—(64 total) PRI 62, PRD 1, PAN 1

Chamber of Deputies: last held on 18 August 1991 (next to be held midyear 1994); results—PRI 53%, PAN 20%, PFCRN 10%, PPS 6%, PARM 7%, PMS (now part of PRD) 4%; seats—(500 total) PRI 320, PAN 89, PRD 41, PFCRN 23, PARM 15, PPS 12

Executive branch: president, Cabinet

Legislative branch: bicameral National Congress (Congreso de la Union) consists of an upper chamber or Senate (Camara de Senadores) and a lower chamber or Chamber of Deputies (Camara de Diputados)

Judicial branch: Supreme Court of Justice (Corte Suprema de Justicia)

Leaders:

Chief of State and Head of Government: President Carlos SALINAS de Gortari (since 1 December 1988)

Member of: AG (observer), CARICOM (observer), CCC, CDB, CG, EBRD, ECLAC, FAO, G-3, G-6, G-11, G-15, G-19, G-24, G-77, GATT, IADB, IAEA, IBRD, ICAO, ICC, ICFTU, IDA, IFAD, IFC, ILO, IMF, IMO, INTELSAT, INTERPOL, IOC, IOM (observer), ISO, ITU, LAES, LAIA, LORCS, NAM (observer), OAS, OPANAL, PCA, RG, UN, UNCTAD, UNESCO, UNIDO, UPU, WCL, WHO, WIPO, WMO, WTO

Diplomatic representation in US:

chief of mission: Ambassador Jorge MONTANO Martinez

chancery: 1911 Pennsylvania Avenue NW, Washington, DC 20006

telephone: (202) 728-1600

consulates general: Chicago, Dallas, Denver, Los Angeles, New Orleans, New York, San Juan (Puerto Rico)

consulates: Albuquerque, Atlanta, Austin, Boston, Brownsville (Texas), Calexico (California), Corpus Christi, Detroit, Fresno (California), Miami, Nogales (Arizona), Philadelphia, Phoenix, St. Louis, Salt Lake City, Seattle

US diplomatic representation:

chief of mission: Ambassador John D. NEGROPONTE, Jr.

embassy: Paseo de la Reforma 305, 06500 Mexico, D.F.

mailing address: P. O. Box 3087, Laredo, TX 78044-3087

telephone: [52] (5) 211-0042

FAX: [52] (5) 511-9980, 208-3373

consulates general: Ciudad Juarez, Guadalajara, Monterrey, Tijuana

consulates: Hermosillo, Matamoros, Mazatlan, Merida, Nuevo Laredo

Flag: three equal vertical bands of green (hoist side), white, and red; the coat of arms (an eagle perched on a cactus with a snake in its beak) is centered in the white band

---

Economy

Overview: Mexico's economy is a mixture of state-owned industrial facilities (notably oil), private manufacturing and services, and both large-scale and traditional agriculture. In the 1980s, Mexico experienced severe economic difficulties: the nation accumulated large external debts as world petroleum prices fell; rapid population growth outstripped the domestic food supply; and inflation, unemployment, and pressures to emigrate became more acute. Growth in national output, however, has recovered, rising from 1.4% in 1988 to 4% in 1990 and 3.6% in 1991 and coming in at 2.6% in 1992. The US is Mexico's major trading partner, accounting for almost three-quarters of its exports and imports. After petroleum, border assembly plants and tourism are the largest earners of foreign exchange. The government, in consultation with international economic agencies, has been implementing programs to stabilize the economy and foster growth. For example, it has privatized more than two-thirds of its state-owned companies (parastatals), including banks. In 1991-92 the government conducted negotiations with the US and Canada on a North American Free Trade Agreement (NAFTA), which was still being discussed by the three countries in early 1993. In January 1993, Mexico replaced its old peso with a new peso, at the rate of 1,000 old to 1 new peso. Notwithstanding the palpable improvements in economic performance in the early 1990s, Mexico faces substantial problems for the remainder of the decade—e.g., rapid population growth, unemployment, and serious pollution, particularly in Mexico City.

National product: GDP—exchange rate conversion—$328 billion (1992 est.)

National product real growth rate: 2.6% (1992)

National product per capita: $3,600 (1992 est.)

Inflation rate (consumer prices): 11.9% (1992)

Unemployment rate: 14%-17% (1991 est.)

Budget: revenues $58.9 billion; expenditures $48.3 billion, including capital expenditures of $6.5 billion (1991); figures do not include state-owned companies

Exports: $27.5 billion (f.o.b., 1992 est.)

commodities: crude oil, oil products, coffee, shrimp, engines, motor vehicles, cotton, consumer electronics

partners: US 74%, Japan 8%, EC 4% (1992 est.)

Imports: $48.1 billion (c.i.f., 1992 est.)

commodities: metal-working machines, steel mill products, agricultural machinery, electrical equipment, car parts for assembly, repair parts for motor vehicles, aircraft, and aircraft parts

partners: US 74%, Japan, 11%, EC 6% (1992)

External debt: $104 billion (1992 est.)

Industrial production: growth rate 5.5% (1991 est.); accounts for 28% of GDP

Electricity: 27,000,000 kW capacity; 120,725 million kWh produced, 1,300 kWh per capita (1992)

Industries: food and beverages, tobacco, chemicals, iron and steel, petroleum, mining, textiles, clothing, motor vehicles, consumer durables, tourism

Agriculture: accounts for 9% of GDP and over 25% of work force; large number of small farms at subsistence level; major food crops—corn, wheat, rice, beans; cash crops—cotton, coffee, fruit, tomatoes; fish catch of 1.4 million metric tons among top 20 nations (1987)

Illicit drugs: illicit cultivation of opium poppy and cannabis continues in spite of active government eradication program; major supplier to the US market; continues as the primary transshipment country for US-bound cocaine from South America

Economic aid: US commitments, including Ex-Im (FY70-89), $3.1 billion; Western (non-US) countries, ODA and OOF bilateral commitments (1970-89), $7.7 billion; Communist countries (1970-89), $110 million

Currency: 1 New Mexican peso (Mex$) = 100 centavos

Exchange rates: market rate of Mexican pesos (Mex$) per US$1—3.100 (January 1993), 3,198 (November 1992), 3,018.4 (1991), 2,812.6 (1990), 2,461.3 (1989), 2,273.1 (1988); note—the new pesos replaced the old pesos on 1 January 1993; 1 new pesos = 1,000 old pesos

Fiscal year: calendar year

Communications

Railroads: 24,500 km total

Highways: 212,000 km total; 65,000 km paved, 30,000 km semipaved or cobblestone, 62,000 km rural roads (improved earth) or roads under construction, 55,000 km unimproved earth roads

Inland waterways: 2,900 km navigable rivers and coastal canals

Pipelines: crude oil 28,200 km; petroleum products 10,150 km; natural gas 13,254 km; petrochemical 1,400 km

Ports: Acapulco, Altamira, Coatzacoalcos, Ensenada, Guaymas, Manzanillo, Mazatlan, Progreso, Puerto Vallarta, Salina Cruz, Tampico, Tuxpan, Veracruz

Merchant marine: 58 ships (1,000 GRT or over) totaling 858,162 GRT/1,278,488 DWT; includes 4 short-sea passenger, 2 cargo, 2 refrigerated cargo, 2 roll-on/roll-off, 31 oil tanker, 4 chemical tanker, 7 liquefied gas, 1 bulk, 5 container

Airports:
total: 1,841
usable: 1,478
with permanent-surface runways: 200
with runways over 3,659 m: 3
with runways 2,440-3,659 m: 35
with runways 1,220-2,439 m: 273

Telecommunications: highly developed system with extensive microwave radio relay links; privatized in December 1990; connected into Central America Microwave System; 6,410,000 telephones; broadcast stations—679 AM, no FM, 238 TV, 22 shortwave; 120 domestic satellite terminals; earth stations—4 Atlantic Ocean INTELSAT and 1 Pacific Ocean INTELSAT

Defense Forces

Branches: National Defense (including Army and Air Force), Navy (including Marines)

Manpower availability: males age 15-49 22,201,567; fit for military service 16,205,926; reach military age (18) annually 1,049,729 (1993 est.)

Defense expenditures: exchange rate conversion—$NA, NA% of GDP

Micronesia, Federated States of

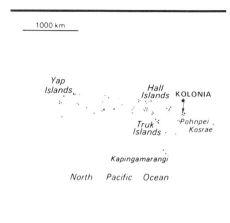

Geography

Location: Oceania, in the North Pacific Ocean, about three-quarters of the way between Hawaii and Indonesia

Map references: Oceania, Southeast Asia, Standard Time Zones of the World

Area:
total area: 702 km²
land area: 702 km²
comparative area: slightly less than four times the size of Washington, DC
note: includes Pohnpei (Ponape), Truk (Chuuk), Yap, and Kosrae

Land boundaries: 0 km

Coastline: 6,112 km

Maritime claims:
exclusive economic zone: 200 nm
territorial sea: 12 nm

International disputes: none

Climate: tropical; heavy year-round rainfall, especially in the eastern islands; located on southern edge of the typhoon belt with occasional severe damage

Terrain: islands vary geologically from high mountainous islands to low, coral atolls; volcanic outcroppings on Pohnpei, Kosrae, and Truk

Natural resources: forests, marine products, deep-seabed minerals

Land use:
arable land: NA%
permanent crops: NA%
meadows and pastures: NA%
forest and woodland: NA%
other: NA%

Irrigated land: NA km²

Environment: subject to typhoons from June to December; four major island groups totaling 607 islands

---

People

Population: 117,588 (July 1993 est.)
Population growth rate: 3.37% (1993 est.)
Birth rate: 28.48 births/1,000 population (1993 est.)

Micronesia, Federated States of
(continued)

Death rate: 6.46 deaths/1,000 population (1993 est.)
Net migration rate: 11.65 migrant(s)/1,000 population (1993 est.)
Infant mortality rate: 37.96 deaths/1,000 live births (1993 est.)
Life expectancy at birth:
total population: 67.45 years
male: 65.49 years
female: 69.44 years (1993 est.)
Total fertility rate: 4.04 children born/woman (1993 est.)
Nationality:
noun: Micronesian(s)
adjective: Micronesian; Kosrae(s), Pohnpeian(s), Trukese, Yapese
Ethnic divisions: nine ethnic Micronesian and Polynesian groups
Religions: Christian (divided between Roman Catholic and Protestant; other churches include Assembly of God, Jehovah's Witnesses, Seventh-Day Adventist, Latter-Day Saints, and the Baha'i Faith)
Languages: English (official and common language), Trukese, Pohnpeian, Yapese, Kosrean
Literacy: age 15 and over can read and write (1980)
total population: 90%
male: 90%
female: 85%
Labor force: NA
by occupation: two-thirds are government employees
note: 45,000 people are between the ages of 15 and 65

Government

Names:
conventional long form: Federated States of Micronesia
conventional short form: none
former: Kosrae, Ponape, Truk, and Yap Districts (Trust Territory of the Pacific Islands)
Abbreviation: FSM
Digraph: FM
Type: constitutional government in free association with the US; the Compact of Free Association entered into force 3 November 1986
Capital: Kolonia (on the island of Pohnpei)
note: a new capital is being built about 10 km southwest in the Palikir valley
Administrative divisions: 4 states; Kosrae, Pohnpei, Chuuk (Truk), Yap
Independence: 3 November 1986 (from the US-administered UN Trusteeship)
Constitution: 10 May 1979
Legal system: based on adapted Trust Territory laws, acts of the legislature, municipal, common, and customary laws
National holiday: Proclamation of the Federated States of Micronesia, 10 May (1979)

Political parties and leaders: no formal parties
Suffrage: 18 years of age; universal
Elections:
President: last held ll May 1991 (next to be held March 1995); results—President Bailey OLTER elected president; Vice-President Jacob NENA
Congress: last held on 5 March 1991 (next to be held March 1993); results—percent of vote NA; seats—(14 total)
Executive branch: president, vice president, Cabinet
Legislative branch: unicameral Congress
Judicial branch: Supreme Court
Leaders:
Chief of State and Head of Government: President Bailey OLTER (since 21 May 1991); Vice President Jacob NENA (since 21 May 1991)
Member of: AsDB, ESCAP, ICAO, SPARTECA, SPC, SPF, UN, UNCTAD, WHO
Diplomatic representation in US:
chief of mission: Ambassador Jesse B. MAREHALAU
chancery: 1725 N St., NW, Washington, DC 20036
telephone: (202) 223-4383
US diplomatic representation:
chief of mission: Ambassador Aurelia BRAZEAL
embassy: address NA, Kolonia
mailing address: P. O. Box 1286, Pohnpei, Federated States of Micronesia 96941
telephone: 691-320-2187
FAX: 691-320-2186
Flag: light blue with four white five-pointed stars centered; the stars are arranged in a diamond pattern

Economy

Overview: Economic activity consists primarily of subsistence farming and fishing. The islands have few mineral deposits worth exploiting, except for high-grade phosphate. The potential for a tourist industry exists, but the remoteness of the location and a lack of adequate facilities hinder development. Financial assistance from the US is the primary source of revenue, with the US pledged to spend $1 billion in the islands in the l990s. Geographical isolation and a poorly developed infrastructure are major impediments to long-term growth.
National product: GNP—purchasing power equivalent—$150 million (1989 est.)
note: GNP numbers reflect US spending
National product real growth rate: NA%
National product per capita: $1,500 (1989 est.)
Inflation rate (consumer prices): NA%
Unemployment rate: NA%

Budget: revenues $165 million; expenditures $115 million, including capital expenditures of $20 million (1988)
Exports: $2.3 million (f.o.b., 1988)
commodities: copra
partners: NA
Imports: $67.7 million (c.i.f., 1988)
commodities: NA
partners: NA
External debt: $NA
Industrial production: growth rate NA%
Electricity: 18,000 kW capacity; 40 million kWh produced, 380 kWh per capita (1990)
Industries: tourism, construction, fish processing, craft items from shell, wood, and pearls
Agriculture: mainly a subsistence economy; black pepper; tropical fruits and vegetables, coconuts, cassava, sweet potatoes, pigs, chickens
Economic aid: under terms of the Compact of Free Association, the US will provide $1.3 billion in grant aid during the period 1986-2001
Currency: US currency is used
Exchange rates: US currency is used
Fiscal year: 1 October-30 September

Communications

Highways: 39 km of paved roads on major islands; also 187 km stone-, coral-, or laterite-surfaced roads
Ports: Colonia (Yap), Truk, Okat and Lelu (Kosrae)
Airports:
total: 6
usable: 5
with permanent-surface runways: 4
with runways over 3,659 m: 0
with runways 2,440-3,659 m: 0
with runways 1,220-2,439 m: 4
Telecommunications: telephone network—960 telephone lines total at Kolonia and Truk; islands interconnected by shortwave radio (used mostly for government purposes); 16,000 radio receivers, 1,125 TV sets (est. 1987); broadcast stations—5 AM, 1 FM, 6 TV, 1 shortwave; 4 Pacific Ocean INTELSAT earth stations

Defense Forces

Note: defense is the responsibility of the US

Midway Islands
(territory of the US)

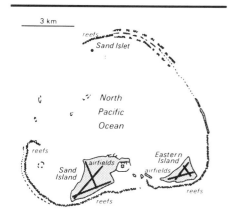

Geography

Location: located in the North Pacific Ocean, 2,350 km west-northwest of Honolulu, about one-third of the way between Honolulu and Tokyo
Map references: Oceania
Area:
total area: 5.2 km²
land area: 5.2 km²
comparative area: about nine times the size of the Mall in Washington, DC
note: includes Eastern Island and Sand Island
Land boundaries: 0 km
Coastline: 15 km
Maritime claims:
contiguous zone: 24 nm
continental shelf: 200 m (depth)
exclusive economic zone: 200 nm
territorial sea: 12 nm
International disputes: none
Climate: tropical, but moderated by prevailing easterly winds
Terrain: low, nearly level
Natural resources: fish, wildlife
Land use:
arable land: 0%
permanent crops: 0%
meadows and pastures: 0%
forest and woodland: 0%
other: 100%
Irrigated land: 0 km²
Environment: coral atoll
Note: closed to the public

People

Population: no indigenous inhabitants; note—there are 453 US military personnel

Government

Names:
conventional long form: none
conventional short form: Midway Islands
Digraph: MQ

Type: unincorporated territory of the US administered by the US Navy, under command of the Barbers Point Naval Air Station in Hawaii and managed cooperatively by the US Navy and the Fish and Wildlife Service of the US Department of the Interior as part of the National Wildlife Refuge System; legislation before Congress in 1990 proposed inclusion of territory within the State of Hawaii
Capital: none; administered from Washington, DC
Flag: the US flag is used

Economy

Overview: The economy is based on providing support services for US naval operations located on the islands. All food and manufactured goods must be imported.
Electricity: supplied by US Military

Communications

Highways: 32 km total
Pipelines: 7.8 km
Ports: Sand Island
Airports:
total: 3
usable: 2
with permanent-surface runways: 1
with runways over 3,659 m: 0
with runways 2,440-3,659 m: 0
with runways 1,220-2,439 m: 1

Defense Forces

Note: defense is the responsibility of the US

Moldova

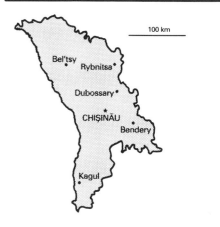

Geography

Location: Eastern Europe, between Ukraine and Romania
Map references: Asia, Europe, Standard Time Zones of the World
Area:
total area: 33,700 km²
land area: 33,700 km²
comparative area: slightly more than twice the size of Hawaii
Land boundaries: total 1,389 km, Romania 450 km, Ukraine 939 km
Coastline: 0 km (landlocked)
Maritime claims: none; landlocked
International disputes: potential dispute with Ukraine over former southern Bessarabian areas; northern Bukovina ceded to Ukraine upon Moldova's incorporation into USSR
Climate: mild winters, warm summers
Terrain: rolling steppe, gradual slope south to Black Sea
Natural resources: lignite, phosphorites, gypsum
Land use:
arable land: 50%
permanent crops: 0%
meadows and pastures: 9%
forest and woodland: 0%
other: 41%
Irrigated land: 2,920 km² (1990)
Environment: heavy use of agricultural chemicals, including banned pesticides such as DDT, has contaminated soil and groundwater; extensive erosion from poor farming methods
Note: landlocked

People

Population: 4,455,645 (July 1993 est.)
Population growth rate: 0.4% (1993 est.)
Birth rate: 16.15 births/1,000 population (1993 est.)
Death rate: 10.01 deaths/1,000 population (1993 est.)

Moldova (continued)

Net migration rate: -2.15 migrant(s)/1,000 population (1993 est.)
Infant mortality rate: 30.8 deaths/1,000 live births (1993 est.)
Life expectancy at birth:
total population: 67.92 years
male: 64.49 years
female: 71.53 years (1993 est.)
Total fertility rate: 2.2 children born/woman (1993 est.)
Nationality:
noun: Moldovan(s)
adjective: Moldovan
Ethnic divisions: Moldovan/Romanian 64.5%, Ukrainian 13.8%, Russian 13%, Gagauz 3.5%, Jewish 1.5%, Bulgarian 2%, other 1.7% (1989 figures)
note: internal disputes with ethnic Russians and Ukrainians in the Dniester region and Gagauz Turks in the south
Religions: Eastern Orthodox 98.5%, Jewish 1.5%, Baptist (only about 1,000 members) (1991)
note: almost all churchgoers are ethnic Moldovan; the Slavic population are not churchgoers
Languages: Moldovan (official); note—virtually the same as the Romanian language, Russian
Literacy: age 9-49 can read and write (1970)
total population: 100%
male: 100%
female: 99%
Labor force: 2.095 million
by occupation: agriculture 34.4%, industry 20.1%, other 45.5% (1985 figures)

Government

Names:
conventional long form: Republic of Moldova
conventional short form: Moldova
local long form: Republica Moldoveneasca
local short form: none
former: Soviet Socialist Republic of Moldova; Moldavia
Digraph: MD
Type: republic
Capital: Chisinau (Kishinev)
Administrative divisions: previously divided into 40 rayons; to be divided into fewer, larger districts at some future point
Independence: 27 August 1991 (from Soviet Union)
Constitution: as of mid-1993 the new constitution had not been adopted; old constitution (adopted NA 1979) is still in effect but has been heavily amended during the past few years
Legal system: based on civil law system; no judicial review of legislative acts; does not accept compulsory ICJ jurisdiction but accepts many UN and CSCE documents

National holiday: Independence Day, 27 August 1991
Political parties and leaders: Christian Democratic Popular Front (formerly Moldovan Popular Front), Ivrie ROSCA, chairman; Yedinstvo Intermovement, V. YAKOVLEV, chairman; Social Democratic Party, Oazul NANTOI, chairman, two other chairmen; Agrarian Democratic Party, Valery CHEBOTARV, leader; Democratic Party, Gheorghe GHIMPU, chairman; Democratic Labor Party, Alexandru ARSENI, chairman
Other political or pressure groups: United Council of Labor Collectives (UCLC), Igor SMIRNOV, chairman; The Ecology Movement of Moldova (EMM), G. MALARCHUK, chairman; The Christian Democratic League of Women of Moldova (CDLWM), L. LARI, chairman; National Christian Party of Moldova (NCPM), D. TODIKE, M. BARAGA, V. NIKU, leaders; The Peoples Movement Gagauz Khalky (GKh), S. GULGAR, leader; The Democratic Party of Gagauzia (DPG), G. SAVOSTIN, chairman; The Alliance of Working People of Moldova (AWPM), G. POLOGOV, president; Christian Alliance for Greater Romania; Women's League; Stefan the Great Movement
Suffrage: 18 years of age; universal
Elections:
President: last held 8 December 1991 (next to be held NA1996); results—Mircea SNEGUR ran unopposed and won 98.17% of vote
Parliament: last held 25 February 1990 (next to be held NA 1995); results—percent of vote by party NA; seats—(350 total) Christian Democratic Popular Front 50; Club of Independent Deputies 25; Agrarian Club 90; Social Democrats 60-70; Russian Conciliation Club 50; 60-70 seats belong to Dniester region deputies who usually boycott Moldovan legislative proceedings; the remaining seats filled by independents; note—until May 1991 was called Supreme Soviet
Executive branch: president, prime minister, Cabinet of Ministers
Legislative branch: unicameral Parliament
Judicial branch: Supreme Court
Leaders:
Chief of State: President Mircea Ivanovich SNEGUR (since 3 September 1990)
Head of Legislature: Chairman of the Parliament Petru LUCINSCHI (since 4 February 1993); Prime Minister Andrei SANGHELI (since 1 July 1992)
Member of: BSEC, CIS, CSCE, EBRD, ECE, IBRD, ICAO, ILO, IMF, NACC, UN, UNCTAD, UNESCO, WHO
Diplomatic representation in US:
chief of mission: Permanent Representative to the UN Tudor PANTIRU (also acts as representative to US)
chancery: NA

telephone: NA
US diplomatic representation:
chief of mission: Ambassador Mary C. PENDLETON
embassy: Strada Alexei Mateevich #103, Chisinau
mailing address: APO AE 09862
telephone: 7-0422-23-37-72 or 23-34-94
FAX: 7-0422-23-34-94
Flag: same color scheme as Romania—3 equal vertical bands of blue (hoist side), yellow, and red; emblem in center of flag is of a Roman eagle of gold outlined in black with a red beak and talons carrying a yellow cross in its beak and a green olive branch in its right talons and a yellow scepter in its left talons; on its breast is a shield divided horizontally red over blue with a stylized ox head, star, rose, and crescent all in black-outlined yellow

Economy

Overview: Moldova, the next-to-smallest of the former Soviet republics in area, is the most densely inhabited. Moldova has a little more than 1% of the population, labor force, capital stock, and output of the former Soviet Union. Living standards have been below average for the European USSR. The country enjoys a favorable climate, and economic development has been primarily based on agriculture, featuring fruits, vegetables, wine, and tobacco. Industry accounts for 20% of the labor force, whereas agriculture employs more than one-third. Moldova has no major mineral resources and has depended on other former Soviet republics for coal, oil, gas, steel, most electronic equipment, machine tools, and major consumer durables such as automobiles. Its industrial and agricultural products, in turn, have been exported to the other republics. Moldova has freed prices on most goods and has legalized private ownership of property. Moldova's near-term economic prospects are dimmed, however, by the difficulties of moving toward a market economy, the political problems of redefining ties to the other former Soviet republics and Romania, and the ongoing separatist movements in the Dniester and Gagauz regions. In 1992, national output fell substantially for the second consecutive year—down 22% in the industrial sector and 20% in agriculture. The decline is mainly attributable to the drop in energy supplies.
National product: GDP $NA
National product real growth rate: -26% (1992)
National product per capita: $NA
Inflation rate (consumer prices): 27% per month (first quarter 1993)
Unemployment rate: 0.7% (includes only officially registered unemployed; also large numbers of underemployed workers)

Budget: revenues $NA; expenditures $NA, including capital expenditures of $NA
Exports: 100 million to outside the successor states of the former USSR (1992)
commodities: foodstuffs, wine, tobacco, textiles and footwear, machinery, chemicals (1991)
partners: Russia, Kazakhstan, Ukraine, Romania
Imports: 100 million from outside the successor states of the former USSR (1992)
commodities: oil, gas, coal, steel machinery, foodstuffs, automobiles, and other consumer durables
partners: Russia, Ukraine, Uzbekistan, Romania
External debt: $100 million (1993 est.)
Industrial production: growth rate −22% (1992)
Electricity: 3,115,000 kW capacity; 11,100 million kWh produced, 2,491 kWh per capita (1992)
Industries: key products (with share of total former Soviet output in parentheses where known): agricultural machinery, foundry equipment, refrigerators and freezers (2.7%), washing machines (5.0%), hosiery (2.0%), refined sugar (3.1%), vegetable oil (3.7%), canned food (8.6%), shoes, textiles
Agriculture: Moldova's principal economic activity; products (shown in share of total output of the former Soviet republics): Grain (1.6%), sugar beets (2.6%), sunflower seed (4.4%), vegetables (4.4%), fruits and berries (9.7%), grapes (20.1%), meat (1.7%), milk (1.4%), eggs (1.4%)
Illicit drugs: illicit producer of opium and cannabis; mostly for CIS consumption; transshipment point for illicit drugs to Western Europe
Economic aid: IMF credit, $18.5 million (1992); EC agricultural credit, $30 million (1992); US commitments, $10 million for grain (1992); World Bank credit, $31 million
Currency: plans to introduce the Moldovan leu in 1993 or 1994, until then retaining Russian ruble as currency
Exchange rates: rubles per US$1—415 (24 December 1992) but subject to wide fluctuations
Fiscal year: calendar year

Communications

Railroads: 1,150 km; does not include industrial lines (1990)
Highways: 20,000 km total; 13,900 km hard-surfaced, 6,100 km earth (1990)
Pipelines: natural gas 310 km (1992)
Ports: none; landlocked
Airports:
total: 26
useable: 15
with permanent-surface runways: 6
with runways over 3,659 m: 0

with runways 2,440-3,659 m: 5
with runways 1,220-2,439 m: 8
Telecommunications: poorly supplied with telephones (as of 1991, 494,000 telephones total, with a density of 111 lines per 1000 persons); 215,000 unsatisfied applications for telephone installations (31 January 1990); connected to Ukraine by landline and to countries beyond the former USSR through the international gateway switch in Moscow

---

Defense Forces

Branches: Ground Forces, Air and Air Defence Force, Security Forces (internal and border troops)
Manpower availability: males age 15-49 1,082,562; fit for military service 859,948; reach military age (18) annually 35,769 (1993 est.)
Defense expenditures: exchange rate conversion—$NA, NA% of GDP

Monaco

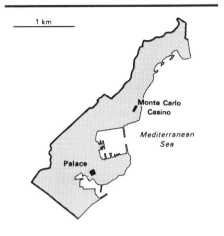

Geography

Location: Western Europe, bordering the Mediterranean Sea, in southern France near the border with Italy
Map references: Europe, Standard Time Zones of the World
Area:
total area: 1.9 km^2
land area: 1.9 km^2
comparative area: about three times the size of the Mall in Washington, DC
Land boundaries: total 4.4 km, France 4.4 km
Coastline: 4.1 km
Maritime claims:
territorial sea: 12 nm
International disputes: none
Climate: Mediterranean with mild, wet winters and hot, dry summers
Terrain: hilly, rugged, rocky
Natural resources: none
Land use:
arable land: 0%
permanent crops: 0%
meadows and pastures: 0%
forest and woodland: 0%
other: 100%
Irrigated land: NA km^2
Environment: almost entirely urban
Note: second smallest independent state in world (after Holy See)

People

Population: 31,008 (July 1993 est.)
Population growth rate: 0.93% (1993 est.)
Birth rate: 10.8 births/1,000 population (1993 est.)
Death rate: 12.32 deaths/1,000 population (1993 est.)
Net migration rate: 10.77 migrant(s)/1,000 population (1993 est.)
Infant mortality rate: 7.3 deaths/1,000 live births (1993 est.)
Life expectancy at birth:
total population: 77.5 years

Monaco (continued)

male: 73.7 years
female: 81.49 years (1993 est.)
Total fertility rate: 1.7 children born/woman (1993 est.)
Nationality:
noun: Monacan(s) or Monegasque(s)
adjective: Monacan or Monegasque
Ethnic divisions: French 47%, Monegasque 16%, Italian 16%, other 21%
Religions: Roman Catholic 95%
Languages: French (official), English, Italian, Monegasque
Literacy:
total population: NA%
male: NA%
female: NA%
Labor force: NA

Government

Names:
conventional long form: Principality of Monaco
conventional short form: Monaco
local long form: Principaute de Monaco
local short form: Monaco
Digraph: MN
Type: constitutional monarchy
Capital: Monaco
Administrative divisions: 4 quarters (quartiers, singular—quartier); Fontvieille, La Condamine, Monaco-Ville, Monte-Carlo
Independence: 1419 (rule by the House of Grimaldi)
Constitution: 17 December 1962
Legal system: based on French law; has not accepted compulsory ICJ jurisdiction
National holiday: National Day, 19 November
Political parties and leaders: National and Democratic Union (UND); Democratic Union Movement (MUD); Monaco Action; Monegasque Socialist Party (PSM)
Suffrage: 25 years of age; universal
Elections:
National Council: last held on 24 January 1988 (next to be held 24 January 1993); results—percent of vote by party NA; seats—(18 total) UND 18
Executive branch: prince, minister of state, Council of Government (cabinet)
Legislative branch: unicameral National Council (Conseil National)
Judicial branch: Supreme Tribunal (Tribunal Supreme)
Leaders:
Chief of State: Prince RAINIER III (since NA November 1949); Heir Apparent Prince ALBERT Alexandre Louis Pierre (born 14 March 1958)
Head of Government: Minister of State Jacques DUPONT (since NA)
Member of: ACCT, CSCE, IAEA, ICAO, IMF (observer), IMO, INMARSAT, INTELSAT, INTERPOL, IOC, ITU, LORCS, UN, UNCTAD, UNESCO, UPU, WHO, WIPO
Diplomatic representation in US:
honorary consulates general: Boston, Chicago, Los Angeles, New Orleans, New York, San Francisco, San Juan (Puerto Rico)
honorary consulates: Dallas, Honolulu, Palm Beach, Philadelphia, and Washington
US diplomatic representation: no mission in Monaco, but the US Consul General in Marseille, France, is accredited to Monaco
Flag: two equal horizontal bands of red (top) and white; similar to the flag of Indonesia which is longer and the flag of Poland which is white (top) and red

Economy

Overview: Monaco, situated on the French Mediterranean coast, is a popular resort, attracting tourists to its casino and pleasant climate. The Principality has successfully sought to diversify into services and small, high-value-added, nonpolluting industries. The state has no income tax and low business taxes and thrives as a tax haven both for individuals who have established residence and for foreign companies that have set up businesses and offices. About 50% of Monaco's annual revenue comes from value-added taxes on hotels, banks, and the industrial sector; about 25% of revenue comes from tourism. Living standards are high, that is, roughly comparable to those in prosperous French metropolitan suburbs.
National product: GDP—exchange rate conversion—$475 million (1991 est.)
National product real growth rate: NA%
National product per capita: $16,000 (1991 est.)
Inflation rate (consumer prices): NA%
Unemployment rate: NEGL%
Budget: revenues $424 million; expenditures $376 million, including capital expenditures of $NA (1991)
Exports: $NA; full customs integration with France, which collects and rebates Monacan trade duties; also participates in EC market system through customs union with France
Imports: $NA; full customs integration with France, which collects and rebates Monacan trade duties; also participates in EC market system through customs union with France
External debt: $NA
Industrial production: growth rate NA%
Electricity: 10,000 kW standby capacity (1992); power imported from France
Agriculture: NA
Economic aid: NA
Currency: 1 French franc (F) = 100 centimes
Exchange rates: French francs (F) per US$1—5.4812 (January 1993), 5.2938 (1992), 5.6421 (1991), 5.4453 (1990), 6.3801 (1989), 5.9569 (1988)
Fiscal year: calendar year

Communications

Railroads: 1.6 km 1.435-meter gauge
Highways: none; city streets
Ports: Monaco
Merchant marine: 1 oil tanker (1,000 GRT or over) totaling 3,268 GRT/4,959 DWT
Airports: 1 usable airfield with permanent-surface runways
Telecommunications: served by cable into the French communications system; automatic telephone system; 38,200 telephones; broadcast stations—3 AM, 4 FM, 5 TV; no communication satellite earth stations

Defense Forces

Note: defense is the responsibility of France

Mongolia

500 km

Geography

Location: East Central Asia, between China and Russia
Map references: Asia, Standard Time Zones of the World
Area:
total area: 1.565 million km²
land area: 1.565 million km²
comparative area: slightly larger than Alaska
Land boundaries: total 8,114 km, China 4,673 km, Russia 3,441 km
Coastline: 0 km (landlocked)
Maritime claims: none; landlocked
International disputes: none
Climate: desert; continental (large daily and seasonal temperature ranges)
Terrain: vast semidesert and desert plains; mountains in west and southwest; Gobi Desert in southeast
Natural resources: oil, coal, copper, molybdenum, tungsten, phosphates, tin, nickel, zinc, wolfram, fluorspar, gold
Land use:
arable land: 1%
permanent crops: 0%
meadows and pastures: 79%
forest and woodland: 10%
other: 10%
Irrigated land: 770 km² (1989)
Environment: harsh and rugged
Note: landlocked; strategic location between China and Russia

People

Population: 2,367,054 (July 1993 est.)
Population growth rate: 2.62% (1993 est.)
Birth rate: 33.41 births/1,000 population (1993 est.)
Death rate: 7.16 deaths/1,000 population (1993 est.)
Net migration rate: 0 migrant(s)/1,000 population (1993 est.)
Infant mortality rate: 44.9 deaths/1,000 live births (1993 est.)

Life expectancy at birth:
total population: 65.77 years
male: 63.53 years
female: 68.13 years (1993 est.)
Total fertility rate: 4.41 children born/woman (1993 est.)
Nationality:
noun: Mongolian(s)
adjective: Mongolian
Ethnic divisions: Mongol 90%, Kazakh 4%, Chinese 2%, Russian 2%, other 2%
Religions: predominantly Tibetan Buddhist, Muslim 4%
note: previously limited religious activity because of Communist regime
Languages: Khalkha Mongol 90%, Turkic, Russian, Chinese
Literacy:
total population: NA%
male: NA%
female: NA%
Labor force: NA
by occupation: primarily herding/agricultural
note: over half the adult population is in the labor force, including a large percentage of women; shortage of skilled labor

Government

Names:
conventional long form: none
conventional short form: Mongolia
local long form: none
local short form: Mongol Uls
former: Outer Mongolia
Digraph: MG
Type: republic
Capital: Ulaanbaatar
Administrative divisions: 18 provinces (aymguud, singular—aymag) and 3 municipalities* (hotuud, singular—hot); Arhangay, Bayanhongor, Bayan-Olgiy, Bulgan, Darhan*, Dornod, Dornogovi, Dundgovi, Dzavhan, Erdenet*, Govi-Altay, Hentiy, Hovd, Hovsgol, Omnogovi, Ovorhangay, Selenge, Suhbaatar, Tov, Ulaanbaatar*, Uvs
Independence: 13 March 1921 (from China)
Constitution: adopted 13 January 1992
Legal system: blend of Russian, Chinese, and Turkish systems of law; no constitutional provision for judicial review of legislative acts; has not accepted compulsory ICJ jurisdiction
National holiday: National Day, 11 July (1921)
Political parties and leaders: Mongolian People's Revolutionary Party (MPRP), Budragchagiin DASH-YONDON, presidium chairman; Mongolian Democratic Party (MDP), Erdenijiyn BAT-UUL, general coordinator; National Progress Party (NPP), S. BYAMBAA and Luusandambyn DASHNYAM, leaders; Social Democratic Party (SDP), BATBAYAR and Tsohiogyyn

ADYASUREN, leaders; Mongolian Independence Party (MIP), D. ZORIGT, leader; United Party of Mongolia (made up of the MDP, SDP, and NPP); Mongolian National Democratic Party (MNDP; merger of the MDP, United Party, Renaissance Party, and PNP), D. GANBOLD
note: opposition parties were legalized in May 1990; additional parties exist: The Mongolian Green Party, The Buddhist Believers' Party, The Republican Party, Mongolian People's Party, and United Herdsmen and Farmers Party (MHFUP), Mongolian Bourgeois Party (BP), Mongolian Private Property Owners Party, Mongolian Workers Party
Suffrage: 18 years of age; universal
Elections:
President: last held 3 September 1990 (next to be held 6 June 1993); results—Punsalmaagiyn OCHIRBAT elected by the People's Great Hural; other candidate Lodongiyn TUDEV (MPRP)
State Great Hural: first time held 28 June 1992 (next to be held NA); results—MPRP 56.9%; seats—(76 total) MPRP 71, MDP/PNP 3, SDP 1, independent 1
note: the People's Small Hural no longer exists
Executive branch: president, vice president, prime minister, first deputy prime minister, cabinet
Legislative branch: unicameral State Great Hural
Judicial branch: Supreme Court serves as appeals court for people's and provincial courts, but to date rarely overturns verdicts of lower courts
Leaders:
Chief of State: President Punsalmaagiyn OCHIRBAT (since 3 September 1990); Vice President Radnaasumbereliyn GONCHIGDORJ (since 7 September 1990)
Head of Government: Prime Minister Putsagiyn JASRAY (since 3 August 1992); First Deputy Prime Minister Puntsagiyn JASRAY (since NA)
Member of: AsDB, CCC, ESCAP, FAO, G-77, IAEA, IBRD, ICAO, IDA, IFC, ILO, IMF, INTERPOL, IOC, ISO, ITU, LORCS, NAM (observer), UN, UNCTAD, UNESCO, UNIDO, UPU, WFTU, WHO, WIPO, WMO, WTO
Diplomatic representation in US:
chief of mission: Ambassador Luvsandorj DAWAGIV
chancery: NA
telephone: (301) 983-1962
FAX: (301) 983-2025
US diplomatic representation:
chief of mission: Ambassador Joseph E. LAKE
embassy: address NA, Ulaanbaatar
mailing address: Ulaanbaatar, c/o American Embassy Beijing, Micro Region II, Big Rind Road; PSC 461, Box 300, FPO AP 96521-0002

Mongolia (continued)

telephone: [976] (1) 329095, 329606
FAX: Telex 080079253 AMEMB MH
Flag: three equal, vertical bands of red (hoist side), blue, and red, centered on the hoist-side red band in yellow is the national emblem ("soyombo"—a columnar arrangement of abstract and geometric representation for fire, sun, moon, earth, water, and the yin-yang symbol)

Economy

Overview: Mongolia's severe climate, scattered population, and wide expanses of unproductive land have constrained economic development. Economic activity traditionally has been based on agriculture and the breeding of livestock—Mongolia has the highest number of livestock per person in the world. In recent years extensive mineral resources have been developed with Soviet support. The mining and processing of coal, copper, molybdenum, tin, tungsten, and gold account for a large part of industrial production. Timber and fishing are also important sectors. In 1992 the Mongolian leadership continued its struggle with severe economic dislocations, mainly attributable to the crumbling of the USSR, by far Mongolia's leading trade and development partner. Moscow cut almost all aid in 1991, and little was provided in 1992. Industry in 1992 was hit hard by energy shortages, mainly due to disruptions in coal production and shortfalls in petroleum imports. By the end of the year, the country was perilously close to a complete shutdown of its centralized energy supply system, due to critical coal shortages. The government is moving away from the Soviet-style, centrally planned economy through privatization and price reform.
National product: GDP—exchange rate conversion—$1.8 billion (1992 est.)
National product real growth rate: -15% (1992 est.)
National product per capita: $800 (1992 est.)
Inflation rate (consumer prices): 325% (1992 est.)
Unemployment rate: 15% (1991 est.)
Budget: deficit of $67 million (1991)
Exports: $347 million (f.o.b., 1991 est.)
commodities: copper, livestock, animal products, cashmere, wool, hides, fluorspar, other nonferrous metals
partners: USSR 75%, China 10%, Japan 4%
Imports: $501 million (f.o.b., 1991 est.)
commodities: machinery and equipment, fuels, food products, industrial consumer goods, chemicals, building materials, sugar, tea
partners: USSR 75%, Austria 5%, China 5%
External debt: $16.8 billion (yearend 1990); 98.6% with USSR

Industrial production: growth rate -15% (1992 est.)
Electricity: 1,248,000 kW capacity; 3,740 million kWh produced, 1,622 kWh per capita (1992)
Industries: copper, processing of animal products, building materials, food and beverage, mining (particularly coal)
Agriculture: accounts for about 20% of GDP and provides livelihood for about 50% of the population; livestock raising predominates (primarily sheep and goats, but also cattle, camels, and horses); crops—wheat, barley, potatoes, forage
Economic aid: about $300 million in trade credits and $34 million in grant aid from USSR and other CEMA countries, plus $7.4 million from UNDP (1990); in 1991, $170 million in grants and technical assistance from Western donor countries, including $30 million from World Bank and $30 million from the IMF; over $200 million from donor countries projected in 1992
Currency: 1 tughrik (Tug) = 100 mongos
Exchange rates: tughriks (Tug) per US$1—40 (1992), 7.1 (1991), 5.63 (1990), 3.00 (1989)
Fiscal year: calendar year

Communications

Railroads: 1,750 km 1.524-meter broad gauge (1988)
Highways: 46,700 km total; 1,000 km hard surface; 45,700 km other surfaces (1988)
Inland waterways: 397 km of principal routes (1988)
Airports:
total: 81
usable: 31
with permanent-surface runways: 11
with runways over 3,659 m: fewer than 5
with runways 2,440-3,659 m: fewer than 20
with runways 1,220-2,439 m: 12
Telecommunications: 63,000 telephones (1989); broadcast stations—12 AM, 1 FM, 1 TV (with 18 provincial repeaters); repeat of Russian TV; 120,000 TVs; 220,000 radios; at least 1 earth station

Defense Forces

Branches: Mongolian People's Army (includes Internal Security Forces and Frontier Guards), Air Force
Manpower availability: males age 15-49 569,135; fit for military service 371,162; reach military age (18) annually 25,406 (1993 est.)
Defense expenditures: exchange rate conversion—$22.8 million of GDP, 1% of GDP (1992)

Montserrat
(dependent territory of the UK)

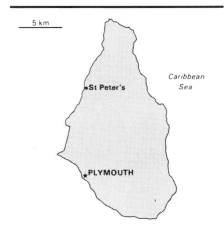

Geography

Location: in the eastern Caribbean Sea, about 400 km southeast of Puerto Rico
Map references: Central America and the Caribbean
Area:
total area: 100 km²
land area: 100 km²
comparative area: about 0.6 times the size of Washington, DC
Land boundaries: 0 km
Coastline: 40 km
Maritime claims:
exclusive fishing zone: 200 nm
territorial sea: 3 nm
International disputes: none
Climate: tropical; little daily or seasonal temperature variation
Terrain: volcanic islands, mostly mountainous, with small coastal lowland
Natural resources: negligible
Land use:
arable land: 20%
permanent crops: 0%
meadows and pastures: 10%
forest and woodland: 40%
other: 30%
Irrigated land: NA km²
Environment: subject to severe hurricanes from June to November
Note: located 400 km east southeast of Puerto Rico in the Caribbean Sea

People

Population: 12,661 (July 1993 est.)
Population growth rate: 0.36% (1993 est.)
Birth rate: 16.35 births/1,000 population (1993 est.)
Death rate: 9.77 deaths/1,000 population (1993 est.)
Net migration rate: -3 migrant(s)/1,000 population (1993 est.)
Infant mortality rate: 11.51 deaths/1,000 live births (1993 est.)
Life expectancy at birth:
total population: 75.76 years

male: 74 years
female: 77.56 years (1993 est.)
Total fertility rate: 2.11 children born/woman (1993 est.)
Nationality:
noun: Montserratian(s)
adjective: Montserratian
Ethnic divisions: black, Europeans
Religions: Anglican, Methodist, Roman Catholic, Pentecostal, Seventh-Day Adventist, other Christian denominations
Languages: English
Literacy: age 15 and over having ever attended school (1970)
total population: 97%
male: 97%
female: 97%
Labor force: 5,100
by occupation: community, social, and personal services 40.5%, construction 13.5%, trade, restaurants, and hotels 12.3%, manufacturing 10.5%, agriculture, forestry, and fishing 8.8%, other 14.4% (1983 est.)

Government

Names:
conventional long form: none
conventional short form: Montserrat
Digraph: MH
Type: dependent territory of the UK
Capital: Plymouth
Administrative divisions: 3 parishes; Saint Anthony, Saint Georges, Saint Peter
Independence: none (dependent territory of the UK)
Constitution: 1 January 1960
Legal system: English common law and statute law
National holiday: Celebration of the Birthday of the Queen (second Saturday of June)
Political parties and leaders: National Progressive Party (NPP) Reuben T. MEADE; People's Liberation Movement (PLM), Noel TUITT; National Development Party (NDP), Bertrand OSBORNE; Independent (IND), Ruby BRAMBLE
Suffrage: 18 years of age; universal
Elections:
Legislative Council: last held on 8 October 1991; results—percent of vote by party NA; seats—(11 total, 7 elected) NPP 4, NDP 1, PLM 1, independent 1
Executive branch: monarch, governor, Executive Council (cabinet), chief minister
Legislative branch: unicameral Legislative Council
Judicial branch: Supreme Court
Leaders:
Chief of State: Queen ELIZABETH II (since 6 February 1952), represented by Governor David TAYLOR (since NA 1990)
Head of Government: Chief Minister Reuben T. MEADE (since October 1991)

Member of: CARICOM, CDB, ECLAC (associate), ICFTU, OECS, WCL
Diplomatic representation in US: none (dependent territory of the UK)
Flag: blue with the flag of the UK in the upper hoist-side quadrant and the Montserratian coat of arms centered in the outer half of the flag; the coat of arms features a woman standing beside a yellow harp with her arm around a black cross

Economy

Overview: The economy is small and open with economic activity centered on tourism and construction. Tourism is the most important sector and accounts for roughly one-fifth of GDP. Agriculture accounts for about 4% of GDP and industry 10%. The economy is heavily dependent on imports, making it vulnerable to fluctuations in world prices. Exports consist mainly of electronic parts sold to the US.
National product: GDP—exchange rate conversion—$73 million (1990 est.)
National product real growth rate: 13.5% (1990 est.)
National product per capita: $5,800 (1990 est.)
Inflation rate (consumer prices): 9% (1991)
Unemployment rate: 3% (1987)
Budget: revenues $12.1 million; expenditures $14.3 million, including capital expenditures of $3.2 million (1988)
Exports: $1.6 million (f.o.b., 1989)
commodities: electronic parts, plastic bags, apparel, hot peppers, live plants, cattle
partners: NA
Imports: $31.0 million (c.i.f., 1989)
commodities: machinery and transportation equipment, foodstuffs, manufactured goods, fuels, lubricants, and related materials
partners: NA
External debt: $2.05 million (1987)
Industrial production: growth rate 8.1% (1986); accounts for 10% of GDP
Electricity: 5,271 kW capacity; 12 million kWh produced, 950 kWh per capita (1992)
Industries: tourism; light manufacturing—rum, textiles, electronic appliances
Agriculture: accounts for 4% of GDP; small-scale farming; food crops—tomatoes, onions, peppers; not self-sufficient in food, especially livestock products
Economic aid: Western (non-US) countries, ODA and OOF bilateral commitments (1970-89), $90 million
Currency: 1 EC dollar (EC$) = 100 cents
Exchange rates: East Caribbean dollars (EC$) per US$1—2.70 (fixed rate since 1976)
Fiscal year: 1 April-31 March

Communications

Highways: 280 km total; about 200 km paved, 80 km gravel and earth
Ports: Plymouth
Airports:
total: 1
usable: 1
with permanent-surface runways 1,036 m: 1
with runways over 3,659 m: 0
with runways 2,440-3,659 m: 0
with runways 1,220-2,439 m: 0
Telecommunications: 3,000 telephones; broadcast stations—8 AM, 4 FM, 1 TV

Defense Forces

Branches: Police Force
Note: defense is the responsibility of the UK

Morocco

Geography

Location: Northern Africa, bordering the Atlantic Ocean and the Mediterranean Sea, between Algeria and Western Sahara
Map references: Africa, Standard Time Zones of the World
Area:
total area: 446,550 km²
land area: 446,300 km²
comparative area: slightly larger than California
Land boundaries: total 2,002 km, Algeria 1,559 km, Western Sahara 443 km
Coastline: 1,835 km
Maritime claims:
contiguous zone: 24 nm
continental shelf: 200 m depth or to depth of exploitation
exclusive economic zone: 200 nm
territorial sea: 12 nm
International disputes: claims and administers Western Sahara, but sovereignty is unresolved; the UN is attempting to hold a referendum; the UN-administered cease-fire has been currently in effect since September 1991; Spain controls five places of sovereignty (plazas de soberania) on and off the coast of Morocco—the coastal enclaves of Ceuta and Melilla which Morocco contests as well as the islands of Penon de Alhucemas, Penon de Velez de la Gomera, and Islas Chafarinas
Climate: Mediterranean, becoming more extreme in the interior
Terrain: mostly mountains with rich coastal plains
Natural resources: phosphates, iron ore, manganese, lead, zinc, fish, salt
Land use:
arable land: 18%
permanent crops: 1%
meadows and pastures: 28%
forest and woodland: 12%
other: 41%
Irrigated land: 12,650 km² (1989 est.)
Environment: northern mountains

geologically unstable and subject to earthquakes; desertification
Note: strategic location along Strait of Gibraltar

People

Population: 27,955,090 (July 1993 est.)
Population growth rate: 2.16% (1993 est.)
Birth rate: 29.23 births/1,000 population (1993 est.)
Death rate: 6.56 deaths/1,000 population (1993 est.)
Net migration rate: -1.13 migrant(s)/1,000 population (1993 est.)
Infant mortality rate: 53.4 deaths/1,000 live births (1993 est.)
Life expectancy at birth:
total population: 67.5 years
male: 65.7 years
female: 69.4 years (1993 est.)
Total fertility rate: 3.96 children born/woman (1993 est.)
Nationality:
noun: Moroccan(s)
adjective: Moroccan
Ethnic divisions: Arab-Berber 99.1%, other 0.7%, Jewish 0.2%
Religions: Muslim 98.7%, Christian 1.1%, Jewish 0.2%
Languages: Arabic (official), Berber dialects, French often the language of business, government, and diplomacy
Literacy: age 15 and over can read and write (1990)
total population: 50%
male: 61%
female: 38%
Labor force: 7.4 million
by occupation: agriculture 50%, services 26%, industry 15%, other 9% (1985)

Government

Names:
conventional long form: Kingdom of Morocco
conventional short form: Morocco
local long form: Al Mamlakah al Maghribiyah
local short form: Al Maghrib
Digraph: MO
Type: constitutional monarchy
Capital: Rabat
Administrative divisions: 37 provinces and 5 municipalities* (wilayas, singular—wilaya); Agadir, Al Hoceima, Azilal, Beni Mellal, Ben Slimane, Boulemane, Casablanca*, Chaouen, El Jadida, El Kelaa des Srarhna, Er Rachidia, Essaouira, Fes, Fes*, Figuig, Guelmim, Ifrane, Kenitra, Khemisset, Khenifra, Khouribga, Laayoune, Larache, Marrakech, Marrakech*, Meknes, Meknes*, Nador, Ouarzazate, Oujda, Rabat-Sale*, Safi, Settat,

Sidi Kacem, Tanger, Tan-Tan, Taounate, Taroudannt, Tata, Taza, Tetouan, Tiznit
Independence: 2 March 1956 (from France)
Constitution: 10 March 1972, revised in September 1992
Legal system: based on Islamic law and French and Spanish civil law system; judicial review of legislative acts in Constitutional Chamber of Supreme Court
National holiday: National Day, 3 March (1961) (anniversary of King Hassan II's accession to the throne)
Political parties and leaders: Morocco has 15 political parties; the major ones are Constitutional Union (UC), Maati BOUABID; National Assembly of Independents (RNI), Ahmed OSMAN; Popular Movement (MP), Mohamed LAENSER; National Popular Movement (MPN), Mahjoubi AHARDANE; Istiqlal, M'Hamed BOUCETTA; Socialist Union of Popular Forces (USFP); National Democratic Party (PND), Mohamed Arsalane EL-JADIDI; Party for Progress and Socialism (PPS), Ali YATA
Suffrage: 21 years of age; universal
Elections:
Chamber of Representatives: last held on 14 September 1984 (were scheduled for September 1990, but postponed until June 1993 when 27 new seats will be added); results—percent of vote by party NA; seats—(306 total, 206 elected) UC 83, RNI 61, MP 47, Istiqlal 41, USFP 36, PND 24, other 14
Executive branch: monarch, prime minister, Council of Ministers (cabinet)
Legislative branch: unicameral Chamber of Representatives (Majlis Nawab)
Judicial branch: Supreme Court
Leaders:
Chief of State: King HASSAN II (since 3 March 1961)
Head of Government: Prime Minister Mohamed KARIM-LAMRANI (since October 1992)
Member of: ABEDA, ACCT (associate), AfDB, AFESD, AL, AMF, AMU, CCC, EBRD, ECA, FAO, G-77, GATT, IAEA, IBRD, ICAO, ICC, ICFTU, IDA, IDB, IFAD, IFC, ILO, IMF, IMO, INTELSAT, INTERPOL, IOC, IOM (observer), ISO, ITU, LORCS, OAS (observer), NAM, OIC, UN, UNAVEM II, UNCTAD, UNESCO, UNHCR, UNIDO, UNOSOM, UPU, WHO, WIPO, WMO, WTO
Diplomatic representation in US:
chief of mission: Ambassador Mohamed BELKHAYAT
chancery: 1601 21st Street NW, Washington, DC 20009;
telephone: (202) 462-7979
consulate general: New York
US diplomatic representation:
chief of mission: (vacant)

embassy: 2 Avenue de Marrakech, Rabat
mailing address: P. O. Box 120, Rabat, or PSC 74, APO AE 09718
telephone: [212] (7) 76-22-65
FAX: [212] (7) 76-56-61
consulate general: Casablanca
Flag: red with a green pentacle (five-pointed, linear star) known as Solomon's seal in the center of the flag; green is the traditional color of Islam

Economy

Overview: The economy had recovered moderately in 1990 because of: the resolution of a trade dispute with India over phosphoric acid sales, a rebound in textile sales to the EC, lower prices for food imports, a sharp increase in worker remittances, increased Arab donor aid, and generous debt rescheduling agreements. Economic performance in 1991 was mixed. A record harvest helped real GDP advance by 4.2%. Inflation accelerated slightly as easier financial policies triggered rapid credit and monetary growth. Despite recovery of domestic demand, import volume growth slowed while export volume was adversely affected by phosphate marketing difficulties. In January 1992, Morocco reached a new 12-month standby arrangement for $129 million with the IMF. In February 1992, the Paris Club rescheduled $1.4 billion of Morocco's commercial debt. This is thought to be Morocco's last rescheduling. By 1993 the Moroccan authorities hope to be in a position to meet all debt service obligations without additional rescheduling. Servicing this large debt, high unemployment, and Morocco's vulnerability to external economic forces remain severe long-term problems. In 1992 Morocco embarked on a program to privatize 112 state-owned companies. A severe winter drought in 1991/92 cut back agricultural output in 1992.
National product: GDP—exchange rate conversion—$28.1 billion (1992 est.)
National product real growth rate: 0% (1992 est.)
National product per capita: $1,060 (1992 est.)
Inflation rate (consumer prices): 6% (1992 est.)
Unemployment rate: 19% (1992 est.)
Budget: revenues $7.5 billion; expenditures $7.7 billion, including capital expenditures of $1.9 billion (1992)
Exports: $4.7 billion (f.o.b., 1992 est.)
commodities: food and beverages 30%, semiprocessed goods 23%, consumer goods 21%, phosphates 17%
partners: EC 58%, India 7%, Japan 5%, former USSR 3%, US 2%
Imports: $7.6 billion (f.o.b., 1992 est.)

commodities: capital goods 24%, semiprocessed goods 22%, raw materials 16%, fuel and lubricants 16%, food and beverages 13%, consumer goods 9%
partners: EC 53%, US 11%, Canada 4%, Iraq 3%, former USSR 3%, Japan 2%
External debt: $20 billion (1991)
Industrial production: growth rate 8.4%; accounts for 27% of GDP (1990)
Electricity: 2,384,000 kW capacity; 8,864 million kWh produced, 317 kWh per capita (1992)
Industries: phosphate rock mining and processing, food processing, leather goods, textiles, construction, tourism
Agriculture: accounts for 16% of GDP, 50% of employment, and 30% of export value; not self-sufficient in food; cereal farming and livestock raising predominate; barley, wheat, citrus fruit, wine, vegetables, olives; fish catch of 491,000 metric tons in 1987
Illicit drugs: illicit producer of hashish; trafficking on the increase for both domestic and international drug markets; shipments of hashish mostly directed to Western Europe; occasional transit point for cocaine from South America destined for Western Europe.
Economic aid: US commitments, including Ex-Im (FY70-89), $1.3 billion and an additional $123.6 million for 1992; Western (non-US) countries, ODA and OOF bilateral commitments (1970-89), $7.5 billion; OPEC bilateral aid (1979-89), $4.8 billion; Communist countries (1970-89), $2.5 billion; $2.8 billion debt canceled by Saudi Arabia (1991); IMF standby agreement worth $13 million; World Bank, $450 million (1991)
Currency: 1 Moroccan dirham (DH) = 100 centimes
Exchange rates: Moroccan dirhams (DH) per US$1—9.207 (February 1993), 8.538 (1992), 8.707 (1991), 8.242 (1990), 8.488 (1989), 8.209 (1988)
Fiscal year: calendar year

Communications

Railroads: 1,893 km 1.435-meter standard gauge (246 km double track, 974 km electrified)
Highways: 59,198 km total; 27,740 km paved, 31,458 km gravel, crushed stone, improved earth, and unimproved earth
Pipelines: crude oil 362 km; petroleum products (abandoned) 491 km; natural gas 241 km
Ports: Agadir, Casablanca, El Jorf Lasfar, Kenitra, Mohammedia, Nador, Safi, Tangier; also Spanish-controlled Ceuta and Melilla
Merchant marine: 50 ships (1,000 GRT or over) totaling 305,758 GRT/484,825 DWT; 10 cargo, 2 container, 11 refrigerated cargo, 6 roll-on/roll-off, 4 oil tanker, 11 chemical tanker, 4 bulk, 2 short-sea passenger

Airports:
total: 73
usable: 65
with permanent-surface runways: 26
with runways over 3,659 m: 2
with runways 2,440-3,659 m: 13
with runways 1,220-2,439 m: 26
Telecommunications: good system composed of wire lines, cables, and microwave radio relay links; principal centers are Casablanca and Rabat; secondary centers are Fes, Marrakech, Oujda, Tangier, and Tetouan; 280,000 telephones (10.5 telephones per 1,000 persons); broadcast stations—20 AM, 7 FM, 26 TV and 26 repeaters; 5 submarine cables; satellite earth stations—2 Atlantic Ocean INTELSAT and 1 ARABSAT; microwave radio relay to Gibraltar, Spain, and Western Sahara; coaxial cable and microwave to Algeria; microwave radio relay network linking Syria, Jordan, Egypt, Libya, Tunisia, Algeria, and Morocco

Defense Forces

Branches: Royal Moroccan Army, Royal Moroccan Navy, Royal Moroccan Air Force, Royal Gendarmerie, Auxiliary Forces
Manpower availability: males age 15-49 6,852,698; fit for military service 4,355,670; reach military age (18) annually 309,666 (1993 est.); limited conscription
Defense expenditures: exchange rate conversion—$1.1 billion, 3.8% of GDP (1993 budget)

Mozambique

Geography

Location: Southern Africa, bordering the Mozambique Channel between South Africa and Tanzania opposite the island of Madagascar
Map references: Africa, Standard Time Zones of the World
Area:
total area: 801,590 km²
land area: 784,090 km²
comparative area: slightly less than twice the size of California
Land boundaries: total 4,571 km, Malawi 1,569 km, South Africa 491 km, Swaziland 105 km, Tanzania 756 km, Zambia 419 km, Zimbabwe 1,231 km
Coastline: 2,470 km
Maritime claims:
exclusive economic zone: 200 nm
territorial sea: 12 nm
International disputes: none
Climate: tropical to subtropical
Terrain: mostly coastal lowlands, uplands in center, high plateaus in northwest, mountains in west
Natural resources: coal, titanium
Land use:
arable land: 4%
permanent crops: 0%
meadows and pastures: 56%
forest and woodland: 20%
other: 20%
Irrigated land: 1,150 km² (1989 est.)
Environment: severe drought and floods occur in south; desertification

People

Population: 16,341,777 (July 1993 est.)
Population growth rate: 6.06% (1993 est.)
Birth rate: 45.35 births/1,000 population (1993 est.)
Death rate: 16.71 deaths/1,000 population (1993 est.)
Net migration rate: 31.95 migrant(s)/1,000 population (1993 est.)

Infant mortality rate: 131.4 deaths/1,000 live births (1993 est.)
Life expectancy at birth:
total population: 48.03 years
male: 46.22 years
female: 49.9 years (1993 est.)
Total fertility rate: 6.31 children born/woman (1993 est.)
Nationality:
noun: Mozambican(s)
adjective: Mozambican
Ethnic divisions: indigenous tribal groups, Europeans about 10,000, Euro-Africans 35,000, Indians 15,000
Religions: indigenous beliefs 60%, Christian 30%, Muslim 10%
Languages: Portuguese (official), indigenous dialects
Literacy: age 15 and over can read and write (1990)
total population: 33%
male: 45%
female: 21%
Labor force: NA
by occupation: 90% engaged in agriculture

Government

Names:
conventional long form: Republic of Mozambique
conventional short form: Mozambique
local long form: Republica Popular de Mocambique
local short form: Mocambique
Digraph: MZ
Type: republic
Capital: Maputo
Administrative divisions: 10 provinces (provincias, singular—provincia); Cabo Delgado, Gaza, Inhambane, Manica, Maputo, Nampula, Niassa, Sofala, Tete, Zambezia
Independence: 25 June 1975 (from Portugal)
Constitution: 30 November 1990
Legal system: based on Portuguese civil law system and customary law
National holiday: Independence Day, 25 June (1975)
Political parties and leaders: Front for the Liberation of Mozambique (FRELIMO), Joaquim Alberto CHISSANO, chairman; formerly a Marxist organization with close ties to the USSR; FRELIMO was the only legal party before 30 November 1990, when the new Constitution went into effect establishing a multiparty system
note: the government plans multiparty elections as early as 1993; 14 parties, including the Liberal Democratic Party of Mozambique (PALMO), the Mozambique National Union (UNAMO), the Mozambique National Movement (MONAMO), and the Mozambique National Resistance

(RENAMO, Alfonso DHLAKAMA, president), have already emerged
Suffrage: 18 years of age; universal
Elections: draft electoral law provides for periodic, direct presidential and Assembly elections
Executive branch: president, prime minister, Cabinet
Legislative branch: unicameral Assembly of the Republic (Assembleia da Republica)
Judicial branch: Supreme Court
Leaders:
Chief of State: President Joaquim Alberto CHISSANO (since 6 November 1986)
Head of Government: Prime Minister Mario da Graca MACHUNGO (since 17 July 1986)
Member of: ACP, AfDB, CCC, ECA, FAO, FLS, G-77, IBRD, ICAO, IDA, IFAD, IFC, ILO, IMF, INMARSAT, IMO, INTELSAT, INTERPOL, IOC, ITU, LORCS, NAM, OAU, SADC, UN, UNCTAD, UNESCO, UNIDO, UPU, WHO, WMO
Diplomatic representation in US:
chief of mission: Ambassador Hipolito PATRICIO
chancery: Suite 570, 1990 M Street NW, Washington, DC 20036
telephone: (202) 293-7146
US diplomatic representation:
chief of mission: Ambassador Townsend B. FRIEDMAN, Jr.
embassy: Avenida Kenneth Kuanda, 193 Maputo
mailing address: P. O. Box 783, Maputo
telephone: [258] (1) 49-27-97, 49-01-67, 49-03-50
FAX: [258] (1) 49-01-14
Flag: three equal horizontal bands of green (top), black, and yellow with a red isosceles triangle based on the hoist side; the black band is edged in white; centered in the triangle is a yellow five-pointed star bearing a crossed rifle and hoe in black superimposed on an open white book

Economy

Overview: One of Africa's poorest countries, Mozambique has failed to exploit the economic potential of its sizable agricultural, hydropower, and transportation resources. Indeed, national output, consumption, and investment declined throughout the first half of the 1980s because of internal disorders, lack of government administrative control, and a growing foreign debt. A sharp increase in foreign aid, attracted by an economic reform policy, resulted in successive years of economic growth in the late 1980s, but aid has declined steadily since 1989. Agricultural output, nevertheless, is at about only 75% of its 1981 level, and grain has to be imported. Industry operates at only

20-40% of capacity. The economy depends heavily on foreign assistance to keep afloat. The continuation of civil strife has dimmed chances of foreign investment, and growth was a mere 0.3% in 1992. Living standards, already abysmally low, fell further in 1991-92.

National product: GDP—exchange rate conversion—$1.75 billion (1992 est.)

National product real growth rate: 0.3% (1992 est.)

National product per capita: $115 (1992 est.)

Inflation rate (consumer prices): 50% (1992 est.)

Unemployment rate: 50% (1989 est.)

Budget: revenues $252 million; expenditures $607 million, including capital expenditures of $NA (1992 est.)

Exports: $162 million (f.o.b., 1991 est.)
commodities: shrimp 48%, cashews 21%, sugar 10%, copra 3%, citrus 3%
partners: US, Western Europe, Germany, Japan

Imports: $899 million (c.i.f., 1991 est.)
commodities: food, clothing, farm equipment, petroleum
partners: US, Western Europe, USSR

External debt: $5.4 billion (1991 est.)

Industrial production: growth rate 5% (1989 est.)

Electricity: 2,270,000 kW capacity; 1,745 million kWh produced, 115 kWh per capita (1991)

Industries: food, beverages, chemicals (fertilizer, soap, paints), petroleum products, textiles, nonmetallic mineral products (cement, glass, asbestos), tobacco

Agriculture: accounts for 50% of GDP and about 90% of exports; cash crops—cotton, cashew nuts, sugarcane, tea, shrimp; other crops—cassava, corn, rice, tropical fruits; not self-sufficient in food

Economic aid: US commitments, including Ex-Im (FY70-89), $350 million; Western (non-US) countries, ODA and OOF bilateral commitments (1970-89), $4.4 billion; OPEC bilateral aid (1979-89), $37 million; Communist countries (1970-89), $890 million

Currency: 1 metical (Mt) = 100 centavos

Exchange rates: meticais (Mt) per US$1—2,74.15 (January 1993), 2,433.34 (1992), 1,434.47 (1991), 929.00 (1990), 800.00 (1989), 528.60 (1988)

Fiscal year: calendar year

Communications

Railroads: 3,288 km total; 3,140 km 1.067-meter gauge; 148 km 0.762-meter narrow gauge; Malawi-Nacala, Malawi-Beira, and Zimbabwe-Maputo lines are subject to closure because of insurgency

Highways: 26,498 km total; 4,593 km paved; 829 km gravel, crushed stone, stabilized soil; 21,076 km unimproved earth

Inland waterways: about 3,750 km of navigable routes

Pipelines: crude oil (not operating) 306 km; petroleum products 289 km

Ports: Maputo, Beira, Nacala

Merchant marine: 4 cargo ships (1,000 GRT or over) totaling 5,686 GRT/9,742 DWT

Airports:
total: 194
usable: 131
with permanent-surface runways: 25
with runways over 3,659 m: 1
with runways 2,440-3,659 m: 4
with runways 1,220-2,439 m: 26

Telecommunications: fair system of troposcatter, open-wire lines, and radio relay; broadcast stations—29 AM, 4 FM, 1 TV; earth stations—2 Atlantic Ocean INTELSAT and 3 domestic Indian Ocean INTELSAT

Defense Forces

Branches: Army, Naval Command, Air and Air Defense Forces, Militia

Manpower availability: males age 15-49 3,675,189; fit for military service 2,110,489 (1993 est.)

Defense expenditures: exchange rate conversion—$118 million, 8% of GDP (1993 est.)

Namibia

Boundary representation is not necessarily authoritative

300 km

Geography

Location: Southern Africa, bordering the South Atlantic Ocean between Angola and South Africa

Map references: Africa, Standard Time Zones of the World

Area:
total area: 824,290 km²
land area: 823,290 km²
comparative area: slightly more than half the size of Alaska

Land boundaries: total 3,935 km, Angola 1,376 km, Botswana 1,360 km, South Africa 966 km, Zambia 233 km

Coastline: 1,489 km

Maritime claims:
contiguous zone: 24 nm
exclusive economic zone: 200 nm
territorial sea: 12 nm

International disputes: short section of boundary with Botswana is indefinite; disputed island with Botswana in the Chobe River; quadripoint with Botswana, Zambia, and Zimbabwe is in disagreement; claim by Namibia to Walvis Bay and 12 offshore islands administered by South Africa; Namibia and South Africa have agreed to jointly administer the area for an interim period; the terms and dates to be covered by joint administration arrangements have not been established at this time, and Namibia will continue to maintain a claim to sovereignty over the entire area; recent dispute with Botswana over uninhabited Kasikili (Sidudu) Island in the Linyanti River

Climate: desert; hot, dry; rainfall sparse and erratic

Terrain: mostly high plateau; Namib Desert along coast; Kalahari Desert in east

Natural resources: diamonds, copper, uranium, gold, lead, tin, lithium, cadmium, zinc, salt, vanadium, natural gas, fish; suspected deposits of oil, natural gas, coal, iron ore

Namibia (continued)

Land use:
arable land: 1%
permanent crops: 0%
meadows and pastures: 64%
forest and woodland: 22%
other: 13%
Irrigated land: 40 km² (1989 est.)
Environment: inhospitable with very limited natural water resources; desertification
Note: Walvis Bay area is an exclave of South Africa in Namibia

People

Population: 1,541,321 (July 1993 est.)
Population growth rate: 3.46% (1993 est.)
Birth rate: 43.77 births/1,000 population (1993 est.)
Death rate: 9.13 deaths/1,000 population (1993 est.)
Net migration rate: 0 migrant(s)/1,000 population (1993 est.)
Infant mortality rate: 63.8 deaths/1,000 live births (1993 est.)
Life expectancy at birth:
total population: 61.2 years
male: 58.57 years
female: 63.91 years (1993 est.)
Total fertility rate: 6.46 children born/woman (1993 est.)
Nationality:
noun: Namibian(s)
adjective: Namibian
Ethnic divisions: black 86%, white 6.6%, mixed 7.4%
note: about 50% of the population belong to the Ovambo tribe and 9% to the Kavangos tribe
Religions: Christian
Languages: English 7% (official), Afrikaans common language of most of the population and about 60% of the white population, German 32%, indigenous languages
Literacy: age 15 and over can read and write (1960)
total population: 38%
male: 45%
female: 31%
Labor force: 500,000
by occupation: agriculture 60%, industry and commerce 19%, services 8%, government 7%, mining 6% (1981 est.)

Government

Names:
conventional long form: Republic of Namibia
conventional short form: Namibia
Digraph: WA
Type: republic
Capital: Windhoek
Administrative divisions: 13 districts; Erango, Hardap, Karas, Khomas, Kunene, Liambezi, Ohanguena, Okarango, Omaheke, Omusat, Oshana, Oshikoto, Otjozondjupa

note: the 26 districts were Bethanien, Boesmanland, Caprivi Oos, Damaraland, Gobabis, Grootfontein, Hereroland Oos, Hereroland Wes, Kaokoland, Karasburg, Karibib, Kavango, Keetmanshoop, Luderitz, Maltahohe, Mariental, Namaland, Okahandja, Omaruru, Otjiwarongo, Outjo, Owambo, Rehoboth, Swakopmund, Tsumeb, Windhoek
Independence: 21 March 1990 (from South African mandate)
Constitution: ratified 9 February 1990
Legal system: based on Roman-Dutch law and 1990 constitution
National holiday: Independence Day, 21 March (1990)
Political parties and leaders: South West Africa People's Organization (SWAPO), Sam NUJOMA; DTA of Namibia (DTA; formerly Democratic Turnhalle Alliance of Namibia), Dirk MUDGE; United Democratic Front (UDF), Justus GAROEB; Action Christian National (ACN), Kosie PRETORIUS; National Patriotic Front (NPF), Moses KATJIUONGUA; Federal Convention of Namibia (FCN), Hans DIERGAARDT; Namibia National Front (NNF), Vekuii RUKORO
Other political or pressure groups: NA
Suffrage: 18 years of age; universal
Elections:
President: last held 16 February 1990 (next to be held March 1995); results—Sam NUJOMA was elected president by the Constituent Assembly (now the National Assembly)
National Assembly: last held on 7-11 November 1989 (next to be held by November 1994); results—percent of vote by party NA; seats—(72 total) SWAPO 41, DTA 21, UDF 4, ACN 3, NNF 1, FCN 1, NPF 1
National Council: last held 30 November-3 December 1992 (next to be held by December 1998); seats—(26 total) SWAPO 19, DTA 6, UDF 1
Executive branch: president, Cabinet
Legislative branch: bicameral legislature consists of an upper house or National Council and a lower house or National Assembly
Judicial branch: Supreme Court
Leaders:
Chief of State and Head of Government: President Sam NUJOMA (since 21 March 1990)
Member of: ACP, C, ECA, FAO, FLS, G-77, IAEA, IBRD, ICAO, IFAD, IFC, ILO, IMF, IOM (observer), ITU, NAM, OAU, SACU, SADC, UN, UNCTAD, UNESCO, UNHCR, UNIDO, UPU, WCL, WFTU, WHO, WIPO, WMO
Diplomatic representation in US:
chief of mission: Ambassador Tuliameni KALOMOH

chancery: 1605 New Hampshire Ave. NW, Washington, DC 20009 (mailing address is PO Box 34738, Washington, DC 20043)
telephone: (202) 986-0540
US diplomatic representation:
chief of mission: Charge d'Affaires Marshall MCCAULEY
embassy: Ausplan Building, 14 Lossen St., Windhoek
mailing address: P. O. Box 9890, Windhoek 9000
telephone: [264] (61) 221-601, 222-675, 222-680
FAX: [264] (61) 229-792
Flag: a large blue triangle with a yellow sunburst fills the upper left section, and an equal green triangle (solid) fills the lower right section; the triangles are separated by a red stripe that is contrasted by two narrow white-edge borders

Economy

Overview: The economy is heavily dependent on the mining industry to extract and process minerals for export. Mining accounts for almost 25% of GDP. Namibia is the fourth-largest exporter of nonfuel minerals in Africa and the world's fifth-largest producer of uranium. Alluvial diamond deposits are among the richest in the world, making Namibia a primary source for gem-quality diamonds. Namibia also produces large quantities of lead, zinc, tin, silver, and tungsten. More than half the population depends on agriculture (largely subsistence agriculture) for its livelihood.
National product: GDP—exchange rate conversion—$2 billion (1992 est.)
National product real growth rate: 2% (1992 est.)
National product per capita: $1,300 (1992 est.)
Inflation rate (consumer prices): 10% (1992) in urban area
Unemployment rate: 25-35% (1992)
Budget: revenues $864 million; expenditures $1,112 million, including capital expenditures of $144 million (FY 92)
Exports: $1.184 billion (f.o.b., 1991)
commodities: diamonds, copper, gold, zinc, lead, uranium, cattle, processed fish, karakul skins
partners: Switzerland, South Africa, Germany, Japan
Imports: $1.238 billion (f.o.b., 1991)
commodities: foodstuffs, petroleum products and fuel, machinery and equipment
partners: South Africa, Germany, US, Switzerland
External debt: about $220 million (1992 est.)
Industrial production: growth rate 4.9% (1991); accounts for 35% of GDP, including mining

Nauru

Electricity: 490,000 kW capacity; 1,290 million kWh produced, 850 kWh per capita (1991)
Industries: meatpacking, fish processing, dairy products, mining (copper, lead, zinc, diamond, uranium)
Agriculture: accounts for 15% of GDP; mostly subsistence farming; livestock raising major source of cash income; crops—millet, sorghum, peanuts; fish catch potential of over 1 million metric tons not being fulfilled, 1988 catch reaching only 384,000 metric tons; not self-sufficient in food
Economic aid: Western (non-US) countries, ODA and OOF bilateral commitments (1970-87), $47.2 million
Currency: 1 South African rand (R) = 100 cents
Exchange rates: South African rand (R) per US$1—3.1576 (May 1993), 2.8497 (1992), 2.7653 (1991), 2.5863 (1990), 2.6166 (1989), 2.2611 (1988)
Fiscal year: 1 April-31 March

Communications

Railroads: 2,341 km 1.067-meter gauge, single track
Highways: 54,500 km; 4,079 km paved, 2,540 km gravel, 47,881 km earth roads and tracks
Ports: Luderitz; primary maritime outlet is Walvis Bay (South Africa)
Airports:
total: 137
usable: 112
with permanent-surface runways: 21
with runways over 3,659 m: 1
with runways 2,440-3,659 m: 4
with runways 1,220-2,439 m: 62
Telecommunications: good urban, fair rural services; radio relay connects major towns, wires extend to other population centers; 62,800 telephones; broadcast stations—4 AM, 40 FM, 3 TV

Defense Forces

Branches: National Defense Force (Army), Police
Manpower availability: males age 15-49 324,599; fit for military service 192,381 (1993 est.)
Defense expenditures: exchange rate conversion—$66 million, 3.4% of GDP (FY92)

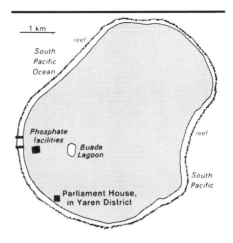

Geography

Location: Oceania, 500 km north-northeast of Papua New Guinea
Map references: Oceania, Standard Time Zones of the World
Area:
total area: 21 km²
land area: 21 km²
comparative area: about one-tenth the size of Washington, DC
Land boundaries: 0 km
Coastline: 30 km
Maritime claims:
exclusive fishing zone: 200 nm
territorial sea: 12 nm
International disputes: none
Climate: tropical; monsoonal; rainy season (November to February)
Terrain: sandy beach rises to fertile ring around raised coral reefs with phosphate plateau in center
Natural resources: phosphates
Land use:
arable land: 0%
permanent crops: 0%
meadows and pastures: 0%
forest and woodland: 0%
other: 100%
Irrigated land: NA km²
Environment: only 53 km south of Equator
Note: Nauru is one of the three great phosphate rock islands in the Pacific Ocean —the others are Banaba (Ocean Island) in Kiribati and Makatea in French Polynesia

People

Population: 9,882 (July 1993 est.)
Population growth rate: 1.42% (1993 est.)
Birth rate: 18.92 births/1,000 population (1993 est.)
Death rate: 5.1 deaths/1,000 population (1993 est.)
Net migration rate: 0.4 migrant(s)/1,000 population (1993 est.)
Infant mortality rate: 40.6 deaths/1,000 live births (1993 est.)

Life expectancy at birth:
total population: 66.68 years
male: 64.3 years
female: 69.18 years (1993 est.)
Total fertility rate: 2.2 children born/woman (1993 est.)
Nationality:
noun: Nauruan(s)
adjective: Nauruan
Ethnic divisions: Nauruan 58%, other Pacific Islander 26%, Chinese 8%, European 8%
Religions: Christian (two-thirds Protestant, one-third Roman Catholic)
Languages: Nauruan (official; a distinct Pacific Island language), English widely understood, spoken, and used for most government and commercial purposes
Literacy:
total population: NA%
male: NA%
female: NA%
Labor force:
by occupation: NA

Government

Names:
conventional long form: Republic of Nauru
conventional short form: Nauru
former: Pleasant Island
Digraph: NR
Type: republic
Capital: no official capital; government offices in Yaren
Administrative divisions: 14 districts; Aiwo, Anabar, Anetan, Anibare, Baiti, Boe, Buada, Denigomodu, Ewa, Ijuw, Meneng, Nibok, Uaboe, Yaren
Independence: 31 January 1968 (from UN trusteeship under Australia, New Zealand, and UK)
Constitution: 29 January 1968
Legal system: own Acts of Parliament and British common law
National holiday: Independence Day, 31 January (1968)
Political parties and leaders: none
Suffrage: 20 years of age; universal and compulsory
Elections:
President: last held 19 November 1992 (next to be held NA November 1995); results—Bernard DOWIYOGO elected by Parliament
Parliament: last held on 14 November 1992 (next to be held NA November 1995); results—percent of vote NA; seats—(18 total) independents 18
Executive branch: president, Cabinet
Legislative branch: unicameral Parliament
Judicial branch: Supreme Court
Leaders:
Chief of State and Head of Government: President Bernard DOWIYOGO (since 12 December 1989)

Nauru *(continued)*

Member of: AsDB, C (special), ESCAP, ICAO, INTERPOL, ITU, SPARTECA, SPC, SPF, UPU
Diplomatic representation in US: there is a Nauruan Consulate in Agana (Guam)
US diplomatic representation: the US Ambassador to Australia is accredited to Nauru
Flag: blue with a narrow, horizontal, yellow stripe across the center and a large white 12-pointed star below the stripe on the hoist side; the star indicates the country's location in relation to the Equator (the yellow stripe) and the 12 points symbolize the 12 original tribes of Nauru

Economy

Overview: Revenues come from the export of phosphates, the reserves of which are expected to be exhausted by the year 2000. Phosphates have given Nauruans one of the highest per capita incomes in the Third World—$10,000 annually. Few other resources exist, so most necessities must be imported, including fresh water from Australia. The rehabilitation of mined land and the replacement of income from phosphates are serious long-term problems. Substantial amounts of phosphate income are invested in trust funds to help cushion the transition.
National product: GNP—exchange rate conversion—$90 million (1989 est.)
National product real growth rate: NA%
National product per capita: $10,000 (1989 est.)
Inflation rate (consumer prices): NA%
Unemployment rate: 0%
Budget: revenues $69.7 million; expenditures $51.5 million, including capital expenditures of $NA (FY86 est.)
Exports: $93 million (f.o.b., 1984)
commodities: phosphates
partners: Australia, NZ
Imports: $73 million (c.i.f., 1984)
commodities: food, fuel, manufactures, building materials, machinery
partners: Australia, UK, NZ, Japan
External debt: $33.3 million
Industrial production: growth rate NA%
Electricity: 14,000 kW capacity; 50 million kWh produced, 5,430 kWh per capita (1990)
Industries: phosphate mining, financial services, coconut products
Agriculture: coconuts; other agricultural activity negligible; almost completely dependent on imports for food and water
Economic aid: Western (non-US) countries (1970-89), $2 million
Currency: 1 Australian dollar ($A) = 100 cents
Exchange rates: Australian dollars ($A) per US$1—1.4837 (January 1993), 1.3600 (1992), 1.2834 (1991), 1.2799 (1990), 1.2618 (1989), 1.2752 (1988)
Fiscal year: 1 July-30 June

Communications

Railroads: 3.9 km; used to haul phosphates from the center of the island to processing facilities on the southwest coast
Highways: about 27 km total; 21 km paved, 6 km improved earth
Ports: Nauru
Merchant marine: 1 bulk ship (1,000 GRT or over) totaling 4,426 GRT/5,750 DWT
Airports:
total: 1
useable: 1
with permanent-surface runways: 1
with runways over 3,659 m: 0
with runways 2,440-3,659 m: 0
with runways 1,220-2,439 m: 1
Telecommunications: adequate local and international radio communications provided via Australian facilities; 1,600 telephones; 4,000 radios; broadcast stations—1 AM, no FM, no TV; 1 Pacific Ocean INTELSAT earth station

Defense Forces

Branches: Directorate of the Nauru Police Force
note: no regular armed forces
Manpower availability: males age 15-49 NA; fit for military service NA
Defense expenditures: $NA—no formal defense structure

Navassa Island
(territory of the US)

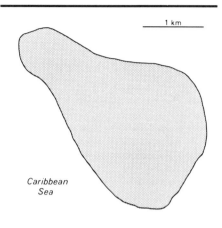

Geography

Location: in the Caribbean Sea, 160 km south of the US Naval Base at Guantanamo (Cuba), between Cuba, Haiti, and Jamaica
Map references: Central America and the Caribbean
Area:
total area: 5.2 km²
land area: 5.2 km²
comparative area: about nine times the size of the Mall in Washington, DC
Land boundaries: 0 km
Coastline: 8 km
Maritime claims:
contiguous zone: 24 nm
continental shelf: 200 m or depth of exploitation
exclusive economic zone: 200 nm
territorial sea: 12 nm
International disputes: claimed by Haiti
Climate: marine, tropical
Terrain: raised coral and limestone plateau, flat to undulating; ringed by vertical white cliffs (9 to 15 meters high)
Natural resources: guano
Land use:
arable land: 0%
permanent crops: 0%
meadows and pastures: 10%
forest and woodland: 0%
other: 90%
Irrigated land: 0 km²
Environment: mostly exposed rock, but enough grassland to support goat herds; dense stands of fig-like trees, scattered cactus
Note: strategic location 160 km south of the US Naval Base at Guantanamo, Cuba

People

Population: uninhabited; note—transient Haitian fishermen and others camp on the island

Nepal

Government

Names:
conventional long form: none
conventional short form: Navassa Island
Digraph: BQ
Type: unincorporated territory of the US administered by the US Coast Guard
Capital: none; administered from Washington, DC

Economy

Overview: no economic activity

Communications

Ports: none; offshore anchorage only

Defense Forces

Note: defense is the responsibility of the US

Geography

Location: South Asia, in the Himalayas, between China and India
Map references: Asia, Standard Time Zones of the World
Area:
total area: 140,800 km²
land area: 136,800 km²
comparative area: slightly larger than Arkansas
Land boundaries: total 2,926 km, China 1,236 km, India 1,690 km
Coastline: 0 km (landlocked)
Maritime claims: none; landlocked
International disputes: none
Climate: varies from cool summers and severe winters in north to subtropical summers and mild winters in south
Terrain: Terai or flat river plain of the Ganges in south, central hill region, rugged Himalayas in north
Natural resources: quartz, water, timber, hydroelectric potential, scenic beauty, small deposits of lignite, copper, cobalt, iron ore
Land use:
arable land: 17%
permanent crops: 0%
meadows and pastures: 13%
forest and woodland: 33%
other: 37%
Irrigated land: 9,430 km² (1989)
Environment: contains eight of world's 10 highest peaks; deforestation; soil erosion; water pollution
Note: landlocked; strategic location between China and India

People

Population: 20,535,466 (July 1993 est.)
Population growth rate: 2.43% (1993 est.)
Birth rate: 37.99 births/1,000 population (1993 est.)
Death rate: 13.66 deaths/1,000 population (1993 est.)
Net migration rate: 0 migrant(s)/1,000 population (1993 est.)

Infant mortality rate: 85.8 deaths/1,000 live births (1993 est.)
Life expectancy at birth:
total population: 51.98 years
male: 51.84 years
female: 52.12 years (1993 est.)
Total fertility rate: 5.33 children born/woman (1993 est.)
Nationality:
noun: Nepalese (singular and plural)
adjective: Nepalese
Ethnic divisions: Newars, Indians, Tibetans, Gurungs, Magars, Tamangs, Bhotias, Rais, Limbus, Sherpas
Religions: Hindu 90%, Buddhist 5%, Muslim 3%, other 2% (1981)
note: only official Hindu state in world, although no sharp distinction between many Hindu and Buddhist groups
Languages: Nepali (official), 20 languages divided into numerous dialects
Literacy: age 15 and over can read and write (1990)
total population: 26%
male: 38%
female: 13%
Labor force: 8.5 million (1991 est.)
by occupation: agriculture 93%, services 5%, industry 2%
note: severe lack of skilled labor

Government

Names:
conventional long form: Kingdom of Nepal
conventional short form: Nepal
Digraph: NP
Type: parliamentary democracy as of 12 May 1991
Capital: Kathmandu
Administrative divisions: 14 zones (anchal, singular and plural); Bagmati, Bheri, Dhawalagiri, Gandaki, Janakpur, Karnali, Kosi, Lumbini, Mahakali, Mechi, Narayani, Rapti, Sagarmatha, Seti
Independence: 1768 (unified by Prithvi Narayan Shah)
Constitution: 9 November 1990
Legal system: based on Hindu legal concepts and English common law; has not accepted compulsory ICJ jurisdiction
National holiday: Birthday of His Majesty the King, 28 December (1945)
Political parties and leaders:
ruling party: Nepali Congress Party (NCP), Party president Krishna Prasad BHATTARAI, Prime Minister Girija Prasad KOIRALA, Ganesh Man SINGH
center: the NDP has two factions: National Democratic Party/Chand (NDP/Chand), Lokendra Bahadur CHAND; and National Democratic Party/Thapa (NDP/Thapa), Surya Bahadur THAPA; Terai Rights Sadbhavana (Goodwill) Party, Gayendra Narayan SINGH

Nepal (continued)

Nepal (continued)

Communist: Communist Party of Nepal/United Marxist and Leninist (CPN/UML), Man Mohan ADIKHARY; United People's Front (UPF), N. K. PRASAI, Lila Mani POKHAREL; Nepal Workers and Peasants Party, leader NA; Rohit Party, N. M. BIJUKCHHE; Democratic Party, leader NA
note: the two factions of the NDP announced a merger in late 1991
Other political or pressure groups: numerous small, left-leaning student groups in the capital; several small, radical Nepalese antimonarchist groups
Suffrage: 18 years of age; universal
Elections:
House of Representatives: last held on 12 May 1991 (next to be held May 1996); results—NCP 38%, CPN/UML 28%, NDP/Chand 6%, UPF 5%, NDP/Thapa 5%, Terai Rights Sadbhavana Party 4%, Rohit 2%, CPN (Democratic) 1%, independents 4%, other 7%; seats—(205 total) NCP 110, CPN/UML 69, UPF 9, Terai Rights Sadbhavana Party 6, NDP/Chand 3, Rohit 2, CPN (Democratic) 2, NDP/Thapa 1, independents 3; note—the new Constitution of 9 November 1990 gave Nepal a multiparty democracy system for the first time in 32 years
Executive branch: monarch, prime minister, Council of Ministers
Legislative branch: bicameral Parliament consists of an upper house or National Council and a lower house or House of Representatives
Judicial branch: Supreme Court (Sarbochha Adalat)
Leaders:
Chief of State: King BIRENDRA Bir Bikram Shah Dev (since 31 January 1972, crowned King 24 February 1985); Heir Apparent Crown Prince DIPENDRA Bir Bikram Shah Dev, son of the King (born 21 June 1971)
Head of Government: Prime Minister Girija Prasad KOIRALA (since 29 May 1991)
Member of: AsDB, CCC, CP, ESCAP, FAO, G-77, IBRD, ICAO, IDA, IFAD, IFC, ILO, IMF, IMO, INTELSAT, INTERPOL, IOC, ITU, LORCS, NAM, SAARC, UN, UNCTAD, UNESCO, UNIDO, UNIFIL, UNPROFOR, UPU, WFTU, WHO, WMO, WTO
Diplomatic representation in US:
chief of mission: Ambassador Yog Prasad UPADHYAYA
chancery: 2131 Leroy Place NW, Washington, DC 20008
telephone: (202) 667-4550
consulate general: New York
US diplomatic representation:
chief of mission: Ambassador Julia Chang BLOCH
embassy: Pani Pokhari, Kathmandu

mailing address: use embassy street address
telephone: [977] (1) 411179 or 412718, 411604, 411613, 413890
FAX: [977] (1) 419963
Flag: red with a blue border around the unique shape of two overlapping right triangles; the smaller, upper triangle bears a white stylized moon and the larger, lower triangle bears a white 12-pointed sun

Economy

Overview: Nepal is among the poorest and least developed countries in the world. Agriculture is the mainstay of the economy, providing a livelihood for over 90% of the population and accounting for 60% of GDP. Industrial activity is limited, mainly involving the processing of agricultural produce (jute, sugarcane, tobacco, and grain). Production of textiles and carpets has expanded recently and accounted for 85% of foreign exchange earnings in FY91. Apart from agricultural land and forests, exploitable natural resources are mica, hydropower, and tourism. Agricultural production in the late 1980s grew by about 5%, as compared with annual population growth of 2.6%. More than 40% of the population is undernourished partly because of poor distribution. The top 10% of the population receives 47% of total income, the bottom 20% less than 5% of the total. Since May 1991, the government has been encouraging trade and foreign investment, e.g., by eliminating business licenses and registration requirements in order to simplify domestic and foreign investment. The government also has been cutting public expenditures by reducing subsides, privatizing state industries, and laying off civil servants. Prospects for foreign trade and investment in the 1990s remain poor, however, because of the small size of the economy, its technological backwardness, and its remoteness.
National product: GDP—exchange rate conversion—$3.4 billion (FY92)
National product real growth rate: 3.1% (FY92)
National product per capita: $170 (FY92)
Inflation rate (consumer prices): 14% (November 1992)
Unemployment rate: 5% (1987); underemployment estimated at 25-40%
Budget: revenues $308.0 million; expenditures $672.0 million, including capital expenditures of $396 million (FY92 est.)
Exports: $313 million (f.o.b., FY92 est.) but does not include unrecorded border trade with India
commodities: carpets, clothing, leather goods, jute goods, grain
partners: US, Germany, India, UK

Imports: $751 million (c.i.f., FY92 est.)
commodities: petroleum products 20%, fertilizer 11%, machinery 10%
partners: India, Singapore, Japan, Germany
External debt: $2.0 billion (FY92 est.)
Industrial production: growth rate 6% (FY91 est.); accounts for 7% of GDP
Electricity: 300,000 kW capacity; 1,000 million kWh produced, 50 kWh per capita (1992)
Industries: small rice, jute, sugar, and oilseed mills; cigarette, textile, carpet, cement, and brick production; tourism
Agriculture: accounts for 60% of GDP and 90% of work force; farm products—rice, corn, wheat, sugarcane, root crops, milk, buffalo meat; not self-sufficient in food, particularly in drought years
Illicit drugs: illicit producer of cannabis for the domestic and international drug markets; probable transit point for heroin from Southeast Asia to the West
Economic aid: US commitments, including Ex-Im (FY70-89), $304 million; Western (non-US) countries, ODA and OOF bilateral commitments (1980-89), $2,230 million; OPEC bilateral aid (1979-89), $30 million; Communist countries (1970-89), $286 million
Currency: 1 Nepalese rupee (NR) = 100 paisa
Exchange rates: Nepalese rupees (NRs) per US$1—43.200 (January 1993), 42.742 (1992), 37.255 (1991), 29.370 (1990), 27.189 (1989), 23.289 (1988)
Fiscal year: 16 July-15 July

Communications

Railroads: 52 km (1990), all 0.762-meter narrow gauge; all in Terai close to Indian border; 10 km from Raxaul to Birganj is government owned
Highways: 7,080 km total (1990); 2,898 km paved, 1,660 km gravel or crushed stone; also 2,522 km of seasonally motorable tracks
Airports:
total: 37
usable: 37
with permanent-surface runways: 5
with runways over 3,659 m: 0
with runways 2,440-3,659 m: 1
with runways 1,220-2,439 m: 8
Telecommunications: poor telephone and telegraph service; fair radio communication and broadcast service; international radio communication service is poor; 50,000 telephones (1990); broadcast stations—88 AM, no FM, 1 TV; 1 Indian Ocean INTELSAT earth station

Defense Forces

Branches: Royal Nepalese Army, Royal Nepalese Army Air Service, Nepalese Police Force

Netherlands

Manpower availability: males age 15-49 4,849,109; fit for military service 2,517,385; reach military age (17) annually 234,060 (1993 est.)
Defense expenditures: exchange rate conversion—$34 million, 2% of GDP (FY91/92)

Geography

Location: Western Europe, bordering the North Sea, between Belgium and Germany
Map references: Europe, Standard Time Zones of the World
Area:
total area: 37,330 km²
land area: 33,920 km²
comparative area: slightly less than twice the size of New Jersey
Land boundaries: total 1,027 km, Belgium 450 km, Germany 577 km
Coastline: 451 km
Maritime claims:
continental shelf: not specified
exclusive fishing zone: 200 nm
territorial sea: 12 nm
International disputes: none
Climate: temperate; marine; cool summers and mild winters
Terrain: mostly coastal lowland and reclaimed land (polders); some hills in southeast
Natural resources: natural gas, petroleum, fertile soil
Land use:
arable land: 26%
permanent crops: 1%
meadows and pastures: 32%
forest and woodland: 9%
other: 32%
Irrigated land: 5,500 km² (1989 est.)
Environment: without an extensive system of dikes and dams, nearly one-half of the total area would be inundated by sea water
Note: located at mouths of three major European rivers (Rhine, Maas or Meuse, Schelde)

People

Population: 15,274,942 (July 1993 est.)
Population growth rate: 0.63% (1993 est.)
Birth rate: 12.81 births/1,000 population (1993 est.)
Death rate: 8.53 deaths/1,000 population (1993 est.)

Net migration rate: 2.06 migrant(s)/1,000 population (1993 est.)
Infant mortality rate: 6.2 deaths/1,000 live births (1993 est.)
Life expectancy at birth:
total population: 77.55 years
male: 74.48 years
female: 80.78 years (1993 est.)
Total fertility rate: 1.59 children born/woman (1993 est.)
Nationality:
noun: Dutchman(men), Dutchwoman(women)
adjective: Dutch
Ethnic divisions: Dutch 96%, Moroccans, Turks, and other 4% (1988)
Religions: Roman Catholic 36%, Protestant 27%, other 6%, unaffiliated 31% (1988)
Languages: Dutch
Literacy: age 15 and over can read and write (1979)
total population: 99%
male: NA%
female: NA%
Labor force: 5.3 million
by occupation: services 50.1%, manufacturing and construction 28.2%, government 15.9%, agriculture 5.8% (1986)

Government

Names:
conventional long form: Kingdom of the Netherlands
conventional short form: Netherlands
local long form: Koninkrijk de Nederlanden
local short form: Nederland
Digraph: NL
Type: constitutional monarchy
Capital: Amsterdam; The Hague is the seat of government
Administrative divisions: 12 provinces (provincien, singular—provincie); Drenthe, Flevoland, Friesland, Gelderland, Groningen, Limburg, Noord-Brabant, Noord-Holland, Overijssel, Utrecht, Zeeland, Zuid-Holland
Dependent areas: Aruba, Netherlands Antilles
Independence: 1579 (from Spain)
Constitution: 17 February 1983
Legal system: civil law system incorporating French penal theory; judicial review in the Supreme Court of legislation of lower order rather than Acts of the States General; accepts compulsory ICJ jurisdiction, with reservations
National holiday: Queen's Day, 30 April (1938)
Political parties and leaders: Christian Democratic Appeal (CDA), Willem van VELZEN; Labor (PvdA), Wim KOK; Liberal (VVD), Frederick BOLKSTEIN; Democrats '66 (D'66), Hans van MIERIO; a host of minor parties
Other political or pressure groups: large multinational firms;

Netherlands (continued)

Federation of Netherlands Trade Union Movement (comprising Socialist and Catholic trade unions) and a Protestant trade union; Federation of Catholic and Protestant Employers Associations; the nondenominational Federation of Netherlands Enterprises; and Interchurch Peace Council (IKV)

Suffrage: 18 years of age; universal

Elections:

First Chamber: last held on 9 June 1991 (next to be held 9 June 1995); results—elected by the country's 12 provincial councils; seats—(75 total) percent of seats by party NA

Second Chamber: last held on 6 September 1989 (next to be held in May 1994); results—CDA 35.3%, PvdA 31.9%, VVD 14.6%, D'66 7.9%, other 10.3%; seats—(150 total) CDA 54, PvdA 49, VVD 22, D'66 12, other 13

Executive branch: monarch, prime minister, vice prime minister, Cabinet, Cabinet of Ministers

Legislative branch: bicameral legislature (Staten Generaal) consists of an upper chamber or First Chamber (Eerste Kamer) and a lower chamber or Second Chamber (Tweede Kamer)

Judicial branch: Supreme Court (De Hoge Raad)

Leaders:

Chief of State: Queen BEATRIX Wilhelmina Armgard (since 30 April 1980); Heir Apparent WILLEM-ALEXANDER, Prince of Orange, son of Queen Beatrix (born 27 April 1967)

Head of Government: Prime Minister Ruud (Rudolph) F. M. LUBBERS (since 4 November 1982); Vice Prime Minister Willem (Wim) KOK (since 2 November 1989)

Member of: AfDB, AG (observer), AsDB, Australia Group, Benelux, BIS, CCC, CE, CERN, COCOM, CSCE, EBRD, EC, ECE, ECLAC, EIB, ESA, ESCAP, FAO, G-10, GATT, IADB, IAEA, IBRD, ICAO, ICC, ICFTU, IDA, IEA, IFAD, IFC, ILO, IMF, IMO, INMARSAT, INTELSAT, INTERPOL, IOC, IOM, ISO, ITU, LORCS, MTCR, NACC, NAM (guest), NATO, NEA, NSG, OAS (observer), OECD, PCA, UN, UNAVEM II, UNCTAD, UNESCO, UNHCR, UNIDO, UNPROFOR, UNTAC, UNTSO, UPU, WCL, WEU, WHO, WIPO, WMO, WTO, ZC

Diplomatic representation in US:

chief of mission: Ambassador Johan Hendrick MEESMAN

chancery: 4200 Linnean Avenue NW, Washington DC 20008

telephone: (202) 244-5300

FAX: (202) 362-3430

consulates general: Chicago, Houston, Los Angeles, Manila (Trust Territories of the Pacific Islands), New York

US diplomatic representation:

chief of mission: (vacant); Charge d'Affaires Thomas H. GEWECKE

embassy: Lange Voorhout 102, The Hague

mailing address: PSC 71, Box 1000, APO AE 09715

telephone: [31] (70) 310-9209

FAX: [31] (70) 361-4688

consulate general: Amsterdam

Flag: three equal horizontal bands of red (top), white, and blue; similar to the flag of Luxembourg, which uses a lighter blue and is longer

---

Economy

Overview: This highly developed and affluent economy is based on private enterprise. The government makes its presence felt, however, through many regulations, permit requirements, and welfare programs affecting most aspects of economic activity. The trade and financial services sector contributes over 50% of GDP. Industrial activity provides about 25% of GDP and is led by the food-processing, oil-refining, and metalworking industries. The highly mechanized agricultural sector employs only 5% of the labor force, but provides large surpluses for export and the domestic food-processing industry. Unemployment and a sizable budget deficit are currently the most serious economic problems. Many of the economic issues of the 1990s will reflect the course of European economic integration.

National product: GDP—purchasing power equivalent—$259.8 billion (1992)

National product real growth rate: 1.6% (1992)

National product per capita: $17,200 (1992)

Inflation rate (consumer prices): 3.5% (1992 est.)

Unemployment rate: 5.3% (1992 est.)

Budget: revenues $109.9 billion; expenditures $122.1 billion, including capital expenditures of $NA (1992 est.)

Exports: $128.5 billion (f.o.b., 1992)

commodities: agricultural products, processed foods and tobacco, natural gas, chemicals, metal products, textiles, clothing

partners: EC 77% (Germany 27%, Belgium-Luxembourg 15%, UK 10%), US 4% (1991)

Imports: $117.7 billion (f.o.b., 1992)

commodities: raw materials and semifinished products, consumer goods, transportation equipment, crude oil, food products

partners: EC 64% (Germany 26%, Belgium-Luxembourg 14%, UK 8%), US 8% (1991)

External debt: none

Industrial production: growth rate 1.6% (1992 est.); accounts for 25% of GDP

Electricity: 22,216,000 kW capacity; 63,500 million kWh produced, 4,200 kWh per capita (1992)

Industries: agroindustries, metal and engineering products, electrical machinery and equipment, chemicals, petroleum, fishing, construction, microelectronics

Agriculture: accounts for 4.6% of GDP; animal production predominates; crops—grains, potatoes, sugar beets, fruits, vegetables; shortages of grain, fats, and oils

Illicit drugs: transit country for illicit narcotics produced in neighboring countries; European producer of illicit amphetamines and other synthetic drugs

Economic aid: donor—ODA and OOF commitments (1970-89), $19.4 billion

Currency: 1 Netherlands guilder, gulden, or florin (f.) = 100 cents

Exchange rates: Netherlands guilders, gulden, or florins (f.) per US$1—1.8167 (January 1993), 1.7585 (1992), 1.8697 (1991), 1.8209 (1990), 2.1207 (1989), 1.9766 (1988)

Fiscal year: calendar year

Communications

Railroads: 2,828 km 1.435-meter standard gauge operated by Netherlands Railways (NS) (includes 1,957 km electrified and 1,800 km double track)

Highways: 108,360 km total; 92,525 km paved (including 2,185 km of limited access, divided highways); 15,835 km gravel, crushed stone

Inland waterways: 6,340 km, of which 35% is usable by craft of 1,000 metric ton capacity or larger

Pipelines: crude oil 418 km; petroleum products 965 km; natural gas 10,230 km

Ports: coastal—Amsterdam, Delfzijl, Den Helder, Dordrecht, Eemshaven, Ijmuiden, Rotterdam, Scheveningen, Terneuzen, Vlissingen; inland—29 ports

Merchant marine: 344 ships (1,000 GRT or over) totaling 2,762,000 GRT/3,675,649 DWT; includes 3 short-sea passenger, 193 cargo, 30 refrigerated cargo, 26 container, 13 roll-on/roll-off, 1 livestock carrier, 11 multifunction large-load carrier, 23 oil tanker, 22 chemical tanker, 10 liquefied gas, 2 specialized tanker, 6 bulk, 4 combination bulk; note—many Dutch-owned ships are also registered on the captive Netherlands Antilles register

Airports:

total: 28

usable: 28

with permanent-surface runways: 20

with runways over 3,659 m: 0

with runways 2,440-3,659 m: 11

with runways 1,220-2,439 m: 6

Telecommunications: highly developed, well maintained, and integrated; extensive redundant system of multiconductor cables, supplemented by microwave radio relay microwave links; 9,418,000 telephones; broadcast stations—3 (3 relays) AM, 12 (39 repeaters) FM, 8 (7 repeaters) TV; 5 submarine cables; 1 communication satellite earth station operating in INTELSAT (1 Indian Ocean and 2 Atlantic Ocean antenna) and EUTELSAT systems; nationwide mobile phone system

Defense Forces

Branches: Royal Netherlands Army, Royal Netherlands Navy (including Naval Air Service and Marine Corps), Royal Netherlands Air Force, Royal Constabulary
Manpower availability: males age 15-49 4,183,167; fit for military service 3,677,445; reach military age (20) annually 104,263 (1993 est.)
Defense expenditures: exchange rate conversion—$7.8 billion, 3% of GDP (1992)

Netherlands Antilles
(part of the Dutch realm)

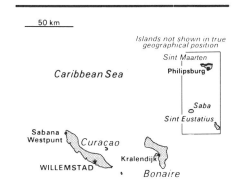

Geography

Location: two island groups—Curacas and Bonaire in the southern Caribbean Sea are about 70 km north of Venezuela near Aruba and the rest of the country is about 800 km to the northeast about one-third of the way between Antigua and Barbuda and Puerto Rico
Map references: Central America and the Caribbean
Area:
total area: 960 km^2
land area: 960 km^2
comparative area: slightly less than 5.5 times the size of Washington, DC
note: includes Bonaire, Curacao, Saba, Sint Eustatius, and Sint Maarten (Dutch part of the island of Saint Martin)
Land boundaries: 0 km
Coastline: 364 km
Maritime claims:
exclusive fishing zone: 12 nm
territorial sea: 12 nm
International disputes: none
Climate: tropical; ameliorated by northeast trade winds
Terrain: generally hilly, volcanic interiors
Natural resources: phosphates (Curacao only), salt (Bonaire only)
Land use:
arable land: 8%
permanent crops: 0%
meadows and pastures: 0%
forest and woodland: 0%
other: 92%
Irrigated land: NA km^2
Environment: Curacao and Bonaire are south of Caribbean hurricane belt, so rarely threatened; Sint Maarten, Saba, and Sint Eustatius are subject to hurricanes from July to October

People

Population: 184,990 (July 1993 est.)
Population growth rate: 0.4% (1993 est.)

Birth rate: 17.23 births/1,000 population (1993 est.)
Death rate: 5.69 deaths/1,000 population (1993 est.)
Net migration rate: -7.57 migrant(s)/1,000 population (1993 est.)
Infant mortality rate: 10.4 deaths/1,000 live births (1993 est.)
Life expectancy at birth:
total population: 75.73 years
male: 73.55 years
female: 78.03 years (1993 est.)
Total fertility rate: 1.99 children born/woman (1993 est.)
Nationality:
noun: Netherlands Antillean(s)
adjective: Netherlands Antillean
Ethnic divisions: mixed African 85%, Carib Indian, European, Latin, Oriental
Religions: Roman Catholic, Protestant, Jewish, Seventh-Day Adventist
Languages: Dutch (official), Papiamento a Spanish-Portuguese-Dutch-English dialect predominates, English widely spoken, Spanish
Literacy: age 15 and over can read and write (1981)
total population: 94%
male: 94%
female: 93%
Labor force: 89,000
by occupation: government 65%, industry and commerce 28% (1983)

Government

Names:
conventional long form: none
conventional short form: Netherlands Antilles
local long form: none
local short form: Nederlandse Antillen
Digraph: NA
Type: part of the Dutch realm; full autonomy in internal affairs granted in 1954
Capital: Willemstad
Administrative divisions: none (part of the Dutch realm)
Independence: none (part of the Dutch realm)
Constitution: 29 December 1954, Statute of the Realm of the Netherlands, as amended
Legal system: based on Dutch civil law system, with some English common law influence
National holiday: Queen's Day, 30 April (1938)
Political parties and leaders: political parties are indigenous to each island
Bonaire: Patriotic Union of Bonaire (UPB), Rudy ELLIS; Democratic Party of Bonaire (PDB), Franklin CRESTIAN
Curacao: National People's Party (PNP), Maria LIBERIA-PETERS; New Antilles Movement (MAN), Domenico Felip Don MARTINA; Workers' Liberation Front (FOL),

Netherlands Antilles (continued)

Wilson (Papa) GODETT; Socialist Independent (SI), George HUECK and Nelson MONTE; Democratic Party of Curacao (DP), Augustin DIAZ; Nos Patria, Chin BEHILIA
Saba: Windward Islands People's Movement (WIPM Saba), Will JOHNSON; Saba Democratic Labor Movement, Vernon HASSELL; Saba Unity Party, Carmen SIMMONDS
Sint Eustatius: Democratic Party of Sint Eustatius (DP-St.E), K. Van PUTTEN; Windward Islands People's Movement (WIPM); St. Eustatius Alliance (SEA), Ralph BERKEL
Sint Maarten: Democratic Party of Sint Maarten (DP-St.M), Claude WATHEY; Patriotic Movement of Sint Maarten (SPA), Vance JAMES
Suffrage: 18 years of age; universal
Elections:
Staten: last held on 16 March 1990 (next to be held March 1994); results—percent of vote by party NA; seats—(22 total) PNP 7, FOL-SI 3, UPB 3, MAN 2, DP-St. M 2, DP 1, SPM 1, WIPM 1, DP-St. E 1, Nos Patria 1; note—the government of Prime Minister Maria LIBERIA-PETERS is a coalition of several parties
Executive branch: Dutch monarch, governor, prime minister, vice prime minister, Council of Ministers (cabinet)
Legislative branch: unicameral legislature (Staten)
Judicial branch: Joint High Court of Justice
Leaders:
Chief of State: Queen BEATRIX Wilhelmina Armgard (since 30 April 1980), represented by Governor General Jaime SALEH (since NA October 1989)
Head of Government: Prime Minister Maria LIBERIA-PETERS (since 17 May 1988, previously served from September 1984 to November 1985)
Member of: CARICOM (observer), ECLAC (associate), ICFTU, INTERPOL, IOC, UNESCO (associate), UPU, WMO, WTO (associate)
Diplomatic representation in US: as an autonomous part of the Netherlands, Netherlands Antillean interests in the US are represented by the Netherlands
US diplomatic representation:
chief of mission: Consul General Bernard J. WOERZ
consulate general: Saint Anna Boulevard 19, Willemstad, Curacao
mailing address: P. O. Box 158, Willemstad, Curacao
telephone: [599] (9) 613066
FAX: [599] (9) 616489
Flag: white with a horizontal blue stripe in the center superimposed on a vertical red band also centered; five white five-pointed stars are arranged in an oval pattern in the center of the blue band; the five stars represent the five main islands of Bonaire, Curacao, Saba, Sint Eustatius, and Sint Maarten

---

Economy

Overview: Tourism, petroleum refining, and offshore finance are the mainstays of the economy. The islands enjoy a high per capita income and a well-developed infrastructure as compared with other countries in the region. Unlike many Latin American countries, the Netherlands Antilles has avoided large international debt. Almost all consumer and capital goods are imported, with the US being the major supplier.
National product: GDP—exchange rate conversion—$1.6 billion (1991 est.)
National product real growth rate: 4% (1991 est.)
National product per capita: $8,700 (1991 est.)
Inflation rate (consumer prices): 4% (1992 est.)
Unemployment rate: 16.4% (1991 est.)
Budget: revenues $209 million; expenditures $232 million, including capital expenditures of $8 million (1992 est.)
Exports: $200 million (f.o.b., 1991)
commodities: petroleum products 98%
partners: US 40%, UK 7%, Guadeloupe 5%
Imports: $1.2 billion (f.o.b., 1991)
commodities: crude petroleum 64%, food, manufactures
partners: Venezuela 42%, US 21%, Netherlands 8%
External debt: $701 million (December 1987)
Industrial production: growth rate NA%
Electricity: 125,000 kW capacity; 365 million kWh produced, 1,980 kWh per capita (1992)
Industries: tourism (Curacao and Sint Maarten), petroleum refining (Curacao), petroleum transshipment facilities (Curacao and Bonaire), light manufacturing (Curacao)
Agriculture: hampered by poor soils and scarcity of water; chief products—aloes, sorghum, peanuts, fresh vegetables, tropical fruit; not self-sufficient in food
Economic aid: Western (non-US) countries, ODA and OOF bilateral commitments (1970-89), $513 million
Currency: 1 Netherlands Antillean guilder, gulden, or florin (NAf.) = 100 cents
Exchange rates: Netherlands Antillean guilders, gulden, or florins (NAf.) per US$1—1.79 (fixed rate since 1989; 1.80 fixed rate 1971-88)
Fiscal year: calendar year

Communications

Highways: 950 km total; 300 km paved, 650 km gravel and earth

Ports: Willemstad, Philipsburg, Kralendijk
Merchant marine: 89 ships (1,000 GRT or over) totaling 781,646 GRT/962,138 DWT; includes 4 passenger, 29 cargo, 14 refrigerated cargo, 7 container, 7 roll-on/roll-off, 12 multifunction large-load carrier, 5 chemical tanker, 6 liquefied gas, 2 bulk, 1 oil tanker, 1 railcar carrier, 1 combination ore/oil; note—all but a few are foreign owned, mostly in the Netherlands
Airports:
total: 5
usable: 4
with permanent-surface runways: 4
with runways over 3,659 m: 0
with runways 2,440-3,659 m: 1
with runways 1,220-2,439 m: 3
Telecommunications: generally adequate facilities; extensive interisland microwave radio relay links; broadcast stations—9 AM, 4 FM, 1 TV; 2 submarine cables; 2 Atlantic Ocean INTELSAT earth stations

---

Defense Forces

Branches: Royal Netherlands Navy, Marine Corps, Royal Netherlands Air Force, National Guard, Police Force
Manpower availability: males age 15-49 48,965; fit for military service 27,531; reach military age (20) annually 1,638 (1993 est.)
Note: defense is responsibility of the Netherlands

New Caledonia

(overseas territory of France)

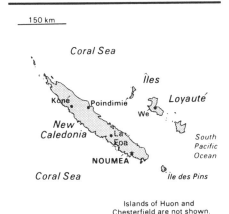

150 km

Coral Sea

Îles Loyauté

Koné Poindimie
We
New Caledonia La South Pacific Ocean
Foa
NOUMEA

Coral Sea Île des Pins

Islands of Huon and Chesterfield are not shown.

Geography

Location: in the South Pacific Ocean, 1,750 km east of Australia
Map references: Oceania
Area:
total area: 19,060 km²
land area: 18,760 km²
comparative area: slightly smaller than New Jersey
Land boundaries: 0 km
Coastline: 2,254 km
Maritime claims:
exclusive economic zone: 200 nm
territorial sea: 12 nm
International disputes: none
Climate: tropical; modified by southeast trade winds; hot, humid
Terrain: coastal plains with interior mountains
Natural resources: nickel, chrome, iron, cobalt, manganese, silver, gold, lead, copper
Land use:
arable land: 0%
permanent crops: 0%
meadows and pastures: 14%
forest and woodland: 51%
other: 35%
Irrigated land: NA km²
Environment: typhoons most frequent from November to March

People

Population: 178,056 (July 1993 est.)
Population growth rate: 1.83% (1993 est.)
Birth rate: 22.7 births/1,000 population (1993 est.)
Death rate: 5.01 deaths/1,000 population (1993 est.)
Net migration rate: 0.58 migrant(s)/1,000 population (1993 est.)
Infant mortality rate: 15.9 deaths/1,000 live births (1993 est.)
Life expectancy at birth:
total population: 73.22 years
male: 69.92 years

female: 76.7 years (1993 est.)
Total fertility rate: 2.67 children born/woman (1993 est.)
Nationality:
noun: New Caledonian(s)
adjective: New Caledonian
Ethnic divisions: Melanesian 42.5%, European 37.1%, Wallisian 8.4%, Polynesian 3.8%, Indonesian 3.6%, Vietnamese 1.6%, other 3%
Religions: Roman Catholic 60%, Protestant 30%, other 10%
Languages: French, 28 Melanesian-Polynesian dialects
Literacy: age 15 and over can read and write (1976)
total population: 91%
male: 91%
female: 90%
Labor force: 50,469 foreign workers for plantations and mines from Wallis and Futuna, Vanuatu, and French Polynesia (1980 est.)
by occupation: NA

Government

Names:
conventional long form: Territory of New Caledonia and Dependencies
conventional short form: New Caledonia
local long form: Territoire des Nouvelle-Caledonie et Dependances
local short form: Nouvelle-Caledonie
Digraph: NC
Type: overseas territory of France since 1956
Capital: Noumea
Administrative divisions: none (overseas territory of France); there are no first-order administrative divisions as defined by the US Government, but there are 3 provinces named Iles Loyaute, Nord, and Sud
Independence: none (overseas territory of France; a referendum on independence will be held in 1998)
Constitution: 28 September 1958 (French Constitution)
Legal system: the 1988 Matignon Accords grant substantial autonomy to the islands; formerly under French law
National holiday: National Day, Taking of the Bastille, 14 July (1789)
Political parties and leaders:
white-dominated Rassemblement pour la Caledonie dans la Republique (RPCR), conservative, Jacques LAFLEUR—affiliated to France's Rassemblement pour la Republique (RPR); Melanesian proindependence Kanaka Socialist National Liberation Front (FLNKS), Paul NEAOUTYINE; Melanesian moderate Kanak Socialist Liberation (LKS), Nidoish NAISSELINE; National Front (FN), extreme right, Guy GEORGE; Caledonie Demain

(CD), right-wing, Bernard MARANT; Union Oceanienne (UO), conservative, Michel HEMA; Front Uni de Liberation Kanak (FULK), proindependence, UREGEI; Union Caledonian (UC), Francois BURCK
Suffrage: 18 years of age; universal
Elections:
French Senate: last held 27 September 1992 (next to be held September 2001); results—percent of vote by party NA; seats—(1 total) RPCR 1
French National Assembly: last held 5 and 12 June 1988 (next to be held 21 and 28 March 1993); results—RPR 83.5%, FN 13.5%, other 3%; seats—(2 total) RPCR 2
Territorial Assembly: last held 11 June 1989 (next to be held 1993); results—RPCR 44.5%, FLNKS 28.5%, FN 7%, CD 5%, UO 4%, other 11%; seats—(54 total) RPCR 27, FLNKS 19, FN 3, other 5; note—election boycotted by FULK
Executive branch: French president, high commissioner, Consultative Committee (cabinet)
Legislative branch: unicameral Territorial Assembly
Judicial branch: Court of Appeal
Leaders:
Chief of State: President Francois MITTERRAND (since 21 May 1981)
Head of Government: High Commissioner and President of the Council of Government Alain CHRISTNACHT (since 15 January 1991)
Member of: ESCAP (associate), FZ, ICFTU, SPC, WMO
Diplomatic representation in US: as an overseas territory of France, New Caledonian interests are represented in the US by France
US diplomatic representation: none (overseas territory of France)
Flag: the flag of France is used

Economy

Overview: New Caledonia has more than 25% of the world's known nickel resources. In recent years the economy has suffered because of depressed international demand for nickel, the principal source of export earnings. Only a negligible amount of the land is suitable for cultivation, and food accounts for about 25% of imports.
National product: GNP—exchange rate conversion—$1 billion (1991 est.)
National product real growth rate: 2.4% (1988)
National product per capita: $6,000 (1991 est.)
Inflation rate (consumer prices): 4.1% (1989)
Unemployment rate: 16% (1989)
Budget: revenues $224.0 million; expenditures $211.0 million, including capital expenditures of $NA (1985)

New Caledonia (continued)

Exports: $671 million (f.o.b., 1989)
commodities: nickel metal 87%, nickel ore
partners: France 52.3%, Japan 15.8%, US 6.4%
Imports: $764 million (c.i.f., 1989)
commodities: foods, fuels, minerals, machines, electrical equipment
partners: France 44.0%, US 10%, Australia 9%
External debt: $NA
Industrial production: growth rate NA%
Electricity: 400,000 kW capacity; 2,200 million kWh produced, 12,790 kWh per capita (1990)
Industries: nickel mining and smelting
Agriculture: large areas devoted to cattle grazing; coffee, corn, wheat, vegetables; 60% self-sufficient in beef
Illicit drugs: illicit cannabis cultivation is becoming a principal source of income for some families
Economic aid: Western (non-US) countries, ODA and OOF bilateral commitments (1970-89), $4,185 million
Currency: 1 CFP franc (CFPF) = 100 centimes
Exchange rates: Comptoirs Francais duPacifique francs (CFPF) per US$1—99.65 (January 1993), 96.24 (1992), 102.57 (1991), 99.00 (1990), 115.99 (1989), 108.30 (1988); note—linked at the rate of 18.18 to the French franc
Fiscal year: calendar year

Communications

Highways: 6,340 km total; only about 10% paved (1987)
Ports: Noumea, Nepoui, Poro, Thio
Airports:
total: 29
usable: 27
with permanent-surface runways: 2
with runways over 3,659 m: 0
with runways 2,440-3,659 m: 1
with runways 1,220-2,439 m: 2
Telecommunications: 32,578 telephones (1987); broadcast stations—5 AM, 3 FM, 7 TV; 1 Pacific Ocean INTELSAT earth station

Defense Forces

Branches: Gendarmerie, Police Force
Note: defense is the responsibility of France

New Zealand

Geography

Location: Oceania, southeast of Australia in the South Pacific Ocean
Map references: Oceania, Standard Time Zones of the World
Area:
total area: 268,680 km²
land area: 268,670 km²
comparative area: about the size of Colorado
note: includes Antipodes Islands, Auckland Islands, Bounty Islands, Campbell Island, Chatham Islands, and Kermadec Islands
Land boundaries: 0 km
Coastline: 15,134 km
Maritime claims:
continental shelf: 200 nm or the edge of continental margin
exclusive economic zone: 200 nm
territorial sea: 12 nm
International disputes: territorial claim in Antarctica (Ross Dependency)
Climate: temperate with sharp regional contrasts
Terrain: predominately mountainous with some large coastal plains
Natural resources: natural gas, iron ore, sand, coal, timber, hydropower, gold, limestone
Land use:
arable land: 2%
permanent crops: 0%
meadows and pastures: 53%
forest and woodland: 38%
other: 7%
Irrigated land: 2,800 km² (1989 est.)
Environment: earthquakes are common, though usually not severe

People

Population: 3,368,774 (July 1993 est.)
Population growth rate: 0.61% (1993 est.)
Birth rate: 15.93 births/1,000 population (1993 est.)
Death rate: 8.11 deaths/1,000 population (1993 est.)

Net migration rate: -1.69 migrant(s)/1,000 population (1993 est.)
Infant mortality rate: 9.1 deaths/1,000 live births (1993 est.)
Life expectancy at birth:
total population: 76.11 years
male: 72.46 years
female: 79.95 years (1993 est.)
Total fertility rate: 2.07 children born/woman (1993 est.)
Nationality:
noun: New Zealander(s)
adjective: New Zealand
Ethnic divisions: European 88%, Maori 8.9%, Pacific Islander 2.9%, other 0.2%
Religions: Anglican 24%, Presbyterian 18%, Roman Catholic 15%, Methodist 5%, Baptist 2%, other Protestant 3%, unspecified or none 9% (1986)
Languages: English (official), Maori
Literacy: age 15 and over can read and write (1980)
total population: 99%
male: NA%
female: NA%
Labor force: 1,603,500 (June 1991)
by occupation: services 67.4%, manufacturing 19.8%, primary production 9.3% (1987)

Government

Names:
conventional long form: none
conventional short form: New Zealand
Abbreviation: NZ
Digraph: NZ
Type: parliamentary democracy
Capital: Wellington
Administrative divisions: 93 counties, 9 districts*, and 3 town districts**; Akaroa, Amuri, Ashburton, Bay of Islands, Bruce, Buller, Chatham Islands, Cheviot, Clifton, Clutha, Cook, Dannevirke, Egmont, Eketahuna, Ellesmere, Eltham, Eyre, Featherston, Franklin, Golden Bay, Great Barrier Island, Grey, Hauraki Plains, Hawera*, Hawke's Bay, Heathcote, Hikurangi**, Hobson, Hokianga, Horowhenua, Hurunui, Hutt, Inangahua, Inglewood, Kaikoura, Kairanga, Kiwitea, Lake, Mackenzie, Malvern, Manaia**, Manawatu, Mangonui, Maniototo, Marlborough, Masterton, Matamata, Mount Herbert, Ohinemuri, Opotiki, Oroua, Otamatea, Otorohanga*, Oxford, Pahiatua, Paparua, Patea, Piako, Pohangina, Raglan, Rangiora*, Rangitikei, Rodney, Rotorua*, Runanga, Saint Kilda, Silverpeaks, Southland, Stewart Island, Stratford, Strathallan, Taranaki, Taumarunui, Taupo, Tauranga, Thames-Coromandel*, Tuapeka, Vincent, Waiapu, Waiheke, Waihemo, Waikato, Waikohu, Waimairi, Waimarino, Waimate, Waimate West, Waimea, Waipa,

Waipawa*, Waipukurau*, Wairarapa South, Wairewa, Wairoa, Waitaki, Waitomo*, Waitotara, Wallace, Wanganui, Waverley**, Westland, Whakatane*, Whangarei, Whangaroa, Woodville

Dependent areas: Cook Islands, Niue, Tokelau

Independence: 26 September 1907 (from UK)

Constitution: no formal, written constitution; consists of various documents, including certain acts of the UK and New Zealand Parliaments; Constitution Act 1986 was to have come into force 1 January 1987, but has not been enacted

Legal system: based on English law, with special land legislation and land courts for Maoris; accepts compulsory ICJ jurisdiction, with reservations

National holiday: Waitangi Day, 6 February (1840) (Treaty of Waitangi established British sovereignty)

Political parties and leaders: National Party (NP; government), James BOLGER; New Zealand Labor Party (NZLP; opposition), Michael MOORE; NewLabor Party (NLP), Jim ANDERTON; Democratic Party, Dick RYAN; New Zealand Liberal Party, Hanmish MACINTYRE and Gilbert MYLES; Green Party, no official leader; Mana Motuhake, Martin RATA; Socialist Unity Party (SUP; pro-Soviet), Kenneth DOUGLAS

note: the New Labor, Democratic, and Mana Motuhake parties formed a coalition called the Alliance Party, Jim ANDERTON, president, in September 1991; the Green Party joined the coalition in May 1992

Suffrage: 18 years of age; universal

Elections:

House of Representatives: last held on 27 October 1990 (next to be held NA November 1993); results—NP 49%, NZLP 35%, Green Party 7%, NLP 5%; seats—(97 total) NP 67, NZLP 29, NLP 1

Executive branch: British monarch, governor general, prime minister, deputy prime minister, Cabinet

Legislative branch: unicameral House of Representatives (commonly called Parliament)

Judicial branch: High Court, Court of Appeal

Leaders:

Chief of State: Queen ELIZABETH II (since 6 February 1952), represented by Governor General Dame Catherine TIZARD (since 12 December 1990)

Head of Government: Prime Minister James BOLGER (since 29 October 1990); Deputy Prime Minister Donald McKINNON (since 2 November 1990)

Member of: ANZUS (US suspended security obligations to NZ on 11 August 1986), APEC, AsDB, Australia Group, C, CCC, CP, COCOM (cooperating country),

EBRD, ESCAP, FAO, GATT, IAEA, IBRD, ICAO, ICFTU, IDA, IEA, IFAD, IFC, ILO, IMF, IMO, INMARSAT, INTELSAT, INTERPOL, IOC, IOM (observer), ISO, ITU, LORCS, MTCR, NAM (guest), OECD, PCA, SPARTECA, SPC, SPF, UN, UNAVEM II, UNCTAD, UNESCO, UNIDO, UNOSOM, UNPROFOR, UNTAC, UNTSO, UPU, WHO, WIPO, WMO

Diplomatic representation in US:

chief of mission: Ambassador Denis Bazely Gordon McLEAN

chancery: 37 Observatory Circle NW, Washington, DC 20008

telephone: (202) 328-4800

consulates general: Los Angeles and New York

US diplomatic representation:

chief of mission: (vacant)

embassy: 29 Fitzherbert Terrace, Thorndon, Wellington

mailing address: P. O. Box 1190, Wellington; PSC 467, Box 1, FPO AP 96531-1001

telephone: [64] (4) 722-068

FAX: [64] (4) 723-537

consulate general: Auckland

Flag: blue with the flag of the UK in the upper hoist-side quadrant with four red five-pointed stars edged in white centered in the outer half of the flag; the stars represent the Southern Cross constellation

---

Economy

Overview: Since 1984 the government has been reorienting an agrarian economy dependent on a guaranteed British market to an open free market economy that can compete on the global scene. The government has hoped that dynamic growth would boost real incomes, broaden and deepen the technological capabilities of the industrial sector, reduce inflationary pressures, and permit the expansion of welfare benefits. The results have been mixed: inflation is down from double-digit levels, but growth was sluggish in 1988-91, and unemployment, always a highly sensitive issue, has exceeded 10% since May 1991. In 1992, growth picked up to 3%, a sign that the new economic approach is beginning to pay off.

National product: GDP—purchasing power equivalent—$49.8 billion (1992)

National product real growth rate: 3% (1992)

National product per capita: $14,900 (1992)

Inflation rate (consumer prices): 2.2% (1991)

Unemployment rate: 10.1% (September 1992)

Budget: revenues $14.0 billion; expenditures $15.2 billion, including capital expenditures of $NA (1992)

Exports: $3.65 billion (f.o.b., FY92)

commodities: wool, lamb, mutton, beef, fruit, fish, cheese, manufactures, chemicals, forestry products

partners: EC 18.3%, Japan 17.9%, Australia 17.5%, US 13.5%, China 3.6%, South Korea 3.1%

Imports: $3.99 billion (f.o.b., FY92)

commodities: petroleum, consumer goods, motor vehicles, industrial equipment

partners: Australia 19.7%, Japan 16.9%, EC 16.9%, US 15.3%, Taiwan 3.0%

External debt: $38.5 billion (September 1992)

Industrial production: growth rate 1.9% (1990); accounts for about 20% of GDP

Electricity: 8,000,000 kW capacity; 31,000 million kWh produced, 9,250 kWh per capita (1992)

Industries: food processing, wood and paper products, textiles, machinery, transportation equipment, banking and insurance, tourism, mining

Agriculture: accounts for about 9% of GDP and about 10% of the work force; livestock predominates—wool, meat, dairy products all export earners; crops—wheat, barley, potatoes, pulses, fruits, vegetables; surplus producer of farm products; fish catch reached a record 503,000 metric tons in 1988

Economic aid: donor—ODA and OOF commitments (1970-89), $526 million

Currency: 1 New Zealand dollar (NZ$) = 100 cents

Exchange rates: New Zealand dollars (NZ$) per US$1—1.9486 (January 1993), 1.8584 (1992), 1.7265 (1991), 1.6750 (1990), 1.6711 (1989), 1.5244 (1988)

Fiscal year: 1 July-30 June

Communications

Railroads: 4,716 km total; all 1.067-meter gauge; 274 km double track; 113 km electrified; over 99% government owned

Highways: 92,648 km total; 49,547 km paved, 43,101 km gravel or crushed stone

Inland waterways: 1,609 km; of little importance to transportation

Pipelines: natural gas 1,000 km; petroleum products 160 km; condensate (liquified petroleum gas—LPG) 150 km

Ports: Auckland, Christchurch, Dunedin, Wellington, Tauranga

Merchant marine: 18 ships (1,000 GRT or over) totaling 182,206 GRT/246,446 DWT; includes 2 cargo, 5 roll-on/roll-off, 1 railcar carrier, 4 oil tanker, 1 liquefied gas, 5 bulk

Airports:

total: 120

usable: 120

with permanent-surface runways: 33

with runways over 3,659 m: 1

with runways 2,440-3,659 m: 2

with runways 1,220-2,439 m: 42

New Zealand (continued)

Telecommunications: excellent international and domestic systems; 2,110,000 telephones; broadcast stations—64 AM, 2 FM, 14 TV; submarine cables extend to Australia and Fiji; 2 Pacific Ocean INTELSAT earth stations

Defense Forces

Branches: New Zealand Army, Royal New Zealand Navy, Royal New Zealand Air Force
Manpower availability: males age 15-49 878,028; fit for military service 741,104; reach military age (20) annually 29,319 (1993 est.)
Defense expenditures: exchange rate conversion—$792 million, 2% of GDP (FY90/91)

Nicaragua

Geography

Location: Central America, between Costa Rica and Honduras
Map references: Central America and the Caribbean, South America
Area:
total area: 129,494 km²
land area: 120,254 km²
comparative area: slightly larger than New York State
Land boundaries: total 1,231 km, Costa Rica 309 km, Honduras 922 km
Coastline: 910 km
Maritime claims:
contiguous zone: 25 nm security zone (status of claim uncertain)
continental shelf: not specified
territorial sea: 200 nm
International disputes: territorial disputes with Colombia over the Archipelago de San Andres y Providencia and Quita Sueno Bank; International Court of Justice (ICJ) referred the maritime boundary question in the Golfo de Fonseca to an earlier agreement in this century and advised that some tripartite resolution among El Salvador, Honduras and Nicaragua likely would be required
Climate: tropical in lowlands, cooler in highlands
Terrain: extensive Atlantic coastal plains rising to central interior mountains; narrow Pacific coastal plain interrupted by volcanoes
Natural resources: gold, silver, copper, tungsten, lead, zinc, timber, fish
Land use:
arable land: 9%
permanent crops: 1%
meadows and pastures: 43%
forest and woodland: 35%
other: 12%
Irrigated land: 850 km² (1989 est.)
Environment: subject to destructive earthquakes, volcanoes, landslides, and occasional severe hurricanes; deforestation; soil erosion; water pollution

People

Population: 3,987,240 (July 1993 est.)
Population growth rate: 2.74% (1993 est.)
Birth rate: 35.61 births/1,000 population (1993 est.)
Death rate: 6.94 deaths/1,000 population (1993 est.)
Net migration rate: -1.25 migrant(s)/1,000 population (1993 est.)
Infant mortality rate: 54.8 deaths/1,000 live births (1993 est.)
Life expectancy at birth:
total population: 63.5 years
male: 60.7 years
female: 66.41 years (1993 est.)
Total fertility rate: 4.48 children born/woman (1993 est.)
Nationality:
noun: Nicaraguan(s)
adjective: Nicaraguan
Ethnic divisions: mestizo 69%, white 17%, black 9%, Indian 5%
Religions: Roman Catholic 95%, Protestant 5%
Languages: Spanish (official)
note: English- and Indian-speaking minorities on Atlantic coast
Literacy: age 15 and over can read and write (1971)
total population: 57%
male: 57%
female: 57%
Labor force: 1.086 million
by occupation: service 43%, agriculture 44%, industry 13% (1986)

Government

Names:
conventional long form: Republic of Nicaragua
conventional short form: Nicaragua
local long form: Republica de Nicaragua
local short form: Nicaragua
Digraph: NU
Type: republic
Capital: Managua
Administrative divisions: 17 departments (departamentos, singular—departamento); Boaco, Carazo, Chinandega, Chontales, Esteli, Granada, Jinotega, Leon, Madriz, Managua, Masaya, Matagalpa, North Atlantic Coast Autonomous Zone (RAAN), Nueva Segovia, Rio San Juan, Rivas, South Atlantic Coast Autonomous Zone (RAAS)
Independence: 15 September 1821 (from Spain)
Constitution: January 1987
Legal system: civil law system; Supreme Court may review administrative acts
National holiday: Independence Day, 15 September (1821)
Political parties and leaders:
ruling coalition: National Opposition Union (UNO) is a 10-party alliance—moderate parties:

National Conservative Party (PNC), Silviano MATAMOROS Lacayo, president; Liberal Constitutionalist Party (PLC), Jose Ernesto SOMARRIBA, Arnold ALEMAN; Christian Democratic Union (UDC), Luis Humberto GUZMAN, Agustin JARQUIN, Azucena FERREY, Roger MIRANDA, Francisco MAYORGA; National Democratic Movement (MDN), Roberto URROZ; National Action Party (PAN), Duilio BALTODANO; hardline parties: Independent Liberal Party (PLI), Wilfredo NAVARRO, Virgilio GODOY Reyes; Social Democratic Party (PSD), Guillermo POTOY, Alfredo CESAR Aguirre, secretary general; Conservative Popular Alliance Party (PAPC), Myriam ARGUELLO; Communist Party of Nicaragua (PCN), Eli ALTAMIRANO Perez; Neo-Liberal Party (PALI), Adolfo GARCIA Esquivel

opposition parties: Sandinista National Liberation Front (FSLN), Daniel ORTEGA; Central American Unionist Party (PUCA), Blanca ROJAS; Democratic Conservative Party of Nicaragua (PCDN), Jose BRENES; Liberal Party of National Unity (PLUIN), Eduardo CORONADO; Movement of Revolutionary Unity (MUR), Francisco SAMPER; Social Christian Party (PSC), Erick RAMIREZ; Revolutionary Workers' Party (PRT), Bonifacio MIRANDA; Social Conservative Party (PSOC), Fernando AGUERRO; Popular Action Movement—Marxist-Leninist (MAP-ML), Isidro TELLEZ; Popular Social Christian Party (PPSC), Mauricio DIAZ

Other political or pressure groups: National Workers Front (FNT) is a Sandinista umbrella group of eight labor unions: Sandinista Workers' Central (CST); Farm Workers Association (ATC); Health Workers Federation (FETASALUD); National Union of Employees (UNE); National Association of Educators of Nicaragua (ANDEN); Union of Journalists of Nicaragua (UPN); Heroes and Martyrs Confederation of Professional Associations (CONAPRO); and the National Union of Farmers and Ranchers (UNAG); Permanent Congress of Workers (CPT) is an umbrella group of four non-Sandinista labor unions: Confederation of Labor Unification (CUS); Autonomous Nicaraguan Workers' Central (CTN-A); Independent General Confederation of Labor (CGT-I); and Labor Action and Unity Central (CAUS); Nicaraguan Workers' Central (CTN) is an independent labor union; Superior Council of Private Enterprise (COSEP) is a confederation of business groups

Suffrage: 16 years of age; universal

Elections:

President: last held on 25 February 1990 (next to be held February 1996); results—Violeta Barrios de CHAMORRO (UNO) 54.7%, Daniel ORTEGA Saavedra (FSLN) 40.8%, other 4.5%

National Assembly: last held on 25 February 1990 (next to be held February 1996); results—UNO 53.9%, FSLN 40.8%, PSC 1.6%, MUR 1.0%; seats—(92 total) UNO 42, FSLN 39, PSC 1, MUR 1, "Centrist" (Dissident UNO) 9

Executive branch: president, vice president, Cabinet

Legislative branch: unicameral National Assembly (Asamblea Nacional)

Judicial branch: Supreme Court (Corte Suprema)

Leaders:

Chief of State and Head of Government: President Violeta Barrios de CHAMORRO (since 25 April 1990); Vice President Virgilio GODOY Reyes (since 25 April 1990)

Member of: BCIE, CACM, ECLAC, FAO, G-77, GATT, IADB, IAEA, IBRD, ICAO, ICFTU, IDA, IFAD, IFC, ILO, IMF, IMO, INTELSAT, INTERPOL, IOC, IOM, ITU, LAES, LAIA (observer), LORCS, NAM, OAS, OPANAL, PCA, UN, UNCTAD, UNESCO, UNHCR, UNIDO, UPU, WCL, WFTU, WHO, WIPO, WMO, WTO

Diplomatic representation in US:

chief of mission: Ambassador Roberto MAYORGA (since January 1993)

chancery: 1627 New Hampshire Avenue NW, Washington, DC 20009

telephone: (202) 939-6570

US diplomatic representation:

chief of mission: Charge d'Affaires Ronald GODARD

embassy: Kilometer 4.5 Carretera Sur., Managua

mailing address: APO AA 34021

telephone: [505] (2) 666010 or 666013, 666015 through 18, 666026, 666027, 666032 through 34

FAX: [505] (2) 666046

Flag: three equal horizontal bands of blue (top), white, and blue with the national coat of arms centered in the white band; the coat of arms features a triangle encircled by the words REPUBLICA DE NICARAGUA on the top and AMERICA CENTRAL on the bottom; similar to the flag of El Salvador, which features a round emblem encircled by the words REPUBLICA DE EL SALVADOR EN LA AMERICA CENTRAL centered in the white band; also similar to the flag of Honduras, which has five blue stars arranged in an X pattern centered in the white band

---

Economy

Overview: Government control of the economy historically has been extensive, although the CHAMORRO government has pledged to greatly reduce intervention. Four private banks have been licensed, and the government has liberalized foreign trade and abolished price controls on most goods. In early 1993, fewer than 50% of the agricultural and industrial firms remain state owned. Sandinista economic policies and the war had produced a severe economic crisis. The foundation of the economy continues to be the export of agricultural commodities, largely coffee and cotton. Farm production fell by roughly 7% in 1989 and 4% in 1990, and remained about even in 1991-92. The agricultural sector employs 44% of the work force and accounts for 15% of GDP and 80% of export earnings. Industry, which employs 13% of the work force and contributes about 25% to GDP, showed a drop of 7% in 1989, fell slightly in 1990, and remained flat in 1991-92; output still is below pre-1979 levels. External debt is one of the highest in the world on a per capita basis. In 1992 the inflation rate was 8%, down sharply from the 766% of 1991.

National product: GDP—exchange rate conversion—$1.7 billion (1992 est.)

National product real growth rate: 0.5% (1992 est.)

National product per capita: $425 (1992 est.)

Inflation rate (consumer prices): 8% (1992)

Unemployment rate: 13% underemployment 50% (1991)

Budget: revenues $347 million; expenditures $499 million, including capital expenditures of $NA million (1991)

Exports: $280 million (f.o.b., 1992 est.)

commodities: coffee, cotton, sugar, bananas, seafood, meat, chemicals

partners: OECD 75%, USSR and Eastern Europe 15%, other 10%

Imports: $720 million (c.i.f., 1992 est.)

commodities: petroleum, food, chemicals, machinery, clothing

partners: Latin America 30%, US 25%, EC 20%, USSR and Eastern Europe 10%, other 15% (1990 est.)

External debt: $10 billion (December 1991)

Industrial production: growth rate NA%; accounts for about 25% of GDP

Electricity: 434,000 kW capacity; 1,118 million kWh produced, 290 kWh per capita (1992)

Industries: food processing, chemicals, metal products, textiles, clothing, petroleum refining and distribution, beverages, footwear

Agriculture: accounts for 15% of GDP and 44% of work force; cash crops—coffee, bananas, sugarcane, cotton; food crops—rice, corn, cassava, citrus fruit, beans; variety of animal products—beef, veal, pork, poultry, dairy; normally self-sufficient in food

Illicit drugs: minor transshipment point for cocaine destined for the US

Nicaragua (continued)

Economic aid: US commitments, including Ex-Im (FY70-89), $294 million; Western (non-US) countries, ODA and OOF bilateral commitments (1970-89), $1,381 million; Communist countries (1970-89), $3.5 billion
Currency: 1 cordoba (C$) = 100 centavos
Exchange rates: cordobas (C$) per US$1—6 (10 January 1993), 25,000,000 (March 1992), 21,354,000 (1991), 15,655 (1989), 270 (1988), 102.60 (1987); note—new gold cordoba issued in 1992
Fiscal year: calendar year

Communications

Railroads: 373 km 1.067-meter narrow gauge, government owned; majority of system not operating; 3 km 1.435-meter gauge line at Puerto Cabezas (does not connect with mainline)
Highways: 25,930 km total; 4,000 km paved, 2,170 km gravel or crushed stone, 5,425 km earth or graded earth, 14,335 km unimproved; Pan-American highway 368.5 km
Inland waterways: 2,220 km, including 2 large lakes
Pipelines: crude oil 56 km
Ports: Corinto, El Bluff, Puerto Cabezas, Puerto Sandino, Rama
Merchant marine: 2 cargo ships (1,000 GRT or over) totaling 2,161 GRT/2,500 DWT
Airports:
total: 226
usable: 151
with permanent-surface runways: 11
with runways over 3,659 m: 0
with runways 2,440-3,659 m: 2
with runways 1,220-2,439 m: 12
Telecommunications: low-capacity radio relay and wire system being expanded; connection into Central American Microwave System; 60,000 telephones; broadcast stations—45 AM, no FM, 7 TV, 3 shortwave; earth stations—1 Intersputnik and 1 Atlantic Ocean INTELSAT

Defense Forces

Branches: Ground Forces, Navy, Air Force
Manpower availability: males age 15-49 911,397; fit for military service 561,448; reach military age (18) annually 44,226 (1993 est.)
Defense expenditures: exchange rate conversion—$40 million, 2.7% of GDP (1992 budget)

Niger

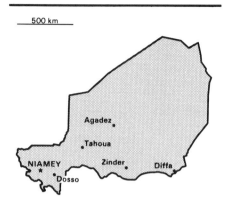

Geography

Location: Western Africa, between Algeria and Nigeria
Map references: Africa, Standard Time Zones of the World
Area:
total area: 1.267 million km²
land area: 1,266,700 km²
comparative area: slightly less than twice the size of Texas
Land boundaries: total 5,697 km, Algeria 956 km, Benin 266 km, Burkina 628 km, Chad 1,175 km, Libya 354 km, Mali 821 km, Nigeria 1,497 km
Coastline: 0 km (landlocked)
Maritime claims: none; landlocked
International disputes: Libya claims about 19,400 km² in northern Niger; demarcation of international boundaries in Lake Chad, the lack of which has led to border incidents in the past, is completed and awaiting ratification by Cameroon, Chad, Niger, and Nigeria; Burkina and Mali are proceeding with boundary demarcation, including the tripoint with Niger
Climate: desert; mostly hot, dry, dusty; tropical in extreme south
Terrain: predominately desert plains and sand dunes; flat to rolling plains in south; hills in north
Natural resources: uranium, coal, iron ore, tin, phosphates
Land use:
arable land: 3%
permanent crops: 0%
meadows and pastures: 7%
forest and woodland: 2%
other: 88%
Irrigated land: 320 km² (1989 est.)
Environment: recurrent drought and desertification severely affecting marginal agricultural activities; overgrazing; soil erosion
Note: landlocked

People

Population: 8,337,352 (July 1993 est.)
Population growth rate: 3.49% (1993 est.)
Birth rate: 57.35 births/1,000 population (1993 est.)
Death rate: 22.44 deaths/1,000 population (1993 est.)
Net migration rate: 0 migrant(s)/1,000 population (1993 est.)
Infant mortality rate: 112.8 deaths/1,000 live births (1993 est.)
Life expectancy at birth:
total population: 44.15 years
male: 42.6 years
female: 45.75 years (1993 est.)
Total fertility rate: 7.35 children born/woman (1993 est.)
Nationality:
noun: Nigerien(s)
adjective: Nigerien
Ethnic divisions: Hausa 56%, Djerma 22%, Fula 8.5%, Tuareg 8%, Beri Beri (Kanouri) 4.3%, Arab, Toubou, and Gourmantche 1.2%, about 4,000 French expatriates
Religions: Muslim 80%, remainder indigenous beliefs and Christians
Languages: French (official), Hausa, Djerma
Literacy: age 15 and over can read and write (1990)
total population: 28%
male: 40%
female: 17%
Labor force: 2.5 million wage earners (1982)
by occupation: agriculture 90%, industry and commerce 6%, government 4%
note: 51% of population of working age (1985)

Government

Names:
conventional long form: Republic of Niger
conventional short form: Niger
local long form: Republique du Niger
local short form: Niger
Digraph: NG
Type: transition government as of November 1991, appointed by national reform conference; scheduled to turn over power to democratically elected government in March 1993
Capital: Niamey
Administrative divisions: 7 departments (departementes, singular—departement); Agadez, Diffa, Dosso, Maradi, Niamey, Tahoua, Zinder
Independence: 3 August 1960 (from France)
Constitution: December 1989 constitution revised November 1991 by National Democratic Reform Conference

Legal system: based on French civil law system and customary law; has not accepted compulsory ICJ jurisdiction
National holiday: Republic Day, 18 December (1958)
Political parties and leaders: National Movement of the Development Society (MNSD-NASSARA), Tandja MAMADOU; Niger Progressive Party—African Democratic Rally (PPN-RDA), Harou KOUKA; Union of Popular Forces for Democracy and Progress (UDFP-SAWABA), Djibo BAKARY; Niger Democratic Union (UDN-SAWABA), Mamoudou PASCAL; Union of Patriots, Democrats, and Progressives (UPDP), Andre SALIFOU; other parties forming
Suffrage: 18 years of age; universal
Elections:
President: President Ali SAIBOU has been in office since December 1989, but the presidency is now a largely ceremonial position
National Assembly: last held 10 December 1989 (next to be held NA); results—MNSD was the only party; seats—(150 total) MNSD 150 (indirectly elected); note—Niger held a national conference from July to November 1991 to decide upon a transitional government and an agenda for multiparty elections
Executive branch: president (ceremonial), prime minister, Cabinet
Legislative branch: unicameral National Assembly
Judicial branch: State Court (Cour d'Etat), Court of Appeal (Cour d'Apel)
Leaders:
Chief of State: President Brig. Gen. Ali SAIBOU (since 14 November 1987); ceremonial post since national conference (1991)
Head of Government: Prime Minister Amadou CHEIFFOU (since NA November 1991)
Member of: ACCT, ACP, AfDB, CCC, CEAO, ECA, ECOWAS, Entente, FAO, FZ, G-77, GATT, IAEA, IBRD, ICAO, IDA, IDB, IFAD, IFC, ILO, IMF, INTELSAT, INTERPOL, IOC, ITU, LORCS, NAM, OAU, OIC, UN, UNCTAD, UNESCO, UNIDO, UPU, WADB, WCL, WHO, WIPO, WMO, WTO
Diplomatic representation in US:
chief of mission: Ambassador Adamou SEYDOU
chancery: 2204 R Street NW, Washington, DC 20008
telephone: (202) 483-4224 through 4227
US diplomatic representation:
chief of mission: Ambassador Jennifer C. WARD
embassy: Avenue des Ambassades, Niamey
mailing address: B. P. 11201, Niamey
telephone: [227] 72-26-61 through 64

FAX: [227] 73-31-67
Flag: three equal horizontal bands of orange (top), white, and green with a small orange disk (representing the sun) centered in the white band; similar to the flag of India, which has a blue spoked wheel centered in the white band

---

Economy

Overview: About 90% of the population is engaged in farming and stock raising, activities that generate almost half the national income. The economy also depends heavily on exploitation of large uranium deposits. Uranium production grew rapidly in the mid-1970s, but tapered off in the early 1980s when world prices declined. France is a major customer, while Germany, Japan, and Spain also make regular purchases. The depressed demand for uranium has contributed to an overall sluggishness in the economy, a severe trade imbalance, and a mounting external debt.
National product: GDP—exchange rate conversion—$2.3 billion (1991 est.)
National product real growth rate: 1.9% (1991 est.)
National product per capita: $290 (1991 est.)
Inflation rate (consumer prices): 1.3% (1991 est.)
Unemployment rate: NA%
Budget: revenues $193 million; expenditures $355 million, including capital expenditures of $106 million (1991 est.)
Exports: $294 million (f.o.b., 1991)
commodities: uranium ore 60%, livestock products 20%, cowpeas, onions
partners: France 77%, Nigeria 8%, Cote d'Ivoire, Italy
Imports: $346 million (c.i.f., 1991)
commodities: primary materials, machinery, vehicles and parts, electronic equipment, cereals, petroleum products, pharmaceuticals, chemical products, foodstuffs
partners: Germany 26%, Cote d'Ivoire 11%, France 5%, Italy 4%, Nigeria 2%
External debt: $1.2 billion (December 1991 est.)
Industrial production: growth rate −2.7% (1991 est.); accounts for 13% of GDP
Electricity: 105,000 kW capacity; 230 million kWh produced, 30 kWh per capita (1991)
Industries: cement, brick, textiles, food processing, chemicals, slaughterhouses, and a few other small light industries; uranium mining began in 1971
Agriculture: accounts for roughly 40% of GDP and 90% of labor force; cash crops—cowpeas, cotton, peanuts; food crops—millet, sorghum, cassava, rice; livestock—cattle, sheep, goats; self-sufficient in food except in drought years

Economic aid: US commitments, including Ex-Im (FY70-89), $380 million; Western (non-US) countries, ODA and OOF bilateral commitments (1970-89), $3,165 million; OPEC bilateral aid (1979-89), $504 million; Communist countries (1970-89), $61 million
Currency: 1 CFA franc (CFAF) = 100 centimes
Exchange rates: Communaute Financiere Africaine francs (CFAF) per US$1—274.06 (January 1993), 264.69 (1992), 282.11 (1991), 272.26 (1990), 319.01 (1989), 297.85 (1988)
Fiscal year: 1 October-30 September

Communications

Highways: 39,970 km total; 3,170 km bituminous, 10,330 km gravel and laterite, 3,470 km earthen, 23,000 km tracks
Inland waterways: Niger River is navigable 300 km from Niamey to Gaya on the Benin frontier from mid-December through March
Airports:
total: 28
usable: 26
with permanent-surface runways: 9
with runways over 3,659 m: 0
with runways 2,440-3,659 m: 2
with runways 1,220-2,439 m: 13
Telecommunications: small system of wire, radiocommunications, and radio relay links concentrated in southwestern area; 14,260 telephones; broadcast stations—15 AM, 5 FM, 18 TV; satellite earth stations—1 Atlantic Ocean INTELSAT, 1 Indian Ocean INTELSAT, and 3 domestic, with 1 planned

---

Defense Forces

Branches: Army, Air Force, Gendarmerie, National Police, Republican Guard
Manpower availability: males age 15-49 1,784,966; fit for military service 961,593; reach military age (18) annually 87,222 (1993 est.)
Defense expenditures: exchange rate conversion—$27 million, 1.3% of GDP (1989)

Nigeria

Geography

Location: Western Africa, bordering the North Atlantic Ocean between Benin and Cameroon
Map references: Africa, Standard Time Zones of the World
Area:
total area: 923,770 km²
land area: 910,770 km²
comparative area: slightly more than twice the size of California
Land boundaries: total 4,047 km, Benin 773 km, Cameroon 1,690 km, Chad 87 km, Niger 1,497 km
Coastline: 853 km
Maritime claims:
continental shelf: 200 m depth or to depth of exploitation
exclusive economic zone: 200 nm
territorial sea: 30 nm
International disputes: demarcation of international boundaries in Lake Chad, the lack of which has led to border incidents in the past, is completed and awaiting ratification by Cameroon, Chad, Niger, and Nigeria; boundary commission, created with Cameroon to discuss unresolved land and maritime boundaries, has not yet convened
Climate: varies; equatorial in south, tropical in center, arid in north
Terrain: southern lowlands merge into central hills and plateaus; mountains in southeast, plains in north
Natural resources: petroleum, tin, columbite, iron ore, coal, limestone, lead, zinc, natural gas
Land use:
arable land: 31%
permanent crops: 3%
meadows and pastures: 23%
forest and woodland: 15%
other: 28%
Irrigated land: 8,650 km² (1989 est.)
Environment: recent droughts in north severely affecting marginal agricultural activities; desertification; soil degradation, rapid deforestation

People

Population: 95,060,430 (July 1993 est.)
Population growth rate: 3.13% (1993 est.)
Birth rate: 43.8 births/1,000 population (1993 est.)
Death rate: 12.85 deaths/1,000 population (1993 est.)
Net migration rate: 0.37 migrant(s)/1,000 population (1993 est.)
Infant mortality rate: 77.3 deaths/1,000 live births (1993 est.)
Life expectancy at birth:
total population: 54.7 years
male: 53.54 years
female: 55.88 years (1993 est.)
Total fertility rate: 6.43 children born/woman (1993 est.)
Nationality:
noun: Nigerian(s)
adjective: Nigerian
Ethnic divisions:
north: Hausa and Fulani
southwest: Yoruba
southeast: Ibos non-Africans 27,000
note: Hausa and Fulani, Yoruba, and Ibos together make up 65% of population
Religions: Muslim 50%, Christian 40%, indigenous beliefs 10%
Languages: English (official), Hausa, Yoruba, Ibo, Fulani
Literacy: age 15 and over can read and write (1990)
total population: 51%
male: 62%
female: 40%
Labor force: 42.844 million
by occupation: agriculture 54%, industry, commerce, and services 19%, government 15%
note: 49% of population of working age (1985)

Government

Names:
conventional long form: Federal Republic of Nigeria
conventional short form: Nigeria
Digraph: NI
Type: military government since 31 December 1983; plans to turn over power to elected civilians in August 1993
Capital: Abuja
note: on 12 December 1991 the capital was officially moved from Lagos to Abuja; many government offices remain in Lagos pending completion of facilities in Abuja
Administrative divisions: 30 states and 1 territory*; Abia, Abuja Capital Territory*, Adamawa, Akwa Ibom, Anambra, Bauchi, Benue, Borno, Cross River, Delta, Edo, Enugu, Imo, Jigawa, Kaduna, Kano, Katsina, Kebbi, Kogi, Kwara, Lagos, Niger, Ogun, Ondo, Osun, Oyo, Plateau, Rivers, Sokoto, Taraba, Yobe

Independence: 1 October 1960 (from UK)
Constitution: 1 October 1979, amended 9 February 1984, revised 1989
Legal system: based on English common law, Islamic law, and tribal law
National holiday: Independence Day, 1 October (1960)
Political parties and leaders: Social Democratic Party (SDP), Alhaji Baba Gana KINGIBE, chairman; National Republican Convention (NRC), Chief Tom IKIMI, chairman
note: these are the only two political parties, and they were established by the government in 1989
Suffrage: 21 years of age; universal
Elections:
President: first presidential elections since the 31 December 1983 coup scheduled for June 1993
Senate: last held 4 July 1992 (next to be held NA 1996); results—percent of vote by party NA; seats—(total 84) SDP 47, NRC 37
House of Representatives: last held 4 July 1992 (next to be held NA 1996); results—percent of vote by party NA; seats—(total 577) SDP 310, NRC 267
Executive branch: president, vice-president, cabinet
Legislative branch: bicameral National Assembly consists of an upper house or Senate and a lower house or House of Representatives
Judicial branch: Supreme Court, Federal Court of Appeal
Leaders:
Chief of State and Head of Government: President and Commander in Chief of Armed Forces Gen. Ibrahim BABANGIDA (since 27 August 1985); Vice-President Admiral (Ret.) Augustus AIKHOMU (since 30 August 1990)
Member of: ACP, AfDB, C, CCC, ECA, ECOWAS, FAO, G-15, G-19, G-24, G-77, GATT, IAEA, IBRD, ICAO, ICC, IDA, IFAD, IFC, ILO, IMO, IMF, INMARSAT, INTELSAT, INTERPOL, IOC, ISO, ITU, LORCS, MINURSO, NAM, OAU, OIC, OPEC, PCA, UN, UNAVEM, UNCTAD, UNESCO, UNHCR, UNIDO, UNIKOM, UNPROFOR, UPU, WCL, WHO, WIPO, WMO, WTO
Diplomatic representation in US:
chief of mission: Ambassador Zubair Mahmud KAZAURE
chancery: 2201 M Street NW, Washington, DC 20037
telephone: (202) 822-1500
consulate general: New York
US diplomatic representation:
chief of mission: Ambassador William L. SWING
embassy: 2 Eleke Crescent, Lagos
mailing address: P. O. Box 554, Lagos

telephone: [234] (1) 610097
FAX: [234] (1) 610257
branch office: Abuja
consulate general: Kaduna
Flag: three equal vertical bands of green (hoist side), white, and green

---

Economy

Overview: Although Nigeria is Africa's leading oil-producing country, it remains poor with a $300 per capita GDP. In 1991-92 massive government spending, much of it to help ensure a smooth transition to civilian rule, ballooned the budget deficit and caused inflation and interest rates to rise. The lack of fiscal discipline forced the IMF to declare Nigeria not in compliance with an 18-month standby facility started in January 1991. Lagos has set ambitious targets for expanding oil production capacity and is offering foreign companies more attractive investment incentives. Government efforts to reduce Nigeria's dependence on oil exports and to sustain noninflationary growth, however, have fallen short because of inadequate new investment funds and endemic corruption. Living standards remain below the level of the early 1980s oil boom.
National product: GDP—exchange rate conversion—$35 billion (1992 est.)
National product real growth rate: 3.6% (1992 est.)
National product per capita: $300 (1992 est.)
Inflation rate (consumer prices): 60% (1992 est.)
Unemployment rate: 28% (1992 est.)
Budget: revenues $9 billion; expenditures $10.8 billion, including capital expenditures of $NA (1992 est.)
Exports: $12.7 billion (f.o.b., 1991)
commodities: oil 95%, cocoa, rubber
partners: EC countries 43%, US 41%
Imports: $7.8 billion (c.i.f., 1991)
commodities: consumer goods, capital equipment, chemicals, raw materials
partners: EC countries 70%, US 16%
External debt: $33.4 billion (1991)
Industrial production: growth rate 5.5% (1991); accounts for 8.5% of GDP
Electricity: 4,740,000 kW capacity; 8,300 million kWh produced, 70 kWh per capita (1991)
Industries: crude oil and mining—coal, tin, columbite; primary processing industries—palm oil, peanut, cotton, rubber, wood, hides and skins; manufacturing industries—textiles, cement, building materials, food products, footwear, chemical, printing, ceramics, steel
Agriculture: accounts for 32% of GDP and half of labor force; inefficient small-scale farming dominates; once a large net exporter of food and now an importer; cash crops—cocoa, peanuts, palm oil, rubber; food crops—corn, rice, sorghum, millet, cassava, yams; livestock—cattle, sheep, goats, pigs; fishing and forestry resources extensively exploited
Illicit drugs: passenger and cargo air hub for West Africa facilitates Nigeria's position as a major transit country for heroin en route from Southeast and Southwest Asia via Africa to Western Europe and North America; increasingly a transit route for cocaine from South America intended for West European and North American markets (some of that cocaine is also consumed in Nigeria)
Economic aid: US commitments, including Ex-Im (FY70-89), $705 million; Western (non-US) countries, ODA and OOF bilateral commitments (1970-89), $3.0 billion; Communist countries (1970-89), $2.2 billion
Currency: 1 naira (N) = 100 kobo
Exchange rates: naira (N) per US$1—19.661 (December 1992), 17.298 (1992), 9.909 (1991), 8.038 (1990), 7.3647 (1989), 4.5370 (1988), 4.0160 (1987)
Fiscal year: calendar year

Communications

Railroads: 3,505 km 1.067-meter gauge
Highways: 107,990 km total 30,019 km paved (mostly bituminous-surface treatment); 25,411 km laterite, gravel, crushed stone, improved earth; 52,560 km unimproved
Inland waterways: 8,575 km consisting of Niger and Benue Rivers and smaller rivers and creeks
Pipelines: crude oil 2,042 km; natural gas 500 km; petroleum products 3,000 km
Ports: Lagos, Port Harcourt, Calabar, Warri, Onne, Sapele
Merchant marine: 28 ships (1,000 GRT or over) totaling 418,046 GRT/664,949 DWT; includes 17 cargo, 1 refrigerated cargo, 1 roll-on/roll-off, 7 oil tanker, 1 chemical tanker, 1 bulk
Airports:
total: 76
usable: 63
with permanent-surface runways: 34
with runways over 3,659 m: 1
with runways 2,440-3,659 m: 15
with runways 1,220-2,439 m: 23
Telecommunications: above-average system limited by poor maintenance; major expansion in progress; radio relay microwave and cable routes; broadcast stations—35 AM, 17 FM, 28 TV; satellite earth stations—2 Atlantic Ocean INTELSAT, 1 Indian Ocean INTELSAT, 20 domestic stations; 1 coaxial submarine cable

---

Defense Forces

Branches: Army, Navy, Air Force, National Guard, paramilitary Police Force
Manpower availability: males age 15-49 21,790,956; fit for military service 12,447,547; reach military age (18) annually 1,297,790 (1993 est.)
Defense expenditures: exchange rate conversion—$172 million, about 1% of GDP (1992)

Niue
(free association with New Zealand)

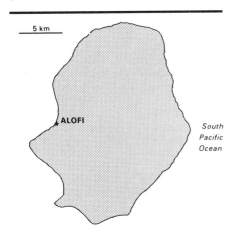

Geography

Location: Oceania, 460 km east of Tonga in the South Pacific Ocean
Map references: Oceania
Area:
total area: 260 km²
land area: 260 km²
comparative area: slightly less than 1.5 times the size of Washington, DC
Land boundaries: 0 km
Coastline: 64 km
Maritime claims:
exclusive economic zone: 200 nm
territorial sea: 12 nm
International disputes: none
Climate: tropical; modified by southeast trade winds
Terrain: steep limestone cliffs along coast, central plateau
Natural resources: fish, arable land
Land use:
arable land: 61%
permanent crops: 4%
meadows and pastures: 4%
forest and woodland: 19%
other: 12%
Irrigated land: NA km²
Environment: subject to typhoons
Note: one of world's largest coral islands

People

Population: 1,977 (July 1993 est.)
Population growth rate: -3.66% (1993 est.)
Birth rate: NA births/1,000 population
Death rate: NA deaths/1,000 population
Net migration rate: NA migrant(s)/1,000 population
Infant mortality rate: NA deaths/1,000 live births
Life expectancy at birth:
total population: NA years
male: NA years
female: NA years
Total fertility rate: NA children born/woman

Nationality:
noun: Niuean(s)
adjective: Niuean
Ethnic divisions: Polynesian (with some 200 Europeans, Samoans, and Tongans)
Religions: Ekalesia Nieue (Niuean Church) 75%—a Protestant church closely related to the London Missionary Society, Morman 10%, other 15% (mostly Roman Catholic, Jehovah's Witnesses, Seventh-Day Adventist)
Languages: Polynesian closely related to Tongan and Samoan, English
Literacy:
total population: NA%
male: NA%
female: NA%
Labor force: 1,000 (1981 est.)
by occupation: most work on family plantations; paid work exists only in government service, small industry, and the Niue Development Board

Government

Names:
conventional long form: none
conventional short form: Niue
Digraph: NE
Type: self-governing territory in free association with New Zealand; Niue fully responsible for internal affairs; New Zealand retains responsibility for external affairs
Capital: Alofi
Administrative divisions: none
Independence: 19 October 1974 (became a self-governing territory in free association with New Zealand on 19 October 1974)
Constitution: 19 October 1974 (Niue Constitution Act)
Legal system: English common law
National holiday: Waitangi Day, 6 February (1840) (Treaty of Waitangi established British sovereignty)
Political parties and leaders: Niue Island Party (NIP), Young VIVIAN
Suffrage: 18 years of age; universal
Elections:
Legislative Assembly: last held on 8 April 1990 (next to be held March 1993); results—percent of vote NA; seats—(20 total, 6 elected) NIP 1, independents 5
Executive branch: British monarch, premier, Cabinet
Legislative branch: unicameral Legislative Assembly
Judicial branch: Appeal Court of New Zealand, High Court
Leaders:
Chief of State: Queen ELIZABETH II (since 6 February 1952), represented by New Zealand Representative John SPRINGFORD (since NA 1974)
Head of Government: Acting Premier Young VIVIAN (since the death of Sir Robert R. REX on 12 December 1992)

Member of: ESCAP (associate), SPARTECA, SPC, SPF
Diplomatic representation in US: none (self-governing territory in free association with New Zealand)
US diplomatic representation: none (self-governing territory in free association with New Zealand)
Flag: yellow with the flag of the UK in the upper hoist-side quadrant; the flag of the UK bears five yellow five-pointed stars—a large one on a blue disk in the center and a smaller one on each arm of the bold red cross

Economy

Overview: The economy is heavily dependent on aid from New Zealand. Government expenditures regularly exceed revenues, with the shortfall made up by grants from New Zealand—the grants are used to pay wages to public employees. The agricultural sector consists mainly of subsistence gardening, although some cash crops are grown for export. Industry consists primarily of small factories to process passion fruit, lime oil, honey, and coconut cream. The sale of postage stamps to foreign collectors is an important source of revenue. The island in recent years has suffered a serious loss of population because of migration of Niueans to New Zealand.
National product: GNP—exchange rate conversion—$2.1 million (1989 est.)
National product real growth rate: NA%
National product per capita: $1,000 (1989 est.)
Inflation rate (consumer prices): 9.6% (1984)
Unemployment rate: NA%
Budget: revenues $5.5 million; expenditures $6.3 million, including capital expenditures of $NA (FY85 est.)
Exports: $175,274 (f.o.b., 1985)
commodities: canned coconut cream, copra, honey, passion fruit products, pawpaw, root crops, limes, footballs, stamps, handicrafts
partners: NZ 89%, Fiji, Cook Islands, Australia
Imports: $3.8 million (c.i.f., 1985)
commodities: food, live animals, manufactured goods, machinery, fuels, lubricants, chemicals, drugs
partners: NZ 59%, Fiji 20%, Japan 13%, Western Samoa, Australia, US
External debt: $NA
Industrial production: growth rate NA%
Electricity: 1,500 kW capacity; 3 million kWh produced, 1,490 kWh per capita (1990)
Industries: tourist, handicrafts, coconut products
Agriculture: coconuts, passion fruit, honey, limes; subsistence crops—taro, yams, cassava (tapioca), sweet potatoes; pigs, poultry, beef cattle

Economic aid: Western (non-US) countries, ODA and OOF bilateral commitments (1970-89), $62 million
Currency: 1 New Zealand dollar (NZ$) = 100 cents
Exchange rates: New Zealand dollars (NZ$) per US$1—1.9486 (January 1993), 1.8584 (1992), 1.7265 (1991), 1.6750 (1990), 1.6711 (1989), 1.5244 (1988)
Fiscal year: 1 April-31 March

Communications

Highways: 123 km all-weather roads, 106 km access and plantation roads
Ports: none; offshore anchorage only
Airports:
total: 1
useable: 1
with permanent-surface runways: 1
with runways over 3,659 m: 0
with runways 2,440-3,659 m: 0
with runways 1,220-2,439 m: 1
Telecommunications: single-line telephone system connects all villages on island; 383 telephones; 1,000 radio receivers (1987 est.); broadcast stations—1 AM, 1 FM, no TV

Defense Forces

Branches: Police Force
Note: defense is the responsibility of New Zealand

Norfolk Island
(territory of Australia)

Geography

Location: Oceania, 1,575 km east of Australia in the South Pacific Ocean
Map references: Oceania
Area:
total area: 34.6 km^2
land area: 34.6 km^2
comparative area: about 0.2 times the size of Washington, DC
Land boundaries: 0 km
Coastline: 32 km
Maritime claims:
exclusive fishing zone: 200 nm
territorial sea: 3 nm
International disputes: none
Climate: subtropical, mild, little seasonal temperature variation
Terrain: volcanic formation with mostly rolling plains
Natural resources: fish
Land use:
arable land: 0%
permanent crops: 0%
meadows and pastures: 25%
forest and woodland: 0%
other: 75%
Irrigated land: NA km^2
Environment: subject to typhoons (especially May to July)

People

Population: 2,665 (July 1993 est.)
Population growth rate: 1.69% (1993 est.)
Birth rate: NA births/1,000 population
Death rate: NA deaths/1,000 population
Net migration rate: NA migrant(s)/1,000 population
Infant mortality rate: NA deaths/1,000 live births
Life expectancy at birth:
total population: NA years
male: NA years
female: NA years
Total fertility rate: NA children born/woman

Nationality:
noun: Norfolk Islander(s)
adjective: Norfolk Islander(s)
Ethnic divisions: descendants of the Bounty mutineers, Australian, New Zealander
Religions: Anglican 39%, Roman Catholic 11.7%, Uniting Church in Australia 16.4%, Seventh-Day Adventist 4.4%, none 9.2%, unknown 16.9%, other 2.4% (1986)
Languages: English (official), Norfolk a mixture of 18th century English and ancient Tahitian
Literacy:
total population: NA%
male: NA%
female: NA%
Labor force: NA

Government

Names:
conventional long form: Territory of Norfolk Island
conventional short form: Norfolk Island
Digraph: NF
Type: territory of Australia
Capital: Kingston (administrative center); Burnt Pine (commercial center)
Administrative divisions: none (territory of Australia)
Independence: none (territory of Australia)
Constitution: Norfolk Island Act of 1957
Legal system: wide legislative and executive responsibility under the Norfolk Island Act of 1979; Supreme Court
National holiday: Pitcairners Arrival Day Anniversary, 8 June (1856)
Political parties and leaders: NA
Suffrage: 18 years of age; universal
Elections:
Legislative Assembly: last held 1989 (held every three years); results—percent of vote by party NA; seats—(9 total) percent of seats by party NA
Executive branch: British monarch, governor general of Australia, administrator, Executive Council (cabinet)
Legislative branch: unicameral Legislative Assembly
Judicial branch: Supreme Court
Leaders:
Chief of State: Queen ELIZABETH II (since 6 February 1952), represented by Administrator A. G. KERR (since NA 1990), who is appointed by the Governor General of Australia
Head of Government: Assembly President and Chief Minister John Terence BROWN (since NA)
Member of: none
Diplomatic representation in US: none (territory of Australia)
US diplomatic representation: none (territory of Australia)

Norfolk Island (continued)

Flag: three vertical bands of green (hoist side), white, and green with a large green Norfolk Island pine tree centered in the slightly wider white band

Economy

Overview: The primary economic activity is tourism, which has brought a level of prosperity unusual among inhabitants of the Pacific Islands. The number of visitors has increased steadily over the years and reached 29,000 in FY89. Revenues from tourism have given the island a favorable balance of trade and helped the agricultural sector to become self-sufficient in the production of beef, poultry, and eggs.
National product: GDP $NA
National product real growth rate: NA%
National product per capita: $NA
Inflation rate (consumer prices): NA%
Unemployment rate: NA%
Budget: revenues $NA; expenditures $4.2 million, including capital expenditures of $400,000 (FY89)
Exports: $1.7 million (f.o.b., FY86)
commodities: postage stamps, seeds of the Norfolk Island pine and Kentia Palm, small quantities of avocados
partners: Australia, Pacific Islands, NZ, Asia, Europe
Imports: $15.6 million (c.i.f., FY86)
commodities: NA
partners: Australia, Pacific Islands, NZ, Asia, Europe
External debt: $NA
Industrial production: growth rate NA%
Electricity: 7,000 kW capacity; 8 million kWh produced, 3,160 kWh per capita (1990)
Industries: tourism
Agriculture: Norfolk Island pine seed, Kentia palm seed, cereals, vegetables, fruit, cattle, poultry
Economic aid: none
Currency: 1 Australian dollar ($A) = 100 cents
Exchange rates: Australian dollars ($A) per US$1—1.4837 (January 1993), 1.3600 (1992), 1.2835 (1991), 1.2799 (1990), 1.2618 (1989), 1.2752 (1988)
Fiscal year: 1 July-30 June

Communications

Highways: 80 km of roads, including 53 km paved; remainder are earth formed or coral surfaced
Ports: none; loading jetties at Kingston and Cascade
Airports:
total: 1
useable: 1
with permanent-surface runways : 1
with runways over 3,659 m: 0
with runways 2,440-3,659 m: 0

with runways 1,220-2,439 m: 1
Telecommunications: 1,500 radio receivers (1982); radio link service with Sydney; 987 telephones (1983); broadcast stations—1 AM, no FM, no TV

Defense Forces

Note: defense is the responsibility of Australia

Northern Mariana Islands
(commonwealth in political union with the US)

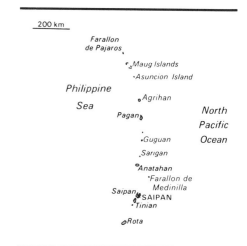

Geography

Location: in the North Pacific Ocean, 5,635 km west-southwest of Honolulu, about three-quarters of the way between Hawaii and the Philippines
Map references: Oceania
Area:
total area: 477 km²
land area: 477 km²
comparative area: slightly more than 2.5 times the size of Washington, DC
note: includes 14 islands including Saipan, Rota, and Tinian
Land boundaries: 0 km
Coastline: 1,482 km
Maritime claims:
contiguous zone: 24 nm
continental shelf: 200 m (depth)
exclusive economic zone: 200 nm
territorial sea: 12 nm
International disputes: none
Climate: tropical marine; moderated by northeast trade winds, little seasonal temperature variation; dry season December. to June, rainy season July to October
Terrain: southern islands are limestone with level terraces and fringing coral reefs; northern islands are volcanic; highest elevation is 471 meters (Mt. Okso' Takpochao on Saipan)
Natural resources: arable land, fish
Land use:
arable land: 5% on Saipan
permanent crops: NA%
meadows and pastures: 19%
forest and woodland: NA%
other: NA%
Irrigated land: NA km²
Environment: active volcanos on Pagan and Agrihan; subject to typhoons (most during August through November)
Note: strategic location in the North Pacific Ocean

People

Population: 48,581 (July 1993 est.)
Population growth rate: 3.04% (1993 est.)
Birth rate: 35.05 births/1,000 population (1993 est.)
Death rate: 4.61 deaths/1,000 population (1993 est.)
Net migration rate: 0 migrant(s)/1,000 population (1993 est.)
Infant mortality rate: 37.96 deaths/1,000 live births (1993 est.)
Life expectancy at birth:
total population: 67.43 years
male: 65.53 years
female: 69.48 years (1993 est.)
Total fertility rate: 2.69 children born/woman (1993 est.)
Nationality:
noun: NA
adjective: NA
Ethnic divisions: Chamorro, Carolinians and other Micronesians, Caucasian, Japanese, Chinese, Korean
Religions: Christian (Roman Catholic majority, although traditional beliefs and taboos may still be found)
Languages: English, Chamorro, Carolinian
note: 86% of population speaks a language other than English at home
Literacy: age NA and over can read and write (1980)
total population: 97%
male: 97%
female: 96%
Labor force: 7,476 total indigenous labor force, 2,699 unemployed; 21,188 foreign workers (1990)
by occupation: NA

Government

Names:
conventional long form: Commonwealth of the Northern Mariana Islands
conventional short form: Northern Mariana Islands
Digraph: CQ
Type: commonwealth in political union with the US; self-governing with locally elected governor, lieutenant governor, and legislature; federal funds to the Commonwealth administered by the US Department of the Interior, Office of Territorial and International Affairs
Capital: Saipan
Administrative divisions: none
Independence: none (commonwealth in political union with the US)
Constitution: Covenant Agreement effective 3 November 1986 and the constitution of the Commonwealth of the Northern Mariana Islands
Legal system: based on US system except for customs, wages, immigration laws, and taxation

National holiday: Commonwealth Day, 8 January (1978)
Political parties and leaders: Republican Party, Governor Lorenzo GUERRERO; Democratic Party, Carlos SHODA, chairman
Suffrage: 18 years of age; universal; indigenous inhabitants are US citizens but do not vote in US presidential elections
Elections:
Governor: last held in NA November 1989 (next to be held NA November 1993); results—Lorenzo I. DeLeon GUERRERO, Republican Party, was elected governor
Senate: last held NA November 1991 (next to be held NA November 1993); results—percent of vote by party NA; seats—(9 total) Republicans 6, Democrats 3
House of Representatives: last held NA November 1991 (next to be held NA November 1993); results—percent of vote by party NA; seats—(18 total) Republicans 10, Democrats 6, Independent 2
US House of Representatives: the Commonwealth does not have a nonvoting delegate in Congress; instead, it has an elected official "resident representative" located in Washington, DC; seats—(1 total) Republican (Juan N. BABAUTA)
Executive branch: US president; governor, lieutenant governor
Legislative branch: bicameral Legislature consists of an upper house or Senate and a lower house or House of Representatives
Judicial branch: Commonwealth Supreme Court, Superior Court, Federal District Court
Leaders:
Chief of State: President William Jefferson CLINTON (since 20 January 1993); Vice President Albert GORE, Jr. (since 20 January 1993)
Head of Government: Governor Lorenzo I. DeLeon GUERRERO (since 9 January 1990); Lieutenant Governor Benjamin T. MANGLONA (since 9 January 1990)
Member of: ESCAP (associate), SPC
Flag: blue with a white five-pointed star superimposed on the gray silhouette of a latte stone (a traditional foundation stone used in building) in the center

Economy

Overview: The economy benefits substantially from financial assistance from the US. The rate of funding has declined as locally generated government revenues have grown. An agreement for the years 1986 to 1992 entitled the islands to $228 million for capital development, government operations, and special programs. A rapidly growing major source of income is the tourist industry, which now employs about 50% of the work force. Japanese tourists predominate. The agricultural sector is made up of cattle ranches and small farms

producing coconuts, breadfruit, tomatoes, and melons. Industry is small scale, mostly handicrafts and light manufacturing.
National product: GNP—purchasing power equivalent—$541 million (1992)
note: GNP numbers reflect US spending
National product real growth rate: NA%
National product per capita: $11,500 (1992)
Inflation rate (consumer prices): 6.5-7.5% (1991 est.)
Unemployment rate: NA%
Budget: revenues $147.0 million; expenditures $127.7 million, including capital expenditures of $NA (1991)
Exports: $263.4 million (f.o.b. 1991 est.)
commodities: manufactured goods, garments, bread, pastries, concrete blocks, light iron work
partners: NA
Imports: $392.4 million (c.i.f. 1991 est.)
commodities: food, construction, equipment, materials
partners: NA
External debt: none
Industrial production: growth rate NA%
Electricity: 25,000 kW capacity; 35 million kWh produced, 740 kWh per capita (1990)
Industries: tourism, construction, light industry, handicrafts
Agriculture: coconuts, fruits, cattle, vegetables
Economic aid: none
Currency: US currency is used
Fiscal year: 1 October-30 September

Communications

Railroads: none
Highways: 381.5 km total; 134.5 km primary, 55 km secondary, 192 km local (1991)
Inland waterways: none
Ports: Saipan, Tinian
Airports:
total: 6
usable: 5
with permanent-surface runways: 3
with runways over 3,659 m: 0
with runways 2,440-3,659 m: 2
with runways 1,220-2,439 m: 2
Telecommunications: broadcast stations—2 AM, 1 FM (1984), 1 TV, 2 cable TV stations; 2 Pacific Ocean INTELSAT earth stations

Defense Forces

Note: defense is the responsibility of the US

Norway

Geography

Location: Northern Europe, bordering the North Atlantic Ocean, west of Sweden
Map references: Arctic Region, Europe, Standard Time Zones of the World
Area:
total area: 324,220 km²
land area: 307,860 km²
comparative area: slightly larger than New Mexico
Land boundaries: total 2,515 km, Finland 729 km, Sweden 1,619 km, Russia 167 km
Coastline: 21,925 km (includes mainland 3,419 km, large islands 2,413 km, long fjords, numerous small islands, and minor indentations 16,093 km)
Maritime claims:
contiguous zone: 10 nm
continental shelf: to depth of exploitation
exclusive economic zone: 200 nm
territorial sea: 4 nm
International disputes: territorial claim in Antarctica (Queen Maud Land); dispute between Denmark and Norway over maritime boundary in Arctic Ocean between Greenland and Jan Mayen is before the Interntional Court of Justice; maritime boundary dispute with Russia over portion of Barents Sea
Climate: temperate along coast, modified by North Atlantic Current; colder interior; rainy year-round on west coast
Terrain: glaciated; mostly high plateaus and rugged mountains broken by fertile valleys; small, scattered plains; coastline deeply indented by fjords; arctic tundra in north
Natural resources: petroleum, copper, natural gas, pyrites, nickel, iron ore, zinc, lead, fish, timber, hydropower
Land use:
arable land: 3%
permanent crops: 0%
meadows and pastures: 0%
forest and woodland: 27%
other: 70%
Irrigated land: 950 km² (1989)

Environment: air and water pollution; acid rain; note—strategic location adjacent to sea lanes and air routes in North Atlantic; one of most rugged and longest coastlines in world; Norway and Turkey only NATO members having a land boundary with Russia
Note: about two-thirds mountains; some 50,000 islands off its much indented coastline

People

Population: 4,297,436 (July 1993 est.)
Population growth rate: 0.41% (1993 est.)
Birth rate: 13.75 births/1,000 population (1993 est.)
Death rate: 10.54 deaths/1,000 population (1993 est.)
Net migration rate: 0.87 migrant(s)/1,000 population (1993 est.)
Infant mortality rate: 6.4 deaths/1,000 live births (1993 est.)
Life expectancy at birth:
total population: 77.16 years
male: 73.79 years
female: 80.73 years (1993 est.)
Total fertility rate: 1.86 children born/woman (1993 est.)
Nationality:
noun: Norwegian(s)
adjective: Norwegian
Ethnic divisions: Germanic (Nordic, Alpine, Baltic), Lapps 20,000
Religions: Evangelical Lutheran 87.8% (state church), other Protestant and Roman Catholic 3.8%, none 3.2%, unknown 5.2% (1980)
Languages: Norwegian (official)
note: small Lapp- and Finnish-speaking minorities
Literacy: age 15 and over can read and write (1976)
total population: 99%
male: NA%
female: NA%
Labor force: 2.004 million (1992)
by occupation: services 39.1%, commerce 17.6%, mining, oil, and manufacturing 16.0%, banking and financial services 7.6%, transportation and communications 7.8%, construction 6.1%, agriculture, forestry, and fishing 5.5% (1989)

Government

Names:
conventional long form: Kingdom of Norway
conventional short form: Norway
local long form: Kongeriket Norge
local short form: Norge
Digraph: NO
Type: constitutional monarchy
Capital: Oslo

Administrative divisions: 19 provinces (fylker, singular—fylke); Akershus, Aust-Agder, Buskerud, Finnmark, Hedmark, Hordaland, More og Romsdal, Nordland, Nord-Trondelag, Oppland, Oslo, Ostfold, Rogaland, Sogn og Fjordane, Sor-Trondelag, Telemark, Troms, Vest-Agder, Vestfold
Dependent areas: Bouvet Island, Jan Mayen, Svalbard
Independence: 26 October 1905 (from Sweden)
Constitution: 17 May 1814, modified in 1884
Legal system: mixture of customary law, civil law system, and common law traditions; Supreme Court renders advisory opinions to legislature when asked; accepts compulsory ICJ jurisdiction, with reservations
National holiday: Constitution Day, 17 May (1814)
Political parties and leaders: Labor Party, Gro Harlem BRUNDTLAND; Conservative Party, Kaci Kullmann FIVE; Center Party, Anne ENGER LAHNSTEIN; Christian People's Party, Kjell Magne BONDEVIK; Socialist Left, Eric SOLHEIM; Norwegian Communist, Ingre IVERSEN; Progress Party, Carl I. HAGEN; Liberal, Odd Einar DORUM; Finnmark List, leader NA
Suffrage: 18 years of age; universal
Elections:
Storting: last held on 11 September 1989 (next to be held 6 September 1993); results—Labor 34.3%, Conservative 22.2%, Progress 13.0%, Socialist Left 10.1%, Christian People's 8.5%, Center Party 6.6%, Finnmark List 0.3%, other 5%; seats—(165 total) Labor 63, Conservative 37, Progress 22, Socialist Left 17, Christian People's 14, Center Party 11, Finnmark List 1
Executive branch: monarch, prime minister, State Council (cabinet)
Legislative branch: unicameral Parliament (Storting) with an Upper Chamber (Lagting) and a Lower Chamber (Odelsting)
Judicial branch: Supreme Court (Hoyesterett)
Leaders:
Chief of State: King HARALD V (since 17 January 1991); Heir Apparent Crown Prince HAAKON MAGNUS (born 20 July 1973)
Head of Government: Prime Minister Gro Harlem BRUNDTLAND (since 3 November 1990)
Member of: AfDB, AsDB, Australia Group, BIS, CBSS, CCC, CE, CERN, COCOM, CSCE, EBRD, ECE, EFTA, ESA, FAO, GATT, IADB, IAEA, IBRD, ICAO, ICC, ICFTU, IDA, IEA, IFAD, IFC, ILO, IMF, IMO, INMARSAT, INTELSAT, INTERPOL, IOC, IOM, ISO, ITU, LORCS, MTCR, NACC, NAM (guest), NATO, NC, NEA, NIB, NSG, OECD, PCA, UN, UNAVEM II, UNCTAD, UNESCO, UNHCR, UNIDO,

UNIFIL, UNIKOM, UNMOGIP, UNOSOM, UNPROFOR, UNTSO, UPU, WHO, WIPO, WMO, ZC

Diplomatic representation in US:
chief of mission: Ambassador Kjeld VIBE
chancery: 2720 34th Street NW, Washington DC 20008
telephone: (202) 333-6000
FAX: (202) 337-0870
consulates general: Houston, Los Angeles, Minneapolis, New York, and San Francisco
consulate: Miami

US diplomatic representation:
chief of mission: (vacant)
embassy: Drammensveien 18, 0244 Oslo 2
mailing address: PSC 69, Box 1000, APO AE 09707
telephone: [47] (2) 44-85-50
FAX: [47] (2) 43-07-77

Flag: red with a blue cross outlined in white that extends to the edges of the flag; the vertical part of the cross is shifted to the hoist side in the style of the Dannebrog (Danish flag)

Economy

Overview: Norway has a mixed economy involving a combination of free market activity and government intervention. The government controls key areas, such as the vital petroleum sector (through large-scale state enterprises) and extensively subsidizes agriculture, fishing, and areas with sparse resources. Norway also maintains an extensive welfare system that helps propel public sector expenditures to slightly more than 50% of the GDP and results in one of the highest average tax burdens in the world (54%). A small country with a high dependence on international trade, Norway is basically an exporter of raw materials and semiprocessed goods, with an abundance of small- and medium-sized firms, and is ranked among the major shipping nations. The country is richly endowed with natural resources—petroleum, hydropower, fish, forests, and minerals—and is highly dependent on its oil sector to keep its economy afloat. Although one of the government's main priorities is to reduce this dependency, this situation is not likely to improve for years to come. The government also hopes to reduce unemployment and strengthen and diversify the economy through tax reform and a series of expansionary budgets. The budget deficit is expected to hit a record 8% of GDP because of welfare spending and bail-outs of the banking system. Unemployment continues at record levels of over 10%—including those in job programs—because of the weakness of the economy outside the oil sector. Overall economic growth is expected to be around

2% in 1993 while inflation is likely to rise slightly to 4%. Oslo, a member of the European Free Trade Area, has applied for EC membership and continues to deregulate and harmonize with EC regulations to prepare for the European Economic Area (EEA)—which creates an EC/EFTA market with free movement of capital, goods, services, and labor—to take effect in late 1993 and its EC bid.

National product: GDP—purchasing power equivalent—$76.1 billion (1992)
National product real growth rate: 2.9% (1992)
National product per capita: $17,700 (1992)
Inflation rate (consumer prices): 2.3% (1992)
Unemployment rate: 5.9% (excluding people in job-training programs) (1992)
Budget: revenues $50.6 billion; expenditures $57.0 billion, including capital expenditures of $NA (1992)
Exports: $35.3 billion (f.o.b., 1992)
commodities: petroleum and petroleum products 37.8%, metals and products 10.7%, natural gas 7.3%, fish 6.6%, chemicals 6.3%, ships 5.4%
partners: EC 67%, Nordic countries 18.2%, developing countries 7.9%, US 5.1%, Japan 1.6% (1992)
Imports: $26.8 billion (c.i.f., 1992)
commodities: machinery, fuels and lubricants, transportation equipment, chemicals, foodstuffs, clothing, ships
partners: EC 48.7%, Nordic countries 26.8%, developing countries 9.3%, US 8.6%, Japan 6.3% (1992)
External debt: $6.5 billion (1992 est.)
Industrial production: growth rate 7.3% (1992)
Electricity: 26,900,000 kW capacity; 111,000 million kWh produced, 25,850 kWh per capita (1992)
Industries: petroleum and gas, food processing, shipbuilding, pulp and paper products, metals, chemicals, timber, mining, textiles, fishing
Agriculture: accounts for 2.6% of GDP and 5.5% of labor force; among world's top 10 fishing nations; livestock output exceeds value of crops; over half of food needs imported; fish catch of 1.76 million metric tons in 1989
Illicit drugs: increasingly used as transshipment point for Latin American cocaine to Europe and gateway for Asian heroin shipped via the CIS and Baltic states for the European market
Economic aid: donor—ODA and OOF commitments (1970-89), $4.4 billion
Currency: 1 Norwegian krone (NKr) = 100 re
Exchange rates: Norwegian kroner (NKr) per US$1—6.8774 (January 1993), 6.2145

(1992), 6.4829 (1991), 6.2597 (1990), 6.9045 (1989), 6.5170 (1988)
Fiscal year: calendar year

Communications

Railroads: 4,223 km 1.435-meter standard gauge; Norwegian State Railways (NSB) operates 4,219 km (2,450 km electrified and 96 km double track); 4 km other
Highways: 79,540 km total; 38,580 km paved; 40,960 km gravel, crushed stone, and earth
Inland waterways: 1,577 km along west coast; 2.4 m draft vessels maximum
Pipelines: refined products 53 km
Ports: Oslo, Bergen, Fredrikstad, Kristiansand, Stavanger, Trondheim
Merchant marine: 829 ships (1,000 GRT or over) totaling 22,312,412 GRT/38,532,109 DWT; includes 13 passenger, 20 short-sea passenger, 106 cargo, 2 passenger-cargo, 19 refrigerated cargo, 15 container, 49 roll-on/roll-off, 23 vehicle carrier, 1 railcar carrier, 174 oil tanker, 91 chemical tanker, 82 liquefied gas, 25 combination ore/oil, 201 bulk, 8 combination bulk; note—the government has created a captive register, the Norwegian International Ship Register (NIS), as a subset of the Norwegian register; ships on the NIS enjoy many benefits of flags of convenience and do not have to be crewed by Norwegians; the majority of ships (777) under the Norwegian flag are now registered with the NIS
Airports:
total: 103
usable: 102
with permanent-surface runways: 63
with runways over 3,659 m: 0
with runways 2,440-3,659 m: 12
with runways 1,220-2,439 m: 16
Telecommunications: high-quality domestic and international telephone, telegraph, and telex services; 2 buried coaxial cable systems; 3,102,000 telephones; broadcast stations—46 AM, 350 private and 143 government FM, 54 (2,100 repeaters) TV; 4 coaxial submarine cables; 3 communications satellite earth stations operating in the EUTELSAT, INTELSAT (1 Atlantic Ocean), MARISAT, and domestic systems

Defense Forces

Branches: Norwegian Army, Royal Norwegian Navy, Royal Norwegian Air Force, Home Guard
Manpower availability: males age 15-49 1,120,744; fit for military service 934,968; reach military age (20) annually 31,903 (1993 est.)
Defense expenditures: exchange rate conversion—$3.8 billion, 3.4% of GDP (1992)

Oman

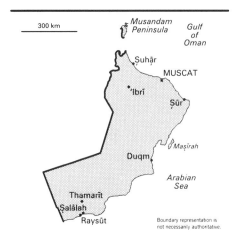

300 km

Musandam Peninsula
Gulf of Oman
Suhār
MUSCAT
Ibrī
Şūr
Maşīrah
Duqm
Arabian Sea
Thamarīt
Şalālah
Raysūt

Boundary representation is not necessarily authoritative.

Geography

Location: Middle East, along the Arabian Sea, between Yemen and the United Arab Emirates
Map references: Middle East, Standard Time Zones of the World
Area:
total area: 212,460 km²
land area: 212,460 km²
comparative area: slightly smaller than Kansas
Land boundaries: total 1,374 km, Saudi Arabia 676 km, UAE 410 km, Yemen 288 km
Coastline: 2,092 km
Maritime claims:
contiguous zone: 24 nm
continental shelf: to be defined
exclusive economic zone: 200 nm
territorial sea: 12 nm
International disputes: no defined boundary with most of UAE; Administrative Line with UAE in far north; a treaty with Yemen to settle the Omani-Yemeni boundary was ratified in December 1992
Climate: dry desert; hot, humid along coast; hot, dry interior; strong southwest summer monsoon (May to September) in far south
Terrain: vast central desert plain, rugged mountains in north and south
Natural resources: petroleum, copper, asbestos, some marble, limestone, chromium, gypsum, natural gas
Land use:
arable land: less than 2%
permanent crops: 0%
meadows and pastures: 5%
forest and woodland: 0%
other: 93%
Irrigated land: 410 km² (1989 est.)
Environment: summer winds often raise large sandstorms and duststorms in interior; sparse natural freshwater resources
Note: strategic location with small foothold on Musandam Peninsula controlling Strait of Hormuz (17% of world's oil production transits this point going from Persian Gulf to Arabian Sea)

People

Population: 1,643,579 (July 1993 est.)
Population growth rate: 3.46% (1993 est.)
Birth rate: 40.56 births/1,000 population (1993 est.)
Death rate: 5.94 deaths/1,000 population (1993 est.)
Net migration rate: 0 migrant(s)/1,000 population (1993 est.)
Infant mortality rate: 38.4 deaths/1,000 live births (1993 est.)
Life expectancy at birth:
total population: 67.32 years
male: 65.47 years
female: 69.27 years (1993 est.)
Total fertility rate: 6.58 children born/woman (1993 est.)
Nationality:
noun: Omani(s)
adjective: Omani
Ethnic divisions: Arab, Balochi, Zanzibari, South Asian (Indian, Pakistani, Bangladeshi)
Religions: Ibadhi Muslim 75%, Sunni Muslim, Shi'a Muslim, Hindu
Languages: Arabic (official), English, Balochi, Urdu, Indian dialects
Literacy:
total population: NA%
male: NA%
female: NA%
Labor force: 430,000
by occupation: agriculture 40% (est.)

Government

Names:
conventional long form: Sultanate of Oman
conventional short form: Oman
local long form: Saltanat Uman
local short form: Uman
Digraph: MU
Type: absolute monarchy with residual UK influence
Capital: Muscat
Administrative divisions: there are no first-order administrative divisions as defined by the US Government, but there are 3 governorates (muhafazah, singular—muhafazat); Musqat, Musandam, Zufar
Independence: 1650 (expulsion of the Portuguese)
Constitution: none
Legal system: based on English common law and Islamic law; ultimate appeal to the sultan; has not accepted compulsory ICJ jurisdiction
National holiday: National Day, 18 November
Political parties and leaders: none

Other political or pressure groups: outlawed Popular Front for the Liberation of Oman (PFLO), based in Yemen
Suffrage: none
Elections: elections scheduled for October 1992
Executive branch: sultan, Cabinet
Legislative branch: unicameral National Assembly
Judicial branch: none; traditional Islamic judges and a nascent civil court system
Leaders:
Chief of State and Head of Government: Sultan and Prime Minister QABOOS bin Sa'id Al Sa'id (since 23 July 1970)
Member of: ABEDA, AFESD, AL, AMF, ESCWA, FAO, G-77, GCC, IBRD, ICAO, IDA, IDB, IFAD, IFC, IMF, IMO, INMARSAT, INTELSAT, INTERPOL, IOC, ISO (correspondent), ITU, NAM, OIC, UN, UNCTAD, UNESCO, UNIDO, UPU, WFTU, WHO, WMO
Diplomatic representation in US:
chief of mission: Ambassador Awadh bin Badr AL-SHANFARI
chancery: 2342 Massachusetts Avenue NW, Washington, DC 20008
telephone: (202) 387-1980 through 1982
US diplomatic representation:
chief of mission: Ambassador David DUNFORD
embassy: address NA, Muscat
mailing address: P. O. Box 50202 Madinat Qaboos, Muscat
telephone: [968] 698-989
FAX: [968] 604-316
Flag: three horizontal bands of white (top, double width), red, and green (double width) with a broad, vertical, red band on the hoist side; the national emblem (a khanjar dagger in its sheath superimposed on two crossed swords in scabbards) in white is centered at the top of the vertical band

Economy

Overview: Economic performance is closely tied to the fortunes of the oil industry. Petroleum accounts for more than 85% of export earnings, about 80% of government revenues, and roughly 40% of GDP. Oman has proved oil reserves of 4 billion barrels, equivalent to about 20 years' supply at the current rate of extraction. Agriculture is carried on at a subsistence level and the general population depends on imported food.
National product: GDP—exchange rate conversion—$10.2 billion (1991)
National product real growth rate: 7.4% (1991)
National product per capita: $6,670 (1991)
Inflation rate (consumer prices): 1.6% (1991)
Unemployment rate: NA%

Budget: revenues $4.1 billion; expenditures $4.8 billion, including capital expenditures of $1 billion (1991)
Exports: $4.9 billion (f.o.b., 1991)
commodities: petroleum 87%, reexports, fish, processed copper, textiles
partners: UAE 30%, Japan 27%, South Korea 10%, Singapore 5%
Imports: $3.0 billion (f.o.b, 1991)
commodities: machinery, transportation equipment, manufactured goods, food, livestock, lubricants
partners: Japan 20%, UAE 19%, UK 19%, US 7%
External debt: $3.1 billion (December 1989 est.)
Industrial production: growth rate 10% (1989), including petroleum sector
Electricity: 1,142,400 kW capacity; 5,100 million kWh produced, 3,200 kWh per capita (1992)
Industries: crude oil production and refining, natural gas production, construction, cement, copper
Agriculture: accounts for 6% of GDP and 40% of the labor force (including fishing); less than 2% of land cultivated; largely subsistence farming (dates, limes, bananas, alfalfa, vegetables, camels, cattle); not self-sufficient in food; annual fish catch averages 100,000 metric tons
Economic aid: US commitments, including Ex-Im (FY70-89), $137 million; Western (non-US) countries, ODA and OOF bilateral commitments (1970-89), $148 million; OPEC bilateral aid (1979-89), $797 million
Currency: 1 Omani rial (RO) = 1,000 baiza
Exchange rates: Omani rials (RO) per US$1—0.3845 (fixed rate since 1986)
Fiscal year: calendar year

Communications

Highways: 26,000 km total; 6,000 km paved, 20,000 km motorable track
Pipelines: crude oil 1,300 km; natural gas 1,030 km
Ports: Mina' Qabus, Mina' Raysut, Mina' al Fahl
Merchant marine: 1 passenger ship (1,000 GRT or over) totaling 4,442 GRT/1,320 DWT
Airports:
total: 138
usable: 130
with permanent-surface runways: 6
with runways over 3,659 m: 1
with runways 2,440-3,659 m: 9
with runways 1,220-2,439 m: 74
Telecommunications: modern system consisting of open-wire, microwave, and radio communications stations; limited coaxial cable; 50,000 telephones; broadcast stations—2 AM, 3 FM, 7 TV; satellite earth stations—2 Indian Ocean INTELSAT, 1 ARABSAT, and 8 domestic

Defense Forces

Branches: Army, Navy, Air Force, Royal Oman Police
Manpower availability: males age 15-49 370,548; fit for military service 210,544; reach military age (14) annually 20,810 (1993 est.)
Defense expenditures: exchange rate conversion—$1.6 billion, 16% of GDP (1993 est.)

Pacific Islands (Palau), Trust Territory of the
(UN trusteeship administered by the US)

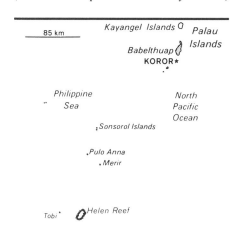

Geography

Location: in the North Pacific Ocean, 850 km southeast of the Philippines
Map references: Oceania
Area:
total area: 458 km²
land area: 458 km²
comparative area: slightly more than 2.5 times the size of Washington, DC
Land boundaries: 0 km
Coastline: 1,519 km
Maritime claims:
contiguous zone: 24 nm
continental shelf: 200 m or depth of exploitation
exclusive fishing zone: 200 nm
territorial sea: 12 nm
International disputes: none
Climate: wet season May to November; hot and humid
Terrain: about 200 islands varying geologically from the high, mountainous main island of Babelthuap to low, coral islands usually fringed by large barrier reefs
Natural resources: forests, minerals (especially gold), marine products, deep-seabed minerals
Land use:
arable land: NA%
permanent crops: NA%
meadows and pastures: NA%
forest and woodland: NA%
other: NA%
Irrigated land: NA km²
Environment: subject to typhoons from June to December; archipelago of six island groups totaling over 200 islands in the Caroline chain
Note: includes World War II battleground of Peleliu and world-famous rock islands

People

Population: 16,071 (July 1993 est.)
Population growth rate: 1.84% (1993 est.)
Birth rate: 22.9 births/1,000 population (1993 est.)

Death rate: 6.61 deaths/1,000 population (1993 est.)

Net migration rate: 2.12 migrant(s)/1,000 population (1993 est.)

Infant mortality rate: 25.07 deaths/1,000 live births (1993 est.)

Life expectancy at birth:
total population: 71.01 years
male: 69.14 years
female: 73.02 years (1993 est.)

Total fertility rate: 2.96 children born/woman (1993 est.)

Nationality:
noun: Palauan(s)
adjective: Palauan

Ethnic divisions: Palauans are a composite of Polynesian, Malayan, and Melanesian races

Religions: Christian (Catholics, Seventh-Day Adventists, Jehovah's Witnesses, the Assembly of God, the Liebenzell Mission, and Latter-Day Saints), Modekngei religion (one-third of the population observes this religion which is indigenous to Palau)

Languages: English (official in all of Palau's 16 states), Sonsorolese (official in the state of Sonsoral), Angaur and Japanese (in the state of Anguar), Tobi (in the state of Tobi), Palauan (in the other 13 states)

Literacy: age 15 and over can read and write (1980)
total population: 92%
male: 93%
female: 91%

Labor force: NA
by occupation: NA

Government

Names:
conventional long form: Trust Territory of the Pacific Islands
conventional short form: none
note: may change to Republic of Palau after independence; the native form of Palau is Belau and is sometimes used incorrectly in English and other languages

Digraph: NQ

Type: UN trusteeship administered by the US
note: constitutional government signed a Compact of Free Association with the US on 10 January 1986, which was never approved in a series of UN-observed plebiscites; until the UN trusteeship is terminated with entry into force of the Compact, Palau remains under US administration as the Palau District of the Trust Territory of the Pacific Islands; administrative authority resides in the Department of the Interior and is exercised by the Assistant Secretary for Territorial and International Affairs through the Palau Office, Trust Territory of the Pacific Islands, J. Victor HOBSON Jr., Director (since 16 December 1990)

Capital: Koror

note: a new capital is being built about 20 km northeast in eastern Babelthuap

Administrative divisions: there are no first-order administrative divisions as defined by the US Government, but there are 16 states: Aimeliik, Airai, Angaur, Kayangel, Koror, Melekeok, Ngaraard, Ngardmau, Ngaremlengui, Ngatpang, Ngchesar, Ngerchelong, Ngiwal, Peleliu, Sonsorol, Tobi

Independence: the last polity remaining under the US-administered UN trusteeship following the departure of the Republic of the Marshall Islands, the Federated States of Micronesia, and the Commonwealth of the Northern Marianas from the trusteeship; administered by the Office of Territorial and International Affairs, US Department of Interior

Constitution: 1 January 1981

Legal system: based on Trust Territory laws, acts of the legislature, municipal, common, and customary laws

National holiday: Constitution Day, 9 July (1979)

Suffrage: 18 years of age; universal

Elections:
President: last held on 4 November 1992 (next to be held NA November 1996); result—Kuniwo NAKAMURA 50.7%, Johnson TORIBIONG 49.3%
Senate: last held 4 November 1992 (next to be held NA November 1996); results—percent of vote by party NA; seats—(14 total); number of seats by party NA
House of Delegates: last held 4 November 1992 (next to be held NA November 1996); results—percent of vote by party NA; seats—(16 total); number of seats by party NA

Executive branch: national president, national vice president

Legislative branch: bicameral Parliament (Olbiil Era Kelulau or OEK) consists of an upper house or Senate and a lower house or House of Delegates

Judicial branch: Supreme Court, National Court, Court of Common Pleas

Leaders:
Chief of State and Head of Government: President Kuniwo NAKAMURA (since 1 January 1993), Vice-President Tommy E. REMENGESAU Jr. (since 1 January 1993)

Member of: ESCAP (associate), SPC, SPF (observer)

Diplomatic representation in US:
administrative officer: Charles UONG,
address: Palau Liaison Office, 444 North Capitol St., N.W., Suite 308, Washington, DC 20001

US diplomatic representation:
director: US Liaison Officer Lloyd W. MOSS
liaison office: US Liaison Office at Top Side, Neeriyas, Koror

mailing address: P.O. Box 6028, Koror, PW 96940
telephone: (680) 488-2920; (680) 488-2911

Flag: light blue with a large yellow disk (representing the moon) shifted slightly to the hoist side

Economy

Overview: The economy consists primarily of subsistence agriculture and fishing. Tourism provides some foreign exchange, although the remote location of Palau and a shortage of suitable facilities has hindered development. The government is the major employer of the work force, relying heavily on financial assistance from the US.

National product: GDP—purchasing power equivalent—$31.6 million (1986)
note: GDP numbers reflect US spending

National product real growth rate: NA%

National product per capita: $2,260 (1986)

Inflation rate (consumer prices): NA%

Unemployment rate: 20% (1986)

Budget: revenues $6.0 million; expenditures $NA, including capital expenditures of $NA (1986)

Exports: $0.5 million (f.o.b., 1986)
commodities: NA
partners: US, Japan

Imports: $27.2 million (c.i.f., 1986)
commodities: NA
partners: US

External debt: about $100 million (1989)

Industrial production: growth rate NA%

Electricity: 16,000 kW capacity; 22 million kWh produced, 1,540 kWh per capita (1990)

Industries: tourism, craft items (shell, wood, pearl), some commercial fishing and agriculture

Agriculture: subsistence-level production of coconut, copra, cassava, sweet potatoes

Economic aid: US commitments, including Ex-Im (FY70-89), $2,560 million; Western (non-US) countries, ODA and OOF bilateral commitments (1970-89), $92 million

Currency: US currency is used

Fiscal year: 1 October-30 September

Communications

Highways: 22.3 km paved, some stone-, coral-, or laterite-surfaced roads (1991)

Ports: Koror

Airports:
total: 3
usable: 3
with permanent-surface runways: 1
with runways over 3,659 m: 0
with runways 2,440-3,659 m: 0
with runways 1,220-2,439 m: 3

Telecommunications: broadcast stations—1 AM, 1 FM, 2 TV; 1 Pacific Ocean INTELSAT earth station

Pacific Ocean

Defense Forces

Note: defense is the responsibility of the US and that will not change when the UN trusteeship terminates if the Compact of Free Association with the US goes into effect

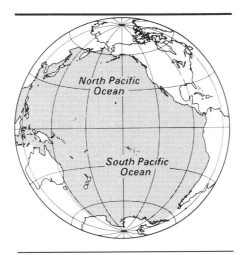

Geography

Location: body of water between the Western Hemisphere and Asia/Australia
Map references: Asia, North America, Oceania, South America, Standard Time Zones of the World
Area:
total area: 165.384 million km^2
comparative area: about 18 times the size of the US; the largest ocean (followed by the Atlantic Ocean, the Indian Ocean, and the Arctic Ocean); covers about one-third of the global surface; larger than the total land area of the world
note: includes Arafura Sea, Banda Sea, Bellingshausen Sea, Bering Sea, Bering Strait, Coral Sea, East China Sea, Gulf of Alaska, Makassar Strait, Philippine Sea, Ross Sea, Sea of Japan, Sea of Okhotsk, South China Sea, Tasman Sea, and other tributary water bodies
Coastline: 135,663 km
International disputes: some maritime disputes (see littoral states)
Climate: the western Pacific is monsoonal—a rainy season occurs during the summer months, when moisture-laden winds blow from the ocean over the land, and a dry season during the winter months, when dry winds blow from the Asian land mass back to the ocean
Terrain: surface currents in the northern Pacific are dominated by a clockwise, warm-water gyre (broad circular system of currents) and in the southern Pacific by a counterclockwise, cool-water gyre; in the northern Pacific sea ice forms in the Bering Sea and Sea of Okhotsk in winter; in the southern Pacific sea ice from Antarctica reaches its northernmost extent in October; the ocean floor in the eastern Pacific is dominated by the East Pacific Rise, while the western Pacific is dissected by deep trenches, including the world's deepest, the 10,924 meter Marianas Trench
Natural resources: oil and gas fields, polymetallic nodules, sand and gravel aggregates, placer deposits, fish
Environment: endangered marine species include the dugong, sea lion, sea otter, seals, turtles, and whales; oil pollution in Philippine Sea and South China Sea; dotted with low coral islands and rugged volcanic islands in the southwestern Pacific Ocean; subject to tropical cyclones (typhoons) in southeast and east Asia from May to December (most frequent from July to October); tropical cyclones (hurricanes) may form south of Mexico and strike Central America and Mexico from June to October (most common in August and September); southern shipping lanes subject to icebergs from Antarctica; occasional El Nino phenomenon occurs off the coast of Peru when the trade winds slacken and the warm Equatorial Countercurrent moves south, killing the plankton that is the primary food source for anchovies; consequently, the anchovies move to better feeding grounds, causing resident marine birds to starve by the thousands because of their lost food source
Note: the major choke points are the Bering Strait, Panama Canal, Luzon Strait, and the Singapore Strait; the Equator divides the Pacific Ocean into the North Pacific Ocean and the South Pacific Ocean; ships subject to superstructure icing in extreme north from October to May and in extreme south from May to October; persistent fog in the northern Pacific from June to December is a hazard to shipping; surrounded by a zone of violent volcanic and earthquake activity sometimes referred to as the Pacific Ring of Fire

Government

Digraph: ZN

Economy

Overview: The Pacific Ocean is a major contributor to the world economy and particularly to those nations its waters directly touch. It provides low-cost sea transportation between East and West, extensive fishing grounds, offshore oil and gas fields, minerals, and sand and gravel for the construction industry. In 1985 over half (54%) of the world's total fish catch came from the Pacific Ocean, which is the only ocean where the fish catch has increased every year since 1978. Exploitation of offshore oil and gas reserves is playing an ever-increasing role in the energy supplies of Australia, New Zealand, China, US, and Peru. The high cost of recovering offshore oil and gas, combined with the wide swings in world prices for oil since 1985, has slowed but not stopped new drillings.
Industries: fishing, oil and gas production

Pacific Ocean *(continued)*

Communications

Ports: Bangkok (Thailand), Hong Kong, Los Angeles (US), Manila (Philippines), Pusan (South Korea), San Francisco (US), Seattle (US), Shanghai (China), Singapore, Sydney (Australia), Vladivostok (Russia), Wellington (NZ), Yokohama (Japan)
Telecommunications: several submarine cables with network nodal points on Guam and Hawaii

Pakistan

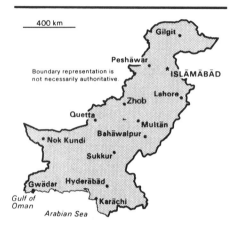

Geography

Location: South Asia, along the Arabian Sea, between India and Afghanistan
Map references: Asia, Standard Time Zones of the World
Area:
total area: 803,940 km²
land area: 778,720 km²
comparative area: slightly less than twice the size of California
Land boundaries: total 6,774 km, Afghanistan 2,430 km, China 523 km, India 2,912 km, Iran 909 km
Coastline: 1,046 km
Maritime claims:
contiguous zone: 24 nm
continental shelf: 200 nm or the edge of continental margin
exclusive economic zone: 200 nm
territorial sea: 12 nm
International disputes: status of Kashmir with India; border question with Afghanistan (Durand Line); water-sharing problems (Wular Barrage) over the Indus with upstream riparian India
Climate: mostly hot, dry desert; temperate in northwest; arctic in north
Terrain: flat Indus plain in east; mountains in north and northwest; Balochistan plateau in west
Natural resources: land, extensive natural gas reserves, limited petroleum, poor quality coal, iron ore, copper, salt, limestone
Land use:
arable land: 26%
permanent crops: 0%
meadows and pastures: 6%
forest and woodland: 4%
other: 64%
Irrigated land: 162,200 km² (1989)
Environment: frequent earthquakes, occasionally severe especially in north and west; flooding along the Indus after heavy rains (July and August); deforestation; soil erosion; desertification; water logging

Note: controls Khyber Pass and Bolan Pass, traditional invasion routes between Central Asia and the Indian Subcontinent

People

Population: 125,213,732 (July 1993 est.)
Population growth rate: 2.87% (1993 est.)
Birth rate: 42.59 births/1,000 population (1993 est.)
Death rate: 12.6 deaths/1,000 population (1993 est.)
Net migration rate: -1.28 migrant(s)/1,000 population (1993 est.)
Infant mortality rate: 103.6 deaths/1,000 live births (1993 est.)
Life expectancy at birth:
total population: 57.11 years
male: 56.54 years
female: 57.72 years (1993 est.)
Total fertility rate: 6.5 children born/woman (1993 est.)
Nationality:
noun: Pakistani(s)
adjective: Pakistani
Ethnic divisions: Punjabi, Sindhi, Pashtun (Pathan), Baloch, Muhajir (immigrants from India and their descendents)
Religions: Muslim 97% (Sunni 77%, Shi'a 20%), Christian, Hindu, and other 3%
Languages: Urdu (official), English (official; lingua franca of Pakistani elite and most government ministries, but official policies are promoting its gradual replacement by Urdu), Punjabi 64%, Sindhi 12%, Pashtu 8%, Urdu 7%, Balochi and other 9%
Literacy: age 15 and over can read and write (1990)
total population: 35%
male: 47%
female: 21%
Labor force: 28.9 million
by occupation: agriculture 54%, mining and manufacturing 13%, services 33%, extensive export of labor (1987 est.)

Government

Names:
conventional long form: Islamic Republic of Pakistan
conventional short form: Pakistan
former: West Pakistan
Digraph: PK
Type: republic
Capital: Islamabad
Administrative divisions: 4 provinces, 1 territory*, and 1 capital territory**; Balochistan, Federally Administered Tribal Areas*, Islamabad Capital Territory**, North-West Frontier, Punjab, Sindh
note: the Pakistani-administered portion of the disputed Jammu and Kashmir region includes Azad Kashmir and the Northern Areas

Independence: 14 August 1947 (from UK)
Constitution: 10 April 1973, suspended 5 July 1977, restored with amendments, 30 December 1985
Legal system: based on English common law with provisions to accommodate Pakistan's stature as an Islamic state; accepts compulsory ICJ jurisdiction, with reservations
National holiday: Pakistan Day, 23 March (1956) (proclamation of the republic)
Political parties and leaders:
government: Pakistan Muslim League-Nawaz (PML-N), Mian Nawaz SHARIF; Jamhoori Watan Party (JWP), Mohammad Akbar Khan BUGTI; Jamiat Ulema-i-Islam (JUI), Fazl-ur-REHMAN and Sami-ul-HAQ; Awami National Party (ANP), Khan Abdul WALI KHAN; Jamiat Ulema-i-Pakistan-Niazi, Maulana Abdul Sattar Khan NIAZI; Pakhtun Khwa Milli Awami Party (PKMAP), Mahmood Khan ACHAKZAI
opposition: Pakistan Peoples Party (PPP), Benazir BHUTTO and Nusrat BHUTTO; Pakistan Muslim League-Chattha (PML-C), Hamid Nasir CHATTHA; Jamaat-i-Islami (JI), Qazi Hussain AHMED; National People's Party (NPP), Ghulam Mustapha JATOI (formerly the PNP); Tehrik-i-Istiqlal (TI), Air Marshal (Ret.) Mohammad ASGHAR KHAN; Tehrik-i-Nifaz-i-Fiqah-i-Jafaria (TNFJ), Agha Hamid Ali MUSAVI; Jamiat Ulema-i-Pakistan-Noorani (JUP-Noorani), Maulana Shah Ahmed NOORANI; Mohajir Quami Mahaz-Haqiqi (MQM-H), Afaq AHMED
Other political or pressure groups:
military remains important political force; ulema (clergy), landowners, industrialists, and small merchants also influential
Suffrage: 21 years of age; universal
Elections:
President: last held on 12 December 1988 (next to be held by NA November 1993); results—Ghulam ISHAQ KHAN was elected by Parliament and the four provincial assemblies
Senate: last held March 1991 (next to be held NA March 1994); seats—(87 total) PML 52, Tribal Area Representatives (nonparty) 8, PPP 5, ANP 5, JWP 4, MQM 3, PNP 2 (name later chaged to NPP), JI 2, JUP 2, JUI 2, PKMAP 1, independent 1
National Assembly: last held on 24 October 1990 (next to be held by October 1995); results—percent of vote by party NA; seats—(217 total) number of seats by party NA; note—President GHULAM ISHAQ Khan dismissed the National Assembly on 18 April 1993; it was reestablished, however, on 26 May 1993 by the Supreme Court, which ruled the dismissal order unconstitutional

Executive branch: president, prime minister, Cabinet
Legislative branch: bicameral Parliament (Majlis-e-Shoora) consists of an upper house or Senate and a lower house or National Assembly
Judicial branch: Supreme Court, Federal Islamic (Shari'at) Court
Leaders:
Chief of State: President Ghulam ISHAQ KHAN (since 13 December 1988)
Head of Government: Prime Minister Mian Nawaz SHARIF (since 6 November 1990); note—President GHULAM ISHAQ Khan dismissed Prime Minister SHARIF on 18 April 1993, but he was reinstated by the Supreme Court on 26 May 1993
Member of: AsDB, C, CCC, CP, ECO, ESCAP, FAO, G-19, G-24, G-77, GATT, IAEA, IBRD, ICAO, ICC, ICFTU, IDA, IDB, IFAD, IFC, ILO, IMF, IMO, INMARSAT, INTELSAT, INTERPOL, IOC, IOM (observer), ISO, ITU, LORCS, MINURSO, NAM, OAS (observer), OIC, PCA, SAARC, UN, UNCTAD, UNESCO, UNHCR, UNIDO, UNIKOM, UNOSOM, UNTAC, UPU, WCL, WFTU, WHO, WIPO, WMO, WTO
Diplomatic representation in US:
chief of mission: (vacant)
chancery: 2315 Massachusetts Avenue NW, Washington, DC 20008
telephone: (202) 939-6200
consulate general: New York
US diplomatic representation:
chief of mission: Ambassador John MONJO
embassy: Diplomatic Enclave, Ramna 5, Islamabad
mailing address: P. O. Box 1048, PSC 1212, Box 2000, Islamabad or APO AE 09812-2000
telephone: [92] (51) 826161 through 79
FAX: [92] (51) 822004
consulates general: Karachi, Lahore
consulate: Peshawar
Flag: green with a vertical white band (symbolizing the role of religious minorities) on the hoist side; a large white crescent and star are centered in the green field; the crescent, star, and color green are traditional symbols of Islam

---

Economy

Overview: Pakistan is a poor Third World country faced with the usual problems of rapidly increasing population, sizable government deficits, and heavy dependence on foreign aid. In addition, the economy must support a large military establishment. A real economic growth rate averaging 5-6% in recent years has helped the country to cope with these problems. Almost all agriculture and small-scale industry is in private hands. In 1990, Pakistan embarked on a sweeping economic liberalization program to boost foreign and domestic private investment and lower foreign aid dependence. The SHARIF government denationalized several state-owned firms and attracted some foreign investment. Pakistan likely will have difficulty raising living standards because of its rapidly expanding population. At the current rate of growth, population would double in 25 years.
National product: GNP—exchange rate conversion—$48.3 billion (FY92 est.)
National product real growth rate: 6.4% (FY92 est.)
National product per capita: $410 (FY92 est.)
Inflation rate (consumer prices): 12.7% (FY91)
Unemployment rate: 10% (FY91 est.)
Budget: revenues $9.4 billion; expenditures $10.9 billion, including capital expenditures of $3.1 billion (FY93 est.)
Exports: $6.8 billion (f.o.b., FY92)
commodities: cotton, textiles, clothing, rice
partners: EC 35%, US 11%, Japan 8% (FY91)
Imports: $9.1 billion (f.o.b., FY92)
commodities: petroleum, petroleum products, machinery, transportation, equipment, vegetable oils, animal fats, chemicals
partners: EC 29%, Japan 13%, US 12% (FY91)
External debt: $16.5 billion (1992 est.)
Industrial production: growth rate 5.7% (FY91); accounts for almost 20% of GNP
Electricity: 10,000,000 kW capacity; 43,000 million kWh produced, 350 kWh per capita (1992)
Industries: textiles, food processing, beverages, construction materials, clothing, paper products, shrimp
Agriculture: 25% of GNP, over 50% of labor force; world's largest contiguous irrigation system; major crops—cotton, wheat, rice, sugarcane, fruits, vegetables; livestock products—milk, beef, mutton, eggs; self-sufficient in food grain
Illicit drugs: illicit producer of opium and hashish for the international drug trade; government eradication efforts on poppy cultivation of limited success; largest producer of Southwest Asian heroin
Economic aid: (including Bangladesh prior to 1972) US commitments, including Ex-Im (FY70-89), $4.5 billion; Western (non-US) countries, ODA and OOF bilateral commitments (1980-89), $9.1 billion; OPEC bilateral aid (1979-89), $2.3 billion; Communist countries (1970-89), $3.2 billion
Currency: 1 Pakistani rupee (PRe) = 100 paisa
Exchange rates: Pakistani rupees (PRs) per US$1—25.904 (January 1993), 25.083 (1992), 23.801 (1991), 21.707 (1990), 20.541 (1989), 18.003 (1988)
Fiscal year: 1 July-30 June

Pakistan *(continued)*

Communications

Railroads: 8,773 km total; 7,718 km broad gauge, 445 km 1-meter gauge, and 610 km less than 1-meter gauge; 1,037 km broad-gauge double track; 286 km electrified; all government owned (1985)
Highways: 101,315 km total (1987); 40,155 km paved, 23,000 km gravel, 29,000 km improved earth, and 9,160 km unimproved earth or sand tracks (1985)
Pipelines: crude oil 250 km; natural gas 4,044 km; petroleum products 885 km (1987)
Ports: Gwadar, Karachi, Port Muhammad bin Qasim
Merchant marine: 29 ships (1,000 GRT or over) totaling 350,916 GRT/530,855 DWT; includes 3 passenger-cargo, 24 cargo, 1 oil tanker, 1 bulk
Airports:
total: 111
usable: 104
with permanent-surface runways: 75
with runways over 3,659 m: 1
with runways 2,440-3,659 m: 31
with runways 1,220-2,439 m: 42
Telecommunications: the domestic telephone system is poor, adequate only for government and business use; about 7 telephones per 1,000 persons; the system for international traffic is better and employs both microwave radio relay and satellites; satellite ground stations—1 Atlantic Ocean INTELSAT and 2 Indian Ocean INTELSAT; broadcast stations—19 AM, 8 FM, 29 TV

Defense Forces

Branches: Army, Navy, Air Force, Civil Armed Forces, National Guard
Manpower availability: males age 15-49 28,657,084; fit for military service 17,585,542; reach military age (17) annually 1,337,352 (1993 est.)
Defense expenditures: exchange rate conversion—$3.2 billion, 6% of GNP (FY91/92)

Palmyra Atoll
(territory of the US)

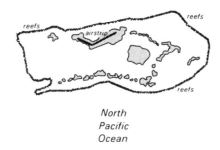

North
Pacific
Ocean

Geography

Location: in the North Pacific Ocean, 1,600 km south-southwest of Honolulu, almost halfway between Hawaii and American Samoa
Map references: Oceania
Area:
total area: 11.9 km^2
land area: 11.9 km^2
comparative area: about 20 times the size of The Mall in Washington, DC
Land boundaries: 0 km
Coastline: 14.5 km
Maritime claims:
contiguous zone: 12 nm
continental shelf: 200 m (depth)
exclusive economic zone: 200 nm
territorial sea: 12 nm
International disputes: none
Climate: equatorial, hot, and very rainy
Terrain: low, with maximum elevations of about 2 meters
Natural resources: none
Land use:
arable land: 0%
permanent crops: 0%
meadows and pastures: 0%
forest and woodland: 100%
other: 0%
Irrigated land: 0 km^2
Environment: about 50 islets covered with dense vegetation, coconut trees, and balsa-like trees up to 30 meters tall

People

Population: uninhabited

Government

Names:
conventional long form: none
conventional short form: Palmyra Atoll
Digraph: LQ
Type: unincorporated territory of the US; privately owned, but administered by the Office of Territorial and International Affairs, US Department of the Interior
Capital: none; administered from Washington, DC

Economy

Overview: no economic activity

Communications

Ports: the main harbor is West Lagoon, which is entered by a channel on the southwest side of the atoll; both the channel and harbor will accommodate vessels drawing 4 meters of water; much of the road and many causeways built during the war are unserviceable and overgrown
Airports:
total: 1
usable: 1
with permanent-surface runways: 1
with runways over 3,659 m: 0
with runways 2,440-3,659 m: 0
with runways 1,220-2,439 m: 1

Defense Forces

Note: defense is the responsibility of the US

Panama

Caribbean Sea

North Pacific Ocean

Geography

Location: extreme southern Central America, between Colombia and Costa Rica
Map references: Central America and the Caribbean, South America, Standard Time Zones of the World
Area:
total area: 78,200 km²
land area: 75,990 km²
comparative area: slightly smaller than South Carolina
Land boundaries: total 555 km, Colombia 225 km, Costa Rica 330 km
Coastline: 2,490 km
Maritime claims:
territorial sea: 200 nm
International disputes: none
Climate: tropical; hot, humid, cloudy; prolonged rainy season (May to January), short dry season (January to May)
Terrain: interior mostly steep, rugged mountains and dissected, upland plains; coastal areas largely plains and rolling hills
Natural resources: copper, mahogany forests, shrimp
Land use:
arable land: 6%
permanent crops: 2%
meadows and pastures: 15%
forest and woodland: 54%
other: 23%
Irrigated land: 320 km² (1989 est.)
Environment: dense tropical forest in east and northwest
Note: strategic location on eastern end of isthmus forming land bridge connecting North and South America; controls Panama Canal that links North Atlantic Ocean via Caribbean Sea with North Pacific Ocean

People

Population: 2,579,047 (July 1993 est.)
Population growth rate: 1.98% (1993 est.)
Birth rate: 25.08 births/1,000 population (1993 est.)

Death rate: 4.94 deaths/1,000 population (1993 est.)
Net migration rate: -0.38 migrant(s)/1,000 population (1993 est.)
Infant mortality rate: 17.2 deaths/1,000 live births (1993 est.)
Life expectancy at birth:
total population: 74.56 years
male: 71.99 years
female: 77.27 years (1993 est.)
Total fertility rate: 2.9 children born/woman (1993 est.)
Nationality:
noun: Panamanian(s)
adjective: Panamanian
Ethnic divisions: mestizo (mixed Indian and European ancestry) 70%, West Indian 14%, white 10%, Indian 6%
Religions: Roman Catholic 85%, Protestant 15%
Languages: Spanish (official), English 14%
note: many Panamanians bilingual
Literacy: age 15 and over can read and write (1990)
total population: 88%
male: 88%
female: 88%
Labor force: 921,000 (1992 est.)
by occupation: government and community services 31.8%, agriculture, hunting, and fishing 26.8%, commerce, restaurants, and hotels 16.4%, manufacturing and mining 9.4%, construction 3.2%, transportation and communications 6.2%, finance, insurance, and real estate 4.3%
note: shortage of skilled labor, but an oversupply of unskilled labor

Government

Names:
conventional long form: Republic of Panama
conventional short form: Panama
local long form: Republica de Panama
local short form: Panama
Digraph: PM
Type: centralized republic
Capital: Panama
Administrative divisions: 9 provinces (provincias, singular—provincia) and 1 territory* (comarca); Bocas del Toro, Chiriqui, Cocle, Colon, Darien, Herrera, Los Santos, Panama, San Blas*, Veraguas
Independence: 3 November 1903 (from Colombia; became independent from Spain 28 November 1821)
Constitution: 11 October 1972; major reforms adopted April 1983
Legal system: based on civil law system; judicial review of legislative acts in the Supreme Court of Justice; accepts compulsory ICJ jurisdiction, with reservations
National holiday: Independence Day, 3 November (1903)

Political parties and leaders:
government alliance: Nationalist Republican Liberal Movement (MOLIRENA), Alfredo RAMIREZ; Authentic Liberal Party (PLA), Arnulfo ESCALONA; Arnulfista Party (PA), Mireya MOSCOSO DE GRUBER
other parties: Christian Democratic Party (PDC), Ricardo ARIAS Calderon; Democratic Revolutionary Party (PRD), Gerardo GONZALEZ; Agrarian Labor Party (PALA), Nestor Tomas GUERRA; Liberal Party (PL), Roberto ALEMAN Zubieta; Doctrinaire Panamenista Party (PPD), Jose Salvador MUNOZ; Papa Egoro Movement, Ruben BLADES; Renovacion Civilista, Manuel BURGOS; Civic Renewal Party (PRC), Tomas HERRERA; National Integration Movement (MINA), Arrigo GUARDIA; National Unity Mission Party (MUN), Jose Manuel PAREDES; Independent Democratic Union Party (UDI), leader NA; Popular Nationalist Party (PNP), leader NA
Other political or pressure groups: National Council of Organized Workers (CONATO); National Council of Private Enterprise (CONEP); Panamanian Association of Business Executives (APEDE); National Civic Crusade; National Committee for the Right to Life; Chamber of Commerce; Panamanian Industrialists Society (SIP); Workers Confederation of the Republic of Panama (CTRP)
Suffrage: 18 years of age; universal and compulsory
Elections:
President: last held on 7 May 1989, annulled but later upheld (next to be held May 1994); results—anti-NORIEGA coalition believed to have won about 75% of the total votes cast
Legislative Assembly: last held on 27 January 1991 (next to be held NA May 1994); results—percent of vote by party NA; seats—(67 total)
progovernment parties: PDC 28, MOLIRENA 15, PA 8, PLA 4
opposition parties: PRD 10, PALA 1, PL 1; note—the PDC went into opposition after President Guillermo ENDARA ousted the PDC from the coalition government in April 1991
Executive branch: president, two vice presidents, Cabinet
Legislative branch: unicameral Legislative Assembly (Asamblea Legislativa)
Judicial branch: Supreme Court of Justice (Corte Suprema de Justicia), 5 superior courts, 3 courts of appeal
Leaders:
Chief of State and Head of Government: President Guillermo ENDARA (since 20 December 1989, elected 7 May 1989); First Vice President Guillermo FORD Boyd (since 24 December 1992); Second Vice President (vacant)

Panama (continued)

Member of: AG (associate), CG, ECLAC, FAO, G-77, IADB, IAEA, IBRD, ICAO, ICFTU, IDA, IFAD, IFC, ILO, IMF, IMO, INMARSAT, INTELSAT, INTERPOL, IOC, IOM, ITU, LAES, LAIA (observer), LORCS, NAM, OAS, OPANAL, PCA, UN, UNCTAD, UNESCO, UNIDO, UPU, WCL, WFTU, WHO, WIPO, WMO, WTO
Diplomatic representation in US:
chief of mission: Ambassador Jaime FORD
chancery: 2862 McGill Terrace NW, Washington, DC 20008
telephone: (202) 483-1407;
note: the status of the consulates general and consulates has not yet been determined
US diplomatic representation:
chief of mission: Ambassador Deane R. HINTON
embassy: Avenida Balboa and Calle 38, Apartado 6959, Panama City 5
mailing address: Box E, APO AA 34002
telephone: (507) 27-1777
FAX: (507) 27-1713
Flag: divided into four, equal rectangles; the top quadrants are white with a blue five-pointed star in the center (hoist side) and plain red, the bottom quadrants are plain blue (hoist side) and white with a red five-pointed star in the center

Economy

Overview: GDP expanded by roughly 8% in 1992, following growth of 9.3% in 1991. The economy thus continues to recover from the crisis that preceded the ouster of Manuel NORIEGA, even though the government's structural adjustment program has been hampered by a lack of popular support and a passive administration. Public investment has been limited as the administration has kept the fiscal deficit below 3% of GDP. Unemployment and economic reform are the two major issues the government must face in 1993-94.
National product: GDP—exchange rate conversion—$6 billion (1992 est.)
National product real growth rate: 8% (1992 est.)
National product per capita: $2,400 (1992 est.)
Inflation rate (consumer prices): 1.8% (1992 est.)
Unemployment rate: 15% (1992 est.)
Budget: revenues $1.8 billion; expenditures $1.9 billion, including capital expenditures of $200 million (1992 est.)
Exports: $486 million (f.o.b., 1992 est.)
commodities: bananas 43%, shrimp 11%, sugar 4%, clothing 5%, coffee 2%
partners: US 38%, Central America and Caribbean, EC (1992 est.)
Imports: $2.0 billion (f.o.b., 1992 est.)
commodities: capital goods 21%, crude oil 11%, foodstuffs 9%, consumer goods, chemicals

partners: US 36%, Japan, EC, Central America and Caribbean, Mexico, Venezuela (1992 est.)
External debt: $5.2 billion (year-end 1992 est.)
Industrial production: growth rate 7.6% (1992 est.); accounts for about 9% of GDP
Electricity: 1,584,000 kW capacity; 4,360 billion kWh produced, 1,720 kWh per capita (1992)
Industries: manufacturing and construction activities, petroleum refining, brewing, cement and other construction material, sugar milling
Agriculture: accounts for 10.5% of GDP (1992 est.), 27% of labor force (1992); crops—bananas, rice, corn, coffee, sugarcane; livestock; fishing; importer of food grain, vegetables
Illicit drugs: major cocaine transshipment point and drug money laundering center
Economic aid: US commitments, including Ex-Im (FY70-89), $516 million; Western (non-US) countries, ODA and OOF bilateral commitments (1970-89), $582 million; Communist countries (1970-89), $4 million
Currency: 1 balboa (B) = 100 centesimos
Exchange rates: balboas (B) per US$1—1.000 (fixed rate)
Fiscal year: calendar year

Communications

Railroads: 238 km total; 78 km 1.524-meter gauge, 160 km 0.914-meter gauge
Highways: 8,530 km total; 2,745 km paved, 3,270 km gravel or crushed stone, 2,515 km improved and unimproved earth
Inland waterways: 800 km navigable by shallow draft vessels; 82 km Panama Canal
Pipelines: crude oil 130 km
Ports: Cristobal, Balboa, Bahia Las Minas
Merchant marine: 3,244 ships (1,000 GRT or over) totaling 51,353,963 GRT/82,138,537 DWT; includes 22 passenger, 26 short-sea passenger, 3 passenger-cargo, 1,091 cargo, 246 refrigerated cargo, 196 container, 63 roll-on/roll-off cargo, 121 vehicle carrier, 9 livestock carrier, 5 multifunction large-load carrier, 403 oil tanker, 180 chemical tanker, 26 combination ore/oil, 121 liquefied gas, 9 specialized tanker, 688 bulk, 34 combination bulk, 1 barge carrier; note—all but 5 are foreign owned and operated; the top 4 foreign owners are Japan 36%, Greece 8%, Hong Kong 8%, and Taiwan 5%; (China owns at least 131 ships, Vietnam 3, Croatia 3, Cuba 4, Cyprus 6, and Russia 16)
Airports:
total: 112
usable: 104
with permanent-surface runways: 39
with runways over 3,659 m: 0
with runways 2,440-3,659 m: 2
with runways 1,220-2,439 m: 15

Telecommunications: domestic and international facilities well developed; connection into Central American Microwave System; 220,000 telephones; broadcast stations—91 AM, no FM, 23 TV; 1 coaxial submarine cable; satellite ground stations—2 Atlantic Ocean INTELSAT

Defense Forces

Branches: the Panamanian Defense Forces (PDF) ceased to exist as a military institution shortly after the United States invaded Panama on 20 December 1989; President ENDARA has restructured the forces, under the new name of Panamanian Public Forces (PPF) and worked to assert civilian control over them; the PPF is divided into the National Police, Maritime Service, and National Air Service; the Judicial Technical Police serve under the Attorney General; the Council of Public Security and National Defense under Menalco SOLIS in the Office of the President is analogous to the US National Security Council; the Institutional Protection Service under Carlos BARES is attached to the presidency
Manpower availability: males age 15-49 671,059; fit for military service 461,471 (1993 est.); no conscription
Defense expenditures: expenditures for the Panamanian Public Forces for internal security amounted to $104.7 million, 1.7% of GDP (1993 est.)

Papua New Guinea

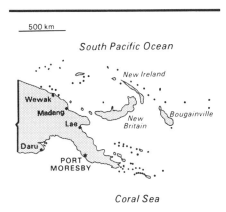

Geography

Location: Southeast Asia, just north of Australia, between Indonesia and the Solomon Islands
Map references: Oceania, Southeast Asia, Standard Time Zones of the World
Area:
total area: 461,690 km²
land area: 451,710 km²
comparative area: slightly larger than California
Land boundaries: total 820 km, Indonesia 820 km
Coastline: 5,152 km
Maritime claims: measured from claimed archipelagic baselines
continental shelf: 200 m depth or to depth of exploitation
exclusive fishing zone: 200 nm
territorial sea: 12 nm
International disputes: none
Climate: tropical; northwest monsoon (December to March), southeast monsoon (May to October); slight seasonal temperature variation
Terrain: mostly mountains with coastal lowlands and rolling foothills
Natural resources: gold, copper, silver, natural gas, timber, oil potential
Land use:
arable land: 0%
permanent crops: 1%
meadows and pastures: 0%
forest and woodland: 71%
other: 28%
Irrigated land: NA km²
Environment: one of world's largest swamps along southwest coast; some active volcanos; frequent earthquakes
Note: shares island of New Guinea with Indonesia

People

Population: 4,100,714 (July 1993 est.)
Population growth rate: 2.32% (1993 est.)

Birth rate: 33.77 births/1,000 population (1993 est.)
Death rate: 10.57 deaths/1,000 population (1993 est.)
Net migration rate: 0 migrant(s)/1,000 population (1993 est.)
Infant mortality rate: 64.9 deaths/1,000 live births (1993 est.)
Life expectancy at birth:
total population: 56.02 years
male: 55.19 years
female: 56.88 years (1993 est.)
Total fertility rate: 4.75 children born/woman (1993 est.)
Nationality:
noun: Papua New Guinean(s)
adjective: Papua New Guinean
Ethnic divisions: Melanesian, Papuan, Negrito, Micronesian, Polynesian
Religions: Roman Catholic 22%, Lutheran 16%, Presbyterian/Methodist/London Missionary Society 8%, Anglican 5%, Evangelical Alliance 4%, Seventh-Day Adventist 1%, other Protestant sects 10%, indigenous beliefs 34%
Languages: English spoken by 1-2%, pidgin English widespread, Motu spoken in Papua region
note: 715 indigenous languages
Literacy: age 15 and over can read and write (1990)
total population: 52%
male: 65%
female: 38%
Labor force: NA

Government

Names:
conventional long form: Independent State of Papua New Guinea
conventional short form: Papua New Guinea
Digraph: PP
Type: parliamentary democracy
Capital: Port Moresby
Administrative divisions: 20 provinces; Central, Chimbu, Eastern Highlands, East New Britain, East Sepik, Enga, Gulf, Madang, Manus, Milne Bay, Morobe, National Capital, New Ireland, Northern, North Solomons, Sandaun, Southern Highlands, Western, Western Highlands, West New Britain
Independence: 16 September 1975 (from UN trusteeship under Australian administration)
Constitution: 16 September 1975
Legal system: based on English common law
National holiday: Independence Day, 16 September (1975)
Political parties and leaders: Papua New Guinea United Party (Pangu Party), Jack GENIA; People's Democratic Movement (PDM), Paias WINGTI; People's Action Party (PAP), Akoka DOI; People's Progress

Party (PPP), Sir Julius CHAN; United Party (UP), Paul TORATO; Papua Party (PP), Galeva KWARARA; National Party (NP), Paul PORA; Melanesian Alliance (MA), Fr. John MOMIS
Suffrage: 18 years of age; universal
Elections:
National Parliament: last held 13-26 June 1992 (next to be held NA 1997); results—percent by party NA; seats—(109 total) Pangu Party 24, PDM 17, PPP 10, PAP 10, independents 30, others 18 (association with political parties is fluid)
Executive branch: British monarch, governor general, prime minister, deputy prime minister, National Executive Council (cabinet)
Legislative branch: unicameral National Parliament (sometimes referred to as the House of Assembly)
Judicial branch: Supreme Court
Leaders:
Chief of State: Queen ELIZABETH II (since 6 February 1952), represented by Governor General Wiwa KOROWI (since NA November 1991)
Head of Government: Prime Minister Paias WINGTI (since 17 July 1992)
Member of: ACP, AsDB, ASEAN (observer), C, CP, ESCAP, FAO, G-77, IBRD, ICAO, ICFTU, IDA, IFAD, IFC, ILO, IMF, IMO, INTELSAT, INTERPOL, IOC, ISO, ITU, LORCS, NAM, SPARTECA, SPC, SPF, UN, UNCTAD, UNESCO, UNIDO, UPU, WFTU, WHO, WMO
Diplomatic representation in US:
chief of mission: Ambassador Margaret TAYLOR
chancery: 3rd floor, 1615 New Hampshire Avenue NW, Washington, DC 20009
telephone: (202) 745-3680
US diplomatic representation:
chief of mission: Ambassador Robert W. FARRAND
embassy: Armit Street, Port Moresby
mailing address: P. O. Box 1492, Port Moresby, or APO AE 96553
telephone: [675] 211-455 or 594, 654
FAX: [675] 213-423
Flag: divided diagonally from upper hoist-side corner; the upper triangle is red with a soaring yellow bird of paradise centered; the lower triangle is black with five white five-pointed stars of the Southern Cross constellation centered

Economy

Overview: Papua New Guinea is richly endowed with natural resources, but exploitation has been hampered by the rugged terrain and the high cost of developing an infrastructure. Agriculture provides a subsistence livelihood for 85% of the population. Mining of numerous

Papua New Guinea (continued)

deposits, including copper and gold, accounts for about 60% of export earnings. Budgetary support from Australia and development aid under World Bank auspices have helped sustain the economy. Robust growth in 1991-92 was led by the mining sector; the opening of a large new gold mine helped the advance.

National product: GDP—exchange rate conversion—$3.4 billion (1992)

National product real growth rate: 8.5% (1992)

National product per capita: $850 (1992)

Inflation rate (consumer prices): 4.5% (1992-93)

Unemployment rate: NA%

Budget: revenues $1.33 billion; expenditures $1.49 billion, including capital expenditures of $NA (1993 est.)

Exports: $1.3 billion (f.o.b., 1990)

commodities: gold, copper ore, coffee, logs, palm oil, cocoa, lobster

partners: FRG, Japan, Australia, UK, Spain, US

Imports: $1.6 billion (c.i.f., 1990)

commodities: machinery and transport equipment, food, fuels, chemicals, consumer goods

partners: Australia, Singapore, Japan, US, New Zealand, UK

External debt: $2.2 billion (April 1991)

Industrial production: growth rate NA%; accounts for 21% of GDP

Electricity: 400,000 kW capacity; 1,600 million kWh produced, 400 kWh per capita (1992)

Industries: copra crushing, palm oil processing, plywood production, wood chip production, mining of gold, silver, and copper, construction, tourism

Agriculture: one-third of GDP; livelihood for 85% of population; fertile soils and favorable climate permits cultivating a wide variety of crops; cash crops—coffee, cocoa, coconuts, palm kernels; other products—tea, rubber, sweet potatoes, fruit, vegetables, poultry, pork; net importer of food for urban centers

Economic aid: US commitments, including Ex-Im (FY70-89), $40.6 million; Western (non-US) countries, ODA and OOF bilateral commitments (1970-89), $6.5 billion; OPEC bilateral aid (1979-89), $17 million

Currency: 1 kina (K) = 100 toea

Exchange rates: kina (K) per US$1—1.0065 (January 1993), 1.0367 (1992), 1.0504 (1991), 1.0467 (1990), 1.1685 (1989), 1.1538 (1988)

Fiscal year: calendar year

Communications

Railroads: none

Highways: 19,200 km total; 640 km paved, 10,960 km gravel, crushed stone, or stabilized-soil surface, 7,600 km unimproved earth

Inland waterways: 10,940 km

Ports: Anewa Bay, Lae, Madang, Port Moresby, Rabaul

Merchant marine: 11 ships (1,000 GRT or over) totaling 20,523 GRT/24,774 DWT; includes 2 cargo, 1 roll-on/roll-off cargo, 5 combination ore/oil, 2 bulk, 1 container

Airports:

total: 504

usable: 457

with permanent-surface runways: 18

with runways over 3,659 m: 0

with runways 2,440-3,659 m: 1

with runways 1,220-2,439 m: 39

Telecommunications: services are adequate and being improved; facilities provide radiobroadcast, radiotelephone and telegraph, coastal radio, aeronautical radio, and international radiocommunication services; submarine cables extend to Australia and Guam; more than 70,000 telephones (1987); broadcast stations—31 AM, 2 FM, 2 TV (1987); 1 Pacific Ocean INTELSAT earth station

Defense Forces

Branches: Papua New Guinea Defense Force (including Army, Navy, Air Force)

Manpower availability: males age 15-49 1,046,929; fit for military service 582,685 (1993 est.)

Defense expenditures: exchange rate conversion—$55 million, 1.8% of GDP (1993 est.)

Paracel Islands

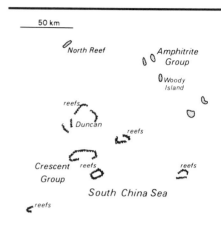

Geography

Location: Southeast Asia, 400 km east of Vietnam in the South China Sea, about one-third of the way between Vietnam and the Philippines

Map references: Asia

Area:

total area: NA km^2

land area: NA km^2

comparative area: NA

Land boundaries: 0 km

Coastline: 518 km

Maritime claims: NA

International disputes: occupied by China, but claimed by Taiwan and Vietnam

Climate: tropical

Terrain: NA

Natural resources: none

Land use:

arable land: 0%

permanent crops: 0%

meadows and pastures: 0%

forest and woodland: 0%

other: 100%

Irrigated land: 0 km^2

Environment: subject to typhoons

People

Population: no indigenous inhabitants; note—there are scattered Chinese garrisons

Government

Names:

conventional long form: none

conventional short form: Paracel Islands

Digraph: PF

Economy

Overview: no economic activity

Paraguay

Communications

Ports: small Chinese port facilities on Woody Island and Duncan Island currently under expansion
Airports: 1 on Woody Island

Defense Forces

Note: occupied by China

Geography

Location: Central South America, between Argentina and Brazil
Map references: South America, Standard Time Zones of the World
Area:
total area: 406,750 km²
land area: 397,300 km²
comparative area: slightly smaller than California
Land boundaries: total 3,920 km, Argentina 1,880 km, Bolivia 750 km, Brazil 1,290 km
Coastline: 0 km (landlocked)
Maritime claims: none; landlocked
International disputes: short section of the boundary with Brazil (just west of Guaira Falls on the Rio Parana) has not been determined
Climate: varies from temperate in east to semiarid in far west
Terrain: grassy plains and wooded hills east of Rio Paraguay; Gran Chaco region west of Rio Paraguay mostly low, marshy plain near the river, and dry forest and thorny scrub elsewhere
Natural resources: hydropower, timber, iron ore, manganese, limestone
Land use:
arable land: 20%
permanent crops: 1%
meadows and pastures: 39%
forest and woodland: 35%
other: 5%
Irrigated land: 670 km² (1989 est.)
Environment: local flooding in southeast (early September to June); poorly drained plains may become boggy (early October to June)
Note: landlocked; buffer between Argentina and Brazil

People

Population: 5,070,856 (July 1993 est.)
Population growth rate: 2.8% (1993 est.)

Birth rate: 32.61 births/1,000 population (1993 est.)
Death rate: 4.58 deaths/1,000 population (1993 est.)
Net migration rate: 0 migrant(s)/1,000 population (1993 est.)
Infant mortality rate: 26.4 deaths/1,000 live births (1993 est.)
Life expectancy at birth:
total population: 72.98 years
male: 71.42 years
female: 74.62 years (1993 est.)
Total fertility rate: 4.37 children born/woman (1993 est.)
Nationality:
noun: Paraguayan(s)
adjective: Paraguayan
Ethnic divisions: mestizo (Spanish and Indian) 95%, white and Indian 5%
Religions: Roman Catholic 90%, Mennonite and other Protestant denominations
Languages: Spanish (official), Guarani
Literacy: age 15 and over can read and write (1990)
total population: 90%
male: 92%
female: 88%
Labor force: 1.641 million (1992 est.)
by occupation: agriculture, industry and commerce, services, government (1986)

Government

Names:
conventional long form: Republic of Paraguay
conventional short form: Paraguay
local long form: Republica del Paraguay
local short form: Paraguay
Digraph: PA
Type: republic
Capital: Asuncion
Administrative divisions: 19 departments (departamentos, singular—departamento); Alto Paraguay, Alto Parana, Amambay, Boqueron, Caaguazu, Caazapa, Canindeyu, Central, Chaco, Concepcion, Cordillera, Guaira, Itapua, Misiones, Neembucu, Nueva Asuncion, Paraguari, Presidente Hayes, San Pedro
Independence: 14 May 1811 (from Spain)
Constitution: 25 August 1967; Constituent Assembly rewrote the Constitution that was promulgated on 20 June 1992
Legal system: based on Argentine codes, Roman law, and French codes; judicial review of legislative acts in Supreme Court of Justice; does not accept compulsory ICJ jurisdiction
National holiday: Independence Days, 14-15 May (1811)
Political parties and leaders: Colorado Party, Blas N. RIQUELME, president; Authentic Radical Liberal Party (PLRA), Domingo LAINO; Christian Democratic Party (PDC), Jose Angel BURRO;

305

Paraguay (continued)

Febrerista Revolutionary Party (PRF), Euclides ACEUEDO; Popular Democratic Party (PDP), Hugo RICHER; National Encounter (EN), Guillermo Caballero VARGAS

Other political or pressure groups: Confederation of Workers (CUT); Roman Catholic Church

Suffrage: 18 years of age; universal and compulsory up to age 60

Elections:

President: last held 1 May 1989 (next to be held 9 May 1993); results—Gen. RODRIGUEZ 75.8%, Domingo LAINO 19.4%

Chamber of Senators: last held 1 May 1989 (next to be held by 9 May 1993); results—percent of vote by party NA; seats—(36 total) Colorado Party 24, PLRA 10, PLR 1, PRF 1

Chamber of Deputies: last held on 1 May 1989 (next to be held by 9 May 1993); results—percent of vote by party NA; seats—(72 total) Colorado Party 48, PLRA 19, PRF 2, PDC 1, other 2

Executive branch: president, Council of Ministers (cabinet), Council of State

Legislative branch: bicameral Congress (Congreso) consists of an upper chamber or Chamber of Senators (Camara de Senadores) and a lower chamber or Chamber of Deputies (Camara de Diputados)

Judicial branch: Supreme Court of Justice (Corte Suprema de Justicia)

Leaders:

Chief of State and Head of Government: President Gen. Andres RODRIGUEZ Pedotti (since 15 May 1989)

Member of: AG (observer), CCC, ECLAC, FAO, G-77, IADB, IAEA, IBRD, ICAO, IDA, IFAD, IFC, ILO, IMF, INTELSAT, INTERPOL, IOC, IOM, ITU, LAES, LAIA, LORCS, MERCOSUR, OAS, OPANAL, PCA, RG, UN, UNCTAD, UNESCO, UNIDO, UPU, WCL, WHO, WIPO, WMO, WTO

Diplomatic representation in US:

chief of mission: Ambassador Juan Esteban Aguirre MARTINEZ

chancery: 2400 Massachusetts Avenue NW, Washington, DC 20008

telephone: (202) 483-6960 through 6962

consulates general: New Orleans and New York

consulate: Houston

US diplomatic representation:

chief of mission: Ambassador Jon David GLASSMAN

embassy: 1776 Avenida Mariscal Lopez, Asuncion

mailing address: C. P. 402, Asuncion, or APO AA 34036-0001

telephone: [595] (21) 213-715

FAX: [595] (21) 213-728

Flag: three equal, horizontal bands of red (top), white, and blue with an emblem centered in the white band; unusual flag in that the emblem is different on each side; the obverse (hoist side at the left) bears the national coat of arms (a yellow five-pointed star within a green wreath capped by the words REPUBLICA DEL PARAGUAY, all within two circles); the reverse (hoist side at the right) bears the seal of the treasury (a yellow lion below a red Cap of Liberty and the words Paz y Justicia (Peace and Justice) capped by the words REPUBLICA DEL PARAGUAY, all within two circles)

Economy

Overview: Agriculture, including forestry, accounts for about 25% of GDP, employs about 45% of the labor force, and provides the bulk of exports. Paraguay lacks substantial mineral or petroleum resources but does have a large hydropower potential. Since 1981 economic performance has declined compared with the boom period of 1976-81, when real GDP grew at an average annual rate of nearly 11%. During the period 1982-86 real GDP fell in three of five years, inflation jumped to an annual rate of 32%, and foreign debt rose. Factors responsible for the erratic behavior of the economy were the completion of the Itaipu hydroelectric dam, bad weather for crops, and weak international commodity prices for agricultural exports. In 1987 the economy experienced a minor recovery because of improved weather conditions and stronger international prices for key agricultural exports. The recovery continued through 1990, on the strength of bumper crops in 1988-89. In a major step to increase its economic activity in the region, Paraguay in March 1991 joined the Southern Cone Common Market (MERCOSUR), which includes Brazil, Argentina, and Uruguay. In 1992, the government, through an unorthodox approach, reduced external debt with both commercial and official creditors by purchasing a sizable amount of the delinquent commercial debt in the secondary market at a substantial discount. The government had paid 100% of remaining official debt arrears to the US, Germany, France, and Spain. All commercial debt arrears have been rescheduled. For the long run, the government must press forward with general, market-oriented economic reforms.

National product: GDP—exchange rate conversion—$7.3 billion (1992 est.)

National product real growth rate: 1.7% (1992 est.)

National product per capita: $1,500 (1992 est.)

Inflation rate (consumer prices): 20% (1992 est.)

Unemployment rate: 10% (1992 est.)

Budget: revenues $1.2 billion; expenditures $1.2 billion, including capital expenditures of $487 million (1991)

Exports: $719 million (f.o.b., 1992)

commodities: cotton, soybean, timber, vegetable oils, coffee, tung oil, meat products

partners: EC 37%, Brazil 25%, Argentina 10%, Chile 6%, US 6%

Imports: $1.33 billion (c.i.f., 1992)

commodities: capital goods 35%, consumer goods 20%, fuels and lubricants 19%, raw materials 16%, foodstuffs, beverages, and tobacco 10%

partners: Brazil 30%, EC 20%, US 18%, Argentina 8%, Japan 7%

External debt: $1.2 billion (1992 est.)

Industrial production: growth rate 5.9% (1989 est.); accounts for 17% of GDP

Electricity: 5,257,000 kW capacity; 16,200 million kWh produced, 3,280 kWh per capita (1992)

Industries: meat packing, oilseed crushing, milling, brewing, textiles, other light consumer goods, cement, construction

Agriculture: accounts for 25% of GDP and 44% of labor force; cash crops—cotton, sugarcane; other crops—corn, wheat, tobacco, soybeans, cassava, fruits, vegetables; animal products—beef, pork, eggs, milk; surplus producer of timber; self-sufficient in most foods

Illicit drugs: illicit producer of cannabis for the international drug trade; important transshipment point for Bolivian cocaine headed for the US and Europe

Economic aid: US commitments, including Ex-Im (FY70-89), $172 million; Western (non-US) countries, ODA and OOF bilateral commitments (1970-89), $1.1 billion

Currency: 1 guarani (G) = 100 centimos

Exchange rates: guaranies (G) per US$—1,637.6 (January 1993), 1,500.3 (1992), 447.5 (March 1992), 1,325.2 (1991), 1,229.8 (1990), 1,056.2 (1989), 550.00 (fixed rate 1986-February 1989)

Fiscal year: calendar year

Communications

Railroads: 970 km total; 440 km 1.435-meter standard gauge, 60 km 1.000-meter gauge, 470 km various narrow gauge (privately owned)

Highways: 21,960 km total; 1,788 km paved, 474 km gravel, and 19,698 km earth

Inland waterways: 3,100 km

Ports: Asuncion, Villeta, Ciudad del Este

Merchant marine: 13 ships (1,000 GRT or over) totaling 16,747 GRT/19,865 DWT; includes 11 cargo, 2 oil tanker; note—1 naval cargo ship is sometimes used commercially

Peru

Airports:
total: 862
usable: 719
with permanent-surface runways: 7
with runways over 3,659 m: 0
with runways 2,440-3,659 m: 4
with runways 1,220-2,439 m: 64
Telecommunications: principal center in Asuncion; fair intercity microwave net; 78,300 telephones; broadcast stations—40 AM, no FM, 5 TV, 7 shortwave; 1 Atlantic Ocean INTELSAT earth station

Defense Forces

Branches: Army, Navy (including Naval Air and Marines), Air Force
Manpower availability: males age 15-49 1,210,171; fit for military service 879,601; reach military age (17) annually 51,361 (1993 est.)
Defense expenditures: exchange rate conversion—$84 million, 1.4% of GDP (1988 est.)

Geography

Location: Western South America, bordering the South Pacific Ocean between Chile and Ecuador
Map references: South America, Standard Time Zones of the World
Area:
total area: 1,285,220 km²
land area: 1.28 million km²
comparative area: slightly smaller than Alaska
Land boundaries: total 6,940 km, Bolivia 900 km, Brazil 1,560 km, Chile 160 km, Colombia 2,900 km, Ecuador 1,420 km
Coastline: 2,414 km
Maritime claims:
territorial sea: 200 nm
International disputes: three sections of the boundary with Ecuador are in dispute
Climate: varies from tropical in east to dry desert in west
Terrain: western coastal plain (costa), high and rugged Andes in center (sierra), eastern lowland jungle of Amazon Basin (selva)
Natural resources: copper, silver, gold, petroleum, timber, fish, iron ore, coal, phosphate, potash
Land use:
arable land: 3%
permanent crops: 0%
meadows and pastures: 21%
forest and woodland: 55%
other: 21%
Irrigated land: 12,500 km² (1989 est.)
Environment: subject to earthquakes, tsunamis, landslides, mild volcanic activity; deforestation; overgrazing; soil erosion; desertification; air pollution in Lima
Note: shares control of Lago Titicaca, world's highest navigable lake, with Bolivia

People

Population: 23,210,352 (July 1993 est.)
Population growth rate: 1.9% (1993 est.)
Birth rate: 26.19 births/1,000 population (1993 est.)

Death rate: 7.15 deaths/1,000 population (1993 est.)
Net migration rate: 0 migrant(s)/1,000 population (1993 est.)
Infant mortality rate: 56.4 deaths/1,000 live births (1993 est.)
Life expectancy at birth:
total population: 65.17 years
male: 63.02 years
female: 67.44 years (1993 est.)
Total fertility rate: 3.22 children born/woman (1993 est.)
Nationality:
noun: Peruvian(s)
adjective: Peruvian
Ethnic divisions: Indian 45%, mestizo (mixed Indian and European ancestry) 37%, white 15%, black, Japanese, Chinese, and other 3%
Religions: Roman Catholic
Languages: Spanish (official), Quechua (official), Aymara
Literacy: age 15 and over can read and write (1990)
total population: 85%
male: 92%
female: 29%
Labor force: 8 million (1992)
by occupation: government and other services 44%, agriculture 37%, industry 19% (1988 est.)

Government

Names:
conventional long form: Republic of Peru
conventional short form: Peru
local long form: Republica del Peru
local short form: Peru
Digraph: PE
Type: republic
Capital: Lima
Administrative divisions: 24 departments (departamentos, singular—departamento) and 1 constitutional province* (provincia constitucional); Amazonas, Ancash, Apurimac, Arequipa, Ayacucho, Cajamarca, Callao*, Cusco, Huancavelica, Huanuco, Ica, Junin, La Libertad, Lambayeque, Lima, Loreto, Madre de Dios, Moquegua, Pasco, Piura, Puno, San Martin, Tacna, Tumbes, Ucayali
note: the 1979 Constitution and legislation enacted from 1987 to 1990 mandate the creation of regions (regiones, singular—region) intended to function eventually as autonomous economic and administrative entities; so far, 12 regions have been constituted from 23 existing departments—Amazonas (from Loreto), Andres Avelino Caceres (from Huanuco, Pasco, Junin), Arequipa (from Arequipa), Chavin (from Ancash), Grau (from Tumbes, Piura), Inca (from Cusco, Madre de Dios, Apurimac), La Libertad (from La Libertad),

Peru *(continued)*

Los Libertadores-Huari (from Ica, Ayacucho, Huancavelica), Mariategui (from Moquegua, Tacna, Puno), Nor Oriental del Maranon (from Lambayeque, Cajamarca, Amazonas), San Martin (from San Martin), Ucayali (from Ucayali); formation of another region has been delayed by the reluctance of the constitutional province of Callao to merge with the department of Lima; because of inadequate funding from the central government, the regions have yet to assume their responsibilities and at the moment coexist with the departmental structure
Independence: 28 July 1821 (from Spain)
Constitution: 28 July 1980 (often referred to as the 1979 Constitution because the Constituent Assembly met in 1979, but the Constitution actually took effect the following year); suspended 5 April 1992; being revised or replaced
Legal system: based on civil law system; has not accepted compulsory ICJ jurisdiction
National holiday: Independence Day, 28 July (1821)
Political parties and leaders: New Majority/Change 90 (Cambio 90), Alberto FUJIMORI; Popular Christian Party (PPC), Luis BEDOYA Reyes; Popular Action Party (AP), Eduardo CALMELL del Solar; Liberty Movement (ML), Luis BUSTAMANTE; American Popular Revolutionary Alliance (APRA), Alan GARCIA; Independent Moralizing Front (FIM), Fernando OLIVERA Vega; National Renewal, Rafael REY; Democratic Coordinator, Jose Barba CAHALLERO; Democratic Left Movement, Gloria HOFLER
Other political or pressure groups: leftist guerrilla groups include Shining Path, Abimael GUZMAN (imprisoned); Tupac Amaru Revolutionary Movement, Nestor SERPA and Victor POLAY (imprisoned)
Suffrage: 18 years of age; universal
Elections:
President: last held on 10 June 1990 (next to be held NA April 1995); results—Alberto FUJIMORI 56.53%, Mario VARGAS Llosa 33.92%, other 9.55%
Democratic Constituent Congress: last held 25 November 1992 (next to be held NA); seats—(80 total) New Majority/Change 90 44, Popular Christian Party 8, Independent Moralization Front 7, Renewal 6, Movement of the Democratic Left 4, Democratic Coordinator 4, others 7; several major parties (American Popular Revolutionary Alliance, Popular Action) did not participate
Executive branch: president, prime minister, Council of Ministers (cabinet)
Legislative branch: unicameral Democratic Constituent Congress (CCD)
Judicial branch: Supreme Court of Justice (Corte Suprema de Justicia)
Leaders:
Chief of State: President Alberto Kenyo FUJIMORI Fujimori (since 28 July 1990)

Head of Government: Prime Minister Oscar DE LA PUENTE Raygada (since 6 April 1992)
Member of: AG, CCC, ECLAC, FAO, G-11, G-15, G-19, G-24, G-77, GATT, IADB, IAEA, IBRD, ICAO, ICFTU, IDA, IFAD, IFC, ILO, IMF, IMO, INMARSAT, INTELSAT, INTERPOL, IOC, IOM, ISO, ITU, LAES, LAIA, LORCS, MINURSO, NAM, OAS, OPANAL, PCA, RG (suspended), UN, UNCTAD, UNESCO, UNIDO, UPU, WCL, WFTU, WHO, WIPO, WMO, WTO
Diplomatic representation in US:
chief of mission: Ambassador Ricardo LUNA
chancery: 1700 Massachusetts Avenue NW, Washington, DC 20036
telephone: (202) 833-9860 through 9869)
consulates general: Chicago, Houston, Los Angeles, Miami, New York, Paterson (New Jersey), San Francisco, and San Juan (Puerto Rico)
US diplomatic representation:
chief of mission: (vacant); Charge d'Affaires Charles H. BRAYSHAW
embassy: corner of Avenida Inca Garcilaso de la Vega and Avenida Espana, Lima
mailing address: P. O. Box 1991, Lima 1, or APO AA 34031
telephone: [51] (14) 33-8000
FAX: [51] (14) 31-6682
Flag: three equal, vertical bands of red (hoist side), white, and red with the coat of arms centered in the white band; the coat of arms features a shield bearing a llama, cinchona tree (the source of quinine), and a yellow cornucopia spilling out gold coins, all framed by a green wreath

---

Economy

Overview: The Peruvian economy is becoming increasingly market oriented, with a large dose of government ownership remaining in mining, energy, and banking. In the 1980s the economy suffered from hyperinflation, declining per capita output, and mounting external debt. Peru was shut off from IMF and World Bank support in the mid-1980s because of its huge debt arrears. An austerity program implemented shortly after the FUJIMORI government took office in July 1990 contributed to a third consecutive yearly contraction of economic activity, but the slide halted late that year, and output rose 2.4% in 1991. After a burst of inflation as the austerity program eliminated government price subsidies, monthly price increases eased to the single-digit level and by December 1991 dropped to the lowest increase since mid-1987. Lima obtained a financial rescue package from multilateral lenders in September 1991, although it faced $14 billion in arrears on its external debt. By

working with the IMF and World Bank on new financial conditions and arrangements, the government succeeded in ending its arrears by March 1993. In 1992, GDP fell by 2.8%, in part because a warmer-than-usual El Nino current resulted in a 30% drop in the fish catch. Meanwhile, revival of growth in GDP continued to be restricted by the large amount of public and private resources being devoted to strengthening internal security.
National product: GDP—exchange rate conversion—$25 billion (1992 est.)
National product real growth rate: -2.8% (1992 est.)
National product per capita: $1,100 (1992 est.)
Inflation rate (consumer prices): 56.7% (1992)
Unemployment rate: 15% (1992 est.); underemployment 70% (1992 est.)
Budget: revenues $2.0 billion; expenditures $2.7 billion, including capital expenditures of $300 million (1992 est.)
Exports: $3.5 billion (f.o.b., 1992)
commodities: copper, fishmeal, zinc, crude petroleum and byproducts, lead, refined silver, coffee, cotton
partners: EC 28%, US 22%, Japan 13%, Latin America 12%, former USSR 2% (1991)
Imports: $4.1 billion (f.o.b., 1992)
commodities: foodstuffs, machinery, transport equipment, iron and steel semimanufactures, chemicals, pharmaceuticals
partners: US 32%, Latin America 22%, EC 17%, Switzerland 6%, Japan 3% (1991)
External debt: $21 billion (December 1992 est.)
Industrial production: growth rate –5% (1992 est.); accounts for almost 24% of GDP
Electricity: 5,042,000 kW capacity; 17,434 million kWh produced, 760 kWh per capita (1992)
Industries: mining of metals, petroleum, fishing, textiles, clothing, food processing, cement, auto assembly, steel, shipbuilding, metal fabrication
Agriculture: accounts for 10% of GDP, about 35% of labor force; commercial crops—coffee, cotton, sugarcane; other crops—rice, wheat, potatoes, plantains, coca; animal products—poultry, red meats, dairy, wool; not self-sufficient in grain or vegetable oil; fish catch of 6.9 million metric tons (1990)
Illicit drugs: world's largest coca leaf producer with about 121,000 hectares under cultivation; source of supply for most of the world's coca paste and cocaine base; at least 85% of coca cultivation is for illicit production; most of cocaine base is shipped to Colombian drug dealers for processing into cocaine for the international drug market

Philippines

Economic aid: US commitments, including Ex-Im (FY70-89), $1.7 billion; Western (non-US) countries, ODA and OOF bilateral commitments (1970-89), $4.3 billion; Communist countries (1970-89), $577 million
Currency: 1 nuevo sol (S/.) = 100 centavos
Exchange rates: nuevo sol (S/. per US$1—1.690 (January 1993), 1.245 (1992), 0.772 (1991), 0.187 (1990), 2.666 (1989), 0.129 (1988)
Fiscal year: calendar year

Communications

Railroads: 1,801 km total; 1,501 km 1.435-meter gauge, 300 km 0.914-meter gauge
Highways: 69,942 km total; 7,459 km paved, 13,538 km improved, 48,945 km unimproved earth
Inland waterways: 8,600 km of navigable tributaries of Amazon system and 208 km Lago Titicaca
Pipelines: crude oil 800 km, natural gas and natural gas liquids 64 km
Ports: Callao, Ilo, Iquitos, Matarani, Talara
Merchant marine: 21 ships (1,000 GRT or over) totaling 194,473 GRT/307,845 DWT; includes 13 cargo, 1 refrigerated cargo, 1 roll-on/roll-off cargo, 2 oil tanker, 4 bulk; note—in addition, 6 naval tankers and 1 naval cargo are sometimes used commercially
Airports:
total: 228
usable: 199
with permanent-surface runways: 37
with runways over 3,659 m: 2
with runways 2,440-3,659 m: 23
with runways 1,220-2,439 m: 46
Telecommunications: fairly adequate for most requirements; nationwide microwave system; 544,000 telephones; broadcast stations—273 AM, no FM, 140 TV, 144 shortwave; satellite earth stations—2 Atlantic Ocean INTELSAT, 12 domestic

Defense Forces

Branches: Army (Ejercito Peruano), Navy (Marina de Guerra del Peru), Air Force (Fuerza Aerea del Peru), National Police
Manpower availability: males age 15-49 6,030,354; fit for military service 4,076,197; reach military age (20) annually 241,336 (1993 est.)
Defense expenditures: exchange rate conversion—$500 million, about 2% of GDP (1991)

Geography

Location: Southeast Asia, between Indonesia and China
Map references: Asia, Oceania, Southeast Asia, Standard Time Zones of the World
Area:
total area: 300,000 km²
land area: 298,170 km²
comparative area: slightly larger than Arizona
Land boundaries: 0 km
Coastline: 36,289 km
Maritime claims: measured from claimed archipelagic baselines
continental shelf: to depth of exploitation
exclusive economic zone: 200 nm
territorial sea: irregular polygon extending up to 100 nm from coastline as defined by 1898 treaty; since late 1970s has also claimed polygonal-shaped area in South China Sea up to 285 nm in breadth
International disputes: involved in a complex dispute over the Spratly Islands with China, Malaysia, Taiwan, Vietnam, and possibly Brunei; claims Malaysian state of Sabah
Climate: tropical marine; northeast monsoon (November to April); southwest monsoon (May to October)
Terrain: mostly mountains with narrow to extensive coastal lowlands
Natural resources: timber, petroleum, nickel, cobalt, silver, gold, salt, copper
Land use:
arable land: 26%
permanent crops: 11%
meadows and pastures: 4%
forest and woodland: 40%
other: 19%
Irrigated land: 16,200 km² (1989 est.)
Environment: astride typhoon belt, usually affected by 15 and struck by five to six cyclonic storms per year; subject to landslides, active volcanoes, destructive earthquakes, tsunami; deforestation; soil erosion; water pollution

People

Population: 68,464,368 (July 1993 est.)
Population growth rate: 1.97% (1993 est.)
Birth rate: 27.9 births/1,000 population (1993 est.)
Death rate: 7.03 deaths/1,000 population (1993 est.)
Net migration rate: -1.19 migrant(s)/1,000 population (1993 est.)
Infant mortality rate: 51.9 deaths/1,000 live births (1993 est.)
Life expectancy at birth:
total population: 65.13 years
male: 62.59 years
female: 67.79 years (1993 est.)
Total fertility rate: 3.45 children born/woman (1993 est.)
Nationality:
noun: Filipino(s)
adjective: Philippine
Ethnic divisions: Christian Malay 91.5%, Muslim Malay 4%, Chinese 1.5%, other 3%
Religions: Roman Catholic 83%, Protestant 9%, Muslim 5%, Buddhist and other 3%
Languages: Pilipino (official; based on Tagalog), English (official)
Literacy: age 15 and over can read and write (1990)
total population: 90%
male: 90%
female: 90%
Labor force: 24.12 million
by occupation: agriculture 46%, industry and commerce 16%, services 18.5%, government 10%, other 9.5% (1989)

Government

Names:
conventional long form: Republic of the Philippines
conventional short form: Philippines
local long form: Republika ng Pilipinas
local short form: Pilipinas
Digraph: RP
Type: republic
Capital: Manila
Administrative divisions: 73 provinces and 61 chartered cities*; Abra, Agusan del Norte, Agusan del Sur, Aklan, Albay, Angeles*, Antique, Aurora, Bacolod*, Bago*, Baguio*, Bais*, Basilan, Basilan City*, Bataan, Batanes, Batangas, Batangas City*, Benguet, Bohol, Bukidnon, Bulacan, Butuan*, Cabanatuan*, Cadiz*, Cagayan, Cagayan de Oro*, Calbayog*, Caloocan*, Camarines Norte, Camarines Sur, Camiguin, Canlaon*, Capiz, Catanduanes, Cavite, Cavite City*, Cebu, Cebu City*, Cotabato*, Dagupan*, Danao*, Dapitan*, Davao City* Davao, Davao del Sur, Davao Oriental, Dipolog*, Dumaguete*, Eastern Samar, General Santos*, Gingoog*, Ifugao, Iligan*, Ilocos Norte, Ilocos Sur, Iloilo, Iloilo City*,

309

Philippines (continued)

Iriga*, Isabela, Kalinga-Apayao, La Carlota*, Laguna, Lanao del Norte, Lanao del Sur, Laoag*, Lapu-Lapu*, La Union, Legaspi*, Leyte, Lipa*, Lucena*, Maguindanao, Mandaue*, Manila*, Marawi*, Marinduque, Masbate, Mindoro Occidental, Mindoro Oriental, Misamis Occidental, Misamis Oriental, Mountain, Naga*, Negros Occidental, Negros Oriental, North Cotabato, Northern Samar, Nueva Ecija, Nueva Vizcaya, Olongapo*, Ormoc*, Oroquieta*, Ozamis*, Pagadian*, Palawan, Palayan*, Pampanga, Pangasinan, Pasay*, Puerto Princesa*, Quezon, Quezon City*, Quirino, Rizal, Romblon, Roxas*, Samar, San Carlos* (in Negros Occidental), San Carlos* (in Pangasinan), San Jose*, San Pablo*, Silay*, Siquijor, Sorsogon, South Cotabato, Southern Leyte, Sultan Kudarat, Sulu, Surigao*, Surigao del Norte, Surigao del Sur, Tacloban*, Tagaytay*, Tagbilaran*, Tangub*, Tarlac, Tawitawi, Toledo*, Trece Martires*, Zambales, Zamboanga*, Zamboanga del Norte, Zamboanga del Sur

Independence: 4 July 1946 (from US)
Constitution: 2 February 1987, effective 11 February 1987
Legal system: based on Spanish and Anglo-American law; accepts compulsory ICJ jurisdiction, with reservations
National holiday: Independence Day, 12 June (1898) (from Spain)
Political parties and leaders: Democratic Filipino Struggle (Laban ng Demokratikong Pilipinas, Laban), Edgardo ESPIRITU; People Power-National Union of Christian Democrats (Lakas ng Edsa, NUCD and Partido Lakas Tao, Lakas/NUCD); Fidel V. RAMOS, President of the Republic, Raul MANGLAPUS, Jose de VENECIA, secretary general; Nationalist People's Coalition (NPC), Eduardo COJUANGCO; Liberal Party, Jovito SALONGA; People's Reform Party (PRP), Miriam DEFENSOR-SANTIAGO; New Society Movement (Kilusan Bagong Lipunan; KBL), Imelda MARCOS; Nacionalista Party (NP), Salvador H. LAUREL, president
Suffrage: 15 years of age; universal
Elections:
President: last held 11 May 1992 (next election to be held NA May 1998); results—Fidel Valdes RAMOS won 23.6% of votes, a narrow plurality
Senate: last held 11 May 1992 (next election to be held NA May 1995); results—LDP 66%, NPC 20%, Lakas-NUCD 8%, Liberal 6%; seats—(24 total) LDP 15, NPC 5, Lakas-NUCD 2, Liberal 1, Independent 1
House of Representatives: last held 11 May 1992 (next election to be held NA May 1995); results—LDP 43.5%; Lakas-NUCD 25%, NPC 23.5%, Liberal 5%, KBL 3%; seats—(200 total) LDP 87, NPC 45, Lakas-NUCD 41, Liberal 15, NP 6, KBL 3, Independent 3

Executive branch: president, vice president, Cabinet
Legislative branch: bicameral Congress (Kongreso) consists of an upper house or Senate (Senado) and a lower house or House of Representatives (Kapulungan Ng Mga Kinatawan)
Judicial branch: Supreme Court
Leaders:
Chief of State and Head of Government: President Fidel Valdes RAMOS (since 30 June 1992); Vice President Joseph Ejercito ESTRADA (since 30 June 1992)
Member of: APEC, AsDB, ASEAN, CCC, CP, ESCAP, FAO, G-24, G-77, GATT, IAEA, IBRD, ICAO, ICFTU, IDA, IFAD, IFC, ILO, IMF, IMO, INMARSAT, INTELSAT, INTERPOL, IOC, IOM, ISO, ITU, LORCS, NAM (observer), UN, UNCTAD, UNESCO, UNHCR, UNIDO, UNTAC, UPU, WCL, WFTU, WHO, WIPO, WMO, WTO
Diplomatic representation in US:
chief of mission: Ambassador Raul RABE
chancery: 1617 Massachusetts Avenue NW, Washington, DC 20036
telephone: (202) 483-1414
consulates general: Agana (Guam), Chicago, Honolulu, Houston, Los Angeles, New York, San Francisco, and Seattle
US diplomatic representation:
chief of mission: (vacant); Charge d'affaires Donald WESTMORE
embassy: 1201 Roxas Boulevard, Manila
mailing address: APO AP 96440
telephone: [63] (2) 521-7116
FAX: [63] (2) 522-4361
consulate general: Cebu
Flag: two equal horizontal bands of blue (top) and red with a white equilateral triangle based on the hoist side; in the center of the triangle is a yellow sun with eight primary rays (each containing three individual rays) and in each corner of the triangle is a small yellow five-pointed star

---

Economy

Overview: Domestic output in this primarily agricultural economy remained the same in 1992 as in 1991. Drought and power supply problems hampered production, while inadequate revenues prevented government pump priming. Despite a flat GDP performance, GNP mustered a small 0.6% expansion, attributable to inflows of workers' remittances combined with smaller foreign interest payments. A marked increase in capital goods imports, particularly power generations equipment, telecommunications equipment, and electronic data processors, contributed to a 20.5% import growth in 1992. Exports rose 11%, led by earnings from the Philippines' two leading manufactures—electronics and garments.

National product: GDP—exchange rate conversion—$54.1 billion (1992 est.)
National product real growth rate: 0.6% (1992 est.)
National product per capita: $860 (1992 est.)
Inflation rate (consumer prices): 8.9% (1992 est.)
Unemployment rate: 9.8% (1992 est.)
Budget: $11.0 billion; expenditures $12.0 billion, including capital expenditures of $NA (1992 est.)
Exports: $9.8 billion (f.o.b., 1992)
commodities: electronics, textiles, coconut oil, copper
partners: US 39%, EC, Japan, ASEAN
Imports: $14.5 billion (f.o.b., 1992)
commodities: raw materials 45%, capital goods 26%, petroleum products 18%
partners: US, Japan, Taiwan, Saudi Arabia
External debt: $29.8 billion (1992)
Industrial production: growth rate −1% (1992 est.); accounts for 34% of GDP
Electricity: 7,850,000 kW capacity; 28,000 million kWh produced, 420 kWh per capita (1992)
Industries: textiles, pharmaceuticals, chemicals, wood products, food processing, electronics assembly, petroleum refining, fishing
Agriculture: accounts for about one-third of GNP and about 45% of labor force; major crops—rice, coconuts, corn, sugarcane, bananas, pineapples, mangos; animal products—pork, eggs, beef; net exporter of farm products; fish catch of 2 million metric tons annually
Illicit drugs: illicit producer of cannabis for the international drug trade; growers are producing more and better quality cannabis despite government eradication efforts
Economic aid: US commitments, including Ex-Im (FY70-89), $3.6 billion; Western (non-US) countries, ODA and OOF bilateral commitments (1970-88), $7.9 billion; OPEC bilateral aid (1979-89), $5 million; Communist countries (1975-89), $123 million
Currency: 1 Philippine peso (P) = 100 centavos
Exchange rates: Philippine pesos (P) per US$1—25.817 (April 1993), 25.512 (1992), 27.479 (1991), 24.311 (1990), 21.737 (1989), 21.095 (1988)
Fiscal year: calendar year

Communications

Railroads: 378 km operable on Luzon, 34% government owned (1982)
Highways: 157,450 km total (1988); 22,400 km paved; 85,050 km gravel, crushed-stone, or stabilized-soil surface; 50,000 km unimproved earth
Inland waterways: 3,219 km; limited to shallow-draft (less than 1.5 m) vessels

Pipelines: petroleum products 357 km
Ports: Cagayan de Oro, Cebu, Davao, Guimaras, Iloilo, Legaspi, Manila, Subic Bay
Merchant marine: 562 ships (1,000 GRT or over) totaling 8,282,936 GRT/13,772,023 DWT; includes 1 passenger, 11 short-sea passenger, 13 passenger-cargo, 155 cargo, 27 refrigerated cargo, 25 vehicle carrier, 9 livestock carrier, 13 roll-on/roll-off cargo, 8 container, 38 oil tanker, 1 chemical tanker, 3 liquefied gas, 1 combination ore/oil, 249 bulk, 8 combination bulk; note—many Philippine flag ships are foreign owned and are on the register for the purpose of long-term bare-boat charter back to their original owners who are principally in Japan and Germany
Airports:
total: 270·
usable: 238
with permanent-surface runways: 73
with runways over 3,659 m: 0
with runways 2,440-3,659 m: 9
with runways 1,220-2,439 m: 57
Telecommunications: good international radio and submarine cable services; domestic and interisland service adequate; 872,900 telephones; broadcast stations—267 AM (including 6 US), 55 FM, 33 TV (including 4 US); submarine cables extended to Hong Kong, Guam, Singapore, Taiwan, and Japan; satellite earth stations—1 Indian Ocean INTELSAT, 2 Pacific Ocean INTELSAT, and 11 domestic

Defense Forces

Branches: Army, Navy (including Coast Guard and Marine Corps), Air Force
Manpower availability: males age 15-49 17,188,695; fit for military service 12,144,278; reach military age (20) annually 716,881 (1993 est.)
Defense expenditures: exchange rate conversion—$915 million, 1.9% of GNP (1991)

Pitcairn Islands
(dependent territory of the UK)

100 km

Sandy
Oeno
Henderson
Ducie
●→ ★ADAMSTOWN
Pitcairn

South Pacific Ocean

Geography

Location: in the South Pacific Ocean, about halfway between Peru and New Zealand
Map references: Oceania
Area:
total area: 47 km²
land area: 47 km²
comparative area: about 0.3 times the size of Washington, DC
Land boundaries: 0 km
Coastline: 51 km
Maritime claims:
exclusive fishing zone: 200 nm
territorial sea: 3 nm
International disputes: none
Climate: tropical, hot, humid, modified by southeast trade winds; rainy season (November to March)
Terrain: rugged volcanic formation; rocky coastline with cliffs
Natural resources: miro trees (used for handicrafts), fish
Land use:
arable land: NA%
permanent crops: NA%
meadows and pastures: NA%
forest and woodland: NA%
other: NA%
Irrigated land: NA km²
Environment: subject to typhoons (especially November to March)

People

Population: 52 (July 1993 est.)
Population growth rate: 0% (1993 est.)
Birth rate: NA births/1,000 population
Death rate: NA deaths/1,000 population
Net migration rate: NA migrant(s)/1,000 population
Infant mortality rate: NA deaths/1,000 live births
Life expectancy at birth:
total population: NA years
male: NA years
female: NA years

Total fertility rate: NA children born/woman
Nationality:
noun: Pitcairn Islander(s)
adjective: Pitcairn Islander
Ethnic divisions: descendants of the Bounty mutineers
Religions: Seventh-Day Adventist 100%
Languages: English (official), Tahitian/English dialect
Literacy:
total population: NA%
male: NA%
female: NA%
Labor force: NA
by occupation: no business community in the usual sense; some public works; subsistence farming and fishing

Government

Names:
conventional long form: Pitcairn, Henderson, Ducie, and Oeno Islands
conventional short form: Pitcairn Islands
Digraph: PC
Type: dependent territory of the UK
Capital: Adamstown
Administrative divisions: none (dependent territory of the UK)
Independence: none (dependent territory of the UK)
Constitution: Local Government Ordinance of 1964
Legal system: local island by-laws
National holiday: Celebration of the Birthday of the Queen, 10 June (1989) (second Saturday in June)
Political parties and leaders: NA
Other political or pressure groups: NA
Suffrage: 18 years of age; universal with three years residency
Elections:
Island Council: last held NA (next to be held NA); results—percent of vote by party NA; seats—(11 total, 5 elected) number of seats by party NA
Executive branch: British monarch, governor, island magistrate
Legislative branch: unicameral Island Council
Judicial branch: Island Court
Leaders:
Chief of State: Queen ELIZABETH II (since 6 February 1952), represented by the Governor and UK High Commissioner to New Zealand David Joseph MOSS (since NA 1990)
Head of Government: Island Magistrate and Chairman of the Island Council Jay WARREN (since NA)
Member of: SPC
Diplomatic representation in US: none (dependent territory of the UK)
US diplomatic representation: none (dependent territory of the UK)

Pitcairn Islands (continued)

Flag: blue with the flag of the UK in the upper hoist-side quadrant and the Pitcairn Islander coat of arms centered on the outer half of the flag; the coat of arms is yellow, green, and light blue with a shield featuring a yellow anchor

Economy

Overview: The inhabitants exist on fishing and subsistence farming. The fertile soil of the valleys produces a wide variety of fruits and vegetables, including citrus, sugarcane, watermelons, bananas, yams, and beans. Bartering is an important part of the economy. The major sources of revenue are the sale of postage stamps to collectors and the sale of handicrafts to passing ships.
National product: GDP $NA
National product real growth rate: NA%
National product per capita: $NA
Inflation rate (consumer prices): NA%
Unemployment rate: NA%
Budget: revenues $430,440; expenditures $429,983, including capital expenditures of $NA (FY87 est.)
Exports: $NA
commodities: fruits, vegetables, curios
partners: NA
Imports: $NA
commodities: fuel oil, machinery, building materials, flour, sugar, other foodstuffs
partners: NA
External debt: $NA
Industrial production: growth rate NA%
Electricity: 110 kW capacity; 0.30 million kWh produced, 5,360 kWh per capita (1990)
Industries: postage stamp sales, handicrafts
Agriculture: based on subsistence fishing and farming; wide variety of fruits and vegetables grown; must import grain products
Economic aid: none
Currency: 1 New Zealand dollar (NZ$) = 100 cents
Exchange rates: New Zealand dollars (NZ$) per US$1—1.9486 (January 1993), 1.8584 (1992), 1.7265 (1991), 1.6750 (1990), 1.6711 (1989), 1.5244 (1988)
Fiscal year: 1 April-31 March

Communications

Railroads: none
Highways: 6.4 km dirt roads
Ports: Bounty Bay
Airports: none
Telecommunications: 24 telephones; party line telephone service on the island; broadcast stations—1 AM, no FM, no TV; diesel generator provides electricity

Defense Forces

Note: defense is the responsibility of the UK

Poland

Boundary representation is not necessarily authoritative.

Geography

Location: Central Europe, between Germany and Belarus
Map references: Asia, Ethnic Groups in Eastern Europe, Europe, Standard Time Zones of the World
Area:
total area: 312,680 km²
land area: 304,510 km²
comparative area: slightly smaller than New Mexico
Land boundaries: total 3,114 km, Belarus 605 km, Czech Republic 658 km, Germany 456 km, Lithuania 91 km, Russia (Kaliningrad Oblast) 432 km, Slovakia 444 km, Ukraine 428 km
Coastline: 491 km
Maritime claims:
exclusive economic zone: 200 nm
territorial sea: 12 nm
International disputes: none
Climate: temperate with cold, cloudy, moderately severe winters with frequent precipitation; mild summers with frequent showers and thundershowers
Terrain: mostly flat plain; mountains along southern border
Natural resources: coal, sulfur, copper, natural gas, silver, lead, salt
Land use:
arable land: 46%
permanent crops: 1%
meadows and pastures: 13%
forest and woodland: 28%
other: 12%
Irrigated land: 1,000 km² (1989 est.)
Environment: plain crossed by a few north flowing, meandering streams; severe air and water pollution in south
Note: historically, an area of conflict because of flat terrain and the lack of natural barriers on the North European Plain

People

Population: 38,519,486 (July 1993 est.)
Population growth rate: 0.35% (1993 est.)

Birth rate: 13.59 births/1,000 population (1993 est.)
Death rate: 9.59 deaths/1,000 population (1993 est.)
Net migration rate: -0.52 migrant(s)/1,000 population (1993 est.)
Infant mortality rate: 13.8 deaths/1,000 live births (1993 est.)
Life expectancy at birth:
total population: 72.2 years
male: 68.14 years
female: 76.51 years (1993 est.)
Total fertility rate: 1.97 children born/woman (1993 est.)
Nationality:
noun: Pole(s)
adjective: Polish
Ethnic divisions: Polish 97.6%, German 1.3%, Ukrainian 0.6%, Belarusian 0.5% (1990 est.)
Religions: Roman Catholic 95% (about 75% practicing), Eastern Orthodox, Protestant, and other 5%
Languages: Polish
Literacy: age 15 and over can read and write (1978)
total population: 98%
male: 99%
female: 98%
Labor force: 15.609 million
by occupation: industry and construction 34.4%, agriculture 27.3%, trade, transport, and communications 16.1%, government and other 22.2% (1991)

Government

Names:
conventional long form: Republic of Poland
conventional short form: Poland
local long form: Rzeczpospolita Polska
local short form: Polska
Digraph: PL
Type: democratic state
Capital: Warsaw
Administrative divisions: 49 provinces (wojewodztwa, singular—wojewodztwo); Biala Podlaska, Bialystok, Bielsko Biala, Bydgoszcz, Chelm, Ciechanow, Czestochowa, Elblag, Gdansk, Gorzow, Jelenia Gora, Kalisz, Katowice, Kielce, Konin, Koszalin, Krakow, Krosno, Legnica, Leszno, Lodz, Lomza, Lublin, Nowy Sacz, Olsztyn, Opole, Ostroleka, Pila, Piotrkow, Plock, Poznan, Przemysl, Radom, Rzeszow, Siedlce, Sieradz, Skierniewice, Slupsk, Suwalki, Szczecin, Tarnobrzeg, Tarnow, Torun, Walbrzych, Warszawa, Wloclawek, Wroclaw, Zamosc, Zielona Gora
Independence: 11 November 1918 (independent republic proclaimed)
Constitution: interim "small constitution" came into effect in December 1992 replacing the Communist-imposed Constitution of 22 July 1952; new democratic Constitution being drafted

Legal system: mixture of Continental (Napoleonic) civil law and holdover Communist legal theory; changes being gradually introduced as part of broader democratization process; limited judicial review of legislative acts; has not accepted compulsory ICJ jurisdiction

National holiday: Constitution Day, 3 May (1791)

Political parties and leaders:

post-Solidarity parties: Democratic Union (UD), Tadeusz MAZOWIECKI; Christian-National Union (ZCHN), Wieslaw CHRZANOWSKI; Centrum (PC), Jaroslaw KACZYNSKI; Liberal-Democratic Congress, Donald TUSK; Peasant Alliance (PL), Gabriel JANOWSKI; Solidarity Trade Union (NSZZ), Marian KRZAKLEWSKI; Union of Labor (UP), Ryszard BUGAJ; Christian-Democratic Party (PCHD), Pawel LACZKOWSKI; Conservative Party, Alexander HALL

non-Communist, non-Solidarity: Confederation for an Independent Poland (KPN), Leszek MOCZULSKI; Polish Economic Program (PPG), Janusz REWINSKI; Christian Democrats (CHD), Andrzej OWSINSKI; German Minority (MN), Henryk KROL; Union of Real Politics (UPR), Janusz KORWIN-MIKKE; Democratic Party (SD), Antoni MACKIEWICZ; Party X, Stanislaw Tyminski

Communist origin or linked: Social Democracy (SDRP, party of Poland), Wlodzimierz Cimoszewicz; Polish Peasants' Party (PSL), Waldemar PAWLAK

Other political or pressure groups: powerful Roman Catholic Church; Solidarity (trade union); All Poland Trade Union Alliance (OPZZ), populist program

Suffrage: 18 years of age; universal

Elections:

president: first round held 25 November 1990, second round held 9 December 1990 (next to be held NA November 1995); results—second round Lech WALESA 74.7%, Stanislaw TYMINSKI 25.3%

Senat: last held 27 October 1991 (next to be held no later than NA October 1995); seats—(100 total)

post-Solidarity bloc: UD 21, NSZZ 11, ZCHN 9, PC 9, Liberal-Democratic Congress 6, PL 7, PCHD 3, other local candidates 11;

non-Communist, non-Solidarity: KPN 4, CHD 1, MN 1, local candidates 5

Communist origin or linked: PSL 8, SLD 4

Sejm: last held 27 October 1991 (next to be held no later than NA October 1995); seats—(460 total)

post-Solidarity bloc: UD 62, ZCHN 49, PC 44, Liberal-Democratic Congress 37, PL 28, NSZZ 27, SP 4, PCHD 4, RDS 1, Krackow Coalition in Solidarity with the President 1, Piast Agreement 1, Bydgoszcz Peasant List 1, Solidarity 80 1

non-Communist, non-Solidarity: KPN 46, PPPP 16, MN 7, CHD 5, Western Union 4, UPR 3, Autonomous Silesia 2, SD 1, Orthodox Election Committee 1, Committee of Women Against Hardships 1, Podhale Union 1, Wielkopolska Group 1, Wielkopolska and Lubuski Inhabitants 1, Party X 3

Communist origin or linked: SLD 60, PSL 48

Executive branch: president, prime minister, Council of Ministers (cabinet)

Legislative branch: bicameral National Assembly (Zgromadzenie Narodowe) consists of an upper house or Senate (Senat) and a lower house or Diet (Sejm)

Judicial branch: Supreme Court

Leaders:

Chief of State: President Lech WALESA (since 22 December 1990)

Head of Government: Prime Minister Hanna SUCHOCKA (since 10 July 1992)

Member of: BIS, CBSS, CCC, CE, CEI, CERN, CSCE, EBRD, ECE, FAO, GATT, IAEA, IBRD, ICAO, ICFTU, IDA, IFC, ILO, IMF, IMO, INMARSAT, INTERPOL, IOC, IOM (observer), ISO, ITU, LORCS, MINURSO, NACC, NAM (guest), NSG, PCA, UN, UNCTAD, UNESCO, UNDOF, UNIDO, UNIFIL, UNIKOM, UNPROFOR, UNTAC, UPU, WCL, WHO, WIPO, WMO, WTO, ZC

Diplomatic representation in US:

chief of mission: Ambassador Kazimierz DZIEWANOWSKI

chancery: 2640 16th Street NW, Washington DC 20009

telephone: (202) 234-3800 through 3802

FAX: (202) 328-6271

consulates general: Chicago, Los Angeles, and New York

US diplomatic representation:

chief of mission: Ambassador Thomas W. SIMONS, Jr.

embassy: Aleje Ujazdowskie 29/31, Warsaw

mailing address: American Embassy Warsaw, Box 5010, Unit 25402, or APO AE 09213-5010

telephone: [48] (2) 628-3041

FAX: [48] (2) 628-8298

consulates general: Krakow, Poznan

Flag: two equal horizontal bands of white (top) and red; similar to the flags of Indonesia and Monaco which are red (top) and white

---

Economy

Overview: Poland is undergoing a difficult transition from a Soviet-style economy—with state ownership and control of productive assets—to a market economy. On January 1, 1990, the new Solidarity-led government implemented shock therapy by slashing subsidies, decontrolling prices, tightening the money supply, stabilizing the foreign exchange rate, lowering import barriers, and restraining state sector wages. As a result, consumer goods shortages and lines disappeared, and inflation fell from 640% in 1989 to 44% in 1992. Western governments, which hold two-thirds of Poland's $48 billion external debt, pledged in 1991 to forgive half of Poland's official debt by 1994. The private sector accounted for 29% of industrial production and nearly half of nonagricultural output in 1992. Production fell in state enterprises, however, and the unemployment rate climbed steadily from virtually nothing in 1989 to 13.6% in December 1992. Poland fell out of compliance with its IMF program by mid-1991, and talks with commercial creditors stalled. The increase in unemployment and the decline in living standards led to strikes in the coal, auto, copper, and railway sectors in 1992. Large state enterprises in the coal, steel, and defense sectors plan to halve employment over the next decade, and the government expects unemployment to reach 3 million (16%) in 1993. A shortfall in tax revenues caused the budget deficit to reach 6% of GDP in 1992, but industrial production began a slow, uneven upturn. In 1993, the government will struggle to win legislative approval for faster privatization and to keep the budget deficit within IMF-approved limits.

National product: GDP—purchasing power equivalent—$167.6 billion (1992 est.)

National product real growth rate: 2% (1992 est.)

National product per capita: $4,400 (1992 est.)

Inflation rate (consumer prices): 44% (1992)

Unemployment rate: 13.6% (December 1992)

Budget: revenues $17.5 billion; expenditures $22.0 billion, including capital expenditures of $1.5 billion (1992 est.)

Exports: $12.8 billion (f.o.b., 1992 est.)

commodities: machinery 22%, metals 16%, chemicals 12%, fuels and power 11%, food 10% (1991)

partners: Germany 28.0%, former USSR 11.7%, UK 8.8%, Switzerland 5.5% (1991)

Imports: $12.9 billion (f.o.b., 1992 est.)

commodities: machinery 38%, fuels and power 20%, chemicals 13%, food 10%, light industry 6% (1991)

partners: Germany 17.4%, former USSR 25.6%, Italy 5.3%, Austria 5.2% (1991)

External debt: $48.5 billion (January 1992); note—Poland's Western government creditors promised in 1991 to forgive 30% of Warsaw's official debt—currently $33 billion—immediately and to forgive another 20% in 1994, if Poland adheres to its IMF program

Poland (continued)

Industrial production: growth rate 3.5% (1992)
Electricity: 31,530,000 kW capacity; 137,000 million kWh produced, 3,570 kWh per capita (1992)
Industries: machine building, iron and steel, extractive industries, chemicals, shipbuilding, food processing, glass, beverages, textiles
Agriculture: accounts for 15% of GDP and 27% of labor force; 75% of output from private farms, 25% from state farms; productivity remains low by European standards; leading European producer of rye, rapeseed, and potatoes; wide variety of other crops and livestock; major exporter of pork products; normally self-sufficient in food
Illicit drugs: illicit producers of opium for domestic consumption and amphetamines for the international market; emerging as a transshipment point for illicit drugs to Western Europe
Economic aid: donor—bilateral aid to non-Communist less developed countries, $2.2 billion (1954-89); the G-24 has pledged $8 billion in grants and credit guarantees to Poland
Currency: 1 zloty (Zl) = 100 groszy
Exchange rates: zlotych (Zl) per US$1—15,879 (January 1993), 13,626 (1992), 10,576 (1991), 9,500 (1990), 1,439.18 (1989), 430.55 (1988)
Fiscal year: calendar year

Communications

Railroads: 26,250 km total; 23,857 km 1.435-meter gauge, 397 km 1.520-meter gauge, 1,996 km narrow gauge; 8,987 km double track; 11,510 km electrified; government owned (1991)
Highways: 360,629 km total (excluding farm, factory and forest roads); 220 km limited access expressways, 45,257 km main highways, 128,775 km regional roads, 186,377 urban or village roads (local traffic); 220,000 km are paved (including all main and regional highways) (1988)
Inland waterways: 3,997 km navigable rivers and canals (1991)
Pipelines: natural gas 4,600 km, crude oil 1,986 km, petroleum products 360 km (1992)
Ports: Gdansk, Gdynia, Szczecin, Swinoujscie; principal inland ports are Gliwice on Kana Gliwice, Wrocaw on the Oder, and Warsaw on the Vistula
Merchant marine: 209 ships (1,000 GRT or over) totaling 2,747,631 GRT/3,992,053 DWT; includes 5 short-sea passenger, 76 cargo, 1 refrigerated cargo, 11 roll-on/roll-off cargo, 9 container, 1 oil tanker, 4 chemical tanker, 101 bulk, 1 passenger; Poland owns 1 ship of 6,333 DWT operating under Liberian registry

Airports:
total: 163
usable: 163
with permanent-surface runways: 100
with runway over 3,659 m: 0
with runways 2,440-3,659 m: 51
with runways 1,220-2,439 m: 95
Telecommunications: severely underdeveloped and outmoded system; cable, open wire and microwave; phone density is 10.5 phones per 100 residents (October 1990); 3.6 million telephone subscribers; exchanges are 86% automatic (1991); broadcast stations—27 AM, 27 FM, 40 (5 Soviet repeaters) TV; 9.6 million TVs; 1 satellite earth station using INTELSAT, EUTELSAT, INMARSAT and Intersputnik

---

Defense Forces

Branches: Army, Navy, Air and Air Defense Force
Manpower availability: males age 15-49 9,914,128; fit for military service 7,774,499; reach military age (19) annually 304,956 (1993 est.)
Defense expenditures: 30.8 trillion zlotych, 1.8% of GNP (1993 est.); note—conversion of defense expenditures into US dollars using the current exchange rate could produce misleading results

Portugal

Geography

Location: Southern Europe, bordering the North Atlantic Ocean west of Spain
Map references: Africa, Europe, Standard Time Zones of the World
Area:
total area: 92,080 km²
land area: 91,640 km²
comparative area: slightly smaller than Indiana
note: includes Azores and Madeira Islands
Land boundaries: total 1,214 km, Spain 1,214 km
Coastline: 1,793 km
Maritime claims:
continental shelf: 200 m depth or to depth of exploitation
exclusive economic zone: 200 nm
territorial sea: 12 nm
International disputes: sovereignty over Timor Timur (East Timor Province) disputed with Indonesia
Climate: maritime temperate; cool and rainy in north, warmer and drier in south
Terrain: mountainous north of the Tagus, rolling plains in south
Natural resources: fish, forests (cork), tungsten, iron ore, uranium ore, marble
Land use:
arable land: 32%
permanent crops: 6%
meadows and pastures: 6%
forest and woodland: 40%
other: 16%
Irrigated land: 6,340 km² (1989 est.)
Environment: Azores subject to severe earthquakes
Note: Azores and Madeira Islands occupy strategic locations along western sea approaches to Strait of Gibraltar

---

People

Population: 10,486,140 (July 1993 est.)
Population growth rate: 0.36% (1993 est.)
Birth rate: 11.59 births/1,000 population (1993 est.)

Death rate: 9.77 deaths/1,000 population (1993 est.)

Net migration rate: 1.8 migrant(s)/1,000 population (1993 est.)

Infant mortality rate: 9.8 deaths/1,000 live births (1993 est.)

Life expectancy at birth:
total population: 74.89 years
male: 71.43 years
female: 78.56 years (1993 est.)

Total fertility rate: 1.45 children born/woman (1993 est.)

Nationality:
noun: Portuguese (singular and plural)
adjective: Portuguese

Ethnic divisions: homogeneous Mediterranean stock in mainland, Azores, Madeira Islands; citizens of black African descent who immigrated to mainland during decolonization number less than 100,000

Religions: Roman Catholic 97%, Protestant denominations 1%, other 2%

Languages: Portuguese

Literacy: age 15 and over can read and write (1990)
total population: 85%
male: 89%
female: 82%

Labor force: 4,605,700
by occupation: services 45%, industry 35%, agriculture 20% (1988)

Government

Names:
conventional long form: Portuguese Republic
conventional short form: Portugal
local long form: Republica Portuguesa
local short form: Portugal

Digraph: PO

Type: republic

Capital: Lisbon

Administrative divisions: 18 districts (distritos, singular—distrito) and 2 autonomous regions* (regioes autonomas, singular—regiao autonoma); Aveiro, Acores (Azores)*, Beja, Braga, Braganca, Castelo Branco, Coimbra, Evora, Faro, Guarda, Leiria, Lisboa, Madeira*, Portalegre, Porto, Santarem, Setubal, Viana do Castelo, Vila Real, Viseu

Dependent areas: Macau (scheduled to become a Special Administrative Region of China on 20 December 1999)

Independence: 1140 (independent republic proclaimed 5 October 1910)

Constitution: 25 April 1976, revised 30 October 1982 and 1 June 1989

Legal system: civil law system; the Constitutional Tribunal reviews the constitutionality of legislation; accepts compulsory ICJ jurisdiction, with reservations

National holiday: Day of Portugal, 10 June

Political parties and leaders: Social Democratic Party (PSD), Anibal CAVACO Silva; Portuguese Socialist Party (PS), Antonio GUTERRES; Party of Democratic Renewal (PRD), Pedro CANAVARRO; Portuguese Communist Party (PCP), Carlos CARVALHAS; Social Democratic Center (CDS), Manuel MONTEIRO; National Solidarity Party, Manuel SERGIO; Center Democratic Party; United Democratic Coalition (CDU; Communists)

Suffrage: 18 years of age; universal

Elections:
President: last held 13 February 1991 (next to be held NA February 1996); results—Dr. Mario Lopes SOARES 70%, Basilio HORTA 14%, Carlos CARVALHAS 13%, Carlos MARQUES 3%
Assembly of the Republic: last held 6 October 1991 (next to be held NA October 1995); results—PSD 50.4%, PS 29.3%, CDU 8.8%, Center Democrats 4.4%, National Solidarity Party 1.7%, PRD 0.6%, other 4.8%; seats—(230 total) PSD 135, PS 72, CDU 17, Center Democrats 5, National Solidarity Party 1

Executive branch: president, Council of State, prime minister, deputy prime minister, Council of Ministers (cabinet)

Legislative branch: unicameral Assembly of the Republic (Assembleia da Republica)

Judicial branch: Supreme Tribunal of Justice (Supremo Tribunal de Justica)

Leaders:
Chief of State: President Dr. Mario Alberto Nobre Lopes SOARES (since 9 March 1986)
Head of Government: Prime Minister Anibal CAVACO SILVA (since 6 November 1985)

Member of: AfDB, Australian Group, BIS, CCC, CE, CERN, COCOM, CSCE, EBRD, EC, ECE, ECLAC, EIB, FAO, GATT, IADB, IAEA, IBRD, ICAO, ICC, ICFTU, IEA, IFAD, IFC, ILO, IMF, IMO, INMARSAT, INTELSAT, INTERPOL, IOC, IOM, ISO, ITU, LAIA (observer), LORCS, MTCR, NACC, NAM (guest), NATO, NEA, NSG, OAS (observer), OECD, PCA, UN, UNCTAD, UNESCO, UNIDO, UNOMOZ, UNPROFOR, UPU, WCL, WEU, WHO, WIPO, WMO, WTO, ZC

Diplomatic representation in US:
chief of mission: Ambassador Francisco Jose Laco Treichler KNOPFLI
chancery: 2125 Kalorama Road NW, Washington DC 20008
telephone: (202) 328-8610
consulates general: Boston, New York, Newark (New Jersey), and San Francisco
consulates: Los Angeles, New Bedford (Massachusetts), and Providence (Rhode Island)

US diplomatic representation:
chief of mission: Ambassador Everett Ellis BRIGGS

embassy: Avenida das Forcas Armadas, 1600 Lisbon
mailing address: PSC 83, APO AE 09726
telephone: [351] (1) 726-6600 or 6659, 8670, 8880
FAX: [351] (1) 726-9109
consulate: Ponta Delgada (Azores)

Flag: two vertical bands of green (hoist side, two-fifths) and red (three-fifths) with the Portuguese coat of arms centered on the dividing line

---

Economy

Overview: Although Portugal has experienced strong growth since joining the EC in 1986—at least 4% each year through 1990—it remains one of the poorest members. To prepare for the European single market, the government is restructuring and modernizing the economy and in 1989 embarked on a major privatization program. As of 1 January 1993, Lisbon has fully liberalized its capital markets and most trade markets. The global slowdown and tight monetary policies to counter inflation caused growth to slow in 1991 and 1992. Growth probably will remain depressed in 1993, but should pick up again in 1994.

National product: GDP—purchasing power equivalent—$93.7 billion (1992)

National product real growth rate: 1.1% (1992)

National product per capita: $9,000 (1992)

Inflation rate (consumer prices): 9% (1992)

Unemployment rate: 5% (1992)

Budget: revenues $27.3 billion; expenditures $33.2 billion, including capital expenditures of $4.5 billion (1991)

Exports: $16.3 billion (f.o.b., 1992 est.)
commodities: cotton textiles, cork and paper products, canned fish, wine, timber and timber products, resin, machinery, appliances
partners: EC 75.4%, other developed countries 12.4%, US 3.8% (1991)

Imports: $26.0 billion (c.i.f., 1992 est.)
commodities: machinery and transport equipment, agricultural products, chemicals, petroleum, textiles
partners: EC 72%, other developed countries 10.9% less developed countries 12.9%, US 3.4%

External debt: $16.9 billion (1992 est.)

Industrial production: growth rate 9.1% (1990); accounts for 40% of GDP

Electricity: 6,624,000 kW capacity; 26,400 million kWh produced, 2,520 kWh per capita (1992)

Industries: textiles and footwear; wood pulp, paper, and cork; metalworking; oil refining; chemicals; fish canning; wine; tourism

Agriculture: accounts for 6.1% of GDP and 20% of labor force; small, inefficient farms; imports more than half of food needs;

Portugal *(continued)*

major crops—grain, potatoes, olives, grapes; livestock sector—sheep, cattle, goats, poultry, meat, dairy products
Illicit drugs: increasingly important gateway country for Latin American cocaine entering the European market
Economic aid: US commitments, including Ex-Im (FY70-89), $1.8 billion; Western (non-US) countries, ODA and OOF bilateral commitments (1970-89), $1.2 billion
Currency: 1 Portuguese escudo (Esc) = 100 centavos
Exchange rates: Portuguese escudos (Esc) per US$1—145.51 (January 1993), 135.00 (1992), 144.48 (1991), 142.55 (1990), 157.46 (1989), 143.95 (1988)
Fiscal year: calendar year

Communications

Railroads: 3,625 km total; state-owned Portuguese Railroad Co. (CP) operates 2,858 km 1.665-meter gauge (434 km electrified and 426 km double track), 755 km 1.000-meter gauge; 12 km (1.435-meter gauge) electrified, double track, privately owned
Highways: 73,661 km total; 61,599 km surfaced (bituminous, gravel, and crushed stone), including 140 km of limited-access divided highway; 7,962 km improved earth; 4,100 km unimproved earth (motorable tracks)
Inland waterways: 820 km navigable; relatively unimportant to national economy, used by shallow-draft craft limited to 300-metric-ton cargo capacity
Pipelines: crude oil 11 km; petroleum products 58 km
Ports: Leixoes, Lisbon, Porto, Ponta Delgada (Azores), Velas (Azores), Setubal, Sines
Merchant marine: 51 ships (1,000 GRT or over) totaling 634,072 GRT/1,130,515 DWT; includes 1 short-sea passenger, 21 cargo, 3 refrigerated cargo, 3 container, 1 roll-on/roll-off cargo, 13 oil tanker, 2 chemical tanker, 5 bulk, 2 liquified gas; note—Portugal has created a captive register on Madeira (MAR) for Portuguese-owned ships that will have the taxation and crewing benefits of a flag of convenience; although only one ship currently is known to fly the Portuguese flag on the MAR register, it is likely that a majority of Portuguese flag ships will transfer to this subregister in a few years
Airports:
total: 64
usable: 62
with permanent-surface runways: 36
with runways over 3,659 m: 2
with runways 2,440-3,659 m: 10
with runways 1,220-2,439 m: 11

Telecommunications: generally adequate integrated network of coaxial cables, open wire and microwave radio relay; 2,690,000 telephones; broadcast stations—57 AM, 66 (22 repeaters) FM, 66 (23 repeaters) TV; 6 submarine cables; 3 INTELSAT earth stations (2 Atlantic Ocean, 1 Indian Ocean), EUTELSAT, domestic satellite systems (mainland and Azores); tropospheric link to Azores

Defense Forces

Branches: Army, Navy (including Marines), Air Force, National Republican Guard, Fiscal Guard, Public Security Police
Manpower availability: males age 15-49 2,696,325; fit for military service 2,188,041; reach military age (20) annually 88,735 (1993 est.)
Defense expenditures: exchange rate conversion—$2.4 billion, 2.9% of GDP (1992)

Puerto Rico
(commonwealth associated with the US)

Geography

Location: in the North Atlantic Ocean, between the Dominican Republic and the Virgin Islands group
Map references: Central America and the Caribbean
Area:
total area: 9,104 km²
land area: 8,959 km²
comparative area: slightly less than three times the size of Rhode Island
Land boundaries: 0 km
Coastline: 501 km
Maritime claims:
contiguous zone: 24 nm
continental shelf: 200 m (depth)
exclusive economic zone: 200 nm
territorial sea: 12 nm
International disputes: none
Climate: tropical marine, mild, little seasonal temperature variation
Terrain: mostly mountains with coastal plain belt in north; mountains precipitous to sea on west coast; sandy beaches along most coastal areas
Natural resources: some copper and nickel, potential for onshore and offshore crude oil
Land use:
arable land: 8%
permanent crops: 9%
meadows and pastures: 41%
forest and woodland: 20%
other: 22%
Irrigated land: 390 km² (1989 est.)
Environment: many small rivers and high central mountains ensure land is well watered; south coast relatively dry; fertile coastal plain belt in north
Note: important location along the Mona Passage—a key shipping lane to the Panama Canal; San Juan is one of the biggest and best natural harbors in the Caribbean

People

Population: 3,797,082 (July 1993 est.)
Population growth rate: 0.13% (1993 est.)

Birth rate: 16.93 births/1,000 population (1993 est.)

Death rate: 7.88 deaths/1,000 population (1993 est.)

Net migration rate: -7.75 migrant(s)/1,000 population (1993 est.)

Infant mortality rate: 14 deaths/1,000 live births (1993 est.)

Life expectancy at birth:

total population: 73.84 years

male: 70.25 years

female: 77.61 years (1993 est.)

Total fertility rate: 2.08 children born/woman (1993 est.)

Nationality:

noun: Puerto Rican(s)

adjective: Puerto Rican

Ethnic divisions: Hispanic

Religions: Roman Catholic 85%, Protestant denominations and other 15%

Languages: Spanish (official), English widely understood

Literacy: age 15 and over can read and write (1980)

total population: 89%

male: 90%

female: 88%

Labor force: 1.17 million (1992)

by occupation: government 20%, manufacturing 14%, trade 17%, construction 5%, communications and transportation 5%, other 39% (1992)

Government

Names:

conventional long form: Commonwealth of Puerto Rico

conventional short form: Puerto Rico

Digraph: QR

Type: commonwealth associated with the US

Capital: San Juan

Administrative divisions: none (commonwealth associated with the US), note: there are 78 municipalities

Independence: none (commonwealth associated with the US)

Constitution: ratified 3 March 1952; approved by US Congress 3 July 1952; effective 25 July 1952

Legal system: based on Spanish civil code

National holiday: US Independence Day, 4 July (1776)

Political parties and leaders: National Republican Party of Puerto Rico, Freddy VALENTIN; Popular Democratic Party (PPD), Rafael HERNANDEZ Colon; New Progressive Party (PNP), Carlos ROMERO Barcelo; Puerto Rican Socialist Party (PSP), Juan MARI Bras and Carlos GALLISA; Puerto Rican Independence Party (PIP), Ruben BERRIOS Martinez; Puerto Rican Communist Party (PCP), leader(s) unknown

Other political or pressure groups: all have engaged in terrorist activities—Armed Forces for National Liberation (FALN); Volunteers of the Puerto Rican Revolution; Boricua Popular Army (also known as the Macheteros); Armed Forces of Popular Resistance

Suffrage: 18 years of age; universal; indigenous inhabitants are US citizens but do not vote in US presidential elections

Elections:

Governor: last held 3 November 1992 (next to be held NA November 1996); results—Pedro ROSSELLO (PND) 50%, Victoria MUNOZ (PPD) 46%, Fernando MARTIN (PIP) 4%

Senate: last held 3 November 1992 (next to be held NA November 1996); results—percent of vote by party NA; seats—(27 total) seats by party NA

US House of Representatives: last held 3 November 1992 (next to be held NA November 1996); results—percent of vote by party NA; seats—(1 total) seats by party NA; note—Puerto Rico elects one representative to the US House of Representatives, Carlos Romero BARCELO

House of Representatives: last held 3 November 1992 (next to be held NA November 1996); results—percent of vote by party NA; seats—(53 total) seats by party NA

Executive branch: US president, US vice president, governor

Legislative branch: bicameral Legislative Assembly consists of an upper house or Senate and a lower house or House of Representatives

Judicial branch: Supreme Court

Leaders:

Chief of State: President William Jefferson CLINTON (since 20 January 1993); Vice President Albert GORE, Jr. (since 20 January 1993)

Head of Government: Governor Pedro ROSSELLO (since NA January 1993)

Member of: CARICOM (observer), ECLAC (associate), FAO (associate), ICFTU, IOC, WCL, WFTU, WHO (associate), WTO (associate)

Diplomatic representation in US: none (commonwealth associated with the US)

Flag: five equal horizontal bands of red (top and bottom) alternating with white; a blue isosceles triangle based on the hoist side bears a large white five-pointed star in the center; design based on the US flag

---

Economy

Overview: Puerto Rico has one of the most dynamic economies in the Caribbean region. Industry has surpassed agriculture as the primary sector of economic activity and income. Encouraged by duty free access to the US and by tax incentives, US firms have invested heavily in Puerto Rico since the 1950s. US minimum wage laws apply.

Important industries include pharmaceuticals, electronics, textiles, petrochemicals, and processed foods. Sugar production has lost out to dairy production and other livestock products as the main source of income in the agricultural sector. Tourism has traditionally been an important source of income for the island, with estimated arrivals of nearly 3 million tourists in 1989.

National product: GNP—purchasing power equivalent—$22.8 billion (1991)

National product real growth rate: 2.2% (FY90)

National product per capita: $6,200 (1991)

Inflation rate (consumer prices): 1.3% (October 1990-91)

Unemployment rate: 17% (1992 est.)

Budget: revenues $5.8 billion; expenditures $5.8 billion, including capital expenditures of $258 million (FY89)

Exports: 20.4 billion (1990)

commodities: pharmaceuticals, electronics, apparel, canned tuna, rum, beverage concentrates, medical equipment, instruments

partners: US 87.8% (1990)

Imports: 16.2 billion (1990)

commodities: chemicals, clothing, food, fish, petroleum products

partners: US 66.6% (1990)

External debt: $NA

Industrial production: growth rate 1.2% (FY92)

Electricity: 5,040,000 kW capacity; 16,100 million kWh produced, 4,260 kWh per capita (1992)

Industries: manufacturing accounts for 55.5% of GDP: manufacturing of pharmaceuticals, electronics, apparel, food products, instruments; tourism

Agriculture: accounts for only 3% of labor force and less than 2% of GDP: crops—sugarcane, coffee, pineapples, plantains, bananas; livestock—cattle, chickens; imports a large share of food needs (1992)

Economic aid: none

Currency: US currency is used

Fiscal year: 1 July-30 June

Communications

Railroads: 96 km rural narrow-gauge system for hauling sugarcane; no passenger railroads

Highways: 13,762 km paved (1982)

Ports: San Juan, Ponce, Mayaguez, Arecibo

Airports:

total: 30

usable: 23

with permanent-surface runways: 19

with runways over 3,659 m: 0

with runways 2,440-3,659 m: 3

with runways 1,220-2,439 m: 5

Telecommunications: modern system, integrated with that of the US by high capacity submarine cable and INTELSAT

Puerto Rico *(continued)*

with high-speed data capability; digital telephone system with about 1 million lines; cellular telephone service; broadcast stations—50 AM, 63 FM, 9 TV; cable television available with US programs (1990)

Defense Forces

Branches: paramilitary National Guard, Police Force
Manpower availability: males age 15-49 830,133; fit for military service NA (1993 est.)
Note: defense is the responsibility of the US

Qatar

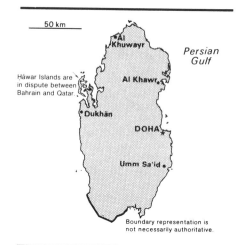

Boundary representation is not necessarily authoritative.

Geography

Location: Middle East, peninsula jutting into the central Persian Gulf, between Iran and Saudi Arabia
Map references: Middle East, Standard Time Zones of the World
Area:
total area: 11,000 km²
land area: 11,000 km²
comparative area: slightly smaller than Connecticut
Land boundaries: total 60 km, Saudi Arabia 60 km
Coastline: 563 km
Maritime claims:
continental shelf: not specified
exclusive economic zone: 200 nm
territorial sea: 12 nm
International disputes: territorial dispute with Bahrain over the Hawar Islands; maritime boundary with Bahrain
Climate: desert; hot, dry; humid and sultry in summer
Terrain: mostly flat and barren desert covered with loose sand and gravel
Natural resources: petroleum, natural gas, fish
Land use:
arable land: 0%
permanent crops: 0%
meadows and pastures: 5%
forest and woodland: 0%
other: 95%
Irrigated land: NA km²
Environment: haze, duststorms, sandstorms common; limited freshwater resources mean increasing dependence on large-scale desalination facilities
Note: strategic location in central Persian Gulf near major petroleum deposits

People

Population: 499,115 (July 1993 est.)
Population growth rate: 2.84% (1993 est.)
Birth rate: 19.61 births/1,000 population (1993 est.)

Death rate: 3.53 deaths/1,000 population (1993 est.)
Net migration rate: 12.36 migrant(s)/1,000 population (1993 est.)
Infant mortality rate: 22.7 deaths/1,000 live births (1993 est.)
Life expectancy at birth:
total population: 72.25 years
male: 69.73 years
female: 74.68 years (1993 est.)
Total fertility rate: 3.88 children born/woman (1993 est.)
Nationality:
noun: Qatari(s)
adjective: Qatari
Ethnic divisions: Arab 40%, Pakistani 18%, Indian 18%, Iranian 10%, other 14%
Religions: Muslim 95%
Languages: Arabic (official), English commonly used as a second language
Literacy: age 15 and over can read and write (1986)
total population: 76%
male: 77%
female: 72%
Labor force: 104,000 85% non-Qatari in private sector (1983)

Government

Names:
conventional long form: State of Qatar
conventional short form: Qatar
local long form: Dawlat Qatar
local short form: Qatar
Digraph: QA
Type: traditional monarchy
Capital: Doha
Administrative divisions: there are no first-order administrative divisions as defined by the US Government, but there are 9 municipalities (baladiyat, singular—baladiyah); Ad Dawhah, Al Ghuwayriyah, Al Jumayliyah, Al Khawr, Al Rayyan, Al Wakrah, Ash Shamal, Jarayan al Batnah, Umm Salal
Independence: 3 September 1971 (from UK)
Constitution: provisional constitution enacted 2 April 1970
Legal system: discretionary system of law controlled by the amir, although civil codes are being implemented; Islamic law is significant in personal matters
National holiday: Independence Day, 3 September (1971)
Political parties and leaders: none
Suffrage: none
Elections:
Advisory Council: constitution calls for elections for part of this consultative body, but no elections have been held; seats—(30 total)
Executive branch: amir, Council of Ministers (cabinet)

Legislative branch: unicameral Advisory Council (Majlis al-Shura)
Judicial branch: Court of Appeal
Leaders:
Chief of State and Head of Government: Amir and Prime Minister KHALIFA bin Hamad Al Thani (since 22 February 1972); Crown Prince HAMAD bin Khalifa Al Thani (appointed 31 May 1977; son of Amir)
Member of: ABEDA, AFESD, AL, AMF, CCC, ESCWA, FAO, G-77, GCC, IAEA, IBRD, ICAO, IDB, IFAD, ILO, IMF, IMO, INMARSAT, INTELSAT, INTERPOL, IOC, ITU, LORCS, NAM, OAPEC, OIC, OPEC, UN, UNCTAD, UNESCO, UNIDO, UPU, WHO, WIPO, WMO
Diplomatic representation in US:
chief of mission: Ambassador 'Abd al-Rahman bin Sa'ud ALTHANI
chancery: Suite 1180, 600 New Hampshire Avenue NW, Washington, DC 20037
telephone: (202) 338-0111
US diplomatic representation:
chief of mission: Ambassador Kenton W. KEITH
embassy: 149 Ali Bin Ahmed St., Farig Bin Omran (opposite the television station), Doha
mailing address: P. O. Box 2399, Doha
telephone: (0974) 864701 through 864703
FAX: (0974) 861669
Flag: maroon with a broad white serrated band (nine white points) on the hoist side

---

Economy

Overview: Oil is the backbone of the economy and accounts for more than 85% of export earnings and roughly 75% of government revenues. Proved oil reserves of 3.3 billion barrels should ensure continued output at current levels for about 25 years. Oil has given Qatar a per capita GDP of about $17,000, comparable to the leading industrial countries. Production and export of natural gas is becoming increasingly important.
National product: GDP—exchange rate conversion—$8.1 billion (1991 est.)
National product real growth rate: 3% (1991 est.)
National product per capita: $17,000 (1991 est.)
Inflation rate (consumer prices): 3% (1990)
Unemployment rate: NA%
Budget: revenues $2.5 billion; expenditures $3.0 billion, including capital expenditures of $440 million (FY92 est.)
Exports: $3.2 billion (f.o.b., 1991)
commodities: petroleum products 85%, steel, fertilizers
partners: Japan 61%, Brazil 6%, South Korea 5%, UAE 4%
Imports: $1.4 billion (f.o.b., 1991 est.)
commodities: machinery and equipment, consumer goods, food, chemicals

partners: France 13%, Japan 12%, UK 11%, Germany 9%
External debt: $1.1 billion (December 1989 est.)
Industrial production: growth rate 0.6% (1987); accounts for 64% of GDP, including oil
Electricity: 1,596,000 kW capacity; 4,818 million kWh produced, 9,655 kWh per capita (1992)
Industries: crude oil production and refining, fertilizers, petrochemicals, steel (rolls reinforcing bars for concrete construction), cement
Agriculture: farming and grazing on small scale, less than 2% of GDP; agricultural area is small and government-owned; commercial fishing increasing in importance; most food imported
Economic aid: donor—pledged $2.7 billion in ODA to less developed countries (1979-88)
Currency: 1 Qatari riyal (QR) = 100 dirhams
Exchange rates: Qatari riyals (QR) per US$1—3.6400 riyals (fixed rate)
Fiscal year: 1 April-31 March

Communications

Highways: 1,500 km total; 1,000 km paved, 500 km gravel or natural surface (est.)
Pipelines: crude oil 235 km, natural gas 400 km
Ports: Doha, Umm Sa'id, Halul Island
Merchant marine: 20 ships (1,000 GRT or over) totaling 390,072 GRT/593,508 DWT; includes 13 cargo, 4 container, 2 oil tanker, 1 refrigerated cargo
Airports:
total: 4
usable: 4
with permanent-surface runways: 1
with runways over 3,659 m: 1
with runways 2,440-3,659 m: 0
with runways 1,220-2,439 m: 2
Telecommunications: modern system centered in Doha; 110,000 telephones; tropospheric scatter to Bahrain; microwave radio relay to Saudi Arabia and UAE; submarine cable to Bahrain and UAE; satellite earth stations—1 Atlantic Ocean INTELSAT, 1 Indian Ocean INTELSAT, 1 ARABSAT; broadcast stations—2 AM, 3 FM, 3 TV

Defense Forces

Branches: Army, Navy, Air Force, Public Security
Manpower availability: males age 15-49 214,977; fit for military service 113,514; reach military age (18) annually 3,578 (1993 est.)
Defense expenditures: exchange rate conversion—$NA, NA%, of GDP

Reunion
(overseas department of France)

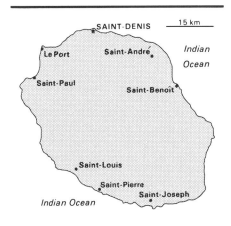

Geography

Location: Southern Africa, in the western Indian Ocean, 750 km east of Madagascar
Map references: World
Area:
total area: 2,510 km²
land area: 2,500 km²
comparative area: slightly smaller than Rhode Island
Land boundaries: 0 km
Coastline: 201 km
Maritime claims:
exclusive economic zone: 200 nm
territorial sea: 12 nm
International disputes: none
Climate: tropical, but moderates with elevation; cool and dry from May to November, hot and rainy from November to April
Terrain: mostly rugged and mountainous; fertile lowlands along coast
Natural resources: fish, arable land
Land use:
arable land: 20%
permanent crops: 2%
meadows and pastures: 4%
forest and woodland: 35%
other: 39%
Irrigated land: 60 km² (1989 est.)
Environment: periodic devastating cyclones

People

Population: 639,622 (July 1993 est.)
Population growth rate: 2.07% (1993 est.)
Birth rate: 25.64 births/1,000 population (1993 est.)
Death rate: 4.94 deaths/1,000 population (1993 est.)
Net migration rate: 0 migrant(s)/1,000 population (1993 est.)
Infant mortality rate: 8.1 deaths/1,000 live births (1993 est.)
Life expectancy at birth:
total population: 73.68 years
male: 70.61 years
female: 76.91 years (1993 est.)

Reunion (continued)

Total fertility rate: 2.81 children born/woman (1993 est.)
Nationality:
noun: Reunionese (singular and plural)
adjective: Reunionese
Ethnic divisions: French, African, Malagasy, Chinese, Pakistani, Indian
Religions: Roman Catholic 94%
Languages: French (official), Creole widely used
Literacy: age 15 and over can read and write (1982)
total population: 69%
male: 67%
female: 74%
Labor force: NA
by occupation: agriculture 30%, industry 21%, services 49% (1981)
note: 63% of population of working age (1983)

Government

Names:
conventional long form: Department of Reunion
conventional short form: Reunion
local long form: none
local short form: Ile de la Reunion
Digraph: RE
Type: overseas department of France
Capital: Saint-Denis
Administrative divisions: none (overseas department of France)
Independence: none (overseas department of France)
Constitution: 28 September 1958 (French Constitution)
Legal system: French law
National holiday: Taking of the Bastille, 14 July (1789)
Political parties and leaders: Rally for the Republic (RPR), Francois MAS; Union for French Democracy (UDF), Gilbert GERARD; Communist Party of Reunion (PCR), Paul VERGES; France-Reunion Future (FRA), Andre THIEN AH KOON; Socialist Party (PS), Jean-Claude FRUTEAU; Social Democrats (CDS); other small parties
Suffrage: 18 years of age; universal
Elections:
General Council: last held 22 March 1991 (next to be held March 1997); results—percent of vote by party NA; seats—(44 total)
Regional Council: last held 28 March 1992 (next to be held NA March 1998); results—UDF 25.6%, PRC 17.9%, PS 10.5%, Independent 30.7%, other 15.3%; seats—(45 total) Independent 17, UDF 14, PRC 9, PS 5
French Senate: last held 24 September 1989 (next to be held NA September 1993); results—percent of vote by party NA;

seats—(3 total) RPR-UDF 1, PS 1, independent 1
French National Assembly: last held 5 and 12 June 1988 (next to be held NA June 1993); results—percent of vote by party NA; seats—(5 total) PCR 2, RPR 1, UDF-CDS 1, FRA 1; note—Reunion elects 3 members to the French Senate and 5 members to the French National Assembly who are voting members
Executive branch: French president, commissioner of the Republic
Legislative branch: unicameral General Council and unicameral Regional Council
Judicial branch: Court of Appeals (Cour d'Appel)
Leaders:
Chief of State: President Francois MITTERRAND (since 21 May 1981)
Head of Government: Commissioner of the Republic Jacques DEWATRE (since NA July 1991)
Member of: FZ
Diplomatic representation in US: as an overseas department of France, Reunionese interests are represented in the US by France
Flag: the flag of France is used

Economy

Overview: The economy has traditionally been based on agriculture. Sugarcane has been the primary crop for more than a century, and in some years it accounts for 85% of exports. The government has been pushing the development of a tourist industry to relieve high unemployment, which recently amounted to one-third of the labor force. The gap in Reunion between the well-off and the poor is extraordinary and accounts for the persistent social tensions. The white and Indian communities are substantially better off than other segments of the population, often approaching European standards, whereas indigenous groups suffer the poverty and unemployment typical of the poorer nations of the African continent. The outbreak of severe rioting in February 1991 illustrates the seriousness of socioeconomic tensions. The economic well-being of Reunion depends heavily on continued financial assistance from France.
National product: GDP—exchange rate conversion—$3.37 billion (1987 est.)
National product real growth rate: 9% (1987 est.)
National product per capita: $6,000 (1987 est.)
Inflation rate (consumer prices): 1.3% (1988)
Unemployment rate: 35% (February 1991)
Budget: revenues $358 million; expenditures $914 million, including capital expenditures of $NA (1986)
Exports: $166 million (f.o.b., 1988)
commodities: sugar 75%, rum and molasses

4%, perfume essences 4%, lobster 3%, vanilla and tea 1%
partners: France, Mauritius, Bahrain, South Africa, Italy
Imports: $1.7 billion (c.i.f., 1988)
commodities: manufactured goods, food, beverages, tobacco, machinery and transportation equipment, raw materials, and petroleum products
partners: France, Mauritius, Bahrain, South Africa, Italy
External debt: $NA
Industrial production: growth rate NA%; about 25% of GDP
Electricity: 245,000 kW capacity; 750 million kWh produced, 1,230 kWh per capita (1991)
Industries: sugar, rum, cigarettes, several small shops producing handicraft items
Agriculture: accounts for 30% of labor force; dominant sector of economy; cash crops—sugarcane, vanilla, tobacco; food crops—tropical fruits, vegetables, corn; imports large share of food needs
Economic aid: Western (non-US) countries, ODA and OOF bilateral commitments (1970-89), $14.8 billion
Currency: 1 French franc (F) = 100 centimes
Exchange rates: French francs (F) per US$1—5.4812 (January 1993), 5.2938 (1992), 5.6421 (1991), 5.4453 (1990), 6.3801 (1989), 5.9569 (1988)
Fiscal year: calendar year

Communications

Highways: 2,800 km total; 2,200 km paved, 600 km gravel, crushed stone, or stabilized earth
Ports: Pointe des Galets
Airports:
total: 2
usable: 2
with permanent-surface runways: 2
with runways over 3,659 m: 0
with runway 2,440-3,659 m: 1
with runway 1,220-2,439 m: 1
Telecommunications: adequate system; modern open-wire and microwave network; principal center Saint-Denis; radiocommunication to Comoros, France, Madagascar; new microwave route to Mauritius; 85,900 telephones; broadcast stations—3 AM, 13 FM, 1 (18 repeaters) TV; 1 Indian Ocean INTELSAT earth station

Defense Forces

Branches: French Forces (including Army, Navy, Air Force, Gendarmerie)
Manpower availability: males age 15-49 167,925; fit for military service 86,764; reach military age (18) annually 5,975 (1993 est.)
Note: defense is the responsibility of France

Romania

Geography

Location: Southeastern Europe, bordering the Black Sea between Bulgaria and the Ukraine

Map references: Ethnic Groups in Eastern Europe, Europe, Standard Time Zones of the World

Area:

total area: 237,500 km²

land area: 230,340 km²

comparative area: slightly smaller than Oregon

Land boundaries: total 2,508 km, Bulgaria 608 km, Hungary 443 km, Moldova 450 km, Serbia and Montenegro 476 km (all with Serbia), Ukraine (north) 362 km, Ukraine (south) 169 km

Coastline: 225 km

Maritime claims:

contiguous zone: 24 nm

continental shelf: 200 m depth or to depth of exploitation

exclusive economic zone: 200 nm

territorial sea: 12 nm

International disputes: none

Climate: temperate; cold, cloudy winters with frequent snow and fog; sunny summers with frequent showers and thunderstorms

Terrain: central Transylvanian Basin is separated from the plain of Moldavia on the east by the Carpathian Mountains and separated from the Walachian Plain on the south by the Transylvanian Alps

Natural resources: petroleum (reserves being exhausted), timber, natural gas, coal, iron ore, salt

Land use:

arable land: 43%

permanent crops: 3%

meadows and pastures: 19%

forest and woodland: 28%

other: 7%

Irrigated land: 34,500 km² (1989 est.)

Environment: frequent earthquakes most severe in south and southwest; geologic structure and climate promote landslides; air pollution in south

Note: controls most easily traversable land route between the Balkans, Moldova, and Ukraine

People

Population: 23,172,362 (July 1993 est.)

Population growth rate: 0.02% (1993 est.)

Birth rate: 13.66 births/1,000 population (1993 est.)

Death rate: 10.17 deaths/1,000 population (1993 est.)

Net migration rate: -3.27 migrant(s)/1,000 population (1993 est.)

Infant mortality rate: 21.2 deaths/1,000 live births (1993 est.)

Life expectancy at birth:

total population: 71.25 years

male: 68.32 years

female: 74.34 years (1993 est.)

Total fertility rate: 1.83 children born/woman (1993 est.)

Nationality:

noun: Romanian(s)

adjective: Romanian

Ethnic divisions: Romanian 89.1%, Hungarian 8.9%, German 0.4%, Ukrainian, Serb, Croat, Russian, Turk, and Gypsy 1.6%

Religions: Romanian Orthodox 70%, Roman Catholic 6% (of which 3% are Uniate), Protestant 6%, unaffiliated 18%

Languages: Romanian, Hungarian, German

Literacy: age 15 and over can read and write (1978)

total population: 98%

male: NA%

female: NA%

Labor force: 10,945,700

by occupation: industry 38%, agriculture 28%, other 34% (1989)

Government

Names:

conventional long form: none

conventional short form: Romania

local long form: none

local short form: Romania

Digraph: RO

Type: republic

Capital: Bucharest

Administrative divisions: 40 counties (judete, singular—judet) and 1 municipality* (municipiu); Alba, Arad, Arges, Bacau, Bihor, Bistrita-Nasaud, Botosani, Braila, Brasov, Bucuresti*, Buzau, Calarasi, Caras-Severin, Cluj, Constanta, Covasna, Dimbovita, Dolj, Galati, Gorj, Giurgiu, Harghita, Hunedoara, Ialomita, Iasi, Maramures, Mehedinti, Mures, Neamt, Olt, Prahova, Salaj, Satu Mare, Sibiu, Suceava, Teleorman, Timis, Tulcea, Vaslui, Vilcea, Vrancea

Independence: 1881 (from Turkey; republic proclaimed 30 December 1947)

Constitution: 8 December 1991

Legal system: former mixture of civil law system and Communist legal theory that increasingly reflected Romanian traditions is being revised

National holiday: National Day of Romania, 1 December (1990)

Political parties and leaders: National Salvation Front (FSN), Petre ROMAN; Democratic National Salvation Front (DNSF), Oliviu GHERMAN; Magyar Democratic Union (UDMR), Geza DOMOKOS; National Liberal Party (PNL), Mircea IONESCU-QUINTUS; National Peasants' Christian and Democratic Party (PNTCD), Corneliu COPOSU; Romanian National Unity Party (PUNR), Gheorghe FUNAR; Socialist Labor Party (PSM), Ilie VERDET; Agrarian Democratic Party of Romania (PDAR), Victor SURDU; The Democratic Convention (CDR), Emil CONSTANTINESCU; Romania Mare Party (PRM), Corneliu Vadim TUDOR

note: there are dozens of smaller parties; although the Communist Party has ceased to exist, small proto-Communist parties, notably the Socialist Labor Party, have been formed

Other political or pressure groups: various human right and professional associations

Suffrage: 18 years of age; universal

Elections:

President: last held 27 September 1992—with runoff between top two candidates on 11 October 1992 (next to be held NA 1998); results—Ion ILIESCU 61.4%, Emil CONSTANTINESCU 38.6%

Senate: last held 27 September 1992 (next to be held NA 1998); results—DFSN 27.5%, CDR 22.5%, FSN 11%, others 39%; seats—(143 total) DFSN 49, CDR 34, FSN 18, PUNR 14, UDMR 12, PRM 6, PDAR 5, PSM 5

House of Deputies: last held 27 September 1992 (next to be held NA 1998); results—DFSN 27.5%, CDR 22.5%, FSN 11%, others 38.5%; seats—(341 total) DFSN 117, CDR 82, FSN 43, PUNR 30, UDMR 27, PRM 16, PSM 13, other 13

Executive branch: president, prime minister, Council of Ministers (cabinet)

Legislative branch: bicameral Parliament consists of an upper house or Senate (Senat) and a lower house or House of Deputies (Adunarea Deputatilor)

Judicial branch: Supreme Court of Justice, Constitutional Court

Leaders:

Chief of State: President Ion ILIESCU (since 20 June 1990, previously President of Provisional Council of National Unity since 23 December 1989)

Head of Government: Prime Minister Nicolae VACAROIU (since November 1992)

Member of: BIS, BSEC, CCC, CSCE,

Romania (continued)

EBRD, ECE, FAO, G-9, G-77, GATT, IAEA, IBRD, ICAO, IFAD, IFC, ILO, IMF, IMO, INMARSAT, INTELSAT, INTERPOL, IOC, IOM (observer), ITU, LORCS, NACC, NAM (guest), NSG, PCA, UN, UNCTAD, UNESCO, UNIDO, UNIKOM, UPU, WHO, WIPO, WMO, WTO, ZC

Diplomatic representation in US:
chief of mission: Ambassador Aurel-Dragos MUNTEANU
chancery: 1607 23rd Street NW, Washington, DC 20008
telephone: (202) 232-4747, 6634, 5693
FAX: (202) 232-4748
US diplomatic representation:
chief of mission: Ambassador John R. DAVIS, Jr.
embassy: Strada Tudor Arghezi 7-9, Bucharest
mailing address: AmConGen (Buch), Unit 25402, APO AE 09213-5260
telephone: [40] (0) 10-40-40
FAX: [40] (0) 12-03-95
Flag: three equal vertical bands of blue (hoist side), yellow, and red; the national coat of arms that used to be centered in the yellow band has been removed; now similar to the flags of Andorra and Chad

Economy

Overview: Industry, which accounts for about one-third of the labor force and generates over half the GDP, suffers from an aging capital plant and persistent shortages of energy. The year 1991 witnessed a 17% drop in industrial production because of energy and input shortages and labor unrest. In recent years the agricultural sector has had to contend with flooding, mismanagement, shortages of inputs, and disarray caused by the dismantling of cooperatives. A shortage of inputs and a severe drought in 1991 contributed to a poor harvest, a problem compounded by corruption and an obsolete distribution system. The new government has instituted moderate land reforms, with more than one-half of cropland now in private hands, and it has liberalized private agricultural output. Private enterprises form an increasingly important portion of the economy largely in services, handicrafts, and small-scale industry. Little progress on large scale privatization has been made since a law providing for the privatization of large state firms was passed in August 1991. Most of the large state firms have been converted into joint-stock companies, but the selling of shares and assets to private owners has been delayed. While the government has halted the old policy of diverting food from domestic consumption to hard currency export markets, supplies remain scarce in some areas. The new government continues

to impose price ceilings on key consumer items. In 1992 the economy muddled along toward the new, more open system, yet output and living standards continued to fall.
National product: GDP—purchasing power equivalent—$63.4 billion (1992 est.)
National product real growth rate: -15% (1992 est.)
National product per capita: $2,700 (1992 est.)
Inflation rate (consumer prices): 200% (1992 est.)
Unemployment rate: 9% (January 1993)
Budget: revenues $19 billion; expenditures $20 billion, including capital expenditures of $2.1 billion (1991 est.)
Exports: $3.5 billion (f.o.b., 1991)
commodities: machinery and equipment 29.3%, fuels, minerals and metals 32.1%, manufactured consumer goods 18.1%, agricultural materials and forestry products 9.0%, other 11.5% (1989)
partners: USSR 27%, Eastern Europe 23%, EC 15%, US 5%, China 4% (1987)
Imports: $5.1 billion (f.o.b., 1991)
commodities: fuels, minerals, and metals 56.0%, machinery and equipment 25.5%, agricultural and forestry products 8.6%, manufactured consumer goods 3.4%, other 6.5% (1989)
partners: Communist countries 60%, non-Communist countries 40% (1987)
External debt: $3 billion (1992)
Industrial production: growth rate –17% (1991 est.); accounts for 48% of GDP
Electricity: 22,500,000 kW capacity; 59,000 million kWh produced, 2,540 kWh per capita (1992)
Industries: mining, timber, construction materials, metallurgy, chemicals, machine building, food processing, petroleum production and refining
Agriculture: accounts for 18% of GDP and 28% of labor force; major wheat and corn producer; other products—sugar beets, sunflower seed, potatoes, milk, eggs, meat, grapes
Illicit drugs: transshipment point for southwest Asian heroin transiting the Balkan route
Economic aid: donor—$4.4 billion in bilateral aid to non-Communist less developed countries (1956-89)
Currency: 1 leu (L) = 100 bani
Exchange rates: lei (L) per US$1—470.10 (January 1993), 307.95 (1992), 76.39 (1991), 22.432 (1990), 14.922 (1989), 14.277 (1988)
Fiscal year: calendar year

Communications

Railroads: 11,275 km total; 10,860 km 1.435-meter gauge, 370 km narrow gauge, 45 km broad gauge; 3,411 km electrified, 3,060 km double track; government owned (1987)

Highways: 72,799 km total; 35,970 km paved; 27,729 km gravel, crushed stone, and other stabilized surfaces; 9,100 km unsurfaced roads (1985)
Inland waterways: 1,724 km (1984)
Pipelines: crude oil 2,800 km, petroleum products 1,429 km, natural gas 6,400 km (1992)
Ports: Constanta, Galati, Braila, Mangalia; inland ports are Giurgiu, Drobeta-Turnu Severin, Orsova
Merchant marine: 249 ships (1,000 GRT or over) totaling 2,882,727 GRT/4,463,879 DWT; includes 1 passenger-cargo, 170 cargo, 2 container, 1 rail-car carrier, 9 roll-on/roll-off cargo, 15 oil tanker, 51 bulk
Airports:
total: 158
usable: 158
with permanent-surface runways: 27
with runways over 3,659 m: 0
with runways 2,440-3,659 m: 21
with runways 1,220-2,439 m: 26
Telecommunications: poor service; about 2.3 million telephone customers; 89% of phone network is automatic; cable and open wire; trunk network is microwave; present phone density is 9.85 per 100 residents; roughly 3,300 villages with no service (February 1990); broadcast stations—12 AM, 5 FM, 13 TV (1990); 1 satellite ground station using INTELSAT

Defense Forces

Branches: Army, Navy, Air and Air Defense Forces, Paramilitary Forces, Civil Defense
Manpower availability: males age 15-49 5,846,332; fit for military service 4,942,746; reach military age (20) annually 185,714 (1993 est.)
Defense expenditures: 137 billion lei, 3% of GDP (1993); note—conversion of defense expenditures into US dollars using the current exchange rate could produce misleading results

Russia

2000 km

Boundary representation is
not necessarily authoritative.

Geography

Location: Europe/North Asia, between
Europe and the North Pacific Ocean
Map references: Asia, Commonwealth of
Independent States—Central Asian States,
Commonwealth of Independent
States—European States, Standard Time
Zones of the World
Area:
total area: 17,075,200 km²
land area: 16,995,800 km²
comparative area: slightly more than 1.8
times the size of the US
Land boundaries: total 20,139 km,
Azerbaijan 284 km, Belarus 959 km, China
(southeast) 3,605 km, China (south) 40 km,
Estonia 290 km, Finland 1,313 km, Georgia
723 km, Kazakhstan 6,846 km, North Korea
19 km, Latvia 217 km, Lithuania
(Kaliningrad Oblast) 227 km, Mongolia
3,441 km, Norway 167 km, Poland
(Kaliningrad Oblast) 432 km, Ukraine 1,576
km
Coastline: 37,653 km
Maritime claims:
continental shelf: 200 m depth or to depth of
exploitation
exclusive economic zone: 200 nm
territorial sea: 12 nm
International disputes: inherited disputes
from former USSR including: sections of the
boundary with China; boundary with Latvia,
Lithuania, and Estonia; Etorofu, Kunashiri,
and Shikotan Islands and the Habomai island
group occupied by the Soviet Union in 1945,
claimed by Japan; maritime dispute with
Norway over portion of the Barents Sea; has
made no territorial claim in Antarctica (but
has reserved the right to do so) and does not
recognize the claims of any other nation
Climate: ranges from steppes in the south
through humid continental in much of
European Russia; subarctic in Siberia to
tundra climate in the polar north; winters
vary from cool along Black Sea coast to
frigid in Siberia; summers vary from warm
in the steppes to cool along Arctic coast

Terrain: broad plain with low hills west of
Urals; vast coniferous forest and tundra in
Siberia; uplands and mountains along
southern border regions
Natural resources: wide natural resource
base including major deposits of oil, natural
gas, coal, and many strategic minerals,
timber
note: formidable obstacles of climate,
terrain, and distance hinder exploitation of
natural resources
Land use:
arable land: NA%
permanent crops: NA%
meadows and pastures: NA%
forest and woodland: NA%
other: NA%
note: agricultural land accounts for 13% of
the total land area
Irrigated land: 61,590 km² (1990)
Environment: despite its size, only a small
percentage of land is arable and much is too
far north for cultivation; permafrost over
much of Siberia is a major impediment to
development; catastrophic pollution of land,
air, water, including both inland waterways
and sea coasts
Note: largest country in the world in terms
of area but unfavorably located in relation to
major sea lanes of the world

People

Population: 149,300,359 (July 1993 est.)
Population growth rate: 0.21% (1993 est.)
Birth rate: 12.73 births/1,000 population
(1993 est.)
Death rate: 11.32 deaths/1,000 population
(1993 est.)
Net migration rate: 0.69 migrant(s)/1,000
population (1993 est.)
Infant mortality rate: 27.6 deaths/1,000
live births (1993 est.)
Life expectancy at birth:
total population: 68.69 years
male: 63.59 years
female: 74.04 years (1993 est.)
Total fertility rate: 1.83 children
born/woman (1993 est.)
Nationality:
noun: Russian(s)
adjective: Russian
Ethnic divisions: Russian 81.5%, Tatar
3.8%, Ukrainian 3%, Chuvash 1.2%, Bashkir
0.9%, Belarusian 0.8%, Moldavian 0.7%,
other 8.1%
Religions: Russian Orthodox, Muslim, other
Languages: Russian, other
Literacy: age 9-49 can read and write
(1970)
total population: 100%
male: 100%
female: 100%
Labor force: 75 million (1993 est.)
by occupation: production and economic
services 83.9%, government 16.1%

Government

Names:
conventional long form: Russian Federation
conventional short form: Russia
local long form: Rossiyskaya Federatsiya
local short form: Rossiya
former: Russian Soviet Federative Socialist
Republic
Digraph: RS
Type: federation
Capital: Moscow
Administrative divisions: 21 autonomous
republics (avtomnykh respublik,
singular—avtomnaya respublika); Adygea
(Maykop), Bashkortostan (Ufa), Buryatia
(Ulan-Ude), Chechenia, Chuvashia
(Cheboksary), Dagestan (Makhachkala),
Gorno-Altay (Gorno-Altaysk), Ingushetia,
Kabardino-Balkaria (Nal'chik), Kalmykia
(Elista), Karachay-Cherkessia (Cherkessk),
Karelia (Petrozavodsk), Khakassia (Abakan),
Komi (Syktyvkar), Mari El (Yoshkar-Ola),
Mordvinia (Saransk), North Ossetia
(Vladikavkaz; formerly Ordzhonikidze),
Tatarstan (Kazan'), Tuva (Kyzyl), Udmurtia
(Izhevsk), Yakutia (Yakutsk); 49 oblasts
(oblastey, singular—oblast'); Amur
(Blagoveshchensk), Arkhangel'sk,
Astrakhan', Belgorod, Bryansk, Chelyabinsk,
Chita, Irkutsk, Ivanovo, Kaliningrad, Kaluga,
Kamchatka (Petropavlovsk-Kamchatskiy),
Kemerovo, Kirov, Kostroma, Kurgan, Kursk,
St. Petersburg (Leningrad), Lipetsk,
Magadan, Moscow, Murmansk, Nizhniy
Novgorod (formerly Gor'kiy), Novgorod,
Novosibirsk, Omsk, Orel, Orenburg, Penza,
Perm', Pskov, Rostov, Ryazan', Sakhalin
(Yuzhno-Sakhalinsk), Samara (formerly
Kuybyshev), Saratov, Smolensk, Sverdlovsk
(Yekaterinburg), Tambov, Tomsk, Tula,
Tver' (formerly Kalinin), Tyumen',
Ul'yanovsk, Vladimir, Volgograd, Vologda,
Voronezh, Yaroslavl'; 6 krays (krayev,
singular—kray); Altay (Barnaul),
Khabarovsk, Krasnodar, Krasnoyarsk,
Primorskiy (Vladivostok), Stavropol'
note: the autonomous republics of Chechenia
and Ingushetia were formerly the automous
republic of Checheno-Ingushetia (the
boundary between Chechenia and Ingushetia
has yet to be determined); the cities of
Moscow and St. Petersburg have oblast
status; an administrative division has the
same name as its administrative center
(exceptions have the administrative center
name following in parentheses); 4 more
administrative divisions may be added
Independence: 24 August 1991 (from
Soviet Union)
Constitution: adopted in 1978; a new
constitution is in the process of being
drafted
Legal system: based on civil law system;
judicial review of legislative acts; does not
accept compulsory ICJ jurisdiction

Russia (continued)

National holiday: Independence Day, June 12

Political parties and leaders:

proreformers: Christian Democratic Party, Aleksandr CHUYEV; Christian Democratic Union of Russia, Aleksandr OGORODNIKOV; Democratic Russia Movement, pro-government faction, Lev PONOMAREV, Gleb YAKUNIN, Vladimir BOKSER; Democratic Russia Movement, radical-liberal faction, Yuriy AFANAS'YEV, Marina SAL'YE; Economic Freedom Party, Konstantin BOROVOY, Svyatoslav FEDOROV; Free Labor Party, Igor' KOROVIKOV; Party of Constitutional Democrats, Viktor ZOLOTAREV; Republican Party of Russia, Vladimir LYSENKO, Vyacheslav SHOSTAKOVSKIY; Russian Democratic Reform Movement, Gavriil POPOV; Social Democratic Party, Boris ORLOV; Social Liberal Party, Vladimir FILIN

moderate reformers: All-Russian Renewal Union (member Civic Union), Arkadiy VOL'SKIY, Aleksandr VLADISLAVLEV; Democratic Party of Russia (member Civic Union), Nikolay TRAVKIN, Valeriy KHOMYAKOV; People's Party of Free Russia (member Civic Union), Aleksandr RUTSKOY, Vasiliy LIPITSKIY; Russian Union of Industrialists and Entrepreneurs, Arkadiy VOL'SKIY, Aleksandr VLADISLAVLEV

antireformers: Communists and neo-Communists have 7 parties—All-Union Communist Party of Bolsheviks, Nina ANDREYEVA; Labor Party, Boris KAGARLITSKIY; Russian Communist Worker's Party, Viktor ANPILOV, Gen. Albert MAKASHOV; Russian Party of Communists, Anatoliy KRYUCHKOV; Socialist Party of Working People, Roy MEDVEDEV; Union of Communists, Aleksey PRIGARIN; Working Russia Movement, Viktor ANPILOV; National Patriots have 6 parties—Constitutional Democratic Party, Mikhail ASTAF'YEV; Council of People and Patriotic Forces of Russia, Gennadiy ZYUGANOV; National Salvation Front, Mikhail ASTAF'YEV, Sergey BABURIN, Vladimir ISAKOV, Il'ya KONSTANTINOV, Aleksandr STERLIGOV; Russian Christian Democratic Movement, Viktor AKSYUCHITS; Russian National Assembly, Aleksandr STERLIGOV; Russian National Union, Sergey BABURIN, Nikolay PAVLOV; extremists have 5 parties—Liberal Democratic Party, Vladimir ZHIRNOVKSKIY; Nashi Movement, Viktor ALKSNIS; National Republican Party of Russia, Nikolay LYSENKO; Russian Party, Viktor KORCHAGIN; Russian National Patriotic Front (Pamyat), Dmitriy VASIL'YEV

Other political or pressure groups: Civic Union, Aleksandr RUTSKOY, Nikolay TRAVKIN, Arkadiy VOL'SKIY, chairmen

Suffrage: 18 years of age; universal

Elections:

President: last held 12 June 1991 (next to be held 1996); results—percent of vote by party NA%

Congress of People's Deputies: last held March 1990 (next to be held 1995); results—percent of vote by party NA%; seats—(1,063 total) number of seats by party NA; election held before parties were formed

Supreme Soviet: last held May 1990 (next to be held 1995); results—percent of vote by party NA%; seats—(252 total) number of seats by party NA; elected from Congress of People's Deputies

Executive branch: president, vice president, Security Council, Presidential Administration, Council of Ministers, Group of Assistants, Council of Heads of Republics

Legislative branch: unicameral Congress of People's Deputies, bicameral Supreme Soviet

Judicial branch: Constitutional Court, Supreme Court

Leaders:

Chief of State: President Boris Nikolayevich YEL'TSIN (since 12 June 1991); Vice President Aleksandr Vladimirovich RUTSKOY (since 12 June 1991); Chairman of the Supreme Soviet Ruslan KHASBULATOV (28 October 1991)

Head of Government: Chairman of the Council of Ministers Viktor Stepanovich CHERNOMYRDIN (since NA December 1992); First Deputy Chairmen of the Council of Ministers Vladimir SHUMEYKO (since 9 June 1992), Oleg LOBW (since NA April 1993), Oleg SOSKOVETS (since NA April 1993)

Member of: BSEC, CBSS, CCC, CERN (observer), CIS, CSCE, EBRD, ECE, ESCAP, IAEA, IBRD, ICAO, ICFTU, IDA, ILO, IMF, IMO, INMARSAT, INTELSAT, INTERPOL, IOC, IOM (observer), ISO, ITU, LORCS, MINURSO, NACC, NSG, OAS (observer), PCA, UN, UNCTAD, UNESCO, UNIDO, UNIKOM, UNPROFOR, UN Security Council, UNTAC, UN Trusteeship Council, UNTSO, UPU, WFTU, WHO, WIPO, WMO, WTO, ZC

Diplomatic representation in US:

chief of mission: Ambassador Vladimir Petrovich LUKIN

chancery: 1125 16th Street NW, Washington, DC 20036

telephone: (202) 628-7551 and 8548

consulates general: New York and San Francisco

US diplomatic representation:

chief of mission: (vacant)

embassy: Ulitsa Chaykovskogo 19/21/23, Moscow

mailing address: APO AE 09721

telephone: [7] (095) 252-2450 through 2459

FAX: [7] (095) 255-9965

consulates: St. Petersburg (formerly Leningrad), Vladivostok

Flag: three equal horizontal bands of white (top), blue, and red

---

Economy

Overview: Russia, a vast country with a wealth of natural resources and a diverse industrial base, continues to experience great difficulties in moving from its old centrally planned economy to a modern market economy. President YEL'TSIN's government made significant strides toward a market economy in 1992 by freeing most prices, slashing defense spending, unifying foreign exchange rates, and launching an ambitious privatization program. At the same time, GDP fell 19%, according to official statistics, largely reflecting government efforts to restructure the economy, shortages of essential imports caused by the breakdown in former Bloc and interstate trade, and reduced demand following the freeing of prices in January. The actual decline, however, may have been less steep, because industrial and agricultural enterprises had strong incentives to understate output to avoid taxes, and official statistics may not have fully captured the output of the growing private sector. Despite the large drop in output, unemployment at yearend stood at an estimated 3%-4% of Russia's 74-million-person labor force; many people, however, are working shortened weeks or are on forced leave. Moscow's financial stabilization program got off to a good start at the beginning of 1992 but began to falter by midyear. Under pressure from industrialists and the Supreme Soviet, the government loosened fiscal policies in the second half. In addition, the Russian Central Bank relaxed its tight credit policy in July at the behest of new Acting Chairman, Viktor GERASHCHENKO. This loosening of financial policies led to a sharp increase in prices during the last quarter, and inflation reached about 25% per month by yearend. The situation of most consumers worsened in 1992. The January price liberalization and a blossoming of private vendors filled shelves across the country with previously scarce food items and consumer goods, but wages lagged behind inflation, making such goods unaffordable for many consumers. Falling real wages forced most Russians to spend a larger share of their income on food and to alter their eating habits. Indeed, many Russians reduced their consumption of higher priced meat, fish, milk, vegetables, and fruit, in favor of more bread and potatoes. As a

324

result of higher spending on food, consumers reduced their consumption of nonfood goods and services. Despite a slow start and some rough going, the Russian government by the end of 1992 scored some successes in its campaign to break the state's stranglehold on property and improve the environment for private businesses. More peasant farms were created than expected; the number of consumers purchasing goods from private traders rose sharply; the portion of the population working in the private sector increased to nearly one-fifth; and the nine-month-long slump in the privatization of small businesses was ended in the fall. Although the output of weapons fell sharply in 1992, most defense enterprises continued to encounter numerous difficulties developing and marketing consumer products, establishing new supply links, and securing resources for retooling. Indeed, total civil production by the defense sector fell in 1992 because of shortages of inputs and lower consumer demand caused by higher prices. Ruptured ties with former trading partners, output declines, and sometimes erratic efforts to move to world prices and decentralize trade—foreign and interstate—took a heavy toll on Russia's commercial relations with other countries. For the second year in a row, foreign trade was down sharply, with exports falling by as much as 25% and imports by 21%. The drop in imports would have been much greater if foreign aid—worth an estimated $8 billion—had not allowed the continued inflow of essential products. Trade with the other former Soviet republics continued to decline, and support for the ruble as a common currency eroded in the face of Moscow's loose monetary policies and rapidly rising prices throughout the region. At the same time, Russia paid only a fraction of the $20 billion due on the former USSR's roughly $80 billion debt; debt rescheduling remained hung up because of a dispute between Russia and Ukraine over division of the former USSR's assets. Capital flight also remained a serious problem in 1992. Russia's economic difficulties did not abate in the first quarter of 1993. Monthly inflation remained at double-digit levels and industrial production continued to slump. To reduce the threat of hyperinflation, the government proposed to restrict subsidies to enterprises; raise interest rates; set quarterly limits on credits, the budget deficit, and money supply growth; and impose temporary taxes and cut spending if budget targets are not met. But many legislators and Central Bank officials oppose various of these austerity measures and failed to approve them in the first part of 1993.

National product: GDP $NA

National product real growth rate: -19% (1992)

National product per capita: $NA

Inflation rate (consumer prices): 25% per month (December 1992)

Unemployment rate: 3%-4% of labor force (1 January 1993 est.)

Budget: revenues $NA; expenditures $NA, including capital expenditures of $NA

Exports: $39.2 billion (f.o.b., 1992)
commodities: petroleum and petroleum products, natural gas, wood and wood products, metals, chemicals, and a wide variety of civilian and military manufactures
partners: Europe

Imports: $35.0 billion (f.o.b., 1992)
commodities: machinery and equipment, chemicals, consumer goods, grain, meat, sugar, semifinished metal products
partners: Europe, North America, Japan, Third World countries, Cuba

External debt: $80 billion (yearend 1992 est.)

Industrial production: growth rate –19% (1992)

Electricity: 213,000,000 KW capacity; 1,014.8 billion kWh produced, 6,824 kWh per capita (1 January 1992)

Industries: complete range of mining and extractive industries producing coal, oil, gas, chemicals, and metals; all forms of machine building from rolling mills to high-performance aircraft and space vehicles; ship-building; road and rail transportation equipment; communications equipment; agricultural machinery, tractors, and construction equipment; electric power generating and transmitting equipment; medical and scientific instruments; consumer durables

Agriculture: grain, sugar beet, sunflower seeds, meat, milk, vegetables, fruits; because of its northern location does not grow citrus, cotton, tea, and other warm climate products

Illicit drugs: illicit producer of cannabis and opium; mostly for domestic consumption; government has active eradication program; used as transshipment point for illicit drugs to Western Europe

Economic aid: US commitments, including Ex-Im (1990-92), $9.0 billion; other countries, ODA and OOF bilateral commitments (1988-92), $91 billion

Currency: 1 ruble (R) = 100 kopeks

Exchange rates: rubles per US$1—415 (24 December 1992) but subject to wide fluctuations

Fiscal year: calendar year

Communications

Railroads: 158,100 km all 1.520-meter broad gauge; 86,800 km in common carrier service, of which 48,900 km are diesel traction and 37,900 km are electric traction; 71,300 km serves specific industry and is not available for common carrier use (31 December 1991)

Highways: 893,000 km total, of which 677,000 km are paved or gravelled and 216,000 km are dirt; 456,000 km are for general use and are maintained by the Russian Highway Corporation (formerly Russian Highway Ministry); the 437,000 km not in general use are the responsibility of various other organizations (formerly ministries); of the 456,000 km in general use, 265,000 km are paved, 140,000 km are gravelled, and 51,000 km are dirt; of the 437,000 km not in general use, 272,000 km are paved or gravelled and 165,000 are dirt (31 December 1991)

Inland waterways: total navigable routes 102,000 km; routes with navigation guides serving the Russian River Fleet 97,300 km (including illumination and light reflecting guides); routes with other kinds of navigational aids 34,300 km; man-made navigable routes 16,900 km (31 December 1991)

Pipelines: crude oil 72,500 km, petroleum products 10,600 km, natural gas 136,000 km (1992)

Ports: coastal—St. Petersburg (Leningrad), Kaliningrad, Murmansk, Petropavlovsk, Arkhangel'sk, Novorossiysk, Vladivostok, Nakhodka, Kholmsk, Korsakov, Magadan, Tiksi, Tuapse, Vanino, Vostochnyy, Vyborg; inland—Astrakhan', Nizhniy Novgorod (Gor'kiy), Kazan', Khabarovsk, Krasnoyarsk, Samara (Kuybyshev), Moscow, Rostov, Volgograd

Merchant marine: 865 ships (1,000 GRT or over) totaling 8,073,954 GRT/11,138,336 DWT; includes 457 cargo, 82 container, 3 multi-function large load carrier, 2 barge carrier, 72 roll-on/roll-off, 124 oil tanker, 25 bulk cargo, 9 chemical tanker, 2 specialized tanker, 16 combination ore/oil, 5 passenger cargo, 18 short-sea passenger, 6 passenger, 28 combination bulk, 16 refrigerated cargo

Airports:
total: 2,550
useable: 964
with permanent surface runways: 565
with runways over 3,659 m: 19
with runways 2,440-3,659 m: 275
with runways 1,220-2,439 m: 426

Telecommunications: NMT-450 analog cellular telephone networks are opertional in Moscow and St. Petersburg; expanding access to international E-mail service via Sprint networks; the inadequacy of Russian telecommunications is a severe handicap to the economy, especially with respect to international connections; total installed telephones 24,400,000, of which in urban areas 20,900,000 and in rural areas 3,500,000; of these, total installed in homes 15,400,000; total pay phones for long distant

Russia (continued)

calls 34,100; telephone density is about 164 telephones per 1,000 persons; international traffic is handled by an inadequate system of satellites, land lines, microwave radio relay and outdated submarine cables; this traffic passes through the international gateway switch in Moscow which carries most of the international traffic for the other countries of the Confederation of Independent States; a new Russian Raduga satellite will soon link Moscow and St. Petersburg with Rome from whence calls will be relayed to destinations in Europe and overseas; satellite ground stations—INTELSAT, Intersputnik, Eutelsat (Moscow), INMARSAT, Orbita; broadcast stations—1,050 AM/FM/SW (reach 98.6% of population), 7,183 TV; receiving sets—54,200,000 TV, 48,800,000 radio receivers; intercity fiberoptic cables installation remains limited

Defense Forces

Branches: Ground Forces, Navy, Air Forces, Air Defense Forces, Strategic Rocket Forces, Command and General Support, Security Forces
note: strategic nuclear units and warning facilities are under joint CIS control; Russian defense forces will be comprised of those ground-, air-, and sea-based conventional assets currently on Russian soil and those still scheduled to be withdrawn from other countries
Manpower availability: males age 15-49 37,092,361; fit for military service 29,253,668; reach military age (18) annually 1,082,115 (1993 est.)
Defense expenditures: $NA, NA% of GDP

Rwanda

Geography

Location: Central Africa, between Tanzania and Zaire
Map references: Africa, Standard Time Zones of the World
Area:
total area: 26,340 km²
land area: 24,950 km²
comparative area: slightly smaller than Maryland
Land boundaries: total 893 km, Burundi 290 km, Tanzania 217 km, Uganda 169 km, Zaire 217 km
Coastline: 0 km (landlocked)
Maritime claims: none; landlocked
International disputes: none
Climate: temperate; two rainy seasons (February to April, November to January); mild in mountains with frost and snow possible
Terrain: mostly grassy uplands and hills; mountains in west
Natural resources: gold, cassiterite (tin ore), wolframite (tungsten ore), natural gas, hydropower
Land use:
arable land: 29%
permanent crops: 11%
meadows and pastures: 18%
forest and woodland: 10%
other: 32%
Irrigated land: 40 km² (1989 est.)
Environment: deforestation; overgrazing; soil exhaustion; soil erosion; periodic droughts
Note: landlocked

People

Population: 8,139,272 (July 1993 est.)
Population growth rate: 2.9% (1993 est.)
Birth rate: 49.92 births/1,000 population (1993 est.)
Death rate: 20.87 deaths/1,000 population (1993 est.)
Net migration rate: 0 migrant(s)/1,000 population (1993 est.)

Infant mortality rate: 119.4 deaths/1,000 live births (1993 est.)
Life expectancy at birth:
total population: 41.23 years
male: 40.2 years
female: 42.28 years (1993 est.)
Total fertility rate: 8.27 children born/woman (1993 est.)
Nationality:
noun: Rwandan(s)
adjective: Rwandan
Ethnic divisions: Hutu 90%, Tutsi 9%, Twa (Pygmoid) 1%
Religions: Roman Catholic 65%, Protestant 9%, Muslim 1%, indigenous beliefs and other 25%
Languages: Kinyarwanda (official), French (official), Kiswahili used in commercial centers
Literacy: age 15 and over can read and write (1990)
total population: 50%
male: 64%
female: 37%
Labor force: 3.6 million
by occupation: agriculture 93%, government and services 5%, industry and commerce 2%
note: 49% of population of working age (1985)

Government

Names:
conventional long form: Republic of Rwanda
conventional short form: Rwanda
local long form: Republika y'u Rwanda
local short form: Rwanda
Digraph: RW
Type: republic; presidential system
note: a new, all-party transitional government is to assume office later this year, replacing the current MRND-dominated coalition
Capital: Kigali
Administrative divisions: 10 prefectures (prefectures, singular—prefecture in French; plural—NA, singular—prefegitura in Kinyarwanda); Butare, Byumba, Cyangugu, Gikongoro, Gisenyi, Gitarama, Kibungo, Kibuye, Kigali, Ruhengeri
Independence: 1 July 1962 (from UN trusteeship under Belgian administration)
Constitution: 18 June 1991
Legal system: based on German and Belgian civil law systems and customary law; judicial review of legislative acts in the Supreme Court; has not accepted compulsory ICJ jurisdiction
National holiday: Independence Day, 1 July (1962)
Political parties and leaders: Republican National Movement for Democracy and Development (MRND), President HABYARIMANA's political movement, remains the dominant party; significant

independent parties include: Democratic Republican Movement (MDR), Faustin TWAGIRAMUNGU; Liberal Party (PL), Justin MUGENZI; Democratic and Socialist Party (PSD), Frederic NZAMURAMBAHO; Coalition for the Defense of the Republic (CDR), Martin BUCYANA; Party for Democracy in Rwanda (PADER), Jean NTAGUNGIRA; Christian Democratic Party (PDL), Nayinzira NEPOMUSCENE

note: formerly a one-party state, Rwanda legalized independent parties in mid-1991; since then, at least 10 new political parties have registered

Other political or pressure groups: since October 1990, Rwanda has been involved in a low-intensity conflict with the Rwandan Patriotic Front/Rwandan Patriotic Army (RPF/RPA)

Suffrage: universal adult at age NA

Elections:

President: last held 19 December 1988 (next to be held NA December 1993); results—President Juvenal HABYARIMANA reelected

National Development Council: last held 19 December 1988 (next to be held NA December 1993); results—MRND was the only party; seats—(70 total) MRND 70

Executive branch: president, prime minister, Council of Ministers (cabinet)

Legislative branch: unicameral National Development Council (Conseil National de Developpement)

Judicial branch: Constitutional Court (consists of the Court of Cassation and the Council of State in joint session)

Leaders:

Chief of State: President Juvenal HABYARIMANA (since 5 July 1973)

Head of Government: Prime Minister Dismas NSENGIYAREMYE (since NA April 1992)

Member of: ACCT, ACP, AfDB, ECA, CCC, CEEAC, CEPGL, FAO, G-77, GATT, IBRD, ICAO, IDA, IFAD, IFC, ILO, IMF, INTELSAT, INTERPOL, IOC, ITU, LORCS, NAM, OAU, UN, UNCTAD, UNESCO, UNIDO, UPU, WCL, WHO, WIPO, WMO, WTO

Diplomatic representation in US:

chief of mission: Ambassador Aloys UWIMANA

chancery: 1714 New Hampshire Avenue NW, Washington, DC 20009

telephone: (202) 232-2882

US diplomatic representation:

chief of mission: Ambassador Robert A. FLATEN

embassy: Boulevard de la Revolution, Kigali

mailing address: B. P. 28, Kigali

telephone: [250] 75601 through 75603

FAX: [250] 72128

Flag: three equal vertical bands of red (hoist side), yellow, and green with a large black letter R centered in the yellow band; uses the popular pan-African colors of Ethiopia; similar to the flag of Guinea, which has a plain yellow band

Economy

Overview: Almost 50% of GDP comes from the agricultural sector; coffee and tea make up 80-90% of total exports. The amount of fertile land is limited, however, and deforestation and soil erosion have created problems. The industrial sector in Rwanda is small, contributing only 17% to GDP. Manufacturing focuses mainly on the processing of agricultural products. The Rwandan economy remains dependent on coffee exports and foreign aid. Weak international prices since 1986 have caused the economy to contract and per capita GDP to decline. A structural adjustment program with the World Bank began in October 1990. An outbreak of insurgency, also in October 1990, has dampened prospects for economic improvement.

National product: GDP—exchange rate conversion—$2.35 billion (1992 est.)

National product real growth rate: 1.3% (1992 est.)

National product per capita: $290 (1992 est.)

Inflation rate (consumer prices): 6% (1992 est.)

Unemployment rate: NA%

Budget: revenues $350 million; expenditures $453.7 million, including capital expenditures of $NA million (1992 est.)

Exports: $66.6 million (f.o.b., 1992 est.)

commodities: coffee 85%, tea, tin, cassiterite, wolframite, pyrethrum

partners: Germany, Belgium, Italy, Uganda, UK, France, US

Imports: $259.5 million (f.o.b., 1992 est.)

commodities: textiles, foodstuffs, machines and equipment, capital goods, steel, petroleum products, cement and construction material

partners: US, Belgium, Germany, Kenya, Japan

External debt: $911 million (1990 est.)

Industrial production: growth rate 1.2% (1988); accounts for 17% of GDP

Electricity: 30,000 kW capacity; 130 million kWh produced, 15 kWh per capita (1991)

Industries: mining of cassiterite (tin ore) and wolframite (tungsten ore), tin, cement, agricultural processing, small-scale beverage production, soap, furniture, shoes, plastic goods, textiles, cigarettes

Agriculture: accounts for almost 50% of GDP and about 90% of the labor force; cash crops—coffee, tea, pyrethrum (insecticide made from chrysanthemums); main food crops—bananas, beans, sorghum, potatoes; stock raising; self-sufficiency declining; country imports foodstuffs as farm production fails to keep up with a 3.8% annual growth in population

Economic aid: US commitments, including Ex-Im (FY70-89), $128 million; Western (non-US) countries, ODA and OOF bilateral commitments (1970-89), $2.0 billion; OPEC bilateral aid (1979-89), $45 million; Communist countries (1970-89), $58 million; note—in October 1990 Rwanda launched a Structural Adjustment Program with the IMF; since September 1991, the EC has given $46 million and the US $25 million in support of this program

Currency: 1 Rwandan franc (RF) = 100 centimes

Exchange rates: Rwandan francs (RF) per US$1—146.34 (January 1993), 133.35 (1992), 125.14 (1991), 82.60 (1990), 79.98 (1989), 76.45 (1988)

Fiscal year: calendar year

Communications

Highways: 4,885 km total; 460 km paved, 1,725 km gravel and/or improved earth, 2,700 km unimproved

Inland waterways: Lac Kivu navigable by shallow-draft barges and native craft

Airports:

total: 8

usable: 7

with permanent-surface runways: 3

with runways over 3,659 m: 0

with runways 2,440-3,659 m: 1

with runways 1,220-2,439 m: 2

Telecommunications: fair system with low-capacity radio relay system centered on Kigali; broadcast stations—2 AM, 1 (7 repeaters) FM, no TV; satellite earth stations—1 Indian Ocean INTELSAT and 1 SYMPHONIE

Defense Forces

Branches: Army (including Air Wing), Gendarmerie

Manpower availability: males age 15-49 1,675,160; fit for military service 853,467 (1993 est.); no conscription

Defense expenditures: exchange rate conversion—$37 million, 1.6% of GDP (1988 est.)

Saint Helena
(dependent territory of the UK)

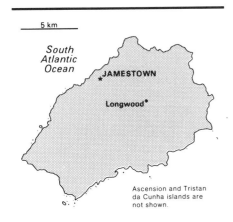

South Atlantic Ocean

JAMESTOWN

Longwood

5 km

Ascension and Tristan da Cunha islands are not shown.

Geography

Location: in the South Atlantic Ocean, 1,920 km west of Angola, about two-thirds of the way between South America and Africa
Map references: Africa
Area:
total area: 410 km²
land area: 410 km²
comparative area: slightly more than 2.3 times the size of Washington, DC
note: includes Ascension, Gough Island, Inaccessible Island, Nightingale Island, and Tristan da Cunha
Land boundaries: 0 km
Coastline: 60 km
Maritime claims:
exclusive fishing zone: 200 nm
territorial sea: 12 nm
International disputes: none
Climate: tropical; marine; mild, tempered by trade winds
Terrain: rugged, volcanic; small scattered plateaus and plains
Natural resources: fish; Ascension is a breeding ground for sea turtles and sooty terns, no minerals
Land use:
arable land: 7%
permanent crops: 0%
meadows and pastures: 7%
forest and woodland: 3%
other: 83%
Irrigated land: NA km²
Environment: very few perennial streams
Note: Napoleon Bonaparte's place of exile and burial; harbors at least 40 species of plants unknown anywhere else in the world

People

Population: 6,720 (July 1993 est.)
Population growth rate: 0.32% (1993 est.)
Birth rate: 9.82 births/1,000 population (1993 est.)
Death rate: 6.67 deaths/1,000 population (1993 est.)

Net migration rate: 0 migrant(s)/1,000 population (1993 est.)
Infant mortality rate: 38.39 deaths/1,000 live births (1993 est.)
Life expectancy at birth:
total population: 74.43 years
male: 72.36 years
female: 76.27 years (1993 est.)
Total fertility rate: 1.16 children born/woman (1993 est.)
Nationality:
noun: Saint Helenian(s)
adjective: Saint Helenian
Ethnic divisions: NA
Religions: Anglican (majority), Baptist, Seventh-Day Adventist, Roman Catholic
Languages: English
Literacy: age 15 and over can read and write (1987)
total population: 98%
male: 97%
female: 98%
Labor force: 2,516
by occupation: professional, technical, and related workers 8.7%, managerial, administrative, and clerical 12.8%, sales people 8.1%, farmer, fishermen, etc. 5.4%, craftspersons, production process workers 14.7%, others 50.3% (1987)

Government

Names:
conventional long form: none
conventional short form: Saint Helena
Digraph: SH
Type: dependent territory of the UK
Capital: Jamestown
Administrative divisions: 1 administrative area and 2 dependencies*; Ascension*, Saint Helena, Tristan da Cunha*
Independence: none (dependent territory of the UK)
Constitution: 1 January 1989
Legal system: NA
National holiday: Celebration of the Birthday of the Queen, 10 June 1989 (second Saturday in June)
Political parties and leaders: Saint Helena Labor Party; Saint Helena Progressive Party
note: both political parties inactive since 1976
Suffrage: NA
Elections:
Legislative Council: last held October 1984 (next to be held NA); results—percent of vote by party NA; seats—(15 total, 12 elected) number of seats by party NA
Executive branch: British monarch, governor commander-in-chief, Executive Council (cabinet)
Legislative branch: unicameral Legislative Council
Judicial branch: Supreme Court

Leaders:
Chief of State: Queen ELIZABETH II (since 6 February 1952)
Head of Government: Governor A. N. HOOLE (since NA)
Member of: ICFTU
Diplomatic representation in US: none (dependent territory of the UK)
US diplomatic representation: none (dependent territory of the UK)
Flag: blue with the flag of the UK in the upper hoist-side quadrant and the Saint Helenian shield centered on the outer half of the flag; the shield features a rocky coastline and three-masted sailing ship

Economy

Overview: The economy depends primarily on financial assistance from the UK. The local population earns some income from fishing, the raising of livestock, and sales of handicrafts. Because there are few jobs, a large proportion of the work force has left to seek employment overseas.
National product: GDP $NA
National product real growth rate: NA%
National product per capita: $NA
Inflation rate (consumer prices): -1.1% (1986)
Unemployment rate: NA%
Budget: revenues $3.2 million; expenditures $2.9 million, including capital expenditures of $NA (1984)
Exports: $23,900 (f.o.b., 1984)
commodities: fish (frozen and salt-dried skipjack, tuna), handicrafts
partners: South Africa, UK
Imports: $2.4 million (c.i.f., 1984)
commodities: food, beverages, tobacco, fuel oils, animal feed, building materials, motor vehicles and parts, machinery and parts
partners: UK, South Africa
External debt: $NA
Industrial production: growth rate NA%
Electricity: 9,800 kW capacity; 10 million kWh produced, 1,390 kWh per capita (1989)
Industries: crafts (furniture, lacework, fancy woodwork), fishing
Agriculture: maize, potatoes, vegetables; timber production being developed; crawfishing on Tristan da Cunha
Economic aid: Western (non-US) countries, ODA and OOF bilateral commitments (1970-89), $198 million
Currency: 1 Saint Helenian pound (£S) = 100 pence
Exchange rates: Saint Helenian pounds (£S) per US$1—0.6527 (January 1993), 0.5664 (1992), 0.5652 (1991), 0.5603 (1990), 0.6099 (1989), 0.5614 (1988); note—the Saint Helenian pound is at par with the British pound
Fiscal year: 1 April-31 March

Saint Kitts and Nevis

Communications

Highways: 87 km paved roads and 20 km earth roads on Saint Helena; 80 km paved roads on Ascension; 2.7 km paved roads on Tristan da Cunha
Ports: Jamestown (Saint Helena), Georgetown (Ascension)
Airports:
total: 1
useable: 1
with permanent-surface runways: 1
with runways over 3,659 m: 0
with runways 2,440-3,659 m: 1
with runways 1,220-2,439 m: 0
Telecommunications: 1,500 radio receivers; broadcast stations—1 AM, no FM, no TV; 550 telephones in automatic network; HF radio links to Ascension, then into worldwide submarine cable and satellite networks; major coaxial submarine cable relay point between South Africa, Portugal, and UK at Ascension; 2 Atlantic Ocean INTELSAT earth stations

Defense Forces

Note: defense is the responsibility of the UK

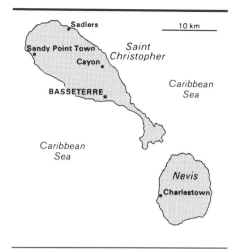

Geography

Location: in the eastern Caribbean Sea, about one-third of the way between Puerto Rico and Trinidad and Tobago
Map references: Central America and the Caribbean, Standard Time Zones of the World
Area:
total area: 269 km²
land area: 269 km²
comparative area: slightly more than 1.5 times the size of Washington, DC
Land boundaries: 0 km
Coastline: 135 km
Maritime claims:
contiguous zone: 24 nm
exclusive economic zone: 200 nm
territorial sea: 12 nm
International disputes: none
Climate: subtropical tempered by constant sea breezes; little seasonal temperature variation; rainy season (May to November)
Terrain: volcanic with mountainous interiors
Natural resources: negligible
Land use:
arable land: 22%
permanent crops: 17%
meadows and pastures: 3%
forest and woodland: 17%
other: 41%
Irrigated land: NA km²
Environment: subject to hurricanes (July to October)

People

Population: 40,407 (July 1993 est.)
Population growth rate: 0.59% (1993 est.)
Birth rate: 23.93 births/1,000 population (1993 est.)
Death rate: 10.39 deaths/1,000 population (1993 est.)
Net migration rate: -7.67 migrant(s)/1,000 population (1993 est.)
Infant mortality rate: 20.5 deaths/1,000 live births (1993 est.)

Life expectancy at birth:
total population: 65.72 years
male: 62.78 years
female: 68.85 years (1993 est.)
Total fertility rate: 2.64 children born/woman (1993 est.)
Nationality:
noun: Kittsian(s), Nevisian(s)
adjective: Kittsian, Nevisian
Ethnic divisions: black African
Religions: Anglican, other Protestant sects, Roman Catholic
Languages: English
Literacy: age 15 and over having ever attended school (1970)
total population: 98%
male: 98%
female: 98%
Labor force: 20,000 (1981)

Government

Names:
conventional long form: Federation of Saint Kitts and Nevis
conventional short form: Saint Kitts and Nevis
former: Federation of Saint Christopher and Nevis
Digraph: SC
Type: constitutional monarchy
Capital: Basseterre
Administrative divisions: 14 parishs; Christ Church Nichola Town, Saint Anne Sandy Point, Saint George Basseterre, Saint George Gingerland, Saint James Windward, Saint John Capesterre, Saint John Figtree, Saint Mary Cayon, Saint Paul Capesterre, Saint Paul Charlestown, Saint Peter Basseterre, Saint Thomas Lowland, Saint Thomas Middle Island, Trinity Palmetto Point
Independence: 19 September 1983 (from UK)
Constitution: 19 September 1983
Legal system: based on English common law
National holiday: Independence Day, 19 September (1983)
Political parties and leaders: People's Action Movement (PAM), Dr. Kennedy SIMMONDS; Saint Kitts and Nevis Labor Party (SKNLP), Dr. Denzil DOUGLAS; Nevis Reformation Party (NRP), Simeon DANIEL; Concerned Citizens Movement (CCM), Vance AMORY
Suffrage: universal adult at age NA
Elections:
House of Assembly: last held 21 March 1989 (next to be held by 21 March 1994); results—percent of vote by party NA; seats—(14 total, 11 elected) PAM 6, SKNLP 2, NRP 2, CCM 1
Executive branch: British monarch, governor general, prime minister, deputy prime minister, Cabinet

Saint Kitts and Nevis *(continued)*

Legislative branch: unicameral House of Assembly
Judicial branch: Eastern Caribbean Supreme Court
Leaders:
Chief of State: Queen ELIZABETH II (since 6 February 1952), represented by Governor General Sir Clement Athelston ARRINDELL (since 19 September 1983, previously Governor General of the Associated State since NA November 1981)
Head of Government: Prime Minister Dr. Kennedy Alphonse SIMMONDS (since 19 September 1983, previously Premier of the Associated State since NA February 1980); Deputy Prime Minister Sydney Earl MORRIS (since NA)
Member of: ACP, C, CARICOM, CDB, ECLAC, FAO, G-77, IBRD, ICFTU, IDA, IFAD, IMF, INTERPOL, OAS, OECS, UN, UNCTAD, UNESCO, UNIDO, UPU, WCL, WHO
Diplomatic representation in US:
chief of mission: Minister-Counselor (Deputy Chief of Mission), Charge d'Affaires ad interim Aubrey Eric HART
chancery: Suite 608, 2100 M Street NW, Washington, DC 20037
telephone: (202) 833-3550
US diplomatic representation: no official presence since the Charge d'Affaires resides in Saint John's (Antigua and Barbuda)
Flag: divided diagonally from the lower hoist side by a broad black band bearing two white five-pointed stars; the black band is edged in yellow; the upper triangle is green, the lower triangle is red

Economy

Overview: The economy has historically depended on the growing and processing of sugarcane and on remittances from overseas workers. In recent years, tourism and export-oriented manufacturing have assumed larger roles.
National product: GDP—exchange rate conversion—$142 million (1991)
National product real growth rate: 6.8% (1991)
National product per capita: $3,500 (1991)
Inflation rate (consumer prices): 4.2% (1991)
Unemployment rate: 12.2% (1990)
Budget: revenues $85.7 million; expenditures $85.8 million, including capital expenditures of $42.4 million (1993)
Exports: $24.6 million (f.o.b., 1990)
commodities: sugar, clothing, electronics, postage stamps
partners: US 53%, UK 22%, Trinidad and Tobago 5%, OECS 5% (1988)
Imports: $103.2 million (f.o.b., 1990)
commodities: foodstuffs, intermediate manufactures, machinery, fuels

partners: US 36%, UK 17%, Trinidad and Tobago 6%, Canada 3%, Japan 3%, OECS 4% (1988)
External debt: $37.2 million (1990)
Industrial production: growth rate 11.8% (1988 est.); accounts for 11% of GDP
Electricity: 15,800 kW capacity; 45 million kWh produced, 1,120 kWh per capita (1992)
Industries: sugar processing, tourism, cotton, salt, copra, clothing, footwear, beverages
Agriculture: accounts for 7% of GDP; cash crop—sugarcane; subsistence crops—rice, yams, vegetables, bananas; fishing potential not fully exploited; most food imported
Illicit drugs: transshipment point for South American drugs destined for the US
Economic aid: US commitments, including Ex-Im (FY85-88), $10.7 million; Western (non-US) countries, ODA and OOF bilateral commitments (1970-89), $67 million
Currency: 1 EC dollar (EC$) = 100 cents
Exchange rates: East Caribbean dollars (EC$) per US$1—2.70 (fixed rate since 1976)
Fiscal year: calendar year

Communications

Railroads: 58 km 0.760-meter gauge on Saint Kitts for sugarcane
Highways: 300 km total; 125 km paved, 125 km otherwise improved, 50 km unimproved earth
Ports: Basseterre (Saint Kitts), Charlestown (Nevis)
Airports:
total: 2
usable: 2
with permanent-surface runways: 2
with runways over 3,659 m: 0
with runways 2,440-3,659 m: 1
with runways 1,220-2,439 m: 0
Telecommunications: good interisland VHF/UHF/SHF radio connections and international link via Antigua and Barbuda and Saint Martin; 2,400 telephones; broadcast stations—2 AM, no FM, 4 TV

Defense Forces

Branches: Royal Saint Kitts and Nevis Police Force, Coast Guard
Manpower availability: NA
Defense expenditures: exchange rate conversion—$NA, NA% of GDP

Saint Lucia

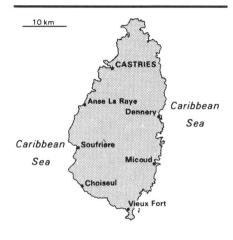

Geography

Location: in the eastern Caribbean Sea, about two-thirds of the way between Puerto Rico and Trinidad and Tobago
Map references: Central America and the Caribbean, South America, Standard Time Zones of the World
Area:
total area: 620 km²
land area: 610 km²
comparative area: slightly less than 3.5 times the size of Washington, DC
Land boundaries: 0 km
Coastline: 158 km
Maritime claims:
contiguous zone: 24 nm
exclusive economic zone: 200 nm
territorial sea: 12 nm
International disputes: none
Climate: tropical, moderated by northeast trade winds; dry season from January to April, rainy season from May to August
Terrain: volcanic and mountainous with some broad, fertile valleys
Natural resources: forests, sandy beaches, minerals (pumice), mineral springs, geothermal potential
Land use:
arable land: 8%
permanent crops: 20%
meadows and pastures: 5%
forest and woodland: 13%
other: 54%
Irrigated land: 10 km² (1989 est.)
Environment: subject to hurricanes and volcanic activity; deforestation; soil erosion

People

Population: 144,337 (July 1993 est.)
Population growth rate: 0.52% (1993 est.)
Birth rate: 23.97 births/1,000 population (1993 est.)
Death rate: 5.91 deaths/1,000 population (1993 est.)
Net migration rate: -12.87 migrant(s)/1,000 population (1993 est.)

Infant mortality rate: 18.7 deaths/1,000 live births (1993 est.)
Life expectancy at birth:
total population: 69.26 years
male: 66.98 years
female: 71.69 years (1993 est.)
Total fertility rate: 2.62 children born/woman (1993 est.)
Nationality:
noun: Saint Lucian(s)
adjective: Saint Lucian
Ethnic divisions: African descent 90.3%, mixed 5.5%, East Indian 3.2%, Caucasian 0.8%
Religions: Roman Catholic 90%, Protestant 7%, Anglican 3%
Languages: English (official), French patois
Literacy: age 15 and over having ever attended school (1980)
total population: 67%
male: 65%
female: 69%
Labor force: 43,800
by occupation: agriculture 43.4%, services 38.9%, industry and commerce 17.7% (1983 est.)

Government

Names:
conventional long form: none
conventional short form: Saint Lucia
Digraph: ST
Type: parliamentary democracy
Capital: Castries
Administrative divisions: 11 quarters; Anse La Raye, Castries, Choiseul, Dauphin, Dennery, Gros Islet, Laborie, Micoud, Praslin, Soufriere, Vieux Fort
Independence: 22 February 1979 (from UK)
Constitution: 22 February 1979
Legal system: based on English common law
National holiday: Independence Day, 22 February (1979)
Political parties and leaders: United Workers' Party (UWP), John COMPTON; Saint Lucia Labor Party (SLP), Julian HUNTE; Progressive Labor Party (PLP), George ODLUM
Suffrage: 18 years of age; universal
Elections:
House of Assembly: last held 27 April 1992 (next to be held by April 1997); results—percent of vote by party NA; seats—(17 total) UWP 11, SLP 6
Executive branch: British monarch, governor general, prime minister, Cabinet
Legislative branch: bicameral Parliament consists of an upper house or Senate and a lower house or House of Assembly
Judicial branch: Eastern Caribbean Supreme Court
Leaders:

Chief of State: Queen ELIZABETH II (since 6 February 1952), represented by Acting Governor General Sir Stanislaus Anthony JAMES (since 10 October 1988)
Head of Government: Prime Minister John George Melvin COMPTON (since 3 May 1982)
Member of: ACCT (associate), ACP, C, CARICOM, CDB, ECLAC, FAO, G-77, IBRD, ICAO, ICFTU, IDA, IFAD, IFC, ILO, IMF, IMO, INTERPOL, LORCS, NAM, OAS, OECS, UN, UNCTAD, UNESCO, UNIDO, UPU, WCL, WHO, WMO
Diplomatic representation in US:
chief of mission: Ambassador Dr. Joseph Edsel EDMUNDS
chancery: Suite 309, 2100 M Street NW, Washington, DC 30037
telephone: (202) 463-7378 or 7379
consulate general: New York
US diplomatic representation: no official presence since the Ambassador resides in Bridgetown (Barbados)
Flag: blue with a gold isosceles triangle below a black arrowhead; the upper edges of the arrowhead have a white border

Economy

Overview: Since 1983 the economy has shown an impressive average annual growth rate of almost 5% because of strong agricultural and tourist sectors. Saint Lucia also possesses an expanding industrial base supported by foreign investment in manufacturing and other activities, such as in data processing. The economy, however, remains vulnerable because the important agricultural sector is dominated by banana production, which is subject to periodic droughts and/or tropical storms.
National product: GDP—exchange rate conversion—$250 million (1991 est.)
National product real growth rate: 2.5% (1991 est.)
National product per capita: $1,650 (1991 est.)
Inflation rate (consumer prices): 6.1% (1991)
Unemployment rate: 16% (1988)
Budget: revenues $131 million; expenditures $149 million, including capital expenditures of $71 million (FY90 est.)
Exports: $105 million (f.o.b., 1991)
commodities: bananas 58%, clothing, cocoa, vegetables, fruits, coconut oil
partners: UK 56%, US 22%,CARICOM 19%
Imports: $267 million (f.o.b., 1991)
commodities: manufactured goods 21%, machinery and transportation equipment 21%, food and live animals, chemicals, fuels
partners: US 34%, CARICOM 17%, UK 14%, Japan 7%, Canada 4%

External debt: $65.7 million (1991 est.)
Industrial production: growth rate 3.5% (1990 est.); accounts for 12% of GDP
Electricity: 32,500 kW capacity; 112 million kWh produced, 740 kWh per capita (1992)
Industries: clothing, assembly of electronic components, beverages, corrugated boxes, tourism, lime processing, coconut processing
Agriculture: accounts for 12% of GDP and 43% of labor force; crops—bananas, coconuts, vegetables, citrus fruit, root crops, cocoa; imports food for the tourist industry
Economic aid: Western (non-US) countries, ODA and OOF bilateral commitments (1970-89), $120 million
Currency: 1 EC dollar (EC$) = 100 cents
Exchange rates: East Caribbean dollars (EC$) per US$1—2.70 (fixed rate since 1976)

Communications

Highways: 760 km total; 500 km paved; 260 km otherwise improved
Ports: Castries, Vieux Fort
Airports:
total: 2
usable: 2
with permanent-surface runways: 2
with runways over 3,659 m: 0
with runways 2,440-3,659 m: 1
with runways 1,220-2,439: 1
Telecommunications: fully automatic telephone system; 9,500 telephones; direct microwave link with Martinique and Saint Vincent and the Grenadines; interisland troposcatter link to Barbados; broadcast stations—4 AM, 1 FM, 1 TV (cable)

Defense Forces

Branches: Royal Saint Lucia Police Force, Coast Guard
Manpower availability: NA
Defense expenditures: exchange rate conversion—$NA, NA% of GDP

Saint Pierre and Miquelon
(territorial collectivity of France)

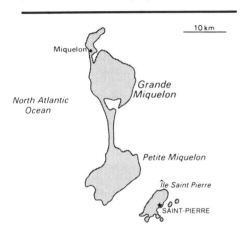

Geography

Location: in the North Atlantic Ocean, 25 km south of Newfoundland (Canada)
Map references: North America
Area:
total area: 242 km²
land area: 242 km²
comparative area: slightly less than 1.5 times the size of Washington, DC
note: includes eight small islands in the Saint Pierre and the Miquelon groups
Land boundaries: 0 km
Coastline: 120 km
Maritime claims:
exclusive economic zone: 200 nm
territorial sea: 12 nm
International disputes: focus of maritime boundary dispute between Canada and France
Climate: cold and wet, with much mist and fog; spring and autumn are windy
Terrain: mostly barren rock
Natural resources: fish, deepwater ports
Land use:
arable land: 13%
permanent crops: 0%
meadows and pastures: 0%
forest and woodland: 4%
other: 83%
Irrigated land: NA km²
Environment: vegetation scanty

People

Population: 6,652 (July 1993 est.)
Population growth rate: 0.79% (1993 est.)
Birth rate: 13.44 births/1,000 population (1993 est.)
Death rate: 6.14 deaths/1,000 population (1993 est.)
Net migration rate: 0.59 migrant(s)/1,000 population (1993 est.)
Infant mortality rate: 12.73 deaths/1,000 live births (1993 est.)
Life expectancy at birth:
total population: 75.19 years
male: 73.56 years
female: 77.16 years (1993 est.)
Total fertility rate: 1.73 children born/woman (1993 est.)
Nationality:
noun: Frenchman(men), Frenchwoman(women)
adjective: French
Ethnic divisions: Basques and Bretons (French fishermen)
Religions: Roman Catholic 98%
Languages: French
Literacy: age 15 and over can read and write (1982)
total population: 99%
male: 99%
female: 99%
Labor force: 2,850 (1988)
by occupation: NA

Government

Names:
conventional long form: Territorial Collectivity of Saint Pierre and Miquelon
conventional short form: Saint Pierre and Miquelon
local long form: Departement de Saint-Pierre et Miquelon
local short form: Saint-Pierre et Miquelon
Digraph: SB
Type: territorial collectivity of France
Capital: Saint-Pierre
Administrative divisions: none (territorial collectivity of France)
Independence: none (territorial collectivity of France; has been under French control since 1763)
Constitution: 28 September 1958 (French Constitution)
Legal system: French law
National holiday: National Day, Taking of the Bastille, 14 July
Political parties and leaders: Socialist Party (PS), Albert PEN; Union for French Democracy (UDF/CDS), Gerard GRIGNON
Suffrage: 18 years of age; universal
Elections:
French President: last held 8 May 1988 (next to be held NA May 1995); results—(second ballot) Jacques CHIRAC 56%, Francois MITTERRAND 44%
French Senate: last held NA September 1986 (next to be held NA September 1995); results—percent of vote by party NA; seats—(1 total) PS 1
French National Assembly: last held 21 and 28 March 1993 (next to be held NA June 1998); results—percent of vote by party NA; seats—(1 total) number of seats by party NA; note—Saint Pierre and Miquelon elects 1 member each to the French Senate and the French National Assembly who are voting members

General Council: last held September-October 1988 (next to be held NA September 1994); results—percent of vote by party NA; seats—(19 total) Socialist and other left-wing parties 13, UDF and right-wing parties 6
Executive branch: French president, commissioner of the Republic
Legislative branch: unicameral General Council
Judicial branch: Superior Tribunal of Appeals (Tribunal Superieur d'Appel)
Leaders:
Chief of State: President Francois MITTERRAND (since 21 May 1981)
Head of Government: Commissioner of the Republic Kamel KHRISSATE (since NA); President of the General Council Marc PLANTEGENET (since NA)
Member of: FZ
Diplomatic representation in US: as a territorial collectivity of France, local interests are represented in the US by France
US diplomatic representation: none (territorial collectivity of France)
Flag: the flag of France is used

Economy

Overview: The inhabitants have traditionally earned their livelihood by fishing and by servicing fishing fleets operating off the coast of Newfoundland. The economy has been declining, however, because the number of ships stopping at Saint Pierre has dropped steadily over the years. In March 1989, an agreement between France and Canada set fish quotas for Saint Pierre's trawlers fishing in Canadian and Canadian-claimed waters for three years. The agreement settles a longstanding dispute that had virtually brought fish exports to a halt. The islands are heavily subsidized by France. Imports come primarily from Canada and France.
National product: GDP—exchange rate conversion—$60 million (1991 est.)
National product real growth rate: NA%
National product per capita: $9,500 (1991 est.)
Inflation rate (consumer prices): NA%
Unemployment rate: 9.6% (1990)
Budget: revenues $18.3 million; expenditures $18.3 million, including capital expenditures of $5.5 million (1989)
Exports: $25.5 million (f.o.b., 1990)
commodities: fish and fish products, fox and mink pelts
partners: US 58%, France 17%, UK 11%, Canada, Portugal
Imports: $87.2 million (c.i.f., 1990)
commodities: meat, clothing, fuel, electrical equipment, machinery, building materials
partners: Canada, France, US, Netherlands, UK

Saint Vincent and the Grenadines

External debt: $NA
Industrial production: growth rate NA%
Electricity: 10,000 kW capacity; 25 million kWh produced, 3,840 kWh per capita (1992)
Industries: fish processing and supply base for fishing fleets; tourism
Agriculture: vegetables, cattle, sheep, pigs for local consumption; fish catch of 20,500 metric tons (1989)
Economic aid: Western (non-US) countries, ODA and OOF bilateral commitments (1970-89), $500 million
Currency: 1 French franc (F) = 100 centimes
Exchange rates: French francs (F) per US$1—5.4812 (January 1993), 5.2938 (1992), 5.6421 (1991), 5.4453 (1990), 6.3801 (1989), 5.9569 (1988)
Fiscal year: calendar year

Communications

Highways: 120 km total; 60 km paved (1985)
Ports: Saint Pierre
Airports:
total: 2
usable: 2
with permanent-surface runways: 2
with runways over 3,659 m: 0
with runways 2,440-3,659 m: 0
with runways 1,220-2,439 m: 1
Telecommunications: 3,601 telephones; broadcast stations—1 AM, 3 FM, no TV; radio communication with most countries in the world; 1 earth station in French domestic satellite system

Defense Forces

Note: defense is the responsibility of France

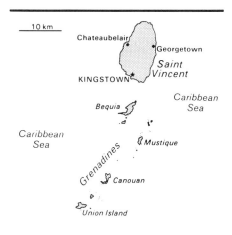

Geography

Location: in the eastern Caribbean Sea about three-fourths of the way between Puerto Rico and Trinidad and Tobago
Map references: Central America and the Caribbean, South America, Standard Time Zones of the World
Area:
total area: 340 km²
land area: 340 km²
comparative area: slightly less than twice the size of Washington, DC
Land boundaries: 0 km
Coastline: 84 km
Maritime claims:
contiguous zone: 24 nm
exclusive economic zone: 200 nm
territorial sea: 12 nm
International disputes: none
Climate: tropical; little seasonal temperature variation; rainy season (May to November)
Terrain: volcanic, mountainous; Soufriere volcano on the island of Saint Vincent
Natural resources: negligible
Land use:
arable land: 38%
permanent crops: 12%
meadows and pastures: 6%
forest and woodland: 41%
other: 3%
Irrigated land: 10 km² (1989 est.)
Environment: subject to hurricanes; Soufriere volcano is a constant threat
Note: some islands of the Grenadines group are administered by Grenada

People

Population: 114,562 (July 1993 est.)
Population growth rate: 0.76% (1993 est.)
Birth rate: 20.86 births/1,000 population (1993 est.)
Death rate: 5.39 deaths/1,000 population (1993 est.)
Net migration rate: -7.92 migrant(s)/1,000 population (1993 est.)

Infant mortality rate: 18.3 deaths/1,000 live births (1993 est.)
Life expectancy at birth:
total population: 71.72 years
male: 70.21 years
female: 73.28 years (1993 est.)
Total fertility rate: 2.16 children born/woman (1993 est.)
Nationality:
noun: Saint Vincentian(s) or Vincentian(s)
adjective: Saint Vincentian or Vincentian
Ethnic divisions: black African descent, white, East Indian, Carib Indian
Religions: Anglican, Methodist, Roman Catholic, Seventh-Day Adventist
Languages: English, French patois
Literacy: age 15 and over having ever attended school (1970)
total population: 96%
male: 96%
female: 96%
Labor force: 67,000 (1984 est.)
by occupation: NA

Government

Names:
conventional long form: none
conventional short form: Saint Vincent and the Grenadines
Digraph: VC
Type: constitutional monarchy
Capital: Kingstown
Administrative divisions: 6 parishes; Charlotte, Grenadines, Saint Andrew, Saint David, Saint George, Saint Patrick
Independence: 27 October 1979 (from UK)
Constitution: 27 October 1979
Legal system: based on English common law
National holiday: Independence Day, 27 October (1979)
Political parties and leaders: New Democratic Party (NDP), James (Son) MITCHELL; Saint Vincent Labor Party (SVLP), Stanley JOHN; United People's Movement (UPM), Adrian SAUNDERS; Movement for National Unity (MNU), Ralph GONSALVES; National Reform Party (NRP), Joel MIGUEL
Suffrage: 18 years of age; universal
Elections:
House of Assembly: last held 16 May 1989 (next to be held NA July 1994); results—percent of vote by party NA; seats—(21 total; 15 elected representatives and 6 appointed senators) NDP 15
Executive branch: British monarch, governor general, prime minister, Cabinet
Legislative branch: unicameral House of Assembly
Judicial branch: Eastern Caribbean Supreme Court
Leaders:
Chief of State: Queen ELIZABETH II (since 6 February 1952), represented by Governor

Saint Vincent and the Grenadines

(continued)

General David JACK (since 29 September 1989)

Head of Government: Prime Minister James F. MITCHELL (since 30 July 1984)

Member of: ACP, C, CARICOM, CDB, ECLAC, FAO, G-77, IBRD, ICAO, ICFTU, IDA, IFAD, IMF, IMO, INTERPOL, IOC, ITU, LORCS, OAS, OECS, UN, UNCTAD, UNESCO, UNIDO, UPU, WCL, WFTU, WHO

Diplomatic representation in US:

chief of mission: Ambassador Kingsley LAYNE

chancery: 1717 Massachusetts Avenue, NW, Suite 102, Washington, DC 20036

telephone: NA

US diplomatic representation: no official presence since the Ambassador resides in Bridgetown (Barbados)

Flag: three vertical bands of blue (hoist side), gold (double width), and green; the gold band bears three green diamonds arranged in a V pattern

Economy

Overview: Agriculture, dominated by banana production, is the most important sector of the economy. The services sector, based mostly on a growing tourist industry, is also important. The economy continues to have a high unemployment rate of 35%-40% because of an overdependence on the weather-plagued banana crop as a major export earner. Government progress toward diversifying into new industries has been relatively unsuccessful.

National product: GDP—exchange rate conversion—$171 million (1992 est.)

National product real growth rate: 3% (1992 est.)

National product per capita: $1,500 (1992 est.)

Inflation rate (consumer prices): 2.3% (1991 est.)

Unemployment rate: 35%-40% (1992 est.)

Budget: revenues $62 million; expenditures $67 million, including capital expenditures of $21 million (FY90 est.)

Exports: $65.7 million (f.o.b., 1991)

commodities: bananas, eddoes and dasheen (taro), arrowroot starch, tennis racquets

partners: UK 43%, CARICOM 37%, US 15%

Imports: $110.7 million (f.o.b., 1991)

commodities: foodstuffs, machinery and equipment, chemicals and fertilizers, minerals and fuels

partners: US 42%, CARICOM 19%, UK 15%

External debt: $50.9 million (1989)

Industrial production: growth rate 0% (1989); accounts for 14% of GDP

Electricity: 16,600 kW capacity; 64 million kWh produced, 555 kWh per capita (1992)

Industries: food processing, cement, furniture, clothing, starch

Agriculture: accounts for 15% of GDP and 60% of labor force; provides bulk of exports; products—bananas, coconuts, sweet potatoes, spices; small numbers of cattle, sheep, hogs, goats; small fish catch used locally

Economic aid: US commitments, including Ex-Im (FY70-87), $11 million; Western (non-US) countries, ODA and OOF bilateral commitments (1970-89), $81 million

Currency: 1 EC dollar (EC$) = 100 cents

Exchange rates: East Caribbean dollars (EC$) per US$1—2.70 (fixed rate since 1976)

Fiscal year: calendar year

Communications

Highways: 1,000 km total; 300 km paved; 400 km improved; 300 km unimproved (est.)

Ports: Kingstown

Merchant marine: 407 ships (1,000 GRT or over) totaling 3,388,427 GRT/5,511,325 DWT; includes 3 passenger, 2 passenger-cargo, 222 cargo, 22 container, 19 roll-on/roll-off cargo, 14 refrigerated cargo, 24 oil tanker, 7 chemical tanker, 4 liquefied gas, 73 bulk, 13 combination bulk, 2 vehicle carrier, 1 livestock carrier, 1 specialized tanker; note—China owns 3 ships; a flag of convenience registry

Airports:

total: 6

usable: 6

with permanent-surface runways: 5

with runways over 3,659 m: 0

with runways 2,440-3,659 m: 0

with runways 1,220-2,439 m: 1

Telecommunications: islandwide fully automatic telephone system; 6,500 telephones; VHF/UHF interisland links from Saint Vincent to Barbados and the Grenadines; new SHF links to Grenada and Saint Lucia; broadcast stations—2 AM, no FM, 1 TV (cable)

Defense Forces

Branches: Royal Saint Vincent and the Grenadines Police Force, Coast Guard

Manpower availability: NA

Defense expenditures: exchange rate conversion—$NA, NA% of GDP

San Marino

Geography

Location: Southern Europe, an enclave in central Italy

Map references: Europe, Standard Time Zones of the World

Area:

total area: 60 km²

land area: 60 km²

comparative area: about 0.3 times the size of Washington, DC

Land boundaries: total 39 km, Italy 39 km

Coastline: 0 km (landlocked)

Maritime claims: none; landlocked

International disputes: none

Climate: Mediterranean; mild to cool winters; warm, sunny summers

Terrain: rugged mountains

Natural resources: building stone

Land use:

arable land: 17%

permanent crops: 0%

meadows and pastures: 0%

forest and woodland: 0%

other: 83%

Irrigated land: NA

Environment: dominated by the Appenines

Note: landlocked; smallest independent state in Europe after the Holy See and Monaco

People

Population: 23,855 (July 1993 est.)

Population growth rate: 1.01% (1993 est.)

Birth rate: 11.32 births/1,000 population (1993 est.)

Death rate: 7.25 deaths/1,000 population (1993 est.)

Net migration rate: 6.08 migrant(s)/1,000 population (1993 est.)

Infant mortality rate: 5.7 deaths/1,000 live births (1993 est.)

Life expectancy at birth:

total population: 81.18 years

male: 77.09 years

female: 85.27 years (1993 est.)

Total fertility rate: 1.54 children born/woman (1993 est.)

Nationality:
noun: Sammarinese (singular and plural)
adjective: Sammarinese
Ethnic divisions: Sammarinese, Italian
Religions: Roman Catholic
Languages: Italian
Literacy: age 14 and over can read and write (1976)
total population: 96%
male: 96%
female: 95%
Labor force: 4,300 (est.)
by occupation: NA

Government

Names:
conventional long form: Republic of San Marino
conventional short form: San Marino
local long form: Repubblica di San Marino
local short form: San Marino
Digraph: SM
Type: republic
Capital: San Marino
Administrative divisions: 9 municipalities (castelli, singular—castello); Acquaviva, Borgo Maggiore, Chiesanuova, Domagnano, Faetano, Fiorentino, Monte Giardino, San Marino, Serravalle
Independence: 301 AD (by tradition)
Constitution: 8 October 1600; electoral law of 1926 serves some of the functions of a constitution
Legal system: based on civil law system with Italian law influences; has not accepted compulsory ICJ jurisdiction
National holiday: Anniversary of the Foundation of the Republic, 3 September
Political parties and leaders: Christian Democratic Party (DCS), Piermarino MENICUCCI; San Marino Democratic Progressive Party (PPDS) formerly San Marino Communist Party (PCS), Gilberto GHIOTTI; San Marino Socialist Party (PSS), Remy GIACOMINI; Unitary Socialst Party (PSU); Democratic Movement (MD), Emilio Della BALDA; San Marino Social Democratic Party (PSDS), Augusto CASALI; San Marino Republican Party (PRS), Cristoforo BUSCARINI
Suffrage: 18 years of age; universal
Elections:
Great and General Council: last held 29 May 1988 (next to be held by NA May 1993); results—percent of vote by party NA; seats—(60 total) DCS 27, PCS 18, PSU 8, PSS 7
Executive branch: two captains regent, Congress of State (cabinet); real executive power is wielded by the secretary of state for foreign affairs and the secretary of state for internal affairs
Legislative branch: unicameral Great and General Council (Consiglio Grande e Generale)

Judicial branch: Council of Twelve (Consiglio dei XII)
Leaders:
Co-Chiefs of State: Captain Regent Patricia BUSIGNANI and Captain Regent Salvatore TONELLI (for the period 1 April—30 September 1993)
Head of Government: Secretary of State Gabriele GATTI (since July 1986)
Member of: CE, CSCE, ECE, ICAO, ICFTU, ILO, IMF, IOC, IOM (observer), ITU, LORCS, NAM (guest), UN, UNCTAD, UNESCO, UPU, WHO, WIPO, WTO
Diplomatic representation in US:
honorary consulates general: Washington and New York
honorary consulate: Detroit
US diplomatic representation: no mission in San Marino, but the Consul General in Florence (Italy) is accredited to San Marino
Flag: two equal horizontal bands of white (top) and light blue with the national coat of arms superimposed in the center; the coat of arms has a shield (featuring three towers on three peaks) flanked by a wreath, below a crown and above a scroll bearing the word LIBERTAS (Liberty)

Economy

Overview: The tourist industry contributes over 50% of GDP. In 1991 over 3.1 million tourists visited San Marino, 2.7 million of whom were Italians. The key industries are wearing apparel, electronics, and ceramics. Main agricultural products are wine and cheeses. The per capita level of output and standard of living are comparable to northern Italy.
National product: GDP—purchasing power equivalent—$465 million (1992 est.)
National product real growth rate: NA%
National product per capita: $20,000 (1992 est.)
Inflation rate (consumer prices): 5% (1992 est.)
Unemployment rate: 3% (1991)
Budget: revenues $NA; expenditures $300 million, including capital expenditures of $NA (1991)
Exports: trade data are included with the statistics for Italy; commodity trade consists primarily of exchanging building stone, lime, wood, chestnuts, wheat, wine, baked goods, hides, and ceramics for a wide variety of consumer manufactures
Imports: see exports
External debt: $NA
Industrial production: growth rate NA%; accounts for 42% of workforce
Electricity: supplied by Italy
Industries: wine, olive oil, cement, leather, textile, tourism
Agriculture: employs 3% of labor force; products—wheat, grapes, maize, olives,

meat, cheese, hides; small numbers of cattle, pigs, horses; depends on Italy for food imports
Economic aid: NA
Currency: Italian currency is used; note—also mints its own coins
Exchange rates: Italian lire (Lit) per US$1—1,482.5 (January 1993), 1,232.4 (1992), 1,240.6 (1991), 1,198.1 (1990), 1,372.1 (1989), 1,301.6 (1988)
Fiscal year: calendar year

Communications

Highways: 104 km
Telecommunications: automatic telephone system completely integrated into Italian system; 11,700 telephones; broadcast services from Italy; microwave and cable links into Italian networks; no communication satellite facilities

Defense Forces

Branches: public security or police force
Manpower availability: all fit men ages 16-60 constitute a militia that can serve as an army
Defense expenditures: $NA, NA% of GDP

Sao Tome and Principe

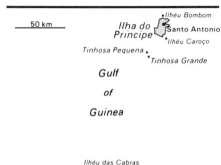

Gulf
of
Guinea

Geography

Location: Western Africa, in the Atlantic Ocean, 340 km off the coast of Gabon straddling the equator
Map references: Africa, Standard Time Zones of the World
Area:
total area: 960 km²
land area: 960 km²
comparative area: slightly less than 5.5 times the size of Washington, DC
Land boundaries: 0 km
Coastline: 209 km
Maritime claims: measured from claimed archipelagic baselines
exclusive economic zone: 200 nm
territorial sea: 12 nm
International disputes: none
Climate: tropical; hot, humid; one rainy season (October to May)
Terrain: volcanic, mountainous
Natural resources: fish
Land use:
arable land: 1%
permanent crops: 20%
meadows and pastures: 1%
forest and woodland: 75%
other: 3%
Irrigated land: NA km²
Environment: deforestation; soil erosion

People

Population: 133,225 (July 1993 est.)
Population growth rate: 2.63% (1993 est.)
Birth rate: 35.39 births/1,000 population (1993 est.)
Death rate: 9.06 deaths/1,000 population (1993 est.)
Net migration rate: 0 migrant(s)/1,000 population (1993 est.)
Infant mortality rate: 64.9 deaths/1,000 live births (1993 est.)
Life expectancy at birth:
total population: 63.02 years
male: 61.19 years

female: 64.9 years (1993 est.)
Total fertility rate: 4.6 children born/woman (1993 est.)
Nationality:
noun: Sao Tomean(s)
adjective: Sao Tomean
Ethnic divisions: mestico, angolares (descendents of Angolan slaves), forros (descendents of freed slaves), servicais (contract laborers from Angola, Mozambique, and Cape Verde), tongas (children of servicais born on the islands), Europeans (primarily Portuguese)
Religions: Roman Catholic, Evangelical Protestant, Seventh-Day Adventist
Languages: Portuguese (official)
Literacy: age 15 and over can read and write (1981)
total population: 57%
male: 73%
female: 42%
Labor force: 21,096 (1981); most of population engaged in subsistence agriculture and fishing; labor shortages on plantations and of skilled workers; 56% of population of working age (1983)

Government

Names:
conventional long form: Democratic Republic of Sao Tome and Principe
conventional short form: Sao Tome and Principe
local long form: Republica Democratica de Sao Tome e Principe
local short form: Sao Tome e Principe
Digraph: TP
Type: republic
Capital: Sao Tome
Administrative divisions: 2 districts (concelhos, singular—concelho); Principe, Sao Tome
Independence: 12 July 1975 (from Portugal)
Constitution: 5 November 1975, approved 15 December 1982
Legal system: based on Portuguese law system and customary law; has not accepted compulsory ICJ jurisdiction
National holiday: Independence Day, 12 July (1975)
Political parties and leaders: Party for Democratic Convergence-Reflection Group (PCD-GR), Daniel Lima Dos Santos DAIO, secretary general; Movement for the Liberation of Sao Tome and Principe (MLSTP), Carlos da GRACA; Christian Democratic Front (FDC), Alphonse Dos SANTOS; Democratic Opposition Coalition (CODO), leader NA; other small parties
Suffrage: 18 years of age; universal
Elections:
President: last held 3 March 1991 (next to be held NA March 1996); results—Miguel TROVOADA was elected without opposition

in Sao Tome's first multiparty presidential election
National People's Assembly: last held 20 January 1991 (next to be held NA January 1996); results—PCD-GR 54.4%, MLSTP 30.5%, CODO 5.2%, FDC 1.5%, other 8.4%; seats—(55 total) PCD-GR 33, MLSTP 21, CODO 1; note—this was the first multiparty election in Sao Tome and Principe
Executive branch: president, prime minister, Council of Ministers (cabinet)
Legislative branch: unicameral National People's Assembly (Assembleia Popular Nacional)
Judicial branch: Supreme Court
Leaders:
Chief of State: President Miguel TROVOADA (since 4 April 1991)
Head of Government: Prime Minister Noberto Jose D'Alva COSTA ALEGRE (since 16 May 1992)
Member of: ACP, AfDB, CEEAC, ECA, FAO, G-77, IBRD, ICAO, IDA, IFAD, ILO, IMF, IMO, INTERPOL, IOM (observer), ITU, LORCS, NAM, OAU, UN, UNCTAD, UNESCO, UNIDO, UPU, WHO, WMO, WTO
Diplomatic representation in US:
chief of mission: Ambassador Joaquim Rafael BRANCO
chancery: (temporary) 801 Second Avenue, Suite 603, New York, NY 10017
telephone: (212) 697-4211
US diplomatic representation: ambassador to Gabon is accredited to Sao Tome and Principe on a nonresident basis and makes periodic visits to the islands
Flag: three horizontal bands of green (top), yellow (double width), and green with two black five-pointed stars placed side by side in the center of the yellow band and a red isosceles triangle based on the hoist side; uses the popular pan-African colors of Ethiopia

Economy

Overview: The economy has remained dependent on cocoa since the country gained independence nearly 15 years ago. Since then, however, cocoa production has gradually deteriorated because of drought and mismanagement, so that by 1987 output had fallen to less than 50% of its former levels. As a result, a shortage of cocoa for export has created a serious balance-of-payments problem. Production of less important crops, such as coffee, copra, and palm kernels, has also declined. The value of imports generally exceeds that of exports by a ratio of 4:1. The emphasis on cocoa production at the expense of other food crops has meant that Sao Tome has to import 90% of food needs. It also has to import all fuels and most manufactured goods. Over the years, Sao Tome has been unable to service its external debt, which amounts to roughly 80% of export earnings. Considerable potential exists for development of a tourist industry, and the government has taken steps to expand facilities in recent years. The government also implemented a Five-Year Plan covering 1986-90 to restructure the economy and reschedule external debt service payments in cooperation with the International Development Association and Western lenders.

National product: GDP—exchange rate conversion—$41.4 million (1992 est.)

National product real growth rate: 1.5% (1992 est.)

National product per capita: $315 (1992 est.)

Inflation rate (consumer prices): 27% (1992 est.)

Unemployment rate: NA%

Budget: revenues $10.2 million; expenditures $36.8 million, including capital expenditures of $22.5 million (1989)

Exports: $5.5 million (f.o.b., 1991 est.)

commodities: cocoa 85%, copra, coffee, palm oil

partners: Germany, Netherlands, China

Imports: $24.5 million (f.o.b., 1991)

commodities: machinery and electrical equipment 54%, food products 23%, other 23%

partners: Portugal, Germany, Angola, China

External debt: $163.6 million (1992)

Industrial production: growth rate 7.1% (1986)

Electricity: 5,000 kW capacity; 10 million kWh produced, 80 kWh per capita (1991)

Industries: light construction, shirts, soap, beer, fisheries, shrimp processing

Agriculture: dominant sector of economy, primary source of exports; cash crops—cocoa (85%), coconuts, palm kernels, coffee; food products—bananas, papaya, beans, poultry, fish; not self-sufficient in food grain and meat

Economic aid: US commitments, including Ex-Im (FY70-89), $8 million; Western (non-US) countries, ODA and OOF bilateral commitments (1970-89), $89 million

Currency: 1 dobra (Db) = 100 centimos

Exchange rates: dobras (Db) per US$1—230 (1992), 260.0 (November 1991), 122.48 (December 1988), 72.827 (1987), 36.993 (1986)

Fiscal year: calendar year

Communications

Highways: 300 km (two-thirds are paved); roads on Principe are mostly unpaved and in need of repair

Ports: Sao Tome, Santo Antonio

Merchant marine: 1 cargo ship (1,000 GRT or over) totaling 1,096 GRT/1,105 DWT

Airports:

total: 2

usable: 2

with permanent-surface runways : 2

with runways over 3,659 m: 0

with runways 2,440-3,659 m: 0

with runways 1,220-2,439 m: 2

Telecommunications: minimal system; broadcast stations—1 AM, 2 FM, no TV; 1 Atlantic Ocean INTELSAT earth station

---

Defense Forces

Branches: Army, Navy, National Police

Manpower availability: males age 15-49 31,326; fit for military service 16,507 (1993 est.)

Defense expenditures: exchange rate conversion—$NA, NA% of GDP

Saudi Arabia

Boundary representation is not necessarily authoritative.

Geography

Location: Middle East, between the Red Sea and the Persian Gulf

Map references: Africa, Middle East, Standard Time Zones of the World

Area:

total area: 1,960,582 km²

land area: 1,960,582 km²

comparative area: slightly less than one-fourth the size of the US

Land boundaries: total 4,415 km, Iraq 814 km, Jordan 728 km, Kuwait 222 km, Oman 676 km, Qatar 60 km, UAE 457 km, Yemen 1,458 km

Coastline: 2,640 km

Maritime claims:

contiguous zone: 18 nm

continental shelf: not specified

territorial sea: 12 nm

International disputes: large section of boundary with Yemen not defined; status of boundary with UAE not final; Kuwaiti ownership of Qaruh and Umm al Maradim Islands is disputed by Saudi Arabia

Climate: harsh, dry desert with great extremes of temperature

Terrain: mostly uninhabited, sandy desert

Natural resources: petroleum, natural gas, iron ore, gold, copper

Land use:

arable land: 1%

permanent crops: 0%

meadows and pastures: 39%

forest and woodland: 1%

other: 59%

Irrigated land: 4,350 km² (1989 est.)

Environment: no perennial rivers or permanent water bodies; developing extensive coastal seawater desalination facilities; desertification

Note: extensive coastlines on Persian Gulf and Red Sea provide great leverage on shipping (especially crude oil) through Persian Gulf and Suez Canal

Saudi Arabia (continued)

People

Population: 17,615,310 (July 1993 est.)
note: the population figure is consistent with a 3.3% growth rate; a 1992 census gives the number of Saudi citizens as 12,304,835 and the number of residents who are not citizens as 4,624,459
Population growth rate: 3.3% (1993 est.)
Birth rate: 38.59 births/1,000 population (1993 est.)
Death rate: 6.05 deaths/1,000 population (1993 est.)
Net migration rate: 0 migrant(s)/1,000 population (1993 est.)
Infant mortality rate: 55.3 deaths/1,000 live births (1993 est.)
Life expectancy at birth:
total population: 67.32 years
male: 65.71 years
female: 69.01 years (1993 est.)
Total fertility rate: 6.7 children born/woman (1993 est.)
Nationality:
noun: Saudi(s)
adjective: Saudi or Saudi Arabian
Ethnic divisions: Arab 90%, Afro-Asian 10%
Religions: Muslim 100%
Languages: Arabic
Literacy: age 15 and over can read and write (1990)
total population: 62%
male: 73%
female: 48%
Labor force: 5 million
by occupation: government 34%, industry and oil 28%, services 22%, agriculture 16%

Government

Names:
conventional long form: Kingdom of Saudi Arabia
conventional short form: Saudi Arabia
local long form: Al Mamlakah al 'Arabiyah as Su'udiyah
local short form: Al 'Arabiyah as Su'udiyah
Digraph: SA
Type: monarchy
Capital: Riyadh
Administrative divisions: 14 emirates (imarat, singular—imarah); Al Bahah, Al Hudud ash Shamaliyah, Al Jawf, Al Madinah, Al Qasim, Al Qurayyat, Ar Riyad, Ash Sharqiyah, 'Asir, Ha'il, Jizan, Makkah, Najran, Tabuk
Independence: 23 September 1932 (unification)
Constitution: none; governed according to Shari'a (Islamic law)
Legal system: based on Islamic law, several secular codes have been introduced; commercial disputes handled by special committees; has not accepted compulsory ICJ jurisdiction

National holiday: Unification of the Kingdom, 23 September (1932)
Political parties and leaders: none allowed
Suffrage: none
Elections: none
Executive branch: monarch and prime minister, crown prince and deputy prime minister, Council of Ministers
Legislative branch: none
Judicial branch: Supreme Council of Justice
Leaders:
Chief of State and Head of Government: King and Prime Minister FAHD bin 'Abd al-'Aziz Al Sa'ud (since 13 June 1982); Crown Prince and Deputy Prime Minister 'ABDALLAH bin 'Abd al-'Aziz Al Sa'ud (half-brother to the King, appointed heir to the throne 13 June 1982)
Member of: ABEDA, AfDB, AFESD, AL, AMF, CCC, ESCWA, FAO, G-19, G-77, GCC, IAEA, IBRD, ICAO, ICC, IDA, IDB, IFAD, IFC, ILO, IMF, IMO, INMARSAT, INTELSAT, INTERPOL, IOC, ISO, ITU, LORCS, NAM, OAPEC, OAS (observer), OIC, OPEC, UN, UNCTAD, UNESCO, UNIDO, UPU, WFTU, WHO, WIPO, WMO
Diplomatic representation in US:
chief of mission: Ambassador BANDAR Bin Sultan
chancery: 601 New Hampshire Avenue NW, Washington, DC 20037
telephone: (202) 342-3800
consulates general: Houston, Los Angeles, and New York
US diplomatic representation:
chief of mission: (vacant); Charge d'Affaires C. David Welch
embassy: Collector Road M, Diplomatic Quarter, Riyadh
mailing address: American Embassy, Unit 61307, Riyadh; International Mail: P. O. Box 94309, Riyadh 11693; or APO AE 09803-1307
telephone: [966] (1) 488-3800
FAX: Telex 406866
consulates general: Dhahran, Jiddah (Jeddah)
Flag: green with large white Arabic script (that may be translated as There is no God but God; Muhammad is the Messenger of God) above a white horizontal saber (the tip points to the hoist side); green is the traditional color of Islam

Economy

Overview: The petroleum sector accounts for roughly 75% of budget revenues, 35% of GDP, and almost all export earnings. Saudi Arabia has the largest reserves of petroleum in the world, ranks as the largest exporter of petroleum, and plays a leading role in OPEC. For the 1990s the government intends to encourage private economic

activity and to foster the gradual process of turning Saudi Arabia into a modern industrial state that retains traditional Islamic values. Four million foreign workers play an important role in the Saudi economy, for example, in the oil and banking sectors.
National product: GDP—exchange rate conversion—$111 billion (1992 est.)
National product real growth rate: 3.6% (1992 est.)
National product per capita: $6,500 (1992 est.)
Inflation rate (consumer prices): 2.5% (1992 est.)
Unemployment rate: 6.5% (1992 est.)
Budget: revenues $45.1 billion; expenditures $52.5 billion, including capital expenditures of $NA (1993 est.)
Exports: $48.2 billion (f.o.b., 1991)
commodities: petroleum and petroleum products 92%
partners: US 21%, Japan 18%, Singapore 6%, France 6%, Korea 5%
Imports: $26.1 billion (f.o.b., 1991)
commodities: food stuffs, manufactured goods, transportation equipment, chemical products, textiles
partners: US 21%, UK 13%, Japan 12%, Germany 8%, France 6%
External debt: $18.9 billion (December 1989 est.)
Industrial production: growth rate −1.1% (1989 est.); accounts for 37% of GDP, including petroleum
Electricity: 28,554,000 kW capacity; 63,000 million kWh produced, 3,690 kWh per capita (1992)
Industries: crude oil production, petroleum refining, basic petrochemicals, cement, two small steel-rolling mills, construction, fertilizer, plastics
Agriculture: accounts for about 10% of GDP, 16% of labor force; subsidized by government; products—wheat, barley, tomatoes, melons, dates, citrus fruit, mutton, chickens, eggs, milk; approaching self-sufficiency in food
Illicit drugs: death penalty for traffickers
Economic aid: donor—pledged $64.7 billion in bilateral aid (1979-89)
Currency: 1 Saudi riyal (SR) = 100 halalas
Exchange rates: Saudi riyals (SR) per US$1—3.7450 (fixed rate since late 1986), 3.7033 (1986)
Fiscal year: calendar year

Communications

Railroads: 1390 km 1.435-meter standard gauge; 448 km are double tracked
Highways: 74,000 km total; 35,000 km paved, 39,000 km gravel and improved earth
Pipelines: crude oil 6,400 km, petroleum products 150 km, natural gas 2,200 km, includes natural gas liquids 1,600 km

Senegal

Ports: Jiddah, Ad Dammam, Ras Tanura, Jizan, Al Jubayl, Yanbu al Bahr, Yanbu al Sinaiyah

Merchant marine: 77 ships (1,000 GRT or over) totaling 860,818 GRT/1,219,345 DWT; includes 1 passenger, 6 short-sea passenger, 11 cargo, 13 roll-on/roll-off cargo, 3 container, 6 refrigerated cargo, 5 livestock carrier, 23 oil tanker, 6 chemical tanker, 1 liquefied gas, 1 specialized tanker, 1 bulk

Airports:
total: 213
usable: 193
with permanent-surface runways: 71
with runways over 3,659 m: 14
with runways 2,440-3,659 m: 36
with runways 1,220-2,439 m: 107

Telecommunications: modern system with extensive microwave and coaxial and fiber optic cable systems; 1,624,000 telephones; broadcast stations—43 AM, 13 FM, 80 TV; microwave radio relay to Bahrain, Jordan, Kuwait, Qatar, UAE, Yemen, and Sudan; coaxial cable to Kuwait and Jordan; submarine cable to Djibouti, Egypt and Bahrain; earth stations—3 Atlantic Ocean INTELSAT, 2 Indian Ocean INTELSAT, 1 ARABSAT, 1 INMARSAT

Defense Forces

Branches: Land Force (Army), Navy, Air Force, Air Defense Force, National Guard, Coast Guard, Frontier Forces, Special Security Force, Public Security Force

Manpower availability: males age 15-49 5,650,492; fit for military service 3,128,620; reach military age (17) annually 140,283 (1993 est.)

Defense expenditures: exchange rate conversion—$16.5 billion, 13% of GDP (1993 budget)

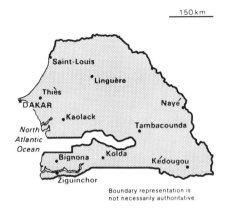

150 km

North Atlantic Ocean

Saint-Louis · Linguère · Thiès · DAKAR · Naye · Kaolack · Tambacounda · Bignona · Kolda · Kédougou · Ziguinchor

Boundary representation is not necessarily authoritative.

Geography

Location: Western Africa, bordering the North Atlantic Ocean between Guinea-Bissau and Mauritania

Map references: Africa, Standard Time Zones of the World

Area:
total area: 196,190 km²
land area: 192,000 km²
comparative area: slightly smaller than South Dakota

Land boundaries: total 2,640 km, The Gambia 740 km, Guinea 330 km, Guinea-Bissau 338 km, Mali 419 km, Mauritania 813 km

Coastline: 531 km

Maritime claims:
contiguous zone: 24 nm
continental shelf: 200 nm or the edge of continental margin
exclusive economic zone: 200 nm
territorial sea: 12 nm

International disputes: short section of the boundary with The Gambia is indefinite; the International Court of Justice (ICJ) on 12 November 1991 rendered its decision on the Guinea-Bissau/Senegal maritime boundary in favor of Senegal - that decision has been rejected by Guinea-Bissau; boundary with Mauritania

Climate: tropical; hot, humid; rainy season (December to April) has strong southeast winds; dry season (May to November) dominated by hot, dry harmattan wind

Terrain: generally low, rolling, plains rising to foothills in southeast

Natural resources: fish, phosphates, iron ore

Land use:
arable land: 27%
permanent crops: 0%
meadows and pastures: 30%
forest and woodland: 31%
other: 12%

Irrigated land: 1,800 km² (1989 est.)

Environment: lowlands seasonally flooded; deforestation; overgrazing; soil erosion; desertification

Note: The Gambia is almost an enclave

People

Population: 8,463,225 (July 1993 est.)
Population growth rate: 3.1% (1993 est.)
Birth rate: 43.42 births/1,000 population (1993 est.)
Death rate: 12.38 deaths/1,000 population (1993 est.)
Net migration rate: 0 migrant(s)/1,000 population (1993 est.)
Infant mortality rate: 77.8 deaths/1,000 live births (1993 est.)
Life expectancy at birth:
total population: 56.01 years
male: 54.59 years
female: 57.48 years (1993 est.)
Total fertility rate: 6.15 children born/woman (1993 est.)
Nationality:
noun: Senegalese (singular and plural)
adjective: Senegalese
Ethnic divisions: Wolof 36%, Fulani 17%, Serer 17%, Toucouleur 9%, Diola 9%, Mandingo 9%, European and Lebanese 1%, other 2%
Religions: Muslim 92%, indigenous beliefs 6%, Christian 2% (mostly Roman Catholic)
Languages: French (official), Wolof, Pulaar, Diola, Mandingo
Literacy: age 15 and over can read and write (1990)
total population: 38%
male: 52%
female: 25%
Labor force: 2.509 million (77% are engaged in subsistence farming; 175,000 wage earners)
by occupation: private sector 40%, government and parapublic 60%
note: 52% of population of working age (1985)

Government

Names:
conventional long form: Republic of Senegal
conventional short form: Senegal
local long form: Republique du Senegal
local short form: Senegal
Digraph: SG
Type: republic under multiparty democratic rule
Capital: Dakar
Administrative divisions: 10 regions (regions, singular—region); Dakar, Diourbel, Fatick, Kaolack, Kolda, Louga, Saint-Louis, Tambacounda, Thies, Ziguinchor
Independence: 20 August 1960 (from France; The Gambia and Senegal signed an agreement on 12 December 1981 that called

for the creation of a loose confederation to be known as Senegambia, but the agreement was dissolved on 30 September 1989)

Constitution: 3 March 1963, last revised in 1991

Legal system: based on French civil law system; judicial review of legislative acts in Supreme Court, which also audits the government's accounting office; has not accepted compulsory ICJ jurisdiction

National holiday: Independence Day, 4 April (1960)

Political parties and leaders: Socialist Party (PS), President Abdou DIOUF; Senegalese Democratic Party (PDS), Abdoulaye WADE; 13 other small uninfluential parties

Other political or pressure groups: students; teachers; labor; Muslim Brotherhoods

Suffrage: 18 years of age; universal

Elections:

President: last held 21 February 1993 (next to be held NA); results—Abdou DIOUF (PS) 58.4%, Abdoulaye WADE (PDS) 32.03%, other 9.57%

National Assembly: last held 28 February 1988 (next to be held NA May 1993); results—PS 71%, PDS 25%, other 4%; seats—(120 total) PS 103, PDS 17

Executive branch: president, prime minister, Council of Ministers (cabinet)

Legislative branch: unicameral National Assembly (Assemblee Nationale)

Judicial branch: Supreme Court (Cour Supreme)

Leaders:

Chief of State: President Abdou DIOUF (since 1 January 1981)

Head of Government: Prime Minister Habib THIAM (since 7 April 1991)

Member of: ACCT, ACP, AfDB, CCC, CEAO, ECA, ECOWAS, FAO, FZ, G-15, G-77, GATT, IAEA, IBRD, ICAO, ICC, IDA, IDB, IFAD, IFC, ILO, IMF, IMO, INTELSAT, INTERPOL, IOC, IOM (observer), ISO (correspondent), ITU, LORCS, NAM, OAU, OIC, PCA, UN, UNAVEM II, UNCTAD, UNESCO, UNIDO, UNIKOM, UNTAC, UPU, WADB, WCL, WFTU, WHO, WIPO, WMO, WTO

Diplomatic representation in US:

chief of mission: Ambassador Ibra Deguene KA

chancery: 2112 Wyoming Avenue NW, Washington, DC 20008

telephone: (202) 234-0540 or 0541

US diplomatic representation:

chief of mission: (vacant); Charge d'Affaires Robert J. KOTT

embassy: Avenue Jean XXIII at the corner of Avenue Kleber, Dakar

mailing address: B. P. 49, Dakar

telephone: [221] 23-42-96 or 23-34-24

FAX: [221] 22-29-91

Flag: three equal vertical bands of green (hoist side), yellow, and red with a small green five-pointed star centered in the yellow band; uses the popular pan-African colors of Ethiopia

---

Economy

Overview: The agricultural sector accounts for about 12% of GDP and provides employment for about 80% of the labor force. About 40% of the total cultivated land is used to grow peanuts, an important export crop. Another principal economic resource is fishing, which brought in about 23% of total foreign exchange earnings in 1990. Mining is dominated by the extraction of phosphate, but production has faltered because of reduced worldwide demand for fertilizers in recent years. Over the past 10 years tourism has become increasingly important to the economy.

National product: GDP—exchange rate conversion—$5.4 billion (1991 est.)

National product real growth rate: 1.2% (1991 est.)

National product per capita: $780 (1991 est.)

Inflation rate (consumer prices): 2% (1990)

Unemployment rate: NA%

Budget: revenues $921 million; expenditures $1,024 million; including capital expenditures of $14 million (FY89 est.)

Exports: $904 million (f.o.b., 1991 est.)

commodities: manufactures 30%, fish products 23%, peanuts 12%, petroleum products 16%, phosphates 9%

partners: France, other EC members, Mali, Cote d'Ivoire, India

Imports: $1.2 billion (c.i.f., 1991 est.)

commodities: semimanufactures 30%, food 27%, durable consumer goods 17%, petroleum 12%, capital goods 14%

partners: France, other EC, Cote d'Ivoire, Nigeria, Algeria, China, Japan

External debt: $2.9 billion (1990)

Industrial production: growth rate 4.7% (1989); accounts for 15% of GDP

Electricity: 215,000 kW capacity; 760 million kWh produced, 100 kWh per capita (1991)

Industries: agricultural and fish processing, phosphate mining, petroleum refining, building materials

Agriculture: major products—peanuts (cash crop), millet, corn, sorghum, rice, cotton, tomatoes, green vegetables; estimated two-thirds self-sufficient in food; fish catch of 354,000 metric tons in 1990

Illicit drugs: increasingly active as a transshipment point for Southwest Asian heroin moving to Europe and North America

Economic aid: US commitments, including Ex-Im (FY70-89), $551 million; Western

(non-US) countries, ODA and OOF bilateral commitments (1970-89), $5.23 billion; OPEC bilateral aid (1979-89), $589 million; Communist countries (1970-89), $295 million

Currency: 1 CFA franc (CFAF) = 100 centimes

Exchange rates: Communaute Financiere Africaine francs (CFAF) per US$1—274.06 (January 1993), 264.69 (1992), 282.11 (1991), 272.26 (1990), 319.01 (1989), 297.85 (1988)

Fiscal year: 1 July-30 June; in January 1993, Senegal will switch to a calendar year

---

Communications

Railroads: 1,034 km 1.000-meter gauge; all single track except 70 km double track Dakar to Thies

Highways: 14,007 km total; 3,777 km paved, 10,230 km laterite or improved earth

Inland waterways: 897 km total; 785 km on the Senegal, 112 km on the Saloum

Ports: Dakar, Kaolack, Foundiougne, Ziguinchor

Merchant marine: 1 bulk ship (1,000 GRT and over) totaling 1,995 GRT/3,775 DWT

Airports:

total: 25

usable: 19

with permanent-surface runways: 10

with runways over 3,659 m: 0

with runways 2,440-3,659 m: 1

with runways 1,220-2,439 m: 15

Telecommunications: above-average urban system, using microwave and cable; broadcast stations—8 AM, no FM, 1 TV; 3 submarine cables; 1 Atlantic Ocean INTELSAT earth station

---

Defense Forces

Branches: Army, Navy, Air Force, Gendarmerie, National Police

Manpower availability: males age 15-49 1,882,551; fit for military service 983,137; reach military age (18) annually 91,747 (1993 est.)

Defense expenditures: exchange rate conversion—$100 million, 2% of GDP (1989 est.)

Serbia and Montenegro

Serbia and Montenegro have asserted the formation of a joint independent state, but this entity has not been formally recognized as a state by the United States.

Note: Serbia and Montenegro have asserted the formation of a joint independent state, but this entity has not been formally recognized as a state by the US; the US view is that the Socialist Federal Republic of Yugoslavia (SFRY) has dissolved and that none of the successor republics represents its continuation

Geography

Location: Southern Europe, bordering the Adriatic Sea, between Bosnia and Herzegovina and Bulgaria
Map references: Ethnic Groups in Eastern Europe, Europe, Standard Time Zones of the World
Area:
total area: 102,350 km²
land area: 102,136 km²
comparative area: slightly larger than Kentucky
note: Serbia has a total area and a land area of 88,412 km² making it slightly larger than Maine; Montenegro has a total area of 13,938 km² and a land area of 13,724 km² making it slightly larger than Connecticut
Land boundaries: total 2,234 km, Albania 287 km (114 km with Serbia; 173 km with Motenegro), Bosnia and Herzegovina 527 km (312 km with Serbia; 215 km with Montenegro), Bulgaria 318 km, Croatia (north) 239 km, Croatia (south) 15 km, Hungary 151 km, Macedonia 221 km, Romania 476 km
note: the internal boundary between Montenegro and Serbia is 211 km
Coastline: 199 km (Montenegro 199 km, Serbia 0 km)
Maritime claims:
territorial sea: 12 nm
International disputes: Sandzak region bordering northern Montenegro and southeastern Serbia—Muslims seeking autonomy; Vojvodina taken from Hungary and awarded to the former Yugoslavia by Treaty of Trianon in 1920; disputes with

Bosnia and Herzegovina and Croatia over Serbian populated areas; Albanian minority in Kosovo seeks independence from Serbian Republic
Climate: in the north, continental climate (cold winter and hot, humid summers with well distributed rainfall); central portion, continental and Mediterranean climate; to the south, Adriatic climate along the coast, hot, dry summers and autumns and relatively cold winters with heavy snowfall inland
Terrain: extremely varied; to the north, rich fertile plains; to the east, limestone ranges and basins; to the southeast, ancient mountain and hills; to the southwest, extremely high shoreline with no islands off the coast; home of largest lake in former Yugoslavia, Lake Scutari
Natural resources: oil, gas, coal, antimony, copper, lead, zinc, nickel, gold, pyrite, chrome
Land use:
arable land: 30%
permanent crops: 5%
meadows and pastures: 20%
forest and woodland: 25%
other: 20%
Irrigated land: NA km²
Environment: coastal water pollution from sewage outlets, especially in tourist-related areas such as Kotor; air pollution around Belgrade and other industrial cities; water pollution along Danube from industrial waste dumped into the Sava which drains into the Danube; subject to destructive earthquakes
Note: controls one of the major land routes from Western Europe to Turkey and the Near East; strategic location along the Adriatic coast

People

Population: 10,699,539 (July 1993 est.)
Population growth rate: NA%
Birth rate: NA births/1,000 population
Death rate: NA deaths/1,000 population
Net migration rate: NA migrant(s)/1,000 population
Infant mortality rate: NA deaths/1,000 live births
Life expectancy at birth:
total population: NA years
male: NA years
female: NA years
Total fertility rate: NA children born/woman
Nationality:
noun: Serb(s) and Montenegrin(s)
adjective: Serbian and Montenegrin
Ethnic divisions: Serbs 63%, Albanians 14%, Montenegrins 6%, Hungarians 4%, other 13%
Religions: Orthodox 65%, Muslim 19%, Roman Catholic 4%, Protestant 1%, other 11%

Languages: Serbo-Croatian 95%, Albanian 5%
Literacy:
total population: NA%
male: NA%
female: NA%
Labor force: 2,640,909
by occupation: industry, mining 40%, agriculture 5% (1990)

Government

Names:
conventional long form: none
conventional short form: Serbia and Montenegro
local long form: none
local short form: Srbija-Crna Gora
Digraph: SR
Type: republic
Capital: Belgrade
Administrative divisions: 2 republics (pokajine, singular—pokajina); and 2 autonomous provinces*; Kosovo*, Montenegro, Serbia, Vojvodina*
Independence: 11 April 1992 (from Yugoslavia)
Constitution: 27 April 1992
Legal system: based on civil law system
National holiday: NA
Political parties and leaders: Serbian Socialist Party (SPS; former Communist Party), Slobodan MILOSEVIC; Serbian Radical Party (SRS), Vojislav SESELJ; Serbian Renewal Party (SPO), Vuk DRASKOVIC; Democratic Party (DS), Dragoljub MICUNOVIC; Democratic Party of Serbia, Vojislav KOSTUNICA; Democratic Party of Socialists (DSSCG), Momir BULATOVIC; People's Party of Montenegro (NS), Novak KILIBARDA; Liberal Alliance of Montenegro, Slavko PEROVIC; Democratic Community of Vojvodina Hungarians (DZVM), Agoston ANDRAS; League of Communists-Movement for Yugoslavia (SK-PJ), Dragan ATANASOVSKI
Other political or pressure groups: Serbian Democratic Movement (DEPOS; coalition of opposition parties)
Suffrage: 16 years of age, if employed; 18 years of age, universal
Elections:
President: Federal Assembly elected Zoran LILIC on 25 June 1993
Chamber of Republics: last held 31 May 1992 (next to be held NA 1996); results—percent of vote by party NA; seats—(40 total; 20 Serbian, 20 Montenegrin)
Chamber of Citizens: last held 31 May 1992 (next to be held NA 1996); results—percent of votes by party NA; seats (138 total; 108 Serbian, 30 Montenegrin)—SPS 73, SRS 33, DSSCG 23, SK-PJ 2, DZVM 2, independents 2, vacant 3

Serbia and Montenegro (continued)

Executive branch: president, vice president, prime minister, deputy prime minister, cabinet

Legislative branch: bicameral Federal Assembly consists of an upper house or Chamber of Republics and a lower house or Chamber of Deputies

Judicial branch: Savezni Sud (Federal Court), Constitutional Court

Leaders:
Chief of State: Zoran LILIC (since 25 June 1993); note—Slobodan MILOSEVIC is president of Serbia (since 9 December 1990); Momir BULATOVIC is president of Montenegro (since 23 December 1990)

Head of Government: Prime Minister Radoje KONTIC (since NA December 1992); Deputy Prime Ministers Jovan ZEBIC (since NA March 1993), Asim TELACEVIC (since NA March 1993), Lovre KOVILJKO (since NA March 1993)

Diplomatic representation in US: US and Serbia and Montenegro do not maintain full diplomatic relations; the Embassy of the former Socialist Federal Republic of Yugoslavia continues to function in the US

US diplomatic representation:
chief of mission: (vacant)
embassy: address NA, Belgrade
mailing address: American Embassy Box 5070, Unit 25402, APO AE 09213-5070
telephone: [38] (11) 645-655
FAX: [38] (11) 645-221

Flag: three equal horizontal bands of blue (top), white, and red

---

Economy

Overview: The swift collapse of the Yugoslav federation has been followed by bloody ethnic warfare, the destabilization of republic boundaries, and the breakup of important interrepublic trade flows. The situation in Serbia and Montenegro remains fluid in view of the extensive political and military strife. Serbia and Montenegro faces major economic problems. First, like the other former Yugoslav republics, it depended on its sister republics for large amounts of foodstuffs, energy supplies, and manufactures. Wide varieties in climate, mineral resources, and levels of technology among the republics accentuate this interdependence, as did the Communist practice of concentrating much industrial output in a small number of giant plants. The breakup of many of the trade links, the sharp drop in output as industrial plants lost suppliers and markets, and the destruction of physical assets in the fighting all have contributed to the economic difficulties of the republics. One singular factor in the economic situation of Serbia and Montenegro is the continuation in office of a Communist government that is primarily interested in political and military mastery, not economic reform. A further complication is the imposition of economic sanctions by the UN.

National product: GDP—exchange rate conversion—$27-37 billion (1992 est.)

National product real growth rate: NA%

National product per capita: $2,500-$3,500 (1992 est.)

Inflation rate (consumer prices): 81% (1991)

Unemployment rate: 25%-40% (1991 est.)

Budget: revenues $NA; expenditures $NA, including capital expenditures of $NA

Exports: $4.4 billion (f.o.b., 1990)
commodities: machinery and transport equipment 29%, manufactured goods 28.5%, miscellaneous manufactured articles 13.5%, chemicals 11%, food and live animals 9%, raw materials 6%, fuels and lubricants 2%, beverages and tobacco 1%
partners: prior to the imposition of sanctions by the UN Security Council trade partners were principally the other former Yugoslav republics; Italy, Germany, other EC, the successor states of the former USSR, East European countries, US

Imports: $6.4 billion (c.i.f., 1990)
commodities: machinery and transport equipment 26%, fuels and lubricants 18%, manufactured goods 16%, chemicals 12.5%, food and live animals 11%, miscellaneous manufactured items 8%, raw materials, including coking coal for the steel industry, 7%, beverages, tobacco, and edible oils 1.5%
partners: prior to the imposition of sanctions by the UN Security Council the trade partners were principally the other former Yugoslav republics; the successor states of the former USSR, EC countries (mainly Italy and Germany), East European countries, US

External debt: $4.2 billion (may assume some part of foreign debt of former Yugoslavia)

Industrial production: growth rate −20% or greater (1991 est.)

Electricity: 8,850,000 kW capacity; 42,000 million kWh produced, 3,950 kWh per capita (1992)

Industries: machine building (aircraft, trucks, and automobiles; armored vehicles and weapons; electrical equipment; agricultural machinery), metallurgy (steel, aluminum, copper, lead, zinc, chromium, antimony, bismuth, cadmium), mining (coal, bauxite, nonferrous ore, iron ore, limestone), consumer goods (textiles, footwear, foodstuffs, appliances), electronics, petroleum products, chemicals, and pharmaceuticals

Agriculture: the fertile plains of Vojvodina produce 80% of the cereal production of the former Yugoslavia and most of the cotton, oilseeds, and chicory; Vojvodina also produces fodder crops to support intensive beef and dairy production; Serbia proper, although hilly, has a well-distributed rainfall and a long growing season; produces fruit, grapes, and cereals; in this area, livestock production (sheep and cattle) and dairy farming prosper; Kosovo produces fruits, vegetables, tobacco, and a small amount of cereals; the mountainous pastures of Kosovo and Montenegro support sheep and goat husbandry; Montenegro has only a small agriculture sector, mostly near the coast where a Mediterranean climate permits the culture of olives, citrus, grapes, and rice

Illicit drugs: NA

Economic aid: NA

Currency: 1 Yugoslav New Dinar (YD) = 100 paras

Exchange rates: Yugoslav New Dinars (YD) per US $1—28.230 (December 1991), 15.162 (1990), 15.528 (1989), 0.701 (1988), 0.176 (1987)

Fiscal year: calendar year

Communications

Railroads: NA

Highways: 46,019 km total (1990); 26,949 km paved, 10,373 km gravel, 8,697 km earth

Inland waterways: NA km

Pipelines: crude oil 415 km, petroleum products 130 km, natural gas 2,110 km

Ports: coastal—Bar; inland—Belgrade

Merchant marine:
Montenegro: 40 ships (1,000 GRT or over) totaling 620,455 GRT/1,024,227 DWT; includes 17 cargo, 5 container, 17 bulk, 1 passenger ship; note—most under Maltese flag except 2 bulk under Panamian flag
Serbia: 4 ships (1,000 GRT or over) totaling 246,631 GRT/451,843 DWT; includes 2 bulk, 2 conbination tanker/ore carrier; note—all under the flag of Saint Vincent and the Grenadines

Airports:
total: 48
useable: 48
with permanent-surface runways: 16
with runways over 3,659 m: 0
with runways 2,440-3,659 m: 6
with runways 1,220-2,439 m: 9

Telecommunications: 700,000 telephones; broadcast stations—26 AM, 9 FM, 18 TV; 2,015,000 radios; 1,000,000 TVs; satellite ground stations—1 Atlantic Ocean INTELSAT

Defense Forces

Branches: People's Army—Ground Forces (internal and border troops), Naval Forces, Air and Air Defense Forces, Frontier Guard, Territorial Defense Force, Civil Defense

Seychelles

Manpower availability: males age 15-49 2,700,485; fit for military service 2,178,128; reach military age (19) annually 83,783 (1993 est.)
Defense expenditures: 245 billion dinars, 4-6% of GDP (1992 est.); note—conversion of defense expenditures into US dollars using the current exchange rate could produce misleading results

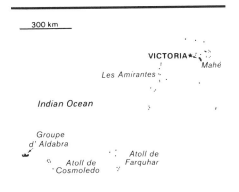

300 km

VICTORIA★
Mahé
Les Amirantes
Indian Ocean
Groupe d' Aldabra
Atoll de Farquhar
Atoll de Cosmoledo

Geography

Location: in the western Indian Ocean northeast of Madagascar
Map references: Africa, Standard Time Zones of the World
Area:
total area: 455 km²
land area: 455 km²
comparative area: slightly more than 2.5 times the size of Washington, DC
Land boundaries: 0 km
Coastline: 491 km
Maritime claims:
continental shelf: 200 nm or the edge of continental margin
exclusive economic zone: 200 nm
territorial sea: 12 nm
International disputes: claims Tromelin Island
Climate: tropical marine; humid; cooler season during southeast monsoon (late May to September); warmer season during northwest monsoon (March to May)
Terrain: Mahe Group is granitic, narrow coastal strip, rocky, hilly; others are coral, flat, elevated reefs
Natural resources: fish, copra, cinnamon trees
Land use:
arable land: 4%
permanent crops: 18%
meadows and pastures: 0%
forest and woodland: 18%
other: 60%
Irrigated land: NA km²
Environment: lies outside the cyclone belt, so severe storms are rare; short droughts possible; no fresh water—catchments collect rain; 40 granitic and about 50 coralline islands

People

Population: 71,494 (July 1993 est.)
Population growth rate: 0.88% (1993 est.)
Birth rate: 22.35 births/1,000 population (1993 est.)

Death rate: 7.12 deaths/1,000 population (1993 est.)
Net migration rate: -6.43 migrant(s)/1,000 population (1993 est.)
Infant mortality rate: 12.1 deaths/1,000 live births (1993 est.)
Life expectancy at birth:
total population: 69.26 years
male: 65.56 years
female: 73.07 years (1993 est.)
Total fertility rate: 2.3 children born/woman (1993 est.)
Nationality:
noun: Seychellois (singular and plural)
adjective: Seychelles
Ethnic divisions: Seychellois (mixture of Asians, Africans, Europeans)
Religions: Roman Catholic 90%, Anglican 8%, other 2%
Languages: English (official), French (official), Creole
Literacy: age 15 and over can read and write (1971)
total population: 58%
male: 56%
female: 60%
Labor force: 27,700 (1985)
by occupation: industry and commerce 31%, services 21%, government 20%, agriculture, forestry, and fishing 12%, other 16% (1985)
note: 57% of population of working age (1983)

Government

Names:
conventional long form: Republic of Seychelles
conventional short form: Seychelles
Digraph: SE
Type: republic
Capital: Victoria
Administrative divisions: 23 administrative districts; Anse aux Pins, Anse Boileau, Anse Etoile, Anse Louis, Anse Royale, Baie Lazare, Baie Sainte Anne, Beau Vallon, Bel Air, Bel Ombre, Cascade, Glacis, Grand' Anse (on Mahe Island), Grand' Anse (on Praslin Island), La Digue, La Riviere Anglaise, Mont Buxton, Mont Fleuri, Plaisance, Pointe Larue, Port Glaud, Saint Louis, Takamaka
Independence: 29 June 1976 (from UK)
Constitution: 5 June 1979
note: new constitution now being drafted by multiparty conference, to take effect in mid-1993
Legal system: based on English common law, French civil law, and customary law
National holiday: Liberation Day, 5 June (1977) (anniversary of coup)
Political parties and leaders: ruling party—Seychelles People's Progressive Front (SPPF), France Albert RENE; Democratic Party (DP), Sir James MANCHAM; Seychelles Party (PS), Wavel RAMKALAWAN;

Seychelles *(continued)*

Seychelles Democratic Movement (MSPD), Jacques HONDOUL; Seychelles Liberal Party (SLP), Ogilvie BERLOUIS
Other political or pressure groups: trade unions; Roman Catholic Church
Suffrage: 17 years of age; universal
Elections:
note: presidential and legislative elections are scheduled to be held once the new, multiparty consitition is ratified later this year
President: last held 9-11 June 1989 (next to be held NA 1993); results—President France Albert RENE reelected without opposition
People's Assembly: last held 5 December 1987 (next to be held mid-1993); results—SPPF was the only legal party; seats—(25 total, 23 elected) SPPF 23
Executive branch: president, Council of Ministers
Legislative branch: unicameral People's Assembly (Assemblee du Peuple)
Judicial branch: Court of Appeal, Supreme Court
Leaders:
Chief of State and Head of Government: President France Albert RENE (since 5 June 1977)
Member of: ACCT, ACP, AfDB, C, ECA, FAO, G-77, IBRD, ICAO, ICFTU, IFAD, IFC, ILO, IMF, IMO, INTERPOL, IOC, NAM, OAU, UN, UNCTAD, UNESCO, UNIDO, UPU, WCL, WHO, WMO, WTO
Diplomatic representation in US:
chief of mission: Second Secretary, Charge d'Affaires ad interim Marc R. MARENGO
chancery: (temporary) 820 Second Avenue, Suite 900F, New York, NY 10017
telephone: (212) 687-9766
US diplomatic representation:
chief of mission: Ambassador Matthew F. MATTINGLY
embassy: 4th Floor, Victoria House, Victoria
mailing address: Victoria House, Box 251, Victoria, Mahe, or Box 148, Unit 62501, APO AE 09815-2501
telephone: (248) 25256
FAX: (248) 25189
Flag: three horizontal bands of red (top), white (wavy), and green; the white band is the thinnest, the red band is the thickest

Economy

Overview: In this small, open, tropical island economy, the tourist industry employs about 30% of the labor force and provides more than 70% of hard currency earnings. In recent years the government has encouraged foreign investment in order to upgrade hotels and other services. At the same time, the government has moved to reduce the high dependence on tourism by promoting the development of farming, fishing, and small-scale manufacturing.

National product: GDP—exchange rate conversion—$350 million (1991 est.)
National product real growth rate: -4.5% (1991 est.)
National product per capita: $5,200 (1991 est.)
Inflation rate (consumer prices): 1.8% (1990 est.)
Unemployment rate: 9% (1987)
Budget: revenues $180 million; expenditures $202 million, including capital expenditures of $32 million (1989)
Exports: $40 million (f.o.b., 1990 est.)
commodities: fish, copra, cinnamon bark, petroleum products (reexports)
partners: France 63%, Pakistan 12%, Reunion 10%, UK 7% (1987)
Imports: $186 million (f.o.b., 1990 est.)
commodities: manufactured goods, food, tobacco, beverages, machinery and transportation equipment, petroleum products
partners: UK 20%, France 14%, South Africa 13%, Yemen 13%, Singapore 8%, Japan 6% (1987)
External debt: $189 million (1991 est.)
Industrial production: growth rate 7% (1987); accounts for 10% of GDP
Electricity: 30,000 kW capacity; 80 million kWh produced, 1,160 kWh per capita (1991)
Industries: tourism, processing of coconut and vanilla, fishing, coir rope factory, boat building, printing, furniture, beverage
Agriculture: accounts for 7% of GDP, mostly subsistence farming; cash crops—coconuts, cinnamon, vanilla; other products—sweet potatoes, cassava, bananas; broiler chickens; large share of food needs imported; expansion of tuna fishing under way
Economic aid: US commitments, including Ex-Im (FY78-89), $26 million; Western (non-US) countries, ODA and OOF bilateral commitments (1978-89), $315 million; OPEC bilateral aid (1979-89), $5 million; Communist countries (1970-89), $60 million
Currency: 1 Seychelles rupee (SRe) = 100 cents
Exchange rates: Seychelles rupees (SRe) per US$1—5.2545 (January 1993), 5.1220 (1992), 5.2893 (1991), 5.3369 (1990), 5.6457 (1989), 5.3836 (1988)
Fiscal year: calendar year

Communications

Highways: 260 km total; 160 km paved, 100 km crushed stone or earth
Ports: Victoria
Merchant marine: 1 refrigerated cargo totaling 1,827 GRT/2,170 DWT
Airports:
total: 14
usable: 14
with permanent-surface runways: 8
with runways over 3,659 m: 0
with runways 2,440-3,659 m: 1
with runways 1,220-2,439 m: 1
Telecommunications: direct radio communications with adjacent islands and African coastal countries; 13,000 telephones; broadcast stations—2 AM, no FM, 2 TV; 1 Indian Ocean INTELSAT earth station; USAF tracking station

Defense Forces

Branches: Army, National Guard, Marines, Coast Guard, Presidential Protection Unit, Police Force
Manpower availability: males age 15-49 18,982; fit for military service 9,710 (1993 est.)
Defense expenditures: exchange rate conversion—$12 million, 4% of GDP (1990 est.)

Sierra Leone

Geography

Location: Western Africa, bordering the North Atlantic Ocean between Guinea and Liberia

Map references: Africa, Standard Time Zones of the World

Area:
total area: 71,740 km²
land area: 71,620 km²
comparative area: slightly smaller than South Carolina

Land boundaries: total 958 km, Guinea 652 km, Liberia 306 km

Coastline: 402 km

Maritime claims:
territorial sea: 200 nm

International disputes: none

Climate: tropical; hot, humid; summer rainy season (May to December); winter dry season (December to April)

Terrain: coastal belt of mangrove swamps, wooded hill country, upland plateau, mountains in east

Natural resources: diamonds, titanium ore, bauxite, iron ore, gold, chromite

Land use:
arable land: 25%
permanent crops: 2%
meadows and pastures: 31%
forest and woodland: 29%
other: 13%

Irrigated land: 340 km² (1989 est.)

Environment: extensive mangrove swamps hinder access to sea; deforestation; soil degradation

People

Population: 4,510,571 (July 1993 est.)

Population growth rate: 2.61% (1993 est.)

Birth rate: 45.47 births/1,000 population (1993 est.)

Death rate: 19.39 deaths/1,000 population (1993 est.)

Net migration rate: 0 migrant(s)/1,000 population (1993 est.)

Infant mortality rate: 145 deaths/1,000 live births (1993 est.)

Life expectancy at birth:
total population: 45.87 years
male: 43.1 years
female: 48.71 years (1993 est.)

Total fertility rate: 6.01 children born/woman (1993 est.)

Nationality:
noun: Sierra Leonean(s)
adjective: Sierra Leonean

Ethnic divisions: 13 native African tribes 99% (Temne 30%, Mende 30%, other 39%), Creole, European, Lebanese, and Asian 1%

Religions: Muslim 30%, indigenous beliefs 30%, Christian 10%, other or none 30%

Languages: English (official; regular use limited to literate minority), Mende principal vernacular in the south, Temne principal vernacular in the north, Krio the language of the re-settled ex-slave population of the Freetown area and is lingua franca

Literacy: age 15 and over can read and write English, Merde, Temne, or Arabic (1990)
total population: 21%
male: 31%
female: 11%

Labor force: 1.369 million (1981 est.)
by occupation: agriculture 65%, industry 19%, services 16% (1981 est.)
note: only about 65,000 wage earners (1985); 55% of population of working age

Government

Names:
conventional long form: Republic of Sierra Leone
conventional short form: Sierra Leone

Digraph: SL

Type: military government

Capital: Freetown

Administrative divisions: 3 provinces and 1 area*; Eastern, Northern, Southern, Western*

Independence: 27 April 1961 (from UK)

Constitution: 1 October 1991; amended September 1991

Legal system: based on English law and customary laws indigenous to local tribes; has not accepted compulsory ICJ jurisdiction

National holiday: Republic Day, 27 April (1961)

Political parties and leaders: status of existing political parties is unknown following 29 April 1992 coup

Suffrage: 18 years of age; universal

Elections: suspended after 29 April 1992 coup; Chairman STRASSER promises multi-party elections sometime within three years

Executive branch: National Provisional Ruling Council

Legislative branch: unicameral House of Representatives (suspended after coup of 29 April 1992)

Judicial branch: Supreme Court (suspended after coup of 29 April 1992)

Leaders:
Chief of State and Head of Government: Chairman of the Supreme Council of State Capt. Valentine E. M. STRASSER (since 29 April 1992)

Member of: ACP, AfDB, C, CCC, ECA, ECOWAS, FAO, G-77, GATT, IAEA, IBRD, ICAO, ICFTU, IDA, IDB, IFAD, IFC, ILO, IMF, IMO, INTERPOL, IOC, ITU, LORCS, NAM, OAU, OIC, UN, UNCTAD, UNESCO, UNIDO, UPU, WCL, WHO, WIPO, WMO, WTO

Diplomatic representation in US:
chief of mission: (vacant)
chancery: 1701 19th Street NW, Washington, DC 20009
telephone: (202) 939-9261

US diplomatic representation:
chief of mission: Ambassador Lauralee M. PETERS
embassy: Walpole and Siaka Stevens Street, Freetown
mailing address: use embassy street address
telephone: [232] (22) 226-481
FAX: [232] (22) 225-471

Flag: three equal horizontal bands of light green (top), white, and light blue

Economy

Overview: The economic and social infrastructure is not well developed. Subsistence agriculture dominates the economy, generating about one-third of GDP and employing about two-thirds of the working population. Manufacturing, which accounts for roughly 10% of GDP, consists mainly of the processing of raw materials and of light manufacturing for the domestic market. Diamond mining provides an important source of hard currency. The economy suffers from high unemployment, rising inflation, large trade deficits, and a growing dependency on foreign assistance. The government in 1990 was attempting to get the budget deficit under control and, in general, to bring economic policy in line with the recommendations of the IMF and the World Bank. Since March 1991, however, military incursions by Liberian rebels in southern and eastern Sierra Leone have severely strained the economy and have undermined efforts to institute economic reforms.

National product: GDP—exchange rate conversion—$1.4 billion (FY92 est.)

National product real growth rate: -1% (FY92 est.)

National product per capita: $330 (FY92 est.)

Inflation rate (consumer prices): 5% (1992)

Unemployment rate: NA%

Sierra Leone (continued)

Budget: revenues $68 million; expenditures $118 million, including capital expenditures of $28 million (FY92 est.)
Exports: $75 million (f.o.b., FY92 est.)
commodities: rutile 50%, bauxite 17%, cocoa 11%, diamonds 3%, coffee 3%
partners: US, UK, Belgium, Germany, other Western Europe
Imports: $62 million (c.i.f., FY92 est.)
commodities: capital goods 40%, food 32%, petroleum 12%, consumer goods 7%, light industrial goods
partners: US, EC countries, Japan, China, Nigeria
External debt: $633 million (FY92 est.)
Industrial production: growth rate NA%
Electricity: 85,000 kW capacity; 185 million kWh produced, 45 kWh per capita (1991)
Industries: mining (diamonds, bauxite, rutile), small-scale manufacturing (beverages, textiles, cigarettes, footwear), petroleum refinery
Agriculture: accounts for over 30% of GDP and two-thirds of the labor force; largely subsistence farming; cash crops—coffee, cocoa, palm kernels; harvests of food staple rice meets 80% of domestic needs; annual fish catch averages 53,000 metric tons
Economic aid: US commitments, including Ex-Im (FY70-89), $161 million; Western (non-US) countries, ODA and OOF bilateral commitments (1970-89), $848 million; OPEC bilateral aid (1979-89), $18 million; Communist countries (1970-89), $101 million
Currency: 1 leone (Le) = 100 cents
Exchange rates: leones (Le) per US$1—552.43 (January 1993), 499.44 (1992), 295.34 (1991), 144.9275 (1990), 58.1395 (1989), 31.2500 (1988)
Fiscal year: 1 July-30 June

Communications

Railroads: 84 km 1.067-meter narrow-gauge mineral line is used on a limited basis because the mine at Marampa is closed
Highways: 7,400 km total; 1,150 km paved, 490 km laterite (some gravel), 5,760 km improved earth
Inland waterways: 800 km; 600 km navigable year round
Ports: Freetown, Pepel, Bonthe
Merchant marine: 1 cargo ship totaling 5,592 GRT/9,107 DWT
Airports:
total: 11
usable: 7
with permanent-surface runways: 4
with runways over 3,659 m: 0
with runways 2,440-3,659 m: 1
with runways 1,220-2,439 m: 3

Telecommunications: marginal telephone and telegraph service; national microwave radio relay system unserviceable at present; 23,650 telephones; broadcast stations—1 AM, 1 FM, 1 TV; 1 Atlantic Ocean INTELSAT earth station

Defense Forces

Branches: Army, Navy, Police, Security Forces
Manpower availability: males age 15-49 983,281; fit for military service 475,855 (1993 est.); no conscription
Defense expenditures: exchange rate conversion—$6 million, 0.7% of GDP (1988 est.)

Singapore

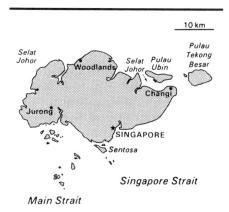

Geography

Location: Southeast Asia, between Malaysia and Indonesia
Map references: Asia, Southeast Asia, Standard Time Zones of the World
Area:
total area: 632.6 km^2
land area: 622.6 km^2
comparative area: slightly less than 3.5 times the size of Washington, DC
Land boundaries: 0 km
Coastline: 193 km
Maritime claims:
exclusive fishing zone: 12 nm
territorial sea: 3 nm
International disputes: two islands in dispute with Malaysia
Climate: tropical; hot, humid, rainy; no pronounced rainy or dry seasons; thunderstorms occur on 40% of all days (67% of days in April)
Terrain: lowland; gently undulating central plateau contains water catchment area and nature preserve
Natural resources: fish, deepwater ports
Land use:
arable land: 4%
permanent crops: 7%
meadows and pastures: 0%
forest and woodland: 5%
other: 84%
Irrigated land: NA km^2
Environment: mostly urban and industrialized
Note: focal point for Southeast Asian sea routes

People

Population: 2,826,331 (July 1993 est.)
Population growth rate: 1.19% (1993 est.)
Birth rate: 17.12 births/1,000 population (1993 est.)
Death rate: 5.25 deaths/1,000 population (1993 est.)
Net migration rate: 0 migrant(s)/1,000 population (1993 est.)

Infant mortality rate: 5.8 deaths/1,000 live births (1993 est.)
Life expectancy at birth:
total population: 75.75 years
male: 73.07 years
female: 78.63 years (1993 est.)
Total fertility rate: 1.89 children born/woman (1993 est.)
Nationality:
noun: Singaporean(s)
adjective: Singapore
Ethnic divisions: Chinese 76.4%, Malay 14.9%, Indian 6.4%, other 2.3%
Religions: Buddhist (Chinese), Atheist (Chinese), Muslim (Malays), Christian, Hindu, Sikh, Taoist, Confucianist
Languages: Chinese (official), Malay (official and national), Tamil (official), English (official)
Literacy: age 15 and over can read and write (1990)
total population: 88%
male: 93%
female: 84%
Labor force: 1,485,800
by occupation: financial, business, and other services 30.2%, manufacturing 28.4%, commerce 22.0%, construction 9.0%, other 10.4% (1990)

Government

Names:
conventional long form: Republic of Singapore
conventional short form: Singapore
Digraph: SN
Type: republic within Commonwealth
Capital: Singapore
Administrative divisions: none
Independence: 9 August 1965 (from Malaysia)
Constitution: 3 June 1959, amended 1965; based on preindependence State of Singapore Constitution
Legal system: based on English common law; has not accepted compulsory ICJ jurisdiction
National holiday: National Day, 9 August (1965)
Political parties and leaders:
government: People's Action Party (PAP), GOH Chok Tong, secretary general
opposition: Workers' Party (WP), J. B. JEYARETNAM; Singapore Democratic Party (SDP), CHIAM See Tong; National Solidarity Party (NSP), leader NA; Barisan Sosialis (BS, Socialist Front), leader NA
Suffrage: 20 years of age; universal and compulsory
Elections:
President: last held 31 August 1989 (next to be held NA August 1993); results—President WEE Kim Wee was reelected by Parliament without opposition

Parliament: last held 31 August 1991 (next to be held 31 August 1996); results—percent of vote by party NA; seats—(81 total) PAP 77, SDP 3, WP 1
Executive branch: president, prime minister, two deputy prime ministers, Cabinet
Legislative branch: unicameral Parliament
Judicial branch: Supreme Court
Leaders:
Chief of State: President WEE Kim Wee (since 3 September 1985)
Head of Government: Prime Minister GOH Chok Tong (since 28 November 1990); Deputy Prime Minister LEE Hsien Loong (since 28 November 1990); Deputy Prime Minister ONG Teng Cheong (since 2 January 1985)
Member of: APEC, AsDB, ASEAN, C, CCC, COCOM (cooperating country), CP, ESCAP, G-77, GATT, IAEA, IBRD, ICAO, ICC, ICFTU, IFC, ILO, IMF, IMO, INMARSAT, INTELSAT, INTERPOL, IOC, ISO, ITU, LORCS, NAM, UN, UNAVEM II, UNCTAD, UNIKOM, UPU, WHO, WIPO, WMO
Diplomatic representation in US:
chief of mission: Ambassador S. R. NATHAN
chancery: 1824 R Street NW, Washington, DC 20009
telephone: (202) 667-7555
US diplomatic representation:
chief of mission: Ambassador Jon M. HUNTSMAN, Jr.
embassy: 30 Hill Street, Singapore 0617
mailing address: FPO AP 96534
telephone: [65] 338-0251
FAX: [65] 338-4550
Flag: two equal horizontal bands of red (top) and white; near the hoist side of the red band, there is a vertical, white crescent (closed portion is toward the hoist side) partially enclosing five white five-pointed stars arranged in a circle

---

Economy

Overview: Singapore has an open entrepreneurial economy with strong service and manufacturing sectors and excellent international trading links derived from its entrepot history. The economy appears to have pulled off a soft landing from the 9% growth rate of the late 1980s, registering higher than expected growth in 1992 while stemming inflation. Economic activity slowed early in 1992, primarily as a result of slackened demand in Singapore's export markets. But after bottoming out in the second quarter, the economy picked up in line with a gradual recovery in the United States. The year's best performers were the construction and financial services industries and manufacturers of computer-related

components. Rising labor costs continue to be a threat to Singapore's competitiveness, but there are indications that productivity is catching up. Government surpluses and the rate of gross national savings remain high. In technology, per capita output, and labor discipline, Singapore is well on its way toward its goal of becoming a developed country.
National product: GDP—exchange rate conversion—$45.9 billion (1992)
National product real growth rate: 5.8% (1992)
National product per capita: $16,500 (1992)
Inflation rate (consumer prices): 2.3% (1992)
Unemployment rate: 2.7% (June 1992)
Budget: revenues $10.4 billion; expenditures $9.4 billion, including capital expenditures of $NA (1993)
Exports: $61.5 billion (f.o.b., 1992)
commodities: computer equipment, rubber and rubber products, petroleum products, telecommunications equipment
partners: US 21%, Malaysia 13%, Hong Kong 8%, Japan 7%, Thailand 6%
Imports: $66.4 billion (f.o.b., 1992)
commodities: aircraft, petroleum, chemicals, foodstuffs
partners: Japan 21%, US 16%, Malaysia 14%, Taiwan 4%
External debt: none; Singapore is a net creditor
Industrial production: growth rate 2.3% (1992); accounts for 28% of GDP
Electricity: 4,860,000 kW capacity; 18,000 million kWh produced, 6,420 kWh per capita (1992)
Industries: petroleum refining, electronics, oil drilling equipment, rubber processing and rubber products, processed food and beverages, ship repair, entrepot trade, financial services, biotechnology
Agriculture: occupies a position of minor importance in the economy; self-sufficient in poultry and eggs; must import much of other food; major crops—rubber, copra, fruit, vegetables
Illicit drugs: transit point for Golden Triangle heroin going to the US, Western Europe, and the Third World; also a major money-laundering center
Economic aid: US commitments, including Ex-Im (FY70-83), $590 million; Western (non-US) countries, ODA and OOF bilateral commitments (1970-89), $1.0 billion
Currency: 1 Singapore dollar (S$) = 100 cents
Exchange rates: Singapore dollars (S$) per US$1—1.6531 (January 1993), 1.6290 (1992), 1.7276 (1991), 1.8125 (1990), 1.9503 (1989), 2.0124 (1988)
Fiscal year: 1 April-31 March

Singapore *(continued)*

Communications

Railroads: 38 km of 1.000-meter gauge
Highways: 2,644 km total (1985)
Ports: Singapore
Merchant marine: 492 ships (1,000 GRT or over) totaling 9,763,511 GRT/15,816,384 DWT; includes 1 passenger-cargo, 125 cargo, 72 container, 7 roll-on/roll-off cargo, 4 refrigerated cargo, 18 vehicle carrier, 1 livestock carrier, 165 oil tanker, 8 chemical tanker, 7 combination ore/oil, 2 specialized tanker, 5 liquefied gas, 74 bulk, 3 combination bulk; note—many Singapore flag ships are foreign owned
Airports:
total: 10
usable: 10
with permanent-surface runways: 10
with runways over 3,659 m: 2
with runways 2,440-3,659 m: 4
with runways 1,220-2,439 m: 3
Telecommunications: good domestic facilities; good international service; good radio and television broadcast coverage; 1,110,000 telephones; broadcast stations—13 AM, 4 FM, 2 TV; submarine cables extend to Malaysia (Sabah and peninsular Malaysia), Indonesia, and the Philippines; satellite earth stations—1 Indian Ocean INTELSAT and 1 Pacific Ocean INTELSAT

Defense Forces

Branches: Army, Navy, Air Force, People's Defense Force, Police Force
Manpower availability: males age 15-49 853,440; fit for military service 629,055 (1993 est.)
Defense expenditures: exchange rate conversion—$1.7 billion, 4% of GDP (1990 est.)

Slovakia

150 km

Geography

Location: Eastern Europe, between Hungary and Poland
Map references: Ethnic Groups in Eastern Europe, Europe, Standard Time Zones of the World
Area:
total area: 48,845 km²
land area: 48,800 km²
comparative area: about twice the size of New Hampshire
Land boundaries: total 1,355 km, Austria 91 km, Czech Republic 215 km, Hungary 515 km, Poland 444 km, Ukraine 90 km
Coastline: 0 km (landlocked)
Maritime claims: none; landlocked
International disputes: Gabcikovo-Nagymaros Dam dispute with Hungary; unresolved property issues with Czech Republic over redistribution of former Czechoslovak federal property; establishment of international border between the Czech Republic and Slovakia
Climate: temperate; cool summers; cold, cloudy, humid winters
Terrain: rugged mountains in the central and northern part and lowlands in the south
Natural resources: brown coal and lignite; small amounts of iron ore, copper and manganese ore; salt; gas
Land use:
arable land: NA%
permanent crops: NA%
meadows and pastures: NA%
forest and woodland: NA%
other: NA%
Irrigated land: NA km²
Environment: severe damage to forests from "acid rain" caused by coal-fired power stations
Note: landlocked

People

Population: 5,375,501 (July 1993 est.)
Population growth rate: 0.51% (1993 est.)

Birth rate: 14.59 births/1,000 population (1993 est.)
Death rate: 9.47 deaths/1,000 population (1993 est.)
Net migration rate: 0 migrant(s)/1,000 population (1993 est.)
Infant mortality rate: 10.8 deaths/1,000 live births (1993 est.)
Life expectancy at birth:
total population: 72.39 years
male: 68.18 years
female: 76.85 years (1993 est.)
Total fertility rate: 1.99 children born/woman (1993 est.)
Nationality:
noun: Slovak(s)
adjective: Slovak
Ethnic divisions: Slovak 85.6%, Hungarian 10.8%, Gypsy 1.5% (the 1992 census figures underreport the Gypsy/Romany community, which could reach 500,000 or more), Czech 1.1%, Ruthenian 15,000, Ukrainian 13,000, Moravian 6,000, German 5,000, Polish 3,000
Religions: Roman Catholic 60.3%, atheist 9.7%, Protestant 8.4%, Orthodox 4.1%, other 17.5%
Languages: Slovak (official), Hungarian
Literacy:
total population: NA%
male: NA%
female: NA%
Labor force: 2.484 million
by occupation: industry 33.2%, agriculture 12.2%, construction 10.3%, communication and other 44.3% (1990)

Government

Names:
conventional long form: Slovak Republic
conventional short form: Slovakia
local long form: Slovenska Republika
local short form: Slovensko
Digraph: LO
Type: parliamentary democracy
Capital: Bratislava
Administrative divisions: 4 departments (departamentos, singular—departamento) Bratislava, Zapadoslovensky, Stredoslovensky, Vychodoslovensky
Independence: 1 January 1993 (from Czechoslovakia)
Constitution: ratified 3 September 1992; fully effective 1 January 1993
Legal system: civil law system based on Austro-Hungarian codes; has not accepted compulsory ICJ jurisdiction; legal code modified to comply with the obligations of Conference on Security and Cooperation in Europe (CSCE) and to expunge Marxist-Leninist legal theory
National holiday: Slovak National Uprising, August 29 (1944)
Political parties and leaders: Hungarian Christian Democratic Movement, Vojtech BUGAR; Christian Democratic Movement,

Jan CARNOGURSKY; Movement for a Democratic Slovakia, Vladimir MECIAR, chairman; Party of the Democratic Left, Peter WEISS, chairman; Slovak National Party, Ludovit CERNAK, chairman; Coexistence, Miklos DURAY, chairman; Party of Conservative Democrats, leader NA

Other political or pressure groups: Green Party; Democratic Party; Social Democratic Party in Slovakia; Movement for Czech-Slovak Accord; Freedom Party; Slovak Christian Union; Hungarian Civic Party

Suffrage: 18 years of age; universal

Elections:

President: last held 8 February 1993 (next to be held NA 1998); results—Michal KOVAC elected by the National Council

National Council: last held 5-6 June 1992 (next to be held NA June 1996); results—Movement for a Democratic Slovakia 37%, Party of the Democratic Left 15%, Christian Democratic Movement 9%, Slovak National Party 8%, Hungarian Christian Democratic Movement/Coexistence 7%; seats—(150 total) Movement for a Democratic Slovakia, 74, Party of the Democratic Left 29, Christian Democratic Movement 18, Slovak National Party 15, Hungarian Christian Democratic Movement/Coexistence 14

Executive branch: president, prime minister, Cabinet

Legislative branch: unicameral National Council (Narodni Rada)

Judicial branch: Supreme Court

Leaders:

Chief of State: President Michal KOVAC (since 8 February 1993)

Head of Government: Prime Minister Vladimir MECIAR (since NA), Deputy Prime Minister Roman KOVAC (since NA)

Member of: BIS, CCC, CE, CEI, CERN, CSCE, EBRD, ECE, FAO, GATT, IAEA, IBRD, ICAO, ICFTU, IDA, IFC, ILO, IMF, IMO, INMARSAT, INTELSAT, INTERPOL, IOC, IOM (observer), ISO, ITU, LORCS, NACC, NAM (guest), NSG, PCA, UN (as of 8 January 1993), UNAVEM II, UNCTAD, UNESCO, UNIDO, UNOSOM, UNPROFOR, UPU, WHO, WIPO, WMO, WTO, ZC

Diplomatic representation in US:

chief of mission: Charge d'Affaires Dr. Milan ERBAN

chancery: 3900 Spring of Freedom Street NW, Washington, DC 20008

telephone: (202) 363-6315 or 6316

US diplomatic representation:

chief of mission: Ambassador Elect Eleanor SUTTER

embassy: Hviczdoslavovo Namestie 4, 81102 Bratislava

mailing address: use embassy street address

telephone: 427 330 861

Flag: three equal horizontal bands of white (top), blue, and red superimposed with a crest with a white double cross on three blue mountains

---

Economy

Overview: The dissolution of Czechoslovakia into two independent states—the Czech Republic and Slovakia—on 1 January 1993 has complicated the task of moving toward a more open and decentralized economy. The old Czechoslovakia, even though highly industrialized by East European standards, suffered from an aging capital plant, lagging technology, and a deficiency in energy and many raw materials. In January 1991, approximately one year after the end of communist control of Eastern Europe, the Czech and Slovak Federal Republic launched a sweeping program to convert its almost entirely state-owned and controlled economy to a market system. In 1991-92 these measures resulted in privatization of some medium- and small-scale economic activity and the setting of more than 90% of prices by the market—but at a cost in inflation, unemployment, and lower output. For Czechoslovakia as a whole inflation in 1991 was roughly 50% and output fell 15%. In 1992 in Slovakia, inflation slowed to an estimated 8.7% and the estimated fall in GDP was a more moderate 7%. In 1993 the government anticipates up to a 7% drop in GDP, with the disruptions from the separation from the Czech lands probably accounting for half the decline; inflation, according to government projections, may rise to 15-20% and unemployment may reach 12-15%. The Slovak government is moving ahead less enthusiastically than the Czech government in the further dismantling of the old centrally controlled economic system. Although the governments of Slovakia and the Czech Republic had envisaged retaining the koruna as a common currency at least in the short run, the two countries ended the currency union in February 1993.

National product: GDP—purchasing power equivalent—$32.1 billion (1992 est.)

National product real growth rate: -7% (1992 est.)

National product per capita: $6,100 (1992 est.)

Inflation rate (consumer prices): 8.7% (1992 est.)

Unemployment rate: 11.3% (1992 est.)

Budget: revenues $NA; expenditures $NA, including capital expenditures of $NA

Exports: $3.6 billion (f.o.b., 1992)

commodities: machinery and transport equipment; chemicals; fuels, minerals, and metals; agricultural products

partners: Czech Republic, CIS republics, Germany, Poland, Austria, Hungary, Italy, France, US, UK

Imports: $3.6 billion (f.o.b., 1992)

commodities: machinery and transport equipment; fuels and lubricants; manufactured goods; raw materials; chemicals; agricultural products

partners: Czech Republic, CIS republics, Germany, Austria, Poland, Switzerland, Hungary, UK, Italy

External debt: $1.9 billion, hard currency indebtedness (December 1992)

Industrial production: growth rate NA%

Electricity: 6,800,000 kW capacity; 24,000 million kWh produced, 4,550 kWh per capita (1992)

Industries: brown coal mining, chemicals, metal-working, consumer appliances, fertilizer, plastics, armaments

Agriculture: largely self-sufficient in food production; diversified crop and livestock production, including grains, potatoes, sugar beets, hops, fruit, hogs, cattle, and poultry; exporter of forest products

Illicit drugs: the former Czechoslavakia was a transshipment point for Southwest Asian heroin and was emerging as a transshipment point for Latin American cocaine (1992)

Economic aid: the former Czechoslovakia was a donor—$4.2 billion in bilateral aid to non-Communist less developed countries (1954-89)

Currency: 1 koruna (Kc) = 100 haleru

Exchange rates: koruny (Kcs) per US$1—28.59 (December 1992), 28.26 (1992), 29.53 (1991), 17.95 (1990), 15.05 (1989), 14.36 (1988)

Fiscal year: calendar year

Communications

Railroads: 3,669 km total (1990)

Highways: 17,650 km total (1990)

Inland waterways: NA km

Pipelines: natural gas 2,700 km; petroleum products NA km

Ports: maritime outlets are in Poland (Gdynia, Gdansk, Szczecin), Croatia (Rijeka), Slovenia (Koper), Germany (Hamburg, Rostock); principal river ports are Komarno on the Danube and Bratislava on the Danube

Merchant marine: the former Czechoslovakia had 22 ships (1,000 GRT or over) totaling 290,185 GRT/437,291 DWT; includes 13 cargo, 9 bulk; may be shared with the Czech Republic

Airports:

total: 34

usable: 34

with permanent-surface runways: 9

with runways over 3,659 m: 0

with runways 2,440-3,659 m: 1

with runways 1,220-2,439 m: 5

Telecommunications: NA

Slovakia *(continued)*

Defense Forces

Branches: Army, Air and Air Defense Forces, Civil Defense, Railroad Units
Manpower availability: males age 15-49 1,407,908; fit for military service 1,082,790; reach military age (18) annually 47,973 (1993 est.)
Defense expenditures: 8.2 billion koruny, NA% of GDP (1993 est.); note—conversion of defense expenditures into US dollars using the current exchange rate could produce misleading results

Slovenia

75 km

Geography

Location: Southern Europe, bordering the Adriatic Sea, between Austria and Croatia
Map references: Ethnic Groups in Eastern Europe, Europe, Standard Time Zones of the World
Area:
total area: 20,296 km²
land area: 20,296 km²
comparative area: slightly larger than New Jersey
Land boundaries: total 999 km, Austria 262 km, Croatia 455 km, Italy 199 km, Hungary 83 km
Coastline: 32 km
Maritime claims:
continental shelf: 200 m depth or to depth of exploitation
territorial sea: 12 nm
International disputes: dispute with Croatia over fishing rights in the Adriatic and over some border areas; the border issue is currently under negotiation; small minority in northern Italy seeks the return of parts of southwestern Slovenia
Climate: Mediterranean climate on the coast, continental climate with mild to hot summers and cold winters in the plateaus and valleys to the east
Terrain: a short coastal strip on the Adriatic, an alpine mountain region adjacent to Italy, mixed mountain and valleys with numerous rivers to the east
Natural resources: lignite coal, lead, zinc, mercury, uranium, silver
Land use:
arable land: 10%
permanent crops: 2%
meadows and pastures: 20%
forest and woodland: 45%
other: 23%
Irrigated land: NA km²
Environment: Sava River polluted with domestic and industrial waste; heavy metals and toxic chemicals along coastal waters; near Koper, forest damage from air pollutants originating at metallurgical and chemical plants; subject to flooding and earthquakes

People

Population: 1,967,655 (July 1993 est.)
Population growth rate: 0.23% (1993 est.)
Birth rate: 11.93 births/1,000 population (1993 est.)
Death rate: 9.6 deaths/1,000 population (1993 est.)
Net migration rate: 0 migrant(s)/1,000 population (1993 est.)
Infant mortality rate: 8.3 deaths/1,000 live births (1993 est.)
Life expectancy at birth:
total population: 74 years
male: 70.08 years
female: 78.13 years (1993 est.)
Total fertility rate: 1.68 children born/woman (1993 est.)
Nationality:
noun: Slovene(s)
adjective: Slovenian
Ethnic divisions: Slovene 91%, Croat 3%, Serb 2%, Muslim 1%, other 3%
Religions: Roman Catholic 96% (including 2% Uniate), Muslim 1%, other 3%
Languages: Slovenian 91%, Serbo-Croatian 7%, other 2%
Literacy:
total population: NA%
male: NA%
female: NA%
Labor force: 786,036
by occupation: agriculture 2%, manufacturing and mining 46%

Government

Names:
conventional long form: Republic of Slovenia
conventional short form: Slovenia
local long form: Republika Slovenije
local short form: Slovenija
Digraph: SI
Type: emerging democracy
Capital: Ljubljana
Administrative divisions: 60 provinces (pokajine, singular—pokajina) Ajdovscina, Brezice, Celje, Cerknica, Crnomelj, Dravograd, Gornja Radgona, Grosuplje, Hrastnik Lasko, Idrija, Ilirska Bistrica, Izola, Jesenice, Kamnik, Kocevje, Koper, Kranj, Krsko, Lenart, Lendava, Litija, Ljubljana-Bezigrad, Ljubljana-Center, Ljubljana-Moste-Polje, Ljubljana-Siska, Ljubljana-Vic-Rudnik, Ljutomer, Logatec, Maribor, Metlika, Mozirje, Murska Sobota, Nova Gorica, Novo Mesto, Ormoz Pesnica, Piran, Postojna, Ptuj, Radlje Ob Dravi, Radovljica, Ravne Na Koroskem, Ribnica, Ruse, Sentjur Pri Celju, Sevnica, Sezana, Skofja Loka, Slovenj Gradec, Slovenska Bistrica, Slovenske Konjice,

Smarje Pri Jelsah, Tolmin, Trbovlje, Trebnje, Trzic, Velenje, Vrhnika, Zagorje Ob Savi, Zalec

Independence: 25 June 1991 (from Yugoslavia)

Constitution: adopted 23 December 1991, effective 23 December 1991

Legal system: based on civil law system

National holiday: Statehood Day, 25 June

Political parties and leaders: Slovene Christian Democratics (SKD), Lozje PETERLE, chairman; Liberal Democratic (LDS), Janez DRNOVSEK, chairman; Social-Democratic Party of Slovenia (SDSS), Joze PUCNIK, chairman; Socialist Party of Slovenia (SSS), Viktor ZAKELJ, chairman; Greens of Slovenia (ZS), Dusan PLUT, chairman; National Democratic, Rajko PIRNAT, chairman; Democratic Peoples Party, Marjan PODOBNIK, chairman; Reformed Socialists (former Communist Party), Ciril RIBICIC, chairman; United List (former Communists and allies); Slovene National Party, leader NA; Democratic Party, Igor BAVCAR; Slovene People's Party (SLS), Ivan OMAN

note: parties have changed as of the December 1992 elections

Other political or pressure groups: none

Suffrage: 16 years of age, if employed; 18 years of age, universal

Elections:

President: last held 6 December 1992 (next to be held NA 1996); results—Milan KUCAN reelected by direct popular vote

State Assembly: last held 6 December 1992 (next to be held NA 1996); results—percent of vote by party NA; seats—(total 90) LDS 22, SKD 15, United List (former Communists and allies) 14, Slovene National Party 12, SN 10, Democratic Party 6, ZS 5, SDSS 4, Hungarian minority 1, Italian minority 1

State Council: will become operational after next election in 1996; in the election of 6 December 1992 40 members were elected to represent local and socio-economic interests

Executive branch: president, prime minister, deputy prime ministers, cabinet

Legislative branch: bicameral National Assembly; consists of the State Assembly and the State Council; note—State Council will become operational after next election

Judicial branch: Supreme Court, Constitutional Court

Leaders:

Chief of State: President Milan KUCAN (since 22 April 1990)

Head of Government: Prime Minister Janez DRNOVSEK (since 14 May 1992)

Member of: CE, CEI, CSCE, EBRD, ECE, IAEA, IBRD, ICAO, ILO, IOM (observer), UN, UNCTAD, UNESCO, UNIDO, UPU, WHO

Diplomatic representation in US:

chief of mission: Ambassador Ernest PETRIC

chancery: (temporary) 1300 19th Street NW, Washington, DC 20036

telephone: (202) 828-1650

US diplomatic representation:

chief of mission: Ambassador E. Allen WENDT

embassy: P.O. Box 254; Cankarjeva 11, 61000 Ljubljana

mailing address: APO AE 09862

telephone: [38] (61) 301-427/472

FAX: [38] (61) 301-401

Flag: three equal horizontal bands of white (top), blue, and red with the Slovenian seal (a shield with the image of Triglav in white against a blue background at the center, beneath it are two wavy blue lines depicting seas and rivers, and around it, there are three six-sided stars arranged in an inverted triangle); the seal is located in the upper hoist side of the flag centered in the white and blue bands

---

Economy

Overview: Slovenia was by far the most prosperous of the former Yugoslav republics, with a per capita income more than twice the Yugoslav average, indeed not far below the levels in neighboring Austria and Italy. Because of its strong ties to Western Europe and the small scale of damage during its fight for independence from Yugoslavia, Slovenia has the brightest prospects among the former Yugoslav republics for economic recovery over the next few years. The dissolution of Yugoslavia, however, has led to severe short-term dislocations in production, employment, and trade ties. For example, overall industrial production fell 10% in 1991; particularly hard hit were the iron and steel, machine-building, chemical, and textile industries. Meanwhile, the continued fighting in other former Yugoslav republics has led to further destruction of long-established trade channels and to an influx of tens of thousands of Croatian and Bosnian refugees. The key program for breaking up and privatizing major industrial firms was established in late 1992. Bright spots for encouraging Western investors are Slovenia's comparatively well-educated work force, its developed infrastructure, and its Western business attitudes, but instability in Croatia is a deterrent. Slovenia in absolute terms is a small economy, and a little Western investment would go a long way.

National product: GDP—purchasing power equivalent—$21 billion (1991 est.)

National product real growth rate: -10% (1991 est.)

National product per capita: $10,700 (1991 est.)

Inflation rate (consumer prices): 2.7% (September 1992)

Unemployment rate: 10% (April 1992)

Budget: revenues $NA; expenditures $NA, including capital expenditures of $NA

Exports: $4.12 billion (f.o.b., 1990)

commodities: machinery and transport equipment 38%, other manufactured goods 44%, chemicals 9%, food and live animals 4.6%, raw materials 3%, beverages and tobacco less than 1%

partners: principally the other former Yugoslav republics, Austria, and Italy

Imports: $4.679 billion (c.i.f., 1990)

commodities: machinery and transport equipment 35%, other manufactured goods 26.7%, chemicals 14.5%, raw materials 9.4%, fuels and lubricants 7%, food and live animals 6%

partners: principally the other former Yugoslav republics, Germany, successor states of the former USSR, US, Hungary, Italy, and Austria

External debt: $2.5 billion

Industrial production: growth rate −1% per month (1991-92 est.)

Electricity: 2,900,000 kW capacity; 10,000 million kWh produced, 5,090 kWh per capita (1992)

Industries: ferrous metallurgy and rolling mill products, aluminum reduction and rolled products, lead and zinc smelting, electronics (including military electronics), trucks, electric power equipment, wood products, textiles, chemicals, machine tools

Agriculture: dominated by stock breeding (sheep and cattle) and dairy farming; main crops—potatoes, hops, hemp, flax; an export surplus in these commodities; Slovenia must import many other agricultural products and has a negative overall trade balance in this sector

Illicit drugs: NA

Economic aid: NA

Currency: 1 tolar (SIT) = 100 NA

Exchange rates: tolars (SIT) per US$1—112 (June 1993), 28 (January 1992)

Fiscal year: calendar year

Communications

Railroads: 1,200 km, 1.435 m gauge (1991)

Highways: 14,553 km total; 10,525 km paved, 4,028 km gravel

Inland waterways: NA

Pipelines: crude oil 290 km, natural gas 305 km

Ports: coastal—Koper

Merchant marine: 22 ships (1,000 GRT or over) totaling 348,784 GRT/596,740 DWT; includes 15 bulk, 7 cargo; all under the flag of Saint Vincent and the Grenadines except for 1 bulk under Liberian flag

Airports:

total: 13

useable: 13

with permanent-surface runways: 5

Slovenia *(continued)*

with runways over 3,659 m: 0
with runways 2,440-3,659 m: 2
with runways 1,220-2,439 m: 4
Telecommunications: 130,000 telephones; broadcast stations—6 AM, 5 FM, 7 TV; 370,000 radios; 330,000 TVs

---

Defense Forces

Branches: Slovene Defense Forces
Manpower availability: males age 15-49 512,186; fit for military service 410,594; reach military age (19) annually 14,970 (1993 est.)
Defense expenditures: 13.5 billion tolars, 4.5% of GDP (1993); note—conversion of the military budget into US dollars using the current exchange rate could produce misleading results

Solomon Islands

Geography

Location: Oceania, just east of Papua New Guinea in the South Pacific Ocean
Map references: Oceania, Standard Time Zones of the World
Area:
total area: 28,450 km²
land area: 27,540 km²
comparative area: slightly larger than Maryland
Land boundaries: 0 km
Coastline: 5,313 km
Maritime claims: measured from claimed archipelagic baselines
exclusive economic zone: 200 nm
territorial sea: 12 nm
International disputes: none
Climate: tropical monsoon; few extremes of temperature and weather
Terrain: mostly rugged mountains with some low coral atolls
Natural resources: fish, forests, gold, bauxite, phosphates
Land use:
arable land: 1%
permanent crops: 1%
meadows and pastures: 1%
forest and woodland: 93%
other: 4%
Irrigated land: NA km²
Environment: subject to typhoons, which are rarely destructive; geologically active region with frequent earth tremors
Note: located just east of Papua New Guinea in the South Pacific Ocean

---

People

Population: 372,746 (July 1993 est.)
Population growth rate: 3.46% (1993 est.)
Birth rate: 39.37 births/1,000 population (1993 est.)
Death rate: 4.76 deaths/1,000 population (1993 est.)
Net migration rate: 0 migrant(s)/1,000 population (1993 est.)

Infant mortality rate: 29 deaths/1,000 live births (1993 est.)
Life expectancy at birth:
total population: 70.13 years
male: 67.73 years
female: 72.65 years (1993 est.)
Total fertility rate: 5.88 children born/woman (1993 est.)
Nationality:
noun: Solomon Islander(s)
adjective: Solomon Islander
Ethnic divisions: Melanesian 93%, Polynesian 4%, Micronesian 1.5%, European 0.8%, Chinese 0.3%, other 0.4%
Religions: Anglican 34%, Roman Catholic 19%, Baptist 17%, United (Methodist/Presbyterian) 11%, Seventh-Day Adventist 10%, other Protestant 5%
Languages: Melanesian pidgin in much of the country is lingua franca, English spoken by 1-2% of population
note: 120 indigenous languages
Literacy:
total population: NA%
male: NA%
female: NA%
Labor force: 23,448 economically active
by occupation: agriculture, forestry, and fishing 32.4%, services 25%, construction, manufacturing, and mining 7.0%, commerce, transport, and finance 4.7% (1984)

Government

Names:
conventional long form: none
conventional short form: Solomon Islands
former: British Solomon Islands
Digraph: BP
Type: parliamentary democracy
Capital: Honiara
Administrative divisions: 7 provinces and 1 town*; Central, Guadalcanal, Honiara*, Isabel, Makira, Malaita, Temotu, Western
Independence: 7 July 1978 (from UK)
Constitution: 7 July 1978
Legal system: common law
National holiday: Independence Day, 7 July (1978)
Political parties and leaders: People's Alliance Party (PAP); United Party (UP), leader NA; Solomon Islands Liberal Party (SILP), Bartholemew ULUFA'ALU; Nationalist Front for Progress (NFP), Andrew NORI; Labor Party (LP), Joses TUHANUKU
Suffrage: 21 years of age; universal
Elections:
National Parliament: last held 22 February 1989 (next to be held 26 May 1993); results—percent of vote by party NA; seats—(38 total) PAP 13, UP 6, NFP 4, SILP 4, LP 2, independents 9
Executive branch: British monarch, governor general, prime minister, Cabinet
Legislative branch: unicameral National Parliament

Judicial branch: High Court
Leaders:
Chief of State: Queen ELIZABETH II (since 6 February 1952), represented by Governor General Sir George LEPPING (since 27 June 1989, previously acted as governor general since 7 July 1988)
Head of Government: Prime Minister Solomon MAMALONI (since 28 March 1989); Deputy Prime Minister Sir Baddeley DEVESI (since NA October 1990)
Member of: ACP, AsDB, C, ESCAP, FAO, G-77, IBRD, ICAO, IDA, IFAD, IFC, ILO, IMF, IMO, IOC, ITU, SPARTECA, SPC, SPF, UN, UNCTAD, UPU, WFTU, WHO, WMO
Diplomatic representation in US:
chief of mission: (vacant); ambassador traditionally resides in Honiara (Solomon Islands)
US diplomatic representation:
chief of mission: Ambassador Robert W. FARRAND
embassy: Mud Alley, Honiara
mailing address: American Embassy, P. O. Box 561, Honiara
telephone: (677) 23890
FAX: (677) 23488
Flag: divided diagonally by a thin yellow stripe from the lower hoist-side corner; the upper triangle (hoist side) is blue with five white five-pointed stars arranged in an X pattern; the lower triangle is green

---

Economy

Overview: About 90% of the population depend on subsistence agriculture, fishing, and forestry for at least part of their livelihood. Agriculture, fishing, and forestry contribute about 70% to GDP, with the fishing and forestry sectors being important export earners. The service sector contributes about 25% to GDP. Most manufactured goods and petroleum products must be imported. The islands are rich in undeveloped mineral resources such as lead, zinc, nickel, and gold. The economy suffered from a severe cyclone in mid-1986 that caused widespread damage to the infrastructure.
National product: GDP—exchange rate conversion—$200 million (1990 est.)
National product real growth rate: 6% (1990 est.)
National product per capita: $600 (1990 est.)
Inflation rate (consumer prices): 14.3% (1991)
Unemployment rate: NA%
Budget: revenues $48 million; expenditures $107 million, including capital expenditures of $45 million (1991 est.)
Exports: $74.2 million (f.o.b., 1991 est.)

commodities: fish 46%, timber 31%, copra 5%, palm oil 5%
partners: Japan 51%, UK 12%, Thailand 9%, Netherlands 8%, Australia 2%, US 2% (1985)
Imports: $87.1 million (c.i.f., 1991 est.)
commodities: plant and machinery 30%, fuel 19%, food 16%
partners: Japan 36%, US 23%, Singapore 9%, UK 9%, NZ 9%, Australia 4%, Hong Kong 4%, China 3% (1985)
External debt: $128 million (1988 est.)
Industrial production: growth rate 0% (1987); accounts for 5% of GDP
Electricity: 21,000 kW capacity; 39 million kWh produced, 115 kWh per capita (1990)
Industries: copra, fish (tuna)
Agriculture: including fishing and forestry, accounts for about 70% of GDP; mostly subsistence farming; cash crops—cocoa, beans, coconuts, palm kernels, timber; other products—rice, potatoes, vegetables, fruit, cattle, pigs; not self-sufficient in food grains; 90% of the total fish catch of 44,500 metric tons was exported (1988)
Economic aid: Western (non-US) countries, ODA and OOF bilateral commitments (1980-89), $250 million
Currency: 1 Solomon Islands dollar (SI$) = 100 cents
Exchange rates: Solomon Islands dollars (SI$) per US$1—3.1211 (January 1993), 2.9281 (1992), 2.7148 (1991), 2.5288 (1990), 2.2932 (1989), 2.0825 (1988)
Fiscal year: calendar year

Communications

Highways: about 2,100 km total (1982); 30 km paved, 290 km gravel, 980 km earth, 800 private logging and plantation roads of varied construction
Ports: Honiara, Ringi Cove
Airports:
total: 30
usable: 29
with permanent-surface runways: 2
with runways over 3,659 m: 0
with runways 2,440-3,659 m: 0
with runways 1,220-2,439 m: 3
Telecommunications: 3,000 telephones; broadcast stations—4 AM, no FM, no TV; 1 Pacific Ocean INTELSAT earth station

---

Defense Forces

Branches: Police Force
Manpower availability: NA
Defense expenditures: exchange rate conversion—$NA, NA% of GDP

Somalia

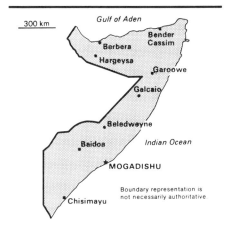

Boundary representation is not necessarily authoritative

Geography

Location: Eastern Africa, bordering the northwestern Indian Ocean, south of the Arabian Peninsula
Map references: Africa, Standard Time Zones of the World
Area:
total area: 637,660 km²
land area: 627,340 km²
comparative area: slightly smaller than Texas
Land boundaries: total 2,366 km, Djibouti 58 km, Ethiopia 1,626 km, Kenya 682 km
Coastline: 3,025 km
Maritime claims:
territorial sea: 200 nm
International disputes: southern half of boundary with Ethiopia is a Provisional Administrative Line; territorial dispute with Ethiopia over the Ogaden; possible claims to Djibouti and parts of Ethiopia and Kenya based on unification of ethnic Somalis
Climate: desert; northeast monsoon (December to February), cooler southwest monsoon (May to October); irregular rainfall; hot, humid periods (tangambili) between monsoons
Terrain: mostly flat to undulating plateau rising to hills in north
Natural resources: uranium and largely unexploited reserves of iron ore, tin, gypsum, bauxite, copper, salt
Land use:
arable land: 2%
permanent crops: 0%
meadows and pastures: 46%
forest and woodland: 14%
other: 38%
Irrigated land: 1,600 km² (1989 est.)
Environment: recurring droughts; frequent dust storms over eastern plains in summer; deforestation; overgrazing; soil erosion; desertification
Note: strategic location on Horn of Africa along southern approaches to Bab el Mandeb and route through Red Sea and Suez Canal

Somalia (continued)

People

Population: 6,514,629 (July 1993 est.)
Population growth rate: 1.35% (1993 est.)
Birth rate: 41.95 births/1,000 population (1993 est.)
Death rate: 28.41 deaths/1,000 population (1993 est.)
Net migration rate: 0 migrant(s)/1,000 population (1993 est.)
Infant mortality rate: 162.7 deaths/1,000 live births (1993 est.)
Life expectancy at birth:
total population: 32.91 years
male: 32.86 years
female: 32.95 years (1993 est.)
Total fertility rate: 6.4 children born/woman (1993 est.)
Nationality:
noun: Somali(s)
adjective: Somali
Ethnic divisions: Somali 85%, Bantu, Arabs 30,000, Europeans 3,000, Asians 800
Religions: Sunni Muslim
Languages: Somali (official), Arabic, Italian, English
Literacy: age 15 and over can read and write (1990)
total population: 24%
male: 36%
female: 14%
Labor force: 2.2 million (very few are skilled laborers)
by occupation: pastoral nomad 70%, agriculture, government, trading, fishing, handicrafts, and other 30%
note: 53% of population of working age (1985)

Government

Names:
conventional long form: none
conventional short form: Somalia
former: Somali Republic
Digraph: SO
Type: none
Capital: Mogadishu
Administrative divisions: 18 regions (plural—NA, singular—gobolka); Awdal, Bakool, Banaadir, Bari, Bay, Galguduud, Gedo, Hiiraan, Jubbada Dhexe, Jubbada Hoose, Mudug, Nugaal, Sanaag, Shabeellaha Dhexe, Shabeellaha Hoose, Sool, Togdheer, Woqooyi Galbeed
Independence: 1 July 1960 (from a merger of British Somaliland, which became independent from the UK on 26 June 1960, and Italian Somaliland, which became independent from the Italian-administered UN trusteeship on 1 July 1960, to form the Somali Republic)
Constitution: 25 August 1979, presidential approval 23 September 1979

Legal system: NA
National holiday: NA
Political parties and leaders: the United Somali Congress (USC) ousted the former regime on 27 January 1991; formerly the only party was the Somali Revolutionary Socialist Party (SRSP), headed by former President and Commander in Chief of the Army Maj. Gen. Mohamed SIAD Barre
Other political or pressure groups: numerous clan and subclan factions are currently vying for power
Suffrage: 18 years of age; universal
Elections:
President: last held 23 December 1986 (next to be held NA); results—President SIAD was reelected without opposition
People's Assembly: last held 31 December 1984 (next to be held NA); results—SRSP was the only party; seats—(177 total, 171 elected) SRSP 171; note—the United Somali Congress (USC) ousted the regime of Maj. Gen. Mohamed SIAD Barre on 27 January 1991; the provisional government has promised that a democratically elected government will be established
Executive branch: president, two vice presidents, prime minister, Council of Ministers (cabinet)
Legislative branch: unicameral People's Assembly (Golaha Shacbiga); non-functioning
Judicial branch: Supreme Court (non-functioning)
Leaders:
Chief of State: Interim President ALI MAHDI Mohamed (since 27 January 1991)
Head of Government: Prime Minister OMAR Arteh Ghalib (since 27 January 1991)
Member of: ACP, AfDB, AFESD, AL, AMF, CAEU, ECA, FAO, G-77, IBRD, ICAO, IDA, IDB, IFAD, IFC, IGADD, ILO, IMF, IMO, INTELSAT, INTERPOL, IOC, IOM (observer), ITU, LORCS, NAM, OAU, OIC, UN, UNCTAD, UNESCO, UNHCR, UNIDO, UPU, WHO, WIPO, WMO
Diplomatic representation in US:
chief of mission: (vacant)
chancery: Suite 710, 600 New Hampshire Avenue NW, Washington, DC 20037
telephone: (202) 342-1575
consulate general: New York
note: Somalian Embassy ceased operations on 8 May 1991
US diplomatic representation: the US Embassy in Mogadishu was evacuated and closed indefinitely in January 1991; United States Liaison Office (USLO) opened in December 1992
Flag: light blue with a large white five-pointed star in the center; design based on the flag of the UN (Italian Somaliland was a UN trust territory)

Economy

Overview: One of the world's poorest and least developed countries, Somalia has few resources. Moreover, much of the economy has been devastated by the civil war. Agriculture is the most important sector, with livestock accounting for about 40% of GDP and about 65% of export earnings. Nomads and seminomads who are dependent upon livestock for their livelihoods make up more than half of the population. Crop production generates only 10% of GDP and employs about 20% of the work force. The main export crop is bananas; sugar, sorghum, and corn are grown for the domestic market. The small industrial sector is based on the processing of agricultural products and accounts for less than 10% of GDP. Greatly increased political turmoil in 1991-92 has resulted in a substantial drop in output, with widespread famine.
National product: $NA
National product real growth rate: NA%
National product per capita: $NA
Inflation rate (consumer prices): NA%
Unemployment rate: NA%
Budget: revenues $NA; expenditures $NA, including capital expenditures of $NA
Exports: $NA
commodities: bananas, livestock, fish, hides, skins
partners: Saudi Arabia, Italy, FRG (1986)
Imports: $NA
commodities: petroleum products, foodstuffs, construction materials
partners: US 13%, Italy, FRG, Kenya, UK, Saudi Arabia (1986)
External debt: $1.9 billion (1989)
Industrial production: growth rate NA%, accounts for NA% of GDP
Electricity: former public power capacity of 75,000 kW is completely shut down by the destruction of the civil war; UN, relief organizations, and foreign military units in Somalia use their own portable power systems
Industries: a few small industries, including sugar refining, textiles, petroleum refining; probably shut down by the widespread destruction during the civil war
Agriculture: dominant sector, led by livestock raising (cattle, sheep, goats); crops—bananas, sorghum, corn, mangoes, sugarcane; not self-sufficient in food; distribution of food disrupted by civil strife; fishing potential largely unexploited
Economic aid: US commitments, including Ex-Im (FY70-89), $639 million; Western (non-US) countries, ODA and OOF bilateral commitments (1970-89), $3.8 billion; OPEC bilateral aid (1979-89), $1.1 billion; Communist countries (1970-89), $336 million
Currency: 1 Somali shilling (So. Sh.) = 100 centesimi

South Africa

Exchange rates: Somali shillings (So. Sh.) per US$1—4,200 (December 1992), 3,800.00 (December 1990), 490.7 (1989), 170.45 (1988), 105.18 (1987), 72.00 (1986)
Fiscal year: calendar year

Communications

Highways: 22,500 km total; including 2,700 km paved, 3,000 km gravel, and 16,800 km improved earth or stabilized soil (1992)
Pipelines: crude oil 15 km
Ports: Mogadishu, Berbera, Chisimayu (Kismaayo), Bender Cassim (Boosaaso)
Merchant marine: 3 ships (1,000 GRT or over) totaling 6,913 GRT/8,718 DWT; includes 2 cargo, 1 refrigerated cargo
Airports:
total: 69.
usable: 48
with permanent-surface runways: 8
with runways over 3,659 m: 2
with runways 2,440-3,659 m: 6
with runways 1,220-2,439 m: 20
Telecommunications: the public telecommunications system was completely destroyed or dismantled by the civil war factions; all relief organizations depend on their own private systems (1993)

Defense Forces

Branches: NA
Manpower availability: males age 15-49 1,596,380; fit for military service 897,660 (1993 est.)
Defense expenditures: exchange rate conversion—$NA, NA% of GDP

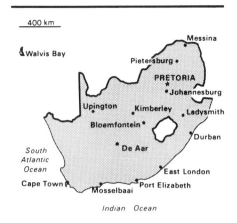

Geography

Location: Southern Africa, at the extreme southern tip of the continent
Map references: Africa, Standard Time Zones of the World
Area:
total area: 1,221,040 km²
land area: 1,221,040 km²
comparative area: slightly less than twice the size of Texas
note: includes Walvis Bay, Marion Island, and Prince Edward Island
Land boundaries: total 4,973 km, Botswana 1,840 km, Lesotho 909 km, Mozambique 491 km, Namibia 1,078 km, Swaziland 430 km, Zimbabwe 225 km
Coastline: 2,881 km
Maritime claims:
continental shelf: 200 m depth or to depth of exploitation
exclusive fishing zone: 200 nm
territorial sea: 12 nm
International disputes: claim by Namibia to Walvis Bay exclave and 12 offshore islands administered by South Africa; South Africa and Namibia have agreed to jointly administer the area for an interim period; the terms and dates to be covered by joint administration arrangements have not been established at this time; and Namibia will continue to maintain a claim to sovereignty over the entire area
Climate: mostly semiarid; subtropical along coast; sunny days, cool nights
Terrain: vast interior plateau rimmed by rugged hills and narrow coastal plain
Natural resources: gold, chromium, antimony, coal, iron ore, manganese, nickel, phosphates, tin, uranium, gem diamonds, platinum, copper, vanadium, salt, natural gas
Land use:
arable land: 10%
permanent crops: 1%
meadows and pastures: 65%
forest and woodland: 3%
other: 21%

Irrigated land: 11,280 km² (1989 est.)
Environment: lack of important arterial rivers or lakes requires extensive water conservation and control measures
Note: Walvis Bay is an exclave of South Africa in Namibia; South Africa completely surrounds Lesotho and almost completely surrounds Swaziland

People

Population: 42,792,804 (July 1993 est.)
Population growth rate: 2.63% (1993 est.)
Birth rate: 33.77 births/1,000 population (1993 est.)
Death rate: 7.65 deaths/1,000 population (1993 est.)
Net migration rate: 0.15 migrant(s)/1,000 population (1993 est.)
Infant mortality rate: 48.3 deaths/1,000 live births (1993 est.)
Life expectancy at birth:
total population: 64.81 years
male: 62.07 years
female: 67.63 years (1993 est.)
Total fertility rate: 4.4 children born/woman (1993 est.)
Nationality:
noun: South African(s)
adjective: South African
Ethnic divisions: black 75.2%, white 13.6%, Colored 8.6%, Indian 2.6%
Religions: Christian (most whites and Coloreds and about 60% of blacks), Hindu (60% of Indians), Muslim 20%
Languages: Afrikaans (official), English (official), Zulu, Xhosa, North Sotho, South Sotho, Tswana, and many other vernacular languages
Literacy: age 15 and over can read and write (1980)
total population: 76%
male: 78%
female: 75%
Labor force: 13.4 million economically active (1990)
by occupation: services 55%, agriculture 10%, industry 20%, mining 9%, other 6%

Government

Names:
conventional long form: Republic of South Africa
conventional short form: South Africa
Abbreviation: RSA
Digraph: SF
Type: republic
Capital: Pretoria (administrative); Cape Town (legislative); Bloemfontein (judicial)
Administrative divisions: 4 provinces; Cape, Natal, Orange Free State, Transvaal; there are 10 homelands not recognized by the US—4 independent (Bophuthatswana, Ciskei, Transkei, Venda) and 6 other

(Gazankulu, Kangwane, KwaNdebele, KwaZulu, Lebowa, QwaQwa)
Independence: 31 May 1910 (from UK)
Constitution: 3 September 1984
Legal system: based on Roman-Dutch law and English common law; accepts compulsory ICJ jurisdiction, with reservations
National holiday: Republic Day, 31 May (1910)
Political parties and leaders:
white political parties and leaders: National Party (NP), Frederik W. DE KLERK (majority party); Conservative Party (CP), leader NA (official opposition party); Democratic Party (DP), Zach DE BEER; Afrikaner Volksunie (AVU), Andries BEYERS
Colored political parties and leaders (see Note): Labor Party (LP), Allan HENDRICKSE (majority party); National Party (NP); Democratic Party (DP); Freedom Party
Indian political parties and leaders: Solidarity, J. N. REDDY (majority party); National People's Party (NPP), Amichand RAJBANSI; Merit People's Party
note: the Democratic Reform Party (DRP) and the United Democratic Party (UDP) were disbanded in May 1991
Other political or pressure groups: African National Congress (ANC), Nelson MANDELA, president; Inkatha Freedom Party (IFP), Mangosuthu BUTHELEZI, president; Pan-Africanist Congress (PAC), Clarence MAKWETU, president
Suffrage: 18 years of age; universal, but voting rights are racially based
Elections:
House of Assembly (whites): last held 6 September 1989 (next to be held by NA March 1995); results—NP 58%, CP 23%, DP 19%; seats—(178 total, 166 elected) NP 103, CP 41, DP 34; note—by February 1992, because of byelections, splits, and defections, changes in number of seats held by parties were as follows: NP 102, CP 36, DP 28, AVU 5, independent 7
House of Representatives (Coloreds): last held 6 September 1989 (next to be held no later than March 1995); results—percent of vote by party NA; seats—(85 total, 80 elected) LP 69, DRP 5, UDP 3, Freedom Party 1, independents 2; note—by October 1992 many representatives had changed their allegiance causing the following changes in seating: NP 44, LP 27, DP 6, Freedom Party 1, independents 6, vacant 1
House of Delegates (Indians): last held 6 September 1989 (next to be held no later than March 1995); results—percent of vote by party NA; seats—(45 total, 40 elected) Solidarity 16, NPP 9, Merit People's Party 3, independents 6, other 6; note - due to delegates changing party affiliation, seating

as of October 1992 is as follows: Solidarity 25, NPP 7, Merit People's Party 2, other 8, independents 3
note: tentative agreement to hold national election open to all races for a 400-seat constitutent assembly on 27 April 1994
Executive branch: state president, Executive Council (cabinet), Ministers' Councils (from the three houses of Parliament)
Legislative branch: tricameral Parliament (Parlement) consists of the House of Assembly (Volksraad; whites), House of Representatives (Raad van Verteenwoordigers; Coloreds), and House of Delegates (Raad van Afgevaardigdes; Indians)
Judicial branch: Supreme Court
Leaders:
Chief of State and Head of Government: State President Frederik Willem DE KLERK (since 13 September 1989)
Member of: BIS, CCC, ECA, GATT, IAEA, IBRD, ICAO (suspended), ICC, IDA, IFC, IMF, INTELSAT, ISO, ITU (suspended), LORCS, SACU, UN, UNCTAD, WFTU, WHO, WIPO, WMO (suspended)
Diplomatic representation in US:
chief of mission: Ambassador Harry SCHWARZ
chancery: 3051 Massachusetts Avenue NW, Washington, DC 20008
telephone: (202) 232-4400
consulates general: Beverly Hills (California), Chicago, Houston, and New York
US diplomatic representation:
chief of mission: Ambassador Princeton N. LYMAN
embassy: Thibault House, 225 Pretorius Street, Pretoria
telephone: [27] (12) 28-4266
FAX: [27] (12) 21-9278
consulates general: Cape Town, Durban, Johannesburg
Flag: actually four flags in one—three miniature flags reproduced in the center of the white band of the former flag of the Netherlands, which has three equal horizontal bands of orange (top), white, and blue; the miniature flags are a vertically hanging flag of the old Orange Free State with a horizontal flag of the UK adjoining on the hoist side and a horizontal flag of the old Transvaal Republic adjoining on the other side

---

Economy

Overview: Many of the white one-seventh of the South African population enjoy incomes, material comforts, and health and educational standards equal to those of Western Europe. In contrast, most of the remaining population suffers from the

poverty patterns of the Third World, including unemployment and lack of job skills. The main strength of the economy lies in its rich mineral resources, which provide two-thirds of exports. Economic developments in the 1990s will be driven partly by the changing relations among the various ethnic groups. The shrinking economy in recent years has absorbed less than 10% of the more than 300,000 workers entering the labor force annually. Local economists estimate that the economy must grow between 5% and 6% in real terms annually to absorb all of the new entrants.
National product: GDP—exchange rate conversion—$115 billion (1992)
National product real growth rate: -2% (1992)
National product per capita: $2,800 (1992)
Inflation rate (consumer prices): 13.9% (1992)
Unemployment rate: 45% (well over 50% in some homeland areas) (1992 est.)
Budget: revenues $28 billion; expenditures $36 billion, including capital expenditures of $3 billion (FY93 est.)
Exports: $23.5 billion (f.o.b., 1992)
commodities: gold 27%, other minerals and metals 20-25%, food 5%, chemicals 3%
partners: Italy, Japan, US, Germany, UK, other EC countries, Hong Kong
Imports: $18.2 billion (f.o.b., 1992)
commodities: machinery 32%, transport equipment 15%, chemicals 11%, oil, textiles, scientific instruments
partners: Germany, Japan, UK, US, Italy
External debt: $18 billion (1992 est.)
Industrial production: growth rate NA%; accounts for about 40% of GDP
Electricity: 46,000,000 kW capacity; 180,000 million kWh produced, 4,100 kWh per capita (1991)
Industries: mining (world's largest producer of platinum, gold, chromium), automobile assembly, metalworking, machinery, textile, iron and steel, chemical, fertilizer, foodstuffs
Agriculture: accounts for about 5% of GDP and 30% of labor force; diversified agriculture, with emphasis on livestock; products—cattle, poultry, sheep, wool, milk, beef, corn, wheat, sugarcane, fruits, vegetables; self-sufficient in food
Economic aid: NA
Currency: 1 rand (R) = 100 cents
Exchange rates: rand (R) per US$1—3.1576 (May 1993), 2.8497 (1992), 2.7563 (1991), 2.5863 (1990), 2.6166 (1989), 2.2611 (1988)
Fiscal year: 1 April-31 March

Communications

Railroads: 20,638 km route distance total; 20,324 km of 1.067-meter gauge trackage (counts double and multiple tracking as single track); 314 km of 610 mm gauge;

substantial electrification of 1.067 meter gauge
Highways: 188,309 km total; 54,013 km paved, 134,296 km crushed stone, gravel, or improved earth
Pipelines: crude oil 931 km, petroleum products 1,748 km, natural gas 322 km
Ports: Durban, Cape Town, Port Elizabeth, Richard's Bay, Saldanha, Mosselbaai, Walvis Bay
Merchant marine: 5 ships (1,000 GRT or over) totaling 213,708 GRT/201,043 DWT; includes 4 container, 1 vehicle carrier
Airports:
total: 899
usable: 713
with permanent-surface runways: 136
with runways over 3,659 m: 5
with runways 2,440-3,659 m: 10
with runways 1,220-2,439 m: 221
Telecommunications: the system is the best developed, most modern, and has the highest capacity in Africa; it consists of carrier-equipped open-wire lines, coaxial cables, radio relay links, fiber optic cable, and radiocommunication stations; key centers are Bloemfontein, Cape Town, Durban, Johannesburg, Port Elizabeth, and Pretoria; over 4,500,000 telephones; broadcast stations—14 AM, 286 FM, 67 TV; 1 submarine cable; satellite earth stations—1 Indian Ocean INTELSAT and 2 Atlantic Ocean INTELSAT

Defense Forces

Branches: South African Defense Force (SADF; including Army, Navy, Air Force, Medical Services), South African Police (SAP)
Manpower availability: males age 15-49 10,294,211; fit for military service 6,279,190; reach military age (18) annually 425,477 (1993 est.); obligation for service in Citizen Force or Commandos begins at 18; black and white volunteers for service in permanent force must be 17; national service obligation for white conscripts is one year; figures include the so-called homelands not recognized by the US
Defense expenditures: exchange rate conversion—$2.9 billion, about 2.5% of GDP (FY93 budget)

South Georgia and the South Sandwich Islands
(dependent territory of the UK)

Geography

Location: in the South Atlantic Ocean, off the south Argentine coast, southeast of the Falkland Islands
Map references: Antarctic Region
Area:
total area: 4,066 km²
land area: 4,066 km²
comparative area: slightly larger than Rhode Island
note: includes Shag Rocks, Clerke Rocks, Bird Island
Land boundaries: 0 km
Coastline: NA km
Maritime claims:
territorial sea: 12 nm
International disputes: administered by the UK, claimed by Argentina
Climate: variable, with mostly westerly winds throughout the year, interspersed with periods of calm; nearly all precipitation falls as snow
Terrain: most of the islands, rising steeply from the sea, are rugged and mountainous; South Georgia is largely barren and has steep, glacier-covered mountains; the South Sandwich Islands are of volcanic origin with some active volcanoes
Natural resources: fish
Land use:
arable land: 0%
permanent crops: 0%
meadows and pastures: 0%
forest and woodland: 0%
other: 100% (largely covered by permanent ice and snow with some sparse vegetation consisting of grass, moss, and lichen)
Irrigated land: 0 km²
Environment: reindeer, introduced early in this century, live on South Georgia; weather conditions generally make it difficult to approach the South Sandwich Islands; the South Sandwich Islands are subject to active volcanism

Note: the north coast of South Georgia has several large bays, which provide good anchorage

---

People

Population: no indigenous population; there is a small military garrison on South Georgia, and the British Antarctic Survey has a biological station on Bird Island; the South Sandwich Islands are uninhabited

---

Government

Names:
conventional long form: South Georgia and the South Sandwich Islands
conventional short form: none
Digraph: SX
Type: dependent territory of the UK
Capital: none; Grytviken on South Georgia is the garrison town
Administrative divisions: none (dependent territory of the UK)
Independence: none (dependent territory of the UK)
Constitution: 3 October 1985
Legal system: English common law
National holiday: Liberation Day, 14 June (1982)
Executive branch: British monarch, commissioner
Legislative branch: none
Judicial branch: none
Leaders:
Chief of State: Queen ELIZABETH II (since 6 February 1952), represented by Commissioner David Everard TATHAM (since August 1992; resident at Stanley, Falkland Islands)

Economy

Overview: Some fishing takes place in adjacent waters. There is a potential source of income from harvesting fin fish and krill. The islands receive income from postage · stamps produced in the UK.
Budget: revenues $291,777; expenditures $451,011, including capital expenditures of $NA (FY88 est.)
Electricity: 900 kW capacity; 2 million kWh produced, NA kWh per capita (1992)

Communications

Highways: NA
Ports: Grytviken on South Georgia
Airports:
total: 5
usable: 5
with permanent-surface runways: 2

with runways over 3,659 m: 0
with runways 2,440-3,659 m: 1
with runways 1,220-2,439 m: 0
Telecommunications: coastal radio station at Grytviken; no broadcast stations

---

Defense Forces

Note: defense is the responsibility of the UK

Spain

Canary Islands, Ceuta, and Melilla are not shown.

Geography

Location: Southwestern Europe, bordering the North Atlantic Ocean and the Mediterranean Sea, between Portugal and France
Map references: Africa, Europe, Standard Time Zones of the World
Area:
total area: 504,750 km²
land area: 499,400 km²
comparative area: slightly more than twice the size of Oregon
note: includes Balearic Islands, Canary Islands, and five places of sovereignty (plazas de soberania) on and off the coast of Morocco—Ceuta, Mellila, Islas Chafarinas, Penon de Alhucemas, and Penon de Velez de la Gomera
Land boundaries: total 1,903.2 km, Andorra 65 km, France 623 km, Gibraltar 1.2 km, Portugal 1,214 km
Coastline: 4,964 km
Maritime claims:
exclusive economic zone: 200 nm
territorial sea: 12 nm
International disputes: Gibraltar question with UK; Spain controls five places of sovereignty (plazas de soberania) on and off the coast of Morocco—the coastal enclaves of Ceuta and Melilla, which Morocco contests, as well as the islands of Penon de Alhucemas, Penon de Velez de la Gomera, and Islas Chafarinas
Climate: temperate; clear, hot summers in interior, more moderate and cloudy along coast; cloudy, cold winters in interior, partly cloudy and cool along coast
Terrain: large, flat to dissected plateau surrounded by rugged hills; Pyrenees in north
Natural resources: coal, lignite, iron ore, uranium, mercury, pyrites, fluorspar, gypsum, zinc, lead, tungsten, copper, kaolin, potash, hydropower
Land use:
arable land: 31%

permanent crops: 10%
meadows and pastures: 21%
forest and woodland: 31%
other: 7%
Irrigated land: 33,600 km² (1989 est.)
Environment: deforestation; air pollution
Note: strategic location along approaches to Strait of Gibraltar

---

People

Population: 39,207,159 (July 1993 est.)
Population growth rate: 0.24% (1993 est.)
Birth rate: 10.88 births/1,000 population (1993 est.)
Death rate: 8.76 deaths/1,000 population (1993 est.)
Net migration rate: 0.24 migrant(s)/1,000 population (1993 est.)
Infant mortality rate: 7 deaths/1,000 live births (1993 est.)
Life expectancy at birth:
total population: 77.51 years
male: 74.22 years
female: 81.04 years (1993 est.)
Total fertility rate: 1.38 children born/woman (1993 est.)
Nationality:
noun: Spaniard(s)
adjective: Spanish
Ethnic divisions: composite of Mediterranean and Nordic types
Religions: Roman Catholic 99%, other sects 1%
Languages: Castilian Spanish, Catalan 17%, Galician 7%, Basque 2%
Literacy: age 15 and over can read and write (1990)
total population: 95%
male: 97%
female: 93%
Labor force: 14.621 million
by occupation: services 53%, industry 24%, agriculture 14%, construction 9% (1988)

Government

Names:
conventional long form: Kingdom of Spain
conventional short form: Spain
local short form: Espana
Digraph: SP
Type: parliamentary monarchy
Capital: Madrid
Administrative divisions: 17 autonomous communities (comunidades autonomas, singular—comunidad autonoma); Andalucia, Aragon, Asturias, Canarias, Cantabria, Castilla-La Mancha, Castilla y Leon, Cataluna, Communidad Valencia, Extremadura, Galicia, Islas Baleares, La Rioja, Madrid, Murcia, Navarra, Pais Vasco
note: there are five places of sovereignty on and off the coast of Morocco (Ceuta, Mellila, Islas Chafarinas, Penon de

Alhucemas, and Penon de Velez de la Gomera) with administrative status unknown
Independence: 1492 (expulsion of the Moors and unification)
Constitution: 6 December 1978, effective 29 December 1978
Legal system: civil law system, with regional applications; does not accept compulsory ICJ jurisdiction
National holiday: National Day, 12 October
Political parties and leaders:
principal national parties, from right to left: Popular Party (PP), Jose Maria AZNAR; Social Democratic Center (CDS), Rafael Calvo ORTEGA; Spanish Socialist Workers Party (PSOE), Felipe GONZALEZ Marquez, secretary general; Socialist Democracy Party (DS), Ricardo Garcia DAMBORENEA; Spanish Communist Party (PCE), Julio ANGUITA; United Left (IU) a coalition of parties including the PCE, a branch of the PSOE, and other small parties, leader NA
chief regional parties: Convergence and Unity (CiU), Jordi PUJOL Saley, in Catalonia; Basque Nationalist Party (PNV), Xabier ARZALLUS; Basque Solidarity (EA), Carlos GARAICOETXEA Urizza; Basque Popular Unity (HB), Jon IDIGORAS; Basque Left (EE), Juan Maria BANDRES; Basque Socialist Party (PSE); coalition of the PSE, EE, and PSOE, Jose Maria BANEGAS; Euskal Ezkerra (EUE), Xabier GURRUTXAGA; Andalusian Party (PA), Pedro PACHECO; Independent Canary Group (AIC), leader NA; Aragon Regional Party (PAR), leader NA; Valencian Union (UV), leader NA
Other political or pressure groups: on the extreme left, the Basque Fatherland and Liberty (ETA) and the First of October Antifascist Resistance Group (GRAPO) use terrorism to oppose the government; free labor unions (authorized in April 1977) include the Communist-dominated Workers Commissions (CCOO); the Socialist General Union of Workers (UGT), and the smaller independent Workers Syndical Union (USO); the Catholic Church; business and landowning interests; Opus Dei; university students
Suffrage: 18 years of age; universal
Elections:
Senate: last held 29 October 1989 (next to be held NA October 1993); results—percent of vote by party NA; seats—(208 total) PSOE 106, PP 79, CiU 10, PNV 4, HB 3, AIC 1, other 5
Congress of Deputies: last held 29 October 1989 (next to be held NA October 1993); results—PSOE 39.6%, PP 25.8%, CDS 9%, IU 9%, CiU 5%, PNV 1.2%, HB 1%, PA 1%, other 8.4%; seats—(350 total) PSOE 175, PP 106, CiU 18, IU 17, CDS 14, PNV 5, HB 4, other 11
Executive branch: monarch, president of

the government (prime minister), deputy prime minister, Council of Ministers (cabinet), Council of State
Legislative branch: bicameral The General Courts or National Assembly (Las Cortes Generales) consists of an upper house or Senate (Senado) and a lower house or Congress of Deputies (Congreso de los Diputados)
Judicial branch: Supreme Court (Tribunal Supremo)
Leaders:
Chief of State: King JUAN CARLOS I (since 22 November 1975)
Head of Government: Prime Minister Felipe GONZALEZ Marquez (since 2 December 1982); Deputy Prime Minister Narcis SERRA y Serra (since 13 March 1991)
Member of: AG (observer), AsDB, Australian Group, BIS, CCC, CE, CERN, COCOM, CSCE, EBRD, AfDB, EC, ECE, ECLAC, EIB, ESA, FAO, G-8, GATT, IADB, IAEA, IBRD, ICAO, ICC, ICFTU, IDA, IEA, IFAD, IFC, ILO, IMF, IMO, INMARSAT, INTELSAT, INTERPOL, IOC, IOM (observer), ISO, ITU, LAIA (observer), LORCS, MTRC, NACC, NAM (guest), NATO, NEA, NSG, OAS (observer), OECD, ONUSAL, PCA, UN, UNAVEM II, UNCTAD, UNESCO, UNIDO, UNOMOZ, UPU, WCL, WEU, WHO, WIPO, WMO, WTO, ZC
Diplomatic representation in US:
chief of mission: Ambassador Jaime De OJEDA y Eiseley
chancery: 2700 15th Street NW, Washington, DC 20009
telephone: (202) 265-0190 or 0191
consulates general: Boston, Chicago, Houston, Los Angeles, Miami, New Orleans, New York, San Francisco, and San Juan (Puerto Rico)
US diplomatic representation:
chief of mission: Ambassador Richard G. CAPEN, Jr.
embassy: Serrano 75, 28006 Madrid
mailing address: PSC 61, APO AE 09642
telephone: [34] (1) 577-4000
FAX: [34] (1) 577-5735
consulate general: Barcelona
consulate: Bilbao
Flag: three horizontal bands of red (top), yellow (double width), and red with the national coat of arms on the hoist side of the yellow band; the coat of arms includes the royal seal framed by the Pillars of Hercules, which are the two promontories (Gibraltar and Ceuta) on either side of the eastern end of the Strait of Gibraltar

---

Economy

Overview: Spain has done well since joining the EC in 1986. Foreign and domestic investments have spurred GDP

growth at an annual average of more than 4% in 1986-91. As of 1 January 1993, Spain has wholly liberalized its trade and capital markets to EC standards, including integrating agriculture two years ahead of schedule. Beginning in 1989, Madrid implemented a tight monetary policy to fight 7% inflation. As a result of this action and the worldwide decline in economic growth, Spain's growth rate declined to 1% in 1992. Spain faces a likely recession in first half 1993. The government expects a recovery in the second half, but this depends on stepped-up growth in Germany and France. The slowdown in growth—along with displacements caused by structural adjustments in preparation for the EC single market—has pushed an already high unemployment rate up to 19%. However, many people listed as unemployed work in the underground economy. If the government can stick to its tough economic policies and push further structural reforms, the economy will emerge stronger at the end of the 1990s.
National product: GDP—purchasing power equivalent—$514.9 billion (1992)
National product real growth rate: 1% (1992)
National product per capita: $13,200 (1992)
Inflation rate (consumer prices): 6% (1992 est.)
Unemployment rate: 19% (yearend 1992)
Budget: revenues $122.9 billion; expenditures $140.2 billion, including capital expenditures of $NA (1992 est.)
Exports: $62 billion (f.o.b., 1992 est.)
commodities: cars and trucks, semifinished manufactured goods, foodstuffs, machinery
partners: EC 71.0%, US 4.9%, other developed countries 7.9% (1991)
Imports: $100 billion (c.i.f., 1992 est.)
commodities: machinery, transport equipment, fuels, semifinished goods, foodstuffs, consumer goods, chemicals
partners: EC 60.0%, US 8.0%, other developed countries 11.5%, Middle East 2.6% (1991)
External debt: $67.5 billion (1992 est.)
Industrial production: growth rate 0.6% (1992 est.)
Electricity: 46,600,000 kW capacity; 157,000 million kWh produced, 4,000 kWh per capita (1992)
Industries: textiles and apparel (including footwear), food and beverages, metals and metal manufactures, chemicals, shipbuilding, automobiles, machine tools, tourism
Agriculture: accounts for about 5% of GDP and 14% of labor force; major products—grain, vegetables, olives, wine grapes, sugar beets, citrus fruit, beef, pork, poultry, dairy; largely self-sufficient in food; fish catch of 1.4 million metric tons is among top 20 nations

Spain (continued)

Illicit drugs: key European gateway country for Latin American cocaine entering the European market
Economic aid: US commitments, including Ex-Im (FY70-87), $1.9 billion; Western (non-US) countries, ODA and OOF bilateral commitments (1970-79), $545.0 million; not currently a recipient
Currency: 1 peseta (Pta) = 100 centimos
Exchange rates: pesetas (Ptas) per US$1—114.59 (January 1993), 102.38 (1992), 103.91 (1991), 101.93 (1990), 118.38 (1989), 116.49 (1988)
Fiscal year: calendar year

Communications

Railroads: 15,430 km total; Spanish National Railways (RENFE) operates 12,691 km (all 1.668-meter gauge, 6,184 km electrified, and 2,295 km double track); FEVE (government-owned narrow-gauge railways) operates 1,821 km (predominantly 1.000-meter gauge, 441 km electrified); privately owned railways operate 918 km (predominantly 1.000-meter gauge, 512 km electrified, and 56 km double track)
Highways: 150,839 km total; 82,513 km national (includes 2,433 km limited-access divided highway, 63,042 km bituminous treated, 17,038 km intermediate bituminous, concrete, or stone block) and 68,326 km provincial or local roads (bituminous treated, intermediate bituminous, or stone block)
Inland waterways: 1,045 km, but of minor economic importance
Pipelines: crude oil 265 km, petroleum products 1,794 km, natural gas 1,666 km
Ports: Algeciras, Alicante, Almeria, Barcelona, Bilbao, Cadiz, Cartagena, Castellon de la Plana, Ceuta, El Ferrol del Caudillo, Puerto de Gijon, Huelva, La Coruna, Las Palmas (Canary Islands), Mahon, Malaga, Melilla, Rota, Santa Cruz de Tenerife, Sagunto, Tarragona, Valencia, Vigo, and 175 minor ports
Merchant marine: 242 ships (1,000 GRT or over) totaling 2,394,175 GRT/4,262,868 DWT; includes 2 passenger, 8 short-sea passenger, 71 cargo, 12 refrigerated cargo, 12 container, 32 roll-on/roll-off cargo, 4 vehicle carrier, 41 oil tanker, 14 chemical tanker, 7 liquefied gas, 3 specialized tanker, 36 bulk
Airports:
total: 105
usable: 99
with permanent-surface runways: 60
with runways over 3,659 m: 4
with runways 2,440-3,659 m: 22
with runways 1,220-2,439 m: 26
Telecommunications: generally adequate, modern facilities; 15,350,464 telephones; broadcast stations—190 AM, 406 (134 repeaters) FM, 100 (1,297 repeaters) TV; 22 coaxial submarine cables; 2 communications satellite earth stations operating in INTELSAT (Atlantic Ocean and Indian Ocean); MARECS, INMARSAT, and EUTELSAT systems; tropospheric links

Defense Forces

Branches: Army, Navy, Air Force, Marines, Civil Guard, National Police, Coastal Civil Guard
Manpower availability: males age 15-49 10,299,960; fit for military service 8,341,046; reach military age (20) annually 338,231 (1993 est.)
Defense expenditures: exchange rate conversion—$9.6 billion, 1.6% of GDP (1992)

Spratly Islands

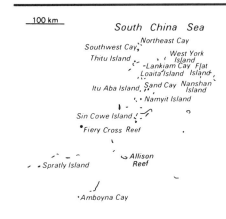

Geography

Location: in the South China Sea, between Vietnam and the Philippines
Map references: Asia, Southeast Asia
Area:
total area: NA km^2 but less than 5 km^2
land area: less than 5 km^2
comparative area: NA
note: includes 100 or so islets, coral reefs, and sea mounts scattered over the South China Sea
Land boundaries: 0 km
Coastline: 926 km
Maritime claims: NA
International disputes: all of the Spratly Islands are claimed by China, Taiwan, and Vietnam; parts of them are claimed by Malaysia and the Philippines; in 1984, Brunei established an exclusive economic zone, which encompasses Louisa Reef, but has not publicly claimed the island
Climate: tropical
Terrain: flat
Natural resources: fish, guano, undetermined oil and natural gas potential
Land use:
arable land: 0%
permanent crops: 0%
meadows and pastures: 0%
forest and woodland: 0%
other: 100%
Irrigated land: 0 km^2
Environment: subject to typhoons; includes numerous small islands, atolls, shoals, and coral reefs
Note: strategically located near several primary shipping lanes in the central South China Sea; serious navigational hazard

People

Population: no indigenous inhabitants; note—there are scattered garrisons

Sri Lanka

Government

Names:
conventional long form: none
conventional short form: Spratly Islands
Digraph: PG

Economy

Overview: Economic activity is limited to commercial fishing; proximity to nearby oil- and gas-producing sedimentary basins suggests the potential for oil and gas deposits, but the region is largely unexplored, and there are no reliable estimates of potential reserves; commercial exploitation has yet to be developed.
Industries: none

Communications

Ports: no natural harbors
Airports:
total: 4
usable: 4
with permanent-surfaced runways: 1
with runways over 3,659 m: 0
with runways 2,440-3,659 m: 0
with runways 1,220-2,439 m: 0

Defense Forces

Note: about 50 small islands or reefs are occupied by China, Malaysia, the Philippines, Taiwan, and Vietnam

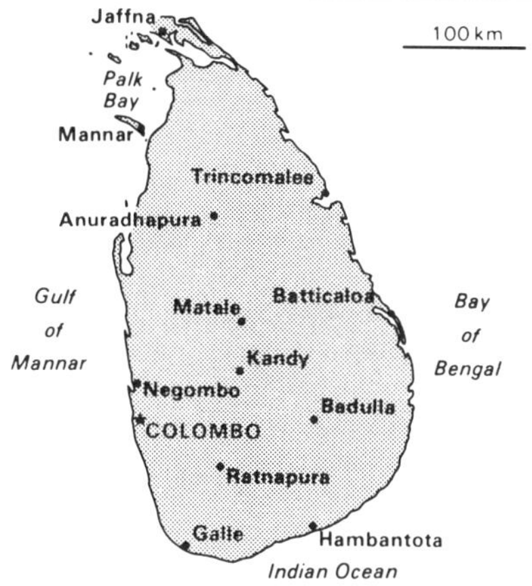

Geography

Location: South Asia, 29 km southeast of India across the Palk Strait in the Indian Ocean
Map references: Asia, Standard Time Zones of the World
Area:
total area: 65,610 km^2
land area: 64,740 km^2
comparative area: slightly larger than West Virginia
Land boundaries: 0 km
Coastline: 1,340 km
Maritime claims:
contiguous zone: 24 nm
continental shelf: 200 nm or the edge of continental margin
exclusive economic zone: 200 nm
territorial sea: 12 nm
International disputes: none
Climate: tropical monsoon; northeast monsoon (December to March); southwest monsoon (June to October)
Terrain: mostly low, flat to rolling plain; mountains in south-central interior
Natural resources: limestone, graphite, mineral sands, gems, phosphates, clay
Land use:
arable land: 16%
permanent crops: 17%
meadows and pastures: 7%
forest and woodland: 37%
other: 23%
Irrigated land: 5,600 km^2 (1989 est.)
Environment: occasional cyclones, tornados; deforestation; soil erosion
Note: strategic location near major Indian Ocean sea lanes

People

Population: 17,838,190 (July 1993 est.)
note: since the outbreak of hostilities between the government and armed Tamil separatists in the mid 1980s, several hundred thousand Tamil civilians have fled the island; as of late 1992, nearly 115,000 were housed in refugee camps in south India, another 95,000 lived outside the Indian camps, and more than 200,000 Tamils have sought political asylum in the West; fewer than 10,000 Tamils have been successfully repatriated to Sri Lanka
Population growth rate: 1.11% (1993 est.)
Birth rate: 18.71 births/1,000 population (1993 est.)
Death rate: 5.84 deaths/1,000 population (1993 est.)
Net migration rate: -1.81 migrant(s)/1,000 population (1993 est.)
Infant mortality rate: 22.8 deaths/1,000 live births (1993 est.)
Life expectancy at birth:
total population: 71.51 years
male: 68.94 years
female: 74.21 years (1993 est.)
Total fertility rate: 2.13 children born/woman (1993 est.)
Nationality:
noun: Sri Lankan(s)
adjective: Sri Lankan
Ethnic divisions: Sinhalese 74%, Tamil 18%, Moor 7%, Burgher, Malay, and Vedda 1%
Religions: Buddhist 69%, Hindu 15%, Christian 8%, Muslim 8%
Languages: Sinhala (official and national language) 74%, Tamil (national language) 18%
note: English is commonly used in government and is spoken by about 10% of the population
Literacy: age 15 and over can read and write (1990)
total population: 88%
male: 93%
female: 84%
Labor force: 6.6 million
by occupation: agriculture 45.9%, mining and manufacturing 13.3%, trade and transport 12.4%, services and other 28.4% (1985 est.)

Government

Names:
conventional long form: Democratic Socialist Republic of Sri Lanka
conventional short form: Sri Lanka
former: Ceylon
Digraph: CE
Type: republic
Capital: Colombo
Administrative divisions: 8 provinces; Central, North Central, North Eastern, North Western, Sabaragamuwa, Southern, Uva, Western
Independence: 4 February 1948 (from UK)
Constitution: 31 August 1978
Legal system: a highly complex mixture of English common law, Roman-Dutch, Muslim, Sinhalese, and customary law;

has not accepted compulsory ICJ jurisdiction
National holiday: Independence and
National Day, 4 February (1948)
Political parties and leaders: United
National Party (UNP), Dingiri Banda
WIJETUNGA; Sri Lanka Freedom Party
(SLFP), Sirimavo BANDARANAIKE; Sri
Lanka Muslim Congress (SLMC), M. H. M.
ASHRAFF; All Ceylon Tamil Congress
(ACTC), Kumar PONNAMBALAM;
People's United Front (MEP, or Mahajana
Eksath Peramuna), Dinesh
GUNAWARDENE; Eelam Democratic Front
(EDF), Edward SEBASTIAN PILLAI; Tamil
United Liberation Front (TULF), leader NA;
Eelam Revolutionary Organization of
Students (EROS), Velupillai
BALAKUMARAN; New Socialist Party
(NSSP, or Nava Sama Samaja Party),
Vasudeva NANAYAKKARA; Lanka
Socialist Party/Trotskyite (LSSP, or Lanka
Sama Samaja Party), Colin R. DE SILVA;
Sri Lanka People's Party (SLMP, or Sri
Lanka Mahajana Party), Ossie
ABEYGUNASEKERA; Communist Party,
K. P. SILVA; Communist Party/Beijing
(CP/B), N. SHANMUGATHASAN;
Democratic United National Front (DUNF),
Lalith ATHULATHMUDALI and Gamini
DISSANAYAKE
note: the United Socialist Alliance (USA)
includes the NSSP, LSSP, SLMP, CP/M, and
CP/B
Other political or pressure groups:
Liberation Tigers of Tamil Eelam (LTTE)
and other smaller Tamil separatist groups;
Janatha Vimukthi Peramuna (JVP or
People's Liberation Front and several other
radical chauvinist Sinhalese groups);
Buddhist clergy; Sinhalese Buddhist lay
groups; labor unions
Suffrage: 18 years of age; universal
Elections:
President: last held 19 December 1988 (next
to be held NA December 1994);
results—Ranasinghe PREMADASA (UNP)
50%, Sirimavo BANDARANAIKE (SLFP)
45%, other 5%; note—following the
assassination of President PREMADASA on
1 May 1993, Prime Minister WIJETUNGA
became acting president; on 7 May 1993, he
was confirmed by a vote of Parliament to
finish out the term of the assassinated
president
Parliament: last held 15 February 1989
(next to be held by NA February 1995);
results—UNP 51%, SLFP 32%, SLMC 4%,
TULF 3%, USA 3%, EROS 3%, MEP 1%,
other 3%; seats—(225 total) UNP 125, SLFP
67, other 33
Executive branch: president, prime
minister, Cabinet
Legislative branch: unicameral Parliament
Judicial branch: Supreme Court
Leaders:

Chief of State: President Dingiri Banda
WIJETUNGA (since 7 May 1993)
Head of Government: Prime Minister Ranil
WICKREMASINGHE (since 7 May 1993)
Member of: AsDB, C, CCC, CP, ESCAP,
FAO, G-24, G-77, GATT, IAEA, IBRD,
ICAO, ICC, ICFTU, IDA, IFAD, IFC, ILO,
IMF, IMO, INMARSAT, INTELSAT,
INTERPOL, IOC, IOM, ISO, ITU, LORCS,
NAM, PCA, SAARC, UN, UNCTAD,
UNESCO, UNIDO, UPU, WCL, WFTU,
WHO, WIPO, WMO, WTO
Diplomatic representation in US:
chief of mission: Ambassador Ananda
GURUGE
chancery: 2148 Wyoming Avenue NW,
Washington, DC 20008
telephone: (202) 483-4025 through 4028
consulate general: New York
US diplomatic representation:
chief of mission: Ambassador Teresita C.
SCHAFFER
embassy: 210 Galle Road, Colombo 3
mailing address: P. O. Box 106, Colombo
telephone: [94] (1) 44-80-07
FAX: [94] (1) 43-73-45
Flag: yellow with two panels; the smaller
hoist-side panel has two equal vertical bands
of green (hoist side) and orange; the other
panel is a large dark red rectangle with a
yellow lion holding a sword, and there is a
yellow bo leaf in each corner; the yellow
field appears as a border that goes around
the entire flag and extends between the two
panels

Economy

Overview: Agriculture, forestry, and fishing
dominate the economy, employing half of
the labor force and accounting for one
quarter of GDP. The plantation crops of tea,
rubber, and coconuts provide about one-third
of export earnings. The economy has been
plagued by high rates of unemployment
since the late 1970s. Economic growth,
which has been depressed by ethnic unrest,
accelerated in 1991-92 as domestic
conditions began to improve and conditions
for foreign investment brightened.
National product: GDP—exchange rate
conversion—$7.75 billion (1992 est.)
National product real growth rate: 4.5%
(1992 est.)
National product per capita: $440 (1992
est.)
Inflation rate (consumer prices): 10%
(1992)
Unemployment rate: 15% (1991 est.)
Budget: revenues $2.0 billion; expenditures
$3.7 billion, including capital expenditures
of $500 million (1992)
Exports: $2.0 billion (f.o.b., 1991)
commodities: textiles and garments, teas,
petroleum products, coconuts, rubber, other

agricultural products, gems and jewelry,
marine products, graphite
partners: US 27.4%, Germany, Japan, UK,
Belgium, Taiwan, Hong Kong, China
Imports: $3.1 billion (c.i.f., 1991)
commodities: food and beverages, textiles
and textile materials, petroleum and
petroleum products, machinery and
equipment
partners: Japan, Iran, US 5.7%, India,
Taiwan, Singapore, Germany, UK
External debt: $5.7 billion (1991 est.)
Industrial production: growth rate 7%
(1991 est.); accounts for 20% of GDP
Electricity: 1,300,000 kW capacity; 3,600
million kWh produced, 200 kWh per capita
(1992)
Industries: processing of rubber, tea,
coconuts, and other agricultural
commodities; cement, petroleum refining,
textiles, tobacco, clothing
Agriculture: accounts for 26% of GDP and
nearly half of labor force; most important
staple crop is paddy rice; other field
crops—sugarcane, grains, pulses, oilseeds,
roots, spices; cash crops—tea, rubber,
coconuts; animal products—milk, eggs,
hides, meat; not self-sufficient in rice
production
Economic aid: US commitments, including
Ex-Im (FY70-89), $1.0 billion; Western
(non-US) countries, ODA and OOF bilateral
commitments (1980-89), $5.1 billion; OPEC
bilateral aid (1979-89), $169 million;
Communist countries (1970-89), $369
million
Currency: 1 Sri Lankan rupee (SLRe) =
100 cents
Exchange rates: Sri Lankan rupees (SLRes)
per US$1—46.342 (January 1993), 43.687
(1992), 41.372 (1991), 40.063 (1990),
36.047 (1989), 31.807 (1988)
Fiscal year: calendar year

Communications

Railroads: 1,948 km total (1990); all
1.868-meter broad gauge; 102 km double
track; no electrification; government owned
Highways: 75,749 km total (1990); 27,637
km paved (mostly bituminous treated),
32,887 km crushed stone or gravel, 14,739
km improved earth or unimproved earth;
several thousand km of mostly unmotorable
tracks (1988 est.)
Inland waterways: 430 km; navigable by
shallow-draft craft
Pipelines: crude oil and petroleum products
62 km (1987)
Ports: Colombo, Trincomalee
Merchant marine: 27 ships (1,000 GRT or
over) totaling 276,074 GRT/443,266 DWT;
includes 12 cargo, 6 refrigerated cargo, 3
container, 3 oil tanker, 3 bulk

Sudan

Airports:
total: 14
usable: 13
with permanent-surface runways: 12
with runways over 3,659 m: 0
with runways 2,440-3,659 m: 1
with runways 1,220-2,439 m: 8
Telecommunications: very inadequate domestic service, good international service; 114,000 telephones (1982); broadcast stations—12 AM, 5 FM, 5 TV; submarine cables extend to Indonesia and Djibouti; 2 Indian Ocean INTELSAT earth stations

Defense Forces

Branches: Army, Navy, Air Force, Police Force
Manpower availability: males age 15-49 4,779,221; fit for military service 3,730,737; reach military age (18) annually 178,032 (1993 est.)
Defense expenditures: exchange rate conversion—$365 million, 4.7% of GDP (1992)

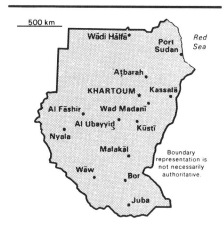

Geography

Location: Northern Africa, along the Red Sea, between Egypt and Ethiopia
Map references: Africa, Standard Time Zones of the World
Area:
total area: 2,505,810 km²
land area: 2.376 million km²
comparative area: slightly more than one-quarter the size of the US
Land boundaries: total 7,697 km, Central African Republic 1,165 km, Chad 1,360 km, Egypt 1,273 km, Ethiopia 2,221 km, Kenya 232 km, Libya 383 km, Uganda 435 km, Zaire 628 km
Coastline: 853 km
Maritime claims:
contiguous zone: 18 nm
continental shelf: 200 m depth or to depth of exploitation
territorial sea: 12 nm
International disputes: administrative boundary with Kenya does not coincide with international boundary; administrative boundary with Egypt does not coincide with international boundary creating the "Hala'ib Triangle," a barren area of 20,580 km², the dispute over this area escalated in 1993
Climate: tropical in south; arid desert in north; rainy season (April to October)
Terrain: generally flat, featureless plain; mountains in east and west
Natural resources: small reserves of petroleum, iron ore, copper, chromium ore, zinc, tungsten, mica, silver
Land use:
arable land: 5%
permanent crops: 0%
meadows and pastures: 24%
forest and woodland: 20%
other: 51%
Irrigated land: 18,900 km² (1989 est.)
Environment: dominated by the Nile and its tributaries; dust storms; desertification
Note: largest country in Africa

People

Population: 28,730,381 (July 1993 est.)
Population growth rate: 2.38% (1993 est.)
Birth rate: 42.65 births/1,000 population (1993 est.)
Death rate: 12.45 deaths/1,000 population (1993 est.)
Net migration rate: -6.4 migrant(s)/1,000 population (1993 est.)
Infant mortality rate: 81.3 deaths/1,000 live births (1993 est.)
Life expectancy at birth:
total population: 53.85 years
male: 53 years
female: 54.73 years (1993 est.)
Total fertility rate: 6.19 children born/woman (1993 est.)
Nationality:
noun: Sudanese (singular and plural)
adjective: Sudanese
Ethnic divisions: black 52%, Arab 39%, Beja 6%, foreigners 2%, other 1%
Religions: Sunni Muslim 70% (in north), indigenous beliefs 25%, Christian 5% (mostly in south and Khartoum)
Languages: Arabic (official), Nubian, Ta Bedawie, diverse dialects of Nilotic, Nilo-Hamitic, Sudanic languages, English
note: program of Arabization in process
Literacy: age 15 and over can read and write (1990)
total population: 27%
male: 43%
female: 12%
Labor force: 6.5 million
by occupation: agriculture 80%, industry and commerce 10%, government 6%
note: labor shortages for almost all categories of skilled employment (1983 est.); 52% of population of working age (1985)

Government

Names:
conventional long form: Republic of the Sudan
conventional short form: Sudan
local long form: Jumhuriyat as-Sudan
local short form: As-Sudan
former: Anglo-Egyptian Sudan
Digraph: SU
Type: military civilian government suspended and martial law imposed after 30 June 1989 coup
Capital: Khartoum
Administrative divisions: 9 states (wilayat, singular—wilayat or wilayah*); A'ali an Nil, Al Wusta*, Al Istiwa'iyah*, Al Khartum, Ash Shamaliyah*, Ash Sharqiyah*, Bahr al Ghazal, Darfur, Kurdufan
Independence: 1 January 1956 (from Egypt and UK)
Constitution: 12 April 1973, suspended following coup of 6 April 1985; interim constitution of 10 October 1985 suspended

Sudan (continued)

following coup of 30 June 1989

Legal system: based on English common law and Islamic law; as of 20 January 1991, the Revolutionary Command Council imposed Islamic law in the six northern states of Al Wusta, Al Khartum, Ash Shamaliyah, Ash Sharqiyah, Darfur, and Kurdufan; the council is still studying criminal provisions under Islamic law; Islamic law will apply to all residents of the six northern states regardless of their religion; some separate religious courts; accepts compulsory ICJ jurisdiction, with reservations

National holiday: Independence Day, 1 January (1956)

Political parties and leaders: none; banned following 30 June 1989 coup

Other political or pressure groups: National Islamic Front, Hasan al-TURABI

Suffrage: none

Elections: none

Executive branch: executive and legislative authority vested in a 10-member Revolutionary Command Council (RCC); chairman of the RCC acts as prime minister; in July 1989, RCC appointed a predominately civilian 22-member cabinet to function as advisers

note: Lt. Gen. BASHIR's military government is dominated by members of Sudan's National Islamic Front, a fundamentalist political organization formed from the Muslim Brotherhood in 1986; front leader Hasan al-TURABI controls Khartoum's overall domestic and foreign policies

Legislative branch: appointed 300-member Transitional National Assembly; note—as announced 1 January 1992 by RCC Chairman BASHIR, the Assembly assumes all legislative authority for Sudan until the eventual, unspecified resumption of national elections

Judicial branch: Supreme Court, Special Revolutionary Courts

Leaders:

Chief of State and Head of Government: Revolutionary Command Council Chairman and Prime Minister Lt. Gen. Umar Hasan Ahmad al-BASHIR (since 30 June 1989); Deputy Chairman of the Command Council and Deputy Prime Minister Maj. Gen. al-Zubayr Muhammad SALIH Ahmed (since 9 July 1989)

Member of: ABEDA, ACP, AfDB, AFESD, AL, AMF, CAEU, CCC, ECA, FAO, G-77, IAEA, IBRD, ICAO, IDA, IDB, IFAD, IFC, IGADD, ILO, IMF, IMO, INTELSAT, INTERPOL, IOC, ISO, ITU, LORCS, NAM, OAU, OIC, PCA, UN, UNCTAD, UNESCO, UNHCR, UNIDO, UPU, WFTU, WHO, WIPO, WMO, WTO

Diplomatic representation in US:

chief of mission: Ambassador 'Abdalla

Ahmad 'ABDALLA

chancery: 2210 Massachusetts Avenue NW, Washington, DC 20008

telephone: (202) 338-8565 through 8570

consulate general: New York

US diplomatic representation:

chief of mission: Ambassador Donald K. PETTERSON

embassy: Shar'ia Ali Abdul Latif, Khartoum

mailing address: P. O. Box 699, Khartoum, or APO AE 09829

telephone: 74700 or 74611

FAX: Telex 22619

Flag: three equal horizontal bands of red (top), white, and black with a green isosceles triangle based on the hoist side

---

Economy

Overview: Sudan is buffeted by civil war, chronic political instability, adverse weather, high inflation, a drop in remittances from abroad, and counterproductive economic policies. The economy is dominated by governmental entities that account for more than 70% of new investment. The private sector's main areas of activity are agriculture and trading, with most private industrial investment predating 1980. The economy's base is agriculture, which employs 80% of the work force. Industry mainly processes agricultural items. Sluggish economic performance over the past decade, attributable largely to declining annual rainfall, has reduced levels of per capita income and consumption. A large foreign debt and huge arrearages continue to cause difficulties. In 1990 the International Monetary Fund took the unusual step of declaring Sudan noncooperative because of its nonpayment of arrearages to the Fund. Despite subsequent government efforts to implement reforms urged by the IMF and the World Bank, the economy remained stagnant in FY91 as entrepreneurs lack the incentive to take economic risks. Growth in 1992 was featured by the recovery of agricultural production in northern Sudan after two years of drought.

National product: GDP—exchange rate conversion—$5.2 billion (FY92 est.)

National product real growth rate: 9% (FY92 est.)

National product per capita: $184 (FY92 est.)

Inflation rate (consumer prices): 150% (FY92 est.)

Unemployment rate: 30% (FY92 est.)

Budget: revenues $1.3 billion; expenditures $2.1 billion, including capital expenditures of $505 million (FY91 est.)

Exports: $315 million (f.o.b., FY92 est.)

commodities: cotton 52%, sesame, gum arabic, peanuts

partners: Western Europe 46%, Saudi Arabia 14%, Eastern Europe 9%, Japan 9%, US 3% (FY88)

Imports: $1.3 billion (c.i.f., FY92 est.)

commodities: foodstuffs, petroleum products, manufactured goods, machinery and equipment, medicines and chemicals, textiles

partners: Western Europe 32%, Africa and Asia 15%, US 13%, Eastern Europe 3% (FY88)

External debt: $15.0 billion (June 1992 est.)

Industrial production: growth rate 4.8%; accounts for 11% of GDP (FY92)

Electricity: 610,000 kW capacity; 905 million kWh produced, 40 kWh per capita (1991)

Industries: cotton ginning, textiles, cement, edible oils, sugar, soap distilling, shoes, petroleum refining

Agriculture: accounts for 35% of GDP and 80% of labor force; water shortages; two-thirds of land area suitable for raising crops and livestock; major products—cotton, oilseeds, sorghum, millet, wheat, gum arabic, sheep; marginally self-sufficient in most foods

Economic aid: US commitments, including Ex-Im (FY70-89), $1.5 billion; Western (non-US) countries, ODA and OOF bilateral commitments (1970-89), $5.1 billion; OPEC bilateral aid (1979-89), $3.1 billion; Communist countries (1970-89), $588 million

Currency: 1 Sudanese pound (£Sd) = 100 piasters

Exchange rates: official rate—Sudanese pounds (£Sd) per US$1—124 (January 1993), 90.1 (March 1992), 5.4288 (1991), 4.5004 (fixed rate since 1987), 2.8121 (1987); note—free market rate 155 (January 1993)

Fiscal year: 1 July-30 June

Communications

Railroads: 5,516 km total; 4,800 km 1.067-meter gauge, 716 km 1.6096-meter-gauge plantation line

Highways: 20,703 km total; 2,000 km bituminous treated, 4,000 km gravel, 2,304 km improved earth, 12,399 km unimproved earth and track

Inland waterways: 5,310 km navigable

Pipelines: refined products 815 km

Ports: Port Sudan, Sawakin

Merchant marine: 5 ships (1,000 GRT or over) totaling 42,277 GRT/59,588 DWT; includes 3 cargo, 2 roll-on/roll-off

Airports:

total: 68

usable: 56

with permanent-surface runways: 10

with runways over 3,659 m: 0

with runways 2,440-3,659 m: 6

with runways 1,220-2,439 m: 30

Suriname

Telecommunications: large, well-equipped system by African standards, but barely adequate and poorly maintained by modern standards; consists of microwave radio relay, cable, radio communications, troposcatter, and a domestic satellite system with 14 stations; broadcast stations—11 AM, 3 TV; satellite earth stations for international traffic—1 Atlantic Ocean INTELSAT and 1 ARABSAT

Defense Forces

Branches: Army, Navy, Air Force, Air Defense Force
Manpower availability: males age 15-49 6,488,864; fit for military service 3,986,084; reach military age (18) annually 301,573 (1993 est.)
Defense expenditures: exchange rate conversion—$339 million, 2.2% of GDP (1989 est.)

Geography

Location: Northern South America, bordering the North Atlantic Ocean between French Guiana and Guyana
Map references: South America, Standard Time Zones of the World
Area:
total area: 163,270 km²
land area: 161,470 km²
comparative area: slightly larger than Georgia
Land boundaries: total 1,707 km, Brazil 597 km, French Guiana 510 km, Guyana 600 km
Coastline: 386 km
Maritime claims:
exclusive economic zone: 200 nm
territorial sea: 12 nm
International disputes: claims area in French Guiana between Litani Rivier and Riviere Marouini (both headwaters of the Lawa); claims area in Guyana between New (Upper Courantyne) and Courantyne/Koetari Rivers (all headwaters of the Courantyne)
Climate: tropical; moderated by trade winds
Terrain: mostly rolling hills; narrow coastal plain with swamps
Natural resources: timber, hydropower potential, fish, shrimp, bauxite, iron ore, and small amounts of nickel, copper, platinum, gold
Land use:
arable land: 0%
permanent crops: 0%
meadows and pastures: 0%
forest and woodland: 97%
other: 3%
Irrigated land: 590 km² (1989 est.)
Environment: mostly tropical rain forest

People

Population: 416,321 (July 1993 est.)
Population growth rate: 1.54% (1993 est.)
Birth rate: 25.85 births/1,000 population (1993 est.)

Death rate: 6.1 deaths/1,000 population (1993 est.)
Net migration rate: -4.33 migrant(s)/1,000 population (1993 est.)
Infant mortality rate: 32.4 deaths/1,000 live births (1993 est.)
Life expectancy at birth:
total population: 69.14 years
male: 66.65 years
female: 71.76 years (1993 est.)
Total fertility rate: 2.85 children born/woman (1993 est.)
Nationality:
noun: Surinamer(s)
adjective: Surinamese
Ethnic divisions: Hindustani (East Indian) 37%, Creole (black and mixed) 31%, Javanese 15.3%, Bush black 10.3%, Amerindian 2.6%, Chinese 1.7%, Europeans 1%, other 1.1%
Religions: Hindu 27.4%, Muslim 19.6%, Roman Catholic 22.8%, Protestant 25.2% (predominantly Moravian), indigenous beliefs 5%
Languages: Dutch (official), English widely spoken, Sranan Tongo (Surinamese, sometimes called Taki-Taki) is native language of Creoles and much of the younger population and is lingua franca among others, Hindi Suriname Hindustani (a variant of Bhoqpuri), Javanese
Literacy: age 15 and over can read and write (1990)
total population: 95%
male: 95%
female: 95%
Labor force: 104,000 (1984)
by occupation: NA

Government

Names:
conventional long form: Republic of Suriname
conventional short form: Suriname
local long form: Republiek Suriname
local short form: Suriname
former: Netherlands Guiana Dutch Guiana
Digraph: NS
Type: republic
Capital: Paramaribo
Administrative divisions: 10 districts (distrikten, singular—distrikt); Brokopondo, Commewijne, Coronie, Marowijne, Nickerie, Para, Paramaribo, Saramacca, Sipaliwini, Wanica
Independence: 25 November 1975 (from Netherlands)
Constitution: ratified 30 September 1987
Legal system: NA
National holiday: Independence Day, 25 November (1975)
Political parties and leaders: The New Front (NF), leader NA, a coalition of four parties (NPS, VHP, KTPI, SPA); Progressive Reform Party (VHP), Jaggernath LACHMON;

National Party of Suriname (NPS), Ronald VENETIAAN; Party of National Unity and Solidarity (KTPI), Willy SOEMITA; Suriname Labor Party (SPA) Fred DARBY; Democratic Alternative '91 (DA '91), Winston JESSURUN, a coalition of four parties (AF, HPP, Pendawa Lima, BEP) formed in January 1991; Alternative Forum (AF), Gerard BRUNINGS, Winston JESSURUN; Reformed Progressive Party (HPP), Panalal PARMESSAR; Party for Brotherhood and Unity in Politics (BEP), Cipriano ALLENDY; Pendawa Lima, Marsha JAMIN; National Democratic Party (NDP), Desire BOUTERSE; Progressive Workers' and Farm Laborers' Union (PALU), Ir Iwan KROLIS, chairman; National Republic Party (PNR), Robin RAVALES

Other political or pressure groups: Surinamese Liberation Army (SLA), Ronnie BRUNSWIJK, Johan "Castro" WALLY; Union for Liberation and Democracy, Kofi AFONGPONG; Saramaccaner Bosneger Angula Movement, Carlos MAASSI; Mandela Bushnegro Liberation Movement, Leendert ADAMS; Tucayana Amazonica, Alex JUBITANA, Thomas SABAJO

Suffrage: 18 years of age; universal

Elections:
President: last held 6 September 1991 (next to be held NA May 1996); results—elected by the National Assembly—Ronald VENETIAAN (NF) 80% (645 votes), Jules WIJDENBOSCH (NDP) 14% (115 votes), Hans PRADE (DA '91) 6% (49 votes)
National Assembly: last held 25 May 1991 (next to be held NA May 1996); results—percent of vote NA; seats—(51 total) NF 30, NDP 10, DA '91 9, Independent 2

Executive branch: president, vice president and prime minister, Cabinet of Ministers, Council of State; note—Commander in Chief of the National Army maintains significant power

Legislative branch: unicameral National Assembly (Assemblee Nationale)

Judicial branch: Supreme Court

Leaders:
Chief of State and Head of Government: President Ronald R. VENETIAAN (since 16 September 1991); Vice President and Prime Minister Jules R. AJODHIA (since 16 September 1991)

Member of: ACP, CARICOM (observer), ECLAC, FAO, GATT, G-77, IADB, IBRD, ICAO, ICFTU, IFAD, ILO, IMF, IMO, INTERPOL, IOC, ITU, LAES, LORCS, NAM, OAS, OPANAL, UN, UNCTAD, UNESCO, UNIDO, UPU, WCL, WHO, WIPO, WMO

Diplomatic representation in US:
chief of mission: Ambassador Willem A. UDENHOUT

chancery: Suite 108, 4301 Connecticut Avenue NW, Washington, DC 20008
telephone: (202) 244-7488 or 7490 through 7492
consulate general: Miami

US diplomatic representation:
chief of mission: Ambassador John (Jack) P. LEONARD
embassy: Dr. Sophie Redmonstraat 129, Paramaribo
mailing address: P. O. Box 1821, Paramaribo
telephone: [597] 472900, 477881, or 476459
FAX: [597] 410025

Flag: five horizontal bands of green (top, double width), white, red (quadruple width), white, and green (double width); there is a large yellow five-pointed star centered in the red band

Economy

Overview: The economy is dominated by the bauxite industry, which accounts for 15% of GDP and about 70% of export earnings. The economy has been in trouble since the Dutch ended development aid in 1982. A drop in world bauxite prices which started in the late 1970s and continued until late 1986 was followed by the outbreak of a guerrilla insurgency in the interior that crippled the important bauxite sector. Although the insurgency has since ebbed and the bauxite sector recovered, a military coup in December 1990 reflected continued political instability and deterred investment and economic reform. High inflation, high unemployment, widespread black market activity, and hard currency shortfalls continue to mark the economy.

National product: GDP—exchange rate conversion—$1.35 billion (1991 est.)

National product real growth rate: -2.5% (1991 est.)

National product per capita: $3,300 (1991 est.)

Inflation rate (consumer prices): 26% (1991)

Unemployment rate: 16.5% (1990)

Budget: revenues $466 million; expenditures $716 million, including capital expenditures of $123 million (1989 est.)

Exports: $417 million (f.o.b., 1992 est.)
commodities: alumina, aluminum, shrimp and fish, rice, bananas
partners: Norway 36%, Netherlands 28%, US 11%, Japan 7%, Brazil 5%, UK 5% (1989)

Imports: $514 million (f.o.b., 1992 est.)
commodities: capital equipment, petroleum, foodstuffs, cotton, consumer goods
partners: US 41%, Netherlands 24%, Trinidad and Tobago 9%, Brazil 4% (1989)

External debt: $138 million (1990 est.)

Industrial production: growth rate −5.0%

(1991 est.); accounts for 27% of GDP

Electricity: 458,000 kW capacity; 2,018 million kWh produced, 4,920 kWh per capita (1992)

Industries: bauxite mining, alumina and aluminum production, lumbering, food processing, fishing

Agriculture: accounts for 10.4% of GDP and 25% of export earnings; paddy rice planted on 85% of arable land and represents 60% of total farm output; other products—bananas, palm kernels, coconuts, plantains, peanuts, beef, chicken; shrimp and forestry products of increasing importance; self-sufficient in most foods

Economic aid: US commitments, including Ex-Im (FY70-83), $2.5 million; Western (non-US) countries, ODA and OOF bilateral commitments (1970-89), $1.5 billion

Currency: 1 Surinamese guilder, gulden, or florin (Sf.) = 100 cents

Exchange rates: Surinamese guilders, gulden, or florins (Sf.) per US$1—1.7850 (fixed rate until October 1992), 25.04 (January 1992)

Fiscal year: calendar year

Communications

Railroads: 166 km total; 86 km 1.000-meter gauge, government owned, and 80 km 1.435-meter standard gauge; all single track

Highways: 8,300 km total; 500 km paved; 5,400 km bauxite gravel, crushed stone, or improved earth; 2,400 km sand or clay

Inland waterways: 1,200 km; most important means of transport; oceangoing vessels with drafts ranging up to 7 m can navigate many of the principal waterways

Ports: Paramaribo, Moengo, Nieuw Nickerie

Merchant marine: 3 ships (1,000 GRT or over) totaling 6,472 GRT/8,914 DWT; includes 2 cargo, 1 container

Airports:
total: 46
usable: 39
with permanent-surface runways: 6
with runways over 3,659 m: 0
with runways 2,440-3,659 m: 1
with runways 1,220-2,439 m: 3

Telecommunications: international facilities good; domestic microwave system; 27,500 telephones; broadcast stations—5 AM, 14 FM, 6 TV, 1 shortwave; 2 Atlantic Ocean INTELSAT earth stations

Defense Forces

Branches: National Army (including Navy which is company-size, small Air Force element), Civil Police

Manpower availability: males age 15-49 111,716; fit for military service 66,429 (1993 est.)

Defense expenditures: exchange rate conversion—$NA, NA% of GDP

Svalbard
(territory of Norway)

Geography

Location: in the Arctic Ocean where the Arctic Ocean, Barents Sea, Greenland Sea, and Norwegian Sea meet, 445 km north of Norway
Map references: Arctic Region, Asia, Standard Time Zones of the World
Area:
total area: 62,049 km²
land area: 62,049 km²
comparative area: slightly smaller than West Virginia
note: includes Spitsbergen and Bjornoya (Bear Island)
Land boundaries: 0 km
Coastline: 3,587 km
Maritime claims:
exclusive fishing zone: 200 nm unilaterally claimed by Norway but not recognized by Russia
territorial sea: 4 nm
International disputes: focus of maritime boundary dispute in the Barents Sea between Norway and Russia
Climate: arctic, tempered by warm North Atlantic Current; cool summers, cold winters; North Atlantic Current flows along west and north coasts of Spitsbergen, keeping water open and navigable most of the year
Terrain: wild, rugged mountains; much of high land ice covered; west coast clear of ice about half the year; fjords along west and north coasts
Natural resources: coal, copper, iron ore, phosphate, zinc, wildlife, fish
Land use:
arable land: 0%
permanent crops: 0%
meadows and pastures: 0%
forest and woodland: 0%
other: 100% (no trees and the only bushes are crowberry and cloudberry)
Irrigated land: NA km²
Environment: great calving glaciers descend to the sea

Note: northernmost part of the Kingdom of Norway; consists of nine main islands; glaciers and snowfields cover 60% of the total area

People

Population: 3,209 (July 1993 est.)
Population growth rate: -2.84% (1993 est.)
Birth rate: NA births/1,000 population
Death rate: NA deaths/1,000 population
Net migration rate: NA migrant(s)/1,000 population
Infant mortality rate: NA deaths/1,000 live births
Life expectancy at birth:
total population: NA years
male: NA years
female: NA years
Total fertility rate: NA children born/woman
Ethnic divisions: Russian 64%, Norwegian 35%, other 1% (1981)
Languages: Russian, Norwegian
Literacy:
total population: NA%
male: NA%
female: NA%
Labor force: NA

Government

Names:
conventional long form: none
conventional short form: Svalbard
Digraph: SV
Type: territory of Norway administered by the Ministry of Industry, Oslo, through a governor (sysselmann) residing in Longyearbyen, Spitsbergen; by treaty (9 February 1920) sovereignty was given to Norway
Capital: Longyearbyen
Independence: none (territory of Norway)
Legal system: NA
National holiday: NA
Leaders:
Chief of State: King HARALD V (since 17 January 1991)
Head of Government: Governor (vacant)
Member of: none
Flag: the flag of Norway is used

Economy

Overview: Coal mining is the major economic activity on Svalbard. By treaty (9 February 1920), the nationals of the treaty powers have equal rights to exploit mineral deposits, subject to Norwegian regulation. Although US, UK, Dutch, and Swedish coal companies have mined in the past, the only companies still mining are Norwegian and Russian. The settlements on Svalbard are essentially company towns. The Norwegian

state-owned coal company employs nearly 60% of the Norwegian population on the island, runs many of the local services, and provides most of the local infrastructure. There is also some trapping of seal, polar bear, fox, and walrus.
Budget: revenues $13.3 million; expenditures $13.3 million, including capital expenditures of $NA (1990)
Electricity: 21,000 kW capacity; 45 million kWh produced, 13,860 kWh per capita (1992)
Currency: 1 Norwegian krone (NKr) = 100 ore
Exchange rates: Norwegian kroner (NKr) per US$1—6.8774 (January 1993), 6.2145 (1992), 6.4829 (1991), 6.2597 (1990), 6.9045 (1989), 6.5170 (1988)

Communications

Ports: limited facilities—Ny-Alesund, Advent Bay
Airports:
total: 4
usable: 4
with permanent-surface runways: 1
with runways over 3,659 m: 0
with runways 2,440-3,659 m: 0
with runways 1,220-2,439 m: 1
Telecommunications: 5 meteorological/radio stations; local telephone service; broadcast stations—1 AM, 1 (2 repeaters) FM, 1 TV; satellite communication with Norwegian mainland

Defense Forces

Note: demilitarized by treaty (9 February 1920)

Swaziland

Geography

Location: Southern Africa, between Mozambique and South Africa
Map references: Africa, Standard Time Zones of the World
Area:
total area: 17,360 km²
land area: 17,200 km²
comparative area: slightly smaller than New Jersey
Land boundaries: total 535 km, Mozambique 105 km, South Africa 430 km
Coastline: 0 km (landlocked)
Maritime claims: none; landlocked
International disputes: none
Climate: varies from tropical to near temperate
Terrain: mostly mountains and hills; some moderately sloping plains
Natural resources: asbestos, coal, clay, cassiterite, hydropower, forests, small gold and diamond deposits, quarry stone, and talc
Land use:
arable land: 8%
permanent crops: 0%
meadows and pastures: 67%
forest and woodland: 6%
other: 19%
Irrigated land: 620 km² (1989 est.)
Environment: overgrazing; soil degradation; soil erosion
Note: landlocked; almost completely surrounded by South Africa

People

Population: 906,932 (July 1993 est.)
Population growth rate: 3.18% (1993 est.)
Birth rate: 43.22 births/1,000 population (1993 est.)
Death rate: 11.41 deaths/1,000 population (1993 est.)
Net migration rate: 0 migrant(s)/1,000 population (1993 est.)
Infant mortality rate: 95.7 deaths/1,000 live births (1993 est.)

Life expectancy at birth:
total population: 55.94 years
male: 51.97 years
female: 60.03 years (1993 est.)
Total fertility rate: 6.16 children born/woman (1993 est.)
Nationality:
noun: Swazi(s)
adjective: Swazi
Ethnic divisions: African 97%, European 3%
Religions: Christian 60%, indigenous beliefs 40%
Languages: English (official; government business conducted in English), siSwati (official)
Literacy: age 15 and over can read and write (1976)
total population: 55%
male: 57%
female: 54%
Labor force: 195,000 (over 60,000 engaged in subsistence agriculture; about 92,000 wage earners—many only intermittently)
by occupation: agriculture and forestry 36%, community and social service 20%, manufacturing 14%, construction 9%, other 21%
note: 15,980 employed in South African gold and coal mines (1991)

Government

Names:
conventional long form: Kingdom of Swaziland
conventional short form: Swaziland
Digraph: WZ
Type: monarchy independent member of Commonwealth
Capital: Mbabane (administrative); Lobamba (legislative)
Administrative divisions: 4 districts; Hhohho, Lubombo, Manzini, Shiselweni
Independence: 6 September 1968 (from UK)
Constitution: none; constitution of 6 September 1968 was suspended on 12 April 1973; a new constitution was promulgated 13 October 1978, but has not been formally presented to the people
Legal system: based on South African Roman-Dutch law in statutory courts, Swazi traditional law and custom in traditional courts; has not accepted compulsory ICJ jurisdiction
National holiday: Somhlolo (Independence) Day, 6 September (1968)
Political parties and leaders: none; banned by the Constitution promulgated on 13 October 1978
Suffrage: none
Elections: direct legislative elections rescheduled for June 1993
Executive branch: monarch, prime minister, Cabinet

Legislative branch: bicameral Parliament is advisory and consists of an upper house or Senate and a lower house or House of Assembly
Judicial branch: High Court, Court of Appeal
Leaders:
Chief of State: King MSWATI III (since 25 April 1986)
Head of Government: Prime Minister Obed Mfanyana DLAMINI (since 12 July 1989)
Member of: ACP, AfDB, C, CCC, ECA, FAO, G-77, IBRD, ICAO, ICFTU, IDA, IFAD, IFC, ILO, IMF, INTELSAT, INTERPOL, IOC, ITU, LORCS, NAM, OAU, PCA, SACU, SADC, UN, UNCTAD, UNESCO, UNIDO, UPU, WHO, WIPO, WMO
Diplomatic representation in US:
chief of mission: Ambassador Absalom Vusani MAMBA
chancery: 3400 International Drive NW, Washington, DC 20008
telephone: (202) 362-6683
US diplomatic representation:
chief of mission: Ambassador Stephen H. ROGERS
embassy: Central Bank Building, Warner Street, Mbabane
mailing address: P. O. Box 199, Mbabane
telephone: [268] 46441 through 46445
FAX: [268] 45959
Flag: three horizontal bands of blue (top), red (triple width), and blue; the red band is edged in yellow; centered in the red band is a large black and white shield covering two spears and a staff decorated with feather tassels, all placed horizontally

Economy

Overview: The economy is based on subsistence agriculture, which occupies most of the labor force and contributes nearly 25% to GDP. Manufacturing, which includes a number of agroprocessing factories, accounts for another quarter of GDP. Mining has declined in importance in recent years; high-grade iron ore deposits were depleted in 1978, and health concerns cut world demand for asbestos. Exports of sugar and forestry products are the main earners of hard currency. Surrounded by South Africa, except for a short border with Mozambique, Swaziland is heavily dependent on South Africa, from which it receives 75% of its imports and to which it sends about half of its exports.
National product: GDP—exchange rate conversion—$700 million (1991 est.)
National product real growth rate: 2.5% (1991 est.)
National product per capita: $800 (1991 est.)
Inflation rate (consumer prices): 13% (1991 est.)

Unemployment rate: NA%
Budget: revenues $342 million; expenditures $410 million, including capital expenditures of $130 million (FY94 est.)
Exports: $575 million (f.o.b., 1991)
commodities: soft drink concentrates, sugar, wood pulp, citrus, canned fruit
partners: South Africa 50% (est.), EC countries, Canada
Imports: $730 million (c.i.f., 1991)
commodities: motor vehicles, machinery, transport equipment, petroleum products, foodstuffs, chemicals
partners: South Africa 75% (est.), Japan, Belgium, UK
External debt: $290 million (1990)
Industrial production: growth rate NA%; accounts for 26% of GDP (1989)
Electricity: 60,000 kW capacity; 155 million kWh produced, 180 kWh per capita (1991)
Industries: mining (coal and asbestos), wood pulp, sugar
Agriculture: accounts for 23% of GDP and over 60% of labor force; mostly subsistence agriculture; cash crops—sugarcane, cotton, maize, tobacco, rice, citrus fruit, pineapples; other crops and livestock—corn, sorghum, peanuts, cattle, goats, sheep; not self-sufficient in grain
Economic aid: US commitments, including Ex-Im (FY70-89), $142 million; Western (non-US) countries, ODA and OOF bilateral commitments (1970-89), $518 million
Currency: 1 lilangeni (E) = 100 cents
Exchange rates: emalangeni (E) per US$1—3.1576 (May 1993), 2.8497 (1992), 2.7563 (1991), 2.5863 (1990), 2.6166 (1989), 2.2611 (1988); note—the Swazi emalangeni is at par with the South African rand
Fiscal year: 1 April-31 March

Communications

Railroads: 297 km (plus 71 km disused), 1.067-meter gauge, single track
Highways: 2,853 km total; 510 km paved, 1,230 km crushed stone, gravel, or stabilized soil, and 1,113 km improved earth
Airports:
total: 23
usable: 21
with permanent-surfaced runways: 1
with runways over 3,659 m: 0
with runways 2,440-3,659 m: 1
with runways 1,220-2,439 m: 1
Telecommunications: system consists of carrier-equipped open-wire lines and low-capacity microwave links; 17,000 telephones; broadcast stations—7 AM, 6 FM, 10 TV; 1 Atlantic Ocean INTELSAT earth station

Defense Forces

Branches: Umbutfo Swaziland Defense Force, Royal Swaziland Police Force
Manpower availability: males age 15-49 197,214; fit for military service 114,097 (1993 est.)
Defense expenditures: exchange rate conversion—$22 million, NA% of GDP (FY93/94)

Sweden

Geography

Location: Northern Europe, bordering the Baltic Sea, between Norway and Finland
Map references: Arctic Region, Asia, Europe, Standard Time Zones of the World
Area:
total area: 449,964 km^2
land area: 410,928 km^2
comparative area: slightly smaller than California
Land boundaries: total 2,205 km, Finland 586 km, Norway 1,619 km
Coastline: 3,218 km
Maritime claims:
continental shelf: 200 m depth or to depth of exploitation
exclusive fishing zone: 200 nm
territorial sea: 12 nm
International disputes: none
Climate: temperate in south with cold, cloudy winters and cool, partly cloudy summers; subarctic in north
Terrain: mostly flat or gently rolling lowlands; mountains in west
Natural resources: zinc, iron ore, lead, copper, silver, timber, uranium, hydropower potential
Land use:
arable land: 7%
permanent crops: 0%
meadows and pastures: 2%
forest and woodland: 64%
other: 27%
Irrigated land: 1,120 km^2 (1989 est.)
Environment: water pollution; acid rain
Note: strategic location along Danish Straits linking Baltic and North Seas

People

Population: 8,730,286 (July 1993 est.)
Population growth rate: 0.58% (1993 est.)
Birth rate: 13.78 births/1,000 population (1993 est.)
Death rate: 10.96 deaths/1,000 population (1993 est.)

Sweden (continued)

Net migration rate: 2.97 migrant(s)/1,000 population (1993 est.)
Infant mortality rate: 5.8 deaths/1,000 live births (1993 est.)
Life expectancy at birth:
total population: 78.08 years
male: 75.3 years
female: 81.02 years (1993 est.)
Total fertility rate: 2.04 children born/woman (1993 est.)
Nationality:
noun: Swede(s)
adjective: Swedish
Ethnic divisions: white, Lapp, foreign born or first-generation immigrants 12% (Finns, Yugoslavs, Danes, Norwegians, Greeks, Turks)
Religions: Evangelical Lutheran 94%, Roman Catholic 1.5%, Pentecostal 1%, other 3.5% (1987)
Languages: Swedish
note: small Lapp- and Finnish-speaking minorities; immigrants speak native languages
Literacy: age 15 and over can read and write (1979)
total population: 99%
male: NA%
female: NA%
Labor force: 4.552 million
by occupation: community, social and personal services 38.3%, mining and manufacturing 21.2%, commerce, hotels, and restaurants 14.1%, banking, insurance 9.0%, communications 7.2%, construction 7.0%, agriculture, fishing, and forestry 3.2% (1991)

Government

Names:
conventional long form: Kingdom of Sweden
conventional short form: Sweden
local long form: Konungariket Sverige
local short form: Sverige
Digraph: SW
Type: constitutional monarchy
Capital: Stockholm
Administrative divisions: 24 provinces (lan, singular and plural); Alvsborgs Lan, Blekinge Lan, Gavleborgs Lan, Goteborgs och Bohus Lan, Gotlands Lan, Hallands Lan, Jamtlands Lan, Jonkopings Lan, Kalmar Lan, Kopparbergs Lan, Kristianstads Lan, Kronobergs Lan, Malmohus Lan, Norrbottens Lan, Orebro Lan, Ostergotlands Lan, Skaraborgs Lan, Sodermanlands Lan, Stockholms Lan, Uppsala Lan, Varmlands Lan, Vasterbottens Lan, Vasternorrlands Lan, Vastmanlands Lan
Independence: 6 June 1809 (constitutional monarchy established)
Constitution: 1 January 1975

Legal system: civil law system influenced by customary law; accepts compulsory ICJ jurisdiction, with reservations
National holiday: Day of the Swedish Flag, 6 June
Political parties and leaders: ruling four-party coalition consists of Moderate Party (conservative), Carl BILDT; Liberal People's Party, Bengt WESTERBERG; Center Party, Olof JOHANSSON; and the Christian Democratic Party, Alf SVENSSON; Social Democratic Party, Ingvar CARLSSON; New Democracy Party, Count Ian WACHTMEISTER; Left Party (VP; Communist), Gudrun SCHYMAN; Communist Workers' Party, Rolf HAGEL; Green Party, no formal leader
Suffrage: 18 years of age; universal
Elections:
Riksdag: last held 15 September 1991 (next to be held NA September 1994); results—Social Democratic Party 37.6%, Moderate Party (conservative) 21.9%, Liberal People's Party 9.1%, Center Party 8.5%, Christian Democrats 7.1%, New Democracy 6.7%, Left Party (Communist) 4.5%, Green Party 3.4%, other 1.2%; seats—(349 total) Social Democratic 138, Moderate Party (conservative) 80, Liberal People's Party 33, Center Party 31, Christian Democrats 26, New Democracy 25, Left Party (Communist) 16; note—the Green Party has no seats in the Riksdag because it received less than the required 4% of the vote
Executive branch: monarch, prime minister, Cabinet
Legislative branch: unicameral parliament (Riksdag)
Judicial branch: Supreme Court (Hogsta Domstolen)
Leaders:
Chief of State: King CARL XVI GUSTAF (since 19 September 1973); Heir Apparent Princess VICTORIA Ingrid Alice Desiree, daughter of the King (born 14 July 1977)
Head of Government: Prime Minister Carl BILDT (since 3 October 1991); Deputy Prime Minister Bengt WESTERBERG (since NA)
Member of: AfDB, AG (observer), AsDB, Australian Group, BIS, CBSS, CCC, CE, CERN, COCOM (cooperating country), CSCE, EBRD, ECE, EFTA, ESA, FAO, G-6, G-8, G-9, G-10, GATT, IADB, IAEA, IBRD, ICAO, ICC, ICFTU, IDA, IEA, IFAD, IFC, ILO, IMF, IMO, INMARSAT, INTERPOL, INTELSAT, IOC, IOM, ISO, ITU, LORCS, MTRC, NAM (guest), NC, NEA, NIB, NSG, OECD, ONUSAL, PCA, UN, UNAVEM II, UNCTAD, UNESCO, UNFICYP, UNHCR, UNIDO, UNIFIL, UNIKOM, UNMOGIP, UNOMOZ, UNPROFOR, UNTSO, UPU, WHO, WIPO, WMO, ZC

Diplomatic representation in US:
chief of mission: Ambassador Carl Henrik LILJEGREN
chancery: Suite 1200 and 715, 600 New Hampshire Avenue NW, Washington, DC 20037
telephone: (202) 944-5600
FAX: (202) 342-1319
consulates general: Chicago, Los Angeles, and New York
US diplomatic representation:
chief of mission: (vacant)
embassy: Strandvagen 101, S-115 89 Stockholm
mailing address: use embassy street address
telephone: [46] (8) 783-5300
FAX: [46] (8) 661-1964
Flag: blue with a yellow cross that extends to the edges of the flag; the vertical part of the cross is shifted to the hoist side in the style of the Dannebrog (Danish flag)

Economy

Overview: Aided by a long period of peace and neutrality during World War I through World War II, Sweden has achieved an enviable standard of living under a mixed system of high-tech capitalism and extensive welfare benefits. It has a modern distribution system, excellent internal and external communications, and a skilled labor force. Timber, hydropower, and iron ore constitute the resource base of an economy that is heavily oriented toward foreign trade. Privately owned firms account for about 90% of industrial output, of which the engineering sector accounts for 50% of output and exports. In the last few years, however, this extraordinarily favorable picture has been clouded by inflation, growing unemployment, and a gradual loss of competitiveness in international markets. Although Prime Minister BILDT'S center-right minority coalition had hoped to charge ahead with free-market-oriented reforms, a skyrocketing budget deficit—almost 13% of GDP in FY94 projections—and record unemployment have forestalled many of the plans. Unemployment in 1993 is forecast at around 7% with another 5% in job training. Continued heavy foreign exchange speculation forced the government to cooperate in late 1992 with the opposition Social Democrats on two crisis packages—one a severe austerity pact and the other a program to spur industrial competitiveness—which basically set economic policy through 1997. In November 1992, Sweden broke its tie to the EC's ECU, and the krona has since depreciated around 2.5% against the dollar. The government hopes the boost in export competitiveness from the depreciation will help lift Sweden out of its 3-year recession. To curb the budget deficit and bolster confidence in the economy, BILDT continues to propose cuts in welfare benefits, subsidies, defense, and foreign aid. Sweden continues to harmonize its economic policies with those of the EC in preparation for concluding its EC membership bid by 1995.

National product: GDP—purchasing power equivalent—$145.6 billion (1992)
National product real growth rate: -1.7% (1992)
National product per capita: $16,900 (1992)
Inflation rate (consumer prices): 2.3% (1992)
Unemployment rate: 5.3% (1992)
Budget: revenues $70.4 billion; expenditures $82.5 billion, including capital expenditures of $NA (FY92)
Exports: $56 billion (f.o.b., 1992)
commodities: machinery, motor vehicles, paper products, pulp and wood, iron and steel products, chemicals, petroleum and petroleum products
partners: EC 55.8% (Germany 15%, UK 9.7%, Denmark 7.2%, France 5.8%), EFTA 17.4% (Norway 8.4%, Finland 5.1%), US 8.2%, Central and Eastern Europe 2.5% (1992)
Imports: $51.7 billion (c.i.f., 1992)
commodities: machinery, petroleum and petroleum products, chemicals, motor vehicles, foodstuffs, iron and steel, clothing
partners: EC 53.6% (Germany 17.9%, UK 6.3%, Denmark 7.5%, France 4.9%), EFTA (Norway 6.6%, Finland 6%), US 8.4%, Central and Eastern Europe 3% (1992)
External debt: $19.5 billion (1992 est.)
Industrial production: growth rate −3.0% (1992)
Electricity: 39,716,000 kW capacity; 142,500 million kWh produced, 16,560 kWh per capita (1992)
Industries: iron and steel, precision equipment (bearings, radio and telephone parts, armaments), wood pulp and paper products, processed foods, motor vehicles
Agriculture: animal husbandry predominates, with milk and dairy products accounting for 37% of farm income; main crops—grains, sugar beets, potatoes; 100% self-sufficient in grains and potatoes; Sweden is about 50% self-sufficient in most products; farming accounted for 1.2% of GDP and 1.9% of jobs in 1990
Illicit drugs: increasingly used as transshipment point for Latin American cocaine to Europe and gateway for Asian heroin shipped via the CIS and Baltic states for the European market
Economic aid: donor—ODA and OOF commitments (1970-89), $10.3 billion
Currency: 1 Swedish krona (SKr) = 100 ore

Exchange rates: Swedish kronor (SKr) per US$1—6.8812 (December 1992), 5.8238 (1992), 6.0475 (1991) 5.9188 (1990), 6.4469 (1989), 6.1272 (1988)
Fiscal year: 1 July-30 June

Communications

Railroads: 12,000 km total; Swedish State Railways (SJ)—10,819 km 1.435-meter standard gauge, 6,955 km electrified and 1,152 km double track; 182 km 0.891-meter gauge; 117 km rail ferry service; privately-owned railways—511 km 1.435-meter standard gauge (332 km electrified) and 371 km 0.891-meter gauge (all electrified)
Highways: 97,400 km total; 51,899 km paved, 20,659 km gravel, 24,842 km unimproved earth
Inland waterways: 2,052 km navigable for small steamers and barges
Pipelines: natural gas 84 km
Ports: Gavle, Goteborg, Halmstad, Helsingborg, Kalmar, Malmo, Stockholm; numerous secondary and minor ports
Merchant marine: 179 ships (1,000 GRT or over) totaling 2,473,769 GRT/3,227,366 DWT; includes 10 short-sea passenger, 29 cargo, 3 container, 43 roll-on/roll-off cargo, 13 vehicle carrier, 2 railcar carrier, 32 oil tanker, 27 chemical tanker, 4 specialized tanker, 2 liquefied gas, 2 combination ore/oil, 10 bulk, 1 combination bulk, 1 refrigerated cargo
Airports:
total: 253
usable: 250
with permanent-surface runways: 139
with runways over 3,659 m: 0
with runways 2,440-3,659 m: 12
with runways 1,220-2,439 m: 94
Telecommunications: excellent domestic and international facilities; 8,200,000 telephones; mainly coaxial and multiconductor cables carry long-distance network; parallel microwave network carries primarily radio, TV and some telephone channels; automatic system; broadcast stations—5 AM, 360 (mostly repeaters) FM, 880 (mostly repeaters) TV; 5 submarine coaxial cables; satellite earth stations—1 Atlantic Ocean INTELSAT and 1 EUTELSAT

Defense Forces

Branches: Swedish Army, Swedish Navy, Swedish Air Force
Manpower availability: males age 15-49 2,156,720; fit for military service 1,884,121; reach military age (19) annually 57,383 (1993 est.)
Defense expenditures: exchange rate conversion—$6.7 billion, 3.8% of GDP (FY92/93)

Switzerland

100 km

Geography

Location: Western Europe, between France and Austria
Map references: Europe, Standard Time Zones of the World
Area:
total area: 41,290 km²
land area: 39,770 km²
comparative area: slightly more than twice the size of New Jersey
Land boundaries: total 1,852 km, Austria 164 km, France 573 km, Italy 740 km, Liechtenstein 41 km, Germany 334 km
Coastline: 0 km (landlocked)
Maritime claims: none; landlocked
International disputes: none
Climate: temperate, but varies with altitude; cold, cloudy, rainy/snowy winters; cool to warm, cloudy, humid summers with occasional showers
Terrain: mostly mountains (Alps in south, Jura in northwest) with a central plateau of rolling hills, plains, and large lakes
Natural resources: hydropower potential, timber, salt
Land use:
arable land: 10%
permanent crops: 1%
meadows and pastures: 40%
forest and woodland: 26%
other: 23%
Irrigated land: 250 km² (1989)
Environment: dominated by Alps
Note: landlocked; crossroads of northern and southern Europe; along with southeastern France and northern Italy, contains the highest elevations in Europe

People

Population: 6,986,621 (July 1993 est.)
Population growth rate: 0.83% (1993 est.)
Birth rate: 12.37 births/1,000 population (1993 est.)
Death rate: 9.24 deaths/1,000 population (1993 est.)

Switzerland *(continued)*

Net migration rate: 5.13 migrant(s)/1,000 population (1993 est.)
Infant mortality rate: 6.6 deaths/1,000 live births (1993 est.)
Life expectancy at birth:
total population: 77.99 years
male: 74.6 years
female: 81.54 years (1993 est.)
Total fertility rate: 1.6 children born/woman (1993 est.)
Nationality:
noun: Swiss (singular and plural)
adjective: Swiss
Ethnic divisions:
total population: German 65% French 18%, Italian 10%, Romansch 1%, other 6%
Swiss nationals: German 74% French 20%, Italian 4%, Romansch 1%, other 1%
Religions: Roman Catholic 47.6%, Protestant 44.3%, other 8.1% (1980)
Languages: German 65%, French 18%, Italian 12%, Romansch 1%, other 4%
note: these are figures for Swiss nationals only— German 74%, French 20%, Italian 4%, Romansch 1%, other 1%
Literacy: age 15 and over can read and write (1980)
total population: 99%
male: NA%
female: NA%
Labor force: 3.31 million (904,095 foreign workers, mostly Italian)
by occupation: services 50%, industry and crafts 33%, government 10%, agriculture and forestry 6%, other 1% (1989)

Government

Names:
conventional long form: Swiss Confederation
conventional short form: Switzerland
local long form: Schweizerische Eidgenossenschaft (German) Confederation Suisse (French) Confederazione Svizzera (Italian)
local short form: Schweiz (German) Suisse (French) Svizzera (Italian)
Digraph: SZ
Type: federal republic
Capital: Bern
Administrative divisions: 26 cantons (cantons, singular—canton in French; cantoni, singular—cantone in Italian; kantone, singular—kanton in German); Aargau, Ausser-Rhoden, Basel-Landschaft, Basel-Stadt, Bern, Fribourg, Geneve, Glarus, Graubunden, Inner-Rhoden, Jura, Luzern, Neuchatel, Nidwalden, Obwalden, Sankt Gallen, Schaffhausen, Schwyz, Solothurn, Thurgau, Ticino, Uri, Valais, Vaud, Zug, Zurich
Independence: 1 August 1291
Constitution: 29 May 1874
Legal system: civil law system influenced by customary law; judicial review of legislative acts, except with respect to federal decrees of general obligatory character; accepts compulsory ICJ jurisdiction, with reservations
National holiday: Anniversary of the Founding of the Swiss Confederation, 1 August (1291)
Political parties and leaders: Free Democratic Party (FDP), Bruno HUNZIKER, president; Social Democratic Party (SPS), Helmut HUBACHER, chairman; Christian Democratic People's Party (CVP), Eva SEGMULLER-WEBER, chairman; Swiss People's Party (SVP), Hans UHLMANN, president; Green Party (GPS), Peter SCHMID, president; Automobile Party (AP), DREYER; Alliance of Independents' Party (LdU), Dr. Franz JAEGER, president; Swiss Democratic Party (SD), NA; Evangelical People's Party (EVP), Max DUNKI, president; Workers' Party (PdA; Communist), Jean SPIELMANN, general secretary; Ticino League, leader NA; Liberal Party (LPS), Gilbert COUTAU, president
Suffrage: 18 years of age; universal
Elections:
Council of States: last held throughout 1991 (next to be held NA 1995); results—percent of vote by party NA; seats—(46 total) FDP 18, CVP 16, SVP 4, SPS 3, LPS 3, LdU 1, Ticino League 1
National Council: last held 20 October 1991 (next to be held NA October 1995); results—percent of vote by party NA; seats—(200 total) FDP 44, SPS 42, CVP 37, SVP 25, GPS 14, LPS 10, AP 8, LdU 6, SD 5, EVP 3, PdA 2, Ticino League 2, other 2
Executive branch: president, vice president, Federal Council (German—Bundesrat, French—Conseil Federal, Italian—Consiglio Federale)
Legislative branch: bicameral Federal Assembly (German—Bundesversammlung, French—Assemblee Federale, Italian—Assemblea Federale) consists of an upper council or Council of States (German—Standerat, French—Conseil des Etats, Italian—Consiglio degli Stati) and a lower council or National Council (German—Nationalrat, French—Conseil National, Italian—Consiglio Nazionale)
Judicial branch: Federal Supreme Court
Leaders:
Chief of State and Head of Government: President Adolf OGI (1993 calendar year; presidency rotates annually); Vice President Otto STICH (term runs concurrently with that of president)
Member of: AfDB, AG (observer), AsDB, Australian Group, BIS, CCC, CE, CERN, COCOM (coopeating country), CSCE, EBRD, ECE, EFTA, ESA, FAO, G-8, G-10, GATT, IADB, IAEA, IBRD, ICAO, ICC, ICFTU, IDA, IEA, IFAD, IFC, ILO, IMF, IMO, INMARSAT, INTELSAT, INTERPOL, IOC, IOM, ISO, ITU, LORCS, MINURSO, MTRC, NAM (guest), NEA, NSG, OAS (observer), OECD, PCA, UN (observer), UNCTAD, UNESCO, UNHCR, UNIDO, UNPROFOR, UNTSO, UPU, WCL, WHO, WIPO, WMO, WTO, ZC
Diplomatic representation in US:
chief of mission: Ambassador Edouard BRUNNER
chancery: 2900 Cathedral Avenue NW, Washington, DC 20008
telephone: (202) 745-7900
FAX: (202) 387-2564
consulates general: Atlanta, Chicago, Houston, Los Angeles, New York, and San Francisco
US diplomatic representation:
chief of mission: Ambassador Joseph B. GILDENHORN
embassy: Jubilaeumstrasse 93, 3005 Bern
mailing address: use embassy street address
telephone: [41] (31) 437-011
FAX: [41] (31) 437-344
branch office: Geneva
consulate general: Zurich
Flag: red square with a bold, equilateral white cross in the center that does not extend to the edges of the flag

Economy

Overview: Switzerland's economy—one of the most prosperous and stable in the world—is nonetheless undergoing a painful adjustment after both the inflationary boom of the late-1980s and the electorate's rejection late last year of membership in the European Economic Area. Stubborn inflation and a soft economy have afflicted Switzerland for more than two years. Despite slow growth in 1991-92, the Swiss central bank has been unable to ease monetary policy because of the threat to the Swiss franc posed by high German interest rates. As a result, unemployment is forecast to rise from 3% in 1992 to more than 4% in 1993, with inflation moving down from 4% to 3%. The voters' rejection in December 1992 of a referendum on membership in the EEA which was supported by most political, business, and financial leaders has raised doubts that the country can maintain its preeminent prosperity and leadership in commercial banking in the 21st century. Despite these problems, Swiss per capita output, general living standards, education and science, health care, and diet remain unsurpassed in Europe. The country has few natural resources except for the scenic natural beauty that has made it a world leader in tourism. Management-labor relations remain generally harmonious.
National product: GDP—purchasing power equivalent—$152.3 billion (1992)
National product real growth rate: -0.6% (1992)

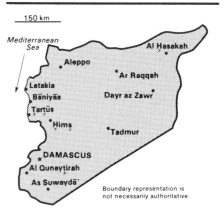

150 km

Mediterranean Sea

Al Hasakah

Aleppo

Ar Raqqah

Latakia
Bāniyās
Tartūs

Dayr az Zawr

Hims

Tadmur

DAMASCUS
Al Qunaytirah

As Suwaydā

Boundary representation is not necessarily authoritative.

National product per capita: $22,300 (1992)
Inflation rate (consumer prices): 4.1% (1992 est.)
Unemployment rate: 3% (1992 est.)
Budget: revenues $24.0 billion; expenditures $23.8 billion, including capital expenditures of $NA (1990)
Exports: $62.2 billion (f.o.b., 1991 est.)
commodities: machinery and equipment, precision instruments, metal products, foodstuffs, textiles and clothing
partners: Western Europe 64% (EC countries 56%, other 8%), US 9%, Japan 4%
Imports: $68.5 billion (c.i.f., 1991 est.)
commodities: agricultural products, machinery and transportation equipment, chemicals, textiles, construction materials
partners: Western Europe 78% (EC countries 71%, other 7%), US 6%
External debt: $NA
Industrial production: growth rate 0.4% (1991 est.)
Electricity: 17,710,000 kW capacity; 56,000 million kWh produced, 8,200 kWh per capita (1992)
Industries: machinery, chemicals, watches, textiles, precision instruments
Agriculture: dairy farming predominates; less than 50% self-sufficient in food; must import fish, refined sugar, fats and oils (other than butter), grains, eggs, fruits, vegetables, meat
Economic aid: donor—ODA and OOF commitments (1970-89), $3.5 billion
Currency: 1 Swiss franc, franken, or franco (SwF) = 100 centimes, rappen, or centesimi
Exchange rates: Swiss francs, franken, or franchi (SwF) per US$1—1.4781 (January 1993), 1.4062 (1992), 1.4340 (1991), 1.3892 (1990), 1.6359 (1989), 1.4633 (1988)
Fiscal year: calendar year

Communications

Railroads: 4,418 km total; 3,073 km are government owned and 1,345 km are nongovernment owned; the government network consists of 2,999 km 1.435-meter standard gauge and 74 km 1.000-meter narrow gauge track; 1,432 km double track, 99% electrified; the nongovernment network consists of 510 km 1.435-meter standard gauge, and 835 km 1.000-meter gauge, 100% electrified
Highways: 62,145 km total (all paved); 18,620 km are canton, 1,057 km are national highways (740 km autobahn), 42,468 km are communal roads
Inland waterways: 65 km; Rhine (Basel to Rheinfelden, Schaffhausen to Bodensee); 12 navigable lakes
Pipelines: crude oil 314 km, natural gas 1,506 km
Ports: Basel (river port)

Merchant marine: 23 ships (1,000 GRT or over) totaling 308,725 GRT/548,244 DWT; includes 5 cargo, 2 roll-on/roll-off cargo, 5 chemical tanker, 2 specialized tanker, 8 bulk, 1 oil tanker
Airports:
total: 66
usable: 65
with permanent-surface runways: 42
with runways over 3,659 m: 2
with runways 2,440-3,659 m: 5
with runways 1,220-2,439 m: 18
Telecommunications: excellent domestic, international, and broadcast services; 5,890,000 telephones; extensive cable and microwave networks; broadcast stations—7 AM, 265 FM, 18 (1,322 repeaters) TV; communications satellite earth station operating in the INTELSAT (Atlantic Ocean and Indian Ocean) system

Defense Forces

Branches: Army (Air Force is part of the Army), Frontier Guards, Fortification Guards
Manpower availability: males age 15-49 1,852,213; fit for military service 1,590,308; reach military age (20) annually 44,124 (1993 est.)
Defense expenditures: exchange rate conversion—$3.5 billion, 1.7% of GDP (1993 est.)

Geography

Location: Middle East, along the Mediterranean Sea, between Turkey and Lebanon
Map references: Africa, Middle East, Standard Time Zones of the World
Area:
total area: 185,180 km²
land area: 184,050 km²
comparative area: slightly larger than North Dakota
note: includes 1,295 km² of Israeli-occupied territory
Land boundaries: total 2,253 km, Iraq 605 km, Israel 76 km, Jordan 375 km, Lebanon 375 km, Turkey 822 km
Coastline: 193 km
Maritime claims:
contiguous zone: 41 nm
territorial sea: 35 nm
International disputes: separated from Israel by the 1949 Armistice Line; Golan Heights is Israeli occupied; Hatay question with Turkey; periodic disputes with Iraq over Euphrates water rights; ongoing dispute over water development plans by Turkey for the Tigris and Euphrates Rivers; Syrian troops in northern Lebanon since October 1976
Climate: mostly desert; hot, dry, sunny summers (June to August) and mild, rainy winters (December to February) along coast
Terrain: primarily semiarid and desert plateau; narrow coastal plain; mountains in west
Natural resources: petroleum, phosphates, chrome and manganese ores, asphalt, iron ore, rock salt, marble, gypsum
Land use:
arable land: 28%
permanent crops: 3%
meadows and pastures: 46%
forest and woodland: 3%
other: 20%
Irrigated land: 6,700 km² (1989)
Environment: deforestation; overgrazing; soil erosion; desertification

Syria *(continued)*

Note: there are 38 Jewish settlements in the Israeli-occupied Golan Heights

People

Population: 14,338,527 (July 1993 est.)
note: in addition, there are at least 14,500 Druze and 14,000 Jewish settlers in the Israeli-occupied Golan Heights (1993 est.)
Population growth rate: 3.76% (1993 est.)
Birth rate: 44.08 births/1,000 population (1993 est.)
Death rate: 6.44 deaths/1,000 population (1993 est.)
Net migration rate: 0 migrant(s)/1,000 population (1993 est.)
Infant mortality rate: 43.9 deaths/1,000 live births (1993 est.)
Life expectancy at birth:
total population: 66.12 years
male: 65.07 years
female: 67.22 years (1993 est.)
Total fertility rate: 6.75 children born/woman (1993 est.)
Nationality:
noun: Syrian(s)
adjective: Syrian
Ethnic divisions: Arab 90.3%, Kurds, Armenians, and other 9.7%
Religions: Sunni Muslim 74%, Alawite, Druze, and other Muslim sects 16%, Christian (various sects) 10%, Jewish (tiny communities in Damascus, Al Qamishli, and Aleppo)
Languages: Arabic (official), Kurdish, Armenian, Aramaic, Circassian, French widely understood
Literacy: age 15 and over can read and write (1990)
total population: 64%
male: 78%
female: 51%
Labor force: 2.951 million (1989)
by occupation: miscellaneous and government services 36%, agriculture 32%, industry and construction 32%;
note—shortage of skilled labor (1984)

Government

Names:
conventional long form: Syrian Arab Republic
conventional short form: Syria
local long form: Al Jumhuriyah al Arabiyah as Suriyah
local short form: Suriyah
former: United Arab Republic (with Egypt)
Digraph: SY
Type: republic under leftwing military regime since March 1963
Capital: Damascus
Administrative divisions: 14 provinces (muhafazat, singular—muhafazah); Al Hasakah, Al Ladhiqiyah, Al Qunaytirah, Ar Raqqah, As Suwayda', Dar'a, Dayr az Zawr, Dimashq, Halab, Hamah, Hims, Idlib, Rif Dimashq, Tartus
Independence: 17 April 1946 (from League of Nations mandate under French administration)
Constitution: 13 March 1973
Legal system: based on Islamic law and civil law system; special religious courts; has not accepted compulsory ICJ jurisdiction
National holiday: National Day, 17 April (1946)
Political parties and leaders: ruling party is the Arab Socialist Resurrectionist (Ba'th) Party; the Progressive National is dominated by Ba'thists but includes independents and members of the Syrian Arab Socialist Party (ASP); Arab Socialist Union (ASU); Syrian Communist Party (SCP); Arab Socialist Unionist Movement; and Democratic Socialist Union Party
Other political or pressure groups: non-Ba'th parties have little effective political influence; Communist party ineffective; conservative religious leaders; Muslim Brotherhood
Suffrage: 18 years of age; universal
Elections:
President: last held 2 December 1991 (next to be held December 1998);
results—President Hafiz al-ASAD was reelected for a fourth seven-year term with 99.98% of the vote
People's Council: last held 22-23 May 1990 (next to be held NA May 1994);
results—Ba'th 53.6%, ASU 3.2%, SCP 3.2%, Arab Socialist Unionist Movement 2.8%, ASP 2%, Democratic Socialist Union Party 1.6%, independents 33.6%;
seats—(250 total) Ba'th 134, ASU 8, SCP 8, Arab Socialist Unionist Movement 7, ASP 5, Democratic Socialist Union Party 4, independents 84; note—the People's Council was expanded to 250 seats total prior to the May 1990 election
Executive branch: president, three vice presidents, prime minister, three deputy prime ministers, Council of Ministers (cabinet)
Legislative branch: unicameral People's Council (Majlis al-Chaab)
Judicial branch: Supreme Constitutional Court, High Judicial Council, Court of Cassation, State Security Courts
Leaders:
Chief of State: President Hafiz al-ASAD (since 22 February 1971 see note); Vice Presidents 'Abd al-Halim KHADDAM, Rif'at al-ASAD, and Muhammad Zuhayr MASHARIQA (since 11 March 1984); note—President ASAD seized power in the November 1970 coup, assumed presidential powers 22 February 1971, and was confirmed as president in the 12 March 1971 national elections

Head of Government: Prime Minister Mahmud ZU'BI (since 1 November 1987); Deputy Prime Minister Lt. Gen. Mustafa TALAS (since 11 March 1984); Deputy Prime Minister Salim YASIN (since NA December 1981); Deputy Prime Minister Rashid AKHTARINI (since 4 July 1992)
Member of: ABEDA, AFESD, AL, AMF, CAEU, CCC, ESCWA, FAO, G-24, G-77, IAEA, IBRD, ICAO, ICC, IDA, IDB, IFAD, IFC, ILO, IMF, IMO, INTELSAT, INTERPOL, IOC, ISO, ITU, LORCS, NAM, OAPEC, OIC, UN, UNCTAD, UNESCO, UNIDO, UNRWA, UPU, WFTU, WHO, WIPO, WMO, WTO
Diplomatic representation in US:
chief of mission: Ambassador Walid MOUALEM
chancery: 2215 Wyoming Avenue NW, Washington, DC 20008
telephone: (202) 232-6313
US diplomatic representation:
chief of mission: Ambassador Christopher W. S. ROSS
embassy: Abu Rumaneh, Al Mansur Street No. 2, Damascus
mailing address: P. O. Box 29, Damascus
telephone: [963] (11) 333052 or 332557, 330416, 332814, 332315, 714108, 337178, 333232
FAX: [963] (11) 718687
Flag: three equal horizontal bands of red (top), white, and black with two small green five-pointed stars in a horizontal line centered in the white band; similar to the flag of Yemen, which has a plain white band and of Iraq, which has three green stars (plus an Arabic inscription) in a horizontal line centered in the white band; also similar to the flag of Egypt, which has a symbolic eagle centered in the white band

Economy

Overview: Syria's state-dominated Ba'thist economy has benefited from the Gulf war, increased oil production, good weather, and economic deregulation. Economic growth averaged nearly 12% annually in 1990-91, buoyed by increased oil production and improved agricultural performance. The Gulf war of early 1991 provided Syria an aid windfall of nearly $5 billion dollars from Arab, European, and Japanese donors. These inflows more than offset Damascus's war-related costs and will help Syria cover some of its debt arrears, restore suspended credit lines, and initiate selected military and civilian purchases. In 1992 the government spurred economic development by loosening controls on domestic and foreign investment while maintaining strict political controls. For the long run, Syria's economy is still saddled with a large number of poorly performing public sector firms and industrial and agricultural productivity is poor. A major long-term concern is the additional drain of upstream Euphrates water by Turkey when its vast dam and irrigation projects are completed by mid-decade.

National product: GDP—exchange rate conversion—$30 billion (1991 est.)

National product real growth rate: 9% (1991 est.)

National product per capita: $2,300 (1991 est.)

Inflation rate (consumer prices): 20% (1992 est.)

Unemployment rate: 5.7% (1989)

Budget: revenues $5.4 billion; expenditures $7.5 billion, including capital expenditures of $2.9 billion (1991 est.)

Exports: $3.5 billion (f.o.b., 1992 est.) *commodities:* petroleum 45%, farm products 11%, textiles, phosphates 5% (1990) *partners:* USSR and Eastern Europe 44%, EC 34%, Arab countries 17%, US/Canada 1% (1990)

Imports: $2.7 billion (f.o.b., 1992 est.) *commodities:* foodstuffs and beverages 21%, machinery 15%, metal and metal products 15%, textiles 7%, petroleum products (1990) *partners:* EC 42%, USSR and Eastern Europe 13%, other Europe 13%, US/Canada 11%, Arab countries 6% (1990)

External debt: $5.3 billion (1990 est.)

Industrial production: growth rate 6% (1991 est.); accounts for 18% of GDP

Electricity: 3,205,000 kW capacity; 11,900 million kWh produced, 830 kWh per capita (1992)

Industries: textiles, food processing, beverages, tobacco, phosphate rock mining, petroleum

Agriculture: accounts for 27% of GDP and one-third of labor force; all major crops (wheat, barley, cotton, lentils, chickpeas) grown mainly on rain-watered land causing wide swings in production; animal products—beef, lamb, eggs, poultry, milk; not self-sufficient in grain or livestock products

Illicit drugs: a transit country for Lebanese and Turkish refined cocaine going to Europe and heroin and hashish bound for the Persian Gulf area

Economic aid: US commitments, including Ex-Im (FY70-81), $538 million; Western (non-US) ODA and OOF bilateral commitments (1970-89) $1.23 billion; OPEC bilateral aid (1979-89), $12.3 billion; former Communist countries (1970-89), $3.3 billion

Currency: 1 Syrian pound (£S) = 100 piasters

Exchange rates: Syrian pounds (£S) per US$1—22.0 (promotional rate since 1991), 22.0 (official rate since 1991), 42.0 (official parallel rate since 1991), 11.2250 (fixed rate 1987-90)

Fiscal year: calendar year

Communications

Railroads: 1,998 km total; 1,766 km standard gauge, 232 km 1.050-meter (narrow) gauge

Highways: 29,000 km total; 670 km expressways; 5,000 km main or national roads; 23,330 km secondary or regional roads (not including municipal roads); 22,680 km of the total is paved (1988)

Inland waterways: 870 km; minimal economic importance

Pipelines: crude oil 1,304 km, petroleum products 515 km

Ports: Tartus, Latakia, Baniyas, Jablah

Merchant marine: 41 ships (1,000 GRT or over) totaling 117,247 GRT/183,607 DWT; includes 36 cargo, 2 vehicle carrier, 3 bulk

Airports:
total: 104
usable: 100
with permanent-surface runways: 24
with runways over 3,659 m: 0
with runways 2,440-3,659 m: 21
with runways 1,220-2,439 m: 3

Telecommunications: fair system currently undergoing significant improvement and digital upgrades, including fiber optic technology; 512,600 telephones (37 telephones per 1,000 persons); broadcast stations—9 AM, 1 FM, 17 TV; satellite earth stations—1 Indian Ocean INTELSAT and 1 Intersputnik; 1 submarine cable; coaxial cable and microwave radio relay to Iraq, Jordan, Lebanon, and Turkey

---

Defense Forces

Branches: Syrian Arab Army, Syrian Arab Navy, Syrian Arab Air Force, Syrian Arab Air Defense Forces

Manpower availability: males age 15-49 3,168,429; fit for military service 1,777,413; reach military age (19) annually 151,102 (1993 est.)

Defense expenditures: exchange rate conversion—$2.2 billion, 6% of GDP (1992)

Entry

follows

Zimbabwe

Tajikistan

Boundary representation is not necessarily authoritative.

150 km

Khudzhand

DUSHANBE

Kurgan-Tyube Kulyab Khorog Murgab

Geography

Location: South Asia, between Uzbekistan and China
Map references: Asia, Commonwealth of Independent States—Central Asian States, Standard Time Zones of the World
Area:
total area: 143,100 km²
land area: 142,700 km²
comparative area: slightly smaller than Wisconsin
Land boundaries: total 3,651 km, Afghanistan 1,206 km, China 414 km, Kyrgyzstan 870 km, Uzbekistan 1,161 km
Coastline: 0 km (landlocked)
Maritime claims: none; landlocked
International disputes: boundary with China under dispute; territorial dispute with Kyrgyzstan on northern boundary in Isfara Valley area; Afghanistan's support to Islamic fighters in Tajikistan's civil war
Climate: midlatitude; semiarid to polar in Pamir Mountains
Terrain: Pamir and Altay Mountains dominate landscape; western Fergana Valley in north, Kafirnigan and Vakhsh Valleys in south or southwest
Natural resources: significant hydropower potential, petroleum, uranium, mercury, brown coal, lead, zinc, antimony, tungsten
Land use:
arable land: 6%
permanent crops: 0%
meadows and pastures: 23%
forest and woodland: 0%
other: 71%
Irrigated land: 6,940 km² (1990)
Environment: NA
Note: landlocked

People

Population: 5,836,140 (July 1993 est.)
Population growth rate: 2.72% (1993 est.)
Birth rate: 35.52 births/1,000 population (1993 est.)

Death rate: 6.87 deaths/1,000 population (1993 est.)
Net migration rate: -1.42 migrant(s)/1,000 population (1993 est.)
Infant mortality rate: 63.6 deaths/1,000 live births (1993 est.)
Life expectancy at birth:
total population: 68.5 years
male: 65.66 years
female: 71.48 years (1993 est.)
Total fertility rate: 4.7 children born/woman (1993 est.)
Nationality:
noun: Tajik(s)
adjective: Tajik
Ethnic divisions: Tajik 64.9%, Uzbek 25%, Russian 3.5% (declining because of emigration), other 6.6%
Religions: Sunni Muslim 80%, Shi'a Muslim 5%
Languages: Tajik (official)
Literacy: age 9-49 can read and write (1970)
total population: 100%
male: 100%
female: 99%
Labor force: 1.938 million
by occupation: agriculture and forestry 43%, industry and construction 22%, other 35% (1990)

Government

Names:
conventional long form: Republic of Tajikistan
conventional short form: Tajikistan
local long form: Respublika i Tojikiston
local short form: none
former: Tajik Soviet Socialist Republic
Digraph: TI
Type: republic
Capital: Dushanbe
Administrative divisions: 2 oblasts (oblastey, singular—oblast') and one autonomous oblast*; Gorno-Badakhshan*; Khatlon, Leninabad (Khudzhand)
note: the rayons around Dushanbe are under direct republic jurisdiction; an oblast usually has the same name as its administrative center (exceptions have the administrative center name following in parentheses)
Independence: 9 September 1991 (from Soviet Union)
Constitution: as of mid-1993, a new constitution had not been formally approved
Legal system: based on civil law system; no judicial review of legislative acts
National holiday: NA
Political parties and leaders: Tajik Democratic Party (TDP), Maksud IKRAMOV, Davia KOUDONAZAROV, Shodmon YUSUPOV; Tajik Socialist Party (TSP), Rakhman NABIYEV, Kakhkhor MAKHKAMOV; Islamic Revival Party

(IRP), Mullah Mukhamedsharif KHIMATZODA, Daviat USMON
Other political or pressure groups: Tajik People's Front
Suffrage: 18 years of age; universal
Elections:
President: last held 27 October 1991 (next to be held NA); results—Rakhman NABIYEV, Communist Party 60%; Davlat KHUDONAZAROV, Democratic Party, Islamic Rebirth Party and Rastokhoz Party 30%
Supreme Soviet: last held 25 February 1990 (next to be held NA); results—Communist Party 99%, other 1%; seats—(230 total) Communist Party 227, other 3
note: in May 1992, the Supreme Soviet was replaced by the transitional 80-member Assembly (Majlis) and in November 1992 Emomili RAKHMANOV, chairman of the Assembly, became Chief of State
Executive branch: president, prime minister, cabinet
Legislative branch: unicameral Assembly (Majlis)
Judicial branch: NA
Leaders:
Chief of State: Acting President and Assembly Chairman Emomili RAKHMANOV (since NA November 1992)
Head of Government: Prime Minister Abdumalik ABULAJANOV (since NA November 1992); First Deputy Prime Minister Tukhtaboy GAFAROV (since NA November 1992)
Member of: CIS, CSCE, EBRD, ECO, ESCAP, NACC, UN, UNCTAD, WHO
Diplomatic representation in US:
chief of mission: NA
chancery: NA
telephone: NA
US diplomatic representation:
chief of mission: Ambassador Stanley T. ESCUDERO
embassy: (temporary) #39 Ainii Street, Dushanbe
mailing address: APO AE 09862
telephone: [7] (3772) 24-82-33
Flag: NA

Economy

Overview: Tajikistan has had the lowest living standards of the CIS republics and now faces the bleakest economic prospects. Agriculture (particularly cotton and fruit growing) is the most important sector, accounting for 38% of employment (1990). Industrial production includes aluminum reduction, hydropower generation, machine tools, refrigerators, and freezers. Throughout 1992 bloody civil disturbances disrupted food imports and several regions became desperately short of basic needs. Hundreds of thousands of people were made homeless

by the strife. In late 1992, one-third of industry was shut down and the cotton crop was only one-half of that of 1991.
National product: GDP $NA
National product real growth rate: -34% (1992 est.)
National product per capita: $NA
Inflation rate (consumer prices): 35% per month (first quarter 1993)
Unemployment rate: 0.4% includes only officially registered unemployed; also large numbers of underemployed workers
Budget: revenues $NA; expenditures $NA, including capital expenditures of $NA
Exports: $100 million to outside successor states of the former USSR (1992)
commodities: aluminum, cotton, fruits, vegetable oil, textiles
partners: Russia, Kazakhstan, Ukraine, Uzbekistan
Imports: $100 million from outside the successor states of the former USSR (1992)
commodities: chemicals, machinery and transport equipment, textiles, foodstuffs
partners: NA
External debt: $650 million (end of 1991 est.)
Industrial production: growth rate −25% (1992 est.)
Electricity: 4,585,000 kW capacity; 16,800 million kWh produced, 2,879 kWh per capita (1992)
Industries: aluminum, zinc, lead, chemicals and fertilizers, cement, vegetable oil, metal-cutting machine tools, refrigerators and freezers
Agriculture: cotton, grain, fruits, grapes, vegetables; cattle, pigs, sheep and goats, yaks
Illicit drugs: illicit producer of cannabis and opium; mostly for CIS consumption; limited government eradication programs; used as transshipment points for illicit drugs from Southwest Asia to Western Europe
Economic aid: $700 million offical and commitments by foreign donors (1992)
Currency: retaining Russian ruble as currency (January 1993)
Exchange rates: rubles per US$1—415 (24 December 1992) but subject to wide fluctuations
Fiscal year: calendar year

Communications

Railroads: 480 km; does not include industrial lines (1990)
Highways: 29,900 km total (1990); 21,400 km hard surfaced, 8,500 km earth
Pipelines: natural gas 400 km (1992)
Airports:
total: 58
useable: 30
with permanent-surface runways: 12
with runways over 3,659 m: 0
with runways 2,440-3,659 m: 4
with runways 1,220-2,439 m: 13
Telecommunications: poorly developed and not well maintained; many towns are not reached by the national network; telephone density in urban locations is about 100 per 1000 persons; linked by cable and microwave to other CIS republics, and by leased connections to the Moscow international gateway switch; satellite earth stations—1 orbita and 2 INTELSAT (TV receive-only; the second INTELSAT earth station provides TV receive-only service from Turkey)

Defense Forces

Branches: Army (being formed), National Guard, Security Forces (internal and border troops)
Manpower availability: males age 15-49 1,313,676; fit for military service 1,079,935; reach military age (18) annually 56,862 (1993 est.)
Defense expenditures: $NA, NA% of GDP

Tanzania

Geography

Location: Eastern Africa, bordering the Indian Ocean between Kenya and Mozambique
Map references: Africa, Standard Time Zones of the World
Area:
total area: 945,090 km²
land area: 886,040 km²
comparative area: slightly larger than twice the size of California
note: includes the islands of Mafia, Pemba, and Zanzibar
Land boundaries: total 3,402 km, Burundi 451 km, Kenya 769 km, Malawi 475 km, Mozambique 756 km, Rwanda 217 km, Uganda 396 km, Zambia 338 km
Coastline: 1,424 km
Maritime claims:
exclusive economic zone: 200 nm
territorial sea: 12 nm
International disputes: boundary dispute with Malawi in Lake Nyasa; Tanzania-Zaire-Zambia tripoint in Lake Tanganyika may no longer be indefinite since it is reported that the indefinite section of the Zaire-Zambia boundary has been settled
Climate: varies from tropical along coast to temperate in highlands
Terrain: plains along coast; central plateau; highlands in north, south
Natural resources: hydropower potential, tin, phosphates, iron ore, coal, diamonds, gemstones, gold, natural gas, nickel
Land use:
arable land: 5%
permanent crops: 1%
meadows and pastures: 40%
forest and woodland: 47%
other: 7%
Irrigated land: 1,530 km² (1989 est.)
Environment: lack of water and tsetse fly limit agriculture; recent droughts affected marginal agriculture; Kilimanjaro is highest point in Africa

Tanzania (continued)

People

Population: 27,286,363 (July 1993 est.)
Population growth rate: 2.56% (1993 est.)
Birth rate: 45.66 births/1,000 population (1993 est.)
Death rate: 19.02 deaths/1,000 population (1993 est.)
Net migration rate: -1.06 migrant(s)/1,000 population (1993 est.)
Infant mortality rate: 110.4 deaths/1,000 live births (1993 est.)
Life expectancy at birth:
total population: 44 years
male: 42.19 years
female: 45.87 years (1993 est.)
Total fertility rate: 6.25 children born/woman (1993 est.)
Nationality:
noun: Tanzanian(s)
adjective: Tanzanian
Ethnic divisions:
mainland: native African 99% (consisting of well over 100 tribes) Asian, European, and Arab 1%
Zanzibar: NA
Religions:
mainland: Christian 40%, Muslim 33%, indigenous beliefs 25%
Zanzibar: Muslim
Languages: Swahili (official; widely understood and generally used for communication between ethnic groups and is used in primary education), English (official; primary language of commerce, administration, and higher education)
note: first language of most people is one of the local languages
Literacy: age 15 and over can read and write (1978)
total population: 46%
male: 62%
female: 31%
Labor force: 732,200 wage earners
by occupation: agriculture 90%, industry and commerce 10% (1986 est.)

Government

Names:
conventional long form: United Republic of Tanzania
conventional short form: Tanzania
former: United Republic of Tanganyika and Zanzibar
Digraph: TZ
Type: republic
Capital: Dar es Salaam
note: some government offices have been transferred to Dodoma, which is planned as the new national capital by the end of the 1990s
Administrative divisions: 25 regions; Arusha, Dar es Salaam, Dodoma, Iringa, Kigoma, Kilimanjaro, Lindi, Mara, Mbeya, Morogoro, Mtwara, Mwanza, Pemba North,

Pemba South, Pwani, Rukwa, Ruvuma, Shinyanga, Singida, Tabora, Tanga, Zanzibar Central/South, Zanzibar North, Zanzibar Urban/West, Ziwa Magharibi
Independence: 26 April 1964 Tanganyika became independent 9 December 1961 (from UN trusteeship under British administration); Zanzibar became independent 19 December 1963 (from UK); Tanganyika united with Zanzibar 26 April 1964 to form the United Republic of Tanganyika and Zanzibar; renamed United Republic of Tanzania 29 October 1964
Constitution: 15 March 1984 (Zanzibar has its own constitution but remains subject to provisions of the union constitution)
Legal system: based on English common law; judicial review of legislative acts limited to matters of interpretation; has not accepted compulsory ICJ jurisdiction
National holiday: Union Day, 26 April (1964)
Political parties and leaders: Chama Chr Mapinduzi (CCM or Revolutionary Party), Ali Hassan MWINYI; Civic United Front (CUF), James MAPALALA; National Committee for Constitutional Reform (NCCK), Mabere MARANDO; Union for Multiparty Democracy (UMD), Abdullah FUNDIKIRA; Democratic Party (DP), Christopher Mtikila
Suffrage: 18 years of age; universal
Elections:
President: last held 28 October 1990 (next to be held NA October 1995); results—Ali Hassan MWINYI was elected without opposition
National Assembly: last held 28 October 1990 (next to be held NA October 1995); results—CCM was the only party; seats—(241 total, 168 elected) CCM 168
Executive branch: president, first vice president and prime minister of the union, second vice president and president of Zanzibar, Cabinet
Legislative branch: unicameral National Assembly (Bunge)
Judicial branch: Court of Appeal, High Court
Leaders:
Chief of State: President Ali Hassan MWINYI (since 5 November 1985); First Vice President John MALECELA (since 9 November 1990); Second Vice President Salmin AMOUR (since 9 November 1990)
Head of Government: Prime Minister John MALECELA (since 9 November 1990)
Member of: ACP, AfDB, C, CCC, EADB, ECA, FAO, FLS, G-6, G-77, GATT, IAEA, IBRD, ICAO, IDA, IFAD, IFC, ILO, IMF, IMO, INTELSAT, INTERPOL, IOC, ISO, ITU, LORCS, NAM, OAU, SADC, UN, UNCTAD, UNESCO, UNHCR, UNIDO, UPU, WCL, WHO, WIPO, WMO, WTO
Diplomatic representation in US:

chief of mission: Ambassador Charles Musama NYIRABU
chancery: 2139 R Street NW, Washington, DC 20008
telephone: (202) 939-6125
US diplomatic representation:
chief of mission: Ambassador Peter Jon DE VOS
embassy: 36 Laibon Road (off Bagamoyo Road), Dar es Salaam
mailing address: P. O. Box 9123, Dar es Salaam
telephone: [255] (51) 66010/13
FAX: [255] (51) 66701
Flag: divided diagonally by a yellow-edged black band from the lower hoist-side corner; the upper triangle (hoist side) is green and the lower triangle is blue

Economy

Overview: Tanzania is one of the poorest countries in the world. The economy is heavily dependent on agriculture, which accounts for about 58% of GDP, provides 85% of exports, and employs 90% of the work force. Industry accounts for 8% of GDP and is mainly limited to processing agricultural products and light consumer goods. The economic recovery program announced in mid-1986 has generated notable increases in agricultural production and financial support for the program by bilateral donors. The World Bank, the International Monetary Fund, and bilateral donors have provided funds to rehabilitate Tanzania's deteriorated economic infrastructure. Growth in 1991-92 featured a pickup in industrial production and a substantial increase in output of minerals led by gold.
National product: GDP—exchange rate conversion—$7.2 billion (1992 est.)
National product real growth rate: 4.5% (1992 est.)
National product per capita: $260 (1992 est.)
Inflation rate (consumer prices): 22% (1992 est.)
Unemployment rate: NA%
Budget: revenues $495 million; expenditures $631 million, including capital expenditures of $118 million (FY90)
Exports: $422 million (f.o.b., 1991)
commodities: coffee, cotton, tobacco, tea, cashew nuts, sisal
partners: FRG, UK, Japan, Netherlands, Kenya, Hong Kong, US
Imports: $1.43 billion (c.i.f., 1991)
commodities: manufactured goods, machinery and transportation equipment, cotton piece goods, crude oil, foodstuffs
partners: FRG, UK, US, Japan, Italy, Denmark
External debt: $6.44 billion (1992)

Industrial production: growth rate 9.3% (1990); accounts for 7% of GDP
Electricity: 405,000 kW capacity; 600 million kWh produced, 20 kWh per capita (1991)
Industries: primarily agricultural processing (sugar, beer, cigarettes, sisal twine), diamond and gold mining, oil refinery, shoes, cement, textiles, wood products, fertilizer
Agriculture: accounts for over 58% of GDP; topography and climatic conditions limit cultivated crops to only 5% of land area; cash crops—coffee, sisal, tea, cotton, pyrethrum (insecticide made from chrysanthemums), cashews, tobacco, cloves (Zanzibar); food crops—corn, wheat, cassava, bananas, fruits, vegetables; small numbers of cattle, sheep, and goats; not self-sufficient in food grain production
Economic aid: US commitments, including Ex-Im (FY70-89), $400 million; Western (non-US) countries, ODA and OOF bilateral commitments (1970-89), $9.8 billion; OPEC bilateral aid (1979-89), $44 million; Communist countries (1970-89), $614 million
Currency: 1 Tanzanian shilling (TSh) = 100 cents
Exchange rates: Tanzanian shillings (TSh) per US$1—325.00 (November 1992), 219.16 (1991), 195.06 (1990), 143.38 (1989), 99.29 (1988), 64.26 (1987)
Fiscal year: 1 July-30 June

Communications

Railroads: 3,555 km total; 960 km 1.067-meter gauge (including the 962 km Tazara Railroad); 2,595 km 1.000-meter gauge, including 6.4 km double track; 115 km of 1.000-meter gauge planned by end of decade
Highways: 81,900 km total, 3,600 km paved; 5,600 km gravel or crushed stone; 72,700 km improved and unimproved earth
Inland waterways: Lake Tanganyika, Lake Victoria, Lake Nyasa
Pipelines: crude oil 982 km
Ports: Dar es Salaam, Mtwara, Tanga, and Zanzibar are ocean ports; Mwanza on Lake Victoria and Kigoma on Lake Tanganyika are inland ports
Merchant marine: 6 ships (1,000 GRT or over) totaling 19,185 GRT/22,916 DWT; includes 2 passenger-cargo, 2 cargo, 1 roll-on/roll-off cargo, 1 oil tanker
Airports:
total: 103
usable: 92
with permanent-surface runways: 12
with runways over 3,659 m: 0
with runways 2,440-3,659 m: 4
with runways 1,220-2,439 m: 40
Telecommunications: fair system operating below capacity; open wire, radio relay, and troposcatter; 103,800 telephones; broadcast stations—12 AM, 4 FM, 2 TV; 1 Indian Ocean and 1 Atlantic Ocean INTELSAT earth station

Defense Forces

Branches: Tanzanian People's Defense Force (TPDF; including Army, Navy, and Air Force), paramilitary Police Field Force Unit, Militia
Manpower availability: males age 15-49 5,835,064; fit for military service 3,375,567 (1993 est.)
Defense expenditures: exchange rate conversion—$NA, NA% of GDP

Thailand

Geography

Location: Southeast Asia, bordering the Gulf of Thailand, between Burma and Cambodia
Map references: Asia, Southeast Asia, Standard Time Zones of the World
Area:
total area: 514,000 km²
land area: 511,770 km²
comparative area: slightly more than twice the size of Wyoming
Land boundaries: total 4,863 km, Burma 1,800 km, Cambodia 803 km, Laos 1,754 km, Malaysia 506 km
Coastline: 3,219 km
Maritime claims:
exclusive economic zone: 200 nm
territorial sea: 12 nm
International disputes: boundary dispute with Laos; unresolved maritime boundary with Vietnam
Climate: tropical; rainy, warm, cloudy southwest monsoon (mid-May to September); dry, cool northeast monsoon (November to mid-March); southern isthmus always hot and humid
Terrain: central plain; eastern plateau (Khorat); mountains elsewhere
Natural resources: tin, rubber, natural gas, tungsten, tantalum, timber, lead, fish, gypsum, lignite, fluorite
Land use:
arable land: 34%
permanent crops: 4%
meadows and pastures: 1%
forest and woodland: 30%
other: 31%
Irrigated land: 42,300 km² (1989 est.)
Environment: air and water pollution; land subsidence in Bangkok area
Note: controls only land route from Asia to Malaysia and Singapore

People

Population: 58,722,437 (July 1993 est.)
Population growth rate: 1.36% (1993 est.)

Thailand (continued)

Birth rate: 19.97 births/1,000 population (1993 est.)

Death rate: 6.33 deaths/1,000 population (1993 est.)

Net migration rate: 0 migrant(s)/1,000 population (1993 est.)

Infant mortality rate: 38.5 deaths/1,000 live births (1993 est.)

Life expectancy at birth:
total population: 68.28 years
male: 65.05 years
female: 71.66 years (1993 est.)

Total fertility rate: 2.16 children born/woman (1993 est.)

Nationality:
noun: Thai (singular and plural)
adjective: Thai

Ethnic divisions: Thai 75%, Chinese 14%, other 11%

Religions: Buddhism 95%, Muslim 3.8%, Christianity 0.5%, Hinduism 0.1%, other 0.6% (1991)

Languages: Thai, English the secondary language of the elite, ethnic and regional dialects

Literacy: age 15 and over can read and write (1990)
total population: 93%
male: 96%
female: 90%

Labor force: 30.87 million
by occupation: agriculture 62%, industry 13%, commerce 11%, services (including government) 14% (1989 est.)

Government

Names:
conventional long form: Kingdom of Thailand
conventional short form: Thailand

Digraph: TH

Type: constitutional monarchy

Capital: Bangkok

Administrative divisions: 73 provinces (changwat, singular and plural); Ang Thong, Buriram, Chachoengsao, Chai Nat, Chaiyaphum, Changwat Mukdahan, Chanthaburi, Chiang Mai, Chiang Rai, Chon Buri, Chumphon, Kalasin, Kamphaeng Phet, Kanchanaburi, Khon Kaen, Krabi, Krung Thep Mahanakhon, Lampang, Lamphun, Loei, Lop Buri, Mae Hong Son, Maha Sarakham, Nakhon Nayok, Nakhon Pathom, Nakhon Phanom, Nakhon Ratchasima, Nakhon Sawan, Nakhon Si Thammarat, Nan, Narathiwat, Nong Khai, Nonthaburi, Pathum Thani, Pattani, Phangnga, Phatthalung, Phayao, Phetchabun, Phetchaburi, Phichit, Phitsanulok, Phra Nakhon Si Ayutthaya, Phrae, Phuket, Prachin Buri, Prachuap Khiri Khan, Ranong, Ratchaburi, Rayong, Roi Et, Sakon Nakhon, Samut Prakan, Samut Sakhon, Samut Songkhram, Sara Buri, Satun, Sing Buri, Sisaket, Songkhla, Sukhothai, Suphan Buri, Surat Thani, Surin,

Tak, Trang, Trat, Ubon Ratchathani, Udon Thani, Uthai Thani, Uttaradit, Yala, Yasothon

Independence: 1238 (traditional founding date; never colonized)

Constitution: 22 December 1978; new constitution approved 7 December 1991; amended 10 June 1992

Legal system: based on civil law system, with influences of common law; has not accepted compulsory ICJ jurisdiction; martial law in effect since 23 February 1991 military coup

National holiday: Birthday of His Majesty the King, 5 December (1927)

Political parties and leaders: Democrat Party (DP), Chuan LIKPHAI; Thai Nation Pary (TNP or Chat Thai Party), Praman ADIREKSAN; National Development Party (NDP or Chat Phattana), Chatchai CHUNHAWAN; New Aspiration Party, Gen. Chawalit YONGCHAIYUT; Phalang Tham (Palang Dharma), Bunchu ROTCHANASATIEN; Social Action Party (SAP), Montri PHONGPHANIT; Liberal Democratic Party (LDP or Seri Tham), Athit URAIRAT; Solidarity Party (SP), Uthai PHIMCHAICHON; Mass Party (Muanchon), Pol. Cpt. Choem YUBAMRUNG; Thai Citizen's Party (Prachakon Thai), Samak SUNTHONWET; People's Party (Ratsadon), Chaiphak SIRIWAT; People's Force Party (Phalang Prachachon), Col. Sophon HANCHAREON

Suffrage: 21 years of age; universal

Elections:
House of Representatives: last held 13 September 1992 (next to be held by NA); results—percent of vote by party NA; seats—(360 total) DP 79, TNP 77, NDP 60, NAP 51, Phalang Tham 47, SAP 22, LDP 8, SP 8, Mass Party 4, Thai Citizen's Party 3, People's Party 1, People's Force Party 0

Executive branch: monarch, prime minister, four deputy prime ministers, Council of Ministers (cabinet), Privy Council

Legislative branch: bicameral National Assembly (Rathasatha) consists of an upper house or Senate (Vuthisatha) and a lower house or House of Representatives (Saphaphoothan-Rajsadhorn)

Judicial branch: Supreme Court (Sarndika)

Leaders:
Chief of State: King PHUMIPHON Adunyadet (since 9 June 1946); Heir Apparent Crown Prince WACHIRALONGKON (born 28 July 1952)
Head of Government: Prime Minister CHUAN Likphai (since 23 September 1992)

Member of: APEC, AsDB, ASEAN, CCC, CP, ESCAP, FAO, G-77, GATT, IAEA, IBRD, ICAO, ICFTU, IDA, IFAD, IFC, ILO, IMF, IMO, INTELSAT, INTERPOL, IOC, IOM, ISO, ITU, LORCS, PCA, UN, UNCTAD, UNESCO, UNHCR, UNIDO, UNIKOM, UNTAC, UPU, WCL, WHO, WIPO, WMO

Diplomatic representation in US:
chief of mission: Ambassador-designate PHIRAPHONG Kasemsi
chancery: 2300 Kalorama Road NW, Washington, DC 20008
telephone: (202) 483-7200
consulates general: Chicago, Los Angeles, and New York

US diplomatic representation:
chief of mission: Ambassador David F. LAMBERTSON
embassy: 95 Wireless Road, Bangkok
mailing address: APO AP 96546
telephone: [66] (2) 252-5040
FAX: [66] (2) 254-2990
consulate general: Chiang Mai
consulates: Songkhla, Udorn

Flag: five horizontal bands of red (top), white, blue (double width), white, and red

Economy

Overview: Thailand's economy recovered rapidly from the political unrest in May 1992 to post an impressive 7% growth rate for the year. Thailand, one of the more advanced developing countries in Asia, depends on exports of manufactures and the development of the service sector to fuel the country's rapid growth. The trade and current account deficits fell in 1992; much of Thailand's recent imports have been for capital equipment suggesting that the export sector is poised for further growth. With foreign investment slowing, Bangkok is working to increase the generation of capital domestically. Prime Minister CHUAN's government—Thailand's fifth government in less than two years—is pledged to continue Bangkok's probusiness policies, and the return of a democratically elected government has improved business confidence. Nevertheless, CHUAN must overcome divisions within his ruling coalition to complete much needed infrastructure development programs if Thailand is to remain an attractive place for business investment. Over the longer-term, Bangkok must produce more college graduates with technical training and upgrade workers' skills to continue its rapid economic development.

National product: GDP—exchange rate conversion—$103 billion (1992 est.)

National product real growth rate: 7% (1992 est.)

National product per capita: $1,800 (1992 est.)

Inflation rate (consumer prices): 4.5% (1992 est.)

Unemployment rate: 4.7% (1992 est.)

Budget: revenues $21.36 billion; expenditures $22.40 billion, including capital expenditures of $6.24 billion (FY93 est.)

Exports: $32.9 billion (f.o.b., 1992)
commodities: machinery and manufactures 76.9%, agricultural products 14.9%, fisheries products 5.9% (1992)
partners: US 21.6%, Japan 18.0%, Singapore 8.7%, Hong Kong 4.8%, Germany 4.4%, Netherlands 4.2%, UK 3.4%, Malaysia, France, China (1992 est.)
Imports: $41.5 billion (c.i.f., 1992)
commodities: capital goods 41.4%, intermediate goods and raw materials 32.8%, consumer goods 10.4%, oil 8.2%
partners: Japan 29.3%, US 11.4%, Singapore 7.6%, Taiwan 5.5%, Germany 5.4%, South Korea 4.6%, Malaysia 4.2%, China 3.3%, Hong Kong 3.3%, UK (1992 est.)
External debt: $33.4 billion (1991)
Industrial production: growth rate 18% (1990); accounts for about 26% of GDP
Electricity: 10,000,000 kW capacity; 43,750 million kWh produced, 760 kWh per capita (1992)
Industries: tourism is the largest source of foreign exchange; textiles and garments, agricultural processing, beverages, tobacco, cement, light manufacturing, such as jewelry; electric appliances and components, integrated circuits, furniture, plastics; world's second-largest tungsten producer and third-largest tin producer
Agriculture: accounts for 12% of GDP and 60% of labor force; leading producer and exporter of rice and cassava (tapioca); other crops—rubber, corn, sugarcane, coconuts, soybeans; except for wheat, self-sufficient in food
Illicit drugs: a minor producer, major illicit trafficker of heroin, particularly from Burma and Laos, and cannabis for the international drug market; eradication efforts have reduced the area of cannabis cultivation and shifted some production to neighboring countries; opium poppy cultivation has been affected by eradication efforts; also a major drug money laundering center
Economic aid: US commitments, including Ex-Im (FY70-89), $870 million; Western (non-US) countries, ODA and OOF bilateral commitments (1970-89), $8.6 billion; OPEC bilateral aid (1979-89), $19 million
Currency: 1 baht (B) = 100 satang
Exchange rates: baht (B) per US$1—25.280 (April 1993), 25.400 (1992), 25.517 (1991), 25.585 (1990), 25.702 (1989), 25.294 (1988)
Fiscal year: 1 October-30 September

Communications

Railroads: 3,940 km 1.000-meter gauge, 99 km double track
Highways: 77,697 km total; 35,855 km paved (including 88 km expressways), 14,092 km gravel or other stabilization, 27,750 km mostly dirt and other (1988)
Inland waterways: 3,999 km principal waterways; 3,701 km with navigable depths of 0.9 m or more throughout the year; numerous minor waterways navigable by shallow-draft native craft
Pipelines: natural gas 350 km, petroleum products 67 km
Ports: Bangkok, Pattani, Phuket, Sattahip, Si Racha
Merchant marine: 169 ships (1,000 GRT or over) totaling 752,055 GRT/1,166,136 DWT; includes 1 short-sea passenger, 91 cargo, 12 container, 40 oil tanker, 9 liquefied gas, 2 chemical tanker, 5 bulk, 6 refrigerated cargo, 2 combination bulk, 1 passenger
Airports:
total: 106
usable: 95
with permanent-surface runways: 51
with runways over 3,659 m: 1
with runways 2,440-3,659 m: 14
with runways 1,220-2,439 m: 28
Telecommunications: service to general public inadequate; bulk of service to government activities provided by multichannel cable and microwave radio relay network; 739,500 telephones (1987); broadcast stations—over 200 AM, 100 FM, and 11 TV in government-controlled networks; satellite earth stations—1 Indian Ocean INTELSAT and 1 Pacific Ocean INTELSAT; domestic satellite system being developed

Defense Forces

Branches: Royal Thai Army, Royal Thai Navy (including Royal Thai Marine Corps), Royal Thai Air Force, Paramilitary Forces
Manpower availability: males age 15-49 16,685,044; fit for military service 10,148,786; reach military age (18) annually 616,042 (1993 est.)
Defense expenditures: exchange rate conversion—$2.6 billion, about 2% of GNP (FY92/93 est.)

Togo

125 km

Bight of Benin

Geography

Location: Western Africa, bordering the North Atlantic Ocean beween Benin and Ghana
Map references: Africa, Standard Time Zones of the World
Area:
total area: 56,790 km²
land area: 54,390 km²
comparative area: slightly smaller than West Virginia
Land boundaries: total 1,647 km, Benin 644 km, Burkina 126 km, Ghana 877 km
Coastline: 56 km
Maritime claims:
exclusive economic zone: 200 nm
territorial sea: 30 nm
International disputes: none
Climate: tropical; hot, humid in south; semiarid in north
Terrain: gently rolling savanna in north; central hills; southern plateau; low coastal plain with extensive lagoons and marshes
Natural resources: phosphates, limestone, marble
Land use:
arable land: 25%
permanent crops: 1%
meadows and pastures: 4%
forest and woodland: 28%
other: 42%
Irrigated land: 70 km² (1989 est.)
Environment: hot, dry harmattan wind can reduce visibility in north during winter; recent droughts affecting agriculture; deforestation

People

Population: 4,104,657 (July 1993 est.)
Population growth rate: 3.61% (1993 est.)
Birth rate: 47.87 births/1,000 population (1993 est.)
Death rate: 11.8 deaths/1,000 population (1993 est.)
Net migration rate: 0 migrant(s)/1,000 population (1993 est.)

Infant mortality rate: 91.3 deaths/1,000 live births (1993 est.)
Life expectancy at birth:
total population: 56.46 years
male: 54.45 years
female: 58.53 years (1993 est.)
Total fertility rate: 6.96 children born/woman (1993 est.)
Nationality:
noun: Togolese (singular and plural)
adjective: Togolese
Ethnic divisions: 37 tribes; largest and most important are Ewe, Mina, and Kabye, European and Syrian-Lebanese under 1%
Religions: indigenous beliefs 70%, Christian 20%, Muslim 10%
Languages: French (official and the language of commerce), Ewe (one of the two major African languages in the south), Mina (one of the two major African languages in the south), Dagomba (one of the two major African languages in the north), Kabye (one of the two major African languages in the north)
Literacy: age 15 and over can read and write (1990)
total population: 43%
male: 56%
female: 31%
Labor force: NA
by occupation: agriculture 78%, industry 22%
note: about 88,600 wage earners, evenly divided between public and private sectors; 50% of population of working age (1985)

Government

Names:
conventional long form: Republic of Togo
conventional short form: Togo
local long form: Republique Togolaise
local short form: none
former: French Togo
Digraph: TO
Type: republic under transition to multiparty democratic rule
Capital: Lome
Administrative divisions: 21 circumscriptions (circonscriptions, singular—circonscription); Amlame (Amou), Aneho (Lacs), Atakpame (Ogou), Badou (Wawa), Bafilo (Assoli), Bassar (Bassari), Dapango (Tone), Kande (Keran), Klouto (Kloto), Pagouda (Binah), Lama-Kara (Kozah), Lome (Golfe), Mango (Oti), Niamtougou (Doufelgou), Notse (Haho), Pagouda, Sotouboua, Tabligbo (Yoto), Tchamba, Nyala, Tchaoudjo, Tsevie (Zio), Vogan (Vo)
note: the 21 units may now be called prefectures (prefectures, singular—prefecture) and reported name changes for individual units are included in parentheses

Independence: 27 April 1960 (from UN trusteeship under French administration)
Constitution: 1980 constitution nullified during national reform conference; transition constitution adopted 24 August 1991; multiparty draft constitution sent to High Council of the Republic for approval in November 1991; adopted by public referendum September 1992
Legal system: French-based court system
National holiday: Independence Day, 27 April (1960)
Political parties and leaders: Rally of the Togolese People (RPT) led by President EYADEMA was the only party until the formation of multiple parties was legalized 12 April 1991; transition regime in place since August 1991
Suffrage: universal adult at age NA
Elections:
President: last held 21 December 1986 (next to be held 1993); results—Gen. EYADEMA was reelected without opposition
National Assembly: last held 4 March 1990; dissolved during national reform conference (next to be held 1993); results—RPT was the only party; seats—(77 total) RPT 77; interim legislative High Council of the Republic (HCR) in place since August 1991
Executive branch: president, prime minister, Council of Ministers (cabinet)
Legislative branch: National Assembly dissolved during national reform conference; 79-member interim High Council for the Republic (HCR) formed to act as legislature during transition to multiparty democracy; legislative elections scheduled to be held in 1993
Judicial branch: Court of Appeal (Cour d'Appel), Supreme Court (Cour Supreme)
Leaders:
Chief of State: President Gen. Gnassingbe EYADEMA (since 14 April 1967)
Head of Government: interim Prime Minister Joseph Kokou KOFFIGOH (since 28 August 1991)
Member of: ACCT, ACP, AfDB, CCC, CEAO (observer), ECA, ECOWAS, Entente, FAO, FZ, G-77, GATT, IBRD, ICAO, ICC, IDA, IFAD, IFC, ILO, IMF, IMO, INTELSAT, INTERPOL, IOC, ITU, LORCS, NAM, OAU, UN, UNCTAD, UNESCO, UNIDO, UPU, WADB, WCL, WHO, WIPO, WMO, WTO
Diplomatic representation in US:
chief of mission: Ambassador Ellom-Kodjo SCHUPPIUS
chancery: 2208 Massachusetts Avenue NW, Washington, DC 20008
telephone: (202) 234-4212 or 4213
US diplomatic representation:
chief of mission: Ambassador Harmon E. KIRBY
embassy: Rue Pelletier Caventou and Rue Vauban, Lome
mailing address: B. P. 852, Lome

telephone: [228] 21-29-91 through 94 and 21-77-17
FAX: [228] 21-79-52
Flag: five equal horizontal bands of green (top and bottom) alternating with yellow; there is a white five-pointed star on a red square in the upper hoist-side corner; uses the popular pan-African colors of Ethiopia

Economy

Overview: The economy is heavily dependent on subsistence agriculture, which accounts for about 33% of GDP and provides employment for 78% of the labor force. Primary agricultural exports are cocoa, coffee, and cotton, which together account for about 30% of total export earnings. Togo is self-sufficient in basic foodstuffs when harvests are normal. In the industrial sector phosphate mining is by far the most important activity, with phosphate exports accounting for about 40% of total foreign exchange earnings. Togo serves as a regional commercial and trade center. The government, over the past decade, with IMF and World Bank support, has been implementing a number of economic reform measures to encourage foreign investment and bring revenues in line with expenditures. Political unrest, including private and public sector strikes throughout 1991 and 1992, has jeopardized the reform program and has disrupted vital economic activity.
National product: GDP—exchange rate conversion—$1.5 billion (1991 est.)
National product real growth rate: 0% (1991 est.)
National product per capita: $400 (1991 est.)
Inflation rate (consumer prices): 0.5% (1991 est.)
Unemployment rate: 2% (1987)
Budget: revenues $284.8 million; expenditures $407 million, including capital expenditures of $NA (1991 est.)
Exports: $512 million (f.o.b., 1991 est.)
commodities: phosphates, cotton, cocoa, coffee
partners: EC 40%, Africa 16%, US 1% (1990)
Imports: $583 million (f.o.b., 1991 est.)
commodities: machinery and equipment, consumer goods, food, chemical products
partners: EC 57%, Africa 17%, US 5%, Japan 4% (1990)
External debt: $1.3 billion (1991)
Industrial production: growth rate 9.0% (1991 est.); accounts for 20% of GDP
Electricity: 179,000 kW capacity; 209 million kWh produced, 60 kWh per capita (1990)
Industries: phosphate mining, agricultural processing, cement, handicrafts, textiles, beverages

Agriculture: accounts for 33% of GDP; cash crops—coffee, cocoa, cotton; food crops—yams, cassava, corn, beans, rice, millet, sorghum; livestock production not significant; annual fish catch, 10,000-14,000 tons

Economic aid: US commitments, including Ex-Im (FY70-90), $142 million; Western (non-US) countries, ODA and OOF bilateral commitments (1970-90), $2 billion; OPEC bilateral aid (1979-89), $35 million; Communist countries (1970-89), $51 million

Currency: 1 CFA franc (CFAF) = 100 centimes

Exchange rates: Communaute Financiere Africaine francs (CFAF) per US$1—274.06 (January 1993), 264.69 (1992), 282.11 (1991), 272.26 (1990), 319.01 (1989), 297.85 (1988)

Fiscal year: calendar year

Communications

Railroads: 570 km 1.000-meter gauge, single track

Highways: 6,462 km total; 1,762 km paved; 4,700 km unimproved roads

Inland waterways: 50 km Mono River

Ports: Lome, Kpeme (phosphate port)

Merchant marine: 2 roll-on/roll-off ships (1,000 GRT or over) totaling 11,118 GRT/20,529 DWT

Airports:
total: 9
usable: 9
with permanent-surface runways: 2
with runways over 3,659 m: 0
with runways 2,440-3,659 m: 2
with runways 1,220-2,439 m: 0

Telecommunications: fair system based on network of radio relay routes supplemented by open wire lines; broadcast stations—2 AM, no FM, 3 (2 relays) TV; satellite earth stations—1 Atlantic Ocean INTELSAT and 1 SYMPHONIE

Defense Forces

Branches: Army, Navy, Air Force, Gendarmerie

Manpower availability: males age 15-49 862,427; fit for military service 452,974 (1993 est.); no conscription

Defense expenditures: exchange rate conversion—$43 million, about 3% of GDP (1989)

Tokelau
(territory of New Zealand)

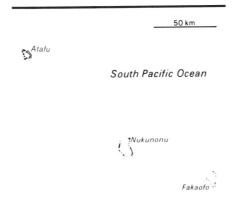

Geography

Location: Oceania, 3,750 km southwest of Honolulu in the South Pacific Ocean, about halfway between Hawaii and New Zealand

Map references: Oceania

Area:
total area: 10 km²
land area: 10 km²
comparative area: about 17 times the size of The Mall in Washington, DC

Land boundaries: 0 km

Coastline: 101 km

Maritime claims:
exclusive economic zone: 200 nm
territorial sea: 12 nm

International disputes: none

Climate: tropical; moderated by trade winds (April to November)

Terrain: coral atolls enclosing large lagoons

Natural resources: negligible

Land use:
arable land: 0%
permanent crops: 0%
meadows and pastures: 0%
forest and woodland: 0%
other: 100%

Irrigated land: NA km²

Environment: lies in Pacific typhoon belt

People

Population: 1,544 (July 1993 est.)

Population growth rate: -1.35% (1993 est.)

Birth rate: NA births/1,000 population

Death rate: NA deaths/1,000 population

Net migration rate: NA migrant(s)/1,000 population

Infant mortality rate: NA deaths/1,000 live births

Life expectancy at birth:
total population: NA years
male: NA years
female: NA years

Total fertility rate: NA children born/woman

Nationality:
noun: Tokelauan(s)
adjective: Tokelauan

Ethnic divisions: Polynesian

Religions: Congregational Christian Church 70%, Roman Catholic 28%, other 2%
note: on Atafu, all Congregational Christian Church of Samoa; on Nukunonu, all Roman Catholic; on Fakaofo, both denominations, with the Congregational Christian Church predominant

Languages: Tokelauan (a Polynesian language), English

Literacy:
total population: NA%
male: NA%
female: NA%

Labor force: NA

Government

Names:
conventional long form: none
conventional short form: Tokelau

Digraph: TL

Type: territory of New Zealand

Capital: none; each atoll has its own administrative center

Administrative divisions: none (territory of New Zealand)

Independence: none (territory of New Zealand)

Constitution: administered under the Tokelau Islands Act of 1948, as amended in 1970

Legal system: British and local statutes

National holiday: Waitangi Day, 6 February (1840) (Treaty of Waitangi established British sovereignty over New Zealand)

Political parties and leaders: NA

Suffrage: NA

Elections: NA

Executive branch: British monarch, administrator (appointed by the Minister of Foreign Affairs in New Zealand), official secretary

Legislative branch: unicameral Council of Elders (Taupulega) on each atoll

Judicial branch: High Court in Niue, Supreme Court in New Zealand

Leaders:
Chief of State: Queen ELIZABETH II (since 6 February 1952)
Head of Government: Administrator Graham ANSELL (since NA 1990); Official Secretary Casimilo J. PEREZ (since NA), Office of Tokelau Affairs; Tokelau's governing Council will elect its first head of government in 1993

Member of: SPC, WHO (associate)

Diplomatic representation in US: none (territory of New Zealand)

US diplomatic representation: none (territory of New Zealand)

Flag: the flag of New Zealand is used

Tokelau (continued)

Economy

Overview: Tokelau's small size, isolation, and lack of resources greatly restrain economic development and confine agriculture to the subsistence level. The people must rely on aid from New Zealand to maintain public services, annual aid being substantially greater than GDP. The principal sources of revenue come from sales of copra, postage stamps, souvenir coins, and handicrafts. Money is also remitted to families from relatives in New Zealand.
National product: GDP—exchange rate conversion—$1.4 million (1988 est.)
National product real growth rate: NA%
National product per capita: $800 (1988 est.)
Inflation rate (consumer prices): NA%
Unemployment rate: NA%
Budget: revenues $430,830; expenditures $2.8 million, including capital expenditures of $37,300 (FY87)
Exports: $98,000 (f.o.b., 1983)
commodities: stamps, copra, handicrafts
partners: NZ
Imports: $323,400 (c.i.f., 1983)
commodities: foodstuffs, building materials, fuel
partners: NZ
External debt: none
Industrial production: growth rate NA%
Electricity: 200 kW capacity; 300,000 kWh produced, 180 kWh per capita (1990)
Industries: small-scale enterprises for copra production, wood work, plaited craft goods; stamps, coins; fishing
Agriculture: coconuts, copra; basic subsistence crops—breadfruit, papaya, bananas; pigs, poultry, goats
Economic aid: Western (non-US) countries, ODA and OOF bilateral commitments (1970-89), $24 million
Currency: 1 New Zealand dollar (NZ$) = 100 cents
Exchange rates: New Zealand dollars (NZ$) per US$1—1.9486 (January 1993), 1.8584 (1992), 1.7265 (1991), 1.6750 (1990), 1.6708 (1989), 1.5244 (1988)
Fiscal year: 1 April-31 March

Communications

Ports: none; offshore anchorage only
Airports: none; lagoon landings by amphibious aircraft from Western Samoa
Telecommunications: radiotelephone service between islands and to Western Samoa

Defense Forces

Note: defense is the responsibility of New Zealand

Tonga

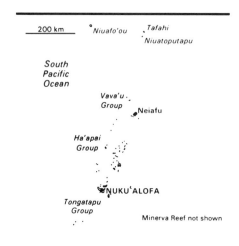

Geography

Location: Oceania, 2,250 km north-northwest of New Zealand, about two-thirds of the way between Hawaii and New Zealand
Map references: Oceania, Standard Time Zones of the World
Area:
total area: 748 km²
land area: 718 km²
comparative area: slightly more than four times the size of Washington, DC
Land boundaries: 0 km
Coastline: 419 km
Maritime claims:
continental shelf: not specified
exclusive economic zone: 200 nm
territorial sea: 12 nm
International disputes: none
Climate: tropical; modified by trade winds; warm season (December to May), cool season (May to December)
Terrain: most islands have limestone base formed from uplifted coral formation; others have limestone overlying volcanic base
Natural resources: fish, fertile soil
Land use:
arable land: 25%
permanent crops: 55%
meadows and pastures: 6%
forest and woodland: 12%
other: 2%
Irrigated land: NA km²
Environment: archipelago of 170 islands (36 inhabited); subject to cyclones (October to April); deforestation

People

Population: 103,949 (July 1993 est.)
Population growth rate: 0.8% (1993 est.)
Birth rate: 25.16 births/1,000 population (1993 est.)
Death rate: 6.75 deaths/1,000 population (1993 est.)
Net migration rate: -10.4 migrant(s)/1,000 population (1993 est.)

Infant mortality rate: 21.38 deaths/1,000 live births (1993 est.)
Life expectancy at birth:
total population: 67.79 years
male: 65.5 years
female: 70.24 years (1993 est.)
Total fertility rate: 3.68 children born/woman (1993 est.)
Nationality:
noun: Tongan(s)
adjective: Tongan
Ethnic divisions: Polynesian, Europeans about 300
Religions: Christian (Free Wesleyan Church claims over 30,000 adherents)
Languages: Tongan, English
Literacy: age 15 and over can read and write (1976)
total population: 57%
male: 60%
female: 60%
Labor force: NA
by occupation: agriculture 70%, mining (600 engaged in mining)

Government

Names:
conventional long form: Kingdom of Tonga
conventional short form: Tonga
former: Friendly Islands
Digraph: TN
Type: hereditary constitutional monarchy
Capital: Nuku'alofa
Administrative divisions: three island groups; Ha'apai, Tongatapu, Vava'u
Independence: 4 June 1970 (from UK)
Constitution: 4 November 1875, revised 1 January 1967
Legal system: based on English law
National holiday: Emancipation Day, 4 June (1970)
Political parties and leaders: Democratic Reform Movement, 'Akilisi POHIVA; Christian Democratic Party, leader NA
Suffrage: all literate, tax-paying males and all literate females over 21
Elections:
Legislative Assembly: last held 14-15 February 1990 (next to be held 3-4 February 1993); results—percent of vote NA; seats—(29 total, 9 elected) 6 proreform, 3 traditionalist
Executive branch: monarch, prime minister, deputy prime minister, Council of Ministers (cabinet), Privy Council
Legislative branch: unicameral Legislative Assembly (Fale Alea)
Judicial branch: Supreme Court
Leaders:
Chief of State: King Taufa'ahau TUPOU IV (since 16 December 1965)
Head of Government: Prime Minister Baron VAEA (since 22 August 1991); Deputy Prime Minister S. Langi KAVALIKU (since 22 August 1991)

Member of: ACP, AsDB, C, ESCAP, FAO, G-77, IBRD, ICAO, ICFTU, IDA, IFAD, IFC, IMF, INTERPOL, IOC, ITU, LORCS, SPARTECA, SPC, SPF, UNCTAD, UNESCO, UNIDO, UPU, WHO
Diplomatic representation in US: Ambassador Sione KITE, resides in London
US diplomatic representation: the US has no offices in Tonga; the ambassador to Fiji is accredited to Tonga and makes periodic visits
Flag: red with a bold red cross on a white rectangle in the upper hoist-side corner

Economy

Overview: The economy's base is agriculture, which employs about 70% of the labor force and contributes 40% to GDP. Coconuts, bananas, and vanilla beans are the main crops and make up two-thirds of exports. The country must import a high proportion of its food, mainly from New Zealand. The manufacturing sector accounts for only 11% of GDP. Tourism is the primary source of hard currency earnings, but the island remains dependent on sizable external aid and remittances to offset its trade deficit.
National product: GDP—exchange rate conversion—$92 million (FY90)
National product real growth rate: 0.4% (FY92 est.)
National product per capita: $900 (FY90)
Inflation rate (consumer prices): 4% (FY92 est.)
Unemployment rate: NA%
Budget: revenues $36.4 million; expenditures $68.1 million, including capital expenditures of $33.2 million (FY91 est.)
Exports: $18.8 million (f.o.b., FY92 est.)
commodities: coconut oil, desiccated coconut, copra, bananas, taro, vanilla beans, fruits, vegetables, fish
partners: Japan 34%, US 17%, Australia 13%, NZ 13% (FY91)
Imports: $68.3 million (c.i.f., FY92 est.)
commodities: food products, machinery and transport equipment, manufactures, fuels, chemicals
partners: NZ 33%, Australia 22%, US 8%, Japan 8% (FY91)
External debt: $47.5 million (FY91)
Industrial production: growth rate 1.7% (FY90); accounts for 11% of GDP
Electricity: 6,000 kW capacity; 8 million kWh produced, 80 kWh per capita (1990)
Industries: tourism, fishing
Agriculture: accounts for 40% of GDP; dominated by coconut, copra, and banana production; vanilla beans, cocoa, coffee, ginger, black pepper
Economic aid: US commitments, including Ex-Im (FY70-89), $16 million; Western (non-US) countries, ODA and OOF bilateral

commitments (1970-89), $258 million
Currency: 1 pa'anga (T$) = 100 seniti
Exchange rates: pa'anga (T$) per US$1—1.3996 (January 1993), 1.3471 (1992), 1.2961 (1991), 1.2809 (1990), 1.2637 (1989), 1.2799 (1988)
Fiscal year: 1 July-30 June

Communications

Highways: 198 km sealed road (Tongatapu); 74 km (Vava'u); 94 km unsealed roads usable only in dry weather
Ports: Nukualofa, Neiafu, Pangai
Merchant marine: 3 ships (1,000 GRT or over) totaling 6,765 GRT/10,597 DWT; includes 1 cargo, 1 roll-on/roll-off cargo, 1 liquefied gas
Airports:
total: 6
usable: 6
with permanent-surface runways: 1
with runways over 3,659 m: 0
with runways 2,440-3,659: 1
with runways 1,220-2,439 m: 1
Telecommunications: 3,529 telephones; 66,000 radios; no TV sets; broadcast stations—1 AM, no FM, no TV; 1 Pacific Ocean INTELSAT earth station

Defense Forces

Branches: Tonga Defense Force, Tonga Maritime Division, Royal Tongan Marines, Royal Tongan Guard, Police
Manpower availability: NA
Defense expenditures: exchange rate conversion—$NA, NA% of GDP

Trinidad and Tobago

Geography

Location: in the extreme southeastern Caribbean Sea, 11 km off the coast of Venezuela
Map references: Central America and the Caribbean, South America, Standard Time Zones of the World
Area:
total area: 5,130 km²
land area: 5,130 km²
comparative area: slightly smaller than Delaware
Land boundaries: 0 km
Coastline: 362 km
Maritime claims:
contiguous zone: 24 nm
continental shelf: 200 nm or the outer edge of continental margin
exclusive economic zone: 200 nm
territorial sea: 12 nm
International disputes: none
Climate: tropical; rainy season (June to December)
Terrain: mostly plains with some hills and low mountains
Natural resources: petroleum, natural gas, asphalt
Land use:
arable land: 14%
permanent crops: 17%
meadows and pastures: 2%
forest and woodland: 44%
other: 23%
Irrigated land: 220 km² (1989 est.)
Environment: outside usual path of hurricanes and other tropical storms

People

Population: 1,313,738 (July 1993 est.)
Population growth rate: 1.1% (1993 est.)
Birth rate: 20.08 births/1,000 population (1993 est.)
Death rate: 6.31 deaths/1,000 population (1993 est.)
Net migration rate: -2.74 migrant(s)/1,000 population (1993 est.)

Infant mortality rate: 16.9 deaths/1,000 live births (1993 est.)

Life expectancy at birth:
total population: 70.53 years
male: 67.91 years
female: 73.22 years (1993 est.)

Total fertility rate: 2.35 children born/woman (1993 est.)

Nationality:
noun: Trinidadian(s), Tobagonian(s)
adjective: Trinidadian, Tobagonian

Ethnic divisions: black 43%, East Indian 40%, mixed 14%, white 1%, Chinese 1%, other 1%

Religions: Roman Catholic 32.2%, Hindu 24.3%, Anglican 14.4%, other Protestant 14%, Muslim 6%, none or unknown 9.1%

Languages: English (official), Hindi, French, Spanish

Literacy: age 15 and over can read and write (1980)
total population: 95%
male: 97%
female: 93%

Labor force: 463,900
by occupation: construction and utilities 18.1%, manufacturing, mining, and quarrying 14.8%, agriculture 10.9%, other 56.2% (1985 est.)

Government

Names:
conventional long form: Republic of Trinidad and Tobago
conventional short form: Trinidad and Tobago

Digraph: TD

Type: parliamentary democracy

Capital: Port-of-Spain

Administrative divisions: 8 counties, 3 municipalities*, and 1 ward**; Arima*, Caroni, Mayaro, Nariva, Port-of-Spain*, Saint Andrew, Saint David, Saint George, Saint Patrick, San Fernando*, Tobago**, Victoria

Independence: 31 August 1962 (from UK)

Constitution: 31 August 1976

Legal system: based on English common law; judicial review of legislative acts in the Supreme Court; has not accepted compulsory ICJ jurisdiction

National holiday: Independence Day, 31 August (1962)

Political parties and leaders: People's National Movement (PNM), Patrick MANNING; United National Congress (UNC), Basdeo PANDAY; National Alliance for Reconstruction (NAR), Carson CHARLES; Movement for Social Transformation (MOTION), David ABDULLAH; National Joint Action Committee (NJAC), Makandal DAAGA

Suffrage: 18 years of age; universal

Elections:
House of Representatives: last held 16 December 1991 (next to be held by December 1996); results—PNM 32%, UNC 13%, NAR 2%; seats—(36 total) PNM 21, UNC 13, NAR 2

Executive branch: president, prime minister, Cabinet

Legislative branch: bicameral Parliament consists of an upper house or Senate and a lower house or House of Representatives

Judicial branch: Court of Appeal, Supreme Court

Leaders:
Chief of State: President Noor Mohammed HASSANALI (since 18 March 1987)
Head of Government: Prime Minister Patrick Augustus Mervyn MANNING (since 17 December 1991)

Member of: ACP, C, CARICOM, CCC, CDB, ECLAC, FAO, G-24, G-77, GATT, IADB, IBRD, ICAO, ICFTU, IDA, IFAD, IFC, ILO, IMF, IMO, INTELSAT, INTERPOL, IOC, ISO, ITU, LAES, LORCS, NAM, OAS, OPANAL, UN, UNCTAD, UNESCO, UNIDO, UPU, WFTU, WHO, WIPO, WMO

Diplomatic representation in US:
chief of mission: Ambassador Corinne BAPTISTE
chancery: 1708 Massachusetts Avenue NW, Washington, DC 20036
telephone: (202) 467-6490
consulate general: New York

US diplomatic representation:
chief of mission: Ambassador Sally G. COWAL
embassy: 15 Queen's Park West, Port-of-Spain
mailing address: P. O. Box 752, Port-of-Spain
telephone: (809) 622-6372 through 6376, 6176
FAX: (809) 628-5462

Flag: red with a white-edged black diagonal band from the upper hoist side

---

Economy

Overview: Trinidad and Tobago's petroleum-based economy has begun to emerge from a lengthy depression in the last few years. The economy fell sharply through most of the 1980s, largely because of the decline in oil prices. This sector accounts for 80% of export earnings and almost 20% of GDP. The government, in response to the oil revenue loss, pursued a series of austerity measures that pushed the unemployment rate as high as 22% in 1988. The economy showed signs of recovery in 1990 and 1991, however, helped along by rising oil prices. Agriculture employs only about 11% of the labor force and produces about 3% of GDP. Since this sector is small, it has been unable to absorb the large numbers of the unemployed. The government currently seeks to diversify its export base.

National product: GDP—exchange rate conversion—$5 billion (1991)

National product real growth rate: 2.6% (1991)

National product per capita: $3,800 (1991)

Inflation rate (consumer prices): 3.8% (1991)

Unemployment rate: 18.5% (1991)

Budget: revenues $1.6 billion; expenditures $1.6 billion, including capital expenditures of $158 million (1993 est.)

Exports: $2.2 billion (f.o.b., 1991)
commodities: includes reexports—petroleum and petroleum products 82%, steel products 9%, fertilizer, sugar, cocoa, coffee, citrus (1988)
partners: US 49%, CARICOM 12%

Imports: $1.7 billion (c.i.f., 1991)
commodities: raw materials and intermediate goods 48%, capital goods 29%, consumer goods 23% (1991)
partners: US 39%, Venezuela 14%, UK 7%, CARICOM 5% (1991)

External debt: $2.4 billion (1991)

Industrial production: growth rate 2.3%, excluding oil refining (1986); accounts for 40% of GDP, including petroleum

Electricity: 1,176,000 kW capacity; 3,480 million kWh produced, 2,680 kWh per capita (1992)

Industries: petroleum, chemicals, tourism, food processing, cement, beverage, cotton textiles

Agriculture: accounts for 3% of GDP; highly subsidized sector; major crops—cocoa, sugarcane; sugarcane acreage is being shifted into rice, citrus, coffee, vegetables; poultry sector most important source of animal protein; must import large share of food needs

Illicit drugs: transshipment point for South American drugs destined for the US

Economic aid: US commitments, including Ex-Im (FY70-89), $373 million; Western (non-US) countries, ODA and OOF bilateral commitments (1970-89), $518 million

Currency: 1 Trinidad and Tobago dollar (TT$) = 100 cents

Exchange rates: Trinidad and Tobago dollars (TT$) per US$1—4.2500 (fixed rate since 1989)

Fiscal year: calendar year

Communications

Railroads: minimal agricultural railroad system near San Fernando

Highways: 8,000 km total; 4,000 km paved, 1,000 km improved earth, 3,000 km unimproved earth

Pipelines: crude oil 1,032 km, petroleum products 19 km, natural gas 904 km

Ports: Port-of-Spain, Pointe-a-Pierre, Scarborough
Merchant marine: 2 cargo ships (1,000 GRT or over) totaling 12,507 GRT/21,923 DWT
Airports:
total: 6
usable: 5
with permanent-surface runways: 2
with runways over 3,659 m: 0
with runways 2,440-3,659 m: 2
with runways 1,220-2,439 m: 1
Telecommunications: excellent international service via tropospheric scatter links to Barbados and Guyana; good local service; 109,000 telephones; broadcast stations—2 AM, 4 FM, 5 TV; 1 Atlantic Ocean INTELSAT earth station

---

Defense Forces

Branches: Trinidad and Tobago Defense Force (including Ground Forces, Coast Guard, and Air Wing), Trinidad and Tobago Police Service
Manpower availability: males age 15-49 351,183; fit for military service 253,084 (1993 est.)
Defense expenditures: exchange rate conversion—$59 million, 1-2% of GDP (1989 est.)

Tromelin Island
(possession of France)

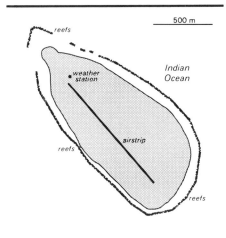

Geography

Location: in the western Indian Ocean, 350 km east of Madagascar and 600 km north of Reunion
Map references: World
Area:
total area: 1 km²
land area: 1 km²
comparative area: about 1.7 times the size of The Mall in Washington, DC
Land boundaries: 0 km
Coastline: 3.7 km
Maritime claims:
contiguous zone: 12 nm
continental shelf: 200 m depth or to depth of exploitation
exclusive economic zone: 200 nm
territorial sea: 12 nm
International disputes: claimed by Madagascar, Mauritius, and Seychelles
Climate: tropical
Terrain: sandy
Natural resources: fish
Land use:
arable land: 0%
permanent crops: 0%
meadows and pastures: 0%
forest and woodland: 0%
other: 100% (scattered bushes)
Irrigated land: 0 km²
Environment: wildlife sanctuary
Note: climatologically important location for forecasting cyclones

---

People

Population: uninhabited

Government

Names:
conventional long form: none
conventional short form: Tromelin Island
local long form: none
local short form: Ile Tromelin

Digraph: TE
Type: French possession administered by Commissioner of the Republic, resident in Reunion
Capital: none; administered by France from Reunion
Independence: none (possession of France)

Economy

Overview: no economic activity

Communications

Ports: none; offshore anchorage only
Airports:
total: 1
usable: 1
with permanent-surface runways: 0
with runways over 3,659 m: 0
with runways 2,440-3,659 m: 0
with runways 1,220-2,439 m: 0
Telecommunications: important meteorological station

Defense Forces

Note: defense is the responsibility of France

Tunisia

Geography

Location: Northern Africa, 144 km from Italy across the Strait of Sicily, between Algeria and Libya
Map references: Africa, Standard Time Zones of the World
Area:
total area: 163,610 km²
land area: 155,360 km²
comparative area: slightly larger than Georgia
Land boundaries: total 1,424 km, Algeria 965 km, Libya 459 km
Coastline: 1,148 km
Maritime claims:
territorial sea: 12 nm
International disputes: maritime boundary dispute with Libya; land boundary disputes with Algeria under discussion
Climate: temperate in north with mild, rainy winters and hot, dry summers; desert in south
Terrain: mountains in north; hot, dry central plain; semiarid south merges into the Sahara
Natural resources: petroleum, phosphates, iron ore, lead, zinc, salt
Land use:
arable land: 20%
permanent crops: 10%
meadows and pastures: 19%
forest and woodland: 4%
other: 47%
Irrigated land: 2,750 km² (1989)
Environment: deforestation; overgrazing; soil erosion; desertification
Note: strategic location in central Mediterranean

People

Population: 8,570,868 (July 1993 est.)
Population growth rate: 1.84% (1993 est.)
Birth rate: 24.24 births/1,000 population (1993 est.)
Death rate: 5.04 deaths/1,000 population (1993 est.)

Net migration rate: -0.79 migrant(s)/1,000 population (1993 est.)
Infant mortality rate: 35.9 deaths/1,000 live births (1993 est.)
Life expectancy at birth:
total population: 72.54 years
male: 70.55 years
female: 74.62 years (1993 est.)
Total fertility rate: 3.02 children born/woman (1993 est.)
Nationality:
noun: Tunisian(s)
adjective: Tunisian
Ethnic divisions: Arab-Berber 98%, European 1%, Jewish less than 1%
Religions: Muslim 98%, Christian 1%, Jewish 1%
Languages: Arabic (official and one of the languages of commerce), French (commerce)
Literacy: age 15 and over can read and write (1990)
total population: 65%
male: 74%
female: 56%
Labor force: 2.25 million
by occupation: agriculture 32%
note: shortage of skilled labor

Government

Names:
conventional long form: Republic of Tunisia
conventional short form: Tunisia
local long form: Al Jumhuriyah at Tunisiyah
local short form: Tunis
Digraph: TS
Type: republic
Capital: Tunis
Administrative divisions: 23 governorates; Beja, Ben Arous, Bizerte, Gabes, Gafsa, Jendouba, Kairouan, Kasserine, Kebili, L'Ariana, Le Kef, Mahdia, Medenine, Monastir, Nabeul, Sfax, Sidi Bou Zid, Siliana, Sousse, Tataouine, Tozeur, Tunis, Zaghouan
Independence: 20 March 1956 (from France)
Constitution: 1 June 1959
Legal system: based on French civil law system and Islamic law; some judicial review of legislative acts in the Supreme Court in joint session
National holiday: National Day, 20 March (1956)
Political parties and leaders: Constitutional Democratic Rally Party (RCD), President BEN ALI (official ruling party); Movement of Democratic Socialists (MDS), Mohammed MOUAADA; five other political parties are legal, including the Communist Party
Other political or pressure groups: the Islamic fundamentalist party, An Nahda (Rebirth), is outlawed
Suffrage: 20 years of age; universal

Elections:
President: last held 2 April 1989 (next to be held NA March 1994); results—Gen. Zine el Abidine BEN ALI was reelected without opposition
Chamber of Deputies: last held 2 April 1989 (next to be held NA April 1994); results—RCD 80.7%, independents/Islamists 13.7%, MDS 3.2%, other 2.4%; seats—(141 total) RCD 141
Executive branch: president, prime minister, Cabinet
Legislative branch: unicameral Chamber of Deputies (Majlis al-Nuwaab)
Judicial branch: Court of Cassation (Cour de Cassation)
Leaders:
Chief of State: President Gen. Zine el Abidine BEN ALI (since 7 November 1987)
Head of Government: Prime Minister Hamed KAROUI (since 26 September 1989)
Member of: ABEDA, ACCT, AfDB, AFESD, AL, AMF, AMU, CCC, ECA, FAO, G-77, GATT, IAEA, IBRD, ICAO, ICC, ICFTU, IDA, IDB, IFAD, IFC, ILO, IMF, IMO, INMARSAT, INTELSAT, INTERPOL, IOC, ISO, ITU, LORCS, MINURSO, NAM, OAPEC (withdrew from active membership in 1986), OAU, OIC, UN, UNCTAD, UNESCO, UNHCR, UNIDO, UNPROFOR, UNTAC, UPU, WHO, WIPO, WMO, WTO
Diplomatic representation in US:
chief of mission: Ambassador Ismail KHELIL
chancery: 1515 Massachusetts Avenue NW, Washington, DC 20005
telephone: (202) 862-1850
US diplomatic representation:
chief of mission: Ambassador John T. McCARTHY
embassy: 144 Avenue de la Liberte, 1002 Tunis-Belvedere
mailing address: use embassy street address
telephone: [216] (1) 782-566
FAX: [216] (1) 789-719
Flag: red with a white disk in the center bearing a red crescent nearly encircling a red five-pointed star; the crescent and star are traditional symbols of Islam

Economy

Overview: The economy depends primarily on petroleum, phosphates, tourism, and exports of light manufactures. Following two years of drought-induced economic decline, the economy came back strongly in 1990-92 as a result of good harvests, continued export growth, and higher domestic investment. High unemployment has eroded popular support for the government, however, and forced Tunis to slow the pace of economic reform. Nonetheless, the government appears committed to implementing its IMF-supported structural adjustment program and to servicing its foreign debt.

National product: GDP—exchange rate conversion—$13.6 billion (1992 est.)
National product real growth rate: 8% (1992 est.)
National product per capita: $1,650 (1992 est.)
Inflation rate (consumer prices): 6% (1992 est.)
Unemployment rate: 15.7% (1992)
Budget: revenues $4.3 billion; expenditures $5.5 billion, including capital expenditures of $NA (1993 est.)
Exports: $3.7 billion (f.o.b., 1992)
commodities: hydrocarbons, agricultural products, phosphates and chemicals
partners: EC countries 74%, Middle East 11%, US 2%, Turkey, former USSR republics
Imports: $6.1 billion (c.i.f., 1992)
commodities: industrial goods and equipment 57%, hydrocarbons 13%, food 12%, consumer goods
partners: EC countries 67%, US 6%, Canada, Japan, Switzerland, Turkey, Algeria
External debt: $7.7 billion (1992 est.)
Industrial production: growth rate 5% (1989); accounts for about 25% of GDP, including petroleum
Electricity: 1,545,000 kW capacity; 5,096 million kWh produced, 600 kWh per capita (1992)
Industries: petroleum, mining (particularly phosphate and iron ore), tourism, textiles, footwear, food, beverages
Agriculture: accounts for 15% of GDP and one-third of labor force; output subject to severe fluctuations because of frequent droughts; export crops—olives, dates, oranges, almonds; other products—grain, sugar beets, wine grapes, poultry, beef, dairy; not self-sufficient in food; fish catch of 99,200 metric tons (1987)
Economic aid: US commitments, including Ex-Im (FY70-89), $730 million; Western (non-US) countries, ODA and OOF bilateral commitments (1970-89), $5.2 billion; OPEC bilateral aid (1979-89), $684 million; Communist countries (1970-89), $410 million
Currency: 1 Tunisian dinar (TD) = 1,000 millimes
Exchange rates: Tunisian dinars (TD) per US$1—0.9931 (February 1993), 0.8844 (1992), 0.9246 (1991), 0.8783 (1990), 0.9493 (1989), 0.8578 (1988)
Fiscal year: calendar year

Communications

Railroads: 2,115 km total; 465 km 1.435-meter (standard) gauge; 1,650 km 1.000-meter gauge

Highways: 17,700 km total; 9,100 km bituminous; 8,600 km improved and unimproved earth
Pipelines: crude oil 797 km, petroleum products 86 km, natural gas 742 km
Ports: Bizerte, Gabes, Sfax, Sousse, Tunis, La Goulette, Zarzis
Merchant marine: 22 ships (1,000 GRT or over) totaling 161,661 GRT/221,959 DWT; includes 1 short-sea passenger, 4 cargo, 2 roll-on/roll-off cargo, 2 oil tanker, 6 chemical tanker, 1 liquefied gas, 6 bulk
Airports:
total: 29
usable: 26
with permanent-surface runways: 13
with runways over 3,659 m: 0
with runways 2,440-3,659 m: 7
with runways 1,220-2,439 m: 7
note: a new airport opened 6 May 1993, length and type of surface NA
Telecommunications: the system is above the African average; facilities consist of open-wire lines, coaxial cable, and microwave radio relay; key centers are Sfax, Sousse, Bizerte, and Tunis; 233,000 telephones (28 telephones per 1,000 persons); broadcast stations—7 AM, 8 FM, 19 TV; 5 submarine cables; satellite earth stations—1 Atlantic Ocean INTELSAT and 1 ARABSAT with back-up control station; coaxial cable and microwave radio relay to Algeria and Libya

---

Defense Forces

Branches: Army, Navy, Air Force, paramilitary forces, National Guard
Manpower availability: males age 15-49 2,164,686; fit for military service 1,244,683; reach military age (20) annually 90,349 (1993 est.)
Defense expenditures: exchange rate conversion—$618 million, 3.7% of GDP (1993 est.)

Turkey

Geography

Location: Southeastern Europe/Southwest Asia, bordering the Mediterranean Sea and Black Sea, between Bulgaria and Iran
Map references: Africa, Europe, Middle East, Standard Time Zones of the World
Area:
total area: 780,580 km²
land area: 770,760 km²
comparative area: slightly larger than Texas
Land boundaries: total 2,627 km, Armenia 268 km, Azerbaijan 9 km, Bulgaria 240 km, Georgia 252 km, Greece 206 km, Iran 499 km, Iraq 331 km, Syria 822 km
Coastline: 7,200 km
Maritime claims:
exclusive economic zone: in Black Sea only—to the maritime boundary agreed upon with the former USSR
territorial sea: 6 nm in the Aegean Sea, 12 nm in the Black Sea and in the Mediterranean Sea
International disputes: complex maritime and air (but not territorial) disputes with Greece in Aegean Sea; Cyprus question; Hatay question with Syria; ongoing dispute with downstream riparians (Syria and Iraq) over water development plans for the Tigris and Euphrates Rivers
Climate: temperate; hot, dry summers with mild, wet winters; harsher in interior
Terrain: mostly mountains; narrow coastal plain; high central plateau (Anatolia)
Natural resources: antimony, coal, chromium, mercury, copper, borate, sulphur, iron ore
Land use:
arable land: 30%
permanent crops: 4%
meadows and pastures: 12%
forest and woodland: 26%
other: 28%
Irrigated land: 22,200 km² (1989 est.)
Environment: subject to severe earthquakes, especially along major river valleys in west; air pollution; desertification

Turkey (continued)

Note: strategic location controlling the Turkish straits (Bosporus, Sea of Marmara, Dardanelles) that link Black and Aegean Seas

People

Population: 60,897,841 (July 1993 est.)
Population growth rate: 2.07% (1993 est.)
Birth rate: 26.62 births/1,000 population (1993 est.)
Death rate: 5.97 deaths/1,000 population (1993 est.)
Net migration rate: 0 migrant(s)/1,000 population (1993 est.)
Infant mortality rate: 52 deaths/1,000 live births (1993 est.)
Life expectancy at birth:
total population: 70.41 years
male: 68.11 years
female: 72.82 years (1993 est.)
Total fertility rate: 3.3 children born/woman (1993 est.)
Nationality:
noun: Turk(s)
adjective: Turkish
Ethnic divisions: Turkish 80%, Kurdish 20% (est.)
Religions: Muslim 99.8% (mostly Sunni), other 0.2% (Christian and Jews)
Languages: Turkish (official), Kurdish, Arabic
Literacy: age 15 and over can read and write (1990)
total population: 81%
male: 90%
female: 71%
Labor force: 20.7 million
by occupation: agriculture 50%, services 35%, industry 15%
note: about 1,800,000 Turks work abroad (1991)

Government

Names:
conventional long form: Republic of Turkey
conventional short form: Turkey
local long form: Turkiye Cumhuriyeti
local short form: Turkiye
Digraph: TU
Type: republican parliamentary democracy
Capital: Ankara
Administrative divisions: 73 provinces (iller, singular—il); Adana, Adiyaman, Afyon, Agri, Aksaray, Amasya, Ankara, Antalya, Artvin, Aydin, Balikesir, Batman, Bayburt, Bilecik, Bingol, Bitlis, Bolu, Burdur, Bursa, Canakkale, Cankiri, Corum, Denizli, Diyarbakir, Edirne, Elazig, Erzincan, Erzurum, Eskisehir, Gaziantep, Giresun, Gumushane, Hakkari, Hatay, Icel, Isparta, Istanbul, Izmir, Kahraman Maras, Karaman, Kars, Kastamonu, Kayseri, Kirikkale, Kirklareli, Kirsehir, Kocaeli, Konya, Kutahya, Malatya, Manisa, Mardin, Mugla, Mus, Nevsehir, Nigde, Ordu, Rize, Sakarya, Samsun, Siirt, Sinop, Sirnak, Sivas, Tekirdag, Tokat, Trabzon, Tunceli, Urfa, Usak, Van, Yozgat, Zonguldak
Independence: 29 October 1923 (successor state to the Ottoman Empire)
Constitution: 7 November 1982
Legal system: derived from various continental legal systems; accepts compulsory ICJ jurisdiction, with reservations
National holiday: Anniversary of the Declaration of the Republic, 29 October (1923)
Political parties and leaders: Correct Way Party (DYP), Suleyman DEMIREL; Motherland Party (ANAP), Mesut YILMAZ; Social Democratic Populist Party (SHP), Erdal INONU; Refah Party (RP), Necmettin ERBAKAN; Democratic Left Party (DSP), Bulent ECEVIT; Nationalist Labor Party (MCP), Alpaslan TURKES; People's Labor Party (HEP), Ahmet TURK; Socialist Unity Party (SBP), Saden AREN; Democratic Center Party (DSP), Bedrettin DALAN; Republican People's Party (CHP), Deniz BAYKAL; Workers' Party (IP), Dogu PERINCEK; National Party (MP), Aykut EDIBALI
Other political or pressure groups: Turkish Confederation of Labor (TURK-IS), Sevket YILMAZ
Suffrage: 21 years of age; universal
Elections:
Grand National Assembly: last held 20 October 1991 (next to be held NA October 1996); results—DYP 27.03%, ANAP 24.01%, SHP 20.75%, RP 16.88%, DSP 10.75%, SBP 0.44%, independent 0.14%; seats—(450 total) DYP 178, ANAP 115, SHP 86, RP 40, MCP 19, DSP 7, other 5
Executive branch: president, Presidential Council, prime minister, deputy prime minister, Cabinet
Legislative branch: unicameral Grand National Assembly (Buyuk Millet Meclisi)
Judicial branch: Court of Cassation
Leaders:
Chief of State: President Suleyman DEMIREL (since 16 May 1993)
Head of Government: Prime Minister Tansu CILLER (since NA June 1993)
Member of: AsDB, BIS, BSEC, CCC, CE, CERN (observer), COCOM, CSCE, EBRD, ECE, ECO, FAO, GATT, IAEA, IBRD, ICAO, ICC, ICFTU, IDA, IDB, IEA, IFAD, IFC, ILO, IMF, IMO, INMARSAT, INTELSAT, INTERPOL, IOC, IOM (observer), ISO, ITU, LORCS, NACC, NATO, NEA, OECD, OIC, PCA, UN, UNCTAD, UNESCO, UNHCR, UNIDO, UNIKOM, UNRWA, UPU, WHO, WIPO, WMO, WTO
Diplomatic representation in US:
chief of mission: Ambassador Nuzhet KANDEMIR
chancery: 1714 Massachusetts Avenue NW, Washington, DC 20036
telephone: (202) 659-8200
consulates general: Chicago, Houston, Los Angeles, and New York
US diplomatic representation:
chief of mission: Ambassador Richard C. BARKLEY
embassy: 110 Ataturk Boulevard, Ankara
mailing address: PSC 88, Box 5000, Ankara, or APO AE 09823
telephone: [90] (4) 426 54 70
FAX: [90] (4) 467-0057 and 0019
consulates general: Istanbul and Izmir
consulate: Adana
Flag: red with a vertical white crescent (the closed portion is toward the hoist side) and white five-pointed star centered just outside the crescent opening

Economy

Overview: After an impressive economic performance through most of the 1980s, Turkey has experienced erratic rates of economic growth since 1988—ranging from a high of 9.2% in 1990 to a low of 0.9% in 1991. Strong consumer demand and increased public investment led the way to a strong 5.9% growth in 1992. Chronic high inflation is Turkey's most serious economic problem, leading to high interest rates and the rapid depreciation of the Turkish lira. The huge public sector deficit—about 12% of GDP—and the Treasury's heavy reliance on Central Bank financing of the deficit are the major causes of Turkish inflation. Meanwhile, wage increases in both the public and private sector have outpaced productivity gains, limited the government's ability to reduce current expenditures, and hindered the return to profitability of many private companies. Agriculture remains an important economic sector, employing about half of the work force, contributing 18% to GDP, and accounting for about 20% of exports. The government has launched a multibillion-dollar development program in the southeastern region, which includes the building of a dozen dams on the Tigris and Euphrates Rivers to generate electric power and irrigate large tracts of farmland. The Turkish economy will probably continue to grow faster than the West European average in 1993, but the shaky coalition government of Prime Minister DEMIREL—which has seen its parliamentary majority shrink from 36 to 11 seats during its first year in power—is unlikely to risk further erosion of its support by implementing the belt-tightening measures necessary to substantially reduce inflation.
National product: GDP—purchasing power equivalent—$219 billion (1992)
National product real growth rate: 5.9% (1992)

National product per capita: $3,670 (1992)
Inflation rate (consumer prices): 70% (1992)
Unemployment rate: 11.1% (1992 est.)
Budget: revenues $40.5 billion; expenditures $46.8 billion, including capital expenditures of $5.5 billion (1993)
Exports: $13.7 billion (f.o.b., 1991)
commodities: manufactured goods 69%, foodstuffs 22%, fuels 2%
partners: EC countries 51%, US 7%, Iran 5%, former USSR 5%
Imports: $21.1 billion (c.i.f., 1991)
commodities: manufactured goods 61%, foodstuffs 8%, fuels 21%
partners: EC countries 44%, US 12%, former USSR 5%
External debt: $48.7 billion (1991)
Industrial production: growth rate 3.2% (1991 est.); accounts for 28% of GDP
Electricity: 14,400,000 kW capacity; 44,000 million kWh produced, 750 kWh per capita (1991)
Industries: textiles, food processing, mining (coal, chromite, copper, boron minerals), steel, petroleum, construction, lumber, paper
Agriculture: accounts for 18% of GDP and employs about half of working force; products—tobacco, cotton, grain, olives, sugar beets, pulses, citrus fruit, variety of animal products; self-sufficient in food most years
Illicit drugs: major transit route for Southwest Asian heroin and hashish to Western Europe and the US via air, land, and sea routes; major Turkish, Iranian, and other international trafficking organizations operate out of Istanbul; laboratories to convert imported morphine base into heroin have sprung up in remote regions of Turkey as well as near Istanbul; government maintains strict controls over areas of legal opium poppy cultivation and output of poppy straw concentrate
Economic aid: US commitments, including Ex-Im (FY70-89), $2.3 billion; Western (non-US) countries, ODA and OOF bilateral commitments (1970-89), $10.1 billion; OPEC bilateral aid (1979-89), $665 million; Communist countries (1970-89), $4.5 billion; note—aid for Persian Gulf war efforts from coalition allies (1991), $4.1 billion; aid pledged for Turkish Defense Fund, $2.5 billion
Currency: 1 Turkish lira (TL) = 100 kurus
Exchange rates: Turkish liras (TL) per US$1—8,814.3 (January 1993), 6,872.4 (1992), 4,171.8 (1991), 2,608.6 (1990), 2,121.7 (1989), 1,422.3 (1988)
Fiscal year: calendar year

Communications

Railroads: 8,429 km 1.435-meter gauge (including 795 km electrified)

Highways: 320,611 km total; 138 km limited access expressways, 31,062 km national (main) roads, 27,853 km regional (secondary) roads, 261,558 km local and municipal roads; 45,526 km of hard surfaced roads (of which about 27,000 km are paved and about 18,500 km are surfaced with gravel or crushed stone) (1988 est.)
Inland waterways: about 1,200 km
Pipelines: crude oil 1,738 km, petroleum products 2,321 km, natural gas 708 km
Ports: Iskenderun, Istanbul, Mersin, Izmir
Merchant marine: 353 ships (1,000 GRT or over) totaling 3,825,274 GRT/6,628,207 DWT; includes 7 short-sea passenger, 1 passenger-cargo, 189 cargo, 1 container, 6 roll-on/roll-off cargo, 2 refrigerated cargo, 1 livestock carrier, 39 oil tanker, 10 chemical tanker, 3 liquefied gas, 9 combination ore/oil, 2 specialized tanker, 80 bulk, 3 combination bulk
Airports:
total: 110
usable: 102
with permanent-surface runways: 65
with runways over 3,659 m: 3
with runways 2,440-3,659 m: 32
with runways 1,220-2,439 m: 26
Telecommunications: fair domestic and international systems; trunk radio relay microwave network; limited open wire network; 3,400,000 telephones; broadcast stations—15 AM; 94 FM; 357 TV; 1 satellite ground station operating in the INTELSAT (2 Atlantic Ocean antennas) and EUTELSAT systems; 1 submarine cable

---

Defense Forces

Branches: Land Forces, Navy (including Naval Air and Naval Infantry), Air Force, Coast Guard, Gendarmerie
Manpower availability: males age 15-49 15,691,874; fit for military service 9,579,453; reach military age (20) annually 604,816 (1993 est.)
Defense expenditures: exchange rate conversion—$5.6 billion, 3.9% of GDP (1992)

Turkmenistan

300 km

Geography

Location: South Asia, bordering the Caspian Sea, between Iran and Uzbekistan
Map references: Asia, Commonwealth of Independent States—Central Asian States, Standard Time Zones of the World
Area:
total area: 488,100 km^2
land area: 488,100 km^2
comparative area: slightly larger than California
Land boundaries: total 3,736 km, Afghanistan 744 km, Iran 992 km, Kazakhstan 379 km, Uzbekistan 1,621 km
Coastline: 0 km
note: Turkmenistan does border the Caspian Sea (1,768 km)
Maritime claims: landlocked, but boundaries in the Caspian Sea with Azerbaijan, Kazakhstan, and Iran will have to be negotiated
International disputes: none
Climate: subtropical desert
Terrain: flat-to-rolling sandy desert with dunes; borders Caspian Sea in west
Natural resources: petroleum, natural gas, coal, sulphur, salt
Land use:
arable land: 3%
permanent crops: 0%
meadows and pastures: 69%
forest and woodland: 0%
other: 28%
Irrigated land: 12,450 km^2 (1990)
Environment: contamination of soil and groundwater with agricultural chemicals, pesticides; salinization, water-logging of soil due to poor irrigation methods
Note: landlocked

People

Population: 3,914,997 (July 1993 est.)
Population growth rate: 2.04% (1993 est.)
Birth rate: 30.91 births/1,000 population (1993 est.)

391

Turkmenistan (continued)

Death rate: 7.6 deaths/1,000 population (1993 est.)

Net migration rate: -2.87 migrant(s)/1,000 population (1993 est.)

Infant mortality rate: 71.2 deaths/1,000 live births (1993 est.)

Life expectancy at birth:
total population: 64.93 years
male: 61.4 years
female: 68.62 years (1993 est.)

Total fertility rate: 3.82 children born/woman (1993 est.)

Nationality:
noun: Turkmen(s)
adjective: Turkmen

Ethnic divisions: Turkmen 73.3%, Russian 9.8%, Uzbek 9%, Kazakhs 2%, other 5.9%

Religions: Muslim 87%, Eastern Orthodox 11%, unknown 2%

Languages: Turkmen 72%, Russian 12%, Uzbek 9%, other 7%

Literacy: age 9-49 can read and write (1970)
total population: 100%
male: 100%
female: 100%

Labor force: 1.542 million
by occupation: agriculture and forestry 42%, industry and construction 21%, other 37% (1990)

Government

Names:
conventional long form: none
conventional short form: Turkmenistan
local long form: Tiurkmenostan Respublikasy
local short form: Turkmanistan
former: Turkmen Soviet Socialist Republic

Digraph: TX

Type: republic

Capital: Ashgabat (Ashkhabad)

Administrative divisions: 5 velayets: Balkan (Nebit Dag), Doshkhovuz (formerly Tashauz), Lebap (Charjev), Mary, Akhal (Ashgabat)
note: all oblasts have the same name as their administrative center except Balkan Oblast, centered at Nebit-Dag

Independence: 27 October 1991 (from the Soviet Union)

Constitution: adopted 18 May 1992

Legal system: based on civil law system

National holiday: Independence Day, 27 October (1991)

Political parties and leaders:
ruling party: Democratic Party (formerly Communist), chairman vacant
opposition: Party for Democratic Development, Durdymurat HOJA-MUHAMMET, chairman ; Agzybirlik, Nurberdy NURMAMEDOV, cochairman, Hubayberdi HALLIYEV, cochairman

Suffrage: 18 years of age; universal

Elections:
President: last held 21 June 1992 (next to be held NA June 1997); results—Saparmurad NIYAZOV 99.5% (ran unopposed)
Majlis: last held 7 January 1990 (next to be held NA 1995); results—percent of vote by party NA; seats—(175 total) elections not officially by party, but Communist Party members won nearly 90% of seats; note—seats to be reduced to 50 at next election

Executive branch: president, prime minister, nine deputy prime ministers, Council of Ministers

Legislative branch: under 1992 constitution there are two parliamentary bodies, a unicameral People's Council (Halk Maslahaty—having more than 100 members and meeting infrequently) and a 50-member unicameral Assembly (Majlis)

Judicial branch: Supreme Court

Leaders:
Chief of State: President Saparmurad NIYAZOV (since NA October 1990)
Head of Government: Prime Minister (vacant); Deputy Prime Ministers Valery G. OCHERTSOV, Orazgeldi AYDOGDYEV, Yagmur OVEZOV, Jourakuli BABAKULIYEV, Matkarim RAJAPOV, Rejep SAPAROV, Boris SHIKHMURADOV (since NA); Chairman of the People's Council Sakhat MURADOV (since NA)

Member of: CIS, CSCE, EBRD, ECO, ESCAP, IBRD, IMF, NACC, UN, UNCTAD

Diplomatic representation in US:
chief of mission: NA
chancery: NA
telephone: NA

US diplomatic representation:
chief of mission: Ambassador Joseph S. HULINGS III
embassy: Yubilenaya Hotel, Ashgabat (Ashkhabad)
mailing address: APO AE 09862
telephone: [7] 36320 24-49-08

Flag: green field, including a vertical stripe on the hoist side, with a claret veritcal stripe in between containing five white, black, and orange carpet guls (an assymetrical design used in producing rugs) associated with five different tribes; a white crescent and five white stars in the upper left corner to the right of the carpet guls

---

Economy

Overview: Like the other 15 former Soviet republics, Turkmenistan faces enormous problems of economic adjustment—to move away from Moscow-based central planning toward a system of decisionmaking by private entrepreneurs, local government authorities, and, hopefully, foreign investors. This process requires wholesale changes in supply sources, markets, property rights, and monetary arrangements. Industry—with 10% of the labor force—is heavily weighted toward the energy sector, which produced 11% of the ex-USSR's gas and 1% of its oil. Turkmenistan ranked second among the former Soviet republics in cotton production, mainly in the irrigated western region, where the huge Karakumskiy Canal taps the Amu Darya. The general decline in national product accelerated in 1992, principally because of inability to obtain spare parts and disputes with customers over the price of natural gas.

National product: GDP $NA

National product real growth rate: -10% (1992 est.)

National product per capita: $NA

Inflation rate (consumer prices): 53% per month (first quarter 1993)

Unemployment rate: 15%-20% (1992 est.)

Budget: revenues $NA; expenditures $NA, including capital expenditures of $NA

Exports: $100 million to outside the successor states of the former USSR (1992)
commodities: natural gas, oil, chemicals, cotton, textiles, carpets
partners: Russia, Ukraine, Uzbekistan

Imports: $100 million from outside the successor states of the former USSR (1992)
commodities: machinery and parts, plastics and rubber, consumer durables, textiles
partners: mostly other than former Soviet Union

External debt: $650 million (end 1991 est.)

Industrial production: growth rate −17% (1992 est.)

Electricity: 2,920,000 kW capacity; 13,100 million kWh produced, 3,079 kWh per capita (1992)

Industries: oil and gas, petrochemicals, fertilizers, food processing, textiles

Agriculture: cotton, fruits, vegetables

Illicit drugs: illicit producer of cannabis and opium; mostly for CIS consumption; limited government eradication program; used as transshipment points for illicit drugs from Southwest Asia to Western Europe

Economic aid: $280 million offical aid commitments by foreign donors (1992)

Currency: retaining Russian ruble as currency; planning to establish own currency, the manat, but no date set (May 1993)

Exchange rates: rubles per US$1—415 (24 December 1992) but subject to wide fluctuations

Fiscal year: calendar year

Communications

Railroads: 2,120 km; does not include industrial lines (1990)

Highways: 23,000 km total; 18,300 km hard surfaced, 4,700 km earth (1990)

Pipelines: crude oil 250 km, natural gas 4,400 km
Ports: inland—Krasnovodsk (Caspian Sea)
Airports:
total: 7
useable: 7
with permanent-surface runways: 4
with runways over 3,659 m: 0
with runways 2,440-3,659 m: 0
with runways 1,220-2,439 m: 4
Telecommunications: poorly developed; only 65 telephones per 1000 persons (1991); linked by cable and microwave to other CIS republics and to other countries by leased connections to the Moscow international gateway switch; a new direct telephone link from Ashgabat (Ashkhabad) to Iran has been established; satellite earth stations—1 Orbita and 1 INTELSAT for TV receive-only service; a newly installed satellite earth station provides TV receiver-only capability for Turkish broadcasts

---

Defense Forces

Branches: National Guard, Republic Security Forces (internal and border troops), Joint Command Turkmenistan/Russia (Ground, Navy or Caspian Sea Flotilla, Air, and Air Defense)
Manpower availability: males age 15-49 933,285; fit for military service 765,824; reach military age (18) annually 39,254 (1993 est.)
Defense expenditures: exchange rate conversion—$NA, NA% of GDP

Turks and Caicos Islands
(dependent territory of the UK)

Geography

Location: in the western North Atlantic Ocean, 190 km north of the Dominican Republic and southeast of The Bahamas
Map references: Central America and the Caribbean
Area:
total area: 430 km^2
land area: 430 km^2
comparative area: slightly less than 2.5 times the size of Washington, DC
Land boundaries: 0 km
Coastline: 389 km
Maritime claims:
exclusive fishing zone: 200 nm
territorial sea: 12 nm
International disputes: none
Climate: tropical; marine; moderated by trade winds; sunny and relatively dry
Terrain: low, flat limestone; extensive marshes and mangrove swamps
Natural resources: spiny lobster, conch
Land use:
arable land: 2%
permanent crops: 0%
meadows and pastures: 0%
forest and woodland: 0%
other: 98%
Irrigated land: NA km^2
Environment: 30 islands (eight inhabited); subject to frequent hurricanes

---

People

Population: 13,137 (July 1993 est.)
Population growth rate: 2.97% (1993 est.)
Birth rate: 14.88 births/1,000 population (1993 est.)
Death rate: 5.17 deaths/1,000 population (1993 est.)
Net migration rate: 20.01 migrant(s)/1,000 population (1993 est.)
Infant mortality rate: 12.7 deaths/1,000 live births (1993 est.)
Life expectancy at birth:
total population: 75.34 years
male: 73.41 years

female: 77.02 years (1993 est.)
Total fertility rate: 2.17 children born/woman (1993 est.)
Nationality:
noun: none
adjective: none
Ethnic divisions: African
Religions: Baptist 41.2%, Methodist 18.9%, Anglican 18.3%, Seventh-Day Adventist 1.7%, other 19.9% (1980)
Languages: English (official)
Literacy: age 15 and over can read and write (1970)
total population: 98%
male: 99%
female: 98%
Labor force: NA
by occupation: majority engaged in fishing and tourist industries; some subsistence agriculture

---

Government

Names:
conventional long form: none
conventional short form: Turks and Caicos Islands
Digraph: TK
Type: dependent territory of the UK
Capital: Grand Turk
Administrative divisions: none (dependent territory of the UK)
Independence: none (dependent territory of the UK)
Constitution: introduced 30 August 1976, suspended in 1986, and a Constitutional Commission is currently reviewing its contents
Legal system: based on laws of England and Wales with a small number adopted from Jamaica and The Bahamas
National holiday: Constitution Day, 30 August (1976)
Political parties and leaders: Progressive National Party (PNP), Washington MISSIC; People's Democratic Movement (PDM), Oswald SKIPPINGS; National Democratic Alliance (NDA), Ariel MISSICK
Suffrage: 18 years of age; universal
Elections:
Legislative Council: last held on 3 April 1991 (next to be held NA); results—percent of vote by party NA; seats—(20 total, 13 elected) PNP 8, PDM 5
Executive branch: British monarch, governor, Executive Council, chief minister
Legislative branch: unicameral Legislative Council
Judicial branch: Supreme Court
Leaders:
Chief of State: Queen ELIZABETH II (since 6 February 1953), represented by Governor Michael J. BRADLEY (since NA 1987)
Head of Government: Chief Minister Washington MISSIC (since NA 1991)

Turks and Caicos Islands
(continued)

Member of: CARICOM (associate), CDB
Diplomatic representation in US: as a dependent territory of the UK, the interests of the Turks and Caicos Islands are represented in the US by the UK
US diplomatic representation: none
Flag: blue with the flag of the UK in the upper hoist-side quadrant and the colonial shield centered on the outer half of the flag; the shield is yellow and contains a conch shell, lobster, and cactus

Economy

Overview: The economy is based on fishing, tourism, and offshore banking. Only subsistence farming—corn, cassava, citrus, and beans—exists on the Caicos Islands, so that most foods, as well as nonfood products, must be imported.
National product: GDP—purchasing power equivalent—$68.5 million (1989 est.)
National product real growth rate: NA%
National product per capita: $5,000 (1989 est.)
Inflation rate (consumer prices): NA%
Unemployment rate: 12% (1992)
Budget: revenues $20.3 million; expenditures $44.0 million, including capital expenditures of $23.9 million (1989)
Exports: $4.1 million (f.o.b., 1987)
commodities: lobster, dried and fresh conch, conch shells
partners: US, UK
Imports: $33.2 million (c.i.f., FY84)
commodities: foodstuffs, drink, tobacco, clothing, manufactures, construction materials
partners: US, UK
External debt: $NA
Industrial production: growth rate NA%
Electricity: 9,050 kW capacity; 11.1 million kWh produced, 860 kWh per capita (1992)
Industries: fishing, tourism, offshore financial services
Agriculture: subsistence farming prevails, based on corn and beans; fishing more important than farming; not self-sufficient in food
Economic aid: Western (non-US) countries, ODA and OOF bilateral commitments (1970-89), $110 million
Currency: US currency is used
Exchange rates: US currency is used
Fiscal year: calendar year

Communications

Highways: 121 km, including 24 km tarmac
Ports: Grand Turk, Salt Cay, Providenciales, Cockburn Harbour
Airports:
total: 7
usable: 7
with permanent-surface runways: 4
with runways over 3,659 m: 0
with runways 2,440-3,659: 0
with runways 1,220-2,439 m: 4
Telecommunications: fair cable and radio services; 1,446 telephones; broadcast stations—3 AM, no FM, several TV; 2 submarine cables; 1 Atlantic Ocean INTELSAT earth station

Defense Forces

Note: defense is the responsibility of the UK

Tuvalu

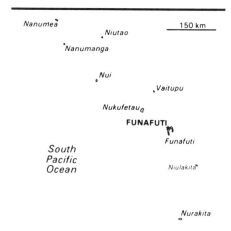

Geography

Location: Oceania, 3,000 km east of Papua New Guinea in the South Pacific Ocean
Map references: Oceania, Standard Time Zones of the World
Area:
total area: 26 km^2
land area: 26 km^2
comparative area: about 0.1 times the size of Washington, DC
Land boundaries: 0 km
Coastline: 24 km
Maritime claims:
contiguous zone: 24 nm
exclusive economic zone: 200 nm
territorial sea: 12 nm
International disputes: none
Climate: tropical; moderated by easterly trade winds (March to November); westerly gales and heavy rain (November to March)
Terrain: very low-lying and narrow coral atolls
Natural resources: fish
Land use:
arable land: 0%
permanent crops: 0%
meadows and pastures: 0%
forest and woodland: 0%
other: 100%
Irrigated land: NA km^2
Environment: severe tropical storms are rare

People

Population: 9,666 (July 1993 est.)
Population growth rate: 1.74% (1993 est.)
Birth rate: 26.79 births/1,000 population (1993 est.)
Death rate: 9.41 deaths/1,000 population (1993 est.)
Net migration rate: 0 migrant(s)/1,000 population (1993 est.)
Infant mortality rate: 26.8 deaths/1,000 live births (1993 est.)

Life expectancy at birth:
total population: 62.64 years
male: 61.27 years
female: 63.82 years (1993 est.)
Total fertility rate: 3.11 children born/woman (1993 est.)
Nationality:
noun: Tuvaluans(s)
adjective: Tuvaluan
Ethnic divisions: Polynesian 96%
Religions: Church of Tuvalu (Congregationalist) 97%, Seventh-Day Adventist 1.4%, Baha'i 1%, other 0.6%
Languages: Tuvaluan, English
Literacy:
total population: NA%
male: NA%
female: NA%
Labor force: NA
by occupation: NA

Government

Names:
conventional long form: none
conventional short form: Tuvalu
former: Ellice Islands
Digraph: TV
Type: democracy; began debating republic status in 1992; referendum expected in 1993
Capital: Funafuti
Administrative divisions: none
Independence: 1 October 1978 (from UK)
Constitution: 1 October 1978
Legal system: NA
National holiday: Independence Day, 1 October (1978)
Political parties and leaders: none
Suffrage: 18 years of age; universal
Elections:
Parliament: last held 28 September 1989 (next to be held by NA September 1993); results - percent of vote NA; seats—(12 total)
Executive branch: British monarch, governor general, prime minister, deputy prime minister, Cabinet
Legislative branch: unicameral Parliament (Palamene)
Judicial branch: High Court
Leaders:
Chief of State: Queen ELIZABETH II (since 6 February 1952), represented by Governor General Toaripi LAUTI (since NA 1992)
Head of Government: Prime Minister Bikenibeu PAENIU (since 16 October 1989); Deputy Prime Minister Dr. Alesana SELUKA (since October 1989)
Member of: ACP, C (special), ESCAP, SPARTECA, SPC, SPF, UNESCO. UPU
Diplomatic representation in US:
chief of mission: (vacant)
US diplomatic representation: none
Flag: light blue with the flag of the UK in the upper hoist-side quadrant; the outer half of the flag represents a map of the country with nine yellow five-pointed stars symbolizing the nine islands

Economy

Overview: Tuvalu consists of a scattered group of nine coral atolls with poor soil. The country has no known mineral resources and few exports. Subsistence farming and fishing are the primary economic activities. The islands are too small and too remote for development of a tourist industry. Government revenues largely come from the sale of stamps and coins and worker remittances. Substantial income is received annually from an international trust fund established in 1987 by Australia, New Zealand, and the UK and supported also by Japan and South Korea.
National product: GNP—exchange rate conversion—$4.6 million (1989 est.)
National product real growth rate: NA%
National product per capita: $530 (1989 est.)
Inflation rate (consumer prices): 3.9% (1984)
Unemployment rate: NA%
Budget: revenues $4.3 million; expenditures $4.3 million, including capital expenditures of $NA (1989)
Exports: $1.0 million (f.o.b., 1983 est.)
commodities: copra
partners: Fiji, Australia, NZ
Imports: $2.8 million (c.i.f., 1983 est.)
commodities: food, animals, mineral fuels, machinery, manufactured goods
partners: Fiji, Australia, NZ
External debt: $NA
Industrial production: growth rate NA%
Electricity: 2,600 kW capacity; 3 million kWh produced, 330 kWh per capita (1990)
Industries: fishing, tourism, copra
Agriculture: coconuts
Economic aid: US commitments, including Ex-Im (FY70-87), $1 million; Western (non-US) countries, ODA and OOF bilateral commitments (1970-89), $101 million
Currency: 1 Tuvaluan dollar ($T) or 1 Australian dollar ($A) = 100 cents
Exchange rates: Tuvaluan dollars ($T) or Australian dollars ($A) per US$1—1.4837 (January 1993), 1.3600 (1992), 1.2835 (1991), 1.2799 (1990), 1.2618 (1989), 1.2752 (1988)
Fiscal year: NA

Communications

Highways: 8 km gravel
Ports: Funafuti, Nukufetau
Merchant marine: 6 ships (1,000 GRT or over) totaling 33,220 GRT/58,518 DWT; includes 1 passenger-cargo, 1 oil tanker, 4 chemical tanker

Airports:
total: 1
useable: 1
with permanent-surface runways: 0
with runways over 3,659 m: 0
with runways 2,440-3,659 m: 0
with runways 1,220-2,439 m: 1
Telecommunications: broadcast stations—1 AM, no FM, no TV; 300 radiotelephones; 4,000 radios; 108 telephones

Defense Forces

Branches: Police Force
Manpower availability: NA
Defense expenditures: exchange rate conversion—$NA, NA% of GNP

Uganda

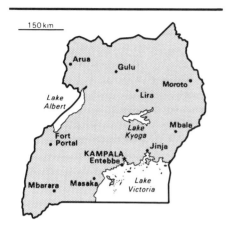

Geography

Location: Eastern Africa, between Kenya and Zaire
Map references: Africa, Standard Time Zones of the World
Area:
total area: 236,040 km²
land area: 199,710 km²
comparative area: slightly smaller than Oregon
Land boundaries: total 2,698 km, Kenya 933 km, Rwanda 169 km, Sudan 435 km, Tanzania 396 km, Zaire 765 km
Coastline: 0 km (landlocked)
Maritime claims: none; landlocked
International disputes: none
Climate: tropical; generally rainy with two dry seasons (December to February, June to August); semiarid in northeast
Terrain: mostly plateau with rim of mountains
Natural resources: copper, cobalt, limestone, salt
Land use:
arable land: 23%
permanent crops: 9%
meadows and pastures: 25%
forest and woodland: 30%
other: 13%
Irrigated land: 90 km² (1989 est.)
Environment: straddles Equator; deforestation; overgrazing; soil erosion
Note: landlocked

People

Population: 19,344,181 (July 1993 est.)
Population growth rate: 2.69% (1993 est.)
Birth rate: 49.86 births/1,000 population (1993 est.)
Death rate: 22.98 deaths/1,000 population (1993 est.)
Net migration rate: 0 migrant(s)/1,000 population (1993 est.)
Infant mortality rate: 112.1 deaths/1,000 live births (1993 est.)

Life expectancy at birth:
total population: 38.4 years
male: 38.09 years
female: 38.71 years (1993 est.)
Total fertility rate: 7.15 children born/woman (1993 est.)
Nationality:
noun: Ugandan(s)
adjective: Ugandan
Ethnic divisions: African 99%, European, Asian, Arab 1%
Religions: Roman Catholic 33%, Protestant 33%, Muslim 16%, indigenous beliefs 18%
Languages: English (official), Luganda, Swahili, Bantu languages, Nilotic languages
Literacy: age 15 and over can read and write (1990)
total population: 48%
male: 62%
female: 35%
Labor force: 4.5 million (est.)
by occupation: agriculture over 80%
note: 50% of population of working age (1983)

Government

Names:
conventional long form: Republic of Uganda
conventional short form: Uganda
Digraph: UG
Type: republic
Capital: Kampala
Administrative divisions: 10 provinces; Busoga, Central, Eastern, Karamoja, Nile, North Buganda, Northern, South Buganda, Southern, Western
Independence: 9 October 1962 (from UK)
Constitution: 8 September 1967, in process of constitutional revision
Legal system: government plans to restore system based on English common law and customary law and reinstitute a normal judicial system; accepts compulsory ICJ jurisdiction, with reservations
National holiday: Independence Day, 9 October (1962)
Political parties and leaders: only party—National Resistance Movement (NRM), Yoweri MUSEVENI
note: the Uganda Patriotic Movement (UPM); Ugandan People's Congress (UPC), Milton OBOTE; Democratic Party (DP), Paul SSEMOGEERE; and Conservative Party (CP), Jeshua NIKHGI continue to exist but are all proscribed from conducting public political activities
Other political or pressure groups: Uganda People's Front (UPF); Uganda People's Christian Democratic Army (UPCDA); Ruwenzori Movement
Suffrage: 18 years of age; universal
Elections:
National Resistance Council: last held 11-28 February 1989 (next to be held by January 1995); results—NRM was the only party;

seats—(278 total, 210 indirectly elected) 210 members elected without party affiliation
Executive branch: president, vice president, prime minister, three deputy prime ministers, Cabinet
Legislative branch: unicameral National Resistance Council
Judicial branch: Court of Appeal, High Court
Leaders:
Chief of State: President Lt. Gen. Yoweri Kaguta MUSEVENI (since 29 January 1986); Vice President Samson Babi Mululu KISEKKA (since NA January 1991)
Head of Government: Prime Minister George Cosmas ADYEBO (since NA January 1991)
Member of: ACP, AfDB, C, CCC, EADB, ECA, FAO, G-77, GATT, IAEA, IBRD, ICAO, ICFTU, IDA, IDB, IFAD, IFC, IGADD, ILO, IMF, INTELSAT, INTERPOL, IOC, IOM, ITU, LORCS, NAM, OAU, OIC, PCA, UN, UNCTAD, UNESCO, UNHCR, UNIDO, UPU, WHO, WIPO, WMO, WTO
Diplomatic representation in US:
chief of mission: Ambassador Stephen Kapimpina KATENTA-APULI
chancery: 5909 16th Street NW, Washington, DC 20011
telephone: (202) 726-7100 through 7102
US diplomatic representation:
chief of mission: Ambassador Johnnie CARSON
embassy: Parliament Avenue, Kampala
mailing address: P. O. Box 7007, Kampala
telephone: [256] (41) 259792, 259793, 259795
Flag: six equal horizontal bands of black (top), yellow, red, black, yellow, and red; a white disk is superimposed at the center and depicts a red-crested crane (the national symbol) facing the staff side

Economy

Overview: Uganda has substantial natural resources, including fertile soils, regular rainfall, and sizable mineral deposits of copper and cobalt. The economy has been devastated by widespread political instability, mismanagement, and civil war since independence in 1962, keeping Uganda poor with a per capita income of about $300. (GDP remains below the levels of the early 1970s, as does industrial production.) Agriculture is the most important sector of the economy, employing over 80% of the work force. Coffee is the major export crop and accounts for the bulk of export revenues. Since 1986 the government has acted to rehabilitate and stabilize the economy by undertaking currency reform, raising producer prices on export crops, increasing prices of petroleum products, and improving civil service wages. The policy changes are especially aimed at dampening inflation, which was running at over 300% in 1987, and boosting production and export earnings. In 1990-92, the economy has turned in a solid performance based on continued investment in the rehabilitation of infrastructure, improved incentives for production and exports, and gradually improving domestic security.

National product: GDP—exchange rate conversion—$6 billion (1992 est.)

National product real growth rate: 4% (1992 est.)

National product per capita: $300 (1992 est.)

Inflation rate (consumer prices): 41.5% (1992 est.)

Unemployment rate: NA%

Budget: revenues $365 million; expenditures $545 million, including capital expenditures of $165 million (FY89 est.)

Exports: $170 million (f.o.b., 1991 est.)
commodities: coffee 97%, cotton, tea
partners: US 25%, UK 18%, France 11%, Spain 10%

Imports: $610 million (c.i.f., 1991 est.)
commodities: petroleum products, machinery, cotton piece goods, metals, transportation equipment, food
partners: Kenya 25%, UK 14%, Italy 13%

External debt: $1.9 billion (1991 est.)

Industrial production: growth rate 7.0% (1990); accounts for 5% of GDP

Electricity: 200,000 kW capacity; 610 million kWh produced, 30 kWh per capita (1991)

Industries: sugar, brewing, tobacco, cotton textiles, cement

Agriculture: mainly subsistence; accounts for 57% of GDP and over 80% of labor force; cash crops—coffee, tea, cotton, tobacco; food crops—cassava, potatoes, corn, millet, pulses; livestock products—beef, goat meat, milk, poultry; self-sufficient in food

Economic aid: US commitments, including Ex-Im (1970-89), $145 million; Western (non-US) countries, ODA and OOF bilateral commitments (1970-89), $1.4 billion; OPEC bilateral aid (1979-89), $60 million; Communist countries (1970-89), $169 million

Currency: 1 Ugandan shilling (USh) = 100 cents

Exchange rates: Ugandan shillings (USh) per US$1—1,217.1 (January 1993), 1.133.8 (1992), 734.0 (1991), 428.85 (1990), 223.1 (1989), 106.1 (1988)

Fiscal year: 1 July-30 June

Communications

Railroads: 1,300 km, 1.000-meter-gauge single track

Highways: 26,200 km total; 1,970 km paved; 5,849 km crushed stone, gravel, and laterite; remainder earth roads and tracks

Inland waterways: Lake Victoria, Lake Albert, Lake Kyoga, Lake George, Lake Edward; Victoria Nile, Albert Nile; principal inland water ports are at Jinja and Port Bell, both on Lake Victoria

Merchant marine: 3 roll-on/roll-off (1,000 GRT or over) totaling 15,091 GRT

Airports:
total: 31
usable: 23
with permanent-surface runways: 5
with runways over 3,659 m: 1
with runways 2,440-3,659 m: 3
with runways 1,220-2,439 m: 11

Telecommunications: fair system with microwave and radio communications stations; broadcast stations—10 AM, no FM, 9 TV; satellite communications ground stations—1 Atlantic Ocean INTELSAT

Defense Forces

Branches: Army, Navy, Air Force

Manpower availability: males age 15-49 4,137,983; fit for military service 2,250,793 (1993 est.)

Defense expenditures: exchange rate conversion—$NA, 15% of budget (FY89/90)

Ukraine

Geography

Location: Eastern Europe, bordering the Black Sea, between Poland and Russia

Map references: Asia, Commonwealth of Independent States—European States, Europe, Standard Time Zones of the World

Area:
total area: 603,700 km²
land area: 603,700 km²
comparative area: slightly smaller than Texas

Land boundaries: total 4,558 km, Belarus 891 km, Hungary 103 km, Moldova 939 km, Poland 428 km, Romania (southwest) 169 km, Romania (west) 362 km, Russia 1,576 km, Slovakia 90 km

Coastline: 2,782 km

Maritime claims: NA

International disputes: potential border disputes with Moldova and Romania in northern Bukovina and southern Odes'ka Oblast'; potential dispute with Moldova over former southern Bessarabian areas; has made no territorial claim in Antarctica (but has reserved the right to do so) and does not recognize the claims of any other nation

Climate: temperate continental; subtropical only on the southern Crimean coast; precipitation disproportionately distributed, highest in west and north, lesser in east and southeast; winters vary from cool along the Black Sea to cold farther inland; summers are warm across the greater part of the country, hot in the south

Terrain: most of Ukraine consists of fertile plains (steppes) and plateaus, mountains being found only in the west (the Carpathians), and in the Crimean Peninsula in the extreme south

Natural resources: iron ore, coal, manganese, natural gas, oil, salt, sulphur, graphite, titanium, magnesium, kaolin, nickel, mercury, timber

Land use:
arable land: 56%
permanent crops: 2%

Ukraine (continued)

meadows and pastures: 12%
forest and woodland: 0%
other: 30%
Irrigated land: 26,000 km^2 (1990)
Environment: air and water pollution, deforestation, radiation contamination around Chornobyl' nuclear power plant
Note: strategic position at the crossroads between Europe and Asia; second largest country in Europe

---

People

Population: 51,821,230 (July 1993 est.)
Population growth rate: 0.06% (1993 est.)
Birth rate: 12.38 births/1,000 population (1993 est.)
Death rate: 12.53 deaths/1,000 population (1993 est.)
Net migration rate: 0.69 migrant(s)/1,000 population (1993 est.)
Infant mortality rate: 21 deaths/1,000 live births (1993 est.)
Life expectancy at birth:
total population: 69.87 years
male: 65.32 years
female: 74.65 years (1993 est.)
Total fertility rate: 1.82 children born/woman (1993 est.)
Nationality:
noun: Ukrainian(s)
adjective: Ukrainian
Ethnic divisions: Ukrainian 73%, Russian 22%, Jewish 1%, other 4%
Religions: Ukrainian Orthodox—Moscow Patriarchate, Ukrainian Orthodox—Kiev Patriarchate, Ukrainian Autocephalous Orthodox, Ukrainian Catholic (Uniate), Protestant, Jewish
Languages: Ukrainian, Russian, Romanian, Polish
Literacy: age 9-49 can read and write (1970)
total population: 100%
male: 100%
female: 100%
Labor force: 25.277 million
by occupation: industry and construction 41%, agriculture and forestry 19%, health, education, and culture 18%, trade and distribution 8%, transport and communication 7%, other 7% (1990)

Government

Names:
conventional long form: none
conventional short form: Ukraine
local long form: none
local short form: Ukrayina
former: Ukrainian Soviet Socialist Republic
Digraph: UP
Type: republic
Capital: Kiev (Kyyiv)

Administrative divisions: 24 oblasts (oblastey, singular—oblast'), 1 autonomous republic* (avtomnaya respublika), and 2 municipalites (singular—misto) with oblast status**; Chernihivs'ka, Cherkas'ka, Chernivets'ka, Dnipropetrovs'ka, Donets'ka, Ivano-Frankivs'ka, Kharkivs'ka, Khersons'ka, Khmel'nyts'ka, Kirovohrads'ka, Kyyiv (Kiev)**, Kyyivs'ka (Kiev), Luhans'ka, L'vivs'ka, Mykolayivs'ka, Odes'ka, Poltavs'ka, Respublika Krym*, Rivnens'ka, Sevastopol'**,Sums'ka, Ternopil's'ka, Vinnyts'ka, Volyns'ka, Zakarpats'ka, Zaporiz'ka, Zhytomyrs'ka
Independence: 1 December 1991 (from Soviet Union)
Constitution: using 1978 pre-independence constitution; new consitution currently being drafted
Legal system: based on civil law system; no judicial review of legislative acts
National holiday: Independence Day, 24 August (1991)
Political parties and leaders: Green Party of Ukraine, Vitaliy KONONOV, leader; Liberal Party of Ukraine, Ihor MERKULOV, chairman; Liberal Democratic Party of Ukraine, Volodymyr KLYMCHUK, chairman; Democratic Party of Ukraine, Volodymyr Oleksandrovych YAVORIVSKIY, chairman; People's Party of Ukraine, Leopol'd TABURYANSKYY, chairman; Peasants' Party of Ukraine, Serhiy DOVGRAN', chairman; Party of Democratic Rebirth of Ukraine, Volodymyr FILENKO, chairman; Social Democratic Party of Ukraine, Yuriy ZBITNEV, chairman; Socialist Party of Ukraine, Oleksandr MOROZ, chairman; Ukrainian Christian Democratic Party, Vitaliy ZHURAVSKYY, chairman; Ukrainian Conservative Republican Party, Stepan KHMARA, chairman; Ukrainian Labor Party, Valentyn LANDIK, chairman; Ukrainian Party of Justice, Mykhaylo HRECHKO, chairman; Ukrainian Peasants' Democratic Party, Serhiy PLACHINDA, chairman; Ukrainian Republican Party, Mykhaylo HORYN', chairman; Ukrainian National Conservative Party, Viktor RADIONOV, chairman
Other political or pressure groups: Ukrainian People's Movement for Restructuring (Rukh); New Ukraine (Nova Ukrayina); Congress of National Democratic Forces
Suffrage: 18 years of age; universal
Elections:
President: last held 1 December 1991 (next to be held NA 1996); results—Leonid KRAVCHUK 61.59%, Vyacheslav CHERNOVIL 23.27%, Levko LUKYANENKO 4.49%, Volodymyr HRYNYOV 4.17%, Iher YUKHNOVSKY 1.74%, Leopold TABURYANSKYY 0.57%, other 4.17%

Supreme Council: last held 4 March 1990 (next scheduled for 1995, may be held earlier in late 1993); results—percent of vote by party NA; seats—(450 total) number of seats by party NA
Executive branch: president, prime minister, cabinet
Legislative branch: unicameral Supreme Council
Judicial branch: being organized
Leaders:
Chief of State: President Leonid Makarovych KRAVCHUK (since 5 December 1991)
Head of Government: Prime Minister Leonid Danilovych KUCHMA (since 13 October 1992); Acting First Deputy Prime Minister Yukhym Leonidovych ZVYAHIL'SKYY (since 11 June 1993) and five deputy prime ministers
Member of: BSEC, CBSS (observer), CIS, CSCE, EBRD, ECE, IAEA, IBRD, ILO, IMF, INMARSAT, IOC, ITU, NACC, PCA, UN, UNCTAD, UNESCO, UNIDO, UNPROFOR, UPU, WHO, WIPO, WMO
Diplomatic representation in US:
chief of mission: Ambassador Oleh Hryhorovych BILORUS
chancery: 3350 M Street NW, Suite 200, Washington, DC 20007
telephone: (202) 333-0606
FAX: (202) 333-0817
US diplomatic representation:
chief of mission: Ambassador Roman POPADIUK
embassy: 10 Vul. Yuria Kotsyubinskovo, 252053 Kiev 53
mailing address: APO AE 09862
telephone: [7] (044) 244-7349
FAX: [7] (044) 244-7350
Flag: two equal horizontal bands of azure (top) and golden yellow represent grainfields under a blue sky

---

Economy

Overview: After Russia, the Ukrainian republic was far and away the most important economic component of the former Soviet Union producing more than three times the output of the next-ranking republic. Its fertile black soil generated more than one fourth of Soviet agricultural output, and its farms provided substantial quantities of meat, milk, grain and vegetables to other republics. Likewise, its well-developed and diversified heavy industry supplied equipment and raw materials to industrial and mining sites in other regions of the former USSR. In 1992 the Ukrainian government liberalized most prices and erected a legal framework for privatizing state enterprises while retaining many central economic controls and continuing subsidies to state production enterprises. In November 1992 the new Prime Minister KUCHMA

launched a new economic reform program promising more freedom to the agricultural sector, faster privatization of small and medium enterprises, and stricter control over state subsidies. Even so, the magnitude of the problems and the slow pace in building new market-oriented institutions preclude a near-term recovery of output to the 1990 level.
National product: GDP $NA
National product real growth rate: -13% (1992 est.)
National product per capita: $NA
Inflation rate (consumer prices): 20%-30% per month (first quarter 1993)
Unemployment rate: NA%
Budget: revenues $NA; expenditures $NA, including capital expenditures of $NA
Exports: $13.5 billion to outside of the successor states of the former USSR (1990)
commodities: coal, electric power, ferrous and nonferrous metals, chemicals, machinery and transport equipment, grain, meat
partners: NA
Imports: $16.7 billion from outside of the successor states of the former USSR (1990)
commodities: machinery and parts, transportation equipment, chemicals, textiles
partners: NA
External debt: $12 billion (1992 est.)
Industrial production: growth rate −9% (1992)
Electricity: 55,882,000 kW capacity; 281,000 million kWh produced, 5,410 kWh per capita (1992)
Industries: coal, electric power, ferrous and nonferrous metals, machinery and transport equipment, chemicals, food-processing (especially sugar)
Agriculture: grain, vegetables, meat, milk, sugar beets
Illicit drugs: illicit producer of cannabis and opium; mostly for CIS consumption; limited government eradication program; used as transshipment points for illicit drugs to Western Europe
Economic aid: $NA
Currency: Ukraine withdrew the Russian ruble from circulation on 12 November 1992 and declared the karbovanets (plural karbovantsi) sole legal tender in Ukrainian markets; Ukrainian officials claim this is an interim move toward introducing a new currency—the hryvnya—possibly in late 1993
Exchange rates: Ukrainian karbovantsi per $US1—3,000 (1 April 1993)
Fiscal year: calendar year

Communications

Railroads: 22,800 km; does not include industrial lines (1990)
Highways: 273,700 km total (1990); 236,400 km hard surfaced, 37,300 km earth

Inland waterways: 1,672 km perennially navigable (Pripyat and Dnipro River)
Pipelines: crude oil 2,010 km, petroleum products 1,920 km, natural gas 7,800 km (1992)
Ports: coastal—Berdyans'k, Illichivs'k Kerch, Kherson, Mariupol' (formerly Zhdanov), Mykolayiv, Odesa, Sevastopol', Pirdenne; inland—Kiev (Kyyiv)
Merchant marine: 394 ships (1,000 GRT or over) totaling 3,952,328 GRT/5,262,161 DWT; includes 234 cargo, 18 container, 7 barge carriers, 55 bulk cargo, 10 oil tanker, 2 chemical tanker, 1 liquefied gas, 12 passenger, 5 passenger cargo, 9 short-sea passenger, 33 roll-on/roll-off, 2 railcar carrier, 1 multi-function-large-load-carrier, 5 refrigerated cargo
Airports:
total: 694
useable: 100
with permanent-surface runways: 111
with runways over 3,659 m: 3
with runways 2,440-3,659 m: 81
with runways 1,220-2,439 m: 78
Telecommunications: international electronic mail system established in Kiev; Ukraine has about 7 million telephone lines (135 telephones for each 1000 persons); as of mid-1992, 650 telephone lines per 1000 persons in Kiev with 15-20 digital switches as of mid-1991; NMT-450 analog cellular network under construction in Kiev; 3.56 million applications for telephones could not be satisfied as of January 1990; international calls can be made via satellite, by landline to other CIS countries, and through the Moscow international switching center on 150 international lines; satellite earth stations employ INTELSAT, INMARSAT, and Intersputnik; fiber optic cable installation (intercity) remains incomplete; new international digital telephone exchange operational in Kiev for direct communication with 167 countries

---

Defense Forces

Branches: Army, Navy, Airspace Defense Forces, Republic Security Forces (internal and border troops), National Guard
Manpower availability: males age 15-49 12,070,775; fit for military service 9,521,697; reach military age (18) annually 365,534 (1993 est.)
Defense expenditures: 544,256 million karbovantsi, NA% of GDP (forecast for 1993); note—conversion of the military budget into US dollars using the current exchange rate could produce misleading results

United Arab Emirates

Boundary representation is not necessarily authoritative.

Geography

Location: Middle East, along the Persian Gulf, between Oman and Saudi Arabia
Map references: Middle East, Standard Time Zones of the World
Area:
total area: 75,581 km²
land area: 75,581 km²
comparative area: slightly smaller than Maine
Land boundaries: total 867 km, Oman 410 km, Saudi Arabia 457 km
Coastline: 1,318 km
Maritime claims:
continental shelf: defined by bilateral boundaries or equidistant line
exclusive economic zone: 200 nm
territorial sea: 3 nm assumed for most of country, 12 nm for Ash Shariqah (Sharjah)
International disputes: location and status of boundary with Saudi Arabia is not final; no defined boundary with most of Oman, but Administrative Line in far north; claims two islands in the Persian Gulf occupied by Iran (Jazireh-ye Tonb-e Bozorg or Greater Tunb, and Jazireh-ye Tonb-e Kuchek or Lesser Tunb); claims island in the Persian Gulf jointly administered with Iran (Jazireh-ye Abu Musa or Abu Musa); in 1992, the dispute over Abu Musa and the Tumb islands became more acute when Iran unilaterally tried to control the entry of third country nationals into the UAE portion of Abu Musa island, Tehran subsequently backed off in the face of significant diplomatic support for the UAE in the region
Climate: desert; cooler in eastern mountains
Terrain: flat, barren coastal plain merging into rolling sand dunes of vast desert wasteland; mountains in east
Natural resources: petroleum, natural gas
Land use:
arable land: 0%
permanent crops: 0%
meadows and pastures: 2%

United Arab Emirates (continued)

forest and woodland: 0%
other: 98%
Irrigated land: 50 km² (1989 est.)
Environment: frequent dust and sand storms; lack of natural freshwater resources being overcome by desalination plants; desertification
Note: strategic location along southern approaches to Strait of Hormuz, a vital transit point for world crude oil

People

Population: 2,657,013 (July 1993 est.)
Population growth rate: 5.06% (1993 est.)
Birth rate: 28.4 births/1,000 population (1993 est.)
Death rate: 3.07 deaths/1,000 population (1993 est.)
Net migration rate: 25.27 migrant(s)/1,000 population (1993 est.)
Infant mortality rate: 22.5 deaths/1,000 live births (1993 est.)
Life expectancy at birth:
total population: 72 years
male: 69.91 years
female: 74.2 years (1993 est.)
Total fertility rate: 4.67 children born/woman (1993 est.)
Nationality:
noun: Emirian(s)
adjective: Emirian
Ethnic divisions: Emirian 19%, other Arab 23%, South Asian 50%, other expatriates (includes Westerners and East Asians) 8% (1982)
note: less than 20% are UAE citizens (1982)
Religions: Muslim 96% (Shi'a 16%), Christian, Hindu, and other 4%
Languages: Arabic (official), Persian, English, Hindi, Urdu
Literacy: age 10 and over can read and write (1980)
total population: 68%
male: 70%
female: 63%
Labor force: 580,000 (1986 est.)
by occupation: industry and commerce 85%, agriculture 5%, services 5%, government 5%
note: 80% of labor force is foreign

Government

Names:
conventional long form: United Arab Emirates
conventional short form: none
local long form: Al Imarat al Arabiyah al Muttahidah
local short form: none
former: Trucial States
Abbreviation: UAE
Digraph: TC
Type: federation with specified powers delegated to the UAE central government

and other powers reserved to member emirates
Capital: Abu Dhabi
Administrative divisions: 7 emirates (imarat, singular—imarah); Abu Zaby (Abu Dhabi), 'Ajman, Al Fujayrah, Ash Shariqah (Sharjah), Dubayy, Ra's al Khaymah, Umm al Qaywayn
Independence: 2 December 1971 (from UK)
Constitution: 2 December 1971 (provisional)
Legal system: secular codes are being introduced by the UAE Government and in several member emirates; Islamic law remains influential
National holiday: National Day, 2 December (1971)
Political parties and leaders: none
Other political or pressure groups: a few small clandestine groups may be active
Suffrage: none
Elections: none
Executive branch: president, vice president, Supreme Council of Rulers, prime minister, deputy prime minister, Council of Ministers
Legislative branch: unicameral Federal National Council (Majlis Watani Itihad)
Judicial branch: Union Supreme Court
Leaders:
Chief of State: President Zayid bin Sultan Al NUHAYYAN, (since 2 December 1971), ruler of Abu Dhabi; Vice President Shaykh Maktum bin Rashid al-MAKTUM (since 8 October 1990), ruler of Dubayy
Head of Government: Prime Minister Shaykh Maktum bin Rashid al-MAKTUM (since 8 October 1990), ruler of Dubayy; Deputy Prime Minister Sultan bin Zayid Al NUHAYYAN (since 20 November 1990)
Member of: ABEDA, AFESD, AL, AMF, CAEU, CCC, ESCWA, FAO, G-77, GCC, IAEA, IBRD, ICAO, IDA, IDB, IFAD, IFC, ILO, IMF, IMO, INMARSAT, INTELSAT, INTERPOL, IOC, ISO (correspondent), ITU, LORCS, NAM, OAPEC, OIC, OPEC, UN, UNCTAD, UNESCO, UNIDO, UPU, WHO, WIPO, WMO, WTO
Diplomatic representation in US:
chief of mission: Ambassador Muhammad bin Husayn Al SHAALI
chancery: Suite 740, 600 New Hampshire Avenue NW, Washington, DC 20037
telephone: (202) 338-6500
US diplomatic representation:
chief of mission: Ambassador William RUGH
embassy: Al-Sudan Street, Abu Dhabi
mailing address: P. O. Box 4009, Abu Dhabi
telephone: [971] (2) 336691, afterhours 338730
FAX: [971] (2) 318441
consulate general: Dubayy (Dubai)
Flag: three equal horizontal bands of green (top), white, and black with a thicker vertical red band on the hoist side

Economy

Overview: The UAE has an open economy with one of the world's highest incomes per capita outside the OECD nations. This wealth is based on oil and gas, and the fortunes of the economy fluctuate with the prices of those commodities. Since 1973, the UAE has undergone a profound transformation from an impoverished region of small desert principalities to a modern state with a high standard of living. At present levels of production, crude oil reserves should last for over 100 years.
National product: GDP—exchange rate conversion—$34.9 billion (1992)
National product real growth rate: NA%
National product per capita: $13,800 (1992)
Inflation rate (consumer prices): 1% (1990 est.)
Unemployment rate: NEGL% (1988)
Budget: revenues $4.3 billion; expenditures $4.8 billion, including capital expenditures of $NA (1993)
Exports: $21.2 billion (f.o.b., 1991 est.)
commodities: crude oil 66%, natural gas, reexports, dried fish, dates
partners: Japan 39%, Singapore 5%, Korea 4%, Iran 4%, India
Imports: $13.9 billion (f.o.b., 1991 est.)
commodities: capital goods, consumer goods, food
partners: Japan 15%, US 10%, UK 9%, Germany 7%, Korea 4%
External debt: $11.0 billion (December 1989 est.)
Industrial production: growth rate 30% (1990 est.); accounts for 56% of GDP, including petroleum
Electricity: 6,090,000 kW capacity; 17,850 million kWh produced, 6,718 kWh per capita (1992)
Industries: petroleum, fishing, petrochemicals, construction materials, some boat building, handicrafts, pearling
Agriculture: accounts for 2% of GDP and 5% of labor force; cash crop—dates; food products—vegetables, watermelons, poultry, eggs, dairy, fish; only 25% self-sufficient in food
Economic aid: donor—pledged $9.1 billion in bilateral aid to less developed countries (1979-89)
Currency: 1 Emirian dirham (Dh) = 100 fils
Exchange rates: Emirian dirhams (Dh) per US$1—3.6710 (fixed rate)
Fiscal year: calendar year

Communications

Highways: 2,000 km total; 1,800 km bituminous, 200 km gravel and graded earth
Pipelines: crude oil 830 km, natural gas, including natural gas liquids, 870 km

United Kingdom

Ports: Al Fujayrah, Khawr Fakkan, Mina' Jabal 'Ali, Mina' Khalid, Mina' Rashid, Mina' Saqr, Mina' Zayid
Merchant marine: 56 ships (1,000 GRT or over) totaling 1,197,306 GRT/2,153,673 DWT; includes 15 cargo, 8 container, 3 roll-on/roll-off, 23 oil tanker, 4 bulk, 1 refrigerated cargo, 1 liquified gas, 1 chemical tanker
Airports:
total: 37
usable: 34
with permanent-surface runways: 20
with runways over 3,659 m: 7
with runways 2,440-3,659 m: 5
with runways 1,220-2,439 m: 5
Telecommunications: modern system consisting of microwave and coaxial cable; key centers are Abu Dhabi and Dubayy; 386,600 telephones; satellite ground stations—1 Atlantic Ocean INTELSAT, 2 Indian Ocean INTELSAT and 1 ARABSAT; submarine cables to Qatar, Bahrain, India, and Pakistan; tropospheric scatter to Bahrain; microwave radio relay to Saudi Arabia; broadcast stations—8 AM, 3 FM, 12 TV

Defense Forces

Branches: Army, Navy, Air Force, Federal Police Force
Manpower availability: males age 15-49 1,008,076; fit for military service 550,965; reach military age (18) annually 15,499 (1993 est.)
Defense expenditures: exchange rate conversion—$1.47 billion, 5.3% of GDP (1989 est.)

Geography

Location: Western Europe, bordering on the North Atlantic Ocean and the North Sea, between Ireland and France
Map references: Europe, Standard Time Zones of the World
Area:
total area: 244,820 km²
land area: 241,590 km²
comparative area: slightly smaller than Oregon
note: includes Rockall and Shetland Islands
Land boundaries: total 360 km, Ireland 360 km
Coastline: 12,429 km
Maritime claims:
continental shelf: as defined in continental shelf orders or in accordance with agreed upon boundaries
exclusive fishing zone: 200 nm
territorial sea: 12 nm
International disputes: Northern Ireland question with Ireland; Gibraltar question with Spain; Argentina claims Falkland Islands (Islas Malvinas); Argentina claims South Georgia and the South Sandwich Islands; Mauritius claims island of Diego Garcia in British Indian Ocean Territory; Rockall continental shelf dispute involving Denmark, Iceland, and Ireland (Ireland and the UK have signed a boundary agreement in the Rockall area); territorial claim in Antarctica (British Antarctic Territory)
Climate: temperate; moderated by prevailing southwest winds over the North Atlantic Current; more than half of the days are overcast
Terrain: mostly rugged hills and low mountains; level to rolling plains in east and southeast
Natural resources: coal, petroleum, natural gas, tin, limestone, iron ore, salt, clay, chalk, gypsum, lead, silica
Land use:
arable land: 29%
permanent crops: 0%
meadows and pastures: 48%
forest and woodland: 9%
other: 14%
Irrigated land: 1,570 km² (1989)
Environment: pollution control measures improving air and water quality; because of heavily indented coastline, no location is more than 125 km from tidal waters
Note: lies near vital North Atlantic sea lanes; only 35 km from France and now being linked by tunnel under the English Channel

---

People

Population: 57,970,200 (July 1993 est.)
Population growth rate: 0.29% (1993 est.)
Birth rate: 13.58 births/1,000 population (1993 est.)
Death rate: 10.87 deaths/1,000 population (1993 est.)
Net migration rate: 0.17 migrant(s)/1,000 population (1993 est.)
Infant mortality rate: 7.4 deaths/1,000 live births (1993 est.)
Life expectancy at birth:
total population: 76.5 years
male: 73.71 years
female: 79.43 years (1993 est.)
Total fertility rate: 1.83 children born/woman (1993 est.)
Nationality:
noun: Briton(s), British (collective pl.)
adjective: British
Ethnic divisions: English 81.5%, Scottish 9.6%, Irish 2.4%, Welsh 1.9%, Ulster 1.8%, West Indian, Indian, Pakistani, and other 2.8%
Religions: Anglican 27 million, Roman Catholic 9 million, Muslim 1 million, Presbyterian 800,000, Methodist 760,000, Sikh 400,000, Hindu 350,000, Jewish 300,000 (1991 est.)
note: the UK does not include a question on religion in its census
Languages: English, Welsh (about 26% of the population of Wales), Scottish form of Gaelic (about 60,000 in Scotland)
Literacy: age 15 and over can read and write (1978)
total population: 99%
male: NA%
female: NA%
Labor force: 28.048 million
by occupation: services 62.8%, manufacturing and construction 25.0%, government 9.1%, energy 1.9%, agriculture 1.2% (June 1992)

Government

Names:
conventional long form: United Kingdom of Great Britain and Northern Ireland
conventional short form: United Kingdom

United Kingdom *(continued)*

Abbreviation: UK
Digraph: UK
Type: constitutional monarchy
Capital: London
Administrative divisions: 47 counties, 7 metropolitan counties, 26 districts, 9 regions, and 3 islands areas
England: 39 counties, 7 metropolitan counties*; Avon, Bedford, Berkshire, Buckingham, Cambridge, Cheshire, Cleveland, Cornwall, Cumbria, Derby, Devon, Dorset, Durham, East Sussex, Essex, Gloucester, Greater London*, Greater Manchester*, Hampshire, Hereford and Worcester, Hertford, Humberside, Isle of Wight, Kent, Lancashire, Leicester, Lincoln, Merseyside*, Norfolk, Northampton, Northumberland, North Yorkshire, Nottingham, Oxford, Shropshire, Somerset, South Yorkshire*, Stafford, Suffolk, Surrey, Tyne and Wear*, Warwick, West Midlands*, West Sussex, West Yorkshire*, Wiltshire
Northern Ireland: 26 districts; Antrim, Ards, Armagh, Ballymena, Ballymoney, Banbridge, Belfast, Carrickfergus, Castlereagh, Coleraine, Cookstown, Craigavon, Down, Dungannon, Fermanagh, Larne, Limavady, Lisburn, Londonderry, Magherafelt, Moyle, Newry and Mourne, Newtownabbey, North Down, Omagh, Strabane
Scotland: 9 regions, 3 islands areas*; Borders, Central, Dumfries and Galloway, Fife, Grampian, Highland, Lothian, Orkney*, Shetland*, Strathclyde, Tayside, Western Isles*
Wales: 8 counties; Clwyd, Dyfed, Gwent, Gwynedd, Mid Glamorgan, Powys, South Glamorgan, West Glamorgan
Dependent areas: Anguilla, Bermuda, British Indian Ocean Territory, British Virgin Islands, Cayman Islands, Falkland Islands, Gibraltar, Guernsey, Hong Kong (scheduled to become a Special Administrative Region of China on 1 July 1997), Jersey, Isle of Man, Montserrat, Pitcairn Islands, Saint Helena, South Georgia and the South Sandwich Islands, Turks and Caicos Islands
Independence: 1 January 1801 (United Kingdom established)
Constitution: unwritten; partly statutes, partly common law and practice
Legal system: common law tradition with early Roman and modern continental influences; no judicial review of Acts of Parliament; accepts compulsory ICJ jurisdiction, with reservations
National holiday: Celebration of the Birthday of the Queen (second Saturday in June)
Political parties and leaders: Conservative and Unionist Party, John MAJOR; Labor Party, John SMITH; Liberal Democrats (LD), Jeremy (Paddy) ASHDOWN; Scottish National Party, Alex SALMOND; Welsh National Party (Plaid Cymru), Dafydd Iwan WIGLEY; Ulster Unionist Party (Northern Ireland), James MOLYNEAUX; Democratic Unionist Party (Northern Ireland), Rev. Ian PAISLEY; Ulster Popular Unionist Party (Northern Ireland), James KILFEDDER; Social Democratic and Labor Party (SDLP, Northern Ireland), John HUME; Sinn Fein (Northern Ireland), Gerry ADAMS
Other political or pressure groups: Trades Union Congress; Confederation of British Industry; National Farmers' Union; Campaign for Nuclear Disarmament
Suffrage: 18 years of age; universal
Elections:
House of Commons: last held 9 April 1992 (next to be held by NA April 1997); results—Conservative 41.9%, Labor 34.5%, Liberal Democratic 17.9%, other 5.7%; seats—(651 total) Conservative 336, Labor 271, Liberal Democratic 20, other 24
Executive branch: monarch, prime minister, Cabinet
Legislative branch: bicameral Parliament consists of an upper house or House of Lords and a lower house or House of Commons
Judicial branch: House of Lords
Leaders:
Chief of State: Queen ELIZABETH II (since 6 February 1952); Heir Apparent Prince CHARLES (son of the Queen, born 14 November 1948)
Head of Government: Prime Minister John MAJOR (since 28 November 1990)
Member of: AfDB, AG (observer), AsDB, Australian Group, BIS, C, CCC, CDB (non-regional), CE, CERN, COCOM, CP, CSCE, EBRD, EC, ECA (associate), ECE, ECLAC, EIB, ESCAP, ESA, FAO, G-5, G-7, G-10, GATT, IADB, IAEA, IBRD, ICAO, ICC, ICFTU, IDA, IEA, IFAD, IFC, ILO, IMF, IMO, INMARSAT, INTELSAT, INTERPOL, IOC, IOM (observer), ISO, ITU, LORCS, MINURSO, MTRC, NACC, NATO, NEA, NSG, OECD, PCA, SPC, UN, UNCTAD, UNFICYP, UNHCR, UNIDO, UNIKOM, UNPROFOR, UNRWA, UN Security Council, UNTAC, UN Trusteeship Council, UPU, WCL, WEU, WHO, WIPO, WMO, ZC
Diplomatic representation in US:
chief of mission: Ambassador Sir Robin RENWICK
chancery: 3100 Massachusetts Avenue NW, Washington, DC 20008
telephone: (202) 462-1340
FAX: (202) 898-4255
consulates general: Atlanta, Boston, Chicago, Cleveland, Houston, Los Angeles, New York, and San Francisco,
consulates: Dallas, Miami, and Seattle
US diplomatic representation:
chief of mission: Ambassador Raymond G. H. SEITZ
embassy: 24/31 Grosvenor Square, London, W.1A1AE
mailing address: PSC 801, Box 40, FPO AE 09498-4040
telephone: [44] (71) 499-9000
FAX: [44] (71) 409-1637
consulates general: Belfast and Edinburgh
Flag: blue with the red cross of Saint George (patron saint of England) edged in white superimposed on the diagonal red cross of Saint Patrick (patron saint of Ireland) which is superimposed on the diagonal white cross of Saint Andrew (patron saint of Scotland); known as the Union Flag or Union Jack; the design and colors (especially the Blue Ensign) have been the basis for a number of other flags including dependencies, Commonwealth countries, and others

---

Economy

Overview: The UK is one of the world's great trading powers and financial centers, and its economy ranks among the four largest in Europe. The economy is essentially capitalistic; over the past thirteen years the ruling Tories have greatly reduced public ownership and contained the growth of social welfare programs. Agriculture is intensive, highly mechanized, and efficient by European standards, producing about 60% of food needs with only 1% of the labor force. The UK has large coal, natural gas, and oil reserves, and primary energy production accounts for 12% of GDP, one of the highest shares of any industrial nation. Services, particularly banking, insurance, and business services, account by far for the largest proportion of GDP while industry continues to decline in importance, now employing only 25% of the work force and generating 21% of GDP. The economy is emerging out of its 3-year recession with only weak recovery expected in 1993. Unemployment is hovering around 10% of the labor force. The government in 1992 adopted a pro-growth strategy, cutting interest rates sharply and removing the pound from the European exchange rate mechanism. Excess industrial capacity probably will moderate inflation which for the first time in a decade is below the EC average. The major economic policy question for Britain in the 1990s is the terms on which it participates in the financial and economic integration of Europe.
National product: GDP—purchasing power equivalent—$920.6 billion (1992)
National product real growth rate: -0.6% (1992)
National product per capita: $15,900 (1992)
Inflation rate (consumer prices): 3.6% (1992)

Unemployment rate: 9.8% (1992)
Budget: revenues $367.6 billion; expenditures $439.3 billion, including capital expenditures of $32.5 billion (FY92 est.)
Exports: $187.4 billion (f.o.b., 1992)
commodities: manufactured goods, machinery, fuels, chemicals, semifinished goods, transport equipment
partners: EC countries 56.7% (Germany 14.0%, France 11.1%, Netherlands 7.9%), US 10.9%
Imports: $210.7 billion (c.i.f., 1992)
commodities: manufactured goods, machinery, semifinished goods, foodstuffs, consumer goods
partners: EC countries 51.7% (Germany 14.9%, France 9.3%, Netherlands 8.4%), US 11.6%
External debt: $16.2 billion (June 1992)
Industrial production: growth rate 0.4% (1992 est.)
Electricity: 99,000,000 kW capacity; 317,000 million kWh produced, 5,480 kWh per capita (1992)
Industries: production machinery including machine tools, electric power equipment, equipment for the automation of production, railroad equipment, shipbuilding, aircraft, motor vehicles and parts, electronics and communications equipment, metals, chemicals, coal, petroleum, paper and paper products, food processing, textiles, clothing, and other consumer goods
Agriculture: accounts for only 1.5% of GDP and 1% of labor force; highly mechanized and efficient farms; wide variety of crops and livestock products produced; about 60% self-sufficient in food and feed needs; fish catch of 665,000 metric tons (1987)
Illicit drugs: increasingly important gateway country for Latin American cocaine entering the European market
Economic aid: donor—ODA and OOF commitments (1970-89), $21.0 billion
Currency: 1 British pound (£) = 100 pence
Exchange rates: British pounds (£) per US$1—0.6527 (January 1993), 0.5664 (1992), 0.5652 (1991), 0.5603 (1990), 0.6099 (1989), 0.5614 (1988)
Fiscal year: 1 April-31 March

Communications

Railroads: UK, 16,914 km total; Great Britain's British Railways (BR) operates 16,584 km 1.435-meter (standard) gauge (including 4,545 km electrified and 12,591 km double or multiple track), several additional small standard-gauge and narrow-gauge lines are privately owned and operated; Northern Ireland Railways (NIR) operates 330 km 1.600-meter gauge (including 190 km double track)

Highways: UK, 362,982 km total; Great Britain, 339,483 km paved (including 2,573 km limited-access divided highway); Northern Ireland, 23,499 km (22,907 paved, 592 km gravel)
Inland waterways: 2,291 total; British Waterways Board, 606 km; Port Authorities, 706 km; other, 979 km
Pipelines: crude oil (almost all insignificant) 933 km, petroleum products 2,993 km, natural gas 12,800 km
Ports: London, Liverpool, Felixstowe, Tees and Hartlepool, Dover, Sullom Voe, Southampton
Merchant marine: 204 ships (1,000 GRT or over) totaling 3,819,719 GRT/4,941,785 DWT; includes 7 passenger, 16 short-sea passenger, 37 cargo, 25 container, 14 roll-on/roll-off, 5 refrigerated cargo, 1 vehicle carrier, 65 oil tanker, 1 chemical tanker, 8 liquefied gas, 1 specialized tanker, 22 bulk, 1 combination bulk, 1 passenger cargo
Airports:
total: 496
usable: 385
with permanent-surface runways: 249
with runways over 3,659 m: 1
with runways 2,440-3,659 m: 37
with runways 1,220-2,439 m: 134
Telecommunications: technologically advanced domestic and international system; 30,200,000 telephones; equal mix of buried cables, microwave and optical-fiber systems; excellent countrywide broadcast systems; broadcast stations—225 AM, 525 (mostly repeaters) FM, 207 (3,210 repeaters) TV; 40 coaxial submarine cables; 5 satellite ground stations operating in INTELSAT (7 Atlantic Ocean and 3 Indian Ocean), INMARSAT, and EUTELSAT systems; at least 8 large international switching centers

Defense Forces

Branches: Army, Royal Navy (including Royal Marines), Royal Air Force
Manpower availability: males age 15-49 14,445,998; fit for military service 12,084,913 (1993 est.); no conscription
Defense expenditures: exchange rate conversion—$42.5 billion, 3.8% of GDP (FY92/93)

United States

Geography

Location: North America, between Canada and Mexico
Map references: North America, Standard Time Zones of the World
Area:
total area: 9,372,610 km²
land area: 9,166,600 km²
comparative area: about half the size of Russia; about three-tenths the size of Africa; about one-half the size of South America (or slightly larger than Brazil); slightly smaller than China; about two and one-half times the size of Western Europe
note: includes only the 50 states and District of Columbia
Land boundaries: total 12,248 km, Canada 8,893 km (including 2,477 km with Alaska), Cuba 29 km (US naval base at Guantanamo), Mexico 3,326 km
Coastline: 19,924 km
Maritime claims:
contiguous zone: 24 nm
continental shelf: 200 m or depth of exploitation
exclusive economic zone: 200 nm
territorial sea: 12 nm
International disputes: maritime boundary disputes with Canada (Dixon Entrance, Beaufort Sea, Strait of Juan de Fuca); US Naval Base at Guantanamo is leased from Cuba and only mutual agreement or US abandonment of the area can terminate the lease; Haiti claims Navassa Island; US has made no territorial claim in Antarctica (but has reserved the right to do so) and does not recognize the claims of any other nation; Republic of Marshall Islands claims Wake Island
Climate: mostly temperate, but tropical in Hawaii and Florida and arctic in Alaska, semiarid in the great plains west of the Mississippi River and arid in the Great Basin of the southwest; low winter temperatures in the northwest are ameliorated occasionally in January and

United States (continued)

February by warm chinook winds from the eastern slopes of the Rocky Mountains

Terrain: vast central plain, mountains in west, hills and low mountains in east; rugged mountains and broad river valleys in Alaska; rugged, volcanic topography in Hawaii

Natural resources: coal, copper, lead, molybdenum, phosphates, uranium, bauxite, gold, iron, mercury, nickel, potash, silver, tungsten, zinc, petroleum, natural gas, timber

Land use:

arable land: 20%

permanent crops: 0%

meadows and pastures: 26%

forest and woodland: 29%

other: 25%

Irrigated land: 181,020 km² (1989 est.)

Environment: pollution control measures improving air and water quality; agricultural fertilizer and pesticide pollution; management of sparse natural water resources in west; desertification; tsunamis, volcanoes, and earthquake activity around Pacific Basin; permafrost in northern Alaska is a major impediment to development

Note: world's fourth-largest country (after Russia, Canada, and China)

People

Population: 258,103,721 (July 1993 est.)

Population growth rate: 1.02% (1993 est.)

Birth rate: 15.48 births/1,000 population (1993 est.)

Death rate: 8.67 deaths/1,000 population (1993 est.)

Net migration rate: 3.41 migrant(s)/1,000 population (1993 est.)

Infant mortality rate: 8.36 deaths/1,000 live births (1993 est.)

Life expectancy at birth:

total population: 75.8 years

male: 72.49 years

female: 79.29 years (1993 est.)

Total fertility rate: 2.05 children born/woman (1993 est.)

Nationality:

noun: American(s)

adjective: American

Ethnic divisions: white 83.4%, black 12.4%, asian 3.3%, native american 0.8% (1992)

Religions: Protestant 56%, Roman Catholic 28%, Jewish 2%, other 4%, none 10% (1989)

Languages: English, Spanish (spoken by a sizable minority)

Literacy: age 15 and over having completed 5 or more years of schooling (1991)

total population: 97.9%

male: 97.9%

female: 97.9%

Labor force: 128.548 million (includes armed forces and unemployed; civilian labor force 126.982 million) (1992)

by occupation: NA

Government

Names:

conventional long form: United States of America

conventional short form: United States

Abbreviation: US or USA

Digraph: US

Type: federal republic; strong democratic tradition

Capital: Washington, DC

Administrative divisions: 50 states and 1 district*; Alabama, Alaska, Arizona, Arkansas, California, Colorado, Connecticut, Delaware, District of Columbia*, Florida, Georgia, Hawaii, Idaho, Illinois, Indiana, Iowa, Kansas, Kentucky, Louisiana, Maine, Maryland, Massachusetts, Michigan, Minnesota, Mississippi, Missouri, Montana, Nebraska, Nevada, New Hampshire, New Jersey, New Mexico, New York, North Carolina, North Dakota, Ohio, Oklahoma, Oregon, Pennsylvania, Rhode Island, South Carolina, South Dakota, Tennessee, Texas, Utah, Vermont, Virginia, Washington, West Virginia, Wisconsin, Wyoming

Dependent areas: American Samoa, Baker Island, Guam, Howland Island, Jarvis Island, Johnston Atoll, Kingman Reef, Midway Islands, Navassa Island, Northern Mariana Islands, Palmyra Atoll, Puerto Rico, Virgin Islands, Wake Island

note: since 18 July 1947, the US has administered the Trust Territory of the Pacific Islands, but recently entered into a new political relationship with three of the four political units; the Northern Mariana Islands is a Commonwealth in political union with the US (effective 3 November 1986); Palau concluded a Compact of Free Association with the US that was approved by the US Congress but to date the Compact process has not been completed in Palau, which continues to be administered by the US as the Trust Territory of the Pacific Islands; the Federated States of Micronesia signed a Compact of Free Association with the US (effective 3 November 1986); the Republic of the Marshall Islands signed a Compact of Free Association with the US (effective 21 October 1986)

Independence: 4 July 1776 (from England)

Constitution: 17 September 1787, effective 4 June 1789

Legal system: based on English common law; judicial review of legislative acts; accepts compulsory ICJ jurisdiction, with reservations

National holiday: Independence Day, 4 July (1776)

Political parties and leaders: Republican Party, Haley BARBOUR, national committee chairman; Jeanie AUSTIN, co-chairman; Democratic Party, David C. WILHELM, national committee chairman; several other groups or parties of minor political significance

Suffrage: 18 years of age; universal

Elections:

President: last held 3 November 1992 (next to be held 5 November 1996); results—William Jefferson CLINTON (Democratic Party) 43.2%, George BUSH (Republican Party) 37.7%, Ross PEROT (Independent) 19.0%, other 0.1%

Senate: last held 3 November 1992 (next to be held 8 November 1994); results—Democratic Party 53%, Republican Party 47%, other NEGL%; seats—(100 total) Democratic Party 57, Republican Party 43

House of Representatives: last held 3 November 1992 (next to be held 8 November 1994); results—Democratic Party 52%, Republican Party 46%, other 2%; seats—(435 total) Democratic Party 258, Republican Party 176, Independent 1

Executive branch: president, vice president, Cabinet

Legislative branch: bicameral Congress consists of an upper house or Senate and a lower house or House of Representatives

Judicial branch: Supreme Court

Leaders:

Chief of State and Head of Government: President William Jefferson CLINTON (since 20 January 1993); Vice President Albert GORE, Jr. (since 20 January 1993)

Member of: AfDB, AG (observer), ANZUS, APEC, AsDB, Australian Group, BIS, CCC, COCOM, CP, CSCE, EBRD, ECE, ECLAC, FAO, ESCAP, G-2, G-5, G-7, G-8, G-10, GATT, IADB, IAEA, IBRD, ICAO, ICC, ICFTU, IDA, IEA, IFAD, IFC, ILO, IMF, IMO, INMARSAT, INTELSAT, INTERPOL, IOC, IOM, ISO, ITU, LORCS, MINURSO, MTCR, NACC, NATO, NEA, NSG, OAS, OECD, PCA, SPC, UN, UNCTAD, UNHCR, UNIDO, UNIKOM, UNRWA, UN Security Council, UNTAC, UN Trusteeship Council, UNTSO, UPU, WCL, WHO, WIPO, WMO, WTO, ZC

Flag: thirteen equal horizontal stripes of red (top and bottom) alternating with white; there is a blue rectangle in the upper hoist-side corner bearing 50 small white five-pointed stars arranged in nine offset horizontal rows of six stars (top and bottom) alternating with rows of five stars; the 50 stars represent the 50 states, the 13 stripes represent the 13 original colonies; known as Old Glory; the design and colors have been the basis for a number of other flags including Chile, Liberia, Malaysia, and Puerto Rico

Economy

Overview: The US has the most powerful, diverse, and technologically advanced economy in the world, with a per capita GDP of $23,400, the largest among major industrial nations. The economy is market oriented with most decisions made by private individuals and business firms and with government purchases of goods and services made predominantly in the marketplace. In 1989 the economy enjoyed its seventh successive year of substantial growth, the longest in peacetime history. The expansion featured moderation in wage and consumer price increases and a steady reduction in unemployment to 5.2% of the labor force. In 1990, however, growth slowed to 1% because of a combination of factors, such as the worldwide increase in interest rates, Iraq's invasion of Kuwait in August, the subsequent spurt in oil prices, and a general decline in business and consumer confidence. In 1991 output fell by 1%, unemployment grew, and signs of recovery proved premature. Growth picked up to 2.1% in 1992. Unemployment, however, remained at nine million, the increase in GDP being mainly attributable to gains in output per worker. Ongoing problems for the 1990s include inadequate investment in economic infrastructure, rapidly rising medical costs, and sizable budget and trade deficits.

National product: GDP—purchasing power equivalent—$5.951 trillion (1992)

National product real growth rate: 2.1% (1992)

National product per capita: $23,400 (1992)

Inflation rate (consumer prices): 3% (1992)

Unemployment rate: 7% (April 1993)

Budget: revenues $1,092 billion; expenditures $1,382 billion, including capital expenditures of $NA (FY92)

Exports: $442.3 billion (f.o.b., 1992)
commodities: capital goods, automobiles, industrial supplies and raw materials, consumer goods, agricultural products
partners: Western Europe 27.3%, Canada 22.1%, Japan 12.1% (1989)

Imports: $544.1 billion (c.i.f., 1992)
commodities: crude oil and refined petroleum products, machinery, automobiles, consumer goods, industrial raw materials, food and beverages
partners: Western Europe 21.5%, Japan 19.7%, Canada 18.8% (1989)

External debt: $NA

Industrial production: growth rate 1.5% (1992 est.); accounts for NA% of GDP

Electricity: 780,000,000 kW capacity; 3,230,000 million kWh produced, 12,690 kWh per capita (1992)

Industries: leading industrial power in the world, highly diversified; petroleum, steel, motor vehicles, aerospace, telecommunications, chemicals, electronics, food processing, consumer goods, lumber, mining

Agriculture: accounts for 2% of GDP and 2.8% of labor force; favorable climate and soils support a wide variety of crops and livestock production; world's second largest producer and number one exporter of grain; surplus food producer; fish catch of 4.4 million metric tons (1990)

Illicit drugs: illicit producer of cannabis for domestic consumption with 1987 production estimated at 3,500 metric tons or about 25% of the available marijuana; ongoing eradication program aimed at small plots and greenhouses has not reduced production

Economic aid: donor—commitments, including ODA and OOF, (FY80-89), $115.7 billion

Currency: 1 United States dollar (US$) = 100 cents

Exchange rates:
British pounds: (£) per US$—0.6527 (January 1993), 0.5664 (1992), 0.5652 (1991), 0.5603 (1990), 0.6099 (1989), 0.5614 (1988)
Canadian dollars: (Can$) per US$—1.2776 (January 1993), 1.2087 (1992), 1.1457 (1991), 1.1668 (1990), 1.1840 (1989), 1.2307 (1988)
French francs: (F) per US$—5.4812 (January 1993), 5.2938 (1992), 5.6421 (1991), 5.4453 (1990), 6.3801 (1989), 5.9569 (1988)
Italian lire: (Lit) per US$—1,482.5 (January 1993), 1,232.4 (1992), 1,240.6 (1991), 1,198.1 (1990), 1.372.1 (1989), 1,301.6 (1988)
Japanese yen: (¥) per US$—125.01 (January 1993), 126.65 (1992), 134.71 (1991), 144.79 (1990), 137.96 (1989), 128.15 (1988)
German deutsche marks: (DM) per US$—1.6158 (January 1993), 1.5617 (1992), 1.6595 (1991), 1.6157 (1990), 1.8800 (1989), 1.7562 (1988)

Fiscal year: 1 October-30 September

Communications

Railroads: 240,000 km of mainline routes, all standard 1.435 meter track, no government ownership (1989)

Highways: 7,599,250 km total; 6,230,000 km state-financed roads; 1,369,250 km federally-financed roads (including 71,825 km interstate limited access freeways) (1988)

Inland waterways: 41,009 km of navigable inland channels, exclusive of the Great Lakes (est.)

Pipelines: petroleum 276,000 km (1991), natural gas 331,000 km (1991)

Ports: Anchorage, Baltimore, Beaumont, Boston, Charleston, Chicago, Cleveland, Duluth, Freeport, Galveston, Hampton Roads, Honolulu, Houston, Jacksonville, Long Beach, Los Angeles, Milwaukee, Mobile, New Orleans, New York, Philadelphia, Portland (Oregon), Richmond (California), San Francisco, Savannah, Seattle, Tampa, Wilmington

Merchant marine: 385 ships (1,000 GRT or over) totaling 12,567,000 GRT/19,511,000 DWT; includes 3 passenger-cargo, 36 cargo, 23 bulk, 169 tanker, 13 tanker tug-barge, 13 liquefied gas, 128 intermodal; in addition, there are 219 government-owned vessels

Airports:
total: 14,177
usable: 12,417
with permanent-surface runways: 4,820
with runways over 3,659 m: 63
with runways 2,440-3,659 m: 325
with runways 1,220-2,439 m: 2,524

Telecommunications: 126,000,000 telephone access lines; 7,557,000 cellular phone subscribers; broadcast stations—4,987 AM, 4,932 FM, 1,092 TV; about 9,000 TV cable systems; 530,000,000 radio sets and 193,000,000 TV sets in use; 16 satellites and 24 ocean cable systems in use; satellite ground stations—45 Atlantic Ocean INTELSAT and 16 Pacific Ocean INTELSAT (1990)

Defense Forces

Branches: Department of the Army, Department of the Navy (including Marine Corps), Department of the Air Force

Manpower availability: males age 15-49 66.826 million; fit for military service NA (1993 est.)

Defense expenditures: exchange rate conversion—$315.5 billion, 5.3% of GDP (1992)

Uruguay

125 km
Boundary representation is not necessarily authoritative.

Geography

Location: Eastern South America, bordering the South Atlantic Ocean between Argentina and Brazil
Map references: South America, Standard Time Zones of the World
Area:
total area: 176,220 km²
land area: 173,620 km²
comparative area: slightly smaller than Washington State
Land boundaries: total 1,564 km, Argentina 579 km, Brazil 985 km
Coastline: 660 km
Maritime claims:
continental shelf: 200 m depth or to depth of exploitation
territorial sea: 200 nm; overflight and navigation permitted beyond 12 nm
International disputes: short section of boundary with Argentina is in dispute; two short sections of the boundary with Brazil are in dispute—Arroyo de la Invernada (Arroio Invernada) area of the Rio Quarai and the islands at the confluence of the Rio Cuareim (Rio Quarai) and the Uruguay
Climate: warm temperate; freezing temperatures almost unknown
Terrain: mostly rolling plains and low hills; fertile coastal lowland
Natural resources: soil, hydropower potential, minor minerals
Land use:
arable land: 8%
permanent crops: 0%
meadows and pastures: 78%
forest and woodland: 4%
other: 10%
Irrigated land: 1,100 km² (1989 est.)
Environment: subject to seasonally high winds, droughts, floods

People

Population: 3,175,050 (July 1993 est.)
Population growth rate: 0.75% (1993 est.)

Birth rate: 17.82 births/1,000 population (1993 est.)
Death rate: 9.52 deaths/1,000 population (1993 est.)
Net migration rate: -0.79 migrant(s)/1,000 population (1993 est.)
Infant mortality rate: 18 deaths/1,000 live births (1993 est.)
Life expectancy at birth:
total population: 73.74 years
male: 70.52 years
female: 77.11 years (1993 est.)
Total fertility rate: 2.46 children born/woman (1993 est.)
Nationality:
noun: Uruguayan(s)
adjective: Uruguayan
Ethnic divisions: white 88%, mestizo 8%, black 4%
Religions: Roman Catholic 66% (less than half adult population attends church regularly), Protestant 2%, Jewish 2%, nonprofessing or other 30%
Languages: Spanish
Literacy: age 15 and over can read and write (1990)
total population: 96%
male: 97%
female: 96%
Labor force: 1.355 million (1991 est.)
by occupation: government 25%, manufacturing 19%, agriculture 11%, commerce 12%, utilities, construction, transport, and communications 12%, other services 21% (1988 est.)

Government

Names:
conventional long form: Oriental Republic of Uruguay
conventional short form: Uruguay
local long form: Republica Oriental del Uruguay
local short form: Uruguay
Digraph: UY
Type: republic
Capital: Montevideo
Administrative divisions: 19 departments (departamentos, singular—departamento); Artigas, Canelones, Cerro Largo, Colonia, Durazno, Flores, Florida, Lavalleja, Maldonado, Montevideo, Paysandu, Rio Negro, Rivera, Rocha, Salto, San Jose, Soriano, Tacuarembo, Treinta y Tres
Independence: 25 August 1828 (from Brazil)
Constitution: 27 November 1966, effective February 1967, suspended 27 June 1973, new constitution rejected by referendum 30 November 1980
Legal system: based on Spanish civil law system; accepts compulsory ICJ jurisdiction
National holiday: Independence Day, 25 August (1828)

Political parties and leaders: National (Blanco) Party, Carlos CAT; Colorado Party, Secretary General (vacant); Broad Front Coalition, Liber SEREGNI Mosquera—includes PSU, PCU, MLN, MRO, PVP; Uruguayan Socialist Party (PSU), Jose Pedro CARDOSO, and; Communist Party (PCU), Marina ARISMENDI; National Liberation Movement (MLN) or Tupamaros, Eleuterio FERNANDEZ Huidobro; Oriental Rvolutionary Movement (MRO), Walter ARTOLA; Party for the Victory of the Poor (PVP), Hugo CORES; New Space Coalition consists of PGP, PDC, and Civic Union, Hugo BATALLA; People's Government Party (PGP), Hugo BATALLA, secretary general; Christian Democratic Party (PDC), Carlos VASSALLO, secretary general; Civic Union, Humberto CIGANDA
Suffrage: 18 years of age; universal and compulsory
Elections:
President: last held 26 November 1989 (next to be held NA November 1994); results—Luis Alberto LACALLE Herrera (Blanco) 37%, Jorge BATLLE Ibanez (Colorado) 29%, Liber SEREGNI Mosquera (Broad Front) 20%
Chamber of Senators: last held 26 November 1989 (next to be held NA November 1994); results—Blanco 40%, Colorado 30%, Broad Front 23% New Space 7%; seats—(30 total) Blanco 12, Colorado 9, Broad Front 7, New Space 2
Chamber of Representatives: last held NA November 1989 (next to be held NA November 1994); results—Blanco 39%, Colorado 30%, Broad Front 22%, New Space 8%, other 1%; seats—(99 total) number of seats by party NA
Executive branch: president, vice president, Council of Ministers (cabinet)
Legislative branch: bicameral General Assembly (Asamblea General) consists of an upper chamber or Chamber of Senators (Camara de Senadores) and a lower chamber or Chamber of Representatives (Camera de Representantes)
Judicial branch: Supreme Court
Leaders:
Chief of State and Head of Government: President Luis Alberto LACALLE (since 1 March 1990); Vice President Gonzalo AGUIRRE Ramirez (since 1 March 1990)
Member of: AG (observer), CCC, ECLAC, FAO, G-11, G-77, GATT, IADB, IAEA, IBRD, ICAO, ICC, IFAD, IFC, ILO, IMF, IMO, INTELSAT, INTERPOL, IOC, IOM, ISO (correspondent), ITU, LAES, LAIA, LORCS, MERCOSUR, NAM (observer), OAS, OPANAL, PCA, RG, UN, UNCTAD, UNESCO, UNIDO, UNIKOM, UNMOGIP, UNOMOZ, UNTAC, UPU, WCL, WFTU, WHO, WIPO, WMO, WTO

Diplomatic representation in US:
chief of mission: Ambassador Eduardo MACGILLYCUDDY
chancery: 1918 F Street NW, Washington, DC 20006
telephone: telephone (202) 331-1313 through 1316
consulates general: Los Angeles, Miami, and New York,
consulate: New Orleans
US diplomatic representation:
chief of mission: Ambassador Richard C. BROWN
embassy: Lauro Muller 1776, Montevideo
mailing address: APO AA 34035
telephone: [598] (2) 23-60-61 or 48-77-77
FAX: [598] (2) 48-86-11
Flag: nine equal horizontal stripes of white (top and bottom) alternating with blue; there is a white square in the upper hoist-side corner with a yellow sun bearing a human face known as the Sun of May and 16 rays alternately triangular and wavy

Economy

Overview: Uruguay is a small economy with favorable climate, good soils, and solid hydropower potential. Economic development has been held back by excessive government regulation of economic detail and 50% to 130% inflation. After several years of sluggish growth, real GDP jumped by about 8% in 1992. The rise is attributable mainly to an increase in Argentine demand for Uruguayan exports, particularly agricultural products and electricity. In a major step toward greater regional economic cooperation, Uruguay in 1991 had joined Brazil, Argentina, and Paraguay in forming the Southern Cone Common Market (Mercosur). A referendum in December 1992 overturned key portions of landmark privatization legislation, dealing a serious blow to President LACALLE's broad economic reform plan.
National product: GDP—exchange rate conversion—$9.8 billion (1992 est.)
National product real growth rate: 8% (1992 est.)
National product per capita: $3,100 (1992 est.)
Inflation rate (consumer prices): 58% (1992 est.)
Unemployment rate: 9% (1992 est.)
Budget: revenues $2.9 billion; expenditures $3.0 billion, including capital expenditures of $388 million (1991)
Exports: $1.7 billion (f.o.b., 1992 est.)
commodities: hides and leather goods 17%, beef 10%, wool 9%, fish 7%, rice 4%
partners: Argentina, Brazil, US, Germany
Imports: $1.7 billion (f.o.b., 1992 est.)
commodities: crude oil, fuels, and lubricants, metals, machinery, transportation equipment, industrial chemicals

partners: Brazil 23%, Argentina 17%, US 10%, EC 27.1% (1990)
External debt: $4.1 billion (1991)
Industrial production: growth rate −1.4% (1990), accounts for almost 25% of GDP
Electricity: 2,168,000 kW capacity; 5,960 million kWh produced, 1,900 kWh per capita (1992)
Industries: meat processing, wool and hides, sugar, textiles, footwear, leather apparel, tires, cement, fishing, petroleum refining, wine
Agriculture: large areas devoted to livestock grazing; wheat, rice, corn, sorghum; self-sufficient in most basic foodstuffs
Economic aid: US commitments, including Ex-Im (FY70-88), $105 million; Western (non-US) countries, ODA and OOF bilateral commitments (1970-89), $420 million; Communist countries (1970-89), $69 million
Currency: 1 new Uruguayan peso (N$Ur) = 100 centesimos
Exchange rates: new Uruguayan pesos (N$Ur) per US$1—3,457.5 (December 1992), 3,026.9 (1992), 2,489 (1991), 1,594 (1990), 805 (1989), 451 (1988), 281 (1987)
Fiscal year: calendar year

Communications

Railroads: 3,000 km, all 1.435-meter (standard) gauge and government owned
Highways: 49,900 km total; 6,700 km paved, 3,000 km gravel, 40,200 km earth
Inland waterways: 1,600 km; used by coastal and shallow-draft river craft
Ports: Montevideo, Punta del Este, Colonia
Merchant marine: 4 ships (1,000 GRT or over) totaling 84,797 GRT/132,296 DWT; includes 1 cargo, 2 container, 1 oil tanker
Airports:
total: 88
usable: 81
with permanent-surface runways: 16
with runways over 3,659 m: 0
with runways 2,440-3,659 m: 2
with runways 1,220-2,439 m: 14
Telecommunications: most modern facilities concentrated in Montevideo; new nationwide microwave network; 337,000 telephones; broadcast stations—99 AM, no FM, 26 TV, 9 shortwave; 2 Atlantic Ocean INTELSAT earth stations

Defense Forces

Branches: Army, Navy (including Naval Air Arm, Coast Guard, Marines), Air Force, Grenadier Guards, Police
Manpower availability: males age 15-49 755,667; fit for military service 613,585 (1993 est.); no conscription
Defense expenditures: exchange rate conversion—$216 million, 2.3% of GDP (1991 est.)

Uzbekistan

300 km

Geography

Location: Central Asia, bordering the Aral Sea, between Kazakhstan and Turkmenistan
Map references: Asia, Commonwealth of Independent States—Central Asian States, Standard Time Zones of the World
Area:
total area: 447,400 km²
land area: 425,400 km²
comparative area: slightly larger than California
Land boundaries: total 6,221 km, Afghanistan 137 km, Kazakhstan 2,203 km, Kyrgyzstan 1,099 km, Tajikistan 1,161 km, Turkmenistan 1,621 km
Coastline: 0 km
note: Uzbekistan does border the Aral Sea (420 km)
Maritime claims: none; landlocked
International disputes: none
Climate: mostly mid latitude desert; semiarid grassland in east
Terrain: mostly flat-to-rolling sandy desert with dunes; Fergana Valley in east surrounded by mountainous Tajikistan and Kyrgyzstan; shrinking Aral Sea in west
Natural resources: natural gas, petroleum, coal, gold, uranium, silver, copper, lead and zinc, tungsten, molybdenum
Land use:
arable land: 10%
permanent crops: 0%
meadows and pastures: 47%
forest and woodland: 0%
other: 43%
Irrigated land: 41,550 km² (1990)
Environment: drying up of the Aral Sea is resulting in growing concentrations of chemical pesticides and natural salts
Note: landlocked

People

Population: 22,127,946 (July 1993 est.)
Population growth rate: 2.17% (1993 est.)
Birth rate: 30.57 births/1,000 population (1993 est.)

407

Uzbekistan (continued)

Death rate: 6.63 deaths/1,000 population (1993 est.)
Net migration rate: -2.2 migrant(s)/1,000 population (1993 est.)
Infant mortality rate: 54.4 deaths/1,000 live births (1993 est.)
Life expectancy at birth:
total population: 68.36 years
male: 65.05 years
female: 71.84 years (1993 est.)
Total fertility rate: 3.78 children born/woman (1993 est.)
Nationality:
noun: Uzbek(s)
adjective: Uzbek
Ethnic divisions: Uzbek 71.4%, Russian 8.3%, Tajik 4.7%, Kazakhs 4.1%, Tartars 2.4% (includes 70% of Crimean Tatars deported during World War II), Karakalpaks 2.1%, other 7%
Religions: Muslim 88% (mostly Sunnis), Eastern Orthodox 9%, other 3%
Languages: Uzbek 85%, Russian 5%, other 10%
Literacy: age 9-49 can read and write (1970)
total population: 100%
male: 100%
female: 100%
Labor force: 7.941 million
by occupation: agriculture and forestry 39%, industry and construction 24%, other 37% (1990)

Government

Names:
conventional long form: Republic of Uzbekistan
conventional short form: Uzbekistan
local long form: Uzbekiston Respublikasi
local short form: none
former: Uzbek Soviet Socialist Republic
Digraph: UZ
Type: republic
Capital: Tashkent (Toshkent)
Administrative divisions: 12 oblasts (oblastey, singular—oblast') and 1 autonomous republic* (avtomnaya respublika); Andizhan, Bukhara, Dzhizak, Fergana, Karakalpakstan* (Nukus), Kashkadar'ya (Karshi), Khorezm (Urgench), Namangan, Navoi, Samarkand, Surkhandar'ya (Termez), Syrdar'ya (Gulistan), Tashkent
note: an administrative division has the same name as its administrative center (exceptions have the administrative center name following in parentheses)
Independence: 31 August 1991 (from Soviet Union)
Constitution: new constitution adopted 8 December 1992
Legal system: evolution of Soviet civil law
National holiday: Independence Day, 1 September (1991)

Political parties and leaders: People's Democratic Party (PDP; formerly Communist Party), Islam A. KARIMOV, chairman; Erk (Freedom) Democratic Party (EDP), Muhammad SOLIKH, chairman
Other political or pressure groups: Birlik (Unity) People's Movement (BPM), Abdul Rakhman PULATOV, chairman; Islamic Rebirth Party (IRP), Abdullah UTAYEV, chairman
Suffrage: 18 years of age; universal
Elections:
President: last held 29 December 1991 (next to be held NA December 1996); results—Islam KARIMOV 86%, Mukhammad SOLIKH 12%, other 2%
Supreme Soviet: last held 18 February 1990 (next to be held NA); results—percent of vote by party NA; seats—(500 total) Communist 450, ERK 10, other 40; note—total number of seats will be reduced to 150 in next election
Executive branch: president, prime minister, cabinet
Legislative branch: unicameral Supreme Soviet
Judicial branch: Supreme Court
Leaders:
Chief of State: President Islam KARIMOV (since NA March 1990)
Head of Government: Prime Minister Abdulkhashim MUTALOV (since 13 January 1992), First Deputy Prime Minister Ismail Hakimovitch DJURABEKOV (since NA); Supreme Soviet Chairman Shavkat Muhitdinovitch YULDASHEV (since NA June 1991)
Member of: CIS, CSCE, EBRD, ECO, ESCAP, IBRD, IDA, IMF, NACC, UN, UNCTAD, WHO
Diplomatic representation in US:
chief of mission: Ambassador Muhammed Babir MALIKOV
chancery: 200 Pennsylvania Avenue NW, Washington, DC 20006
telephone: (202) 778-0107
FAX: (202) 861-0472
US diplomatic representation:
chief of mission: Ambassador Henry L. CLARKE
embassy: 55 Chelanzanskaya, Tashkent
mailing address: APO AE 09862
telephone: [7] (3712) 77-14-07
Flag: three equal horizontal bands of blue (top), white, and green separated by red fimbriations with a crescent moon and 12 stars in the upper hoist-side quadrant

Economy

Overview: Although Uzbekistan accounted for only 3.4% of total Soviet output, it produced two-thirds of the USSR's cotton and ranks as the fourth largest global producer. Moscow's push for ever-increasing

amounts of cotton had included massive irrigation projects which caused extensive environmental damage to the Aral Sea and rivers of the republic. Furthermore, the lavish use of chemical fertilizers has caused extensive pollution and widespread health problems. Recently the republic has sought to encourage food production at the expense of cotton. The small industrial sector specializes in such items as agricultural machinery, mineral fertilizers, vegetable oil, and bridge cranes. Uzbekistan also has some important natural resources including gold (about 30% of former Soviet production), uranium, and natural gas. The Uzbek Government has encouraged some land reform but has shied away from other aspects of economic reform. Output and living standards continued to fall in 1992 largely because of the cumulative impact of disruptions in supply that have followed the dismemberment of the USSR.
National product: GDP $NA
National product real growth rate: -10% (1992)
National product per capita: $NA
Inflation rate (consumer prices): at least 17% per month (first quarter 1993)
Unemployment rate: 0.1% includes only officially registered unemployed; there are also large numbers of underemployed workers
Budget: revenues $NA; expenditures $NA, including capital expenditures of $NA
Exports: $900 million to outside the successor states of the former USSR (1992)
commodities: cotton, gold, textiles, chemical and mineral fertilizers, vegetable oil
partners: Russia, Ukraine, Eastern Europe
Imports: $900 million from outside the successor states of the former USSR (1992)
commodities: machinery and parts, consumer durables, grain, other foods
partners: principally other former Soviet republics
External debt: $2 billion (end 1991 est.)
Industrial production: growth rate –6%
Electricity: 11,950,000 kW capacity; 50,900 million kWh produced, 2,300 kWh per capita (1992)
Industries: chemical and mineral fertilizers, vegetable oil, textiles
Agriculture: cotton, with much smaller production of grain, fruits, vegetables, and livestock
Illicit drugs: illicit producers of cannabis and opium; mostly for CIS consumption; limited government eradication programs; used as transshipment points for illicit drugs to Western Europe
Economic aid: $950 million official aid commitments by foreign donors (1992)
Currency: retaining Russian ruble as currency (January 1993)

Vanuatu

Exchange rates: rubles per US$1—415 (24 December 1992) but subject to wide fluctuations
Fiscal year: calendar year

Communications

Railroads: 3,460 km; does not include industrial lines (1990)
Highways: 78,400 km total; 67,000 km hard-surfaced, 11,400 km earth (1990)
Pipelines: crude oil 250 km, petroleum products 40 km, natural gas 810 km (1992)
Ports: none; landlocked
Airports:
totol: 265
useable: 74
with permanent-surface runways: 30
with runways over 3,659 m: 2
with runways 2,440-3,659 m: 20
with runways 1,220-2,439 m: 19
Telecommunications: poorly developed; NMT-450 analog cellular network established in Tashkent; 1.4 million telephone lines with 7.2 lines per 100 persons (1992); linked by landline or microwave with CIS member states and by leased connection via the Moscow international gateway switch to other countries; satellite earth stations—Orbita and INTELSAT (TV receive only); new intelsat earth station provides TV receive only capability for Turkish broadcasts; new satellite ground station also installed in Tashkent for direct linkage to Tokyo.

Defense Forces

Branches: Army, National Guard, Republic Security Forces (internal and border troops)
Manpower availability: males age 15-49 5,214,075; fit for military service 4,272,398; reach military age (18) annually 218,916 (1993 est.)
Defense expenditures: exchange rate conversion—$NA, NA% of GDP

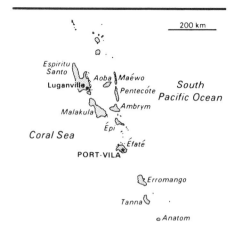

Geography

Location: Oceania, 5,750 km southwest of Honolulu in the South Pacific Ocean, about three-quarters of the way between Hawaii and Australia
Map references: Oceania, Standard Time Zones of the World
Area:
total area: 14,760 km^2
land area: 14,760 km^2
comparative area: slightly larger than Connecticut
note: includes more than 80 islands
Land boundaries: 0 km
Coastline: 2,528 km
Maritime claims: measured from claimed archipelagic baselines
contiguous zone: 24 nm
continental shelf: 200 nm or the edge of continental margin
exclusive economic zone: 200 nm
territorial sea: 12 nm
International disputes: none
Climate: tropical; moderated by southeast trade winds
Terrain: mostly mountains of volcanic origin; narrow coastal plains
Natural resources: manganese, hardwood forests, fish
Land use:
arable land: 1%
permanent crops: 5%
meadows and pastures: 2%
forest and woodland: 1%
other: 91%
Irrigated land: NA km^2
Environment: subject to tropical cyclones or typhoons (January to April); volcanism causes minor earthquakes

People

Population: 165,876 (July 1993 est.)
Population growth rate: 2.36% (1993 est.)
Birth rate: 33.16 births/1,000 population (1993 est.)

Death rate: 9.57 deaths/1,000 population (1993 est.)
Net migration rate: 0 migrant(s)/1,000 population (1993 est.)
Infant mortality rate: 69.9 deaths/1,000 live births (1993 est.)
Life expectancy at birth:
total population: 58.8 years
male: 57.11 years
female: 60.58 years (1993 est.)
Total fertility rate: 4.47 children born/woman (1993 est.)
Nationality:
noun: Ni-Vanuatu (singular and plural)
adjective: Ni-Vanuatu
Ethnic divisions: indigenous Melanesian 94%, French 4%, Vietnamese, Chinese, Pacific Islanders
Religions: Presbyterian 36.7%, Anglican 15%, Catholic 15%, indigenous beliefs 7.6%, Seventh-Day Adventist 6.2%, Church of Christ 3.8%, other 15.7%
Languages: English (official), French (official), pidgin (known as Bislama or Bichelama)
Literacy: age 15 and over can read and write (1979)
total population: 53%
male: 57%
female: 48%
Labor force: NA
by occupation: NA

Government

Names:
conventional long form: Republic of Vanuatu
conventional short form: Vanuatu
former: New Hebrides
Digraph: NH
Type: republic
Capital: Port-Vila
Administrative divisions: 11 island councils; Ambrym, Aoba/Maewo, Banks/Torres, Efate, Epi, Malakula, Paama, Pentecote, Santo/Malo, Shepherd, Tafea
Independence: 30 July 1980 (from France and UK)
Constitution: 30 July 1980
Legal system: unified system being created from former dual French and British systems
National holiday: Independence Day, 30 July (1980)
Political parties and leaders: Vanuatu Party (VP), Donald KALPOKAS; Union of Moderate Parties (UMP), Serge VOHOR; Melanesian Progressive Party (MPP), Barak SOPE; National United Party (NUP), Walter LINI; Tan Union Party (TUP), Vincent BOULEKONE; Nagriamel Party, Jimmy STEVENS; Friend Melanesian Party, leader NA
Suffrage: 18 years of age; universal
Elections:
Parliament: last held 2 December 1991 (next to be held by November 1995);

Vanuatu (continued)

note—after election, a coalition was formed by the Union of Moderate Parties and the National United Party to form new government on 16 December 1991; seats—(46 total) UMP 19; NUP 10; VP 10; MPP 4; TUP 1; Nagriamel 1; Friend 1
Executive branch: president, prime minister, deputy prime minister, Council of Ministers (cabinet)
Legislative branch: unicameral Parliament; note—the National Council of Chiefs advises on matters of custom and land
Judicial branch: Supreme Court
Leaders:
Chief of State: President Frederick TIMAKATA (since 30 January 1989)
Head of Government: Prime Minister Maxime CARLOT KORMAN (since 16 December 1991); Deputy Prime Minister Sethy REGENVANU (since 17 December 1991)
Member of: ACCT, ACP, AsDB, C, ESCAP, FAO, G-77, IBRD, ICAO, IDA, IFC, IMF, IMO, IOC, ITU, NAM, SPARTECA, SPC, SPF, UN, UNCTAD, UNIDO, UPU, WHO, WMO
Diplomatic representation in US: Vanuatu does not have a mission in Washington
US diplomatic representation: the ambassador to Papua New Guinea is accredited to Vanuatu
Flag: two equal horizontal bands of red (top) and green with a black isosceles triangle (based on the hoist side) all separated by a black-edged yellow stripe in the shape of a horizontal Y (the two points of the Y face the hoist side and enclose the triangle); centered in the triangle is a boar's tusk encircling two crossed namele leaves, all in yellow

Economy

Overview: The economy is based primarily on subsistence farming which provides a living for about 80% of the population. Fishing and tourism are the other mainstays of the economy. Mineral deposits are negligible; the country has no known petroleum deposits. A small light industry sector caters to the local market. Tax revenues come mainly from import duties.
National product: GDP—exchange rate conversion—$142 million (1988 est.)
National product real growth rate: 6% (1990)
National product per capita: $900 (1988 est.)
Inflation rate (consumer prices): 5% (1990)
Unemployment rate: NA%
Budget: revenues $90 million; expenditures $103 million, including capital expenditures of $45 million (1989 est.)
Exports: $15.6 million (f.o.b., 1990 est.)
commodities: copra 59%, cocoa 11%, meat 9%, fish 8%, timber 4%

partners: Netherlands, Japan, France, New Caledonia, Belgium
Imports: $60.4 million (f.o.b., 1990 est.)
commodities: machines and vehicles 25%, food and beverages 23%, basic manufactures 18%, raw materials and fuels 11%, chemicals 6%
partners: Australia 36%, Japan 13%, NZ 10%, France 8%, Fiji 8%
External debt: $30 million (1990 est.)
Industrial production: growth rate NA%; accounts for about 10% of GDP
Electricity: 17,000 kW capacity; 30 million kWh produced, 180 kWh per capita (1990)
Industries: food and fish freezing, wood processing, meat canning
Agriculture: accounts for 40% of GDP; export crops—coconuts, cocoa, coffee, fish; subsistence crops—taro, yams, coconuts, fruits, vegetables
Economic aid: Western (non-US) countries, ODA and OOF bilateral commitments (1970-89), $606 million
Currency: 1 vatu (VT) = 100 centimes
Exchange rates: vatu (VT) per US$1—120.77 (January 1993), 113.39 (1992), 111.68 (1991), 116.57 (1990), 116.04 (1989), 104.43 (1988)
Fiscal year: calendar year

Communications

Railroads: none
Highways: 1,027 km total; at least 240 km sealed or all-weather roads
Ports: Port-Vila, Luganville, Palikoulo, Santu
Merchant marine: 125 ships (1,000 GRT or over) totaling 2,121,819 GRT/3,193,942 DWT; includes 23 cargo, 16 refrigerated cargo, 6 container, 11 vehicle carrier, 1 livestock carrier, 6 oil tanker, 2 chemical tanker, 3 liquefied gas, 54 bulk, 1 combination bulk, 1 passenger, 1 short-sea passenger; note—a flag of convenience registry
Airports:
total: 31
usable: 31
with permanent-surface runways: 2
with runways over 3,659 m: 0
with runways 2,440-3,659 m: 1
with runways 1,220-2,439 m: 2
Telecommunications: broadcast stations—2 AM, no FM, no TV; 3,000 telephones; 1 Pacific Ocean INTELSAT ground station

Defense Forces

Branches: Vanuatu Police Force (VPF), paramilitary Vanuatu Mobile Force (VMF)
note: no military forces
Manpower availability: males age 15-49 NA; fit for military service NA
Defense expenditures: exchange rate conversion—$NA, NA% of GDP

Venezuela

400 km

Caribbean Sea

Boundary representation is not necessarily authoritative.

Geography

Location: Northern South America, bordering the Caribbean Sea between Colombia and Guyana
Map references: South America, Standard Time Zones of the World
Area:
total area: 912,050 km²
land area: 882,050 km²
comparative area: slightly more than twice the size of California
Land boundaries: total 4,993 km, Brazil 2,200 km, Colombia 2,050 km, Guyana 743 km
Coastline: 2,800 km
Maritime claims:
contiguous zone: 15 nm
continental shelf: 200 m depth or to depth of exploitation
exclusive economic zone: 200 nm
territorial sea: 12 nm
International disputes: claims all of Guyana west of the Essequibo river; maritime boundary dispute with Colombia in the Gulf of Venezuela
Climate: tropical; hot, humid; more moderate in highlands
Terrain: Andes mountains and Maracaibo lowlands in northwest; central plains (llanos); Guyana highlands in southeast
Natural resources: petroleum, natural gas, iron ore, gold, bauxite, other minerals, hydropower, diamonds
Land use:
arable land: 3%
permanent crops: 1%
meadows and pastures: 20%
forest and woodland: 39%
other: 37%
Irrigated land: 2,640 km² (1989 est.)
Environment: subject to floods, rockslides, mudslides; periodic droughts; increasing industrial pollution in Caracas and Maracaibo
Note: on major sea and air routes linking North and South America

People

Population: 20,117,687 (July 1993 est.)
Population growth rate: 2.22% (1993 est.)
Birth rate: 26.37 births/1,000 population (1993 est.)
Death rate: 4.69 deaths/1,000 population (1993 est.)
Net migration rate: 0.48 migrant(s)/1,000 population (1993 est.)
Infant mortality rate: 28.9 deaths/1,000 live births (1993 est.)
Life expectancy at birth:
total population: 72.69 years
male: 69.76 years
female: 75.77 years (1993 est.)
Total fertility rate: 3.14 children born/woman (1993 est.)
Nationality:
noun: Venezuelan(s)
adjective: Venezuelan
Ethnic divisions: mestizo 67%, white 21%, black 10%, Indian 2%
Religions: nominally Roman Catholic 96%, Protestant 2%
Languages: Spanish (official), Indian dialects spoken by about 200,000 Amerindians in the remote interior
Literacy: age 15 and over can read and write (1990)
total population: 88%
male: 87%
female: 90%
Labor force: 5.8 million
by occupation: services 56%, industry 28%, agriculture 16% (1985)

Government

Names:
conventional long form: Republic of Venezuela
conventional short form: Venezuela
local long form: Republica de Venezuela
local short form: Venezuela
Digraph: VE
Type: republic
Capital: Caracas
Administrative divisions: 21 states (estados, singular—estado), 1 territory* (territorio), 1 federal district** (distrito federal), and 1 federal dependence*** (dependencia federal); Amazonas*, Anzoategui, Apure, Aragua, Barinas, Bolivar, Carabobo, Cojedes, Delta Amacuro, Dependencias Federales***, Distrito Federal**, Falcon, Guarico, Lara, Merida, Miranda, Monagas, Nueva Esparta, Portuguesa, Sucre, Tachira, Trujillo, Yaracuy, Zulia
note: the federal dependence consists of 11 federally controlled island groups with a total of 72 individual islands
Independence: 5 July 1811 (from Spain)
Constitution: 23 January 1961
Legal system: based on Napoleonic code; judicial review of legislative acts in Cassation Court only; has not accepted compulsory ICJ jurisdiction
National holiday: Independence Day, 5 July (1811)
Political parties and leaders: Social Christian Party (COPEI), Hilarion CARDOZO, president, and Jose CURIEL, secretary general (acting); Democratic Action (AD), Humberto CELLI, president, and Luis ALFARO Ucero, secretary general; Movement Toward Socialism (MAS), Argelia LAYA, president, and Freddy MUNOZ, secretary general; The Radical Cause (La Causa R), Pablo Medina, secretary general
Other political or pressure groups: FEDECAMARAS, a conservative business group; Venezuelan Confederation of Workers (labor organization dominated by the Democratic Action); VECINOS groups
Suffrage: 18 years of age; universal
Elections:
President: last held 4 December 1988 (next to be held 5 December 1993); results—Carlos Andres PEREZ (AD) 54.6%, Eduardo FERNANDEZ (COPEI) 41.7%, other 3.7%; note—President Carlos Andres PEREZ suspended pending trial on corruption charges
Senate: last held 4 December 1988 (next to be held 5 December 1993); results—percent of vote by party NA; seats—(49 total) AD 23, COPEI 22, other 4; note—3 former presidents (1 from AD, 2 from COPEI) hold lifetime senate seats
Chamber of Deputies: last held 4 December 1992 (next to be held 5 December 1993); results—AD 43.7%, COPEI 31.4%, MAS 10.3%, other 14.6%; seats—(201 total) AD 97, COPEI 67, MAS 18, other 19
Executive branch: president, Council of Ministers (cabinet)
Legislative branch: bicameral Congress of the Republic (Congreso de la Republica) consists of an upper chamber or Senate (Senado) and a lower chamber or Chamber of Deputies (Camara de Diputados)
Judicial branch: Supreme Court of Justice (Corte Suprema de Justicia)
Leaders:
Chief of State and Head of Government: Interim President Ramon Jose VELASQUEZ (since 5 June 1993); note—President Carlos Andres PEREZ suspended pending trial on corruption charges
Member of: AG, CARICOM (observer), CDB, CG, ECLAC, FAO, G-3, G-11, G-15, G-19, G-24, G-77, GATT, IADB, IAEA, IBRD, ICAO, ICC, ICFTU, IFAD, IFC, ILO, IMF, IMO, INTELSAT, INTERPOL, IOC, IOM, ISO, ITU, LAES, LAIA, LORCS, MINURSO, NAM, OAS, ONUSAL, OPANAL, OPEC, PCA, RG, UN, UNCTAD, UNESCO, UNHCR, UNIDO, UNIKOM, UNPROFOR, UPU, WCL, WFTU, WHO, WIPO, WMO, WTO
Diplomatic representation in US:
chief of mission: Ambassador Simon Alberto CONSALVI Bottaro
chancery: 1099 30th Street NW, Washington, DC 20007
telephone: (202) 342-2214
consulates general: Baltimore, Boston, Chicago, Houston, Miami, New Orleans, New York, Philadelphia, San Francisco, and San Juan (Puerto Rico)
US diplomatic representation:
chief of mission: Ambassador Michael Martin SKOL
embassy: Avenida Francisco de Miranda and Avenida Principal de la Floresta, Caracas
mailing address: P. O. Box 62291, Caracas 1060-A, or APO AA 34037
telephone: [58] (2) 285-2222
FAX: [58] (2) 285-0336
consulate: Maracaibo
Flag: three equal horizontal bands of yellow (top), blue, and red with the coat of arms on the hoist side of the yellow band and an arc of seven white five-pointed stars centered in the blue band

Economy

Overview: Petroleum is the backbone of the economy, accounting for 23% of GDP, 70% of central government revenues, and 82% of export earnings in 1992. President PEREZ introduced an economic readjustment program when he assumed office in February 1989. Lower tariffs and the removal of price controls, a free market exchange rate, and market-linked interest rates threw the economy into confusion, causing an 8% decline in GDP in 1989. However, the economy recovered part way in 1990 and grew by 10.4% in 1991 and 7.3% in 1992, led by the non-petroleum sector.
National product: GDP—exchange rate conversion—$57.8 billion (1992 est.)
National product real growth rate: 7.3% (1992 est.)
National product per capita: $2,800 (1992 est.)
Inflation rate (consumer prices): 32% (1992 est.)
Unemployment rate: 8.4% (1992 est.)
Budget: revenues $13.2 billion; expenditures $13.1 billion, including capital expenditures of $NA (1992)
Exports: $14.0 billion (f.o.b., 1992 est.)
commodities: petroleum 82%, bauxite and aluminum, iron ore, agricultural products, basic manufactures
partners: US 50.7%, Europe 13.7%, Japan 4.0% (1989)
Imports: $12.4 billion (f.o.b., 1992 est.)
commodities: foodstuffs, chemicals, manufactures, machinery and transport equipment

Venezuela (continued)

partners: US 44%, FRG 8.0%, Japan 4%, Italy 7%, Canada 2% (1989)
External debt: $27.1 billion (1992)
Industrial production: growth rate 11.9% (1992 est.); accounts for 25% of GDP, including petroleum
Electricity: 21,130,000 kW capacity; 58,541 million kWh produced, 2,830 kWh per capita (1992)
Industries: petroleum, iron-ore mining, construction materials, food processing, textiles, steel, aluminum, motor vehicle assembly
Agriculture: accounts for 6% of GDP and 16% of labor force; products—corn, sorghum, sugarcane, rice, bananas, vegetables, coffee, beef, pork, milk, eggs, fish; not self-sufficient in food other than meat
Illicit drugs: illicit producer of cannabis and coca leaf for the international drug trade on a small scale; however, large quantities of cocaine transit the country from Colombia; important money-laundering hub
Economic aid: US commitments, including Ex-Im (FY70-86), $488 million; Communist countries (1970-89), $10 million
Currency: 1 bolivar (Bs) = 100 centimos
Exchange rates: bolivares (Bs) per US$1—80.18 (January 1993), 68.38 (1992), 56.82 (1991), 46.90 (1990), 34.68 (1989), 14.50 (fixed rate 1987-88)
Fiscal year: calendar year

Communications

Railroads: 542 km total; 363 km 1.435-meter standard gauge all single track, government owned; 179 km 1.435-meter gauge, privately owned
Highways: 77,785 km total; 22,780 km paved, 24,720 km gravel, 14,450 km earth roads, and 15,835 km unimproved earth
Inland waterways: 7,100 km; Rio Orinoco and Lago de Maracaibo accept oceangoing vessels
Pipelines: crude oil 6,370 km; petroleum products 480 km; natural gas 4,010 km
Ports: Amuay Bay, Bajo Grande, El Tablazo, La Guaira, Puerto Cabello, Puerto Ordaz
Merchant marine: 56 ships (1,000 GRT or over) totaling 837,375 GRT/1,344,795 DWT; includes 1 short-sea passenger, 1 passenger cargo, 19 cargo, 2 container, 4 roll-on/roll-off, 18 oil tanker, 1 chemical tanker, 2 liquefied gas, 6 bulk, 1 vehicle carrier, 1 combination bulk
Airports:
total: 360
usable: 331
with permanent-surface runways: 133
with runways over 3,659 m: 0
with runways 2,440-3,659 m: 15
with runways 1,220-2,439 m: 87

Telecommunications: modern and expanding; 1,440,000 telephones; broadcast stations—181 AM, no FM, 59 TV, 26 shortwave; 3 submarine coaxial cables; satellite ground stations—1 Atlantic Ocean INTELSAT and 3 domestic

Defense Forces

Branches: National Armed Forces (Fuerzas Armadas Nacionales, FAN) includes—Ground Forces or Army (Fuerzas Terrestres or Ejercito), Naval Forces (Fuerzas Navales or Armada), Air Forces (Fuerzas Aereas or Aviacion), Armed Forces of Cooperation or National Guard (Fuerzas Armadas de Cooperacion or Guardia Nacional)
Manpower availability: males age 15-49 5,192,107; fit for military service 3,769,441; reach military age (18) annually 221,043 (1993 est.)
Defense expenditures: exchange rate conversion—$1.95 billion, 4% of GDP (1991)

Vietnam

Geography

Location: Southeast Asia, bordering the South China Sea, between Laos and the Philippines
Map references: Asia, Southeast Asia, Standard Time Zones of the World
Area:
total area: 329,560 km²
land area: 325,360 km²
comparative area: slightly larger than New Mexico
Land boundaries: total 3,818 km, Cambodia 982 km, China 1,281 km, Laos 1,555 km
Coastline: 3,444 km (excludes islands)
Maritime claims:
contiguous zone: 24 nm
continental shelf: 200 nm or the edge of continental margin
exclusive economic zone: 200 nm
territorial sea: 12 nm
International disputes: maritime boundary with Cambodia not defined; involved in a complex dispute over the Spratly Islands with China, Malaysia, Philippines, Taiwan, and possibly Brunei; unresolved maritime boundary with Thailand; maritime boundary dispute with China in the Gulf of Tonkin; Paracel Islands occupied by China but claimed by Vietnam and Taiwan
Climate: tropical in south; monsoonal in north with hot, rainy season (mid-May to mid-September) and warm, dry season (mid-October to mid-March)
Terrain: low, flat delta in south and north; central highlands; hilly, mountainous in far north and northwest
Natural resources: phosphates, coal, manganese, bauxite, chromate, offshore oil deposits, forests
Land use:
arable land: 22%
permanent crops: 2%
meadows and pastures: 1%
forest and woodland: 40%
other: 35%

Irrigated land: 18,300 km² (1989 est.)
Environment: occasional typhoons (May to January) with extensive flooding

People

Population: 71,787,608 (July 1993 est.)
Population growth rate: 1.85% (1993 est.)
Birth rate: 27.99 births/1,000 population (1993 est.)
Death rate: 7.92 deaths/1,000 population (1993 est.)
Net migration rate: -1.56 migrant(s)/1,000 population (1993 est.)
Infant mortality rate: 46.4 deaths/1,000 live births (1993 est.)
Life expectancy at birth:
total population: 65.1 years
male: 63.08 years
female: 67.25 years (1993 est.)
Total fertility rate: 3.45 children born/woman (1993 est.)
Nationality:
noun: Vietnamese (singular and plural)
adjective: Vietnamese
Ethnic divisions: Vietnamese 85-90%, Chinese 3%, Muong, Thai, Meo, Khmer, Man, Cham
Religions: Buddhist, Taoist, Roman Catholic, indigenous beliefs, Islamic, Protestant
Languages: Vietnamese (official), French, Chinese, English, Khmer, tribal languages (Mon-Khmer and Malayo-Polynesian)
Literacy: age 15 and over can read and write (1990)
total population: 88%
male: 92%
female: 84%
Labor force: 32.7 million
by occupation: agricultural 65%, industrial and service 35% (1990 est.)

Government

Names:
conventional long form: Socialist Republic of Vietnam
conventional short form: Vietnam
local long form: Cong Hoa Chu Nghia Viet Nam
local short form: Viet Nam
Abbreviation: SRV
Digraph: VM
Type: Communist state
Capital: Hanoi
Administrative divisions: 50 provinces (tinh, singular and plural), 3 municipalities* (thanh pho, singular and plural); An Giang, Ba Ria-Vung Tau, Bac Thai, Ben Tre, Binh Dinh, Binh Thuan, Can Tho, Cao Bang, Dac Lac, Dong Nai, Dong Thap, Gia Lai, Ha Bac, Ha Giang, Ha Noi*, Ha Tay, Ha Tinh, Hai Hung, Hai Phong*, Ho Chi Minh*, Hoa Binh, Khanh Hoa, Kien Giang, Kon Tum, Lai Chau, Lam Dong, Lang Son, Lao Cai, Long An, Minh Hai, Nam Ha, Nghe An, Ninh Binh, Ninh Thuan, Phu Yen, Quang Binh, Quang Nam-Da Nang, Quang Ngai, Quang Ninh, Quang Tri, Soc Trang, Son La, Song Be, Tay Ninh, Thai Binh, Thanh Hoa, Thua Thien, Tien Giang, Tra Vinh, Tuyen Quang, Vinh Long, Vinh Phu, Yen Bai
Independence: 2 September 1945 (from France)
Constitution: NA April 1992
Legal system: based on Communist legal theory and French civil law system
National holiday: Independence Day, 2 September (1945)
Political parties and leaders: only party—Vietnam Communist Party (VCP), DO MUOI, general secretary
Suffrage: 18 years of age; universal
Elections:
National Assembly: last held 19 July 1992 (next to be held NA July 1997); results—VCP is the only party; seats—(395 total) VCP or VCP-approved 395
Executive branch: president, prime minister, three deputy prime ministers
Legislative branch: unicameral National Assembly (Quoc-Hoi)
Judicial branch: Supreme People's Court
Leaders:
Chief of State: President Le Duc ANH (since 23 September 1992)
Head of Government: Prime Minister Vo Van KIET (since 9 August 1991); First Deputy Prime Minister Phan Van KHAI (since 10 August 1991); Deputy Prime Minister Nguyen KHANH (since NA February 1987); Deputy Prime Minister Tran Duc LUONG (since NA February 1987)
Member of: ACCT, AsDB, ESCAP, FAO, G-77, IAEA, IBRD, ICAO, IDA, IFAD, IFC, ILO, IMF, IMO, INTELSAT, INTERPOL, IOC, IOM (observer), ISO, ITU, LORCS, NAM, UN, UNCTAD, UNESCO, UNIDO, UPU, WCL, WFTU, WHO, WIPO, WMO, WTO
Diplomatic representation in US: none
US diplomatic representation: none
Flag: red with a large yellow five-pointed star in the center

Economy

Overview: Vietnam has made significant progress in recent years moving away from the planned economic model and toward a more effective market-based economic system. Most prices are now fully decontrolled and the Vietnamese currency has been effectively devalued and floated at world market rates. In addition, the scope for private sector activity has been expanded, primarily through decollectivization of the agricultural sector and introduction of laws giving legal recognition to private business. Despite such positive indicators, the country's economic turnaround remains tenuous. Nearly three-quarters of export earnings are generated by only two commodities, rice and crude oil. Meanwhile, industrial production stagnates, burdened by uncompetitive state-owned enterprises the government is unwilling or unable to privatize. Unemployment looms as the most serious problem with over 25% of the workforce without jobs and population growth swelling the ranks of the unemployed yearly.
National product: GNP—exchange rate conversion—$16 billion (1992 est.)
National product real growth rate: 7.4% (1992 est.)
National product per capita: $230 (1992 est.)
Inflation rate (consumer prices): 15%-20% (1992 est.)
Unemployment rate: 25% (1992 est.)
Budget: revenues $1.7 billion; expenditures $1.9 billion, including capital expenditures of $NA (1990)
Exports: $2.3 billion (f.o.b., 1992)
commodities: agricultural and handicraft products, coal, minerals, crude oil, ores, seafood
partners: Japan, Singapore, Thailand, Hong Kong, Taiwan
Imports: $1.9 billion (c.i.f., 1992)
commodities: petroleum products, steel products, railroad equipment, chemicals, medicines, raw cotton, fertilizer, grain
partners: Japan, Singapore, Thailand
External debt: $16.8 billion (1990 est.)
Industrial production: growth rate 15% (1992); accounts for 30% of GNP
Electricity: 3,300,000 kW capacity; 9,000 million kWh produced, 130 kWh per capita (1992)
Industries: food processing, textiles, machine building, mining, cement, chemical fertilizer, glass, tires, oil
Agriculture: accounts for half of GNP; paddy rice, corn, potatoes make up 50% of farm output; commercial crops (rubber, soybeans, coffee, tea, bananas) and animal products 50%; since 1989 self-sufficient in food staple rice; fish catch of 943,100 metric tons (1989 est.)
Economic aid: US commitments, including Ex-Im (FY70-74), $3.1 billion; Western (non-US) countries, ODA and OOF bilateral commitments (1970-89), $2.9 billion; OPEC bilateral aid (1979-89), $61 million; Communist countries (1970-89), $12.0 billion
Currency: 1 new dong (D) = 100 xu
Exchange rates: new dong (D) per US$1—10,800 (November 1992), 8,100 (July 1991), 7,280 (December 1990), 3,996 (March 1990), 2,047 (1988), 225 (1987); note—1985-89 figures are end of year
Fiscal year: calendar year

Vietnam *(continued)*

Communications

Railroads: 3,059 km total; 2,454
1.000-meter gauge, 151 km 1.435-meter
(standard) gauge, 230 km dual gauge (three
rails), and 224 km not restored to service
after war damage
Highways: 85,000 km total; 9,400 km
paved, 48,700 km gravel or improved earth,
26,900 km unimproved earth (est.)
Inland waterways: 17,702 km navigable;
more than 5,149 km navigable at all times
by vessels up to 1.8 meter draft
Pipelines: petroleum products 150 km
Ports: Da Nang, Haiphong, Ho Chi Minh
City
Merchant marine: 99 ships (1,000 GRT or
over) totaling 460,712 GRT/739,246 DWT;
includes 84 cargo, 3 refrigerated cargo, 1
roll-on/roll-off, 8 oil tanker, 3 bulk
Airports:
total: 100
usable: 100
with permanent-surface runways: 50
with runways over 3,659 m: 0
with runways 2,440-3,659 m: 10
with runways 1,220-2,439 m: 20
Telecommunications: the inadequacies of
the obsolete switching equipment and cable
system is a serious constraint on the
business sector and on economic growth,
and restricts access to the international links
that Vietnam has established with most
major countries; the telephone system is not
generally available for private use (25
telephones for each 10,000 persons); 3
satellite earth stations; broadcast
stations—NA AM, 288 FM; 36 (77
repeaters) TV; about 2,500,000 TV receivers
and 7,000,000 radio receivers in use (1991)

Defense Forces

Branches: Ground, Navy (including Naval
Infantry), Air Force
Manpower availability: males age 15-49
17,835,536; fit for military service
11,338,880; reach military age (17) annually
771,792 (1993 est.)
Defense expenditures: exchange rate
conversion—$NA, NA% of GNP

Virgin Islands
(territory of the US)

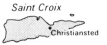

Geography

Location: in the eastern Caribbean Sea,
about 110 km east and southeast of Puerto
Rico
Map references: Central America and the
Caribbean
Area:
total area: 352 km²
land area: 349 km²
comparative area: slightly less than twice
the size of Washington, DC
Land boundaries: 0 km
Coastline: 188 km
Maritime claims:
contiguous zone: 24 nm
continental shelf: 200 m or depth of
exploitation
exclusive economic zone: 200 nm
territorial sea: 12 nm
International disputes: none
Climate: subtropical, tempered by easterly
tradewinds, relatively low humidity, little
seasonal temperature variation; rainy season
May to November
Terrain: mostly hilly to rugged and
mountainous with little level land
Natural resources: sun, sand, sea, surf
Land use:
arable land: 15%
permanent crops: 6%
meadows and pastures: 26%
forest and woodland: 6%
other: 47%
Irrigated land: NA km²
Environment: rarely affected by hurricanes;
subject to frequent severe droughts, floods,
earthquakes; lack of natural freshwater
resources
Note: important location along the Anegada
Passage—a key shipping lane for the
Panama Canal; Saint Thomas has one of the
best natural, deepwater harbors in the
Caribbean

People

Population: 98,130 (July 1993 est.)
Population growth rate: -0.76% (1993 est.)
Birth rate: 20.26 births/1,000 population
(1993 est.)
Death rate: 5.2 deaths/1,000 population
(1993 est.)
Net migration rate: -22.64 migrant(s)/1,000
population (1993 est.)
Infant mortality rate: 12.54 deaths/1,000
live births (1993 est.)
Life expectancy at birth:
total population: 75.29 years
male: 73.6 years
female: 77.2 years (1993 est.)
Total fertility rate: 2.64 children
born/woman (1993 est.)
Nationality:
noun: Virgin Islander(s)
adjective: Virgin Islander; US citizens
Ethnic divisions: West Indian (45% born in
the Virgin Islands and 29% born elsewhere
in the West Indies) 74%, US mainland 13%,
Puerto Rican 5%, other 8%; black 80%,
white 15%, other 5%; Hispanic origin 14%
Religions: Baptist 42%, Roman Catholic
34%, Episcopalian 17%, other 7%
Languages: English (official), Spanish,
Creole
Literacy:
total population: NA%
male: NA%
female: NA%
Labor force: 45,500 (1988)
by occupation: tourism 70%

Government

Names:
conventional long form: Virgin Islands of the
United States
conventional short form: Virgin Islands
Digraph: VQ
Type: organized, unincorporated territory of
the US administered by the Office of
Territorial and International Affairs, US
Department of the Interior
Capital: Charlotte Amalie
Administrative divisions: none (territory of
the US)
Constitution: Revised Organic Act of 22
July 1954
Legal system: based on US
National holiday: Transfer Day, 31 March
(1917) (from Denmark to US)
Political parties and leaders: Democratic
Party, Marilyn STAPLETON; Independent
Citizens' Movement (ICM), Virdin C.
BROWN; Republican Party, Charlotte-Poole
DAVIS
Suffrage: 18 years of age; universal
Elections:
Governor: last held 6 November 1990 (next
to be held November 1994);

results—Governor Alexander FARRELLY (Democratic Party) 56.5% defeated Juan LUIS (independent) 38.5%
Senate: last held 3 November 1992 (next to be held 2 November 1994); results—percent of vote by party NA; seats—(15 total) number of seats by party NA
US House of Representatives: last held 3 November 1992 (next to be held 2 November 1994); results—Ron DE LUGO reelected as delegate; seats—(1 total); seat by party NA; note—the Virgin Islands elect one representative to the US House of Representatives
Executive branch: US president, popularly elected governor and lieutenant governor
Legislative branch: unicameral Senate
Judicial branch:
US District Court: handles civil matters over $50,000, felonies (persons 15 years of age and over), and federal cases
Territorial Court: handles civil matters up to $50,000, small claims, juvenile, domestic, misdemeanors, and traffic cases
Leaders:
Chief of State: President William Jefferson CLINTON (since 20 January 1993); Vice President Albert GORE, Jr. (since 20 January 1993)
Head of Government: Governor Alexander A. FARRELLY (since 5 January 1987); Lieutenant Governor Derek M. HODGE (since 5 January 1987)
Member of: ECLAC (associate), IOC
Diplomatic representation in US: none (territory of the US)
Flag: white with a modified US coat of arms in the center between the large blue initials V and I; the coat of arms shows an eagle holding an olive branch in one talon and three arrows in the other with a superimposed shield of vertical red and white stripes below a blue panel

Economy

Overview: Tourism is the primary economic activity, accounting for more than 70% of GDP and 70% of employment. The manufacturing sector consists of textile, electronics, pharmaceutical, and watch assembly plants. The agricultural sector is small, most food being imported. International business and financial services are a small but growing component of the economy. One of the world's largest petroleum refineries is at Saint Croix.
National product: GDP—purchasing power equivalent—$1.2 billion (1987)
National product real growth rate: NA%
National product per capita: $11,000 (1987)
Inflation rate (consumer prices): NA%
Unemployment rate: 3.7% (1992)

Budget: revenues $364.4 million; expenditures $364.4 million, including capital expenditures of $NA (FY90)
Exports: $2.8 billion (f.o.b., 1990)
commodities: refined petroleum products
partners: US, Puerto Rico
Imports: $3.3 billion (c.i.f., 1990)
commodities: crude oil, foodstuffs, consumer goods, building materials
partners: US, Puerto Rico
External debt: $NA
Industrial production: growth rate 12%; accounts for NA% of GDP
Electricity: 380,000 kW capacity; 565 million kWh produced, 5,710 kWh per capita (1992)
Industries: tourism, petroleum refining, watch assembly, rum distilling, construction, pharmaceuticals, textiles, electronics
Agriculture: truck gardens, food crops (small scale), fruit, sorghum, Senepol cattle
Economic aid: Western (non-US) countries, ODA and OOF bilateral commitments (1970-89), $42 million
Currency: US currency is used
Fiscal year: 1 October-30 September

Communications

Highways: 856 km total
Ports: Saint Croix—Christiansted, Frederiksted; Saint Thomas—Long Bay, Crown Bay, Red Hook; Saint John—Cruz Bay
Airports:
total: 2
usable: 2
with permanent-surface runways : 2
with runways over 3,659 m: 0
with runways 2,440-3,659 m: 0
with runways 1,220-2,439 m: 2
note: international airports on Saint Thomas and Saint Croix
Telecommunications: modern telephone system using fiber-optic cable, submarine cable, microwave radio, and satellite facilities; 58,931 telephones; 98,000 radios; 63,000 TV sets in use; broadcast stations—4 AM, 8 FM, 4 TV (1988)

Defense Forces

Note: defense is the responsibility of the US

Wake Island
(territory of the US)

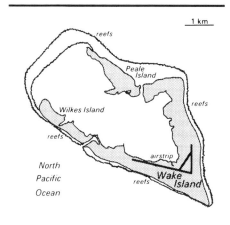

Geography

Location: in the North Pacific Ocean, 3,700 km west of Honolulu, about two-thirds of the way between Hawaii and the Northern Mariana Islands
Map references: Oceania
Area:
total area: 6.5 km²
land area: 6.5 km²
comparative area: about 11 times the size of The Mall in Washington, DC
Land boundaries: 0 km
Coastline: 19.3 km
Maritime claims:
contiguous zone: 24 nm
continental shelf: 200 m or depth of exploitation
exclusive economic zone: 200 nm
territorial sea: 12 nm
International disputes: claimed by the Republic of the Marshall Islands
Climate: tropical
Terrain: atoll of three coral islands built up on an underwater volcano; central lagoon is former crater, islands are part of the rim; average elevation less than 4 meters
Natural resources: none
Land use:
arable land: 0%
permanent crops: 0%
meadows and pastures: 0%
forest and woodland: 0%
other: 100%
Irrigated land: 0 km²
Environment: subject to occasional typhoons
Note: strategic location in the North Pacific Ocean; emergency landing location for transpacific flights

People

Population: no indigenous inhabitants; note—there are 302 US Air Force personnel, civilian weather service personnel, and US and Thai contractors; population peaked

Wake Island (continued)

about 1970 with over 1,600 persons during the Vietnam conflict

Government

Names:
conventional long form: none
conventional short form: Wake Island
Digraph: WQ
Type: unincorporated territory of the US administered by the US Air Force (under an agreement with the US Department of Interior) since 24 June 1972
Capital: none; administered from Washington, DC
Independence: none (territory of the US)
Flag: the US flag is used

Economy

Overview: Economic activity is limited to providing services to US military personnel and contractors located on the island. All food and manufactured goods must be imported.
Electricity: supplied by US military

Communications

Ports: none; because of the reefs, there are only two offshore anchorages for large ships
Airports:
total: 1
usable: 1
with permanent-surface runways: 1
with runways over 3,659 m: 0
with runways 2,440-3,659 m: 1
with runways 1,220-2,439 m: 0
Telecommunications: underwater cables to Guam and through Midway to Honolulu; 1 Autovon circuit off the Overseas Telephone System (OTS); Armed Forces Radio/Television Service (AFRTS) radio and television service provided by satellite; broadcast stations—1 AM, no FM, no TV
Note: formerly an important commercial aviation base, now used only by US military and some commercial cargo planes

Defense Forces

Note: defense is the responsibility of the US

Wallis and Futuna
(overseas territory of France)

Geography

Location: in the South Pacific Ocean, 4,600 km southwest of Honolulu, about two-thirds of the way from Hawaii to New Zealand
Map references: Oceania
Area:
total area: 274 km²
land area: 274 km²
comparative area: slightly larger than Washington, DC
note: includes Ile Uvea (Wallis Island), Ile Futuna (Futuna Island), Ile Alofi, and 20 islets
Land boundaries: 0 km
Coastline: 129 km
Maritime claims:
exclusive economic zone: 200 nm
territorial sea: 12 nm
International disputes: none
Climate: tropical; hot, rainy season (November to April); cool, dry season (May to October)
Terrain: volcanic origin; low hills
Natural resources: negligible
Land use:
arable land: 5%
permanent crops: 20%
meadows and pastures: 0%
forest and woodland: 0%
other: 75%
Irrigated land: NA km²
Environment: both island groups have fringing reefs

People

Population: 14,175 (July 1993 est.)
Population growth rate: 1.15% (1993 est.)
Birth rate: 26.42 births/1,000 population (1993 est.)
Death rate: 5.38 deaths/1,000 population (1993 est.)
Net migration rate: -9.5 migrant(s)/1,000 population (1993 est.)
Infant mortality rate: 27.59 deaths/1,000 live births (1993 est.)

Life expectancy at birth:
total population: 71.2 years
male: 70.54 years
female: 71.9 years (1993 est.)
Total fertility rate: 3.34 children born/woman (1993 est.)
Nationality:
noun: Wallisian(s), Futunan(s), or Wallis and Futuna Islanders
adjective: Wallisian, Futunan, or Wallis and Futuna Islander
Ethnic divisions: Polynesian
Religions: Roman Catholic
Languages: French, Wallisian (indigenous Polynesian language)
Literacy: all ages can read and write (1969)
total population: 50%
male: 50%
female: 51%
Labor force: NA
by occupation: agriculture, livestock, and fishing 80%, government 4% (est.)

Government

Names:
conventional long form: Territory of the Wallis and Futuna Islands
conventional short form: Wallis and Futuna
local long form: Territoire des Iles Wallis et Futuna
local short form: Wallis et Futuna
Digraph: WF
Type: overseas territory of France
Capital: Mata-Utu (on Ile Uvea)
Administrative divisions: none (overseas territory of France)
Independence: none (overseas territory of France)
Constitution: 28 September 1958 (French Constitution)
Legal system: French legal system
Political parties and leaders: Rally for the Republic (RPR); Union Populaire Locale (UPL); Union Pour la Democratie Francaise (UDF); Lua kae tahi (Giscardians); Mouvement des Radicaux de Gauche (MRG)
Suffrage: 18 years of age; universal
Elections:
Territorial Assembly: last held 15 March 1987 (next to be held NA March 1992); results—percent of vote by party NA; seats—(20 total) RPR 7, UPL 5, UDF 4, UNF 4
French Senate: last held 24 September 1989 (next to be held by NA September 1998); results—percent of vote by party NA; seats—(1 total) RPR 1
French National Assembly: last held 21 and 28 March 1992 (next to be held by NA September 1996); results—percent of vote by party NA; seats—(1 total) MRG 1
Executive branch: French president, chief administrator; note—there are three traditional kings with limited powers

Legislative branch: unicameral Territorial Assembly (Assemblee Territoriale)
Judicial branch: none; justice generally administered under French law by the chief administrator, but the three traditional kings administer customary law and there is a magistrate in Mata-Utu
Leaders:
Chief of State: President Francois MITTERRAND (since 21 May 1981)
Head of Government: Chief Administrator Robert POMMIES (since 26 September 1990)
Member of: FZ, SPC
Diplomatic representation in US: as an overseas territory of France, local interests are represented in the US by France
US diplomatic representation: none (overseas territory of France)
Flag: the flag of France is used

Economy

Overview: The economy is limited to traditional subsistence agriculture, with about 80% of the labor force earning its livelihood from agriculture (coconuts and vegetables), livestock (mostly pigs), and fishing. About 4% of the population is employed in government. Revenues come from French Government subsidies, licensing of fishing rights to Japan and South Korea, import taxes, and remittances from expatriate workers in New Caledonia. Wallis and Futuna imports food, fuel, clothing, machinery, and transport equipment, but its exports are negligible, consisting of copra and handicrafts.
National product: GDP—exchange rate conversion—$25 million (1991 est.)
National product real growth rate: NA%
National product per capita: $1,500 (1991 est.)
Inflation rate (consumer prices): NA%
Unemployment rate: NA%
Budget: revenues $2.7 million; expenditures $2.7 million, including capital expenditures of $NA (1983)
Exports: negligible
commodities: copra, handicrafts
partners: NA
Imports: $13.3 million (c.i.f., 1984)
commodities: foodstuffs, manufactured goods, transportation equipment, fuel
partners: France, Australia, New Zealand
External debt: $NA
Industrial production: growth rate NA%
Electricity: 1,200 kW capacity; 1 million kWh produced, 70 kWh per capita (1990)
Industries: copra, handicrafts, fishing, lumber
Agriculture: dominated by coconut production, with subsistence crops of yams, taro, bananas, and herds of pigs and goats
Economic aid: Western (non-US) countries,

ODA and OOF bilateral commitments (1970-89), $118 million
Currency: 1 CFP franc (CFPF) = 100 centimes
Exchange rates: Comptoirs Francais du Pacifique francs (CFPF) per US$1—99.65 (January 1993), 96.24 (1992), 102.57 (1991), 99.0 (1990), 115.99 (1989), 108.30 (1988); note—linked at the rate of 18.18 to the French franc
Fiscal year: NA

Communications

Highways: 100 km on Ile Uvea, 16 km sealed; 20 km earth surface on Ile Futuna
Inland waterways: none
Ports: Mata-Utu, Leava
Airports:
total: 2
useable: 2
with permanent-surface runways: 1
with runways over 3,659 m: 0
with runways 2,440-3,659 m: 0
with runways 1,220-2,439 m: 1
Telecommunications: 225 telephones; broadcast stations—1 AM, no FM, no TV

Defense Forces

Note: defense is the responsibility of France

West Bank

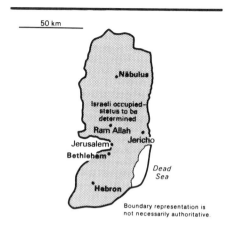

Boundary representation is not necessarily authoritative.

Note: The war between Israel and the Arab states in June 1967 ended with Israel in control of the West Bank and the Gaza Strip, the Sinai, and the Golan Heights. As stated in the 1978 Camp David Accords and reaffirmed by President Bush's post-Gulf crisis peace initiative, the final status of the West Bank and the Gaza Strip, their relationship with their neighbors, and a peace treaty between Israel and Jordan are to be negotiated among the concerned parties. Camp David further specifies that these negotiations will resolve the respective boundaries. Pending the completion of this process, it is US policy that the final status of the West Bank and the Gaza Strip has yet to be determined. In the view of the US, the term West Bank describes all of the area west of the Jordan River under Jordanian administration before the 1967 Arab-Israeli war. However, with respect to negotiations envisaged in the framework agreement, it is US policy that a distinction must be made between Jerusalem and the rest of the West Bank because of the city's special status and circumstances. Therefore, a negotiated solution for the final status of Jerusalem could be different in character from that of the rest of the West Bank.

Geography

Location: Middle East, between Jordan and Israel
Map references: Middle East
Area:
total area: 5,860 km²
land area: 5,640 km²
comparative area: slightly larger than Delaware
note: includes West Bank, East Jerusalem, Latrun Salient, Jerusalem No Man's Land, and the northwest quarter of the Dead Sea, but excludes Mt. Scopus
Land boundaries: total 404 km, Israel 307 km, Jordan 97 km
Coastline: 0 km (landlocked)

West Bank (continued)

Maritime claims: none; landlocked
International disputes: Israeli occupied with status to be determined
Climate: temperate, temperature and precipitation vary with altitude, warm to hot summers, cool to mild winters
Terrain: mostly rugged dissected upland, some vegetation in west, but barren in east
Natural resources: negligible
Land use:
arable land: 27%
permanent crops: 0%
meadows and pastures: 32%
forest and woodland: 1%
other: 40%
Irrigated land: NA km²
Environment: highlands are main recharge area for Israel's coastal aquifers
Note: landlocked; there are 175 Jewish settlements in the West Bank and 14 Israeli-built Jewish neighborhoods in East Jerusalem

People

Population: 1,404,114 (July 1993 est.)
note: in addition, there are 102,000 Jewish settlers in the West Bank and 134,000 in East Jerusalem (1993 est.)
Population growth rate: 2.9% (1993 est.)
Birth rate: 33.78 births/1,000 population (1993 est.)
Death rate: 5.32 deaths/1,000 population (1993 est.)
Net migration rate: 0.52 migrant(s)/1,000 population (1993 est.)
Infant mortality rate: 35.4 deaths/1,000 live births (1993 est.)
Life expectancy at birth:
total population: 69.93 years
male: 68.48 years
female: 71.46 years (1993 est.)
Total fertility rate: 4.37 children born/woman (1993 est.)
Nationality:
noun: NA
adjective: NA
Ethnic divisions: Palestinian Arab and other 88%, Jewish 12%
Religions: Muslim 80% (predominantly Sunni), Jewish 12%, Christian and other 8%
Languages: Arabic, Hebrew spoken by Israeli settlers, English widely understood
Literacy:
total population: NA%
male: NA%
female: NA%
Labor force: NA
by occupation: small industry, commerce, and business 29.8%, construction 24.2%, agriculture 22.4%, service and other 23.6% (1984)
note: excluding Israeli Jewish settlers

Government

Note: The West Bank is currently governed by Israeli military authorities and Israeli civil administration. It is US policy that the final status of the West Bank will be determined by negotiations among the concerned parties. These negotiations will determine how the area is to be governed.
Names:
conventional long form: none
conventional short form: West Bank
Digraph: WG

Economy

Overview: Economic progress in the West Bank has been hampered by Israeli military administration and the effects of the Palestinian uprising (intifadah). Industries using advanced technology or requiring sizable investment have been discouraged by a lack of local capital and restrictive Israeli policies. Capital investment consists largely of residential housing, not productive assets that would enable local firms to compete with Israeli industry. A major share of GNP is derived from remittances of workers employed in Israel and Persian Gulf states, but such transfers from the Gulf dropped dramatically after Iraq invaded Kuwait in August 1990. In the wake of the Persian Gulf crisis, many Palestinians have returned to the West Bank, increasing unemployment, and export revenues have plunged because of the loss of markets in Jordan and the Gulf states. Israeli measures to curtail the intifadah also have pushed unemployment up and lowered living standards. The area's economic outlook remains bleak.
National product: GNP—exchange rate conversion—$1.3 billion (1990 est.)
National product real growth rate: -10% (1990 est.)
National product per capita: $1,200 (1990 est.)
Inflation rate (consumer prices): 11% (1991 est.)
Unemployment rate: 15% (1990 est.)
Budget: revenues $31.0 million; expenditures $36.1 million, including capital expenditures of $NA (FY88)
Exports: $150 million (f.o.b., 1988 est.)
commodities: NA
partners: Jordan, Israel
Imports: $410 million (c.i.f., 1988 est.)
commodities: NA
partners: Jordan, Israel
External debt: $NA
Industrial production: growth rate 1% (1989); accounts for about 4% of GNP
Electricity: power supplied by Israel
Industries: generally small family businesses that produce cement, textiles, soap, olive-wood carvings, and

mother-of-pearl souvenirs; the Israelis have established some small-scale modern industries in the settlements and industrial centers
Agriculture: accounts for about 15% of GNP; olives, citrus and other fruits, vegetables, beef, and dairy products
Economic aid: NA
Currency: 1 new Israeli shekel (NIS) = 100 new agorot; 1 Jordanian dinar (JD) = 1,000 fils
Exchange rates: new Israeli shekels (NIS) per US$1—2.6480 (November 1992), 2.2791 (1991), 2.0162 (1990), 1.9164 (1989), 1.5989 (1988), 1.5946 (1987); Jordanian dinars (JD) per US$1—0.6890 (January 1993), 0.6797 (1992), 0.6808 (1991), 0.6636 (1990), 0.5704 (1989), 0.3709 (1988)
Fiscal year: calendar year (since 1 January 1992)

Communications

Highways: small road network, Israelis developing east-west axial highways to service new settlements
Airports:
total: 2
usable: 2
with permanent-surface runways: 2
with runways over 3,659 m: 0
with runways 2,440-3,659 m: 0
with runways 1,220-2,439 m: 1
Telecommunications: open-wire telephone system currently being upgraded; broadcast stations—no AM, no FM, no TV

Defense Forces

Branches: NA
Manpower availability: males age 15-49 NA; fit for military service NA
Defense expenditures: exchange rate conversion—$NA, NA% of GDP

Western Sahara

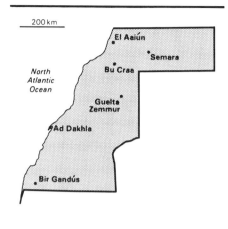

Geography

Location: Northern Africa, along the Atlantic Ocean, between Morocco and Mauritania
Map references: Africa, Standard Time Zones of the World
Area:
total area: 266,000 km²
land area: 266,000 km²
comparative area: slightly smaller than Colorado
Land boundaries: total 2.046 km, Algeria 42 km, Mauritania 1,561 km, Morocco 443 km
Coastline: 1,110 km
Maritime claims: contingent upon resolution of sovereignty issue
International disputes: claimed and administered by Morocco, but sovereignty is unresolved and the UN is attempting to hold a referendum on the issue; the UN-administered cease-fire has been currently in effect since September 1991
Climate: hot, dry desert; rain is rare; cold offshore air currents produce fog and heavy dew
Terrain: mostly low, flat desert with large areas of rocky or sandy surfaces rising to small mountains in south and northeast
Natural resources: phosphates, iron ore
Land use:
arable land: 0%
permanent crops: 0%
meadows and pastures: 19%
forest and woodland: 0%
other: 81%
Irrigated land: NA km²
Environment: hot, dry, dust/sand-laden sirocco wind can occur during winter and spring; widespread harmattan haze exists 60% of time, often severely restricting visibility; sparse water and arable land

People

Population: 206,629 (July 1993 est.)
Population growth rate: 2.52% (1993 est.)

Birth rate: 47.54 births/1,000 population (1993 est.)
Death rate: 19.57 deaths/1,000 population (1993 est.)
Net migration rate: -2.79 migrant(s)/1,000 population (1993 est.)
Infant mortality rate: 155.5 deaths/1,000 live births (1993 est.)
Life expectancy at birth:
total population: 44.88 years
male: 43.98 years
female: 46.06 years (1993 est.)
Total fertility rate: 7.01 children born/woman (1993 est.)
Nationality:
noun: Sahrawi(s), Sahraoui(s)
adjective: Sahrawian, Sahraouian
Ethnic divisions: Arab, Berber
Religions: Muslim
Languages: Hassaniya Arabic, Moroccan Arabic
Literacy:
total population: NA%
male: NA%
female: NA%
Labor force: 12,000
by occupation: animal husbandry and subsistence farming 50%

Government

Names:
conventional long form: none
conventional short form: Western Sahara
Digraph: WI
Type: legal status of territory and question of sovereignty unresolved; territory contested by Morocco and Polisario Front (Popular Front for the Liberation of the Saguia el Hamra and Rio de Oro), which in February 1976 formally proclaimed a government in exile of the Sahrawi Arab Democratic Republic (SADR); territory partitioned between Morocco and Mauritania in April 1976, with Morocco acquiring northern two-thirds; Mauritania, under pressure from Polisario guerrillas, abandoned all claims to its portion in August 1979; Morocco moved to occupy that sector shortly thereafter and has since asserted administrative control; the Polisario's government in exile was seated as an OAU member in 1984; guerrilla activities continued sporadically, until a UN-monitored cease-fire was implemented 6 September 1991
Capital: none
Administrative divisions: none (under de facto control of Morocco)
Leaders: none
Member of: none
Diplomatic representation in US: none
US diplomatic representation: none

Economy

Overview: Western Sahara, a territory poor in natural resources and having little rainfall, has a per capita GDP of roughly $300. Pastoral nomadism, fishing, and phosphate mining are the principal sources of income for the population. Most of the food for the urban population must be imported. All trade and other economic activities are controlled by the Moroccan Government.
National product: GDP—exchange rate conversion—$60 million (1991 est.)
National product real growth rate: NA%
National product per capita: $300 (1991 est.)
Inflation rate (consumer prices): NA%
Unemployment rate: NA%
Budget: revenues $NA; expenditures $NA, including capital expenditures of $NA
Exports: $8 million (f.o.b., 1982 est.)
commodities: phosphates 62%
partners: Morocco claims and administers Western Sahara, so trade partners are included in overall Moroccan accounts
Imports: $30 million (c.i.f., 1982 est.)
commodities: fuel for fishing fleet, foodstuffs
partners: Morocco claims and administers Western Sahara, so trade partners are included in overall Moroccan accounts
External debt: $NA
Industrial production: growth rate NA%
Electricity: 60,000 kW capacity; 79 million kWh produced, 425 kWh per capita (1989)
Industries: phosphate mining, fishing, handicrafts
Agriculture: limited largely to subsistence agriculture; some barley is grown in nondrought years; fruit and vegetables are grown in the few oases; food imports are essential; camels, sheep, and goats are kept by the nomadic natives; cash economy exists largely for the garrison forces
Economic aid: NA
Currency: 1 Moroccan dirham (DH) = 100 centimes
Exchange rates: Moroccan dirhams (DH) per US$1—9.034 (January 1993), 8.538 (1992), 8.707 (1991), 8.242 (1990), 8.488 (1989), 8.209 (1988)
Fiscal year: NA

Communications

Highways: 6,200 km total; 1,450 km surfaced, 4,750 km improved and unimproved earth roads and tracks
Ports: El Aaiun, Ad Dakhla
Airports:
total: 14
usable: 14
with permanent-surface runways: 3
with runways over 3,659 m: 0
with runways 2,440-3,659 m: 3
with runways 1,220-2,439 m: 5

Western Sahara (continued)

Telecommunications: sparse and limited system; tied into Morocco's system by microwave radio relay, troposcatter, and 2 Atlantic Ocean INTELSAT earth stations linked to Rabat, Morocco; 2,000 telephones; broadcast stations—2 AM, no FM, 2 TV

Defense Forces

Branches: NA
Manpower availability: NA
Defense expenditures: exchange rate conversion—$NA, NA% of GDP

Western Samoa

Geography

Location: Oceania, 4,300 km southwest of Honolulu in the South Pacific Ocean, about halfway between Hawaii and New Zealand
Map references: Oceania, Standard Time Zones of the World
Area:
total area: 2,860 km²
land area: 2,850 km²
comparative area: slightly smaller than Rhode Island
Land boundaries: 0 km
Coastline: 403 km
Maritime claims:
exclusive economic zone: 200 nm
territorial sea: 12 nm
International disputes: none
Climate: tropical; rainy season (October to March), dry season (May to October)
Terrain: narrow coastal plain with volcanic, rocky, rugged mountains in interior
Natural resources: hardwood forests, fish
Land use:
arable land: 19%
permanent crops: 24%
meadows and pastures: 0%
forest and woodland: 47%
other: 10%
Irrigated land: NA km²
Environment: subject to occasional typhoons; active volcanism

People

Population: 199,652 (July 1993 est.)
Population growth rate: 2.37% (1993 est.)
Birth rate: 33 births/1,000 population (1993 est.)
Death rate: 6.17 deaths/1,000 population (1993 est.)
Net migration rate: -3.14 migrant(s)/1,000 population (1993 est.)
Infant mortality rate: 38.6 deaths/1,000 live births (1993 est.)
Life expectancy at birth:
total population: 67.58 years

male: 65.19 years
female: 70.08 years (1993 est.)
Total fertility rate: 4.28 children born/woman (1993 est.)
Nationality:
noun: Western Samoan(s)
adjective: Western Samoan
Ethnic divisions: Samoan 92.6%, Euronesians 7% (persons of European and Polynesian blood), Europeans 0.4%
Religions: Christian 99.7% (about half of population associated with the London Missionary Society; includes Congregational, Roman Catholic, Methodist, Latter Day Saints, Seventh-Day Adventist)
Languages: Samoan (Polynesian), English
Literacy: age 15 and over can read and write (1971)
total population: 97%
male: 97%
female: 97%
Labor force: 38,000
by occupation: agriculture 22,000 (1987 est.)

Government

Names:
conventional long form: Independent State of Western Samoa
conventional short form: Western Samoa
Digraph: WS
Type: constitutional monarchy under native chief
Capital: Apia
Administrative divisions: 11 districts; A'ana, Aiga-i-le-Tai, Atua, Fa'asaleleaga, Gaga'emauga, Gagaifomauga, Palauli, Satupa'itea, Tuamasaga, Va'a-o-Fonoti, Vaisigano
Independence: 1 January 1962 (from UN trusteeship administered by New Zealand)
Constitution: 1 January 1962
Legal system: based on English common law and local customs; judicial review of legislative acts with respect to fundamental rights of the citizen; has not accepted compulsory ICJ jurisdiction
National holiday: National Day, 1 June
Political parties and leaders: Human Rights Protection Party (HRPP), TOFILAU Eti, chairman; Samoan National Development Party (SNDP), TAPUA Tamasese Efi, chairman
Suffrage: 21 years of age; universal, but only matai (head of family) are able to run for the Legislative Assembly
Elections:
Legislative Assembly: last held 5 April 1991 (next to be held by NA 1996); results—percent of vote by party NA; seats—(47 total) HRPP 28, SNDP 18, independents 1
Executive branch: chief, Executive Council, prime minister, Cabinet
Legislative branch: unicameral Legislative Assembly (Fono)

Judicial branch: Supreme Court, Court of Appeal

Leaders:

Chief of State: Chief Susuga Malietoa TANUMAFILI II (Co-Chief of State from 1 January 1962 until becoming sole Chief of State on 5 April 1963)

Head of Government: Prime Minister TOFILAU Eti Alesana (since 7 April 1988)

Member of: ACP, AsDB, C, ESCAP, FAO, G-77, IBRD, ICFTU, IDA, IFAD, IFC, IMF, IOC, ITU, LORCS, SPARTECA, SPC, SPF, UN, UNCTAD, UNESCO, UPU, WHO

Diplomatic representation in US:

chief of mission: Ambassador-designate Neroni SLADE

chancery: (temporary) suite 510, 1155 15th Street NW, Washington, DC 20005

telephone: (202) 833-1743

US diplomatic representation:

chief of mission: the ambassador to New Zealand is accredited to Western Samoa

embassy: address NA, Apia

mailing address: P.O. Box 3430, Apia

telephone: (685) 21-631

FAX: (685) 22-030

Flag: red with a blue rectangle in the upper hoist-side quadrant bearing five white five-pointed stars representing the Southern Cross constellation

---

Economy

Overview: Agriculture employs more than half of the labor force, contributes 50% to GDP, and furnishes 90% of exports. The bulk of export earnings comes from the sale of coconut oil and copra. The economy depends on emigrant remittances and foreign aid to support a level of imports several times export earnings. Tourism has become the most important growth industry, and construction of the first international hotel is under way.

National product: GDP—exchange rate conversion—$115 million (1990)

National product real growth rate: -4.5% (1990 est.)

National product per capita: $690 (1990)

Inflation rate (consumer prices): 15% (1990)

Unemployment rate: NA%

Budget: revenues $95.3 million; expenditures $95.4 million, including capital expenditures of $41 million (FY92)

Exports: $9 million (f.o.b., 1990)

commodities: coconut oil and cream 54%, taro 12%, copra 9%, cocoa 3%

partners: NZ 28%, American Samoa 23%, Germany 22%, US 6% (1990)

Imports: $75 million (c.i.f., 1990)

commodities: intermediate goods 58%, food 17%, capital goods 12%

partners: New Zealand 41%, Australia 18%, Japan 13%, UK 6%, US 6%

External debt: $83 million (December 1990 est.)

Industrial production: growth rate –4% (1990 est.); accounts for 14% of GDP

Electricity: 29,000 kW capacity; 45 million kWh produced, 240 kWh per capita (1990)

Industries: timber, tourism, food processing, fishing

Agriculture: accounts for 50% of GDP; coconuts, fruit (including bananas, taro, yams)

Economic aid: US commitments, including Ex-Im (FY70-89), $18 million; Western (non-US) countries, ODA and OOF bilateral commitments (1970-89), $306 million; OPEC bilateral aid (1979-89), $4 million

Currency: 1 tala (WS$) = 100 sene

Exchange rates: tala (WS$) per US$1—2.5681 (January 1993), 2.4655 (1992), 2.3975 (1991), 2.3095 (1990), 2.2686 (1989), 2.0790 (1988)

Fiscal year: calendar year

Communications

Highways: 2,042 km total; 375 km sealed; 1,667 km mostly gravel, crushed stone, or earth

Ports: Apia

Merchant marine: 1 roll-on/roll-off ship (1,000 GRT or over) totaling 3,838 GRT/5,536 DWT

Airports:

total: 3

usable: 3

with permanent-surface runways: 1

with runways over 3,659 m: 0

with runways 2,440-3,659 m: 1

with runways 1,220-2,439 m: 0

Telecommunications: 7,500 telephones; 70,000 radios; broadcast stations—1 AM, no FM, no TV; 1 Pacific Ocean INTELSAT ground station

---

Defense Forces

Branches: Department of Police and Prisons

Manpower availability: males age 15-49 NA; fit for military service NA

Defense expenditures: exchange rate conversion—$NA, NA% of GDP

World

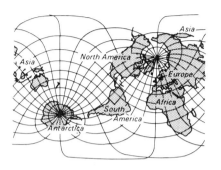

Geography

Map references: Standard Time Zones of the World

Area:

total area: 510.072 million km^2

land area: 148.94 million km^2

water area: 361.132 million km^2

comparative area: land area about 16 times the size of the US

note: 70.8% of the world is water, 29.2% is land

Land boundaries: the land boundaries in the world total 250,883.64 km (not counting shared boundaries twice)

Coastline: 356,000 km

Maritime claims:

contiguous zone: 24 nm claimed by most but can vary

continental shelf: 200 m depth claimed by most or to the depth of exploitation, others claim 200 nm or to the edge of the continental margin

exclusive fishing zone: 200 nm claimed by most but can vary

exclusive economic zone: 200 nm claimed by most but can vary

territorial sea: 12 nm claimed by most but can vary

note: boundary situations with neighboring states prevent many countries from extending their fishing or economic zones to a full 200 nm; 42 nations and other areas that are landlocked include Afghanistan, Andorra, Armenia, Austria, Azerbaijan, Belarus, Bhutan, Bolivia, Botswana, Burkina, Burundi, Central African Republic, Chad, Czech Republic, Ethiopia, Holy See (Vatican City), Hungary, Kazakhstan, Kyrgyzstan, Laos, Lesotho, Liechtenstein, Luxembourg, Macedonia, Malawi, Mali, Moldova, Mongolia, Nepal, Niger, Paraguay, Rwanda, San Marino, Slovakia, Swaziland, Switzerland, Tajikistan, Turkmenistan, Uganda, Uzbekistan, West Bank, Zambia, Zimbabwe

Climate: two large areas of polar climates separated by two rather narrow temperate zones from a wide equatorial band of tropical to subtropical climates

Terrain: highest elevation is Mt. Everest at 8,848 meters and lowest depression is the Dead Sea at 392 meters below sea level; greatest ocean depth is the Marianas Trench at 10,924 meters

Natural resources: the rapid using up of nonrenewable mineral resources, the depletion of forest areas and wetlands, the extinction of animal and plant species, and the deterioration in air and water quality (especially in Eastern Europe and the former USSR) pose serious long-term problems that governments and peoples are only beginning to address

Land use:

arable land: 10%

permanent crops: 1%

meadows and pastures: 24%

forest and woodland: 31%

other: 34%

Irrigated land: NA km^2

Environment: large areas subject to severe weather (tropical cyclones), natural disasters (earthquakes, landslides, tsunamis, volcanic eruptions), overpopulation, industrial disasters, pollution (air, water, acid rain, toxic substances), loss of vegetation (overgrazing, deforestation, desertification), loss of wildlife resources, soil degradation, soil depletion, erosion

---

People

Population: 5,554,552,453 (July 1993 est.)

Population growth rate: 1.6% (1993 est.)

Birth rate: 25 births/1,000 population (1993 est.)

Death rate: 9 deaths/1,000 population (1993 est.)

Infant mortality rate: 66 deaths/1,000 live births (1993 est.)

Life expectancy at birth:

total population: 62 years

male: 60 years

female: 64 years (1993 est.)

Total fertility rate: 3.2 children born/woman (1993 est.)

Literacy: age 15 and over can read and write (1990 est.)

combined: 74%

male: 81%

female: 67%

Labor force: 2.24 billion (1992)

by occupation: NA

Government

Digraph: XX

Administrative divisions: 265 sovereign nations, dependent areas, other, and miscellaneous entries

Legal system: varies by individual country; 182 are parties to the United Nations International Court of Justice (ICJ or World Court)

Economy

Overview: Real global output—gross world product (GWP)—rose one-half of 1% in 1992, with results varying widely among regions and countries. Average growth of 1.5% in the GDP of industrialized countries (62% of GWP in 1992) and average growth of 5% in the GDP of less developed countries (30% of GWP) were offset by a further 15-20% drop in the GDP of the former Soviet-East European area (now only 8% of GWP). The United States accounted for 23% of GWP in 1992; the 12-member European Community, which established a single internal market on 1 January 1993, accounted for another 23%, and Japan accounted for 10%. These are the three "economic superpowers" presumably destined to compete for mastery in international markets on into the 21st century. In general, growth in the industrialized countries was sluggish in 1992, with unemployment typically at 7-11%. As for the less developed countries, China, India, and the Four Dragons—South Korea, Taiwan, Hong Kong, and Singapore—posted good records; however, many other countries, especially in Africa, suffered bitterly from drought, rapid population growth, and civil strife. The continued plunge in production in practically all the former Warsaw Pact economies strained the political and social fabric of these newly independent nations, in particular in Russia. The addition of nearly 100 million people each year to an already overcrowded globe is exacerbating the problems of pollution, desertification, underemployment, epidemics, and famine. Because of their own internal problems, the industrialized countries have inadequate resources to deal effectively with the poorer areas of the world, which, at least from the economic point of view, are becoming further marginalized. (For the specific economic problems of each country, see the individual country entries in this volume.)

National product: GWP (gross world product)—purchasing power equivalent—$25.6 trillion (1992 est.)

National product real growth rate: 0.5% (1992 est.)

National product per capita: $4,600 (1992 est.)

Inflation rate (consumer prices):

developed countries: 5% (1992 est.)

developing countries: 50% (1992 est.)

note: these figures vary widely in individual cases

Unemployment rate: developed countries typically 7-11%; developing countries, extensive unemployment and underemployment (1992)

Exports: $3.64 trillion (f.o.b., 1992 est.)

commodities: the whole range of industrial and agricultural goods and services

partners: in value, about 75% of exports from the developed countries

Imports: $3.82 trillion (c.i.f., 1992 est.)

commodities: the whole range of industrial and agricultural goods and services

partners: in value, about 75% of imports by the developed countries

External debt: $1.0 trillion for less developed countries (1992 est.)

Industrial production: growth rate –1% (1992 est.)

Electricity: 2,864,000,000 kW capacity; 11,450,000 million kWh produced, 2,150 kWh per capita (1990)

Industries: industry worldwide is dominated by the onrush of technology, especially in computers, robotics, telecommunications, and medicines and medical equipment; most of these advances take place in OECD nations; only a small portion of non-OECD countries have succeeded in rapidly adjusting to these technological forces, and the technological gap between the industrial nations and the less-developed countries continues to widen; the rapid development of new industrial (and agricultural) technology is complicating already grim environmental problems

Agriculture: the production of major food crops has increased substantially in the last 20 years; the annual production of cereals, for instance, has risen by 50%, from about 1.2 billion metric tons to about 1.8 billion metric tons; production increases have resulted mainly from increased yields rather than increases in planted areas; while global production is sufficient for aggregate demand, about one-fifth of the world's population remains malnourished, primarily because local production cannot adequately provide for large and rapidly growing populations, which are too poor to pay for food imports; conditions are especially bad in Africa where drought in recent years has intensified the consequences of overpopulation

Economic aid: NA

Communications

Railroads: 239,430 km of narrow gauge track; 710,754 km of standard gauge track; 251,153 km of broad gauge track; includes about 190,000 to 195,000 km of electrified routes of which 147,760 km are in Europe, 24,509 km in the Far East, 11,050 km in Africa, 4,223 km in South America, and only 4,160 km in North America; fastest

Yemen

speed in daily service is 300 km/hr attained by France's SNCF TGV-Atlantique line
Ports: Mina al Ahmadi (Kuwait), Chiba, Houston, Kawasaki, Kobe, Marseille, New Orleans, New York, Rotterdam, Yokohama
Merchant marine: 23,943 ships (1,000 GRT or over) totaling 397,225,000 GRT/652,025,000 DWT; includes 347 passenger-cargo, 12,581 freighters, 5,473 bulk carriers, and 5,542 tankers (January 1992)

Defense Forces

Branches: ground, maritime, and air forces at all levels of technology
Defense expenditures: $1.0 trillion, 4% of total world output; decline of 5-10% (1991 est.)

300 km

Boundary representation is not necessarily authoritative.

Geography

Location: Middle East, along the Red Sea and the Arabian Sea, south of Saudi Arabia
Map references: Africa, Middle East, Standard Time Zones of the World
Area:
total area: 527,970 km^2
land area: 527,970 km^2
comparative area: slightly larger than twice the size of Wyoming
note: includes Perim, Socotra, the former Yemen Arab Republic (YAR or North Yemen), and the former People's Democratic Republic of Yemen (PDRY or South Yemen)
Land boundaries: total 1,746 km, Oman 288 km, Saudi Arabia 1,458 km
Coastline: 1,906 km
Maritime claims:
contiguous zone: 18 nm in the North 24 nm in the South
continental shelf: 200 m depth in the North 200 nm in the South or to the edge of the continental margin
exclusive economic zone: 200 nm
territorial sea: 12 nm
International disputes: undefined section of boundary with Saudi Arabia; Administrative Line with Oman; a treaty with Oman to settle the Yemeni-Omani boundary was ratified in December 1992
Climate: mostly desert; hot and humid along west coast; temperate in western mountains affected by seasonal monsoon; extraordinarily hot, dry, harsh desert in east
Terrain: narrow coastal plain backed by flat-topped hills and rugged mountains; dissected upland desert plains in center slope into the desert interior of the Arabian Peninsula
Natural resources: petroleum, fish, rock salt, marble, small deposits of coal, gold, lead, nickel, and copper, fertile soil in west
Land use:
arable land: 6%
permanent crops: 0%
meadows and pastures: 30%

forest and woodland: 7%
other: 57%
Irrigated land: 3,100 km^2 (1989 est.)
Environment: subject to sand and dust storms in summer; scarcity of natural freshwater resources; overgrazing; soil erosion; desertification
Note: controls Bab el Mandeb, the strait linking the Red Sea and the Gulf of Aden, one of world's most active shipping lanes

People

Population: 10,742,395 (July 1993 est.)
Population growth rate: 3.31% (1993 est.)
Birth rate: 51 births/1,000 population (1993 est.)
Death rate: 15.37 deaths/1,000 population (1993 est.)
Net migration rate: -2.56 migrant(s)/1,000 population (1993 est.)
Infant mortality rate: 115.6 deaths/1,000 live births (1993 est.)
Life expectancy at birth:
total population: 50.94 years
male: 49.83 years
female: 52.11 years (1993 est.)
Total fertility rate: 7.27 children born/woman (1993 est.)
Nationality:
noun: Yemeni(s)
adjective: Yemeni
Ethnic divisions: predominantly Arab; Afro-Arab concentrations in coastal locations; South Asians in southern regions; small European communities in major metropolitan areas; 60,000 (est.) Somali refugees encamped near Aden
Religions: Muslim (including Sha'fi, Sunni, and Zaydi Shi'a), Jewish, Christian, Hindu
Languages: Arabic
Literacy: age 15 and over can read and write (1990)
total population: 38%
male: 53%
female: 26%
Labor force:
North: NA
by occupation: agriculture and herding 70%, expatriate laborers 30% (est.)
South: 477,000
by occupation: agriculture 45.2%, services 21.2%, construction 13.4%, industry 10.6%, commerce and other 9.6% (1983)

Government

Names:
conventional long form: Republic of Yemen
conventional short form: Yemen
local long form: Al Jumhuriyah al Yamaniyah
local short form: Al Yaman
Digraph: YM
Type: republic

Yemen *(continued)*

Capital: Sanaa

Administrative divisions: 17 governorates (muhafazat, singular—muhafazah); Abyan, 'Adan, Al Bayda', Al Hudaydah, Al Jawf, Al Mahrah, Al Mahwit, Dhamar, Hadramawt, Hajjah, Ibb, Lahij, Ma'rib, Sa'dah, San'a', Shabwah, Ta'izz
note: there may be a new capital district of San'a'

Independence: 22 May 1990 Republic of Yemen was established on 22 May 1990 with the merger of the Yemen Arab Republic {Yemen (Sanaa) or North Yemen} and the Marxist-dominated People's Democratic Republic of Yemen {Yemen (Aden) or South Yemen}; previously North Yemen had become independent on NA November 1918 (from the Ottoman Empire) and South Yemen had become independent on 30 November 1967 (from the UK)

Constitution: 16 April 1991

Legal system: based on Islamic law, Turkish law, English common law, and local customary law; does not accept compulsory ICJ jurisdiction

National holiday: Proclamation of the Republic, 22 May (1990)

Political parties and leaders: General People's Congress, 'Ali 'Abdallah SALIH; Yemeni Socialist Party (YSP; formerly South Yemen's ruling party—a coalition of National Front, Ba'th, and Communist Parties), Ali Salim al-BIDH; Yemen Grouping for Reform or Islaah, Abdallah Husayn AHMAR

Other political or pressure groups: conservative tribal groups; Muslim Brotherhood; Islamist parties; pro-Iraqi Ba'thists; Nasirists

Suffrage: 18 years of age; universal

Elections:
House of Representatives: last held NA (next to be held 27 April 1993); results—percent of vote NA; seats—(301); number of seats by party NA; note—the 301 members of the new House of Representatives come from North Yemen's Consultative Assembly (159 members), South Yemen's Supreme People's Council (111 members), and appointments by the New Presidential Council (31 members)

Executive branch: five-member Presidential Council (president, vice president, two members from northern Yemen and one member from southern Yemen), prime minister

Legislative branch: unicameral House of Representatives

Judicial branch: Supreme Court

Leaders:
Chief of State and Head of Government: President 'Ali 'Abdallah SALIH (since 22 May 1990, the former president of North Yemen); Vice President Ali Salim al-BIDH (since 22 May 1990); Presidential Council Member Salim Salih MUHAMMED; Presidential Council Member Kadi Abdul-Karim al-ARASHI; Presidential Council Member Abdul-Aziz ABDUL-GHANI; Prime Minister Haydar Abu Bakr al-'ATTAS (since 22 May 1990, the former president of South Yemen)

Member of: ACC, AFESD, AL, AMF, CAEU, ESCWA, FAO, G-77, IBRD, ICAO, IDA, IDB, IFAD, IFC, ILO, IMF, IMO, INTELSAT, INTERPOL, IOC, ITU, LORCS, NAM, OIC, UN, UNCTAD, UNESCO, UNIDO, UPU, WFTU, WHO, WIPO, WMO, WTO

Diplomatic representation in US:
chief of mission: Ambassador Muhsin Ahmad al-AYNI
chancery: Suite 840, 600 New Hampshire Avenue NW, Washington, DC 20037
telephone: (202) 965-4760 or 4761
consulate general: Detroit
consulate: San Francisco

US diplomatic representation:
chief of mission: Ambassador Arthur H. HUGHES
embassy: Dhahr Himyar Zone, Sheraton Hotel District, Sanaa
mailing address: P. O. Box 22347 Sanaa or Sanaa, Department of State, Washington, DC 20521-6330
telephone: [967] (2) 238-842 through 238-852
FAX: [967] (2) 251-563

Flag: three equal horizontal bands of red (top), white, and black; similar to the flag of Syria which has two green stars and of Iraq which has three green stars (plus an Arabic inscription) in a horizontal line centered in the white band; also similar to the flag of Egypt which has a symbolic eagle centered in the white band

---

Economy

Overview: Whereas the northern city Sanaa is the political capital of a united Yemen, the southern city Aden, with its refinery and port facilities, is the economic and commercial capital. Future economic development depends heavily on Western-assisted development of promising oil resources. Former South Yemen's willingness to merge stemmed partly from the steady decline in Soviet economic support. The low level of domestic industry and agriculture have made northern Yemen dependent on imports for virtually all of its essential needs. Large trade deficits have been compensated for by remittances from Yemenis working abroad and by foreign aid. Once self-sufficient in food production, northern Yemen has become a major importer. Land once used for export crops—cotton, fruit, and vegetables—has been turned over to growing qat, a mildly narcotic shrub chewed by Yemenis which has no significant export market. Oil export revenues started flowing in late 1987 and boosted 1988 earnings by about $800 million. Economic growth in former South Yemen has been constrained by a lack of incentives, partly stemming from centralized control over production decisions, investment allocation, and import choices.

National product: GDP—exchange rate conversion—$8 billion (1992 est.)

National product real growth rate: NA%

National product per capita: $775 (1992 est.)

Inflation rate (consumer prices): 100% (December 1992)

Unemployment rate: 30% (December 1992)

Budget: revenues $NA, expenditures $NA, including capital expenditures of $NA

Exports: $908 million (f.o.b., 1990 est.)
commodities: crude oil, cotton, coffee, hides, vegetables, dried and salted fish
partners: US, EC countries, South Korea, Saudi Arabia

Imports: $2.1 billion (f.o.b., 1990 est.)
commodities: textiles and other manufactured consumer goods, petroleum products, sugar, grain, flour, other foodstuffs, cement, machinery, chemicals
partners: Japan, Saudi Arabia, Australia, EC countries, China, Russia, US

External debt: $5.75 billion (December 1989 est.)

Industrial production: growth rate NA%, accounts for 18% of GDP

Electricity: 714,000 kW capacity; 1,224 million kWh produced, 120 kWh per capita (1992)

Industries: crude oil production and petroleum refining; small-scale production of cotton textiles and leather goods; food processing; handicrafts; small aluminum products factory; cement

Agriculture: accounted for 26% of GDP; products—grain, fruits, vegetables, qat (mildly narcotic shrub), coffee, cotton, dairy, poultry, meat, fish; not self-sufficient in grain

Economic aid: US commitments, including Ex-Im (FY70-89), $389 million; Western (non-US) countries, ODA and OOF bilateral commitments (1970-89), $2.0 billion; OPEC bilateral aid (1979-89), $3.2 billion; Communist countries (1970-89), $2.4 billion

Currency: Yemeni rial (new currency); 1 North Yemeni riyal (YR) = 100 fils; 1 South Yemeni dinar (YD) = 1,000 fils
note: following the establishment of the Republic of Yemen on 22 May 1990, the North Yemeni riyal and the South Yemeni dinar are to be replaced with a new Yemeni rial

Exchange rates: Yemeni rials per US$1—12.0 (official); 30-40 (unofficial) (est.); North Yemeni riyals (YR) per

Zaire

US$1—12.1000 (June 1992), 12.0000 (1991), 9.7600 (1990), 9.7600 (January 1989), 9.7717 (1988), 10.3417 (1987); South Yemeni dinars (YD) per US$1—0.3454 (fixed rate)

note: following the establishment of the Republic of Yemen on 22 May 1990, the North Yemeni riyal and the South Yemeni dinar are to be replaced with a new Yemeni rial

Fiscal year: calendar year

Communications

Highways: 15,500 km total; 4,000 km paved, 11,500 km natural surface (est.)
Pipelines: crude oil 644 km, petroleum products 32 km
Ports: Aden, Al Hudaydah, Al Khalf, Al Mukalla, Mocha, Nishtun, Ra's Kathib, Salif
Merchant marine: 3 ships (1,000 GRT or over) totaling 4,309 GRT/6,568 DWT; includes 2 cargo, 1 oil tanker
Airports:
total: 45
usable: 39
with permanent-surface runways: 10
with runways over 3,659 m: 0
with runways 2,440-3,659 m: 18
with runways 1,220-2,439 m: 11
Telecommunications: since unification in 1990, efforts are still being made to create a national domestic civil telecommunications network; the network consists of microwave radio relay, cable and troposcatter; 65,000 telephones (est.); broadcast stations—4 AM, 1 FM, 10 TV; satellite earth stations—2 Indian Ocean INTELSAT, 1 Atlantic Ocean INTELSAT, 1 Intersputnik, 2 ARABSAT; microwave radio relay to Saudi Arabia, and Djibouti

---

Defense Forces

Branches: Army, Navy, Air Force, Police
Manpower availability: males age 15-49 2,060,124; fit for military service 1,172,633; reach military age (14) annually 133,727 (1993 est.)
Defense expenditures: exchange rate conversion—$762 million, 10% of GDP (1992)

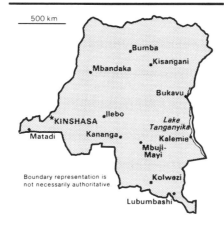

500 km

KINSHASA

Bumba
Kisangani
Mbandaka
Bukavu
Ilebo
Lake Tanganyika
Matadi
Kananga
Kalemie
Mbuji-Mayi
Kolwezi
Lubumbashi

Boundary representation is not necessarily authoritative

Geography

Location: Central Africa, between Congo and Zambia
Map references: Africa, Standard Time Zones of the World
Area:
total area: 2,345,410 km²
land area: 2,267,600 km²
comparative area: slightly more than one-quarter the size of US
Land boundaries: total 10,271 km, Angola 2,511 km, Burundi 233 km, Central African Republic 1,577 km, Congo 2,410 km, Rwanda 217 km, Sudan 628 km, Uganda 765 km, Zambia 1,930 km
Coastline: 37 km
Maritime claims:
exclusive fishing zone: 200 nm
territorial sea: 12 nm
International disputes: Tanzania-Zaire-Zambia tripoint in Lake Tanganyika may no longer be indefinite since it is reported that the indefinite section of the Zaire-Zambia boundary has been settled; long section with Congo along the Congo River is indefinite (no division of the river or its islands has been made)
Climate: tropical; hot and humid in equatorial river basin; cooler and drier in southern highlands; cooler and wetter in eastern highlands; north of Equator—wet season April to October, dry season December to February; south of Equator—wet season November to March, dry season April to October
Terrain: vast central basin is a low-lying plateau; mountains in east
Natural resources: cobalt, copper, cadmium, petroleum, industrial and gem diamonds, gold, silver, zinc, manganese, tin, germanium, uranium, radium, bauxite, iron ore, coal, hydropower potential
Land use:
arable land: 3%
permanent crops: 0%
meadows and pastures: 4%

forest and woodland: 78%
other: 15%
Irrigated land: 100 km² (1989 est.)
Environment: dense tropical rain forest in central river basin and eastern highlands; periodic droughts in south
Note: straddles Equator; very narrow strip of land that controls the lower Congo River and is only outlet to South Atlantic Ocean

---

People

Population: 41,345,738 (July 1993 est.)
Population growth rate: 3.2% (1993 est.)
Birth rate: 48.43 births/1,000 population (1993 est.)
Death rate: 16.91 deaths/1,000 population (1993 est.)
Net migration rate: 0.52 migrant(s)/1,000 population (1993 est.)
Infant mortality rate: 113.2 deaths/1,000 live births (1993 est.)
Life expectancy at birth:
total population: 47.26 years
male: 45.45 years
female: 49.12 years (1993 est.)
Total fertility rate: 6.7 children born/woman (1993 est.)
Nationality:
noun: Zairian(s)
adjective: Zairian
Ethnic divisions: over 200 African ethnic groups, the majority are Bantu; four largest tribes - Mongo, Luba, Kongo (all Bantu), and the Mangbetu-Azande (Hamitic) make up about 45% of the population
Religions: Roman Catholic 50%, Protestant 20%, Kimbanguist 10%, Muslim 10%, other syncretic sects and traditional beliefs 10%
Languages: French, Lingala, Swahili, Kingwana, Kikongo, Tshiluba
Literacy: age 15 and over can read and write (1990)
total population: 72%
male: 84%
female: 61%
Labor force: 15 million (13% of the labor force is wage earners; 51% of the population is of working age)
by occupation: agriculture 75%, industry 13%, services 12% (1985)

Government

Names:
conventional long form: Republic of Zaire
conventional short form: Zaire
local long form: Republique du Zaire
local short form: Zaire
former: Belgian Congo Congo/Leopoldville Congo/Kinshasa
Digraph: CG
Type: republic with a strong presidential system
Capital: Kinshasa

Administrative divisions: 10 regions (regions, singular—region) and 1 town* (ville); Bandundu, Bas-Zaire, Equateur, Haut-Zaire, Kasai-Occidental, Kasai-Oriental, Kinshasa*, Maniema, Nord-Kivu, Shaba, Sud-Kivu

Independence: 30 June 1960 (from Belgium)

Constitution: 24 June 1967, amended August 1974, revised 15 February 1978; amended April 1990; new constitution to be put to referendum in 1993

Legal system: based on Belgian civil law system and tribal law; has not accepted compulsory ICJ jurisdiction

National holiday: Anniversary of the Regime (Second Republic), 24 November (1965)

Political parties and leaders: sole legal party until January 1991—Popular Movement of the Revolution (MPR); other parties include Union for Democracy and Social Progress (UDPS), Etienne TSHISEKEDI wa Mulumba; Democratic Social Christian Party (PDSC), Joseph ILEO; Union of Federalists and Independent Republicans (UFERI), NGUZ a Karl-I-Bond; Unified Lumumbast Party (PALU), leader NA

Suffrage: 18 years of age; universal and compulsory

Elections:
President: last held 29 July 1984 (next to be scheduled by High Council, the opposition-controlled transition legislature); results—President MOBUTU was reelected without opposition
Legislative Council: last held 6 September 1987 (next to be scheduled by High Council); results—MPR was the only party; seats—(210 total) MPR 210; note—MPR still holds majority of seats but some deputies have joined other parties

Executive branch: president, prime minister, Executive Council (cabinet)

Legislative branch: unicameral National Parliament; anti-Mobutu opposition claims National Parliament replaced by High Council

Judicial branch: Supreme Court (Cour Supreme)

Leaders:
Chief of State: President Marshal MOBUTU Sese Seko Kuku Ngbendu wa Za Banga (since 24 November 1965)
Head of Government: Interim Prime Minister Faustin BIRINDWA (since 18 March 1993)

Member of: ACCT, ACP, AfDB, CCC, CEEAC, CEPGL, ECA, FAO, G-19, G-24, G-77, GATT, IAEA, IBRD, ICAO, ICC, IDA, IFAD, IFC, ILO, IMF, IMO, INTELSAT, INTERPOL, IOC, ITU, LORCS, NAM, OAU, PCA, UN, UNCTAD, UNESCO, UNHCR, UNIDO, UPU, WCL, WFTU, WHO, WIPO, WMO, WTO

Diplomatic representation in US:
chief of mission: Ambassador TATANENE Manata
chancery: 1800 New Hampshire Avenue NW, Washington, DC 20009
telephone: (202) 234-7690 or 7691

US diplomatic representation:
chief of mission: Deputy Chief of Mission John YATES
embassy: 310 Avenue des Aviateurs, Kinshasa
mailing address: APO AE 09828
telephone: [243] (12) 21532, 21628
FAX: [243] (12) 21232
consulate general: Lubumbashi (closed and evacuated in October 1991 because of the poor security situation)

Flag: light green with a yellow disk in the center bearing a black arm holding a red flaming torch; the flames of the torch are blowing away from the hoist side; uses the popular pan-African colors of Ethiopia

Economy

Overview: In 1992, Zaire's formal economy continued to disintegrate. While meaningful economic figures are difficult to come by, Zaire's hyperinflation, the largest government deficit ever, and plunging mineral production have made the country one of the world's poorest. Most formal transactions are conducted in hard currency as indigenous banknotes have lost almost all value, and a barter economy now flourishes in all but the largest cities. Most individuals and families hang on grimly through subsistence farming and petty trade. The government has not been able to meet its financial obligations to the International Momentary Fund or put in place the financial measures advocated by the IMF. Although short-term prospects for improvement are dim, improved political stability would boost Zaire's long-term potential to effectively exploit its vast wealth of mineral and agricultural resources.

National product: GDP—exchange rate conversion—$9.2 billion (1992, at 1990 exchange rate)

National product real growth rate: -6% (1992 est.)

National product per capita: $235 (1992, at 1990 exchange rate)

Inflation rate (consumer prices): 35-40% per month (1992 est.)

Unemployment rate: NA%

Budget: revenues $NA, expenditures $NA, including capital expenditures of $NA

Exports: $1.5 billion (f.o.b., 1992 est.)
commodities: copper, coffee, diamonds, cobalt, crude oil
partners: US, Belgium, France, Germany, Italy, UK, Japan, South Africa

Imports: $1.2 billion (f.o.b., 1992 est.)

commodities: consumer goods, foodstuffs, mining and other machinery, transport equipment, fuels
partners: South Africa, US, Belgium, France, Germany, Italy, Japan, UK

External debt: $9.2 billion (May 1992 est.)

Industrial production: growth grate NA%

Electricity: 2,580,000 kW capacity; 6,000 million kWh produced, 160 kWh per capita (1991)

Industries: mining, mineral processing, consumer products (including textiles, footwear, and cigarettes), processed foods and beverages, cement, diamonds

Agriculture: cash crops—coffee, palm oil, rubber, quinine; food crops—cassava, bananas, root crops, corn

Illicit drugs: illicit producer of cannabis, mostly for domestic consumption

Economic aid: US commitments, including Ex-Im (FY70-89), $1.1 billion; Western (non-US) countries, ODA and OOF bilateral commitments (1970-89), $6.9 billion; OPEC bilateral aid (1979-89), $35 million; Communist countries (1970-89), $263 million; except for humanitarian aid to private organizations, no US assistance was given to Zaire in 1992

Currency: 1 zaire (Z) = 100 makuta

Exchange rates: zaire (Z) per US$1—2,000,000 (January 1993), 15,587 (1991), 719 (1990), 381 (1989), 187 (1988), 112 (1987)

Fiscal year: calendar year

Communications

Railroads: 5,254 km total; 3,968 km 1.067-meter gauge (851 km electrified); 125 km 1.000-meter gauge; 136 km 0.615-meter gauge; 1,025 km 0.600-meter gauge; limited trackage in use because of civil strife

Highways: 146,500 km total; 2,800 km paved, 46,200 km gravel and improved earth; 97,500 unimproved earth

Inland waterways: 15,000 km including the Congo, its tributaries, and unconnected lakes

Pipelines: petroleum products 390 km

Ports: Matadi, Boma, Banana

Merchant marine: 1 passenger cargo ship (1,000 GRT or over) totaling 15,489 GRT/13,481 DWT

Airports:
total: 281
usable: 235
with permanent-surface runways: 25
with runways over 3,659 m: 1
with runways 2,440-3,659 m: 6
with runways 1,220-2,439 m: 73

Telecommunications: barely adequate wire and microwave service; broadcast stations—10 AM, 4 FM, 18 TV; satellite earth stations—1 Atlantic Ocean INTELSAT, 14 domestic

Zambia

Defense Forces

Branches: Army, Navy, Air Force, paramilitary National Gendarmerie, Civil Guard, Special Presidential Division
Manpower availability: males age 15-49 8,879,731; fit for military service 4,521,768 (1993 est.)
Defense expenditures: exchange rate conversion—$49 million, 0.8% of GDP (1988)

300 km

Boundary representation is not necessarily authoritative.

Geography

Location: Southern Africa, between Zaire and Zimbabwe
Map references: Africa, Standard Time Zones of the World
Area:
total area: 752,610 km²
land area: 740,720 km²
comparative area: slightly larger than Texas
Land boundaries: total 5,664 km, Angola 1,110 km, Malawi 837 km, Mozambique 419 km, Namibia 233 km, Tanzania 338 km, Zaire 1,930 km, Zimbabwe 797 km
Coastline: 0 km (landlocked)
Maritime claims: none; landlocked
International disputes: quadripoint with Botswana, Namibia, and Zimbabwe is in disagreement; Tanzania-Zaire-Zambia tripoint in Lake Tanganyika may no longer be indefinite since it is reported that the indefinite section of the Zaire-Zambia boundary has been settled
Climate: tropical; modified by altitude; rainy season (October to April)
Terrain: mostly high plateau with some hills and mountains
Natural resources: copper, cobalt, zinc, lead, coal, emeralds, gold, silver, uranium, hydropower potential
Land use:
arable land: 7%
permanent crops: 0%
meadows and pastures: 47%
forest and woodland: 27%
other: 19%
Irrigated land: 320 km² (1989 est.)
Environment: deforestation; soil erosion; desertification
Note: landlocked

People

Population: 8,926,099 (July 1993 est.)
Population growth rate: 2.96% (1993 est.)
Birth rate: 46.53 births/1,000 population (1993 est.)

Death rate: 16.88 deaths/1,000 population (1993 est.)
Net migration rate: -0.05 migrant(s)/1,000 population (1993 est.)
Infant mortality rate: 83.9 deaths/1,000 live births (1993 est.)
Life expectancy at birth:
total population: 45.56 years
male: 44.97 years
female: 46.16 years (1993 est.)
Total fertility rate: 6.75 children born/woman (1993 est.)
Nationality:
noun: Zambian(s)
adjective: Zambian
Ethnic divisions: African 98.7%, European 1.1%, other 0.2%
Religions: Christian 50-75%, Muslim and Hindu 24-49%, indigenous beliefs 1%
Languages: English (official)
note: about 70 indigenous languages
Literacy: age 15 and over can read and write (1990)
total population: 73%
male: 81%
female: 65%
Labor force: 2.455 million
by occupation: agriculture 85%, mining, manufacturing, and construction 6%, transport and services 9%

Government

Names:
conventional long form: Republic of Zambia
conventional short form: Zambia
former: Northern Rhodesia
Digraph: ZA
Type: republic
Capital: Lusaka
Administrative divisions: 9 provinces; Central, Copperbelt, Eastern, Luapula, Lusaka, Northern, North-Western, Southern, Western
Independence: 24 October 1964 (from UK)
Constitution: NA August 1991
Legal system: based on English common law and customary law; judicial review of legislative acts in an ad hoc constitutional council; has not accepted compulsory ICJ jurisdiction
National holiday: Independence Day, 24 October (1964)
Political parties and leaders: Movement for Multiparty Democracy (MMD), Frederick CHILUBA; United National Independence Party (UNIP), Kebby MUSOKATWANE; United Democratic Party, Enoch KAVINDELE
Suffrage: 18 years of age; universal
Elections:
President: last held 31 October 1991 (next to be held mid-1995); results—Frederick CHILUBA 84%, Kenneth KAUNDA 16%
National Assembly: last held 31 October 1991 (next to be held mid-1995);

427

Zambia (continued)

results—percent of vote by party NA; seats—(150 total) MMD 125, UNIP 25
Executive branch: president, Cabinet
Legislative branch: unicameral National Assembly
Judicial branch: Supreme Court
Leaders:
Chief of State and Head of Government: President Frederick CHILUBA (since 31 October 1991)
Member of: ACP, AfDB, C, CCC, ECA, FAO, FLS, G-19, G-77, GATT, IAEA, IBRD, ICAO, IDA, IFAD, IFC, ILO, IMF, INTELSAT, INTERPOL, IOC, IOM, ITU, LORCS, NAM, OAU, SADC, UN, UNCTAD, UNESCO, UNIDO, UNOMOZ, UPU, WCL, WHO, WIPO, WMO, WTO
Diplomatic representation in US:
chief of mission: Ambassador Dunstan KAMONA
chancery: 2419 Massachusetts Avenue NW, Washington, DC 20008
telephone: (202) 265-9717 through 9721
US diplomatic representation:
chief of mission: Ambassador Gordon L. STREEB
embassy: corner of Independence Avenue and United Nations Avenue, Lusaka
mailing address: P. O. Box 31617, Lusaka
telephone: [260-1] 228-595, 228-601, 228-602, 228-603
FAX: [260-1] 251-578
Flag: green with a panel of three vertical bands of red (hoist side), black, and orange below a soaring orange eagle, on the outer edge of the flag

---

Economy

Overview: The economy has been in decline for more than a decade with falling imports and growing foreign debt. Economic difficulties stem from a chronically depressed level of copper production and ineffective economic policies. In 1991 real GDP fell by 2% and in 1992 by 3% more. An annual population growth of more than 3% has brought a decline in per capita GDP of 50% over the past decade. A high inflation rate has also added to Zambia's economic woes in recent years, as well as severe drought in the crop year 1991/92.
National product: GDP—exchange rate conversion—$4.7 billion (1992 est.)
National product real growth rate: -3% (1992 est.)
National product per capita: $550 (1992 est.)
Inflation rate (consumer prices): 170% (1992 est.)
Unemployment rate: NA%
Budget: revenues $665 million; expenditures $767 million, including capital expenditures of $300 million (1991 est.)
Exports: $1.0 billion (f.o.b., 1992 est.)

commodities: copper, zinc, cobalt, lead, tobacco
partners: EC countries, Japan, South Africa, US, India
Imports: $1.2 billion (c.i.f., 1992 est.)
commodities: machinery, transportation equipment, foodstuffs, fuels, manufactures
partners: EC countries, Japan, Saudi Arabia, South Africa, US
External debt: $7.6 billion (1991)
Industrial production: growth rate -2% (1991); accounts for 50% of GDP
Electricity: 2,775,000 kW capacity; 12,000 million kWh produced, 1,400 kWh per capita (1991)
Industries: copper mining and processing, construction, foodstuffs, beverages, chemicals, textiles, and fertilizer
Agriculture: accounts for 17% of GDP and 85% of labor force; crops—corn (food staple), sorghum, rice, peanuts, sunflower, tobacco, cotton, sugarcane, cassava; cattle, goats, beef, eggs
Economic aid: US commitments, including Ex-Im (1970-89), $4.8 billion; Western (non-US) countries, ODA and OOF bilateral commitments (1970-89), $4.8 billion; OPEC bilateral aid (1979-89), $60 million; Communist countries (1970-89), $533 million
Currency: 1 Zambian kwacha (ZK) = 100 ngwee
Exchange rates: Zambian kwacha (ZK) per US$1—178.5714 (August 1992), 61.7284 (1991), 28.9855 (1990), 12.9032 (1989), 8.2237 (1988), 8.8889 (1987)
Fiscal year: calendar year

Communications

Railroads: 1,266 km, all 1.067-meter gauge; 13 km double track
Highways: 36,370 km total; 6,500 km paved, 7,000 km crushed stone, gravel, or stabilized soil; 22,870 km improved and unimproved earth
Inland waterways: 2,250 km, including Zambezi and Luapula Rivers, Lake Tanganyika
Pipelines: crude oil 1,724 km
Ports: Mpulungu (lake port)
Airports:
total: 116
usable: 104
with permanent-surface runways: 13
with runways over 3,659 m: 1
with runways 2,440-3,659 m: 4
with runways 1,220-2,439 m: 22
Telecommunications: facilities are among the best in Sub-Saharan Africa; high-capacity microwave connects most larger towns and cities; broadcast stations—11 AM, 5 FM, 9 TV; satellite earth stations—1 Indian Ocean INTELSAT and 1 Atlantic Ocean INTELSAT

Defense Forces

Branches: Army, Air Force, Police, paramilitary
Manpower availability: males age 15-49 1,810,442; fit for military service 949,878 (1993 est.)
Defense expenditures: exchange rate conversion—$45 million, 1% of GDP (1992 est.)

Zimbabwe

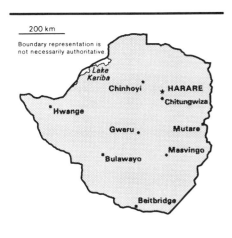

200 km
Boundary representation is not necessarily authoritative

Geography

Location: Southern Africa, between South Africa and Zambia
Map references: Africa, Standard Time Zones of the World
Area:
total area: 390,580 km²
land area: 386,670 km²
comparative area: slightly larger than Montana
Land boundaries: total 3,066 km, Botswana 813 km, Mozambique 1,231 km, South Africa 225 km, Zambia 797 km
Coastline: 0 km (landlocked)
Maritime claims: none; landlocked
International disputes: quadripoint with Botswana, Namibia, and Zambia is in disagreement
Climate: tropical; moderated by altitude; rainy season (November to March)
Terrain: mostly high plateau with higher central plateau (high veld); mountains in east
Natural resources: coal, chromium ore, asbestos, gold, nickel, copper, iron ore, vanadium, lithium, tin, platinum group metals
Land use:
arable land: 7%
permanent crops: 0%
meadows and pastures: 12%
forest and woodland: 62%
other: 19%
Irrigated land: 2,200 km² (1989 est.)
Environment: recurring droughts; floods and severe storms are rare; deforestation; soil erosion; air and water pollution
Note: landlocked

People

Population: 10,837,772 (July 1993 est.)
Population growth rate: 1.32% (1993 est.)
Birth rate: 38.16 births/1,000 population (1993 est.)
Death rate: 17.68 deaths/1,000 population (1993 est.)

Net migration rate: -7.27 migrant(s)/1,000 population (1993 est.)
Infant mortality rate: 75.3 deaths/1,000 live births (1993 est.)
Life expectancy at birth:
total population: 42.82 years
male: 41.2 years
female: 44.49 years (1993 est.)
Total fertility rate: 5.26 children born/woman (1993 est.)
Nationality:
noun: Zimbabwean(s)
adjective: Zimbabwean
Ethnic divisions: African 98% (Shona 71%, Ndebele 16%, other 11%), white 1%, mixed and Asian 1%
Religions: syncretic (part Christian, part indigenous beliefs) 50%, Christian 25%, indigenous beliefs 24%, Muslim and other 1%
Languages: English (official), Shona, Sindebele
Literacy: age 15 and over can read and write (1990)
total population: 67%
male: 74%
female: 60%
Labor force: 3.1 million
by occupation: agriculture 74%, transport and services 16%, mining, manufacturing, construction 10% (1987)

Government

Names:
conventional long form: Republic of Zimbabwe
conventional short form: Zimbabwe
former: Southern Rhodesia
Digraph: ZI
Type: parliamentary democracy
Capital: Harare
Administrative divisions: 8 provinces; Manicaland, Mashonaland Central, Mashonaland East, Mashonaland West, Masvingo (Victoria), Matabeleland North, Matabeleland South, Midlands
Independence: 18 April 1980 (from UK)
Constitution: 21 December 1979
Legal system: mixture of Roman-Dutch and English common law
National holiday: Independence Day, 18 April (1980)
Political parties and leaders: Zimbabwe African National Union-Patriotic Front (ZANU-PF), Robert MUGABE; Zimbabwe African National Union-Sithole (ZANU-S), Ndabaningi SITHOLE; Zimbabwe Unity Movement (ZUM), Edgar TEKERE; Democratic Party (DP), Emmanuel MAGOCHE; Forum Party, Enock DUMBUTSHENA
Suffrage: 18 years of age; universal
Elections:

Executive President: last held 28-30 March 1990 (next to be held NA March 1996); results—Robert MUGABE 78.3%, Edgar TEKERE 21.7%
Parliament: last held 28-30 March 1990 (next to be held NA March 1995); results—percent of vote by party NA; seats—(150 total, 120 elected) ZANU-PF 117, ZUM 2, ZANU-S 1
Executive branch: executive president, 2 vice presidents, Cabinet
Legislative branch: unicameral Parliament
Judicial branch: Supreme Court
Leaders:
Chief of State and Head of Government: Executive President Robert Gabriel MUGABE (since 31 December 1987); Co-Vice President Simon Vengai MUZENDA (since 31 December 1987); Co-Vice President Joshua M. NKOMO (since 6 August 1990)
Member of: ACP, AfDB, C, CCC, ECA, FAO, FLS, G-15, G-77, GATT, IAEA, IBRD, ICAO, IDA, IFAD, IFC, ILO, IMF, INTELSAT, INTERPOL, IOC, IOM (observer), ITU, LORCS, NAM, OAU, PCA, SADC, UN, UNAVEM II, UNCTAD, UNESCO, UNIDO, UNOSOM, UPU, WCL, WHO, WIPO, WMO, WTO
Diplomatic representation in US:
chief of mission: Counselor (Political Affairs), Head of Chancery, Ambassador-designate Amos Bernard Muvengwa MIDZI
chancery: 1608 New Hampshire Avenue NW, Washington, DC 20009
telephone: (202) 332-7100
US diplomatic representation:
chief of mission: Ambassador Edward Gibson LANPHER
embassy: 172 Herbert Chitapo Avenue, Harare
mailing address: P. O. Box 3340, Harare
telephone: [263] (4) 794-521
FAX: [263] (4) 796-488
Flag: seven equal horizontal bands of green, yellow, red, black, red, yellow, and green with a white equilateral triangle edged in black based on the hoist side; a yellow Zimbabwe bird is superimposed on a red five-pointed star in the center of the triangle

Economy

Overview: Agriculture employs three-fourths of the labor force and supplies almost 40% of exports. The manufacturing sector, based on agriculture and mining, produces a variety of goods and contributes 35% to GDP. Mining accounts for only 5% of both GDP and employment, but supplies of minerals and metals account for about 40% of exports. Wide fluctuations in agricultural production over the past six years have resulted in an uneven growth

Zimbabwe (continued)

rate, one that on average has matched the 3% annual increase in population. Helped by an IMF/World Bank structural adjustment program, output rose 3.5% in 1991. A severe drought in 1991/92 caused the economy to contract by about 10% in 1992.

National product: GDP—exchange rate conversion—$6.2 billion (1992 est.)

National product real growth rate: -10% (1992 est.)

National product per capita: $545 (1992 est.)

Inflation rate (consumer prices): 45% (1992 est.)

Unemployment rate: at least 35% (1993 est.)

Budget: revenues $2.7 billion; expenditures $3.3 billion, including capital expenditures of $330 million (FY91)

Exports: $1.5 billion (f.o.b., 1992 est.)
commodities: agricultural 35% (tobacco 20%, other 15%), manufactures 20%, gold 10%, ferrochrome 10%, cotton 5%
partners: UK 14%, Germany 11%, South Africa 10%, Japan 7%, US 5% (1991)

Imports: $1.8 billion (c.i.f., 1992 est.)
commodities: machinery and transportation equipment 37%, other manufactures 22%, chemicals 16%, fuels 15%
partners: UK 15%, Germany 9%, South Africa 5%, Botswana 5%, US 5%, Japan 5% (1991)

External debt: $3.9 billion (March 1993 est.)

Industrial production: growth rate 5% (1991 est.); accounts for 38% of GDP

Electricity: 3,650,000 kW capacity; 8,920 million kWh produced, 830 kWh per capita (1991)

Industries: mining, steel, clothing and footwear, chemicals, foodstuffs, fertilizer, beverage, transportation equipment, wood products

Agriculture: accounts for 13% of GDP and employs 74% of population; 40% of land area divided into 4,500 large commercial farms and 42% in communal lands; crops—corn (food staple), cotton, tobacco, wheat, coffee, sugarcane, peanuts; livestock—cattle, sheep, goats, pigs; self-sufficient in food

Economic aid: US commitments, including Ex-Im (FY80-89), $389 million; Western (non-US) countries, ODA and OOF bilateral commitments (1970-89), $2.6 billion; OPEC bilateral aid (1979-89), $36 million; Communist countries (1970-89), $134 million

Currency: 1 Zimbabwean dollar (Z$) = 100 cents

Exchange rates: Zimbabwean dollars (Z$) per US$1—6.3532 (February 1993), 5.1046 (1992), 3.4282 (1991), 2.4480 (1990), 2.1133 (1989), 1.8018 (1988)

Fiscal year: 1 July-30 June

Communications

Railroads: 2,745 km 1.067-meter gauge (including 42 km double track, 355 km electrified)

Highways: 85,237 km total; 15,800 km paved, 39,090 km crushed stone, gravel, stabilized soil; 23,097 km improved earth; 7,250 km unimproved earth

Inland waterways: Lake Kariba is a potential line of communication

Pipelines: petroleum products 212 km

Airports:
total: 485
usable: 403
with permanent-surface runways: 22
with runways over 3,659 m: 2
with runways 2,440-3,659 m: 3
with runways 1,220-2,439 m: 29

Telecommunications: system was once one of the best in Africa, but now suffers from poor maintenance; consists of microwave links, open-wire lines, and radio communications stations; 247,000 telephones; broadcast stations—8 AM, 18 FM, 8 TV; 1 Atlantic Ocean INTELSAT earth station

Defense Forces

Branches: Zimbabwe National Army, Air Force of Zimbabwe, Zimbabwe Republic Police (including Police Support Unit, Paramilitary Police), People's Militia

Manpower availability: males age 15-49 2,315,461; fit for military service 1,436,671 (1993 est.)

Defense expenditures: exchange rate conversion—$412.4 million, about 6% of GDP (FY91 est.)

Taiwan

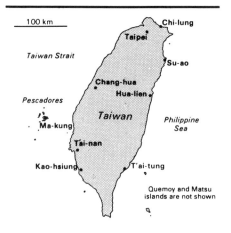

Geography

Location: East Asia, off the southeastern coast of China, between Japan and the Philippines

Map references: Asia, Oceania, Southeast Asia

Area:
total area: 35,980 km²
land area: 32,260 km²
comparative area: slightly larger than Maryland and Delaware combined
note: includes the Pescadores, Matsu, and Quemoy

Land boundaries: 0 km

Coastline: 1,448 km

Maritime claims:
exclusive economic zone: 200 nm
territorial sea: 12 nm

International disputes: involved in complex dispute over the Spratly Islands with China, Malaysia, Philippines, Vietnam, and possibly Brunei; Paracel Islands occupied by China, but claimed by Vietnam and Taiwan; Japanese-administered Senkaku-shoto (Senkaku Islands/Diaoyu Tai) claimed by China and Taiwan

Climate: tropical; marine; rainy season during southwest monsoon (June to August); cloudiness is persistent and extensive all year

Terrain: eastern two-thirds mostly rugged mountains; flat to gently rolling plains in west

Natural resources: small deposits of coal, natural gas, limestone, marble, and asbestos

Land use:
arable land: 24%
permanent crops: 1%
meadows and pastures: 5%
forest and woodland: 55%
other: 15%

Irrigated land: NA km²

Environment: subject to earthquakes and typhoons

People

Population: 21,091,663 (July 1993 est.)
Population growth rate: 1% (1993 est.)
Birth rate: 15.88 births/1,000 population (1993 est.)
Death rate: 5.54 deaths/1,000 population (1993 est.)
Net migration rate: -0.38 migrant(s)/1,000 population (1993 est.)
Infant mortality rate: 5.7 deaths/1,000 live births (1993 est.)
Life expectancy at birth:
total population: 75.04 years
male: 71.84 years
female: 78.39 years (1993 est.)
Total fertility rate: 1.81 children born/woman (1993 est.)
Nationality:
noun: Chinese (singular and plural)
adjective: Chinese
Ethnic divisions: Taiwanese 84%, mainland Chinese 14%, aborigine 2%
Religions: mixture of Buddhist, Confucian, and Taoist 93%, Christian 4.5%, other 2.5%
Languages: Madarin Chinese (official), Taiwanese (Min), Hakka dialects
Literacy: age 15 and over can read and write (1980)
total population: 86%
male: 93%
female: 79%
Labor force: 7.9 million
by occupation: industry and commerce 53%, services 22%, agriculture 15.6%, civil administration 7% (1989)

Administration

Names:
conventional long form: none
conventional short form: Taiwan
local long form: none
local short form: T'ai-wan
Digraph: TW
Type: multiparty democratic regime; opposition political parties legalized in March, 1989
Capital: Taipei
Administrative divisions: some of the ruling party in Taipei claim to be the government of all China; in keeping with that claim, the central administrative divisions include 2 provinces (sheng, singular and plural) and 2 municipalities* (shih, singular and plural)—Fu-chien (some 20 offshore islands of Fujian Province including Quemoy and Matsu), Kao-hsiung*, T'ai-pei*, and Taiwan (the island of Taiwan and the Pescadores islands); the more commonly referenced administrative divisions are those of Taiwan Province—16 counties (hsien, singular and plural), 5 municipalities* (shih, singular and plural), and 2 special municipalities** (chuan-shih,

singular and plural); Chang-hua, Chia-i, Chia-i*, Chi-lung*, Hsin-chu, Hsin-chu*, Hua-lien, I-lan, Kao-hsiung, Kao-hsiung**, Miao-li, Nan-t'ou, P'eng-hu, P'ing-tung, T'ai-chung, T'ai-chung*, T'ai-nan, T'ai-nan*, T'ai-pei, T'ai-pei**, T'ai-tung, T'ao-yuan, and Yun-lin; the provincial capital is at Chung-hsing-hsin-ts'un
note: Taiwan uses the Wade-Giles system for romanization
Constitution: 25 December 1947, presently undergoing revision
Legal system: based on civil law system; accepts compulsory ICJ jurisdiction, with reservations
National holiday: National Day, 10 October (1911) (Anniversary of the Revolution)
Political parties and leaders: Kuomintang (KMT, Nationalist Party), LI Teng-hui, chairman; Democratic Progressive Party (DPP); China Social Democratic Party (CSDP); Labor Party (LP)
Other political or pressure groups: Taiwan independence movement, various environmental groups
note: debate on Taiwan independence has become acceptable within the mainstream of domestic politics on Taiwan; political liberalization and the increased representation of the opposition Democratic Progressive Party in Taiwan's legislature have opened public debate on the island's national identity; advocates of Taiwan independence, both within the DPP and the ruling Kuomintang, oppose the ruling party's traditional stand that the island will eventually unify with mainland China; the aims of the Taiwan independence movement include establishing a sovereign nation on Taiwan and entering the UN; other organizations supporting Taiwan independence include the World United Formosans for Independence and the Organization for Taiwan Nation Building
Suffrage: 20 years of age; universal
Elections:
President: last held 21 March 1990 (next to be held NA March 1996); results—President LI Teng-hui was reelected by the National Assembly
Vice President: last held 21 March 1990 (next to be held NA March 1996); results—LI Yuan-zu was elected by the National Assembly
Legislative Yuan: last held 19 December 1992 (next to be held near the end of 1995); results—KMT 60%, DPP 31%, independents 9%; seats—(304 total, 161 elected) KMT 96, DPP 50, independents 15
National Assembly: first National Assembly elected in November 1946 with a supplementary election in December 1986; second and present National Assembly elected in December 1991; seats—403 total, KMT 318, DPP 75, other 10; (next election to be held in 1997)

Executive branch: president, vice president, premier of the Executive Yuan, vice premier of the Executive Yuan, Executive Yuan
Legislative branch: unicameral Legislative Yuan and unicameral National Assembly
Judicial branch: Judicial Yuan
Leaders:
Chief of State: President LI Teng-hui (since 13 January 1988); Vice President LI Yuan-zu (since 20 May 1990)
Head of Government: Premier (President of the Executive Yuan) LIEN Chan (since 23 February 1993); Vice Premier (Vice President of the Executive Yuan) HSU Li-teh (since 23 February 1993)
Member of: expelled from UN General Assembly and Security Council on 25 October 1971 and withdrew on same date from other charter-designated subsidiary organs; expelled from IMF/World Bank group April/May 1980; seeking to join GATT; attempting to retain membership in INTELSAT; suspended from IAEA in 1972, but still allows IAEA controls over extensive atomic development, APEC, AsDB, ICC, ICFTU, IOC
Diplomatic representation in US: none; unofficial commercial and cultural relations with the people of the US are maintained through a private instrumentality, the Coordination Council for North American Affairs (CCNAA) with headquarters in Taipei and field offices in Washington and 10 other US cities
US diplomatic representation: unofficial commercial and cultural relations with the people of Taiwan are maintained through a private institution, the American Institute in Taiwan (AIT), which has offices in Taipei at #7, Lane 134, Hsiu Yi Road, Section 3, telephone [886] (2) 709-2000, and in Kao-hsiung at #2 Chung Cheng 3d Road, telephone [886] (7) 224-0154 through 0157, and the American Trade Center at Room 3207 International Trade Building, Taipei World Trade Center, 333 Keelung Road Section 1, Taipei 10548, telephone [886] (2) 720-1550
Flag: red with a dark blue rectangle in the upper hoist-side corner bearing a white sun with 12 triangular rays

Economy

Overview: Taiwan has a dynamic capitalist economy with considerable government guidance of investment and foreign trade and partial government ownership of some large banks and industrial firms. Real growth in GNP has averaged about 9% a year during the past three decades. Export growth has been even faster and has provided the impetus for industrialization. Agriculture contributes about 4% to GNP, down from 35% in 1952. Taiwan currently ranks as number 13 among major trading countries. Traditional labor-intensive industries are steadily being replaced with more capital- and technology-intensive industries. Taiwan has become a major investor in China, Thailand, Indonesia, the Philippines, and Malaysia. The tightening of labor markets has led to an influx of foreign workers, both legal and illegal.

National product: GNP—purchasing power equivalent—$209 billion (1992 est.)

National product real growth rate: 6.7% (1992 est.)

National product per capita: $10,000 (1992 est.)

Inflation rate (consumer prices): 4.4% (1992 est.)

Unemployment rate: 1.6% (1992 est.)

Budget: revenues $30.3 billion; expenditures $30.1 billion, including capital expenditures of $NA (FY91 est.)

Exports: $82.4 billion (f.o.b., 1992 est.)
commodities: electrical machinery 18.5%, textiles 14.7%, general machinery and equipment 17.7%, footwear 4.5%, foodstuffs 1.1%, plywood and wood products 1.1% (1992 est.)
partners: US 29.1%, Hong Kong 18.7%, EC countries 17.1% (1992 est.)

Imports: $72.1 billion (c.i.f., 1992 est.)
commodities: machinery and equipment 15.8%, chemicals 10.0%, crude oil 4.2%, foodstuffs 2.1% (1992 est.)
partners: Japan 30.3%, US 21.9%, EC countries 17.1% (1992 est.)

External debt: $620 million (1992 est.)

Industrial production: growth rate 6.5% (1992 est.); accounts for more than 40% of GDP

Electricity: 18,382,000 kW capacity; 98,500 million kWh produced, 4,718 kWh per capita (1992)

Industries: electronics, textiles, chemicals, clothing, food processing, plywood, sugar milling, cement, shipbuilding, petroleum refining

Agriculture: accounts for 4% of GNP and 16% of labor force (includes part-time farmers); heavily subsidized sector; major crops—vegetables, rice, fruit, tea; livestock—hogs, poultry, beef, milk; not self-sufficient in wheat, soybeans, corn; fish catch increasing, reached 1.4 million metric tons in 1988

Illicit drugs: an important heroin transit point; also a major drug money laundering center

Economic aid: US, including Ex-Im (FY46-82), $4.6 billion; Western (non-US) countries, ODA and OOF bilateral commitments (1970-89), $500 million

Currency: 1 New Taiwan dollar (NT$) = 100 cents

Exchange rates: New Taiwan dollars per US$1—25.125 (1992 est.), 25.748 (1991), 27.108 (1990), 26.407 (1989) 28.589 (1988), 31.845 (1987)

Fiscal year: 1 July-30 June

Communications

Railroads: about 4,600 km total track with 1,075 km common carrier lines and 3,525 km industrial lines; common carrier lines consist of the 1.067-meter gauge 708 km West Line and the 367 km East Line; a 98.25 km South Link Line connection was completed in late 1991; common carrier lines owned by the government and operated by the Railway Administration under Ministry of Communications; industrial lines owned and operated by government enterprises

Highways: 20,041 km total; 17,095 km bituminous or concrete pavement, 2,371 km crushed stone or gravel, 575 km graded earth

Pipelines: petroleum products 615 km, natural gas 97 km

Ports: Kao-hsiung, Chi-lung (Keelung), Hua-lien, Su-ao, T'ai-tung

Merchant marine: 223 ships (1,000 GRT or over) totaling 6,761,609 GRT/9,375,677 DWT; includes 1 passenger-cargo, 43 cargo, 11 refrigerated cargo, 85 container, 19 oil tanker, 2 combination ore/oil, 1 specialized tanker, 57 bulk, 1 roll-on/roll-off, 2 combination bulk, 1 chemical tanker

Airports:
total: 40
usable: 38
with permanent-surface runways: 36
with runways over 3,659 m: 3
with runways 2,440-3,659 m: 16
with runways 1,220-2,439 m: 7

Telecommunications: best developed system in Asia outside of Japan; 7,800,000 telephones; extensive microwave radio relay links on east and west coasts; broadcast stations—91 AM, 23 FM, 15 TV (13 repeaters); 8,620,000 radios; 6,386,000 TVs (5,680,000 color, 706,000 monochrome); satellite earth stations—1 Pacific Ocean INTELSAT and 1 Indian Ocean INTELSAT; submarine cable links to Japan (Okinawa), the Philippines, Guam, Singapore, Hong Kong, Indonesia, Australia, Middle East, and Western Europe

Defense Forces

Branches: General Staff, Ministry of National Defense, Army, Navy (including Marines), Air Force, Coastal Patrol and Defense Command, Armed Forces Reserve Command, Military Police Command

Manpower availability: males age 15-49 6,095,857; fit for military service 4,731,172 (1993 est.); about 184,740 currently reach military age (19) annually

Defense expenditures: exchange rate conversion—$10.9 billion, 5.4% of GNP (FY93/94 est.)

Appendix A:

The United Nations System

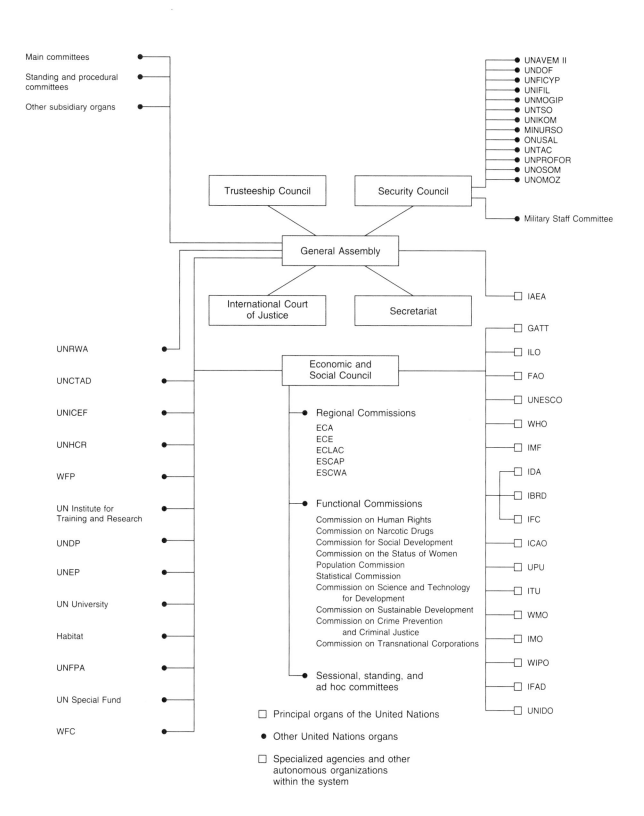

Main committees

Standing and procedural committees

Other subsidiary organs

UNAVEM II
UNDOF
UNFICYP
UNIFIL
UNMOGIP
UNTSO
UNIKOM
MINURSO
ONUSAL
UNTAC
UNPROFOR
UNOSOM
UNOMOZ

Military Staff Committee

Trusteeship Council

Security Council

General Assembly

International Court of Justice

Secretariat

IAEA

GATT
ILO
FAO
UNESCO
WHO
IMF
IDA
IBRD
IFC
ICAO
UPU
ITU
WMO
IMO
WIPO
IFAD
UNIDO

UNRWA

UNCTAD

UNICEF

UNHCR

WFP

UN Institute for Training and Research

UNDP

UNEP

UN University

Habitat

UNFPA

UN Special Fund

WFC

Economic and Social Council

Regional Commissions
ECA
ECE
ECLAC
ESCAP
ESCWA

Functional Commissions
Commission on Human Rights
Commission on Narcotic Drugs
Commission for Social Development
Commission on the Status of Women
Population Commission
Statistical Commission
Commission on Science and Technology for Development
Commission on Sustainable Development
Commission on Crime Prevention and Criminal Justice
Commission on Transnational Corporations

Sessional, standing, and ad hoc committees

☐ Principal organs of the United Nations

● Other United Nations organs

☐ Specialized agencies and other autonomous organizations within the system

Based on chart from the *UN Chronicle*

Appendix B:

Abbreviations for International Organizations and Groups

A	ABEDA	Arab Bank for Economic Development in Africa
	ACC	Arab Cooperation Council
	ACCT	Agence de Cooperation Culturelle et Technique; see Agency for Cultural and Technical Cooperation
	ACP	African, Caribbean, and Pacific Countries
	AfDB	African Development Bank
	AFESD	Arab Fund for Economic and Social Development
	AG	Andean Group
	AL	Arab League
	ALADI	Asociacion Latinoamericana de Integracion; see Latin American Integration Association (LAIA)
	AMF	Arab Monetary Fund
	AMU	Arab Maghreb Union
	ANZUS	Australia-New Zealand-United States Security Treaty
	APEC	Asia Pacific Economic Cooperation
	AsDB	Asian Development Bank
	ASEAN	Association of Southeast Asian Nations
B	BAD	Banque Africaine de Developpement; see African Development Bank (AfDB)
	BADEA	Banque Arabe de Developpement Economique en Afrique; see Arab Bank for Economic Development in Africa (ABEDA)
	BCIE	Banco Centroamericano de Integracion Economico; see Central American Bank for Economic Integration (BCIE)
	BDEAC	Banque de Developppment des Etats de l'Afrique Centrale; see Central African States Development Bank (BDEAC)
	Benelux	Benelux Economic Union
	BID	Banco Interamericano de Desarrollo; see Inter-American Development Bank (IADB)
	BIS	Bank for International Settlements
	BOAD	Banque Ouest-Africaine de Developpement; see West African Development Bank (WADB)
	BSEC	Black Sea Economic Cooperation Zone
C	C	Commonwealth
	CACM	Central American Common Market
	CAEU	Council of Arab Economic Unity
	CARICOM	Caribbean Community and Common Market
	CBSS	Council of the Baltic Sea States
	CCC	Customs Cooperation Council
	CDB	Caribbean Development Bank
	CE	Council of Europe
	CEAO	Communaute Economique de l'Afrique de l'Ouest; see West African Economic Community (CEAO)
	CEEAC	Communaute Economique des Etats de l'Afrique Centrale; see Economic Community of Central African States (CEEAC)
	CEI	Central European Initiative
	CEMA	Council for Mutual Economic Assistance; also known as CMEA or Comecon; abolished 1 January 1991

		CEPGL	Communaute Economique des Pays des Grands Lacs; see Economic Community of the Great Lakes Countries (CEPGL)
		CERN	Conseil Europeen pour la Recherche Nucleaire; see European Organization for Nuclear Research (CERN)
		CG	Contadora Group
		CIS	Commonwealth of Independent States
		CMEA	Council for Mutual Economic Assistance (CEMA); also known as Comecon; abolished 1 January 1991
		COCOM	Coordinating Committee on Export Controls
		Comecon	Council for Mutual Economic Assistance (CEMA); also known as CMEA; abolished 1 January 1991
		CP	Colombo Plan
		CSCE	Conference on Security and Cooperation in Europe
D		DC	developed country
E		EADB	East African Development Bank
		EBRD	European Bank for Reconstruction and Development
		EC	European Community
		ECA	Economic Commission for Africa
		ECAFE	Economic Commission for Asia and the Far East; see Economic and Social Commission for Asia and the Pacific (ESCAP)
		ECE	Economic Commission for Europe
		ECLA	Economic Commission for Latin America; see Economic Commission for Latin America and the Caribbean (ECLAC)
		ECLAC	Economic Commission for Latin America and the Caribbean
		ECO	Economic Cooperation Organization
		ECOSOC	Economic and Social Council
		ECOWAS	Economic Community of West African States
		ECWA	Economic Commission for Western Asia; see Economic and Social Commission for Western Asia (ESCWA)
		EFTA	European Free Trade Association
		EIB	European Investment Bank
		Entente	Council of the Entente
		ESA	European Space Agency
		ESCAP	Economic and Social Commission for Asia and the Pacific
		ESCWA	Economic and Social Commission for Western Asia
F		FAO	Food and Agriculture Organization
		FLS	Front Line States
		FZ	Franc Zone
G		G-2	Group of 2
		G-3	Group of 3
		G-5	Group of 5
		G-6	Group of 6 (not to be confused with the Big Six)
		G-7	Group of 7
		G-8	Group of 8
		G-9	Group of 9
		G-10	Group of 10
		G-11	Group of 11
		G-15	Group of 15
		G-19	Group of 19
		G-24	Group of 24
		G-30	Group of 30

	G-33	Group of 33	
	G-77	Group of 77	
	GATT	General Agreement on Tariffs and Trade	
	GCC	Gulf Cooperation Council	
H	Habitat	Commission on Human Settlements	
I	IADB	Inter-American Development Bank	
	IAEA	International Atomic Energy Agency	
	IBEC	International Bank for Economic Cooperation	
	IBRD	International Bank for Reconstruction and Development	
	ICAO	International Civil Aviation Organization	
	ICC	International Chamber of Commerce	
	ICEM	Intergovernmental Committee for European Migration; see International Organization for Migration (IOM)	
	ICFTU	International Confederation of Free Trade Unions	
	ICJ	International Court of Justice	
	ICM	Intergovernmental Committee for Migration; see International Organization for Migration (IOM)	
	ICRC	International Committee of the Red Cross	
	IDA	International Development Association	
	IDB	Islamic Development Bank	
	IEA	International Energy Agency	
	IFAD	International Fund for Agricultural Development	
	IFC	International Finance Corporation	
	IGADD	Inter-Governmental Authority on Drought and Development	
	IIB	International Investment Bank	
	ILO	International Labor Organization	
	IMCO	Intergovernmental Maritime Consultative Organization; see International Maritime Organization (IMO)	
	IMF	International Monetary Fund	
	IMO	International Maritime Organization	
	INMARSAT	International Maritime Satellite Organization	
	INTELSAT	International Telecommunications Satellite Organization	
	INTERPOL	International Criminal Police Organization	
	IOC	International Olympic Committee	
	IOM	International Organization for Migration	
	ISO	International Organization for Standardization	
	ITU	International Telecommunication Union	
L	LAES	Latin American Economic System	
	LAIA	Latin American Integration Association	
	LAS	League of Arab States; see Arab League (AL)	
	LDC	less developed country	
	LLDC	least developed country	
	LORCS	League of Red Cross and Red Crescent Societies	
M	MERCOSUR	Mercado Comun del Cono Sur; see Southern Cone Common Market	
	MINURSO	United Nations Mission for the Referendum in Western Sahara	
	MTCR	Missile Technology Control Regime	
N	NACC	North Atlantic Cooperation Council	
	NAM	Nonaligned Movement	

	NATO	North Atlantic Treaty Organization
	NC	Nordic Council
	NEA	Nuclear Energy Agency
	NIB	Nordic Investment Bank
	NIC	newly industrializing country; see newly industrializing economy (NIE)
	NIE	newly industrializing economy
	NSG	Nuclear Suppliers Group
O	OAPEC	Organization of Arab Petroleum Exporting Countries
	OAS	Organization of American States
	OAU	Organization of African Unity
	OECD	Organization for Economic Cooperation and Development
	OECS	Organization of Eastern Caribbean States
	OIC	Organization of the Islamic Conference
	ONUSAL	United Nations Observer Mission in El Salvador
	OPANAL	Organismo para la Proscripcion de las Armas Nucleares en la America Latina y el Caribe; see Agency for the Prohibition of Nuclear Weapons in Latin America and the Caribbean
	OPEC	Organization of Petroleum Exporting Countries
P	PCA	Permanent Court of Arbitration
R	RG	Rio Group
S	SAARC	South Asian Association for Regional Cooperation
	SACU	Southern African Customs Union
	SADC	Southern African Development Community
	SELA	Sistema Economico Latinoamericana; see Latin American Economic System (LAES)
	SPARTECA	South Pacific Regional Trade and Economic Cooperation Agreement
	SPC	South Pacific Commission
	SPF	South Pacific Forum
U	UDEAC	Union Douaniere et Economique de l'Afrique Centrale; see Central African Customs and Economic Union (UDEAC)
	UN	United Nations
	UNAVEM II	United Nations Angola Verification Mission
	UNCTAD	United Nations Conference on Trade and Development
	UNDOF	United Nations Disengagement Observer Force
	UNDP	United Nations Development Program
	UNEP	United Nations Environment Program
	UNESCO	United Nations Educational, Scientific, and Cultural Organization
	UNFICYP	United Nations Force in Cyprus
	UNFPA	United Nations Fund for Population Activities; see UN Population Fund (UNFPA)
	UNHCR	United Nations Office of the High Commissioner for Refugees
	UNICEF	United Nations Children's Fund
	UNIDO	United Nations Industrial Development Organization
	UNIFIL	United Nations Interim Force in Lebanon
	UNIKOM	United Nations Iraq-Kuwait Observation Mission
	UNMOGIP	United Nations Military Observer Group in India and Pakistan
	UNOMOZ	United Nations Operation in Mozambique
	UNOSOM	United Nations Operation in Somalia

	UNPROFOR	United Nations Protection Force	
	UNRWA	United Nations Relief and Works Agency for Palestine Refugees in the Near East	
	UNTAC	United Nations Transitional Authority in Cambodia	
	UNTSO	United Nations Truce Supervision Organization	
	UPU	Universal Postal Union	
	USSR/EE	USSR/Eastern Europe	
W	WADB	West African Development Bank	
	WCL	World Confederation of Labor	
	WEU	Western European Union	
	WFC	World Food Council	
	WFP	World Food Program	
	WFTU	World Federation of Trade Unions	
	WHO	World Health Organization	
	WIPO	World Intellectual Property Organization	
	WMO	World Meteorological Organization	
	WP	Warsaw Pact (members met 1 July 1991 to dissolve the alliance)	
	WTO	World Tourism Organization	
Z	ZC	Zangger Committee	

Note: Not all international organizations and groups have abbreviations.

Appendix C:

International Organizations and Groups

advanced developing countries	another term for those less developed countries (LDCs) with particularly rapid industrial development; see newly industrializing economies (NIEs)
African, Caribbean, and Pacific Countries (ACP) *established*—1 April 1976 *aim*—members have a preferential economic and aid relationship with the EC	*members*—(69) Angola, Antigua and Barbuda, The Bahamas, Barbados, Belize, Benin, Botswana, Burkina, Burundi, Cameroon, Cape Verde, Central African Republic, Chad, Comoros, Congo, Cote d'Ivoire, Djibouti, Dominica, Dominican Republic, Equatorial Guinea, Ethiopia, Fiji, Gabon, The Gambia, Ghana, Grenada, Guinea, Guinea-Bissau, Guyana, Haiti, Jamaica, Kenya, Kiribati, Lesotho, Liberia, Madagascar, Malawi, Mali, Mauritania, Mauritius, Mozambique, Namibia, Niger, Nigeria, Papua New Guinea, Rwanda, Saint Kitts and Nevis, Saint Lucia, Saint Vincent and the Grenadines, Sao Tome and Principe, Senegal, Seychelles, Sierra Leone, Solomon Islands, Somalia, Sudan, Suriname, Swaziland, Tanzania, Togo, Tonga, Trinidad and Tobago, Tuvalu, Uganda, Vanuatu, Western Samoa, Zaire, Zambia, Zimbabwe
African Development Bank (AfDB), also known as Banque Africaine de Developpement (BAD) *established*—4 August 1963 *aim*—to promote economic and social development	*regional members*—(50) Algeria, Angola, Benin, Botswana, Burkina, Burundi, Cameroon, Cape Verde, Central African Republic, Chad, Comoros, Congo, Cote d'Ivoire, Djibouti, Egypt, Equatorial Guinea, Ethiopia, Gabon, The Gambia, Ghana, Guinea, Guinea-Bissau, Kenya, Lesotho, Liberia, Libya, Madagascar, Malawi, Mali, Mauritania, Mauritius, Morocco, Mozambique, Niger, Nigeria, Rwanda, Sao Tome and Principe, Senegal, Seychelles, Sierra Leone, Somalia, Sudan, Swaziland, Tanzania, Togo, Tunisia, Uganda, Zaire, Zambia, Zimbabwe *nonregional members*—(25) Argentina, Austria, Belgium, Brazil, Canada, China, Denmark, Finland, France, Germany, India, Italy, Japan, South Korea, Kuwait, Netherlands, Norway, Portugal, Saudi Arabia, Spain, Sweden, Switzerland, UK, US, Yugoslavia
Agence de Cooperation Culturelle et Technique (ACCT)	see Agency for Cultural and Technical Cooperation (ACCT)
Agency for Cultural and Technical Cooperation (ACCT) *note*—acronym from Agence de Cooperation Culturelle et Technique *established*—21 March 1970 *aim*—to promote cultural and technical cooperation among French-speaking countries	*members*—(31) Belgium, Benin, Burkina, Burundi, Canada, Central African Republic, Chad, Comoros, Congo, Cote d'Ivoire, Djibouti, Dominica, Equatorial Guinea, France, Gabon, Guinea, Haiti, Luxembourg, Madagascar, Mali, Mauritius, Monaco, Niger, Rwanda, Senegal, Seychelles, Togo, Tunisia, Vanuatu, Vietnam, Zaire *associate members*—(7) Cameroon, Egypt, Guinea-Bissau, Laos, Mauritania, Morocco, Saint Lucia *participating governments*—(2) New Brunswick (Canada), Quebec (Canada)
Agency for the Prohibition of Nuclear Weapons in Latin America and the Caribbean (OPANAL) *note*—acronym from Organismo para la Proscripcion de las Armas Nucleares en la America Latina y el Caribe (OPANAL) *established*—14 February 1967 *aim*—to encourage the peaceful uses of atomic energy and prohibit nuclear weapons	*members*—(26) Antigua and Barbuda, The Bahamas, Barbados, Bolivia, Brazil, Chile, Colombia, Costa Rica, Dominica, Dominican Republic, Ecuador, El Salvador, Grenada, Guatemala, Haiti, Honduras, Jamaica, Mexico, Nicaragua, Panama, Paraguay, Peru, Suriname, Trinidad and Tobago, Uruguay, Venezuela
Andean Group (AG) *established*—26 May 1969 *effective*—16 October 1969 *aim*—to promote harmonious development through economic integration	*members*—(5) Bolivia, Colombia, Ecuador, Peru, Venezuela *associate member*—(1) Panama *observers*—(26) Argentina, Australia, Austria, Belgium, Brazil, Canada, Costa Rica, Denmark, Egypt, Finland, France, Germany, India, Israel, Italy, Japan, Mexico, Netherlands, Paraguay, Spain, Sweden, Switzerland, UK, US, Uruguay, Yugoslavia

Note: The Socialist Federal Republic of Yugoslavia (SFRY) has dissolved, and ceases to exist. None of the successor states of the former Yugoslavia, including Serbia and Montenegro, have been permitted to participate solely on the basis of the membership of the former Yugoslavia in the United Nations General Assembly and Economic and Social Council and their subsidiary bodies and in various United Nations Specialized Agencies. The United Nations, however, permits the seat and nameplate of the SFRY to remain, permits the SFRY mission to continue to function, and continues to fly the flag of the former Yugoslavia. For a variety of reasons, a number of other organizations have not yet taken action with regard to the membership of the former Yugoslavia. The *The World Factbook* therefore continues to list Yugoslavia under international organizations where the SFRY seat remains or where no action has yet been taken.

Arab Bank for Economic Development in Africa (ABEDA) *note*—also known as Banque Arabe de Developpement Economique en Afrique (BADEA) *established*—18 February 1974 *effective*—16 September 1974 *aim*—to promote economic development	*members*—(17 plus the Palestine Liberation Organization) Algeria, Bahrain, Egypt, Iraq, Jordan, Kuwait, Lebanon, Libya, Mauritania, Morocco, Oman, Qatar, Saudi Arabia, Sudan, Syria, Tunisia, UAE, Palestine Liberation Organization; note—these are all the members of the Arab League except Djibouti, Somalia, and Yemen
Arab Cooperation Council (ACC) *established*—16 February 1989 *aim*—to promote economic cooperation and integration, possibly leading to an Arab Common Market	*members*—(4) Egypt, Iraq, Jordan, Yemen
Arab Fund for Economic and Social Development (AFESD) *established*—16 May 1968 *aim*—to promote economic and social development	*members*—(20 plus the Palestine Liberation Organization) Algeria, Bahrain, Djibouti, Egypt (suspended from 1979 to 1988), Iraq, Jordan, Kuwait, Lebanon, Libya, Mauritania, Morocco, Oman, Qatar, Saudi Arabia, Somalia, Sudan, Syria, Tunisia, UAE, Yemen, Palestine Liberation Organization
Arab League (AL) *note*—also known as League of Arab States (LAS) *established*—22 March 1945 *aim*—to promote economic, social, political, and military cooperation	*members*—(20 plus the Palestine Liberation Organization) Algeria, Bahrain, Djibouti, Egypt, Iraq, Jordan, Kuwait, Lebanon, Libya, Mauritania, Morocco, Oman, Qatar, Saudi Arabia, Somalia, Sudan, Syria, Tunisia, UAE, Yemen, Palestine Liberation Organization
Arab Maghreb Union (AMU) *established*—17 February 1989 *aim*—to promote cooperation and integration among the Arab states of northern Africa	*members*—(5) Algeria, Libya, Mauritania, Morocco, Tunisia
Arab Monetary Fund (AMF) *established*—27 April 1976 *effective*—2 February 1977 *aim*—to promote Arab cooperation, development, and integration in monetary and economic affairs	*members*—(19 plus the Palestine Liberation Organization) Algeria, Bahrain, Egypt, Iraq, Jordan, Kuwait, Lebanon, Libya, Mauritania, Morocco, Oman, Qatar, Saudi Arabia, Somalia, Sudan, Syria, Tunisia, UAE, Yemen, Palestine Liberation Organization
Asia Pacific Economic Cooperation (APEC) *established*—NA November 1989 *aim*—to promote trade and investment in the Pacific basin	*members*—(15) all ASEAN members (Brunei, Indonesia, Malaysia, Philippines, Singapore, Thailand) plus Australia, Canada, China, Hong Kong, Japan, South Korea, NZ, Taiwan, US
Asian Development Bank (AsDB) *established*—19 December 1966 *aim*—to promote regional economic cooperation	*regional members*—(36) Afghanistan, Australia, Bangladesh, Bhutan, Burma, Cambodia, China, Cook Islands, Fiji, Hong Kong, India, Indonesia, Japan, Kiribati, South Korea, Laos, Malaysia, Maldives, Marshall Islands, Federated States of Micronesia, Mongolia, Nauru, Nepal, NZ, Pakistan, Papua New Guinea, Philippines, Singapore, Solomon Islands, Sri Lanka, Taiwan, Thailand, Tonga, Vanuatu, Vietnam, Western Samoa *nonregional members*—(16) Austria, Belgium, Canada, Denmark, Finland, France, Germany, Italy, Netherlands, Norway, Spain, Sweden, Switzerland, Turkey, UK, US
Asociacion Latinoamericana de Integracion (ALADI)	see Latin American Integration Association (LAIA)

Association of Southeast Asian Nations (ASEAN) *established*—9 August 1967 *aim*—to encourage regional economic, social, and cultural cooperation among the non-Communist countries of Southeast Asia	*members*—(6) Brunei, Indonesia, Malaysia, Philippines, Singapore, Thailand *observer*—(1) Papua New Guinea
Australia Group *established*—1984 *aim*—to consult on and coordinate export controls related to chemical and biological weapons	*members*—(25) Argentina, Australia, Austria, Belgium, Canada, Denmark, Finland, France, Germany, Greece, Hungary, Iceland, Ireland, Italy, Japan, Luxembourg, Netherlands, NZ, Norway, Portugal, Spain, Sweden, Switzerland, UK, US *observer*—(1) Singapore
Australia—New Zealand—United States Security Treaty (ANZUS) *established*—1 September 1951 *effective*—29 April 1952 *aim*—to implement a trilateral mutual security agreement, although the US suspended security obligations to NZ on 11 August 1986	*members*—(3) Australia, NZ, US
Banco Centroamericano de Integracion Economico (BCIE)	see Central American Bank for Economic Integration (BCIE)
Banco Interamericano de Desarrollo (BID)	see Inter-American Development Bank (IADB)
Bank for International Settlements (BIS) *established*—20 January 1930 *effective*—17 March 1930 *aim*—to promote cooperation among central banks in international financial settlements	*members*—(30) Australia, Austria, Belgium, Bulgaria, Canada, Czech Republic, Denmark, Finland, France, Germany, Greece, Hungary, Iceland, Ireland, Italy, Japan, Netherlands, Norway, Poland, Portugal, Romania, Slovakia, South Africa, Spain, Sweden, Switzerland, Turkey, UK, US, Yugoslavia
Banque Africaine de Developpement (BAD)	see African Development Bank (AfDB)
Banque Arabe de Developpement Economique en Afrique (BADEA)	see Arab Bank for Economic Development in Africa (ABEDA)
Banque de Developpement des Etats de l'Afrique Centrale (BDEAC)	see Central African States Development Bank (BDEAC)
Banque Ouest-Africaine de Developpement (BOAD)	see West African Development Bank (WADB)
Benelux Economic Union (Benelux) *note*—acronym from Belgium, Netherlands, and Luxembourg *established*—3 February 1958 *effective*—1 November 1960 *aim*—to develop closer economic cooperation and integration	*members*—(3) Belgium, Luxembourg, Netherlands
Big Seven *note*—membership is the same as the Group of 7 *established*—NA *aim*—to discuss and coordinate major economic policies	*members*—(7) Big Six (Canada, France, Germany, Italy, Japan, UK) plus the US

Big Six

note—not to be confused with the Group of 6

established—NA

aim—to foster economic cooperation

members—(6) Canada, France, Germany, Italy, Japan, UK

Black Sea Economic Cooperation Zone (BSEC)

established—25 June 1992

aim—to enhance regional stability through economic cooperation

members—(11) Albania, Armenia, Azerbaijan, Bulgaria, Georgia, Greece, Moldova, Romania, Russia, Turkey, Ukraine

---

Caribbean Community and Common Market (CARICOM)

established—4 July 1973

effective—1 August 1973

aim—to promote economic integration and development, especially among the less developed countries

members—(13) Antigua and Barbuda, The Bahamas, Barbados, Belize, Dominica, Grenada, Guyana, Jamaica, Montserrat, Saint Kitts and Nevis, Saint Lucia, Saint Vincent and the Grenadines, Trinidad and Tobago

associate members—(2) British Virgin Islands, Turks and Caicos Islands

observers—(10) Anguilla, Bermuda, Cayman Islands, Dominican Republic, Haiti, Mexico, Netherlands Antilles, Puerto Rico, Suriname, Venezuela

---

Caribbean Development Bank (CDB)

established—18 October 1969

effective—26 January 1970

aim—to promote economic development and cooperation

regional members—(20) Anguilla, Antigua and Barbuda, The Bahamas, Barbados, Belize, British Virgin Islands, Cayman Islands, Colombia, Dominica, Grenada, Guyana, Jamaica, Mexico, Montserrat, Saint Kitts and Nevis, Saint Lucia, Saint Vincent and the Grenadines, Trinidad and Tobago, Turks and Caicos Islands, Venezuela

nonregional members—(5) Canada, France, Germany, Italy, UK

Cartagena Group

see Group of 11

Central African Customs and Economic Union (UDEAC)

note—acronym from Union Douaniere et Economique de l'Afrique Centrale

established—8 December 1964

effective—1 January 1966

aim—to promote the establishment of a Central African Common Market

members—(6) Cameroon, Central African Republic, Chad, Congo, Equatorial Guinea, Gabon

---

Central African States Development Bank (BDEAC)

note—acronym from Banque de Developpement des Etats de l'Afrique Centrale

established—3 December 1975

aim—to provide loans for economic development

members—(9) Cameroon, Central African Republic, Chad, Congo, Equatorial Guinea, France, Gabon, Germany, Kuwait

---

Central American Bank for Economic Integration (BCIE)

note—acronym from Banco Centroamericano de Integracion Economico

established—13 December 1960

aim—to promote economic integration and development

members—(5) Costa Rica, El Salvador, Guatemala, Honduras, Nicaragua

Central American Common Market (CACM)	*members*—(5) Costa Rica, El Salvador, Guatemala, Honduras, Nicaragua
established—13 December 1960	
effective—3 June 1961	
aim—to promote establishment of a Central American Common Market	
Central European Initiative (CEI)	*members*—(10) Austria, Bosnia and Herzegovina, Croatia, Czech Republic, Hungary, Italy, Poland, Slovakia, Slovenia, Yugoslavia
note—evolved from the Hexagonal Group	
established—July 1991	
aim—to form an economic and political cooperation group for the region between the Adriatic and the Baltic Seas	
centrally planned economies	a term applied mainly to the traditionally Communist states that looked to the former USSR for leadership; most are now evolving toward more democratic and market-oriented systems; also known formerly as the Second World or as the Communist countries; through the 1980s, this group included Albania, Bulgaria, Cambodia, China, Cuba, Czechoslovakia, GDR, Hungary, North Korea, Laos, Mongolia, Poland, Romania, USSR, Vietnam, Yugoslavia
Colombo Plan (CP)	*members*—(26) Afghanistan, Australia, Bangladesh, Bhutan, Burma, Cambodia, Canada, Fiji, India, Indonesia, Iran, Japan, South Korea, Laos, Malaysia, Maldives, Nepal, NZ, Pakistan, Papua New Guinea, Philippines, Singapore, Sri Lanka, Thailand, UK, US
established—1 July 1951	
aim—to promote economic and social development in Asia and the Pacific	
Commission for Social Development	*members*—(32) selected on a rotating basis from all regions
established—21 June 1946 as the Social Commission, renamed 29 July 1966	
aim—Economic and Social Council organization dealing with social development programs of UN	
Commission on Human Rights	*members*—(53) selected on a rotating basis from all regions
established—18 February 1946	
aim—Economic and Social Council organization dealing with human rights programs of UN	
Commission on Human Settlements (Habitat)	*members*—(58) selected on a rotating basis from all regions
established—12 October 1978	
aim—Economic and Social Council organization assisting in solving human settlement problems of UN	
Commission on Narcotic Drugs	*members*—(53) selected on a rotating basis from all regions with emphasis on producing and processing countries
established—16 February 1946	
aim—Economic and Social Council organization dealing with illicit drugs programs of UN	
Commission on the Status of Women	*members*—(32) selected on a rotating basis from all regions
established—21 June 1946	
aim—Economic and Social Council organization dealing with women's rights goals of UN	

Commonwealth (C) *established*—31 December 1931 *aim*—voluntary association that evolved from the British Empire and that seeks to foster multinational cooperation and assistance	*members*—(48) Antigua and Barbuda, Australia, The Bahamas, Bangladesh, Barbados, Belize, Botswana, Brunei, Canada, Cyprus, Dominica, The Gambia, Ghana, Grenada, Guyana, India, Jamaica, Kenya, Kiribati, Lesotho, Malawi, Malaysia, Maldives, Malta, Mauritius, Namibia, NZ, Nigeria, Pakistan, Papua New Guinea, Saint Kitts and Nevis, Saint Lucia, Saint Vincent and the Grenadines, Seychelles, Sierra Leone, Singapore, Solomon Islands, Sri Lanka, Swaziland, Tanzania, Tonga, Trinidad and Tobago, Uganda, UK, Vanuatu, Western Samoa, Zambia, Zimbabwe *special members*—(2) Nauru, Tuvalu
Commonwealth of Independent States (CIS) *established*—8 December 1991 *effective*—21 December 1991 *aim*—to coordinate intercommonwealth relations and to provide a mechanism for the orderly dissolution of the USSR	*members*—(11) Armenia, Azerbaijan, Belarus, Kazakhstan, Kyrgyzstan, Moldova, Russia, Tajikistan, Turkmenistan, Ukraine, Uzbekistan
Communaute Economique de l'Afrique de l'Ouest (CEAO)	see West African Economic Community (CEAO)
Communaute Economique des Etats de l'Afrique Centrale (CEEAC)	see Economic Community of Central African States (CEEAC)
Communaute Economique des Pays des Grands Lacs (CEPGL)	see Economic Community of the Great Lakes Countries (CEPGL)
Communist countries	traditionally the Marxist-Leninist states with authoritarian governments and command economies based on the Soviet model; most of the successor states are no longer Communist; see centrally planned economies
Conference on Security and Cooperation in Europe (CSCE) *established*—NA November 1972 *aim*—discusses issues of mutual concern and reviews implementation of the Helsinki Agreement	*members*—(53) Albania, Armenia, Austria, Azerbaijan, Belarus, Belgium, Bosnia and Herzegovina, Bulgaria, Canada, Croatia, Cyprus, Czech Republic, Denmark, Estonia, Finland, France, Georgia, Germany, Greece, Holy See, Hungary, Iceland, Ireland, Italy, Kazakhstan, Kyrgyzstan, Latvia, Liechtenstein, Lithuania, Luxembourg, Malta, Moldova, Monaco, Netherlands, Norway, Poland, Portugal, Romania, Russia, San Marino, Slovakia, Slovenia, Spain, Sweden, Switzerland, Tajikistan, Turkey, Turkmenistan, Ukraine, UK, US, Uzbekistan, Yugoslavia *observer*—(1) Japan
Conseil Europeen pour la Recherche Nucleaire (CERN)	see European Organization for Nuclear Research (CERN)
Contadora Group (CG)	was established 5 January 1983 (on the Panamanian island of Contadora) to reduce tensions and conflicts in Central America but evolved into the Rio Group (RG); members included Colombia, Mexico, Panama, Venezuela
Cooperation Council for the Arab States of the Gulf	see Gulf Cooperation Council (GCC)
Coordinating Committee on Export Controls (COCOM) *established*—NA 1949 *aim*—to control the export of strategic products and technical data from member countries to proscribed destinations	*members*—(17) Australia, Belgium, Canada, Denmark, France, Germany, Greece, Italy, Japan, Luxembourg, Netherlands, Norway, Portugal, Spain, Turkey, UK, US cooperating countries *cooperating countries*—(8) Austria, Finland, Ireland, South Korea, NZ, Singapore, Sweden, Switzerland
Council for Mutual Economic Assistance (CEMA)	also known as CMEA or Comecon, was established 25 January 1949 to promote the development of socialist economies and was abolished 1 January 1991; members included Afghanistan (observer), Albania (had not participated since 1961 break with USSR), Angola (observer), Bulgaria, Cuba, Czechoslovakia, Ethiopia (observer), GDR, Hungary, Laos (observer), Mongolia, Mozambique (observer), Nicaragua (observer), Poland, Romania, USSR, Vietnam, Yemen (observer), Yugoslavia (associate)
Council of Arab Economic Unity (CAEU) *established*—3 June 1957 *effective*—30 May 1964 *aim*—to promote economic integration among Arab nations	*members*—(11 plus the Palestine Liberation Organization) Egypt, Iraq, Jordan, Kuwait, Libya, Mauritania, Somalia, Sudan, Syria, UAE, Yemen, Palestine Liberation Organization

Council of the Baltic Sea States (CBSS) *established*—5 March 1992 *aim*—to promote cooperation among the Baltic Sea states in the areas of aid to new democratic institutions, economic development, humanitarian aid, energy and the environment, cultural and education, and transportation and communication	*members*—(10) Denmark, Estonia, Finland, Germany, Latvia, Lithuania, Norway, Poland, Russia, Sweden *observers*—(2) Belarus, Ukraine
Council of Europe (CE) *established*—5 May 1949 *effective*—3 August 1949 *aim*—to promote increased unity and quality of life in Europe	*members*—(29) Austria, Belgium, Bulgaria, Cyprus, Czech Republic, Denmark, Finland, France, Germany, Greece, Hungary, Iceland, Ireland, Italy, Liechtenstein, Luxembourg, Malta, Netherlands, Norway, Poland, Portugal, San Marino, Slovakia, Slovenia, Spain, Sweden, Switzerland, Turkey, UK
Council of the Entente (Entente) *established*—29 May 1959 *aim*—to promote economic, social, and political coordination	*members*—(5) Benin, Burkina, Cote d'Ivoire, Niger, Togo
Customs Cooperation Council (CCC) *established*—15 December 1950 *aim*—to promote international cooperation in customs matters	*members*—(114) Algeria, Angola, Argentina, Australia, Austria, The Bahamas, Bangladesh, Belgium, Bermuda, Botswana, Brazil, Bulgaria, Burkina, Burma, Burundi, Cameroon, Canada, Central African Republic, Chile, China, Congo, Cote d'Ivoire, Cuba, Cyprus, Czech Republic, Denmark, Egypt, Ethiopia, Finland, France, Gabon, The Gambia, Germany, Ghana, Greece, Guatemala, Guinea, Guyana, Haiti, Hong Kong, Hungary, Iceland, India, Indonesia, Iran, Iraq, Ireland, Israel, Italy, Jamaica, Japan, Jordan, Kenya, South Korea, Lebanon, Lesotho, Liberia, Libya, Luxembourg, Madagascar, Malawi, Malaysia, Mali, Malta, Mauritania, Mauritius, Mexico, Mongolia, Morocco, Mozambique, Nepal, Netherlands, NZ, Niger, Nigeria, Norway, Pakistan, Paraguay, Peru, Philippines, Poland, Portugal, Qatar, Romania, Russia, Rwanda, Saudi Arabia, Senegal, Sierra Leone, Singapore, Slovakia, South Africa, Spain, Sri Lanka, Sudan, Swaziland, Sweden, Switzerland, Syria, Tanzania, Thailand, Togo, Trinidad and Tobago, Tunisia, Turkey, Uganda, UAE, UK, US, Uruguay, Yugoslavia, Zaire, Zambia, Zimbabwe
developed countries (DCs)	the top group in the comprehensive but mutually exclusive hierarchy of developed countries (DCs), former USSR/Eastern Europe (former USSR/EE), and less developed countries (LDCs); includes the market-oriented economies of the mainly democratic nations in the Organization for Economic Cooperation and Development (OECD), Bermuda, Israel, South Africa, and the European ministates; also known as the First World, high-income countries, the North, industrial countries; generally have a per capita GNP/GDP in excess of $10,000 although some OECD countries and South Africa have figures well under $10,000 and two of the excluded OPEC countries have figures of more than $10,000; the 34 DCs are: Andorra, Australia, Austria, Belgium, Bermuda, Canada, Denmark, Faroe Islands, Finland, France, Germany, Greece, Holy See, Iceland, Ireland, Israel, Italy, Japan, Liechtenstein, Luxembourg, Malta, Monaco, Netherlands, NZ, Norway, Portugal, San Marino, South Africa, Spain, Sweden, Switzerland, Turkey, UK, US
developing countries	an imprecise term for the less developed countries with growing economies; see less developed countries (LDCs)
East African Development Bank (EADB) *established*—6 June 1967 *effective*—1 December 1967 *aim*—to promote economic development	*members*—(3) Kenya, Tanzania, Uganda
Economic and Social Commission for Asia and the Pacific (ESCAP) *established*—28 March 1947 as Economic Commission for Asia and the Far East (ECAFE) *aim*—to promote economic development as a regional commission for the UN's Economic and Social Council	*members*—(46) Afghanistan, Australia, Azerbaijan, Bangladesh, Bhutan, Brunei, Burma, Cambodia, China, Fiji, France, India, Indonesia, Iran, Japan, Kiribati, North Korea, South Korea, Kyrgyzstan, Laos, Malaysia, Maldives, Marshall Islands, Federated States of Micronesia, Mongolia, Nauru, Nepal, Netherlands, NZ, Pakistan, Papua New Guinea, Philippines, Russia, Singapore, Solomon Islands, Sri Lanka, Tajikistan, Thailand, Tonga, Turkmenistan, Tuvalu, UK, US, Vanuatu, Vietnam, Western Samoa *associate members*—(10) American Samoa, Cook Islands, French Polynesia, Guam, Hong Kong, Macau, New Caledonia, Niue, Northern Mariana Islands, Trust Territory of the Pacific Islands (Palau)

Economic and Social Commission for Western Asia (ESCWA) *established*—9 August 1973 as Economic Commission for Western Asia (ECWA) *aim*—to promote economic development as a regional commission for the UN's Economic and Social Council	*members*—(12 and the Palestine Liberation Organization) Bahrain, Egypt, Iraq, Jordan, Kuwait, Lebanon, Oman, Qatar, Saudi Arabia, Syria, UAE, Yemen, Palestine Liberation Organization
Economic and Social Council (ECOSOC) *established*—26 June 1945 *effective*—24 October 1945 *aim*—to coordinate the economic and social work of the UN; includes five regional commissions (see Economic Commission for Africa, Economic Commission for Europe, Economic Commission for Latin America and the Caribbean, Economic and Social Commission for Asia and the Pacific, Economic and Social Commission for Western Asia) and six functional commissions (see Commission for Social Development, Commission on Human Rights, Commission on Narcotic Drugs, Commission on the Status of Women, Population Commission, Statistical Commission, Commission on Science and Technology for Development, Commission on Sustainable Development, Commission on Crime Prevention and Criminal Justice, and Commission on Transnational Corporations)	*members*—(54) selected on a rotating basis from all regions
Economic Commission for Africa (ECA) *established*—29 April 1958 *aim*—to promote economic development as a regional commission of the UN's Economic and Social Council	*members*—(52) Algeria, Angola, Benin, Botswana, Burkina, Burundi, Cameroon, Cape Verde, Central African Republic, Chad, Comoros, Congo, Cote d'Ivoire, Djibouti, Egypt, Equatorial Guinea, Ethiopia, Gabon, The Gambia, Ghana, Guinea, Guinea-Bissau, Kenya, Lesotho, Liberia, Libya, Madagascar, Malawi, Mali, Mauritania, Mauritius, Morocco, Mozambique, Namibia, Niger, Nigeria, Rwanda, Sao Tome and Principe, Senegal, Seychelles, Sierra Leone, Somalia, South Africa (suspended), Sudan, Swaziland, Tanzania, Togo, Tunisia, Uganda, Zaire, Zambia, Zimbabwe *associate members*—(2) France, UK
Economic Commission for Asia and the Far East (ECAFE)	see Economic and Social Commission for Asia and the Pacific (ESCAP)
Economic Commission for Europe (ECE) *established*—28 March 1947 *aim*—to promote economic development as a regional commission of the UN's Economic and Social Council	*members*—(44) Albania, Austria, Belarus, Belgium, Bosnia and Herzegovina, Bulgaria, Canada, Croatia, Cyprus, Czech Republic, Denmark, Estonia, Finland, France, Germany, Greece, Hungary, Iceland, Ireland, Israel, Italy, Latvia, Lichtenstein, Lithuania, Luxembourg, Malta, Moldova, Netherlands, Norway, Poland, Portugal, Romania, Russia, San Marino, Slovakia, Slovenia, Spain, Sweden, Switzerland, Turkey, Ukraine, UK, US, Yugoslavia
Economic Commission for Latin America (ECLA)	see Economic Commission for Latin America and the Caribbean (ECLAC)
Economic Commission for Latin America and the Caribbean (ECLAC) *established*—25 February 1948 as Economic Commission for Latin America (ECLA) *aim*—to promote economic development as a regional commission of the UN's Economic and Social Council	*members*—(41) Antigua and Barbuda, Argentina, The Bahamas, Barbados, Belize, Bolivia, Brazil, Canada, Chile, Colombia, Costa Rica, Cuba, Dominica, Dominican Republic, Ecuador, El Salvador, France, Grenada, Guatemala, Guyana, Haiti, Honduras, Italy, Jamaica, Mexico, Netherlands, Nicaragua, Panama, Paraguay, Peru, Portugal, Saint Kitts and Nevis, Saint Lucia, Saint Vincent and the Grenadines, Spain, Suriname, Trinidad and Tobago, UK, US, Uruguay, Venezuela *associate members*—(6) Aruba, British Virgin Islands, Montserrat, Netherlands Antilles, Puerto Rico, Virgin Islands
Economic Commission for Western Asia (ECWA)	see Economic and Social Commission for Western Asia (ESCWA)

Economic Community of Central African States (CEEAC)—acronym from Communaute Economique des Etats de l'Afrique Centrale *established*—18 October 1983 *aim*—to promote regional economic cooperation and establish a Central African Common Market	*members*—(10) Burundi, Cameroon, Central African Republic, Chad, Congo, Equatorial Guinea, Gabon, Rwanda, Sao Tome and Principe, Zaire *observer*—(1) Angola
Economic Community of the Great Lakes Countries (CEPGL) *note*—acronym from Communaute Economique des Pays des Grands Lacs *established*—26 September 1976 *aim*—to promote regional economic cooperation and integration	*members*—(3) Burundi, Rwanda, Zaire
Economic Community of West African States (ECOWAS) *established*—28 May 1975 *aim*—to promote regional economic cooperation	*members*—(17) Benin, Burkina, Cape Verde, Cote d'Ivoire, Equatorial Guinea, The Gambia, Ghana, Guinea, Guinea-Bissau, Liberia, Mali, Mauritania, Niger, Nigeria, Senegal, Sierra Leone, Togo
Economic Cooperation Organization (ECO) *established*—1985 *aim*—to promote regional cooperation in trade, transportation, communications, tourism, cultural affairs, and economic development	*members*—(10) Afghanistan, Azerbaijan, Iran, Kazakhstan, Kyrgyzstan, Pakistan, Tajikistan, Turkey, Turkmenistan, Uzbekistan
European Bank for Reconstruction and Development (EBRD) *established*—15 April 1991 *aim*—to facilitate the transition of seven centrally planned economies in Europe (Bulgaria, former Czechoslovakia, Hungary, Poland, Romania, former USSR, and former Yugoslavia) to market economies by committing 60% of its loans to privatization	*members*—(58) Albania, Armenia, Australia, Austria, Azerbaijan, Belgium, Bulgaria, Canada, Cyprus, Czech Republic, Denmark, European Community (EC), Egypt, European Investment Bank (EIB), Estonia, Finland, France, Georgia, Germany, Greece, Hungary, Iceland, Ireland, Israel, Italy, Japan, Kazakhstan, South Korea, Kyrgyzstan, Latvia, Liechtenstein, Lithuania, Luxembourg, Macedonia, Malta, Mexico, Moldova, Morocco, Netherlands, NZ, Norway, Poland, Portugal, Russia, Romania, Slovakia, Slovenia, Spain, Sweden, Switzerland, Tajikistan, Turkey, Turkmenistan, Ukraine, UK, US, Uzbekistan, Yugoslavia; note—includes all 24 members of the OECD and the EC as an institution
European Community (EC) *established*—8 April 1965 *effective*—1 July 1967 *aim*—to integrate the European Atomic Energy Community (Euratom), the European Coal and Steel Community (ESC), and the European Economic Community (EEC or Common Market); the EC plans to establish a completely integrated common market and an eventual federation of Europe	*members*—(12) Belgium, Denmark, France, Germany, Greece, Ireland, Italy, Luxembourg, Netherlands, Portugal, Spain, UK
European Free Trade Association (EFTA) *established*—4 January 1960 *effective*—3 May 1960 *aim*—to promote expansion of free trade	*members*—(7) Austria, Finland, Iceland, Leichtenstein, Norway, Sweden, Switzerland

European Investment Bank (EIB) *established*—25 March 1957 *effective*—1 January 1958 *aim*—to promote economic development of the EC	*members*—(12) Belgium, Denmark, France, Germany, Greece, Ireland, Italy, Luxembourg, Netherlands, Portugal, Spain, UK
European Organization for Nuclear Research (CERN) *note*—acronym retained from the predecessor organization Conseil Europeen pour la Recherche Nucleaire *established*—1 July 1953 *effective*—29 September 1954 *aim*—to foster nuclear research for peaceful purposes only	*members*—(19) Austria, Belgium, Czech Republic, Denmark, Finland, France, Germany, Greece, Hungary, Italy, Netherlands, Norway, Poland, Portugal, Slovakia, Spain, Sweden, Switzerland, UK *observers*—(6) EC, Israel, Russia, Turkey, United Nations Educational, Scientific, and Cultural Organization (UNESCO), Yugoslavia
European Space Agency (ESA) *established*—31 July 1973 *effective*—1 May 1975 *aim*—to promote peaceful cooperation in space research and technology	*members*—(13) Austria, Belgium, Denmark, France, Germany, Ireland, Italy, Netherlands, Norway, Spain, Sweden, Switzerland, UK *associate member*—(1) Finland *cooperating state*—(1) Canada
First World	another term for countries with advanced, industrialized economies; this term is fading from use; see developed countries (DCs)
Food and Agriculture Organization (FAO) *established*—16 October 1945 *aim*—UN specialized agency to raise living standards and increase availability of agricultural products	*members*—(162) Afghanistan, Albania, Algeria, Angola, Antigua and Barbuda, Argentina, Australia, Austria, The Bahamas, Bahrain, Bangladesh, Barbados, Belgium, Belize, Benin, Bhutan, Bolivia, Botswana, Brazil, Brunei, Bulgaria, Burkina, Burma, Burundi, Cambodia, Cameroon, Canada, Cape Verde, Central African Republic, Chad, Chile, China, Colombia, Comoros, Congo, Costa Rica, Cote d'Ivoire, Cuba, Cyprus, Czech Republic, Denmark, Djibouti, Dominica, Dominican Republic, Ecuador, EC, Egypt, El Salvador, Equatorial Guinea, Estonia, Ethiopia, Fiji, Finland, France, Gabon, The Gambia, Germany, Ghana, Greece, Grenada, Guatemala, Guinea, Guinea-Bissau, Guyana, Haiti, Honduras, Hungary, Iceland, India, Indonesia, Iran, Iraq, Ireland, Israel, Italy, Jamaica, Japan, Jordan, Kenya, North Korea, South Korea, Kuwait, Laos, Latvia, Lebanon, Lesotho, Liberia, Libya, Lithuania, Luxembourg, Madagascar, Malawi, Malaysia, Maldives, Mali, Malta, Mauritania, Mauritius, Mexico, Mongolia, Morocco, Mozambique, Namibia, Nepal, Netherlands, NZ, Nicaragua, Niger, Nigeria, Norway, Oman, Pakistan, Panama, Papua New Guinea, Paraguay, Peru, Philippines, Poland, Portugal, Qatar, Romania, Rwanda, Saint Kitts and Nevis, Saint Lucia, Saint Vincent and the Grenadines, Sao Tome and Principe, Saudi Arabia, Senegal, Seychelles, Sierra Leone, Slovakia, Solomon Islands, Somalia, Spain, Sri Lanka, Sudan, Suriname, Swaziland, Sweden, Switzerland, Syria, Tanzania, Thailand, Togo, Tonga, Trinidad and Tobago, Tunisia, Turkey, Uganda, UAE, UK, US, Uruguay, Vanuatu, Venezuela, Vietnam, Western Samoa, Yemen, Yugoslavia, Zaire, Zambia, Zimbabwe *associate member*—(1) Puerto Rico
Former USSR/Eastern Europe (former USSR/EE)	the middle group in the comprehensive but mutually exclusive hierarchy of developed countries (DCs), former USSR/Eastern Europe (former USSR/EE), and less developed countries (LDCs); these countries are in political and economic transition and may well be grouped differently in the near future; this group of 27 countries includes Albania, Armenia, Azerbaijan, Belarus, Bosnia and Herzegovina, Bulgaria, Croatia, Czech Republic, Estonia, Georgia, Hungary, Kazakhstan, Kyrgyzstan, Latvia, Lithuania, Macedonia. Moldova, Poland, Romania, Russia, Serbia and Montenegro, Slovakia, Slovenia, Tajikistan, Turkmenistan, Ukraine, Uzbekistan
Four Dragons	the four small Asian less developed countries (LDCs) that have experienced unusually rapid economic growth; also known as the Four Tigers; this group includes Hong Kong, South Korea, Singapore, Taiwan
Four Tigers	another term for the Four Dragons; see Four Dragons
Franc Zone (FZ) *established*—NA *aim*—to form a monetary union among countries whose currencies are linked to the French franc	*members*—(15) Benin, Burkina, Cameroon, Central African Republic, Chad, Comoros, Congo, Cote d'Ivoire, Equatorial Guinea, France, Gabon, Mali, Niger, Senegal, Togo; note—France includes metropolitan France, the four overseas departments of France (French Guiana, Guadeloupe, Martinique, Reunion), the two territorial collectivities of France (Mayotte, Saint Pierre and Miquelon), and the three overseas territories of France (French Polynesia, New Caledonia, Wallis and Futuna)

Front Line States (FLS) *established*—NA *aim*—to achieve black majority rule in South Africa	*members*—(7) Angola, Botswana, Mozambique, Namibia, Tanzania, Zambia, Zimbabwe
General Agreement on Tariffs and Trade (GATT) *established*—30 October 1947 *effective*—1 January 1948 *aim*—to promote the expansion of international trade on a nondiscriminatory basis	*members*—(104) Antigua and Barbuda, Argentina, Australia, Austria, Bangladesh, Barbados, Belgium, Belize, Benin, Bolivia, Botswana, Brazil, Burkina, Burma, Burundi, Cameroon, Canada, Central African Republic, Chad, Chile, Colombia, Congo, Costa Rica, Cote d'Ivoire, Cuba, Cyprus, Czech Republic, Denmark, Dominican Republic, Egypt, El Salvador, Finland, France, Gabon, The Gambia, Germany, Ghana, Greece, Guatemala, Guyana, Haiti, Hong Kong, Hungary, Iceland, India, Indonesia, Ireland, Israel, Italy, Jamaica, Japan, Kenya, South Korea, Kuwait, Lesotho, Luxembourg, Macau, Madagascar, Malawi, Malaysia, Maldives, Malta, Mauritania, Mauritius, Mexico, Morocco, Netherlands, NZ, Nicaragua, Niger, Nigeria, Norway, Pakistan, Peru, Philippines, Poland, Portugal, Romania, Rwanda, Senegal, Sierra Leone, Singapore, Slovakia, South Africa, Spain, Sri Lanka, Suriname, Sweden, Switzerland, Tanzania, Thailand, Togo, Trinidad and Tobago, Tunisia, Turkey, Uganda, UK, US, Uruguay, Venezuela, Yugoslavia, Zaire, Zambia, Zimbabwe
Group of 2 (G-2) *established*—informal term that came into use about 1986 *aim*—bilateral economic cooperation between the two most powerful economic giants	*members*—(2) Japan, US
Group of 3 (G-3) *established*—NA October 1990 *aim*—mechanism for policy coordination	*members*—(3) Colombia, Mexico, Venezuela
Group of 5 (G-5) *established*—22 September 1985 *aim*—the five major non-Communist economic powers	*members*—(5) France, Germany, Japan, UK, US
Group of 6 (G-6) *note*—not to be confused with the Big Six *established*—22 May 1984 *aim*—to achieve nuclear disarmament	*members*—(6) Argentina, Greece, India, Mexico, Sweden, Tanzania
Group of 7 (G-7) *note*—membership is the same as the Big Seven *established*—22 September 1985 *aim*—the seven major non-Communist economic powers	*members*—(7) Group of 5 (France, Germany, Japan, UK, US) plus Canada and Italy
Group of 8 (G-8) *established*—NA October 1975 *aim*—the developed countries (DCs) that participated in the Conference on International Economic Cooperation (CIEC), held in several sessions between NA December 1975 and 3 June 1977	*members*—(8) Australia, Canada, EC (as one member), Japan, Spain, Sweden, Switzerland, US
Group of 9 (G-9) *established*—NA *aim*—informal group that meets occasionally on matters of mutual interest	*members*—(9) Austria, Belgium, Bulgaria, Denmark, Finland, Hungary, Romania, Sweden, Yugoslavia

Group of 10 (G-10) *note*—also known as the Paris Club *established*—NA October 1962 *aim*—wealthiest members of the IMF who provide most of the money to be loaned and act as the informal steering committee; name persists in spite of the addition of Switzerland on NA April 1984	*members*—(11) Belgium, Canada, France, Germany, Italy, Japan, Netherlands, Sweden, Switzerland, UK, US
Group of 11 (G-11) *note*—also known as the Cartagena Group *established*—22 June 1984, in Cartagena, Colombia *aim*—forum for largest debtor nations in Latin America	*members*—(11) Argentina, Bolivia, Brazil, Chile, Colombia, Dominican Republic, Ecuador, Mexico, Peru, Uruguay, Venezuela
Group of 15 (G-15) *note*—byproduct of the Non-Aligned Movement *established*—1989 *aim*—to promote economic cooperation among developing nations; to act as the main political organ for the Non-Aligned Movement	*members*—(15) Algeria, Argentina, Brazil, Egypt, India, Indonesia, Jamaica, Malaysia, Mexico, Nigeria, Peru, Senegal, Venezuela, Yugoslavia, Zimbabwe
Group of 19 (G-19) *established*—NA October 1975 *aim*—the less developed countries (LDCs) that participated in the Conference on International Economic Cooperation (CIEC) held in several sessions between NA December 1975 and 3 June 1977	*members*—(19) Algeria, Argentina, Brazil, Cameroon, Egypt, India, Indonesia, Iran, Iraq, Jamaica, Mexico, Nigeria, Pakistan, Peru, Saudi Arabia, Venezuela, Yugoslavia, Zaire, Zambia
Group of 24 (G-24) *established*—NA January 1972 *aim*—to promote the interests of developing countries in Africa, Asia, and Latin America within the IMF	*members*—(24) Algeria, Argentina, Brazil, Colombia, Cote d'Ivoire, Egypt, Ethiopia, Gabon, Ghana, Guatemala, India, Iran, Lebanon, Mexico, Nigeria, Pakistan, Peru, Philippines, Sri Lanka, Syria, Trinidad and Tobago, Venezuela, Yugoslavia, Zaire
Group of 30 (G-30) *established*—NA 1979 *aim*—to discuss and propose solutions to the world's economic problems	*members*—(30) informal group of 30 leading international bankers, economists, financial experts, and businessmen organized by Johannes Witteveen (former managing director of the IMF)
Group of 33 (G-33) *established*—NA 1987 *aim*—to promote solutions to international economic problems	*members*—(33) leading economists from 13 countries

Group of 77 (G-77) *established*—NA October 1967 *aim*—to promote economic cooperation among developing countries; name persists in spite of increased membership	*members*—(127 plus the Palestine Liberation Organization) Afghanistan, Algeria, Angola, Antigua and Barbuda, Argentina, The Bahamas, Bahrain, Bangladesh, Barbados, Belize, Benin, Bhutan, Bolivia, Botswana, Brazil, Brunei, Burkina, Burma, Burundi, Cambodia, Cameroon, Cape Verde, Central African Republic, Chad, Chile, Colombia, Comoros, Congo, Costa Rica, Cote d'Ivoire, Cuba, Cyprus, Djibouti, Dominica, Dominican Republic, Ecuador, Egypt, El Salvador, Equatorial Guinea, Ethiopia, Fiji, Gabon, The Gambia, Ghana, Grenada, Guatemala, Guinea, Guinea-Bissau, Guyana, Haiti, Honduras, India, Indonesia, Iran, Iraq, Jamaica, Jordan, Kenya, North Korea, South Korea, Kuwait, Laos, Lebanon, Lesotho, Liberia, Libya, Madagascar, Malawi, Malaysia, Maldives, Mali, Malta, Mauritania, Mauritius, Mexico, Mongolia, Morocco, Mozambique, Namibia, Nepal, Nicaragua, Niger, Nigeria, Oman, Pakistan, Panama, Papua New Guinea, Paraguay, Peru, Philippines, Qatar, Romania, Rwanda, Saint Kitts and Nevis, Saint Lucia, Saint Vincent and the Grenadines, Sao Tome and Principe, Saudi Arabia, Senegal, Seychelles, Sierra Leone, Singapore, Solomon Islands, Somalia, Sri Lanka, Sudan, Suriname, Swaziland, Syria, Tanzania, Thailand, Togo, Tonga, Trinidad and Tobago, Tunisia, Uganda, UAE, Uruguay, Vanuatu, Venezuela, Vietnam, Western Samoa, Yemen, Yugoslavia, Zaire, Zambia, Zimbabwe, Palestine Liberation Organization
Gulf Cooperation Council (GCC) *note*—also known as the Cooperation Council for the Arab States of the Gulf *established*—25-26 May 1981 *aim*—to promote regional cooperation in economic, social, political, and military affairs	*members*—(6) Bahrain, Kuwait, Oman, Qatar, Saudi Arabia, UAE
Habitat	Commission on Human Settlements
Hexagonal Group	see Central European Initiative (CEI)
high-income countries	another term for the industrialized countries with high per capita GNPs/GDPs; see developed countries (DCs)
industrial countries	another term for the developed countries; see developed countries (DCs)
Inter-American Development Bank (IADB) *note*—also known as Banco Interamericano de Desarrollo (BID) *established*—8 April 1959 *effective*—30 December 1959 *aim*—to promote economic and social development in Latin America	*members*—(44) Argentina, Austria, The Bahamas, Barbados, Belgium, Bolivia, Brazil, Canada, Chile, Colombia, Costa Rica, Denmark, Dominican Republic, Ecuador, El Salvador, Finland, France, Germany, Guatemala, Guyana, Haiti, Honduras, Israel, Italy, Jamaica, Japan, Mexico, Netherlands, Nicaragua, Norway, Panama, Paraguay, Peru, Portugal, Spain, Suriname, Sweden, Switzerland, Trinidad and Tobago, UK, US, Uruguay, Venezuela, Yugoslavia
Inter-Governmental Authority on Drought and Development (IGADD) *established*—NA January 1986 *aim*—to promote cooperation on drought-related matters	*members*—(6) Djibouti, Ethiopia, Kenya, Somalia, Sudan, Uganda
International Atomic Energy Agency (IAEA) *established*—26 October 1956 *effective*—29 July 1957 *aim*—to promote peaceful uses of atomic energy	*members*—(115) Afghanistan, Albania, Algeria, Argentina, Australia, Austria, Bangladesh, Belarus, Belgium, Bolivia, Brazil, Bulgaria, Burma, Cambodia, Cameroon, Canada, Chile, China, Colombia, Costa Rica, Cote d'Ivoire, Cuba, Cyprus, Czech Republic, Denmark, Dominican Republic, Ecuador, Egypt, El Salvador, Estonia, Ethiopia, Finland, France, Gabon, Germany, Ghana, Greece, Guatemala, Haiti, Holy See, Hungary, Iceland, India, Indonesia, Iran, Iraq, Ireland, Israel, Italy, Jamaica, Japan, Jordan, Kenya, North Korea, South Korea, Kuwait, Lebanon, Liberia, Libya, Liechtenstein, Luxembourg, Madagascar, Malaysia, Mali, Mauritius, Mexico, Monaco, Mongolia, Morocco, Namibia, Netherlands, NZ, Nicaragua, Niger, Nigeria, Norway, Pakistan, Panama, Paraguay, Peru, Philippines, Poland, Portugal, Qatar, Romania, Russia, Saudi Arabia, Senegal, Sierra Leone, Singapore, Slovakia, Slovenia, South Africa, Spain, Sri Lanka, Sudan, Sweden, Switzerland, Syria, Tanzania, Thailand, Tunisia, Turkey, Uganda, Ukraine, UAE, UK, US, Uruguay, Venezuela, Vietnam, Yugoslavia, Zaire, Zambia, Zimbabwe
International Bank for Economic Cooperation (IBEC)	established in 22 October 1963; aim was to promote economic cooperation and development; members were Bulgaria, Cuba, Czechoslovakia, East Germany, Hungary, Mongolia, Poland, Romania, USSR, Vietnam; now it is a Russian bank with a new charter

International Bank for Reconstruction and Development (IBRD) *note*—also known as the World Bank *established*—22 July 1944 *effective*—27 December 1945 *aim*—UN specialized agency that initially promoted economic rebuilding after World War II and now provides economic development loans	*members*—(174) Afghanistan, Albania, Algeria, Angola, Antigua and Barbuda, Argentina, Armenia, Australia, Austria, Azerbaijan, The Bahamas, Bahrain, Bangladesh, Barbados, Belarus, Belgium, Belize, Benin, Bhutan, Bolivia, Botswana, Brazil, Bulgaria, Burkina, Burma, Burundi, Cameroon, Canada, Cape Verde, Central African Republic, Chad, Chile, China, Colombia, Comoros, Congo, Costa Rica, Cote d'Ivoire, Cyprus, Czech Republic, Denmark, Djibouti, Dominica, Dominican Republic, Ecuador, Egypt, El Salvador, Equatorial Guinea, Estonia, Ethiopia, Fiji, Finland, France, Gabon, The Gambia, Georgia, Germany, Ghana, Greece, Grenada, Guatemala, Guinea, Guinea-Bissau, Guyana, Haiti, Honduras, Hungary, Iceland, India, Indonesia, Iran, Iraq, Ireland, Israel, Italy, Jamaica, Japan, Jordan, Kazakhstan, Kenya, Kiribati, South Korea, Kuwait, Kyrgyzstan, Laos, Latvia, Lebanon, Lesotho, Liberia, Libya, Lithuania, Luxembourg, Madagascar, Malawi, Malaysia, Maldives, Mali, Malta, Marshall Islands, Mauritania, Mauritius, Mexico, Moldova, Mongolia, Morocco, Mozambique, Namibia, Nepal, Netherlands, New Zealand, Nicaragua, Niger, Nigeria, Norway, Oman, Pakistan, Panama, Papua New Guinea, Paraguay, Peru, Philippines, Poland, Portugal, Qatar, Romania, Russia, Rwanda, Saint Kitts and Nevis, Saint Lucia, Saint Vincent and the Grenadines, Sao Tome and Principe, Saudi Arabia, Senegal, Seychelles, Sierra Leone, Singapore, Slovakia, Solvenia, Solomon Islands, Somalia, South Africa, Spain, Sri Lanka, Sudan, Suriname, Swaziland, Sweden, Switzerland, Syria, Tanzania, Thailand, Togo, Tonga, Trinidad and Tobago, Tunisia, Turkey, Turkmenistan, Uganda, Ukraine, UAE, UK, US, Uruguay, Uzbekistan, Vanuatu, Venezuela, Vietnam, Western Samoa, Yemen, Yugoslavia, Zaire, Zambia, Zimbabwe
International Chamber of Commerce (ICC) *established*—NA 1919 *aim*—to promote free trade and private enterprise and to represent business interests at national and international levels	*members*—(58 national councils) Argentina, Australia, Austria, Belgium, Brazil, Burkina, Cameroon, Canada, Colombia, Cote d'Ivoire, Cyprus, Denmark, Ecuador, Egypt, Finland, France, Gabon, Germany, Greece, Iceland, India, Indonesia, Iran, Ireland, Israel, Italy, Japan, Jordan, South Korea, Lebanon, Luxembourg, Madagascar, Mexico, Morocco, Netherlands, Nigeria, Norway, Pakistan, Portugal, Saudi Arabia, Senegal, Singapore, South Africa, Spain, Sri Lanka, Sweden, Switzerland, Syria, Taiwan, Togo, Tunisia, Turkey, UK, US, Uruguay, Venezuela, Yugoslavia, Zaire
International Civil Aviation Organization (ICAO) *established*—7 December 1944 *effective*—4 April 1947 *aim*—UN specialized agency to promote international cooperation in civil aviation	*members*—(173) Afghanistan, Albania, Algeria, Angola, Antigua and Barbuda, Argentina, Armenia, Australia, Austria, The Bahamas, Bahrain, Bangladesh, Barbados, Belgium, Belize, Benin, Bhutan, Bolivia, Botswana, Brazil, Brunei, Bulgaria, Burkina, Burma, Burundi, Cambodia, Cameroon, Canada, Cape Verde, Central African Republic, Chad, Chile, China, Colombia, Comoros, Congo, Cook Islands, Costa Rica, Cote d'Ivoire, Croatia, Cuba, Cyprus, Czech Republic, Denmark, Djibouti, Dominican Republic, Ecuador, Egypt, El Salvador, Equatorial Guinea, Estonia, Ethiopia, Fiji, Finland, France, Gabon, The Gambia, Germany, Ghana, Greece, Grenada, Guatemala, Guinea, Guinea-Bissau, Guyana, Haiti, Honduras, Hungary, Iceland, India, Indonesia, Iran, Iraq, Ireland, Israel, Italy, Jamaica, Japan, Jordan, Kenya, Kiribati, North Korea, South Korea, Kuwait, Laos, Latvia, Lebanon, Lesotho, Liberia, Libya, Lithuania, Luxembourg, Macedonia, Madagacar, Malawi, Malaysia, Maldives, Mali, Malta, Marshall Islands, Mauritania, Mauritius, Mexico, Federated States of Micronesia, Moldova, Monaco, Mongolia, Morocco, Mozambique, Namibia, Nauru, Nepal, Netherlands, NZ, Nicaragua, Niger, Nigeria, Norway, Oman, Pakistan, Panama, Papua New Guinea, Paraguay, Peru, Philippines, Poland, Portugal, Qatar, Romania, Russia, Rwanda, Saint Lucia, Saint Vincent and the Grenadines, San Marino, Sao Tome and Principe, Saudi Arabia, Senegal, Seychelles, Sierra Leone, Singapore, Slovakia, Slovenia, Solomon Islands, Somalia, South Africa (suspended), Spain, Sri Lanka, Sudan, Suriname, Swaziland, Sweden, Switzerland, Syria, Tanzania, Thailand, Togo, Tonga, Trinidad and Tobago, Tunisia, Turkey, Uganda, UAE, UK, US, Uruguay, Vanuatu, Venezuela, Vietnam, Yemen, Yugoslavia, Zaire, Zambia, Zimbabwe
International Committee of the Red Cross (ICRC) *established*—NA 1863 *aim*—to provide humanitarian aid in wartime	*members*—(25 individuals) all Swiss nationals
International Confederation of Free Trade Unions (ICFTU) *established*—NA December 1949 *aim*—to promote the trade union movement	*members*—(144 national organizations in the following 104 areas) Antigua and Barbuda, Argentina, Australia, Austria, The Bahamas, Bangladesh, Barbados, Basque Country, Belgium, Bermuda, Botswana, Brazil, Bulgaria, Burkina, Canada, Central African Republic, Chad, Chile, Colombia, Costa Rica, Curacao, Cyprus, Czech Republic, Denmark, Dominica, Dominican Republic, Ecuador, El Salvador, Estonia, Falkland Islands, Fiji, Finland, France, French Polynesia, The Gambia, Germany, Greece, Grenada, Guatemala, Guyana, Holy See, Honduras, Hong Kong, Iceland, India, Indonesia, Israel, Italy, Jamaica, Japan, Kiribati, South Korea, Lebanon, Lesotho, Liberia, Luxembourg, Madagascar, Malawi, Malaysia, Malta, Mauritius, Mexico, Montserrat, Morocco, Netherlands, New Caledonia, NZ, Nicaragua, Norway, Pakistan, Panama, Papua New Guinea, Peru, Philippines, Poland, Portugal, Puerto Rico, Russia, Saint Helena, Saint Kitts and Nevis, Saint Lucia, Saint Vincent and the Grenadines, San Marino, Seychelles, Sierra Leone, Singapore, Slovakia, Spain, Sri Lanka, Suriname, Swaziland, Sweden, Switzerland, Taiwan, Thailand, Tonga, Trinidad and Tobago, Tunisia, Turkey, Uganda, UK, US, Venezuela, Western Samoa

International Court of Justice (ICJ) *note*—also known as the World Court *established*—26 June 1945 *effective*—24 October 1945 *aim*—primary judicial organ of the UN	*members*—(15 judges) elected by the General Assembly and Security Council to represent all principal legal systems
International Criminal Police Organization (INTERPOL) *established*—13 June 1956 *aim*—to promote international cooperation between criminal police authorities	*members*—(159) Albania, Algeria, Andorra, Angola, Antigua and Barbuda, Argentina, Aruba, Australia, Austria, The Bahamas, Bahrain, Bangladesh, Barbados, Belgium, Belize, Benin, Bolivia, Botswana, Brazil, Brunei, Bulgaria, Burkina, Burma, Burundi, Cambodia, Cameroon, Canada, Cape Verde, Central African Republic, Chad, Chile, China, Colombia, Congo, Costa Rica, Cote d'Ivoire, Cuba, Cyprus, Czech Republic, Denmark, Djibouti, Dominica, Dominican Republic, Ecuador, Egypt, Equatorial Guinea, Ethiopia, Fiji, Finland, France, Gabon, The Gambia, Germany, Ghana, Greece, Grenada, Guatemala, Guinea, Guyana, Haiti, Honduras, Hungary, Iceland, India, Indonesia, Iran, Iraq, Ireland, Israel, Italy, Jamaica, Japan, Jordan, Kenya, Kiribati, South Korea, Kuwait, Laos, Lebanon, Lesotho, Liberia, Libya, Liechtenstein, Lithuania, Luxembourg, Madagascar, Malawi, Malaysia, Maldives, Mali, Malta, Marshall Islands, Mauritania, Mauritius, Mexico, Monaco, Mongolia, Morocco, Mozambique, Nauru, Nepal, Netherlands, Netherlands Antilles, NZ, Nicaragua, Niger, Nigeria, Norway, Oman, Pakistan, Panama, Papua New Guinea, Paraguay, Peru, Philippines, Poland, Portugal, Qatar, Romania, Russia, Rwanda, Saint Kitts and Nevis, Saint Lucia, Saint Vincent and the Grenadines, Sao Tome and Principe, Saudi Arabia, Senegal, Seychelles, Sierra Leone, Singapore, Slovakia, Somalia, Spain, Sri Lanka, Sudan, Suriname, Swaziland, Sweden, Switzerland, Syria, Tanzania, Thailand, Togo, Tonga, Trinidad and Tobago, Tunisia, Turkey, Uganda, UAE, UK, US, Uruguay, Venezuela, Vietnam, Yemen, Yugoslavia, Zaire, Zambia, Zimbabwe *subbureaus*—(5) American Samoa, Bermuda, Cayman Islands, Gibraltar, Hong Kong
International Development Association (IDA) *established*—26 January 1960 *effective*—24 September 1960 *aim*—UN specialized agency and IBRD affiliate that provides economic loans for low income countries	*members*—(147); Part I—(23 more economically advanced countries) Australia, Austria, Belgium, Canada, Denmark, Finland, France, Germany, Iceland, Ireland, Italy, Japan, Kuwait, Luxembourg, Netherlands, NZ, Norway, South Africa, Sweden, Switzerland, UAE, UK, US Part II—(124 less developed nations) Afghanistan, Albania, Algeria, Angola, Argentina, Bangladesh, Belize, Benin, Bhutan, Bolivia, Botswana, Brazil, Burkina, Burma, Burundi, Cambodia, Cameroon, Cape Verde, Central African Republic, Chad, Chile, China, Colombia, Comoros, Congo, Costa Rica, Cote d'Ivoire, Cyprus, Czech Republic, Djibouti, Dominica, Dominican Republic, Ecuador, Egypt, El Salvador, Equatorial Guinea, Ethiopia, Fiji, Gabon, The Gambia, Ghana, Greece, Grenada, Guatemala, Guinea, Guinea-Bissau, Guyana, Haiti, Honduras, Hungary, India, Indonesia, Iran, Iraq, Israel, Jordan, Kazakhstan, Kenya, Kiribati, South Korea, Kyrgyzstan, Laos, Latvia, Lebanon, Lesotho, Liberia, Libya, Madagascar, Malawi, Malaysia, Maldives, Mali, Mauritania, Mauritius, Mexico, Mongolia, Morocco, Mozambique, Nepal, Nicaragua, Niger, Nigeria, Oman, Pakistan, Panama, Papua New Guinea, Paraguay, Peru, Philippines, Poland, Russia, Rwanda, Saint Kitts and Nevis, Saint Lucia, Saint Vincent and the Grenadines, Sao Tome and Principe, Saudi Arabia, Senegal, Sierra Leone, Slovakia, Solomon Islands, Somalia, Spain, Sri Lanka, Sudan, Swaziland, Syria, Tanzania, Thailand, Togo, Tonga, Trinidad and Tobago, Tunisia, Turkey, Uganda, Uzbekistan, Vanuatu, Vietnam, Western Samoa, Yemen, Yugoslavia, Zaire, Zambia, Zimbabwe
International Energy Agency (IEA) *established*—15 November 1974 *aim*—established by the OECD to promote cooperation on energy matters, especially emergency oil sharing and relations between oil consumers and oil producers	*members*—(21) Australia, Austria, Belgium, Canada, Denmark, Germany, Greece, Ireland, Italy, Japan, Luxembourg, Netherlands, NZ, Norway, Portugal, Spain, Sweden, Switzerland, Turkey, UK, US

International Finance Corporation (IFC)

established—25 May 1955

effective—20 July 1956

aim—UN specialized agency and IBRD affiliate that helps private enterprise sector in economic development

members—(149) Afghanistan, Albania, Algeria, Angola, Antigua and Barbuda, Argentina, Australia, Austria, The Bahamas, Bangladesh, Barbados, Belgium, Belize, Benin, Bolivia, Botswana, Brazil, Bulgaria, Burkina, Burma, Burundi, Cameroon, Canada, Cape Verde, Central African Republic, Chile, China, Colombia, Comoros, Congo, Costa Rica, Cote d'Ivoire, Cyprus, Czech Republic, Denmark, Djibouti, Dominica, Dominican Republic, Ecuador, Egypt, El Salvador, Equatorial Guinea, Ethiopia, Fiji, Finland, France, Gabon, The Gambia, Germany, Ghana, Greece, Grenada, Guatemala, Guinea, Guinea-Bissau, Guyana, Haiti, Honduras, Hungary, Iceland, India, Indonesia, Iran, Iraq, Ireland, Israel, Italy, Jamaica, Japan, Jordan, Kenya, Kiribati, South Korea, Kuwait, Laos, Lebanon, Lesotho, Liberia, Libya, Luxembourg, Madagascar, Malawi, Malaysia, Maldives, Mali, Marshall Islands, Mauritania, Mauritius, Mexico, Mongolia, Morocco, Mozambique, Namibia, Nepal, Netherlands, NZ, Nicaragua, Niger, Nigeria, Norway, Oman, Pakistan, Panama, Papua New Guinea, Paraguay, Peru, Philippines, Poland, Portugal, Romania, Rwanda, Saint Lucia, Saudi Arabia, Senegal, Seychelles, Sierra Leone, Singapore, Slovakia, Solomon Islands, Somalia, South Africa, Spain, Sri Lanka, Sudan, Swaziland, Sweden, Switzerland, Syria, Tanzania, Thailand, Togo, Tonga, Trinidad and Tobago, Tunisia, Turkey, Uganda, UAE, UK, US, Uruguay, Vanuatu, Venezuela, Vietnam, Western Samoa, Yemen, Yugoslavia, Zaire, Zambia, Zimbabwe

International Fund for Agricultural Development (IFAD)

established—NA November 1974

aim—UN specialized agency that promotes agricultural development

members—(147);

Category I—(21 industrialized aid contributors) Australia, Austria, Belgium, Canada, Denmark, Finland, France, Germany, Greece, Ireland, Italy, Japan, Luxembourg, Netherlands, NZ, Norway, Spain, Sweden, Switzerland, UK, US

Category II—(12 petroleum-exporting aid contributors) Algeria, Gabon, Indonesia, Iran, Iraq, Kuwait, Libya, Nigeria, Qatar, Saudi Arabia, UAE, Venezuela

Category III—(114 aid recipients) Afghanistan, Albania, Angola, Antigua and Barbuda, Argentina, Bangladesh, Barbados, Belize, Benin, Bhutan, Bolivia, Botswana, Brazil, Burkina, Burma, Burundi, Cambodia, Cameroon, Cape Verde, Central African Republic, Chad, Chile, China, Colombia, Comoros, Congo, Costa Rica, Cote d'Ivoire, Cuba, Cyprus, Djibouti, Dominica, Dominican Republic, Ecuador, Egypt, El Salvador, Equatorial Guinea, Ethiopia, Fiji, The Gambia, Ghana, Grenada, Guatemala, Guinea, Guinea-Bissau, Guyana, Haiti, Honduras, India, Israel, Jamaica, Jordan, Kenya, North Korea, South Korea, Laos, Lebanon, Lesotho, Liberia, Madagascar, Malawi, Malaysia, Maldives, Mali, Malta, Mauritania, Mauritius, Mexico, Morocco, Mozambique, Namibia, Nepal, Nicaragua, Niger, Oman, Pakistan, Panama, Papua New Guinea, Paraguay, Peru, Philippines, Portugal, Romania, Rwanda, Saint Kitts and Nevis, Saint Lucia, Saint Vincent and the Grenadines, Sao Tome and Principe, Senegal, Seychelles, Sierra Leone, Solomon Islands, Somalia, Sri Lanka, Sudan, Suriname, Swaziland, Syria, Tanzania, Thailand, Togo, Tonga, Trinidad and Tobago, Tunisia, Turkey, Uganda, Uruguay, Vietnam, Western Samoa, Yemen, Yugoslavia, Zaire, Zambia, Zimbabwe

International Investment Bank (IIB)

established on 7 July 1970; to promote economic development; members were Bulgaria, Cuba, Czechoslovakia, East Germany, Hungary, Mongolia, Poland, Romania, USSR, Vietnam; now it is a Russian bank with a new charter

International Labor Organization (ILO)

established—11 April 1919 (affiliated with the UN 14 December 1946)

aim—UN specialized agency concerned with world labor issues

members—(158) Afghanistan, Algeria, Angola, Antigua and Barbuda, Argentina, Australia, Austria, Azerbaijan, The Bahamas, Bahrain, Bangladesh, Barbados, Belarus, Belgium, Belize, Benin, Bolivia, Botswana, Brazil, Bulgaria, Burkina, Burma, Burundi, Cambodia, Cameroon, Canada, Cape Verde, Central African Republic, Chad, Chile, China, Colombia, Comoros, Congo, Costa Rica, Cote d'Ivoire, Cuba, Cyprus, Czech Republic, Denmark, Djibouti, Dominica, Dominican Republic, Ecuador, Egypt, El Salvador, Equatorial Guinea, Estonia, Ethiopia, Fiji, Finland, France, Gabon, Germany, Ghana, Greece, Grenada, Guatemala, Guinea, Guinea-Bissau, Guyana, Haiti, Honduras, Hungary, Iceland, India, Indonesia, Iran, Iraq, Ireland, Israel, Italy, Jamaica, Japan, Jordan, Kenya, South Korea, Kuwait, Kyrgyzstan, Laos, Latvia, Lebanon, Lesotho, Liberia, Libya, Lithuania, Luxembourg, Madagascar, Malawi, Malaysia, Mali, Malta, Mauritania, Mauritius, Mexico, Moldova, Mongolia, Morocco, Mozambique, Namibia, Nepal, Netherlands, NZ, Nicaragua, Niger, Nigeria, Norway, Pakistan, Panama, Papua New Guinea, Paraguay, Peru, Philippines, Poland, Portugal, Qatar, Romania, Russia, Rwanda, Saint Lucia, San Marino, Sao Tome and Principe, Saudi Arabia, Senegal, Seychelles, Sierra Leone, Singapore, Slovakia, Slovenia, Solomon Islands, Somalia, Spain, Sri Lanka, Sudan, Suriname, Swaziland, Sweden, Switzerland, Syria, Tanzania, Thailand, Togo, Trinidad and Tobago, Tunisia, Turkey, Uganda, Ukraine, UAE, UK, US, Uruguay, Venezuela, Vietnam, Yemen, Yugoslavia, Zaire, Zambia, Zimbabwe

International Maritime Organization (IMO) *note*—name changed from Intergovernmental Maritime Consultative Organization (IMCO) on 22 May 1982 *established*—17 March 1958 *aim*—UN specialized agency concerned with world maritime affairs	*members*—(138) Algeria, Angola, Antigua and Barbuda, Argentina, Australia, Austria, The Bahamas, Bahrain, Bangladesh, Barbados, Belgium, Belize, Benin, Bolivia, Brazil, Brunei, Bulgaria, Burma, Cambodia, Cameroon, Canada, Cape Verde, Chile, China, Colombia, Congo, Costa Rica, Cote d'Ivoire, Croatia, Cuba, Cyprus, Czech Republic, Denmark, Djibouti, Dominica, Dominican Republic, Ecuador, Egypt, El Salvador, Equatorial Guinea, Estonia, Ethiopia, Fiji, Finland, France, Gabon, The Gambia, Germany, Ghana, Greece, Guatemala, Guinea, Guinea-Bissau, Guyana, Haiti, Honduras, Hungary, Iceland, India, Indonesia, Iran, Iraq, Ireland, Israel, Italy, Jamaica, Japan, Jordan, Kenya, North Korea, South Korea, Kuwait, Lebanon, Liberia, Libya, Luxembourg, Madagascar, Malawi, Malaysia, Maldives, Malta, Mauritania, Mauritius, Mexico, Monaco, Morocco, Mozambique, Nepal, Netherlands, NZ, Nicaragua, Nigeria, Norway, Oman, Pakistan, Panama, Papua New Guinea, Peru, Philippines, Poland, Portugal, Qatar, Romania, Russia, Saint Lucia, Saint Vincent and the Grenadines, Sao Tome and Principe, Saudi Arabia, Senegal, Seychelles, Sierra Leone, Singapore, Slovakia, Solomon Islands, Somalia, Spain, Sri Lanka, Sudan, Suriname, Sweden, Switzerland, Syria, Tanzania, Thailand, Togo, Trinidad and Tobago, Tunisia, Turkey, UAE, UK, US, Uruguay, Vanuatu, Venezuela, Vietnam, Yemen, Yugoslavia, Zaire *associate members*—(2) Hong Kong, Macau
International Maritime Satellite Organization (INMARSAT) *established*—3 September 1976 *effective*—26 July 1979 *aim*—to provide worldwide communications for maritime and other applications	*members*—(66) Algeria, Argentina, Australia, Bahrain, Belarus, Belgium, Brazil, Bulgaria, Cameroon, Canada, Chile, China, Colombia, Cuba, Cyprus, Czech Republic, Denmark, Egypt, Finland, France, Gabon, Germany, Greece, Iceland, India, Indonesia, Iran, Iraq, Israel, Italy, Japan, South Korea, Kuwait, Liberia, Malaysia, Malta, Monaco, Mozambique, Netherlands, NZ, Nigeria, Norway, Oman, Pakistan, Panama, Peru, Philippines, Poland, Portugal, Qatar, Romania, Russia, Saudi Arabia, Singapore, Slovakia, Spain, Sri Lanka, Sweden, Switzerland, Tunisia, Turkey, Ukraine, UAE, UK, US, Yugoslavia
International Monetary Fund (IMF) *established*—22 July 1944 *effective*—27 December 1945 *aim*—UN specialized agency concerned with world monetary stability and economic development	*members*—(175) Afghanistan, Albania, Algeria, Angola, Antigua and Barbuda, Argentina, Armenia, Australia, Austria, Azerbaijan, The Bahamas, Bahrain, Bangladesh, Barbados, Belarus, Belgium, Belize, Benin, Bhutan, Bolivia, Botswana, Brazil, Bulgaria, Burkina, Burma, Burundi, Cambodia, Cameroon, Canada, Cape Verde, Central African Republic, Chad, Chile, China, Colombia, Comoros, Congo, Costa Rica, Cote d'Ivoire, Cyprus, Czech Republic, Denmark, Djibouti, Dominica, Dominican Republic, Ecuador, Egypt, El Salvador, Equatorial Guinea, Estonia, Ethiopia, Fiji, Finland, France, Gabon, The Gambia, Georgia, Germany, Ghana, Greece, Grenada, Guatemala, Guinea, Guinea-Bissau, Guyana, Haiti, Honduras, Hungary, Iceland, India, Indonesia, Iran, Iraq, Ireland, Israel, Italy, Jamaica, Japan, Jordan, Kazakhstan, Kenya, Kiribati, South Korea, Kuwait, Kyrgyzstan, Laos, Latvia, Lebanon, Lesotho, Liberia, Libya, Lithuania, Luxembourg, Macedonia, Madagascar, Malawi, Malaysia, Maldives, Mali, Malta, Marshall Islands, Mauritania, Mauritius, Mexico, Moldova, Mongolia, Morocco, Mozambique, Namibia, Nepal, Netherlands, NZ, Nicaragua, Niger, Nigeria, Norway, Oman, Pakistan, Panama, Papua New Guinea, Paraguay, Peru, Philippines, Poland, Portugal, Qatar, Romania, Russia, Rwanda, Saint Kitts and Nevis, Saint Lucia, Saint Vincent and the Grenadines, San Marino, Sao Tome and Principe, Saudi Arabia, Senegal, Seychelles, Sierra Leone, Singapore, Slovakia, Solomon Islands, Somalia, South Africa, Spain, Sri Lanka, Sudan, Suriname, Swaziland, Sweden, Switzerland, Syria, Tanzania, Thailand, Togo, Tonga, Trinidad and Tobago, Tunisia, Turkey, Turkmenistan, Uganda, Ukraine, UAE, UK, US, Uruguay, Uzbekistan, Vanuatu, Venezuela, Vietnam, Western Samoa, Yemen, Yugoslavia, Zaire, Zambia, Zimbabwe *observers*—(3) Holy See, North Korea, Monaco
International Olympic Committee (IOC) *established*—23 June 1894 *aim*—to promote the Olympic ideals and administer the Olympic games: 1992 Winter Olympics in Albertville, France (8-23 February); 1992 Summer Olympics in Barcelona, Spain (25 July-9 August); 1994 Winter Olympics in Lillehammer; Norway (12-27 February); 1996 Summer Olympics in Atlanta, United States (20 July-4 August); 1998 Winter Olympics in Nagano, Japan (date NA)	*members*—(168) Afghanistan, Albania, Algeria, American Samoa, Andorra, Angola, Antigua and Barbuda, Argentina, Aruba, Australia, Austria, The Bahamas, Bahrain, Bangladesh, Barbados, Belarus, Belgium, Belize, Benin, Bermuda, Bhutan, Bolivia, Botswana, Brazil, British Virgin Islands, Brunei, Bulgaria, Burkina, Burma, Cameroon, Canada, Cayman Islands, Central African Republic, Chad, Chile, China, Colombia, Congo, Cook Islands, Costa Rica, Cote d'Ivoire, Cuba, Cyprus, Czech Republic, Denmark, Djibouti, Dominican Republic, Ecuador, Egypt, El Salvador, Equatorial Guinea, Ethiopia, Fiji, Finland, France, Gabon, The Gambia, Germany, Ghana, Greece, Grenada, Guam, Guatemala, Guinea, Guyana, Haiti, Honduras, Hong Kong, Hungary, Iceland, India, Indonesia, Iran, Iraq, Ireland, Israel, Italy, Jamaica, Japan, Jordan, Kenya, North Korea, South Korea, Kuwait, Laos, Lebanon, Lesotho, Liberia, Libya, Liechtenstein, Luxembourg, Madagascar, Malawi, Malaysia, Maldives, Mali, Malta, Mauritania, Mauritius, Mexico, Monaco, Mongolia, Morocco, Mozambique, Nepal, Netherlands, Netherlands Antilles, NZ, Nicaragua, Niger, Nigeria, Norway, Oman, Pakistan, Panama, Papua New Guinea, Paraguay, Peru, Philippines, Poland, Portugal, Puerto Rico, Qatar, Romania, Russia, Rwanda, Saint Vincent and the Grenadines, San Marino, Saudi Arabia, Senegal, Seychelles, Sierra Leone, Singapore, Slovakia, Solomon Islands, Somalia, Spain, Sri Lanka, Sudan, Suriname, Swaziland, Sweden, Switzerland, Syria, Taiwan, Tanzania, Thailand, Togo, Tonga, Trinidad and Tobago, Tunisia, Turkey, Uganda, Ukraine, UAE, UK, US, Uruguay, Vanuatu, Venezuela, Vietnam, Virgin Islands, Western Samoa, Yemen, Yugoslavia, Zaire, Zambia, Zimbabwe

International Organization for Migration (IOM)—established as Provisional Intergovernmental Committee for the Movement of Migrants from Europe; renamed Intergovernmental Committee for European Migration (ICEM) on 15 November 1952; renamed Intergovernmental Committee for Migration (ICM) in November 1980; current name adopted 14 November 1989 *established*—5 December 1951 *aim*—to facilitate orderly international emigration and immigration	*members*—(46) Angola, Argentina, Australia, Austria, Bangladesh, Belgium, Bolivia, Canada, Chile, Colombia, Costa Rica, Cyprus, Denmark, Dominican Republic, Ecuador, Egypt, El Salvador, Finland, France, Germany, Greece, Guatemala, Honduras, Hungary, Israel, Italy, Kenya, South Korea, Luxembourg, Netherlands, Nicaragua, Norway, Panama, Paraguay, Peru, Philippines, Portugal, Sri Lanka, Sweden, Switzerland, Thailand, Uganda, US, Uruguay, Venezuela, Zambia *observers*—(41) Albania, Belize, Brazil, Bulgaria, Cape Verde, Croatia, Czech Republic, Federation of Ethnic Communities' Council of Australia Inc., Ghana, Guinea-Bissau, Holy See, India, Indonesia, Japan, Japan International Friendship and Welfare Foundation, Jordan, Latvia, Malta, Mexico, Morocco, Namibia, NZ, Niwano Peace Foundation, Pakistan, Partnership with the Children of the Third World, Poland, Presiding Bishop's Fund for World Relief/Episcopal Church Refuge Council of Australia, Romania, Russia, San Marino, Sao Tome and Principe, Senegal, Slovakia, Slovenia, Somalia, Spain, Turkey, UK, Vietnam, Yugoslavia, Zimbabwe
International Organization for Standardization (ISO) *established*—NA February 1947 *aim*—to promote the development of international standards	*members*—(73 national standards organizations) Albania, Algeria, Argentina, Australia, Austria, Bangladesh, Belgium, Brazil, Bulgaria, Canada, Chile, China, Colombia, Cote d'Ivoire, Cuba, Cyprus, Czech Republic, Denmark, Egypt, Ethiopia, Finland, France, Germany, Ghana, Greece, Hungary, India, Indonesia, Iran, Iraq, Ireland, Israel, Italy, Jamaica, Japan, Kenya, North Korea, South Korea, Malaysia, Mexico, Mongolia, Morocco, Netherlands, NZ, Nigeria, Norway, Pakistan, Papua New Guinea, Peru, Philippines, Poland, Portugal, Russia, Saudi Arabia, Singapore, Slovakia, South Africa, Spain, Sri Lanka, Sudan, Sweden, Switzerland, Syria, Tanzania, Thailand, Trinidad and Tobago, Tunisia, Turkey, UK, US, Venezuela, Vietnam, Yugoslavia *correspondent members*—(14) Bahrain, Barbados, Brunei, Guinea, Hong Kong, Iceland, Jordan, Kuwait, Malawi, Mauritius, Oman, Senegal, UAE, Uruguay
International Red Cross and Red Crescent Movement *established*—NA 1928 *aim*—to promote worldwide humanitarian aid through the International Committee of the Red Cross (ICRC) in wartime, and League of Red Cross and Red Crescent Societies (LORCS) in peacetime	*members*—(9) 2 representatives from ICRC, 2 from LORCS, and 5 from national societies elected by the international conference of the International Red Cross and Red Crescent Movement
International Telecommunication Union (ITU) *established*—9 December 1932 *effective*—1 January 1934 *affiliated with the UN*—15 November 1947 *aim*—UN specialized agency concerned with world telecommunications	*members*—(168) Afghanistan, Albania, Algeria, Angola, Antigua and Barbuda, Argentina, Australia, Austria, Azerbaijan, The Bahamas, Bahrain, Bangladesh, Barbados, Belarus, Belgium, Belize, Benin, Bhutan, Bolivia, Botswana, Brazil, Brunei, Bulgaria, Burkina, Burma, Burundi, Cambodia, Cameroon, Canada, Cape Verde, Central African Republic, Chad, Chile, China, Colombia, Comoros, Congo, Costa Rica, Cote d'Ivoire, Cuba, Cyprus, Czech Republic, Denmark, Djibouti, Dominican Republic, Ecuador, Egypt, El Salvador, Equatorial Guinea, Ethiopia, Fiji, Finland, France, Gabon, The Gambia, Germany, Ghana, Greece, Grenada, Guatemala, Guinea, Guinea-Bissau, Guyana, Haiti, Holy See, Honduras, Hungary, Iceland, India, Indonesia, Iran, Iraq, Ireland, Israel, Italy, Jamaica, Japan, Jordan, Kenya, Kiribati, North Korea, South Korea, Kuwait, Laos, Latvia, Lebanon, Lesotho, Liberia, Libya, Liechtenstein, Lithuania, Luxembourg, Madagascar, Malawi, Malaysia, Maldives, Mali, Malta, Mauritania, Mauritius, Mexico, Monaco, Mongolia, Morocco, Mozambique, Namibia, Nauru, Nepal, Netherlands, NZ, Nicaragua, Niger, Nigeria, Norway, Oman, Pakistan, Panama, Papua New Guinea, Paraguay, Peru, Philippines, Poland, Portugal, Qatar, Romania, Russia, Rwanda, Saint Vincent and the Grenadines, San Marino, Sao Tome and Principe, Saudi Arabia, Senegal, Sierra Leone, Singapore, Slovakia, Solomon Islands, Somalia, South Africa (suspended), Spain, Sri Lanka, Sudan, Suriname, Swaziland, Sweden, Switzerland, Syria, Tanzania, Thailand, Togo, Tonga, Trinidad and Tobago, Tunisia, Turkey, Uganda, Ukraine, UAE, UK, US, Uruguay, Vanuatu, Venezuela, Vietnam, Western Samoa, Yemen, Yugoslavia, Zaire, Zambia, Zimbabwe
International Telecommunications Satellite Organization (INTELSAT) *established*—20 August 1971 *effective*—12 February 1973 *aim*—to develop and operate a global commercial telecommunications satellite system	*members*—(125) Afghanistan, Algeria, Angola, Argentina, Australia, Austria, Azerbaijan, The Bahamas, Bangladesh, Barbados, Belgium, Benin, Bhutan, Bolivia, Brazil, Burkina, Cameroon, Canada, Cape Verde, Central African Republic, Chad, Chile, China, Colombia, Congo, Costa Rica, Cote d'Ivoire, Cyprus, Czech Republic, Denmark, Dominican Republic, Ecuador, Egypt, El Salvador, Ethiopia, Fiji, Finland, France, Gabon, Germany, Ghana, Greece, Guatemala, Guinea, Haiti, Holy See, Honduras, Iceland, India, Indonesia, Iran, Iraq, Ireland, Israel, Italy, Jamaica, Japan, Jordan, Kenya, South Korea, Kuwait, Lebanon, Libya, Liechtenstein, Luxembourg, Madagascar, Malawi, Malaysia, Mali, Mauritania, Mauritius, Mexico, Monaco, Morocco, Mozambique, Nepal, Netherlands, NZ, Nicaragua, Niger, Nigeria, Norway, Oman, Pakistan, Panama, Papua New Guinea, Paraguay, Peru, Philippines, Portugal, Qatar, Romania, Russia, Rwanda, Saudi Arabia, Senegal, Singapore, Slovakia, Somalia, South Africa, Spain, Sri Lanka, Sudan, Swaziland, Sweden, Switzerland, Syria, Tanzania, Thailand, Togo, Trinidad and Tobago, Tunisia, Turkey, Uganda, UAE, UK, US, Uruguay, Venezuela, Vietnam, Yemen, Yugoslavia, Zaire, Zambia, Zimbabwe

Islamic Development Bank (IDB) *established*—15 December 1973 *aim*—to promote Islamic economic aid and social development	*members*—(44 plus the Palestine Liberation Organization) Afghanistan (suspended), Algeria, Azerbaijan, Bahrain, Bangladesh, Benin, Brunei, Burkina, Cameroon, Chad, Comoros, Djibouti, Egypt, Gabon, The Gambia, Guinea, Guinea-Bissau, Indonesia, Iran, Iraq, Jordan, Kuwait, Lebanon, Libya, Malaysia, Maldives, Mali, Mauritania, Morocco, Niger, Oman, Pakistan, Qatar, Saudi Arabia, Senegal, Sierra Leone, Somalia, Sudan, Syria, Tunisia, Turkey, Uganda, UAE, Yemen, Palestine Liberation Organization
Latin American Economic System (LAES) *note*—also known as Sistema Economico Latinoamericana (SELA) *established*—17 October 1975 *aim*—to promote economic and social development through regional cooperation	*members*—(26) Argentina, Barbados, Bolivia, Brazil, Chile, Colombia, Costa Rica, Cuba, Dominican Republic, Ecuador, El Salvador, Grenada, Guatemala, Guyana, Haiti, Honduras, Jamaica, Mexico, Nicaragua, Panama, Paraguay, Peru, Suriname, Trinidad and Tobago, Uruguay, Venezuela
Latin American Integration Association (LAIA) *note*—also known as Asociacion Latinoamericana de Integracion (ALADI) *established*—12 August 1980 *effective*—18 March 1981 *aim*—to promote freer regional trade	*members*—(11) Argentina, Bolivia, Brazil, Chile, Colombia, Ecuador, Mexico, Paraguay, Peru, Uruguay, Venezuela *observers*—(16) Commission of the European Communities, Costa Rica, Cuba, Dominican Republic, El Salvador, Guatemala, Honduras, Inter-American Development Bank, Italy, Nicaragua, Organization of American States, Panama, Portugal, Spain, United Nations Development Program, United Nations Economic Commission for Latin America and the Caribbean
League of Arab States (LAS)	see Arab League (AL)
League of Red Cross and Red Crescent Societies (LORCS) *established*—5 May 1919 *aim*—to provide humanitarian aid in peacetime	*members*—(148) Afghanistan, Albania, Algeria, Angola, Argentina, Australia, Austria, The Bahamas, Bahrain, Bangladesh, Barbados, Belgium, Belize, Benin, Bolivia, Botswana, Brazil, Bulgaria, Burkina, Burma, Burundi, Cambodia, Cameroon, Canada, Cape Verde, Central African Republic, Chad, Chile, China, Colombia, Congo, Costa Rica, Cote d'Ivoire, Cuba, Czech Republic, Denmark, Djibouti, Dominica, Dominican Republic, Ecuador, Egypt, El Salvador, Ethiopia, Fiji, Finland, France, The Gambia, Germany, Ghana, Greece, Grenada, Guatemala, Guinea, Guinea-Bissau, Guyana, Haiti, Honduras, Hungary, Iceland, India, Indonesia, Iran, Iraq, Ireland, Italy, Jamaica, Japan, Jordan, Kenya, North Korea, South Korea, Kuwait, Laos, Lebanon, Lesotho, Liberia, Libya, Liechtenstein, Luxembourg, Madagascar, Malawi, Malaysia, Mali, Mauritania, Mauritius, Mexico, Monaco, Mongolia, Morocco, Mozambique, Nepal, Netherlands, NZ, Nicaragua, Niger, Nigeria, Norway, Pakistan, Panama, Papua New Guinea, Paraguay, Peru, Philippines, Poland, Portugal, Qatar, Romania, Russia, Rwanda, Saint Lucia, Saint Vincent and the Grenadines, San Marino, Sao Tome and Principe, Saudi Arabia, Senegal, Sierra Leone, Singapore, Slovakia, Somalia, South Africa, Spain, Sri Lanka, Sudan, Suriname, Swaziland, Sweden, Switzerland, Syria, Tanzania, Thailand, Togo, Tonga, Trinidad and Tobago, Tunisia, Turkey, Uganda, UAE, UK, US, Uruguay, Venezuela, Vietnam, Western Samoa, Yemen, Yugoslavia, Zaire, Zambia, Zimbabwe *associate members*—(2) Equatorial Guinea, Gabon
least developed countries (LLDCs)	that subgroup of the less developed countries (LDCs) initially identified by the UN General Assembly in 1971 as having no significant economic growth, per capita GNPs/GDPs normally less than $500, and low literacy rates; also known as the undeveloped countries. The 42 LLDCs are: Afghanistan, Bangladesh, Benin, Bhutan, Botswana, Burkina, Burma, Burundi, Cape Verde, Central African Republic, Chad, Comoros, Djibouti, Equatorial Guinea, Eritrea, Ethiopia, The Gambia, Guinea, Guinea-Bissau, Haiti, Kiribati, Laos, Lesotho, Malawi, Maldives, Mali, Mauritania, Mozambique, Nepal, Niger, Rwanda, Sao Tome and Principe, Sierra Leone, Somalia, Sudan, Tanzania, Togo, Tuvalu, Uganda, Vanuatu, Western Samoa, Yemen

less developed countries (LDCs)	the bottom group in the comprehensive but mutually exclusive hierarchy of developed countries (DCs), former USSR/Eastern Europe (former USSR/EE), and less developed countries (LDCs); mainly countries with low levels of output, living standards, and technology; per capita GNPs/GDPs are generally below $5,000 and often less than $1,000; however, the group also includes a number of countries with high per capita incomes, areas of advanced technology, and rapid rates of growth; includes the advanced developing countries, developing countries, Four Dragons (Four Tigers), least developed countries (LLDCs), low-income countries, middle-income countries, newly industrializing economies (NIEs), the South, Third World, underdeveloped countries, undeveloped countries; the 175 LDCs are: Afghanistan, Algeria, American Samoa, Angola, Anguilla, Antigua and Barbuda, Argentina, Aruba, The Bahamas, Bahrain, Bangladesh, Barbados, Belize, Benin, Bhutan, Bolivia, Botswana, Brazil, British Virgin Islands, Brunei, Burkina, Burma, Burundi, Cambodia, Cameroon, Cape Verde, Cayman Islands, Central African Republic, Chad, Chile, China, Christmas Island, Cocos Islands, Colombia, Comoros, Congo, Cook Islands, Costa Rica, Cote d'Ivoire, Cuba, Cyprus, Djibouti, Dominica, Dominican Republic, Ecuador, Egypt, El Salvador, Equatorial Guinea, Eritrea, Ethiopia, Falkland Islands, Fiji, French Guiana, French Polynesia, Gabon, The Gambia, Gaza Strip, Ghana, Gibraltar, Greenland, Grenada, Guadeloupe, Guam, Guatemala, Guernsey, Guinea, Guinea-Bissau, Guyana, Haiti, Honduras, Hong Kong, India, Indonesia, Iran, Iraq, Jamaica, Jersey, Jordan, Kenya, Kiribati, North Korea, South Korea, Kuwait, Laos, Lebanon, Lesotho, Liberia, Libya, Macau, Madagascar, Malawi, Malaysia, Maldives, Mali, Isle of Man, Marshall Islands, Martinique, Mauritania, Mauritius, Mayotte, Mexico, Federated States of Micronesia, Mongolia, Montserrat, Morocco, Mozambique, Namibia, Nauru, Nepal, Netherlands Antilles, New Caledonia, Nicaragua, Niger, Nigeria, Niue, Norfolk Island, Northern Mariana Islands, Oman, Trust Territory of the Pacific Islands (Palau), Pakistan, Panama, Papua New Guinea, Paraguay, Peru, Philippines, Pitcairn Islands, Puerto Rico, Qatar, Reunion, Rwanda, Saint Helena, Saint Kitts and Nevis, Saint Lucia, Saint Pierre and Miquelon, Saint Vincent and the Grenadines, Sao Tome and Principe, Saudi Arabia, Senegal, Seychelles, Sierra Leone, Singapore, Solomon Islands, Somalia, Sri Lanka, Sudan, Suriname, Swaziland, Syria, Taiwan, Tanzania, Thailand, Togo, Tokelau, Tonga, Trinidad and Tobago, Tunisia, Turks and Caicos Islands, Tuvalu, UAE, Uganda, Uruguay, Vanuatu, Venezuela, Vietnam, Virgin Islands, Wallis and Futuna, West Bank, Western Sahara, Western Samoa, Yemen, Zaire, Zambia, Zimbabwe
low-income countries	another term for those less developed countries with below-average per capita GNPs/GDPs; see less developed countries (LDCs)
London Suppliers Group	see Nuclear Suppliers Group (NSG)
Mercado Comun del Cono Sur (MERCOSUR)	see Southern Cone Common Market
middle-income countries	another term for those less developed countries with above-average per capita GNPs/GDPs; see less developed countries (LDCs)
Missile Technology Control Regime (MTCR) *established*—April 1987 *aim*—to arrest missile proliferation by controlling the export of key missile technologies and equipment	*members*—(24) Australia, Austria, Belgium, Canada, Denmark, Finland, France, Germany, Greece, Hungary, Iceland, Ireland, Italy, Japan, Luxembourg, Netherlands, NZ, Norway, Portugal, Spain, Sweden, Switzerland, UK, US
newly industrializing countries (NICs)	former term for the newly industrializing economies; see newly industrializing economies (NIEs)
newly industrializing economies (NIEs)	that subgroup of the less developed countries (LDCs) that has experienced particularly rapid industrialization of their economies; formerly known as the newly industrializing countries (NICs); also known as advanced developing countries; usually includes the Four Dragons (Hong Kong, South Korea, Singapore, Taiwan) plus Brazil and Mexico

Nonaligned Movement (NAM)

established—1-6 September 1961

aim—to establish political and military cooperation apart from the traditional East or West blocs

members—(102 plus the Palestine Liberation Organization) Afghanistan, Algeria, Angola, The Bahamas, Bahrain, Bangladesh, Barbados, Belize, Benin, Bhutan, Bolivia, Botswana, Burkina, Burundi, Cambodia, Cameroon, Cape Verde, Central African Republic, Chad, Colombia, Comoros, Congo, Cote d'Ivoire, Cuba, Cyprus, Djibouti, Ecuador, Egypt, Equatorial Guinea, Ethiopia, Gabon, The Gambia, Ghana, Grenada, Guatemala, Guinea, Guinea-Bissau, Guyana, India, Indonesia, Iran, Iraq, Jamaica, Jordan, Kenya, North Korea, Kuwait, Laos, Lebanon, Lesotho, Liberia, Libya, Madagascar, Malawi, Malaysia, Maldives, Mali, Malta, Mauritania, Mauritius, Mongolia, Morocco, Mozambique, Namibia, Nepal, Nicaragua, Niger, Nigeria, Oman, Pakistan, Panama, Papua New Guinea, Peru, Qatar, Rwanda, Saint Lucia, Sao Tome and Principe, Saudi Arabia, Senegal, Seychelles, Sierra Leone, Singapore, Somalia, Sri Lanka, Sudan, Suriname, Swaziland, Syria, Tanzania, Togo, Trinidad and Tobago, Tunisia, Uganda, UAE, Vanuatu, Venezuela, Vietnam, Yemen, Yugoslavia, Zaire, Zambia, Zimbabwe, Palestine Liberation Organization

observers—(19) African National Congress, Afro-Asian Solidarity Organization, Antigua and Barbuda, Arab League, Brazil, China, Costa Rica, Dominica, El Salvador, Islamic Conference, Kanaka Socialist National Liberation Front (New Caledonia), Mexico, Mongolia, Organization of African Unity, Pan Africanist Congress of Azania, Philippines, Socialist Party of Puerto Rico, UN, Uruguay

guests—(21) Australia, Austria, Bulgaria, Canada, Czech Republic, Dominican Republic, Finland, Germany, Greece, Hungary, Netherlands, NZ, Norway, Poland, Portugal, Romania, San Marino, Slovakia, Spain, Sweden, Switzerland

---

Nordic Council (NC)

established—16 March 1952

effective—12 February 1953

aim—to promote regional economic, cultural, and environmental cooperation

members—(5) Denmark, Finland, Iceland, Norway, Sweden; note—Denmark includes Faroe Islands and Greenland

---

Nordic Investment Bank (NIB)

established—4 December 1975

effective—1 June 1976

aim—to promote economic cooperation and development

members—(5) Denmark, Finland, Iceland, Norway, Sweden

North

a popular term for the rich industrialized countries generally located in the northern portion of the Northern Hemisphere; the counterpart of the South; see developed countries (DCs)

North Atlantic Cooperation Council (NACC)—an extension of NATO

established—8 November 1991

effective—20 December 1991

aim—to form a forum to discuss cooperation concerning mutual political and security issues

members—(38) Albania, Armenia, Azerbaijan, Belarus, Belgium, Bulgaria, Canada, Czech Republic, Denmark, Estonia, France, Georgia, Germany, Greece, Hungary, Iceland, Italy, Kyrgyzstan, Latvia, Lithuania, Luxembourg, Moldova, Netherlands, Norway, Poland, Portugal, Romania, Russia, Slovakia, Spain, Tajikistan, Turkey, Turkmenistan, Ukraine, UK, US, Uzbekistan, Yugoslavia

---

North Atlantic Treaty Organization (NATO)

established—17 September 1949

aim—to promote mutual defense and cooperation

members—(16) Belgium, Canada, Denmark, France, Germany, Greece, Iceland, Italy, Luxembourg, Netherlands, Norway, Portugal, Spain, Turkey, UK, US

Nuclear Energy Agency (NEA)

established—NA 1958

aim—associated with OECD, seeks to promote the peaceful uses of nuclear energy

members—(23) Australia, Austria, Belgium, Canada, Denmark, Finland, France, Germany, Greece, Iceland, Ireland, Italy, Japan, Luxembourg, Netherlands, Norway, Portugal, Spain, Sweden, Switzerland, Turkey, UK, US

Nuclear Suppliers Group (NSG) *note*—also known as the London Suppliers Group *established*—1974 *aim*—to establish guidelines on exports of enrichment and processing plant assistance and nuclear exports to countries of proliferation concern and regions of conflict and instability	*members*—(28) Australia, Austria, Belgium, Bulgaria, Canada, Czech Republic, Denmark, Finland, France, Germany, Greece, Hungary, Ireland, Italy, Japan, Luxembourg, Netherlands, Norway, Poland, Portugal, Romania, Russia, Slovakia, Spain, Sweden, Switzerland, UK, US
Organismo para la Proscripcion de las Armas Nucleares en la America Latina y el Caribe (OPANAL)	see Agency for the Prohibition of Nuclear Weapons in Latin America and the Caribbean (OPANAL)
Organization for Economic Cooperation and Development (OECD) *established*—14 December 1960, effective 30 September 1961 *aim*—to promote economic cooperation and development	*members*—(24) Australia, Austria, Belgium, Canada, Denmark, Finland, France, Germany, Greece, Iceland, Ireland, Italy, Japan, Luxembourg, Netherlands, NZ, Norway, Portugal, Spain, Sweden, Switzerland, Turkey, UK, US *special members*—(2) EC, Yugoslavia
Organization of African Unity (OAU) *established*—25 May 1963 *aim*—to promote unity and cooperation among African states	*members*—(52) Algeria, Angola, Benin, Botswana, Burkina, Burundi, Cameroon, Cape Verde, Central African Republic, Chad, Comoros, Congo, Cote d'Ivoire, Djibouti, Egypt, Equatorial Guinea, Eritrea, Ethiopia, Gabon, The Gambia, Ghana, Guinea, Guinea-Bissau, Kenya, Lesotho, Liberia, Libya, Madagascar, Malawi, Mali, Mauritania, Mauritius, Mozambique, Namibia, Niger, Nigeria, Rwanda, Sahrawi Arab Democratic Republic, Sao Tome and Principe, Senegal, Seychelles, Sierra Leone, Somalia, Sudan, Swaziland, Tanzania, Togo, Tunisia, Uganda, Zaire, Zambia, Zimbabwe
Organization of American States (OAS) *established*—30 April 1948 *effective*—13 December 1951 *aim*—to promote peace and security as well as economic and social development	*members*—(35) Antigua and Barbuda, Argentina, The Bahamas, Barbados, Belize, Bolivia, Brazil, Canada, Chile, Colombia, Costa Rica, Cuba (excluded from formal participation since 1962), Dominica, Dominican Republic, Ecuador, El Salvador, Grenada, Guatemala, Guyana, Haiti, Honduras, Jamaica, Mexico, Nicaragua, Panama, Paraguay, Peru, Saint Kitts and Nevis, Saint Lucia, Saint Vincent and the Grenadines, Suriname, Trinidad and Tobago, US, Uruguay, Venezuela *observers*—(24) Algeria, Austria, Belgium, Cyprus, EC, Egypt, Equatorial Guinea, Finland, France, Germany, Greece, Holy See, Israel, Italy, Japan, South Korea, Morocco, Netherlands, Pakistan, Portugal, Russia, Saudi Arabia, Spain, Switzerland,
Organization of Arab Petroleum Exporting Countries (OAPEC) *established*—9 January 1968 *aim*—to promote cooperation in the petroleum industry	*members*—(11) Algeria, Bahrain, Egypt, Iraq, Kuwait, Libya, Qatar, Saudi Arabia, Syria, Tunisia (withdrew from active membership in 1986), UAE
Organization of Eastern Caribbean States (OECS) *established*—18 June 1981 *effective*—4 July 1981 *aim*—to promote political, economic, and defense cooperation	*members*—(7) Antigua and Barbuda, Dominica, Grenada, Montserrat, Saint Kitts and Nevis, Saint Lucia, Saint Vincent and the Grenadines *associate member*—(1) British Virgin Islands
Organization of Petroleum Exporting Countries (OPEC) *established*—14 September 1960 *aim*—to coordinate petroleum policies	*members*—(12) Algeria, Gabon, Indonesia, Iran, Iraq, Kuwait, Libya, Nigeria, Qatar, Saudi Arabia, UAE, Venezuela
Organization of the Islamic Conference (OIC) *established*—22-25 September 1969 *aim*—to promote Islamic solidarity and cooperation in economic, social, cultural, and political affairs	*members*—(47 plus the Palestine Liberation Organization) Afghanistan (suspended), Albania, Algeria, Azerbaijan, Bahrain, Bangladesh, Benin, Brunei, Burkina, Cameroon, Chad, Comoros, Djibouti, Egypt, Gabon, The Gambia, Guinea, Guinea-Bissau, Indonesia, Iran, Iraq, Jordan, Kazakhstan, Kuwait, Lebanon, Libya, Malaysia, Maldives, Mali, Mauritania, Morocco, Niger, Nigeria, Oman, Pakistan, Qatar, Saudi Arabia, Senegal, Sierra Leone, Somalia, Sudan, Syria, Tunisia, Turkey, Uganda, UAE, Yemen, Palestine Liberation Organization *observer*—(1) Turkish-Cypriot administered area of Cyprus

Paris Club	see Group of 10
Permanent Court of Arbitration (PCA) *established*—NA 1899 *aim*—to facilitate the settlement of international disputes	*members*—(78) Argentina, Australia, Austria, Belarus, Belgium, Bolivia, Brazil, Bulgaria, Burkina, Cambodia, Cameroon, Canada, Chile, China, Colombia, Cuba, Czech Republic, Denmark, Dominican Republic, Ecuador, Egypt, El Salvador, Fiji, Finland, France, Germany, Greece, Guatemala, Haiti, Honduras, Hungary, Iceland, India, Iran, Iraq, Israel, Italy, Japan, Jordan, Kyrgyzstan, Laos, Lebanon, Luxembourg, Malta, Mauritius, Mexico, Netherlands, NZ, Nicaragua, Nigeria, Norway, Pakistan, Panama, Paraguay, Peru, Poland, Portugal, Romania, Russia, Senegal, Slovakia, Spain, Sri Lanka, Sudan, Swaziland, Sweden, Switzerland, Thailand, Turkey, Uganda, Ukraine, UK, US, Uruguay, Venezuela, Yugoslavia, Zaire, Zimbabwe
Population Commission *established*—3 October 1946 *aim*—Economic and Social Council organization dealing with population matters of importance to the UN	*members*—(27) selected on a rotating basis from all regions
Rio Group (RG) *established*—NA 1988 *aim*—a consultation mechanism on regional Latin American issues	*members*—(11) Argentina, Bolivia, Brazil, Chile, Colombia, Ecuador, Mexico, Paraguay, Peru (suspended), Uruguay, Venezuela; note—Panama was expelled in 1988; Peru was suspended after April 1992 coup
Second World	another term for the traditionally Marxist-Leninist states with authoritarian governments and command economies based on the Soviet model; the term is fading from use; see centrally planned economies
socialist countries	in general, countries in which the government owns and plans the use of the major factors of production; note—the term is sometimes used incorrectly as a synonym for Communist countries
South	a popular term for the poorer, less industrialized countries generally located south of the developed countries; the counterpart of the North; see less developed countries (LDCs)
South Asian Association for Regional Cooperation (SAARC) *established*—8 December 1985 *aim*—to promote economic, social, and cultural cooperation	*members*—(7) Bangladesh, Bhutan, India, Maldives, Nepal, Pakistan, Sri Lanka
South Pacific Commission (SPC) *established*—6 February 1947 *effective*—29 July 1948 *aim*—to promote regional cooperation in economic and social matters	*members*—(27) American Samoa, Australia, Cook Islands, Fiji, France, French Polynesia, Guam, Kiribati, Marshall Islands, Federated States of Micronesia, Nauru, New Caledonia, NZ, Niue, Northern Mariana Islands, Trust Territory of the Pacific Islands (Palau), Papua New Guinea, Pitcairn Islands, Solomon Islands, Tokelau, Tonga, Tuvalu, UK, US, Vanuatu, Wallis and Futuna, Western Samoa
South Pacific Forum (SPF) *established*—5 August 1971 *aim*—to promote regional cooperation in political matters	*members*—(15) Australia, Cook Islands, Fiji, Kiribati, Marshall Islands, Federated States of Micronesia, Nauru, NZ, Niue, Papua New Guinea, Solomon Islands, Tonga, Tuvalu, Vanuatu, Western Samoa *observer*—(1) Trust Territory of the Pacific Islands (Palau)
South Pacific Regional Trade and Economic Cooperation Agreement (SPARTECA) *established*—NA 1981 *aim*—to redress unequal trade relationship of Australia and New Zealand with small island economies in Pacific region	*members*—(15) Australia, Cook Islands, Fiji, Kiribati, Marshall Islands, Federated States of Micronesia, Nauru, NZ, Niue, Papua New Guinea, Solomon Islands, Tonga, Tuvalu, Vanuatu, Western Samoa
Southern African Customs Union (SACU) *established*—11 December 1969 *aim*—to promote free trade and cooperation in customs matters	*members*—(9) Bophuthatswana, Botswana, Ciskei, Lesotho, Namibia, South Africa, Swaziland, Transkei, Venda

Southern African Development Community (SADC)	*members*—(10) Angola, Botswana, Lesotho, Malawi, Mozambique, Namibia, Swaziland, Tanzania, Zambia, Zimbabwe
note—evolved from the Southern African Development Coordination Conference (SADCC)	
established—17 August 1992	
aim—to promote regional economic development and integration	
Southern Cone Common Market (MERCOSUR)	*members*—(4) Argentina, Brazil, Paraguay, Uruguay
established—26 March 1991	
aim—regional economic cooperation	
Statistical Commission	*members*—(25) selected on a rotating basis from all regions
established—21 June 1946	
aim—Economic and Social Council organization dealing with development and standardization of national statistics of interest to the UN	
Third World	another term for the less developed countries; the term is fading from use; see less developed countries (LDCs)
underdeveloped countries	refers to those less developed countries with the potential for above-average economic growth; see less developed countries (LDCs)
undeveloped countries	refers to those extremely poor less developed countries (LDCs) with little prospect for economic growth; see least developed countries (LLDCs)
Union Douaniere et Economique de l'Afrique Centrale (UDEAC)	see Central African Customs and Economic Union (UDEAC)
United Nations (UN)	*members*—(182 excluding Yugoslavia) Afghanistan, Albania, Algeria, Angola, Antigua and Barbuda, Argentina, Armenia, Australia, Austria, Azerbaijan, The Bahamas, Bahrain, Bangladesh, Barbados, Belarus, Belgium, Belize, Benin, Bhutan, Bolivia, Bosnia and Herzegovina, Botswana, Brazil, Brunei, Bulgaria, Burkina, Burma, Burundi, Cambodia, Cameroon, Canada, Cape Verde, Central African Republic, Chad, Chile, China, Colombia, Comoros, Congo, Costa Rica, Cote d'Ivoire, Croatia, Cuba, Cyprus, Czech Republic, Denmark, Djibouti, Dominica, Dominican Republic, Ecuador, Egypt, El Salvador, Equatorial Guinea, Eritrea, Estonia, Ethiopia, Fiji, Finland, France, Gabon, The Gambia, Georgia, Germany, Ghana, Greece, Grenada, Guatemala, Guinea, Guinea-Bissau, Guyana, Haiti, Honduras, Hungary, Iceland, India, Indonesia, Iran, Iraq, Ireland, Israel, Italy, Jamaica, Japan, Jordan, Kazakhstan, Kenya, North Korea, South Korea, Kuwait, Kyrgyzstan, Laos, Latvia, Lebanon, Lesotho, Liberia, Libya, Liechtenstein, Lithuania, Luxembourg, Macedonia, Madagascar, Malawi, Malaysia, Maldives, Mali, Malta, Marshall Islands, Mauritania, Mauritius, Mexico, Federated States of Micronesia, Moldova, Monaco, Mongolia, Morocco, Mozambique, Namibia, Nepal, Netherlands, NZ, Nicaragua, Niger, Nigeria, Norway, Oman, Pakistan, Panama, Papua New Guinea, Paraguay, Peru, Philippines, Poland, Portugal, Qatar, Romania, Russia, Rwanda, Saint Kitts and Nevis, Saint Lucia, Saint Vincent and the Grenadines, San Marino, Sao Tome and Principe, Saudi Arabia, Senegal, Seychelles, Sierra Leone, Singapore, Slovakia, Slovenia, Solomon Islands, Somalia, South Africa, Spain, Sri Lanka, Sudan, Suriname, Swaziland, Sweden, Syria, Tajikistan, Tanzania, Thailand, Togo, Trinidad and Tobago, Tunisia, Turkey, Turkmenistan, Uganda, Ukraine, UAE, UK, US, Uruguay, Uzbekistan, Vanuatu, Venezuela, Vietnam, Western Samoa, Yemen, Yugoslavia, Zaire, Zambia, Zimbabwe; note—all UN members are represented in the General Assembly
established—26 June 1945	
effective—24 October 1945	
aim—to maintain international peace and security and to promote cooperation involving economic, social, cultural and humanitarian problems	
	observers—(2 and the Palestine Liberation Organization) Holy See, Switzerland, Palestine Liberation Organization
United Nations Angola Verification Mission (UNAVEM II)	*members*—(25) Algeria, Argentina, Brazil, Canada, Congo, Czech Republic, Egypt, Guinea-Bissau, Hungary, India, Ireland, Jordan, Malaysia, Morocco, Netherlands, NZ, Nigeria, Norway, Senegal, Singapore, Slovakia, Spain, Sweden, Yugoslavia, Zimbabwe
note—successor to original UNAVEM	
established—20 December 1988	
aim—established by the UN Security Council to verify the withdrawal of Cuban troops from Angola	

United Nations Children's Fund (UNICEF)	*members*—(41) selected on a rotating basis from all regions
note—acronym retained from the predecessor organization UN International Children's Emergency Fund	
established—11 December 1946	
aim—to help establish child health and welfare services	
United Nations Conference on Trade and Development (UNCTAD)	*members*—(186) all UN members plus Holy See, Switzerland, Tonga
established—30 December 1964	
aim—to promote international trade	
United Nations Development Program (UNDP)	*members*—(48) selected on a rotating basis from all regions
established—22 November 1965	
aim—to provide technical assistance to stimulate economic and social development	
United Nations Disengagement Observer Force (UNDOF)	*members*—(4) Austria, Canada, Finland, Poland
established—31 May 1974	
aim—established by the UN Security Council to observe the 1973 Arab-Israeli ceasefire	
United Nations Educational, Scientific, and Cultural Organization (UNESCO)	*members*—(172) Afghanistan, Albania, Algeria, Angola, Antigua and Barbuda, Argentina, Armenia, Australia, Austria, Azerbaijan, The Bahamas, Bahrain, Bangladesh, Barbados, Belarus, Belgium, Belize, Benin, Bhutan, Bolivia, Botswana, Brazil, Bulgaria, Burkina, Burma, Burundi, Cambodia, Cameroon, Canada, Cape Verde, Central African Republic, Chad, Chile, China, Colombia, Comoros, Congo, Cook Islands, Costa Rica, Cote d'Ivoire, Croatia, Cuba, Cyprus, Czech Republic, Denmark, Djibouti, Dominica, Dominican Republic, Ecuador, Egypt, El Salvador, Equatorial Guinea, Estonia, Ethiopia, Fiji, Finland, France, Gabon, The Gambia, Georgia, Germany, Ghana, Greece, Grenada, Guatemala, Guinea, Guinea-Bissau, Guyana, Haiti, Honduras, Hungary, Iceland, India, Indonesia, Iran, Iraq, Ireland, Israel, Italy, Jamaica, Japan, Jordan, Kazakhstan, Kenya, Kiribati, North Korea, South Korea, Kuwait, Kyrgyzstan, Laos, Latvia, Lebanon, Lesotho, Liberia, Libya, Lithuania, Luxembourg, Madagascar, Malawi, Malaysia, Maldives, Mali, Malta, Mauritania, Mauritius, Mexico, Moldova, Monaco, Mongolia, Morocco, Mozambique, Namibia, Nepal, Netherlands, NZ, Nicaragua, Niger, Nigeria, Norway, Oman, Pakistan, Panama, Papua New Guinea, Paraguay, Peru, Philippines, Poland, Portugal, Qatar, Romania, Russia, Rwanda, Saint Kitts and Nevis, Saint Lucia, Saint Vincent and the Grenadines, San Marino, Sao Tome and Principe, Saudi Arabia, Senegal, Seychelles, Sierra Leone, Slovakia, Slovenia, Somalia, Spain, Sri Lanka, Sudan, Suriname, Swaziland, Sweden, Switzerland, Syria, Tanzania, Thailand, Togo, Tonga, Trinidad and Tobago, Tunisia, Turkey, Tuvalu, Uganda, Ukraine, UAE, Uruguay, Venezuela, Vietnam, Western Samoa, Yemen, Yugoslavia, Zaire, Zambia, Zimbabwe
established—16 November 1945	
effective—4 November 1946	
aim—to promote cooperation in education, science, and culture	
	associate members—(3) Aruba, British Virgin Islands, Netherlands Antilles
United Nations Environment Program (UNEP)	*members*—(58) selected on a rotating basis from all regions
established—15 December 1972	
aim—to promote international cooperation on all environmental matters	
United Nations Force in Cyprus (UNFICYP)	*members*—(7) Austria, Canada, Denmark, Finland, Ireland, Sweden, UK
established—4 March 1964	
aim—established by the UN Security Council to serve as a peacekeeping force beween Greek Cypriots and Turkish Cypriots in Cyprus	

United Nations General Assembly	members—(183) all UN members are represented in the General Assembly
established—26 June 1945	
effective—24 October 1945	
aim—primary deliberative organ in the UN	

United Nations Industrial Development Organization (UNIDO)	members—(160) Afghanistan, Albania, Algeria, Angola, Argentina, Armenia, Australia, Austria, The Bahamas, Bahrain, Bangladesh, Barbados, Belarus, Belgium, Belize, Benin, Bhutan, Bolivia, Botswana, Brazil, Bulgaria, Burkina, Burma, Burundi, Cameroon, Canada, Cape Verde, Central African Republic, Chad, Chile, China, Colombia, Comoros, Congo, Costa Rica, Cote d'Ivoire, Croatia, Cuba, Cyprus, Czech Republic, Denmark, Djibouti, Dominica, Dominican Republic, Ecuador, Egypt, El Salvador, Equatorial Guinea, Ethiopia, Fiji, Finland, France, Gabon, The Gambia, Germany, Ghana, Greece, Grenada, Guatemala, Guinea, Guinea-Bissau, Guyana, Haiti, Honduras, Hungary, India, Indonesia, Iran, Iraq, Ireland, Israel, Italy, Jamaica, Japan, Jordan, Kenya, North Korea, South Korea, Kuwait, Laos, Latvia, Lebanon, Lesotho, Liberia, Libya, Lithuania, Luxembourg, Madagascar, Malawi, Malaysia, Maldives, Mali, Malta, Mauritania, Mauritius, Mexico, Mongolia, Morocco, Mozambique, Namibia, Nepal, Netherlands, NZ, Nicaragua, Niger, Nigeria, Norway, Oman, Pakistan, Panama, Papua New Guinea, Paraguay, Peru, Philippines, Poland, Portugal, Qatar, Romania, Russia, Rwanda, Saint Kitts and Nevis, Saint Lucia, Saint Vincent and the Grenadines, Sao Tome and Principe, Saudi Arabia, Senegal, Seychelles, Sierra Leone, Slovakia, Slovenia, Somalia, Spain, Sri Lanka, Sudan, Suriname, Swaziland, Sweden, Switzerland, Syria, Tanzania, Thailand, Togo, Tonga, Trinidad and Tobago, Tunisia, Turkey, Uganda, Ukraine, UAE, UK, US, Uruguay, Vanuatu, Venezuela, Vietnam, Yemen, Yugoslavia, Zaire, Zambia, Zimbabwe
established—17 November 1966	
effective—1 January 1967	
aim—UN specialized agency that promotes industrial development especially among the members	

United Nations Interim Force in Lebanon (UNIFIL)	members—(10) Fiji, Finland, France, Ghana, Ireland, Italy, Nepal, Norway, Poland, Sweden
established—19 March 1978	
aim—established by the UN Security Council to confirm the withdrawal of Israeli forces, restore peace, and reestablish Lebanese authority in southern Lebanon	

United Nations Iraq-Kuwait Observation Mission (UNIKOM)	members—(34) Argentina, Austria, Bangladesh, Canada, Chile, China, Denmark, Fiji, Finland, France, Ghana, Greece, Hungary, India, Indonesia, Ireland, Italy, Kenya, Malaysia, Nigeria, Norway, Pakistan, Poland, Romania, Russia, Senegal, Singapore, Sweden, Thailand, Turkey, UK, US, Uruguay, Venezuela
established—NA 1991	
aim—established by the UN Security Council to observe and monitor the demilitarized zone established between Iraq and Kuwait	

United Nations Military Observer Group in India and Pakistan (UNMOGIP)	members—(8) Belgium, Chile, Denmark, Finland, Italy, Norway, Sweden, Uruguay
established—13 August 1948	
aim—established by the UN Security Council to observe the 1949 India-Pakistan ceasefire	

United Nations Mission for the Referendum in Western Sahara (MINURSO)	members—(25) Argentina, Australia, Austria, Bangladesh, Canada, China, Egypt, France, Ghana, Greece, Guinea, Ireland, Italy, Kenya, Malaysia, Nigeria, Pakistan, Peru, Poland, Russia, Switzerland, Tunisia, UK, US, Venezuela
established—NA 1990	
aim—established by the UN Security Council to supervise the referendum in Western Sahara	

United Nations Observer Mission in El Salvador (ONUSAL)	members—(9) Brazil, Canada, Colombia, Ecuador, India, Ireland, Spain, Sweden, Venezuela
established—NA 1991	
aim—established by the UN Security Council to verify ceasefire arrangments and to monitor the maintenance of public order pending the organization of a new National Civil Police	

United Nations Office of the High Commissioner for Refugees (UNHCR) *established*—3 December 1949 *effective*—1 January 1951 *aim*—to try to ensure the humanitarian treatment of refugees and find permanent solutions to refugee problems	*members*—(46) Algeria, Argentina, Australia, Austria, Belgium, Brazil, Canada, China, Colombia, Denmark, Ethiopia, Finland, France, Germany, Greece, Holy See, Hungary, Iran, Israel, Italy, Japan, Lebanon, Lesotho, Madagascar, Morocco, Namibia, Netherlands, Nicaragua, Nigeria, Norway, Pakistan, Philippines, Somalia, Sudan, Sweden, Switzerland, Tanzania, Thailand, Tunisia, Turkey, Uganda, UK, US, Venezuela, Yugoslavia, Zaire
United Nations Operation in Mozambique (UNOMOZ) *established*—NA 1992 *aim*—established by the UN Security Council to supervise the ceasefire	*members*—(18) Argentina, Bangladesh, Botswana, Brazil, Canada, Cape Verde, Egypt, Guinea-Bissau, Hungary, India, Italy, Japan, Malaysia, Portugal, Spain, Sweden, Uruguay, Zambia
United Nations Operation in Somalia (UNOSOM) *established*—NA 1992 *aim*—established by the UN Security Council to facilitate an immediate cessation of hostilities, to maintain a ceasefire to promote a political settlement, and to provide urgent humanitarian assistance	*members*—(17) Australia, Austria, Bangladesh, Belgium, Canada, Czech Republic, Egypt, Fiji, Finland, Indonesia, Jordan, Morocco, NZ, Norway, Pakistan, Slovakia, Zimbabwe
United Nations Population Fund (UNFPA) *note*—acronym retained from predecessor organization UN Fund for Population Activities *established*—NA July 1967 *aim*—to promote assistance in dealing with population problems	*members*—(51) selected on a rotating basis from all regions
United Nations Protection Force (UN-PROFOR) *established*—NA 1992 *aim*—established by the UN Security Council to create conditions for peace and security required for the negotiation of an overall settlement of the "Yugoslav" crisis	*members*—(31) Argentina, Bangladesh, Belgium, Brazil, Canada, Colombia, Czech Republic, Denmark, Egypt, Finland, France, Ghana, Ireland, Jordan, Kenya, Luxembourg, Nepal, Netherlands, NZ, Nigeria, Norway, Poland, Portugal, Russia, Slovakia, Sweden, Switzerland, Tunisia, Ukraine, UK, Venezuela
United Nations Relief and Works Agency for Palestine Refugees in the Near East (UNRWA) *established*—8 December 1949 *aim*—to provide assistance to Palestinian refugees	*members*—(10) Belgium, Egypt, France, Japan, Jordan, Lebanon, Syria, Turkey, UK, US
United Nations Secretariat *established*—26 June 1945 *effective*—24 October 1945 *aim*—primary administrative organ of the UN	*member*—Secretary General appointed for a five-year term by the General Assembly on the recommendation of the Security Council
United Nations Security Council *established*—26 June 1945 *effective*—24 October 1945 *aim*—to maintain international peace and security	*permanent members*—(5) China, France, Russia, UK, US *nonpermanent members*—(10) elected for two-year terms by the UN General Assembly; Austria (1991-92), Belgium (1991-92), Cape Verde (1992-93), Ecuador (1991-92), Hungary (1992-93), India (1991-92), Japan (1992-93), Morocco (1992-93), Venezuela (1992-93), Zimbabwe (1991-92)

United Nations Transitional Authority in Cambodia (UNTAC) *established*—NA 1992 *aim*—established by the UN Security Council to contribute to the restoration and maintenance of peace and to the holding of free elections	*members*—(31) Algeria, Argentina, Australia, Austria, Bangladesh, Belgium, Bulgaria, Cameroon, Canada, Chile, China, Congo, France, Germany, Ghana, India, Indonesia, Ireland, Malaysia, Netherlands, NZ, Pakistan, Philippines, Poland, Russia, Senegal, Thailand, Tunisia, UK, US, Uruguay
United Nations Truce Supervision Organization (UNTSO) *established*—NA May 1948 *aim*—initially established by the UN Security Council to supervise the 1948 Arab-Israeli ceasefire and subsequently extended to work in the Sinai, Lebanon, Jordan, Afghanistan, and Pakistan	*members*—(19) Argentina, Australia, Austria, Belgium, Canada, Chile, China, Denmark, Finland, France, Ireland, Italy, Netherlands, NZ, Norway, Russia, Sweden, Switzerland, US
United Nations Trusteeship Council *established*—26 June 1945 *effective*—24 October 1945 *aim*—to supervise the administration of the UN trust territories; only one of the original 11 trusteeships remains—the Trust Territory of the Pacific Islands (Palau)	*members*—(5) China, France, Russia, UK, US
Universal Postal Union (UPU) *established*—9 October 1874, affiliated with the UN 15 November 1947 *effective*—1 July 1948 *aim*—UN specialized agency that promotes international postal cooperation	*members*—(178) Afghanistan, Albania, Algeria, Angola, Argentina, Armenia, Australia, Austria, The Bahamas, Bahrain, Bangladesh, Barbados, Belarus, Belgium, Belize, Benin, Bhutan, Bolivia, Botswana, Brazil, Brunei, Bulgaria, Burkina, Burma, Burundi, Cambodia, Cameroon, Canada, Cape Verde, Central African Republic, Chad, Chile, China, Colombia, Comoros, Congo, Costa Rica, Cote d'Ivoire, Croatia, Cuba, Cyprus, Czech Republic, Denmark, Djibouti, Dominica, Dominican Republic, Ecuador, Egypt, El Salvador, Equatorial Guinea, Estonia, Ethiopia, Fiji, Finland, France, Gabon, The Gambia, Germany, Ghana, Greece, Grenada, Guatemala, Guinea, Guinea-Bissau, Guyana, Haiti, Holy See, Honduras, Hungary, Iceland, India, Indonesia, Iran, Iraq, Ireland, Israel, Italy, Jamaica, Japan, Jordan, Kazakhstan, Kenya, Kiribati, North Korea, South Korea, Kuwait, Laos, Latvia, Lebanon, Lesotho, Liberia, Libya, Liechtenstein, Lithuania, Luxembourg, Madagascar, Malawi, Malaysia, Maldives, Mali, Malta, Mauritania, Mauritius, Mexico, Monaco, Mongolia, Morocco, Mozambique, Namibia, Nauru, Nepal, Netherlands, Netherlands Antilles, NZ, Nicaragua, Niger, Nigeria, Norway, Oman, Overseas Territories of the UK, Pakistan, Panama, Papua New Guinea, Paraguay, Peru, Philippines, Poland, Portugal, Qatar, Romania, Russia, Rwanda, Saint Kitts and Nevis, Saint Lucia, Saint Vincent and the Grenadines, San Marino, Sao Tome and Principe, Saudi Arabia, Senegal, Seychelles, Sierra Leone, Singapore, Slovakia, Slovenia, Solomon Islands, Somalia, Spain, Sri Lanka, Sudan, Suriname, Swaziland, Sweden, Switzerland, Syria, Tanzania, Thailand, Togo, Tonga, Trinidad and Tobago, Tunisia, Turkey, Tuvalu, Uganda, Ukraine, UAE, UK, US, Uruguay, Vanuatu, Venezuela, Vietnam, Western Samoa, Yemen, Yugoslavia, Zaire, Zambia, Zimbabwe
Warsaw Pact (WP)	was established 14 May 1955 to promote mutual defense; members met 1 July 1991 to dissolve the alliance; member states at the time of dissolution were Bulgaria, Czechoslovakia, Hungary, Poland, Romania, and the USSR; earlier members included East Germany and Albania
West African Development Bank (WADB) *note*—also known as Banque Ouest-Africaine de Developpement (BOAD) *established*—14 November 1973 *aim*—to promote economic development and integration	*members*—(7) Benin, Burkina, Cote d'Ivoire, Mali, Niger, Senegal, Togo
West African Economic Community (CEAO) *note*—acronym from Communaute Economique de l'Afrique de l'Ouest *established*—3 June 1972 *aim*—to promote regional economic development	*members*—(7) Benin, Burkina, Cote d'Ivoire, Mali, Mauritania, Niger, Senegal *observers*—(2) Guinea, Togo

Western European Union (WEU)	*members*—(9) Belgium, France, Germany, Italy, Luxembourg, Netherlands, Portugal, Spain, UK
established—23 October 1954	*associate member*—(1) Iceland
effective—6 May 1955	*observer*—(1) Greece
aim—mutual defense and progressive political unification	

World Bank	see International Bank for Reconstruction and Development (IBRD)

World Bank Group	includes International Bank for Reconstruction and Development (IBRD), International Development Association (IDA), and International Finance Corporation (IFC)

World Confederation of Labor (WCL) *established*—19 June 1920 as the International Federation of Christian Trade Unions (IFCTU), renamed 4 October 1968 *aim*—to promote the trade union movement	*members*—(94 national organizations) Algeria, Angola, Antigua and Barbuda, Argentina, Aruba, Austria, Bangladesh, Belgium, Belize, Benin, Bolivia, Bonaire Island, Botswana, Brazil, Burkina, Cameroon, Canada, Cape Verde, Central African Republic, Chad, Chile, Colombia, Costa Rica, Cote d'Ivoire, Cuba, Curacao, Cyprus, Dominica, Dominican Republic, Ecuador, El Salvador, France, French Guiana, Gabon, The Gambia, Ghana, Grenada, Guadaloupe, Guatemala, Guinea, Guyana, Haiti, Honduras, Hong Kong, Indonesia, Italy, Jamaica, Kenya, Lesotho, Liechtenstein, Luxembourg, Madagascar, Malaysia, Mali, Martinique, Mauritius, Mexico, Montserrat, Namibia, Netherlands, Nicaragua, Niger, Nigeria, Pakistan, Panama, Paraguay, Peru, Philippines, Poland, Portugal, Puerto Rico, Rwanda, Saint Kitts and Nevis, Saint Lucia, Saint Martin, Saint Vincent and the Grenadines, Senegal, Seychelles, Sierra Leone, Spain, Sri Lanka, Suriname, Switzerland, Tanzania, Thailand, Togo, UK, US, Uruguay, Venezuela, Vietnam, Zaire, Zambia, Zimbabwe

World Court	see International Court of Justice (ICJ)

World Federation of Trade Unions (WFTU) *established*—NA 1945 *aim*—to promote the trade union movement	*members*—(67) Afghanistan, Angola, Argentina, Australia, Austria, Bahrain, Bangladesh, Bolivia, Brazil, Burkina, Cambodia, Chile, Colombia, Congo, Costa Rica, Cuba, Cyprus, Dominican Republic, Ecuador, El Salvador, Ethiopia, France, The Gambia, Guatemala, Guinea-Bissau, Guyana, Haiti, Honduras, India, Indonesia, Iran, Iraq, Jamaica, Japan, Jordan, North Korea, Kuwait, Laos, Lebanon, Madagascar, Mauritius, Mongolia, Namibia, Nepal, Nicaragua, Oman, Pakistan, Panama, Papua New Guinea, Peru, Philippines, Puerto Rico, Russia, Saint Vincent and the Grenadines, Saudi Arabia, Senegal, Solomon Islands, South Africa, Sri Lanka, Sudan, Syria, Trinidad and Tobago, Uruguay, Venezuela, Vietnam, Yemen, Zaire

World Food Council (WFC) *established*—17 December 1974 *aim*—ECOSOC organization that studies world food problems and recommends solutions	*members*—(36) selected on a rotating basis from all regions

World Food Program (WFP) *established*—24 November 1961 *aim*—ECOSOC organization that provides food aid to assist in development or disaster relief	*members*—(42) selected on a rotating basis from all regions

World Health Organization (WHO) *established*—22 July 1946 *effective*—7 April 1948 *aim*—UN specialized agency concerned with health matters	*members*—(180) Afghanistan, Albania, Algeria, Angola, Antigua and Barbuda, Argentina, Armenia, Australia, Austria, The Bahamas, Bahrain, Bangladesh, Barbados, Belarus, Belgium, Belize, Benin, Bhutan, Bolivia, Bosnia and Herzegovina, Botswana, Brazil, Brunei, Bulgaria, Burkina, Burma, Burundi, Cambodia, Cameroon, Canada, Cape Verde, Central African Republic, Chad, Chile, China, Colombia, Comoros, Congo, Cook Islands, Costa Rica, Cote d'Ivoire, Croatia, Cuba, Cyprus, Czech Republic, Denmark, Djibouti, Dominica, Dominican Republic, Ecuador, Egypt, El Salvador, Equatorial Guinea, Ethiopia, Fiji, Finland, France, Gabon, The Gambia, Georgia, Germany, Ghana, Greece, Grenada, Guatemala, Guinea, Guinea-Bissau, Guyana, Haiti, Honduras, Hungary, Iceland, India, Indonesia, Iran, Iraq, Ireland, Israel, Italy, Jamaica, Japan, Jordan, Kenya, Kiribati, North Korea, South Korea, Kuwait, Kyrgyzstan, Laos, Latvia, Lebanon, Lesotho, Liberia, Libya, Lithuania, Luxembourg, Madagascar, Malawi, Malaysia, Maldives, Mali, Malta, Marshall Islands, Mauritania, Mauritius, Mexico, Federated States of Micronesia, Moldova, Monaco, Mongolia, Morocco, Mozambique, Namibia, Nepal, Netherlands, NZ, Nicaragua, Niger, Nigeria, Norway, Oman, Pakistan, Panama, Papua New Guinea, Paraguay, Peru, Philippines, Poland, Portugal, Qatar, Romania, Russia, Rwanda, Saint Kitts and Nevis, Saint Lucia, Saint Vincent and the Grenadines, San Marino, Sao Tome and Principe, Saudi Arabia, Senegal, Seychelles, Sierra Leone, Singapore, Slovakia, Slovenia, Solomon Islands, Somalia, South Africa, Spain, Sri Lanka, Sudan, Suriname, Swaziland, Sweden, Switzerland, Syria, Tajikistan, Tanzania, Thailand, Togo, Tonga, Trinidad and Tobago, Tunisia, Turkey, Uganda, Ukraine, UAE, UK, US, Uruguay, Uzbekistan, Vanuatu, Venezuela, Vietnam, Western Samoa, Yemen, Yugoslavia, Zaire, Zambia, Zimbabwe *associate members*—(2) Puerto Rico, Tokelau

World Intellectual Property Organization (WIPO)

established—14 July 1967

effective—26 April 1970

aim—UN specialized agency concerned with the protection of literary, artistic, and scientific works

members—(133) Albania, Algeria, Angola, Argentina, Australia, Austria, The Bahamas, Bangladesh, Barbados, Belarus, Belgium, Benin, Brazil, Bulgaria, Burkina, Burundi, Cameroon, Canada, Central African Republic, Chad, Chile, China, Colombia, Congo, Costa Rica, Cote d'Ivoire, Cuba, Cyprus, Czech Republic, Denmark, Dominican Republic, Ecuador, Egypt, El Salvador, Fiji, Finland, France, Gabon, The Gambia, Germany, Ghana, Greece, Guatemala, Guinea, Guinea-Bissau, Haiti, Holy See, Honduras, Hungary, Iceland, India, Indonesia, Iran, Iraq, Ireland, Israel, Italy, Jamaica, Japan, Jordan, Kenya, North Korea, South Korea, Lebanon, Lesotho, Liberia, Libya, Liechtenstein, Lithuania, Luxembourg, Madagascar, Malawi, Malaysia, Mali, Malta, Mauritania, Mauritius, Mexico, Monaco, Mongolia, Morocco, Namibia, Netherlands, NZ, Nicaragua, Niger, Nigeria, Norway, Pakistan, Panama, Paraguay, Peru, Philippines, Poland, Portugal, Qatar, Romania, Russia, Rwanda, Saudi Arabia, Senegal, Sierra Leone, Singapore, Slovakia, Somalia, South Africa, Spain, Sri Lanka, Sudan, Suriname, Swaziland, Sweden, Switzerland, Syria, Tanzania, Thailand, Togo, Trinidad and Tobago, Tunisia, Turkey, Uganda, Ukraine, UAE, UK, US, Uruguay, Venezuela, Vietnam, Yemen, Yugoslavia, Zaire, Zambia, Zimbabwe

World Meteorological Organization (WMO)

established—11 October 1947

effective—4 April 1951

aim—specialized UN agency concerned with meteorological cooperation

members—(162) Afghanistan, Albania, Algeria, Angola, Antigua and Barbuda, Argentina, Australia, Austria, The Bahamas, Bahrain, Bangladesh, Barbados, Belarus, Belgium, Belize, Benin, Bolivia, Botswana, Brazil, British Caribbean Territories, Brunei, Bulgaria, Burkina, Burma, Burundi, Cambodia, Cameroon, Canada, Cape Verde, Central African Republic, Chad, Chile, China, Colombia, Comoros, Congo, Costa Rica, Cote d'Ivoire, Cuba, Czech Republic, Denmark, Djibouti, Dominica, Dominican Republic, Ecuador, Egypt, El Salvador, Ethiopia, Fiji, Finland, France, French Polynesia, Gabon, The Gambia, Germany, Ghana, Greece, Guatemala, Guinea, Guinea-Bissau, Guyana, Haiti, Honduras, Hong Kong, Hungary, Iceland, India, Indonesia, Iran, Iraq, Ireland, Israel, Italy, Jamaica, Japan, Jordan, Kenya, North Korea, South Korea, Kuwait, Laos, Lebanon, Lesotho, Liberia, Libya, Luxembourg, Macedonia, Madagascar, Malawi, Malaysia, Maldives, Mali, Malta, Mauritania, Mauritius, Mexico, Mongolia, Morocco, Mozambique, Namibia, Nepal, Netherlands, Netherlands Antilles, New Caledonia, NZ, Nicaragua, Niger, Nigeria, Norway, Oman, Pakistan, Panama, Papua New Guinea, Paraguay, Peru, Philippines, Poland, Portugal, Qatar, Romania, Russia, Rwanda, Saint Lucia, Sao Tome and Principe, Saudi Arabia, Senegal, Seychelles, Sierra Leone, Singapore, Slovakia, Solomon Islands, Somalia, South Africa (suspended), Spain, Sri Lanka, Sudan, Suriname, Swaziland, Sweden, Switzerland, Syria, Tanzania, Thailand, Togo, Trinidad and Tobago, Tunisia, Turkey, Uganda, Ukraine, UAE, UK, US, Uruguay, Vanuatu, Venezuela, Vietnam, Yemen, Yugoslavia, Zaire, Zambia, Zimbabwe

World Tourism Organization (WTO)

established—2 January 1975

aim—promote tourism as a means of contributing to economic development, international understanding, and peace

members—(110) Afghanistan, Algeria, Angola, Argentina, Austria, Bangladesh, Belgium, Benin, Bolivia, Brazil, Bulgaria, Burkina, Burundi, Cambodia, Cameroon, Canada, Chad, Chile, China, Colombia, Congo, Cote d'Ivoire, Cuba, Cyprus, Czech Republic, Dominican Republic, Ecuador, Egypt, Ethiopia, Finland, France, Gabon, The Gambia, Germany, Ghana, Greece, Grenada, Guinea, Guinea-Bissau, Haiti, Hungary, India, Indonesia, Iran, Iraq, Israel, Italy, Jamaica, Japan, Jordan, Kenya, North Korea, South Korea, Kuwait, Laos, Lebanon, Lesotho, Libya, Madagascar, Malawi, Malaysia, Maldives, Mali, Malta, Mauritania, Mauritius, Mexico, Mongolia, Morocco, Nepal, Netherlands, Nicaragua, Niger, Nigeria, Pakistan, Panama, Paraguay, Peru, Philippines, Poland, Portugal, Romania, Russia, Rwanda, San Marino, Sao Tome and Principe, Senegal, Seychelles, Sierra Leone, Slovakia, Spain, Sri Lanka, Sudan, Switzerland, Syria, Tanzania, Togo, Tunisia, Turkey, Uganda, UAE, US, Uruguay, Venezuela, Vietnam, Yemen, Yugoslavia, Zaire, Zambia, Zimbabwe

associate members—(4) Aruba, Macau, Netherlands Antilles, Puerto Rico

observer—(1) Holy See

Zangger Committee (ZC)

established—early 1970s

aim—to establish guidelines for the export control provisions of the nuclear Non-Proliferation Treaty

members—(28) Australia, Austria, Belgium, Bulgaria, Canada, Czech Republic, Denmark, Finland, France, Germany, Greece, Hungary, Ireland, Italy, Japan, Luxembourg, Netherlands, Norway, Poland, Portugal, Romania, Russia, Slovakia, Spain, Sweden, Switzerland, UK, US

Appendix D:

Weights and Measures

Mathematical Notation	Mathematical Power	Name
	10^{18} or 1,000,000,000,000,000, 000	one quintillion
	10^{15} or 1,000,000,000,000,000	one quadrillion
	10^{12} or 1,000,000,000,000	one trillion
	10^{9} or 1,000,000,000	one billion
	10^{6} or 1,000,000	one million
	10^{3} or 1,000	one thousand
	10^{2} or 100	one hundred
	10^{1} or 10	ten
	10^{0} or 1	one
	10^{-1} or 0.1	one tenth
	10^{-2} or 0.01	one hundredth
	10^{-3} or 0.001	one thousandth
	10^{-6} or 0.000 001	one millionth
	10^{-9} or 0.000 000 001	one billionth
	10^{-12} or 0.000 000 000 001	one trillionth
	10^{-15} or 0.000 000 000 000 001	one quadrillionth
	10^{-18} or 0.000 000 000 000 000 001	one quintillionth

Metric Interrelationships

Conversions from a multiple or submultiple to the basic units of meters, liters, or grams can be done using the table. For example, to convert from kilometers to meters, multiply by 1,000 (9.26 kilometers equals 9,260 meters) or to convert from meters to kilometers, multiply by 0.001 (9,260 meters equals 9.26 kilometers)

Prefix	Symbol	Length, weight, or capacity	Area	Volume
exa	E	10^{18}	10^{36}	10^{54}
peta	P	10^{15}	10^{30}	10^{45}
tera	T	10^{12}	10^{24}	10^{36}
giga	G	10^{9}	10^{18}	10^{27}
mega	M	10^{6}	10^{12}	10^{18}
hectokilo	hk	10^{5}	10^{10}	10^{15}
myria	ma	10^{4}	10^{8}	10^{12}
kilo	k	10^{3}	10^{6}	10^{9}
hecto	h	10^{2}	10^{4}	10^{6}
basic unit	—	1 meter, 1 gram, 1 liter	1 meter^2	1 meter^3
deci	d	10^{-1}	10^{-2}	10^{-3}
centi	c	10^{-2}	10^{-4}	10^{-6}
milli	m	10^{-3}	10^{-6}	10^{-9}
decimilli	dm	10^{-4}	10^{-8}	10^{-12}
centimilli	cm	10^{-5}	10^{-10}	10^{-15}
micro	u	10^{-6}	10^{-12}	10^{-18}
nano	n	10^{-9}	10^{-18}	10^{-27}
pico	p	10^{-12}	10^{-24}	10^{-36}
femto	f	10^{-15}	10^{-30}	10^{-45}
atto	a	10^{-18}	10^{-36}	10^{-54}

Equivalents

Unit	Metric Equivalent	US Equivalent
acre	0.404 685 64 hectares	43,560 feet^2
acre	4,046,856 4 meters^2	4,840 yards ^{2}
acre	0.004 046 856 4 kilometers^2	0.001 562 5 miles^2, statute
are	100 meters^2	119.599 yards^2
barrel (petroleum, US)	158.987 29 liters	42 gallons
(proof spirits, US)	151.416 47 liters	40 gallons
(beer, US)	117.347 77 liters	31 gallons
bushel	35.239 07 liters	4 pecks
cable	219.456 meters	120 fathoms
chain (surveyor's)	20.116 8 meters	66 feet
cord (wood)	3.624 556 meters^3	128 feet^3
cup	0.236 588 2 liters	8 ounces, liquid (US)
degrees, celsius	(water boils at 100° degrees C, freezes at 0° C)	multiply by 1.8 and add 32 to obtain °F
degrees, fahrenheit	subtract 32 and divide by 1.8 to obtain °C	(water boils at 212 °F, freezes at 32 °F)
dram, avoirdupois	1.771 845 2 grams	0.0625 5 ounces, avoirdupois
dram, troy	3.887 934 6 grams	0.125 ounces, troy
dram, liquid (US)	3.696 69 milliliters	0.125 ounces, liquid
fathom	1.828 8 meters	6 feet
foot	30.48 centimeters	12 inches
foot	0.304 8 meters	0.333 333 3 yards
foot	0.000 304 8 kilometers	0.000 189 39 miles, statute
foot^2	929.030 4 centimeters^2	144 inches^2
foot	2 0.092 903 04 meters^2	0.111 111 1 yards^2
foot^3	28.316 846 592 liters	7.480 519 gallons
foot^3	0.028 316 847 meters^3	1,728 inches^3
furlong	201.168 meters	220 yards
gallon, liquid (US)	3.785 411 784 liters	4 quarts, liquid
gill (US)	118.294 118 milliliters	4 ounces, liquid
grain	64.798 91 milligrams	0.002 285 71 ounces, advp.
gram	1,000 milligrams	0.035 273 96 ounces, advp.
hand (height of horse)	10.16 centimeters	4 inches
hectare	10,000 meters^2	2.471 053 8 acres
hundredweight, long	50.802 345 kilograms	112 pounds, avoirdupois
hundredweight, short	45.359 237 kilograms	100 pounds, avoirdupois
inch	2.54 centimeters	0.083 333 33 feet
inch^2	6.451 6 centimeters^2	0.006 944 44 feet^2
inch^3	16.387 064 centimeters^3	0.000 578 7 feet^3
inch^3	16.387 064 milliliters	0.029 761 6 pints, dry
inch^3	16.387 064 milliliters	0.034 632 0 pints, liquid
kilogram	0.001 tons, metric	2.204 623 pounds, avoirdupois
kilometer	1,000 meters	0.621 371 19 miles, statute
kilometer^2	100 hectares	247.105 38 acres
kilometer^2	1,000,000 meters^2	0.386 102 16 miles^2, statute
knot (1 nautical mi/hr)	1.852 kilometers/hour	1.151 statute miles/hour
league, nautical	5.559 552 kilometers	3 miles, nautical
league, statute	4.828.032 kilometers	3 miles, statute
link (surveyor's)	20.116 8 centimeters	7.92 inches

Equivalents

Unit	Metric Equivalent	US Equivalent
liter	0.001 meters^3	61.023 74 inches^3
liter	0.1 dekaliter	0.908 083 quarts, dry
liter	1,000 milliliters	1.056 688 quarts, liquid
meter	100 centimeters	1.093 613 yards
meter^2	10,000 centimeters^2	1.195 990 yards^2
meter^3	1,000 liters	1.307 951 yards^3
micron	0.000 001 meter	0.000 039 4 inches
mil	0.025 4 millimeters	0.001 inch
mile, nautical	1.852 kilometers	1.150 779 4 miles, statute
mile^2, nautical	3.429 904 kilometers^2	1.325 miles^2, statute
mile, statute	1.609 344 kilometers	5,280 feet or 8 furlongs
mile^2, statute	258.998 811 hectares	640 acres or 1 section
mile^2, statute	2.589 988 11 kilometers^2	0.755 miles^2, nautical
minim (US)	0.061 611 52 milliliters	0.002 083 33 ounces, liquid
ounce, avoirdupois	28.349 523 125 grams	437.5 grains
ounce, liquid (US)	29.573 53 milliliters	0.062 5 pints, liquid
ounce, troy	31.103 476 8 grams	480 grains
pace	76.2 centimeters	30 inches
peck	8.809 767 5 liters	8 quarts, dry
pennyweight	1.555 173 84 grams	24 grains
pint, dry (US)	0.550 610 47 liters	0.5 quarts, dry
pint, liquid (US)	0.473 176 473 liters	0.5 quarts, liquid
point (typographical)	0.351 459 8 millimeters	0.013 837 inches
pound, avoirdupois	453.592 37 grams	16 ounces, avourdupois
pound, troy	373.241 721 6 grams	12 ounces, troy
quart, dry (US)	1.101 221 liters	2 pints, dry
quart, liquid (US)	0.946 352 946 liters	2 pints, liquid
quintal	100 kilograms	220.462 26 pounds, avdp.
rod	5.029 2 meters	5.5 yards
scruple	1.295 978 2 grams	20 grains
section (US)	2.589 988 1 kilometers^2	1 mile^2, statute or 640 acres
span	22.86 centimeters	9 inches
stere	1 meter^3	1.307 95 yards^3
tablespoon	14.786 76 milliliters	3 teaspoons
teaspoon	4.928 922 milliliters	0.333 333 tablespoons
ton, long or deadweight	1,016.046 909 kilograms	2,240 pounds, avoirdupois
ton, metric	1,000 kilograms	2,204.623 pounds, avoirdupois
ton, metric	1,000 kilograms	32,150.75 ounces, troy
ton, register	2.831 684 7 meters^3	100 feet^3
ton, short	907.184 74 kilograms	2,000 pounds, avoirdupois
township (US)	93.239 572 kilometers^2	36 miles^2, statute
yard	0.914 4 meters	3 feet
yard^2	0.836 127 36 meters^2	9 feet^2
yard^3	0.764 554 86 meters^3	27 feet^3
yard^3	764.554 857 984 liters	201.974 gallons

Appendix E:

Cross-Reference List of Geographic Names

This list indicates where various names including all United States Foreign Service Posts, alternate names, former names, and political or geographical portions of larger entities can be found in *The World Factbook*. Spellings are not necessarily those approved by the United States Board on Geographic Names (BGN). Alternate names are included in parentheses; additional information is included in brackets.

	Name	Entry in *The World Factbook*
A	Abidjan [US Embassy]	Cote d'Ivoire
	Abu Dhabi [US Embassy]	United Arab Emirates
	Abuja [US Embassy Branch Office]	Nigeria
	Acapulco [US Consular Agency]	Mexico
	Accra [US Embassy]	Ghana
	Adamstown	Pitcairn Islands
	Adana [US Consulate]	Turkey
	Addis Ababa [US Embassy]	Ethiopia
	Adelaide [US Consular Agency]	Australia
	Adelie Land (Terre Adelie) [claimed by France]	Antarctica
	Aden	Yemen
	Aden, Gulf of	Indian Ocean
	Admiralty Islands	Papua New Guinea
	Adriatic Sea	Atlantic Ocean
	Aegean Islands	Greece
	Aegean Sea	Atlantic Ocean
	Afars and Issas, French Territory of the (F.T.A.I.)	Djibouti
	Agalega Islands	Mauritius
	Agana	Guam
	Aland Islands	Finland
	Alaska	United States
	Alaska, Gulf of	Pacific Ocean
	Aldabra Islands	Seychelles
	Alderney	Guernsey
	Aleutian Islands	United States
	Alexander Island	Antarctica
	Alexandria [US Consulate General]	Egypt
	Algiers [US Embassy]	Algeria
	Alhucemas, Penon de	Spain
	Alma-Ata (Almaty)	Kazakhstan
	Almaty (Alma-Ata) [US Embassy]	Kazakhstan
	Alofi	Niue
	Alphonse Island	Seychelles
	Amami Strait	Pacific Ocean
	Amindivi Islands	India
	Amirante Isles	Seychelles
	Amman [US Embassy]	Jordan
	Amsterdam [US Consulate General]	Netherlands
	Amsterdam Island (Ile Amsterdam)	French Southern and Antarctic Lands
	Amundsen Sea	Pacific Ocean
	Amur	China; Russia
	Andaman Islands	India
	Andaman Sea	Indian Ocean
	Andorra la Vella	Andorra
	Anegada Passage	Atlantic Ocean
	Anglo-Egyptian Sudan	Sudan
	Anjouan	Comoros
	Ankara [US Embassy]	Turkey
	Annobon	Equatorial Guinea
	Antananarivo [US Embassy]	Madagascar
	Antipodes Islands	New Zealand

Name	Entry in *The World Factbook*
Antwerp [US Consulate General]	Belgium
Aozou Strip [claimed by Libya]	Chad
Apia [US Embassy]	Western Samoa
Aqaba, Gulf of	Indian Ocean
Arabian Sea	Indian Ocean
Arafura Sea	Pacific Ocean
Argun	China; Russia
Ascension Island	Saint Helena
Ashgabat (Ashkhabad)	Turkmenistan
Ashkhabad [US Embassy]	Turkmenistan
Asmara [US Embassy]	Eritrea
Asmera (see Asmara)	Eritrea
Assumption Island	Seychelles
Asuncion [US Embassy]	Paraguay
Asuncion Island	Northern Mariana Islands
Atacama	Chile
Athens [US Embassy]	Greece
Attu	United States
Auckland [US Consulate General]	New Zealand
Auckland Islands	New Zealand
Australes Iles (Iles Tubuai)	French Polynesia
Avarua	Cook Islands
Axel Heiberg Island	Canada
Azores	Portugal
Azov, Sea of	Atlantic Ocean

B		
	Bab el Mandeb	Indian Ocean
	Babuyan Channel	Pacific Ocean
	Babuyan Islands	Philippines
	Baffin Bay	Arctic Ocean
	Baffin Island	Canada
	Baghdad [US Embassy temporarily suspended; US Interests Section located in Poland's embassy in Baghdad]	Iraq
	Baku [US Embassy]	Azerbaijan
	Baky (Baku)	Azerbaijan
	Balabac Strait	Pacific Ocean
	Balearic Islands	Spain
	Balearic Sea (Iberian Sea)	Atlantic Ocean
	Bali [US Consular Agency]	Indonesia
	Bali Sea	Indian Ocean
	Balintang Channel	Pacific Ocean
	Balintang Islands	Philippines
	Balleny Islands	Antarctica
	Balochistan	Pakistan
	Baltic Sea	Atlantic Ocean
	Bamako [US Embassy]	Mali
	Banaba (Ocean Island)	Kiribati
	Bandar Seri Begawan [US Embassy]	Brunei
	Banda Sea	Pacific Ocean
	Bangkok [US Embassy]	Thailand
	Bangui [US Embassy]	Central African Republic
	Banjul [US Embassy]	Gambia, The
	Banks Island	Canada
	Banks Islands (Iles Banks)	Vanuatu
	Barcelona [US Consulate General]	Spain
	Barents Sea	Arctic Ocean
	Barranquilla [US Consulate]	Colombia
	Bashi Channel	Pacific Ocean
	Basilan Strait	Pacific Ocean
	Bass Strait	Indian Ocean

Name	Entry in *The World Factbook*
Basse-Terre	Gaudeloupe
Basseterre	Saint Kitts and Nevis
Batan Islands	Philippines
Basutoland	Lesotho
Bavaria (Bayern)	Germany
Beagle Channel	Atlantic Ocean
Bear Island (Bjornoya)	Svalbard
Beaufort Sea	Arctic Ocean
Bechuanaland	Botswana
Beijing [US Embassy]	China
Beirut [US Embassy]	Lebanon
Belau	Pacific Islands, Trust Territory of the
Belem [US Consular Agency]	Brazil
Belep Islands (Iles Belep)	New Caledonia
Belfast [US Consulate General]	United Kingdom
Belgian Congo	Zaire
Belgrade [US Embassy; US does not maintain full diplomatic relations with Serbia and Montenegro]	Serbia and Montenegro
Belize City [US Embassy]	Belize
Belle Isle, Strait of	Atlantic Ocean
Bellingshausen Sea	Pacific Ocean
Belmopan	Belize
Belorussia	Belarus
Bengal, Bay of	Indian Ocean
Bering Sea	Pacific Ocean
Bering Strait	Pacific Ocean
Berkner Island	Antarctica
Berlin [US Branch Office]	Germany
Berlin, East	Germany
Berlin, West	Germany
Bern [US Embassy]	Switzerland
Bessarabia	Romania; Moldova
Bijagos, Arquipelago dos	Guinea-Bissau
Bikini Atoll	Marshall Islands
Bilbao [US Consulate]	Spain
Bioko	Equatorial Guinea
Biscay, Bay of	Atlantic Ocean
Bishkek [Interim Chancery]	Kyrgyzstan
Bishop Rock	United Kingdom
Bismarck Archipelago	Papua New Guinea
Bismarck Sea	Pacific Ocean
Bissau [US Embassy]	Guinea-Bissau
Bjornoya (Bear Island)	Svalbard
Black Rock	Falkland Islands (Islas Malvinas)
Black Sea	Atlantic Ocean
Bloemfontein	South Africa
Boa Vista	Cape Verde
Bogota [US Embassy]	Colombia
Bombay [US Consulate General]	India
Bonaire	Netherlands Antilles
Bonifacio, Strait of	Atlantic Ocean
Bonin Islands	Japan
Bonn [US Embassy]	Germany
Bophuthatswana	South Africa
Bora-Bora	French Polynesia
Bordeaux [US Consulate General]	France
Borneo	Brunei; Indonesia; Malaysia
Bornholm	Denmark
Bosporus	Atlantic Ocean
Bothnia, Gulf of	Atlantic Ocean
Bougainville Island	Papua New Guinea

Name	Entry in *The World Factbook*
Bougainville Strait	Pacific Ocean
Bounty Islands	New Zealand
Brasilia [US Embassy]	Brazil
Bratislava [US Embassy]	Slovakia
Brazzaville [US Embassy]	Congo
Bridgetown [US Embassy]	Barbados
Brisbane [US Consulate]	Australia
British East Africa	Kenya
British Guiana	Guyana
British Honduras	Belize
British Solomon Islands	Solomon Islands
British Somaliland	Somalia
Brussels [US Embassy, US Mission to European Communities, US Mission to the North Atlantic Treaty Organization (USNATO)]	Belgium
Bucharest [US Embassy]	Romania
Budapest [US Embassy]	Hungary
Buenos Aires [US Embassy]	Argentina
Bujumbura [US Embassy]	Burundi
Burnt Pine	Norfolk Island
Byelorussia	Belarus

C

Name	Entry in *The World Factbook*
Cabinda	Angola
Cabot Strait	Atlantic Ocean
Caicos Islands	Turks and Caicos Islands
Cairo [US Embassy]	Egypt
Calcutta [US Consulate General]	India
Calgary [US Consulate General]	Canada
California, Gulf of	Pacific Ocean
Campbell Island	New Zealand
Canal Zone	Panama
Canary Islands	Spain
Canberra [US Embassy]	Australia
Cancun [US Consular Agency]	Mexico
Canton (Guangzhou)	China
Canton Island	Kiribati
Cape Town [US Consulate General]	South Africa
Caracas [US Embassy]	Venezuela
Cargados Carajos Shoals	Mauritius
Caroline Islands	Micronesia, Federated States of; Pacific Islands, Trust Territory of the
Caribbean Sea	Atlantic Ocean
Carpentaria, Gulf of	Pacific Ocean
Casablanca [US Consulate General]	Morocco
Castries	Saint Lucia
Cato Island	Australia
Cayenne	French Guiana
Cebu [US Consulate General]	Philippines
Celebes	Indonesia
Celebes Sea	Pacific Ocean
Celtic Sea	Atlantic Ocean
Central African Empire	Central African Republic
Ceuta	Spain
Ceylon	Sri Lanka
Chafarinas, Islas	Spain
Chagos Archipelago (Oil Islands)	British Indian Ocean Territory
Channel Islands	Guernsey; Jersey
Charlotte Amalie	Virgin Islands
Chatham Islands	New Zealand
Cheju-do	Korea, South
Cheju Strait	Pacific Ocean
Chengdu [US Consulate General]	China

Name	Entry in *The World Factbook*
Chesterfield Islands (Iles Chesterfield)	New Caledonia
Chiang Mai [US Consulate General]	Thailand
Chihli, Gulf of (Bo Hai)	Pacific Ocean
China, People's Republic of	China
China, Republic of	Taiwan
Chisinau [US Embassy]	Moldova
Choiseul	Solomon Islands
Christchurch [US Consular Agency]	New Zealand
Christmas Island [Indian Ocean]	Australia
Christmas Island [Pacific Ocean] (Kiritimati)	Kiribati
Chukchi Sea	Arctic Ocean
Ciskei	South Africa
Ciudad Juarez [US Consulate General]	Mexico
Cochabamba [US Consular Agency]	Bolivia
Coco, Isla del	Costa Rica
Cocos Islands	Cocos (Keeling) Islands
Colombo [US Embassy]	Sri Lanka
Colon [US Consular Agency]	Panama
Colon, Archipielago de (Galapagos Islands)	Ecuador
Commander Islands (Komandorskiye Ostrova)	Russia
Conakry [US Embassy]	Guinea
Congo (Brazzaville)	Congo
Congo (Kinshasa)	Zaire
Congo (Leopoldville)	Zaire
Con Son Islands	Vietnam
Cook Strait	Pacific Ocean
Copenhagen [US Embassy]	Denmark
Coral Sea	Pacific Ocean
Corn Islands (Islas del Maiz)	Nicaragua
Corsica	France
Cosmoledo Group	Seychelles
Cotonou [US Embassy]	Benin
Crete	Greece
Crooked Island Passage	Atlantic Ocean
Crozet Islands (Iles Crozet)	French Southern and Antarctic Lands
Curacao [US Consulate General]	Netherlands Antilles
Cusco [US Consular Agency]	Peru
Czechoslovakia	Czech Republic; Slovakia
D Dahomey	Benin
Daito Islands	Japan
Dakar [US Embassy]	Senegal
Daman (Damao)	India
Damascus [US Embassy]	Syria
Danger Atoll	Cook Islands
Danish Straits	Atlantic Ocean
Danzig (Gdansk)	Poland
Dao Bach Long Vi	Vietnam
Dardanelles	Atlantic Ocean
Dar es Salaam [US Embassy]	Tanzania
Davis Strait	Atlantic Ocean
Deception Island	Antarctica
Denmark Strait	Atlantic Ocean
D'Entrecasteaux Islands	Papua New Guinea
Devon Island	Canada
Dhahran [US Consulate General]	Saudi Arabia
Dhaka [US Embassy]	Bangladesh
Diego Garcia	British Indian Ocean Territory
Diego Ramirez	Chile
Diomede Islands	Russia [Big Diomede]; United States [Little Diomede]
Diu	India
Djibouti [US Embassy]	Djibouti

Name	Entry in *The World Factbook*
Dodecanese	Greece
Dodoma	Tanzania
Doha [US Embassy]	Qatar
Douala [US Consulate]	Cameroon
Douglas	Man, Isle of
Dover, Strait of	Atlantic Ocean
Drake Passage	Atlantic Ocean
Dubai (Dubayy) [US Consulate General]	United Arab Emirates
Dublin [US Embassy]	Ireland
Durango [US Consular Agency]	Mexico
Durban [US Consulate General]	South Africa
Dushanbe [Interim Chancery]	Tajikistan
Dusseldorf [US Consulate General]	Germany
Dutch East Indies	Indonesia
Dutch Guiana	Suriname

	Name	Entry in *The World Factbook*
E	East China Sea	Pacific Ocean
	Easter Island (Isla de Pascua)	Chile
	Eastern Channel (East Korea Strait or Tsushima Strait)	Pacific Ocean
	East Germany (German Democratic Republic)	Germany
	East Korea Strait (Eastern Channel or Tsushima Strait)	Pacific Ocean
	East Pakistan	Bangladesh
	East Siberian Sea	Arctic Ocean
	East Timor (Portuguese Timor)	Indonesia
	Edinburgh [US Consulate General]	United Kingdom
	Elba	Italy
	Ellef Ringnes Island	Canada
	Ellesmere Island	Canada
	Ellice Islands	Tuvalu
	Elobey, Islas de	Equatorial Guinea
	Enderbury Island	Kiribati
	Enewetak Atoll (Eniwetok Atoll)	Marshall Islands
	England	United Kingdom
	English Channel	Atlantic Ocean
	Eniwetok Atoll	Marshall Islands
	Epirus, Northern	Albania; Greece
	Essequibo [claimed by Venezuela]	Guyana
	Etorofu	Russia [de facto]

	Name	Entry in *The World Factbook*
F	Farquhar Group	Seychelles
	Fernando de Noronha	Brazil
	Fernando Po (Bioko)	Equatorial Guinea
	Finland, Gulf of	Atlantic Ocean
	Florence [US Consulate General]	Italy
	Florida, Straits of	Atlantic Ocean
	Formosa	Taiwan
	Formosa Strait (Taiwan Strait)	Pacific Ocean
	Fort-de-France [US Consulate General]	Martinique
	Frankfurt am Main [US Consulate General]	Germany
	Franz Josef Land	Russia
	Freetown [US Embassy]	Sierra Leone
	French Cameroon	Cameroon
	French Indochina	Cambodia; Laos; Vietnam
	French Guinea	Guinea
	French Sudan	Mali
	French Territory of the Afars and Issas (F.T.A.I.)	Djibouti
	French Togo	Togo
	Friendly Islands	Tonga
	Frunze (Bishkek)	Kyrgyzstan
	Fukuoka [US Consulate]	Japan
	Funafuti	Tuvalu
	Funchal [US Consular Agency]	Portugal

Name	Entry in *The World Factbook*
Fundy, Bay of	Atlantic Ocean
Futuna Islands (Hoorn Islands)	Wallis and Futuna

G

Name	Entry in *The World Factbook*
Gaborone [US Embassy]	Botswana
Galapagos Islands (Archipielago de Colon)	Ecuador
Galleons Passage	Atlantic Ocean
Gambier Islands (Iles Gambier)	French Polynesia
Gaspar Strait	Indian Ocean
Geneva [Branch Office of the US Embassy, US Mission to European Office of the UN and Other International Organizations]	Switzerland
Genoa [US Consulate General]	Italy
George Town [US Consular Agency]	Cayman Islands
Georgetown [US Embassy]	Guyana
German Democratic Republic (East Germany)	Germany
German Federal Republic of (West Germany)	Germany
Gibraltar	Gibraltar
Gibraltar, Strait of	Atlantic Ocean
Gilbert Islands	Kiribati
Goa	India
Gold Coast	Ghana
Golan Heights	Syria
Good Hope, Cape of	South Africa
Goteborg	Sweden
Gotland	Sweden
Gough Island	Saint Helena
Grand Banks	Atlantic Ocean
Grand Cayman	Cayman Islands
Grand Turk [US Consular Agency]	Turks and Caicos Islands
Great Australian Bight	Indian Ocean
Great Belt (Store Baelt)	Atlantic Ocean
Great Britain	United Kingdom
Great Channel	Indian Ocean
Greater Sunda Islands	Brunei; Indonesia; Malaysia
Green Islands	Papua New Guinea
Greenland Sea	Arctic Ocean
Grenadines, Northern	Saint Vincent and the Grenadines
Grenadines, Southern	Grenada
Guadalajara [US Consulate General]	Mexico
Guadalcanal	Solomon Islands
Guadalupe, Isla de	Mexico
Guangzhou [US Consulate General]	China
Guantanamo [US Naval Base]	Cuba
Guatemala [US Embassy]	Guatemala
Gubal, Strait of	Indian Ocean
Guinea, Gulf of	Atlantic Ocean
Guayaquil [US Consulate General]	Ecuador

H

Name	Entry in *The World Factbook*
Ha'apai Group	Tonga
Habomai Islands	Russia [de facto]
Hague, The [US Embassy]	Netherlands
Haifa [US Consular Agency]	Israel
Hainan Dao	China
Halifax [US Consulate General]	Canada
Halmahera	Indonesia
Hamburg [US Consulate General]	Germany
Hamilton [US Consulate General]	Bermuda
Hanoi	Vietnam
Harare [US Embassy]	Zimbabwe
Hatay	Turkey
Havana [US post not maintained, representation by US Interests Section (USINT) of the Swiss Embassy]	Cuba
Hawaii	United States

Name	Entry in *The World Factbook*
Heard Island	Heard Island and McDonald Islands
Helsinki [US Embassy]	Finland
Hermosillo [US Consulate]	Mexico
Hispaniola	Dominican Republic; Haiti
Hokkaido	Japan
Hong Kong [US Consulate General]	Hong Kong
Honiara [US Consulate]	Solomon Islands
Honshu	Japan
Hormuz, Strait of	Indian Ocean
Horn, Cape (Cabo de Hornos)	Chile
Horne, Iles de	Wallis and Futuna
Horn of Africa	Ethiopia; Somalia
Hudson Bay	Arctic Ocean
Hudson Strait	Arctic Ocean

I Inaccessible Island	Saint Helena
Indochina	Cambodia; Laos; Vietnam
Inner Mongolia (Nei Mongol)	China
Ionian Islands	Greece
Ionian Sea	Atlantic Ocean
Irian Jaya	Indonesia
Irish Sea	Atlantic Ocean
Islamabad [US Embassy]	Pakistan
Islas Malvinas	Falkland Islands (Islas Malvinas)
Istanbul [US Consulate General]	Turkey
Italian Somaliland	Somalia
Ivory Coast	Cote d'Ivoire
Iwo Jima	Japan
Izmir [US Consulate General]	Turkey

J Jakarta [US Embassy]	Indonesia
Jamestown	Saint Helena
Japan, Sea of	Pacific Ocean
Java	Indonesia
Java Sea	Indian Ocean
Jeddah [US Consulate General]	Saudi Arabia
Jerusalem [US Consulate General]	Israel; West Bank
Johannesburg [US Consulate General]	South Africa
Juan de Fuca, Strait of	Pacific Ocean
Juan Fernandez, Isla de	Chile
Juventud, Isla de la (Isle of Youth)	Cuba

K Kabul [US Embassy now closed]	Afghanistan
Kaduna [US Consulate General]	Nigeria
Kalimantan	Indonesia
Kamchatka Peninsula (Poluostrov Kamchatka)	Russia
Kampala [US Embassy]	Uganda
Kampuchea	Cambodia
Karachi [US Consulate General]	Pakistan
Kara Sea	Arctic Ocean
Karimata Strait	Indian Ocean
Kathmandu [US Embassy]	Nepal
Kattegat	Atlantic Ocean
Kauai Channel	Pacific Ocean
Keeling Islands	Cocos (Keeling) Islands
Kerguelen, Iles	French Southern and Antarctic Lands
Kermadec Islands	New Zealand
Khabarovsk	Russia
Khartoum [US Embassy]	Sudan
Khmer Republic	Cambodia
Khuriya Muriya Islands (Kuria Muria Islands)	Oman
Khyber Pass	Pakistan
Kiel Canal (Nord-Ostsee Kanal)	Atlantic Ocean

Name	Entry in *The World Factbook*
Kiev [US Embassy]	Ukraine
Kigali [US Embassy]	Rwanda
Kingston [US Embassy]	Jamaica
Kingston	Norfolk Island
Kingston	Saint Vincent and the Grenadines
Kinshasa [US Embassy]	Zaire
Kirghiziya	Kyrgyzstan
Kiritimati (Christmas Island)	Kiribati
Kishinev (Chisinau)	Moldova
Kithira Strait	Atlantic Ocean
Kodiak Island	United States
Kola Peninsula (Kol'skiy Poluostrov)	Russia
Kolonia [US Embassy]	Micronesia, Federated States of
Korea Bay	Pacific Ocean
Korea, Democratic People's Republic of	Korea, North
Korea, Republic of	Korea, South
Korea Strait	Pacific Ocean
Koror [US Liaison Office]	Pacific Islands, Trust Territory of
Kosovo	Serbia and Montenegro
Kowloon	Hong Kong
Krakow [US Consulate General]	Poland
Kuala Lumpur [US Embassy]	Malaysia
Kunashiri (Kunashir)	Russia [de facto]
Kuril Islands	Russia [de facto]
Kuwait [US Embassy]	Kuwait
Kwajalein Atoll	Marshall Islands
Kyushu	Japan
Kyyiv (Kiev)	Ukraine

L	Labrador	Canada

Name	Entry in *The World Factbook*
Laccadive Islands	India
Laccadive Sea	Indian Ocean
La Coruna [US Consular Agency]	Spain
Lagos [US Embassy]	Nigeria
Lahore [US Consulate General]	Pakistan
Lakshadweep	India
La Paz [US Embassy]	Bolivia
La Perouse Strait	Pacific Ocean
Laptev Sea	Arctic Ocean
Las Palmas [US Consular Agency]	Spain
Lau Group	Fiji
Leipzig [US Consulate General]	Germany
Leningrad (see Saint Petersburg)	Russia
Lesser Sunda Islands	Indonesia
Leyte	Philippines
Liancourt Rocks [claimed by Japan]	Korea, South
Libreville [US Embassy]	Gabon
Ligurian Sea	Atlantic Ocean
Lilongwe [US Embassy]	Malawi
Lima [US Embassy]	Peru
Lincoln Sea	Arctic Ocean
Line Islands	Kiribati; Palmyra Atoll
Lisbon [US Embassy]	Portugal
Ljubljana [US Embassy]	Slovenia
Lobamba	Swaziland
Lombok Strait	Indian Ocean
Lome [US Embassy]	Togo
London [US Embassy]	United Kingdom
Longyearbyen	Svalbard
Lord Howe Island	Australia
Louisiade Archipelago	Papua New Guinea

Name	Entry in *The World Factbook*
Loyalty Islands (Iles Loyaute)	New Caledonia
Luanda [US Liaison Office]	Angola
Lubumbashi [US Consulate General closed since October 1991]	Zaire
Lusaka [US Embassy]	Zambia
Luxembourg [US Embassy]	Luxembourg
Luzon	Philippines
Luzon Strait	Pacific Ocean
Lyon [US Consulate General]	France

M	Macao	Macau
	Macedonia	Bulgaria
	Macquarie Island	Australia
	Madeira Islands	Portugal
	Madras [US Consulate General]	India
	Madrid [US Embassy]	Spain
	Magellan, Strait of	Atlantic Ocean
	Maghreb	Algeria, Libya, Mauritania, Morocco, Tunisia
	Mahe Island	Seychelles
	Maiz, Islas del (Corn Islands)	Nicaragua
	Majorca (Mallorca)	Spain
	Majuro [US Embassy]	Marshall Islands
	Makassar Strait	Pacific Ocean
	Malabo [US Embassy]	Equatorial Guinea
	Malacca, Strait of	Indian Ocean
	Malaga [US Consular Agency]	Spain
	Malagasy Republic	Madagascar
	Male [US post not maintained, representation from Colombo, Sri Lanka]	Maldives
	Mallorca (Majorca)	Spain
	Malpelo, Isla de	Colombia
	Malta Channel	Atlantic Ocean
	Malvinas, Islas	Falkland Islands (Islas Malvinas)
	Mamoutzou	Mayotte
	Managua [US Embassy]	Nicaragua
	Manama [US Embassy]	Bahrain
	Manaus [US Consular Agency]	Brazil
	Manchukuo	China
	Manchuria	China
	Manila [US Embassy]	Philippines
	Manipa Strait	Pacific Ocean
	Mannar, Gulf of	Indian Ocean
	Manua Islands	American Samoa
	Maputo [US Embassy]	Mozambique
	Maracaibo [US Consulate]	Venezuela
	Marcus Island (Minami-tori-shima)	Japan
	Mariana Islands	Guam; Northern Mariana Islands
	Marion Island	South Africa
	Marmara, Sea of	Atlantic Ocean
	Marquesas Islands (Iles Marquises)	French Polynesia
	Marseille [US Consulate General]	France
	Martin Vaz, Ilhas	Brazil
	Mas a Tierra (Robinson Crusoe Island)	Chile
	Mascarene Islands	Mauritius; Reunion
	Maseru [US Embassy]	Lesotho
	Matamoros [US Consulate]	Mexico
	Mata Utu	Wallis and Futuna
	Mazatlan [US Consulate]	Mexico
	Mbabane [US Embassy]	Swaziland
	McDonald Islands	Heard Island and McDonald Islands
	Medan [US Consulate]	Indonesia

Name	Entry in *The World Factbook*
Mediterranean Sea	Atlantic Ocean
Melbourne [US Consulate General]	Australia
Melilla	Spain
Mensk (Minsk)	Belarus
Merida [US Consulate]	Mexico
Messina, Strait of	Atlantic Ocean
Mexico [US Embassy]	Mexico
Mexico, Gulf of	Atlantic Ocean
Milan [US Consulate General]	Italy
Minami-tori-shima	Japan
Mindanao	Philippines
Mindoro Strait	Pacific Ocean
Minicoy Island	India
Minsk [US Embassy]	Belarus
Mogadishu [US Liaison Office]	Somalia
Moldovia	Moldova
Mombasa [US Consulate]	Kenya
Monaco	Monaco
Mona Passage	Atlantic Ocean
Monrovia [US Embassy]	Liberia
Montego Bay [US Consular Agency]	Jamaica
Montenegro	Serbia and Montenegro
Monterrey [US Consulate General]	Mexico
Montevideo [US Embassy]	Uruguay
Montreal [US Consulate General, US Mission to the International Civil Aviation Organization (ICAO)]	Canada
Moravian Gate	Czech Republic
Moroni [US Embassy]	Comoros
Mortlock Islands	Micronesia, Federated States of
Moscow [US Embassy]	Russia
Mozambique Channel	Indian Ocean
Mulege [US Consular Agency]	Mexico
Munich [US Consulate General]	Germany
Musandam Peninsula	Oman; United Arab Emirates
Muscat [US Embassy]	Oman
Muscat and Oman	Oman
Myanma, Myanmar	Burma

N	Naha [US Consulate General]	Japan
	Nairobi [US Embassy]	Kenya
	Nampo-shoto	Japan
	Naples [US Consulate General]	Italy
	Nassau [US Embassy]	Bahamas, The
	Natuna Besar Islands	Indonesia
	N'Djamena [US Embassy]	Chad
	Netherlands East Indies	Indonesia
	Netherlands Guiana	Suriname
	Nevis	Saint Kitts and Nevis
	New Delhi [US Embassy]	India
	Newfoundland	Canada
	New Guinea	Indonesia; Papua New Guinea
	New Hebrides	Vanuatu
	New Siberian Islands	Russia
	New Territories	Hong Kong
	New York, New York [US Mission to the United Nations (USUN)]	United States
	Niamey [US Embassy]	Niger
	Nice [US Consular Agency]	France
	Nicobar Islands	India
	Nicosia [US Embassy]	Cyprus
	Nightingale Island	Saint Helena
	North Atlantic Ocean	Atlantic Ocean

Name	Entry in *The World Factbook*
North Channel	Atlantic Ocean
Northeast Providence Channel	Atlantic Ocean
Northern Epirus	Albania; Greece
Northern Grenadines	Saint Vincent and the Grenadines
Northern Ireland	United Kingdom
Northern Rhodesia	Zambia
North Island	New Zealand
North Korea	Korea, North
North Pacific Ocean	Pacific Ocean
North Sea	Atlantic Ocean
North Vietnam	Vietnam
Northwest Passages	Arctic Ocean
North Yemen (Yemen Arab Republic)	Yemen
Norwegian Sea	Atlantic Ocean
Nouakchott [US Embassy]	Mauritania
Noumea	New Caledonia
Nuku' alofa	Tonga
Novaya Zemlya	Russia
Nuevo Laredo [US Consulate]	Mexico
Nuuk (Godthab)	Greenland
Nyasaland	Malawi
O	
Oahu	United States
Oaxaca [US Consular Agency]	Mexico
Ocean Island (Banaba)	Kiribati
Ocean Island (Kure Island)	United States
Ogaden	Ethiopia; Somalia
Oil Islands (Chagos Archipelago)	British Indian Ocean Territory
Okhotsk, Sea of	Pacific Ocean
Okinawa	Japan
Oman, Gulf of	Indian Ocean
Ombai Strait	Pacific Ocean
Oporto [US Consulate]	Portugal
Oran [US Consulate]	Algeria
Oranjestad	Aruba
Oresund (The Sound)	Atlantic Ocean
Orkney Islands	United Kingdom
Osaka-Kobe [US Consulate General]	Japan
Oslo [US Embassy]	Norway
Otranto, Strait of	Atlantic Ocean
Ottawa [US Embassy]	Canada
Ouagadougou [US Embassy]	Burkina
Outer Mongolia	Mongolia
P	
Pagan	Northern Mariana Islands
Pago Pago	American Samoa
Palau	Pacific Islands, Trust Territory of the
Palawan	Philippines
Palermo [US Consulate General]	Italy
Palk Strait	Indian Ocean
Palma de Mallorca [US Consular Agency]	Spain
Pamirs	China; Tajikistan
Panama [US Embassy]	Panama
Panama Canal	Panama
Panama, Gulf of	Pacific Ocean
Papeete	French Polynesia
Paramaribo [US Embassy]	Suriname
Parece Vela	Japan
Paris [US Embassy, US Mission to the Organization for Economic Cooperation and Development (OECD), US Observer Mission at the UN Educational, Scientific, and Cultural Organization (UNESCO)]	France
Pascua, Isla de (Easter Island)	Chile

Name	Entry in *The World Factbook*
Passion, Ile de la	Clipperton Island
Pashtunistan	Afghanistan; Pakistan
Peking (Beijing)	China
Pemba Island	Tanzania
Pentland Firth	Atlantic Ocean
Perim	Yemen
Perouse Strait, La	Pacific Ocean
Persian Gulf	Indian Ocean
Perth [US Consulate General]	Australia
Pescadores	Taiwan
Peshawar [US Consulate]	Pakistan
Peter I Island	Antarctica
Philip Island	Norfolk Island
Philippine Sea	Pacific Ocean
Phnom Penh [US Embassy]	Cambodia
Phoenix Islands	Kiribati
Pines, Isle of (Isla de la Juventud)	Cuba
Piura [US Consular Agency]	Peru
Pleasant Island	Nauru
Plymouth	Montserrat
Ponape (Pohnpei)	Micronesia
Ponta Delgada [US Consulate]	Portugal
Port-au-Prince [US Embassy]	Haiti
Port Louis [US Embassy]	Mauritius
Port Moresby [US Embassy]	Papua New Guinea
Porto Alegre [US Consulate]	Brazil
Port-of-Spain [US Embassy]	Trinidad and Tobago
Porto-Novo	Benin
Port Said [US Consular Agency]	Egypt
Portuguese Guinea	Guinea-Bissau
Portuguese Timor (East Timor)	Indonesia
Port-Vila	Vanuatu
Poznan [US Consulate General]	Poland
Prague [US Embassy]	Czech Republic
Praia [US Embassy]	Cape Verde
Pretoria [US Embassy]	South Africa
Pribilof Islands	United States
Prince Edward Island	Canada
Prince Edward Islands	South Africa
Prince Patrick Island	Canada
Principe	Sao Tome and Principe
Puerto Plata [US Consular Agency]	Dominican Republic
Puerto Vallarta [US Consular Agency]	Mexico
Pusan [US Consulate]	Korea, South
P'yongyang	Korea, North
Q Quebec [US Consulate General]	Canada
Queen Charlotte Islands	Canada
Queen Elizabeth Islands	Canada
Queen Maud Land [claimed by Norway]	Antarctica
Quito [US Embassy]	Ecuador
R Rabat [US Embassy]	Morocco
Ralik Chain	Marshall Islands
Rangoon [US Embassy]	Burma
Ratak Chain	Marshall Islands
Recife [US Consulate]	Brazil
Redonda	Antigua and Barbuda
Red Sea	Indian Ocean
Revillagigedo Island	United States
Revillagigedo Islands	Mexico
Reykjavik [US Embassy]	Iceland

Name	Entry in *The World Factbook*
Rhodes	Greece
Rhodesia	Zimbabwe
Rhodesia, Northern	Zambia
Rhodesia, Southern	Zimbabwe
Riga [US Embassy]	Latvia
Rio de Janeiro [US Consulate General]	Brazil
Rio de Oro	Western Sahara
Rio Muni	Equatorial Guinea
Riyadh [US Embassy]	Saudi Arabia
Road Town	British Virgin Islands
Robinson Crusoe Island (Mas a Tierra)	Chile
Rocas, Atol das	Brazil
Rockall [disputed]	United Kingdom
Rodrigues	Mauritius
Rome [US Embassy, US Mission to the UN Agencies for Food and Agriculture (FODAG)]	Italy
Roncador Cay	Colombia
Roosevelt Island	Antarctica
Roseau	Dominica
Ross Dependency [claimed by New Zealand]	Antarctica
Ross Island	Antarctica
Ross Sea	Antarctica
Rota	Northern Mariana Islands
Rotuma	Fiji
Ryukyu Islands	Japan

	Name	Entry in *The World Factbook*
S	Saba	Netherlands Antilles
	Sabah	Malaysia
	Sable Island	Canada
	Sahel	Burkina, Cape Verde, Chad, The Gambia, Guinea-Bissau, Mali, Mauritania, Niger, Senegal
	Saigon (Ho Chi Minh City)	Vietnam
	Saint Brandon	Mauritius
	Saint Christopher and Nevis	Saint Kitts and Nevis
	Saint-Denis	Reunion
	Saint George's [US Embassy]	Grenada
	Saint George's Channel	Atlantic Ocean
	Saint Heliar	Jersey
	Saint John's [US Embassy]	Antigua and Barbuda
	Saint Lawrence, Gulf of	Atlantic Ocean
	Saint Lawrence Island	United States
	Saint Lawrence Seaway	Atlantic Ocean
	Saint Martin	Guadeloupe
	Saint Martin (Sint Maarten)	Netherlands Antilles
	Saint Paul Island	Canada
	Saint Paul Island	United States
	Saint Paul Island (Ile Saint-Paul)	French Southern and Antarctic Lands
	Saint Peter and Saint Paul Rocks (Penedos de Sao Pedro e Sao Paulo)	Brazil
	Saint Peter Port	Guernsey
	Saint Petersburg [US Consulate]	Russia
	Saint-Pierre	Saint Pierre and Miquelon
	Saint Vincent Passage	Atlantic Ocean
	Saipan	Northern Mariana Islands
	Sakhalin Island (Ostrov Sakhalin)	Russia
	Sala y Gomez, Isla	Chile
	Salisbury (Harare)	Zimbabwe
	Salvador de Bahia [US Consular Agency]	Brazil
	Salzburg [US Consulate General]	Austria
	Sanaa [US Embassy]	Yemen
	San Ambrosio	Chile
	San Andres y Providencia, Archipielago	Colombia

Name	Entry in *The World Factbook*
San Bernardino Strait	Pacific Ocean
San Felix, Isla	Chile
San Jose [US Embassy]	Costa Rica
San Juan	Puerto Rico
San Luis Potosi [US Consular Agency]	Mexico
San Marino	San Marino
San Miguel Allende [US Consular Agency]	Mexico
San Salvador [US Embassy]	El Salvador
Santa Cruz [US Consular Agency]	Bolivia
Santa Cruz Islands	Solomon Islands
Santiago [US Embassy]	Chile
Santo Domingo [US Embassy]	Dominican Republic
Sao Luis [US Consular Agency]	Brazil
Sao Paulo [US Consulate General]	Brazil
Sao Pedro e Sao Paulo, Penedos de	Brazil
Sao Tome	Sao Tome and Principe
Sapporo [US Consulate General]	Japan
Sapudi Strait	Indian Ocean
Sarajevo	Bosnia and Herzegovina
Sarawak	Malaysia
Sardinia	Italy
Sargasso Sea	Atlantic Ocean
Sark	Guernsey
Scotia Sea	Atlantic Ocean
Scotland	United Kingdom
Scott Island	Antarctica
Senyavin Islands	Micronesia, Federated States of
Seoul [US Embassy]	Korea, South
Serbia	Serbia and Montenegro
Serrana Bank	Colombia
Serranilla Bank	Colombia
Settlement, The	Christmas Island
Severnaya Zemlya (Northland)	Russia
Seville [US Consular Agency]	Spain
Shag Island	Heard Island and McDonald Islands
Shag Rocks	Falkland Islands (Islas Malvinas)
Shanghai [US Consulate General]	China
Shenyang [US Consulate General]	China
Shetland Islands	United Kingdom
Shikoku	Japan
Shikotan (Shikotan-to)	Japan
Siam	Thailand
Sibutu Passage	Pacific Ocean
Sicily	Italy
Sicily, Strait of	Atlantic Ocean
Sikkim	India
Sinai	Egypt
Singapore [US Embassy]	Singapore
Singapore Strait	Pacific Ocean
Sinkiang (Xinjiang)	China
Sint Eustatius	Netherlands Antilles
Sint Maarten (Saint Martin)	Netherlands Antilles
Skagerrak	Atlantic Ocean
Skopje	Macedonia
Society Islands (Iles de la Societe)	French Polynesia
Socotra	Yemen
Sofia [US Embassy]	Bulgaria
Solomon Islands, northern	Papua New Guinea
Solomon Islands, southern	Solomon Islands
Soloman Sea	Pacific Ocean

Name	Entry in *The World Factbook*
Songkhla [US Consulate]	Thailand
Sound, The (Oresund)	Atlantic Ocean
South Atlantic Ocean	Atlantic Ocean
South China Sea	Pacific Ocean
Southern Grenadines	Grenada
Southern Rhodesia	Zimbabwe
South Georgia	South Georgia and the South Sandwich Islands
South Island	New Zealand
South Korea	Korea, South
South Orkney Islands	Antarctica
South Pacific Ocean	Pacific Ocean
South Sandwich Islands	South Georgia and the South Sandwich Islands
South Shetland Islands	Antarctica
South Tyrol	Italy
South Vietnam	Vietnam
South-West Africa	Namibia
South Yemen (People's Democratic Republic of Yemen)	Yemen
Soviet Union	Armenia, Azerbaijan, Belarus, Estonia, Georgia, Kazakhstan, Kyrgyzstan, Latvia, Lithuania, Moldova, Russia, Tajikistan, Turkmenistan, Ukraine, Uzbekistan
Spanish Guinea	Equatorial Guinea
Spanish Sahara	Western Sahara
Spitsbergen	Svalbard
Stanley	Falkland Islands (Islas Malvinas)
Stockholm [US Embassy]	Sweden
Strasbourg [US Consulate General]	France
Stuttgart [US Consulate General]	Germany
Suez, Gulf of	Indian Ocean
Sulu Archipelago	Philippines
Sulu Sea	Pacific Ocean
Sumatra	Indonesia
Sumba	Indonesia
Sunda Islands (Soenda Isles)	Indonesia; Malaysia
Sunda Strait	Indian Ocean
Surabaya [US Consulate]	Indonesia
Surigao Strait	Pacific Ocean
Surinam	Suriname
Suva [US Embassy]	Fiji
Swains Island	American Samoa
Swan Islands	Honduras
Sydney [US Consulate General]	Australia
T Tahiti	French Polynesia
Taipei	Taiwan
Taiwan Strait	Pacific Ocean
Tallin [US Embassy]	Estonia
Tampico [US Consular Agency]	Mexico
Tanganyika	Tanzania
Tangier	Morocco
Tarawa	Kiribati
Tartar Strait	Pacific Ocean
Tashkent [US Embassy]	Uzbekistan
Tasmania	Australia
Tasman Sea	Pacific Ocean
Taymyr Peninsula (Poluostrov Taymyra)	Russia
Tegucigalpa [US Embassy]	Honduras
Tehran [US post not maintained, representation by Swiss Embassy]	Iran
Tel Aviv [US Embassy]	Israel
Terre Adelie (Adelie Land) [claimed by France]	Antarctica
Thailand, Gulf of	Pacific Ocean
Thessaloniki [US Consulate General]	Greece

Name	Entry in *The World Factbook*
Thimphu	Bhutan
Thurston Island	Antarctica
Tibet (Xizang)	China
Tibilisi (Tbilisi) [US Embassy]	Georgia
Tierra del Fuego	Argentina; Chile
Tijuana [US Consulate General]	Mexico
Timor	Indonesia
Timor Sea	Indian Ocean
Tinian	Northern Mariana Islands
Tiran, Strait of	Indian Ocean
Tirane [US Embassy]	Albania
Tobago	Trinidad and Tobago
Tokyo [US Embassy]	Japan
Tonkin, Gulf of	Pacific Ocean
Toronto [US Consulate General]	Canada
Torres Strait	Pacific Ocean
Torshavn	Faroe Islands
Toshkent (Tashkent)	Uzbekistan
Transjordan	Jordan
Transkei	South Africa
Transylvania	Romania
Trieste [US Consular Agency]	Italy
Trindade, Ilha de	Brazil
Tripoli [US post not maintained, representation by Belgian Embassy]	Libya
Tristan da Cunha Group	Saint Helena
Trobriand Islands	Papua New Guinea
Trucial States	United Arab Emirates
Truk Islands	Micronesia
Tsugaru Strait	Pacific Ocean
Tuamotu Islands (Iles Tuamotu)	French Polynesia
Tubuai Islands (Iles Tubuai)	French Polynesia
Tunis [US Embassy]	Tunisia
Turin	Italy
Turkish Straits	Atlantic Ocean
Turkmeniya	Turkmenistan
Turks Island Passage	Atlantic Ocean
Tyrol, South	Italy
Tyrrhenian Sea	Atlantic Ocean

U		
	Udorn [US Consulate]	Thailand
	Ulaanbaatar [US Embassy]	Mongolia
	Ullung-do	Korea, South
	Unimak Pass [strait]	Pacific Ocean
	Union of Soviet Socialist Republics	Armenia, Azerbaijan, Belarus, Estonia, Georgia, Kazakhstan, Kyrgyzstan, Latvia, Lithuania, Moldova, Russia, Tajikistan, Turkmenistan, Ukraine, Uzbekistan
	United Arab Republic	Egypt; Syria
	Upper Volta	Burkina
	USSR	Armenia, Azerbaijan, Belarus, Estonia, Georgia, Kazakhstan, Kyrgyzstan, Latvia, Lithuania, Moldova, Russia, Tajikistan, Turkmenistan, Ukraine, Uzbekistan

V		
	Vaduz [US post not maintained, representation from Zurich, Switzerland]	Liechtenstein
	Vakhan Corridor (Wakhan)	Afghanistan
	Valencia [US Consular Agency]	Spain
	Valletta [US Embassy]	Malta
	Valley, The	Anguilla
	Vancouver [US Consulate General]	Canada
	Vancouver Island	Canada
	Van Diemen Strait	Pacific Ocean
	Vatican City [US Embassy]	Holy See

Name	Entry in *The World Factbook*
Velez de la Gomera, Penon de	Spain
Venda	South Africa
Veracruz [US Consular Agency]	Mexico
Verde Island Passage	Pacific Ocean
Victoria [US Embassy]	Seychelles
Vienna [US Embassy, US Mission to International Organizations in Vienna (UNVIE)]	Austria
Vientiane [US Embassy]	Laos
Vilnius [US Embassy]	Lithuania
Vladivostok [US Consulate]	Russia
Volcano Islands	Japan
Vostok Island	Kiribati
Vrangelya, Ostrov (Wrangel Island)	Russia

W	Wakhan Corridor (now Vakhan Corridor)	Afghanistan
	Wales	United Kingdom
	Walvis Bay	South Africa
	Warsaw [US Embassy]	Poland
	Washington, DC [The Permanent Mission of the USA to the Organization of American States (OAS)]	United States
	Weddell Sea	Atlantic Ocean
	Wellington [US Embassy]	New Zealand
	Western Channel (West Korea Strait)	Pacific Ocean
	West Germany (Federal Republic of Germany)	Germany
	West Island	Cocos (Keeling) Islands
	West Korea Strait (Western Channel)	Pacific Ocean
	West Pakistan	Pakistan
	Wetar Strait	Pacific Ocean
	White Sea	Arctic Ocean
	Willemstad	Netherlands Antilles
	Windhoek [US Embassy]	Namibia
	Windward Passage	Atlantic Ocean
	Winnipeg [US Consular Agency]	Canada
	Wrangel Island (Ostrov Vrangelya)	Russia [de facto]

Y	Yamoussoukro	Cote d'Ivoire
	Yaounde [US Embassy]	Cameroon
	Yap Islands	Micronesia
	Yellow Sea	Pacific Ocean
	Yemen (Aden) [People's Democratic Republic of Yemen]	Yemen
	Yemen Arab Republic	Yemen
	Yemen, North [Yemen Arab Republic]	Yemen
	Yemen (Sanaa) [Yemen Arab Republic]	Yemen
	Yemen, People's Democratic Republic of	Yemen
	Yemen, South [People's Democratic Republic of Yemen]	Yemen
	Yerevan [US Embassy]	Armenia
	Youth, Isle of (Isla de la Juventud)	Cuba
	Yucatan Channel	Atlantic Ocean
	Yugoslavia	Bosnia and Herzegovina, Croatia, Macedonia, Serbia and Montenegro, Slovenia

Z	Zagreb [US Embassy]	Croatia
	Zanzibar	Tanzania
	Zurich [US Consulate General]	Switzerland